Chemistry in Context

Applying Chemistry to Society

A Project of the American Chemic

Ninth Edition

Chemistry in Context

Applying Chemistry to Society

Bradley D. Fahlman
Central Michigan University

Kathleen L. Purvis-Roberts
Claremont McKenna, Pitzer, and Scripps Colleges

John S. Kirk
Carthage College

Anne K. Bentley
Lewis & Clark College

Patrick L. Daubenmire
Loyola University Chicago

Jamie P. Ellis
Ithaca College

Michael T. Mury
All Saints Academy

A Project of the American Chemical Society

CHEMISTRY IN CONTEXT: APPLYING CHEMISTRY TO SOCIETY, NINTH EDITION

This book is printed on acid-free paper.

5 6 7 8 9 LWI 21 20 19 18

ISBN 978-1-259-63814-5

MHID 1-259-63814-6

Chief Product Officer, SVP Products & Markets: *G. Scott Virkler*
Vice President, General Manager, Products & Markets: *Marty Lange*
Vice President, Content Design & Delivery: *Betsy Whalen*
Managing Director: *Thomas Timp*
Director of Chemistry: *David Spurgeon, Ph.D.*
Director, Product Development: *Rose Koos*
Product Developer: *Jodi Rhomberg*
Marketing Manager: *Matthew Garcia*
Market Development Manager: *Tamara Hodge*
Director of Digital Content: *Shirley Hino, Ph.D.*
Digital Product Developer: *Joan Weber*
Digital Product Anaylst: *Patrick Diller*
Director, Content Design & Delivery: *Linda Avenarius*
Program Manager: *Lora Neyens*
Content Project Managers: *Sherry Kane / Tammy Juran*
Buyer: *Laura M. Fuller*
Designer: *Tara McDermott*
Content Licensing Specialists: *Carrie Burger / Lori Slattery*
Cover Image: © Ingram Publishing/SuperStock (landfill); © Image Source/Getty Images (smoke stacks); © Johan Swanepoel/Shutterstock (finger print); © Echo/Getty Images (store clerk); © William Leaman/Alamy (spider web); © payless images/123RF (recycle bin); © McGraw-Hill Higher Education (periodic table)
Compositor: *Aptara®, Inc.*
Typeface: *10/12 STIX Mathjax Main*
Printer: *LSC Communications*

Library of Congress Cataloging-in-Publication Data

Names: Fahlman, Bradley D. | American Chemical Society.
Title: Chemistry in context : applying chemistry to society.
Description: Ninth edition / Bradley D. Fahlman, Central Michigan University [and six others]. | New York, NY : McGraw-Hill Education, [2018] | Previous edition: chemistry in context : applying chemistry to society / Catherine H. Middlecamp (New York, NY : McGraw-Hill Education, 2015). | "A project of the American Chemical Society."
Identifiers: LCCN 2016044871 | ISBN 9781259638145 (alk. paper) | ISBN 1259638146 (alk. paper)
Subjects: LCSH: Biochemistry. | Environmental chemistry. | Geochemistry.
Classification: LCC QD415 .C482 2018 | DDC 540—dc23 LC record available at https://lccn.loc.gov/2016044871

Logo applies to the text stock only

mheducation.com/highered

Brief Contents

Contents

© Thinkstock/Index Stock RF

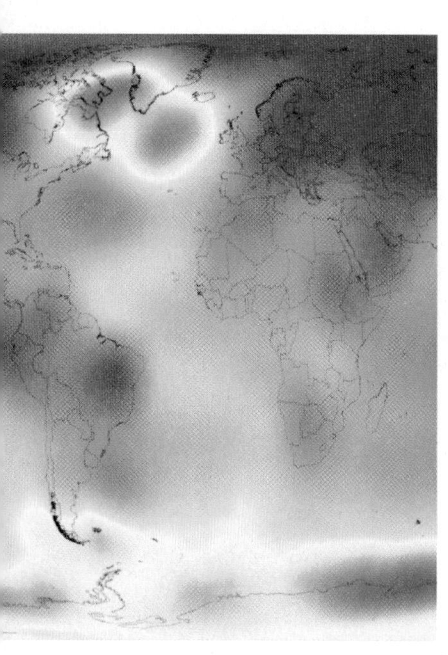

Source: NASA/Scientific Visualization Studio/Goddard Space Flight Center

© Bignai/Shutterstock.com

Chapter 11

© Stock Footage, Inc./Getty Images

Chapter 12

Chapter 13

Chapter 14

Preface

Climate change. Water contamination. Air pollution. Food shortages. These and other societal issues are regularly featured in the media. However, did you know that chemistry plays a crucial role in addressing these challenges? A knowledge of chemistry is also essential to improve the quality of our lives. For instance, faster electronic devices, stronger plastics, and more effective medicines and vaccines all rely on the innovations of chemists throughout the world. With our world so dependent on chemistry, it is unfortunate that most chemistry textbooks do not provide significant details regarding real-world applications. Enter *Chemistry in Context*—"the book that broke the mold." Since its inception in 1993, *Chemistry in Context* has focused on the presentation of chemistry fundamentals within a contextual framework.

So, what is "context," and how will this make your study of chemistry more interesting and relevant?

Context! This word is derived from the Latin word meaning "to weave." Hence, *Chemistry in Context* weaves together connections between chemistry and society. In the absence of societal issues, there could be no *Chemistry in Context*. Similarly, without teachers and students who are willing (and brave enough) to engage in these issues, there could be no *Chemistry in Context*. As the "Central Science," chemistry is woven into the fabric of practically every issue that our society faces today.

Context! Do you enjoy good stories about the world in which you live? If so, look inside this book for stories that intrigue, challenge, and possibly even motivate you to act in new or different ways. In almost all contexts—local, regional, and global—parts of these stories are still unfolding. The ways in which you and others make choices today will determine the nature of the stories told in the future.

Context! Are you aware that using a real-world context to engage people is a high-impact practice backed up by research on how people learn? *Chemistry in Context* offers real-world contexts through which to engage learners on multiple levels: personal, societal, and global. Given the rapidly changing nature of these contexts, *Chemistry in Context* also offers teachers the opportunity to become learners alongside their students.

Sustainability—The Ultimate Context

Global sustainability is not just a challenge. Rather, it is *the* defining challenge of our century. Accordingly, the ninth edition of *Chemistry in Context* continues to focus on this challenge, both as a topic worth studying and as a problem worth solving. As a topic, sustainability provides an important source of content for students to master. For example, the tragedy of the commons, the Triple Bottom Line, and the concept of cradle-to-cradle are all part of this essential content. As a problem worth solving, sustainability generates new questions for students to ask—ones that help them to imagine and achieve a sustainable future. For example, students will find questions about the risks and benefits of acting (or not acting) to reduce emissions of greenhouse gases.

Incorporating sustainability requires more than a casual rethinking of the curriculum. Unlike most general chemistry texts, *Chemistry in Context* is context rich. In essence, you can think of our coverage as a "Citizens First" approach that is context-driven, rather than the content-driven "Atoms First" approach used in many general chemistry curricula. Thus, unlike any other textbook, we provide interesting real-world scenarios about energy, materials, food, water, and health in order to convey essential chemistry content alongside the key concepts of sustainability.

Green chemistry, a means to sustainability, continues to be an important theme in *Chemistry in Context.* As in previous editions, examples of green chemistry are highlighted in each chapter. In this new edition, we provide even more examples. This expanded coverage offers the reader a better sense of the need for, and importance of, greening our chemical processes.

Updates to Existing Content

People sometimes ask us, "Why do you release new editions so often?" Indeed, we are on a fast publishing cycle, turning out a new version every three years. We do this because the content in *Chemistry in Context* is time sensitive.

The ninth edition of *Chemistry in Context* represents a significant update, which is reflected by a change in cover art from previous editions. We now feature new contexts: portable electronics (Chapter 1) and "kitchen" chemistry (Chapter 10). A third new context, forensics, represents the final capstone chapter of the textbook (Chapter 14), and is written as a "whodunit" storyline. Concepts from all of the previous 13 chapters are woven into the story, which takes students through the process of investigating crime scenes and employing appropriate techniques for evidence collection and analyses.

All other chapters have been extensively revised in order to improve the flow of topics while incorporating new scientific developments, changes in policies, energy trends, and current world events. Some highlights of updates to *Chemistry in Context,* 9e, include:

- Chapter 2 (air quality) and Chapter 4 (climate change): updated data and environmental contexts, policies, and regulations are woven throughout each chapter.
- Chapter 3 (radiation from the Sun): more details are provided regarding the role of nanoparticles in sunscreen formulations.
- Chapter 5 (energy from combustion): more details are given for the properties of fuels, and contextual comparisons are provided for various energy values. New information regarding current oil reserves is included, as well as the processes involved to obtain fossil fuels from underground reservoirs, including fracking. A thorough discussion of London dispersion intermolecular forces is also provided.
- Chapter 6 (alternative energies): the original chapter placement has been moved to immediately follow the hydrocarbon-fuel chapter. More details regarding solar, wind, and thermoelectric sources of energy are now included.
- Chapter 7 (energy storage): new details are provided regarding supercapacitors *versus* batteries for electric vehicle applications.
- Chapter 8 (water quality): discussions of water contamination issues from Flint, Michigan, and Durango, Colorado, are included, as well as more details regarding acid–base equilibria.
- Chapter 9 (polymers): updated statistics and new information regarding plastics recycling are provided.
- Chapter 11 (nutrition): an introduction to issues in food safety and food security are included.
- Chapter 12 (health and medicine): this heavily revised chapter now includes new details regarding the role of equilibria on the health of our bodies and the processes involved in modern drug design.
- Chapter 13 (genetics): additional information and references are added regarding GMOs, as well as more details on how synthetic insulin is produced via genetic engineering.

Each chapter has available online, an introductory video that introduce the overall topic to be discussed, with a "Reflection" activity for students to ponder before reading the chapter. This is immediately followed by a new section "The Big Picture", which clearly identifies the main questions that are addressed in the chapter. Every chapter then concludes with a "Learning Outcomes" section that outlines the important concepts that were introduced, with citations to their particular section(s).

A number of interactive simulations are also included in various chapters. The digital edition of *Chemistry in Context,* 9e, features embedded videos and activities, whereas the print version provides these experiences via pointing to the **Connect** website. Relative to previous editions, more activities are woven throughout each chapter that direct students to search the Internet to find appropriate data or reports in order to draw their own conclusions regarding current worldwide issues.

Teaching and Learning in Context

This new edition of *Chemistry in Context* continues with the organizational scheme used in previous editions. However, a new introductory chapter focusing on portable electronics is used to introduce the periodic table, elements, and compounds. Subsequent chapters delve into other real-world themes that provide a foundation of chemistry concepts that is built upon in later chapters.

A variety of embedded in-chapter question types—"Skill Building" (basic review, more traditional, "Scientific Practices" (critical thinking), and "You Decide" (analytical reasoning—also includes questions that directly use the Internet. The questions are plentiful and varied. They range from simpler practice exercises focusing on traditional chemical principles to those requiring more thorough analysis and integration of applications. Some of the questions are the basis for small group work, class discussions, or individual projects. These activities will afford students the opportunity to explore interests, as time permits, beyond the core topics.

Web-based activities found on the **Connect** site are integrated throughout the text. These web-based activities help students develop critical thinking and analytical problem-solving skills based on real-time information.

Many chapters include a figure that "comes alive" through interactivity. This feature resides on the **Connect** site and can be assigned by the instructor.

Chemistry in Context, 9e—A Team Effort

Once again, we have the pleasure of offering our readers a new edition of *Chemistry in Context.* But the work is not done by just one individual; rather, it is the work of a talented team. The ninth edition builds on the legacy of prior author teams led by Cathy Middlecamp, A. Truman Schwartz, Conrad L. Stanitski, and Lucy Pryde Eubanks, all leaders in the chemical education community.

This new edition was prepared by Bradley Fahlman, Kathleen Purvis-Roberts, John Kirk, Anne Bentley, Patrick Daubenmire, Jamie Ellis, and Michael Mury. The accompanying laboratory manual was extensively revised by Jennifer Tripp and Lallie McKenzie. Each author brought their own experiences and expertise to the project, which helped to greatly expand the depth and breadth of the contexts in order to reach a variety of audiences. Stephanie Ryan and Jaclyn Trate also did an amazing job with writing solutions to all in-chapter activities, which were greatly expanded from previous editions.

At the American Chemical Society, leadership was provided by Mary Kirchhoff, Director of the Education Division. She supported the writing team, cheering on its efforts to "connect the dots" between chemistry contexts and the underlying fundamental chemistry content. Terri Taylor, Assistant Director for K–12 Science at the American Chemical Society, provided superior support throughout the project, with great insights regarding the effective use of *CiC* in the classroom. Former production manager, Michael Mury, and current production manager, Emily Bones, were also instrumental in the successful completion of this edition. Michael was able to effectively bring together all of the parties involved—the author team, the publisher, and the American Chemical Society, which was no small feat. Emily's attention to detail and extensive experience in the classroom significantly improved the flow and readability of this edition. The introductory videos for each chapter were completed by an extremely talented videographer at the American Chemical Society, Janali Thompson. Input from Terri Taylor, Kevin McCue, and Adam Dylewski at ACS was also instrumental in achieving professional-quality videos in record time.

The many pedagogical improvements offered in *CiC,* 9e were greatly assisted through input from an Editorial Advisory Board: Renee Cole (University of Iowa), Max Houck (Forensic and Intelligence Services, LLC), Andy Jorgensen (University of Toledo), Steve Keller (University of Missouri-Columbia), Resa Kelly (San Jose State University), Kasi Kiehlbaugh (University of Arizona), Peter Mahaffy (King's University), and Ted Picciotto (Portland Community College). The feedback obtained from this exceptional group substantially improved the quality of the completed work.

The McGraw-Hill team was superb in all aspects of this project, with special thanks to Jodi Rhomberg and Sherry Kane for shepherding the project to the finish line. Marty Lange (Vice President and General Manager), Thomas Timp (Managing Director), David Spurgeon, PhD (Director of Chemsitry), Rose Koos (Director of Development), Shirley Hino, PhD (Director of Digital Content Development), Matthew Garcia (Marketing Manager), Tami Hodge (Director of Marketing), and Jodi Rhomberg (Senior Product Developer), Sherry Kane and Tammy Juran (Content Project Managers), Carrie Burger and Lori Slattery (Content Licensing Specialists), Tara McDermott (Designer), Laura Fuller (Buyer), Patrick Diller (Digital Product Analyst) and Lora Neyens (Program Manager).

The author team truly benefited from the expertise of a wider community. We would like to thank the following individuals who wrote and/or reviewed learning-goal-oriented content for **LearnSmart**.

David G. Jones, *Vistamar School*
Adam I. Keller, *Columbus State Community College*
Margaret Ruth Leslie, *Kent State University*
Peter de Lijser, *California State University—Fullerton*

Input from instructors teaching this course is invaluable to the development of each new edition. Our thanks and gratitude go out to the instructors from the following institutions who participated in *Chemistry in Context* workshops:

American River College
Arizona Agribusiness & Equine Center
Arizona State University
Baruch College
Benito Juarez Community Academy
Bluegrass Community & Technical College
Bronx Community College
Butler University
Cerritos College
Chandler-Gibert Community College
Claremont McKenna, Pitzer & Scripps Colleges
Clemson University
College of DuPage
College of the Canyons
Columbia Secondary School
Delta College
DePaul University
Durham Public Schools
Eastern Maine Community College
Florida International University—Biscayne Bay
Florida Southern College
Florida SouthWestern State College
Florida State College at Jacksonville
Gateway Technical College
Georgia Gwinnett College
Georgia Southwestern State University
Harold Washington College
Hueneme High School
J.D. Clement Early College High School
Johns Hopkins University
LaGuardia Community College
Lake Michigan College
Lake–Sumter State College
Lancaster High School
Merrimack College
Misericordia University
Montgomery College
Moorpark College
Neosho County Community College
New Jersey City University
Norco College
North Hennepin Community College
Northern Virginia Community College
Ohlone College
Oklahoma State University—Oklahoma City
Ozarks Technical Community College
Payson High School
Penn State Altoona
Phoenix College
Plymouth State University
Rock Valley College
Scottsdale Community College

Socorro High School
Southlands Christian Schools
Southwestern College
St. John Fisher College
St. Louis Community College
St. Xavier's College (India)
Suffolk County Community College
SUNY Oneonta
SUNY Plattsburgh
Texas Woman's University
Truckee Meadows Community College
University of Abuja (Nigeria)
University of Baltimore
University of Central Florida
University of Southern Indiana
University of Tennessee
University of Toledo
University of Wisconsin—Milwaukee
Warren County R-III School District
Washington College

We are very excited by the new contexts and features provided in this edition. As you explore these contexts, we hope that your study of the underlying fundamental chemistry concepts will become more relevant in your life. We believe that the chemistry contexts and content provided in this edition, alongside the interactive and thought-provoking activities embedded throughout, will make you think differently about the world around you and the challenges we face. The solutions to current and future societal problems will require an interdisciplinary approach. Whether you decide to continue your studies in chemistry, or transition to other fields of study, we believe that the critical thinking skills fostered in *Chemistry in Context*, 9e will be of value to all of your future endeavors.

Sincerely, on behalf of the author team,

Bradley D. Fahlman

Senior Author and Editor-in-Chief
August, 2016

Required=Results

©Getty Images/iStockphoto

McGraw-Hill Connect®
Learn Without Limits

Connect is a teaching and learning platform that is proven to deliver better results for students and instructors.

Connect empowers students by continually adapting to deliver precisely what they need, when they need it, and how they need it, so your class time is more engaging and effective.

Connect's Impact on Retention Rates, Pass Rates, and Average Exam Scores

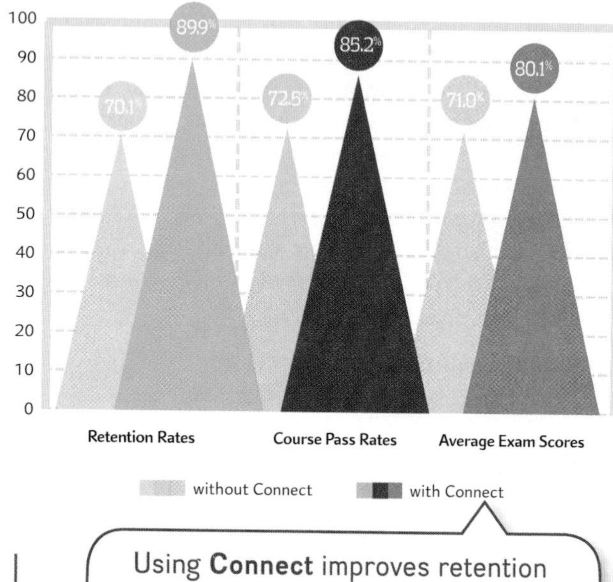

without Connect   with Connect

73% of instructors who use **Connect** require it; instructor satisfaction **increases** by 28% when **Connect** is required.

Using **Connect** improves retention rates by **19.8%**, passing rates by **12.7%**, and exam scores by **9.1%**.

Analytics

Connect Insight®

Connect Insight is Connect's new one-of-a-kind visual analytics dashboard that provides at-a-glance information regarding student performance, which is immediately actionable. By presenting assignment, assessment, and topical performance results together with a time metric that is easily visible for aggregate or individual results, Connect Insight gives the user the ability to take a just-in-time approach to teaching and learning, which was never before available. Connect Insight presents data that helps instructors improve class performance in a way that is efficient and effective.

Impact on Final Course Grade Distribution

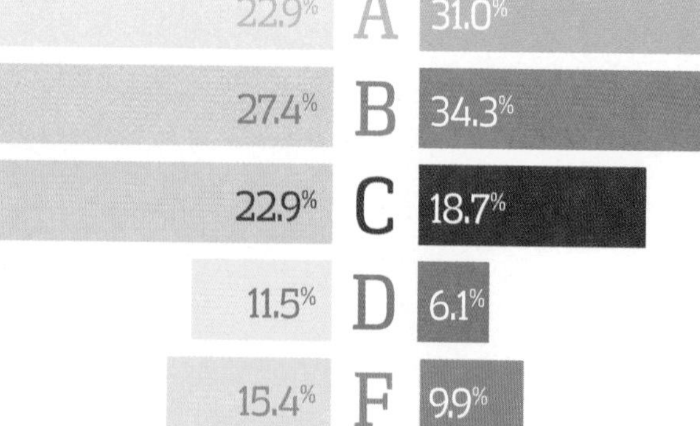

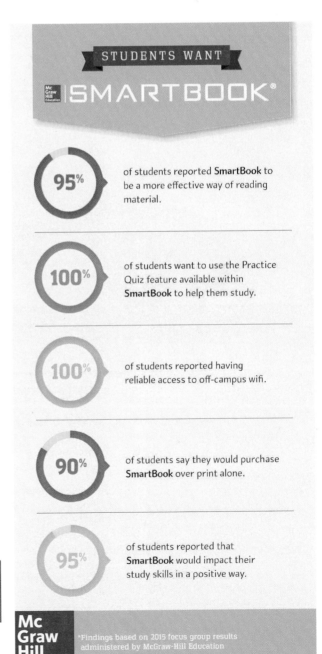

CHAPTER 1

Portable Electronics: The Periodic Table in the Palm of Your Hand

© LifestyleVideoFootage/Shutterstock.com

REFLECTION

What's in Your Cell Phone?

As you will see in this chapter, chemistry plays a central role in controlling the properties of electronic devices.

a. List some desirable attributes of a cell phone, and some that you would like to see in the future.
b. The majority of materials that comprise your cell phone may be classified as metals, plastics, or glass. Using the Web as a resource, describe where these materials come from (both the region(s) of the world where they are produced, and the raw materials used in their fabrication).
c. Cite two elements that combine to form a substance important to your cell phone.
d. What is the expected lifespan of your cell phone?

The Big Picture

In this chapter, you will explore the following questions:

- What are the different components in your portable electronic device made from?
- How does the periodic table of elements guide us in the design of your device?
- How does the touchscreen on your portable electronic device work?
- What role do metals play in electronic devices?
- What are rocks, and how do we isolate and purify metals from these natural sources?
- How is ordinary sand converted into silicon—the fundamental component of processor chips?
- How is sand converted into glass, and how can its structure be modified for crack-resistant screens?
- What are the environmental implications of fabricating and recycling your portable electronic devices?

Introduction

Email, phone calls, texts, tweets, and, of course, Facebook. Our modern society demands constant contact during busy days filled with meetings, classes, travel, and social activities. The tablet or cell phone you hold in your hand is a combination of a variety of materials that have been carefully crafted to give you special capabilities you can't live without.

In order to satisfy the ever-rigorous demands of today's consumer, the latest portable electronics must be lightweight, thin, durable, multifunctional, and easily synced with computers and next-generation wearable devices. Such complex designs are only possible by putting together the elements of the periodic table in many different ways to form materials with the above physical properties that we need or desire.

In this chapter, you will learn about the various components that make up your cell phone, tablet, or other portable electronic device. Perhaps most importantly, you will discover where these components came from and what happens to them after their lifetime is finished. Throughout this book, you will see that the world around us may be described by various length scales. Let's now begin our discovery into the submicroscopic depths of your electronic device. You will never look at your cell phone the same way again ...

Your Turn 1.1 Scientific Practices How Small?!

The smallest building blocks inside your cell phone are about *1000 times* smaller than the diameter of a human hair fiber!

a. What is a typical diameter of an individual hair fiber?

b. Using the answer found in question **a**, how many hair-fiber widths would it take to span the length your cell phone?

1.1 What's the Matter with Materials? A Survey of the Periodic Table

It's wintertime and you need to respond to an urgent text on your smartphone. You touch the screen with a gloved finger and get no response. The hassle of removing your gloves and risking frostbite, just to operate your cell phone or tablet, is an all-too-common occurrence for those who live in cold climates. However, there are now a variety of commercially available gloves that use a special thread or have pads sewn into them, which allow a user to seamlessly control their touchscreen device. Most smartphones and tablets will also respond to a special pen-like object known as a stylus. Nevertheless, this begs the question: Why are touchscreens so restrictive in responding to only a small number of stimuli?

Your Turn 1.2 Scientific Practices Touchscreen Response

Taking care not to damage your screen, use a variety of materials to touch the screen of your portable electronic device. In addition to your finger, items that may be used include a paper clip, plastic pen, key, battery, fabrics, pencil lead, sponge (wet and dry), pencil eraser, coin, glass marble, paper, cardboard, or any other items. Did any of these materials other than your finger cause a response? We will revisit this question later in the chapter.

The properties of a device are governed by what it is made of—its **composition**. What compositions are required for a touchscreen to be transparent, crack-resistant, and touch-sensitive? This is no minor feat, and requires scientists to constantly explore the world around them in order to select the most appropriate constituents.

Chemistry is the branch of science that focuses on the composition, structure, properties, and changes of matter.

Everything around you—the air you breathe, the water you drink, and the mobile device in your hand—is defined as **matter**. Matter is considered to be anything that occupies space and has a mass. This consists of solids, liquids, gases, or plasmas that exist as either pure substances or mixtures (Figure 1.1).

Plasmas are seen in superheated conditions, such as a lightning strike.

For instance, in dissolving sugar in water, both the solid sugar and liquid water are considered pure substances—each composed of a single substance. The mixing together of these separate pure substances will result in a **homogeneous mixture**,

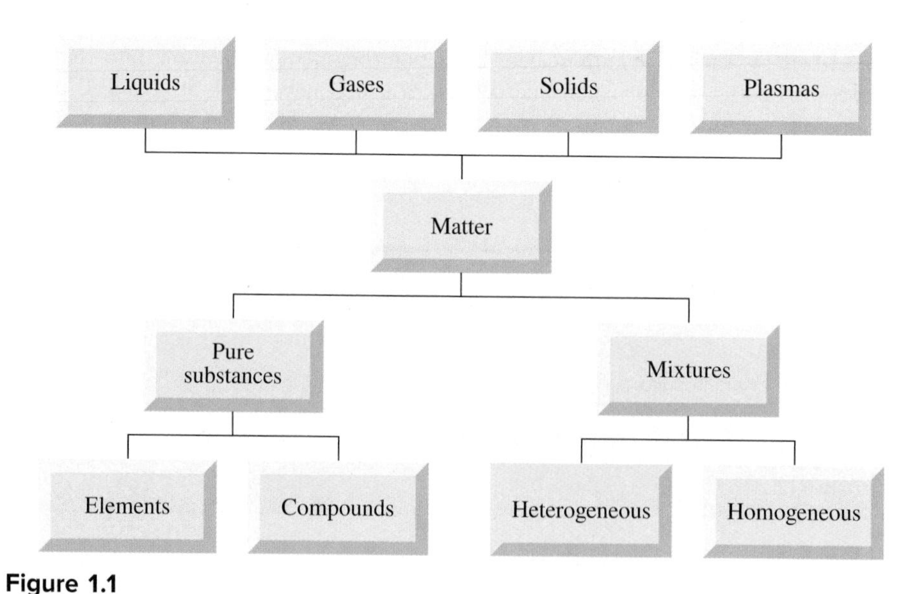

Figure 1.1

A classification scheme for matter.

which will be uniform in composition throughout. Quite often, a homogeneous mixture is referred to as a **solution**. If we take a few spoonfuls of a sugar solution, each one would contain the same ratio of sugar and water. In contrast, if one digs up a handful of soil, you will discover a complicated mixture of sand, particles of varying shapes and colors, liquid water within the pores, and perhaps even some resident earthworms. This is known as a **heterogeneous mixture**, because it is not uniform in composition throughout. That is, the relative amounts of sand, dirt, rocks, etc., will vary from one handful to the next.

As we will see shortly, the smallest building blocks of matter are known as **atoms**. An **element** is composed of many atoms of the same type. Every day, we take for granted the use of pure elements such as copper in household pipes, aluminum in home exteriors, lithium in batteries, and carbon in pencil nibs. In contrast, a **compound** is a pure substance that is made up of two or more different types of atoms in a fixed, characteristic chemical combination. Reconsidering a sugar solution, water (H_2O) is a compound consisting of oxygen and hydrogen atoms. Sugar ($C_{12}H_{22}O_{11}$) is also a compound, but instead contains carbon, hydrogen, and oxygen atoms. Even though the types of atoms in compounds and elements are identical, they are bonded to one another in a different manner within each substance. For instance, the oxygen atoms in sugar are exactly identical to the oxygen atoms that comprise elemental oxygen gas (O_2). However, it would take a chemical reaction to break apart the atoms within sugar to return the oxygen atoms to their elemental form—gaseous oxygen.

Chemical symbols are one- or two-letter abbreviations for the elements. These symbols, established by international agreement, are used throughout the world. Some of them make immediate sense to those who speak English or related languages. For example, oxygen is O, aluminum is Al, lithium is Li, and silicon is Si. However, other symbols have their origin in other languages, such as some metals that were discovered by ancient civilizations and given Latin names long ago. For example, argentum (Ag) is silver, ferrum (Fe) is iron, plumbum (Pb) is lead, and hydrargyrum (Hg) is mercury.

Chemical symbols sometimes also are referred to as atomic symbols.

Elements have been named for properties, planets, places, and people. Hydrogen (H) means "water former," because hydrogen and oxygen gases burn in a flame to form the compound water (H_2O). Neptunium (Np) and plutonium (Pu) were named after two planets in our solar system. Berkelium (Bk) and californium (Cf) honor the University of California, Berkeley, lab in which they were first produced. Flerovium (Fl) and livermorium (Lv) were both named after the laboratories in which the elements were discovered. Only a few atoms of each have been produced by nuclear fusion reactions.

Did You Know? Pluto was discovered in 1930, and for over 75 years was considered a planet. However, in 2006, Pluto was reclassified as a dwarf planet. Regardless of this reclassification, the name plutonium still appears in the periodic table.

Plutonium can fuel both nuclear reactors and nuclear bombs. See Chapter 6 for details.

It is fitting that Russian chemist Dmitri Mendeleev (1834–1907) has his own element (Mendelevium, Md), because the most common way of arranging the elements—the periodic table—reflects the system he developed. This is an orderly arrangement of all the elements based on similarities in their reactivities and properties.

About 90 elemental substances occur naturally on Earth and, as far as we know, elsewhere in the universe. The other two dozen or so elements, including those most recently discovered, have been created from existing elements through nuclear reactions. Plutonium is probably the best known of the synthetic elements, although it does occur in trace amounts in nature.

Four new elements were recognized in 2015 after being discovered years earlier. Elements 113, 115, 117, and 118 have been named Nihonium (Nh; one of two ways to say Japan in Japanese), Moscovium (Mc; to recognize a laboratory in Moscow, Russia), Tennessine (Ts; to recognize laboratories in Tennessee in the U.S.), and Oganesson (Og; to recognize the Russian nuclear physicist Yuri Oganessian), respectively.

Among all known elements, the vast majority are solids at room temperature. At room temperature, nitrogen ($N_2(g)$), oxygen ($O_2(g)$), argon ($Ar(g)$), and eight other elements are gases; in contrast, only bromine ($Br_2(l)$) and mercury ($Hg(l)$) are liquids.

Throughout the text, we will use italics to indicate the phase of the substance; *(s)* indicates a solid, *(l)* a liquid, and *(g)* a gas. In Section 1.4, we will describe why only some elements need a "2" subscript.

The modern periodic table shown in Figure 1.2 lists the elements by number. The green shading indicates the *metals*, which represent most of the periodic table. These elements are usually solid at room temperature, shiny in appearance, may be permanently deformed without breaking or cracking, and are effective conductors of electricity and heat. Ancient civilizations used some metallic elements (iron, copper, tin, lead, gold, and silver) for weaponry, currency, and decoration. Today, the cases of portable electronic devices sometimes employ the metal aluminum, and the circuitry that powers the device utilizes metals such as gold, copper, and tin.

1 1A	2 2A	3 3B	4 4B	5 5B	6 6B	7 7B	8 8B	9 8B	10 8B	11 1B	12 2B	13 3A	14 4A	15 5A	16 6A	17 7A	18 8A
Hydrogen 1 **H** 1.008																	Helium 2 **He** 4.003
Lithium 3 **Li** 6.941	Beryllium 4 **Be** 9.012											Boron 5 **B** 10.81	Carbon 6 **C** 12.01	Nitrogen 7 **N** 14.01	Oxygen 8 **O** 16.00	Fluorine 9 **F** 19.00	Neon 10 **Ne** 20.18
Sodium 11 **Na** 22.99	Magnesium 12 **Mg** 24.31											Aluminum 13 **Al** 26.98	Silicon 14 **Si** 28.09	Phosphorus 15 **P** 30.97	Sulfur 16 **S** 32.07	Chlorine 17 **Cl** 35.45	Argon 18 **Ar** 39.95
Potassium 19 **K** 39.10	Calcium 20 **Ca** 40.08	Scandium 21 **Sc** 44.96	Titanium 22 **Ti** 47.88	Vanadium 23 **V** 50.94	Chromium 24 **Cr** 52.00	Manganese 25 **Mn** 54.94	Iron 26 **Fe** 55.85	Cobalt 27 **Co** 58.93	Nickel 28 **Ni** 58.69	Copper 29 **Cu** 63.55	Zinc 30 **Zn** 65.39	Gallium 31 **Ga** 69.72	Germanium 32 **Ge** 72.61	Arsenic 33 **As** 74.92	Selenium 34 **Se** 78.96	Bromine 35 **Br** 79.90	Krypton 36 **Kr** 83.80
Rubidium 37 **Rb** 85.47	Strontium 38 **Sr** 87.62	Yttrium 39 **Y** 88.91	Zirconium 40 **Zr** 91.22	Niobium 41 **Nb** 92.91	Molybdenum 42 **Mo** 95.94	Technetium 43 **Tc** (98)	Ruthenium 44 **Ru** 101.1	Rhodium 45 **Rh** 102.9	Palladium 46 **Pd** 106.4	Silver 47 **Ag** 107.9	Cadmium 48 **Cd** 112.4	Indium 49 **In** 114.8	Tin 50 **Sn** 118.7	Antimony 51 **Sb** 121.8	Tellurium 52 **Te** 127.6	Iodine 53 **I** 126.9	Xenon 54 **Xe** 131.3
Cesium 55 **Cs** 132.9	Barium 56 **Ba** 137.3	Lanthanum 57 **La** 138.9	Hafnium 72 **Hf** 178.5	Tantalum 73 **Ta** 180.9	Tungsten 74 **W** 183.8	Rhenium 75 **Re** 186.2	Osmium 76 **Os** 190.2	Iridium 77 **Ir** 192.2	Platinum 78 **Pt** 195.1	Gold 79 **Au** 197.0	Mercury 80 **Hg** 200.6	Thallium 81 **Tl** 204.4	Lead 82 **Pb** 207.2	Bismuth 83 **Bi** 209.0	Polonium 84 **Po** (209)	Astatine 85 **At** (210)	Radon 86 **Rn** (222)
Francium 87 **Fr** (223)	Radium 88 **Ra** (226)	Actinium 89 **Ac** (227)	Rutherfordium 104 **Rf** (261)	Dubnium 105 **Db** (262)	Seaborgium 106 **Sg** (266)	Bohrium 107 **Bh** (264)	Hassium 108 **Hs** (277)	Meitnerium 109 **Mt** (268)	Darmstadtium 110 **Ds** (281)	Roentgenium 111 **Rg** (280)	Copernicium 112 **Cn** (285)	Ununtrium 113 **Uut** (284)	Flerovium 114 **Fl** (289)	Ununpentium 115 **Uup** (288)	Livermorium 116 **Lv** (293)	Ununseptium 117 **Uus** (294)	Ununoctium 118 **Uuo** (294)

Cerium 58 **Ce** 140.1	Praseodymium 59 **Pr** 140.9	Neodymium 60 **Nd** 144.2	Promethium 61 **Pm** (145)	Samarium 62 **Sm** 150.4	Europium 63 **Eu** 152.0	Gadolinium 64 **Gd** 157.3	Terbium 65 **Tb** 158.9	Dysprosium 66 **Dy** 162.5	Holmium 67 **Ho** 164.9	Erbium 68 **Er** 167.3	Thulium 69 **Tm** 168.9	Ytterbium 70 **Yb** 173.0	Lutetium 71 **Lu** 175.0
Thorium 90 **Th** 232.0	Protactinium 91 **Pa** 231.0	Uranium 92 **U** 238.0	Neptunium 93 **Np** (237)	Plutonium 94 **Pu** (244)	Americium 95 **Am** (243)	Curium 96 **Cm** (247)	Berkelium 97 **Bk** (247)	Californium 98 **Cf** (251)	Einsteinium 99 **Es** (252)	Fermium 100 **Fm** (257)	Mendelevium 101 **Md** (258)	Noblelium 102 **No** (259)	Lawrencium 103 **Lr** (262)

Metals

Metalloids

Nonmetals

Figure 1.2

The periodic table of elements, showing the locations of metals, metalloids, and nonmetals.

Did You Know? Lothar Meyer (1830–1895), a German chemist, also developed a periodic table at the same time as Mendeleev. Interestingly, both periodic tables were developed independently, but were nearly identical to each other.

Far fewer in number are the *nonmetals*—elements that may be in gaseous, liquid, or solid states at room temperature. The nonmetals are characterized by poor conductivity of heat or electricity, and those in the solid state cannot be deformed without cracking or breaking. A mere eight elements fall into a category known as *metalloids*—elements that lie between metals and nonmetals in the periodic table, and whose properties do not fall cleanly into either category. As a reflection of their intermediate electrical conductivity relative to metals and nonmetals, the metalloids are also often called *semimetals* or *semiconductors*. The metalloid element silicon serves as the key component in all integrated circuits, known as *chips*, that are at the heart of all electronic devices.

Your Turn 1.3 Scientific Practices The Periodic Table Inside Your Cell Phone

Survey the periodic table shown above. Which elements do you think are found in your cell phone?

The elements in the periodic table fall into vertical columns called **groups**. Groups serve to organize elements according to important properties they have in common, and are numbered from left to right. Some groups are given names as well. For example, the metals in the first two columns, Groups 1 and 2, are referred to as the *alkali metals* and *alkaline earth metals*, respectively. Compounds containing metals from either of these groups will give rise to alkaline conditions in soil and water. Additionally, the alkaline earths are mostly responsible for the hard water found in some vicinities.

The nonmetals in Group 17 are known as *halogens*, which include fluorine, chlorine, bromine, and iodine. The final column, Group 18, consists of the *noble gases*—inert elements that undergo few, if any, chemical reactions. You may recognize helium as the noble gas used to make balloons rise, because it is less dense than air. Radon is a noble gas that is radioactive, a characteristic that distinguishes it from the other elements in Group 18.

Chapter 6 will provide more details about radioactivity.

1.2 Atomic Legos—Atoms as Building Blocks for Matter

Elements are made up of *atoms*—the smallest building block that can exist as a stable, independent entity. The word atom comes from the Greek word for "uncuttable." Although today it is possible to "split" atoms using specialized processes, atoms remain indivisible by ordinary chemical or mechanical means.

Atoms are extremely small. Because they are so tiny, we need colossal numbers of them in order to see, touch, or weigh them. For example, a *single drop of water* contains about 5.3×10^{21} atoms. To put this into perspective, this is roughly a trillion times greater than the 7 billion people on Earth—almost enough to give each person a trillion atoms!

Notice a particular format, called *scientific notation*, for '5.3×10^{21} atoms' was used. In decimal notation, that number of atoms would be written as 5,300,000,000,000,000,000,000. More details regarding scientific notation will be provided in Section 1.8.

Although individual atoms are infinitesimally small, we have technology capable of moving them into desired positions and imaging them on a surface. As incredible as this sounds, scientists at Ohio University were able to assemble atoms on a silver surface to create a smiley face (Figure 1.3). **Nanotechnology** refers to the manipulation of matter with at least one dimension sized between 1–100 nanometers, where 1 nanometer (nm) = 1×10^{-9} m. Whereas individual atoms and small molecules are sized in the sub-nanoscopic range, larger biomolecules such as DNA, hemoglobin, and most viruses are nanoscopic in size. Numerous components found in consumer products such as cosmetics, sunscreens, and paints are sized within the nano-regime. The smiley face shown in Figure 1.3 is only a few nanometers tall and wide. At this size, about 250 million smileys could fit on a cross section of a human hair!

Chapter 3 will describe the types of nanoparticles used in sunscreens, as well as their overall benefits and hazards.

In order to convert a quantity into a different unit, a conversion factor must be used. For instance, the conversion of 12 m to nm would be:

$$(12\ \cancel{\text{m}}) \times \left(\frac{1 \times 10^{9}\ \text{nm}}{1\ \cancel{\text{m}}}\right) = 1.2 \times 10^{10}\ \text{nm}$$

When a unit is converted from one form to another, it is often referred to as *dimensional analysis*.

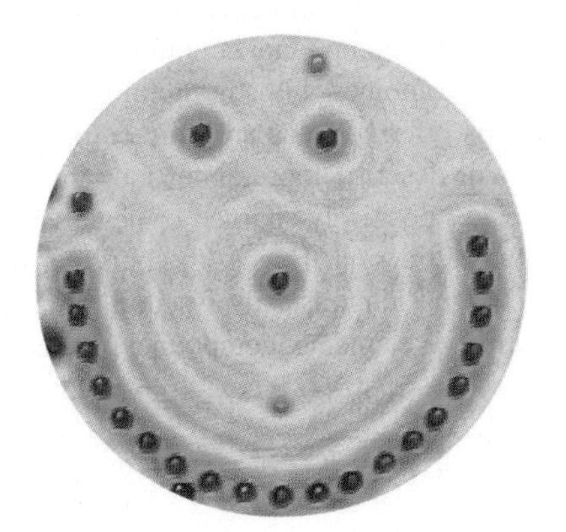

Figure 1.3

A nano-sized smiley face formed by the arrangement of individual silver atoms on a surface, as imaged with a scanning tunneling microscope.

© *Saw-Wai Hla/Hla Group/Ohio University*

Your Turn 1.4 Scientific Practices Unit Conversions

In **Your Turn 1.1**, you discovered the extremely small dimensions of an individual hair fiber. Let's now explore other length scales that are in the macroscopic world around us, and the invisible micro- and sub-microscopic worlds that comprise our cell phones.

a. List some examples of macroscopic objects in your surroundings with dimensions (length, width, height, diameter, etc.) on the order of: (i) millimeters, (ii) centimeters, and (iii) meters.
b. Describe the dimensions (length, width, height) of your cell phone or tablet using the three units described in question **a**. Express your answers in standard decimal notation.

1.3 Compounding the Complexity—From Elements to Compounds

Did You Know? Chemists in the late 18th century isolated what they thought were pure Group 2 elements, which were found to be insoluble in water and resistant to heating. The term "earth" was historically used to describe these characteristic properties. However, these chemists had instead isolated *compounds* of the Group 2 elements, such as calcium oxide (CaO) and magnesium oxide (MgO). Years later, it was discovered that pure alkaline earth metals have drastically different properties than these compounds, such as extreme reactivity with water and rapid burning in air with a brilliant-colored flame.

Using the concept of atoms, we can better explain the terms element and compound that are so prevalent in the language of chemistry. Elements are made up of only one kind of atom. For example, the element carbon is made up of carbon atoms only. By contrast, compounds are made up of two or more different kinds of atoms that are chemically bonded to one another. For instance, the compound aluminum oxide (Al_2O_3) contains both aluminum and oxygen atoms in a 2:3 ratio. Silicon dioxide (SiO_2) is made up of silicon and oxygen atoms.

A **chemical formula** is a symbolic way to represent the elementary composition of a substance. It reveals both the elements present (by chemical symbols) and the atomic ratio of those elements (by the subscripts). For example, in the compound CO_2, the elements C and O are present in a ratio of one carbon atom for every two oxygen atoms. Similarly, H_2O indicates two hydrogen atoms for each oxygen atom. Note that when an atom occurs only once, such as the O in H_2O or the C in CO_2, the subscript of "1" is omitted.

So what about the term **molecule** that is so pervasive in chemistry vocabulary? Are molecules the same as compounds? Are elements also considered molecules? The definition of compounds and molecules is quite similar—both being the combination of more than one atom in a specific spatial arrangement. However, only molecules may feature a single type of atom. For instance, water (H_2O) is considered *both* a compound and a molecule, because it is composed of two different types of atoms—hydrogen and oxygen. In contrast, ozone (O_3) is best referred to as a molecule, but is *not* considered a compound because only oxygen is present.

At this juncture, it would be tempting to say that all compounds could also be defined as molecules (*e.g.*, H_2O, CO_2, SO_2). This is indeed the case for compounds composed of two or more nonmetals, which are commonly denoted as **molecular compounds**. However, this is not accurate if the compound contains a metal and nonmetal. For instance, when the metal sodium combines with the nonmetal chlorine, the compound NaCl is formed. This substance is referred to as an **ionic compound** and should not be designated as a molecule. We will describe more about ions in Section 1.7; however, at this stage, consider ions to be either positively charged or negatively charged species that are held together by their mutual attraction. Hence, the building blocks for these types of compounds are oppositely charged ions instead of neutral atoms. Figure 1.4 provides a summarizing definition scheme for elements, compounds, molecules, and atoms.

Your Turn 1.5 Skill Building Classification of Matter

Use the classification scheme shown in **Figure 1.4** to categorize the following:

a. Your cell phone **b.** Aluminum foil **c.** Red wine
d. Chlorine gas **e.** Stainless steel **f.** Table salt
g. Sugar

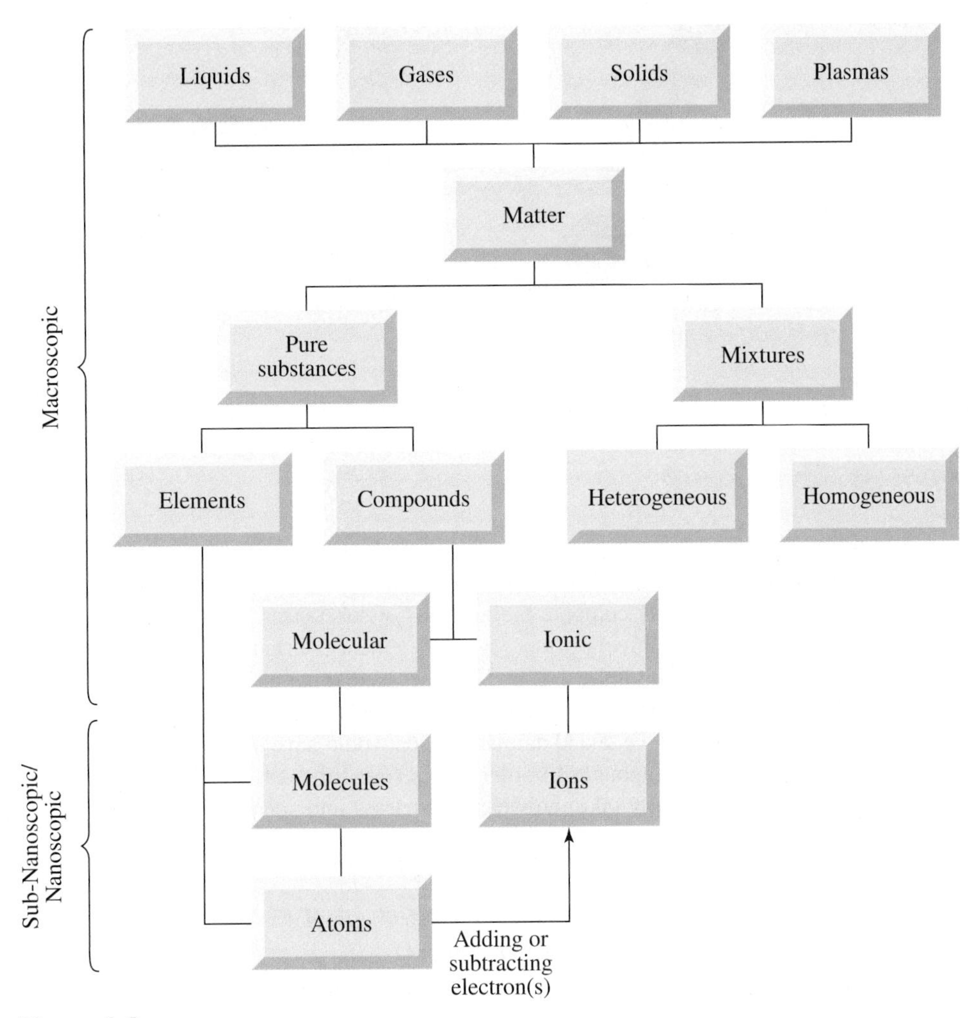

Figure 1.4

An explicit classification scheme for matter, showing the difference between elements and the two types of compounds: ionic and molecular. The formation of ions from atoms will be discussed in Section 1.7.

Although 118 elements exist, over 20 million compounds have been isolated, identified, and characterized. Some are very familiar, naturally occurring substances such as water, table salt, and sucrose (*i.e.*, table sugar). Many known compounds are chemically synthesized by people across our planet. You might be wondering how 20 million compounds could possibly be formed from so few elements. But consider that over 1 million words in the English language can be formed from only 26 letters.

For example, iron and oxygen can combine in a number of different ways. Anyone who has driven extensively on salty roads during the winter has observed the compound Fe_2O_3, or rust, on the metal sides or undercarriages of cars. Pure samples of this compound will contain 69.9% iron and 30.1% oxygen atoms by mass. Thus, 100 grams of rust will always consist of 70 grams of iron atoms and 30 grams of oxygen atoms, which are chemically combined to form this particular compound. These values never vary, no matter where the rust is found. Every compound exhibits a constant characteristic chemical composition.

A small paper clip weighs about a gram.

However, iron atoms may also combine with oxygen atoms to form a different compound, Fe_3O_4, which is referred to as magnetite. A pure sample of Fe_3O_4 contains 72.4% iron atoms and 27.6% oxygen atoms by mass. You might be wondering that if the formula of magnetite contains a 3:4 Fe:O atomic ratio, why isn't the composition expressed as 43% Fe (that is, $\frac{3\text{ Fe atoms}}{7\text{ atoms total}}$) and 57% O (that is, $\frac{4\text{ O atoms}}{7\text{ atoms total}}$)? Similarly, why doesn't the compound Fe_2O_3 above have 40% Fe (that is, $\frac{2\text{ Fe atoms}}{5\text{ atoms total}}$) and 60% O (that is, $\frac{3\text{ O atoms}}{5\text{ atoms total}}$)? If iron and oxygen atoms had the same masses, these calculations would exactly describe the composition of each compound. However, if you compare the weight of a piece of iron relative to a similar-sized piece of aluminum, the iron will be much heavier. Hence, every

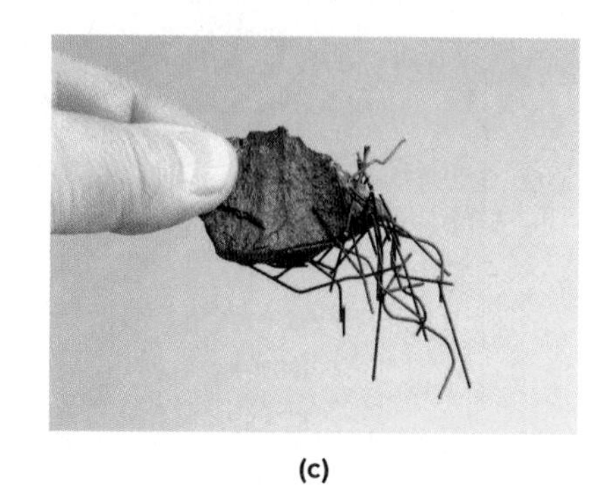

(a) (b) (c)

Figure 1.5

A comparison of the relative magnetism of various iron-containing solids. Shown are pure iron (Fe) filings **(a)**, rust, $Fe_2O_3(s)$ **(b)**, and magnetite, $Fe_3O_4(s)$, picking up bits of iron wire **(c)**.

(a) and (b): © GIPhotoStock/Science Source; (c): © sciencephotos/Alamy Stock Photo

element has a different mass, and so chemists use a number known as a **mole** (commonly abbreviated as *mol*) to easily compare the amounts of substances.

The unit of moles is used a lot in chemistry; many chapters of *CiC* will present applications for this essential concept.

The molar mass, in units of g/mol, of each element is listed below each symbol in the periodic table shown in Figure 1.2. For instance, one mole of iron atoms would have a mass of 55.85 g, but a mole of oxygen atoms would only measure 16.00 g. Accordingly, for compounds that contain Fe and O atoms, the iron would have the greatest contribution to the total mass of the compound. In order to determine the correct mass percent of Fe and O in $Fe_2O_3(s)$, we simply take the weighted masses into account:

$$\frac{\text{mass of two Fe atoms}}{\text{total mass of Fe atoms + O atoms}} = \frac{(55.85 \text{ g/mol Fe})(2)}{(55.85 \text{ g/mol})(2) + (16.00 \text{ g/mol})(3)} \times 100\% = 69.94\% \text{ Fe}$$

$$\frac{\text{mass of three O atoms}}{\text{total mass of Fe atoms + O atoms}} = \frac{(16.00 \text{ g/mol O})(3)}{(55.85 \text{ g/mol})(2) + (16.00 \text{ g/mol})(3)} \times 100\% = 30.06\% \text{ O}$$

Did You Know? Another compound, $FeO(s)$, may be formed between iron and oxygen, and is often referred to as black rust. This compound is also magnetic, but to a much lesser extent than magnetite. It is readily converted to familiar red rust, $Fe_2O_3(s)$, in the presence of air.

Even though the ratio of iron atoms to oxygen atoms is similar for both iron compounds, they will exhibit very different properties. As shown in Figure 1.5, not only are the colors of each iron oxide compound different, but they also vary significantly in their densities, melting points, and magnetic properties. In fact, Fe_3O_4 is the most magnetic naturally occurring mineral on our planet, whereas rust is nonmagnetic. The black stripe across the back of a credit card contains small particles of magnetite that are used to encode your personal identification details, your account number, and the routing number for the banking institution.

Your Turn 1.6 Skill Building Atomic Percentages

For each of the following compounds, calculate their atomic percentages. Report your answers to two decimal places.

a. TiO_2 **b.** MnO_2 **c.** CuO

Your Turn 1.7 You Decide A Mystery Solid

You discover an unknown white solid at the bottom of an aluminum container, and are able to determine the atomic percentages of aluminum, oxygen, and hydrogen using a variety of experimental techniques. How would you decide whether the compound is alumina (Al_2O_3), boehmite ($AlO(OH)$), or gibbsite ($Al(OH)_3$)?

1.4 | What Makes Atoms Tick? Atomic Structure

Atoms, though still indivisible by chemical or mechanical means, contain a nucleus—a minuscule and highly dense center composed of protons and neutrons. Whereas protons are positively charged particles, neutrons are electrically neutral particles. Both species have almost exactly the same mass and together they account for almost all of the mass of an atom. Outside the nucleus are the electrons that define the boundary of the atom. An electron has a mass much smaller than that of a proton or neutron. In addition, electrons carry an electrical charge equal in magnitude to that of a proton, but opposite in sign. Therefore, in any electrically neutral atom, the number of electrons equals the number of protons. The properties of these particles are summarized in Table 1.1. Atoms are held together in part by the attraction of the negative charge of the electrons to the positive charge of the protons in the nucleus.

The number of protons in the nucleus (the **atomic number**) determines the identity of the atom. For example, all hydrogen (H) nuclei contain one proton; hence, hydrogen has an atomic number of 1. Similarly, all helium (He) nuclei contain two protons and have an atomic number of 2. As seen in the periodic table shown in Figure 1.2, the atomic number increases for each successive element in the periodic table. For example, the nucleus of element 92 (U, uranium) contains 92 protons. Since atoms are neutral, they must contain the same number of negatively charged electrons as protons. Accordingly, a H atom will contain one proton and one electron, whereas a He atom will contain two protons and two electrons (Figure 1.6).

The **mass number** refers to the number of protons and neutrons residing in the nucleus. For instance, the mass number of hydrogen is 1, which indicates that there is one proton and no neutrons. However, helium has a mass number of 4, which means there are two protons and two neutrons in the nucleus (Figure 1.6).

In Chapter 6, we will describe atoms with varying numbers of neutrons, known as *isotopes*, which have applications for nuclear energy.

Table 1.1 Properties of Subatomic Particles

Particle	Relative Charge	Relative Mass	Actual Mass, kg
proton	+1	1	1.67×10^{-27}
neutron	0	1	1.67×10^{-27}
electron	−1	0*	9.11×10^{-31}

*This value is zero when rounded to the nearest whole number. The electron does indeed have mass, though very small!

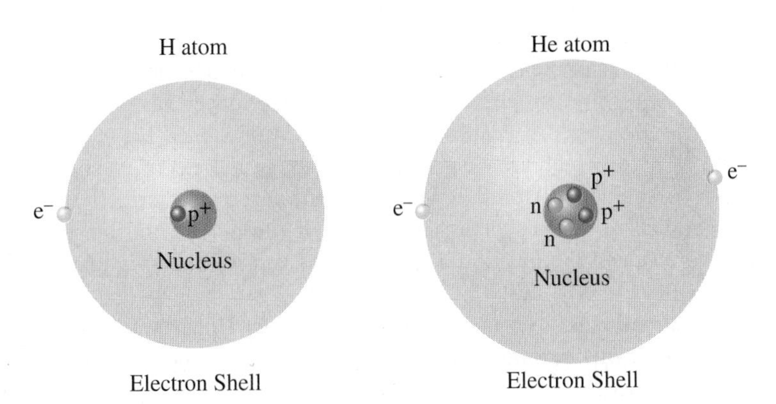

Figure 1.6
Comparison of the atomic structures for hydrogen and helium, showing the location of protons (p^+), neutrons (n), and electrons (e^-).

© demarcomedia/Shutterstock.com

© Malcolm Fife/age fotostock RF

(top) The combination of copper's conductive properties and relatively low price make it the primary material used in electrical wiring. (bottom) Aluminum is used in overhead electrical transmission lines because it is lighter than copper and less expensive.

Your Turn 1.8 Skill Building Atomic Structure

Determine the number of protons and electrons in each of the following atoms:

a. Ga **b.** Sn **c.** Pb **d.** Fe

Determine the number of protons, neutrons, and electrons in each of the following atoms:

a. H (mass number of 2)
b. Cr (mass number of 52)
c. Al (mass number of 27)
d. As (mass number of 75)

The conductivity of a material is dependent on its three-dimensional (3-D) structure and mobility of electrons. Electricity is basically the movement of charge. Hence, the conductivity of a material is related to the ability of electrons to move from one atom to another. The more easily electrons are able to move, the more conductive the material becomes. Metallic solids have an ordered 3-D structure with plentiful electrons that are not tightly bound to distinct metal atoms. This allows for extremely effective electrical conductivity.

What are some materials that you know to be electrically conductive? Copper is conductive. Other metals such as aluminum, silver, and gold are all conductive, too. In fact, among the 100+ elements in the periodic table, the metals are the most electrically conductive. It makes sense that when manufacturers create products requiring the conduction of electricity, they will most often use a metal.

1.5 One-Touch Surfing: How Do Touchscreens Work?

If you have experienced a shock by touching a metal object after sliding your feet across a carpet, you realize that the human body is a conductor of **electricity**—a type of energy that can build up in one place (*static electricity*), or the flow of electrons from one location to another (*current electricity*, as mentioned in the previous section).

If a material enables the flow of electricity, it is said to be electrically *conductive*. Some examples of electrically conductive materials are metals such as copper, silver, and aluminum. On the other hand, a material that does not allow electricity to flow through it is referred to as electrically *insulating*. Materials such as concrete, wood, most plastics, and glass are electrically insulating materials. A third category of materials, known as *semiconductors*, are intermediate between metals and insulators in their transport of electrons. Examples of semiconductors include metalloid elements such as silicon and germanium, as well as compounds formed between elements from Groups 13 and 15 (*e.g.,* GaAs, InSb, etc.), Groups 12 and 16 (*e.g.,* ZnSe, CdS, etc.), or combinations of other Groups.

The flow of electrons via electrical conductivity is analogous to the flow of heat via thermal conductivity. Metals are used for cookware because they effectively transfer heat from the stove to the food in the pot or pan. Likewise, metals are used as conduits for electricity because they transport electrons effectively from one location to another.

Most modern touchscreens contain two layers of narrow, electrically conductive wires placed on top of a glass surface (Figure 1.7). Each layer has parallel wires that form a two-dimensional grid. Because the wires are sandwiched between glass and a protective film (both of which are electrically insulating), the electrical current flowing in the wires is isolated and stored within this multilayer structure. If the screen is touched by a conductive object such as your finger, the uniformity of the stored electrical field is distorted, and the location of the touch is determined by a controller in the processing chip.

Since touchscreens rely on the conductivity of your finger on the surface of the glass screen to create this current flow, the device will not respond to objects that are nonconductive, such as a plastic pen or your gloved finger. However, as noted earlier, one may wear special gloves that contain a conductive pad to allow one to operate touchscreens. The pad comprises a conductive thread that acts as a conduit for electrical charge between your finger and the touchscreen.

Most devices also use computer algorithms to ignore contact points that are relatively small and that can give rise to false signals from objects other than your finger.

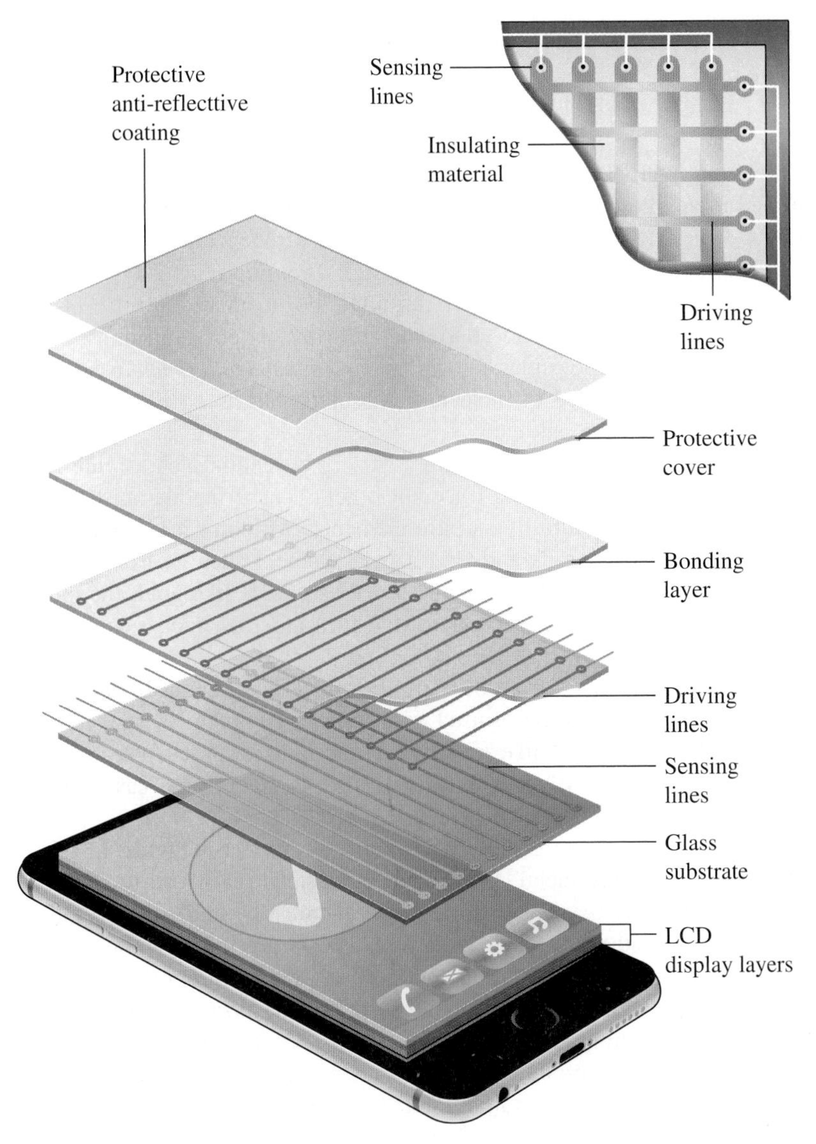

Figure 1.7

Diagram of the structure of a touchscreen.

In multi-touch screens, more than one group of lines will have a significant current flow, telling the device that more than one finger is in contact with the surface. When you swipe your finger across the surface, the location of the greatest current flow moves with your finger. A microprocessor within the device keeps track of the locations over time, allowing you to interact with applications or programs on the touchscreen device.

With an understanding of how touchscreens work, let's now focus on the transformations of matter that are needed to create the materials that make up our electronic devices.

Your Turn 1.9 You Decide Touchscreens Revisited: A Homemade Stylus

Earlier in the chapter, you experimented with a variety of materials to determine whether their contact with a touchscreen would cause a response. Based on the discussion of how touchscreens work and the materials that you observed to cause a touchscreen response, how could you design a stylus using common materials you have at home?

1.6 A Look at the Elements in Their Natural States

The buckminsterfullerene allotrope of carbon, also known as a buckyball, resembles a soccer ball with a spherical shape composed of both six-membered and five-membered rings of carbon atoms.

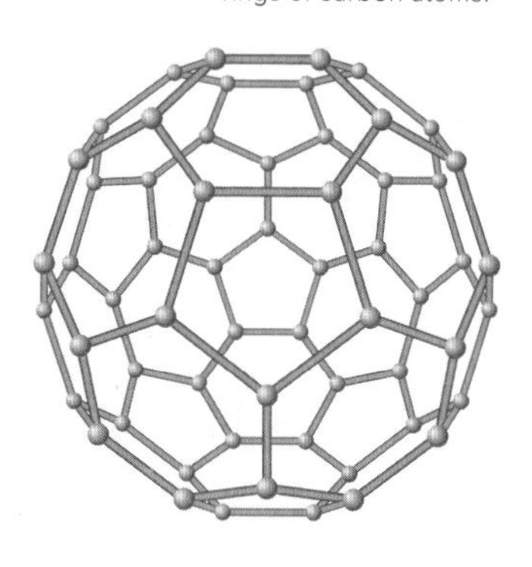

The different ways that atoms are arranged to form the bulk, macroscopic elements are referred to as **allotropes**. Some elements are made of diatomic molecules, or molecules that feature two identical atoms such as hydrogen ($H_2(g)$), nitrogen ($N_2(g)$), oxygen ($O_2(g)$), fluorine ($F_2(g)$), chlorine ($Cl_2(g)$), bromine ($Br_2(l)$), and iodine ($I_2(s)$). Other elements are composed of larger sub-units. For instance, sulfur typically exists as eight-membered rings, S_8, phosphorus as an array of four-atom units, P_4, and one form of carbon as C_{60} molecules known as *buckminsterfullerene*. In contrast, other allotropes are not composed of molecular sub-units, but as an infinite 3-D array of atoms (*e.g.*, graphite, diamond, etc.).

Portable electronic devices contain a variety of metals such as aluminum, copper, nickel, lithium, tin, lead, and traces of others. These metals must be extremely pure, but are not found naturally as pure elements. Wouldn't it be great if we could simply dig into our backyards and find a pure element such as iron, aluminum, or even carbon? With the exception of some precious metals like gold, the metallic elements do not exist in nature in their pure states. Instead, they must be obtained from compounds.

Let's look at one metal from the periodic table—aluminum. In addition to holding our beverages, aluminum (Al) is used extensively in automobiles since it is extremely lightweight and will not rust like iron does. Relevant to this chapter, some portable electronic devices such as iPhones and iPads also use Al for the case, which makes these gadgets extremely lightweight and highly recyclable.

Did You Know? While the atomic symbols are international, element names are not. In countries other than the United States and Canada, such as Great Britain, aluminum is commonly spelled *aluminium*.

Even though aluminum is readily found in Earth's crust (Figure 1.8), it does not exist in nature as the pure metal. Many elements instead react readily with a common gas in our atmosphere, oxygen, to form more chemically stable compounds. Consequently, aluminum and many other metallic elements are found within **rocks**, which are heterogeneous mixtures of solid compounds known as **minerals**. Considering the elemental makeup of Earth's crust, it is no surprise that most rocks are complex mixtures of oxygen-containing minerals designated as *oxides*. The combination of oxygen with silicon in minerals results in *silicates*, whereas aluminum and oxygen minerals are known as *aluminates*. As you might imagine, silicon, aluminum, and oxygen atoms might all form some of the compounds found in a rock formation, which is known as an *aluminosilicate* mineral.

There are also very large reserves of un-mined bauxite in other countries such as Guinea, Australia, Jamaica, and Vietnam. Aluminum may also be produced from clay, oil shales, coal wastes, and other non-bauxitic natural resources throughout the world.

To visualize the structure of rocks, let's consider an image of an aluminum-containing rock formation known as *bauxite*, found mostly in Australia, Guinea, China, Indonesia, and Brazil. In the cross-section image of bauxite shown in Figure 1.9, you can see a variety of solids in a random distribution. This is known as a heterogeneous mixture, because the composition of bauxite is not uniform throughout. By contrast, if you were to pick out one of the individual grains within bauxite, you would get a mineral with a defined composition. Each of these mineral grains is classified as a homogeneous mixture, because it has a defined and unchanging composition throughout its structure.

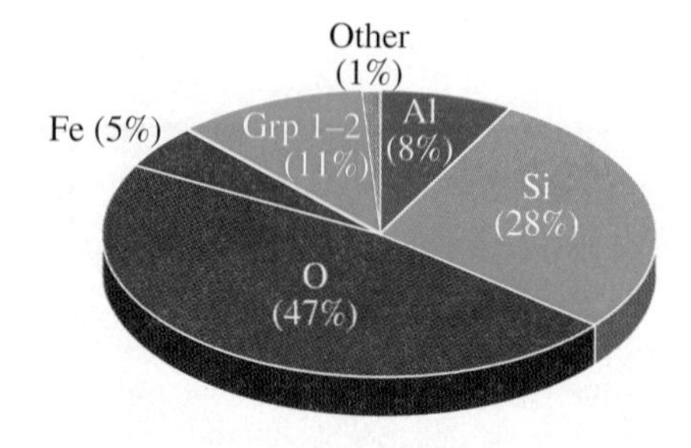

Figure 1.8

Atomic composition of the minerals comprising Earth's crust.

Figure 1.9

Cross-section image of bauxite, illustrating the complex heterogeneous mixture of homogeneous minerals. For instance, in addition to iron and titanium oxides, bauxite contains grains of gibbsite (composition: $Al(OH)_3$) and boehmite (composition: $AlO(OH)$)—each with a defined composition.

© Doug Sherman/Geofile

Your Turn 1.10 Scientific Practices Minerals

In addition to aluminum, other metals such as scandium and yttrium, known as rare earth metals, are also found in cell phones and most electronic devices. What are the natural sources of these metals, and in which parts of the world are these minerals found?

Mixtures are typically described in terms of *relative* (%-based) concentrations of the components. If the mixture comprises solids, one can simply define the concentrations by mass. For instance, consider the example below:

200.0 g of bauxite, composed of 100.0 g gibbsite, 50.5 g boehmite, and 49.5 g iron oxide.

The relative concentrations are:

50.00% gibbsite (*i.e.,* 100.0 g gibbsite ÷ 200.0 g total mixture × 100%)

25.3% boehmite (*i.e.,* 50.5 g boehmite ÷ 200.0 g total mixture × 100%)

24.7% iron oxide (*i.e.,* 49.5 g iron oxide ÷ 200.0 g total mixture × 100%)

$$\text{Total} = 50.00\% + 25.3\% + 24.7\% = 100.0\%$$

(the total must equal 100% if we have taken into account all of the components)

The concentrations above are often designated as percent by weight, wt%, or %(w/w), which indicates that the weight of an individual component is being compared to the total weight of the sample.

Although you can see that ***mass*** and ***weight*** are different entities, we often use them interchangeably. For instance, we say that a person weighs "170 lb" or "77 kg"—units of mass instead of weight. Even though this is not strictly correct, it is an acceptable approximation for practical purposes, as long as we are describing objects on Earth's surface, a place where the gravitational force is relatively constant. http://education.ssc.nasa.gov/mvw_intro_video.asp
Source: NASA

Your Turn 1.11 Skill Building Significant Figures, Part 1

In the calculations above, the answers were reported to a specific number of *significant figures,* a description of the uncertainty of a particular measuring device. For instance, a mass of "one gram" that has been determined using a balance with a precision of ±0.01 g should be reported as 1.00 g. If we incorrectly report this mass as 1.0 g, we have understated the precision of the measuring device. In contrast, if we report the mass as 1.000 g, we have overstated its precision since the last decimal place indicates the level of uncertainty in its measurement.

In counting the number of significant figures for a measurement, follow these rules:

1. All non-zero digits are significant. For example, the number 1.55 g would have 3 significant figures.

2. All zeroes embedded between non-zero digits are significant. For instance, 1.003 mL would have 4 significant figures.
3. Trailing zeroes that follow a non-zero digit(s) and decimal point are significant. For example, the number 1.000 g would have 4 significant figures. The number 3.0 mL would have 2 significant figures.
4. Leading zeroes placed ahead of non-zero digits are not significant. For instance, 0.00032 g would only have 2 significant figures. The number 0.00305 mL would have 3 significant figures, since only the leading zeroes are not significant.

For each of the values below, determine the number of significant figures.

a. 100.0 mL **b.** 60.1 g **c.** 0.0001 L **d.** 1.003 g

Your Turn 1.12 Skill Building Significant Figures, Part 2

In reporting values to the proper number of significant figures, it is important whether the measurements are being added/subtracted or multiplied/divided. In reporting an answer that has been calculated from measured values, use the following rules:

1. For *addition and subtraction,* the answer should contain the smallest number of *decimal places* among numbers that are being added or subtracted. For instance, 1.003 g + 0.2 g + 0.001 g = 1.204 g. However, based on the smallest number of decimal places (1 decimal place for 0.2 g), the answer should be reported as 1.2 g.
2. For *multiplication and division,* the answer should contain the smallest number of *significant figures* among numbers that are being multiplied or divided. For example, 1.002 cm × 0.005 cm = 0.0050 cm^2. However, based on the smallest number of significant figures (1 sig. fig. for 0.005 cm), the answer should be reported as 0.005 cm^2, or in scientific notation as we will see later in Section 1.8.

For each of the following, report the answer to the correct number of significant figures. Remember to also include the correct unit for each of the calculations.

a. 5.0 g ÷ 0.31 mL **b.** 15.0 m × 0.003 m **c.** 1.003 g + 0.01 g **d.** 1.000 mL − 0.1 mL

1.7 Chemical Rock-'n-Roll: How Do We Obtain Pure Metals from Complex Rocks?

A rock formation that contains a considerable concentration of a desired metal is known as an *ore*. This term is generally used for rocks that can be practically obtained for mining. Even though there are mineral deposits worth millions of dollars, it may not always be economically feasible to extract the metal due to its remote location and/or exorbitant cost of processing. For instance, mineral deposits worth $100 million in a remote part of northern Canada would simply be called rock instead of ore if it would cost more than $100 million to mine and process the deposit!

It is no small undertaking to convert ore into pure elements, requiring many purification steps. In order to understand the specific reactions involved, we need to understand how electrons may be gained or lost by atoms. In Section 1.4 you discovered that atoms consist of a nucleus (protons and neutrons) that is surrounded by electrons. In chemistry, we are most concerned about the electrons, since they are farthest from the nucleus and may be more easily transferred to a neighboring atom. When an electron is lost by an atom, the atom undergoes **oxidation**. On the other hand, if an atom picks up an electron from a neighboring atom, the atom gaining the electron has undergone **reduction**. An easy way to remember this is to think of the mnemonic OIL RIG—*Oxidation Is Loss; Reduction Is Gain*. As you will see in Chapter 7, the combination of both a reduction and an oxidation process is known as a *redox* reaction. Redox reactions are essential to the operation of batteries.

As you saw in Section 1.1, the elements in Group 18 (noble gases) of the periodic table are the least likely to react with other elements. All other elements try to reach the same number of electrons as Group 18 by adding, removing, or sharing electrons. For now, let's look at how many electrons are involved for various groups.

Since electrons are negatively charged, the addition of electrons to a neutral (zero-charged) atom will generate a negatively charged *ion*:

$$\mathrm{A} + e^- \longrightarrow \mathrm{A}^-$$
$$\mathrm{A} + 2e^- \longrightarrow \mathrm{A}^{2-}$$
$$\text{in general: } \mathrm{A} + ne^- \longrightarrow \mathrm{A}^{n-}$$

Likewise, if you remove an electron from a neutral atom (oxidation), the remaining ion will now become positively charged:

$$\mathrm{A} \longrightarrow \mathrm{A}^+ + e^- \qquad (\textit{i.e.}, \mathrm{A} - e^- \longrightarrow \mathrm{A}^+)$$
$$\mathrm{A} \longrightarrow \mathrm{A}^{2+} + 2e^-$$
$$\text{in general: } \mathrm{A} \longrightarrow \mathrm{A}^{n+} + ne^-$$

If you look at the periodic table in Figure 1.3, you can see that the Group 17 elements are one column (*i.e.*, one electron) away from the stable noble gases. Accordingly, they will likely add a single electron to their atoms to achieve the same number of electrons as their neighboring noble gas. That is why chlorine (Cl), fluorine (F), and others in Group 17 tend to become Cl^-, F^-, etc., because they have added one electron. We would say that they tend to undergo *reduction*—the addition of an electron. Since oxygen, sulfur, and others in Group 16 are *two* columns away from the noble gases, they will add *two* electrons to become O^{2-}, S^{2-}, etc. How many electrons will the nonmetals in Group 15 add? You guessed it—three—to become N^{3-} or P^{3-}, for instance.

Now that we have talked about reduction, let's look at oxidation. If you consider the Group 1 element Li, it would have to add seven electrons to become like its role model, the noble gas neon (Ne)! However, it is highly improbable to add so many electrons during reduction. Instead, atoms in this group can more easily lose one electron to share the electron count of a noble gas. Because the Group 1 elements lose a single electron, they form Li^+, Na^+, etc. Likewise, the Group 2 elements can lose two electrons to become Be^{2+}, Mg^{2+}, etc. It is not so straightforward for the next set of elements, known as the *transition metals*. At the moment, we will skip over Groups 3–12, and focus on Group 13. This group continues the trend, and can lose three electrons to become B^{3+}, Al^{3+}, etc. You might be wondering about the next group, the Group 14 elements. These are midway between the above scenarios, and can either lose four electrons or gain four to achieve the noble gas configuration.

Consider the analogy of the exits nearest you on an airplane. During the pre-flight announcement, the flight attendant always states that your nearest exit may be either in front of or behind you. Likewise, the best route for an atom to reach the electron count of a noble gas may be either to lose one or more electrons, or add electrons.

Returning to bauxite, the various minerals found in the ore are aluminum compounds that contain Al^{3+} ions. We saw earlier that oxygen typically has a charge of O^{2-}. Chemical compounds will generally have a net zero, or neutral charge. In order to determine the ratios of Al and O ions within the compound, we have to consider how they can be added together so that the compound has zero net charge:

The presence of an element or compound in an expression without any charge has a superscript that implies there is a zero or neutral charge; *i.e.*, "Al_xO_y" = $Al_xO_y^0$.

$$\mathrm{x\ Al^{3+} + y\ O^{2-} \longrightarrow Al_xO_y}$$

Because Al_xO_y has to have an overall charge of zero (it is a neutral compound),

$$\mathrm{x = 2 \text{ and } y = 3}\ (\textit{i.e.}, \text{ you have } (2) \times (+3) = +6 \text{ and } (3) \times (-2) = -6)$$
$$\textit{i.e.}, 2\ [\mathrm{Al^{3+}}] \text{ and } 3\ [\mathrm{O^{2-}}]$$
$$\text{or } [\mathrm{Al^{3+}}]_2[\mathrm{O^{2-}}]_3$$
$$= \mathrm{Al_2O_3}$$

Your Turn 1.13 Skill Building Practice with Charges

For each of the following compounds, provide the correct ionic charges.

a. ZnO **b.** $CuCl_2$ **c.** CaS **d.** TiO_2

The compound Al_2O_3 is a solid known as *aluminum oxide*, or *alumina*. The first step in bauxite processing is to convert all of the various aluminum-containing minerals to alumina. In order to obtain pure aluminum metal from alumina, the aluminum ions must be converted into neutral atoms (*i.e.*, Al). For this to happen, the aluminum ions (Al^{3+}) must be *reduced* by adding three electrons via an electric current in a specialized high-temperature reaction cell:

$$Al^{3+} + 3e^- \longrightarrow Al(s) \qquad \text{(reduction half-reaction)}$$

Because the electrons used for reduction must come from somewhere, we need to consider the other reaction involved in the overall redox process. That is, the negatively charged *oxide* ions in alumina are *oxidized*, generating electrons that are used to reduce the aluminum ions:

$$2\,O^{2-} \longrightarrow O_2(g) + 4e^- \qquad \text{(oxidation half-reaction)}$$

Since the reduction and oxidation reactions each only describe half of the overall redox reaction, they are denoted as **half-reactions**. Notice that for each of the reduction and oxidation half-reactions, there is a conservation of charge and mass. That is, both sides of each equation have equal numbers of atoms/ions, and the overall charge of both sides is equal. As you can see above, it is also customary to include subscripts that indicate the phase of the substances involved in the reaction. For instance, the reduction process generates solid aluminum metal, *(s)*, whereas the oxidation process forms oxygen gas, *(g)*.

Producing one ton (2000 lb or 907 kg) of aluminum metal requires 4 tons of dried bauxite, or 2 tons of pure alumina. This same reduction process is not limited to aluminum production, but can be used to obtain almost any metal from its ore. Even though these reactions are complicated to set up and usually require very high temperatures, the fundamental chemistry is quite straightforward. Simply adding something that is negative (electrons) to something that is positive (*e.g.*, Al^{3+} ions in Al_2O_3) will result in neutral metal atoms with no overall charge.

Your Turn 1.14 You Decide Designing Your Own Cell Phone

You have an unlimited budget and are brainstorming ideas about the next design for a cell phone that will revolutionize the world.

a. Describe the properties that are most desirable for your design.
b. Which chemical elements are incorporated in your design and why did you choose them?

1.8 Your Cell Phone Started with a Day at the Beach: From Sand to Silicon

As seen in the previous activity, we often focus on properties such as weight and durability and take for granted the processing speeds of our electronic devices. For instance, simply touching the icon for a weather app instantly displays the temperature and weather conditions for our part of the world. Such rapid computational speeds would not be possible without continual improvement of the heart of any electronic device—the microprocessor, known as the *chip.* All microprocessors, whether they control your laptop or desktop computer, your coffee maker, or your cell phone, contain the element silicon (Si).

One of the most intriguing applications of chemistry occurs when ordinary sand is transformed into the ultra-high-purity silicon that is used in every electronic device on the planet. Analogous to aluminum and most other metals, due to the high concentration of oxygen in Earth's crust (Figure 1.8), silicon doesn't exist in nature as the pure element. Instead, this element is found as a compound containing Si and O atoms, known as *silica* (SiO_2). This is the chemical composition of most types of sand, in which silicon may be thought to exist as a positively charged ion (Si^{4+}) and oxygen as its usual O^{2-} ion:

The structures of molecular compounds such as SiO_2, CO, PCl_5, and others are not actually composed of discrete ions. However, for redox reactions, it is useful to assign formal charges to their atoms, known as their *oxidation states*.

By definition, the atoms within elemental forms (H_2, N_2, Cu, etc.) are given an oxidation state of zero.

$$[Si^{4+}] \text{ and } 2\ [O^{2-}]$$

$$\text{charge balance: } +4 - 4 = 0 \text{ (neutral charge)}$$

(the +4 oxidation state of Si is balanced by the total charge of −4 for the two oxygens)

$$\text{or} \quad [Si^{4+}][O^{2-}]_2 = SiO_2$$

Your Turn 1.15 Skill Building Oxidation States of Molecular Compounds and Elements

For each of the following, denote the oxidation states of all atoms:

a. P_2O_5 **b.** Cl_2 **c.** Zn **d.** CO_2 **e.** SF_6

In order to remove the oxygen and produce pure Si, a procedure similar to the one that produced aluminum from Al_2O_3 may be used. This process consists of passing an electrical current through a high-temperature cell to produce solid Si and O_2 gas. However, it is more common that carbon is first added to sand to reduce the silicon:

$$SiO_2(s) + 2\ C(s) \xrightarrow{>1000°C} Si(s) + 2\ CO(g)$$

In fact, the carbon used for this reaction is similar to the charcoal briquettes used for backyard grilling! This carbon-based reduction process is used as the first purification step for many metals.

Did You Know? Carbon is not able to be employed for aluminum processing. Unfortunately, aluminum is much too reactive, and will form a compound with carbon known as *aluminum carbide*.

The silicon generated by the reaction above is called metallurgical-grade Si with a purity of 95–98%—not yet pure enough to be used for electronics applications. This concentration implies that for every 100 atoms of silicon, there are 2–5 atoms of impurities such as phosphorus (P), boron (B), carbon (C), oxygen (O), and a variety of metals. Even though this seems like a very low amount of impurities, metallurgical-grade Si has too much variation in physical properties to be used in electronic circuits. In fact, in order to be used for electronics applications, the silicon must have a purity of at least 99.9999999%, which is known as 9N (9 nines). Some companies today even produce Si with a purity of 99.9999999999%, or 12N!

Although we could also indicate the total impurity concentration by a percentage, it would be an extremely small number in this context—only 0.0000001% for 9N silicon, or 0.0000000001% for 12N silicon. Instead of using zeros for very small (or large) numbers, it is most convenient to use **scientific notation** (Figure 1.10). This simply consists of moving a decimal point an appropriate number of digits, and indicating this shift either by negative exponents (moving the decimal to the right for small numbers) or positive exponents (moving the decimal to the left for large numbers). The 9N and 12N impurity concentrations listed above can be represented using scientific notation as $1 \times 10^{-7}\%$ and $1 \times 10^{-10}\%$, respectively.

Engineering notation is a special type of scientific notation where the exponents are listed in multiples of 3, such as 10^{-3}, 10^{-6}, 10^{-9}, etc. This is especially useful for converting between metric units; for instance: 1 kilogram = 10^3 g; 1 mg = 1×10^{-3} g, 1 µg = 1×10^{-6} g; 1 ng = 1×10^{-9} g, and so on.

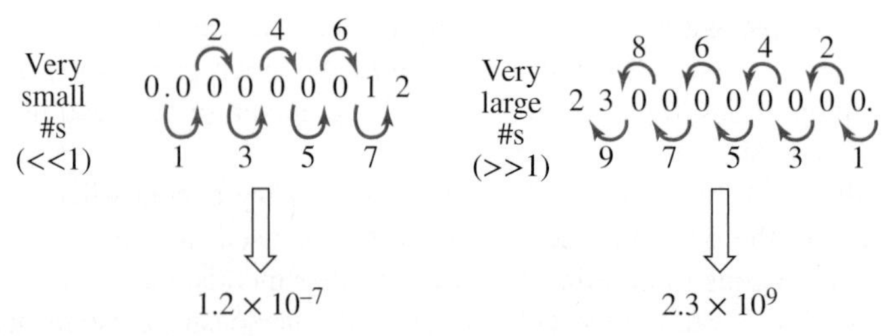

Figure 1.10

Examples of converting numbers into scientific notation for small and large numbers. The number in front of the exponent must be between 1 and 9.99.

Your Turn 1.16 Skill Building Scientific Notation

a. Express the current U.S. national debt and world population in scientific notation.
b. Express your answers from **Your Turn 1.4b** in scientific notation.

More explicitly, this astounding level of purity implies that only one foreign (non-Si) atom may be present for every billion or trillion atoms of silicon. You will see other examples of ppb, ppt, and parts per million (ppm) units throughout this book when describing relatively low concentrations of substances such as air and water pollutants.

As an alternative to using small numbers with many zeros or negative exponents, it is often preferred to designate such low concentrations as one *part per billion* (1 ppb) or *part per trillion* (ppt). An interesting way to visualize these small concentrations is to think of stacking yellow tennis balls (representing Si atoms) from your front step to the surface of the Moon. For silicon of 9N purity, if you replace only six of the yellow balls with red ones, that would represent the maximum number of impurities that are allowed for electronics applications. To put the 12N purity in perspective, the maximum number of impurities would correspond to a single red ball within 170 separate stacks of yellow tennis balls from Earth to the Moon (Figure 1.11)!

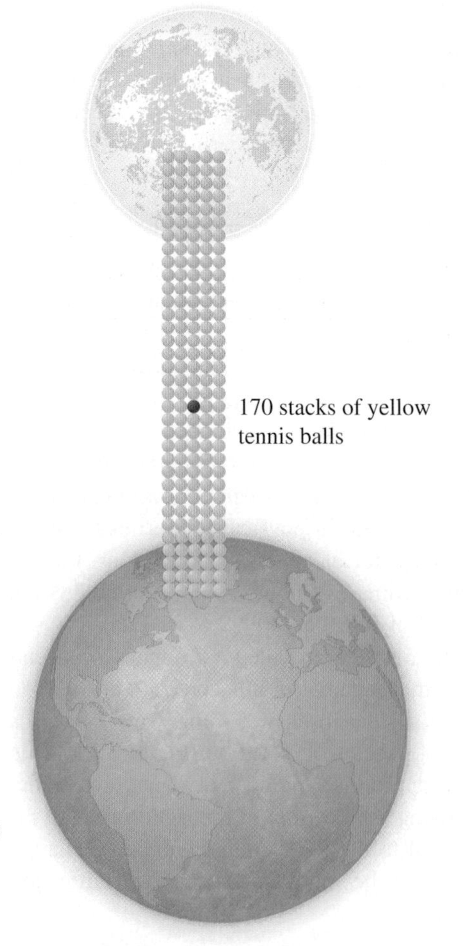

Figure 1.11

An illustration representing the 12N purity of silicon used in some electronics applications.

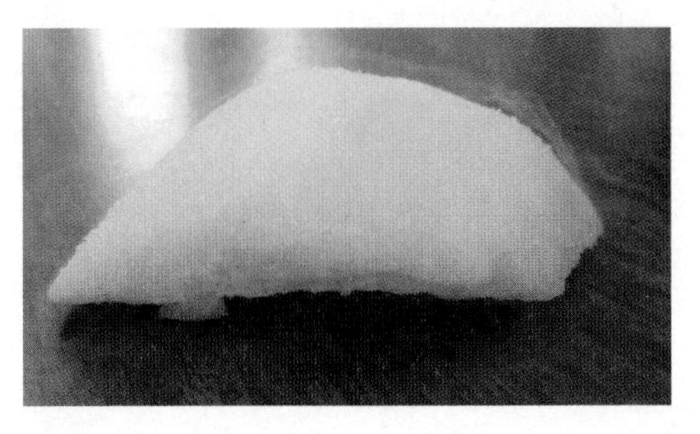

Figure 1.12

The direct transformation of solid carbon dioxide (dry ice—$CO_2(s)$), to gaseous carbon dioxide ($CO_2(g)$), a process known as sublimation. The photo on the right shows what happens when dry ice (–78 °C) is placed into water at room temperature. As the dry ice sublimes, the gaseous carbon dioxide being released is still quite cold. This causes water in the air to condense in the cold carbon dioxide gas as it rises above the water. This is a similar process to cloud formation that occurs in the cold upper atmosphere by condensation of water vapor.

Both: © Bradley D. Fahlman

In order to achieve the desired Si purity of 9N or 12N, the next step is to have the metallurgical-grade silicon react with hydrogen chloride gas, $HCl(g)$, to form a compound that is easily decomposed. You have probably heard about HCl as an aqueous solution—hydrochloric acid, which is sold in hardware stores as *muriatic acid*. Although the natural state of HCl is a gas, it is most often bubbled into water to form a liquid solution that is easier and safer to work with. The reaction of hydrogen chloride gas with silicon is:

$$Si(s) + 3\ HCl(g) \xrightarrow{300°C} SiHCl_3(g) + H_2(g)$$

The final step consists of decomposing $SiHCl_3$ at high temperatures to produce ultra-high-purity Si:

$$2\ SiHCl_3(g) \xrightarrow{1150°C} Si(s) + 2\ HCl(g) + SiCl_4(g)$$

These reactions are performed in very large reactors, and the atmosphere must be free of oxygen, nitrogen, and other gases that could preferentially react with the Si that is formed. In order to save costs, the gaseous HCl and $SiCl_4$ that are formed as side products are recycled and then reused in earlier steps. If you look closely at the reaction, you will also notice that $SiHCl_3$ is a gas that directly forms solid silicon. The process of converting a gas directly to a solid is known as **vapor deposition**. This process is the opposite of **sublimation**, in which a solid is converted into a gas—what happens when dry ice ($CO_2(s)$) is allowed to warm up (Figure 1.12).

Throughout this book, we will denote reactions in the manner shown here, with substances known as **reactants** on the left side of an arrow, and **products** listed on the right side of an arrow. Although there are many types of chemical reactions, each one will have this framework, which simply indicates reactants are transformed into products—each with differing structures and properties.

The various stages of Si processing, from high-purity silicon crystals to the final computer chip, are illustrated in Figure 1.13. This process occurs over hundreds of steps taking place within specialized rooms known as *clean rooms* and utilize high temperatures, ultra-high-purity environments, and a variety of liquid and gaseous chemicals. To prevent contamination by dust particles that would render the chip inoperable, clean rooms make extensive use of stainless steel, sloped surfaces to avoid dust accumulation, and perforated floors and special ceiling tiles to promote air circulation. Prior to entering the clean room, personnel must cover their clothing

Did You Know? The billion-dollar silicon processing facilities used to make computer chips are often referred to as *fabs*.

Figure 1.13

Processing of silicon to produce computer chips. Shown are: **(a)** ultra-pure silicon produced from sand; **(b)** a cylindrical piece of purified silicon known as an *ingot* produced via *zone refining,* and silicon wafers produced by slicing the ingot into thin slices; **(c)** a variety of clean-room processes used to fabricate integrated circuits by workers; **(d)** a final silicon wafer with many computer chips known as dies, and equipment used to test the chips; **(e)** chip packaging equipment and a final computer chip ready to be placed into an electronic device.

with a white "bunny suit" that has non-lint and anti-static properties (Figure 1.14). To enter the clean room, the worker must also walk over a sticky pad and pass through strong bursts of air (referred to as an air shower) to remove dust particles from shoes and clothing. The ratings of clean rooms range from Class 1 to Class 10,000—an indication of the number of particles per cubic meter. As a familiar reference, in uncontrolled environments such as a typical home or office, the particle count is 35 million per cubic meter!

Although computer chips are now comparable in size to a single grain of rice (Figure 1.15), they still contain billions of individual components known as *transistors* that are used to perform the operations needed by our computers and portable electronic devices. Figure 1.16 puts this scale into perspective by sequentially zooming into

Figure 1.14

Technicians work inside a clean room at Sanan Optoelectronics Co., Ltd. in Tianjin, China.

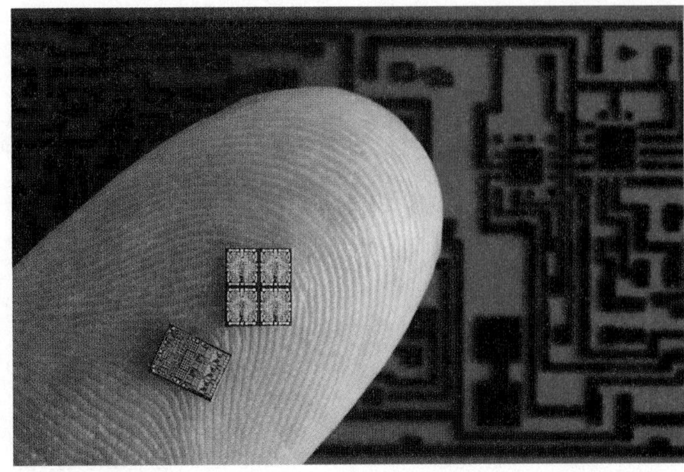

Figure 1.15

Computer processing chips placed onto a single fingertip.

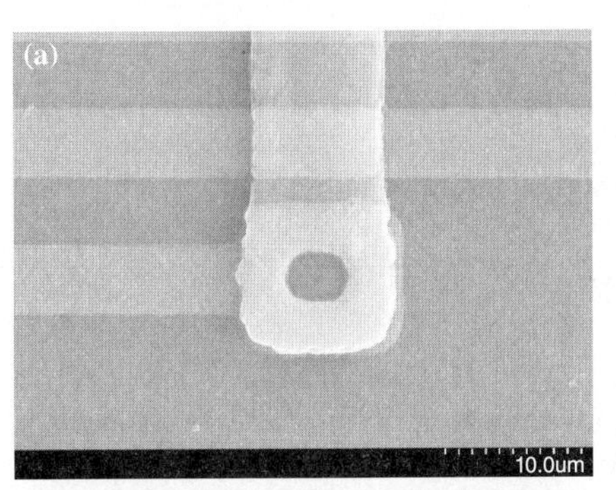

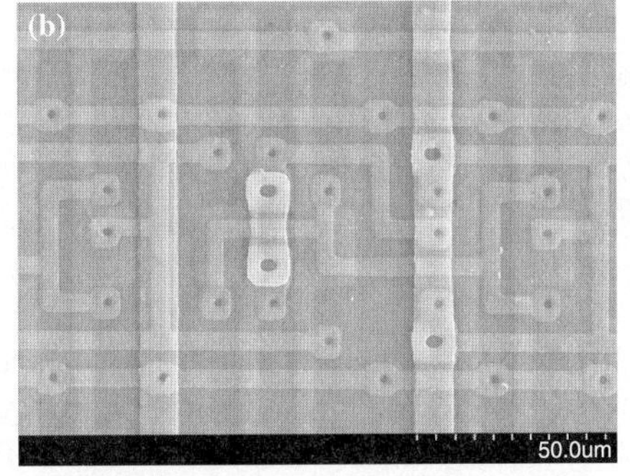

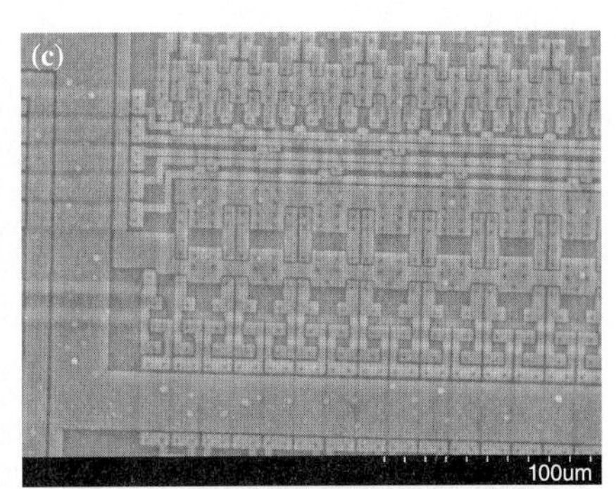

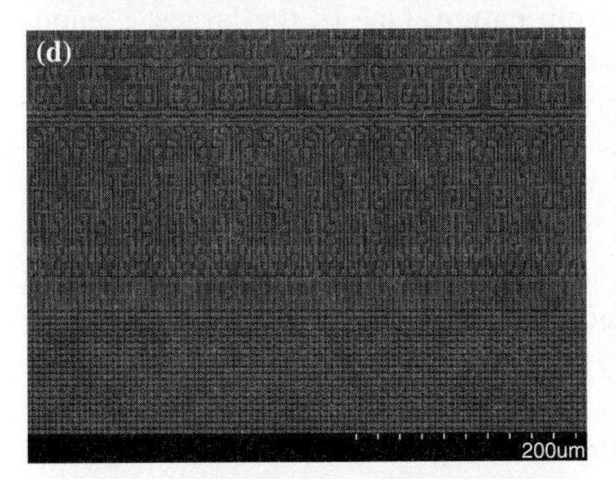

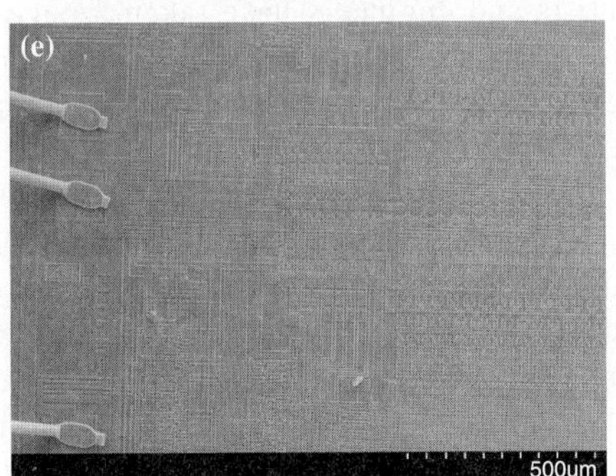

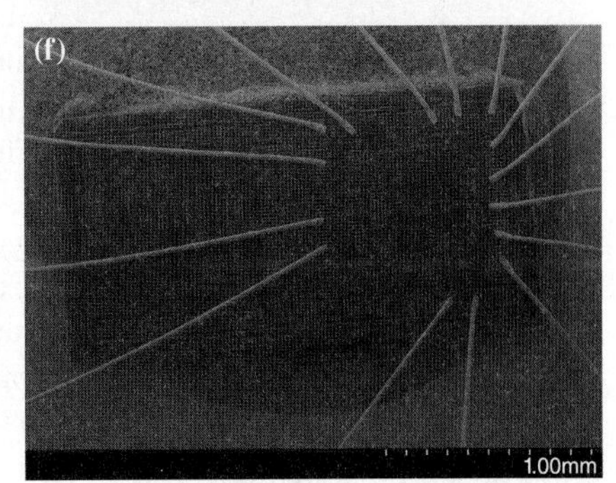

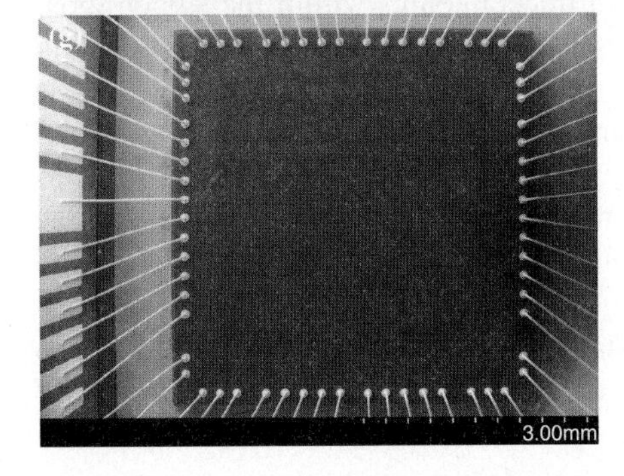

Figure 1.16

A comparative perspective of an integrated circuit. The scale bar in each image roughly corresponds to the: **(a)** diameter of a cloud water droplet; **(b)** diameter of mold spores; **(c)** diameter of a human hair fiber; **(d)** diameter of common beach sand; **(e)** thickness of a human cornea; **(f)** diameter of a pinhead; **(g)** diameter of a pupil a computer chip to reveal the variety of complex micro- and nano-sized architectures. Indeed, the chip that runs a computer or a cell phone is truly an engineering marvel that would not be possible without the numerous chemical reactions that occur during chip processing. Perhaps most astonishing, all of this is possible through the conversion of ordinary sand into silicon!

1.9 More Fun at the Beach: From Sand to Glass

So far, we have talked about the metals and semiconductors used in a portable electronic device. However, those of us who have peered through the cracked screens on our cell phones are very familiar with another component in most portable electronic devices—glass. You might be surprised to learn that sand is not only used to fabricate high-purity silicon, but also the crystal-clear transparent glass that we interact with on mobile devices.

Your Turn 1.17 Scientific Practices Light-Matter Interactions

Using a laser pointer, predict and then determine whether light will be transmitted, reflected, or absorbed during its contact with a/an:

a. Glass window
b. LCD screen
c. Plasma TV screen
d. Concrete sidewalk
e. Asphalt road
f. Ceramic plate
g. Cotton shirt

When you consider the front surface of your mobile device, what kind of properties should it have? Qualities that you may look for include transparency, scratch resistance, and shatter resistance. To find a material with these properties, scientists and engineers have taken a page from nature. One of the largest components of Earth's crust is *silica* (*i.e.*, silicon dioxide, $SiO_2(s)$), which is found in many different forms. These forms vary by composition and structure, with each having different properties.

At the atomic level, silica consists of repeating linkages between silicon and oxygen in a dense, spider web-like structure. There are some naturally formed silicon dioxide structures with very well-ordered structures at the atomic level. This ordered structure is called a **crystal**. Pure crystallized silicon dioxide is known as *quartz*, a clear and colorless mineral that is the primary component of sand (Figure 1.17). When small amounts of other elements are present in the crystal, it can give the mineral some color. For example, the yellow color of citrine and the purple color of amethyst are from different forms of iron that are present in trace amounts within the silicon dioxide crystal (Figure 1.18).

Figure 1.17
Light microscope image of sand taken from Big Talbot Island, Florida, illustrating the individual crystals of silica.

In contrast to well-ordered quartz, the structure of glass is disordered on the atomic level with a random array of silicon and oxygen linkages throughout the solid (Figure 1.19). What a tangled web we weave with glass! Disordered materials such as this are called **amorphous** solids. Although relatively brittle compared to crystalline silicon dioxide, glass has the ability to be molten in a fluid-like state and worked into different shapes for various purposes.

Silica glass is made from heating ordinary sand to a high enough temperature to melt it (Figure 1.20), then cooling the liquid until it hardens to a glass. A variety of

Figure 1.18

Photos of **(a)** citrine, and **(b)** amethyst—forms of quartz with iron impurities that give it varying colors.

(a): © TinaImages/Shutterstock.com; (b): © Alexander Hoffmann/Shutterstock.com

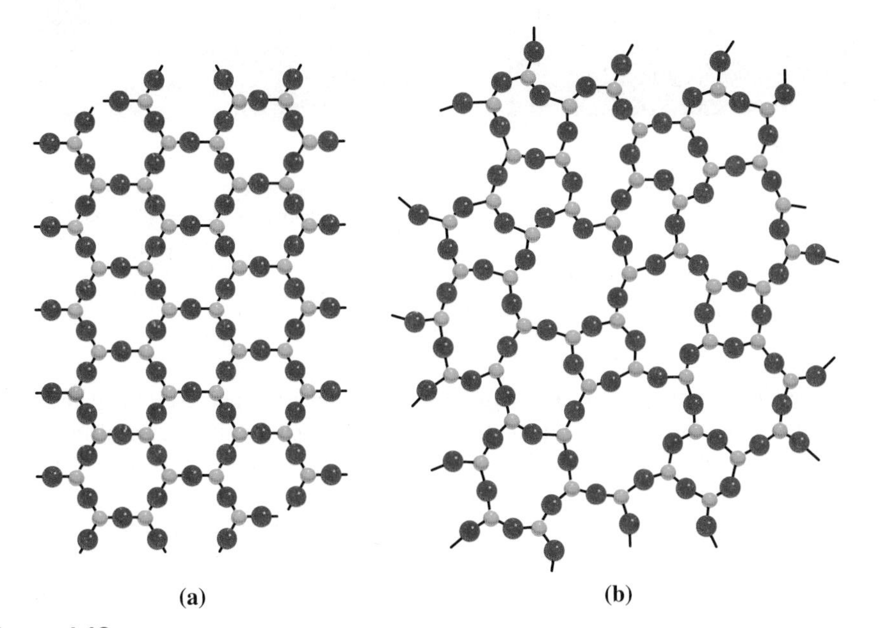

Figure 1.19

Molecular representations of **(a)** crystalline quartz, and **(b)** amorphous glass.

(a-b): © McGraw-Hill Education

Figure 1.20

Molten sand being poured from a ceramic crucible.

Source: Photo Courtesy of the University of Wisconsin-Stout Archives and Area Research Center

additives may be mixed with the silica raw material before melting, which can give the resulting glass a wide range of properties. For instance, the Pyrex™ glass used for cookware not only contains silicon and oxygen atoms, but also boron (B) and traces of other metals such as sodium (Na), aluminum (Al), and potassium (K). The addition of these elements to glass greatly improves its thermal properties by limiting the extent of expansion at high temperatures, or contraction at low temperatures, to reduce its likelihood of cracking.

As you can imagine, temperatures required for melting pure silicon dioxide are very high—in excess of 1700 °C! Adding a type of material called *flux* lowers the melting point by breaking some of the linkages between silicon and oxygen atoms. Common fluxes are salts such as sodium carbonate (Na_2CO_3), calcium carbonate ($CaCO_3$), and magnesium carbonate ($MgCO_3$). In addition to lowering the melting temperature of the glass, these additives make the molten glass less viscous and easier to work into the intricate shapes that you often see in glass artwork.

The lower the viscosity of a liquid, the easier it will flow when being poured from a vessel. For instance, water has much less viscosity than molasses or honey, and will therefore flow much easier.

As you saw in the previous activity, when light shines onto a piece of silicon dioxide, whether it is crystalline quartz or amorphous glass, it mostly passes straight through the material. This means that the material is *transparent*. Whenever there are

Figure 1.21

Stained glass windows of St. Chappelle in Paris, France.

© *John Kirk*

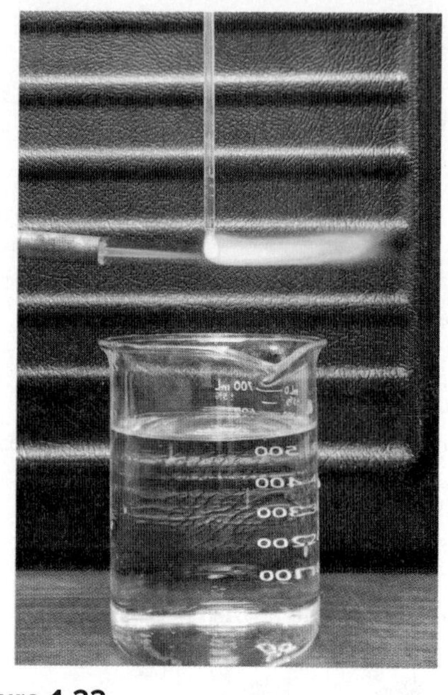

Figure 1.22

Photo showing the formation of a Prince Rupert's drop from quickly cooling a drop of molten glass, demonstrating its high mechanical strength.

© *Bradley D. Fahlman*

Did You Know? Mendeleev's experience in the family glass works fueled his interest in the elements and his search for a periodic arrangement.

A crystal of smoky quartz.

© *Albert Russ/Shutterstock.com*

differences in the structure or the composition at the microscopic level, the path of light through the material is altered, potentially making it opaque. Pure crystalline quartz, having the same structure throughout, certainly is transparent. However, if there are imperfections or impurities present, such as in smoky quartz and milky quartz, the mineral becomes opaque. Although amorphous glass has a variation in its atomic-level structure distributed randomly throughout the entire material, it will still allow light to pass through the material giving it transparency. Of course, over the past several millennia, glassworkers have discovered quite a few additives that give glass some color or make the glass opaque. Beautiful examples of this are the stained glass windows commonly found in Europe's many cathedrals, such as those of St. Chappelle in Paris, France, shown in Figure 1.21. Some of this stained glass is colored red by the inclusion of nanoparticles of gold!

If you ever roughhoused in your family's living room when you were young or played softball close to parked cars, you probably know that glass can be quite fragile. How could this material be useful for a device that has the potential to be dropped and broken? Much research has gone into improving the strength and scratch resistance of glass. In the 17th century, it was discovered that quickly cooling a drop of molten glass in cold water results in a hardened drop that could withstand a hammer blow (Figure 1.22). However, a small amount of force to the long tail of these so-called *Prince Rupert's drops* would cause the entire glass piece to shatter explosively into small fragments. The strength of the material comes from the quick hardening of the outer portion of the glass, freezing it into place while the inside of the drop is still cooling. As we'll discuss later, cooling down an object tends to shrink its size. Because the outer surface of the drop is locked in place, there is a lot of internal stress in the drop as the interior of the glass tries to pull the outer surface inward. Many types of tempered glass have been heat-treated to behave very similarly to these drops. Since heat-strengthened glass tends to shatter into very small pieces when broken, it is often laminated or coated with a thin layer of plastic. For instance, when an automobile windshield is broken, the pieces are quite small and tend to stick together, thus resulting in fewer severe injuries from large pieces of glass (Figure 1.23).

Figure 1.23

Photo of a broken windshield, showing the retainment of smaller glass fragments by the plastic film coating.

© *Esa Hiltula/Alamy RF*

Figure 1.24

Illustration of the pressure felt by each foot of a bottom elephant, if 100,000 four-ton elephants were stacked on top of each other—certainly, an impossible task!

© Bradley D. Fahlman

In addition to heat treatment, chemical treatments can also strengthen glass. This is precisely the technique that has been used by Corning Corp. to fabricate Gorilla Glass™—the tough scratch-resistant glass that is used in a wide variety of mobile device screens, including cell phones, tablets, and laptop computers. This glass is theoretically able to withstand a pressure of 10 GPa, which is equivalent to the pressure exerted by a stack of 100,000 elephants (Figure 1.24)! This incredible strength is achieved by submerging the glass into a bath of molten potassium nitrate (KNO_3). As shown in Figure 1.25, potassium ions from the bath will replace some of the smaller sodium ions close to the surface of the glass. This results in the same types of stresses on the surface of the glass as found in Prince Rupert's drops. Gorilla Glass™ screens are scratch/shatter-resistant when dropped, but are not scratch/shatter-proof.

The unit GPa (gigapascals) refers to 1×10^9 Pa—1,000,000 times greater than kPa and 1,000 times greater than MPa.

Corning and other companies are actively researching the next generation of materials for mobile device screens. These materials include not just amorphous materials like glass, but crystalline materials, too. Sapphire "glass" is one of the potential

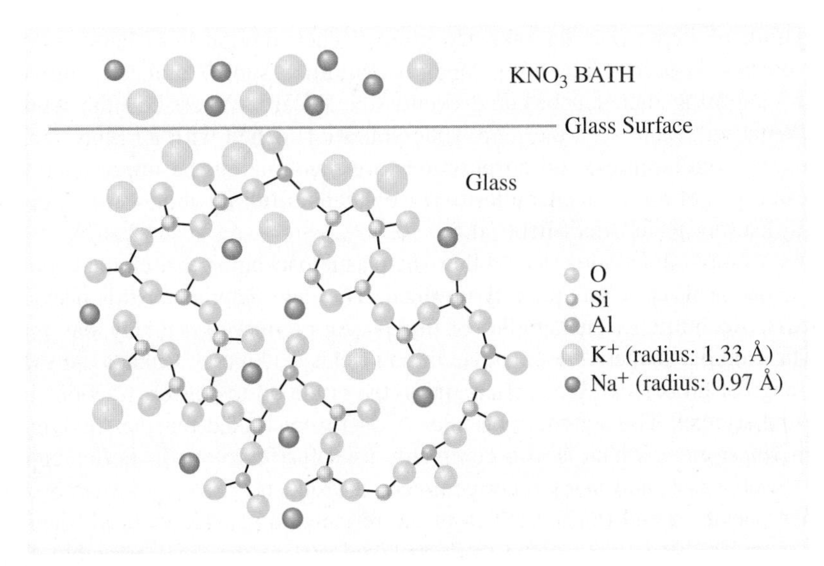

Figure 1.25

Structural schematic of Gorilla Glass™, in which sodium ions are replaced with larger potassium ions.

replacements. Sapphire is a natural gemstone that is harder than quartz. In fact, sapphire is the second-hardest material known after diamond. Sapphire is composed of aluminum oxide, the same substance discussed in Section 1.7. As a comparison, the crystal structure of aluminum oxide is three times harder than Gorilla Glass™.

The **density** of a material refers to its mass/volume ratio. Lightweight materials such as aluminum and plastics that are used for portable electronics will have a relatively low density, whereas building materials such as steel, concrete, and others used for bridges and buildings, generally have comparatively high densities.

Synthetic sapphire requires heating fine aluminum oxide powder at extremely high temperatures—as high as 1800 °C! The crystal growth process is very slow, taking more than two weeks to grow a single large crystal. Once formed, it is cut into its final size and shape by a diamond saw or a laser. While sapphire is extremely hard, it is denser than glass. This means that for the same size and thickness of material, sapphire will weigh more. The weight of sapphire is 67% heavier than an equivalent size piece of glass. The production of synthetic sapphire is also more costly and slower than glass; however, further development of production methods are bringing the cost down. Sapphire is already used for surfaces that see a lot of wear-and-tear such as checkout scanners, airplane windows, and high-end watches. With improvements in production, we may be seeing many more sapphire screens on mobile devices in the very near future.

Your Turn 1.18 Scientific Practices Density

a. An unknown metal was found to have a mass of 424 g. By water displacement, the volume of the solid was determined to be 47.8 mL. Identify the metal based on these known densities: gold, 19.3 g/mL; copper, 8.86 g/mL; bronze, 9.87 g/mL.

b. Why is there an increase in the use of aluminum-based frames in automobiles in place of iron/steel-based frames?

1.10 From Cradle to Grave: The Life Cycle of a Cell Phone

With the increase of active cell phones in the world outpacing the growth of the human population, it is essential that we understand the environmental impacts that their production, use, and disposal have on our planet. The expression *cradle-to-grave* is an approach to analyzing the life cycle of an item, starting with the raw materials from which it came and ending with its ultimate disposal.

Think of items that we take for granted each day such as batteries, plastic water bottles, T-shirts, cleaning supplies, running shoes, and, of course, cell phones—anything that you buy and eventually discard. Where did the item come from? What will happen to the item when you are finished with it? More than ever, individuals, communities, and corporations are recognizing the importance of asking these types of questions. Cradle-to-grave means thinking about *every* step in the process, leading to its final disposal.

As a simple illustration, let's follow the plastic packaging that cradled your shiny new cell phone when it was proudly unveiled. The raw material for this packaging is petroleum. Accordingly, the "cradle" of this plastic product most likely was crude oil somewhere on our planet—for example, the oil fields of Alberta, Canada. At the refinery, a range of processes were carried out on the crude oil to convert fractions into the compound styrene. The styrene molecules (C_8H_8) were linked together (polymerized) to form *polystyrene*, which is also commonly used for Styrofoam™ coffee cups, CD/DVD "jewel" cases, and many other commercial products. The polystyrene packaging was then packaged and transported from the refinery in Canada or the United States (burning jet or diesel fuels—other refinery products) to the final assembly plant in China or Taiwan.

However, what was the fate of this packaging material after you removed the new cell phone? This is not really a cradle-to-grave scenario, but rather cradle-to-your-trash—definitely several steps short of any graveyard. The term *grave* describes wherever an item eventually ends up. Unlike other types of plastics that may be easily recycled, polystyrene is not accepted in most plastic recycling bins. As a result, this type of plastic is the principle component of landfills, urban litter, and marine debris where it begins a presumed 1000-year cycle of slow decomposition into carbon dioxide and water, as well as potentially toxic substances.

Cradle-to-a-grave-somewhere-on-the-planet is a poorly planned scenario for plastic packaging. If the polystyrene waste instead was to serve as the starting material for a new product, or creatively reused in its native state, we then would have a more sustainable situation. **Cradle-to-cradle**, a term that emerged in the 1970s, refers to a responsible use of materials in which the end of the life cycle of one item dovetails with the beginning of the life cycle of another, so that everything is reused rather than disposed of as waste. When considering the most responsible end-use of a product, one should consider the **three pillars of sustainability**:

In Chapter 9, we will examine the main classes of plastics, their applications, and a variety of recycle-and-reuse scenarios.

- Environmental—pollution prevention, natural resource use
- Social—better quality of life for all members of society
- Economic—fair distribution and efficient allocation of resources

As you would expect, the life cycle of a cell phone—an assemblage of many different types of materials from varying parts of the world—would be much more complex than that of its packaging materials. Among the materials comprising a cell phone, 40% are metals, 40% are plastics, and 20% are ceramics and glass. Properties of metals such as electrical and thermal conductivity, durability, and malleability (ability to be bent into complex shapes) are exploited for the circuit board, battery, and touch-sensitive screen. In contrast, the lightweight, inexpensive, and moldable properties of plastics are well suited for the protective case and LCD screen. Ceramics and glass exhibit brittleness and are electrically insulating. Glass is most often used for the outer screen to protect the underlying LCD display, whereas ceramics are used within the circuit board, speaker, and antenna.

The circuit board is the "brain" of the phone, controlling multiple functionalities. The circuit board consists of analog-to-digital (and vice versa) chips, flash memory and ROM (storage) chips, and the microprocessor that controls the keyboard and screen functions. Common metals employed in the circuit board include copper, gold, lead, nickel, zinc, beryllium, tantalum, and others in trace amounts.

So, how much energy is required to fabricate such a complex design? After all, electronic devices are getting smaller/thinner and more efficient (Figure 1.26), which means less energy will be required to produce them, correct? In fact, it's just the opposite,

The term *energy* is a transferable property of matter. While energy may be transferred from one object to another, it cannot be created or destroyed.

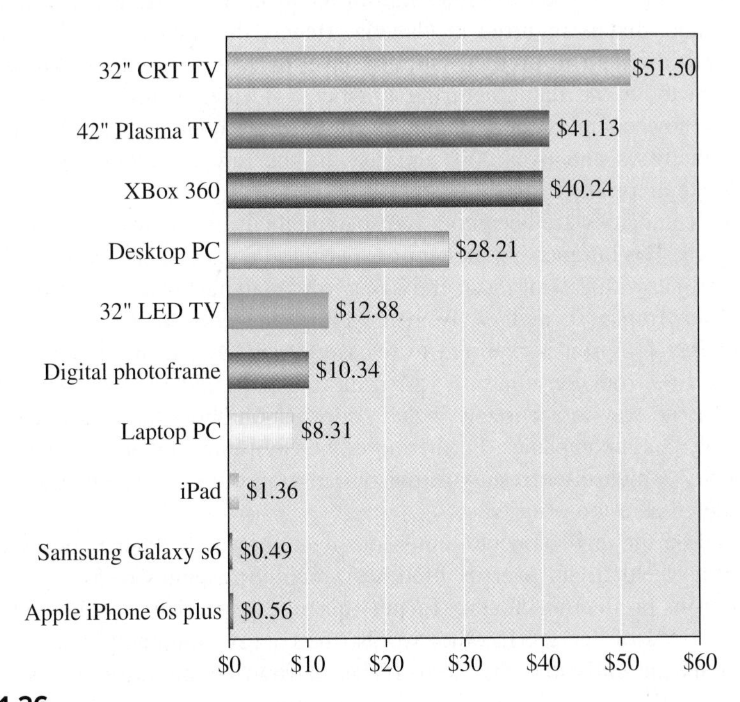

Figure 1.26

Comparison of annual operating costs for various electronic devices. Annual costs are based on an average U.S. residential electricity rate of $0.12/kWh.

with their production from raw materials accounting for more than 90% of the energy consumed over their lifetime! This is not the case with low-tech products such as light bulbs, vacuum cleaners, and ovens that consume much more energy over their lifetimes than was spent for their fabrication. Automobiles used to be in the same "low-tech" category, controlled by analog devices; however, microprocessors now monitor and control every aspect of modern vehicles from the fuel injection system to tailpipe emissions. The increased energy consumption during the production of high-tech devices is primarily because:

- More diverse materials are needed, which requires greater costs for mining and purification, as well as the manufacturing of ceramics and plastics.
- Microprocessors (computer chips) originate from the energy-intensive conversion of sand into ultra-high-purity silicon, and must go through hundreds of complex steps required to fabricate the integrated circuits.
- Complex devices require many hours of design with teams of people using multiple high-speed computers that run continuously for 24 hours a day, 7 days a week.

While it is quite easy to determine how much energy an electronic device consumes during its operation, it is very difficult to calculate the energy used in its fabrication. For instance, a new cell phone begins its production many years before it is released, in the hands of engineers who plan out its features and design the complex architecture and computer chips that it will employ. It's hard to estimate how much energy this initiative will consume, because it involves the electricity to power the buildings and laboratories used for research and development. Administrators and members of the sales force also use electricity in their offices and consume fossil fuels during their extensive travel.

Overlooking these pre-manufacturing activities simplifies the situation somewhat, but we still have the problem of globalization. That is, the silicon employed for the computer chips may be purified in Michigan, the circuit board built in California, the lithium for the battery mined and purified in Chile, and the plastics synthesized in China. Some variability in these locations depends on the company's supply chain, which will vary dramatically between electronics companies. The amount of energy required to mine lithium metal in South America would be very different than what is required in Canada. Hence, this makes a general life-cycle analysis very difficult to predict with any level of accuracy without knowing more information about the manufacturing practices of each materials supplier. As an example of how complex the situation is for a single company, Apple has 18 final assembly facilities and over 200 suppliers of the raw materials and components needed for their product lines.

The unit "kgCO_2e" (kg of equivalent CO_2) refers to the relative emissions of greenhouse gases (carbon dioxide, methane (CH_4), and/or nitrous oxide (N_2O)) per unit of fuel that is consumed.

More companies are becoming transparent about the environmental footprint of their products. For instance, Apple reports that the iPad is responsible for 220 kgCO_2e over its lifetime, with 75% of those emissions from manufacturing, 19% from consumer use, 5% from transport, and 1% from recycling. In contrast, the iPhone 6 with its smaller energy footprint is reported to release 80 kgCO_2e, with 84% generated from production, 10% from consumer use, 5% from transport, and 1% from recycling. However, there is no way to accurately include information about the energy consumption of the supply chain companies. Furthermore, the environmental standards of countries differ greatly, which often results in outsourcing to countries where sustainability is not considered as a top priority.

The unit MJ is a megajoule, or 1×10^6 joules. As you will see in Chapter 5, a joule is the standard unit of energy, and corresponds to the energy required to lift a small apple (with a mass of 100 grams) vertically through one meter of air. A megajoule (MJ) corresponds to the kinetic energy of a one-tonne (1000 kg) vehicle moving at 100 mph (160 km/h).

Based on the environmental emissions data above, an iPhone consumes a total of 152 kWh of electricity over its lifetime, which corresponds to 546 MJ of energy (464 MJ from production alone). To put this in perspective, a gallon of gasoline contains 131 MJ of energy. In other words, the energy contained in four gallons of gasoline (and the emissions that were released from its combustion) was needed to fabricate a single iPhone. While this may not seem too significant, bear in mind that there are currently over 7 billion cell phones in use on the planet, with approximately 2 billion new phones sold every year. Further, there are significantly more tablets, laptops, and other electronic devices that each require more energy to produce than cell phones. In fact, a 27″ iMac computer requires approximately 3,500 MJ of electrical energy (or 480 $kgCO_2e$) to fabricate—7.5 times more energy than an iPhone. Folding in consumer use, transportation, and recycling, the iMac will consume a total of 7,200 MJ of electrical energy (980 $kgCO_2e$) over its lifetime!

The environmental impact numbers we have discussed thus far only deal with the direct fabrication, use, and recycling of electronic devices. However, the full life cycle of a device also includes many other energy-intensive activities that are needed to extract, refine, and transport the raw materials from various parts of the world to the central fabrication facility (Figure 1.27). How much energy does it take to extract lithium metal from an ore in Chile? It depends on how difficult the ore is to reach, and what specific techniques the company uses to break apart the ore, extract the metal, and then refine/purify the metal once it is removed. The same may be said about other components of the phone such as the outer screen. Whereas Samsung doesn't expend much energy in attaching the glass to the case in its final assembly

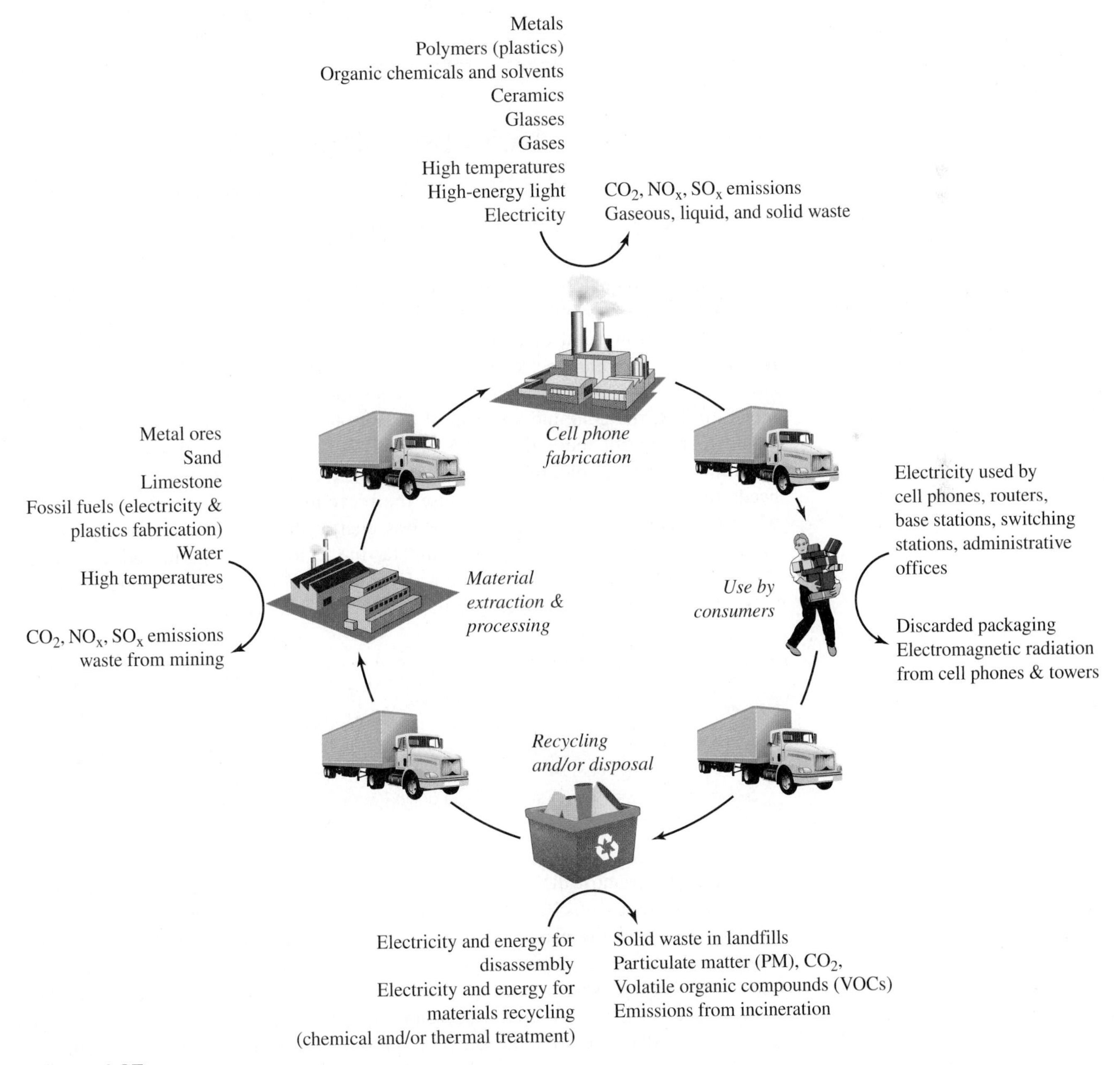

Figure 1.27

The life cycle of a cell phone. The years spent for its design and marketing are not included.

plant, how much energy did the glass manufacturer consume to convert sand into a high-strength glass, and then ship large crates of the material to China for final assembly? The answers to these questions are not easily obtainable, and illustrate just how complicated it is to determine the full environmental impact of a high-tech device in our globalized society.

Your Turn 1.19 Scientific Practices Energy Requirements

Other than charging, what are some other energy requirements of your cell phone?

Your Turn 1.20 Scientific Practices Smartphone Usage

Considering how energy-intensive it is to fabricate cell phones, do you think increasing smartphone usage could cause a decrease in the overall energy consumption in our planet? Explain.

1.11 Howdy Neighbor, May We Borrow a Few Metals? The Importance of Recycling and Protecting Our Supply Chains

Did You Know? In 2015, Apple reported the recovery of over 61 million pounds of steel, aluminum, and other metals from old Mac computers. Of this total, over 2,200 pounds of gold was recovered, which corresponds to around $40 million!

Although approximately 2 billion new cell phones are purchased each year, over 90% of these phones will either collect dust at home or be sent to landfills after their owners grow tired of them. A sparse 3% will be recycled, while 7% will be re-sold. However, did you know that each cell phone contains about 300 mg of silver and 30 mg of gold, which is 50 times more concentrated than its ore in a mine? In fact, the gold and silver used in cell phones sold this year alone are estimated to be worth more than $2.5 billion! Who would have thought our urban landfills are virtual goldmines? Needless to say, the process of recycling electronics needs to be further developed, because the recycling of metals from electronics—while not easy—requires significantly less energy than mining and purifying the metal from its ore. Some companies are starting to focus on this initiative, such as the Brussels-based company Umicore. Even automakers are developing in-house recycling programs for their electronic devices and batteries.

Your Turn 1.21 You Decide Recycling

An aluminum mining company has claimed that it is less expensive and energy intensive to extract Al from ore instead of recycling aluminum cans. Consider the costs and energy sources involved in both processes, and decide whether this claim is valid.

Perhaps the most difficult step in electronics recycling is to remove the metals from the device itself. This process consists of boiling the circuit boards in solvents to remove the plastics and then leaching out the metals with strong acids. However, if one is not careful, groundwater could become contaminated with heavy metals and organic waste, possibly contributing to an increase of cancers and other life-threatening illnesses in the surrounding communities. Unfortunately, these recycling practices are often outsourced to developing countries where environmental regulations are not established and proper safety precautions are not adopted for workers.

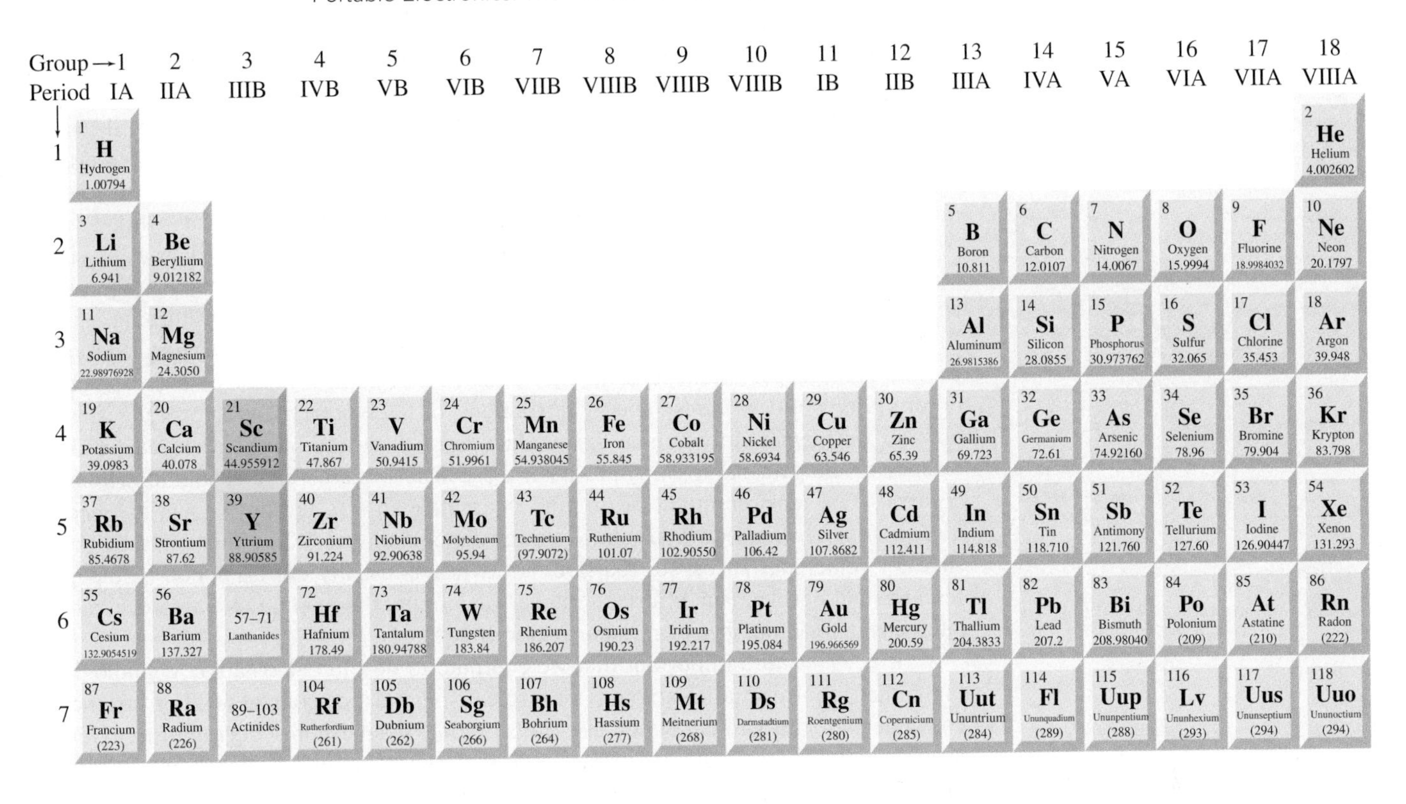

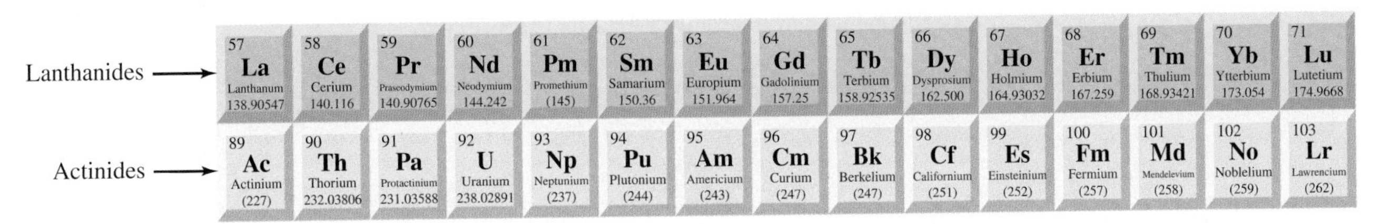

Figure 1.28

Periodic table of the elements. The positions of the rare earth metals are highlighted in blue.

Other than the precious metals of silver, gold, and platinum, another class of metals that are increasingly important for our society are the "rare earth" metals (Figure 1.28). These elements are employed for many applications that we rely on every day such as vehicle catalytic converters and fluorescent lighting, as well as memory chips, rechargeable batteries, magnets, and speakers found inside cell phones and portable electronic devices. The military also uses a variety of rare earths for night-vision goggles, advanced weaponry, GPS equipment, batteries, and advanced electronics.

China is the world's leading producer of rare earth metals (Figure 1.29), but is also an increasing consumer for the finished electronic products. Over 90% of the

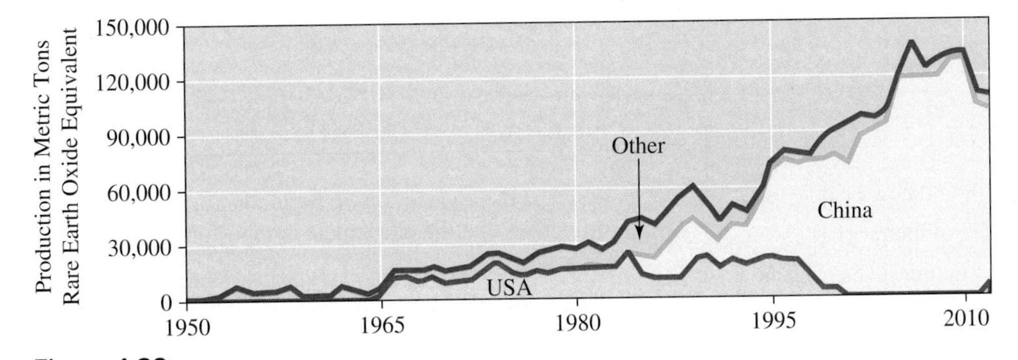

Figure 1.29

Illustration of the dominance of China in the mining of the rare earth metals.

world's supply of rare earth elements are exported from China, who also holds more than 50% of the world's total reserve of these metals.

Indeed, advanced technology can be thought of as a double-edged sword for a society. We are fortunate to enjoy the benefits of faster, lighter, and more powerful portable devices. As a result of this technology, we are more accessible in our businesses, more quickly connected with our social circles, and better able to navigate while away from computers. However, we also open ourselves up to a heightened level of risk associated with the availability of our source materials. Once a society progresses beyond antiquated devices and fully adopts new technology, there is no turning back. But what if a key raw material needed to fabricate our cell phones and electronic devices is no longer available? This may arise from a number of causes such as natural disasters, political unrest, energy restrictions, or trade barriers. Whatever the cause, how do we manage to continue the production of electronic devices needed for our businesses and personal lives? And what if these materials are also essential for national security? One answer might be to find alternative raw materials, known as *earth-abundant materials*, that would have similar functionalities. While this could be possible for some substances, for the rare earths this is usually not an option. Although research and development efforts are underway around the world, there are no suitable alternatives for a number of rare earth elements.

It is even more important that we continue to develop alternative technologies that either require smaller amounts of rare earth metals, or none at all. For instance, consumers in the United States are transitioning from fluorescent lighting, which uses a relatively large amount of rare earths, to more energy-efficient light-emitting diodes (LEDs). Although the components responsible for light generation in LEDs, known as *phosphors*, may also be composed of rare earth oxides, these elements are present in a lesser amount than is required for fluorescent lights.

Conclusions

It is hard for many to imagine life without the use of a smartphone or portable electronic device. The latest weather report, our favorite music, and the answers to life's most difficult questions are now only a touch away. Without the role of chemistry, we would not be able to acquire the elements and compounds that comprise our modern electronic devices. Indeed, the chemical transformations of rocks and minerals into pure Si and metals are required for virtually all aspects of our modern lifestyles.

However, there are limited global reserves for some elements used in portable electronics, such as the rare earths. As the world scurries to find more sources for the rare earths—even looking on the ocean floors—we can more easily acquire these and other low-abundant materials that have already been mined. This can be realized by simply developing low-cost (and environmentally friendly) recycling protocols for the used electronic devices sitting in our drawers at home or those discarded in urban landfills.

The next chapter will describe how our manufacturing and end-use practices for electronics affect the very air we breathe. We will move beyond the clean room, where a trace of oxygen will cause problems with computer chip fabrication, to the real world that needs oxygen in order to sustain human life.

Learning Outcomes

The numbers in parentheses indicate the sections within the chapter where these outcomes were discussed.

Having studied this chapter, you should now be able to:

- classify and compare the states of matter (1.1)
- describe how manipulation of matter influences its properties (1.1)
- define chemistry (1.1)
- describe the connection between macroscopic properties and the particulate composition of matter (1.2, 1.3, 1.4)
- classify metals, nonmetals, and metalloids in terms of electrical conductivity, indicate their location on the periodic table, and predict some components of a portable electronic device they would be most suited for (1.1)

- select an appropriate unit of measurement based on the scale of the object (1.2)
- convert among different units (1.2)
- describe the differences between atoms, molecules, elements, and compounds, and give examples for each (1.2, 1.3)
- describe the differences between ionic and molecular compounds, and calculate the atomic percentages of various compounds (1.3)
- distinguish between mixtures and pure substances, and categorize matter into these two classifications (1.1, 1.6)
- define mixtures, and classify them as either heterogeneous or homogeneous (1.1, 1.6)
- explain the physical and chemical transformations involved in the fabrication and recycling of portable electronic devices (throughout the chapter)
- illustrate the structure of an atom, including the neutron, electron, and proton and compare the relative locations, charges, and masses of the subatomic particles (1.4)
- evaluate how the subatomic particles govern the identity of the elements and their placement in the periodic table (1.4)
- define electrical and thermal conductivity and describe their relationship (1.5)
- describe and diagram how a touchscreen works (1.5)
- describe the composition of Earth's crust, in terms of relative concentrations of its components (1.6)
- determine the correct number of significant figures for measured and calculated values (1.6)
- describe how metals and silicon are separated from ore (1.7)
- define oxidation and reduction, and illustrate how atoms become positively and negatively charged ions (1.7)
- write the formulas of simple ionic compounds (1.7)
- convert numbers between decimal form and scientific notation (1.8).
- measure, calculate, and compare different densities of materials (1.9)
- explain why transparency is important for electronic displays (1.9)
- predict and compare the way light interacts with different types of matter (1.9)
- explain how we can alter materials to change or enhance properties such as durability or transparency (1.9)
- identify and select materials based on their properties (throughout the chapter)
- define energy and describe its role in the fabrication, use, and recycling of portable electronic devices (1.10)
- define the three pillars of sustainability, and relate these principles to the fabrication, use, and recycling of portable electronic devices (1.10)
- distinguish cradle-to-grave from cradle-to-cradle, and predict some environmental, economic, and social impacts of both philosophies (1.10)
- identify sources and possible alternatives for low-abundant materials used in portable electronics (1.11)

Questions

The end-of-chapter questions are grouped in three ways:

- **Emphasizing Essentials** questions give you the opportunity to practice fundamental skills. They are similar to the *Skill Building* exercises in the chapter.
- **Concentrating on Concepts** questions are more difficult and may relate to societal issues. They are similar to the *Scientific Practices* activities in the chapter.
- **Exploring Extensions** questions challenge you to go beyond the information presented in the text. They are similar to the *You Decide* activities in the chapter.

Appendix 5 contains the answers to questions with numbers in blue.

Questions marked with this icon relate to green chemistry.

Emphasizing Essentials

1. In these diagrams, two different types of atoms are represented by color and size. Characterize each sample as an element, a compound, or a mixture. Explain your reasoning.

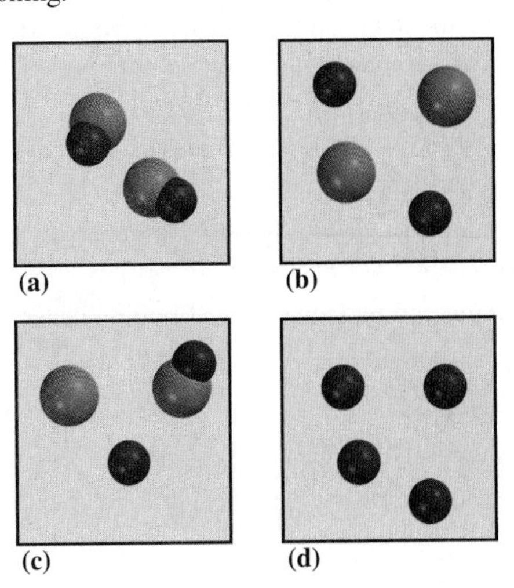

2. From the solids, liquids, or gases that are present in your favorite room or office, list three homogeneous mixtures and three heterogeneous mixtures. Also, provide the names and symbols or chemical formulas of any elements or compounds, respectively.
3. Convert the diameter of the period at the end of this sentence into nanometers.
4. Express each of these numbers in scientific notation.
   a. 1500 m, the distance of a foot race
   b. 0.0000000000958 m, the distance between O and H atoms in a water molecule
   c. 0.0000075 m, the diameter of a red blood cell
5. Express 1 m in terms of cm, μm, and nm. Use proper scientific notation in your answers.
6. Consider this portion of the periodic table and the groups shaded on it.

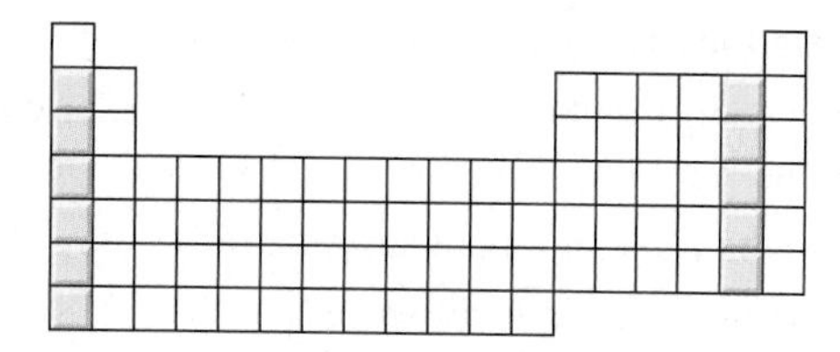

   a. What is the group number for each shaded region?
   b. Name the elements that make up each group.
   c. Give a general characteristic of the elements in each of these groups.
7. Consider the following blank periodic table.

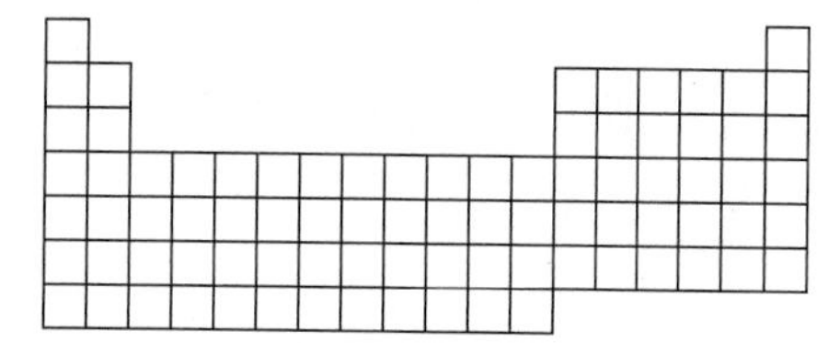

   a. Locate the region of the periodic table in which metals are found.
   b. Common metals include iron, magnesium, aluminum, sodium, potassium, and silver. Give the chemical symbol for each.
   c. Give the name and chemical symbol for five nonmetals (elements that are not in your shaded region).
8. Classify each of these substances as an element, a compound, or a mixture.
   a. a sample of "laughing gas" (dinitrogen monoxide, also called nitrous oxide)
   b. steam coming from a pan of boiling water
   c. a bar of deodorant soap
   d. a sample of copper
   e. a cup of mayonnaise
   f. the helium filling a balloon
9. Draw the structures and describe the properties for two allotropes of sulfur. How are these fabricated?
10. Provide the number of protons, neutrons, and electrons for an aluminum atom with a mass number of 27. How do these numbers change once an Al atom is oxidized to form an Al^{3+} ion? Provide the numbers of protons, neutrons, and electrons for a S^{2-} ion with a mass number of 32.
11. Classify each of the following compounds as molecular or ionic.
    a. KBr b. P_2O_5 c. SO_3
    d. $SrCl_2$ e. XeF_4
12. Calculate the atomic percentages for each of the following compounds.
    a. HfO_2 b. $BeCl_2$ c. $Ti(OH)_4$
    d. FeO e. SiO_2 f. $B(OH)_3$
13. For the following molecules, list the number and type of atoms that each contain.
    a. CO_2 b. H_2S
    c. NO_2 d. SiO_2
14. The density of a mystery solid is 1.14 g/cm^3. Will this float or sink in pure water? Explain.
15. What are the oxidation states of the metals in the following compounds?
    a. CuO b. Al_2O_3
    c. $FeCl_3$ d. Mn_2O_7

Concentrating on Concepts

16. In the text, we illustrated the 12N purity of silicon in terms of colored tennis balls. Provide illustrations of your own for 9N and 12N purities.
17. The processor chips in portable and desktop electronics are composed of tiny switches, known as *transistors*. What are the smallest dimensions of the transistors used in current processors? Relate these dimensions in terms of nm and km.
18. What is meant by "Moore's Law" and is this still valid?
19. Describe how aluminum metal is isolated from its natural ore, as well as the processes involved in its purification.
20. The use of sapphire for the screens of portable electronic devices will soon become prevalent. Compare the physical properties, molecular structures, and fabrication techniques for glass and sapphire.
21. Glass is generally thought to be an electrical insulator. However, is it possible to fabricate "conductive glass"? Explain.
22. Using a molecular perspective, describe the formation of *Prince Rupert's drops* and their violent implosion when the droplet tail is fractured.
23. Describe some components of your cell phone that are in units of cm, mm, μm and nm.
24. List three metals that are currently used in cell phones that have a natural abundance in Earth's crust of < 50 ppm.

25. Can you fabricate high-purity silicon for use in portable electronic devices from plentiful sea sand? Explain.
26. List some waste products generated from the fabrication of high-purity silicon.
27. Critique the accuracy of the following statement: "As cell phones become smaller in size and less expensive, their impact on the environment will increase."
28. Evaluate the current portable electronics industry in terms of the three pillars of sustainability. For each pillar, provide a letter-grade rating and suggest three possibilities for improvement.

Exploring Extensions

29. The crystal structures of many gemstones are based on SiO_2 and Al_2O_3 frameworks. Considering that pure silica and alumina are white solids, describe the origin of the diverse colors exhibited by gemstones.
30. It can be said that "impurities affect the physical properties of most crystalline solids." Explain.
31. "Smart glass" that becomes opaque with a flip of a switch is now being used in businesses and hotel rooms across the world. Describe how glass can transition from transparent to opaque with the passage of electrical current.
32. Describe some procedures that have been used to recycle the metals found in cell phones.
33. Cell phone companies have advertised "superior toxic substance removal" from their products. Which elements have been removed and where were these located within cell phones?
34. Find a precedent for soil and water pollution that arose from the improper recycling of electronic devices. How could these situations have been prevented?
35. Provide a cradle-to-cradle strategy for the recycling of processor "chips" found in portable electronics.
36. Draw a flowchart that illustrates the reactions required to convert SiO_2 sand into high-purity silicon. What happens to the waste products that are generated in each step? How sustainable is this process?
37. In this chapter, we described the reactions required to convert SiO_2 sand into high-purity Si. Compare and contrast the *Czochralski* (CZ) and *float zone* processes to fabricate long cylinders of the high-purity silicon, known as *ingots*. Which technique is more energy intensive? How are these Si ingots used in the fabrication of processor chips?
38. Using Internet resources, perform a life-cycle analysis for your cell phone. Try to be as detailed as possible for two scenarios: cradle-to-grave and cradle-to-cradle.
39. Describe some environmental impacts that are involved during the design, research and development, and marketing phases of cell phones before their ultimate production and release to consumers?
40. Compare and contrast the steps, associated costs, and energy use required to extract aluminum from ore vs. recycling, and rate these practices based on their overall efficiency and sustainability.
41. Consider the image below that shows the increasing global demand for rare earth metals. Calculate the percentage increases in demand for China, Japan/NE Asia, USA, and the rest of the world between 2012 and 2016. Due to rising prices of the rare earths and limited global supplies, more countries are evaluating recycling programs to extract and reuse these elements from existing devices. What devices contain rare-earth metals? Based on the number of these devices sold annually, their average lifetimes, and assuming that 100% of available devices are recycled with 100% recovery of the metals, could the U.S. meet its current demand through recycling efforts alone? Explain.

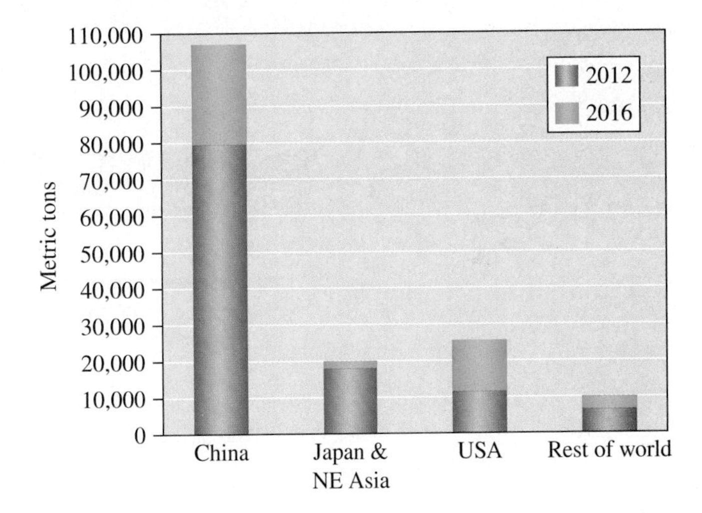

CHAPTER 2

The Air We Breathe

© Milmotion/Getty Images

REFLECTION

The Components of Air

The air we breathe is composed of a wide variety of substances.

a. Identify three indoor and three outdoor sources that emit chemicals into the air around you.

b. Briefly describe how each of these chemicals might affect your health.

The Big Picture

In this chapter, we will answer the following questions:

- What is air? What are the components that make up the air we breathe?
- How does the composition of air change from place to place?
- What are the impurities in air and how did they get there?
- What are the health implications of inhaling certain impurities?
- How do we determine if the air is safe to breathe?
- Are there harmful components in the air you breathe indoors?
- Are there ways we can prevent or limit contaminants from polluting our atmosphere?

Introduction

Not unlike our portable electronic devices discussed in Chapter 1, we take the very air we breathe for granted and trust that it will always be there. However, what makes up this air that we breathe? Why is it necessary for life? Are there components in the air that can harm us? Together with earth, fire, and water, the ancient Greeks named air as a basic element of nature. Hundreds of years later, chemists experimented to learn more about the composition of air, and found that it is made of matter—a gaseous mixture containing elements, molecules, and particles. Today, we can view air in Earth's atmosphere from outer space. And daily, just like the ancients, we can also peer up through the air to catch a glimpse of the twinkling stars at night.

Our atmosphere completely surrounds us, acting as a thin, invisible veil that separates us from outer space. This chapter describes the atmospheric gases that support life on Earth. The next chapter describes the ozone in the stratosphere that protects us from harmful ultraviolet radiation emitted by the Sun. And the chapter that follows describes the greenhouse gases in our atmosphere that protect us from the bitter cold of outer space. Truly, our atmosphere is a resource beyond price.

This chapter also describes how, by our actions, we humans have altered the composition of the atmosphere. Many of these changes have occurred as a result of industrial processing that is needed to fuel our increasing drive for advanced technology (Chapter 1). However, with over 7 billion humans on the planet and counting, it is important to realize that our individual actions—however insignificant they may seem—may have lasting repercussions for our environment. The next activity invites you to think about how our lifestyles, both individually and collectively, can change the air we breathe.

Your Turn 2.1 Scientific Practices Footprints in the Air

Hiking boot treads, asphalt pavement, corn fields—each of these is an example of a "ground print" left by humans because each one alters the lay of the land. Similarly, our activities leave "air prints" that alter the composition of our atmosphere.

Identify three indoor and three outdoor sources that emit chemicals into the air around you. For each of these sources, describe whether they (1) hurt the air quality, (2) improve the air quality, or (3) have some effect, but you don't know what it is.

2.1 | Why Do We Breathe?

Take a breath! Automatically and unconsciously, you do this thousands of times each day. No one has to tell you to breathe. You just do it! Although a doctor or nurse may have encouraged your first breath, nature then took over and you began doing it unconsciously. Even if you were to hold your breath in a moment of fear or suspense, you soon would involuntarily gasp a lungful of that invisible stuff we call air. Indeed, you could survive only minutes without a fresh supply.

Your Turn 2.2 Scientific Practices Take a Breath

What total volume of air do you inhale (and exhale) in a typical day? Figure this out. First, determine how much air you exhale in a single "normal" breath. Then, determine how many breaths you take per minute. Finally, calculate how much air you exhale per day. Describe how you made your estimate, provide your data, and list any factors you believe may have affected the accuracy of your answer.

How much did you estimate that you breathe in a day? Typically, an adult breathes more than 11,000 liters (about 3,000 gallons) of air per day. The value would be even higher if you had spent the day on a bike trail or hiking in the mountains. Are you surprised by how much air you actually breathe?

We breathe in air because it keeps us alive. The air around us contains oxygen, which is essential to our survival. In the process called **respiration**, we take in oxygen in order to help metabolize the foods we eat. Sugar and oxygen are transformed into carbon dioxide and water, and energy is made available to carry out other essential processes in our bodies. With each breath, we inhale air to obtain oxygen, and we exhale carbon dioxide (and small amounts of water) into the atmosphere.

2.2 | Defining the Invisible: What Is Air?

Air is matter, a collection of gases mixed together in various proportions. You know by now that it contains oxygen, but is that all it contains?

Your Turn 2.3 Scientific Practices What's in a Breath?

Take a breath. What are you breathing in? Exhale. What are you breathing out?
Hint: Is there an "ideal" atmosphere in which you might breathe? If so, describe it.

Two types of mixtures were discussed in Section 1.1.

Although you can't tell by looking, the air you are breathing is not a single pure substance, such as only oxygen (O_2). Rather, it is a **mixture**—the physical interaction of two or more pure substances present in variable amounts. In this section, we focus on the pure substances that are in air: nitrogen (N_2), oxygen (O_2), argon (Ar), carbon dioxide (CO_2), and water (H_2O). All are colorless, odorless gases that are invisible to the eye and undetectable to the nose.

These components came to be in our atmosphere across essentially three stages of development. In the first stage, very light and very fast substances, hydrogen (H_2) and helium (He), left Earth's mass shortly after its formation and dissipated quickly into space. In the second stage, during Earth's early development, numerous volcanoes spewed out water (H_2O), ammonia (NH_3), and carbon dioxide (CO_2). Compared to hydrogen and helium, these gases are relatively heavier and slower, and so, they began to settle around Earth. As small photosynthesizing organisms took in carbon dioxide and released oxygen, other animals took in the oxygen and released carbon dioxide. A sort of balance between those two substances began to be reached. Finally, the ammonia in the air was eventually transformed by intense sunlight into nitrogen (N_2) and hydrogen (H_2) gases. The lighter hydrogen drifted into space and the inert nitrogen (N_2) hung around.

The result of this formation today is a multilayered atmosphere (Figure 2.1). The lowest layer, where we live, is called the *troposphere* and accounts for 75% of the mass of the entire atmosphere. The next layer is the *stratosphere*, followed by the *mesosphere* and *thermosphere*. The last (farthest from Earth) is called the *exosphere*. Temperature fluctuations exist across these levels, and overall pressure decreases the farther away from Earth due to the lower amounts of gases at each higher level. The composition of the mixture that we call "air" depends on where you are. Because exhaled air is a slightly different mixture than inhaled air (Table 2.1), we at least temporarily change the air around us when we breathe.

Given that the air is a collection of gases, because of their nature, gas particles can move around and "through" one another. This means that trace amounts of substances can be carried in the air. Often, we can detect these by smell. For example, in areas of France, the scent of lavender may permeate the air outside, whereas in the mountainous

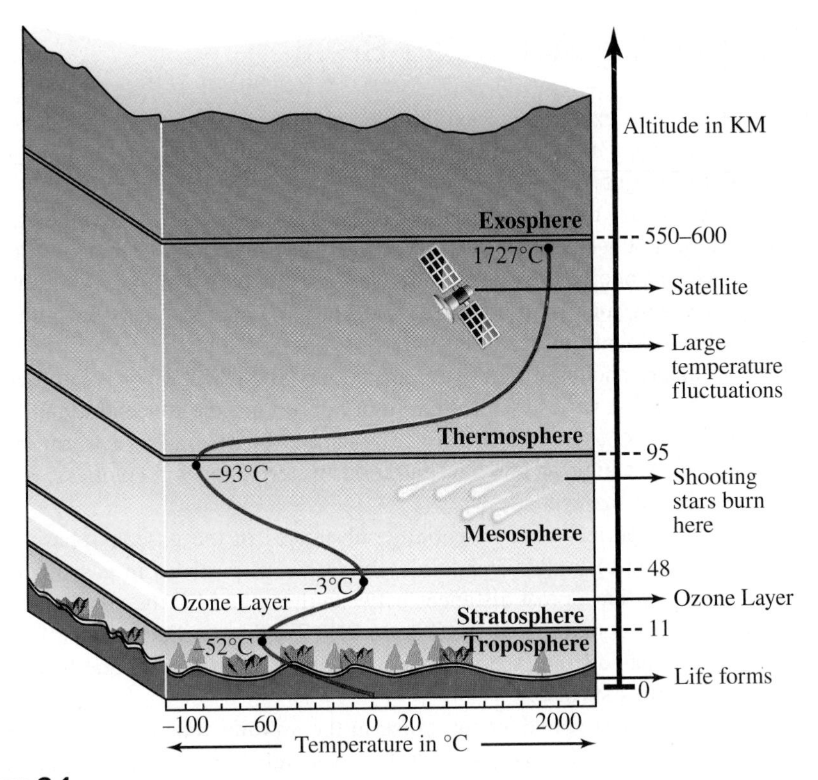

Figure 2.1
Regions of Earth's atmosphere

Table 2.1 Typical Composition of Inhaled and Exhaled Air

Substance	Inhaled Air (%)*	Exhaled Air (%)*
Nitrogen (N_2)	78.0	78.0
Oxygen (O_2)	21.0	16.0
Argon (Ar)	0.9	0.9
Carbon dioxide (CO_2)	0.04	4.0
Water (H_2O)	Variable	Variable

*In unit of percent by volume, %(v/v)

areas of the United States, pine may fill your nostrils. Indoors, the aroma of freshly brewed coffee may beckon you to the kitchen, or outside, the smell of the sea may invite you to the sands of a beach. Of course, other less desirable substances can be released in the air, too, such as those from fuel combustion, landfills, or cow flatulence.

Your Turn 2.4 Scientific Practices Your Nose Knows

The air is different in a pine forest, a bakery, an Italian restaurant, and a dairy barn. Blindfolded, you could smell the difference. Our noses alert us to the fact that air contains trace quantities of many substances.

a. Name three indoor and three outdoor smells that indicate small quantities of chemicals are present in the air.

b. Our noses warn us to avoid certain things. Give three examples of when a smell indicates a hazard or something to be avoided.

2.3 | You Are What You Breathe

Hopefully, the previous exercise has you thinking about how changes to the components of air can cause changes in you—in your mood and even in your health. In other words, when the composition of air changes, its properties also change.

Using a pie chart and a bar graph, Figure 2.3 represents the composition of air. Regardless of how we present the data, the air you breathe is primarily nitrogen and oxygen. More specifically, the composition of air by volume is about 78% nitrogen, 21% oxygen, and 1% other gases. **Percent** *(%)* means "parts per hundred." In this case, the parts are either molecules or atoms.

The percentages shown in Figure 2.3 are for *dry* air. Water vapor is not included, because its concentration varies by location. In dry desert air, the concentration of water vapor can be close to 0%. In contrast, it can reach 5% by volume in a warm tropical rain forest. Whether at high or low concentration, water vapor is a colorless, odorless gas that is invisible to the eye.

But wait—you may ask: "Why can fog and clouds be seen?" These formations are actually not composed of water vapor. Rather, they consist of tiny droplets of liquid water or crystals of ice (Figure 2.2).

So, what we experience are essentially substances in the gaseous phase in our atmosphere. Compared to liquids and solids, the atoms or particles in gaseous phases are more free-flowing, have great distances between them, and fill the volume and take the shape of anything containing them (Figure 2.4). Earth's gravitational pull is what actually keeps these gases from escaping our atmosphere, creating a virtual "container" of gases around Earth's surface.

Nitrogen is the most abundant substance in the air, and constitutes about 78% of what we breathe. This gas is colorless, odorless, and relatively unreactive, passing in and out of our lungs unchanged (Table 2.1). Although nitrogen is essential for life and is part of all living things, its form in the atmosphere is not usable to lifeforms. Most plants and animals obtain their needs from altered or alternative sources of nitrogen in our atmosphere.

Even though oxygen is less abundant than nitrogen in our atmosphere, it plays a key role on our planet. Oxygen is absorbed into our blood via the lungs, and reacts with the foods we eat to release the energy needed to power chemical processes within our bodies. It is necessary for many other chemical reactions as well, including

Figure 2.2

Clouds consist of minuscule droplets of water that remain suspended because of upward air currents. Clouds can weigh millions of pounds.

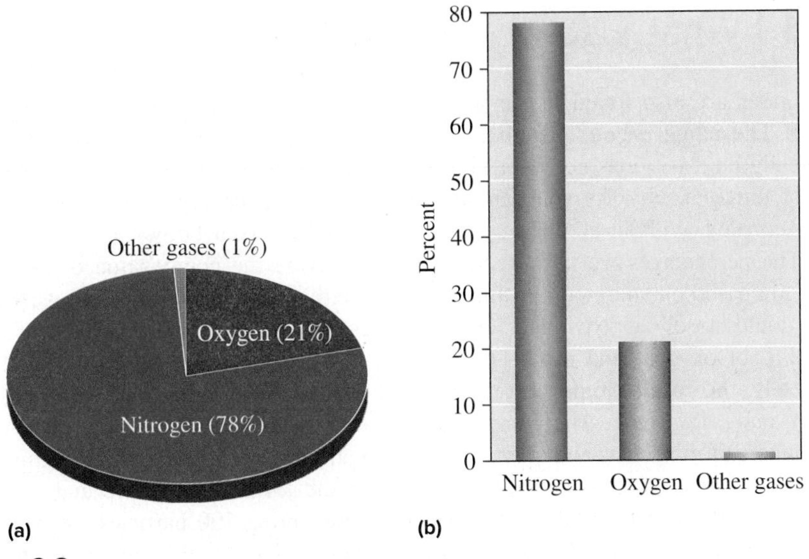

Figure 2.3

The composition of dry air by volume represented as a: **(a)** pie chart, and **(b)** bar graph.

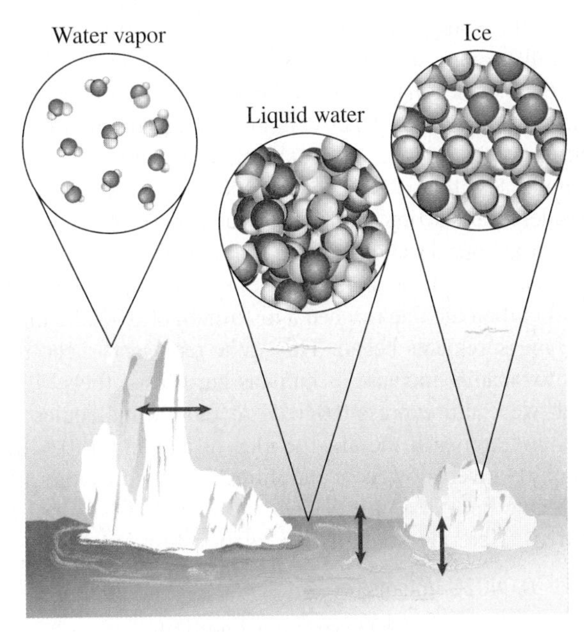

Figure 2.4

Molecular arrangements of water molecules in their different states.

combustion (burning) and oxidation (*e.g.,* rusting). Largely due to its presence in water molecules (H_2O), oxygen is the most abundant element (by mass) in the human body. As we discussed in Chapter 1, its presence in many rocks and minerals also makes it the most abundant element in Earth's crust.

Your Turn 2.5 Scientific Practices More Oxygen ... ?

We live in an atmosphere of 21% oxygen. A match burns in less than a minute, a fireplace consumes a small pine log in about 20 minutes, and we exhale about 15 times a minute. Life on Earth would be *very* different if the oxygen concentration were twice as high. List at least four ways Earth or our lives might be different with such an increased amount of oxygen.

2.4 | What Else Is in the Air?

Other gases are also found in our atmosphere. Argon, for example, is about 0.9% of the air. The name *argon,* meaning "lazy" in Greek, reflects the fact that argon is chemically inert—it does not want to react with another substance (much like the inert form of nitrogen, N_2). As you can see from Table 2.1, the argon that you inhale is simply exhaled, without adding any benefits or harm to your body.

The percentages we have been using to describe the composition of the atmosphere are based on volume—the amount of space that each gas occupies. If we wanted to, we could closely approximate 100 liters (L) of dry air by combining 78 L of nitrogen, 21 L of oxygen, and nearly 1 L of argon. Since the separate gases would mix completely, the result would correspond to a mixture containing 78% nitrogen, 21% oxygen, and ~1% argon. The composition of air can also be represented in terms of the numbers of molecules and atoms present. Equal volumes of gases will contain equal numbers of particles, provided the gases are at the same temperature and pressure. Thus, if you were able to take a sample of air containing 100 particles, 78 would be nitrogen molecules, 21 would be oxygen molecules, and 1 would be an argon atom. In other words, when we say that air is 21% oxygen, we mean that there are 21 molecules of oxygen per 100 of the total molecules and atoms in the air.

Although the sum of percentages of nitrogen, oxygen, and argon appear to make up our atmosphere, there are many other components there as well—substances in trace amounts, but still present nonetheless. Some of these make us feel good; for example, breathing in the molecules that make up the scent of fresh bread may trigger a refreshing smile to some faces. However, other components can be harmful to our health, such as the vehicle emissions in an urban environment. These molecules are smaller in concentration—for example, less than 1 molecule in 100. No matter where you live, each lungful of air you inhale contains tiny amounts of substances other than nitrogen and oxygen. Many are present at concentrations less than 1%, or less than one part per hundred. Such is the case with carbon dioxide, a gas you both inhale and exhale. In our atmosphere, the concentration of carbon dioxide reached a maximum of 0.0402% in 2015, the highest value recorded since measurements began. This value represents a recent exponential rise that now continues to steadily increase as humans burn fossil fuels.

Carbon dioxide is also considered a *greenhouse gas,* which means its properties cause heat from the sun to remain near Earth's surface rather than dissipate back to space. The production and consequences of carbon dioxide and other greenhouse gases will be described in more detail in Chapter 4.

Although we could express 0.0402% as 0.0402 molecules of carbon dioxide per 100 molecules and atoms in the air, the idea of a fraction (0.0402) of a molecule is a bit strange. For relatively low concentrations, it is more convenient to use parts per million (ppm). One ppm is a unit of concentration 10,000 times smaller than 1% (one part per hundred). Here are some useful relationships:

0.0402% means:

0.0402 parts per hundred
0.402 parts per thousand
4.02 parts per ten thousand
40.2 parts per hundred thousand
402 parts per million

Do not misunderstand. A low concentration does not necessarily mean low impact. Small concentrations of some substances can have a great impact, and carbon dioxide is such a substance. Even at 402 ppm in air, it contributes to increasing global temperatures and climate change.

For instance, out of a sample of air containing 1,000,000 molecules and atoms, 402 of them will be carbon dioxide molecules. The carbon dioxide concentration is denoted as 402 ppm.

Your Turn 2.6 You Decide Really One Part per Million?

Some say that a part per million is the same as one second in nearly 12 days. Is this an accurate analogy? How about one step (~2.5 feet) in a 568-mile journey? What about 4 drops (20 drops ~1 mL) of ink in a 55-gallon barrel of water? Check the validity of these analogies, explaining your reasoning. Then, come up with an analogy or two of your own.

Figure 2.5

Photographs taken from the same vantage point on different days in Beijing, China.

Some of the substances found in low concentrations are also considered air pollutants. Although these are found everywhere on Earth, they are more likely to be more concentrated in metropolitan areas. When large numbers of people do certain activities, like cooking meals over open fires or driving combustion engine vehicles, they tend to pollute the air. For example, Figure 2.5 shows two days of varying pollution levels in Beijing, China. Other large cities such as Los Angeles, Mexico City, Mumbai, and Santiago, Chile, often have dirty air as well. Human activities leave "air prints," both indoors and out.

Certain gases contribute to air pollution at the surface of Earth. One of these gases, carbon monoxide (CO), is odorless; others—ozone (O_3), sulfur dioxide (SO_2), and nitrogen dioxide (NO_2)—have characteristic odors. All can be hazardous to your health, even at concentrations well below 1 ppm.

Your Turn 2.7 Skill Building Practice with Parts per Million

a. In some countries, the limit for the average concentration of carbon monoxide in an 8-hour period is set at 9 ppm. Express this amount as a percentage.

b. Exhaled air typically contains about 78% nitrogen. Express this concentration in parts per million.

2.5 | Home Sweet Home: The Troposphere

About 75% of our air, by mass, is in the **troposphere**, the lowest region of the atmosphere in which we live that lies directly above the surface of Earth. *Tropos* is Greek for "turning" or "changing." The troposphere contains the air currents and turbulent storms that turn and mix our air. This is one feature that explains why our atmosphere can have varying concentrations at different locations.

The warmest air in the troposphere usually lies at ground level because the Sun's rays penetrate the air and primarily heat the ground, which, when reflected from the surface, warms the air above it. Cooler air is found higher up, a phenomenon you may have observed if you have hiked or driven to higher elevations. However, *air inversions* occur when cooler air gets trapped beneath warmer air due to weather conditions in an area. Air pollutants can also accumulate in the cooler air of an inversion layer, especially if the layer remains stationary for an extended period. This often occurs in cities surrounded by mountains, such as Salt Lake City, Utah, in the United States (Figure 2.6).

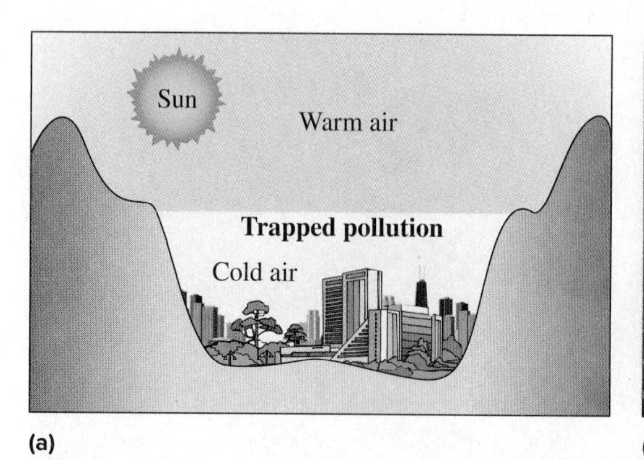

(a)

(b)

Figure 2.6

(a) An air inversion can trap pollution. **(b)** An air inversion, trapping a smoggy layer of air over Salt Lake City, Utah.

(b): © AP Photo/The Salt Lake Tribune, Steve Griffin

One hundred years ago, Earth was home to fewer than 2 billion people. We are now over the 7 billion mark, with the majority of people living in urban regions. This growth in population has been accompanied by a massive growth in both the consumption of resources and the production of waste. The waste that we stash in our atmosphere is called *air pollution.* To better understand the characteristics of air pollutants, we first must understand the chemistry of the atmosphere in general.

2.6 I Can "See" You! Visualizing the Particles in the Air

Chemists typically use three viewpoints to study and understand matter (Figure 2.7). One is the macroscopic view, which consists of viewing matter through the lens of senses, observations, and measurements. Characteristics that can be described in this viewpoint are properties such as color, odor, chemical reactivity, density, etc. However, we can also describe matter using symbols. As we have seen in Chapter 1, these descriptions use letters and numbers within chemical formulas to represent samples of matter (H_2O for water,

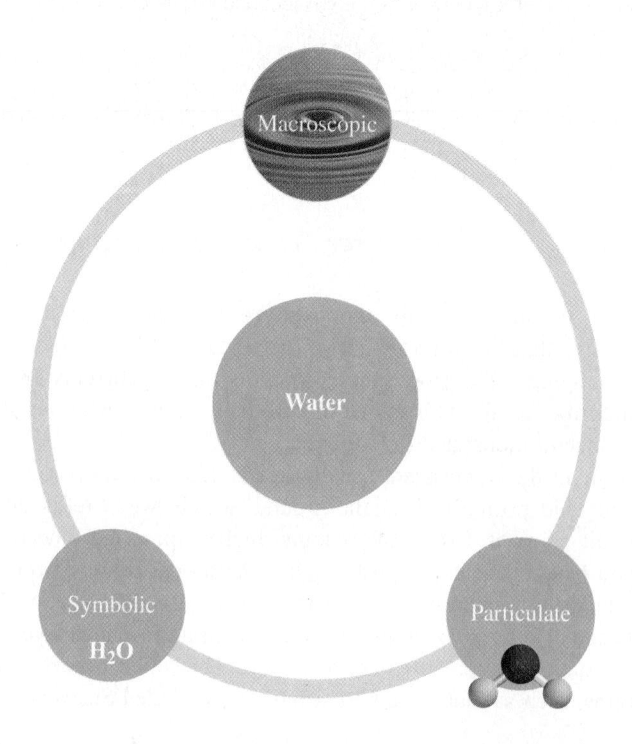

Figure 2.7

Three viewpoints of water.

© Mario7/Shutterstock.com

for example). We can also use symbols within equations to describe various physical relationships of matter (*e.g.*, d = m/V for the relationship of density, mass, and volume, respectively). The third view of matter is the particulate view. In this view, we "see" or imagine what the actual particles, atoms, or molecules look like, and how they might interact. For instance, the water molecule contains two hydrogen atoms (*represented in white*) combined with one oxygen atom (*represented in red*), as seen in Figure 2.7.

We now apply these concepts to the mixture known as air. Some of its components are elemental substances: nitrogen and oxygen exist as diatomic molecules (N_2 and O_2), while argon and helium exist as single, uncombined atoms (Ar and He). Other components, most notably water vapor (H_2O) and carbon dioxide (CO_2), are compounds. In carbon dioxide, the carbon and oxygen atoms are *not* present as separate entities. Rather, the atoms are chemically combined to form a carbon dioxide **molecule**, in which the two atoms are held together by a chemical bond in a certain spatial arrangement (Figure 2.8). More specifically, two oxygen atoms (*red*) are combined with one carbon atom (*black*) to form a carbon dioxide molecule. We'll describe what the double lines between the carbon and oxygen atoms represent later in the text.

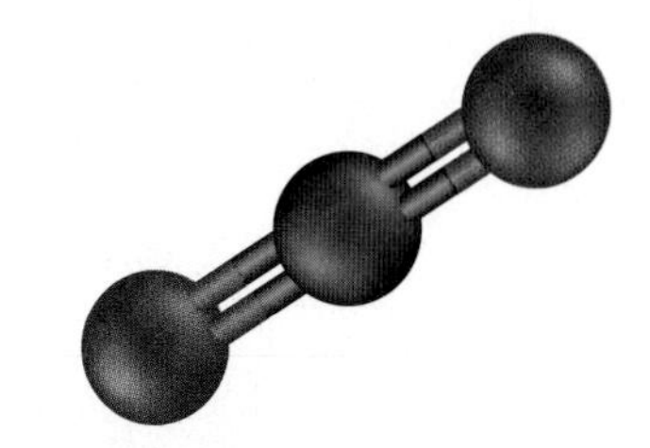

Figure 2.8
Molecular representation of a carbon dioxide (CO_2) molecule, showing a central carbon atom (black) chemically bonded to two oxygen atoms (red).

Atoms are commonly color-coded in molecular structure representations in the following way:

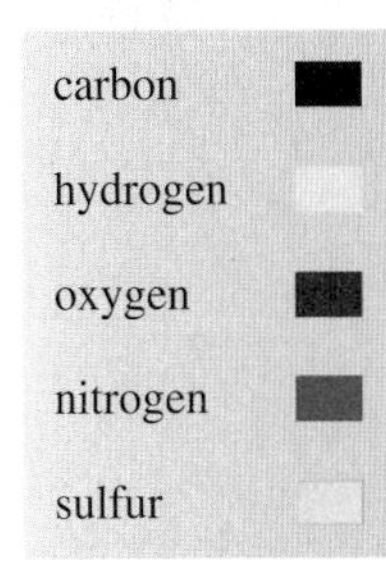

2.7 A Chemical Meet & Greet—Naming Molecular Compounds

If chemical symbols are the alphabet of chemistry, then chemical formulas are the words. The language of chemistry, like any other language, has rules of spelling and syntax. In this section, we help you to "speak chemistry" using chemical formulas and names. As you'll see, each name corresponds uniquely to one chemical formula. However, chemical formulas are *not* unique and may correspond to more than one name.

Right now, we will focus on the chemical names and formulas (known as *nomenclature*) of the compounds relating to the air you breathe. However, rules and practices for naming and symbolizing other categories of chemicals will be shared later in the text. We have already named some of the pure substances found in air, including carbon monoxide, carbon dioxide, sulfur dioxide, ozone, water vapor, and nitrogen dioxide. Although it may not be apparent, this list includes two types of names: systematic and common.

Systematic names for compounds follow a reasonably straightforward set of rules. Here are the rules for compounds composed of two or more nonmetals, such as carbon dioxide (CO_2) and carbon monoxide (CO):

- Name each element in the chemical formula, modifying the name of the second element to end in *-ide*. For example, oxygen becomes *oxide*, and sulfur becomes *sulfide*.
- Use prefixes to indicate the numbers of atoms in the chemical formula (Table 2.2). For example, *di-* means 2, and thus the name carbon *di*oxide means two oxygen atoms for each carbon atom.

Table 2.2 Prefixes Corresponding to the Number of Atoms for Molecular Compounds

Number of Atoms	Prefix	Number of Atoms	Prefix
1	*mono-*	6	*hexa-*
2	*di-*	7	*hepta-*
3	*tri-*	8	*octa-*
4	*tetra-*	9	*nona-*
5	*penta-*	10	*deca-*

- If there is only one atom for the first element in the chemical formula, omit the prefix *mono-*. For example, CO is carbon monoxide, not monocarbon monoxide.

If instead you are writing a chemical formula from a name, remember that the subscript of 1 is not shown in chemical formulas—it is understood. Thus, the chemical formula for carbon dioxide is CO_2 and *not* C_1O_2. Similarly, carbon monoxide is CO and *not* C_1O_1. The next activity gives you a chance to practice.

Did You Know? Two of these have a different common name. NO is also known as nitric oxide and N_2O is known as nitrous oxide (or sometimes just "nitrous"). Nitrous oxide is used as a propellant in ReddiWip® whipped cream and as the source of extra oxygen in modified race cars.

Your Turn 2.8 Skill Building Writing Symbols and Naming Oxides

a. Write chemical formulas for nitrogen monoxide, nitrogen dioxide, dinitrogen monoxide, and dinitrogen tetraoxide.
b. Give the names for SO_2 and SO_3.

Some names ("common names") do not follow a set of rules. Water (H_2O) is one example—why isn't this called dihydrogen monoxide? Although this would make sense and is actually an accurate name for the compound, water was given its name long before anybody knew anything about hydrogen and oxygen. Chemists, being reasonable folks, did not rename water. Likewise, O_3 goes more often by its common name, ozone (trioxygen would be more official), and NH_3 goes by ammonia (its formal name is nitrogen trihydride). Common names cannot be figured out by simply looking at the chemical formula. You have to know them or look them up.

In the next two sections, we explore the connection between air quality and the fuels we burn. Accordingly, we also need to introduce the names of several **hydrocarbons**—compounds composed entirely of hydrogen and carbon atoms. Hydrocarbons follow a very different set of naming rules from the ones just presented.

Methane (CH_4) is the smallest hydrocarbon. Other small hydrocarbons include ethane, propane, and butane. Although methane may not appear to be a systematic name, it indeed is one; *meth-* means 1; in this case, 1 carbon atom. Similarly, *eth-* means 2 carbon atoms, and C_2H_6 is ethane. *Prop-* means 3 carbon atoms, and *but-* means 4. So, propane is C_3H_8, and butane is C_4H_{10}. Just as *mono-, di-, tri-,* and *tetra-* are used to count, so are *meth-, eth-, prop-,* and *but-*. The suffix *-ane* tells us something specific about the ratio of carbon atoms to hydrogen atoms in the molecule; in particular, having the general formula C_xH_{2x+2}. Other ratios exist, and so there are also other suffixes, which we will explain in Chapter 5.

These new prefixes listed in Table 2.3 are very versatile. They can be used not only at the beginning of chemical names, but also within them to represent groups of chains of carbon and hydrogen atoms.

Table 2.3 Names of Hydrocarbons Based on the Number of Carbon Atoms

Chemical Formula	Number of Carbon Atoms	Compound Name	Chemical Formula	Number of Carbon Atoms	Compound Name
CH_4	1	Methane	C_6H_{14}	6	Hexane
C_2H_6	2	Ethane	C_7H_{16}	7	Heptane
C_3H_8	3	Propane	C_8H_{18}	8	Octane
C_4H_{10}	4	Butane	C_9H_{20}	9	Nonane
C_5H_{12}	5	Pentane	$C_{10}H_{22}$	10	Decane

Your Turn 2.9 Skill Building Mother Eats Peanut Butter

Many generations of chemistry students have used the memory aid "mother eats peanut butter" to remember *meth–*, *eth–*, *prop–*, *but–*. Use this, or another memory aid of your choice, to tell how many carbon atoms are in each of these compounds.

a. Ethanol (a component of some beverages and a gasoline additive)
b. Methylene chloride (a component of paint strippers and a possible indoor air pollutant)
c. Propane (the major component in liquid petroleum gas, LPG)

Hydrocarbon molecules can contain 50 carbon atoms or more—each with a distinctive chemical formula and name. At least with the smaller molecules, we can use the prefixes shown in Table 2.4. For example, octane molecules contain 8 carbon atoms and decane contains 10 carbon atoms. You now have the ability to name some simple nonmetal compounds, including short-chain hydrocarbons. Now, let's look at other aspects of the components of air.

2.8 The Dangerous Few: A Look at Air Pollutants

Why might we want to be concerned about the concentrations of carbon monoxide, ozone, sulfur dioxide, nitrogen dioxide, or any other component in the air? Even at concentrations well below 1 ppm, some of these can be hazardous to your health. For instance, consider the following:

- **Carbon monoxide** *(CO)* has earned the nickname "the silent killer" because it has no color, taste, or smell. When you inhale carbon monoxide, it passes into your bloodstream and then interferes with the ability of your hemoglobin to carry oxygen. If you breathe carbon monoxide, at first you may feel dizzy and nauseous or get a headache—symptoms that could easily be mistaken for another illness. Continued exposure, however, can make you extremely ill or kill you. Both automobile exhaust and charcoal fires are sources of carbon monoxide. Propane camping stoves (Figure 2.9) can be another.
- **Ozone** (O_3) has a sharp odor that you may have detected around electric motors or welding equipment. Even at very low concentrations, ozone can reduce your lung function. The symptoms you experience may include chest pain, coughing, sneezing, or lung congestion. Ozone also mottles the leaves of crops and yellows pine needles (Figure 2.10). Here on Earth's surface, ozone is definitely a harmful pollutant. However, at high altitudes, it plays an essential role in screening out harmful ultraviolet (UV) radiation, as you will learn in the next chapter.
- **Sulfur dioxide** (SO_2) has a sharp, unpleasant odor. If you inhale sulfur dioxide, it dissolves in the moist tissue of your lungs to form an acid. The elderly, the young, and individuals with emphysema or asthma are most susceptible to sulfur dioxide poisoning. At present, sulfur dioxide in the air comes primarily from the burning of coal. For example, the 1952 London smog that eventually killed over 10,000 people was in part caused by the SO_2 emissions from coal-fired stoves. The causes of death included acute respiratory distress, heart failure (from preexisting conditions), and asphyxiation. Even those who survived had permanent lung damage, despite their attempts to protect themselves from smog exposure (Figure 2.11).

Figure 2.9

A propane camping stove.

Figure 2.10

The impact of ozone on pine needles.

Did You Know? The acid formed from sulfur dioxide in the air is a diluted form of the same acid found inside your automobile battery, known as *sulfuric acid*. The properties of various acids and bases will be described in more detail in Chapter 8.

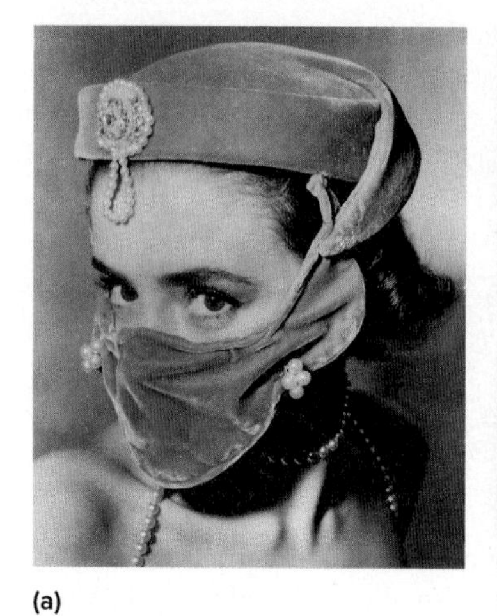

(a)

(b)

Figure 2.11

Masks worn for smog protection **(a)** in 1950's London, and **(b)** in modern-day Beijing.

(a): © Keystone Pictures USA/Alamy Stock Photo; (b): © ChinaFotoPress via Getty Images

Did You Know? The word "smog" (smoke and fog) originates from this infamous event, in which the cool, damp conditions of London, England, caused toxic smoke and fog to combine into a deadly atmosphere.

- **Nitrogen oxides** *(NO_x)*. Nitrogen dioxide (NO_2) has a characteristic brown color and is the primary visible component of urban smog, as shown in Figure 2.6b. Like sulfur dioxide, it can combine with the moist tissue in your lungs to produce an acid. In our atmosphere, nitrogen dioxide is produced from nitrogen monoxide (NO; common name: nitric oxide), another pollutant that is a colorless gas. Nitrogen monoxide is formed from the reaction of N_2 and O_2 in the air from anything that is hot, including vehicle engines and coal-fired power plants. Nitrogen oxides, NO and NO_2, can also form naturally in grain silos and can injure or kill farmers who may inadvertently inhale the gases.
- **Particulate matter** *(PM)* is a complex mixture of tiny solid particles and microscopic liquid droplets, and is the least understood of the air pollutants that we have listed. Particulate matter is classified by size rather than composition, and its size is larger relative to the individual molecules we have described thus far (Figure 2.12). The size of the particles are inversely correlated with the

The size range of particulate matter is larger than the nanomaterials discussed in Chapter 1.

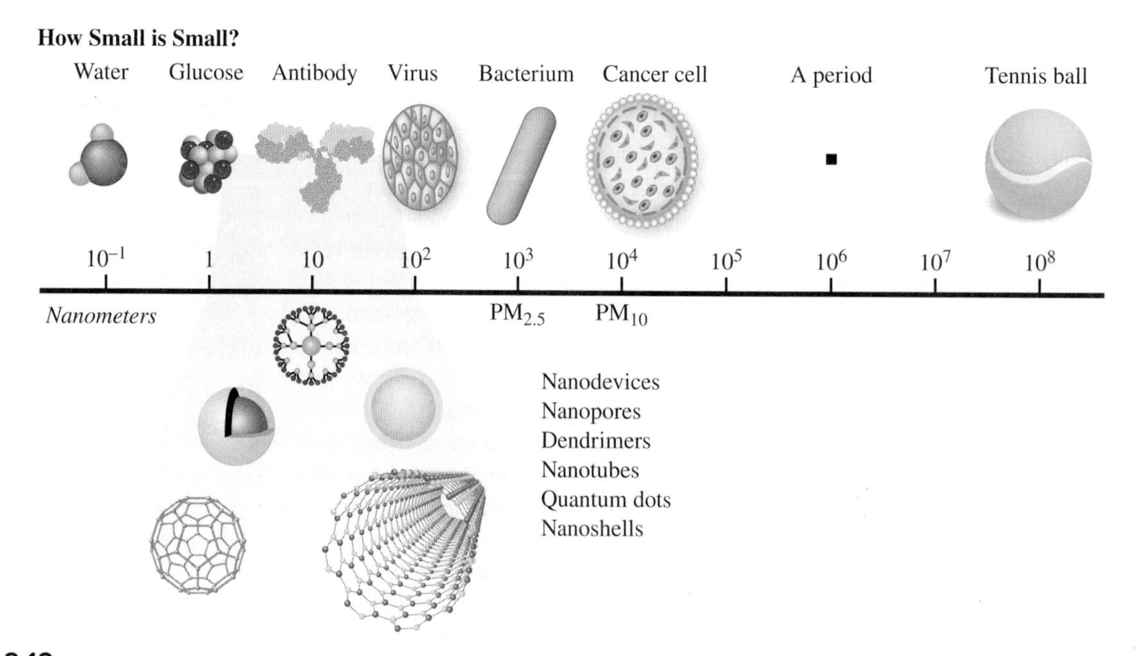

Figure 2.12

Comparison of common length scales of interest to chemistry. A micron (also known as a micrometer, μm) is 1000 times larger than a nanometer (nm).

severity of the health consequence. PM_{10} includes particles with an average diameter of 10 μm (1.0×10^4 nm), a length on the order of 4×10^{-4} (0.0004) inches or one-fifth the width of a human hair. $PM_{2.5}$ is a subset of PM_{10} and includes particles with an average diameter of 2.5 μm (2.5×10^3 nm) or less. These tinier and more deadly particles are sometimes called *fine particulates.* Particulate matter originates from many sources, including vehicle engines, coal-burning power plants, wildfires, and blowing dust. Sometimes, particulate matter is visible as soot or smoke (Figure 2.13), but the two types described here, PM_{10} and $PM_{2.5}$, are too tiny to see. These particles, when inhaled, go deep into your lungs and cause irritation. The smallest particles pass from your lungs into your bloodstream and can cause heart disease.

Figure 2.13

A 2012 wildfire near Denver, Colorado. This fire is releasing particulate matter, some of which is visible as soot.

We end this section with a fact that may surprise you. All of the air pollutants that we just listed can occur naturally! For example, a wildfire (Figure 2.13) produces particulate matter and carbon monoxide, lightning produces ozone and nitrogen oxides, and volcanoes release sulfur dioxide. Because they are the same chemicals, whether released from natural or human sources, the pollutants have the same hazards. Their levels or concentrations in the air are what is significant. What are the risks to your health? We now turn to this topic.

Your Turn 2.10 Scientific Practices What's in a Polluted Breath?

Take another breath. What components of air are you breathing in?

Hint: Exhale. What are you breathing out? Due to the production of portable electronics (discussed in Chapter 1) and other consumer products, our air is not as clean as it once was. How do you think the pollutants created from the portable electronics production process could impact your health?

2.9 Are You Feeling Lucky? Assessing the Risk of Air Pollutants

Risk is a part of living. Although we cannot avoid risk, we still try to minimize it. For example, certain practices are illegal because they carry risks that are judged to be unacceptable. Other activities carry high risks, and we label them as such. For example, cigarette packages carry a warning label regarding lung cancer. Wine bottles carry warnings about birth defects and about operating machinery under the influence of alcohol. The absence of a warning, however, does not guarantee safety. The risk may be too low to label, it may be obvious or unavoidable, or it may be far outweighed by other benefits.

Warnings are just that. They do not mean that somebody *will* be affected. Rather, they report the likelihood of an adverse outcome. Let's say that the odds of dying from a vehicle accident are one in a million for each 30,000 miles traveled. On average, this means that one person out of every million traveling 30,000 miles would die in an accident. Such a prediction is not simply a guess, but the result of **risk assessment**—the process of evaluating scientific data and making predictions in an organized manner about the probabilities of an outcome.

When is it risky to breathe the air? Fortunately, existing air quality standards can offer you guidance. We say *guidance,* because standards are set through a complex interaction of scientists, medical experts, governmental agencies, and politicians. People may not necessarily agree on which standards are reasonable and safe, and standards may change over time as new scientific knowledge is generated (or as political decisions change).

In the United States, national air quality standards were first established in 1970, as a result of the Clean Air Act. If pollutant levels are below these standards, "presumably" the air is healthy to breathe. We say "presumably" because air quality standards usually become stricter over time. If you look worldwide, you will find that air quality regulations vary both in their strictness and in the degree to which they are enforced.

The risks presented by an air pollutant are a function of both **toxicity**, the intrinsic health hazard of a substance, and **exposure**, the amount of the substance encountered. Toxicities are difficult to accurately assess for many reasons, including that it is unethical to run experiments in which people are exposed to harmful substances on purpose. Even if data were available, we would still have to determine the levels of risk that are acceptable for different groups of people. In spite of these complexities, government agencies have succeeded in establishing limits of exposure for the major air pollutants. Table 2.4 shows the National Ambient Air Quality Standards established by the U.S. Environmental Protection Agency (EPA). Here, **ambient air** refers to the air surrounding us, usually meaning the outside air. As our knowledge grows, we modify these standards. For example, in 2006, these standards were made more stringent for $PM_{2.5}$, because additional scientific studies showed that inhaling increased levels of fine particulate matter was damaging to human health. Similarly, in 2015, the standards were lowered for ozone, and in 2010 a new standard for nitrogen dioxide was added.

Table 2.4 U.S. Ambient Air Quality Standards

Pollutant	Standard (ppm)	Approximate Equivalent Concentration ($\mu g/m^3$)
carbon monoxide		
1-h average	35	40,000
8-h average	9	10,000
nitrogen dioxide		
1-h average	0.100	200
Annual average	0.053	100
ozone		
8-h average	0.070	140
particulates		
PM_{10}, 24-h average	–	150
$PM_{2.5}$, 24-h average	–	35
$PM_{2.5}$, annual average	–	15
sulfur dioxide		
1-h average	0.075	210
3-h average	0.50	1,300

Exposure is far more straightforward to assess relative to toxicity, because exposure depends on factors that we can measure more easily. These include:

- *Concentration in the air.* The more toxic the pollutant, the lower its concentration must be set. Concentrations are expressed either as parts per million (ppm) or as micrograms per cubic meter (μg/m^3), as shown in Table 2.4. Earlier, we used the prefix *micro-* with micrometers (μm), meaning a millionth of a meter (10^{-6} m). Similarly, one microgram (μg) is a millionth of a gram (g), or 10^{-6} g.
- *Length of time.* Higher concentrations of a pollutant can be tolerated only briefly. A pollutant may have several standards, each for a different length of time.
- *Rate of breathing.* People breathe at a higher rate during physical activity such as running. If the air quality is poor, reducing activity is one way to reduce exposure.

Suppose you collect an air sample on a city street. An analysis shows that it contains 5000 μg of carbon monoxide (CO) per cubic meter of air. Is this concentration of CO harmful to breathe? We can use Table 2.4 to answer this question. Two standards are reported for carbon monoxide, one for a 1-hour exposure and another for an 8-hour exposure. The 1-hour exposure is set at a higher level because a higher concentration can be tolerated for a short time. Since the analyzed CO concentration of 5000 μg/m^3 is less than both the 1-h and 8-h exposure limits, the air quality is considered safe to breathe.

Table 2.4 also allows us to assess the relative toxicities of pollutants. For example, we can compare the 8-hour average exposure standards for carbon monoxide and ozone: 9 ppm *vs.* 0.070 ppm. Doing some quick math, this indicates that ozone is about 120 times more hazardous to breathe than carbon monoxide! Nonetheless, carbon monoxide still can be exceedingly dangerous. As the "silent killer," it may impair your judgment before you recognize the danger.

Your Turn 2.11 You Decide Estimating Toxicities

a. Which pollutant in **Table 2.4** is likely to be the most toxic? Exclude particulate matter. Share a reason for your decision.

b. Examine the particulate matter standards. Earlier, we stated that "fine particles," $PM_{2.5}$, are more deadly than the coarser ones, PM_{10}. Do the values in **Table 2.4** support this claim? Why or why not?

Although the standards for air pollutants are expressed in parts per million, the concentrations of sulfur dioxide and nitrogen dioxide could conveniently be reported in parts per billion (ppb), meaning one part out of one billion, or 1000 times less concentrated than one part per million.

$$\textit{sulfur dioxide } 0.075 \textit{ ppm} = 75 \textit{ ppb}$$

$$\textit{nitrogen dioxide } 0.100 \textit{ ppm} = 100 \textit{ ppb}$$

As these values reveal, converting from parts per million to parts per billion involves moving the decimal point three places to the right. Even though making this change of units for SO_2 and NO_2 may create a more convenient number, when reporting concentrations across a variety of chemicals in the air, it is beneficial to have a common unit for direct comparison.

Your Turn 2.12 Skill Building Living Downwind

Sulfur dioxide (SO_2) is released in the air when copper ore is smelted to make copper metal. Let's assume that a woman living downwind of a smelter inhaled 44 μg of SO_2 in an hour.

a. If she inhaled 625 liters (0.625 m^3) of air per hour, would she exceed the 1-hr average for the U.S. National Ambient Air Quality Standards for SO_2? Support your answer with a calculation.

b. If she were exposed at this rate for three hours, would she exceed the 3-hr average?

Established in 1948, the World Health Organization (WHO) has the authority to direct and coordinate responses to world health concerns within the United Nations. Part of their charge is to set norms and standards for public health, which led to the development of air pollution guidelines. You will discover how these guidelines differ from those established by the U.S. EPA in the next activity.

Your Turn 2.13 You Decide Difference in Standards

Conduct some research to determine the WHO Air Pollution Guidelines. Compare the pollutants identified in **Table 2.4** by both the U.S. EPA and WHO. Select one pollutant and name three differences between the EPA standards and the WHO guidelines for your chosen pollutant. Which agency has set the more stringent standard?

To end this section, we note that our *perception* of a risk also plays an important role. For example, the risks of traveling by car far exceed those of flying. Each day in the United States, more than 100 people die in automobile accidents. Yet some people avoid taking flights because of their fear of a plane crash. Similarly, some people fear living inland near a dormant volcano. Yet, as some extreme hurricanes have demonstrated, living in a coastal area can be a far riskier proposition. Whether perceived as a risk or not, air pollution presents real hazards, both to present and future generations. In the next section, we offer you the tools to assess these hazards.

2.10 Is It Safe to Leave My House? Air Quality Monitoring and Reporting

Depending on where you live, you will breathe air of varying quality. Some locations always have healthy air, others have air of moderate quality, and still others have unhealthy air much of the time. As we will see, these differences arise from a number of factors such as population, regional activities, geographical features, prevailing weather patterns, and the activities of people in neighboring regions.

To improve air quality, many nations have enacted legislation. For example, we already have cited the U.S. Clean Air Act (1970) that led to the establishment of air quality standards. Like many environmental laws, this one focused on limiting our exposure to hazardous substances. It has been named as a "command and control law" or an "end of the pipe solution," because it tries to limit the spread of hazardous substances or clean them up after the fact.

The Pollution Prevention Act (1990) was a significant piece of legislation that followed the Clean Air Act. It focused on *preventing* the formation of hazardous substances, stating that "pollution should be prevented or reduced at the source whenever feasible." The language shift is significant. Rather than cleaning up pollutants, people should not produce them in the first place! With the Pollution Prevention Act, it became national policy to employ practices that reduce, or ideally eliminate, pollutants at their source.

Your Turn 2.14 You Decide The Logic of Prevention

"Take off your muddy shoes at the door—I'm not going to clean the carpet after you!" may be a phrase you have heard from a parent in your past. It is a common-sense practice. List three "common sense" examples that prevent air pollution rather than cleaning it up after the fact.

Hint: Revisit Your Turn 2.1 on "air prints."

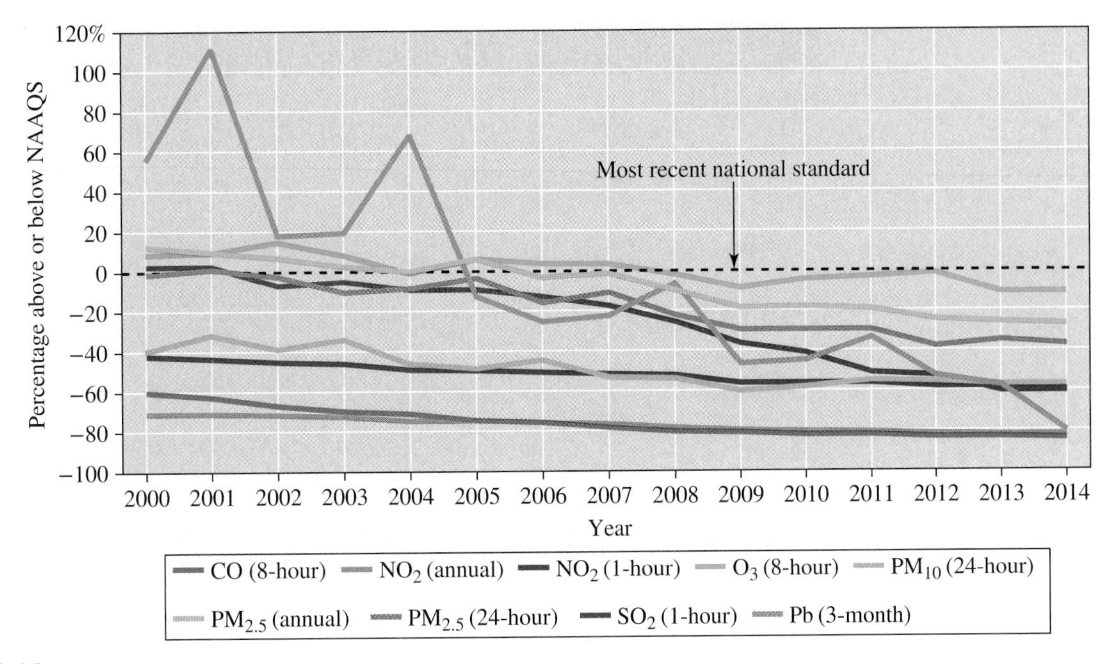

Figure 2.14

U.S. average levels of air pollutants (at selected sites) compared with national ambient air quality standards, 2000–2014.

Source: EPA

The recent decrease in the concentration of air pollutants in the United States has been dramatic (Figure 2.14). Some improvements occurred through a combination of laws and regulations, such as the ones we just mentioned. Others stemmed from local decisions. For example, a community may have built a new public transportation system, or an industry may have installed more modern equipment. Still others occurred because of the ingenuity of chemists, most notably via a set of practices called "green chemistry," which will be described in the final section of this chapter.

Although air quality may have improved, on average, people in some metropolitan areas breathe air that contains unhealthy levels of pollutants. To illustrate this, check the data for the United States presented in Table 2.5. The label "unhealthy" means just that. As we described earlier, air pollutants are the perpetrators of biological mischief. To help you more quickly assess the hazards, the U.S. EPA developed the color-coded Air Quality Index (AQI) shown in Table 2.6. This index is scaled from 1–500, with the value of 100 pegged to the national standard for the pollutant. Green or yellow (<100) indicates air of good or moderate quality. Orange (100–150) indicates that the air has become unhealthy for some groups. Red, purple, or maroon (>150) indicates that the air is unhealthy for *everybody* to breathe.

Figure 2.15 shows the air quality forecast for ozone and particulates around New Year's Day in 2016 in Phoenix, Arizona. The pollutant of concern was $PM_{2.5}$ from wood smoke and fireworks. People were advised to "limit or even avoid outdoor exertion such as jogging or riding bicycles." With the development of inexpensive and small sensors, it is now even possible to monitor the air quality of your location in real time via your cell phone.[1]

The AQI score does not include units because it is a number created to communicate potential health impact based on the concentrations of four pollutants. The pollutants included in the calculation are particulate matter, carbon monoxide, sulfur dioxide, and ozone.

Your Turn 2.15 You Decide Air Quality Indices

Examine **Tables 2.5** and **2.6**, as well as **Figure 2.15**. Look up pollution levels for where you live on the AirNow website. What is the air quality for your location? How does the AQI affect your planned activities for the day?

[1] http://www.kurzweilai.net/monitoring-air-pollution-on-smart-phones

Table 2.5 Air Quality Data for Selected U.S. Metropolitan Areas

Metropolitan Area	# Of Unhealthy Days/Year*	
	O_3	$PM_{2.5}$
Boston	0	8
Chicago	10	0
Cleveland	10	1
Houston	21	0
Los Angeles	43	0
Phoenix	11	4
Pittsburgh	14	1
Sacramento	35	13
Seattle	2	0
Washington, DC	21	2

*Days that exceed the U.S. EPA standards for the particular pollutant

Table 2.6 Levels for the Air Quality Index (AQI)

When the AQI is in this range:	... air quality conditions are:	... as symbolized by this color:
0–50	Good	Green
51–100	Moderate	Yellow
101–150	Unhealthy for sensitive groups	Orange
151–200	Unhealthy	Red
201–300	Very unhealthy	Purple
301–500	Hazardous	Maroon

FORECAST DATE AIR POLLUTANT	YESTERDAY WED 12/30/2015	TODAY THURS 12/31/2015	TOMORROW FRI 1/1/2016	EXTENDED SAT 1/2/2016
O_3	43	41	36	34
PM_{10}	43	54	109	30
$PM_{2.5}$	75	107	203	59

Figure 2.15

Air quality forecast for Phoenix, Arizona, December 30, 2015 through January 2, 2016, using the colors as defined in Table 2.6.

2.11 | The Origin of Pollutants: Who's to Blame?

Life on Earth bears the stamp of oxygen. Compounds containing oxygen occur in the atmosphere, in your body, and in the rocks and soils of the planet. Why? The answer is that many different elements combine chemically with oxygen. One such element is carbon. You were already introduced to the compound carbon monoxide, CO, a pollutant listed in Table 2.4. Fortunately, CO is relatively rare in our atmosphere. In contrast, carbon dioxide, CO_2, is far more abundant (about 400 ppm). Even so, at this concentration, CO_2 plays an important role as a greenhouse gas. In this section, we explain how both CO_2 and CO are emitted into our atmosphere.

As you know, humans exhale CO_2 with each breath. Breathing is one natural source of CO_2 in our atmosphere, but it is not a significant cause of the recent increase in atmospheric levels. Carbon dioxide is also produced when humans burn fuels. **Combustion** is the chemical process of burning; that is, the rapid reaction of fuel with oxygen to release energy in the form of heat and light. When carbon-containing compounds burn, the carbon combines with oxygen to produce carbon dioxide (CO_2). When the oxygen supply is limited, carbon monoxide (CO) is likely to form as well.

Combustion is a type of **chemical reaction**, a process whereby substances described as reactants are transformed into different substances called products. A **chemical equation** is a representation of a chemical reaction using chemical formulas. To students, a chemical equation is probably better known as "the thing with an arrow in it." Chemical equations are the sentences in the language of chemistry. They are made up of chemical symbols (corresponding to letters) that are often combined in the formulas of compounds (the words of chemistry). Like a sentence, a chemical equation conveys information—in this case, about the chemical change taking place. A chemical equation must also obey some of the same constraints that apply to a mathematical equation.

At the most fundamental level, a chemical equation is a qualitative description of this process:

$$\text{Reactants} \longrightarrow \text{Products}$$

By convention, the reactants are always written on the left and the products on the right. The arrow represents a chemical transformation and can be read as "converted to."

The combustion of carbon (charcoal) to produce carbon dioxide, as shown in Figure 2.16, can be represented in several ways. One is with chemical names:

$$\text{carbon} + \text{oxygen} \longrightarrow \text{carbon dioxide}$$

Another more common way is to use chemical formulas:

$$C + O_2 \longrightarrow CO_2 \qquad \textbf{[2.1]}$$

This compact symbolic statement conveys a good deal of information. It might sound something like this: "One atom of the element carbon reacts with one molecule of the element oxygen to yield one molecule of carbon dioxide." Using black for carbon and red for oxygen, we can also represent the molecules and atoms involved using images, as shown in Figure 2.16.

Atoms are neither created nor destroyed in a chemical reaction. The elements present do not change their identities when converted from reactants to products, although they may change the way their atoms are bonded to one another. This relationship is known as the **law of conservation of matter and mass**. That is, in a chemical reaction, the mass of the reactants consumed equals the mass of the products formed.

Chemical equations follow this law and are similar to a mathematical expression in that the number and kind of each atom on the left side of the arrow *must* equal those on the right:

$$\text{Left side: 1 C and 2 O} \longrightarrow \text{Right side: 1 C and 2 O}$$

If they do not equal, adjustments may be made only to the amounts of chemicals in the reaction, *never* to the subscripts of the chemicals themselves. This would change the chemical compound to a new substance.

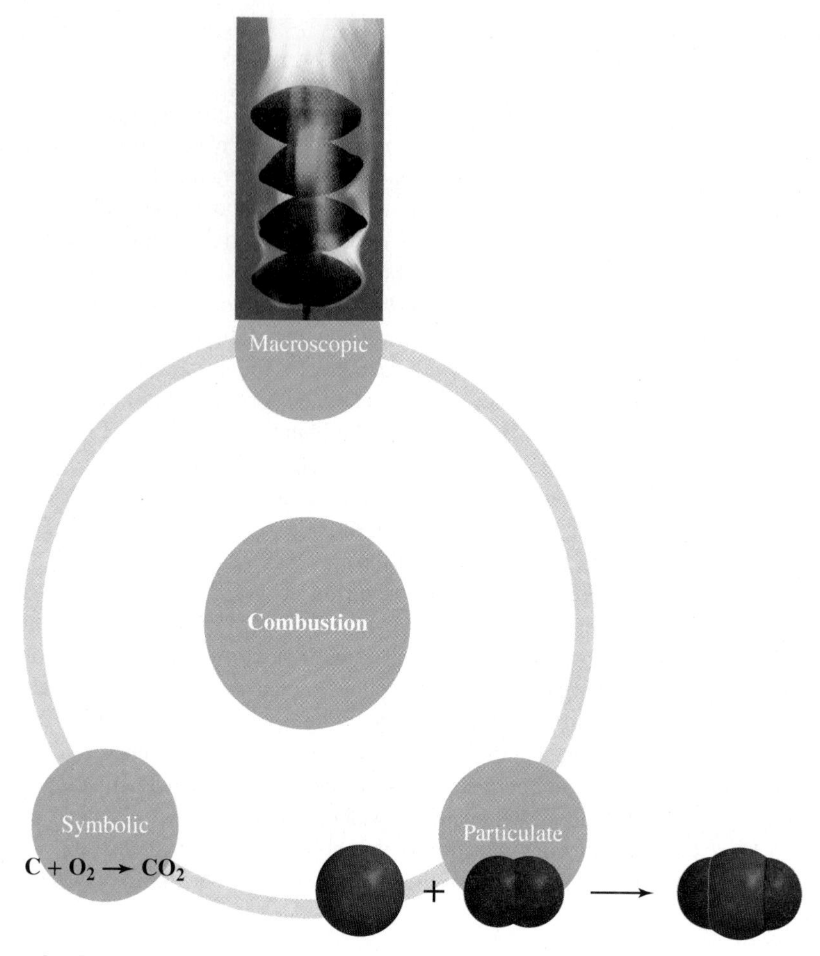

Figure 2.16

Three representations of charcoal burning in air.

(photo): © McGraw-Hill Education. Bob Coyle, photographer

Let's look at another example. Using yellow for sulfur, we can represent how sulfur burns in oxygen to produce the air pollutant sulfur dioxide:

$$S + O_2 \longrightarrow SO_2 \qquad \textbf{[2.2]}$$

This equation is balanced—the same number and types of atoms are present on each side of the arrow. These atoms, however, were rearranged. This is what a chemical reaction is all about!

We designate the physical states when this information is particularly important; otherwise, for simplicity, we will omit them.

It is possible to pack even more information into a chemical equation by specifying the physical states of the reactants and products. As mentioned in Section 1.1, solid is designated by the subscript *(s)*, a liquid by *(l)*, and a gas by *(g)*. Because carbon and sulfur are solids, and oxygen, carbon dioxide, and sulfur dioxide are gases at ordinary temperatures and pressures, Equations 2.1 and 2.2 may become:

$$C(s) + O_2(g) \longrightarrow CO_2(g)$$
$$S(s) + O_2(g) \longrightarrow SO_2(g)$$

Equation 2.1 describes the combustion of pure carbon in an ample supply of oxygen, but this is not always the case. If the oxygen supply is limited, carbon monoxide may be one of the products. Let's take the extreme case in which CO is the sole product:

$$C + O_2 \longrightarrow CO \text{ (unbalanced equation)} \qquad \textbf{[2.3a]}$$

This equation is not balanced because there are 2 oxygen atoms on the left, but only 1 on the right. You might be tempted to balance the equation by simply adding an additional oxygen atom to the right side. However, once we write the *correct* chemical formulas for the reactants and products, we cannot change them. We may only use whole-number coefficients (or occasionally fractional ones) in front of the given chemical formulas to make the adjustments. In simple cases like this, the coefficients can be found by trial and error. If we place a 2 in front of CO, it signifies two molecules of carbon monoxide. This balances the oxygen atoms:

$$C + O_2 \longrightarrow 2\ CO \text{ (still not balanced)} \qquad \textbf{[2.3b]}$$

But now the carbon atoms do not balance. Fortunately, this is easily corrected by placing a 2 in front of the carbon on the left side of the equation:

$$2\ C + O_2 \longrightarrow 2\ CO \text{ (balanced equation)} \qquad \textbf{[2.3c]}$$

From a molecular perspective, this corresponds to:

By comparing Equations 2.1 and 2.3c, you can see that more O_2 is required to produce CO_2 from carbon than is needed to produce CO. This matches the conditions we stated for the formation of carbon monoxide; namely, that the supply of oxygen was limited. Consult Table 2.7 for some tips about balancing chemical equations.

Table 2.7 Characteristics of Chemical Equations

Always Conserved
Identity of atoms in reactants = identity of atoms in products
Number of atoms of each element in reactants = number of atoms of each element in products
Mass of all reactants = mass of all products
May Change
Number of molecules in reactants may differ from the number in products
Physical states (*s*, *l*, or *g*) of reactants may differ from those of products

You may be surprised to learn the origin of the air pollutant nitrogen monoxide. It comes from the nitrogen and oxygen found in the air! Both oxygen and nitrogen are safe substances, but can react to create something quite dangerous. These two gases chemically combine in the presence of something very hot, such as an automobile engine or a forest fire:

$$N_2(g) + O_2(g) \xrightarrow{\text{high temperature}} NO(g) \text{(unbalanced equation)} \qquad \textbf{[2.4a]}$$

The equation is not balanced, as 2 oxygen atoms are on the left side, but only 1 is on the right. The same is true for nitrogen atoms. Balancing the equation simply consists of placing a 2 in front of NO, which supplies 2 N and 2 O atoms on the right:

$$N_2(g) + O_2(g) \xrightarrow{\text{high temperature}} 2\ NO(g) \text{(balanced equation)} \qquad \textbf{[2.4b]}$$

high temperature

Your Turn 2.16 Skill Building Chemical Equations

Balance these chemical equations and draw representations of all reactants and products, analogous to **Equation 2.4b.**

a. $H_2 + O_2 \longrightarrow H_2O$ **b.** $N_2 + O_2 \longrightarrow NO_2$

2.12 More Oxygen, Please: The Effect of Combustion on Air Quality

As we mentioned earlier, hydrocarbons are compounds composed exclusively of hydrogen and carbon. The hydrocarbons we use today are primarily obtained from crude oil. Methane (CH_4), the simplest hydrocarbon, is the primary component of natural gas. Both gasoline and kerosene are mixtures of many long-chain hydrocarbons.

Given an ample supply of oxygen, hydrocarbon fuels burn completely. You may hear this called "complete combustion." In essence, all of the carbon atoms in the hydrocarbon molecule combine with O_2 molecules from the air to form CO_2. Similarly, all the hydrogen atoms combine with O_2 to form H_2O. For example, below is the chemical equation for the complete combustion of methane. This equation is your first peek at why burning carbon-based fuels releases carbon dioxide into the atmosphere.

$$CH_4 + O_2 \longrightarrow CO_2 + H_2O \text{ (unbalanced equation)} \qquad \textbf{[2.5a]}$$

Note that O appears in *both* products: CO_2 and H_2O. To balance the equation, start with an element that appears in *only one substance* on each side of the arrow. In this case, both H and C qualify. No coefficients need to be changed for carbon, because both sides contain 1 C atom. Balance the H atoms by placing a 2 in front of the H_2O:

$$CH_4 + O_2 \longrightarrow CO_2 + 2\ H_2O \text{ (still not balanced)} \qquad \textbf{[2.5b]}$$

Balance the oxygen atoms last. Four O atoms are on the right side and 2 O atoms are on the left, so we need 2 O_2 molecules to balance the equation:

$$CH_4 + 2\ O_2 \longrightarrow CO_2 + 2\ H_2O \text{ (balanced equation)} \qquad \textbf{[2.5c]}$$

A nice feature of chemical equations is that simply counting the number of each type of atom on both sides of the arrow tells you if it is balanced. Here, the equation is balanced because each side has 1 C atom, 4 H atoms, and 4 O atoms:

+ + high temperature + +

Most automobiles run on the complex mixture of hydrocarbons we call gasoline. Octane, C_8H_{18}, is one of the pure substances in this mixture. With sufficient oxygen, octane burns completely to form carbon dioxide and water:

$$2\ C_8H_{18} + 25\ O_2 \longrightarrow 16\ CO_2 + 18\ H_2O \qquad \textbf{[2.6]}$$

Both products travel from the engine out the exhaust pipe and into the air. Are these combustion products visible? Usually not. Water, in the form of water vapor, and carbon dioxide are both colorless gases. But if you happen to be outside on a winter day, the water vapor condenses to form clouds of tiny ice crystals that you can see. Occasionally, the frozen vapor gets trapped in an inversion layer and forms an ice fog (Figure 2.17).

Figure 2.17

A winter ice fog in Fairbanks, Alaska.

With less oxygen, the hydrocarbon mixture we call gasoline burns incompletely ("incomplete combustion"). Water is still produced together with both CO_2 and CO. The extreme case occurs when only carbon monoxide is formed, as is shown here for the incomplete combustion of octane:

$$2\,C_8H_{18} + 17\,O_2 \longrightarrow 16\,CO + 18\,H_2O \quad [2.7]$$

Compare the coefficient of 17 for O_2 in Equation 2.7 with that of 25 for O_2 in Equation 2.6. Less oxygen is needed for incomplete combustion, as CO contains less oxygen than CO_2.

Your Turn 2.17 Skill Building Is It Balanced?

Demonstrate that **Equations 2.6** and **2.7** are balanced by counting the number of atoms of each element on both sides of the arrow.

What is the actual mixture of products formed when gasoline is burned in your car? This is not a simple question, because the products vary with the fuel, the engine, and its operating conditions. It is safe to say that gasoline burns primarily to form H_2O and CO_2. However, some CO is also produced. The amounts of CO and CO_2 that go out the tailpipe indicate how efficiently the car burns the fuel, which in turn indicates how well the engine is tuned. Some regions of the United States monitor auto emissions with a probe that detects CO (Figure 2.18). The CO concentrations in the exhaust are

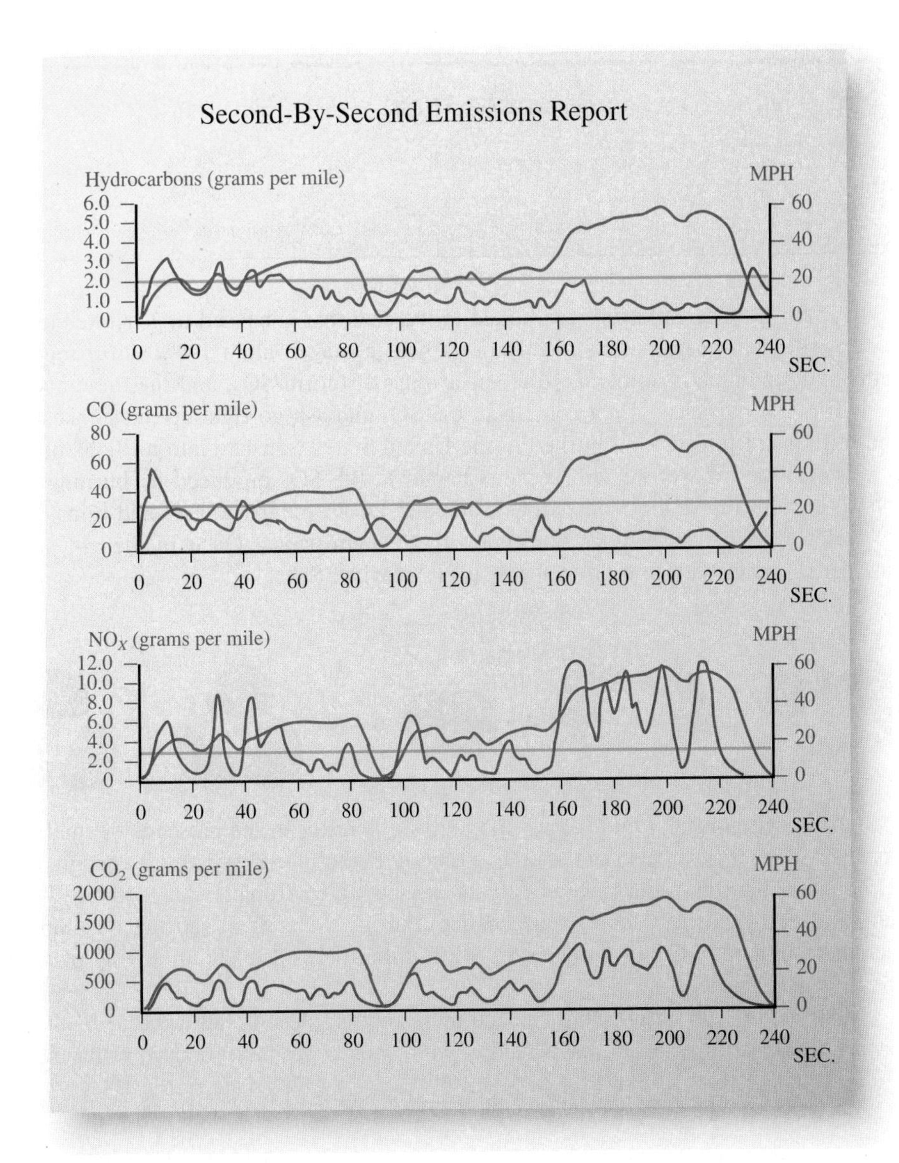

Figure 2.18

A U.S. auto emissions report. The blue line shows the change in engine speed; the red line shows the change in emissions. Any emissions below the green line are in the acceptable range.

© Data from Cathy Middlecamp

compared with established standards—for example, currently 4.2 grams per mile in the state of California. If the vehicle fails the emissions test, it must be serviced to meet at least the minimum emission standards.

Your Turn 2.18 Skill Building Auto Emissions Report

a. **Figure 2.18** reports NO_x emissions in grams per mile. NO_x is a way to collectively represent the oxides of nitrogen. If x = 1 and x = 2, write the corresponding chemical formulas. Also give the chemical names.
b. NO is the primary oxide of nitrogen emitted. What is the source of this compound? *Hint:* Revisit Equation 2.4.
c. The green line is missing on the CO_2 graph, but present on the others. Provide an explanation for this apparent oversight.

2.13 | Air Pollutants: Direct Sources

In this section, we examine two major sources of air pollutants: motor vehicles and coal-fired plants that generate electricity. These sources emit SO_2, CO, NO, and PM directly, and we will revisit each of these pollutants in turn. We also digress to discuss VOCs (volatile organic compounds), pollutants that are not regulated but are still intimately connected with the ones that are.

Your Turn 2.19 Scientific Practices Tailpipe Gases

What comes out of the tailpipe of an automobile? Start your list now and build it as you work through this section.
Hint: Some of the air that enters the engine also comes out the tailpipe.

Sulfur dioxide emissions are linked to the coal that is burned to generate electric power. Although coal consists mostly of carbon, it may contain 1–3% sulfur together with small amounts of minerals. The sulfur burns to form SO_2, and the minerals end up as fine ash particles. If not contained, the SO_2 and ash go right up the smokestack. The millions of tons of coal burned in the United States translate into millions of tons of waste in the air. As we will see in Chapter 8, the SO_2 produced by burning coal can dissolve in the water droplets of clouds and fall to the ground as acid rain.

However, the story does not end with SO_2 emissions. Once in the air, sulfur dioxide can react with oxygen to form sulfur trioxide, SO_3:

$$2\,SO_2 + O_2 \longrightarrow 2\,SO_3 \qquad [2.8]$$

Although normally quite slow, this reaction is faster in the presence of small ash particles. The particles also aid another process. If the humidity is high enough, they help condense water vapor into an aerosol of tiny water droplets. Aerosols are liquid and solid particles that remain suspended in the air rather than settling out. Smoke, such as from a campfire or a cigarette, is a familiar aerosol made up of tiny particles of solids and liquids.

The aerosol of concern here is made up of tiny droplets of sulfuric acid, H_2SO_4. It forms because sulfur trioxide reacts readily with water droplets to produce sulfuric acid:

$$H_2O + SO_3 \longrightarrow H_2SO_4 \qquad [2.9]$$

If inhaled, the droplets of the sulfuric acid aerosol are small enough to become trapped in the lung tissue and cause severe damage.

The good news? Sulfur dioxide emissions in the United States are declining (Figure 2.14). For example, in 1985, approximately 20 million tons of SO_2 was emitted from the burning of coal. Today, the value is closer to 11 million tons. This impressive decrease can be credited to the Clean Air Act of 1970 that mandated many reductions, including those from coal-fired electric power plants. More stringent regulations were established in the Clean Air Act Amendments and the Pollution Prevention Act of 1990. For example, gasoline and diesel fuel both once contained small amounts of sulfur, but the allowable amounts were drastically lowered in 1993 and in 2006, respectively.

Did You Know? In the U.S. and Canada, diesel fuel with <15 ppm sulfur content is currently required for all on-road vehicles. However, the European Union and countries such as Australia and China have adopted more stringent regulations for diesel fuel, with an allowable sulfur content of <10 ppm.

Your Turn 2.20 Scientific Practices SO_2 from the Mining Industry

Burning coal is not the only source of sulfur dioxide. As you saw in **Your Turn 2.12,** smelting is another. Silver and copper metal can be produced from their sulfide ores. Write the balanced chemical equations for the following reactions:

a. Silver sulfide (Ag_2S) is heated in air to produce silver and sulfur dioxide.
b. Copper sulfide (CuS) is heated in air to produce copper and sulfur dioxide.

Hint: Reacting with oxygen in air at elevated temperatures.

With more than 250 million vehicles (and over 300 million people), the United States has more vehicles per capita than any other nation. Do these vehicles emit sulfur dioxide? Fortunately, the answer is no since cars have internal combustion engines primarily fueled by gasoline. We already mentioned that the combustion of hydrocarbons in gasoline produces—at best—carbon dioxide and water vapor (Equation 2.6). Since gasoline contains little or no sulfur, its combustion produces little or no sulfur dioxide. Nonetheless, each tailpipe puffs out its share of air pollutants. In addition to carbon dioxide and water, the ubiquitous automobile adds to the atmospheric concentrations of carbon monoxide, volatile organic compounds, nitrogen oxides, and particulate matter. We discuss each of these in turn.

Carbon monoxide pollution comes primarily from automobiles. But think in terms of all the tailpipes out there, not just those attached to cars. Some are attached to heavy trucks, SUVs, and the three m's: motorcycles, minibikes, and mopeds. Others are on equipment such as farm tractors, bulldozers, motor boats, and lawnmowers. The tailpipes attached to all gasoline and diesel engines emit carbon monoxide.

Your Turn 2.21 Scientific Practices Other Tailpipes

Visit "Nonroad Engines, Equipment, and Vehicles" on the EPA's website to answer the following:

a. The text mentioned tractors, bulldozers, and boats. Name five other engine-powered machines or vehicles that do not run on roads.
b. Select a machine or vehicle of interest to you. How are emissions from its engine being reduced? What is the time scale for the reduction?

Cars not only emit carbon in the form of carbon monoxide but also in the form of unburned and partially burned hydrocarbons. This leads us to the topic of VOCs, volatile organic compounds. A *volatile* substance readily passes into the vapor phase; that is, it evaporates easily. Gasoline and nail polish remover are both volatile. If you were to spill either of these, the puddle would soon evaporate. When you apply varnish to a surface, you can smell the volatile compounds that evaporate with each brush stroke. An **organic compound** always contains carbon, almost always contains hydrogen, and may contain other

In contrast to organic compounds, inorganic compounds are those that contain little to no carbon atoms.

elements such as oxygen and nitrogen. Organic compounds include methane and octane—two examples of hydrocarbons that were mentioned earlier. However, organics also include alcohol and sugar, compounds that contain oxygen in addition to carbon and hydrogen.

Nitrogen monoxide and nitrogen dioxide are collectively known as NO_x, as mentioned in Your Turn 2.18.[2] NO_2 is brown in color, giving smog its characteristic brownish tinge. Recall that N_2 and O_2 combine to produce NO, which is a colorless gas (Equation 2.4). But what is the origin of NO_2? Here is a balanced equation that appears to be a likely candidate, which proceeds most rapidly in the presence of VOCs:

$$2\ NO + O_2 \longrightarrow 2\ NO_2 \qquad [2.10]$$

Volatile organic compounds (VOCs) are carbon-containing compounds that pass easily into the vapor phase. They originate from a variety of sources. For example, you can smell naturally occurring VOCs in a spruce or pine forest. VOCs from tailpipes are not so pleasant, because they are vapors of incompletely burned gasoline molecules or fragments of these molecules. The exhaust gas still contains oxygen, as not all of it is consumed in the engine. Catalytic converters utilize this oxygen to burn VOCs to form carbon dioxide and water. Section 2.14 describes the connection between VOCs and ozone formation. However, right now, we want to connect VOCs with the formation of NO_2, which is formed by more complex pathways.

Equation 2.11 shows a reaction that predominates in urban settings, where you are likely to find NO. In some cities, this actually lowers the ground-level ozone concentrations along highways congested with vehicles emitting NO:

$$NO + O_3 \longrightarrow NO_2 + O_2 \qquad [2.11]$$

To further complicate things, on a sunny day, some of the NO_2 converts back into NO, as we will see in the next section. Again, this is why people refer to NO_x, rather than to either NO or NO_2.

You might have heard of "free radicals" in the context of skin creams and other anti-aging cosmetics. The term *free radical* means that an atom or molecule has an odd number of electrons (indicated by a single dot by the molecular formula), which makes it extremely reactive toward neighboring species. Electrons like to pair up to form stable molecules and atoms; hence, radical species are very reactive to anything that can be used to sequester an additional electron. More applications for free radicals will be provided in Chapters 3 and 9.

The conversion of NO to NO_2 connects to the breakdown of VOCs in the air. A new player is involved, the reactive hydroxyl radical •OH. This reactive species is present in tiny amounts in air, polluted or otherwise.

$$\begin{aligned} VOC + \cdot OH &\longrightarrow A \\ A + O_2 &\longrightarrow A' \\ A' + NO &\longrightarrow A'' + NO_2 \end{aligned} \qquad [2.12]$$

Here, A, A', and A" represent reactive molecules that can form in the air from •OH and VOCs. The bottom line? Atmospheric chemistry is complex and involves many players. You have met some of them, including NO, NO_2, O_2, O_3, VOCs, and •OH.

Despite early claims from the auto industry that it would be impossible (or too costly) to meet new emissions standards, the industry is curbing emissions by improving engine designs and gasoline formulations. In particular, the United States has had limited success in curbing NO_x emissions. As we'll see in the next section, this also means limited success in curbing ozone. Nonetheless, given the increasing number of vehicles, any decrease in NO_x and CO emissions is impressive.

Based on measurements by the EPA at over 250 sites in the United States, the average CO concentration has decreased almost 60% since 1980. If wildfires are excluded, today's levels are the lowest reported in three decades. The decrease is due to several factors, including improved engine design, computerized sensors that better adjust the fuel–oxygen mixture, and most importantly, the requirement that all cars manufactured since the mid-1970s have catalytic converters (Figure 2.19). Catalytic converters reduce the amount of carbon monoxide in the exhaust stream by catalyzing the combustion of CO to CO_2. They also lower NO_x emissions by catalyzing the conversion of nitrogen oxides back to N_2 and O_2, the two atmospheric gases

[2] For more details regarding NO_x and other air pollutants, see: http://www.epa.gov/criteria-air-pollutants

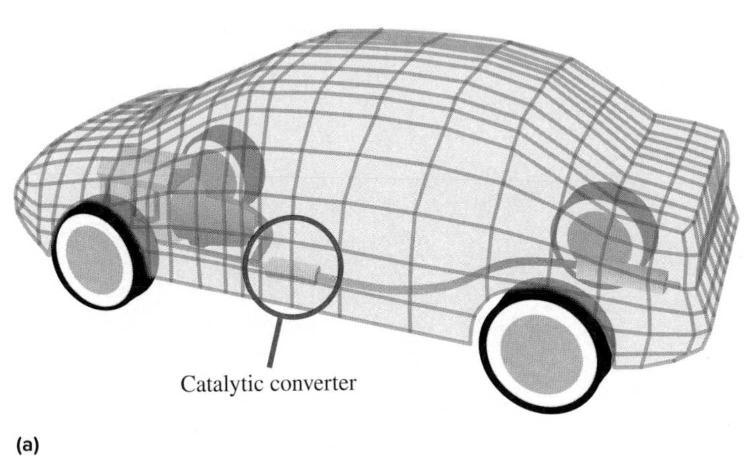

(a)

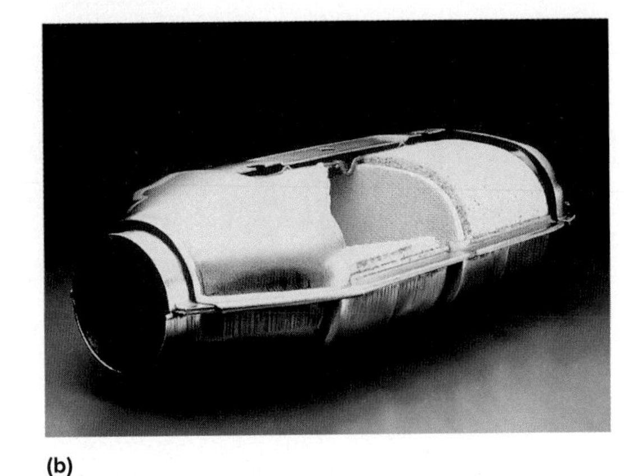

(b)

Figure 2.19

(a) Location of a catalytic converter in a car. **(b)** Cutaway view of a catalytic converter. Metals such as platinum and rhodium serve as catalysts and form a coating on the surface of ceramic beads.

Source: © Corning Incorporated

that formed them. In general, a **catalyst** is a chemical substance that participates in a chemical reaction and influences its rate, without itself undergoing permanent change. Catalytic converters typically use expensive metals such as platinum and rhodium as catalysts.

Your Turn 2.22 You Decide Save Money with the Way You Drive

Burning less gasoline equates to fewer tailpipe emissions. Which driving practices conserve fuel? Which practices expend it more than necessary? Think about the behavior of motorists on highways, city streets, and in parking lots. For each of these venues, list at least three ways that drivers could burn less gasoline.

Hint: Consider how you accelerate, coast, idle, brake, and park.

Particulate matter comes in a range of sizes, but only the tiny particles (PM_{10} and $PM_{2.5}$) are regulated as pollutants. Particles of this size can penetrate deeply into your lungs, pass into your bloodstream, and inflame your cardiovascular system. In terms of regulation, particles are the new pollutant on the block. Data collection in the United States for PM_{10} and $PM_{2.5}$ started in 1990 and 1999, respectively. In 2006, the daily air quality standard for $PM_{2.5}$ was lowered from 65 to 35 $\mu g/m^3$ because these particles proved to be more hazardous to human health than originally thought.

Particulate matter has many different sources. In the summer, wildfires may raise the concentration of particulate matter to a hazardous level. In the winter, wood stoves may produce exactly the same effect. At any time of the year in almost any urban environment, older diesel engines on trucks and buses emit clouds of black smoke. Diesel engines on tractors similarly can pollute. Construction sites, mining operations, and the unpaved roads that serve them also loft tiny particles of dust and dirt into the atmosphere. Particulate matter can even form right in the atmosphere. For example, the compound ammonia, used in agriculture, is a major player in forming ammonium sulfate and ammonium nitrate in the air, both $PM_{2.5}$.

Given all these sources, particulate matter has proven a tough pollutant to control. Even so, the EPA reported a decrease of 35% in the annual $PM_{2.5}$ concentrations from 2000 to 2014. However, some of the sites monitored still showed an increase in particle pollution. Again, what you breathe depends very much on where you live.

Your Turn 2.23 Scientific Practices Particles Where You Live

Shown is a map of the continental U.S. that shows $PM_{2.5}$ data for January 1, 2016.

a. In terms of air quality, what do the green, yellow, orange, and red colors indicate?
b. Which groups of people are most sensitive to particulate matter?
c. Visit "State of the Air," a website posted by the American Lung Association. How many days a year does your state have "orange days" and "red days" for particulate pollution? What is the difference?

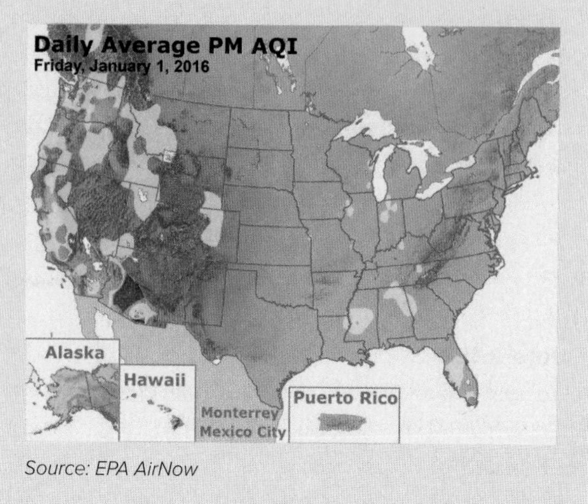

Source: EPA AirNow

2.14 | Ozone: A Secondary Pollutant

Ozone is a bad actor in the troposphere. Even at very low concentrations, it reduces lung function in healthy people who are exercising outdoors. Ozone also damages crops and the leaves of trees. But ozone does not come out of a tailpipe, and is not produced when coal is burned. How is it produced? Before we fill you in on the details, perform this activity.

Your Turn 2.24 Scientific Practices Ozone Around the Clock

Ozone concentrations vary during the day, as shown in **Figure 2.20.**

a. Near which cities is the air hazardous to one or more groups?
Hint: Refer back to the color-coded AQI (see Table 2.6).
b. At about what time does the ozone level peak?
c. Can moderate levels (shown in yellow) of ozone exist in the absence of sunlight? Assume sunrise occurs around 6 am and sunset about 8 pm.

This activity raises several related questions. Why is ozone more prevalent in some areas than others? What role does sunlight play in ozone production? We now address these.

Unlike the pollutants described in Section 2.13, ozone is a **secondary pollutant**. It is produced from chemical reactions involving one or more other pollutants. For ozone, these other pollutants are VOCs and NO_2. Recall that NO, rather than NO_2, comes directly out of a tailpipe (or a smokestack). But, over time and in the presence of VOCs and ·OH, NO in the atmosphere is converted to NO_2.

Nitrogen dioxide meets several fates in the atmosphere. The one of most interest to us occurs when the Sun is high in the sky. The energy provided by sunlight splits one of the bonds in the NO_2 molecule:

$$NO_2 \xrightarrow{Sunlight} NO + O \qquad [2.13]$$

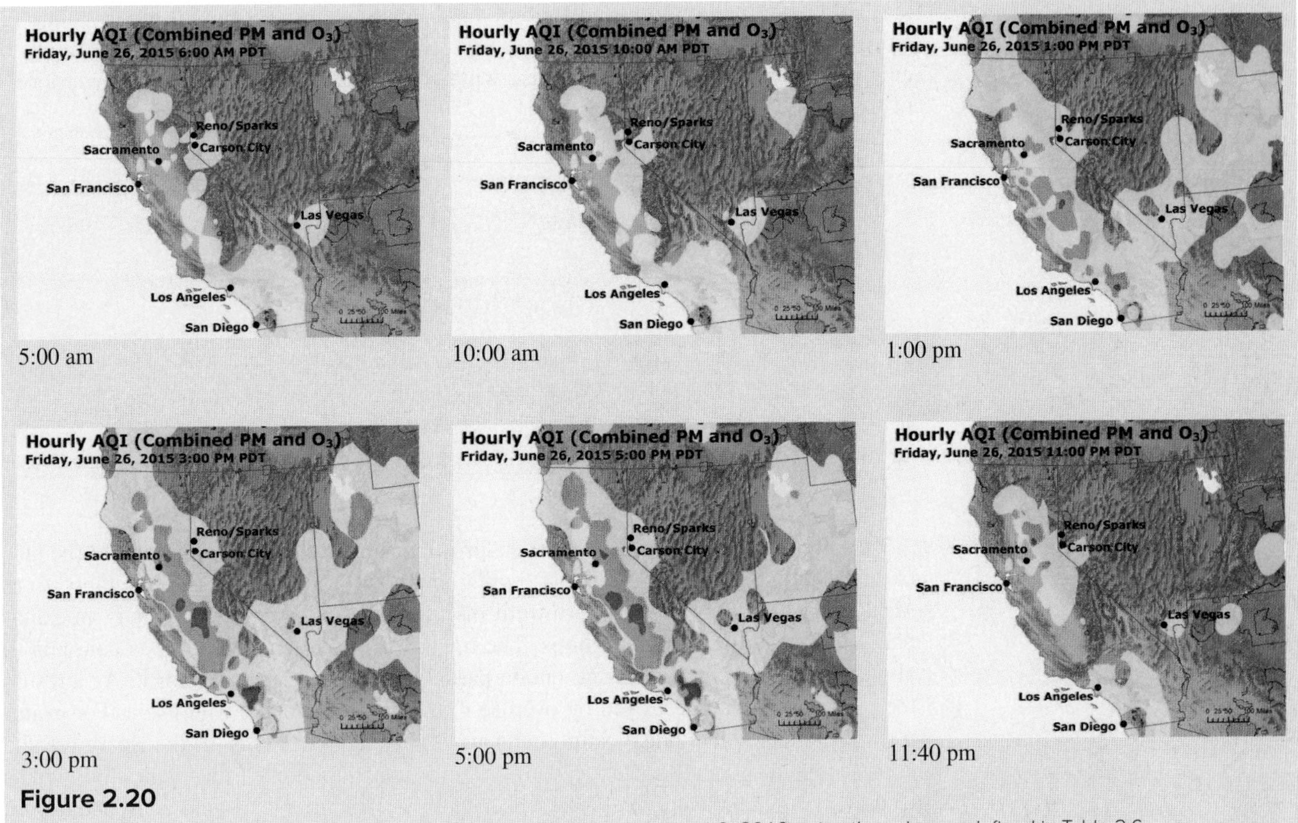

Figure 2.20

Air quality forecast for Phoenix, Arizona, December 30, 2015 through January 2, 2016, using the colors as defined in Table 2.6.

Source: EPA AirNow

The oxygen atoms produced can react with oxygen molecules to produce ozone:

$$O + O_2 \longrightarrow O_3 \qquad [2.14]$$

This explains why ozone formation requires sunlight. Sunlight first splits NO_2 to release O atoms. These in turn react with O_2 to form O_3. Thus, once the Sun goes down, the ozone concentrations drop off sharply (Figure 2.20). What happened to the ozone? In just a matter of hours, the ozone molecules react with many things, including animal and plant tissue.

Note that Equation 2.14 contains three different forms (*allotropes*) of elemental oxygen: O, O_2, and O_3. All three are found in nature, but O_2 is the least reactive and by far the most abundant, constituting about one-fifth of the air we breathe. As you will see in the next chapter, our atmosphere naturally contains tiny amounts of protective ozone in the stratosphere.

Did You Know? Oxygen atoms also exist in our upper atmosphere and are even more reactive than ozone.

Your Turn 2.25 Skill Building Ozone Summary

Summarize what you have learned about ozone formation by developing your own way to arrange these chemicals sequentially and in relation to one another: O, O_2, O_3, VOCs, NO, and NO_2. Chemicals may appear as many times as you like. You also may wish to include sunlight.

Since sunlight is involved in ozone formation, the concentration of ground-level ozone varies with weather, season, and latitude. High levels of O_3 are much more likely to occur on long, sunny summer days—especially in congested urban areas. Stagnant air also favors the buildup of air pollution. For example, revisit the air quality data for cities shown in Table 2.5. Ozone was usually the culprit responsible for pollution in cities with sunny climates. In contrast, windy and rainy cities usually have lower levels of ozone.

Your Turn 2.26 Scientific Practices Ozone and You

The AirNow website, courtesy of the EPA, provides a wealth of information about ground-level ozone levels in the United States.

a. Let's say that the ozone level is "orange," actually a common occurrence in many U.S. cities during the summer months. Does air of this quality affect you if you have no health concerns but are actively exercising outdoors?

b. How does the air quality in your state compare with others?

Although the Air Quality Index maps often stop at the U.S. border due to the lack of international data (Figure 2.21), pollution knows no boundaries! Pollutants from Mexico are easily transported northward into the U.S., and vice versa. This is an example of the **tragedy of the commons**. The tragedy arises when a resource is common to all and used by many, but has no one in particular who is responsible for it. As a result, the resource may be destroyed by overuse to the detriment of all who use it. For example, we cannot lay individual claim to the air—it belongs to all of us. If the air we breathe has waste dumped into it, this leads to an unhealthy situation for everyone. Individuals whose activities have little or no effect on the air still suffer the same consequences as those who pollute. The costs are shared by all. In later chapters, we will see other examples of the tragedy of the commons that relate to water, energy, and food.

Air pollution, once primarily a local concern, is now a serious international issue. Many cities worldwide have high ozone levels. Couple motor vehicles with a sunny location anywhere on the planet, and you are likely to find unacceptable levels of ozone. Some places, however, are worse than others. London with its cool, foggy days typically has low ozone levels. In contrast, ozone is a serious problem in Mexico City.

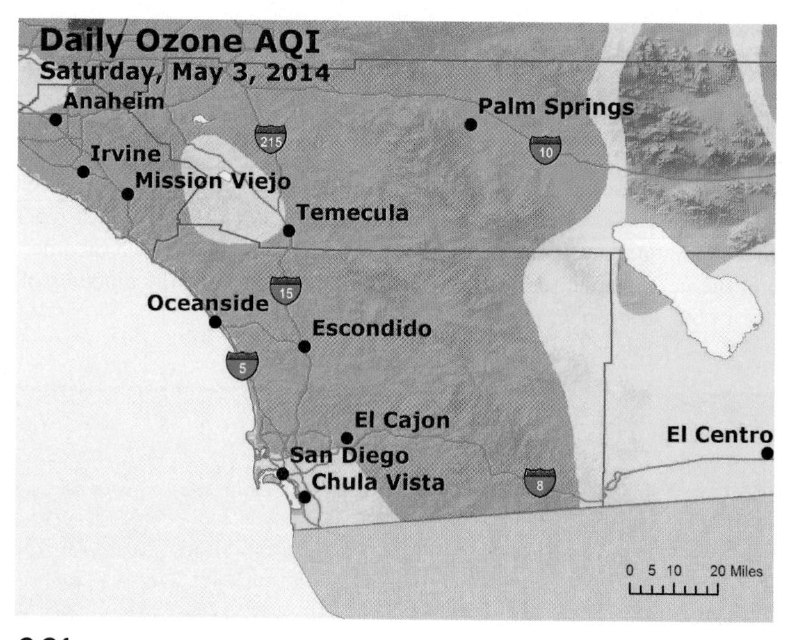

Figure 2.21

Tropospheric ozone map for the Orange County and San Diego, California, area on May 3, 2014 (purple: very unhealthy; red: unhealthy; orange: unhealthy for sensitive groups; yellow: moderate).

Source: EPA AirNow

It is quite ironic that ozone attacks rubber, thereby damaging the tires of the vehicles that led to its production in the first place! Should you park your car indoors in the garage to minimize possible rubber damage? In fact, should you park yourself indoors if the levels of ozone outside are unhealthy? The next section speaks to the quality of *indoor* air.

2.15 Are We Really Safe from Polluted Air by Staying Indoors?

In *The Wizard of Oz,* Dorothy hugged her dog Toto and exclaimed, "There's no place like home!" Of course she was right, but when it comes to air quality, your home may not always be the best place to be. In fact, the levels of air pollution indoors may far exceed those existing outside. Given that most of us sleep, work, study, and play indoors, we should learn about the air in the place we call home.

Indoor air may contain up to a thousand substances at low levels. If you are in a room where somebody is smoking, add another thousand or so. Indoor air contains some familiar culprits: VOCs, NO, NO_2, SO_2, CO, ozone, and PM. These pollutants are present either because they came in with the outside air, or because they were generated right inside your dwelling.

Let's begin our discussion with the question, should you move indoors to escape the ozone present outside? In general, if a pollutant is highly reactive, it does not persist long enough to be transported indoors. Thus, for highly reactive molecules such as O_3, NO_2, and SO_2, you should expect lower levels indoors. Indeed, this is the case, because indoor air is typically 10–30% lower in ozone concentration relative to outdoor air. Similarly, sulfur dioxide and nitrogen dioxide levels are lower indoors, although the decrease is not as dramatic as that for ozone.

Carbon monoxide is a different story. As a relatively unreactive pollutant, CO has an appreciably long atmospheric lifetime to move freely in and out of buildings through doors, windows, or a ventilation system. The same is true for some VOCs, but not for the more reactive ones such as those that give pine forests their scent. If you want to inhale the delicious volatile compounds emitted from the bark of Ponderosa pines, it is best to remain outside near the trees.

Some pollutants are trapped by filters in the heating or cooling system of the building. For example, many air-handling systems contain filters that remove larger-sized particulate matter and pollen. As a result, people who suffer from seasonal allergies can find relief indoors. Similarly, those near a wildfire can go inside to escape some of the irritating smoke particles. However, gas molecules such as O_3, CO, NO_2, and SO_2 are *not* trapped by the filters used in most ventilation systems.

These days, many buildings are constructed with an eye to increasing their energy efficiency. This is a win–win situation, lowering your heating bills while lowering the amounts of pollutants generated in producing the heat. But there can be a downside. A building that is airtight with a limited intake of fresh air may have unhealthy levels of indoor air pollutants. Therefore, what appeared initially to be a benefit (better energy efficiency) can turn into a higher risk (increased pollutant levels). In some cases, poor ventilation can cause indoor pollutants to reach hazardous levels, creating a condition known as "sick building syndrome." Clearly, this is an undesirable outcome. Today, architects and builders are finding ways to make buildings more energy efficient, while still maintaining an effective air exchange.

Even with good ventilation, indoor activities can compromise air quality. For example, tobacco smoke is a serious indoor air pollutant, containing over a thousand chemical substances. Nicotine is one that you may recognize; others include benzene and formaldehyde. Taken as a whole, tobacco smoke is **carcinogenic**, meaning it's capable of causing cancer.

Combustion of carbon-containing materials also can generate carbon monoxide and nitrogen oxides. For example, carbon monoxide from cigarette smoking in bars can reach 50 ppm, a value well within the unhealthy range. In cigarette smoke, NO_2 levels can exceed 50 ppb. Fortunately, smokers take puffs rather than constantly breathing cigarette smoke.

Your Turn 2.27 You Decide Indoor Cigar Party

In 2007, student researchers attended a cigar party and trade show in Times Square, New York. They carried concealed detectors and found levels of 1943 μg of particulate matter per cubic meter of air inside the ballroom.[3]

a. The news article did not report whether the particulate matter was $PM_{2.5}$ or PM_{10}. For either, does the value they measured exceed the air quality standards in the United States?
Hint: See Table 2.5.

b. Assume that the students were measuring $PM_{2.5}$. What are the health implications?

People also burn candles, perhaps to soften the lighting or set a mood. However, candles deplete the oxygen in a room. They also can produce soot, carbon monoxide, and VOCs. Similarly, people may burn incense in their homes for one reason or another. Researchers who studied incense burning in churches in Europe found that the pollutants in smoke from incense and candles may be more toxic than the emissions originating from vehicle engines.[4]

Burning candles or incense can generate fumes more rapidly than they can be removed by your ventilation system or by the breezes that pass through open windows. The next activity gives you the opportunity to further investigate sources of indoor pollutants.

Your Turn 2.28 Scientific Practices Indoor Activities

Name 10 activities that add pollutants or VOCs to indoor air. To get you started, two are pictured in **Figure 2.22.** Remember that some pollutants have no detectable odors.

Figure 2.22
Examples of activities that can cause indoor air pollution.
Source: (left): © Image Source/Corbis RF; (right): © Digital Vision/Getty Images RF

[3] http://www.nytimes.com/2007/11/22/nyregion/22cigar.html?_r=0

[4] https://www.sciencenews.org/article/holy-smoke-burning-incense-candles-pollute-air-churches

Figure 2.23

A home radon test kit.

© Kieth Eng, 2008

Figure 2.24

A cookstove used for cooking and/or heat also gives off significant air pollution.

© Joerg Boethling/Alamy Stock Photo

One naturally occurring air pollutant of concern is radon, a noble gas (Group 18). Radon is a special case of indoor air pollution, because it occurs naturally in tiny amounts and usually is no problem. However, it may reach hazardous levels in basements, mines, and caves. Like all noble gases, radon is colorless, odorless, tasteless, and chemically unreactive. But unlike the others, it is radioactive. Radon is generated in the decay series of uranium, another naturally occurring radioactive element. Since uranium occurs at a concentration of about 4 ppm in all the rocks of our planet, radon is ubiquitous. Depending on how your apartment or residence hall is constructed, the radon produced from uranium-containing rocks may find entry into the basement. Radon is the second-leading cause of lung cancer worldwide behind tobacco smoke. As is the case for other pollutants, the threshold for danger can be estimated but is not precisely known. Radon test kits, such as the one shown in Figure 2.23, are used to measure the radon concentration in living spaces.

We will describe the chemistry associated with radioactive elements in Chapter 6.

The examples of indoor air pollution mentioned in this section already are of concern around the world. However, around 3 billion people on the planet still cook with solid fuels (wood, crop waste, animal dung, and coal) in their homes over open fires and simple stoves (Figure 2.24). This occurs primarily in developing countries, but is not limited to these locations exclusively. Increased concentrations of particulate matter cause a range of health effects, including excess pneumonia, strokes, heart disease, lung cancer, and more. Organizations such as the World Health Organization are in the process of designing more efficient cookstoves that provide more heat and less pollution. Indoors or out, we need to breathe healthy air; with each breath, we inhale a truly prodigious number of molecules and atoms.

2.16 | Is There a Sustainable Way Forward?

Air pollution provides the first context in which we can discuss **sustainability**—making decisions with a concern not only for today's outcomes, but also for the needs of future generations. It makes sense to avoid actions that produce pollutants that can compromise our health and well-being. This is the logic behind the Pollution Prevention Act of 1990, legislation that calls for preventing pollution, rather than cleaning pollution up after it is produced.

The Pollution Prevention Act provided the impetus for **green chemistry**, a set of key ideas to guide all in the chemical community, including teachers and students. Green chemistry is "benign by design." It calls for designing chemical products and processes that reduce or eliminate the use or generation of hazardous substances.

In the broadest sense, this entails the use of our principles and knowledge to make products and processes better for humans, economics, and the environment.

Begun under the EPA Design for the Environment Program, green chemistry reduces pollution through the design or redesign of chemical processes. The goal is to use less energy, create less waste, use fewer resources, and use renewable resources. Green chemistry is a tool for achieving sustainability, rather than an end in itself.

Innovative "green" chemical methods have already decreased or eliminated toxic substances used or created in chemical manufacturing processes. Some examples include:

- Plastics synthesized from renewable sources instead of typical fossil fuel-derived precursors
- Paints that contain fewer volatile organic compounds
- Cheaper and less wasteful ways to produce pharmaceuticals, pesticides, and consumer products, such as contact lenses and disposable diapers
- Limiting or eliminating the use of organic solvents for dry cleaning and electronics fabrication processes
- Removing arsenic from the touchscreens of portable electronics

Some of the research chemists and chemical engineers who developed these methods have received Presidential Green Chemistry Challenge Awards.[5] Begun in 1995, these presidential-level awards recognize chemists for their innovations on behalf of a less polluted world.

Conclusions

Nobody wants dirty air. It makes you sick, reduces the quality of your life, and may hasten your death. However, the problem is that many people have become so accustomed to breathing dirty air that they don't notice it. One concept describing this is **shifting baselines**. This refers to the idea that what people expect as "normal" on our planet has changed over time, especially with regard to ecosystems. Burning eyes and breathing disorders have become so common that we have forgotten that they once were not. We have become accustomed to living in *megacities*, urban areas with 10 million people or more, such as Tokyo, New York City, Mexico City, and Mumbai. Pollutants such as wood smoke, car exhaust, and industrial emissions often are generated in populated areas, and thus they are concentrated in the troposphere around megacities.

Clearly, we have some problems! Your knowledge of chemistry can lead you to make better choices to deal with these problems, both as an individual and in your local community. The air we breathe affects both our health and the health of the planet. Our atmosphere contains the essentials for life, including two elements (oxygen and nitrogen) and two compounds (water and carbon dioxide). Our very existence on this planet depends on having a large supply of relatively clean, unpolluted air.

But the air you breathe may be polluted with carbon monoxide, ozone, sulfur dioxide, and the oxides of nitrogen. Emergency room visits correlate with bad air quality. So do shortness of breath, scratchy throats, and stinging eyes. The pollutants that cause us harm are, for the most part, relatively simple chemical substances. They are largely produced as consequences of our dependence on coal for electricity production in power plants, gasoline in internal combustion engines, and the fuels we burn to heat and cook.

Over the past 30 years, governmental regulations, industry initiatives, and modern technology have reduced pollutant levels. Both catalytic converters on cars and emissions controls on smokestacks have been important players. But it makes more sense not to generate "people fumes" in the first place. Here is where green chemistry plays an important role. By designing new processes that do not produce air pollutants, we do not later have to clean them up.

Indoors or out, the oxygen-laden air we breathe is very close to the surface of Earth. However, Earth's atmosphere extends upward for considerable distance and contains other gases that also are essential for life on this planet. Chapters 3 and 4 will describe two of these: stratospheric ozone and carbon dioxide. We will see that our human footprints and "air prints" on planet Earth connect in surprising ways to both of these gases.

[5] For an updated list of clever Green Chemistry strategies, see: http://www2.epa.gov/green-chemistry/presidential-green-chemistry-challenge-winners

Learning Outcomes

The numbers in parentheses indicate the sections within the chapter where these outcomes were discussed.

Having studied this chapter, you should now be able to:

- describe what happens to the air we breathe in when we take a breath (2.1, 2.2)
- identify the components of the air that keep us alive and healthy (2.2, 2.3)
- outline the structure of the atmosphere in terms of the different regions and their temperatures, pressures, and compositions (2.2, 2.5)
- relate how the composition of the atmosphere is essential to life (2.3)
- distinguish among states of matter on a molecular level (2.3)
- explain how gases differ in properties relative to condensed phases (2.3)
- convert among concentration units (2.4)
- express pollutant concentrations using appropriate scientific notation and units (2.4)
- identify the primary pollutants from industrial, residential, and transportation sources (2.4, 2.8, 2.11)
- create, use, and interpret particulate diagrams for atoms and molecules (2.6)
- write formulas and names for molecular compounds (2.7)
- describe some health implications of inhaling pollutants (throughout the chapter)
- identify the sources of the "criteria pollutants" (2.8)
- evaluate a concentration in terms of its threat in polluting the air (2.9)
- describe and calculate concentrations of pollutants in the atmosphere (2.9, 2.10)
- list the major air pollutants and explain why EPA has identified them as problematic (2.9, 2.10)
- interpret tables and graphs and use EPA and WHO data to identify the concentrations of gases and PM that are harmful to health (2.10)
- describe how agencies such as EPA set their regulatory limits (2.10)
- define and apply the law of conservation of matter and mass (2.11)
- define a chemical reaction (2.11)
- relate a description of a chemical reaction in words, and write balanced chemical reactions that form the gaseous pollutants (2.11)
- recognize that some components present in small quantities can have big effects on Earth and its people (throughout the chapter)
- evaluate the health consequences of indoor and outdoor air pollution (sections throughout the chapter)
- define volatility and VOCs (2.13)
- summarize the role of sunlight in the production of ozone, and why ozone concentrations vary by location (2.14)
- define and identify sources and potential health impacts for the indoor pollutants of smoke, VOCs, and ozone (2.15)
- design possible solutions for indoor pollution (2.15)
- devise and rate the effectiveness of some methods for the remediation of atmospheric pollution (2.16)
- describe legislation related to controlling atmospheric pollution, and evaluate their successes and shortcomings (2.16)
- explain the role of "green chemistry" practices in reducing pollution (2.16)

Questions

Emphasizing Essentials

1. a. Calculate the volume of air in liters that you might inhale (and exhale) while you are sleeping for 7.5 hours. Assume that each breath has a volume of about 0.5 L, and that you are breathing 10 times per minute.

   b. From this calculation, you can see that breathing exposes you to a large volume of air. Name five things you can do to improve the quality of the air you and others breathe.

2. Some of the gases found in the troposphere are Rn, CO_2, CO, O_2, Ar, and N_2.

   a. Rank them in order of their abundance in the troposphere.

   b. For which of these gases is it convenient to express concentration in parts per million?

   c. Which of these gases is/are currently regulated as an air pollutant where you live?

   d. Which of these gases is/are found in Group 8 of the periodic table, the noble gases?

3. Identify three sources of particulate matter found in air. Explain the difference between $PM_{2.5}$ and PM_{10} in terms of size and health effects.

4. a. The concentration of argon in air is approximately 0.934%. Express this value in ppm.

   b. The air exhaled from the lungs of a smoker has a concentration of 20–50 ppm CO. In contrast, air exhaled by nonsmokers is 0–2 ppm CO. Express each concentration as a percent.

c. On a very humid day, the water vapor concentration in the air might be 8,500 ppm. Express this as a percent.

d. A sample of air taken from Antarctica was found to contain 8 ppm water vapor. Express this as a percent.

5. Gases found in the atmosphere in small amounts include Xe, N_2O, and CH_4.

a. What information does each chemical formula convey about the number and types of atoms present?

b. Write the names of these gases.

6. Hydrocarbons are important fuels that we burn primarily for energy.

a. What is a hydrocarbon?

b. Rank the following hydrocarbons by the number of carbons they contain from smallest to largest: pentane, ethane, octane, hexane, propane.

c. We suggested "mother eats peanut butter" as a memory aid for the names of the first four hydrocarbons by number of carbon atoms. Propose a new one that includes pent-, the prefix that indicates five carbon atoms.

7. Air contains trace amounts of substances. Define trace amounts and give two examples of possible trace substances found in air.

8. If you had a sample of 500 particles of air, how many of these particles would be nitrogen, oxygen, and argon?

9. Count the atoms on both sides of the equation to demonstrate that these equations are balanced.

a. $2\ C_3H_8(g) + 7\ O_2(g) \longrightarrow 6\ CO(g) + 8\ H_2O(l)$

b. $2\ C_8H_{18}(g) + 25\ O_2(g) \longrightarrow 16\ CO_2(g) + 18\ H_2O(l)$

10. Consider this representation of the reaction between nitrogen and hydrogen to form ammonia (NH_3).

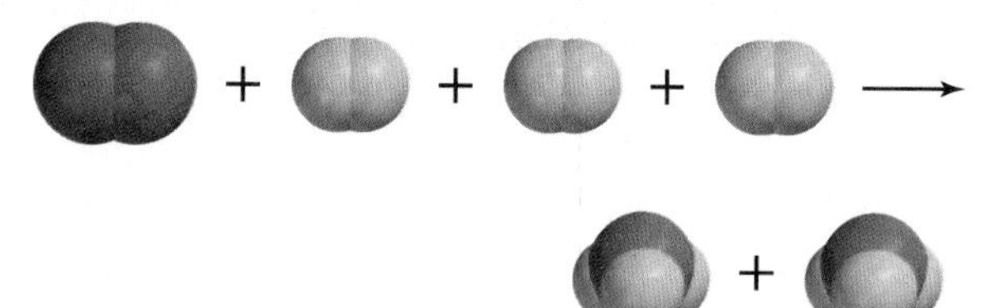

a. Are the masses of reactants and products the same? Explain.

b. Are the numbers of molecules of reactants and of products the same? Explain.

c. Are the total number of atoms in the reactants and the total number of atoms in the products the same? Explain.

11. Write balanced chemical equations to represent these reactions.
Hint: Nitrogen and oxygen are both diatomic molecules.

a. Nitrogen reacts with oxygen to form nitrogen monoxide.

b. Ozone decomposes into oxygen and atomic oxygen (O).

c. Sulfur reacts with oxygen to form sulfur trioxide.

12. These questions relate to the combustion of hydrocarbons.

a. LPG (liquid petroleum gas) is mostly propane, C_3H_8. Balance this equation.

$$C_3H_8(g) + O_2(g) \longrightarrow CO_2(g) + H_2O(g)$$

b. Cigarette lighters burn butane, C_4H_{10}. Write a balanced equation for the combustion of butane.

c. With a limited supply of oxygen, both propane and butane can burn incompletely to form carbon monoxide. Write balanced equations for both reactions.

13. Balance the following equations in which ethane burns in oxygen.

a. $C_2H_6(g) + O_2(g) \longrightarrow C(s) + H_2O(g)$

b. $C_2H_6(g) + O_2(g) \longrightarrow CO(g) + H_2O(g)$

c. $C_2H_6(g) + O_2(g) \longrightarrow CO_2(g) + H_2O(g)$

d. Explain why the coefficients for oxygen vary, depending on whether C, CO, or CO_2 is formed.

14. Air is an example of a mixture. Provide two more examples of mixtures and explain why they are mixtures.

15. When a person inhales air, 21% of that air is oxygen. However, only 16% of the air exhaled is oxygen. Provide at least two reasons for why there is a decrease in the amount of oxygen exhaled.

16. What is the troposphere and why is it important?

17. Dry ice is solid carbon dioxide. Represent carbon dioxide in the following three ways: symbolic, particulate, and macroscopic.

18. Name the following nitrogen-containing compounds: NO_2, N_2O, NO, NCl_3, and N_2O_4.

19. Name the following compounds: CCl_4, SO_3, Cl_2O_6, and P_4S_3.

20. A carbon monoxide detector will go off if the concentration of CO is 400 ppm or greater for a period of 4–15 minutes.

a. Express 400 ppm as a percent.

b. Why is carbon monoxide considered to be an air pollutant?

c. What are the health effects resulting from long-term exposure to carbon monoxide?

21. Sulfur dioxide and nitrogen dioxide are considered to be air pollutants.

a. Where would you most likely find these pollutants?

b. Which of these pollutants is more toxic?

c. Express 0.045 ppm nitrogen dioxide in ppb.

22. Nail polish remover containing acetone was spilled in a room 6 m × 5 m × 3 m. Measurements indicated that 3,600 mg of acetone evaporated. Calculate the acetone concentration in micrograms per cubic meter.

Concentrating on Concepts

23. "Air prints" were mentioned in Your Turn 2.1. Examine these two photographs. The first picture is from a waterfront café on the Greek Island of Hydra, and the other is above a busy street in Tianjin, China. List three ways in which each photo shows the air print of humans. **Hint:** Some may not be visible but rather implied by the photograph.

(both): © Bradley D. Fahlman

24. The EPA AirNow website states that "Air quality directly affects our quality of life." Demonstrate the wisdom of this statement for two air pollutants of your choice.

25. In Your Turn 2.2, you calculated the volume of air exhaled in a day. How does this volume compare with the volume of air in your chemistry classroom? Show your calculations.
Hint: Think ahead about the most convenient unit to use for measuring or estimating the dimensions of your classroom.

26. According to Table 2.1 the percentage of carbon dioxide in inhaled air is lower than it is in exhaled air. How can you account for this relationship?

27. A headline from the *Anchorage Daily News* in Alaska (January, 17, 2008): "Family in car overcome by carbon monoxide. Fire department saves five after slide into snow bank."
    a. If your car is in the snow bank and the engine is running, CO may accumulate inside your car. Normally, however, CO does not accumulate in the car. Explain.
    b. Why didn't the occupants detect the CO?

28. Consider how life on Earth would change if the concentration of oxygen were cut in half. Give two examples of things that would be affected.

29. Explain why CO is named the "silent killer." Select two other pollutants for which this name would not apply and explain why not.

30. Undiluted cigarette smoke may contain 2–3% CO.
    a. How many parts per million is this? How many parts per billion?
    b. How does this value compare with the National Ambient Air Quality Standards for CO in both a 1-hour and an 8-hour period?
    c. Propose a reason why smokers do not die from carbon monoxide poisoning.

31. In the Northern Hemisphere, the ozone season runs from about May 1 to October 1. Why are ozone levels typically not reported in the winter months?

32. A certain city has an ozone reading of 0.13 ppm for 1 hour, and the permissible limit is 0.12 for that time. You have the choice of reporting that the city has exceeded the ozone limit by 0.01 ppm or saying that it has exceeded the limit by 8%. Compare these two methods of reporting.

33. Here are ozone air quality data for Atlanta, Georgia, from August 1–10, 2015. The primary pollutant was ozone:

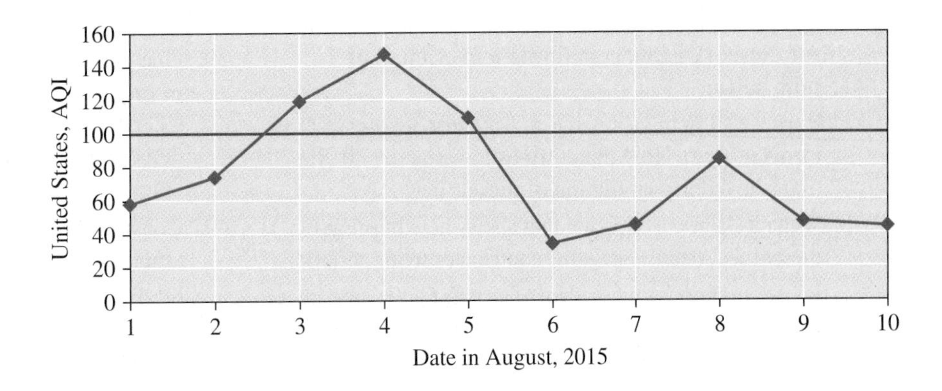

a. In general, which groups of people are the most sensitive to ozone?

b. The U.S. Environmental Protection Agency determined that air rated above 100 is hazardous for some or all groups. For the data shown, how many days was the air hazardous?

c. Ozone levels drop off sharply at night. Explain why.

d. During the daytime, the ozone dropped off sharply after August 5. Propose two different reasons that could account for this observation.

e. Data for the month of December is not shown. Would you expect the ozone levels to be higher or lower than those in August? Explain.

34. Here are air quality data for the last week of 2015, in Beijing, China, for air quality based on the primary pollutant of $PM_{2.5}$.

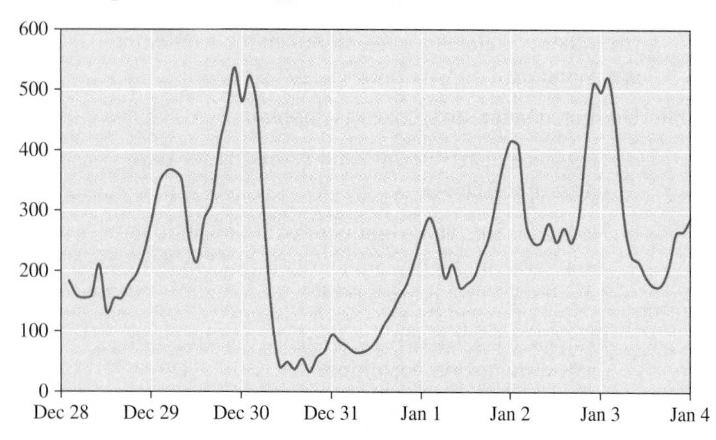

a. In general, which groups of people are the most sensitive to particulate matter?

b. The U.S. Environmental Protection Agency determined that air rated between 51–100 is acceptable except for particularly sensitive groups. For the data shown, when was the air acceptable or better?

c. The levels of PM do not necessarily drop off at night the way they do for ozone. Explain.

d. The levels of particulate matter can increase sharply. Propose two different reasons that could account for this observation.

35. Prior to 1990, diesel fuel in the United States could contain as much as 2% sulfur. New regulations have changed this, and today most diesel fuel is ultra-low sulfur diesel (ULSD) containing a maximum of 15 ppm sulfur.

a. Express 15 ppm as a percent. Likewise, express 2% in terms of ppm. How many times lower is the ULSD than the older formulation of diesel fuel?

b. Write a chemical equation that shows how burning diesel fuel containing sulfur contributes to air pollution.

c. Diesel fuel contains the hydrocarbon $C_{12}H_{26}$. Write a chemical equation that shows how burning diesel adds carbon dioxide to the atmosphere.

d. Comment on burning diesel fuel as a sustainable practice, both in terms of how things have improved and in terms of where they still need to go.

36. Here is a U.S. map with the peak ozone data for August 17, 2015:

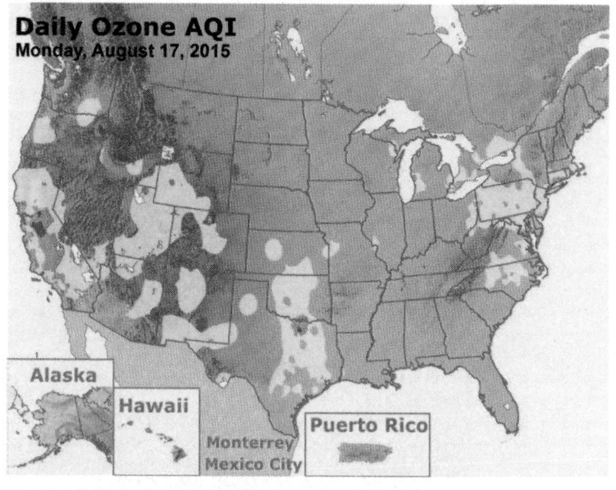

Source: EPA AirNow

a. These data are typical in that most of the ozone pollution is expected in California, Denver, Texas, the Midwest, and the East Coast. Why is ozone pollution so high in these regions of the country?

b. The midwest typically has high ozone levels in the summer, but not on this particular day. Offer a possible explanation.

c. Why are inland areas in California, such as the Sacramento Valley, likely to have worse air quality than the California coast?

37. Look up the air quality data for ozone in two different cities, one hot and dry, and the other cooler and more rainy. Account for any differences you find. **Hint:** Use *State of the Air*, a website posted by the American Lung Association or the AirNow website.

38. At certain times of the year, inhabitants of the beautiful city of Santiago, Chile, breathe some of the worst air on the planet.

a. Driving private cars has been severely restricted in Santiago. How specifically does this improve air quality?

b. Although the pollution of Santiago is comparable to that in other cities, its air quality is much worse. Suggest geographical features that might be responsible.

39. One can purchase a carbon monoxide monitor that immediately sounds an alarm if the concentration of CO reaches a threshold. In contrast, most radon detection systems sample the air over a period of time before an alarm sounds. Why the difference?

40. Consumers now can purchase paints that emit only low amounts of VOCs. However, these consumers may not know why it matters to purchase this paint.

a. What would you print on the label of a paint can to make the point that a low-VOC paint is a good idea?

b. We apply paint to many outdoor surfaces, such as buildings, bridges, and fence posts. Comment on the environmental effects of the VOCs that these paints emit.

41. a. Explain why running outdoors (as opposed to sitting outdoors) increases your exposure to pollutants.

    b. Running indoors at home can decrease your exposure to some pollutants, but may increase your exposure to others. Explain.

42. Select a profession of your choice, possibly the one you intend to pursue. Name at least one way that a person in this profession could have a positive effect on air quality.

Exploring Extensions

43. "Air pollution is a diffuse problem, the shared fault of many emitters. It is a classic example of the tragedy of the commons." (Source: Introduction to Air in California by David Carle, 2006.) Explain the phrase "tragedy of the commons," and how air pollution is a classic example.

44. Mercury, another serious air pollutant, is not described in this chapter. If you were a textbook author, what would you include about mercury emissions? How would you connect mercury emissions to the sustainable use of resources? Write several paragraphs in a style that would match that of this textbook. Be sure to reference your information sources.

45. The EPA oversees the Presidential Green Chemistry Challenge Awards. Use the EPA website to find the most recent winners of the award. Pick one winner and summarize in your own words the green chemistry advance that merited the award.

46. Here are two scanning electron micrograph images of particulate matter, courtesy of the National Science Foundation and researchers at Arizona State University. The first is of a soil particle and the second of a rubber particle, and each is about 10 μm in diameter.

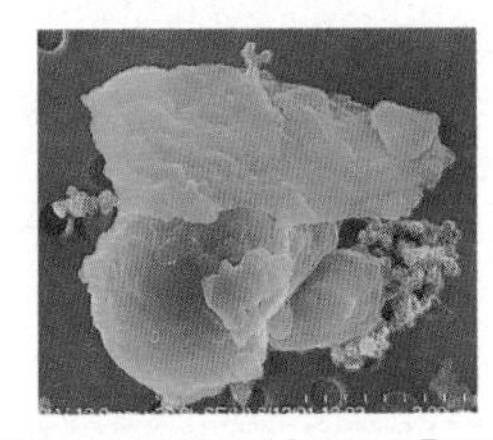

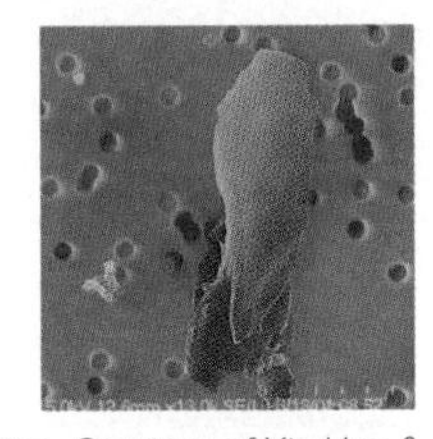

(both) Source: National Science Foundation, Courtesy of Xin Hua & James Anderson

    a. Suggest a likely source of the rubber particle. Name two other substances that might contribute PM to the air.

    b. The soil particle is composed mainly of silicon and oxygen. What other elements are commonly present in the rocks and minerals in Earth's crust?

    c. What about these photographs suggests that these particles would inflame your blood vessels?

47. Ultrafine particles have diameters less than 0.1 μm. In terms of their sources and health effects, how do these particles compare with $PM_{2.5}$ and PM_{10}? Use the Internet to locate the most up-to-date information. Be sure to reference your information sources.

48. Consider this graph that shows the effects of carbon monoxide inhalation on humans.

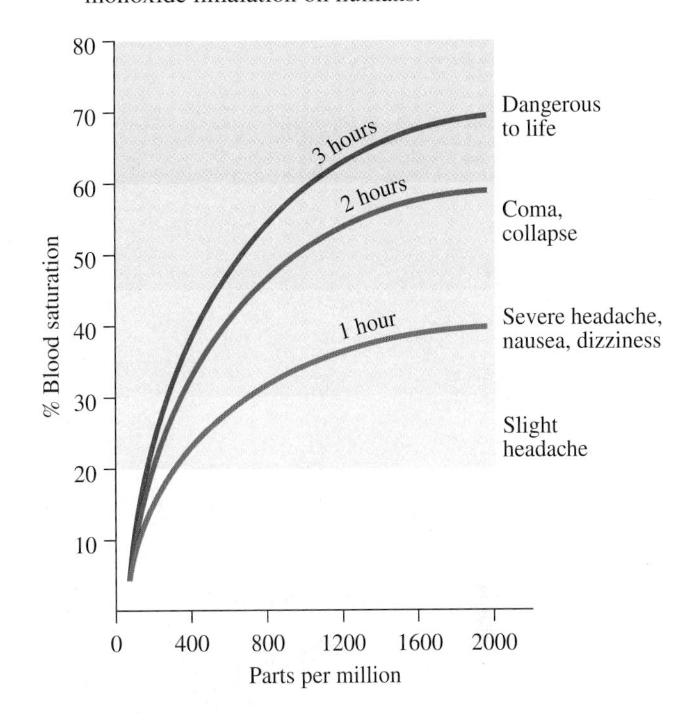

    a. Both the amount of exposure and the duration of exposure have an effect on CO toxicity in humans. Use the graph to explain why.

    b. Use the information in this graph to prepare a statement to include with a home carbon monoxide detection kit about the health hazards of carbon monoxide gas.

49. You may have admired the beauty of hardwood floors. Polyurethane is the finish of choice for floors because it is more durable than varnishes and shellacs. Until recently, polyurethane was always an oil-based paint. But recently, the Bayer Corporation developed a water-based polyurethane that reduces the amount of VOCs by 50–90%. In 2000, Bayer was awarded a Presidential Green Chemistry Challenge award for this development. Prepare a summary of this work. Also check stores to see if any water-based polyurethanes are available in your area.

50. Homeowners eager to improve their energy bills will use spray-polyurethane foam to add insulation to their attic spaces.

    a. While safe when the reaction is complete, many insulation foams release isocyanate during and immediately after installation. What are the hazards of this volatile organic compound?

    b. Professor Richard Wool at the University of Delaware developed bio-based substitutes for high-performance materials, including isocyanate foams. For his innovative work, he won a Presidential Green Chemistry Challenge Award in 2013. Prepare a summary of his accomplishments.

CHAPTER 3 Radiation from the Sun

REFLECTION

Protection from the Sun

If you were outdoors on a sunny day, list the ways that you would protect yourself from sunlight. Rank them from what you think is most effective to least effective.

The Big Picture

In this chapter, we will answer the following questions:

- How is sunlight measured?
- What types of radiation are more energetic than others?
- What is the link between sun exposure and skin cancer?
- How does Earth's atmosphere naturally protect us from the Sun?
- Can this source of natural protection be eliminated, and if so, can it be restored?
- How do sunscreens and sunblocks work?

Introduction

Although many people love to be outside on a sunny day, it is important for us to understand both the positive and negative implications of the Sun's radiation on Earth. On the positive side, the Sun's radiation provides energy to grow food that you enjoy, forms the important vitamin D in your bloodstream, and maintains Earth's temperature at a livable level. However, the Sun's radiation causes premature weathering of materials and health problems such as sunburns, skin cancer, and damage to your eyes.

Your Turn 3.1 You Decide Making Decisions

In this chapter, we will examine exposure to sunlight and the means of protection. In order to guide our exploration, complete the following activity:

Over a three-day period, record the types of sun exposure you experience throughout the day. Note the time of day you are exposed to sunlight, the amount of your skin exposed, your location, and the length of time you are exposed to the Sun.

In this chapter, we will examine the Sun's radiation by investigating its composition and effects on our health. You may be aware that applying sunscreen protects your skin against sunlight in the summertime. However, did you know that our atmosphere also provides a natural source of protection against these harmful rays? We will expand upon the discussion of the atmosphere from Chapter 2 to pinpoint the components in our atmosphere that provide this natural shielding. We will also describe the chemistry of consumer sunscreen formulations. Throughout this chapter, you will learn how chemistry plays an important role in shielding us from the harmful effects of radiation.

Let's start by answering the question: What *is* radiation?

3.1 Dissecting the Sun: The Electromagnetic Spectrum

Imagine you are relaxing or working in the sun on a clear summer day. The Sun is bright, your skin becomes warm, and your skin may eventually darken—although hopefully the damage does not cause a sunburn! Have you given much thought regarding the composition of the Sun's radiation? What do you think is in a sunray? What is radiation? Before we discuss these questions, complete the following activity.

Your Turn 3.2 You Decide Sun Damage

To get you started thinking about the Sun and the damage it can cause, discuss the following questions with a classmate:

a. What is emitted from the Sun?
b. Why do you think exposure to the Sun can cause damage?
c. What kinds of damage can the Sun cause?
d. What are some of the positive effects of the Sun?

You may have heard the term "radiation" used in several contexts: radiation from the Sun, radiation from nuclear reactions, microwave radiation, or radiation used for medical treatments/diagnoses. In this chapter, we will focus on radiation from the Sun; however, future chapters will address the other types of radiation.

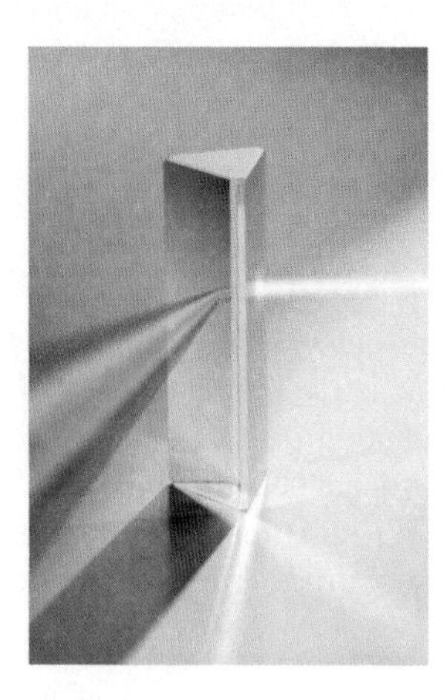

Figure 3.1

Separation of visible light into its components via a diffraction crystal.

© *Thinkstock/Index Stock RF*

Radiation is often designated as *electromagnetic (EM)*. This term indicates that the waves are composed of both electric and magnetic fields.

You may have come across the acronym ROY G. BIV to remember the colors in a spectrum: **R**ed, **O**range, **Y**ellow, **G**reen, **B**lue, **I**ndigo, **V**iolet. Indigo is the most controversial color on this list. Scientists are not in total agreement about its placement in the spectrum.

One type of **radiation** is energy emitted by a hot object such as the Sun. This energy is then absorbed by other objects such as humans, Earth, plants, animals, etc. The energy from the Sun takes forms that we can feel (*e.g.,* heat) and see (*e.g.,* visible light). However, there are other forms that are actually invisible to the naked eye. One of these is ultraviolet (UV) radiation. As we look at the characteristics of radiation, we will use visible light as an example.

Light is continually emitted by the Sun and, after some time, reaches Earth. Have you ever observed a rainbow or the colors generated by a prism or crystal? Prisms and raindrops can separate the visible light reaching Earth into a spectrum of colors (Figure 3.1). The general spectrum we see contains the colors red, orange, yellow, green, blue, indigo, and violet.

Each color of the visible light spectrum represents a particular range of wavelengths of radiation. The word *wavelength* correctly suggests that light travels through space like waves travel through a body of water (Figure 3.2). However, while water waves travel exclusively through water, electromagnetic radiation can travel as a wave

Figure 3.2

While water waves travel through water, electromagnetic radiation can travel as a wave through empty space.

© *McGraw-Hill Education/Brian Kanof*

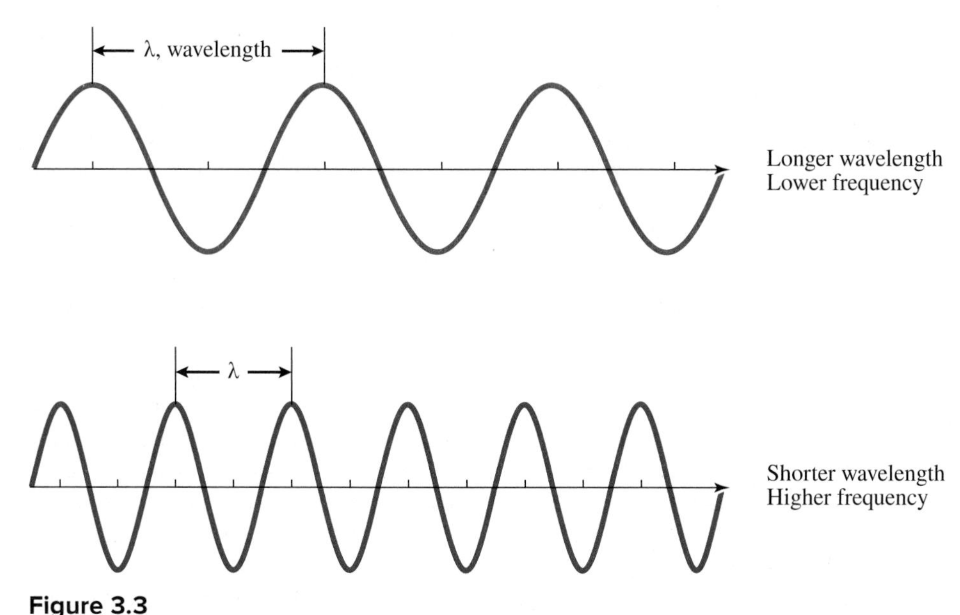

Figure 3.3

Illustration of wavelength and frequency for a wave.

through air or even through empty space. **Wavelength** (Greek: lambda, λ) simply refers to the distance between successive peaks in a wave and is reported in units of length (Figure 3.3). Sometimes, scientists are more interested in how many waves pass a fixed point in a certain amount of time—the **frequency** (Greek: nu, ν) of the wave. Frequency is also illustrated in Figure 3.3. In this Figure, two different waves are represented, with the assumption that both waves are traveling at the same speed across the page. The top wave has a longer wavelength and a lower frequency, whereas the bottom wave has a shorter wavelength and a higher frequency.

As seen above, wavelength and frequency are inversely proportional. This means that as either wavelength or frequency increases, the other decreases. We can show this relationship by using a simple equation (Equation 3.1), in which c is a constant known as the **speed of light**. The value of c is 3.00×10^8 m/s and represents the maximum velocity that light is able to travel through air.

$$\text{frequency } (\nu) = \frac{\text{speed of light (c)}}{\text{wavelength } (\lambda)} \qquad \textbf{[3.1]}$$

Did You Know? The speed of light is best defined as the maximum speed of light in a *vacuum*. That is, an atmosphere that is free of gas molecules that would impede the propagation of light. It should be noted that the exact value for the speed of light in air *vs.* vacuum is actually 870 μm/s slower: 2.99705×10^8 m/s *vs.* 2.99792×10^8 m/s, respectively. However, since the value of c is so large, the speed of light in air is still the same at three significant figures of accuracy (3.00×10^8 m/s).

Your Turn 3.3 Scientific Practices Wavelength and Frequency

Using **Equation 3.1**, answer the following questions:

a. What is the frequency (in Hz) of green light, with a wavelength of 525 nm?

b. How many waves of green light would pass a fixed point in 1 minute? How about in 1 hour?

c. What is meant by the *amplitude* of a wave? Do changes in amplitude affect the wavelength or frequency of the wave? Explain.

The velocity or speed of a wave is not limited to visible light. In fact, we can determine the speed of any wave if we know its frequency and wavelength:

$$\text{velocity (v)} = \text{wavelength } (\lambda) \times \text{frequency } (\nu) \qquad \textbf{[3.2]}$$

If we evaluate Equations 3.1 and 3.2, we can identify the units for each of these parameters. For instance, if the speed of light is in units of m/s, and wavelength is in m, then:

$$\text{frequency } (\nu) = \frac{\cancel{m}/s}{\cancel{m}} = 1/s \text{ or } s^{-1} \text{ (also known as Hertz, Hz)}$$

The combination of wavelength (m) and frequency (s^{-1}) generates the velocity (or speed) of any wave in units of $m{\cdot}s^{-1}$ or m/s. Although this is referred to as a *standard* unit, you are probably more familiar with other units of speed such as km/h or miles per hour (mph). These are known as *derived* units, and may be found once conversion factors are known. For instance, to convert the speed of a wave from m/s to km/h, we would need two conversion factors:

i) there are 1000 meters (m) in 1 kilometer (km)
ii) there are 3600 seconds (s) in one hour (h); that is, 60 seconds/min × 60 min/h

Accordingly, we could perform the conversion of 3.00×10^8 m/s to km/h as:

$$\left(\frac{3.00 \times 10^8\ \cancel{m}}{1\cancel{s}}\right) \times \left(\frac{1\ km}{1000\ \cancel{m}}\right) \times \left(\frac{3600\ \cancel{s}}{1\ h}\right) = 1.08 \times 10^9\ km/h$$

The human eye can only detect a very small range of wavelengths—namely those between 4×10^{-7} m and 7×10^{-7} m. You will recognize from Chapter 1 that these numbers are in **scientific notation**. However, most scientists find it more convenient to represent the values as 400×10^{-9} m and 700×10^{-9} m, respectively. Take a few moments to convince yourself that the numbers are the same. You may remember from your studies of the metric system that the prefix for 10^{-9} means *nano- (n)*, or one-billionth. One nanometer is one-billionth the length of a meter. In mathematical relationships, this is:

$$1\ nm = \frac{1}{1{,}000{,}000{,}000}\ m = \frac{1}{1 \times 10^9}\ m = 1 \times 10^{-9}\ m$$

Applying this relationship to the visible wavelengths, we can use dimensional analysis to convert from m to nm:

$$700 \times 10^{-9}\ \cancel{m} \times \frac{1\ nm}{1 \times 10^{-9}\ \cancel{m}} = 700\ nm$$

Therefore, we can represent the visible range of wavelengths as 400 nm to 700 nm.

Now that we have used visible light to define many aspects of radiation, let's examine the entire spectrum of sunlight referred to as the **electromagnetic (EM) spectrum** (Figure 3.4). This spectrum can be thought of as a continuum of waves, from gamma rays that have very small wavelengths—comparable to the diameter of an atomic nucleus (approx. $10^{-14} - 10^{-12}$ m)—to radio waves that have large wavelengths of the same magnitude as the items in our macroscopic world (*ca.* 1–100 m).

Your Turn 3.4 Skill Building Analyzing a Rainbow

Water droplets in a rainbow act as prisms to separate visible light into its colors.

a. In **Figure 3.4**, which color has the longest wavelength? The highest frequency?
b. Green light has a wavelength of 500 nm. Express this value in meters.

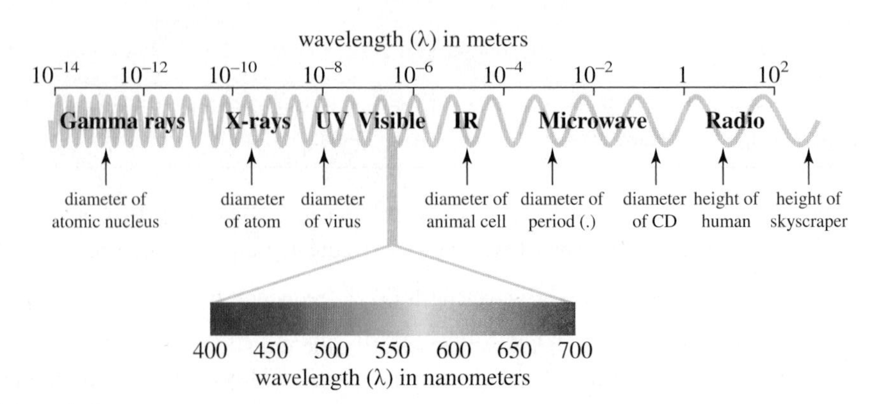

For an interactive illustration of the visible spectrum, see **Connect**.

Figure 3.4

The electromagnetic spectrum, highlighting the visible region. Note that 10^x is equivalent to 1×10^x.

Source: U.S. EPA

Throughout *Chemistry in Context*, we will examine several regions of the electromagnetic spectrum. Beyond the red side of the visible spectrum, with longer wavelengths than red light, lies the infrared (IR) region. Although you cannot see this radiation, you can feel its warming effect. Microwaves (at even longer wavelengths) are also invisible to the naked eye, and are used for radar detection and heating our food. The length of a microwave is on the sub-millimeter to centimeter scale—on average, about the diameter of the period that follows this sentence. Finally, at the longest range of wavelengths are radio waves, which are transmitted through the air to facilitate your cell phone conversations.

Beyond the violet side of the visible region are ultraviolet, X-ray, and gamma radiation. The two shortest wavelength ranges in the EM spectrum are exhibited by gamma rays that accompany nuclear radiation, and X-rays used in medical diagnoses. The wavelength of X-rays are on the Angstrom (Å) scale, or 1×10^{-10} m, which is comparable to the diameter of individual atoms.

Your Turn 3.5 Skill Building Relative Wavelengths

Consider these four types of radiant energy from the electromagnetic spectrum: infrared, microwave, ultraviolet, visible.

Hint: See Figure 3.4.

a. Arrange them in order of increasing wavelength.
b. Approximately how many times longer is a wavelength associated with a radio wave than one associated with an X–ray?

Even though the Sun emits many types of radiation, not all of it reaches Earth's surface with equal intensities. In this context, intensity relates to the amount of radiation that impacts a certain area of Earth. Figure 3.5 displays a graph of the relative intensity of solar radiation at Earth's surface as a function of the wavelength of the radiation.

Your Turn 3.6 Scientific Practices Energy from the Sun

Examine the distribution of energy from the Sun shown in **Figure 3.5**.

a. Which type of electromagnetic radiation comprises the greatest portion of the energy from the Sun?
b. Which type of radiation is the most intense?

The graph shows that approximately 53% of the total energy emitted by the Sun reaches Earth's surface as infrared radiation, which is a major source of heat for our planet. We will describe the consequences of this phenomenon in Chapter 4. Although only approximately 8% of the sunlight reaching Earth is in the form of UV radiation, we will see in the next two sections that this is the most harmful type of radiation that showers our skin.

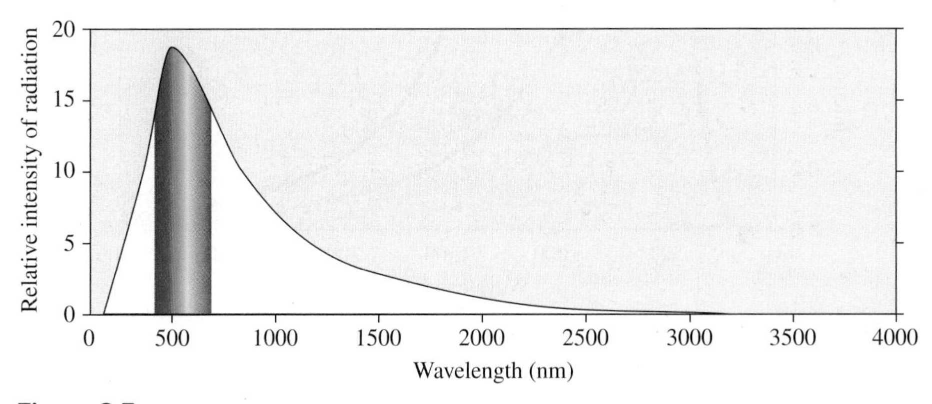

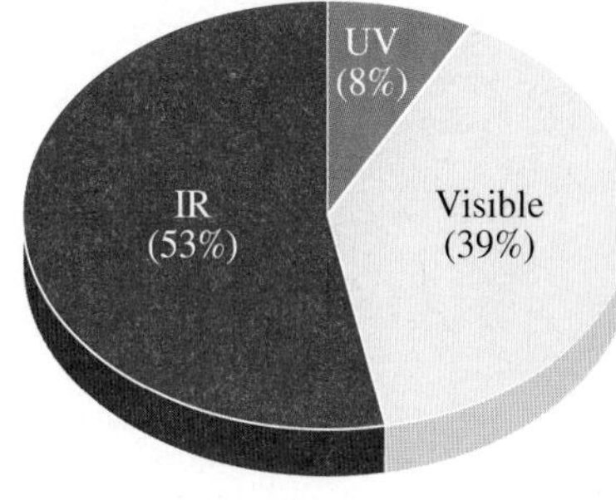

Figure 3.5

Composition of the Sun's radiation reaching the surface of Earth.

3.2 | The Personalities of Radiation

In the early 1900s, scientists were puzzled by radiation distribution profiles similar to the one shown in Figure 3.5. In particular, the prevailing theory predicted that as the frequency of emitted radiation increased, its intensity would also increase. This meant that heated objects such as an oven, a flame, or even the Sun were predicted to emit a continuum of high-intensity radiation, which would become more intense as it approached the low-wavelength UV region of the electromagnetic spectrum. However, as shown in Figure 3.6, this behavior is not observed experimentally.

Did You Know? This failure of the classical theory was referred to as "the UV catastrophe"!

The German physicist Max Planck (1858–1947) was particularly interested in this problem and worked to explain the disparity between experiment and theory. In particular, he noted that the peaks in the radiation profiles shown in Figure 3.6 shift from the IR to visible regions with increasing temperature. To illustrate this phenomenon, let's consider the flame from a propane torch. As the temperature increases, its color shifts from red/orange to yellow and eventually blue (Figure 3.7). If we measure the temperature of the various colored flames, we can see that the hottest is blue (> 2000 °C), and the coolest is red (~800 °C). Since temperature is a measure of thermal energy, this observation implies that the *energy* of the flame is related to its wavelength. Hence, in addition to wavelengths and frequencies, each region of the electromagnetic spectrum also has a range of energies.

However, since only certain colors are observed in flames of different temperatures, a heated object will emit only certain energies of radiation, rather than a continuum. When energy is distributed in this noncontinuous way, it is said to be **quantized**. As an analogy, consider a staircase as representing various energies of electromagnetic radiation (Figure 3.8). Just as our feline friends are only able to sit on individual steps, and not the space between levels, the Sun emits radiation with only certain energies.

Up to now, we have described radiation in terms of waves. Although this description is well-established and very useful, there is another convenient way to describe radiation. In 1921, Albert Einstein (1879–1955) won the Nobel Prize in physics for suggesting that this quantized radiation be viewed as a collection of individual bundles of energy called **photons**. Photons can be considered "particles of light"; however, they do not have the same properties as other particles you may know. One major difference is that photons do not have a mass. The ideas of quantized energy and photons form the basis *of quantum theory,* which studies the behavior of atoms, molecules, and subatomic particles by examining their energies.

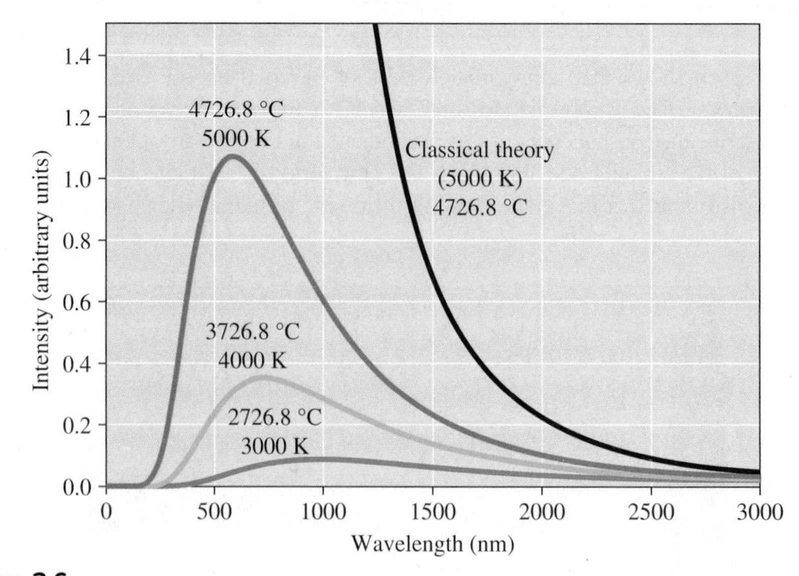

Figure 3.6

Comparison of the radiation emission profiles predicted by the early 20th century theory (black line), and the experimentally observed emission profiles at varying temperatures (colored lines). Note that the temperatures are shown in Kelvin (K), with their corresponding temperatures in °C. The conversion between K and °C is: K = °C + 273.15.

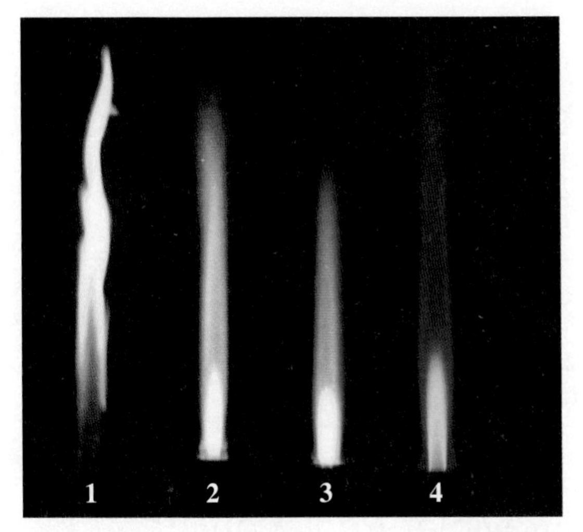

Figure 3.7

Comparison of the colors of Bunsen burner flames based on their temperatures. The oxygen:fuel ratio is increasing for flames 1–4, which results in an increasingly hotter flame.

Source: Arthur Jan Fijałkowski (via Wikimedia: https://commons.wikimedia.org/wiki/File:Bunsen_burner_flame_types.jpg)

Figure 3.8

An analogy for quantization, illustrating that radiation consists of only certain values (steps) of energy.

© Bradley D. Fahlman

Although scientists have discovered the particulate nature of energy, the wave description is still useful and relevant. Both are valid descriptions for energy. This twofold nature of radiation is known as **wave-particle duality**, and it may seem to defy common sense. After all, how can light be described as waves and particles at the same time? To add to the confusion, this description is not just limited to electromagnetic radiation, but to all forms of matter. For instance, Figure 3.9 illustrates how water can be visualized as having both wave-like and particle-like properties.

These two views are linked in a relationship that is one of the most important equations in modern science (Equation 3.3). This equation will also assist us as we look at the Sun's radiation throughout this chapter:

$$E = \frac{hc}{\lambda} \quad [3.3]$$

where: E = the energy of a single photon in units of joules, J

h = Planck's constant (6.626×10^{-34} J·s)

c = 3.00×10^{8} m/s

The above equation demonstrates that the energy of radiation is inversely proportional to its wavelength, λ. Therefore, as the wavelength of radiation gets shorter, the energy of its photons increases. This fact will be very important as we look at natural and chemical protection from the Sun's radiation.

As the size of an object gets larger, it becomes more particle-like rather than wave-like. For instance, an apple falling from a tree behaves as a pure particle, being described by its mass, velocity, and momentum, rather than by its wavelength and frequency. In contrast, subatomic species, such as electrons, are best described by their wave-like attributes.

Photon self-identity issues

© Nick Kim/CartoonStock

Figure 3.9

An illustration of the wave-particle duality of matter—a drop of water (particle-like) suspended above liquid water (wave-like).

© Trout55/Getty Images RF

Your Turn 3.7 Skill Building Wavelength, Frequency, and Energy

Return to **Figure 3.4**, and calculate the amount of energy present in a wave of red light and a wave of blue light. Are they the same amount of energy? Why or why not?

In the not-too-distant past, many people used film exposure to take pictures. A "dark" room was a place where this film was developed into the actual images based on how light exposed the film during the taking of the pictures. These dark rooms used red lights during the time of developing the film because a special dye was added to the film that was not affected by light with a wavelength between 620 and 750 nm. Do you think these rooms could have also used blue lights? Why or why not?

3.3 | The ABCs of Ultraviolet Radiation

Refer back to Figure 3.4 to see the full electromagnetic spectrum.

We know the colors of visible light by their names: red, blue, yellow, and so on. Similarly, we call ultraviolet light by different names. Admittedly, these names are not as colorful: UVA, UVB, and UVC. The UVA region lies closest to the violet region of visible light and is the lowest in energy; you may know it as "black light." In contrast, UVC has the highest energy and lies next to the X-ray region of the electromagnetic spectrum. Table 3.1 shows the characteristics of the different regions of UV light.

Your Turn 3.8 Skill Building The ABCs of Solar UV

a. Arrange UVA, UVB, and UVC in order of increasing frequency.
b. Is the order for increasing energy the same as for frequency? Explain.
c. Should you use a sunscreen that claims to only protect against UVC? Explain.

UVC is used to sterilize medical equipment by killing all bacteria on the surface.

With their short wavelengths and high energies, we are fortunate that most UVB and all UVC radiation is absorbed by our atmosphere, and consequently does not reach Earth's surface. Due to their extreme energies, these regions of the electromagnetic spectrum would cause significant damage to biological tissue if allowed to come into contact with humans, animals, or plant life. But how are these wavelengths absorbed by the atmosphere? As we noted in Chapter 2, about 21% of the atmosphere consists of oxygen, O_2. Photons with wavelengths of 242 nm or shorter have sufficient energy to break the bond in an O_2 molecule. Since these wavelengths are found in the UVC region, this radiation is absorbed by reacting with the atmospheric oxygen molecules.

If oxygen was the only molecule that absorbed UV light from the Sun, Earth's surface and the creatures that live on it would still be subjected to damaging ultraviolet radiation of wavelengths longer than 242 nm (*i.e.*, UVB and partial UVC). It is at

Table 3.1 Types of UV Radiation

Type	Wavelength	Relative Energy	Comments
UVA	320–400 nm	Lowest energy	Reaches Earth's surface in the greatest quantity and penetrates farthest into the skin.
UVB	280–320 nm	Medium energy	Most UVB is absorbed by ozone (O_3) in the stratosphere. UVB damages the outermost layer of skin.
UVC	200–280 nm	Highest energy	Although UVC radiation is very harmful, it is completely absorbed by O_2 and O_3 in the stratosphere.

these lower energies that ozone (O_3) plays an important protective role. You may remember from Chapter 2 that ground-level ozone is a harmful component of smog. However, ozone in the upper atmosphere is beneficial, and protects humans from ultraviolet radiation. Because the atoms in oxygen and ozone are connected with bonds of different strengths, the O_3 molecule is more easily broken apart than O_2. Accordingly, UV light of a lower energy (longer wavelength) is sufficient to separate the atoms in O_3. In particular, photons of wavelength 320 nm or shorter carry enough energy to break the O–O bond in ozone. Because all UVB and UVC light is used to break the bonds in oxygen and ozone molecules, effectively none of this harmful radiation reaches Earth's surface. Of course, this is only the case if the ozone layer is completely intact. But what if a hole develops in this protective sheath? We will discuss if and how this is possible and the implications of this worst-case scenario later in this chapter.

See **Connect** to find out where UV light comes from, and explore how the atmosphere protects us from UV light and other harmful solar radiation.

Your Turn 3.9 Skill Building Energy and Wavelength

We just stated that it takes photons in the UVC range (≤ 242 nm) to break the bond in O_2. The bonds in O_3 are somewhat weaker than those in O_2, so lower-energy photons (≤ 320 nm) can break those bonds. How much more energy does a 242-nm photon have compared to a 320-nm photon?

3.4 The Biological Effects of Ultraviolet Radiation

The Sun bombards Earth with countless photons, which are essentially indivisible packages of energy. The atmosphere, the planet's surface, and Earth's living matter all absorb these photons. Radiation in the infrared (IR) region of the spectrum warms Earth and its oceans. The cells in our retinas are tuned to the wavelengths of visible light, which trigger a series of complex chemical reactions that ultimately lead to sight. Compared with animals, green plants capture most of their photons in an even narrower region of the visible spectrum (corresponding to red light). **Photosynthesis** is the process through which green plants (including algae) and some bacteria capture the energy of sunlight to produce glucose ($C_6H_{12}O_6$) and oxygen from carbon dioxide and water.

Photons in the UV region of the electromagnetic spectrum have enough energy to remove electrons from neutral molecules, leaving them positively charged. Even shorter UV wavelength photons break bonds, causing molecules to come apart. In living things, such changes disrupt cells and create the potential for genetic defects and cancers. The interaction of UV radiation with chemical bonds is shown schematically in Figure 3.10. It is an irony of nature that this interaction of radiation with matter explains both the damage ultraviolet radiation can cause and the atmospheric mechanism that protects us from it. We will examine both of these effects in more detail in the next two sections. First, we'll see what happens to the skin when it is exposed to UV radiation.

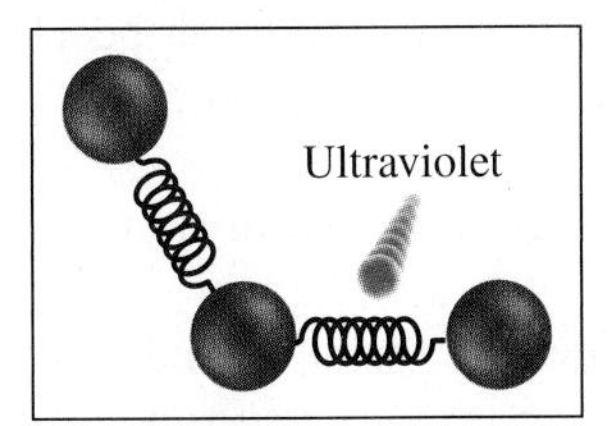

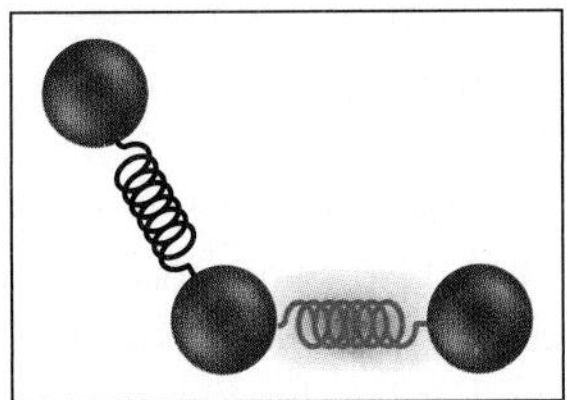
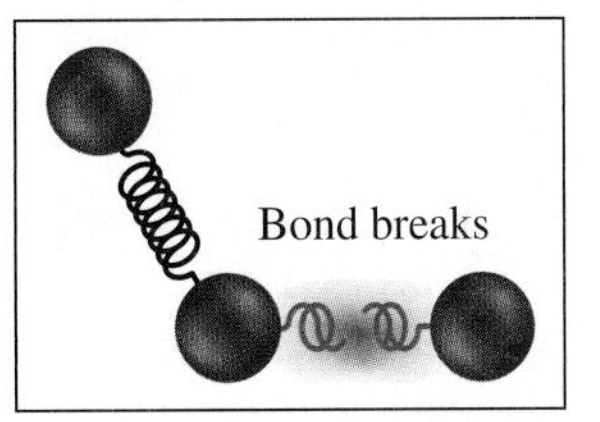

Figure 3.10

Ultraviolet radiation is able to break some, but not all, chemical bonds. Bonds are represented as springs that hold the atoms together but allow the atoms to move relative to each other.

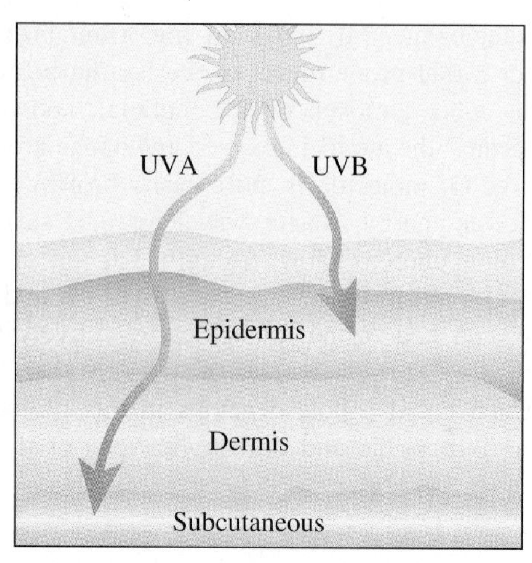

Figure 3.11

UVA and UVB radiation penetrate the layers of the skin to different degrees.

Source: U.S. Surgeon General's report, July 2014

The energy of UV radiation is approximately 10 million times greater than the energy emitted by your favorite radio station. However, did you realize that whether or not your radio is turned on, you are continuously bombarded by radio waves? Your body cannot detect them, but your radio can. The energy associated with each of the radio photons is very low and not sufficient to produce a local increase in the concentration of the skin pigment, **melanin**, as happens with exposure to ultraviolet radiation.

Melanin provides a natural protection against the harmful effects of ultraviolet radiation.

Figure 3.11 demonstrates how UVA and UVB penetrate into your skin. It is rather counterintuitive that UVA radiation, which has a *lower* energy than UVB, actually penetrates *deeper* into the skin. This occurs because the energy of the UVB radiation is aligned well with the energy needed to break chemical bonds; thus, it is rapidly absorbed at the surface of the skin. UVA, on the other hand, does not have enough energy to break bonds, so it travels farther into the body before its energy is absorbed.

When this radiation hits your skin and is absorbed, it sets off a chain of events. First, the energy from the UV photons is absorbed by skin cells. Lower-energy UVA light may remove electrons from molecules such as water, creating **free radicals** and reactive oxygen species. UVB light, with more energy, can cause some of the chemical bonds to break in nearby molecules. DNA molecules in the skin cells can be damaged, either by the free radicals generated by UVA absorption or directly by the UVB radiation. This damage stimulates the production and release of melanin, the chemical compound that gives skin its color. In all cases, skin darkens as a result. Most of the time, your body repairs the damage or the cell dies.

DNA stands for deoxyribonucleic acid, the molecule that stores genetic information. You will learn more about DNA in Chapter 13.

However, another outcome is possible. Although bonds can break in many different molecules, those broken in the DNA molecules of a skin cell are the most serious. Why? Because this damage may cause the DNA to be mutated in a way that leads to skin cancer. Important points about skin cancer include:

- Most skin cancers are linked to the exposure to sunlight.
- Although skin cancer can appear at any age, it is more common in older people.
- Skin cancer can develop many years after repeated, excessive exposure has stopped.
- The UVA in sunlight at Earth's surface is most strongly linked with skin cancers. However, UVB may also play a role.
- Cancer can arise in different types of skin cells. Those in the basal and squamous cells are common, but seldom fatal. In contrast, cancers in the melanocyte cells (melanomas) are more deadly.
- Melanoma is the third-most common cancer diagnosed in young adults between the ages of 15 and 39.

Did You Know? Squamous skin cells are found in the skin's outer layer, the epidermis, and are constantly shed as new ones form. Basal skin cells, on the other hand, are located in the lower part of the epidermis and constantly divide to form new cells, which eventually replace the squamous cells that wear off the skin's surface.

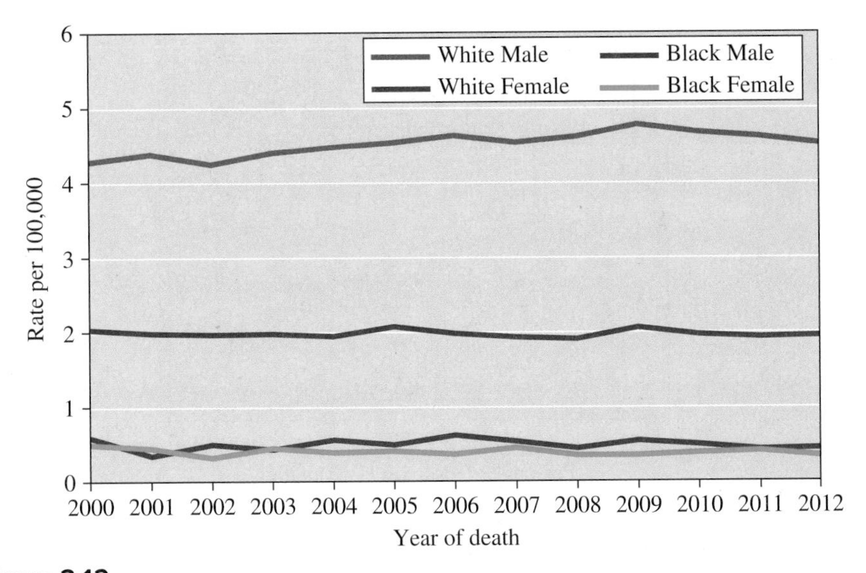

Figure 3.12

Melanoma mortality rates for people (all ages) in the United States, 2000–2012.

Source: National Cancer Institute, SEER Fast Stats, 2012

Skin cancer rates are rising in all countries, despite increased awareness of the dangers of exposure to UV radiation. Figure 3.12 presents the trends in skin cancer in the United States. Although everyone is susceptible to skin cancer, you can see that it is more common for whites, with the highest rates for white males. For all ethnicities, the rate of new melanoma diagnoses in the U.S. has tripled over the past 35 years.

Your Turn 3.10 You Decide Melanoma

a. Using a reliable source, find data for melanoma cases that have occurred over the past 40 years in your country. Did you find diagnosis or mortality rates? Do they trend in the same direction? Look up at least three other forms of cancer. Are the mortality rates for those cancers rising, dropping, or staying the same?

b. Why do the highest incidences of skin cancer occur for white males?

c. Based on what you have read thus far, how might you alter your current behavior regarding sun exposure? How might you advise friends and/or relatives?

Currently, your risk of developing skin cancer is related to a complex blend of chemistry, physics, biology, geography, and human psychology. Factors include your altitude and latitude, how well you protect yourself from the Sun when its rays are the most harsh, whether or not you use indoor tanning beds, and how well you respond to public health campaigns for early detection of skin cancer. Our genetic makeup is another important factor that we are unable to change.

For more about genetics, see Chapter 13.

Some amount of UV radiation can actually be beneficial. UVB light stimulates the production of vitamin D, which is important for healthy bones and immune systems. However, most people acquire sufficient levels of vitamin D through the food they eat, so we don't need to be very concerned about getting it from the Sun. Figure 3.13 illustrates the rather tenuous relationship between UV exposure and our health. While we do need some UV radiation, most people can get enough by spending just a few minutes outside daily.

Your Turn 3.11 You Decide Exposure

Examine the curve shown in **Figure 3.13**. Can you think of another example of something that is required for good health, but dangerous to health in high quantities?

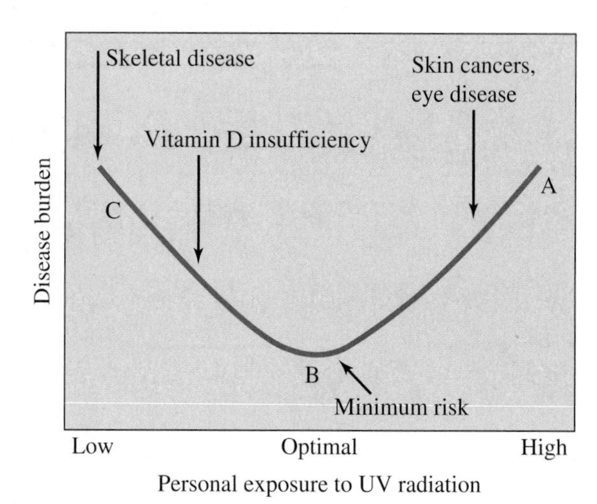

Figure 3.13

The delicate balance of "optimal" exposure to UV radiation and health.

How do you know whether the UV levels are at a dangerous level on any given day? Most developed countries provide a *UV Index* forecast, which uses computer models to predict the risk of sunburn from overexposure to UV light from the Sun. Important factors considered by most models include the ozone concentration in the upper atmosphere, elevation, and cloud cover. Daily UV Index values range from 0 (during nighttime) to 15 (extremely high risk) and are color-coded for ease of interpretation (Figure 3.14). These values also are accompanied by suggestions to help protect your eyes and skin from UV damage as shown in Table 3.2.

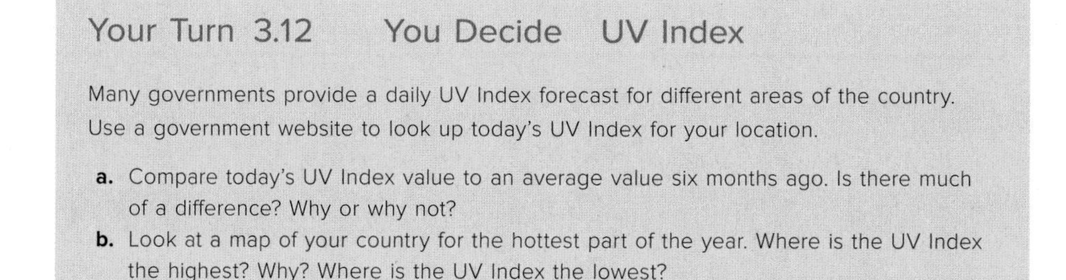

Your Turn 3.12 You Decide UV Index

Many governments provide a daily UV Index forecast for different areas of the country. Use a government website to look up today's UV Index for your location.

a. Compare today's UV Index value to an average value six months ago. Is there much of a difference? Why or why not?

b. Look at a map of your country for the hottest part of the year. Where is the UV Index the highest? Why? Where is the UV Index the lowest?

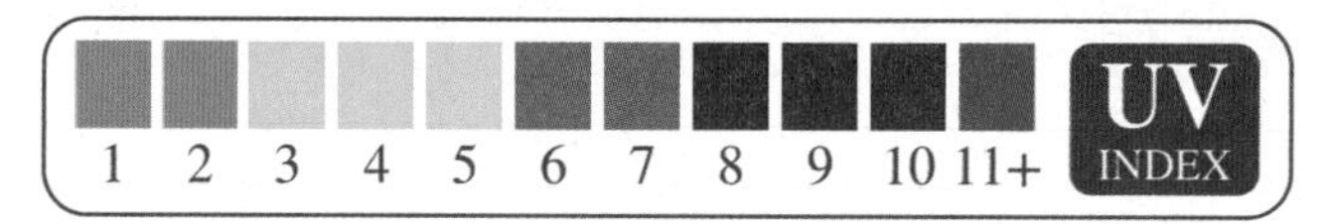

Figure 3.14

The color-coded UV Index scale.

Table 3.2 Suggestions Based on the UV Index Scale

Exposure Category	Index	Tips to Avoid Harmful Exposure to UV
LOW	0–2	If you burn easily, cover up and use sunscreen.
MODERATE	3–5	Stay in the shade when the rays of the Sun are strongest.
HIGH	6–7	Cover up, wear a hat and sunglasses, and use sunscreen. Reduce exposure between 10 am and 4 pm.
VERY HIGH	8–10	Be especially careful if you are outside on sand, snow, or water, as these surfaces reflect UV, increasing your exposure. Minimize exposure between 10 am and 4 pm.
EXTREME	11+	Take full precautions against sunburn. Unprotected skin can burn in minutes. Avoid the Sun between 10 am and 4 pm.

Although the UV Index focuses on skin damage, this is not the only biological effect of UV radiation. Your eyes can be damaged as well. For example, all people, no matter what the pigmentation level of their skin, are susceptible to retinal damage caused by UV exposure. Another effect is cataracts, a clouding of the lens of the eye that can be caused by excessive exposure to UVB radiation (Figure 3.15). However, just as proper clothing and sunscreen can cut down on skin damage, we should also protect our eyes by wearing optical-quality sunglasses capable of blocking at least 99% of UVA and UVB.

In 2015, the U.S. Food and Drug Administration (FDA) proposed a nationwide ban on tanning bed use by people under the age of 18. The U.S. Centers for Disease Control and Prevention (CDC) also warns that the use of tanning booths, tanning beds, and sun lamps is dangerous, particularly for young people. The reason is straightforward: these booths use primarily UVA radiation, the same dangerous wavelengths that are emitted by the Sun. The CDC points out that the concept of "getting a base tan" is actually a response to skin injury and is not as smart as people may think. Rather, it is wise to protect your skin from the Sun.

Normal Eye
Eye affected by cataracts

Figure 3.15

Illustration of cataracts, which can be caused by exposure to UV radiation.

Your Turn 3.13 Scientific Practices Tanning Beds

What are the laws concerning tanning beds in your state or country? Currently, at least 44 U.S. states and many countries restrict tanning in some way for those under the age of 18. Look up the rules where you live.

3.5 | The Atmosphere as Natural Protection

Without Earth's atmosphere above us, ultraviolet radiation from the Sun would inflict tremendous damage on the planet. In this section, we will look at how the atmosphere protects us from the most harmful radiation from the Sun. In particular, this invisible shield contains an especially active ingredient for protection—the ozone layer.

Your Turn 3.14 Scientific Practices The Ozone Layer

a. Had you heard of the ozone layer before this course? If so, in which context? What do you know about the ozone layer?
b. Where is the ozone layer located? Which molecules make up the ozone layer?

In the previous activity, you were able to assess your current knowledge about the ozone layer. What exactly is ozone? Let's look more in depth at this portion of the atmosphere.

If you have ever been near a sparking electric motor or in a severe lightning storm, you may have smelled ozone. Its odor is unmistakable but difficult to describe. Some compare it to that of chlorine gas; others think the odor reminds them of freshly mown grass. Appropriately enough, the name *ozone* comes from a Greek word meaning "to smell."

The ozone and oxygen molecules differ by only one atom (Figure 3.16). As we will see, this difference in molecular structure translates to significant differences in

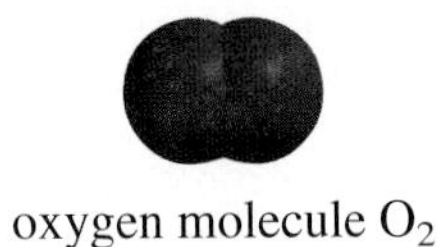

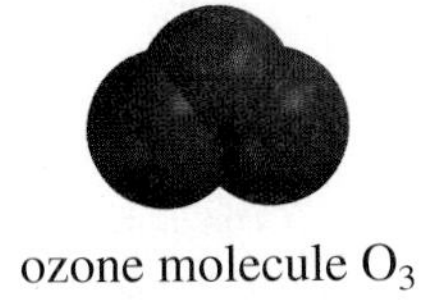

Figure 3.16

Space-filling models of oxygen and ozone molecules

chemical properties. One difference is that ozone is far more chemically reactive than O_2. Ozone can be used to kill microorganisms in water and to bleach paper pulp and fabrics. At one time, ozone was even advocated as a deodorizer for air. This use only makes sense, however, if nobody breathes the air during the deodorizing process! In contrast, you safely can (and must) breathe oxygen constantly.

Ozone forms both naturally and as a result of human activities, as you saw in Chapter 2. Given the high reactivity of ozone, it does not usually persist for long periods of time. If it were not for the fact that ozone is formed anew on our planet, you would not find it except as a curiosity in the chemistry lab.

Ozone can be formed from oxygen, but this process requires energy. A simple chemical equation summarizes the process:

$$\text{Energy} + 3\,O_2 \longrightarrow 2\,O_3 \quad \textbf{[3.4]}$$

Equation 3.4 helps explain why ozone is formed from oxygen in the presence of a high-energy electrical discharge, whether from an electric spark or lightning.

Ozone is reasonably rare in the *troposphere*, the region of the atmosphere closest to Earth's surface (Figure 3.17). Here, only somewhere between 20 and 100 ozone molecules typically occur for each billion molecules and atoms that make up the air. This equates to 20–100 ppb, a unit we first saw in Section 1.8. In contrast, the *stratosphere*, farther away from Earth's surface is mostly where ozone filters some types of ultraviolet light from the Sun. The concentration of ozone in this region is several orders of magnitude greater than in the troposphere, but still very low. As an upper limit, about 12,000 ozone molecules are present for every billion molecules and atoms of gases that make up the atmosphere at this level.

Most of the ozone on our planet, about 90% of the total, is found in the stratosphere. The term *ozone layer* refers to a designated region in the stratosphere with maximum ozone concentration. Figure 3.18 shows the relative ozone concentrations within the troposphere and stratosphere.

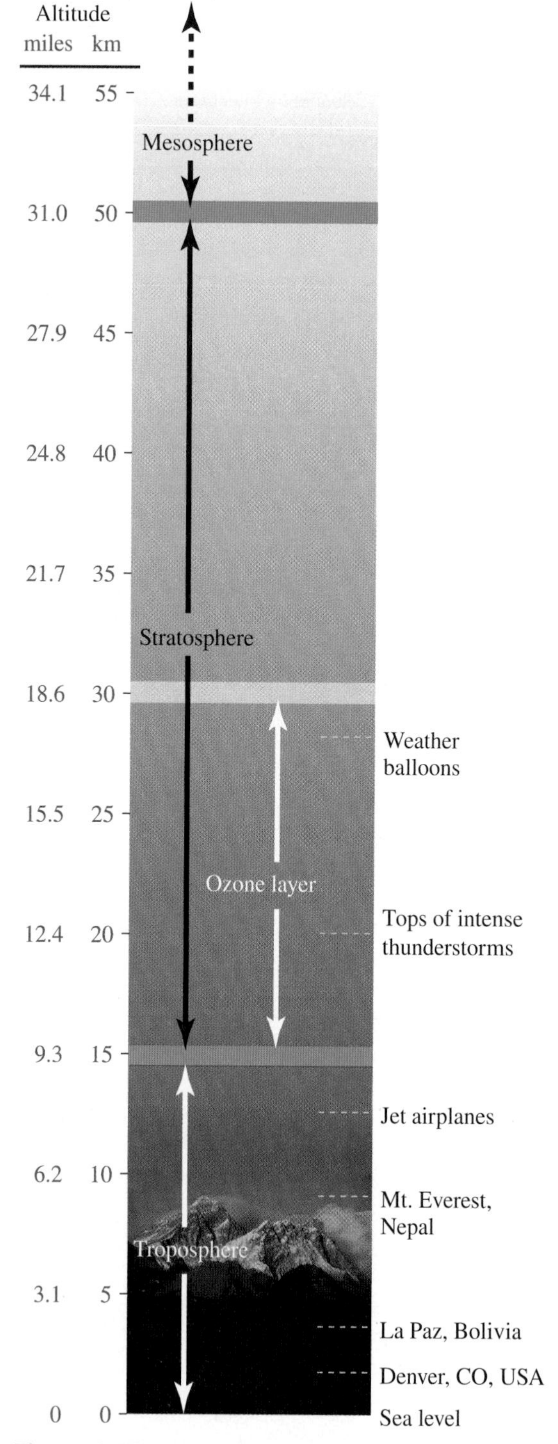

Figure 3.17

The regions of our atmosphere, showing where the ozone layer is situated.

© Galen Rowell/Corbis

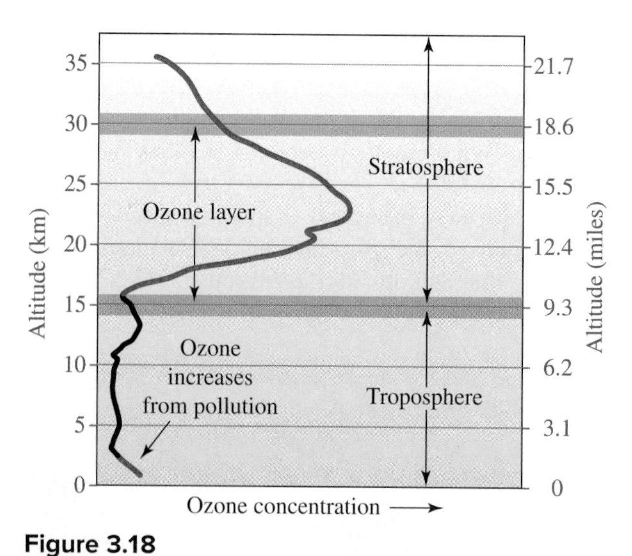

Figure 3.18

The comparative concentrations of ozone within the troposphere and stratosphere.

Your Turn 3.15 You Decide Ozone Concentrations

Use **Figure 3.18** and values given in the text to answer these questions:

a. What is the approximate altitude of maximum ozone concentration?
b. What is the maximum number of ozone molecules per billion molecules and atoms of all types found in the stratosphere?
c. What is the maximum number of ozone molecules per billion molecules and atoms of all types found in ambient air just meeting the EPA limit for an 8-hr average?

Because the range of altitudes is so broad, the concept of an "ozone layer" can be a little misleading. No thick, fluffy blanket of ozone exists in the stratosphere. At altitudes of the maximum ozone concentration, the atmosphere is very thin, so the total amount of ozone is surprisingly small. If all the O_3 in the atmosphere could be isolated and brought to the average pressure and temperature at Earth's surface (1.0 atm and 15 °C), the resulting layer of gas would have a thickness of less than 0.5 cm, or about 0.2 inch. This is a very small amount of matter. Yet, this ozone shield protects the surface of Earth and its inhabitants from the harmful effects of ultraviolet radiation.

A common unit for pressure is atmospheres, atm. At sea level, the normal atmospheric pressure at Earth's surface is 1.0 atm. The effect of elevation on atmospheric pressure will be discussed in Section 10.5.

3.6 Counting Molecules: How Can We Measure the Ozone Concentration?

G.M.B. Dobson (1889–1975), a scientist at Oxford University, invented the first instrument used to quantitatively measure the concentration of ozone in our atmosphere. This technique considers the total amount of ozone in a vertical column of air of known volume (Figure 3.19). The determination can be done from Earth's surface by measuring the amount of UV radiation reaching a detector. The lower the intensity of the radiation, the greater the amount of ozone in the column. Therefore, it is fitting that the unit of such measurements is named for him, referred to as *Dobson units* (DU).

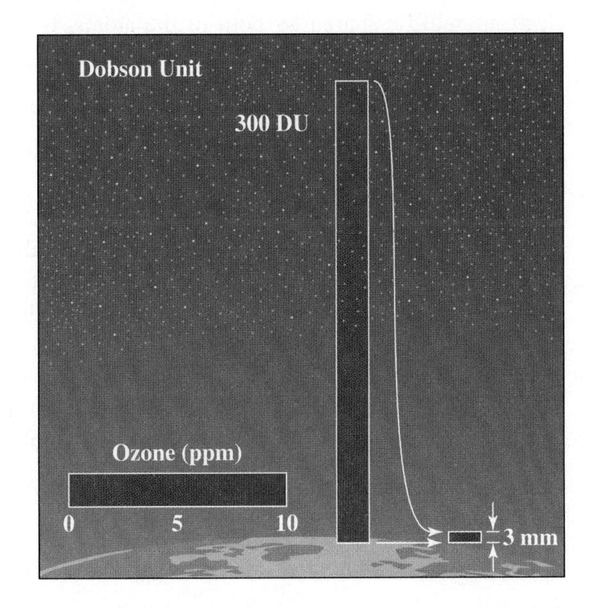

Figure 3.19

An illustration of the definition of the Dobson unit. If all the ozone over a certain area is compressed at 0 °C to form a 3-mm thick slab with a pressure of 1 atmosphere (atm), it is defined as 300 Dobson units (DU). If a 5-mm thick slab is produced, the ozone concentration will be 500 DU, and so on.

Figure 3.20

A photo of Earth from Space Shuttle *Columbia*, showing the various regions of our lower and upper atmospheres as a visible haze above Earth's surface.

Source: Earth Sciences and Image Analysis Laboratory/Johnson Space Center/NASA

Your Turn 3.16 Scientific Practices This Year's Ozone Hole

Citizens and scientists alike can examine the data for the ozone hole over Antarctica. What is happening this year? NASA posts the data on its website. For the most recent year:

a. What is the area of the hole? How does this compare with recent years?
b. What is the lowest reading observed for ozone? Again, how does this compare?

Scientists continue to measure and evaluate ozone levels using ground observations, weather balloons, and high-flying aircraft. However, since the 1970s, measurements of total-column ozone have also been made from the top of the atmosphere. Satellite-mounted detectors record the intensity of the UV radiation scattered by the upper atmosphere. The results are then related to the amount of O_3 present.

The Space Shuttle *Columbia* tested a new approach for monitoring ozone. Rather than looking directly downward toward Earth from a satellite, the equipment aboard the Shuttle looked sideways through the thin blue haze that rises above the denser regions of the troposphere (Figure 3.20). This region is known as Earth's "limb," and thus this new technique is known as *limb viewing*. Reliable information can be gathered at each level of the atmosphere, allowing scientists to better understand the chemistry taking place in the lower regions of the stratosphere.

3.7 | How Does Ozone Decompose in UV Light?

The process by which ozone protects us from damaging solar radiation involves the interaction of matter and energy from the Sun. In order to understand these reactions, we must first consider the molecular models, which provide information regarding how the atoms are connected within each molecule.

Let's begin with the simplest molecule, hydrogen (H_2). As seen in Chapter 1, each hydrogen atom has one electron. Once two hydrogen atoms become bonded together, the two electrons will be shared by both of the atoms. If we represent each electron by a dot, the two separate hydrogen atoms look something like this:

H • and • H

Bringing the two atoms together yields a molecule that can be represented this way:

H**:**H

Since each atom effectively has a share in both electrons, the resulting H_2 molecule has a lower energy than the sum of the energies of two separate H atoms, and is therefore more stable. The two electrons that are shared constitute a **covalent bond**. Appropriately, the name *covalent* implies "shared strength."

Lewis structures, also called dot structures, can be drawn for many simple nonionic compounds and molecules by following a set of straightforward steps. We first illustrate the procedure with hydrogen fluoride, HF, another simple molecular compound.

The column number for Group A elements in the periodic table (see Figure 1.2) denotes the number of valence electrons; *i.e.*, Group 2 elements have 2 valence electrons; Group 15 have 5 valence electrons.

Step 1. Note the number of outer electrons (also known as **valence electrons**) contributed by each of the atoms.

1 H atom × 1 valence electron per atom = 1 valence electron
1 F atom × 7 valence electrons per atom = 7 valence electrons

Step 2. Add the valence electrons contributed by the individual atoms to obtain the total number of valence electrons available:

1 + 7 = 8 total valence electrons

Step 3. Arrange the valence electrons in pairs. Then, distribute them so as to maximize stability by giving each atom a share in enough electrons to fully fill its outermost energy level:

2 electrons in the case of hydrogen, 8 electrons for most other atoms

$$\mathrm{H{:}\ddot{\underset{\cdot\cdot}{F}}{:}}$$

Notice that we surrounded the F atom with 8 dots, organized into 4 pairs. The pair of dots between the H and the F atoms represents the bond that unites the hydrogen and fluorine atoms. The other 3 pairs of electrons are not shared with other atoms; as such, they are called non-bonding electrons, or **lone pairs**.

A **single covalent bond** is formed when 2 electrons (one pair) are shared between two atoms. A line is typically used to represent the two electrons in the bond:

$$\mathrm{H{-}\ddot{\underset{\cdot\cdot}{F}}{:}}$$

Sometimes, the lone pairs are removed from a Lewis structure, simplifying it even more. The result is called a **structural formula**, a representation of how the atoms in a molecule are connected:

$$\mathrm{H{-}F}$$

Recall that the single line in a Lewis structure represents one pair of shared electrons. These two electrons, plus the 6 electrons in the lone pairs, mean that the fluorine atom is associated with 8 valence electrons—whether or not the electrons are explicitly shown. However, the hydrogen atom has no additional electrons other than the single pair it shares with fluorine. It is at maximum capacity with 2 electrons. The fact that electrons in many molecules are arranged so that every atom (except hydrogen) shares 8 electrons is called the **octet rule**.

Beryllium and the elements in Group 13 will typically have less than eight electrons. In addition, elements beyond the second row of the periodic table *may* have more than eight electrons surrounding them.

As another example, consider the Cl_2 molecule—the diatomic form of elemental chlorine. From the periodic table, we can see that chlorine, like fluorine, is in Group 17, which means that its atoms each have 7 valence electrons. Using the scheme given earlier for HF, we first count and add up the valence electrons for Cl_2:

2 Cl atoms × 7 valence electrons per atom = 14 total valence electrons

For the Cl_2 molecule to exist, a bond must connect the two atoms. The remaining 12 electrons constitute 6 lone pairs, distributed in such a way as to give each chlorine atom a share in 8 electrons (2 bonding electrons and 3 lone pairs). Here is the Lewis structure:

$$\mathrm{{:}\ddot{\underset{\cdot\cdot}{Cl}}{-}\ddot{\underset{\cdot\cdot}{Cl}}{:}}$$

So far, we have dealt only with molecules having just two atoms. However, the octet rule also applies to larger molecules. Let's use a water molecule, H_2O, as an example. Just as with two-atom molecules, first tally the valence electrons:

2 H atoms × 1 valence electron per atom = 2 valence electrons

1 O atom × 6 valence electrons per atom = 6 valence electrons

Total = 8 valence electrons

In molecules like water that have a single atom bonded to two or more atoms of a different element (or elements), *the single atom is usually the central one.* Since oxygen is the "single atom" in H_2O, we place it in the center of the Lewis structure. Each of the H atoms bonds to the O atom, using 2 electrons each. The remaining 4 electrons go on the O atom as 2 lone pairs:

$$\mathrm{H{:}\ddot{\underset{\cdot\cdot}{O}}{:}H}$$

A quick count confirms that the O atom is surrounded by 8 electrons, so it satisfies the octet rule. Alternatively, we could use lines for the single bonds:

$$\mathrm{H{-}\ddot{\underset{\cdot\cdot}{O}}{-}H}$$

Chapter 4 will describe the process used to predict the shapes of molecules based on their Lewis structures.

Chemical formulas show the types and ratio of the atoms present in a compound. In contrast, Lewis structures indicate *how* the atoms are connected, and show the presence of lone pairs of electrons, if present. Note that Lewis structures do *not* directly reveal the shape of a molecule. For example, from the Lewis structure, it might appear that the water molecule is linear. In fact, the molecule is bent, which has an enormous effect on its properties.

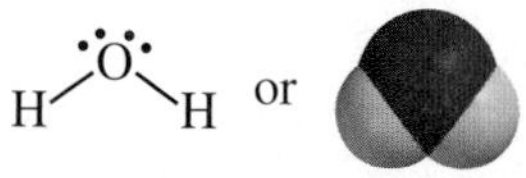

Did You Know? Methane is the primary component of *natural gas*, used to heat our homes and fuel some city buses.

Another molecule to consider is methane, CH_4. Again, we begin by tallying the valence electrons:

$$4\ H\ atoms \times 1\ valence\ electron\ per\ atom = 4\ valence\ electrons$$
$$1\ C\ atom \times 4\ valence\ electrons\ per\ atom = 4\ valence\ electrons$$
$$Total = 8\ valence\ electrons$$

The central carbon atom is surrounded by the 8 electrons, giving carbon an octet of electrons. In the Lewis structure, each H atom uses two of the electrons to bond with the C atom, for a total of 4 single covalent bonds:

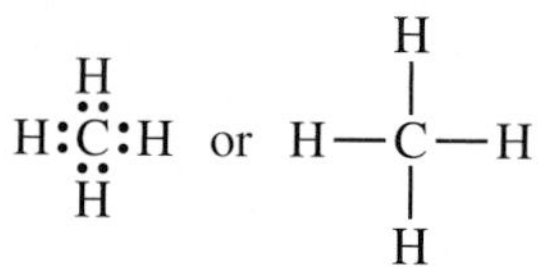

Remember that H can only accommodate one pair of electrons. The next activity gives you the opportunity to practice with other molecules.

Your Turn 3.17 Skill Building Lewis Structures for Diatomic Molecules

Draw the Lewis structure for each molecule.

a. HBr
b. Br_2

In some structures, single covalent bonds do not allow the atoms to follow the octet rule. Consider, for example, the O_2 molecule. Here, we have 12 valence electrons to distribute, six from each of the oxygen atoms. As seen below, there are not enough electrons to satisfy the octet rule for both atoms if only one pair is held in common:

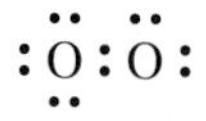

However, the octet rule can be satisfied for all atoms if a lone pair becomes shared by two atoms:

A covalent bond consisting of two pairs of shared electrons is called a **double bond**, which is shorter, stronger, and requires more energy to break than a single bond between the same atoms. A double bond is represented by four dots, or by two lines:

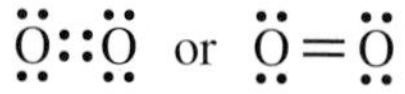

A **triple bond** is a covalent linkage made up of three pairs of shared electrons. For the same atoms, triple bonds are even shorter, stronger, and harder to break than double bonds. Since electrons are negatively charged, and atomic nuclei are positively charged, you may think of electrons as a "glue" that holds the atomic nuclei together. The more electrons shared between two atoms, the more "glue," which results in a much stronger bond between the two atoms.

For an example of a triple bond, consider the nitrogen molecule, N_2. Each nitrogen atom (Group 15) contributes 5 valence electrons, for a total of 10. These 10 electrons can be distributed in accordance with the octet rule if six of them (three pairs) are shared between the two atoms, leaving four to form lone pairs, one on each nitrogen atom:

$$:N\vdots\vdots N: \quad \text{or} \quad :N\equiv N:$$

The ozone molecule introduces another structural feature. We start again with tallying the valence electrons. Each of the 3 oxygen atoms contributes 6 valence electrons, for a total of 18. However, these 18 electrons can be arranged in two ways—each way allows all oxygens to satisfy the octet rule:

$$\underset{\textbf{a}}{\ddot{O}::\ddot{O}:\ddot{O}:} \qquad \underset{\textbf{b}}{:\ddot{O}:\ddot{O}::\ddot{O}}$$

Structures **a** and **b** predict that the molecule should contain one single bond and one double bond. In structure **a**, the double bond is shown to the left of the central atom, whereas in **b** it is shown to the right. But experiments reveal that the two bonds in the O_3 molecule are identical, being intermediate between the length and strength of a single and double bond. Structures **a** and **b** are called **resonance forms**, Lewis structures that represent hypothetical extremes of electron arrangements in a molecule. For example, no single resonance form represents the electron arrangement in the ozone molecule. Rather, the actual structure is something like a hybrid of the two resonance forms, **a** and **b**. A double-headed arrow linking the different forms is used to represent the resonance phenomenon:

$$\ddot{O}=\ddot{O}-\ddot{O}: \longleftrightarrow :\ddot{O}-\ddot{O}=\ddot{O}$$

Your Turn 3.18 Skill Building More Lewis Structures

Draw the Lewis structure for each of these molecules.

a. hydrogen sulfide (H_2S)
b. dichlorodifluoromethane (CCl_2F_2)

Your Turn 3.19 Skill Building Lewis Structures with Multiple Bonds

Draw the Lewis structure for each compound. Indicate resonance forms, if appropriate.

a. carbon monoxide (CO)
b. sulfur dioxide (SO_2)
c. sulfur trioxide (SO_3)

It should be noted that like water, the O_3 molecule is not linear as the simple Lewis structure seems to indicate. Remember that Lewis structures tell us only what is connected to what and do not necessarily show the shape of the molecule. The O_3 molecule is actually bent, as in this representation:

$$\ddot{O}{=}\ddot{O}{-}\ddot{O}: \longleftrightarrow :\ddot{O}{-}\ddot{O}{=}\ddot{O} \quad \text{(bent)}$$

Although you will learn about predicting molecular shapes in the next chapter, at this point you only need to know *how* the bonding in O_2 and O_3 molecules relates to their interaction with sunlight. Let's now look at how this chemistry allows for the oxygen–ozone protective screen.

Earlier, we described how UV radiation reacts with both oxygen and ozone. Here are the corresponding equations, taking into account their Lewis structures and geometries as well as the photon that triggers the reaction:

$$:\ddot{\text{O}}{=}\ddot{\text{O}}: \xrightarrow[\lambda\, <\, 242\ \text{nm}]{\text{UV photon}} :\ddot{\text{O}}: + :\ddot{\text{O}}: \qquad [3.5]$$

$$\text{O}{=}\ddot{\text{O}}{-}\ddot{\text{O}}: \xrightarrow[\lambda\, <\, 320\ \text{nm}]{\text{UV photon}} :\ddot{\text{O}}{=}\ddot{\text{O}}: + :\ddot{\text{O}}: \qquad [3.6]$$

Equations 3.5 and 3.6 describe a key set of chemical reactions that occur in the stratosphere. Note that the energy required to break an O=O double bond is higher (*i.e.*, requires radiation with a shorter wavelength) than that required to break an O–O single bond.

Everyday, 300 million (3×10^8) tons of stratospheric O_3 forms, and an equal mass decomposes. New matter is neither created nor destroyed, but merely changes its chemical form. Consequently, the overall concentration of ozone remains constant in this natural cycle. This process is an example of a **steady state**, a condition in which a dynamic system is in balance, so that there is no net change in concentration of the major species involved. A steady state arises when a number of chemical reactions, typically competing reactions, balance one another. Stratospheric ozone is formed and decomposed by a natural cycle of steady-state reactions. In the next section, we will consider what happens when something disturbs the steady state of the cycle, and thus leads to the destruction of our protective stratospheric ozone.

As an illustration, consider the water level at the top of Niagara Falls. Even though millions of gallons of water spill continually over a cliff, a steady upstream flow of water constantly replenishes the waterfall, resulting in a constant water level.

Your Turn 3.20 You Decide The Ozone Layer

Based on the reactions and molecules you examined in this section, if left undisturbed, do you think the ozone concentration in the atmosphere would be uniform across the globe at all times? Why or why not?

3.8 | How Safe Is Our Protective Ozone Layer?

Since 1926, the stratospheric O_3 concentrations have been measured at the Swiss Meteorological Institute. These measurements show that the natural concentration of stratospheric O_3 is not uniform across the globe.

Did You Know? Switzerland holds the record for the longest continuous set of ozone-level measurements.

On average, the total stratospheric ozone concentration increases as one approaches either pole. In previous sections, we showed how UV radiation affects the production and destruction of O_2 and O_3. Ozone production increases with the intensity of the radiation striking the stratosphere, which in turn depends on the angle of Earth with respect to the Sun, and the distance between the Sun and Earth (Figure 3.21). At the equator, the period of highest intensity occurs at the equinox (March and October)

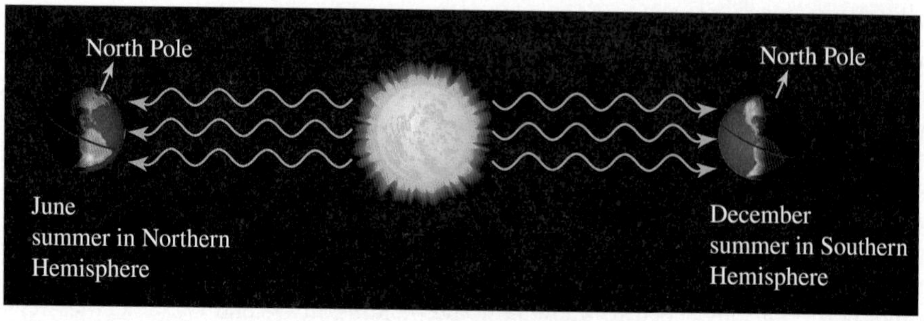

Figure 3.21

Illustration of Earth's tilt and the seasonal variation of the Sun's intensity.

when the Sun is directly overhead. Outside the tropics, the Sun is never directly overhead, so the maximum intensity occurs at the summer solstice (June in the Northern Hemisphere, December in the Southern). The angle of Earth with respect to the Sun dominates both ozone production, and the seasons. There is a slight (~7%) increase in solar intensity reaching Earth in early January, when Earth is nearest the Sun, compared with July, when Earth is farthest away. The wind patterns in the stratosphere cause other variations in ozone concentrations—some on a seasonal basis, and others over a longer cycle.

Extraordinary images of Earth are color-coded to clearly show stratospheric ozone concentrations. The dark blue and purple regions indicate where the lowest concentrations of O_3 are observed. A value of 250–270 DU is typical at the equator. As one moves away from the equator, values range between 300 and 350 DU, with seasonal variations. At the highest northern latitudes, values can be as high as 400 DU.

Your Turn 3.21 Scientific Practices The Ozone Hole

Visit NASA's Ozone Hole Watch website and click on the "Ozone Maps" tab.

a. How is the map of Earth oriented? Why do you think this orientation was chosen?
b. Use the color key at the bottom of the map to determine whether the ozone levels are currently at a healthy level.
c. Use the month tabs to page through a year's worth of maps. In which month(s) does the ozone hole appear?
d. On the site, find a video demonstrating the ozone hole in that month from 1979 to the present. Describe the trends you observe.

As you probably saw in the previous activity, there is reason for concern regarding the thinning of ozone (the "ozone hole") that occurs seasonally over the South Pole (Figure 3.22). These changes were so pronounced that, when the British monitoring team at Halley Bay in Antarctica first observed it in 1985, they thought their instruments were malfunctioning! The area over which ozone levels are reported to be less than 220 DU is usually considered to be the "hole." From the mid-1990s on, the annual size of the ozone hole has nearly equaled the total area of the North American continent, and in some cases exceeded it.

Check out the dramatic decline in stratospheric ozone levels observed near the South Pole shown in Figure 3.23. In recent years, the minimum has been around 100 DU. Keep in mind that seasonal variation has always occurred in the ozone concentration

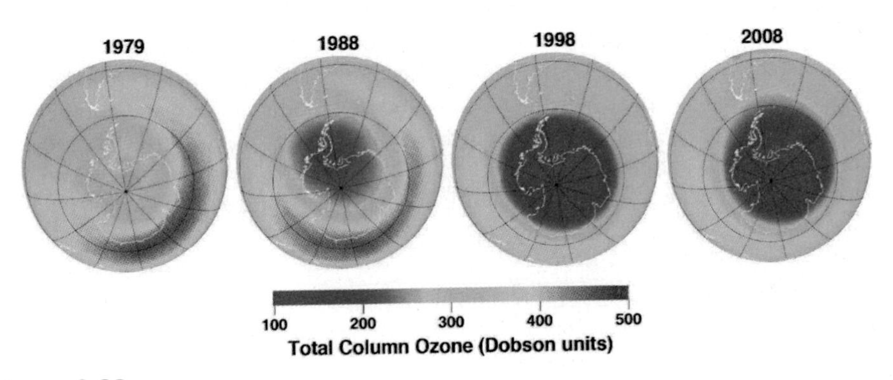

Figure 3.22

Map of the total ozone concentration and size of the ozone hole over Antarctica for a three-decade timespan.

Source: Images and data courtesy NASA Ozone Watch

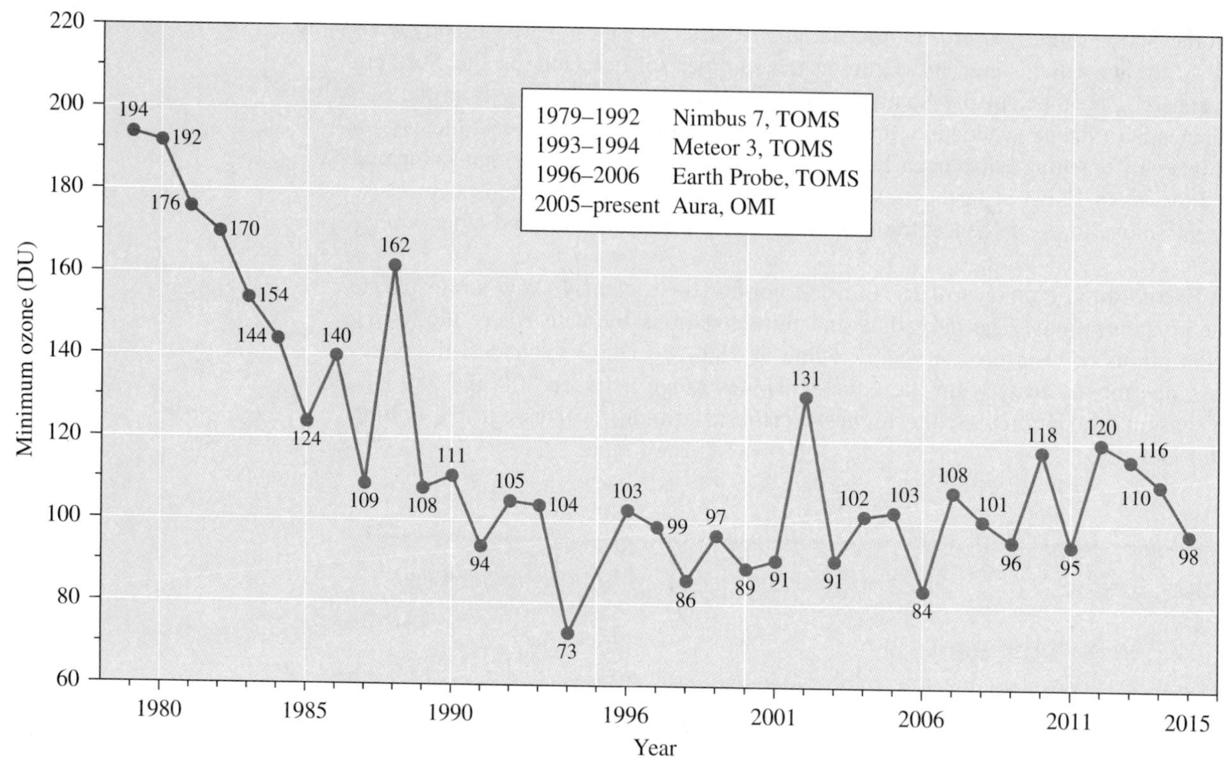

Figure 3.23

The lowest values, in Dobson Units, recorded each Spring (1979–2015) for the stratospheric ozone concentration above Antarctica. *Note:* The high value in 2002 was due to an early breakdown of the vortex that isolates polar from mid-latitude air. No data were acquired in 1995.

Source: NASA Ozone Watch

over the South Pole, with a minimum in late September or early October—the Antarctic spring. Unprecedented, however, is the striking decrease in this minimum that has been observed in recent decades.

Your Turn 3.22 Scientific Practices The Ozone Hole

How would you describe the trend of data present in **Figure 3.23**? Based on the patterns you observe, what would you predict for the data for years 2016–2019? How might you verify or reject your predictions?

The major natural cause of ozone destruction, wherever it takes place around the globe, is a series of reactions involving water vapor and its breakdown products. The great majority of H_2O molecules that evaporate from the oceans and lakes fall back to Earth's surface as rain or snow. But a few molecules reach the stratosphere, where the H_2O concentration is about 5 ppm. At this altitude, photons of UV radiation trigger the dissociation of water molecules into hydrogen (H·) and hydroxyl (·OH) free radicals. A **free radical** is a highly reactive chemical species with one or more unpaired electron(s). An unpaired electron is often indicated with a single dot:

$$H_2O \xrightarrow{\text{photon}} H\cdot + \cdot OH$$

Because of the unpaired electron, free radicals are highly reactive. Thus, the H· and ·OH radicals participate in many reactions, including some that ultimately convert O_3 to O_2. This is the most efficient mechanism for destroying ozone at altitudes above 50 km.

Water molecules and their breakdown products are not the only agents responsible for natural ozone destruction. Another is the free radical ·NO, also called nitrogen monoxide. Most of the ·NO in the stratosphere is of natural origin. It is formed from dinitrogen monoxide, N_2O, a naturally occurring compound that is produced in the soil and oceans by microorganisms. Natural air currents cause N_2O to gradually drift upward from the troposphere to the stratosphere, where it reacts with oxygen atoms to produce ·NO. Little can or should be done to control this process. It is part of a natural cycle involving nitrogen compounds. However, as we will now see, human activities also play a significant role in ozone destruction.

Your Turn 3.23 Scientific Practices Free Radicals

Consider the reaction between the hydroxyl free radical and the molecule methane (CH_4). The three-step process (shown below) occurs in the upper atmosphere. Rewrite these equations using full Lewis structures and identify all of the free radicals present throughout the steps:

$$OH + CH_4 \longrightarrow H_2O + CH_3$$
$$CH_3 + O_2 \longrightarrow OOCH_3$$
$$OOCH_3 + NO \longrightarrow OCH_3 + NO_2$$

Your Turn 3.24 You Decide Free Radicals in the Atmosphere

Based on information you read about in **Chapters 2** and **3**, do you think having free radicals in the atmosphere is beneficial or detrimental to the healthy functioning of the atmosphere? Explain.

3.9 Chemistry to the ~~Rescue~~ Detriment? Human Roles in the Destruction of the Ozone Layer

Even when the effects of water, nitrogen oxides, and other naturally occurring compounds are included in stratospheric models, the measured ozone concentration is still lower than predicted. Measurements worldwide indicate that the ozone concentration has been decreasing over the past 40 years. There is a good deal of fluctuation in the data, but the trend is clear. The stratospheric ozone concentration at mid-latitudes (60° south to 60° north) has decreased by more than 8% in some cases. These changes cannot be correlated with changes in the intensity of solar radiation, so we must look elsewhere for an explanation. Thus, it is time to turn our attention to the culprit: chemicals that were once quite prevalent in aerosol spray cans and air conditioners.

A major human cause of stratospheric ozone depletion was uncovered through the masterful scientific sleuthing of F. Sherwood Rowland (1927–2012), Mario Molina (b. 1943), and Paul Crutzen (b. 1933). They analyzed vast quantities of atmospheric data and studied hundreds of chemical reactions. As with most scientific investigations, some uncertainties remained. Nonetheless, their evidence pointed to an unlikely group of compounds: the chlorofluorocarbons.

For their work, Rowland, Molina, and Crutzen jointly won the 1995 Nobel Prize in chemistry.

As the name implies, **chlorofluorocarbons (CFCs)** are compounds composed of the elements chlorine, fluorine, and carbon only. Fluorine and chlorine are members of Group 17, the halogens. In their elemental state, most of the halogens are diatomic molecules, but only fluorine and chlorine are gases. Fluorine is extremely reactive and cannot even be contained in a glass receptacle. In contrast, CFC molecules are highly unreactive. To get started with CFCs, let's examine two examples:

```
      :F̤̈:                        :F̤̈:
       |                           |
:C̤̈l—C—C̤̈l:        and       :C̤̈l—C—F̤̈:
       |                           |
      :C̤̈l:                       :C̤̈l:
```

CCl_3F
trichlorofluoromethane
Freon-11

CCl_2F_2
dichlorodifluoromethane
Freon-12

Note how the names of the compounds show the connection of CFCs to methane, CH_4. The prefixes *di-* and *tri-* specify the number of chlorine and fluorine atoms that substitute for hydrogen atoms of methane. These two CFCs are known by their trade names, Freon-11 and Freon-12. You may also hear them called CFC-11 and CFC-12, respectively, following a naming scheme developed in the 1930s by chemists at DuPont.

CFCs do not occur in nature; we humans synthesized them for a variety of uses. This is an important verification point in the debate over the role of CFCs in stratospheric ozone depletion. As we saw previously, other contributors to the destruction of ozone, such as the ·OH and ·NO free radicals, are formed in the atmosphere both from natural sources and human activities.

Halons are close cousins of CFCs and are also inert, nontoxic compounds. They contain bromine in addition to fluorine or chlorine that surround a carbon atom. Like CFCs, no hydrogens are in halons. For example, here is the Lewis structure for bromotrifluoromethane, $CBrF_3$, also known as Halon-1301:

```
    ..
   :Br:
    |
 ..    ..
:F—C—F:
 ..    ..
    |
   :F:
    ..
```

Halons are used as fire suppressants. They are especially helpful when a fire hose or sprinkler system would be inappropriate—for instance, in libraries (especially rare book rooms), grease fires (where water might spread the fire), chemical stockrooms (where some chemicals react with water), and aircraft (where hosing down the cockpit would definitely be a bad idea!). Unfortunately, these compounds are as harmful as CFCs when it comes to ozone depletion.

The introduction of CCl_2F_2 as a refrigerant gas in the 1930s was hailed as a great triumph of chemistry and an important advance in consumer safety. It replaced ammonia (NH_3) or sulfur dioxide (SO_2), two toxic and corrosive refrigerant gases. In many respects, CCl_2F_2 was (and still is) an ideal substitute. It is nontoxic, odorless, colorless, and does not burn. In fact, the CCl_2F_2 molecule is so stable that it does not react with much of anything!

Given the desirable nontoxic properties of CFCs, they soon were put to other uses. For example, CCl_3F was often blown into mixtures to make foams for cushions and foamed insulation. Other CFCs served as propellants in aerosol spray cans and as nontoxic solvents for oil and grease.

For better or worse, the synthesis of CFCs has had a major effect on our lives. Because these compounds are nontoxic, nonflammable, cheap, and widely available, they revolutionized air conditioning, making it readily accessible for homes, office buildings, shops, schools, and automobiles. Beginning in the 1960s and 1970s, CFCs helped spur the growth of cities in hot and humid parts of the world. In effect, a major demographic shift occurred because of CFC-based technology that transformed the economy and business potential of entire regions of the globe.

Ironically, the very property that makes CFCs ideal for so many applications—their chemical inertness—ended up causing harm to our atmosphere. The C–Cl and C–F bonds in the CFCs are so strong that the molecules are virtually indestructible. For example, it has been estimated that an average CCl_2F_2 molecule can persist in the atmosphere for 120 years before it meets its final decomposing fate. In contrast, it only takes about five years for atmospheric wind currents to bring molecules up to the stratosphere, which is exactly where some of the CFC molecules ended up.

What happens to CFCs in the stratosphere? As altitude increases, the concentrations of oxygen and ozone decrease, but the intensity of UV radiation increases. The high-energy UVC radiation can break C–Cl bonds. Here is the chemical reaction that releases chlorine atoms from dichlorodifluoromethane:

$$\mathrm{F{-}\overset{\displaystyle Cl}{\underset{\displaystyle F}{C}}{-}Cl} \xrightarrow[\lambda\ \leq\ 220\ \mathrm{nm}]{\text{UV photon}} \mathrm{F{-}\overset{\displaystyle Cl}{\underset{\displaystyle F}{C}}\cdot} + \cdot\mathrm{Cl}$$

See **Connect** to discover the effects that EM radiation has on CFCs and ozone molecules.

This reaction actually produces two free radicals. Through several reactions, the chlorine radical reacts with ozone and forms oxygen gas. Earlier (Equation 3.4), we mentioned that energy was required for this reaction; in this case, the unstable chlorine radical essentially provides that energy:

$$Cl\cdot + O_3 \longrightarrow ClO\cdot + O_2 \quad [3.7]$$

$$ClO\cdot + O \longrightarrow Cl\cdot + O_2 \quad [3.8]$$

As you can see, the complex interaction of ozone with atomic chlorine provides a pathway for the destruction of ozone. We will explore these pathways in the next activity.

Your Turn 3.25 Scientific Practices Ozone Concentration

Based on the graph in **Figure 3.24**, make a claim about the relationship between concentrations of stratospheric ozone and stratospheric chlorine. In what way do **Equations 3.7** and **3.8** verify the data trends in the figure?

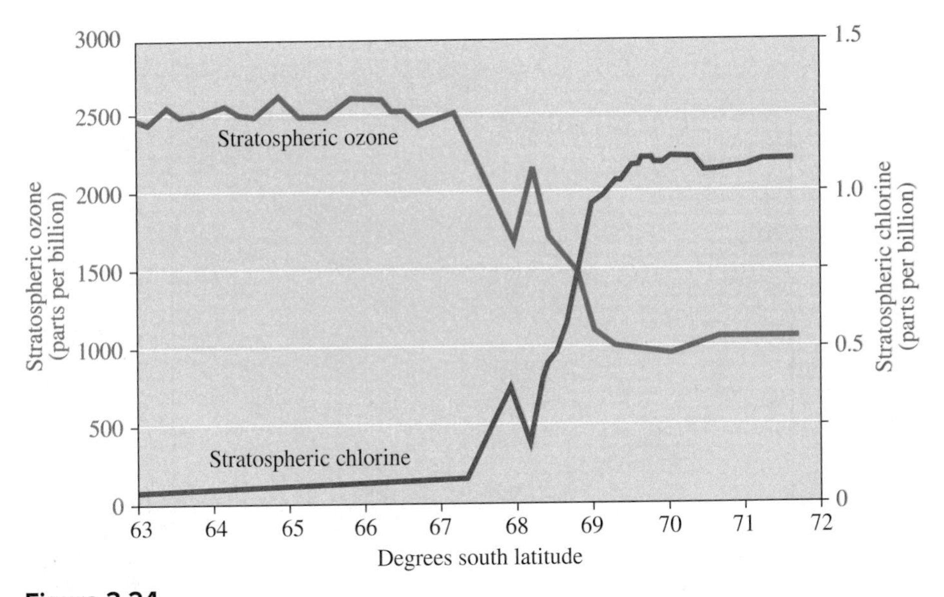

Figure 3.24

Antarctic stratospheric concentrations of ozone and chlorine, from a flight into the Antarctic ozone hole, 1987.

Source: United Nations Environment Programme (UNEP)

Notice that Cl· appears both as a reactant in Equation 3.7 and as a product in Equation 3.8. This means that Cl· is both consumed *and* regenerated in the cycle of ozone depletion, with no net change in its concentration. Such behavior is characteristic of a **catalyst**, a chemical substance that participates in a chemical reaction and influences its speed without undergoing permanent change. Atomic chlorine acts catalytically by being regenerated and recycled to remove more ozone molecules. On average, a single atom can catalyze the destruction of as many as 1×10^5 ozone molecules before it is carried back to the lower atmosphere by winds.

Not all of the chlorine implicated in stratospheric ozone destruction comes from CFCs. Natural sources such as seawater and volcanoes produce other chlorinated carbon compounds. However, most chlorine from natural sources is in water-soluble forms. Therefore, any natural chlorine-containing substances are washed out of the atmosphere by rainfall long before they can reach the stratosphere. Of particular significance are the data gathered by NASA and international researchers, which have established that high concentrations of HCl (hydrogen chloride) and HF (hydrogen fluoride) always occur together in the stratosphere. Although some of the HCl might conceivably arise from a variety of natural sources, the only reasonable origin of stratospheric concentrations of HF is CFCs.

Ozone-depleting gases are present throughout the stratosphere. Furthermore, as a result of global wind patterns, CFCs are present in comparable abundance in lower parts of the atmosphere over *both* hemispheres. Why, then, have the greatest losses of stratospheric ozone occurred over Antarctica? Furthermore, given that more ozone-depleting gases are emitted in the Northern Hemisphere, why are their effects felt most strongly in the Southern Hemisphere?

A special set of conditions exists in Antarctica; the lower stratosphere over the South Pole is the coldest spot on Earth. From June to September (Antarctic winter), the winds that circulate around the South Pole form a vortex that prevents warmer air from entering the region. As a result, the temperature may drop to as low as −90 °C. Under these conditions, **polar stratospheric clouds (PSCs)** can form. These thin clouds are composed of tiny ice crystals formed from the small amount of water vapor present in the stratosphere. The chemical reactions that occur on the surface of these ice crystals convert molecules that do not deplete ozone, such as HCl (mentioned previously), to more reactive species that do, such as Cl_2.

It has been shown that the compound $ClONO_2$ may be converted to the reactive HOCl species on the surface of ice crystals, which also contributes to ozone depletion alongside Cl_2.

Neither HOCl nor Cl_2 cause any harm in the dark of winter. But when sunlight returns to the South Pole in late September, the light splits HOCl and Cl_2 to release

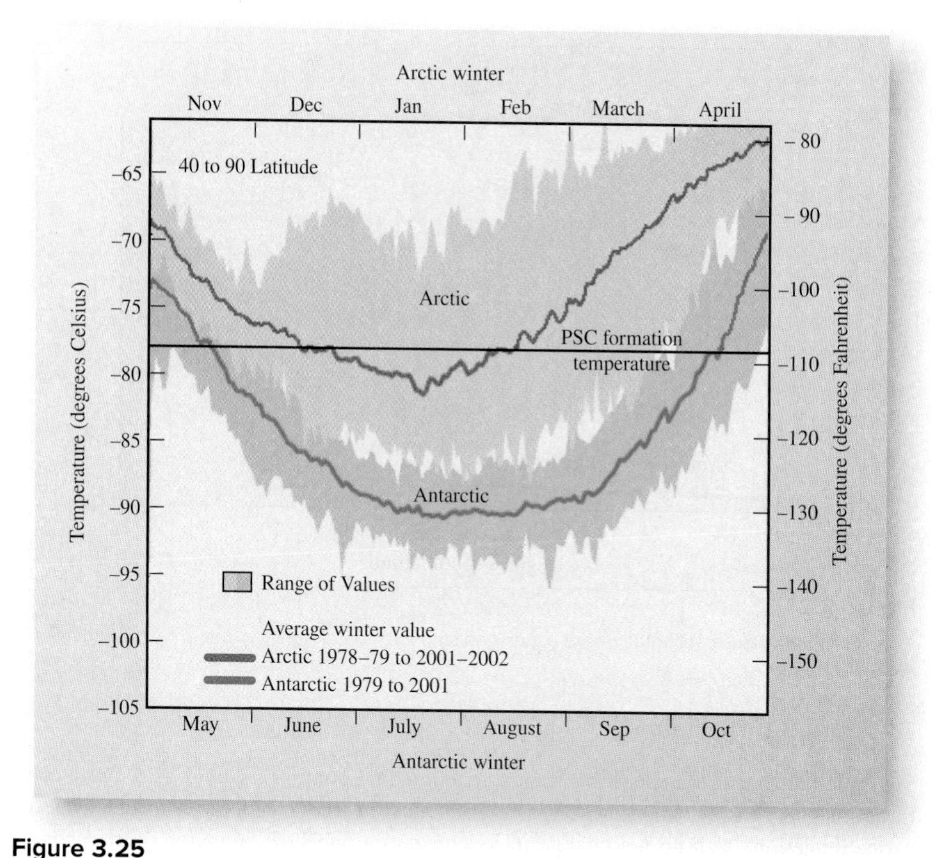

Figure 3.25

Minimum air temperatures in the polar lower stratospheres.

Source: Scientific Assessment of Ozone Depletion: 2002, World Meteorological Organization, UNEP

chlorine atoms. Given this increase in Cl•, a species that destroys vast quantities of ozone, the hole starts to form. Notice the conditions required: extreme cold, a circular wind pattern (vortex), enough time for ice crystals to form and provide a surface for the reactions, and darkness followed by rapidly increasing levels of sunlight. Figure 3.25 shows the seasonal variation, and compares the minimum temperatures above the Arctic and the Antarctic.

As you can see in the figure, the necessary conditions for PSC formation more often are found in Antarctica. Changes in ozone concentrations above Antarctica closely follow the seasonal temperatures. Typically, rapid ozone depletion takes place during spring at the South Pole, which is from September to early November. As the sunlight warms the stratosphere, the polar stratospheric clouds dissipate, thereby releasing Cl• species that attack ozone molecules. However, as the spring season draws to a close, air from lower latitudes flows into the polar region, replenishing the depleted ozone levels. By the end of November, the hole is largely refilled. It turns out that ozone depletion in the Northern Hemisphere is not nearly as severe as it is in the Southern. The difference stems mainly from the fact that the air above the North Pole is not as cold. Even so, polar stratospheric clouds have been repeatedly observed in the Arctic.

What are the global effects of this ozone hole? Decreased stratospheric ozone over the South Pole leads to increased UVB levels reaching Earth. In turn, skin cancer rates have increased in Australia and southern Chile. Furthermore, Australian scientists believe that wheat, sorghum, and pea crop yields have decreased as a result of increased UV radiation. Similar effects are also being felt in southern Chile in the area around Punta Arenas and on the island of Tierra del Fuego at the southernmost tip of South America. Chile's health minister has warned the 120,000 residents of Punta Arenas not to be out in the Sun during the noon hours in the spring, when ozone depletion is greatest.

3.10 Where Do We Go from Here: Can the Ozone Hole Be Restored?

Once people understood the role of CFCs in ozone destruction, they responded with surprising speed. Individual countries took the first steps. For example, the use of CFCs in spray cans was banned in the United States and Canada in 1978; their use as foaming agents for plastics was discontinued in 1990. The problem of CFC production and subsequent release, however, spanned the globe and required global cooperation to address.

The world came together and ratified a treaty (the Montreal Protocol) to eliminate the production of CFCs worldwide. The following activities will look at some of the effects from this treaty, and then we will look at replacements for CFCs—after all, air conditioners and refrigerators are still in existence!

Your Turn 3.26 You Decide Save the Ozone Layer

The cartoon below, **Figure 3.26**, is making a statement about the ozone layer. Based on what you have learned in this chapter, what do you think about the message of this cartoon?

Figure 3.26

A satirical cartoon about the use of CFCs and the ozone layer.

Source: UNEP

Your Turn 3.27 Scientific Practices The Global Response

Examine **Figures 3.27** and **3.28**. Do you think there is substantial evidence that the ozone "hole" is being repaired? Describe how this evidence either supports or does not support effective strategies for protecting the ozone layer.

In finding suitable replacements for CFCs, no one advocated a return to toxic gases such as ammonia and sulfur dioxide in home refrigeration units. Similarly, no one advocated giving up air conditioning entirely. Instead, chemists sought to prepare new compounds that were similar to nontoxic CFCs but without their long-term effects on stratospheric ozone.

Any substitute for a CFC should minimize three undesirable properties: toxicity, flammability, and atmospheric lifetime. At the same time, it should preserve a boiling

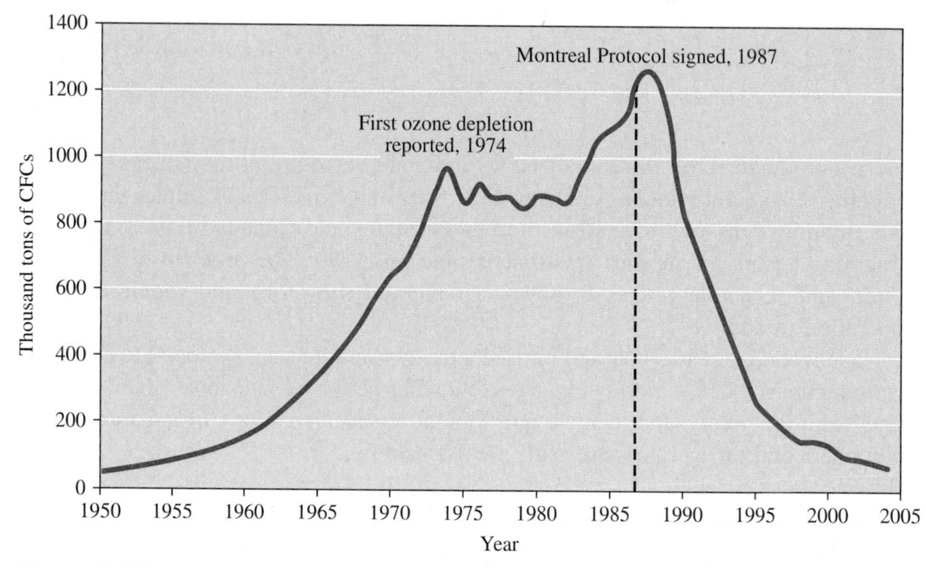

Figure 3.27

Global production of CFCs, 1950–2004.

Source: UNEP

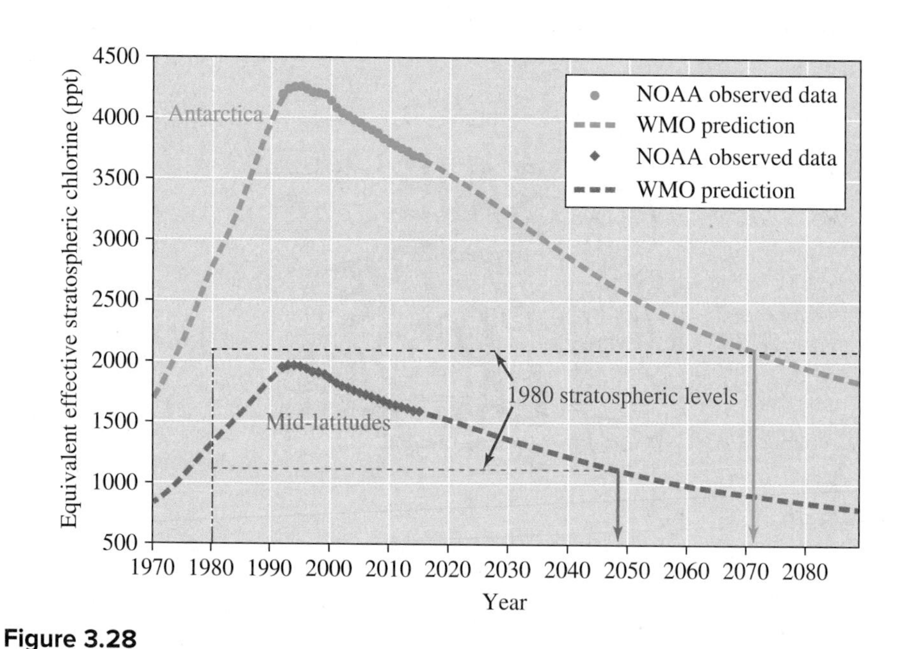

Figure 3.28

Past and projected future changes in reactive halogen concentrations in the atmosphere. The down-pointing arrows mark the estimated dates when concentrations of stratospheric halogen will return to levels present in the 1980s.

Source: The NOAA Ozone Depleting Gas Index: Guiding Recovery of the Ozone Layer; May, 2015

point that is compatible with those of existing refrigerant gases, typically in the range of −10 to −40 °C. Indeed, obtaining a substitute is a delicate balancing act!

One strategy for reducing the atmospheric lifetime of a CFC is to replace one of its C–Cl bonds with a C–H bond. Unlike C–Cl bonds, the C–H bond is susceptible to attack by the hydroxyl radical (·OH) and therefore breaks down more quickly in the lower atmosphere. However, the substitution of a hydrogen atom increases the flammability of the molecule—clearly, an undesirable side effect. The introduction of a lighter atom such as hydrogen also decreases the boiling point, which necessitated re-engineering some of the equipment. Even so, this strategy produced some very useful substitutes. When a hydrogen atom is substituted for one of the chlorine atoms of a CFC, the result is a *hydrochlorofluorocarbon (HCFC)*, a compound of hydrogen, chlorine, fluorine, and carbon (and no other elements). For example, the substitution of a chlorine atom in CCl_2F_2 with hydrogen produces $CHClF_2$:

CCl_2F_2
CFC-12 or R-12

$CHClF_2$
HCFC-22 or R-22

The naming convention for CFCs will be explained in end-of-chapter problem #55.

HCFC-22 has an atmospheric lifetime of about 12 years, compared with 110 years (or so) for the CFC-12, also called R-12 or Freon-12. Because $CHClF_2$ is largely broken down in the lower atmosphere, it does not accumulate in the stratosphere. As a result, its ozone-depleting potential is about 5% that of CCl_2F_2—definitely a step in the right direction.

However, hydrochlorofluorocarbons still contain chlorine and will contribute to ozone depletion. Even so, HCFCs offered considerable improvements over CFCs. At one point, $CHClF_2$ (also called R-22) was the most widely used HCFC. It is suitable for many applications, including air conditioners and as a blowing agent to produce lightweight fast-food containers from plastic "foam."

The phaseout of CFCs and the continuing development of alternative materials were accompanied by major economic concerns. At its peak, the annual worldwide market for CFCs reached $2 billion, the tip of a very large financial iceberg. In the United States alone, CFCs were used in or used to produce goods valued at about $28 billion per year. Although the conversion to CFC replacements has been accompanied by some additional costs in retooling equipment, the overall effect on the U.S. economy has been minimal. Furthermore, the conversions provided a market opportunity for innovative syntheses, based on the key ideas in green chemistry (Chapter 2) to produce environmentally benign substances—a win for both current and future generations.

When first utilized, CFCs appeared to be ideal refrigerant gases. But unexpectedly, they turned out to be the party responsible for destroying Earth's protective ozone layer. When HCFCs replaced CFCs, these were only an interim solution because they also affected the ozone layer, though to a lesser extent. Most HCFCs are scheduled to be phased out by 2030. Currently, HCFCs are no longer being manufactured in developed nations. Although this should lower the concentrations of HCFCs in the atmosphere, as of 2010 the concentration of $CHClF_2$ (R-22) was still increasing. Given the high demand for R-22 worldwide, including home air-conditioning unit use, this is not surprising. Higher demand has also brought higher prices. For example, during the exceptionally hot summer months of 2012, R-22 was reportedly in short supply in the United States and selling for several times its usual price.

With HCFCs on their way out, what is replacing them? *Hydrofluorocarbons (HFCs)*, compounds of hydrogen, fluorine, and carbon (and no other elements), seemed to be the likely candidates because they are similar compounds, but without any chlorine. Here are two examples:

C_2HF_5
pentafluoroethane
HFC-125 (R-125)

and

CH_2F_2
difluoromethane
HFC-32 (R-32)

Neither of these molecules has been implicated in the depletion of stratospheric ozone, and neither has an excessively long atmospheric lifetime.

The switch from HCFCs to HFCs is currently underway. In some cases, a blend of HFCs is substituted for R-22, rather than dropping in a single HFC. Among the most widely used is R-410a, a blend of C_2HF_5 and CH_2F_2. Even so, using the blend requires retooling the equipment so that it can run smoothly with it. Newer designs for air conditioners are engineered from the start to use R-410a rather than its HCFC predecessor, R-22. Another compound, R-134a, with chemical formula $C_2H_2F_4$, is widely used in home refrigerators and in the air conditioners of automobiles.

But with HFCs, another unintended consequence emerges. HFCs are greenhouse gases, which play a role in global climate change! Actually, so were the HCFCs and CFCs they replaced. Like carbon dioxide, HFCs absorb infrared radiation, trap heat in the atmosphere, and contribute to global warming (Chapter 4). In particular, R-32 is of interest because it is a by-product in the synthesis of R-22, currently one of the world's most widely used coolants, as mentioned above. Thus, both in the short haul and over the long one, HFCs are problematic as replacements. So, where do we go from here? Are there any better alternatives?

If you got bogged down with CFCs, HCFCs, and HFCs, prepare yourself for more alphabet soup. One of the newest classes of refrigerant gases is HFOs; that is, hydrofluoroolefins. First, let's take a moment to focus on the name.

- **hydro** means that these compounds contain C–H bonds, just as do *hydro*fluorocarbons and *hydro*chlorofluorocarbons.
- **fluoro** means that these compounds contain C–F bonds, just as do hydro*fluoro*carbons and hydrochloro*fluoro*carbons.
- **olefin** means that these compounds contain C=C double bonds.

Putting these three pieces together, here is the structural formula for HFO-1234yf, an example of an HFO:

$$H_2C{=}C(F){-}CF_3$$

Although it absorbs infrared radiation, the presence of the reactive C=C double bond shortens its atmospheric lifetime. Therefore, it does not persist long in the atmosphere. The compound is somewhat flammable, given the C–H bonds that it contains.

Your Turn 3.28 Skill Building Composition, Structure, and Properties

Look closely at the three molecules presented in this section: HCFC-22, HFC-32, and HFO-1234yf. Recall both the desirable and undesirable properties of refrigerants. Based on the chemical structures of these molecules, describe what components you think make them have similar properties, and what components are present in each molecule that create different properties.

Two other refrigerant gases deserve note. One is R-744, better known as carbon dioxide. This gas was used in refrigeration systems in the 1800s, but suffered from the disadvantage that high pressures are needed to compress it—sometimes more than 100 times atmospheric pressure. It was largely replaced, first by ammonia, and then by CFCs. Although today there is renewed interest in using CO_2 as a refrigerant gas, no revival has yet occurred.

The naming scheme for hydrocarbons was introduced in Section 2.7.

The second is another naturally occurring compound, propane, a small hydrocarbon (C_3H_8) that we use for outdoor grilling. Although it is inexpensive and nontoxic, like all hydrocarbons, it is flammable. As a refrigerant gas, propane has properties similar to R-22. With the advent of the HFOs, also more flammable than their earlier counterparts, propane is poised to make a comeback.

As we end this section, an analogy comes to mind, one that Nobel Laureate Mario Molina employed at a 2011 symposium on ozone depletion and global climate change. He commented that some perceive science as a house of cards: If one part is disturbed, the whole house crumbles. He suggested that a better metaphor is a jigsaw puzzle of a kitten, noting that even if pieces are missing, you still can see the kitten. With stratospheric ozone depletion, this was the case. Even with a few key pieces missing, the picture was still discernible. Although with the help of chemistry one can see the picture, clearly there are still missing pieces. Ultimately, the debate among governments and their citizens about how best to protect the stratospheric ozone layer will determine the outcome in the global political arena.

We have seen that chemistry and scientists played a central role in tackling a major issue in the 20th century. Complete the last activity in this section to review this information.

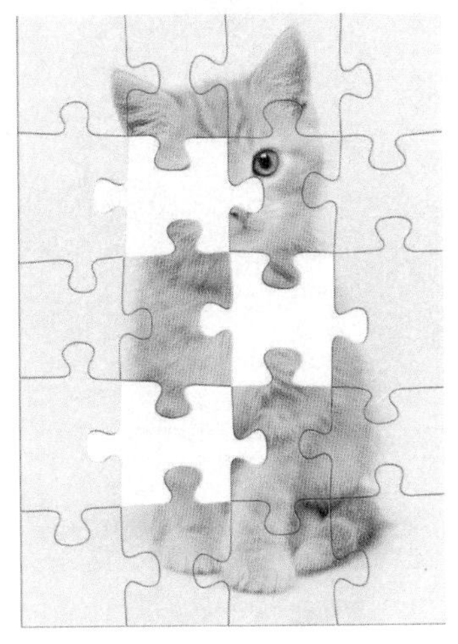

© Stockbyte/Getty Images RF

Your Turn 3.29 Scientific Practices So Where Are We Now?

Imagine you are in a coffee shop with a few friends and the conversation you are having turns to the idea of the ozone layer. One friend says, "I really don't think there is much we can or need to do about the ozone hole. The things that happen to the ozone layer are just part of nature's cycles." Based on what you have learned so far in this chapter, how would you respond to this friend?

3.11 How Do Sunscreens Work?

In addition to our natural source of protection via the ozone layer, there are some simple ways to protect ourselves from UV radiation. The most obvious is to avoid exposure to sunlight when it is at its strongest, especially between 10 am and 4 pm. We can also wear clothing that minimizes skin exposure; there are several brands of clothing that promise to provide extra UV protection. Be aware that sand, water, and snow all reflect sunlight, and your risk of sunburn increases with gains in altitude. Even cloudy days can be risky, as 80% of the Sun's radiation passes through the cloud cover. If it is not possible to stay out of the sun, then doctors and public health workers recommend you apply a sunblock or sunscreen to scatter or absorb UV radiation.

Mineral-based sunblock formulations contain relatively large particles of the compounds zinc oxide (ZnO) or titanium dioxide (TiO_2). Sunblock products physically block the light from reaching your skin cells, much as tightly woven clothing would. You might have seen examples such as the white opaque sunblock cream used by lifeguards ("lifeguard nose") at a pool or beach. However, the metal oxides may also be present in nanoparticulate form, with dimensions often less than 100 nm in diameter. In comparison, recall from Chapter 1 that a human hair is tens of thousands of nanometers across, while a red blood cell is a few thousand nanometers in diameter. In the previous chapter, we saw that particulate matter pollutants are categorized as PM_{10} or $PM_{2.5}$, depending on the size of the particulates. However, PM is on the *micrometer* (or micron, μm) scale, which is 1000 times larger than a nanometer.

Sunblock creams physically block the Sun's UV light, while sunscreens contain chemical and/or physical ingredients that absorb UV light.

Remember that a nanometer is one-billionth of a meter.
1 m = 1000 mm
1 mm = 1000 μm
1 μm = 1000 nm

Your Turn 3.30 Skill Building Particulate Classification

An individual dust particle is 6 μm in diameter.

a. Would this particle be classified as PM_{10} or $PM_{2.5}$?
b. What is the diameter of this particle in nm?

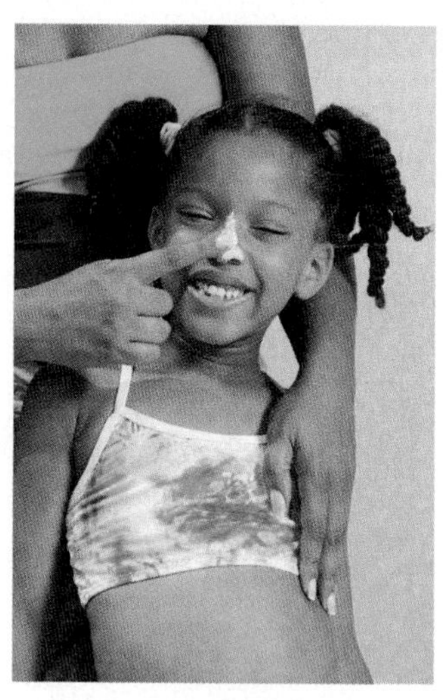

© Image Source/Alamy RF

Since the nanoparticles of ZnO and TiO_2 in sunscreens are so tiny, they do not scatter as much light as the larger particles used in opaque sunblocks. As a result, nanoparticle-based sunscreens are transparent, definitely a cosmetic plus to those who wear them. The nanoparticle products spread more evenly, are cost-effective, and are extremely effective at absorbing, scattering, and reflecting UV radiation. Unlike particulate matter in the air, which can become lodged in the lungs and endanger our health, the nanoparticles used in sunscreens are often suspended in a lotion, so they cannot be inhaled. However, as we will see next, it is important to realize that wearing a sunblock or sunscreen does not mean you are free of risk from damaging UV rays.

Most sunscreens are applied as a lotion or cream; they are also available in aerosol sprays, which are considered less effective. Sunscreens labeled as "broad spectrum" contain chemical compounds that absorb UVB to some extent together with others that absorb UVA. The American Academy of Dermatology recommends a sunscreen with a sun protection factor (SPF) of at least 30. The SPF rating refers to the length of time a sunscreen will protect your skin from reddening/burning, as compared to how long it would take to burn without sunscreen protection. However, these ratings assume that a dosage of 2 mg/cm^2 is applied to your skin. This roughly corresponds to 1 teaspoon (or 5 mL) for your face, and a whopping 1 ounce (30 mL—approximately two tablespoons) for a single application to your body. Research has shown that users typically use much less sunscreen per application, resulting in an *actual* SPF that is 20–50% of the value expected from the product label. Furthermore, it is recommended that you re-apply sunscreen every two hours, regardless of SPF, to ensure the best protection—another guideline often overlooked by beachgoers.

Your Turn 3.31 You Decide Sunscreens

A sunscreen's SPF is a rough indication of how much longer a person can remain in the sun while using the sunscreen, as compared to without any protection. Recently, sunscreens with SPF as high as 70 have been introduced.

a. Estimate how long you think it would take your skin to burn at noon on a midsummer day, assuming you use no sunscreen.
b. If you buy a sunscreen advertising SPF 70, calculate how long you might be able to stay in the sun without burning.
c. What assumptions are being made in predicting an SPF? Why are some dermatologists concerned about the SPF 70 sunscreens?

Your Turn 3.32 You Decide SPF Ratings

Sunscreens with an SPF of 15 filter out 93% of incoming UV radiation, while SPF 30 sunscreens filter out 97% and SPF 50 filter out 98%.

a. Is an SPF 30 sunscreen twice as "good" as an SPF 15 sunscreen?
b. How much additional protection do you get by choosing to use an SPF 50 sunscreen over an SPF 30?

Earlier in the chapter, we saw the absorption of UV light by ozone and oxygen as a consequence of their reactivity. However, the absorption by physical and chemical sunscreen components is not due to reactions with UV light, but rather is simply due to their structures. As we saw earlier, when atoms are bonded together to form

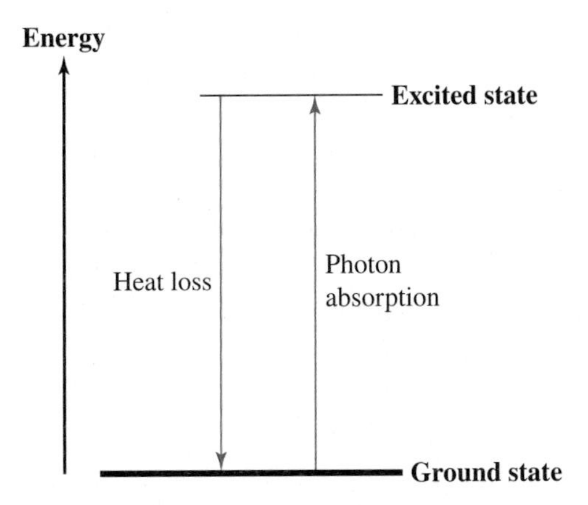

Figure 3.29

Illustration of the absorption of light of different wavelengths by electrons in a lower-energy level (ground state) into a higher-energy excited state. The electrons return to the lowest level through loss of heat to the surrounding solid.

molecules, their valence electrons are shared between the two atoms to form a covalent bond. When trillions of atoms come together to form a solid, as is the case with the nanoparticle components in sunscreens, the electrons are housed in a very complicated structure of quantized energy levels that allows for light of certain energies to be absorbed. When the light is absorbed, the energy is transferred to electrons in lower-energy levels, which then become excited and move into higher-energy levels. Eventually, electrons lose this excess energy and relax back to their original energy levels, with the release of heat to the surrounding solid (Figure 3.29). The absorption properties of the solid depend on its composition as well as the size of the particulates. For instance, solids of identical composition but with particulate diameters of 20 nm versus 50 nm will absorb radiation of different wavelengths. Based on extensive studies, ZnO appears to have the best broadband protection against both UVA and UVB radiation.

While some are concerned that nanoparticles may present a health risk if they are absorbed into the bloodstream through the skin, studies to date indicate that nanoparticles in sunscreen products do not penetrate beyond the epidermis, the first layer of skin. Both consumers and government agencies continue to call for further studies to better quantify the risks. As shown in Figure 3.30, a variety of organic-based chemicals may also be incorporated within sunscreen formulations to either enhance or supplant the use of nanoparticles. However, these ingredients also pose their own set of potential safety risks. Although the health implications of using a sunscreen are not yet completely understood, the risk of too much exposure to the Sun is already very well-known, and sunscreens play an important role in protecting us from UV radiation.

Your Turn 3.33 You Decide Sunscreen Labels

The U.S. Food and Drug Administration (FDA) introduced new regulations concerning sunscreen labels in 2012. With other students, gather a collection of sunscreen products and compare the labels.

a. What claims are made by the companies on the front of the products?
b. Look at the composition information required by the FDA on the back of the product. Which active ingredients are used? For each sunscreen product, determine whether it is organic-based or mineral-based.

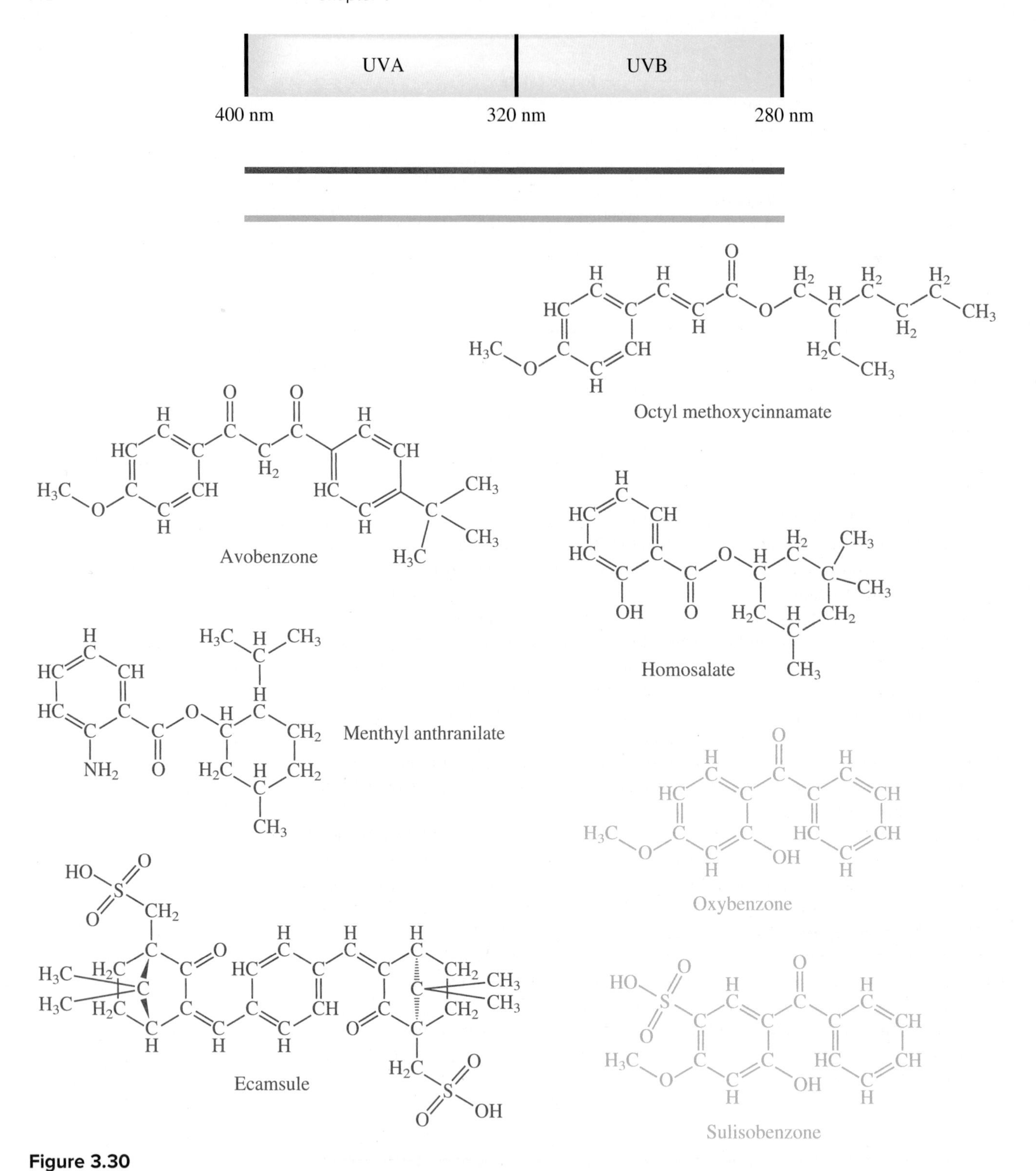

Figure 3.30

Chemical structures and UVA and UVB blocking ranges of some organic-based active ingredients in sunscreens.

Your Turn 3.34 Scientific Practices UV Interactions with Matter

Investigate the two simulations in **Connect** that illustrate the interactions of ultraviolet radiation with nanoparticles, a component of some sunscreens. Explain the influences of nanoparticle size and density on the overall effectiveness of a mineral-based sunscreen.

Conclusions

Most people enjoy being outside, especially when the weather is sunny and warm. We may bask in the sun, play games, go swimming, or climb a mountain, and some jobs require workers to be outside all day.

The Sun's rays are not all the same—in fact, the Sun is sending a range of energies to Earth. We feel some of this energy as heat and see some of it as color. But the invisible ultraviolet energy we receive from the Sun is cause for concern. Ultraviolet radiation contains enough energy to damage human skin. The effects of exposure to too much UV radiation can appear quickly, as in the case with a sunburn. Or, the effects can take years to become evident, as with wrinkles and skin cancer. In this chapter, we learned about the damage ultraviolet radiation can cause when it hits our skin.

Luckily for life on Earth, the ozone found in the stratosphere far above us protects us from the majority of harmful rays. However, advancements in technology may not always act in our best interest. In the 20th century, common refrigerant gases began depleting the ozone in the protective layer. Fortunately, governments around the world acted relatively quickly to pass the Montreal Protocol, a treaty that limited the production of these harmful gases. Over the 25 years since the treaty took effect, the ozone hole has stabilized, and scientists now predict a full recovery by the year 2050. Indeed, chemistry is often a two-edged sword, being used in this case study to defend the very natural source of UV protection that it almost destroyed.

The ozone layer does not absorb all of the UV radiation from the Sun, so people must still take precautions to protect their skin. Chemical- and mineral-based sunscreens and sunblocks, together with protective clothing and headgear, are the most effective means of limiting UV exposure.

Learning Outcomes

The numbers in parentheses indicate the sections within the chapter where these outcomes were discussed.

Having studied this chapter, you should now be able to:

- categorize radiation in terms of electromagnetic waves, subatomic particles, and/or photons (3.1, 3.2)
- identify the forms of radiation present in the electromagnetic spectrum (3.1)
- distinguish forms of radiation based on wavelength, frequency, and energy (3.1, 3.2)
- calculate the energy of radiation at different wavelengths and frequencies (3.2)
- sketch the emission spectrum from the Sun and label each portion of the spectrum (3.3)
- model how radiation interacts with matter (3.3)
- explain why some forms of radiation are harmful to humans, while others are not (3.4)
- differentiate the three regions of UV radiation based on their energy, impact on human molecules, and wavelength (3.4)
- use your understanding to predict how UV radiation can affect other organisms and materials (3.4)
- summarize how Earth's atmosphere, skin, and shade naturally protect us from UV light (3.4, 3.5)
- identify which gases in the upper atmosphere protect us from UV (over)exposure (3.5)
- describe where ozone is found in the atmosphere, and compare its effects at the different levels of the atmosphere (3.5)
- describe why the interaction of ozone with UV results in protection from radiation (3.5)
- describe how the ozone concentration is measured and how it led to discovery of the "ozone hole" (3.6, 3.8)
- define and apply the law of conservation of mass (3.6)
- define covalent bonding as the sharing of electrons (3.7)
- illustrate a covalent bond between two atoms (3.7)
- draw and interpret Lewis structures using the octet rule (3.7)
- predict the number of valence electrons for an element based on its position in the periodic table (3.7)
- examine the relationship between bond strength and the magnitude of energy required for dissociation, and predict the stability of molecules irradiated by various regions of the EM spectrum (3.7)
- using data, explain the relationship between the increase in use of CFCs and ozone depletion (3.8)
- discuss why CFCs were chosen for industrial applications, and what alternatives are being used to help restore the ozone layer (3.9, 3.10)
- summarize the role of free radical chemistry in the formation of the ozone hole (3.9)
- explain and illustrate how chemical compounds may be used to prevent overexposure to sunlight (3.11)
- convert among various length scales and relate these length scales to bulk and micro/nanoscale objects (3.11)
- summarize what governs a molecule's ability to absorb certain regions of the EM spectrum (sections throughout the chapter)
- describe risk assessment in choosing the best sunscreens or sunblocks for protection from sunlight (3.11)

Questions

Emphasizing Essentials

1. How does ozone differ from oxygen in its chemical formula? In its properties?
2. The text states that the odor of ozone can be detected in concentrations as low as 10 ppb. Would you be able to smell ozone in either of these air samples?
   a. 0.118 ppm of ozone, a concentration reached in an urban area
   b. 25 ppm of ozone, a concentration measured in the stratosphere
3. A journalist wrote "Hovering 10 miles above the South Pole is a sprawling patch of stratosphere with disturbingly low levels of radiation-absorbing ozone."
   a. How big is this sprawling patch?
   b. Is the figure of 10 miles correct? Express this value in kilometers.
   c. What type of radiation does ozone absorb?
4. It has been suggested that the term *ozone screen* would be a better descriptor than *ozone layer* to describe ozone in the stratosphere. What are the advantages and disadvantages to each term?
5. Describe three differences between air in the troposphere and the stratosphere. In your answer, consider material from both Chapter 2 and Chapter 3.
6. a. What is a Dobson unit?
   b. Does a reading of 320 DU or 275 DU indicate more total column ozone overhead?
7. Using the periodic table as a guide, specify the number of protons and electrons in a neutral atom of each of these elements.
   a. oxygen (O) b. magnesium (Mg)
   c. nitrogen (N) d. sulfur (S)
8. Consider this representation of a periodic table.

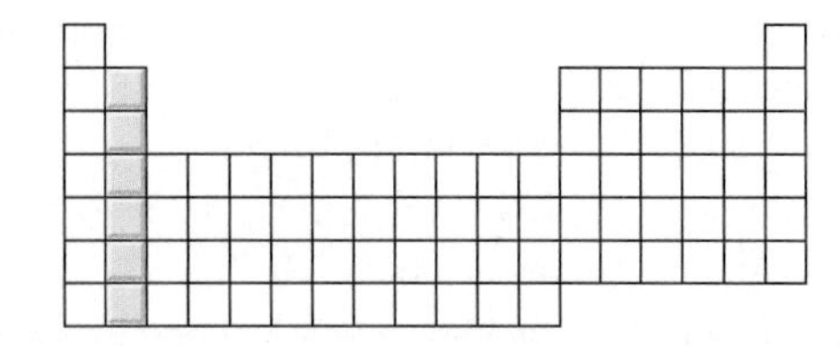

   a. What is the group number of the shaded column?
   b. Which elements make up this group?
   c. What is the number of electrons for a neutral atom of each element in this group?
   d. What is the number of outer electrons for a neutral atom of each element of this group?
9. Give the name and symbol for the element with this number of protons.
   a. 2 b. 19 c. 29
10. Draw the Lewis structure for each of these atoms.
   a. calcium b. chlorine
   c. nitrogen d. helium
11. Assuming that the octet rule applies, draw the Lewis structure for each of these molecules.
   a. CCl_4 (carbon tetrachloride, a substance formerly used as a cleaning agent)
   b. H_2O_2 (hydrogen peroxide, a mild disinfectant; the atoms are bonded in this order: H–O–O–H)
   c. H_2S (hydrogen sulfide, a gas with the unpleasant odor of rotten eggs)
   d. N_2 (nitrogen gas, the major component of the atmosphere)
   e. HCN (hydrogen cyanide, a molecule found in space and a poisonous gas)
   f. N_2O (nitrous oxide, "laughing gas"; the atoms are bonded N–N–O)
   g. CS_2 (carbon disulfide, used to kill rodents; the atoms are bonded S–C–S)
12. Several oxygen species play important chemical roles in the stratosphere, including oxygen atoms, oxygen molecules, ozone molecules, and hydroxyl radicals. Draw Lewis structures for each.
13. Consider these two waves representing different parts of the electromagnetic spectrum. How do they compare in terms of:

Wave 1 Wave 2

   a. wavelength
   b. frequency
   c. speed of travel
14. Use Figure 3.4 to specify the region of the electromagnetic spectrum where radiation of each of the following wavelengths is found.
   Hint: Change each wavelength to meters before making the comparison.
   a. 2.0 cm b. 50 um
   c. 400 nm d. 150 mm
15. What determines the color of light? Describe the differences between orange and violet light.

16. Arrange the wavelengths in question 14 in order of *increasing* energy. Which wavelength possesses the most energetic photons?

17. Does all light travel at the same speed in a vacuum? Explain your reasoning.

18. Arrange these types of radiation in order of *increasing* energy per photon: gamma rays, infrared radiation, radio waves, visible light.

19. The microwaves in home microwave ovens have a frequency of $2.45 \times 10^9\ s^{-1}$. Is this radiation more or less energetic than radio waves? Than X-rays?

20. Ultraviolet radiation is categorized as UVA, UVB, or UVC. Arrange these types in order of increasing:

a. wavelength b. potential for biological damage

c. energy

21. Calculate the wavelength, in nanometers, of the following wave frequencies:

a. $6.79 \times 10^{14}\ s^{-1}$ b. 4.44×10^{12} Hz

22. The distance from Earth to the Sun is about 1.50×10^8 km. How long does it take light from the Sun to travel to Earth?

23. Draw Lewis structures for any two different CFCs.

24. CFCs were used in hair sprays, refrigerators, air conditioners, and plastic foams. Which properties of CFCs made them desirable for these uses?

25. a. Can a molecule that contains hydrogen be classified as a CFC?

b. What is the difference between an HCFC and an HFC?

26. a. Most CFCs are based either on the structure of methane, CH_4, or ethane, C_2H_6. Use structural formulas to represent these two compounds.

b. Substituting both chlorine atoms and fluorine atoms for all of the hydrogen atoms on a methane molecule, you obtain CFCs. How many possibilities exist?

c. Which of the substituted CFC compounds in part b has been the most successful?

d. Why weren't all of these compounds equally successful?

27. The following free radicals all play a role in catalyzing ozone depletion reactions: Cl•, •NO_2, ClO•, and •OH.

a. Count the number of outer electrons available and then draw a Lewis structure for each free radical.

b. What characteristic is shared by these species that makes them so reactive?

28. a. How were the original measurements of increases in chlorine monoxide and the stratospheric ozone depletion over the Antarctic obtained?

b. How are these measurements made today?

29. Which graph shows how measured increases in UVB radiation correlate with percent reduction in the concentration of ozone in the stratosphere over the South Pole? Explain.

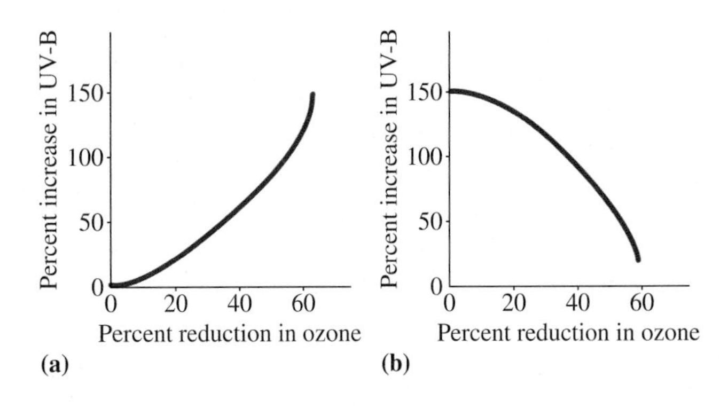

Concentrating on Concepts

30. The EPA has used the slogan "*Ozone: Good Up High, Bad Nearby*" in some of its publications for the general public. Explain the message.

31. Nobel Laureate F. Sherwood Rowland referred to the ozone layer as the Achilles heel of our atmosphere. Explain the metaphor.

32. In the abstract of a talk he gave in 2007, Nobel Laureate F. Sherwood Rowland wrote "Solar UV radiation creates an ozone layer in the atmosphere which in turn completely absorbs the most energetic fraction of this radiation."

a. What is the most energetic UV fraction? **Hint:** See Figure 2.4.

b. How does solar UV radiation "create an ozone layer"?

33. What are some of the reasons that the solution to ozone depletion proposed in this Sydney Harris cartoon will not work?

AEROSOL PRODUCTS
"OH, FOR PETE'S SAKE, LET'S JUST GET SOME OZONE AND SEND IT BACK UP THERE!"

Source: ScienceCartoonsPlus.com. Reprinted with permission.

34. "*We risk solving one global environmental problem while possibly exacerbating another unless other alternatives can be found.*" The date of this quote by a U.S. official is 2009, and the context is phasing out the use of HCFCs.

a. What compounds were HCFCs being replaced with in 2009?

b. What is the risk of this replacement?

35. It is possible to write three resonance structures for ozone, not just the two shown in the text. Verify that all three structures satisfy the octet rule and offer an explanation of why the triangular structure is not reasonable.

$$O{=}O{-}O \longleftrightarrow O{-}O{=}O \longleftrightarrow \text{(triangular } O_3\text{)}$$

36. The average length of an O–O single bond is 132 pm. The average length of an O=O double bond is 121 pm. What do you predict the O–O bond lengths will be in ozone? Will they all be the same? Explain your predictions.

37. More ozone is found in the stratosphere than the troposphere. Why is this?

38. Describe why ozone is more reactive than oxygen gas.

39. Consider the Lewis structures for SO_2. How do they compare with the Lewis structures for ozone?

40. Even if you have skin with little pigment, you cannot get a tan from standing in front of a radio. Why?

41. The morning newspaper reports a UV Index Forecast of 6.5. Given the amount of pigment in your skin, how might this affect how you plan your daily activities?

42. All the reports of the damage caused by UV radiation focus on UVA and UVB radiation. Why isn't there more attention paid to the damaging effects of UVC radiation?

43. If all 3.3×10^8 tons of stratospheric ozone that are formed every day are also destroyed every day, how is it possible for stratospheric ozone to offer any protection from UV radiation?

44. How does the chemical inertness of CCl_2F_2 (Freon-12) relate to the usefulness and the problems associated with this compound?

45. Chlorine is a catalyst in chemical reactions involving ozone in the atmosphere. Name two other catalysts found in the atmosphere.

46. Explain how the small changes in Cl· concentrations (measured in parts per billion) can cause the much larger changes in O_3 concentrations (measured in parts per million).

Exploring Extensions

47. DVD players use a laser with a wavelength of 650 nm (a red light) to read the information stored on the discs. Blu-ray players use a 405 nm wavelength laser (blue light). Single-layer DVDs can store 4.7 GB of data compared to 25 GB stored on a Blu-ray disc. Why do Blu-ray discs store so much more data than DVDs?

48. Development of the stratospheric ozone hole has been most dramatic over Antarctica. What set of conditions exist over Antarctica that help to explain why this area is well-suited to studying changes in stratospheric ozone concentration? Are these same conditions not operating in the Arctic? Explain.

49. The free radical CF_3O· is produced during the decomposition of HFC-134a.

a. Propose a Lewis structure for this free radical.

b. Offer a possible reason why CF_3O· does not cause ozone depletion.

50. Consider this graph that shows the atmospheric abundance of bromine-containing gases from 1950 to 2100.

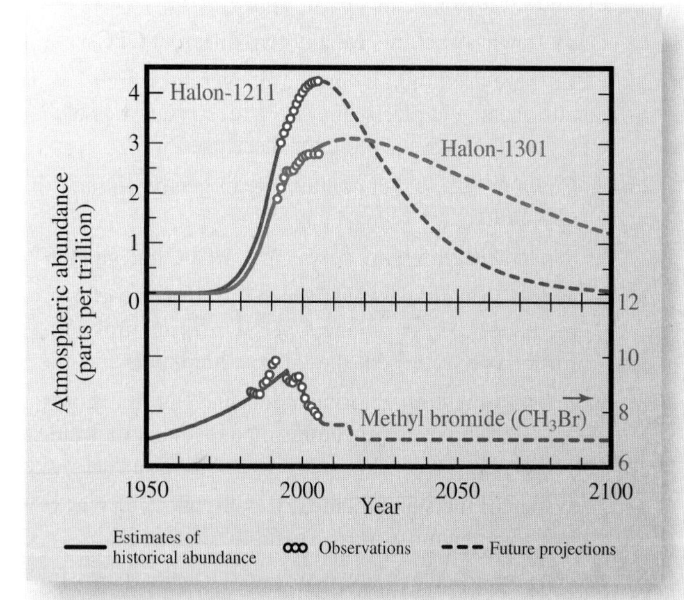

a. Halon-1301 is $CBrF_3$ and Halon-1211 is $CClBrF_2$. Why were these compounds once manufactured?

b. Compare the patterns for Halon-1211 and Halon-1301. Why doesn't Halon-1301 drop off as quickly?

c. In 2005, methyl bromide was phased out in the United States except for critical uses. Why is its future use predicted as a straight line, rather than tailing off? Section 2.13 discussed the role of nitrogen monoxide (NO) in forming photochemical smog. What role, if any, does NO play in stratospheric ozone depletion? Are NO sources the same in the stratosphere as in the troposphere?

51. Resonance structures can be used to explain the bonding in charged groups of atoms as well as in neutral molecules, such as ozone. The nitrate ion, NO_3^-, has one additional electron plus the outer electrons contributed by nitrogen and oxygen atoms. That extra electron gives the ion its charge. Draw the resonance structures, verifying that each obeys the octet rule.

52. Some places like China have an issue with smog, a fog created in the reaction of sunlight with nitrogen oxides and carbon compounds released from cars or burning coal. How does smog affect the amount of UV radiation received in these areas? How does it affect the breathing conditions?

53. Ozone depletion has been a concern due to measurements indicating the worldwide ozone concentration has decreased over the past 20 years. Could this issue be solved by releasing ozone into the atmosphere? Why or why not?

54. Although oxygen exists as O_2 and O_3, nitrogen exists only as N_2. Propose an explanation for these facts. **Hint:** Try drawing a Lewis structure for N_3.

55. The chemical formulas for a CFC, such as CFC-11 (CCl_3F), can be figured out from its code number by adding 90 to it to get a three-digit number. For example, with CFC-11 you get 90 + 11 = 101. The first digit is the # of C atoms, the second is the # of H atoms, and the third is the # of F atoms. Accordingly, CCl_3F has 1 C atom, no H atoms, and 1 F atom. All remaining bonds are assumed to be chlorine.

a. What is the chemical formula for CFC-12?

b. What is the code number for CCl_4?

c. Does this "90" method work for HCFCs? Use HCFC-22 ($CHClF_2$) in explaining your answer.

d. Does this method work for halons? Use Halon-1301 (CF_3Br) in explaining your answer.

56. Many different types of ozone generators ("ozonators") are on the market for sanitizing air, water, and even food. They are often sold with a slogan such as this one from a pool store. "Ozone, the world's most powerful sanitizer!"

a. What claims are made for ozonators intended to purify air?

b. What risks are associated with these devices?

57. The effect a chemical substance has on the ozone layer is measured by a value called its *ozone-depleting potential,* ODP. This is a numerical scale that estimates the lifetime potential stratospheric ozone that could be destroyed by a given mass of the substance. All values are relative to CFC-11, which has an ODP defined as equal to 1.0. Use those facts to answer these questions.

a. Name two factors that affect the ODP value of a compound, and explain the reason for each one.

b. Most CFCs have ODP values ranging from 0.6 to 1.0. What range do you expect for HCFCs? Explain your reasoning.

c. What ODP values do you expect for HFCs? Explain your reasoning.

58. Cooking with an electric stove can have a negative effect on the environment, just as one that uses natural gas can, even though gas is not being burned directly. Why is this? **Hint:** Where does the electricity used to power the stove come from?

59. One mechanism that helps break down ozone in the Antarctic region involves the BrO· free radical. Once formed, it reacts with ClO· to form BrCl and O_2. BrCl, in turn, reacts with sunlight to break into Cl· and Br·, both of which react with O_3 and form O_2.

a. Represent this information with a set of equations.

b. What is the net equation for this cycle?

60. Polar stratospheric clouds (PSCs) play an important role in stratospheric ozone depletion.

a. Why do PSCs form more often over Antarctica than in the Arctic?

b. Reactions occur more quickly on the surface of PSCs than in the atmosphere. One such reaction is that of two species that do not deplete ozone, hydrogen chloride and chlorine nitrate ($ClONO_2$), which react to produce a chlorine molecule and nitric acid (HNO_3). Write the chemical equation.

c. The chlorine molecule produced does not deplete ozone either. However, when sunlight returns to the Antarctic in the springtime, it is converted to a species that does. Show how this happens using a chemical equation.

61. Recent experimental evidence indicates that ClO· initially reacts to form Cl_2O_2.

a. Predict a reasonable Lewis structure for this molecule. Assume the order of atom linkage is Cl–O–O–Cl.

b. What effect does this evidence have on understanding the mechanism for the catalytic destruction of ozone by ClO·?

CHAPTER 4 Climate Change

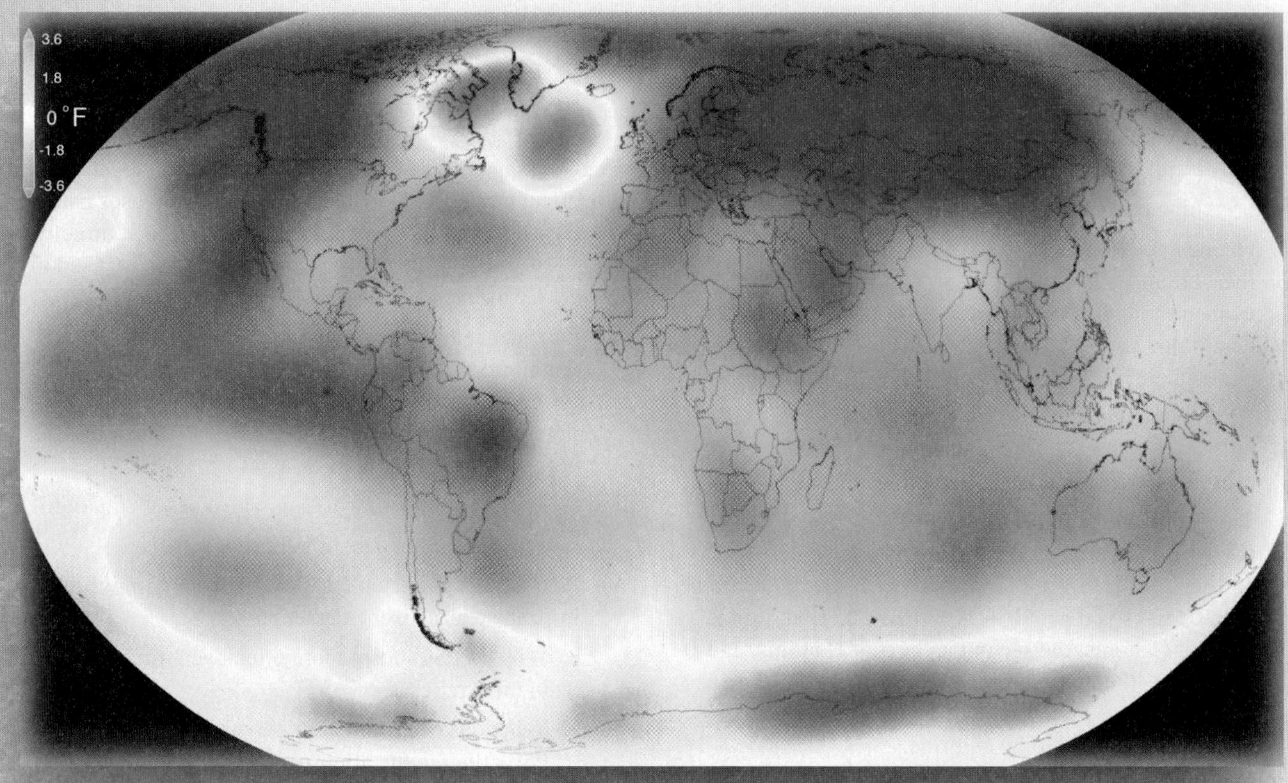

Source: NASA/Scientific Visualization Studio/Goddard Space Flight Center

REFLECTION

Climate Change

The controversial topic of climate change has been widely portrayed in the media. Answer these questions based on your current knowledge.

a. Do you think the warming effect of greenhouse gases is a positive effect or a negative effect? Why?
b. Where does the carbon dioxide in the air come from?
c. What is climate change?
d. Is climate change occurring now? Why or why not?

The Big Picture

In this chapter, you will explore the following questions:

- What are the sources of carbon on Earth?
- How does carbon move between reservoirs, and how do scientists measure this?
- What are "greenhouse gases," and what are their positive and negative effects?
- What are the global consequences of climate change?
- How do our climate trends differ from the past?
- How can my daily actions affect the global environment?

Introduction

Now that we know what the public is thinking about climate change and the greenhouse effect, let's ponder what it will be like 100 years in the future ...

As you maneuver your hovercraft to visit different locations around Earth, think about two different scenarios:

i. Earth has increased in temperature by 6 °C (10.8 °F), on average. Heat waves are more common, which leads to an increase in wildfires and heat-related illnesses. The parched Western United States never did recover from the drought that started in 2000. Lake Mead, created when the Hoover Dam was built in the early 1930s, has completely dried up, resulting in water shortages to large population centers such as Los Angeles, Phoenix, and Las Vegas. Many people have had to move from these locations due to the lack of potable water and the extreme heat. In Greenland and Antarctica, the volume of glaciers has decreased rapidly over the last 100 years. The melting of glaciers has resulted in a sea level rise of 4 meters (approximately 13 feet), and as you fly over low-lying land such as tropical islands, lower Manhattan, Miami, and Bangladesh, you notice that sea water is encroaching on population centers and forcing people and businesses to move. More intense hurricanes and cyclones are occurring due to warming ocean waters, so the people living in these low-lying areas are impacted even more.

Source: GNP Archives/USGS

Source: Karen Holzer/USGS

ii. As you travel to Glacier National Park in Montana, some of the 115 glaciers that had disappeared in 2015 have started to return. When you visit the small island nation of the Maldives off the coast of India, you observe that the sea waters encroaching on the islands have started to recede, and the government no longer has to think about purchasing large tracts of land to move their 345,000 citizens to the mainland (Figure 4.1). The coral reefs around the Maldives have started to recover from the excess warmth and acidity of ocean water. You observe that agricultural yields in Africa have doubled due to increased precipitation, and although hurricanes and cyclones still occur, their intensity is not as strong, causing less destruction.

Figure 4.1

In October 2009, the President of the Maldives and his cabinet had an underwater meeting to call attention to the threat of sea level rise to the archipelago nation.

© AP Photo/Mohammed Seeneen

Which world would you choose? In this chapter, we will discuss the chemistry behind climate change. We will present data about the climate, and help you analyze the data to come up with your own conclusions about what is happening on Earth. Think about what you know now. What have you read or heard about Earth's climate? What causes warming and cooling effects? Has our climate really changed in the past 50 years, and what can we expect in the next 50 years?

Your Turn 4.1 You Decide Is It Really Possible?

Think about the two scenarios presented in the Introduction. What would cause each of those situations? Are there actions by humans that could influence them? Explain your reasoning.

Keep your answers to Your Turn 4.1 and the Reflection in mind as we investigate possible reasons for climate change and how chemistry plays an important role. We will start by examining a very important element.

4.1 | Carbon, Carbon Everywhere!

If you look on the periodic table at the element with atomic number 6, you will find carbon—one of the most important elements for life on Earth. Chemists consider carbon so important that there is a branch of chemistry devoted to studying carbon and its compounds. *Organic chemistry* is the field that works at understanding and producing carbon-based compounds. You may have seen carbon in its elemental form as a diamond or as pencil "lead" (actually graphite), but there are millions of compounds that have carbon atoms in them. A few examples include sugar ($C_{12}H_{22}O_{11}$), fingernail polish remover (acetone, C_3H_6O), or one compound we will examine extensively, carbon dioxide (CO_2).

Your Turn 4.2 You Decide What Do You Know About the "Big C"?

In chemistry, substances containing carbon atoms play a crucial role in understanding life and interactions on Earth. What do you know about carbon compounds currently? Why do you think it is an important element? Provide some examples of other compounds that contain carbon.

As chemists, it is important to understand where we can find species containing carbon atoms on Earth. As shown in Figure 4.2, compounds with carbon atoms are found in several places, referred to as *reservoirs*. In the next two sections, we will describe this figure in detail, but for now you only need to pay attention to the locations of carbon-containing compounds.

You may notice that one reservoir is the atmosphere. Many of the carbon atoms here are in the form of $CO_2(g)$ (~400 ppm), $CH_4(g)$ (~1.8 ppm), and $CO(g)$ (trace amounts as an air pollutant). A second reservoir for carbon is carbonate-containing rocks. A third reservoir for carbon is plants and animals where the carbon atoms are combined with oxygen, hydrogen, nitrogen, and other elements to form carbohydrates, proteins, and lipids.

We will examine carbohydrates, proteins, and lipids in more detail in Chapter 11.

In Section 4.2, we will further examine how different substances containing carbon atoms cycle through nature and we will interpret "the global carbon cycle" (Figure 4.2) more fully. However, you may notice that carbon atoms rarely show up in their elemental forms. It is generally found in compounds, mixed in fixed ratios with other elements. Combined with other elements, carbon atoms appear in both molecular and ionic compounds. Chemists use different conventions for naming molecular and ionic compounds. In this section, we will describe the naming of ionic compounds in detail since this will help us identify the solid forms of compounds that contain carbon atoms.

See Section 2.7 for more details on naming molecular compounds. These types of compounds are sometimes denoted as *covalent compounds* since their atoms are held together by covalent bonds.

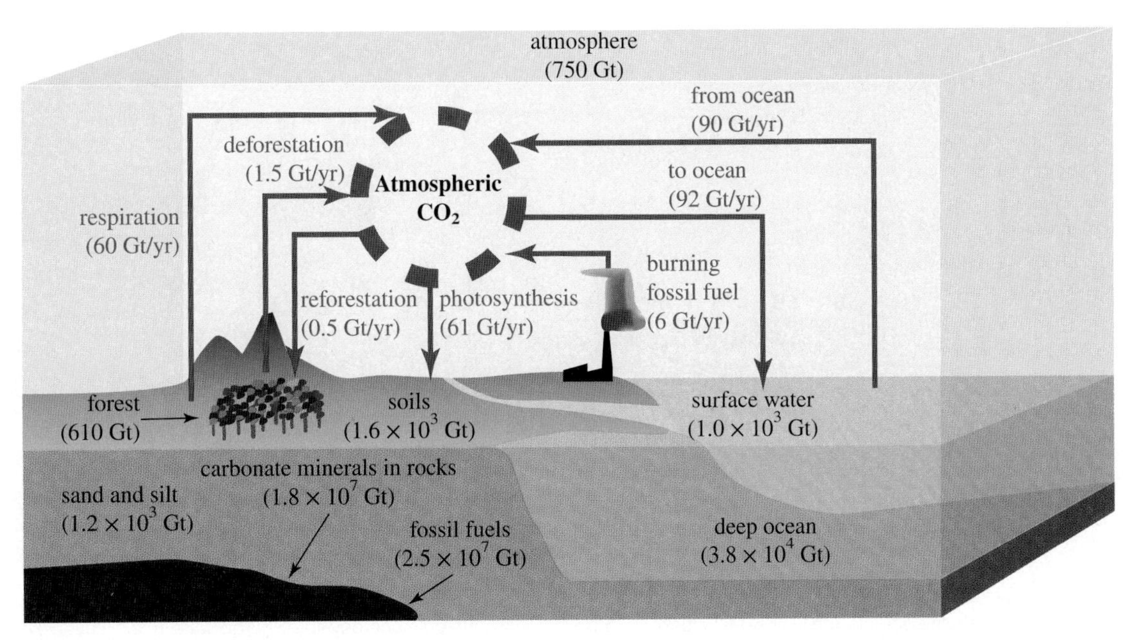

Figure 4.2

The global carbon cycle. The numbers show the quantity of carbon, expressed in gigatonnes (Gt), that is stored in various carbon reservoirs (*black*) or moving through the system per year (*red*).

Source: Purves, Orians, Heller, and Sadava, Life, The Science of Biology, *5th edition, 1998, p. 1186. Reprinted with permission of Sinauer Associates, Inc.*

The carbon atoms found in rocks are generally part of an ionic compound. In Section 1.7, we discussed the chemical formulas of simple ionic compounds, formed from positive ions (cations) and negative ions (anions). Here is another example to refresh your memory. Knowing that boron atoms tend to lose three electrons to form B^{3+}, you can write the chemical formula of the ionic compound formed from B^{3+} and O^{2-} as B_2O_3. Here, a 2:3 ratio of ions is needed so that the overall electric charge on the compound is zero. Note that it is *not* correct to write the chemical formula as "$B_2^{3+}O_3^{2-}$".

A gigatonne (Gt) is a billion metric tons, or about 2.2 trillion pounds. For comparison, a fully loaded Boeing 747 jet weighs about 800,000 lb. It would take nearly 3 million 747s to have a total mass of 1 Gt.

You may have heard several ionic compounds called by their names, including sodium chloride, aluminum oxide, and potassium chloride. Observe the pattern: name the cation first, then the anion, and modify the name to end in the suffix *-ide*. Thus, CaO is calcium oxide, with each ion named for its element and oxygen modified to read oxide. Similarly, NaI is sodium iodide, and KCl is potassium chloride.

The elements presented thus far formed only one type of ion. Group 1 and 2 elements only form ions with charges of 1+ and 2+, respectively. The halogens typically form ions with a 1– charge. Hence, the chemical formula for lithium bromide is LiBr. The ratio of 1:1 is understood because lithium atoms only form Li^+ ions and bromine atoms only form Br^- ions. The prefixes *mono-*, *di-*, *tri-*, and *tetra-* are *not* used when naming ionic compounds such as these. There is no need to call LiBr "monolithium monobromide." Likewise, $MgBr_2$ is magnesium bromide, not magnesium dibromide. Magnesium atoms *only* form Mg^{2+} ions, so the ratio of 1:2 is understood and does not need to be explicitly stated as done when naming molecular compounds.

But some elements form more than one ion, as you can see in Figure 4.3. Prefixes still are not used, but now the charge on the ion must be specified using a Roman numeral. Consider copper compounds, for example. If your instructor asks you to head down to the stockroom and grab some "copper oxide," what do you do? You should ask if they want either copper(I) oxide or copper(II) oxide, because copper ions form different ions. Similarly, as you saw in Chapter 1, iron ions can form different oxides. Two forms are FeO (formed from Fe^{2+}) and Fe_2O_3 (commonly called rust and formed from Fe^{3+}). The names for FeO and Fe_2O_3 are iron(II) oxide and iron(III) oxide, respectively. Note the space after, but not before, the parentheses enclosing the Roman numeral.

Prefixes such as *di-* and *tri-* generally are not used when naming ionic compounds. Roman numerals are used with the name of the cation if it has more than one possible charge.

As a final comparison, the name of $CuCl_2$ is copper(II) chloride, but the name of $MgCl_2$ is magnesium chloride. Magnesium atoms only form one ion (Mg^{2+}), whereas copper atoms can form two ions (Cu^+ and Cu^{2+}).

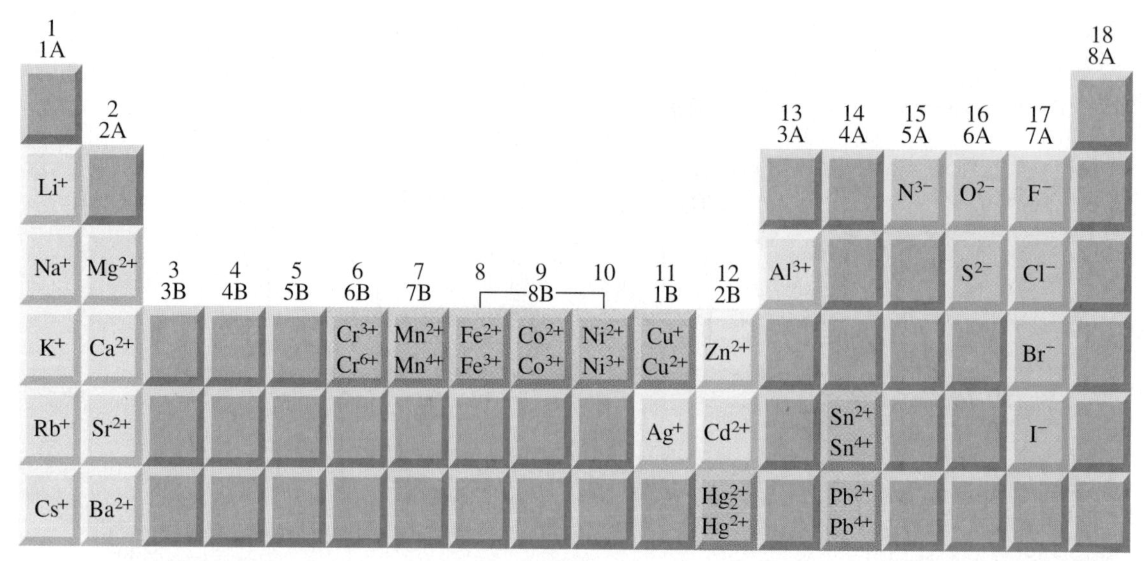

Figure 4.3

Common ions formed from their elements. Ions in **green** (cations) or **blue** (anions) have only one possible charge. Ions in red (cations) have more than one possible charge.

Your Turn 4.3 Skill Building Forming Ionic Compounds

Each pair of elements forms one or more ionic compounds. For each, write all of the possible chemical formulas and names.

a. Ca and S
b. F and K
c. Mn and O
d. Cl and Al
e. Co and Br

One or both of the ions in an ionic compound can be a **polyatomic ion**, two or more atoms covalently bonded together that have an overall positive or negative charge. An example is the carbonate anion, CO_3^{2-}, with three oxygen atoms covalently bonded to a central carbon atom. Table 4.1 lists common polyatomic ions. Most are anions, but polyatomic cations also are possible, as in the case of the ammonium ion, NH_4^+. Note that some elements (carbon, sulfur, and nitrogen) form more than one polyatomic anion with oxygen.

The rules for naming ionic compounds containing polyatomic ions are similar to those for ionic compounds with two monatomic ions. Consider calcium carbonate, an ionic compound that is also known as chalk and found in the rock portion of the carbon cycle. When you see $CaCO_3$, you should interpret this chemical formula as a compound

Table 4.1 Common Polyatomic Ions

Name	Formula	Name	Formula
acetate	$C_2H_3O_2^-$	nitrite	NO_2^-
bicarbonate*	HCO_3^-	phosphate	PO_4^{3-}
carbonate	CO_3^{2-}	sulfate	SO_4^{2-}
hydroxide	OH^-	sulfite	SO_3^{2-}
hypochlorite	ClO^-	ammonium	NH_4^+
nitrate	NO_3^-		

*Also called hydrogen carbonate

that contains the ions calcium (Ca^{2+}) and carbonate (CO_3^{2-}) in a 1:1 ratio. As is true for all ionic compounds, the name of the cation is given first.

Another compound that is formed from a polyatomic anion is aluminum sulfate, $Al_2(SO_4)_3$. The subscript "3" applies to the *entire* SO_4^{2-} ion that is enclosed in parentheses. Accordingly, this compound contains three sulfate ions. Similarly, in the ionic compound ammonium sulfide, $(NH_4)_2S$, the NH_4^+ is enclosed in parentheses. The subscript "2" indicates that there are two ammonium ions for each sulfide ion. Note that the charges are *not* shown in the chemical formula; they are assumed to be there. In some cases, though, the polyatomic ion is *not* enclosed in parentheses. Parentheses are omitted when the subscript of the polyatomic ion would be 1. Nonetheless, you still have to "read" the chemical formula of $AlPO_4$ as containing the phosphate ion, and you have to "read" NH_4Cl as containing the ammonium ion. The next three activities will help you practice naming compounds that have polyatomic ions.

The subscripts in a chemical formula for an ionic compound are not arbitrary, and are needed to ensure that the positive and negative charges of the ions cancel. This will result in an overall zero charge for the entire compound.

Your Turn 4.4 Skill Building Polyatomic Ions I

Write the chemical formula for the ionic compound formed from each pair of ions.

a. Na^+ and SO_4^{2-}
b. OH^- and Mg^{2+}
c. Al^{3+} and $C_2H_3O_2^-$
d. CO_3^{2-} and K^+

Your Turn 4.5 Skill Building Polyatomic Ions II

Name each of these compounds.

a. KNO_3
b. $(NH_4)_2SO_4$
c. $FeSO_4$
d. $NaHCO_3$
e. $Mg_3(PO_4)_2$

Your Turn 4.6 Skill Building Polyatomic Ions III

Write the chemical formula for each of these compounds.

a. sodium hypochlorite (a type of bleach)
b. magnesium carbonate (found in limestone rocks)
c. ammonium nitrate (fertilizer)
d. calcium hydroxide (an agent to remove impurities in water)

Now that we know the location of carbon on Earth and are able to name compounds, especially those that contain carbon, let's look at how carbon cycles through Earth.

4.2 Where Did All the Carbon Atoms Go?

In the previous section, we identified that carbon atoms can show up on Earth in the atmosphere ($CO_2(g)$, $CO(g)$, $CH_4(g)$), as carbonate minerals in rocks, and in plants and animals (proteins, carbohydrates, lipids). However, carbon atoms are always on the move. Through processes such as combustion, photosynthesis, and sedimentation, carbon atoms move from one reservoir to another. It is estimated that an average carbon atom has been recycled from sediment, through the mobile reservoirs of Earth, and back to sediment more than 20 times over Earth's history. Perhaps some of the carbon atoms in your body once belonged to a dinosaur or Julius Caesar! Carbon dioxide gas in the air today may have been released from campfires burning more than a thousand years ago. As you reexamine Figure 4.2, take note that all of the processes are happening simultaneously. However, they may not all occur at the same rate.

It is important to know how carbon atoms move throughout Earth. For example, the slow transformation of carbon from living organisms into fossil fuels millions of years ago is of great importance to us. However, today's transfer of carbon back into the atmosphere by burning fossil fuels will affect future generations, who must deal with the consequences of increased rates of burning these fuels. The next activity gives you a closer look at reservoirs of carbon-containing compounds, and the processes that move carbon atoms among them.

Your Turn 4.7 Scientific Practices Understanding the Carbon Cycle

Using **Figure 4.2,** answer the following questions.

a. Which processes *add* carbon atoms (in the form of CO_2) to the atmosphere?
b. Which processes *remove* carbon atoms from the atmosphere?
c. What are the two largest reservoirs of carbon atoms?
d. Which parts of the carbon cycle are most influenced by the activities of humans?
e. Why do you think **Figure 4.2** is called the "carbon cycle"?

Considering the above activity, you will notice that the carbon cycle is a dynamic system, consisting of both natural addition and removal mechanisms. Respiration adds carbon dioxide to the atmosphere and photosynthesis removes it:

$C_6H_{12}O_6$ is the formula for glucose, a simple sugar.

$$\text{Respiration: } 6\ O_2 + C_6H_{12}O_6 \longrightarrow 6\ CO_2 + 6\ H_2O \quad \textbf{[4.1]}$$

$$\text{Photosynthesis: } 6\ CO_2 + 6\ H_2O \longrightarrow 6\ O_2 + C_6H_{12}O_6 \quad \textbf{[4.2]}$$

Did You Know? There was a major methane leak from a faulty well in a natural gas storage facility near Los Angeles from October 2015 through February 2016. This caused the release of more than 87,000 metric tons of methane into the atmosphere, which was estimated to be 25% of the state of California's daily methane emissions.

As members of the animal kingdom, we *Homo sapiens* participate in the carbon cycle along with our fellow creatures. As is true for any animal, we inhale and exhale, ingest and excrete, live and die. Human civilization, however, relies on processes that put many more carbon atoms into the atmosphere rather than on processes that remove them (Figure 4.4). Widespread burning of coal, petroleum, and natural gas for electricity production, transportation, and home heating all transfer carbon atoms from the largest underground carbon reservoir into the atmosphere.

Another human influence on CO_2 emissions is deforestation by burning, a practice that releases about 1.5 Gt of carbon into the atmosphere each year. It is estimated that forested land the size of two football fields is lost every second of every day from the rainforests of the world. Although firm numbers are elusive, Brazil is the country with the greatest annual loss of rainforest acreage; more than 5.4 million acres of Amazon rainforest is vanishing each year. Trees, very efficient absorbers of carbon dioxide, are removed from the carbon cycle through deforestation. If the wood is burned, vast quantities of CO_2 are generated; if it is left to decay, that process also

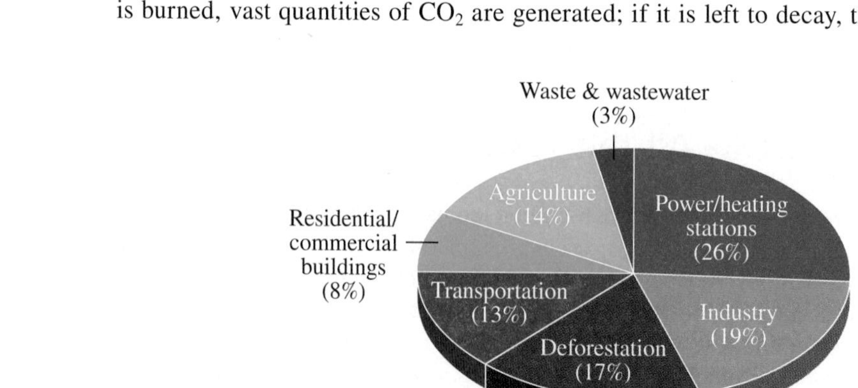

Figure 4.4

Global carbon dioxide emissions.

Source: EPA

releases carbon dioxide, but more slowly. Even if the lumber is harvested for construction purposes and the land is replanted with cultivated crops, the loss in carbon dioxide-absorbing capacity may approach 80%.

Your Turn 4.8 Scientific Practices Rainforests

Search the Internet to find a map of the current locations for rainforests in the world. Then, find a map from approximately 20 years ago. What differences do you notice? Predict what you think may occur in 20 years?

The oceans both absorb and emit carbon dioxide. We will learn more about this in Chapter 8, but it is important to know that gases such as CO_2 and solids such as Group 1 ionic compounds can both dissolve in water. Imagine for a moment that you place a pot of water on the stove. If you add table salt to the water, it dissolves. When a solid or gas dissolves in a liquid, a **solution** is formed. The substance that is dissolved is called the **solute** and the substance that does the dissolving is called the **solvent**. Therefore, saltwater is a type of *aqueous* solution with the salt being the solute, and water acting as the solvent.

As mentioned, gas can also dissolve in a liquid. Carbonated beverages are an example of this. Carbon dioxide is dissolved in the aqueous solvent to create the fizziness we expect from these beverages. What have you seen happen when sodas warm up? In many cases, more bubbling occurs. When bubbling occurs, carbon dioxide is coming out of the solution in a process known as **degasification**. The same process can occur in our oceans when the water temperature rises. During degasification, some of the carbon dioxide originally dissolved in the ocean is released into the atmosphere. With much lower CO_2 concentrations than your favorite soft drink, the oceans represent a large sink for absorbing CO_2 gas emitted into the atmosphere.

Have you ever been in a cave or seen a rock formation? Some of these structures are as a result of a solid solute dropping out of solution, similar to **crystallization**. When a solid "falls" out of solution, we call this **precipitation** and the solid is called a **precipitate**. Changes in temperature can affect precipitation. Can you think of other factors that influence precipitation? We will discuss this in more detail in Chapter 8, but an example to keep in mind is coral reefs. Coral reefs form by precipitation reactions at specific conditions. When those conditions change, coral reefs can either grow larger or dissolve—do either of these situations pose an issue?

Precipitation has two meanings—either the formation of a solid from a solution, or falling rain, snow, or hail in the atmosphere.

The total quantity of carbon-containing substances released by the human activities of deforestation and burning fossil fuels is about 7.5 Gt per year. About half of this is eventually recycled into the oceans and the biosphere; however, carbon dioxide is not always removed as quickly as the rate of addition to the atmosphere. Much of the CO_2 stays in the atmosphere, adding between 3.1 and 3.5 Gt of carbon per year to the existing base of 750 Gt noted in Figure 4.2. We are concerned primarily with the relatively rapid *increase* in atmospheric carbon dioxide, because *excess* CO_2 can affect Earth's climate and the health of Earth's oceans. Therefore, it would be useful to know the mass (Gt) of CO_2 added to the atmosphere each year. In other words, what mass of CO_2 contains 3.3 Gt of carbon, the midpoint between 3.1 and 3.5 Gt? Answering this question requires that we return to some more fundamental and quantitative aspects of chemistry.

Be careful—sometimes "Gt of carbon" is used, while other times you will see "Gt of carbon dioxide." Scientists most often use "Gt of carbon" because they are talking about the carbon cycle.

4.3 | Quantifying Carbon—First Stop: Mass

To solve the problem posed in the previous section, we need to know how the mass of C atoms in a sample of CO_2 gas is related to the mass of CO_2 molecules. Regardless of the source of CO_2, its chemical formula is always the same. The mass percent of C atoms in CO_2 is also unwavering; therefore, we must calculate the mass percent of C in CO_2, based on the formula of the compound. As you work through the next two sections, keep in mind that we are seeking a value for that percentage.

The approach requires using the masses of all the atoms involved. But this raises an important question: How much does an individual atom weigh? The mass of an atom is mainly due to the neutrons and protons in the nucleus. As you learned in Chapter 1, varying elements differ in mass because their atoms differ in composition. Instead of using the absolute masses of individual atoms, chemists have found it convenient to employ relative masses—in other words, to relate the masses of all atoms to some convenient standard. The internationally accepted mass standard is carbon-12, the isotope that makes up 98.90% of all carbon atoms. Carbon-12 (C-12) atoms have a mass number of 12 because each atom has a nucleus consisting of 6 protons and 6 neutrons.

To see the effect of varying the isotopic ratio of an element on its average atomic weight, check out the simulation in **Connect**.

However, the periodic table in the text shows that the atomic mass of carbon is 12.01, not 12.00. This is not an error; it reflects the fact that carbon exists naturally as three isotopes. Although C-12 predominates, 1.10% of carbon is C-13, the isotope with 6 protons and 7 neutrons. In addition, natural carbon contains a trace of C-14, the isotope with 6 protons and 8 neutrons. The tabulated mass value of 12.01 is called the **atomic mass**, a weighted average that takes into consideration the masses and percent of the natural abundances of all naturally occurring isotopes of carbon. This isotopic distribution and average mass of 12.01 characterize carbon obtained from any natural source—a graphite ("lead") pencil, a tank of gasoline, a loaf of bread, a lump of limestone, or your body. Review Table 4.2 for a summary of the naturally occurring isotopes of carbon.

You may notice that no units are used with the atomic masses. These are expressed in *unified atomic mass units* (u), which equals 1.67×10^{-27} kg.

Table 4.2 Isotopes of Carbon

Isotope	Mass Number	Relative Percent	Contribution to Atomic Mass*
C-12	12	98.90%	11.868
C-13	13	1.10%	0.143
C-14	14	~0.001%	0.0001
			Avg. atomic mass of C = 11.868 + 0.143 + 0.0001 = 12.011

* Calculated as the relative percent × mass number. For instance, 98.90% of the C-12 isotope is 0.9890 × 12 = 11.868.

Your Turn 4.9 Skill Building Isotopes of Nitrogen

Nitrogen (N_2) is an important element in the atmosphere and biological systems. It consists of two naturally occurring isotopes: N-14 and N-15.

a. Use the periodic table to find the atomic number and atomic mass of nitrogen atoms.
b. What is the number of protons, neutrons, and electrons in a neutral atom of N-14?
c. Compare your answers for part **b** with those for a neutral atom of N-15.
d. Given the atomic mass of nitrogen, which isotope has the greatest natural abundance? Refer to **Connect** for the simulation to confirm your answer.

Having reviewed the meaning of isotopes, we return to the matter at hand—the masses of atoms, and particularly the atoms in CO_2. Not surprisingly, it is difficult to weigh a single atom because of its extremely small mass. A typical laboratory balance can detect a minimum mass of 0.1 mg; this corresponds to 5,000,000,000,000,000,000 carbon atoms, or 5×10^{18} carbon atoms. A unified atomic mass unit (u) is far too small to measure in a conventional chemistry laboratory. Rather, the gram is the chemist's mass unit of choice. Therefore, scientists use exactly 12 g of carbon-12 as the reference for the atomic masses of all the elements. We define atomic mass as the mass (in grams) of the same number of atoms that are found in exactly 12 g of carbon-12. This number of atoms is, of course, *very* large; in fact, **Avogadro's number** is 602,000,000,000,000,000,000,000. It is more compactly written in scientific

This important chemical number is named after an Italian scientist with the impressive name of Count Lorenzo Romano Amadeo Carlo Avogadro di Quaregna e di Ceretto (1776–1856). (His friends called him Amadeo.)

notation as 6.02×10^{23}. This is the incredible number of atoms in exactly 12 g of carbon-12, no more than a tablespoon of soot!

Avogadro's number counts a collection of atoms, like the term *dozen* counts a collection of eggs. It does not matter if the eggs are large or small, brown or white, "organic" or not. In all cases, if there are 12 eggs, they are counted as a dozen. However, a dozen ostrich eggs has a greater mass than a dozen quail eggs. Figure 4.5 illustrates this point with a half-dozen tennis and a half-dozen golf balls. Like atoms of different elements, the masses of tennis and golf balls differ even when the number of units is the same.

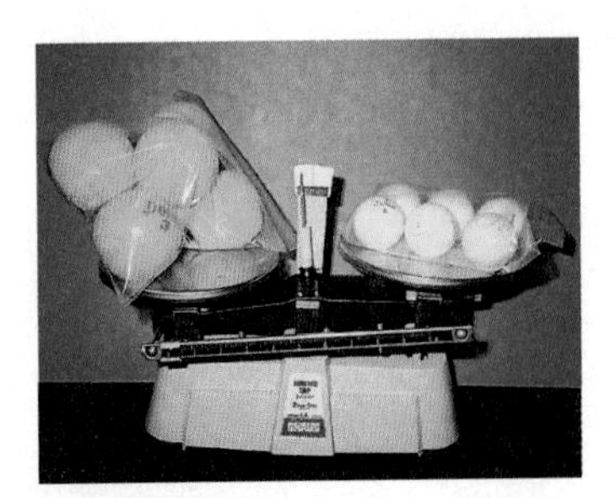

Figure 4.5

Comparative masses of tennis balls versus golf balls. The six tennis balls have a greater mass than six golf balls.

© Conrad Stanitski

Your Turn 4.10 You Decide Marshmallows and Pennies

Avogadro's number is so large that the only way to visualize it is through analogies. For example, one Avogadro's number of regular-sized marshmallows, 6.02×10^{23} of them, would cover the surface of the United States to a depth of 650 miles. Or, if you are more impressed by money than marshmallows, assume 6.02×10^{23} pennies were distributed evenly among the approximately 7 billion inhabitants of Earth. Every man, woman, and child could spend $1 million every hour, day and night, and half of those pennies would still be left unspent by the time each person passes away. Can these fantastic claims be correct? Check one or both, showing your reasoning. Come up with an analogy of your own.

Knowledge of Avogadro's number and the atomic mass of any element permit us to calculate the average mass of an individual atom of that element. Thus, the mass of 6.02×10^{23} oxygen atoms is 16.00 g, the atomic mass from the periodic table. To find the average mass of just one oxygen atom, we must divide the mass of the large collection of atoms by the size of the collection. In chemist's terms, this means dividing the atomic mass by Avogadro's number. Fortunately, calculators help make this job quick and easy:

$$\frac{16.00 \text{ g oxygen}}{6.02 \times 10^{23} \text{ oxygen atoms}} = 2.66 \times 10^{-23} \text{ grams per oxygen atom}$$

This very small mass confirms once again why chemists do not generally work with small numbers of atoms. We manipulate trillions at a time. Therefore, practitioners of this art need to measure matter with a sort of chemist's dozen—a very large one, indeed.

Your Turn 4.11 Skill Building Calculating the Mass of Atoms

Follow these instructions for the three examples:

i. Predict whether the value will be large or small.
ii. Calculate the value.
iii. Do your calculations match your predictions? Think about whether your predictions were reasonable.

a. The average mass in grams of an individual atom of carbon.
b. The mass in grams of 5 trillion carbon atoms.
c. The mass in grams of 6×10^{15} carbon atoms.

4.4 Quantifying Carbon—Next Stop: Molecules and Moles

Chemists have another way of communicating the number of atoms, molecules, or other small particles present. This is to use the term **mole** (mol), defined as an Avogadro's number of objects. The term is derived from the Latin word to "heap," or "pile up." Thus, 1 mol of carbon atoms is 6.02×10^{23} C atoms and 1 mole of aluminum atoms

When used with a number, *mol* is an abbreviation for *mole*.

There are 2 moles of oxygen atoms, O, in every mole of oxygen molecules, O_2.

Did You Know? Some chemists celebrate "mole day" every year on October 23 (10^{23}), between 6:02 AM and 6:02 PM. Let's just hope they don't try to eat a mole of ice cream during their celebration!

is 6.02×10^{23} Al atoms. In fact, 1 mol of people would be 6.02×10^{23} people! So, how many atoms will comprise 1 mol of oxygen gas? Before you answer 6.02×10^{23} oxygen atoms, remember that oxygen gas is made up of O_2 *molecules*. Hence, for 6.02×10^{23} O_2 *molecules,* there are two times Avogadro's number, 1.20×10^{24} of O atoms.

As you already know from previous chapters, chemical formulas and equations are written in terms of atoms and molecules. For example, reconsider Equation 4.3 for the complete combustion of carbon in oxygen:

$$C(s) + O_2(g) \longrightarrow CO_2(g) \quad \textbf{[4.3]}$$

This equation tells us that one *atom* of carbon combines with one *molecule* of oxygen to yield one *molecule* of carbon dioxide. Thus, it reflects the ratio in which the particles interact. It is equally correct to say that 10 C atoms react with 10 O_2 molecules (20 O atoms) to form 10 CO_2 molecules. Or, putting the reaction on a grander scale, we can say 6.02×10^{23} C atoms react with 6.02×10^{23} O_2 molecules (1.20×10^{24} O atoms) to yield 6.02×10^{23} CO_2 molecules! The last statement is equivalent to saying: "one *mole* of carbon plus one *mole* of oxygen yields one *mole* of carbon dioxide." Thus, the numbers of *atoms and molecules* taking part in a reaction are proportional to the numbers of *moles* of the same substances. The ratio of two oxygen atoms to one carbon atom remains the same regardless of the number of carbon dioxide molecules, as summarized in Table 4.3.

In the laboratory and the factory, the quantity of matter required for a reaction is often measured by mass. The mole is a practical way to relate the number of particles to the more easily measured mass. The **molar mass** is the mass of Avogadro's number, or one mole, of whatever particles are specified. For example, from the periodic table we can see that the mass of one mole of carbon atoms, rounded to the nearest tenth of a gram, is 12.0 g. A mole of oxygen atoms has a mass of 16.0 g. But we can also speak of a mole of O_2 molecules. Because there are two oxygen atoms in each oxygen molecule, there are two moles of oxygen atoms in each mole of molecular oxygen, O_2. Consequently, the molar mass of O_2 is 32.0 g, twice the molar mass of O.

The same logic for determining the molar mass of the element O_2 applies to compounds, such as carbon dioxide. The formula of carbon dioxide, CO_2, reveals that each molecule contains one carbon atom and two oxygen atoms. Scaling up by 6.02×10^{23}, we can say that each mole of CO_2 consists of 1 mol of C atoms and 2 mol of O atoms (Table 4.3). But, remember that we are interested in the molar mass of carbon dioxide, which we obtain by adding the molar mass of carbon atoms to twice the molar mass of oxygen atoms:

$$\begin{aligned}
1 \text{ mol } CO_2 &= 1 \text{ mol C} + 2 \text{ mol O} \\
&= \left(1 \cancel{\text{mol C}} \times \frac{12.0 \text{ g C}}{1 \cancel{\text{mol C}}}\right) + \left(2 \cancel{\text{mol O}} \times \frac{16.0 \text{ g O}}{1 \cancel{\text{mol O}}}\right) \\
&= 12.0 \text{ g C} + 32.0 \text{ g O} \\
1 \text{ mol } CO_2 &= 44.0 \text{ g } CO_2
\end{aligned}$$

The term *atomic weight* has long been used by chemists; however, many prefer to use the equivalent term *relative atomic mass.*

This procedure is used in chemical calculations, where molar mass is an important property. Some examples are included in the next activity. In every case, you multiply the number of moles of each atom by its corresponding standard atomic weight (in grams/mole), and add the result.

Table 4.3 Ways to Interpret a Chemical Equation

C	+	O_2	$\longrightarrow$	CO_2
1 atom		1 molecule		1 molecule
6.02×10^{23} atoms		6.02×10^{23} molecules		6.02×10^{23} molecules
1 mol		1 mol		1 mol

Your Turn 4.12 Skill Building Molar Mass

Calculate the molar mass of each of the following gases that are found in the atmosphere.

a. O_3 (ozone)
b. N_2O (dinitrogen monoxide or nitrous oxide)
c. CCl_3F (Freon-11; trichlorofluoromethane)

We started out on this mathematical excursion so that we could calculate the mass of CO_2 produced from burning 3.3 Gt of carbon. We now have all the pieces assembled. Out of every 44.0 g (1 mole) of CO_2, 12.0 g is C. This mass ratio holds for all samples of CO_2 and we can use it to calculate the mass of C in any known mass of CO_2. More to the point, we can use it to calculate the mass of CO_2 released by burning any known mass of carbon. It only depends on how we arrange the ratio. The C:CO_2 mass ratio is $\frac{12.0 \text{ g C}}{44.0 \text{ g } CO_2}$, but it is equally true that the CO_2:C ratio is $\frac{44.0 \text{ g } CO_2}{12.0 \text{ g C}}$.

Further, we could calculate the grams of C in 100.0 g CO_2 by setting up the relationship in this manner:

As first seen in Chapter 1, this method is referred to as "dimensional analysis" or "unit analysis."

$$100.0 \cancel{\text{g } CO_2} \times \frac{12.0 \text{ g C}}{44.0 \cancel{\text{g } CO_2}} = 27.3 \text{ g C}$$

The fact that there are 27.3 g of carbon in 100.0 g of carbon dioxide is equivalent to saying that the mass percent of C in CO_2 is 27.3%. Note that carrying along the units "g CO_2" and "g C" helps you do the calculation correctly. The unit "g CO_2" can be canceled, and you are left with "g C," because one unit is in the numerator and the other is in the denominator. Keeping track of the units, and canceling where appropriate, are useful strategies in solving many problems.

The concept of mass percent was introduced in Section 1.3.

Your Turn 4.13 Skill Building Mass Ratios and Percents

a. Calculate the mass ratio of S atoms in SO_2 molecules.
b. Find the mass percent of S atoms in SO_2 molecules.
c. Calculate the mass ratio and the mass percent of N atoms in N_2O molecules.

To find the mass of CO_2 that contains 3.3 Gt of C atoms, we use a similar approach. We could convert 3.3 Gt to grams, but it is not necessary. As long as we use the same mass unit for C atoms and CO_2 molecules, the same numerical ratio holds. Compared with our last calculation, this problem has one important difference in how we use the ratio. We are solving for the mass of CO_2 molecules, not the mass of C atoms. Look carefully at the units this time:

$$3.3 \cancel{\text{Gt C}} \times \frac{44.0 \text{ Gt } CO_2}{12.0 \cancel{\text{Gt C}}} = 12 \text{ Gt } CO_2$$

Once again, the units cancel and we are left with Gt of CO_2 molecules.

Our burning question of the mass of CO_2 molecules added to the atmosphere each year from the combustion of fossil fuels has finally been answered: 12 Gt! Along the way, we demonstrated the problem-solving power of chemistry and introduced five important ideas: atomic mass, Avogadro's number, mole, molar mass, and mass percent. The next few activities provide opportunities to practice your skill with these essential concepts.

Your Turn 4.14 Skill Building SO_2 from Volcanoes

a. It is estimated that volcanoes globally release about 1.9×10^7 t (19 million metric tons) of SO_2 per year. Calculate the mass of sulfur atoms in this amount of SO_2 molecules.
b. If 1.42×10^8 t of SO_2 is released per year by fossil fuel combustion, calculate the mass of sulfur atoms in this amount of SO_2 molecules.

If you know how to apply these ideas, you have the ability to critically evaluate media reports about releases of carbon or CO_2 (and other substances), and judge their accuracy. One can either take such statements on faith, or check their accuracy by applying mathematics to the relevant chemical concepts. Obviously, there is insufficient time to check every assertion, but we hope that you develop questioning and critical attitudes toward all statements about chemistry and society, including those found in this book.

Your Turn 4.15 You Decide Checking Carbon from Cars

A clean-burning automobile engine emits about 5 lb of C atoms in the form of CO_2 molecules for every gallon of gasoline it consumes. The average American car is driven about 12,000 miles per year. Using this information, check the statement that the average American car releases its own weight in carbon into the atmosphere each year. List the assumptions you make to solve this problem. Compare your list and your answer with those of your classmates.

4.5 Why Does It Matter Where Carbon Atoms End Up?

Now that we can quantify the amounts of carbon in different reservoirs, we need to look at what effect carbon-containing material can have in these reservoirs. To start this discussion, we need to understand Earth's energy balance and how it is heated and cooled. The energy to heat Earth comes mainly from the Sun; however, this is not the entire story. Based on Earth's distance from the Sun and the amount of solar radiation the Sun emits, the average temperature on Earth should be −18 °C (0 °F) and the oceans should be frozen year round. Thankfully, this is not true. Earth's average temperature is currently around 15 °C (59 °F).

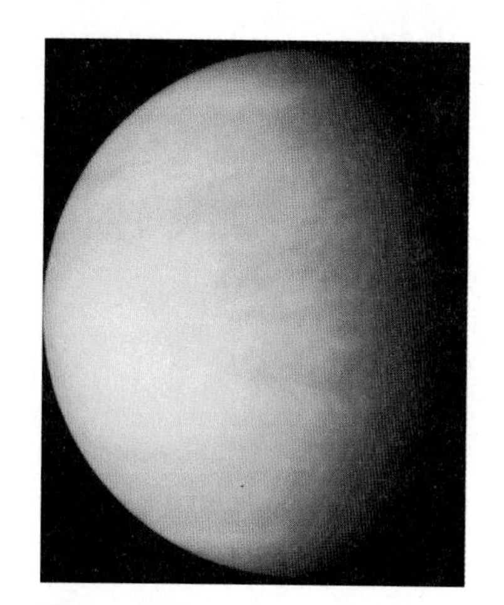

Figure 4.6

Venus, as photographed by the *Galileo* spacecraft.

Source: Galileo Spacecraft, JPL, NASA

Venus (Figure 4.6) is another planet whose temperature is inconsistent with its distance from the Sun. Considered by some to be the brightest and most beautiful body in the night sky, Venus has an average temperature of about 450 °C (840 °F). Based simply on its distance from the Sun, however, the average temperature on Venus should be 100 °C, the boiling point of water. What do Earth and Venus have in common that would explain these discrepancies? They both have an atmosphere. To see the role that our atmosphere plays, we now examine what happens when solar radiation reaches Earth.

The energy processes that contribute to Earth's energy balance appear in Figure 4.7. Earth receives nearly all of its energy from the Sun (orange arrows), primarily in the form of ultraviolet, visible, and infrared radiation. Some of this incoming radiation is reflected back into space (blue arrows) by the dust and aerosol particles suspended in our atmosphere (25%). Other parts of this incoming radiation are reflected by the surface of Earth itself, especially those regions white with snow or sea ice (6%). Thus, 31% of the radiation received from the Sun is reflected.

The remaining 69% of the radiation from the Sun is absorbed, either by the atmosphere (23%) or by land masses and oceans (46%). We can account for all of the Sun's radiation by adding the reflected and absorbed radiation: 31% + 69% = 100%.

Your Turn 4.16 Skill Building Light from the Sun Refresher

In **Chapter 3**, we discussed the various components of the electromagnetic spectrum. Consider these three types of radiant energy emitted by the Sun: infrared (IR), ultraviolet (UV), and visible.

a. Arrange them in order of increasing wavelength.
b. Arrange them in order of increasing energy.

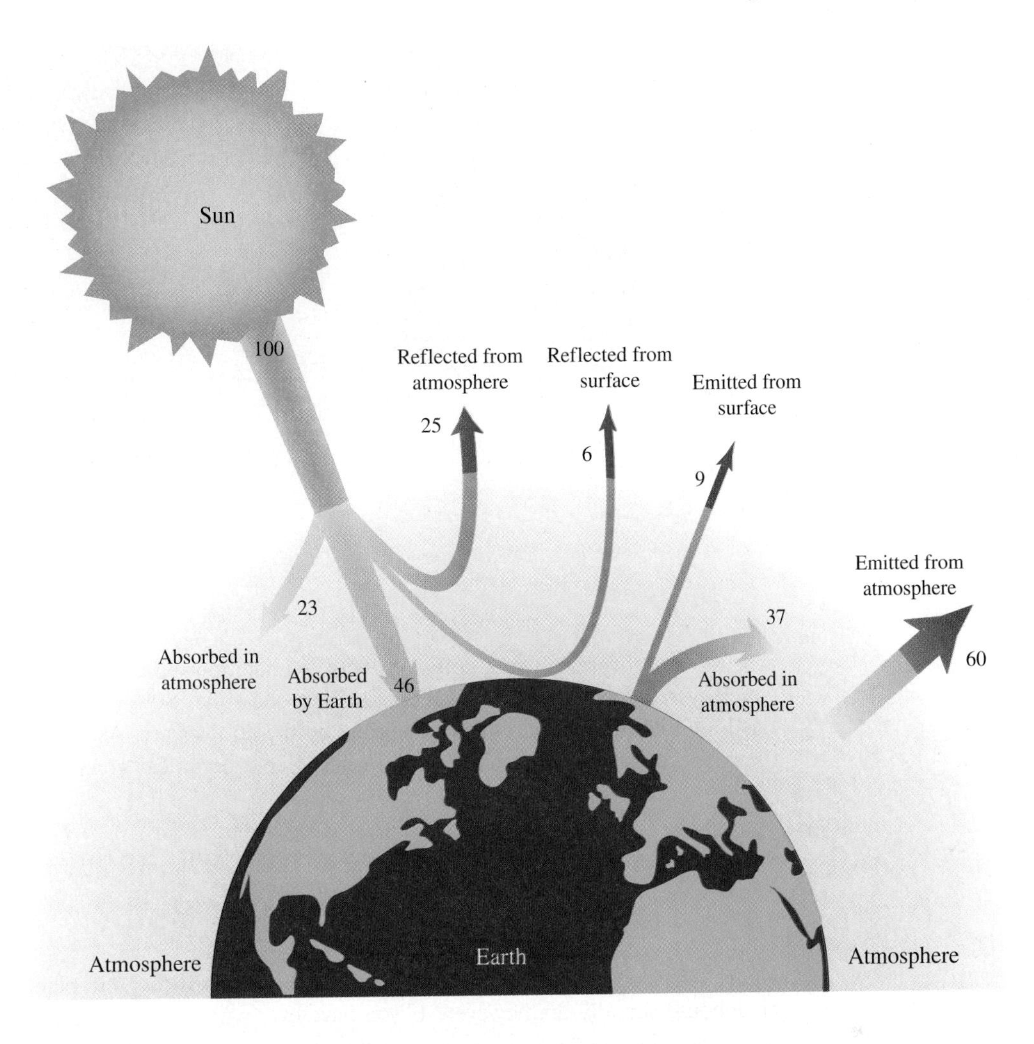

Figure 4.7

Earth's energy balance. Orange represents a mixture of wavelengths of electromagnetic radiation; shorter wavelengths of radiation are shown in blue, longer ones in red. The values are given in percentages of the total incoming solar radiation.

In order to maintain Earth's energy balance, all of the radiation that is absorbed from the Sun must eventually go back into space. Figure 4.7 shows us that:

- 46% of the Sun's radiation is absorbed by Earth.
- Earth re-emits all of the radiation it absorbs, but at a longer wavelength in the form of IR radiation.
  - Part of what Earth emits escapes into space (9%).
  - The remainder is absorbed by the atmosphere (37%).
- 54% of the Sun's radiation is either absorbed by the atmosphere (23%), reflected by the atmosphere (25%), or reflected by Earth's surface (6%).

The 60% of the radiation that is absorbed by the atmosphere, either directly from the Sun (23%) or from Earth's surface (37%), eventually is emitted into space to complete the energy balance.

Your Turn 4.17 Scientific Practices Building Your Own Energy Diagram

Figure 4.7 is not easy to fully comprehend. Use the figure and the description of Earth's energy balance presented in this section to devise your own model describing Earth's energy balance. Compare your model with the models of your classmates.

Again, of the 46% of the Sun's radiation that is absorbed and eventually emitted by Earth, 37% is absorbed in the atmosphere prior to its emission into space. This

Figure 4.8

A typical botanical greenhouse.

process of absorption causes collisions between neighboring atmospheric gas molecules, which warms up Earth's atmosphere. At any one time, 80% (or 37 ÷ 46 × 100%) of Earth's emitted radiation will be absorbed by the atmosphere. As we hope you can see, the gases in Earth's atmosphere hold in heat like a botanical greenhouse (Figure 4.8)!

4.6 Warming by Greenhouse Gases: Good, Bad, or a Little of Both?

Figure 4.8 shows a typical greenhouse that may be used to grow plants. What does a greenhouse actually do, however? If you have ever parked a car with its windows closed on a sunny day, you probably have experienced firsthand how a greenhouse can trap heat. The car, with its glass windows, operates much the same way as a greenhouse for growing plants. The glass windows transmit visible and UV light from the Sun. This energy is absorbed by the car's interior, particularly by any dark surfaces. Some of this radiant energy is then re-emitted as longer wavelength IR radiation (heat). Unlike visible light, infrared light is not easily transmitted through the glass windows and so becomes "trapped" inside the car. When you re-enter the vehicle, a blast of hot air greets you. The temperature inside of a car can exceed 49 °C (120 °F) in the summer in certain climates! Although the physical barrier of the windows is not an exact analogy to Earth's atmosphere, the effect of warming the car's interior is similar to the warming of Earth.

The **greenhouse effect** is the natural process by which atmospheric gases trap a major portion (about 80%) of the infrared radiation radiated by Earth. Again, Earth's higher-than-expected average annual temperature of 15 °C (59 °F) is a result of the heat-trapping gases in our atmosphere. The atmosphere of Venus acts in a similar manner; however, it traps even more heat. This is because Venus's atmosphere is made up of nearly 96% carbon dioxide, which, as we will see, is a far greater concentration than that in Earth's atmosphere.

Water vapor is the most abundant greenhouse gas in our atmosphere and increases with increasing temperature of Earth's atmosphere. However, the amount of water vapor is primarily due to natural processes.

Carbon dioxide, which is present in the atmospheres of both Earth and Venus, is a greenhouse gas. **Greenhouse gases** (GHGs) are those gases capable of absorbing and emitting IR radiation, thereby warming the atmosphere. In addition to carbon dioxide, other examples of GHGs are water vapor, methane, nitrous oxide, ozone, and chlorofluorocarbons. The presence of these gases is essential in keeping our planet at habitable temperatures.

In our energy balance discussion, we showed that 80% of Earth's absorbed solar radiation is emitted into the atmosphere. The exchange of energy among Earth, atmosphere, and space results in a steady state and a continuous average temperature of Earth. However, the increase in concentration of greenhouse gases that is taking place today is changing the energy balance and causing changes in the temperature of the planet. The term **enhanced greenhouse effect** refers to the process in which atmospheric gases trap and return *more than* 80% of the heat energy radiated by Earth. An increase in the concentration of greenhouse gases will mean that more than 80% of the radiated energy will be returned to Earth's surface, with an accompanying increase in average global temperature. The term **global warming**, which you likely have heard, is frequently used to describe the increase in average global temperatures that results from an enhanced greenhouse effect.

Why is the amount of greenhouse gases in the atmosphere increasing? One explanation considers **anthropogenic** influences due to human activities such as industry, transportation, mining, and agriculture. These activities require carbon-based fuels, which produce carbon dioxide when burned. In the late 19th century, Swedish scientist Svante Arrhenius (1859–1927) considered the problems that increased industrialization might cause by building up CO_2 in the atmosphere. He calculated that doubling the concentration of CO_2 would result in an increase of 5–6 °C in the average temperature of the planet's surface. But, how are we adding CO_2 to the atmosphere?

Your Turn 4.18 You Decide Evaporating Coal Mines

Writing in the *London, Edinburgh, and Dublin Philosophical Magazine*, Arrhenius described the phenomenon: "We are evaporating our coal mines into the air." Although the statement was effective in grabbing attention in 1898, what process do you think he really was referring to in discussing the amount of CO_2 being added to the air? Explain your reasoning.

Your Turn 4.19 You Decide Only Anthropogenic?

We have just discussed how anthropogenic influences have changed the concentration of CO_2 in Earth's atmosphere. Are there other reasons why the concentration of CO_2 has changed? Explain. We will continue to investigate this throughout the chapter.

Your Turn 4.20 Scientific Practices The Earth as a Greenhouse

Draw a diagram or model that describes how Earth functions as a greenhouse. Explain your representation in words.

Your Turn 4.21 You Decide Revisiting the Greenhouse Question

As we started this chapter, we asked the question: Is supplying Earth's atmosphere with greenhouse gases a good or a bad thing?

How would you answer this question now? Explain.

Now that we have learned about the greenhouse effect and how it influences Earth, we need to determine what makes molecules such as CO_2 and water vapor greenhouse gases.

4.7 How Do You Recognize a "Greenhouse Gas"?

Carbon dioxide, water, and methane are greenhouse gases; in contrast, nitrogen and oxygen are not. Why the difference? The answer relates in part to how many atoms a molecule has and its molecular shape. In this section, we will put your knowledge of Lewis structures to work to predict shapes of molecules. In the subsequent section, we will connect these shapes to molecular vibrations, which can help us explain the difference between greenhouse gases and non-greenhouse gases.

Remember, the atmosphere is composed of 78% nitrogen and 21% oxygen gases.

In Chapter 3, you used Lewis structures to predict how electrons are arranged in atoms and molecules. Shape was not the primary consideration. What is the shape for diatomic molecules such as O_2 and N_2? Here, the shape is unambiguous because the molecule must be linear:

:N⋮⋮N: or :N≡N: or N≡N

Ö::Ö or Ö=Ö or O=O

Exceptions to the normal pattern of four pairs of electrons around each atom can be found. For instance, H and He will have a maximum of two electrons, and atoms from Group 13 will usually have less than an octet. In addition, atoms from the 3rd period and below are able to "expand their octet," and may have more than eight electrons.

Even though different geometries are possible with larger molecules, Lewis structures can still be useful models to help us understand the experimental shapes of molecules. Therefore, the first step in predicting the shape of a molecule is to draw its Lewis structure. Each atom (except hydrogen and helium) is usually associated with four pairs of electrons, known as the *octet rule*. Some molecules may include non-bonding lone pair electrons, but all molecules must contain some bonding electrons or they would not be molecules!

Opposite charges attract and like charges repel. Negatively charged electrons are attracted to a positively charged nucleus. However, the electrons all have the same charge and are found as far from each other in space as possible, while still maintaining their attraction to the positively charged nucleus. Groups of negatively charged electrons repel one another. Therefore, the experimental shapes of molecules can be rationalized by identifying the arrangements where the mutually repelling regions of electron density are as far apart as possible. We illustrate the procedure for understanding the experimental shape of a molecule with methane (CH_4), a greenhouse gas.

1. **Determine the number of valence electrons associated with each atom in the molecule**. The carbon atom (Group IVA or 14) has four valence electrons; each of the four hydrogen atoms contributes one electron. This gives $4 + (4 \times 1)$, or 8 valence electrons.
2. **Draw a Lewis structure in which you arrange the valence electrons in pairs to include 8 electrons around the central atom**. This may require single, double, or triple bonds. For the methane molecule, use the eight valence electrons to form four single bonds (four electron pairs) around the central carbon atom. This is the Lewis structure:

```
   H              H
   ..             |
 H:C:H   or   H—C—H
   ..             |
   H              H
```

   Although this structure seems to imply that the CH_4 molecule is flat, it is not. In fact, the methane molecule is *tetrahedral*, as we will see in the next step.
3. **Assume that the most stable molecular shape has the bonding electron pairs as far apart as possible**. (*Note:* In other molecules, we also need to consider non-bonding electrons, but CH_4 has none.) The four bonding electron pairs around the carbon atom in CH_4 repel one another, and in their most stable arrangement they are as far from one another as possible. As a result, the four hydrogen atoms are also as far from one another as possible. This shape is *tetrahedral,* because the hydrogen atoms correspond to the corners of a *tetrahedron*, a four-cornered geometric shape with four equal triangular sides, sometimes called a triangular pyramid.

Figure 4.9

The legs and the shaft of a music stand approximate the geometry of the bonds in a tetrahedral molecule such as methane.

© Mark Hall/Taxi/Getty Images

One way to describe the shape of a CH_4 molecule is by analogy to the base of a folding music stand. The four C–H bonds correspond to the three evenly spaced legs and the vertical shaft of the stand (Figure 4.9). The angle between each pair of bonds is 109.5°. The tetrahedral shape of a CH_4 molecule has been experimentally confirmed. Indeed, it is one of the most common atomic arrangements in nature, particularly in carbon-containing molecules.

Your Turn 4.22 Skill Building Methane—Flat or Tetrahedral?

a. If the methane molecule were flat, as the two-dimensional Lewis structure seems to predict, what would the H–C–H bond angle be?

b. Offer a reason why the tetrahedral shape, not the two-dimensional flat shape, is more advantageous for this molecule.
Hint: Think about the position of electrons around the carbon atom.

c. Consider the part of the music stand circled in yellow, shown in **Figure 4.9.** In the analogy of shape using a music stand, where would the carbon atom be located? Where would each of the hydrogen atoms be?

Chemists represent molecules in several ways. The simplest, of course, is the chemical formula itself. In the case of methane, this is simply CH_4. Another is the Lewis structure, but again this is only a two-dimensional representation that gives information about the valence electrons. Figure 4.10 shows these two representations, as well as two others that are three-dimensional in appearance. One has a wedge-shaped line that represents a bond coming out of the paper toward the reader. The dashed wedge in the same structural formula represents a bond pointing away from the reader. The two solid lines lie in the plane of the paper. The other, a space-filling model, was drawn with a molecular modeling program. Space-filling models enclose the volume occupied by electrons in an atom or molecule. Seeing and manipulating physical models, either in the classroom or laboratory, can help you visualize the structure of molecules.

However, not all valence electrons reside in bonding pairs. In some molecules, the central atom has non-bonding electron pairs, also called *lone pairs*. For example, Figure 4.11 shows the ammonia molecule, in which nitrogen follows the normal pattern of being surrounded by eight electrons; in this case, there are three bonding pairs and one lone pair of electrons.

A lone pair of electrons occupies greater space than a bonding pair of electrons. Consequently, the lone pair repels the bonding pairs more strongly than the bonding pairs repel one another. This enhanced repulsion forces the bonding pairs closer to one another, creating a H–N–H angle slightly less than the predicted 109.5° associated with a regular tetrahedron. The experimental value of 107.3° is close to the tetrahedral angle, again indicating that our model is reasonably reliable.

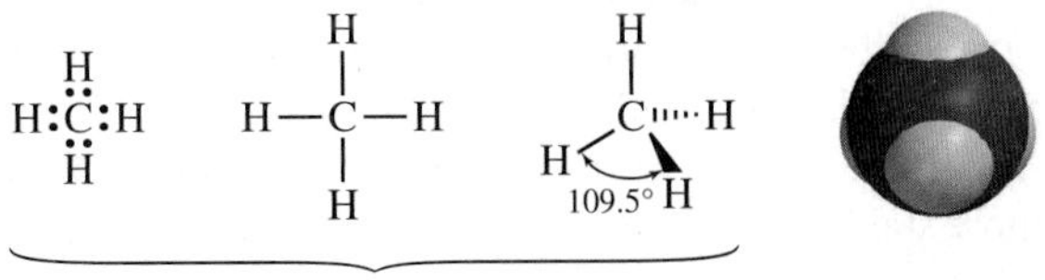

109.5°

(a) (b)

Figure 4.10

Representations of molecules of methane, CH_4. Shown are the: **(a)** Lewis structures and structural formula, and **(b)** space-filling model.

107.3°

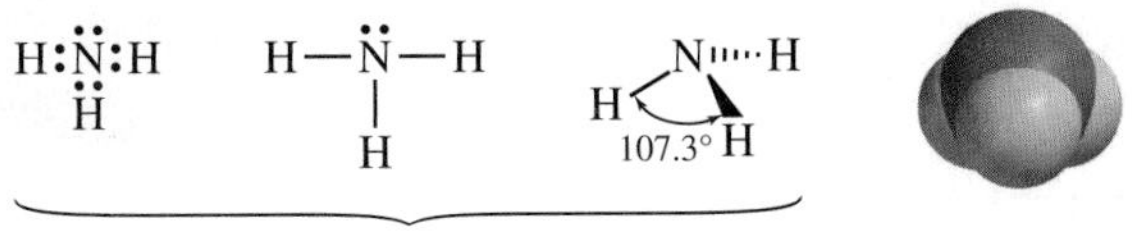

(a) (b)

Figure 4.11

Representations of an ammonia molecule, NH_3. Shown are the **(a)** Lewis structures and structural formula, and **(b)** space-filling model.

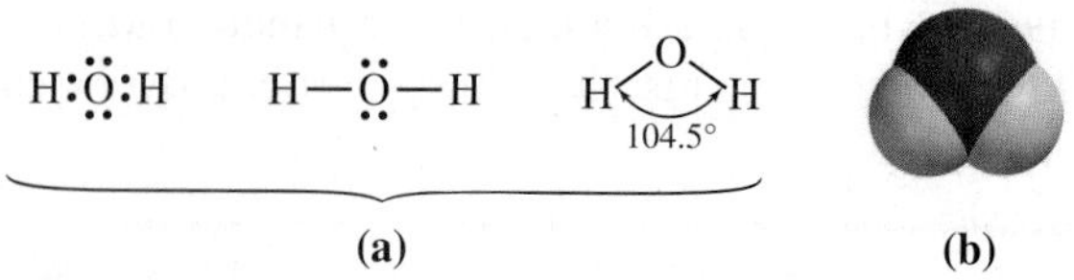

$H{:}\ddot{\underset{..}{O}}{:}H$ $\quad$ $H-\ddot{\underset{..}{O}}-H$ $\quad$ H–O–H 104.5°

(a) (b)

Figure 4.12

Representations of a water molecule, H_2O. Shown are the: **(a)** Lewis structures and structural formulas, and **(b)** space-filling model.

The shape of a molecule is described by its arrangement of atoms (Table 4.4). The hydrogen atoms of a NH_3 molecule form a triangle with the nitrogen atom above them at the top of the pyramid. Thus, a molecule of ammonia is said to have a *trigonal pyramidal* shape. Going back to the analogy of the folding music stand (Figure 4.9), you could expect to find hydrogen atoms at the tip of each leg of the music stand. This places the nitrogen atom at the intersection of the legs with the shaft, with the non-bonding electron pair corresponding to the shaft of the stand.

The water molecule is *bent*, illustrating another shape. There are eight valence electrons on the central oxygen atom: one from each of the two hydrogen atoms, plus six from the oxygen atom (Group VIA or 16). These eight electrons are distributed in two bonding and two lone pairs of electrons (Figure 4.12a).

If these four pairs of electrons were arranged as far apart as possible, we might predict the H–O–H bond angle to be 109.5°, the same as the H–C–H bond angle in a molecule methane. However, unlike a methane molecule, a water molecule has two non-bonding pairs of electrons. The repulsion between the two non-bonding pairs causes the bond angle to be less than 109.5°. Experiments indicate a value of approximately 104.5°.

Your Turn 4.23 Skill Building Predicting Molecular Shapes, Part 1

Using **Table 4.4** and the strategies just described, sketch the shape for each of these molecules.

a. CCl_4 (carbon tetrachloride)
b. CCl_2F_2 (Freon-12, dichlorodifluoromethane)
c. H_2S (hydrogen sulfide)

We already looked at the structures of several molecules important for understanding the chemistry of climate change. What about the Lewis structure of the carbon dioxide molecule? With a total of 16 valence electrons, the C atom contributes four electrons, and six come from each of the two oxygen atoms. If only single bonds were involved, each atom would not be surrounded by eight electrons. However, this becomes possible if the central carbon atom forms a double bond with each of the two oxygen atoms, thus sharing four electrons with each oxygen atom.

So, what is the shape of the CO_2 molecule? Again, groups of electrons repel one another, and the most stable configuration provides the farthest separation of the negative charges. In this case, the groups of electrons are the double bonds, and these are farthest apart with an O=C=O bond angle of 180°. The model predicts that all three atoms in a CO_2 molecule will be in a straight line, and that the molecule will be *linear*. In fact, this is the case, as shown in Figure 4.13.

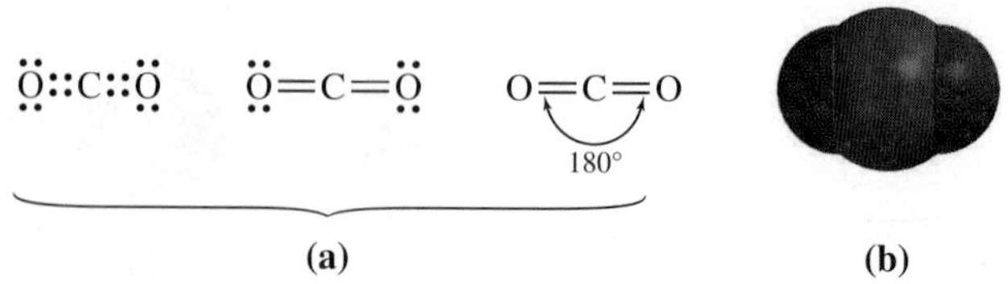

$\ddot{\underset{..}{O}}{::}C{::}\ddot{\underset{..}{O}}$ $\quad$ $\ddot{\underset{..}{O}}=C=\ddot{\underset{..}{O}}$ $\quad$ O=C=O 180°

(a) (b)

Figure 4.13

Representations of a carbon dioxide molecule, CO_2. Shown are the **(a)** Lewis structures and structural formula, and **(b)** the space-filling model.

Table 4.4 Common Molecular Geometries

# of Bonded Atoms to the Central Atom	# of Non-bonding Electron Pairs on Central Atom	Geometry	Space-Filling Model
2	0	linear	CO_2 
2	2	bent	H_2O 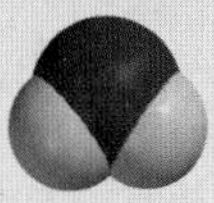
3	0	trigonal planar	BCl_3 
3	1	trigonal pyramid	NH_3 
4	0	tetrahedral	CH_4 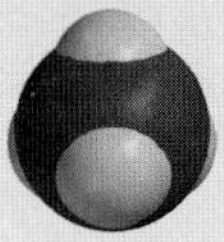

We applied the idea of electron-pair repulsion to molecules in which there are four groups of electrons (CH_4, NH_3, and H_2O), and two groups of electrons (CO_2) surrounding the central atom. Electron-pair repulsion also applies reasonably well to molecules that include three, five, or six groups of electrons. In most molecules, the electrons and atoms are still arranged to maximize the separation of the electrons. This logic accounts for the bent shape we associated with the ozone molecule in Section 3.7.

The Lewis structure for the ozone (O_3) molecule with its 18 valence electrons contains a single bond and a double bond, and the central oxygen atom carries a non-bonding lone pair of electrons. Thus, the central O atom has three groups of electrons: the pair that makes up the single bond, the two pairs that constitute the double bond, and the lone pair. These three groups of electrons repel one another, and the minimum energy of the molecule corresponds to their farthest separation. This occurs when the electron groups are all in the same plane, and at an angle of about 120° from one another. We predict, therefore, that the O_3 molecule should be bent, and the angle made by the three atoms should be approximately 120°. Experiments show the bond angle to be 117°, just slightly smaller than the prediction (Figure 4.14). The non-bonding electron pair on the central oxygen atom occupies a greater volume than the bonding pairs of electrons, causing a greater repulsion force responsible for the slightly smaller bond angle.

The O_3 molecule is best represented by two equivalent *resonance* forms. Revisit this topic in Chapter 3.

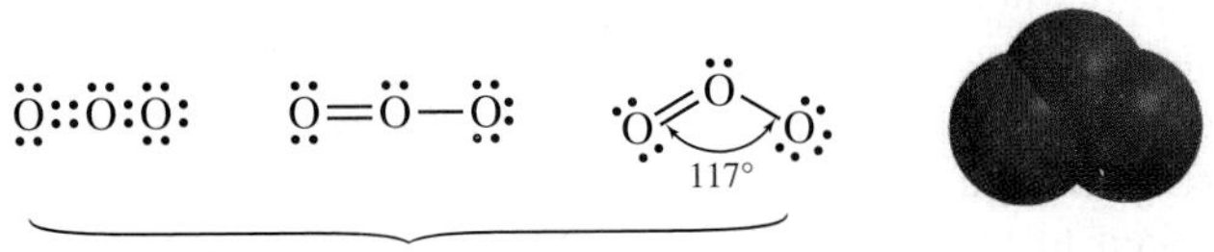

117°

(a) (b)

Figure 4.14

Representations of an ozone molecule, O_3. Shown are the: **(a)** Lewis structures and structural formula for one resonance form, and **(b)** the space-filling model.

Your Turn 4.24 Skill Building Predicting Molecular Shapes, Part 2

Using **Table 4.4** and the strategies just described, predict and sketch the shape of molecules of SO_2 (sulfur dioxide) and SO_3 (sulfur trioxide).

Hint: Because S and O are in the same group on the periodic table, the structures for SO_2 and O_3 will be closely related.

As promised, in this section we helped you see that molecules have different shapes, ones that can be predicted. In the next section, we return to our story of greenhouse gases, putting your knowledge of molecular shapes to work to explore why all gases are not greenhouse gases.

4.8 How Do Greenhouse Gases Work?

How do greenhouse gases trap heat, keeping our planet at more or less comfortable temperatures? In part, the answer lies in how molecules respond to photons of energy. This topic is complex, but we will provide the basics to understand how the greenhouse gases in our atmosphere trap heat. At the same time, we'll reveal why some gases do *not* trap heat.

We begin this topic by revisiting the interaction of UV light with molecules, as discussed in Chapter 3 in relation to the ozone layer. You saw that high-energy photons (UVC) could break the covalent bonds in O_2, and that photons of lower energy (UVB) could break the bonds in O_3. Put another way, both ozone and oxygen molecules can absorb UV radiation. When this absorption occurs, an oxygen-to-oxygen bond is broken.

A simulation related to the interaction of EM radiation with a CFC molecule is found in **Connect**.

Fortunately, infrared (IR) photons do not contain enough energy to cause chemical bonds to break. Instead, a photon of IR radiation can only add sufficient energy to the molecule to cause vibrations. Depending on the molecular structure, only certain vibrations are possible. In order for a photon to be absorbed, the energy of the incoming photon must correspond exactly to the vibrational energy of the molecule. This means that different molecules absorb IR radiation at different wavelengths, and thus vibrate at different energies.

We illustrate these ideas with a model of the CO_2 molecule, representing the atoms as balls and the covalent bonds as springs. As illustrated in Figure 4.15, every CO_2 molecule constantly vibrates in four distinct ways. The arrows indicate the direction of motion of each atom during each vibration, and the atoms move forward and backward along the arrows. Vibrations **(a)** and **(b)** are stretching vibrations. In vibration **(a)**, the central carbon atom is stationary and the oxygen atoms move back and forth (stretch) in opposite directions away from the central atom. Alternatively, the oxygen atoms can move in the same direction and the carbon atom in the opposite direction (vibration **(b)**). Vibrations **(c)** and **(d)** look very much alike. In both cases, the molecule bends from its normal linear shape. The bending counts as two vibrations because it occurs in either of two possible planes. In vibration **(c)**, the molecule is shown bending up and down in the plane of the paper on which the diagram is printed, whereas in vibration **(d)**, the molecule is shown bending out of the plane of the paper.

Recall that energy and wavelength are inversely related.

If you have ever examined a spring, you probably noticed that more energy is required to stretch it than to bend it. Similarly, more energy is required to stretch a CO_2 molecule than to bend it. This means that more energetic photons, those with shorter wavelengths, are needed to cause stretching vibrations, **(a)** or **(b)**, than to cause bending vibrations, **(c)** or **(d)**. For example, absorption of IR radiation with a wavelength of 15.0 micrometers (μm) adds energy to the bending vibrations, **(c)** and **(d)**. When that occurs, the atoms move farther from their equilibrium positions and move

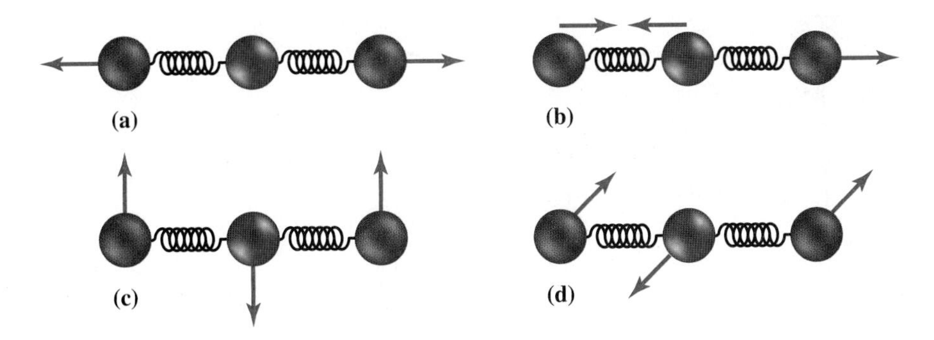

Figure 4.15

Illustration of molecular vibrations in carbon dioxide, CO_2. Each spring represents a C=O double bond. Vibrations **(a)** and **(b)** are stretching vibrations, whereas **(c)** and **(d)** are bending vibrations.

faster (on average) than they do normally. For the same thing to happen with vibration **(b)**, higher-energy radiation having a wavelength of 4.3 μm is required. This is considered an asymmetric stretch. Together, vibrations **(b)**, **(c)**, and **(d)** account for the greenhouse properties of carbon dioxide since all of these result from the absorption of IR radiation (heat).

In contrast, direct absorption of IR radiation does not add energy to vibration **(a)**. In a CO_2 molecule, the average concentration of electrons is greater on the oxygen atoms than on the carbon atom. The property of **electronegativity** (EN) described here is a measure of an atom's ability to attract bonded electrons. Elements with the greatest electronegativity lie in the top-right of the periodic table (toward F, which has the greatest EN). In contrast, the lowest electronegativity is found for the elements in the bottom-left of the periodic table (toward Francium, Fr).

The greater electronegativity of oxygen versus carbon means that the oxygen atoms carry a partial negative charge (δ^-) relative to the carbon atom, which carries a partial positive charge (δ^+). As the bonds stretch, the positions of the electrons change, thereby changing the charge distribution in the molecule. Because of the linear shape and symmetry of CO_2, the changes in charge distribution during the symmetric stretch (vibration **(a)**) cancel, and no infrared absorption occurs.

Your Turn 4.25 Skill Building How Will a Molecule Stretch?

For each molecule below, determine its shape, and then determine where in the molecule any asymmetric stretches will occur.

a. NO_2 **b.** O_3
c. CH_4 **d.** NH_3

The infrared (heat) energy that molecules absorb can be measured with an instrument called an **infrared spectrometer**. Infrared radiation is passed through a sample of the compound to be studied, in this case gaseous CO_2. A detector measures the amount of radiation, at various wavelengths, which is transmitted by the sample. High transmission means low absorbance, and vice versa. This information is displayed graphically, where the relative intensity of the transmitted radiation is plotted versus wavelength. The result is the infrared spectrum of the compound. Figure 4.16 shows the infrared spectrum of gaseous CO_2.

Spectroscopy is the field of study that examines matter by passing electromagnetic energy through a sample.

The infrared spectrum shown in Figure 4.16 was acquired using a laboratory sample of CO_2 gas, but the same absorption takes place in the atmosphere. Gaseous molecules of CO_2 that absorb specific wavelengths of infrared energy experience different fates. Some hold that extra energy for a brief time, and then re-emit it in all directions as heat. Others collide with atmospheric molecules like N_2 and O_2, and can transfer some of the absorbed energy to those molecules, also as heat. Through

You learned about the role of CFCs in ozone depletion in Chapter 3, but an entirely different process is at work there.

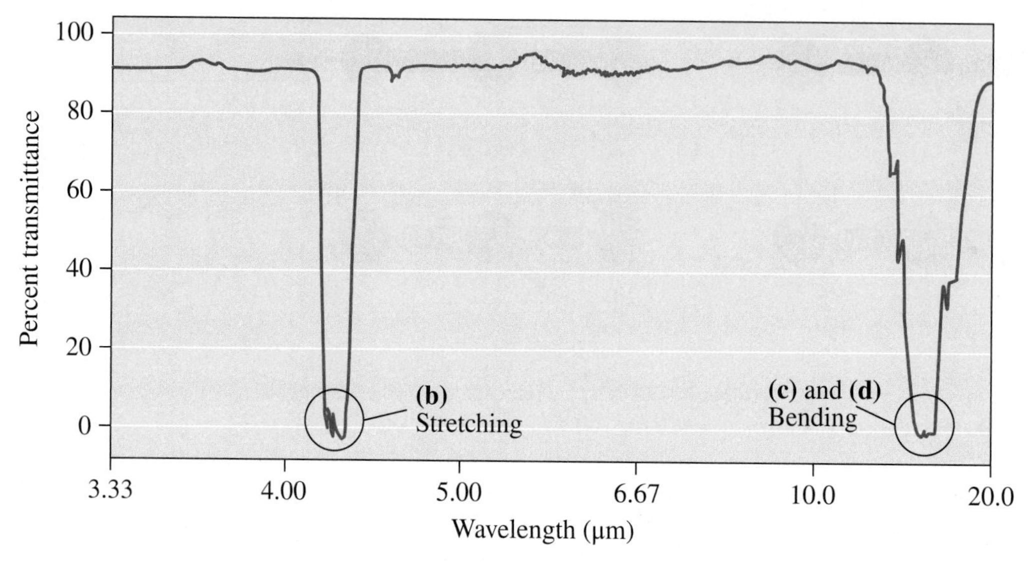

Figure 4.16

Infrared spectrum of carbon dioxide gas. The letters **(b)**, **(c)**, and **(d)** refer to the molecular vibrations shown in Figure 4.15.

both of these processes, CO_2 "traps" some of the infrared radiation emitted by Earth, keeping our planet comfortably warm. This is what makes CO_2 a greenhouse gas.

Any molecule that has three or more atoms will be able to absorb photons of IR radiation and behave as a greenhouse gas. Water is by far the most important gas in maintaining Earth's temperature, followed by carbon dioxide. Figure 4.17 shows the IR spectrum of gaseous H_2O, which displays strong absorption bands. However, methane, nitrous oxide, ozone, and chlorofluorocarbons (such as CCl_3F) also strongly absorb IR radiation and help retain planetary heat.

Interactive comparative IR absorption profiles for a variety of molecules are found in **Connect**.

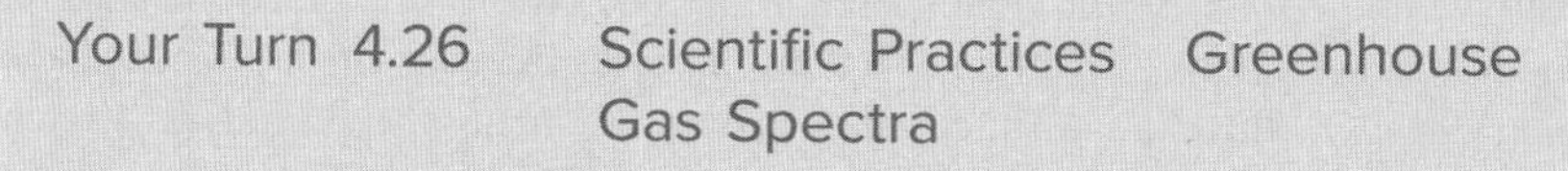

Your Turn 4.26 Scientific Practices Greenhouse Gas Spectra

a. Use **Figure 4.17** to estimate the wavelengths corresponding to the strongest IR absorbance for water vapor.

b. Which wavelengths do you predict represent bending vibrations, and which represent stretching? Explain the basis of your predictions.

Hint: Compare the IR spectrum of H_2O with that of CO_2.

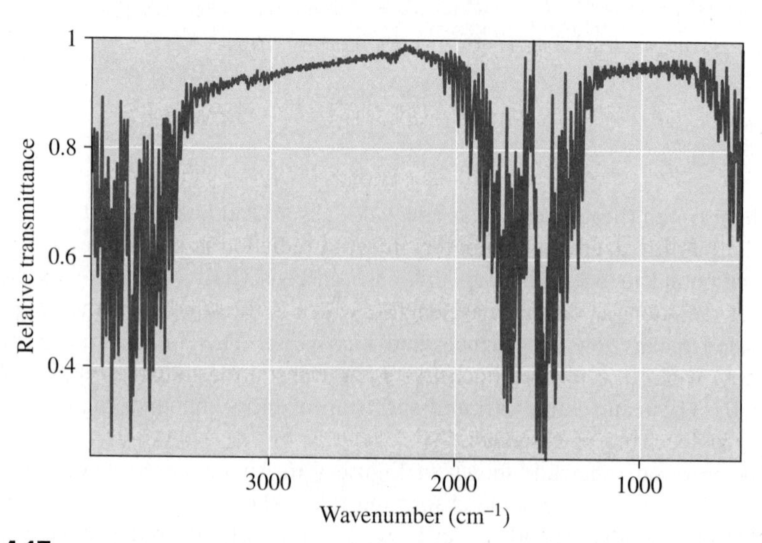

Figure 4.17

Infrared spectrum of water vapor.

Source: NIST Chemistry WebBook (http://webbook.nist.gov/chemistry)

Your Turn 4.27 Scientific Practices Greenhouse Gas Spectra

We have shown the IR spectra of CO_2 and water vapor in **Figures 4.16** and **4.17**, respectively. Identify another gas that is labeled a greenhouse gas. Perform an Internet search to find the IR spectrum for this gas. What are the similarities and differences among your spectra, and those of CO_2 and water vapor? Next, choose a gas that is known not to be a greenhouse gas. Perform the same comparison. What is unique about the spectra of greenhouse gases?

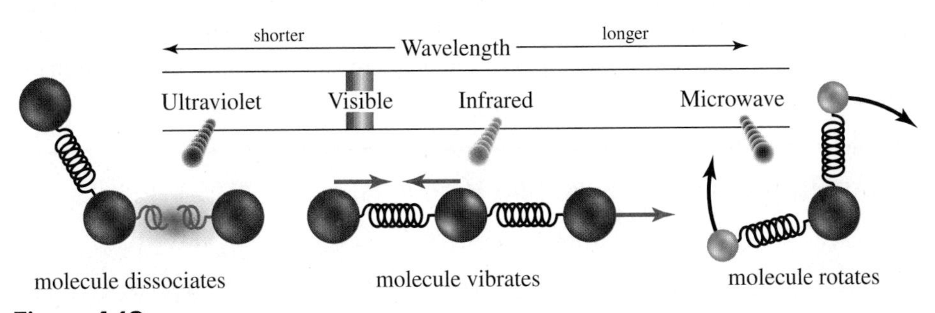

Figure 4.18

Molecular response to types of radiation.

Diatomic gases that are made of two identical atoms, such as N_2 and O_2, are not greenhouse gases. Although molecules consisting of two identical atoms do vibrate, the overall electric charge distribution does not change during these vibrations. Hence, these molecules cannot be greenhouse gases. Earlier, we discussed this lack of change in the overall electric charge distribution as the reason why the symmetric stretching vibration in Figure 4.16 was not responsible for the greenhouse gas behavior of CO_2.

So far, you have encountered two ways that molecules respond to radiation. Highly energetic photons with high frequencies and short wavelengths (such as UV radiation) can break bonds within molecules. The less energetic photons (such as IR radiation) cause an increase in molecular vibrations. Both processes are depicted in Figure 4.18, which also includes another response of molecules to radiant energy that is probably more familiar to you. Longer wavelengths than those in the IR range have only enough energy to cause molecules to rotate faster.

For example, microwave ovens generate electromagnetic radiation that causes water molecules within food to spin faster. The radiation generated in such a device is of relatively long wavelengths, about a centimeter. Thus, the energy per photon is quite low. As the H_2O molecules absorb the photons and spin more rapidly, the resulting friction cooks your food, warms up the leftovers, or heats your coffee. The same region of the spectrum is used for radar. Beams of microwave radiation are sent out from a generator. When the beams strike an object such as an airplane, the microwaves bounce back and are detected by a sensor.

A simulation of an ozone molecule interacting with various regions of the EM spectrum is found in **Connect**.

Your Turn 4.28 You Decide Revisiting the Greenhouse Effect

Now that you have learned the identity of greenhouse gases and how those molecules can cause atmospheric warming, let's revisit some of our initial questions. Provide responses to the questions below. Indicate whether your thoughts have changed, and if so, why they have changed.

a. Is atmospheric warming a good or a bad thing?
b. Does the greenhouse effect cause climate change?
c. Is climate change occurring?

4.9 | How Can We Learn from Our Past?

Now that we have examined reservoirs of carbon and determined how the shapes of molecules can influence their ability to be greenhouse gases, we need to look at what has happened to the climate on Earth in the past. An important distinction needs to be made between the terms **climate** and **weather**. *Weather* includes the daily high and low temperatures, the drizzles and downpours, the blizzards and heat waves, and the fall breezes and hot summer winds—all of which have relatively short durations. In contrast, *climate* describes regional temperatures, humidity, winds, rain, and snowfall over decades, not days. And while the weather varies on a daily basis, our climate has stayed relatively uniform over the last 10,000 years. The values quoted for the "average global temperature" are but one measure of climate phenomena. The key point is that relatively small changes in average global temperature can have huge effects on many aspects of our climate.

Your Turn 4.29 Scientific Practices "Climate" Versus "Weather"

List three similarities and three differences between climate and weather. Is the term "global warming" a description of climate or weather phenomena? Explain.

Over the past 4.5 billion years—the approximate age of Earth—both Earth's climate and its atmosphere have varied widely. Earth's climate has been directly affected by periodic changes in the shape of Earth's orbit and the tilt of Earth's axis. Such changes are thought to be responsible for the ice ages that have occurred regularly during the past million years. Even the Sun itself has changed; its energy output half a billion years ago was 25–30% less than it is today. In addition, changes in atmospheric greenhouse gas concentrations affect Earth's energy balance, and hence its climate. Carbon dioxide was once 20 times more prevalent in the atmosphere than it is today. However, that level was lowered as CO_2 dissolved in the oceans or was incorporated into rocks such as limestone. The biological process of photosynthesis also radically altered the composition of our atmosphere by removing CO_2 and producing oxygen. Certain geological events like volcanic eruptions add millions of tons of CO_2 and other gases to the atmosphere.

Although these natural phenomena will continue to influence Earth's atmosphere and its climate in the coming years, we must also assess the role that human activities are playing. With the development of modern industry and transportation, humans have moved huge quantities of carbon from terrestrial sources like coal, oil, and natural gas into the atmosphere in the form of CO_2 over a relatively short period of time. To evaluate the influence humans are having on the atmosphere, and hence on any climate change, it is important to investigate the fate of this large, rapid influx of carbon dioxide. Indeed, CO_2 concentrations in the atmosphere have increased significantly in the past half-century. The best direct measurements are taken from the Mauna Loa Observatory in Hawaii, as displayed in Figure 4.19. The red zigzag line shows the average monthly concentrations, with a small increase each April, followed by a small decrease in October. The black line is a 12-month moving average. Notice the steady increase in average annual values from 315 ppm in 1960 to more than 400 ppm today.

Your Turn 4.30 You Decide What is Causing CO_2 to Increase?

Figure 4.19 shows that the amount of carbon dioxide in Earth's atmosphere is increasing. Make a list of the reasons why this could be occurring over this time frame, and provide an explanation for the increase.

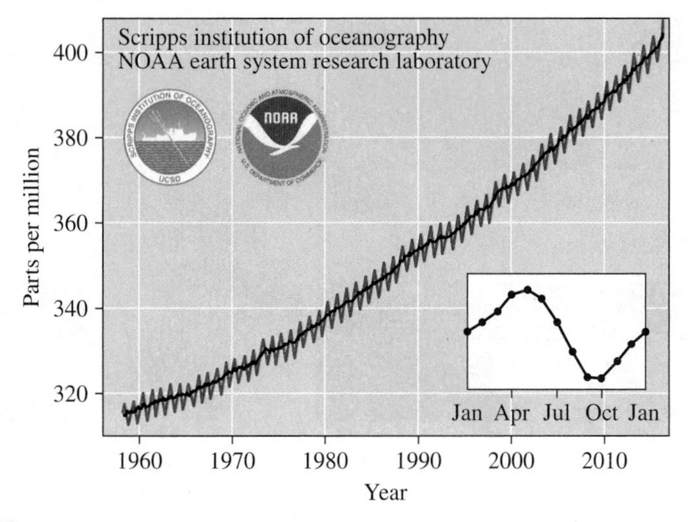

Figure 4.19

Carbon dioxide concentrations from 1958 to 2015, as measured at Mauna Loa, Hawaii. Inset: one year of the monthly variations.

Source: NOAA

Your Turn 4.31 Scientific Practices What About the Other Greenhouse Gases?

Visit the Mauna Loa Observatory website and view the data on at least three other greenhouse gases we have discussed. How does the change in concentration over time of each of these gases compare to carbon dioxide? What other information from this data do you feel is important?

Your Turn 4.32 Skill Building The Cycles of Mauna Loa

a. Calculate the percent increase in CO_2 concentration during the last 50 years.
b. Estimate the variation in ppm of CO_2 within any given year.
c. On average, the CO_2 concentrations are higher each April than each October. Why is this?

How can we obtain data about the composition of our atmosphere farther back in time? Much relevant information comes from the analysis of ice core samples. Regions on the planet that have permanent snow cover contain preserved histories of the atmosphere, buried in layers of ice. Figure 4.20a shows a dramatic example of annual ice layers from the Peruvian Andes. The oldest ice on the planet is located in Antarctica, and scientists have been drilling and collecting ice core samples there for more than 50 years (Figure 4.20b). Air bubbles trapped in the ice (Figure 4.20c) provide a vertical timeline of the history of concentrations of trace atmospheric gases in the atmosphere; the deeper you drill, the farther back in time you go.

Relatively shallow ice core data show that for the first 800 years of the last millennium, the CO_2 concentration was relatively constant at about 280 ppm. Figure 4.21 combines the Mauna Loa data with data from ice cores in Antarctica from the last three decades. Beginning in the early 1800s, CO_2 and other greenhouse gases began accumulating in the atmosphere at an ever-increasing rate, corresponding to the Industrial Revolution and the accompanying combustion of **fossil fuels** that powered that transformation.

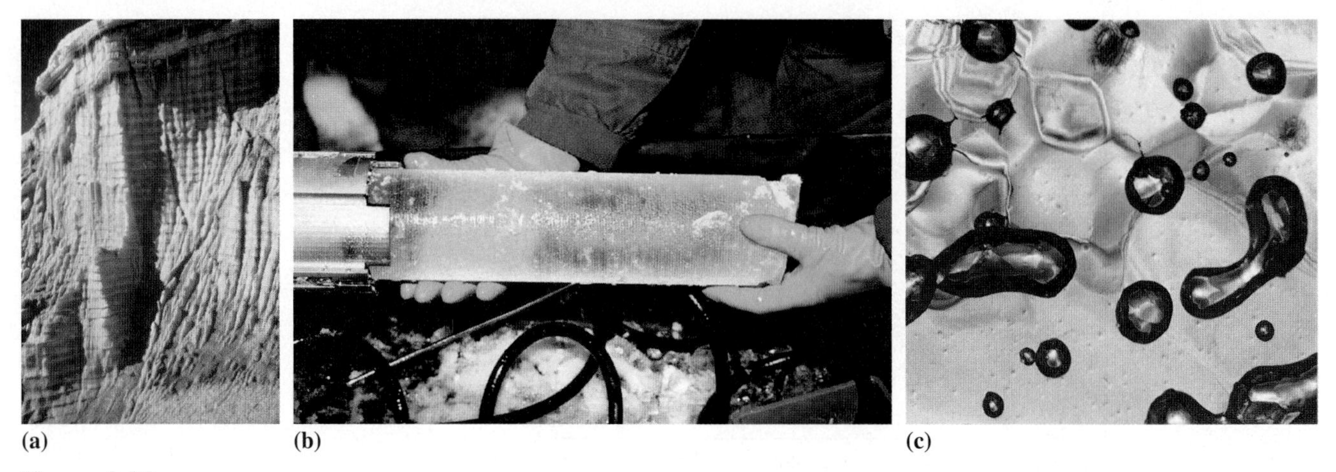

Figure 4.20

(a) Quelccaya ice cap (Peruvian Andes) showing the annual layers. **(b)** Ice core that can be used to determine changes in concentrations of greenhouse gases over time. **(c)** Microscopic air bubbles in ice.

a: © Lonnie G. Thompson, Ohio State University; b: © Vin Morgan/AFP/Getty Images; c: © W. Berner, 1978, PhD Thesis University of Bern, Switzerland. (D. Lüthi, M.Le Floch, B. Bereiter, T. Blunier, J. -M. Bamola, U. siegenthaler, D. Raynaud, J. Jouzel, H. Fischer, K. Kawamura, and T. F. Stocker, High-resolution carbon dioxide concentration record 640,000-800,000 years before present Nature ***2008****, 453, 379-382 (DOI: 10.1038/nature06949).*

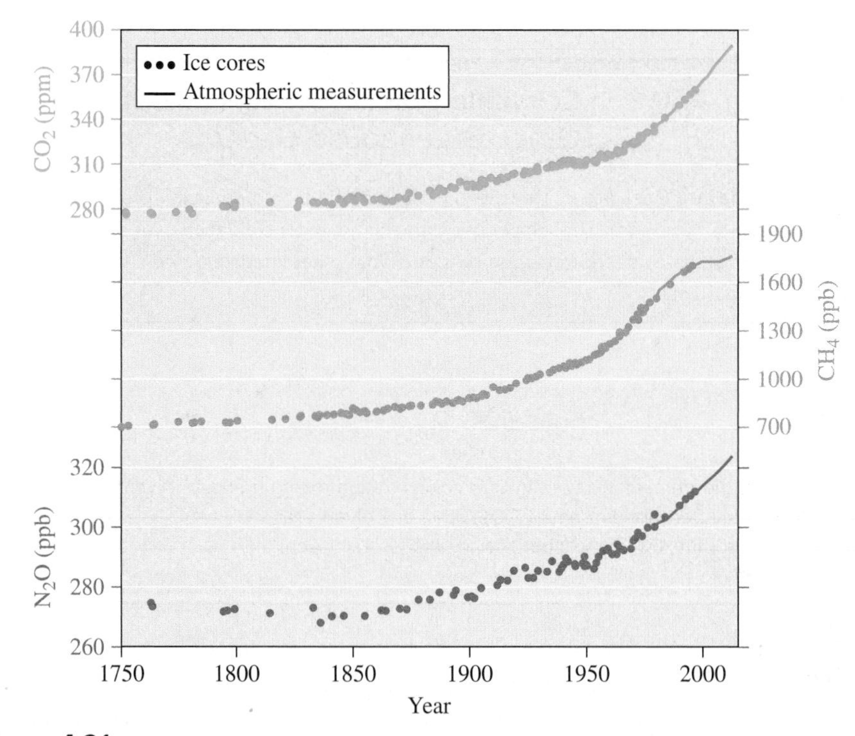

Figure 4.21

Carbon dioxide and other GHG concentrations over the last three centuries as measured from Antarctic ice cores and atmospheric measurements from the Mauna Loa Observatory.

Source: Intergovernmental Panel on Climate Change (IPCC) Synthesis Report, 2014.

Your Turn 4.33 You Decide Checking the Facts on CO_2 Increases

a. A recent government report states that the atmospheric level of CO_2 has increased 30% since 1860. Use the data in **Figure 4.21** to evaluate this statement.

b. A global warming skeptic states that the percent increase in the atmospheric level of CO_2 since 1957 has been only about half as great as the percent increase from 1860 to the present. Comment on the accuracy of that statement and how it could affect potential greenhouse gas emissions policy.

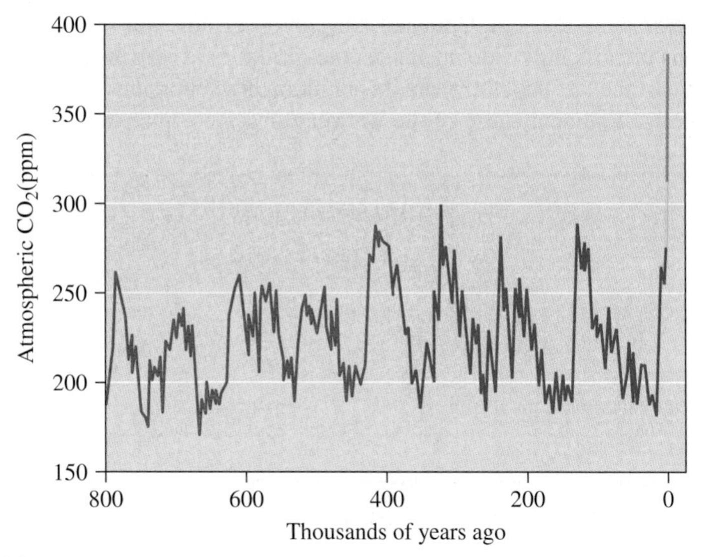

Figure 4.22

Atmospheric carbon dioxide concentrations over the last eight glacial cycles to modern times. Shown are composite datasets from ice cores in Antarctica (EPICA Dome C **(red)**, Vostok **(blue)**, Law Dome **(yellow)**, and atmospheric measurements from Mauna Loa **(orange)**).

What about further back in time? Drilling by a team of Russian, French, and U.S. scientists at varying locations in Antarctica yielded ice cores taken from the snows of 800 millennia. The atmospheric carbon dioxide concentrations going back over 800,000 years are shown in Figure 4.22.

Most obvious from the graph are the periodic cycles of high and low carbon dioxide concentrations, which occur roughly in 100,000-year intervals. Two important conclusions can be drawn from these data. First, the current atmospheric CO_2 concentration is about 100 ppm *higher* than at any time in the last million years. Also during that time, never has the CO_2 concentration risen as rapidly as it is rising today.

Your Turn 4.34 You Decide Do You Agree?

a. In the previous paragraph, we made two conclusions about the data presented in **Figure 4.22.** Do you agree with our conclusions? Explain.
b. What do you suspect is causing the changes we mentioned in our conclusions?

What about the global temperature? Measurements indicate that during the past 135 years or so, the average temperature of the planet has increased about 0.8 °C (1.4 °F). Ten of the warmest years since 1880 have occurred in the past 17 years. Some scientists correctly point out that a century or two is relatively an instant in the 4.5-billion-year history of our planet. They caution restraint in reading too much into short-term temperature fluctuations. Short-term changes in atmospheric circulation patterns like El Niño and La Niña events are certainly implicated in some of the observed temperature anomalies.

Although the trend in temperatures over the last 50 years generally follows an increase in carbon dioxide concentrations, the temperature data from year to year are much less consistent. Whether the temperature increase is a consequence of the increased CO_2 concentration cannot be determined with absolute certainty.[1] This being said,

[1] Climate Change 2014 Synthesis Report, Summary for Policy Makers, p. 4: http://www.ipcc.ch/pdf/assessment-eport/ar5/syr/AR5_SYR_FINAL_SPM.pdf

Connect contains an interactive simulation describing the role of selected greenhouse gases on climate change over the past 800,000 years.

the Intergovernmental Panel on Climate Change report from 2014 says that atmospheric concentrations of carbon dioxide, methane, and nitrous oxide are at the highest concentration in at least the last 800,000 years. In addition, these emissions are extremely likely to have been the dominant cause of the warming observed since the mid-20th century.

Your Turn 4.35 Scientific Practices Temperature Change over Time

We have presented information about how the global temperature has changed over time. Use the Internet to find some graphs that show the temperature change (make sure to use reputable sources!). Do our assertions match the graph(s) you found? What other conclusions can you make from the graph(s)?

Your Turn 4.36 You Decide Cause Versus Correlation

We stated that the global temperature has increased as the level of CO_2 has increased. However, we cannot make the statement that one causes the other. Why can we not say this? Although there is a *correlation*, we cannot prove *causation*. Search the Internet to find three other cases where there is a correlation but no causation.

It is important to realize that an increase in the average global temperature does not mean that across the globe every day is now 0.6 °C warmer than it was in 1970. Many regions on Earth have experienced just a little warming and others have cooled. Yet there are other regions, particularly in the higher latitudes, that have experienced greater-than-average warming. Warming is most drastic in the Arctic, where, not surprisingly, much of the tangible effects of climate change have already been observed.

Your Turn 4.37 Scientific Practices Temperature Changes

Search the Internet to find data on the temperature changes over the last 100 years in the area where you live. Has the temperature changed much? Now search for a map of the world that shows temperature change. Which areas have changed the most in the last 100 years? Which have changed the least? How does this compare to your area?

As we will see in more detail in Section 6.1, isotope symbols are often written using a superscript designation. For example, ^{2}H for deuterium (H-2) and ^{1}H for hydrogen (H-1).

Ice cores can also provide data for estimating temperatures further back in time because of the hydrogen isotopes found in the frozen water. Water molecules containing the most abundant form of hydrogen atoms, ^{1}H, are lighter than those that contain deuterium atoms, ^{2}H. The lighter H_2O molecules evaporate just a bit more readily than the heavier ones. As a result, there is more ^{1}H than ^{2}H in the water vapor of the atmosphere than in the oceans. Likewise, the heavier H_2O molecules in the atmosphere condense just a bit more readily than the lighter ones. Therefore, snow that condenses from atmospheric water vapor is enriched in ^{2}H. The degree of enrichment depends on temperature. The ratio of ^{2}H:^{1}H in water molecules in the ice core can be measured and used to estimate the temperature at the time the snow fell.

Connect contains more information about *isotope ratio mass spectrometry (IRMS)* that is used to evaluate our planet's climate history.

Another use of isotopes to determine the effect of atmospheric changes is carbon. The radioactive isotope carbon-14, although present only in trace amounts, provides direct evidence that the combustion of fossil fuels is the *predominant* cause of the rise in atmospheric CO_2 concentrations over the past 150 years. In all living things, only 1 out of 10^{12} carbon atoms is ^{14}C, which is radioactive. A plant or animal constantly exchanges CO_2 with the environment and this maintains a constant ^{14}C concentration in the organism. However, when the organism dies, the biochemical processes that exchange carbon stop functioning and the ^{14}C is no longer replenished. This means that after the death of the

We will discuss radioactive isotopes, known as radioisotopes, in more detail in Chapter 6.

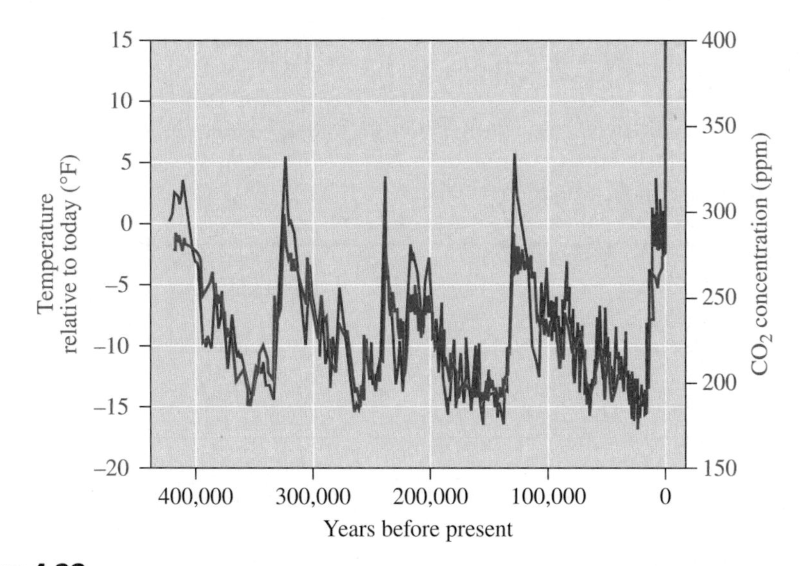

Figure 4.23

Carbon dioxide concentration **(blue)** and global temperatures **(red)** over the last 400,000 years from ice core data.

Source: Environmental Defense Fund

organism, the concentration of ^{14}C decreases with time because it undergoes radioactive decay to form ^{14}N. Coal, oil, and natural gas are remnants of plant life that died hundreds of millions of years ago. Hence, in fossil fuels and in the carbon dioxide released when fossil fuels burn, the level of ^{14}C is essentially zero. Careful measurements show that the concentration of ^{14}C in atmospheric CO_2 has recently decreased. This strongly suggests that the origin of the added CO_2 is the burning of fossil fuels, a human activity.

When we look back into the past, we see that the global temperature has undergone fairly regular cycles, matching the highs and lows in CO_2 concentration quite remarkably (Figure 4.23). Other data show that periods of high temperature have also been characterized by high atmospheric concentrations of methane and nitrous oxide, two other significant greenhouse gases. The accuracy of these data do not allow an assignment of cause and effect. Evidence other than a simple correlation of the data supports causation, but does not prove it definitively. What is clear, however, is that the current CO_2 and methane levels are much higher than any time in the last million years. Notice that the variation from hottest to coldest is only about 11 °C (20 °F). However, this represents the difference between the moderate climate we have today and ice covering much of northern North America and Eurasia, as was the case during the last glacial maximum 20,000 years ago.

Over the past million years, Earth has experienced 10 major periods of glacier activity and 40 minor ones. Without question, mechanisms other than human-caused greenhouse gas concentrations are involved in the periodic fluctuations of global climate. Some of this temperature variation is caused by minor changes in Earth's orbit that affect the distance from Earth to the Sun, and the angle at which sunlight strikes the planet. However, this hypothesis cannot fully explain the observed temperature fluctuations. Orbital effects are most likely coupled with terrestrial events such as changes in reflectivity, cloud cover, and airborne dust, as well as CO_2 and CH_4 concentrations. The feedback mechanisms that couple these effects together are complicated and not completely understood, but it is likely that the effects from each are *additive*. In other words, the existence of natural climate cycles doesn't preclude the effects on global climate caused by human activities since the Industrial Revolution.

We are a long way from the hothouse of Venus, but we face difficult decisions. These decisions will be better informed with an understanding of the mechanisms by which greenhouse gases interact with electromagnetic radiation to create the greenhouse effect that we discussed earlier. How can we use the past to help us make decisions for the future?

4.10 | Can We Predict the Future?

"Prediction is very difficult, especially about the future." Niels Bohr (1885–1962), one of the foremost contributors to our modern view of the atom, spoke these words years ago. His words still hold true today!

Your Turn 4.38 You Decide Sun Skeptics

Some people have stated that changes in the Sun, such as increased solar flares, are causing global climate change. Look at Internet sources, such as Skeptical Science's website, for your research. What are your thoughts?

Although admittedly a difficult task, we still need to make predictions. To this end, in 1988, the United Nations Environment Programme and the World Meteorological Organization teamed up to establish the UN Intergovernmental Panel on Climate Change (IPCC). The IPCC was charged with assembling and assessing the climate change data, including socioeconomic data. Thousands of international scientists were involved in this review. In their fifth report, published in 2014, the vast majority of scientists agreed on several key points:

- Human influence on Earth's climate is clear, and human-caused emissions of greenhouse gases are the highest in history.
- Human activities (primarily the combustion of fossil fuels and deforestation) are responsible for atmospheric and ocean warming, lower concentrations of ice and snow on the planet, and sea level rise.
- Continued emission of greenhouse gases will result in further warming and long-lasting change in Earth's climate system. This will cause an increased likelihood of severe, pervasive, and irreversible impacts for both ecosystems and people.

The challenge, however, is to understand current climate change well enough to *predict* future changes, and by doing so, to determine the decrease in emissions required to minimize harmful changes. To make predictions, scientists work with models. They design computer models of the oceans and the atmosphere that take into account the ability of each to absorb heat, as well as to circulate and transport matter (Figure 4.24). If that weren't difficult enough, the models must also include astronomical, meteorological, geological, and biological factors, ones that are often incompletely understood. Human influences, such as population, industrialization levels, and pollution emissions must also be included. Dr. Michael Schlesinger, who directs climate research at the University of Illinois, remarked: "If you were going to pick a planet to model, this is the *last* planet you would choose."

Figure 4.24

Climate scientists use computer simulations to understand future climate change.

Source: Lawrence Berkeley National Laboratory/U.S. Dept. of Energy

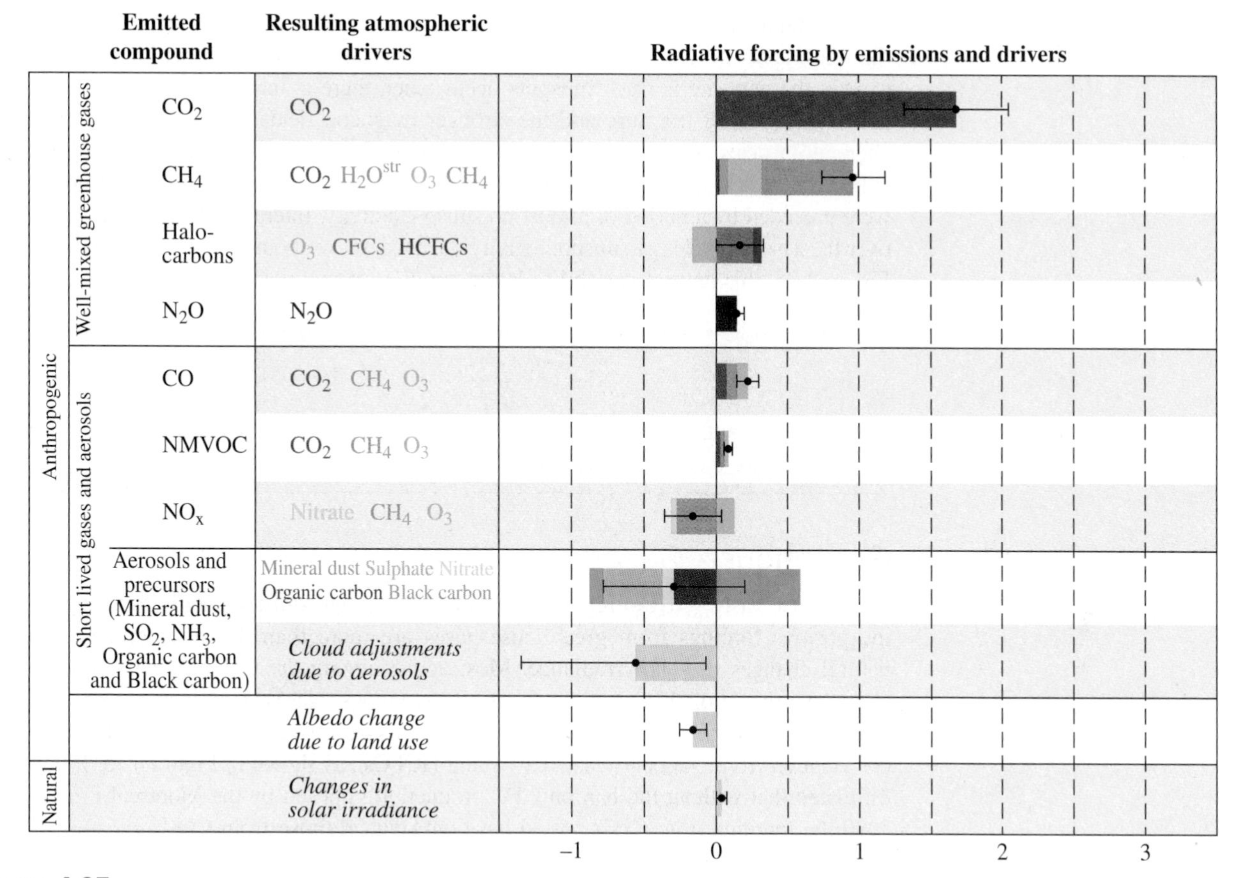

Figure 4.25

Selected radiative forcings of climate from 1750 to 2011. The units of the x-axis are watts per square meter (W/m^2), the light energy hitting a square meter of Earth's surface every second. NMVOC stands for non-metheane volatile organic compounds–typically other types of hydrocarbons.

Source: Adapted from Climate Change 2013: The Physical Science Basis. *Contribution of Working Group I to the Fourth Assessment Report of the Intergovernmental Panel on Climate Change.*

Climate scientists call the factors (both natural and anthropogenic) that influence the balance of Earth's incoming and outgoing radiation **radiative forcings**. Negative forcings have a cooling effect, while positive forcings have a warming effect. The primary forcings used in climate models are solar irradiance ("solar brightness"), greenhouse gas concentrations, land use, and aerosols. The effects of these forcings on Earth's energy balance are summarized in Figure 4.25, with GHGs shown to create a warming effect, and short-lived gases and aerosols mostly yielding a cooling effect. Each forcing has an error bar associated with it; the larger the error bar, the more uncertain the value. Let's consider some of these factors in more detail.

Solar Irradiance

We can directly observe the natural seasonal variations in sunlight intensity. In the higher latitudes, temperatures are warmer in the summer. Compared to winter months, the Sun is higher in the sky and stays up longer. Across the globe, these variations essentially cancel, because when it is winter in the Northern Hemisphere, it is summer in the Southern Hemisphere.

Subtle periodic changes occur in the brightness of the Sun. Earth's orbit oscillates slightly over a 100,000-year period, changing its shape. In addition, the magnitude of the tilt of Earth's axis and the direction of that tilt both change over the course of several tens of thousands of years. These small changes will affect the amount of solar radiation that hits Earth. However, neither of these variations occurs on a time scale short enough to explain the warming since the Industrial Revolution.

Additionally, sunspots occur in large numbers about every 11 years. You might think that dark spots on the Sun would mean a smaller amount of radiation hitting Earth, but exactly the opposite is true. Sunspots occur when there is increased magnetic activity in the outer layers of the Sun, and the stronger magnetic fields stir up a larger amount of charged particles that emit radiation. Notably, the 17th and 18th centuries were sometimes called the "Little Ice Age" because of the below-average temperatures in Europe, which were preceded by a period of almost no sunspot activity. Interestingly, the solar brightness over the 11-year cycles of sunspot activity varies only by about 0.1%. As you can see from Figure 4.25, this natural variability is the *smallest* of any positive forcing listed.

Your Turn 4.39 Skill Building Radiation from the Sun

Sunlight strikes Earth continually. Which types of light are emitted by the Sun? Which one makes up the largest percentage of sunlight? ***Hint:*** Refer back to **Figure 4.7**.

Greenhouse Gases

Greenhouse gases (GHGs) represent the most dominant **anthropogenic forcings**. In fact, the positive forcings from greenhouse gases are more than 30 times greater than the natural changes in solar irradiance. Most active among the GHGs is CO_2, constituting about two-thirds of the warming from all greenhouse gases. However, as we explained earlier, methane, nitrous oxide, and other gases also contribute. Notice the relatively small contribution from "halocarbons" (CFCs and HCFCs), as shown in Figure 4.25. It has been estimated that without the ban on CFC production imposed by the Montreal Protocol, by 1990 the forcings from CFCs would have outweighed those from CO_2!

Land Use

The *reflection coefficient* known as albedo is a dimensionless quantity. It is reported on a scale from zero (no reflection from a perfectly black surface) to 1 (perfect reflection from a white surface).

Changes in land use drive climate change because this alters the amount of incoming solar radiation that is absorbed by the surface of Earth. The ratio of electromagnetic radiation *reflected* from a surface relative to the amount of radiation *incident* on it is called the **albedo**. In short, albedo is a measure of the reflectivity of a surface. The albedo of Earth's surface varies between about 0.1 and 0.9, as you can see from the values listed in Table 4.5. You will notice that the higher the number, the more reflective the surface.

As the seasons change, so does the albedo of Earth. When a snow-covered area melts, the albedo decreases and more sunlight is absorbed, creating a positive feedback loop and additional warming. This effect will lead to greater increases in average temperature observed in the Arctic, where the amount of sea ice and permanent snow cover is decreasing. Similarly, when glaciers retreat and expose darker rock, the albedo decreases, causing further warming.

Table 4.5 Albedo Values for Different Ground Covers

Surface	Range of Albedo
fresh snow	0.80–0.90
old/melting snow	0.40–0.80
desert sand	0.40
grassland	0.25
deciduous trees	0.15–0.18
coniferous forest	0.08–0.15
tundra	0.20
ocean	0.07–0.10

Human activity can also change Earth's albedo, most notably through deforestation in the tropics. The crops we plant reflect more incoming light than does the dark green foliage of the rain forests, causing an increase in the albedo that results in a cooling effect. In addition, sunlight is more consistent in the tropics, so changes in land use at low latitudes produce greater effects than changes in the polar regions. The conversion of a tropical rain forest to crop and pastureland has more than offset the decrease in the amount of sea ice and snow cover near the poles. Therefore, the changes in Earth's albedo have caused a net cooling effect.

Your Turn 4.40 Skill Building White Roofs, Green Roofs

a. In 2009, U.S. Energy Secretary Steven Chu suggested that painting roofs white would be one way to combat global warming. Explain the reasoning behind this course of action.
b. The idea of "green roofs" is also attracting attention. Planting gardens on rooftops has benefits in addition to those of white roofs. But such gardens also have limitations. Explain.

Aerosols

Because they are a complex class of materials, **aerosols**—suspensions of small solid particles in a gas or liquid—have a correspondingly complex effect on climate. Many natural sources of aerosols exist, including dust storms, ocean spray, forest fires, and volcanic eruptions, remember the discussion of aerosols or particulate matter in Chapter 2. Human activity also can release aerosols into the environment in the form of smoke, soot, and SO_x from coal combustion.

The effect of aerosols on climate is probably the least-well understood of the forcings listed in Figure 4.25. Tiny aerosol particles (< 4 μm) are efficient at scattering incoming solar radiation. Other aerosols absorb incoming radiation, and still other particles both scatter and absorb. Both processes decrease the amount of radiation available for absorption by greenhouse gases. In a dramatic example, the 1991 eruption of Mt. Pinatubo in the Philippines spewed over 20 million tons of SO_2 into the atmosphere (Figure 4.26). In addition to providing spectacular sunsets for several months, the sulfur dioxide transformed to sulfate aerosols in the atmosphere, which caused temperatures around the world to drop slightly and therefore had a cooling effect (negative forcing). The results provided the climate modelers a mini-control experiment. The most reliable models were able to reproduce the cooling effect caused by the eruption.

In addition to a direct cooling effect, aerosol particles can serve as nuclei for the condensation of water droplets, and hence promote cloud formation. Clouds reflect incoming solar radiation, although the effects of increased cloud cover are more complex than this. Therefore, in both direct and indirect ways, aerosols counter the warming effects of greenhouse gases.

Given the complexity inherent in all the forcings we have just described, you can appreciate that assembling these forcings into a climate model is no easy task. Furthermore, once a model has been built, scientists have difficulty assessing its validity. However, scientists do have one trick up their sleeves. They can test climate models with known data sets as a means to tease apart the contributions of different forcings. For example, we know the temperature data of the 20th century. In Figure 4.27, the black lines represent the known data. Next, examine the green bands. These represent temperature ranges that were predicted by the climate model using *only* natural forcings. As you can see, the natural forcings do not map well onto the actual temperatures. Finally, examine the pink bands to see that when anthropogenic forcings are included, the temperature increases of the 20th century can be accurately reproduced. So, although the last 30 years of warming were *influenced* by natural factors, the actual temperatures cannot be accounted for without including the effects of human activities.

Figure 4.26

Photo of the aerosol smoke clouds emitted from the Mt. Pinatubo eruption in 1991.

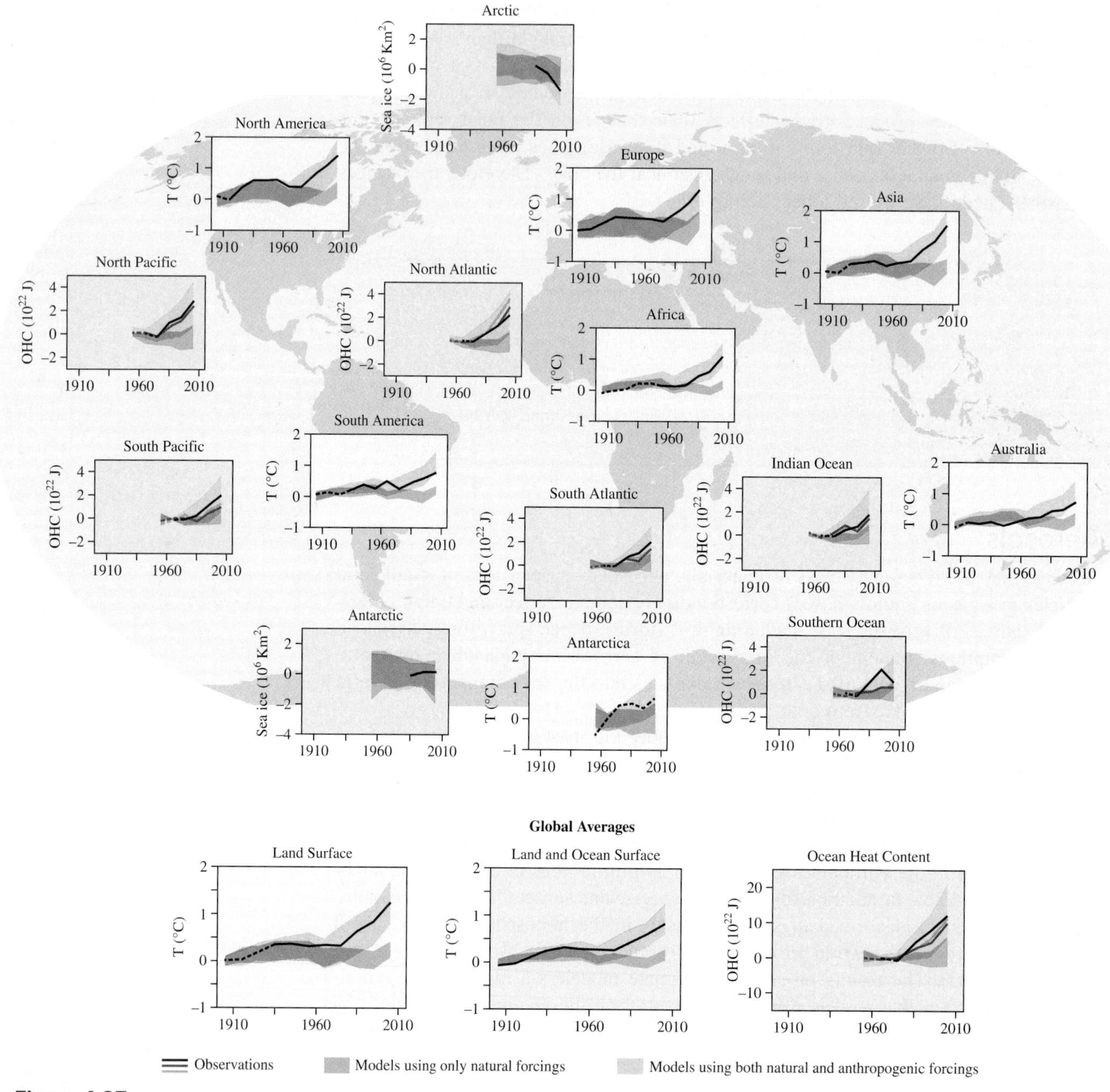

Figure 4.27

Comparison of observed and simulated climate change based on three large-scale indicators in the atmosphere.

Source: Adapted from Climate Change 2013: The Physical Science Basis. *Contribution of Working Group I to the Fourth Assessment Report of the Intergovernmental Panel on Climate Change.*

Your Turn 4.41 Scientific Practices Assessing Climate Models

Between 1950 and 2005, the climate models that used natural forcings only (green bands in **Figure 4.27**) showed an overall cooling effect, and thus did not match the observed temperatures.

a. Name the forcings included in the models that only included natural forcings.

b. List two additional forcings included in the models that more accurately re-create the temperatures of the 20th century (pink bands in **Figure 4.27**).

The magnitude of future emissions, and hence the magnitude of future warming, depends on many factors. As you might expect, one is population. As of 2015, the global population stood at about 7.3 billion. Assuming that there will be more people on the planet in the future, we humans are likely to have a larger **carbon footprint**, an estimate of the amount of CO_2 and other greenhouse gas emissions in a given time frame, usually a year. Having more people to feed, clothe, house, and transport will require the consumption of more energy. In turn, this translates to more CO_2 emissions, at least if the rate of using current fuels continues to grow. In addition, scientists who create climate models have to include values for two factors: (1) the rate of economic growth, and (2) the rate of development of "green" (less carbon-intensive) energy sources. Again, as you might expect, both are difficult to predict.

So what, if anything, can computer models tell us about Earth's future climate? Given the uncertainties we have listed, hundreds of different projected temperature scenarios for the 21st century are possible. Most predictions show that the temperature will increase. With some amount of future warming virtually ensured, we now turn our discussion to the consequences of climate change.

Find a radiation balance necessary for life on a planet using the interactive simulation found in **Connect**.

4.11 | A Look at Our Future World

Considering even the most extreme predictions of warming described in the last section, you may be thinking, "So what?" After all, the temperature changes predicted by most models are only a few degrees. At any single spot on the planet, the temperature fluctuates several times that amount daily. It is important to remember a statement we mentioned earlier. The key point is that relatively small changes in the average global temperature can have huge effects on many aspects of our climate.

In addition to modeling various future temperature scenarios, the 2014 IPCC report estimated the likelihood of various consequences. The report employed descriptive terms ("judgmental estimates of confidence") to help both policy makers and the general public better understand the inherent uncertainty of the data. Subsequent updates to the IPCC reports will use these terms, along with their assigned probabilities, which are found in Table 4.6.

Table 4.6 Judgmental Estimates of Confidence

Term	Probability That a Result Is True
virtually certain	99–100%
extremely likely	95–100%
very likely	90–100%
likely	66–100%
more likely than not	50–100%
about as likely as not	33–66%
unlikely	0–33%
very unlikely	0–10%
extremely unlikely	0–5%
exceptionally unlikely	0–1%

Source: Adapted from Climate Change 2013: The Physical Science Basis. *Contribution of Working Group I to the Fifth Assessment Report of the Intergovernmental Panel on Climate Change.*

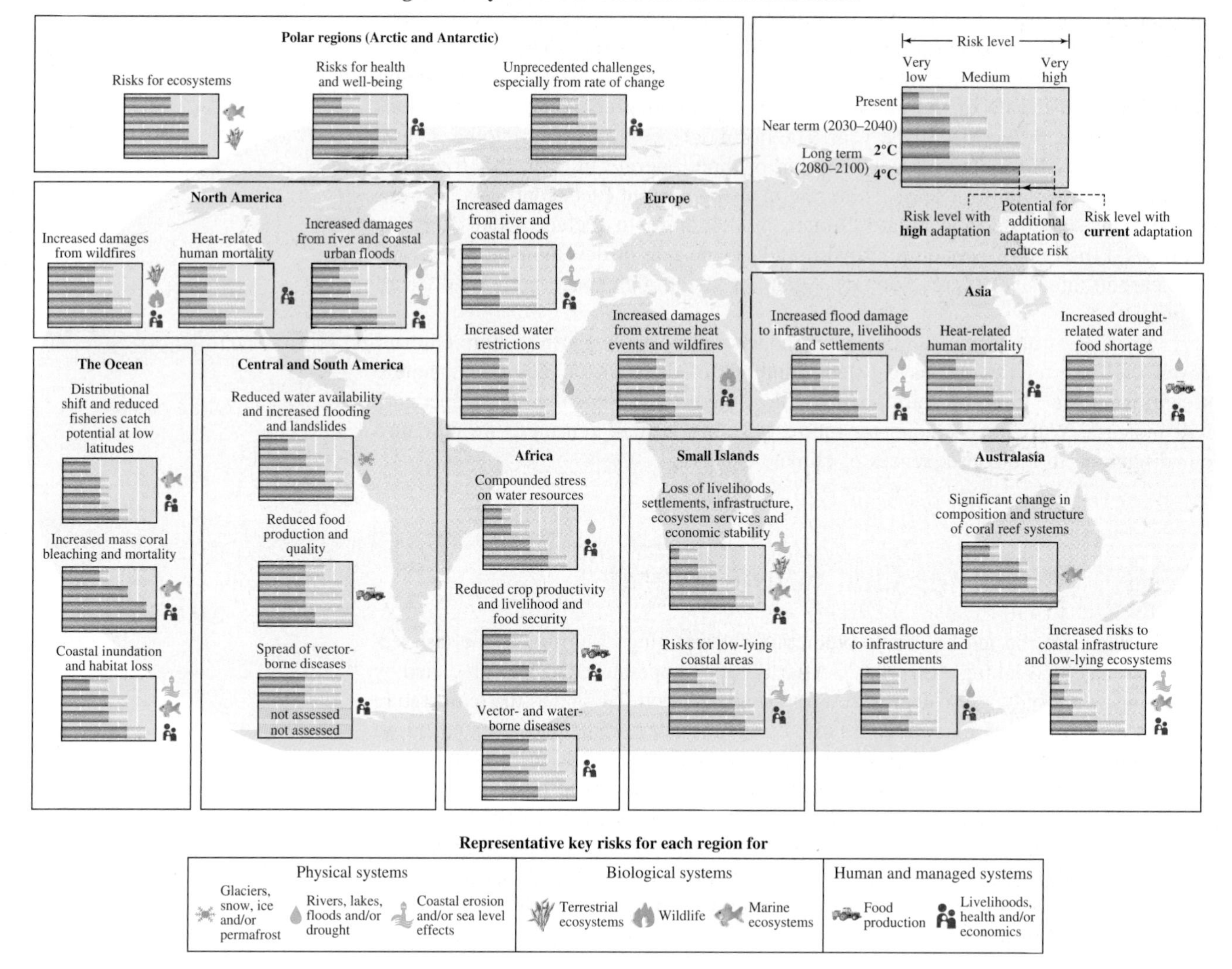

Figure 4.28

Global patterns of impacts attributed to climate change.

Source: Intergovernmental Panel on Climate Change 2014: Impacts, Adaptation, and Vulnerability, Summary for Policymakers

Conclusions from the 2014 IPCC report are illustrated in Figure 4.28. The evidence for climate change impacts has been studied most comprehensively for natural systems. For example, many species have changed migration patterns, seasonal activities, and habitats due to the impact of climate change. Impacts on human systems have also been observed. Table 4.7 provides some other conclusions from the IPCC related to the impacts of global climate change.

Many scientific organizations, including the American Association for the Advancement of Science (AAAS) and the American Chemical Society (ACS), have also recognized the threats posed by climate change. In an open letter to United States senators, these organizations cited sea level rise, more extreme weather events, increased water scarcity, and disturbances of local ecosystems as likely eventualities of a warmer planet.

To conclude this section, we describe these and other outcomes we might expect, including sea ice disappearance, more extreme weather, changes in ocean chemistry, loss of biodiversity, and harm to human health.

Table 4.7 IPCC Conclusions, 2014

Virtually Certain

- The upper ocean (0–700 m) warmed from 1971 to 2010.
- There will be more frequent hot and fewer cold temperature extremes over most land areas on daily and seasonal timescales.
- The extent of near-surface permafrost at high northern latitudes will be reduced.
- Global mean sea level rise will continue for many centuries beyond 2100.

Extremely Likely

- Human-caused emissions are the main factor causing warming since 1951.

Very Likely

- Anthropogenic influences, particularly GHGs and stratospheric ozone depletion, have led to a detectable observed pattern of tropospheric warming and a corresponding cooling in the lower stratosphere since 1961. These influences also have contributed to Arctic sea ice loss since 1979.
- Since the mid-20th century, the number of cold days and nights has decreased and the number of warm days and nights has increased on the global scale.
- Heat waves will occur with a higher frequency and longer duration.
- Extreme precipitation events over most mid-latitude land masses and over wet tropical regions will become more intense and more frequent.

Likely

- Anthropogenic forcings have made a substantial contribution to surface temperature increases since the mid-20th century.
- Anthropogenic influences have affected the global water cycle and the retreat of glaciers since 1960.
- Over the second half of the 20th century, there are more land regions where the number of heavy precipitation events has increased than where it has decreased.
- The ocean warmed at depths of 700–2000 m from 1957 to 2009, and depths of 3000 m to the bottom for the period 1992 to 2005.
- Tropical oxygen-minimum zones have expanded in recent decades.
- The global mean surface warming is in the range of 0.5 to 1.3 °C over the period 1951 to 2010.

Unlikely

- Relative to 1850–1900, the global surface temperature change for the end of the 21st century (2081–2100) will exceed 2 °C.

Your Turn 4.42 You Decide IPCC Report

For each of the conclusions presented in **Table 4.7,** write a brief summary of your thoughts on the conclusion. Do you support the IPCC findings? Which areas do you still have questions about?

Sea Ice Disappearance

As we mentioned earlier, the temperatures in the Arctic are rising faster than anywhere else on Earth. One result is that sea ice is shrinking (Figure 4.29). A record low for ice cover was set in September 2012. Summer sea ice has declined about 40% from when satellites started tracking ice coverage in the late 1970s. A new analysis that uses both computer models and data from actual conditions in the Arctic region forecasts that most of the Arctic sea ice will be gone in 30 years. Not only would significant populations of wildlife be endangered, but the accompanying decrease in albedo would lead to even more warming.

Sea Level Rise

Warmer temperatures result in an increase in sea level. This increase occurs primarily because as water warms, it expands. A smaller effect is caused by the influx of

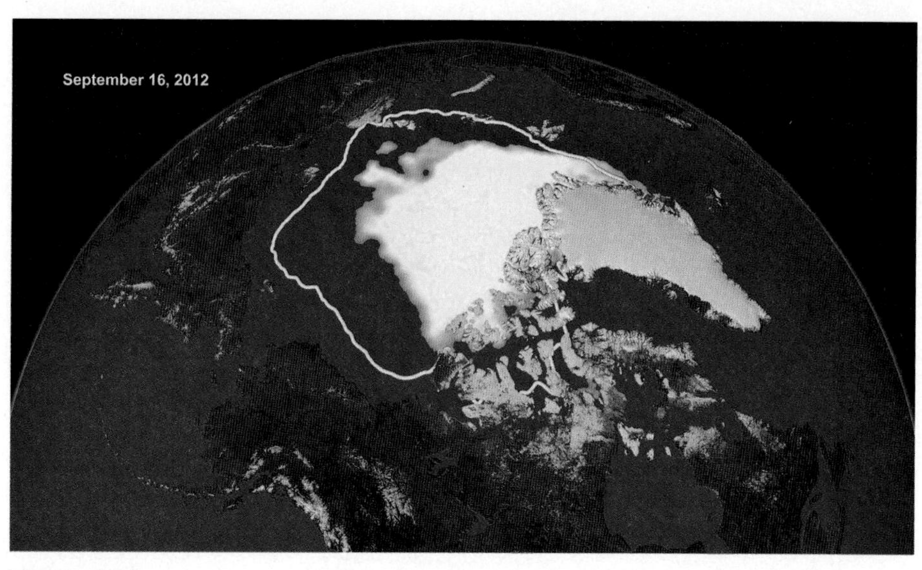

Figure 4.29

The extent of Arctic ice in September 2012 in comparison to the 30-year average sea ice minimum (yellow line).

Source: NASA GSFC Scientific Visualization Studio

freshwater into the ocean from land-based glacier runoff. According to a 2008 study published in the journal *Nature*, the increase was about 1.5 mm each year (a total of about 7.5 cm over the past 50 years) between 1961 and 2003.[2] However, the increases are not seen uniformly across the globe. In addition, they are influenced by regional weather patterns. Even so, these small increases in sea levels can cause erosion in coastal areas and the stronger storm surges associated with hurricanes and cyclones.

Your Turn 4.43 You Decide External Costs

The consequences described earlier and later on are examples of what are known as external costs. These costs are not reflected in the price of a commodity, such as the price of a gallon of gasoline or a ton of coal, but nevertheless take a toll on the environment. The external costs of burning fossil fuels often are shared by those who emit very little carbon dioxide, such as the people of the island nation of Maldives. Although a rise of sea level of just a few millimeters may not seem like much, the effects could be catastrophic for nations that lie close to sea level. Use the resources of the Internet to investigate how the people of Maldives are preparing for rising sea levels. Also comment on how this is an example of the *tragedy of the commons*.

More Extreme Weather

An increase in the average global temperature could cause more extreme weather, including storms, floods, and droughts. In the Northern Hemisphere, the summers are predicted to be drier and the winters wetter. Over the past several decades, more frequent wildfires and floods have occurred on every continent. The severity (although not the frequency) of cyclones and hurricanes may also be increasing. These tropical storms extract their energy from the oceans; a warmer ocean provides more energy to feed the storms.

Changes in Ocean Chemistry

"Over the past 200 years, the oceans have absorbed approximately 550 billion tons of CO_2 from the atmosphere, or about a third of the total amount of anthropogenic emissions over that period ..." reports Richard A. Feely, a senior scientist with the National Pacific Marine Environmental Laboratory in Seattle. Scientists estimate that 1 million tons of CO_2 are absorbed into the oceans every hour of every day! In their role as carbon

[2] *Nature* **2008,** *453,* 1090-1093 (DOI: 10.1038/nature07080).

absorbers, the world's oceans have prevented some of the warming that carbon dioxide would have caused had it remained in the atmosphere. However, this absorption has come with a cost. Critical changes are already occurring in the oceans (Figure 4.30), as we will further explore in Chapter 8. For example, carbon dioxide is slightly soluble in water and dissolves to form carbonic acid. This alters the acidity of the ocean and changes the concentration of carbon-containing species such as carbonic acid, bicarbonate, and carbonate ions. As a consequence it is difficult for marine organisms to maintain the integrity of their shells and skeletons. The increase in carbon dioxide concentrations in the atmosphere is putting entire marine ecosystems at risk.

Figure 4.30
Coral bleaching in Maldives, Indian Ocean, Asia

© Helmut Corneli/imagebroker.net/ SuperStock

Your Turn 4.44 You Decide Plankton and You

Plankton are microscopic plant- and animal-like creatures found in both salt- and freshwater systems. Many plankton species have shells made of calcium carbonate that are weakened by more acidic environments. Although humans do not eat plankton, many other marine organisms do. Construct a food chain to show the link between plankton and humans. What might be the impact of ocean acidification on the food chain that you drew?

Loss of Biodiversity

Climate change is already affecting plant, insect, and animal species around the world. Species as diverse as the California starfish, alpine herbs, and Checkerspot butterflies (Figure 4.31) have all exhibited changes in either their ranges or their habits. Dr. Richard P. Alley, a Pennsylvania State University expert on past climate shifts, sees particular significance in the fact that animals and plants that rely on each other will not necessarily change ranges or habits at the same rate. Referring to affected species, he said, "You'll have to change what you eat, or rely on fewer things to eat, or travel farther to eat, all of which have costs." In extreme cases, those costs can cause the extinction of species. Currently, the rate of extinction worldwide is nearly 1000 times greater than at any time during the last 65 million years! A 2004 report in the journal *Nature* projects that about 20% of the plants and animals considered will face extinction by 2050, even under the most optimistic climate forecasts.[3] Mass extinction is primarily driven by human development of the environment and only secondarily by climate change.

Figure 4.31
Many different species of Checkerspot butterflies exist. This one is found in parts of Wisconsin.

© www.wisconsinbutterflies.org

Vulnerability of Freshwater Resources

Like polar and sea ice, glaciers in many parts of the world are shrinking due to increased average temperatures (Figure 4.32, on the next page). Billions of people rely on glacier runoff for both drinking water and crop irrigation. The 2013 report of the IPCC predicts that a 1 °C increase in global temperature corresponds to more than half a billion people experiencing water shortages that they have not known before. The redistribution of freshwater also has implications for food production. Drought and high temperatures could reduce crop yields in the American Midwest, but the growing range might extend farther into Canada. However, soil types further north may not be appropriate to support the same level of food production. It is also possible that some desert regions could get sufficient rain to become arable. One region's loss may well become another region's gain, but it is too early to tell.

Human Health

We may all be losers in a warmer world. In 2000, the World Health Organization attributed more than 150,000 premature deaths worldwide to the effects of climate change. Those effects included more frequent and severe heat waves, increased droughts in already water-stressed regions, and infectious diseases in regions where they had not occurred before. Further increases in average temperatures are expected to expand the geographical range of mosquitoes, tsetse flies, and other disease-carrying insects. The result could be a significant upturn in illnesses such as malaria, yellow and dengue fevers, and sleeping sickness in new areas, including Asia, Europe, and the United States.

Now that we have examined some rather concerning possibilities for our future, let's examine what we can do now to try to prevent these scenarios.

[3] *Nature* **2004,** *427,* 145-148 (DOI: 10.1038/nature02121).

Figure 4.32

A view of the Athabasca Glacier in Jasper National Park, Alberta, Canada (photo taken in 2012).

© Mary Caperton Morton

4.12 Action Plans to Prevent Future Global Catastrophes—Who and How?

The debate over climate change has shifted in the last 20 years. Today's scientific data makes it clear that Earth's climate is changing. For example, measurements of higher surface and ocean temperatures, retreating glaciers and sea ice, and rising sea levels are unequivocal. In addition, the carbon isotopic ratio found in atmospheric CO_2 leaves little doubt that human activity is responsible for much of the observed warming. However, at issue is what we *can* do and what we *should* do about the changes that are occurring.

You already may know how to estimate the gas mileage for a vehicle. Likewise, you can estimate how many Calories you consume. How might you estimate how much of Earth's natural capital it takes to support the way in which you live? Clearly, this is far more difficult. Fortunately, other scientists already have grappled with how to do the math. They base the calculations on the way in which a person lives coupled with the available renewable resources needed to sustain this lifestyle.

Consider the metaphor of a footprint. You can see the footprints that you leave in sand or snow. You also can see the muddy tracks that your boots leave on the kitchen floor. Similarly, one might argue that your life leaves a footprint on planet Earth. To understand this footprint, you need to think in units of hectares or acres. A hectare is a bit more than twice the area of an acre. The ecological footprint is a means of estimating the amount of biologically productive space (land and water) necessary to support a particular standard of living or lifestyle.

Your Turn 4.45 Skill Building Footprint Calculations

Investigate some websites that calculate your personal carbon and ecological footprints.

a. For each site, list the name, the sponsor, and the information requested in order to calculate the footprints.

b. Does the information requested differ from site to site? If so, report the differences.

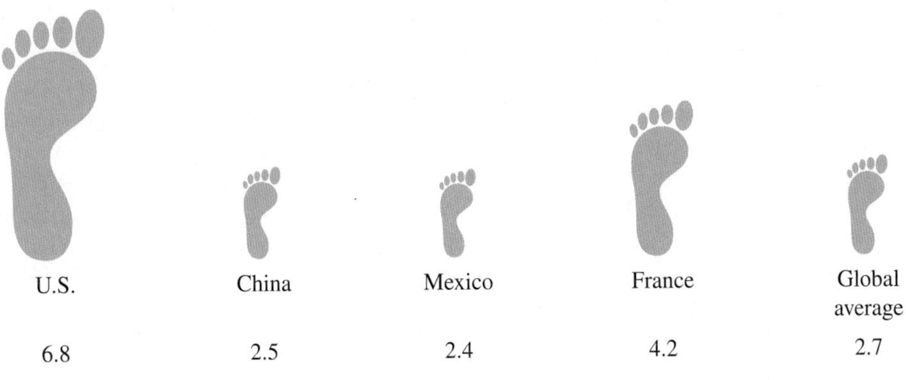

Figure 4.33

A comparison of ecological footprints, in global hectares per person.

Source: National Footprint Accounts, 2015 edition (based on 2011 data)

For the average U.S. citizen, the ecological footprint was estimated in 2011 to be about 6.8 hectares (17 acres). In other words, if you live in the United States, on average it requires 6.8 hectares of land to provide the resources to feed you, clothe you, transport you, and give you a dwelling with the creature comforts to which you are accustomed. The people of the United States have relatively big feet, as you can see in Figure 4.33. The world average in 2011 was estimated to be 2.7 hectares per person.

How much biologically productive land and water is available on our planet? We can estimate this by including regions such as croplands and fishing zones, and omitting regions such as deserts and ice caps. Currently, the value is estimated at about 12 billion hectares (roughly 30 billion acres) of land, water, and sea surface. This turns out to be about a quarter of Earth's surface. Is this enough to sustain everybody on the planet with the lifestyle that people in the United States have? The next activity allows you to see for yourself.

Your Turn 4.46 Scientific Practices Your Personal Share of the Planet

As stated earlier, an estimated 12 billion hectares (about 30 billion acres) of biologically productive land, water, and sea are available on our planet.

a. Find the current estimate for the world population. Cite your source.
b. Use this estimate, together with the estimate for biologically productive land, to calculate the amount of land theoretically available for each person in the world.

Why is this important? We have been exceeding Earth's ability to meet our demands since the 1970s. A nation whose people have an average footprint greater than about 1.7 hectares is exceeding the "carrying capacity" of Earth. Using the United States as an example, let's do one more calculation to see by how much.

Your Turn 4.47 Scientific Practices How Many Earths?

In 2011, the United States had an ecological footprint of about 6.8 hectares (about 17 acres) per person.

a. Find an estimate of the current population of the United States. Cite your source.
b. Calculate the amount of biologically productive land that the United States currently requires for this population.
c. What percentage is this amount of the biologically productive space that is available on our planet?

Energy is essential for every human endeavor. Personally, you obtain the energy you need by eating and metabolizing food. As a community or nation, we meet our energy needs in a variety of ways, including burning coal, petroleum, and natural gas. The combustion of these carbon-based fuels produces several waste products, including carbon dioxide. The countries with large populations and those that are highly industrialized tend to burn the largest quantities of fuels, and as a result, emit the most CO_2. According to the Carbon Dioxide Information Analysis Center (CDIAC) of Oak Ridge National Laboratory, in 2011, the top CO_2 emitters were China, the United States, India, the Russian Federation, and Japan. Which other nations rank high on the list? The next activity shows you how to find out.

Your Turn 4.48 Scientific Practices Carbon Emissions by Nation

CDIAC publishes a list of the top 20 nations for CO_2 emissions.

a. From what you already know, predict any five of the nations (in addition to those listed in the previous paragraph) that are on this list. Check how accurate your predictions were by using the Internet.

b. How would these rankings change if they were listed per capita?

John Holdren, the senior advisor to President Obama, summarized our options in dealing with climate change with three words: mitigation, adaptation, and suffering. "Basically, if we do less mitigation and adaptation, we're going to do a lot more suffering," he concluded. But who will be responsible for the mitigation? Who will be forced to adapt? Who will bear the brunt of the suffering? It is likely that significant disagreements will arise regarding answers to these questions. But we can agree that any practical solution must be global in nature, and include a complicated mix of risk perception, societal values, politics, and economics.

Climate mitigation is any action taken to permanently eliminate or reduce the long-term risk and hazards of climate change to human life, property, or the environment. The most obvious strategy for minimizing anthropogenic climate change is to reduce the amount of CO_2 and other greenhouse gases emitted into the atmosphere in the first place. Take a look back at Figure 4.4. It is difficult to imagine curtailing any of these "necessities" to any great extent. Therefore, decreasing our energy consumption will not be easy, at least in the short term. The simplest and least-expensive approach is to improve energy efficiency, which is one of the key ideas in green chemistry. Due to the inefficiencies associated with energy production, saving energy on the consumer end multiplies its effect on the production end three to five times. However, relying on the individual consumers worldwide to buy climate-friendly goods and do climate-friendly things will not be sufficient to hold CO_2 emissions below dangerous levels.

Your Turn 4.49 Skill Building Trees as Carbon Sinks

An average-sized tree absorbs 25 to 50 pounds of carbon dioxide each year. In the United States, the average annual per capita CO_2 emission is 19 tons.

a. How many new trees would be required to absorb the annual CO_2 emissions for an average U.S. citizen?

b. What percentage of annual global emissions from burning fossil fuels could be absorbed by 12 billion trees?

Hint: Refer back to Figure 4.2.

Regardless of any potential decreases in future emissions, some effects of climate change are unavoidable. As mentioned previously, many of the CO_2 molecules emitted today will remain in the atmosphere for centuries. **Climate adaptation** refers to the ability of a system to adjust to climate change (including climate variability and extremes) to moderate potential damage, to take advantage of opportunities, or to cope with the consequences. Some adaptive methods include developing new crop varieties and shoring up or constructing new coastline defense systems for low-lying countries and islands. The further spread of infectious diseases could be minimized by enhanced public health systems. Many of these strategies are win–win situations that would benefit societies even in the absence of climate change challenges.

Compared with the scientific consensus on understanding the role greenhouse gases play in Earth's climate, there is much less agreement among governments regarding what actions should be taken to limit greenhouse gas emissions. One outcome from the Earth Summit held in 1992 in Rio de Janeiro was the Framework Convention on Climate Change. The goal of this international treaty was "to achieve stabilization of greenhouse gas concentrations in the atmosphere at a low enough level to prevent dangerous anthropogenic interference with the climate system." Not only was this treaty nonbinding, but also there was no agreement about what "dangerous anthropogenic interference" meant, or what level of greenhouse gas emissions would be necessary to avoid it.

In 1997, the first international treaty imposing legally binding limits on greenhouse gas emissions was written by nearly 10,000 participants from 161 countries gathered in Kyoto, Japan. The result has come to be known as the Kyoto Protocol. Binding emission targets based on 1990 levels were set for 38 developed nations to reduce their emissions of six greenhouse gases. The gases regulated include carbon dioxide, methane, nitrous oxide, hydrofluorocarbons (HFCs), perfluorocarbons (PFCs), and sulfur hexafluoride. The United States was expected to reduce emissions to 7% below its 1990 levels, the European Union nations 8%, and Canada and Japan 6% by 2012.

Your Turn 4.50 You Decide The British Experience

The British Labour Party in 1997, under the leadership of Tony Blair, committed to cut British greenhouse gas emissions 20% by 2010. This is significantly more than the 12.5% required by the Kyoto treaty. Did Britain meet its goal? Research this question and write a short report on the British experience in reducing greenhouse gases. Have other countries been able to reduce their emissions significantly since 1997?

Although the treaty went into effect in 2005 (when ratified by the Russian Federation), the United States never opted to participate. One reason was the belief that meeting the reduction requirements set by the protocol would cause serious harm to the U.S. economy. Another reason for not ratifying the protocol was concern about the lack of emissions limitations on developing nations, mainly China and India; those countries are expected to show the most dramatic increases in carbon dioxide emissions in the coming years. The administration of President George W. Bush argued that such unequal burdens between developed and developing countries would be economically disastrous to the United States.

The United States has also resisted domestic legislation to restrict CO_2 emissions on similar economic grounds. Voluntary reduction programs implemented during the early 2000s proved insufficient to reduce emissions for a variety of reasons. One problem is that fossil fuels are too inexpensive. A second problem is that any mitigation measures entail significant upfront costs, and just as importantly, the cost of mitigation is not known with certainty, making it difficult for corporations to plan effectively. The world's current energy infrastructure cost $15 trillion to develop and distribute, and reducing carbon dioxide emissions will mean replacing much of that infrastructure. A final problem lies in the fact that the benefits of emissions reductions will not be felt for decades because of the long residence time of CO_2 molecules in the atmosphere.

The Paris Agreement in 2015 aimed to reach global peaking of greenhouse gas emissions, in order to keep the average global temperature well below 2 °C above pre-industrial levels.

Connect contains a simulation that explores the summative effects of various technologies on their carbon output.

Now, 20 years after the first Earth Summit, scientific consensus is beginning to focus on determining what levels of CO_2 are considered "dangerous." At the United Nations Climate Conference in 2007, participating scientists concluded that greenhouse gas emissions need to peak by about 2020, and then be reduced to well below half of current levels by 2050. In absolute terms, that means that annual global emissions must be decreased by about 9 billion tons. To give you a scale of the magnitude of this goal, reducing emissions by 1 billion tons requires one of the following changes:

- Cutting energy usage in the world's buildings by 20–25% below business-as-usual.
- Having *all* cars get 60 mpg instead of 30 mpg.
- Capturing and sequestering carbon dioxide at 800 coal-burning power plants.
- Replacing 700 large coal-burning power plants with nuclear, wind, or solar power.

Clearly, implementation of any one of those will not be accomplished on a purely voluntary basis. In the United States and elsewhere, there is a burgeoning realization that laws and regulations are needed to reduce greenhouse gas emissions. One example is a "cap-and-trade" system, such as the one that has been successful in reducing the emission of oxides of both sulfur and nitrogen in the United States. The "trade" part of the cap-and-trade system works through a system of allowances. Companies are assigned allowances that authorize the emission of a certain quantity of CO_2, either during the current year or any year thereafter. At the end of a year, each company must have sufficient allowances to cover its actual emissions. If it has extra allowances, it can trade or sell them to another company that might have exceeded their emissions limit. If a company has insufficient allowances, it must purchase them. The "cap" is enforced by creating only a certain number of allowances each year.

Here's an example of how cap-and-trade works (Figure 4.34). Without emission restrictions, Plant A emits 600 tons of CO_2 and Plant B emits 400 tons. To get under the imposed cap, they are required to reduce their combined emissions by 300 tons (30%). One way to accomplish this is for each to reduce their own emissions by 30%, each accruing the associated costs. It is likely, however, that one of the plants (Plant B) would be more efficient in their emissions reductions, and lower their emissions below the prescribed 30%. In that case, Plant A can purchase some unused emissions permits from Plant B, at a cost less than that required for Plant A to comply with the

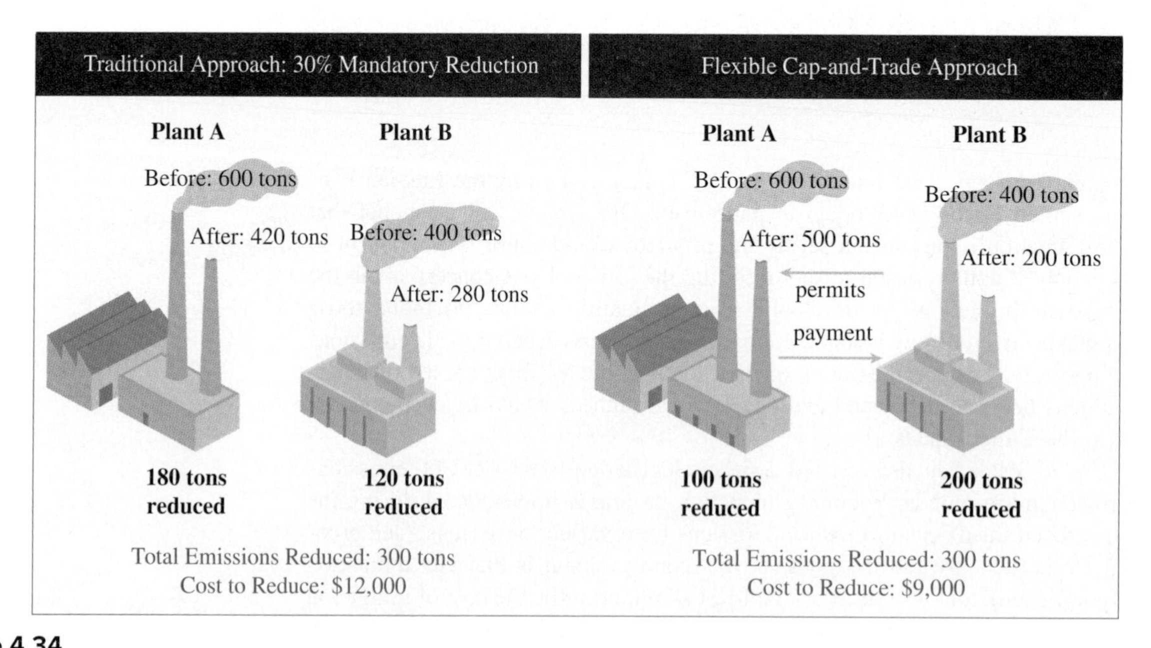

Figure 4.34

The emissions cap-and-trade concept.

Source: EPA, Clearing the Air, The Facts About Capping and Trading Emissions, 2002, p. 3

30% emissions reduction. The *overall* emissions reductions are then arrived at in the most financially beneficial way for both plants.

The cap-and-trade system has some possible disadvantages, including a potentially volatile market for the emissions permits. Energy providers might experience wide, often unpredictable swings in their energy costs. Those swings would result in large fluctuations in consumer costs. As an alternative to cap-and-trade, some advocate a carbon tax instead of a cap-and-trade program. Instead of limiting emissions and letting the market decide how "best" to comply, a carbon tax simply increases the cost of burning fossil fuels. Placing an additional cost based on the amount of carbon contained in a certain quality of fuel is intended to make alternative energy sources more competitive in the near term. Of course, levying a tax on carbon fuels or emissions will mean higher prices for consumers as well.

Your Turn 4.51 You Decide Climate Change Insurance?

Mitigation of climate change can be seen as a risk–benefit scenario. As such, uncertainty about future effects may discourage governments from taking financially costly actions. Another way of tackling climate change is to view it as a risk-management problem, analogous to the reasons we buy insurance. Having car insurance doesn't reduce the likelihood of being involved in an accident, but it can limit the costs if an accident should occur. How might the insurance analogy fit in with climate change actions and policies?

Although the U.S. federal government has been slow to produce binding climate change legislation, individual states have taken matters into their own hands. The 10 northeastern states that make up the Regional Greenhouse Gas Initiative (RGGI) signed the first U.S. cap-and-trade program for carbon dioxide. The program began by capping emissions at current levels in 2009 and then reducing emissions 10% by 2019. The Midwestern Regional Greenhouse Gas Reduction Accord developed a multisector cap-and-trade system to help meet a long-term target of 60–80% below current emissions levels. Western Climate Initiative states, as well as British Columbia and Manitoba (the first participating jurisdictions outside of the United States), agreed to mandatory emissions reporting, as well as regional efforts to accelerate development of renewable energy technologies.

More locally, the U.S. Mayors Climate Protection Agreement included 227 cities committed to cutting emissions to meet the targets of the Kyoto Protocol. The cities represented include some of the largest in the Northeast, the Great Lakes region, and West Coast, and their mayors represent some 44 million people.

Your Turn 4.52 You Decide A Drop in the Bucket?

Critics suggest that actions made by individual states or countries, even if successful, cannot possibly have a significant effect on global emissions of greenhouse gases. Proponents for immediate action, such as NASA climate scientist James Hansen, take a different approach. "China and India have the most to lose from uncontrolled climate change because they have huge populations living near sea level. Conversely then, they also have the most to gain from reduced local air pollution. They must be a part of the solution to global warming, and I believe they will be if developed nations such as the United States take the appropriate first steps..." After studying this chapter, which side do you fall on? Explain.

Your Turn 4.53 You Decide The "Big Question" Revisited

Answer these questions based on your current knowledge.

a. Do you feel the greenhouse effect and warming of the atmosphere is good or bad? Why or why not?
b. Where does the carbon dioxide in the air come from?
c. What is climate change?
d. Is climate change occurring? Why or why not?

Compare your answers to those from the start of the chapter. How have your answers changed?

Conclusions

© *Fuse/Getty Images RF*

This group of chimpanzees contributed minimally, or not at all, to global climate change and likely are not discussing the issue. However, they must adapt to the changes that will occur. Unlike humans, chimpanzees, along with plants and other animals, don't argue with each other about whether climate is changing. They just attempt to adapt to the ever-changing world, which can affect their way of life including their access to food, water, and habitat. For example, as the climate changes, food availability shifts, thus forcing animals such as the chimpanzee to adapt in order to obtain enough calories to survive. The changes also affect their habitat, with variations due to differing weather patterns.

Like much of the planet, the salt water in the oceans has no voice, but it still responds to climate change and has a story to tell. In colder climates, it quietly freezes to form sea ice when temperatures drop. And perhaps more noisily, this ice breaks up with the return of warmer temperatures in the spring. This freeze–thaw cycle has been occurring for thousands of years, gradually shifting to form more or less ice as the temperatures on Earth have shifted. In recent years, however, the freeze–thaw cycle has been more pronounced and the waters in the Arctic have been free of ice for longer periods of time. Might carbon dioxide be the culprit of changes witnessed in the Arctic? As a greenhouse gas, carbon dioxide plays a role in keeping our planet comfortably warm and able to support life, but there can be too much of a good thing. John Holdren has said several times, "Global warming is a misnomer, because it implies something that is gradual, something that is uniform, something that is quite possibly benign. What we are experiencing with climate change is none of those things."

The first assertion is that global warming isn't gradual. By this he means that in comparison with the past, the climate changes we are seeing today are occurring much more rapidly. Natural climate changes are part of our planet's history. Glaciers, for example, have advanced and retreated numerous times, and global temperatures have been both much higher and much lower than the temperatures we currently experience. But the geologic evidence indicates these past changes occurred over millennia, not decades as they occur today. So Holdren is correct. Global warming is not gradual, at least not in comparison with the geologic time frames of the past.

Second, he asserts that global warming does not occur uniformly across the globe. Holdren is right again. To date, the most dramatic effects have been observed at the poles. These include quickly receding glaciers, shrinking sea ice, and melting permafrost. So far, the more densely populated lower latitudes have experienced far smaller effects from climate change.

His third assertion, that global warming might not be benign, is the most difficult to assess. The issue is complicated in part because we cannot predict with certainty which aspects of our planet global warming will affect and to what degree. It is further complicated because we cannot easily understand why only a few degrees of warming might be catastrophic.

As evidenced by Holdren's points, global climate change is an extremely complicated phenomenon. Like it or not, we are in the midst of conducting a planetwide experiment, one that will test our ability to sustain both our economic development and our environment.

In the next chapter, you will learn about various types of fossil fuels and the details of their combustion. Although these are the culprit reactions responsible for the increased concentration of CO_2 in the atmosphere, we rely on the burning of these fuels for electricity, transportation, and warmth. However, as you will learn, more sustainable fuel options are beginning to be used, including a sustainable use for CO_2 in oil recovery projects.

Learning Outcomes

The numbers in parentheses indicate the sections within the chapter where these outcomes were discussed.

Having studied this chapter, you should now be able to:

- name and identify carbon-containing compounds (4.1)
- illustrate, interpret, and predict sources of carbon using carbon cycle diagrams (4.1)
- identify where carbon is located on Earth (4.1)
- write formulas and names of ionic compounds and transition metal compounds (4.1)
- name and identify the charges for polyatomic ions (4.1)
- summarize photosynthesis, combustion, and respiration using chemical reactions (4.2)
- describe mole–mass relationships, and relate this relationship to real-world reactions (4.3)
- convert among grams, moles, and number of molecules, ions, or atoms using mole–mass relationships (4.3)
- calculate the average atomic mass based on the relative percentages of isotopes (4.3)
- use molar masses to calculate the amount of carbon in gas molecules and ionic compounds (4.4)
- estimate the amounts of carbon in various carbon reservoirs (4.4)
- outline the path(s) of incoming and outgoing radiation in Earth's atmosphere (4.5)
- diagram the greenhouse effect and explain how it influences the temperature on Earth (4.6)
- describe the characteristics of a greenhouse gas (4.6)
- construct and use Lewis structures to predict molecular shape (4.7)
- describe how IR radiation interacts with molecules and can lead to asymmetrical stretching and bending (4.8)
- model and explain how radiation is released from molecules and further warms Earth (4.8)
- interpret graphs to make claims about climate conditions in the past (4.9)
- explain the processes for collecting historical and current climate data and assess the reliability of data (4.9)
- distinguish between weather and climate (4.9)
- differentiate between causation and correlation (4.9)
- recognize that elements can exist as different isotopes and identify that the presence of isotopes allow scientists to determine age (4.3, 4.9)
- make predictions based on trends and models (4.10)
- differentiate between observation and inference in relation to future climate predictions (4.10)
- relate scientific processes to the prediction of Earth's future climates (4.10)
- evaluate conclusions from a scientific report on climate change and interpret how data supported those conclusions (4.9, 4.10, 4.11)
- estimate the possible global consequences of climate change and describe factors that can mitigate the severity of these potential consequences (4.11)
- identify factors that influence individual carbon footprints (4.12)
- devise actions that can lower individual carbon footprints (4.12)
- predict how changes in individual carbon footprints, cities, and nations can mitigate climate change consequences (4.12)

Questions

Emphasizing Essentials

1. The chapter concluded with a quote from John Holdren: "Global warming is a misnomer, because it implies something that is gradual, something that is uniform, something that is quite possibly benign. What we are experiencing with climate change is none of those things." Use examples to:
   a. explain why climate change is not uniform.
   b. explain why it is not gradual, at least in comparison to how quickly social and environmental systems can adjust.
   c. explain why it probably will not be benign.
2. The surface temperatures of both Venus and Earth are warmer than would be expected on the basis of their respective distances from the Sun. Why is this so?
3. Using the analogy of a greenhouse to understand the energy radiated by Earth, of what are the "windows" of Earth's greenhouse made? In what ways is the analogy not precisely correct?
4. Consider the photosynthetic conversion of CO_2 and H_2O to form glucose, $C_6H_{12}O_6$, and O_2.
   a. Write the balanced equation.
   b. Is the number of each type of atom on either side of the equation the same?
   c. Is the number of molecules on either side of the equation the same? Explain.
5. Describe the difference between climate and weather.
6. a. It is estimated that 29 megajoules per square meter (MJ/m^2) of energy comes to the top of our atmosphere from the Sun each day, but only 17 MJ/m^2 reaches the surface. What happens to the rest?
   b. Under steady-state conditions, how much energy would leave the top of the atmosphere?
7. Consider Figure 4.23.
   a. How does the present concentration of CO_2 in the atmosphere compare with its concentration 20,000 years ago? With its concentration 120,000 years ago?
   b. How does the present temperature of the atmosphere compare with the 1950–1980 mean temperature? With the temperature 20,000 years ago? How does each of these values compare with the average temperature 120,000 years ago?
   c. Do your answers to parts **a** and **b** indicate causation, correlation, or no relation? Explain.
8. Understanding Earth's energy balance is essential to understanding the issue of global warming. For example, the solar energy striking Earth's surface averages 168 watts per square meter (W/m^2), but the energy leaving Earth's surface averages 390 W/m^2. Why isn't Earth cooling rapidly?
9. Explain each of these observations.
   a. A car parked in a sunny location may become hot enough to endanger the lives of pets or small children left in it.
   b. Clear winter nights tend to be colder than cloudy ones.
   c. A desert shows much wider daily temperature variation than a moist environment.
   d. People wearing dark clothing in the summertime put themselves at a greater risk of heatstroke than those wearing white clothing.
10. Construct a methane molecule (CH_4) from a molecular model kit (or use Styrofoam™ balls or gumdrops to represent the atoms and toothpicks to represent the bonds). Demonstrate that the hydrogen atoms would be farther from one another in a tetrahedral arrangement than if they all were in the same plane (square planar arrangement).
11. Draw the Lewis structure and name the molecular geometry for each molecule.
    a. H_2S
    b. OCl_2 (oxygen is the central atom)
    c. N_2O (nitrogen is the central atom)
12. Draw the Lewis structure and name the molecular geometry for these molecules.
    a. PF_3
    b. HCN (carbon is the central atom)
    c. CF_2Cl_2 (carbon is the central atom)
13. a. Draw the Lewis structure for methanol (wood alcohol), H_3COH.
    b. Based on this structure, predict the H–C–H bond angle. Explain your reasoning.
    c. Based on this structure, predict the H–O–C bond angle. Explain your reasoning.
14. a. Draw the Lewis structure for ethene (ethylene), H_2CCH_2, a small hydrocarbon with a C=C double bond.
    b. Based on this structure, predict the H–C–H bond angle. Explain your reasoning.
    c. Sketch the molecule showing the predicted bond angles.

15. Three different modes of vibration of a water molecule are shown. Which of these modes of vibration contributes to the greenhouse effect? Explain.

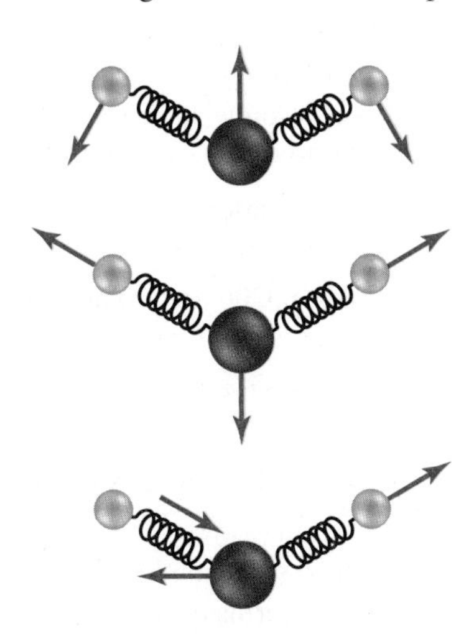

16. If a carbon dioxide molecule interacts with certain photons in the IR region, the vibrational motions of the atoms are increased. For CO_2, the major wavelengths of absorption occur at 4.26 micrometers (μm) and 15.00 micrometers (μm).
   a. What is the energy corresponding to each of these IR photons?
   b. What happens to the energy in the vibrating CO_2 species?
17. Water vapor and carbon dioxide are greenhouse gases, but N_2 and O_2 are not. Explain.
18. Explain how each of these relates to global climate change.
   a. volcanic eruptions
   b. CFCs in the stratosphere
19. Termites possess enzymes that allow them to break down cellulose into glucose, $C_6H_{12}O_6$, and then metabolize the glucose into CO_2 and CH_4.
   a. Write a balanced equation for the metabolism of glucose into CO_2 and CH_4.
   b. What mass of CO_2, in grams, could one termite produce in one year if it metabolized 1.0 mg glucose in one day?
20. Consider Figure 4.4.
   a. Which sector has the highest CO_2 emission from fossil fuel combustion?
   b. What alternatives exist for each of the major sectors of CO_2 emissions?
21. Silver has an atomic number of 47.
   a. Give the number of protons, neutrons, and electrons in a neutral atom of the most common isotope, Ag-107.
   b. How do the numbers of protons, neutrons, and electrons in a neutral atom of Ag-109 compare with those of Ag-107?
22. Silver only has two naturally occurring isotopes: Ag-107 and Ag-109. Why isn't the average atomic mass of silver given on the periodic table simply 108?
23. a. Calculate the average mass in grams of an individual atom of silver.
   b. Calculate the mass in grams of 10 trillion silver atoms.
   c. Calculate the mass in grams of 5.00×10^{45} silver atoms.
24. Calculate the molar mass of these compounds. Each plays a role in atmospheric chemistry.
   a. H_2O
   b. CCl_2F_2 (Freon-12)
   c. N_2O
25. a. Calculate the mass percent of chlorine in CCl_3F (Freon-11).
   b. Calculate the mass percent of chlorine in CCl_2F_2 (Freon-12).
   c. What is the maximum mass of chlorine that could be released in the stratosphere by 100 g of each compound?
   d. How many atoms of chlorine correspond to the masses calculated in part **c**?
26. The total mass of carbon in living systems is estimated to be 7.5×10^{17} g. Given that the total mass of carbon on Earth is estimated to be 7.5×10^{22} g, what is the ratio of carbon atoms in living systems to the total carbon atoms on Earth? Report your answer in percent and in ppm.
27. Other than atmospheric concentration, what two other properties are included in the calculation of the global warming potential for a substance?

Concentrating on Concepts

28. Give the number of protons, neutrons, and electrons in each of these neutral atoms.
   a. oxygen-18 ($^{18}_{8}O$)
   b. sulfur-35 ($^{35}_{16}S$)
   c. uranium-239 ($^{239}_{82}U$)
   d. bromine-82 ($^{82}_{35}Br$)
   e. neon-19 ($^{19}_{10}Ne$)
   f. radium-226 ($^{226}_{88}Ra$)
29. Give the symbol showing the atomic number and the mass number for the isotope that has:
   a. 9 protons and 10 neutrons (used in nuclear medicine).
   b. 26 protons and 30 neutrons (the most stable isotope of this element).
   c. 86 protons and 136 neutrons (the radioactive gas found in some homes).

30. John Holdren, quoted in the conclusion of the chapter, suggests that we use the term *global climatic disruption* rather than *global warming*. After studying this chapter, do you agree with his suggestion? Explain.

31. The Arctic has been called "our canary in the coal mine for climate impacts that will affect us all."
   a. What does the phrase "canary in the coal mine" mean?
   b. Explain why the Arctic serves as a canary in a coal mine.
   c. The melting of the tundra accelerates changes elsewhere. Give one reason why.

32. Do you think the comment made in the cartoon is justified? Explain.

Pepper . . . and Salt

"This winter has lowered my concerns about global warming . . . ."

Source: The Wall Street Journal. By permission of Cartoon Features Syndicate.

33. Given that direct measurements of Earth's atmospheric temperature over the last several thousands of years are not available, how can scientists estimate past fluctuations in the temperature?

34. A friend tells you about a newspaper story that stated, "The greenhouse effect poses a serious threat to humanity." What is your reaction to that statement? What would you tell your friend?

35. Over the last 20 years, about 120 billion tons of CO_2 has been emitted from the burning of fossil fuels, yet the amount of CO_2 in the atmosphere has risen by only about 80 billion tons. Explain.

36. Carbon dioxide gas and water vapor both absorb IR radiation. Do they also absorb visible radiation? Offer some evidence based on your everyday experiences to help explain your answer.

37. How would the energy required to cause IR-absorbing vibrations in CO_2 change if the carbon and oxygen atoms were connected by single rather than double bonds?

38. Explain why water in a glass cup is quickly warmed in a microwave oven, but the glass cup itself warms much more slowly, if at all.

39. Ethanol, C_2H_5OH, can be produced from sugars and starches in crops such as corn or sugarcane. The ethanol is used as a gasoline additive, and when burned it combines with O_2 to form H_2O and CO_2.
   a. Write a balanced equation for the complete combustion of C_2H_5OH.
   b. How many moles of CO_2 are produced from each mole of C_2H_5OH completely burned?
   c. How many moles of O_2 are required to burn 10 mol of C_2H_5OH?

40. Why is the atmospheric lifetime of a greenhouse gas important?

41. Compare and contrast stratospheric ozone depletion and climate change in terms of the chemical species involved, the type of radiation involved, and the predicted environmental consequences.

42. Explain the term *radiative forcings* to someone unfamiliar with climate modeling.

43. It is estimated that Earth's ruminants, such as cattle and sheep, produce 73 million metric tons of CH_4 each year. How many metric tons of carbon are present in this mass of CH_4?

44. Nine of the ten warmest years since 1880 have occurred since the year 2000. Does this *prove* that the enhanced greenhouse effect (global warming) is taking place? Explain.

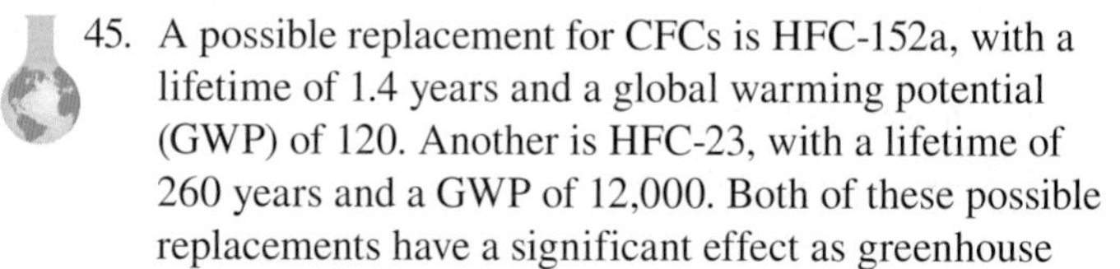

45. A possible replacement for CFCs is HFC-152a, with a lifetime of 1.4 years and a global warming potential (GWP) of 120. Another is HFC-23, with a lifetime of 260 years and a GWP of 12,000. Both of these possible replacements have a significant effect as greenhouse gases and are regulated under the Kyoto Protocol.
   a. Based on the given information, which appears to be the better replacement? Consider only the potential for global warming.
   b. What other considerations are there in choosing a replacement?

46. The emissions of CO_2 from fossil fuel burning can be reported in different ways. For example, the Carbon Dioxide Information Analysis Center (CDIAC) reported in 2009 that China, the United States, and India ranked highest among world nations:

Ranking	Nation	Metric tons of CO_2
#1	China (mainland)	2,096,295
#2	United States	1,445,204
#3	India	539,794

a. Would the rankings change if expressed on a per capita basis? If so, which nation would rank first?

b. CDIAC reports the per capita rankings on the basis of metric tons of carbon emitted, rather than metric tons of CO_2. Qatar leads the world in per capita emissions at 12.01 metric tons of carbon. Would this value be higher or lower if expressed on the basis of metric tons of CO_2 emitted? Explain.

47. Compare and contrast a cap-and-trade system with a carbon tax.

48. When Arrhenius first theorized the role of atmospheric greenhouses, he calculated that doubling the concentration of CO_2 would result in an increase of 5–6 °C in the average global temperature. How far off was he from the current IPCC modeling?

Exploring Extensions

49. Former vice president Al Gore writes in his 2006 book and film, *An Inconvenient Truth*: "We can no longer afford to view global warming as a political issue—rather, it is the biggest moral challenge facing our global civilization."

a. Do you believe that global warming is a moral issue? If so, why?

b. Do you believe that global warming is a political issue? If so, why?

50. China's growing economy is fueled largely by its dependence on coal, described as China's "double-edged sword." Coal is both the new economy's "black gold" and the "fragile environment's dark cloud."

a. What are some of the consequences of dependence on high-sulfur coal?

b. Sulfur pollution from China may slow global warming, but only temporarily. Explain.

c. What other country is rapidly stepping up its construction of coal-fired power plants and is expected to have a larger population than China by the year 2030?

51. The Quino checkerspot butterfly is an endangered species with a small range in northern Mexico and southern California. Evidence reported in 2003 indicates that the range of this species is even smaller than previously thought.

a. Propose an explanation why this species is being pushed north, out of Mexico.

b. Propose an explanation why this species is being pushed south, out of southern California.

c. Propose a plan to prevent further harm to this endangered species.

52. Data taken over time reveal an increase in CO_2 in the atmosphere. The large increase in the combustion of hydrocarbons since the Industrial Revolution is often cited as a reason for the increasing levels of CO_2. However, an increase in water vapor has *not* been observed during the same period. Remembering the general equation for the combustion of a hydrocarbon, does the difference in these two trends *disprove* any connection between human activities and global warming? Explain your reasoning.

53. In the energy industry, 1 standard cubic foot (SCF) of natural gas contains 1196 mol of methane (CH_4) at 15.6 °C (60 °F). Refer to Appendix 1 for conversion factors.

a. How many moles of CO_2 could be produced by the complete combustion of 1 SCF of natural gas?

b. How many kilograms of CO_2 could be produced?

c. How many metric tons of CO_2 could be produced?

54. An international conference on climate change was held in Paris, France, in November 2015. Write a brief summary of the outcomes of this conference.

55. A solar oven is a low-tech, low-cost device for focusing sunlight to cook food. How might solar ovens help mitigate global warming? Which regions of the world would benefit most from using this technology?

56. In 2005, the European Union adopted a cap-and-trade policy for carbon dioxide. Write a short report on the outcomes of this policy, both in terms of the economic result and the effect it has had on European greenhouse gas emissions.

CHAPTER 5

Energy from Combustion

REFLECTION

What Does It Take?

a. **How much fuel does it take to move you?**
Imagine you are going to take a road trip across the United States, from New York City to Los Angeles. The distance of this trip is 4,460 km and the vehicle you will use gets 30 miles per gallon of gasoline. How much gasoline would you need for this trip?

b. **How much alternative fuel does it take to move you?**
Now, imagine that the car you are using has the capability of also using an alternative, renewable fuel such as biodiesel or ethanol. Using ethanol is less efficient than using gasoline. In other words, when using ethanol, you can only achieve 20 miles per gallon. How many gallons of ethanol would you need to cover your entire trip?

c. **How much land does it take to make the corn used to produce the ethanol for your trip?**
Currently, the ethanol used for fuel is predominately made from corn. Assume it takes 26.1 pounds of corn to make one gallon of ethanol, and one acre of land to produce 7,110 pounds of corn. How much land is needed to produce sufficient fuel for your trip?

The Big Picture

In this chapter, you will explore the following questions:

- What are fuels?
- How is electricity generated?
- What are the positive and negative consequences of burning carbon-based fuels?
- How much heat is released when fuels are burned?
- What are biofuels, and what are the benefits of using them as a transportation fuel?

Introduction

Since the beginning of recorded history, fire has been central to our society as a source of heat, light, and security. Our modern fuels, the substances we burn or combust, are available in many different forms. We use coal in power plants to generate electricity. We use gasoline to run our cars. We use natural gas or heating oil to warm our homes. We use propane, charcoal, or wood to cook our food at a summer barbecue. We might even use wax to provide light for a romantic candlelit dinner. In each of these cases, *using* fuels means *burning* them. This process of combustion causes a difference in energy between the reactants and products of combustion, and that energy is released as light and heat—a flame!

In this chapter, we will describe fuels and their characteristics, including promising new fuels to help us meet our future energy needs. We begin by describing the properties of fuels and what happens when fuels are burned. We then take you inside a power plant to illustrate how the combustion of fuels is converted to electricity. Since

there are many types of fuels, we will also describe how their varying compositions affect not only heat output, but also the gaseous products generated from their combustion. Let's begin our analysis by answering the following questions: What is a fuel, and what happens when it burns?

5.1 Fossil Fuels: A Prehistoric Fill-Up at the Gas Station

A **fuel** is any solid, liquid, or gaseous substance that may be combusted (burned) to produce heat or work. Sources of fuel date back to prehistoric times, where solids such as grass and straw were burned for heat. The use of coal as a fuel actually dates back to ancient civilizations, where it was used to isolate copper from ore in northeastern China as early as 1000 BC. However, the Industrial Revolution in the late 18th century sparked the large-scale use of coal for steam engines and steelmaking. The development of drilling technology for oil wells in the mid-19th century in the U.S. gave rise to the petroleum industry and mass consumption of petroleum products for transportation, electricity, heating, and even plastics fabrication. Currently, the world's energy needs are provided by burning fossil fuels, coal, and/or oil. With so many choices of fuels available to us, what are some desirable properties of fuels?

A whimsical electric meter cover in Ithaca, New York, illustrating the common notion that fossil fuels were derived from dinosaurs.

© Jamie Ellis, Ithaca College

Your Turn 5.1 You Decide The "Best" Fuel?

Consider the various types of fuels that have been used since ancient times.

a. What properties make a particular fuel desirable or undesirable?
b. What makes a good match between a fuel and its intended use?
c. With the above considerations in mind, propose some reasons why industrialized nations shifted from using wood to coal, and then eventually to petroleum and natural gas as the chief sources of fuel?
d. What fuel properties will be needed to take the place of petroleum in the future?

As you may have seen in the previous activity, a fuel is considered valuable if it ignites easily at a low temperature and produces a large quantity of heat during its combustion. In addition, fuels should be inexpensively isolated and have properties that allow for their safe and efficient storage or transport. Lastly, a desirable fuel should leave little residue behind after being burned, and produce by-products that are not harmful to human health or the environment. Unfortunately, no fuel satisfies all of these conditions.

Humans currently use coal, petroleum products (*e.g.*, gasoline, diesel, propane, etc.), and natural gas as our primary sources of fuel. Contrary to popular belief, these so-called **fossil fuels** are not the prehistoric remains of dinosaurs. In fact, most of the fossil fuels we use today were formed from decaying plant life that flourished millions of years before the first dinosaurs appeared.

The process of photosynthesis exhibited by green plants, including primeval plant life, involves the capture of sunlight to produce glucose and oxygen from carbon dioxide and water:

$$6\ CO_2(g) + 6\ H_2O(l) \longrightarrow C_6H_{12}O_6(aq) + 6\ O_2(g) \qquad [5.1]$$

Did You Know? The discovery of intact dinosaur fossils during oil and natural gas exploration—some found at the top of a coal or oil field—is evidence that dinosaurs were not the primary source of fossil fuels.

The process of decaying matter releases energy and reverses this process, producing $CO_2(g)$ and $H_2O(l)$. However, under certain conditions, the carbon-containing compounds that make up the organism only *partially* decompose. In the prehistoric past, vast quantities of plant and animal life became buried beneath layers of sediment in swamps or at the bottom of the oceans. Oxygen failed to reach the decaying material, thus showing the decomposition process. The temperature and pressure increased as additional layers of mud and rock covered the buried remnants, causing additional chemical reactions to occur. Over time, the plants that once captured the Sun's rays were transformed into the substances we call coal, petroleum, and natural gas. In a very real sense, these fossils may be thought of as ancient solar energy (sunshine) stored in solid, liquid, and gaseous states.

Your Turn 5.2 Scientific Practices Steamy Compost

Want to recycle and reuse plant and animal material? Start a compost pile. Under the right weather conditions, steam can be seen rising from a pile of compost. Explain this observation.

Yes, today's plant life will become tomorrow's fossil fuels. But this will not occur in a time frame useful to humans. It is staggering to realize that the amount of fossil fuels we will have consumed in just a few centuries took nature several millennia to produce. Over geological periods of time—millions and millions of years—changes in temperature and pressure transformed dead plant and animal life into valuable forms, such as coal or crude oil.

Considering how long it takes to convert plant life to fossil fuels, the rate at which we are burning coal, petroleum, and natural gas is not sustainable, at least in terms of having enough of it available to meet current and future energy needs. Perhaps you are somewhat skeptical of this claim. Fossil fuels may appear to be in adequate supply, because new deposits are always being found and extraction technologies are continually improving to capture more of it. But even if the supply of fossil fuels is infinite (it is not), sustainability involves more than just availability.

Your Turn 5.3 You Decide Where are the Fossil Fuels?

Consult the interactive trends found in the International Energy Data link on the U.S. Energy Information Administration (EIA) website.

a. Which countries are the world's largest producers of coal, petroleum, and natural gas?
b. Do these countries also have the greatest reserves of these fossil fuels?
c. Propose reasons why the rankings for **a** and **b** are not identical. For instance, Canada is ranked #3 in petroleum resources, but only has a production ranking of #5; in contrast, the United States is ranked #11 for world petroleum reserves, but is the #1 producer.
d. Have the trends in production and reserve capacity of particular countries changed over the last 10 years? Propose some socioeconomic factors that may explain these trends.

Your Turn 5.4 You Decide Will We Run Out of Fossil Fuels?

Using a variety of Internet sources, predict some likely time frames that will exhaust our available supply of nonrenewable coal, petroleum, and natural gas. What factors may worsen or improve these predictions?

Burning fossil fuels for energy fails to meet the criteria of sustainability in two ways. First, the fuels themselves are nonrenewable. Once gone, they cannot be replaced—at least within a useful timescale. Second, the waste products of combustion have adverse effects on our environment, both today and in the future. Chapter 4 described how the concentration of atmospheric $CO_2(g)$—a greenhouse gas—has risen dramatically since the beginning of the Industrial Revolution. These increases will continue to affect our climate for generations to come. Burning coal also releases pollutants such as soot, carbon monoxide, mercury, and the oxides of sulfur and nitrogen. These emissions affect us right now because they lower the quality of our air, acidify our rain, aggravate existing health conditions, and accelerate global climate change. Before we consider what we can do about this, let's first consider what happens to a fossil fuel when it is burned and why it provides so much of our current energy needs.

5.2 Burn, Baby! Burn! The Process of Combustion

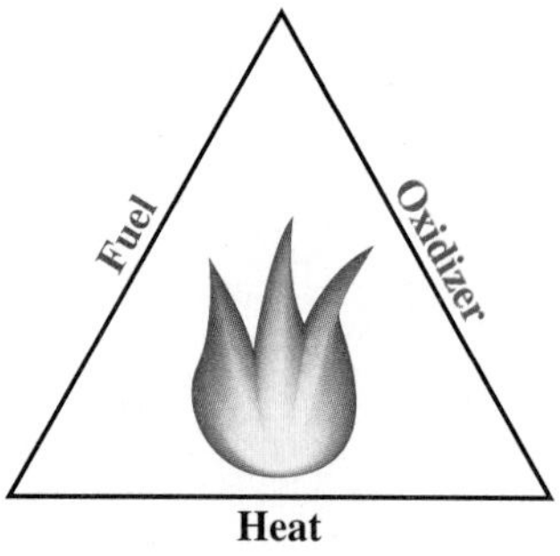

Figure 5.1

The fire triangle, illustrating the three components required for combustion.

There are three necessary requirements to generate a fire—a source of heat, a fuel, and an oxidizer (Figure 5.1). When these components are combined, a chemical reaction takes place that releases a variety of by-products and a significant amount of heat. Once a fire is generated, the heat or ignition source is no longer needed. The fire will continue to burn until either the oxygen or fuel source is removed. For instance, fire blankets are used to extinguish a fire by preventing available oxygen from reacting with the fuel source.

In a sealed environment, the available oxygen in the room will eventually be consumed by reacting with the fuel, and the fire will subside. However, inadvertently opening a door and exposing the dying fire to a fresh source of oxygen results in a dangerously explosive situation, referred to as a *backdraft.*

The oxygen ($O_2(g)$) in our atmosphere is the most typical oxidizing agent; however, other sources may be used, such as ozone ($O_3(g)$), hydrogen peroxide ($H_2O_2(l)$), nitrous oxide ($N_2O(g)$), or many other oxygen-containing compounds. Although much less common, it is also possible to have a combustion reaction with a halogen serving as the oxidizing agent, such as fluorine ($F_2(g)$), chlorine ($Cl_2(g)$), bromine ($Br_2(l)$), or even halogen-containing compounds.

Regardless of the specific source of fuel or oxidizing agent, the general chemical reaction is the same (Equation 5.2). As in all chemical reactions, the species on the left-hand side of the chemical equation are referred to as **reactants** (*i.e.,* fuel and oxidizer molecules for a combustion reaction), whereas those appearing on the right-hand side are known as **products**:

$$\text{Fuel} + \text{Oxidizer} \xrightarrow{heat} \text{Products} \qquad [5.2]$$

The identity of the products will differ, depending on the fuel and oxidizer used for combustion. However, the chemical makeup of these products is rarely straightforward.

The great majority of fuels are **hydrocarbons**, compounds made up only of the elements hydrogen and carbon. A few basic rules for chemical bonding can help you create order out of the seeming chaos of hydrocarbons. One is the **octet rule** introduced in Chapter 3. The C atoms in hydrocarbons form chemical bonds that share eight valence electrons. For example, in methane (CH_4), the primary component of natural gas, the central C atom is surrounded by eight electrons that are arranged to form four covalent bonds:

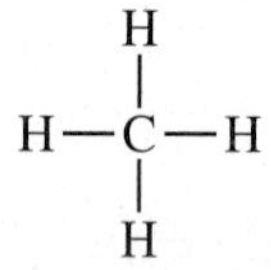

```
    H
    |
H — C — H
    |
    H
```

Another useful rule is that carbon forms four bonds in hydrocarbons. One possibility is four single bonds, as in methane. Another possibility is one double and two single bonds around each carbon, again for a total of four bonds, as in ethene (also known as *ethylene*):

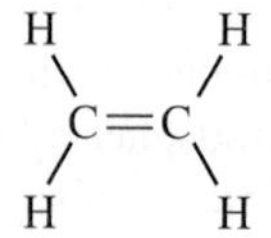

```
H       H
 \     /
  C = C
 /     \
H       H
```

As we mentioned in Chapter 1, chemical formulas, such as CH_4, C_2H_4, or C_8H_{18}, indicate the kinds and numbers of atoms present in a molecule, but do not show how the atoms are connected. To get this level of detail, you need a **structural formula**. For example, the structural formula for *n*-butane, C_4H_{10}, a hydrocarbon used to fuel lighters and camp stoves, is shown below. The *n* in the chemical name stands for *normal*, meaning that the carbon atoms are in a straight chain:

```
    H   H   H   H
    |   |   |   |
H — C — C — C — C — H
    |   |   |   |
    H   H   H   H
```

As we discussed in Section 4.7, each of the carbon atoms features a tetrahedral arrangement of C–H and C–C bonds.

A drawback to structural formulas is that they take up a lot of space on the page. To convey the same information more compactly, **condensed structural formulas** are often used in which some bonds are not shown. In these forms, the structural formula is understood to contain an appropriate number of bonds. Here are two condensed structural formulas for *n*-butane, the second more "condensed" than the first:

$$CH_3—CH_2—CH_2—CH_3 \qquad CH_3CH_2CH_2CH_3$$

Although the H atoms in these structures appear to be part of the chain of C atoms, it is understood that they are not, and instead are connected to the prior carbon atom.

If a hydrocarbon fuel is burned, the only products that will be generated upon complete combustion are carbon dioxide and water vapor. As an example, consider the combustion of *propane*, C_3H_8:

The names of simple hydrocarbons were provided in Section 2.7.

$$C_3H_8(g) + 5\ O_2(g) \longrightarrow 3\ CO_2(g) + 4\ H_2O(g) \qquad \textbf{[5.3]}$$

However, there may also be some carbon monoxide ($CO(g)$) and/or carbon soot generated due to *incomplete* combustion. For combustion of propane, this would occur if the molar ratio of O_2:C_3H_8 would be less than 5:1, referred to as oxygen-deficient conditions.

Propane, coal, and most other fuels also contain sulfur, which results in sulfur oxide emissions (SO_x, where x = 2 or 3). Furthermore, nitrogen oxides (NO_x, where x = 1 or 2) are often generated through the high-temperature reaction of $N_2(g)$ and $O_2(g)$ during combustion. As we will see throughout this chapter, a great diversity of products are possible from the combustion of a fuel.

Recall the role of NO_x emissions in the formation of harmful ground-level ozone, described in Chapter 2.

Your Turn 5.5 Skill Building Practice with Combustion Reactions

For each of the fuels below, write the balanced combustion reaction.

a. Glucose, sugar ($C_6H_{12}O_6$)
b. Methane, natural gas (CH_4)
c. Butane, fuel in lighters (C_4H_{10})

Hint: Go back to Chapter 2 for information on balancing equations.

Your Turn 5.6 Scientific Practices Combustion in Air

The balanced equation presented in **Equation 5.3** describes the combustion of propane in pure oxygen. However, this combustion in real life occurs in air, which is predominantly composed of nitrogen. How would you write a balanced equation to illustrate the combustion of propane in air? Assume that the nitrogen is *inert*—that is, it doesn't react with propane or oxygen.

5.3 | What Is "Energy"?

Creating fire was a critical discovery that promoted the survival of our early Paleolithic ancestors. The energy generated from burning fuels keeps us warm and powers our automobiles, lights, and appliances. In Chapter 3, you explored a form of energy related to UV radiation and its effect on human health. However, what exactly is energy, and where does it come from?

Energy is a fundamental property of our universe, referring to the ability or capacity of matter to do work or to produce change. For instance, the combustion of fuel in a car drives an engine that turns the wheels. Alternatively, energy is given off as heat and light when fireworks are ignited, or from nuclear reactions occurring in the Sun.

Electromagnetic (EM) radiation results from the movement of electric and magnetic fields. Hence, it is best classified as a form of kinetic energy.

Although there are many different forms of energy, there are only two general types. **Kinetic energy** is the energy of motion, which includes the movement of atoms and molecules, as well as our own activities such as walking, climbing, or running. In contrast, **potential energy** is stored energy, or the energy of position; that is, position with respect to another object. Because different fuels have different compositions and arrangements of atoms, they hold different amounts of potential energy. The magnitude of this energy is very much dependent on the types of bonds present in the fuel molecules. During combustion reactions, the relative positions of the atoms and molecules in the fuel change, which alters its potential energy.

Forces in nature and energy are closely related. A bicycle at the top of a hill has more potential energy than the same bicycle when it reaches the bottom of the hill because of its change in position. As it moves from the top to the bottom of the hill, the gravitational force exerted on it changes. At the particle or molecular level, the amounts of energy are connected to electrostatic forces—the attractions and repulsions of charged particles (*e.g.*, protons, electrons, and ions). Most often, atoms and molecules are driven toward a low-energy state, which is characterized by a particular balance between attractive and repulsive forces. Even when we describe a covalent bond as the sharing of electrons between two atoms, we still recognize that there is a certain balance of attractions and repulsions of the charged subatomic particles "holding" the atoms together (Figure 5.2). During combustion, these interactions are disrupted and new interactions occur, which changes the energy states of the components. If the products are of a lower energy than the reactants, this energy difference must go somewhere and is transformed into heat and light energy—*voilà*, a flame!

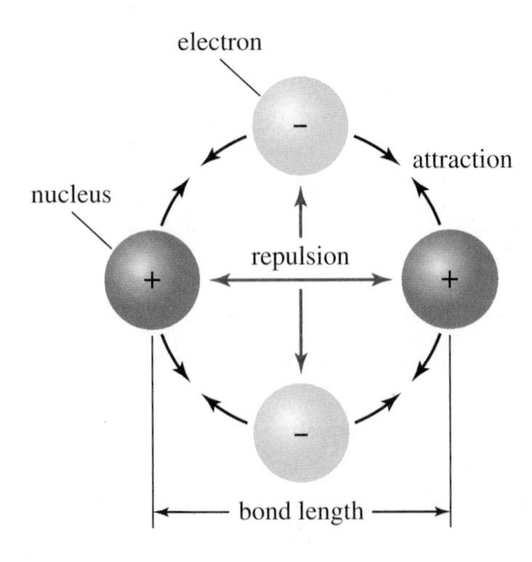

Figure 5.2

Illustration of a covalent bond as the optimized balance of attractive and repulsive forces between neighboring atoms.

Your Turn 5.7 Skill Building Amounts of Potential Energy

For each reaction below, decide whether the reactants or products have more potential energy.

a. The detonation of TNT, an explosive.
b. The synthesis of hydrazine, a rocket fuel.
c. The decomposition of a dead plant material.
d. The building of new muscle in the human body.

Heat is the kinetic energy that flows from a hotter object to a colder one. When two bodies are in contact, heat always flows from the object at the higher temperature to one at a lower temperature. **Temperature** is a measure of the average kinetic energy of the atoms and/or molecules present in a substance. Everything around us is at some temperature—hot, cold, or lukewarm. An object is "cold" when its atoms and molecules move more slowly, on average, relative to an object that is "warm." Therefore, for the temperature of an object to increase, the kinetic energy of its atoms and molecules must increase.

Heat cannot be measured directly. However, changes in heat can usually be detected as changes in temperature.

Although the concepts of temperature and heat are related, they are not identical. Your bottle of water and the Pacific Ocean may be at the same temperature, but the ocean contains and can transfer far more heat than the bottle of water. This is because the number of water molecules in the Pacific Ocean is exceedingly larger than in the bottle of water. Indeed, bodies of water can affect the climate of an entire region as a consequence of their ability to absorb and transfer large amounts of heat.

5.4 How Hot Is "Hot"? Measuring Energy Changes

The ability of a substance to provide heat energy makes it a good fuel. This section will describe how we measure and express the energy released from the combustion of a fuel. The **calorie** *(cal)* was introduced with the metric system in the late 18th century and was defined as the amount of heat necessary to raise the temperature of one gram of water by one degree Celsius. When Calorie is capitalized, as in a nutritional Calorie, it generally means kilocalorie on the metric scale. The values tabulated on food package labels and in cookbooks are, in fact, kilocalories:

$$1 \text{ kilocalorie (kcal)} = 1000 \text{ calories (cal)} = 1 \text{ Calorie (Cal)}$$

The modern system of units uses the **joule** *(J)*, a unit of energy equal to 239 cal (or 0.239 Cal). One joule (1 J) is approximately equal to the energy required to raise a 100-g apple to a height of 1 m against the force of gravity. In terms of calorie equivalents, 1 cal = 4.184 J. On a more personal basis, each beat of the human heart requires about 1 J of energy. Figure 5.3 provides a contextual comparison of various energy magnitudes, in terms of joules.

Your Turn 5.8 Skill Building Energy Calculations

a. A slice of pizza contains 217 kcal (217 Cal). Express this value in kilojoules.
b. Calculate the number of 1-kg books you could lift to a shelf 2 m off the floor with the amount of energy from metabolizing one slice of pizza.

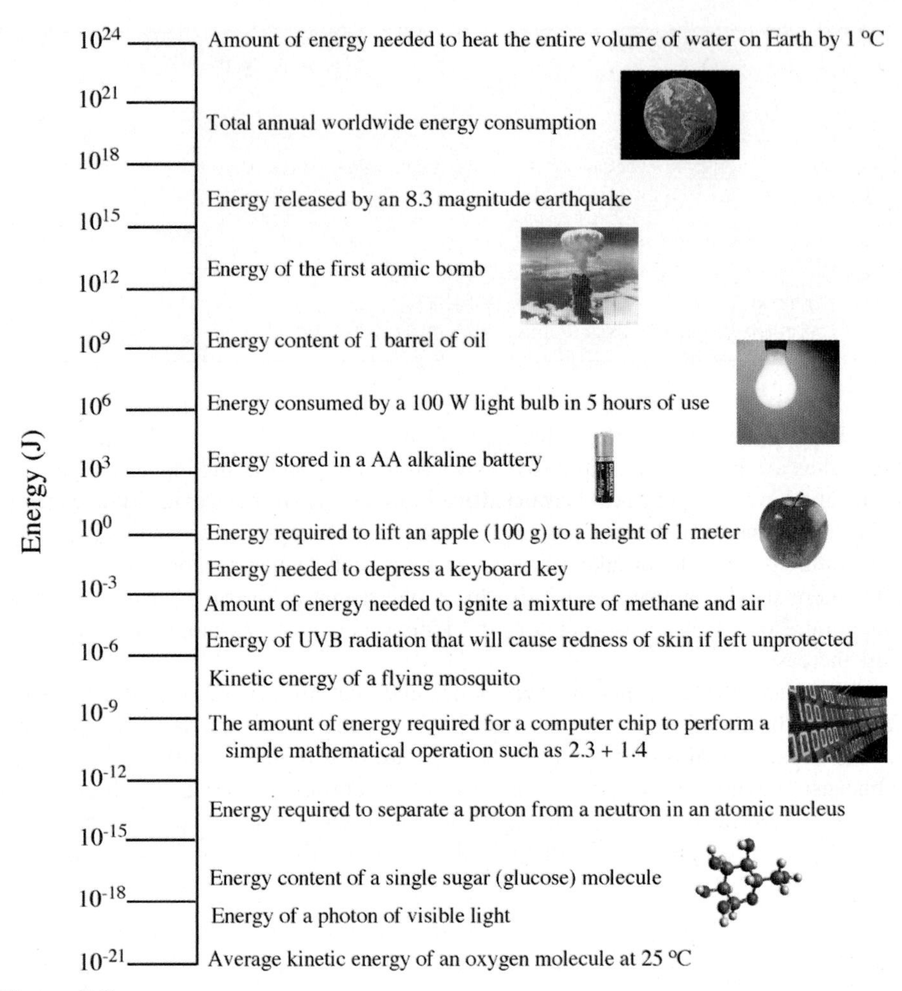

Figure 5.3

A contextual comparison of various energy magnitudes.

(Earth): © Studio Photogram/Alamy RF; (Bomb): Source Library of Congress Prints & Photographs Division [LC-USZ62-36452]; (Bulb): © Ingram Publishing/SuperStock RF; (Battery): © Jeffrey B. Banke/Shutterstock.com; (Apple): © lynx/iconotec.com/Glow Images RF; (Binary): © Mmaxer/Shutterstock.com

So, how is food a fuel? After all, we don't combust when we eat! The process of combustion is actually an oxidation process—a reaction with oxygen. Carbohydrates (sugars) and fats are categories of biomolecules that provide our bodies with energy. They do so because when they react with oxygen, their products have less potential energy than the reactants. This energy difference is both transformed into usable energy in the body, and dissipated as heat throughout the body.

Your Turn 5.9 You Decide Checking Assumptions

A simplifying assumption was made in doing the calculations in part **b** of the preceding activity.

a. What was the assumption, and is it reasonable?

b. Based on this assumption, is your answer too high or too low? Explain your reasoning.

The **calorimeter** is a device used to experimentally measure the quantity of heat energy released in a combustion reaction. Figure 5.4 shows a schematic representation of a calorimeter. To use it, you introduce a known mass of fuel and an excess of oxygen into the heavy-walled stainless steel container. The container is then sealed and submerged in a bucket of water. The reaction is initiated with a

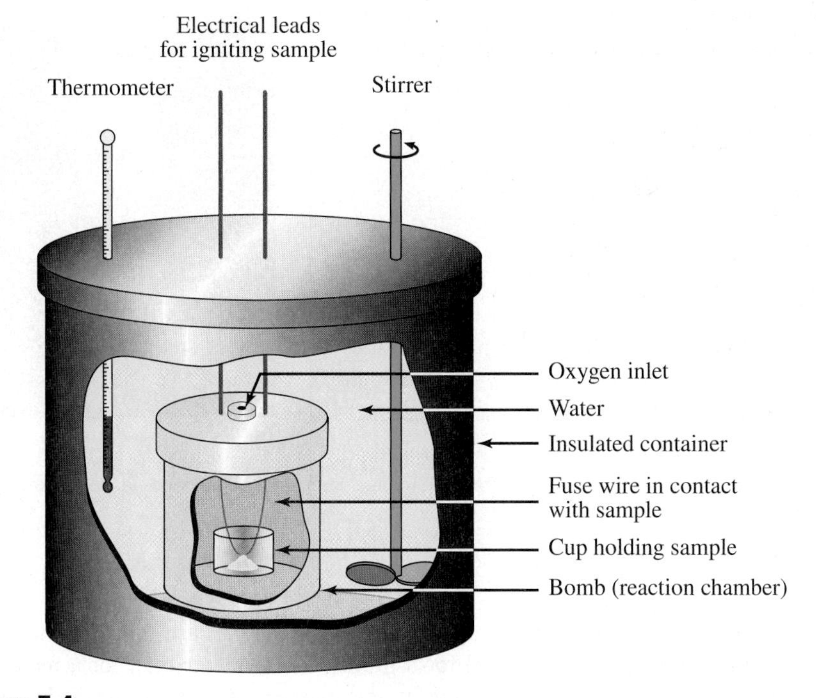

Figure 5.4

Schematic drawing of a calorimeter. The reaction chamber is often referred to as a "bomb" due to large pressures that are created inside a sealed container during combustion.

spark, and the heat evolved by the reaction flows from the container to the water and the rest of the apparatus. As a consequence, the temperature of the entire calorimeter system increases. The quantity of heat given off by the reaction can be calculated from this temperature rise, by using the known heat-absorbing properties of the calorimeter and the amount of water it contains. The greater the temperature increase (measured in °C), the greater the quantity of energy evolved from the reaction (measured in J).

Experimental measurements of this sort are the source of most of the tabulated values of heats of combustion. As the name suggests, the **heat of combustion** is the quantity of heat energy given off when a specified amount of a substance burns in oxygen. Heats of combustion are typically reported in units of kilojoules per mole (kJ/mol), kilojoules per gram (kJ/g), kilocalories per mole (kcal/mol), or kilocalories per gram (kcal/g). For example, the experimentally determined heat of combustion of methane is 802.3 kJ/mol. This means that 802.3 kJ of heat is given off when 1 mole of $CH_4(g)$ reacts with 2 moles of $O_2(g)$ to form 1 mole of $CO_2(g)$ and 2 moles of $H_2O(g)$:

$$CH_4(g) + 2\ O_2(g) \longrightarrow CO_2(g) + 2\ H_2O(g) + 802.3\ \text{kJ} \qquad \textbf{[5.4]}$$

Burning methane is analogous to water flowing from the top of a waterfall. Initially in a state of higher potential energy, the water drops down to one of lower potential energy. The potential energy is converted into kinetic energy, which is then released when the water hits the rocks below. Similarly, when methane is burned, energy is released when the atoms from the reactants change their interactions and "fall" to a state of lower potential energy as they are transformed into products. Figure 5.5 is a schematic representation of this process. The downward arrow indicates that the energy associated with 1 mole of $CO_2(g)$ and 2 moles of $H_2O(g)$ is less than the energy associated with 1 mole of $CH_4(g)$ and 2 moles of $O_2(g)$.

The combustion of methane is **exothermic**, a term applied to any chemical or physical change accompanied by the release of heat. In this reaction, the energy difference is −802.3 kJ. The negative sign attached to the energy change for all exothermic reactions signifies the decrease in potential energy going from reactants to products. Not surprisingly, the total amount of energy released depends on the amount of fuel burned.

For an exothermic reaction, one can consider the evolved heat as a product, as shown in Equation 5.4.

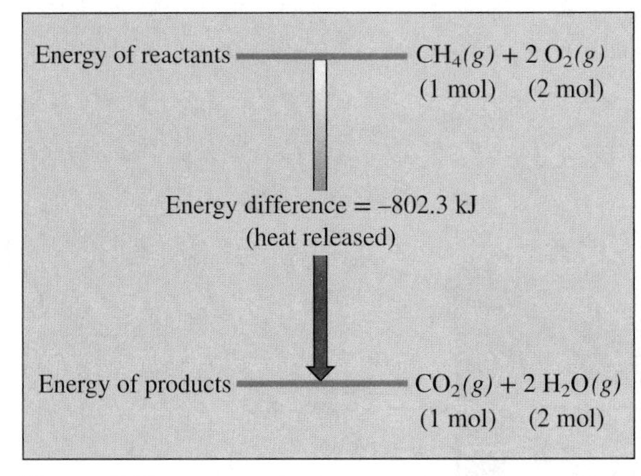

Figure 5.5

Comparison of a waterfall to the energy difference in the combustion of methane—an exothermic reaction.

© Ingram Publishing/SuperStock RF

For a more straightforward comparison of fuels, we can use this value to calculate the number of kilojoules released for a gram of fuel, rather than for a mole. The molar mass of CH_4, calculated from the atomic masses of carbon and hydrogen, is 16.0 g/mol. We then can calculate the heat of combustion per gram of methane:

$$\frac{802.3\ \text{kJ}}{1\ \cancel{\text{mol CH}_4}} \times \frac{1\ \cancel{\text{mol CH}_4}}{16.0\ \text{g CH}_4} = 50.1\ \text{kJ/g CH}_4$$

As fuels go, this is a high heat of combustion! In terms of heat energy released during combustion, the more energy released, the better the fuel. Figure 5.6 compares the energy differences (in kJ/g) of several different fuels. We can make some interesting generalizations based on the chemical formulas of the fuels. First, the fuels with the highest heats of combustion are hydrocarbons. Second, as the ratio of hydrogen-to-carbon decreases, the heat of combustion decreases. And third, as the amount of oxygen in the fuel molecule increases, the heat of combustion decreases.

Nuclear fuels will be discussed in Chapter 6.

It is no coincidence that the molecular structures of the most viable fuels, other than uranium-containing nuclear fuels, are composed largely of carbon and hydrogen atoms. These atoms are easily oxidized to form stable (low potential energy) carbon dioxide and water molecules, resulting in an overall release of energy from the reacting fuel and oxygen.

methane
$CH_4 + 2\ O_2$
Energy difference = –50.1 kJ/g
$CO_2 + 2\ H_2O$

octane
$C_8H_{18} + 25/2\ O_2$
Energy difference = –44.4 kJ/g
$8\ CO_2 + 9\ H_2O$

coal
$C + O_2$
Energy difference = –32.8 kJ/g
CO_2

ethanol
$C_2H_5OH + 3\ O_2$
Energy difference = –26.8 kJ/g
$2\ CO_2 + 3\ H_2O$

wood (glucose)
$C_6H_{12}O_6 + 6\ O_2$
Energy difference = –14.1 kJ/g
$6\ CO_2 + 6\ H_2O$

Figure 5.6

Energy differences (in kJ/g) for the combustion of methane (CH_4), *n*-octane (C_8H_{18}), coal (assumed to be pure carbon), ethanol (C_2H_5OH), and wood (presumed to be glucose). Carbon dioxide and water are formed in the gas phase.

Your Turn 5.10 Scientific Practices Coal Versus Ethanol

On the basis of their chemical composition, explain why ethanol and coal have very different chemical formulas but similar heats of combustion.

Not all chemical reactions are exothermic; some reactions *absorb* energy as they occur. We discussed two important examples in earlier chapters. One is the decomposition of O_3 to yield O_2 and O, which requires the input of energy in the form of high-energy (UVB and UVC) photons. The other reaction is the combination of N_2 and O_2 to yield two molecules of NO, which requires a high-temperature environment. Both of these reactions are **endothermic**, the term applied to any chemical or physical change that absorbs energy and creates products of a higher potential energy state. A chemical reaction is endothermic when the potential energy of the products is *higher* than that of the reactants. The convention for representing an endothermic reaction is to place a positive sign (or no sign at all) in front of the energy value and unit; for example, +29 J means that the reaction has absorbed 29 J of energy.

For an endothermic reaction, one can consider the absorbed heat as a reactant, as shown in Equation 5.5.

Photosynthesis is endothermic. This process requires the absorption of 2800 kJ of sunlight per mole of glucose ($C_6H_{12}O_6$) formed, or 15.5 kJ per gram. The complete process involves many steps, but the overall reaction can be described with this equation:

$$2800\text{ kJ} + 6\,CO_2(g) + 6\,H_2O(l) \xrightarrow{\text{chlorophyll}} \underset{\text{glucose}}{C_6H_{12}O_6(s)} + 6\,O_2(g) \qquad [5.5]$$

This reaction requires the participation of the green pigment chlorophyll. The chlorophyll molecule absorbs energy from the photons of visible sunlight and uses this energy to drive the photosynthetic process, an energetically uphill reaction. Photosynthesis plays an essential role in the carbon cycle, because it removes CO_2 from the atmosphere.

A general summary of endothermic and exothermic reactions is as follows (Figure 5.7):

i) Heat added to reactants > heat evolved by formation of products: *endothermic* (*e.g.,* baking bread, producing sugar by photosynthesis)

ii) Heat added to reactants < heat evolved with products: *exothermic* (*e.g.,* combustion of fuels)

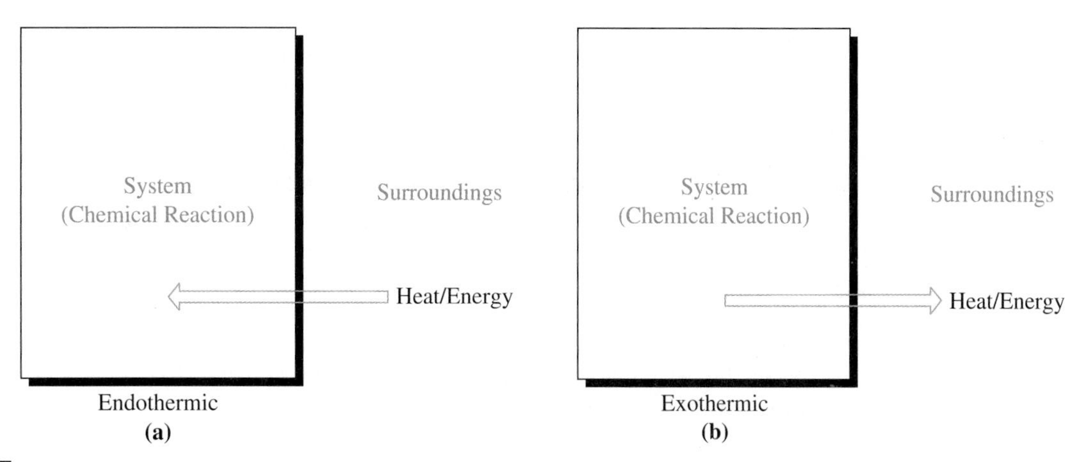

Figure 5.7

An illustration of the difference between an **(a)** endothermic process, and **(b)** exothermic process. The term *system* is used to denote the reaction taking place, whereas the surroundings are everything else outside the reactants (glass, countertop, room, etc.)

Your Turn 5.11 Scientific Practices Do-It-Yourself Hot and Cold Packs

You hurt your ankle while jogging in your favorite 5K race. Luckily, your friend comes to the rescue with a room temperature pack from the first aid closet; one snap and you have a soothing cold pack. But, how does this work? Therapeutic hot and cold packs sold in pharmacies or supermarkets consist of isolated compartments containing water and a salt. Once the divider between the two compartments is broken, the salt and water are allowed to mix and the pack gets either hot or cold.

The term "salt" does not always refer to table salt, NaCl, but to an entire class of ionic compounds that completely dissolve in water.

a. Explain which type of reaction (exothermic or endothermic) would be needed to make a hot pack and cold pack?

b. Obtain a sample of as many of the following salts as possible:

- calcium chloride ($CaCl_2$—available at hardware or retail stores; salt used as a sidewalk de-icer)
- water softener salt (mostly NaCl—available at hardware or retail stores)
- sodium chloride (NaCl—available at grocery stores; ordinary table salt)
- ammonium chloride (NH_4Cl—available at hardware, retail, or landscaping stores; active ingredient in 34-0-0 fertilizer)
- potassium chloride (KCl—known as "Morton Lite," available at hardware or retail stores)
- sodium bicarbonate ($NaHCO_3$—known as baking soda, available at grocery or retail stores)

34-0-0 fertilizer may not be widely available due to local regulations. Goggles should be worn to protect your eyes from chemical spills or splashes.

c. Place 50 mL of water into separate Styrofoam™ cups (one for each salt), and record the initial temperature of the water using a thermometer. Record the temperature changes that occur when 1 tablespoon of a salt is dissolved in water.

d. For those reactions in which a temperature change is observed, which correspond to endothermic processes, and which correspond to exothermic processes?

e. Which salts would be the most effective choices for hot and cold packs? Are there any other factors that should be considered in making your final decisions?

5.5 Hyperactive Fuels: How Is Energy Released during Combustion?

For a review of covalent bonds see Section 3.7.

As you have seen in previous chapters, molecular compounds are composed of atoms that are bonded together by covalent bonds. Chemical reactions involve the breaking and forming of these bonds. Energy is required to break bonds, just as energy is required to break chains or tear paper. In contrast, forming chemical bonds releases energy. The overall energy change associated with a chemical reaction depends on the net difference of the energy needed to break bonds, and the energy released when bonds form.

For example, consider the combustion of hydrogen. Hydrogen is desirable as a fuel because, compared with other fuels, it releases a large amount of energy when it burns:

$$2\ H_2(g) + O_2(g) \longrightarrow 2\ H_2O(g) \qquad 249\ \text{kJ/mol or } 125\ \text{kJ/g} \qquad [5.6]$$

To calculate the energy released from the combustion of hydrogen to form water vapor, let us assume that all the bonds in the reactant molecules are broken, and then the individual atoms are recombined to form the products. In fact, the reaction does not occur this way, but we are interested in only the relative states of reactants and products, not the mechanistic details.

The covalent bond energies given in Table 5.1 provide the numbers needed to compute the energy difference between reactants and products. **Bond energy** is the amount of energy that must be absorbed to break a specific chemical bond. Since energy must be absorbed, breaking bonds is an endothermic process, and all the bond energies in Table 5.1 are positive. The values are expressed in kJ/mol of bonds broken.

Table 5.1 Covalent Bond Energies (in kJ/mol)

	H	C	N	O	S	F	Cl	Br	I
Single Bonds									
H	436								
C	416	356							
N	391	285	160						
O	467	336	201	146					
S	347	272	—	—	226				
F	566	485	272	190	326	158			
Cl	431	327	193	205	255	255	242		
Br	366	285	—	234	213	—	217	193	
I	299	213	—	201	—	—	209	180	151
Multiple Bonds									
C=C	598			C=N	616		C=O*	803	
C≡C	813			C≡N	866		C≡O	1073	
N=N	418			O=O	498				
N≡N	946								

*In CO_2

Note that the atoms appear both across the top and down the left side of the table. The number at the intersection of any row and column is the energy (in kJ) needed to break a mole of the covalent bonds between the two atoms. For example, the energy required to break 1 mole of C–H bonds is 416 kJ. Similarly, the energy to break 1 mole of N≡N triple bonds is 946 kJ, not three times 160 kJ.

Your Turn 5.12 You Decide O_3 Versus O_2

As noted in **Chapter 3**, ozone absorbs UV radiation of wavelengths less than 320 nm, while oxygen requires higher-energy electromagnetic radiation with wavelengths less than 242 nm. Use the bond energies in **Table 5.1** plus information about the resonance structures of O_3 from **Chapter 3** to explain this difference.

To determine whether the overall reaction is endothermic or exothermic, we need to keep track of whether energy is absorbed or released. To do this, we indicate when energy is absorbed with a positive sign. This is the energy absorbed when the bond is broken. Forming a bond releases energy, and the sign is negative. For example, when 1 mole of O=O double bonds is broken, the energy change is +498 kJ, and when 1 mole of O=O double bonds is formed, the energy change is −498 kJ.

Now we are finally ready to apply these concepts and conventions to the burning of hydrogen gas, $H_2(g)$. The next equation shows the Lewis structures of the species involved so that we can count the bonds that need to be broken and formed:

Lewis structures include all lone pairs, whereas structural formulas typically omit these non-bonding electrons.

$$2\,\mathrm{H{-}H} + \ddot{\underset{\cdot\cdot}{\mathrm{O}}}{=}\ddot{\underset{\cdot\cdot}{\mathrm{O}}} \longrightarrow 2\,\mathrm{H{-}\ddot{\underset{\cdot\cdot}{O}}{-}H} \quad [5.7]$$

Remember that chemical equations can be read in terms of moles. Both Equations 5.6 and 5.7 indicate "2 moles of H_2 plus 1 mole of O_2 yields 2 moles of H_2O."

To use bond energies, we need to count the number of moles of bonds involved. Here is a summary:

Molecule	Bonds per Molecule	Moles in Reaction	Moles of Bonds	Bond Process	Energy per Bond	Total Energy
H–H	1	2	1 × 2 = 2	breaking	+436 kJ	2 × (+436 kJ) = +872 kJ
O=O	1	1	1 × 1 = 1	breaking	+498 kJ	1 × (+498 kJ) = +498 kJ
H–O–H	2	2	2 × 2 = 4	forming	−467 kJ	4 × (−467 kJ) = −1868 kJ
					Total:	**−498 kJ**

From the last column, we can see that the overall energy change in breaking bonds (872 kJ + 498 kJ = 1370 kJ) and forming new ones (−1868 kJ) results in a net energy change of −498 kJ.

This calculation is diagrammed in Figure 5.8. The energy of the reactants, two H_2 molecules and one O_2 molecule, is set at zero—an arbitrary but convenient value. The green arrows pointing upward signify energy absorbed to break the bonds in the reactant molecules and form four H atoms and two O atoms. The red arrow on the right pointing downward represents energy released as these atoms bond to form the product H_2O molecules. The shorter red arrow corresponds to the net energy change of −498 kJ, signifying that the overall combustion reaction is strongly exothermic. The products are lower in energy than the reactants, so the energy change is negative. The net result is the release of energy, mostly in the form of heat.

The energy change we just calculated from bond energies, −498 kJ for burning 2 mol of hydrogen, compares favorably with the experimentally determined value when all of the species are gases. This agreement justifies the use of our rather unrealistic model of analysis: that all the bonds in the reactant molecules are first broken and then all the bonds in the product molecules are formed. The energy change that accompanies a chemical reaction depends only on the energy *difference* between the products and the reactants, not on the particular process, mechanism, or individual steps that connect

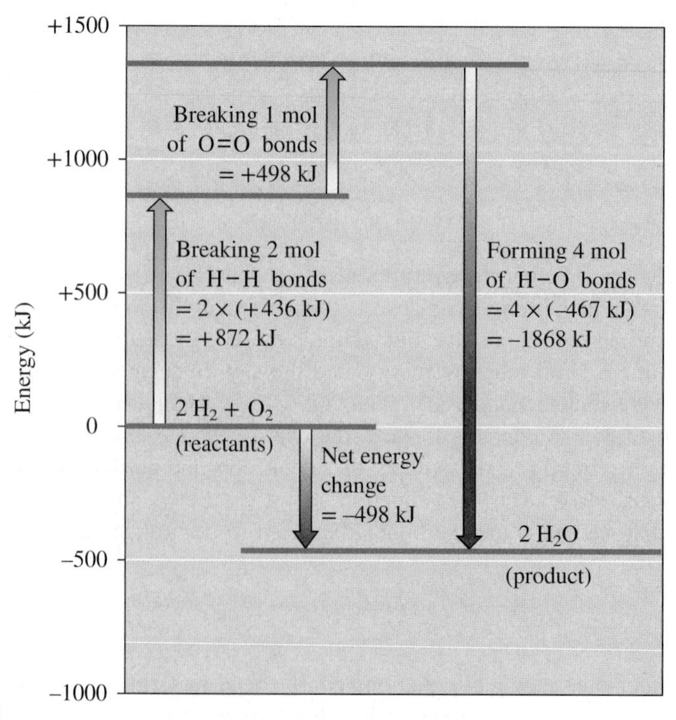

Figure 5.8

The energy changes during the combustion of hydrogen to form water vapor.

the two. This is an extremely powerful idea when doing calculations related to energy changes in reactions.

Not all calculations will result in the same close agreement with experimental data. One possible source of error is that the bond energies listed in Table 5.1 apply only to gases. Hence, calculations are most accurate if all reactants and products are in the gaseous state. Moreover, tabulated bond energies (other than C=O in Table 5.1) are listed as average values, which are based on many different types of molecules. The strength of a bond depends on the overall structure of the molecule in which it is found; in other words, what else the atoms are bonded to. Thus, the strength of an O–H bond is slightly different in HOH (H_2O), HOOH (H_2O_2), and H_3COH. Nevertheless, the procedure illustrated here is a useful way of estimating energy changes in a range of reactions. The approach also helps illustrate the relationship between bond strength and chemical energy.

Extended bond energy tables include details on the exact molecule and its physical state.

This analysis also helps clarify why products of combustion reactions (such as H_2O or CO_2) cannot be used as fuels. There are no substances into which these compounds can be converted that have stronger bonds and that are lower in potential energy. Therefore, their conversion into something else is not favorable in terms of heat energy. Bottom line: You cannot run a car on its exhaust fumes; however, it would be a remarkable and very beneficial chemical recycling discovery to do so!

After you get some practice with calculations in the next activity, we will describe how the combustion of fuels generates the power we need to survive, and just how efficient is the overall process.

Your Turn 5.13 Skill Building Heat of Combustion for Ethyne

Use the bond energies in **Table 5.1** to calculate the heat of combustion for ethyne, C_2H_2, also called acetylene. Report your answer both in kilojoules per mole (kJ/mol) C_2H_2 and kilojoules per gram (kJ/g) C_2H_2. Here is the balanced chemical equation:

$$2\,H{-}C{\equiv}C{-}H + 5\,\ddot{O}{=}\ddot{O} \longrightarrow 4\,\ddot{O}{=}C{=}\ddot{O} + 2\,H{-}\ddot{O}{-}H$$

Hint: The coefficient for acetylene in the chemical equation is 2. The heat of combustion is for 1 mole.

5.6 | Fossil Fuels and Electricity

Beyond using fossil fuels for our direct transportation and heating needs, about 70% of the electricity generated in the United States comes from their combustion—primarily from coal. But how do electrical power plants "produce" electricity, and what really goes on inside them? Our task in this section is to take a closer look at the energy transformations in a power plant.

The first step in producing electricity from coal is to burn it. Examine the photographs in Figure 5.9. You can almost feel the heat from the burning coal! In the coal beds of the boilers, the temperature can reach 650 °C. To generate this level of heat, the power plant burns a train car load of coal every few hours.

We will describe the properties of coal and its combustion in more detail in Section 5.8.

The second step in producing electricity is to use the heat released from combustion to boil water—usually in a closed, high-pressure system (Figure 5.10). The elevated pressure serves two purposes: It raises the boiling point of the water above 100 °C, and it compresses the resulting water vapor. The hot high-pressure steam is then directed toward a steam turbine.

The third and final step generates electricity. As the steam expands and cools, it rushes past the turbine, causing it to spin. The shaft of the turbine is connected to a large coil of wire that rotates within a magnetic field, and the turning of this coil generates an electric current. Meanwhile, the water vapor leaves the turbine and continues to cycle through the system. It passes through a condenser, where a stream of cooling water carries

Did You Know? Nuclear power plants, which will be described in Chapter 6, operate on a similar principle—the use of superheated steam to turn a turbine. The difference in both cases is simply the fuel used to heat the water into steam.

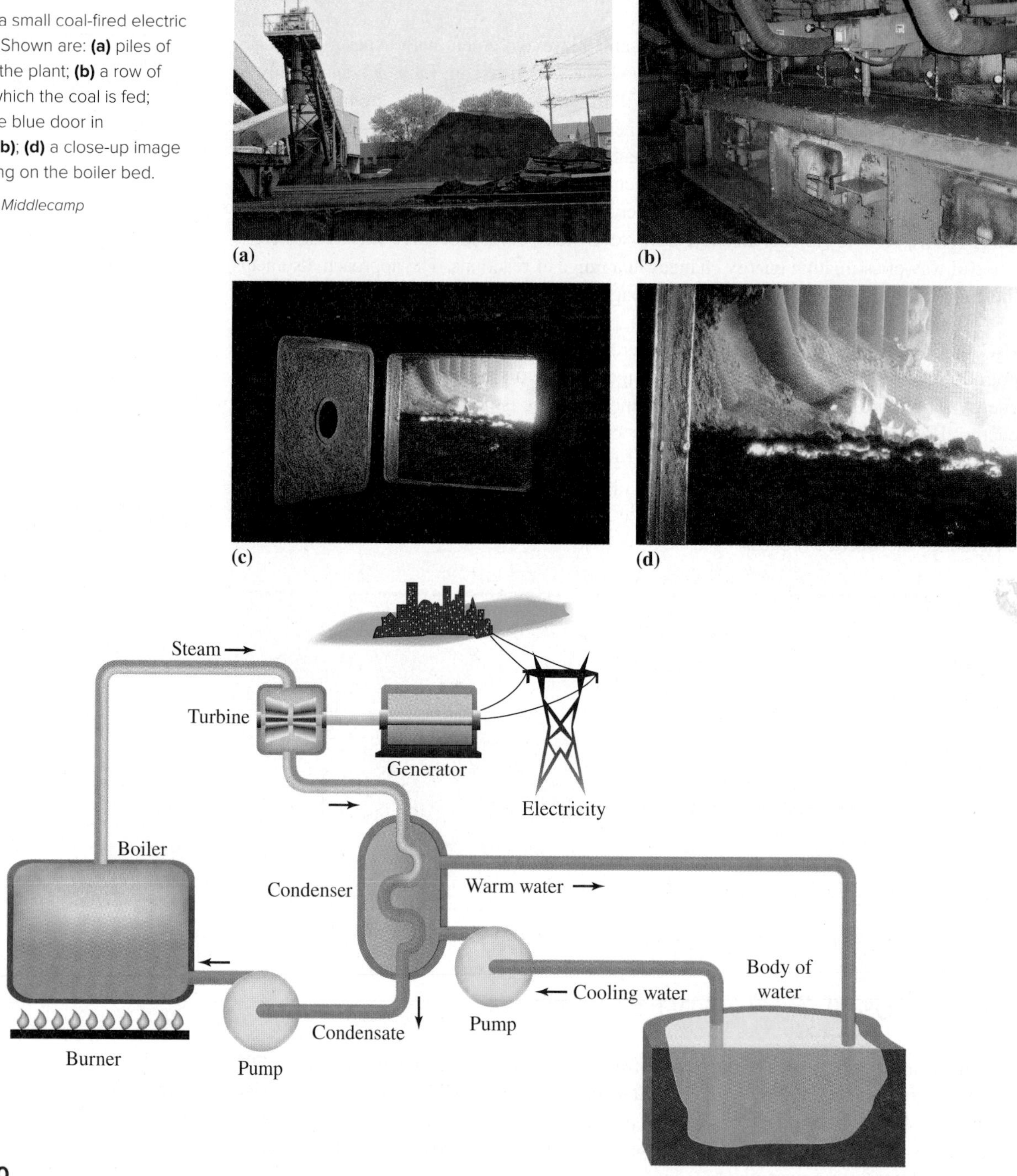

Figure 5.9

Photos from a small coal-fired electric power plant. Shown are: **(a)** piles of coal outside the plant; **(b)** a row of boilers into which the coal is fed; **(c)** behind the blue door in photograph **(b)**; **(d)** a close-up image of coal burning on the boiler bed.

(a–d): © Cathy Middlecamp

Figure 5.10

Diagram of an electric power plant illustrating the conversion of energy from the combustion of fuels to electricity. Components are not illustrated to scale.

away the remainder of the heat energy originally acquired from the fuel. The condensed water then re-enters the boiler, ready to resume the energy transfer cycle.

When fuel molecules combust, their potential energy is converted into heat, which is then absorbed by the water in the boiler. As the water molecules absorb the heat, they move faster and faster in all directions—*i.e.,* their *kinetic* energy increases until they can escape contact with one another and vaporize to steam. As pressurized steam, water molecules have a tremendous amount of kinetic energy, which is transferred to the turbine that is spun into motion. The kinetic (motion) energy of the water molecules has been transformed to mechanical energy in the turbine. The generator then turns and converts the mechanical energy (a form of kinetic energy) in the turbine into **electrical energy**, which is another form of kinetic energy that we discussed in Chapter 1. The various energy transformation steps are summarized in Figure 5.11.

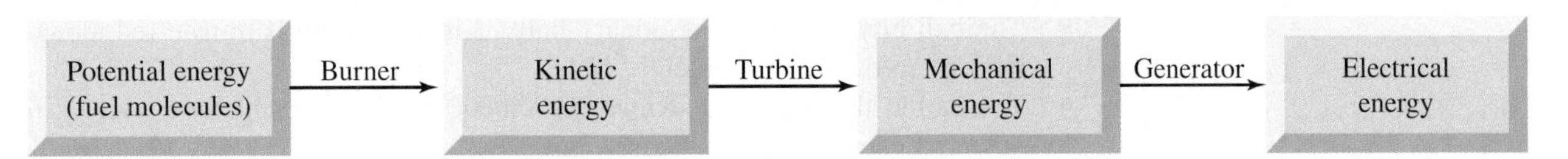

Figure 5.11

Summary of the energy transformations in a fossil fuel-powered electric power plant.

Your Turn 5.14 Scientific Practices Energy Conversion

Although power plants require several steps to transform potential energy into electrical energy, other devices do this more simply. For example, a battery converts chemical energy to electrical energy in one step. List three other devices that convert energy from one form to another. For each one, name the types of energy and show the path of energy transformations involved.

Revisit the processes of combustion and photosynthesis. Overall, energy is released during combustion, but in photosynthesis energy is absorbed by the reaction. The relationship between the two hints at a cycle, as shown in Figure 5.12. Over time, the products of combustion could again be returned through photosynthesis to a high potential energy state. The **First Law of Thermodynamics**, also called the Law of Conservation of Energy, states that energy is neither created nor destroyed. It implies that although the *forms* of energy change, the total amount of energy before and after any transformation remains the same. The solar energy that is transformed to chemical potential energy during photosynthesis is then released as heat and light during combustion.

Remember that even our energy-rich hydrocarbon fossil fuels originally came from energy investments in photosynthesis.

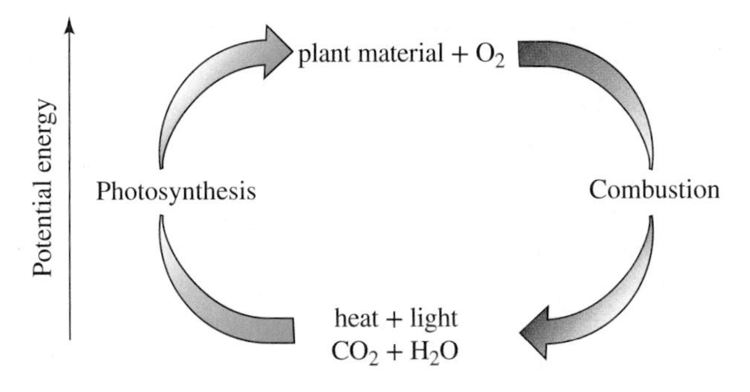

Figure 5.12

The energy relationship between photosynthesis and combustion.

5.7 | How Efficient Is a Power Plant?

In the First Law of Thermodynamics, we are assured that the total energy of the universe is conserved. If this is true, how can we ever experience an energy crisis? To be sure, no new energy is created during combustion, but none is destroyed either. Although we may not be able to win, can we at least break even? The question is not as facetious as it might sound. Disappointedly, we *cannot* break even. In burning coal, natural gas, and petroleum, we always convert at least some of the energy in the fuels into forms that we cannot easily use or recover.

You may have seen the desktop toy known as a Newton's cradle (Figure 5.13). Like a power plant, but much simpler, this device transforms potential energy (position or stored energy) into kinetic energy (motion energy). Here's how it works:

- A ball is lifted at one end. The movement of the ball in the upward direction increases the potential energy of the ball.
- The ball is released and falls back toward its starting point. The potential energy is converted into kinetic energy.

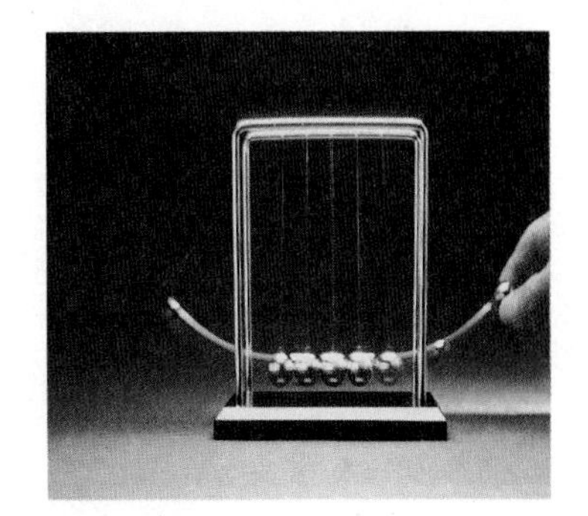

Figure 5.13

A Newton's cradle.

- The ball hits the row of stationary balls. Kinetic energy is transferred along the row to the ball at the other end.
- The ball at the other end swings up. Kinetic energy is gradually converted into potential energy as the ball rises and slows.
- The ball at this end begins to fall and the process repeats.

With each successive cycle, however, each ball does not rise quite as high as the previous one. Eventually, the balls all come to rest at their original positions.

Why do they stop moving? Where did their energy go? Is this a violation of the First Law of Thermodynamics? In fact, it is *not* a violation. In each collision, some of the energy is transformed to vibrational energy (we hear sound) and some becomes heat. If we could measure precisely enough, we would observe the balls heating up slightly. This heat is then transferred to the surrounding atoms and molecules in the air, thus increasing their kinetic energy. The First Law of Thermodynamics does not specify whether kinetic or potential energy is conserved. Instead, this law governs the conservation of the total energy, or the sum of the two. Therefore, when all the balls finally come to rest, the energy of the universe has been conserved. All the energy you initially put into the system has ultimately been dissipated through random motion of the atoms and molecules in the surrounding air.

These same principles can be used to explain why no electric power plant, no matter how well designed, can completely convert one type of energy into another. In spite of the best engineers and the most competent green chemists, inefficiency is inevitable. This inefficiency is caused by the transformation of energy into useless heat. Overall, the % net efficiency is given by the ratio of the electrical energy produced, to the energy supplied by the fuel:

$$\%\text{ net efficiency} = \frac{\text{electrical energy produced}}{\text{heat from fuel}} \times 100 \qquad \textbf{[5.8]}$$

Newer boiler systems and advanced turbine technologies have pushed the efficiencies of each step in Figure 5.11 to 90% or better. So, you might be surprised to learn that the net efficiencies of most fossil fuel power plants are only between 35 and 50%. Why so low?

The problem is that not all of the heat energy from fuel combustion in the boilers can be converted into electricity. Consider, for example, the high-temperature steam that initially spins the turbines. As the steam transfers energy to the turbines, the kinetic energy of the steam decreases, it cools, and its pressure drops. It isn't long before the steam does not have enough energy to spin the turbines anymore. Yet, the production of this "unused" steam still requires a significant amount of energy—energy that is not converted into electricity.

Power plants using very high-temperature steam (600 °C) have efficiencies at the high-end of the range. In fact, the efficiency goes up as the difference between the steam temperature and the temperature outside the plant increases. Of course, there is a trade-off. Higher-temperature steam means higher pressures, thus requiring the use of improved construction materials in order to be able to withstand such extreme conditions.

Now consider the case of electrical home heating, sometimes advertised as being "clean and efficient." Assume that electricity from a coal-burning power plant (net efficiency of 37%) is used to heat a house. If the house requires 3.5×10^7 kJ of energy for heat annually—a typical value for a city in a cooler climate—how much coal would be burned?

To answer this question, we need a value for the energy content of the coal. Let's assume that the combustion of this particular coal releases 29 kJ/g. Remember that only 37% of the energy released by burning the coal is available to heat the house. We now can calculate the total annual quantity of heat that we need by burning coal at the power plant:

energy generated at plant × efficiency = energy required to heat house

energy generated at plant $\times 0.37 = 3.5 \times 10^7$ *kJ*

energy generated at plant $= 9.5 \times 10^7$ *kJ*

In these calculations, note that we expressed the percent efficiency in decimal form. We now take into account that each gram of coal burned yields 29 kJ:

$$9.5 \times 10^7 \text{ kJ} \times \left(\frac{1 \text{ g coal}}{29 \text{ kJ}}\right) = 3.3 \times 10^6 \text{ g coal}$$

This shows that 3.3×10^6 g of coal must be burned each year at the power plant in order to furnish the needed 3.5×10^7 kJ of energy to heat one house.

The above calculation assumed a net efficiency of 37% at the coal plant. Higher efficiencies would mean that less fuel would have to be burned to generate the same amount of energy, and that less carbon dioxide and other pollutants would be emitted. The next activity explores these connections.

Your Turn 5.15 Scientific Practices Comparing Power Plants

Consider two coal-fired power plants that generate 5.0×10^{12} J of electricity daily. Plant A has an overall net efficiency of 38%. Plant B, a proposed replacement, would operate at higher temperatures with an overall net efficiency of 46%. The grade of coal used releases 30 kJ of heat per gram. Assume that coal is pure carbon.

a. If 1000 kg of coal costs $30, what is the difference in daily fuel costs for the two plants?

b. How many fewer grams of CO_2 are emitted daily by Plant B, assuming complete combustion?

Cars and trucks also convert energy from one form to another. The internal combustion engine uses the gaseous combustion products (CO_2 and H_2O) to push a series of pistons, thus converting the potential energy of the gasoline or diesel fuel into mechanical energy. Other mechanisms eventually transform that mechanical energy into the kinetic energy of the vehicle's motion. Internal combustion engines are even less efficient than coal-fired power plants. Only about 15% of the energy released by the combustion of the gasoline is actually used to move the vehicle. Much of the energy is dissipated as waste heat, including about 60% lost from the internal combustion engine alone.

Your Turn 5.16 Scientific Practices Transportation Inefficiency

a. List some of the energy losses that take place when driving a car. Use the Internet to verify and expand your list, if necessary.

b. Given the assumption that only 15% of the energy from fuel combustion is used to move the vehicle, estimate the percent used to move the passengers.

To bring this section to a close, we ask you to revisit the Newton's cradle. You would never expect the balls at rest to start knocking into one another on their own, right? For this to occur, all the heat energy dissipated when the balls were colliding would have to be gathered back together. The inability of a Newton's cradle to start up on its own relates to another concept—entropy. **Entropy** is a measure of how much energy gets dispersed in a given process. The **Second Law of Thermodynamics** has many versions, the most general of which is that the entropy, or randomness, of the universe is constantly increasing. The Newton's cradle provides an example of the Second Law of Thermodynamics. When we lift one of the balls of the Newton's cradle, we add potential energy. After the balls knock for a while and come to rest, this potential energy has become transformed into the chaotic (and hence more random and dispersed) motion of heat energy, and never the other way around. The entropy of the universe has increased.

Do you find it difficult to visualize how energy can disperse? If so, here is an analogy that might help. Imagine you are sitting in the middle of a large auditorium and someone down in the front breaks a bottle of perfume. You don't smell anything at first, because it takes time for the molecules of the perfume to diffuse to where you're sitting. This process of **diffusion** is predicted by the Second Law of Thermodynamics. When the perfume molecules disperse from the smaller volume of the bottle into the larger volume of the auditorium, the energy of the molecules gets dispersed as well. As with the Newton's cradle, the end result is an increase in the entropy of the universe. It is extremely unlikely that all of the perfume molecules would suddenly gather in one corner of the room. Rather, once dispersed, they stay dispersed unless energy is expended to recollect them.

In the same way, it is essentially impossible for the Newton's cradle to begin to move on its own after the energy originally added was dissipated as heat. Although it may not be as obvious, the Second Law of Thermodynamics also explains the inability of a power plant or an auto engine to convert energy from one type to another with 100% efficiency.

Your Turn 5.17 Scientific Practices More Entropy Examples

An input of energy can be used to decrease entropy "locally." Even so, energy expended in one place requires a net increase in entropy elsewhere in the universe.

a. Consider the energy input from burning coal. The entropy of the universe increased elsewhere. Give an example of how it could have increased.

b. Consider the decrease in entropy that occurs when somebody arranges the socks in a drawer. What must have accompanied this decrease in entropy?

5.8 | Power from Ancient Plants: Coal

About two centuries ago, the Industrial Revolution began the great exploitation of fossil fuels that continues today. In the early 1800s, wood was the major energy source in the United States. Coal turned out to be an even better energy source than wood, because it yielded more heat per gram. By the 1960s, most coal was used for generating electricity, and today the electrical power sector accounts for 92% of all U.S. coal consumption (Figure 5.14).

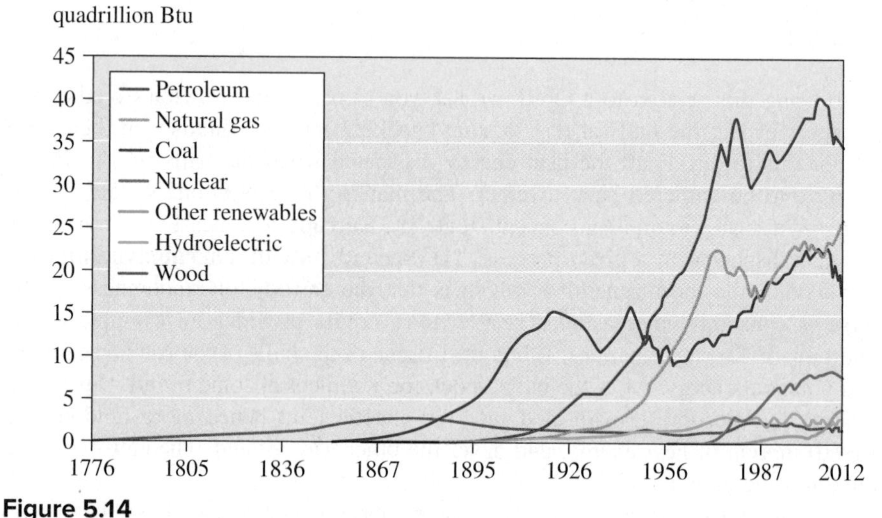

Figure 5.14

History of U.S. energy consumption by source, 1776–2012.

Your Turn 5.18 Scientific Practices Changing Fuel Patterns

Use **Figure 5.14** and Internet resources to:

a. Describe two ways in which fuel consumption in the United States has changed over time. Propose reasons for the changes.
b. Estimate the fraction of total energy that is produced by the burning of coal.

How do the trends you identified in **a** compare to other countries in Europe, Asia, or Central America? Explain why differences exist for consumption trends in these countries, if applicable.

Although coal is often assumed to be composed of pure carbon, it is a more complex mixture, containing small amounts of many other elements. An approximate chemical formula for coal is $C_{135}H_{96}O_9NS$, which corresponds to a carbon content of about 85% by mass. The smaller amounts of hydrogen, oxygen, nitrogen, and sulfur come from the ancient plant material and other substances present when the plants were buried. In addition, some samples of coal typically contain trace amounts of silicon, sodium, calcium, aluminum, nickel, copper, zinc, arsenic, lead, and mercury.

Did You Know? Despite the common belief that diamonds are formed from coal, this has rarely (if at all) taken place. Coal is found at depths less than 2 miles (3.2 km) from Earth's surface. However, the high-temperature and high-pressure reactions necessary for diamond formation occur in limited zones of Earth's mantle—approximately 90 miles (150 km) below Earth's surface.

Your Turn 5.19 Scientific Practices Coal Calculations

a. Assuming the composition of coal can be approximated by the formula $C_{135}H_{96}O_9NS$, calculate the mass of carbon (in tons) in 1.5 million tons of coal. This quantity of coal might be burned by a typical power plant in 1 year.
b. Compute the amount of energy (in kJ) released by burning this mass of coal. Assume the process releases 30 kJ/g of coal. Useful conversion factors: 1 ton = 2000 lb and 1 pound = 454 g.
c. What mass of CO_2 would be formed by the complete combustion of 1.5 million tons of this coal?
Hint: In the balanced chemical equation, assume a mole ratio of coal:CO_2 of 1:135.

Coal occurs in varying grades; however, all are better fuels than wood because they contain a higher percentage of carbon and a lower percentage of oxygen. For example, burning 1 mole of C to produce CO_2 yields about 40% more energy than is obtained from burning 1 mole of CO to produce CO_2.

Figure 5.15 shows a structural comparison of various types of coal. Soft lignite, or brown coal, is the lowest grade. The plant matter from which it originated underwent the least amount of change, and its chemical composition is similar to that of wood or peat (Table 5.2). Consequently, the amount of energy released when lignite is burned is only slightly greater than that of wood.

Table 5.2 Energy Content of U.S. Coals

Type of Coal	State of Origin	% Carbon	% Moisture	% Ash[1]	Energy Content[2] (kJ/g)
Anthracite	Pennsylvania	85–98	<15	10–20	30–35
Bituminous	Maryland	45–85	2–15	3–12	26–35
Sub-bituminous	Washington	35–45	10–45	<10	20–30
Lignite	North Dakota	25–35	30–60	10–50	9–19
Peat	Mississippi	10–25	40–85	2–15	3–13

[1] Ash consists of inorganic matter from Earth's crust (iron, aluminum oxides, clay, silicon dioxide, limestone (calcium carbonate, $CaCO_3$), and trace elements).

[2] For comparison, the energy content of wood ranges from 10–14 kJ/g, depending on the type.

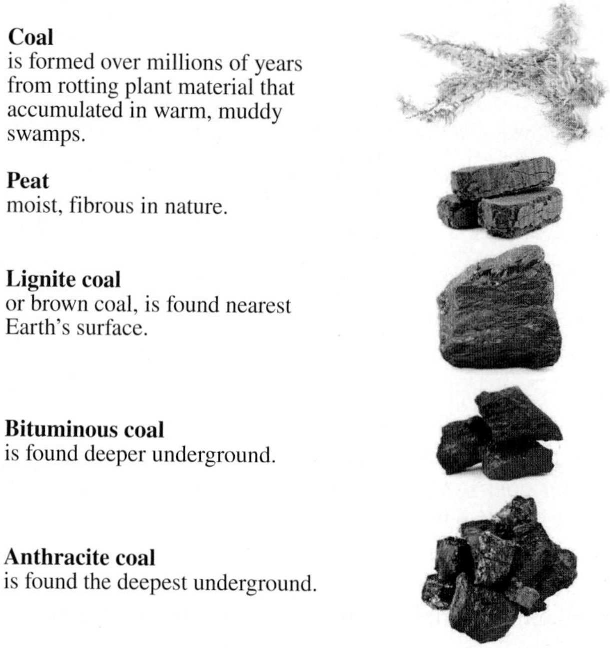

Coal
is formed over millions of years from rotting plant material that accumulated in warm, muddy swamps.

Peat
moist, fibrous in nature.

Lignite coal
or brown coal, is found nearest Earth's surface.

Bituminous coal
is found deeper underground.

Anthracite coal
is found the deepest underground.

Figure 5.15

A comparison of the three principal grades of coal and partially decayed plant material known as peat.

Sources (top-bottom): © Kuttelvaserova Stuchelova/Shutterstock.com; © Perutskyi Petro/Shutterstock.com; © Jiang Hongyan/Shutterstock.com; © Swapan Photography/Shutterstock.com; © Madlen/Shutterstock.com

The higher grades of coal, bituminous and anthracite, have been exposed to higher pressures and temperatures for longer periods of time. During that process, they lost more oxygen and moisture to become harder and more dense, and exhibit a higher crystallinity. These grades of coal contain a higher percentage of carbon than lignite. As a hard, black solid, anthracite has a relatively high carbon content and a low sulfur content. Both of these qualities make it the most desirable grade of coal that results in fewer harmful sulfur oxide emissions. Unfortunately, the deposits of anthracite are relatively small, and the supply of it in the United States is almost exhausted.

Although coal is available across the globe and remains a widely used fuel, it has serious drawbacks—the first of which relates to underground mining that is both dangerous and expensive. Mine safety has dramatically improved in the United States. However, during the past century, more than 100,000 workers have been killed by accidents, cave-ins, fires, explosions, and poisonous gases. Many more have been incapacitated by respiratory diseases. Worldwide, the picture is far worse.

Did You Know? Many streams and rivers in Appalachia suffer from high levels of pollution due to decades of mining operations.

A second drawback is the environmental harm caused by coal mining. When groundwater infiltrates abandoned mine shafts, or comes in contact with sulfur-rich rock often associated with coal deposits, it becomes acidified. This acidic mine drainage also dissolves excessive amounts of iron and aluminum, making the water uninhabitable for many fish species and placing drinking water sources at risk for many communities.

When coal deposits lie sufficiently close to the surface, safer mining techniques are possible, but there are still environmental costs. One technique, called *mountaintop mining*, is most common in West Virginia and eastern Kentucky. This process calls for scraping away the overlying vegetation, and then blasting off the top several hundred feet of a mountain to reveal the underlying coal seam. Mountaintop mining creates massive quantities of rubble (referred to as "overburden") that are often disposed of by dumping the debris into nearby river valleys. In 2005, the U.S. EPA estimated that more than 700 miles of Appalachian streams were completely buried as a result of mountaintop mining between 1985 and 2001. Furthermore, increased sediments and mineral content in the surrounding water systems has adversely affected many aquatic ecosystems. As if these environmental issues were not serious enough, the airborne coal-dust particulates generated by this practice have recently been implicated in a number of health issues such as lung cancer and birth defects.

A third drawback is that coal is a dirty fuel. It is, of course, physically dirty; however, the issue here is the dirty combustion products. Soot from countless coal fires in cities in the 19th and early 20th centuries blackened both buildings and lungs. In fact, during the mid-1940s, the burning of coal in steel mills near Pittsburgh, Pennsylvania, generated enough atmospheric pollution to mask the mid-day sun (Figure 5.16).

Figure 5.16

A photo taken in 1940 at 10:55 am in Pittsburgh, Pennsylvania, showing the mid-day sun completely shielded by smoke and soot from the burning of coal in nearby steel mills and power plants.

When coal is burned, microscopic particulates known as *fly ash* are generated in the exhaust gas. In the past, this ash was released to the environment; however, more stringent regulations in recent years have resulted in extremely low particulate emissions. A series of filters and precipitators are now used in most coal-fired power plants to capture the ash before it rises up the chimney (or "stack"). Coal combustion also results in larger/heavier particles known as *bottom ash*, which are more easily collected at the power plant. It is estimated that 92 million tons of coal ash are generated from coal-fired power plants each year in the United States.

Although coal is mostly carbon, its ash has very little carbon remaining. The composition of coal ash is similar to that found in Earth's crust, featuring oxides of silicon, aluminum, iron, and calcium, as well as smaller concentrations of magnesium, potassium, sodium, titanium, and sulfur. Coal contains only minor amounts (50–200 ppb) of mercury, lead, and cadmium; however, these toxic elements may become concentrated in the ash and may leach into the environment from landfills or storage sites. Furthermore, the ash left on site also presents a storage hazard. For example, Figure 5.17 shows the devastation caused by millions of gallons of fly ash sludge that spilled down a valley when the retaining walls of a storage pond failed.

In order to prevent environmental harm and enhance sustainability, there has been much recent interest in reusing coal ash waste for consumer products. For instance, the addition of ash in concrete is shown to improve its durability, chemical resistance, and shrinkage during hardening. Other products such as carpet backing, fire and heat protection devices, automobile and marine bodies, insulation, paints, and even toothpaste may contain microscopic fly ash in their formulations.

Figure 5.17

A photo taken in December of 2008 showing homes near Knoxville, Tennessee, buried by 300 million gallons of coal sludge.

A final drawback may ultimately be the most serious—burning of coal produces gases that contribute to acid rain (NO_x and SO_x), as well as global warming (CO_2). In fact, coal combustion produces more CO_2 per kilojoule of heat released than either petroleum or natural gas. In recent years, coal has been ranked as the world's fastest-growing fossil fuel source, with an explosion of production and use in China and India.

Your Turn 5.20 Scientific Practices Coal Emissions

In the United States, coal-burning power plants are responsible for two-thirds of the sulfur dioxide emissions and one-fifth of the nitrogen monoxide emissions.

a. Why does burning coal produce sulfur dioxide? Name another source of SO_2 in the atmosphere.

b. Why does burning coal produce nitrogen monoxide? Name two other sources of NO.

Because of these drawbacks, and given that coal reserves are relatively plentiful in the United States, significant research efforts are underway to improve coal technologies. Although it may sound like an oxymoron, "clean coal" is promoted by its supporters as an important step toward decreasing our reliance on petroleum imports and reducing air pollution. The term "clean coal technology" actually encompasses a variety of methods that aim to increase the efficiency of coal-fired power plants while decreasing harmful emissions.

What does the future hold for the dirtiest of the fossil fuels? The answer depends on where you live. Figure 5.18 compares coal consumption in different regions of the

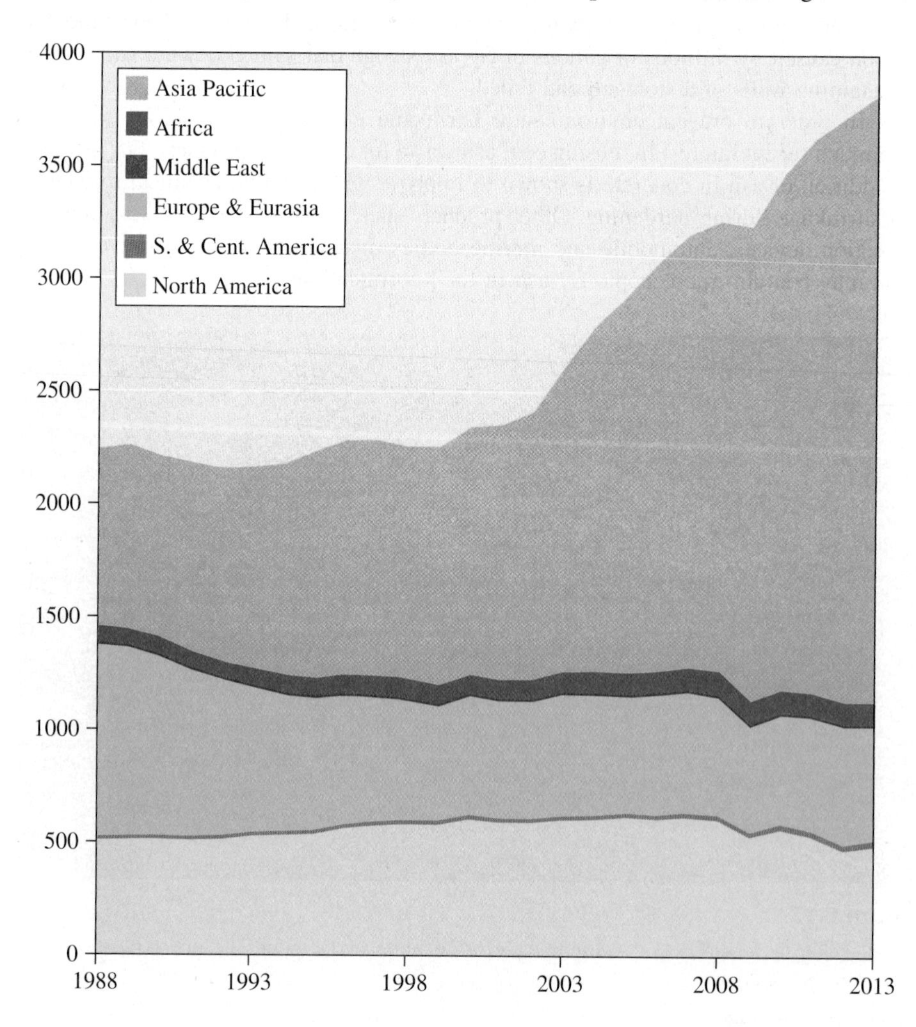

Figure 5.18

World consumption of coal by region, 1988–2013, in million tonnes oil equivalent.

Source: BP Statistical Review of World Energy, June 2014

globe between 1988 and 2013. Although most regions showed modest changes, the use of coal in Asia is skyrocketing. On one hand this makes sense, because China has enormous coal reserves to fuel its rapid growth. But on the other, coal burning (by any nation) clearly does not meet the criteria for sustainability.

Your Turn 5.21 You Decide Clean Coal?

In 2011, a newspaper columnist remarked that the idea of clean coal "remains a distant dream."

a. List three factors that contribute to making coal a dirty fuel.
b. Now that it is several years later, has the dream of clean coal come any closer to being realized? Select a format and use it to argue your case either way.

5.9 From Steam Engines to Sports Cars: The Shift from Coal to Oil

Throughout the late 19th century, coal was still the most heavily used energy source in the United States, with some modest oil production efforts limited to Pennsylvania. However, in 1901, the modern oil industry in the U.S. was forever changed. Drilling into a hill composed of a giant salt dome in southeast Texas released an eruption of oil to heights greater than 150 feet (Figure 5.19). The so-called *Lucas Spindletop gusher* was not brought under control for 9 days, which resulted in a loss of 850,000 barrels of oil—by today's prices, around $30,000,000! The ensuing Texas oil boom caused oil prices to drop from $2 to $0.03 per barrel, and eventually led to the U.S. becoming one of the world's leading producers of crude oil.

Due primarily to energy demands caused by the automobile, 1950 marked the first year that petroleum surpassed coal as the major energy source in the United States. The reasons are relatively easy to understand. Unlike coal, petroleum has the distinct advantage of being a liquid, making it easily pumped to the surface and transported via pipelines to refineries. Moreover, petroleum yields about 40–60% more energy per gram than coal. A typical value for petroleum is 48 kJ/g as compared to 30 kJ/g for a high-grade coal.

Depending on its origin, crude oil ranges from a clear golden viscous fluid to a black tarry liquid. The distinctive foul odor of crude oil pumped from wells or transported by tankers is mostly due to sulfur-containing compounds such as hydrogen sulfide (H_2S) and others that feature –SH groups bonded to carbon atoms (*e.g.*, CH_3CH_2SH).

Today, the top five oil producers in the world are: 1. United States, 2. Saudi Arabia, 3. The Russian Federation, 4. China, and 5. Canada.

Did You Know? Technological advances in drilling and transportation starting in the late 1860s, helped lay the groundwork for petroleum to outpace coal as a major energy source in the U.S. The required infrastructure needed to develop and expand exploration, refining, and transportation of crude oil and its derived products was spearheaded by businessman John D. Rockefeller (1839–1937), who formed the Standard Oil Company in 1870.

Carbon compounds that contain –SH groups are known as *mercaptans*.

Figure 5.19
Photo of the Lucas Spindletop gusher on January 10, 1901 (left), and the widespread exploration of the oil field a year later (above).

(both): © & Courtesy of Texas Energy Museum

5.10 Squeezing Oil from Rock: How Long Can This Continue?

A barrel of oil is equivalent to 159 L (42 gal), and 7 1/3 barrels have a mass of 1000 kg.

As you saw earlier in Your Turns 5.3 and 5.4, oil is a fossil fuel with a limited total supply. However, there are still plentiful oil reserves located throughout the world (Figure 5.20). Globally, there are 1.7 trillion barrels (bbl) of proven oil reserves, with 87 million barrels produced per day. Accordingly, the potential threat to our societal dependence is not the *quantity* of fossil fuels remaining, but rather the *rate* at which we can extract them. In the mid-1950s, the average global oil consumption was about 8 million barrels per day, with a production of 15 million barrels. Today, we use over 91 million barrels per day, but only produce 87 million barrels. The world oil production to consumption ratio is not fixed, but may tip in either direction depending on geopolitical factors and the success of crude oil exploration and recovery efforts (Figure 5.21).

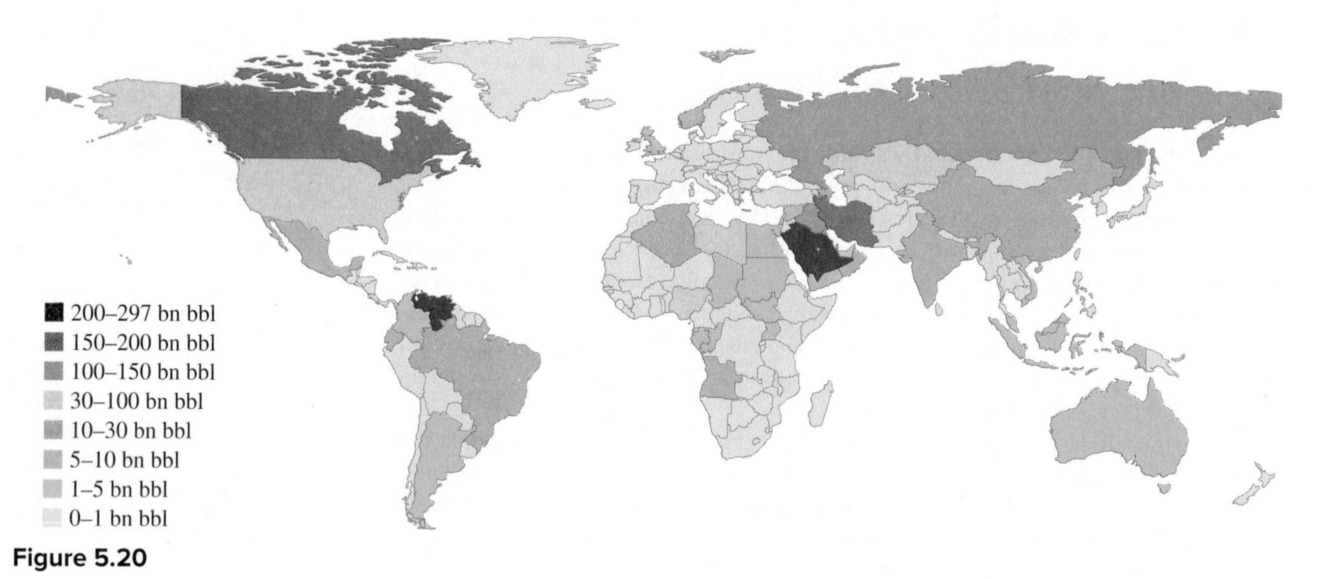

Figure 5.20

Worldwide proven oil reserves.

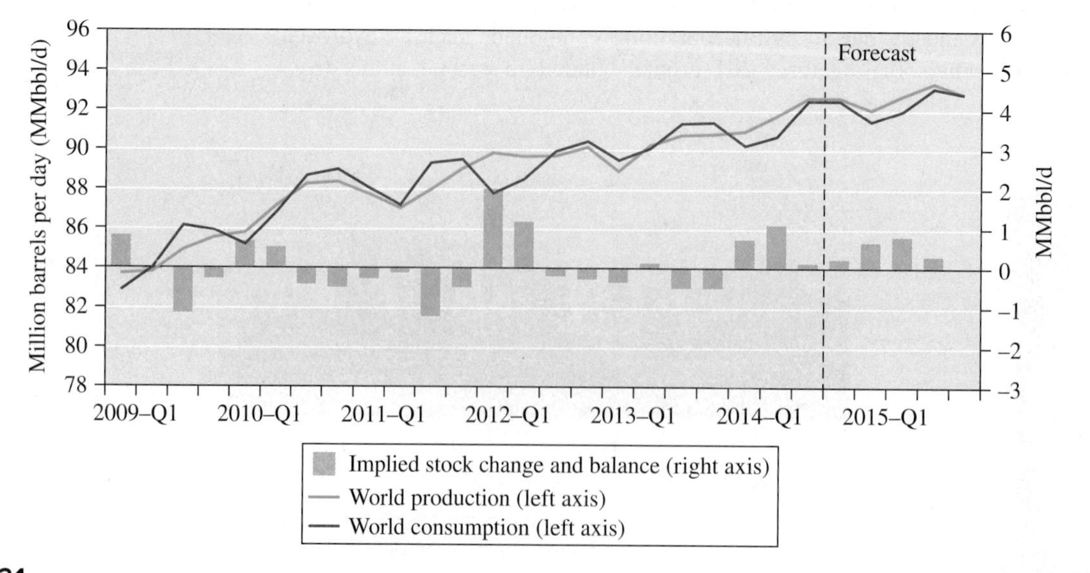

Figure 5.21

The world oil production to consumption ratio, 2009–2015.

Your Turn 5.22 Skill Building How Much Is a Barrel?

The petroleum industry uses barrels/day as a standard measure of consumption and production. However, we often speak in terms of the weight of fuels, such as tons, or the volume, such as gallons (gal) or liters (L). Using Appendix 1 or your favorite unit conversion website or app, convert the current world consumption and production figures for crude oil into tons, gallons, and liters.

Contrary to popular belief, oil is not found in underground pools, but is found within the pores of geologic rock formations such as sandstone, tarry oil sands, and shales (Figure 5.22). As you might expect, a higher degree of porosity within the rock usually means a higher potential to store oil. However, porosity is only part of the story. In order for a rock formation to yield sufficient quantities of oil or gas, the pores must be interconnected, something known as the *permeability* of the reservoir. If the pore spaces are isolated from one another, oil will not be able to flow, regardless of whether there is appreciable porosity and oil content.

A great deal of time and money are spent in identifying the likely locations of oil-rich rock formations, known as *reservoirs*. These reservoirs can exist on land or under the ocean, at varying depths from the surface. Using sound waves, scientists can first determine whether underground rocks are likely to contain oil or gas deposits. From there, an exploratory well is drilled, and core samples are taken to the surface where they are examined to determine whether oil is contained within the pores. Due to this time-consuming process, the oil we use today comes from fields that were discovered decades ago. A recent large oil discovery was at Kashagan in the Caspian Sea and is expected to yield more than 10 billion barrels over its lifetime. Although discovered in 2000, Kashagan oil only went into production in 2013.

An oil reservoir is under pressure from the weight of millions of tons of overlying rock, as well as Earth's natural heat—both of which expand the gases present in the pores of the rock. When an oil well first penetrates this reservoir, the pressure is released, which pushes oil from the pores of rock up to the surface through the well. Complicated valve systems known as *Christmas trees* are now used to control the pressure inside the well, to prevent an uncontrolled gusher such as the previously discussed Lucas Spindletop. This spontaneous flow of oil to the surface may continue for days or years, but will eventually slow/stop due to a loss of pressure. When this happens, well pumps must be used, and deposits of natural gas may also be separated from oil and injected back into the reservoir to increase the pressure and keep oil flowing. Eventually, however, these strategies will not be sufficient to maintain the flow of oil.

In order to recover even more oil, injection wells must be drilled to pump water into the oil reservoir. The pressurized water washes some additional oil from the rock pores and pushes it to the surface in a process known as *secondary recovery* (also known as "waterflooding"). However, the majority of the original oil still remains bound within the reservoir rock formation. In fact, only 20–25% of the original oil deposit will be released during primary drilling, with an additional 5–10% being obtained through secondary recovery. That leaves about 65–70% of the original oil still remaining within the reservoir. To put this into perspective, for every barrel of oil produced, two barrels are left behind. The challenge that threatens our future access to petroleum products, such as gasoline, is to find economical and sustainable strategies to extract more of this trapped oil.

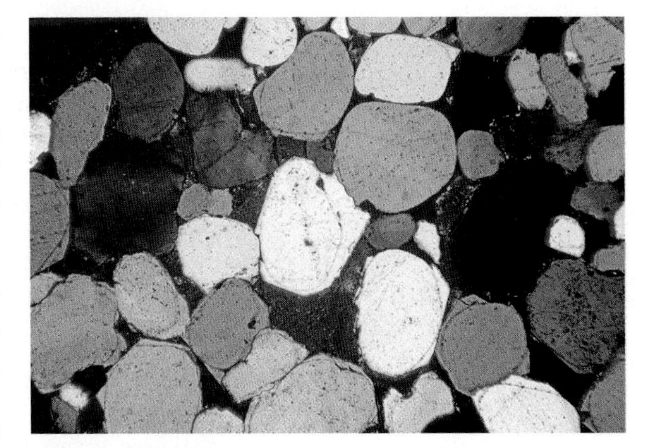

Figure 5.22

Microscopic image of a sandstone reservoir rock, showing the locations of trapped oil deposits within its pores and grain boundaries (cracks). Water and natural gas are often also present within the pores of these rocks.

One strategy, carbon dioxide enhanced oil recovery (CO_2-EOR), is currently attracting the most interest because it

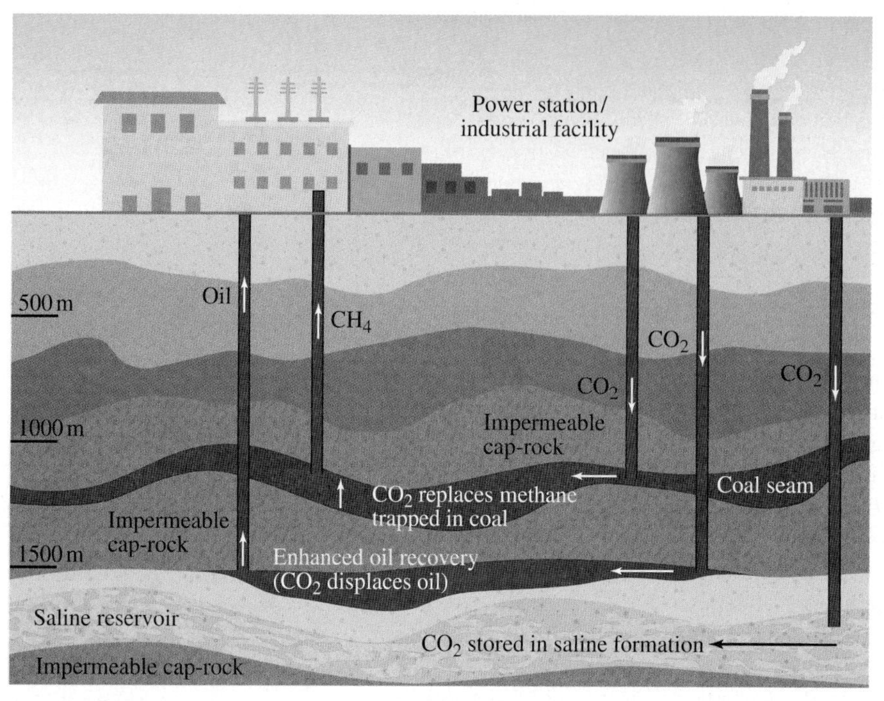

Figure 5.23

Schematic of a CO_2-EOR strategy to use CO_2 that has been sequestered from a fossil fuel-fired power plant for oil recovery.

provides a sustainable use for a greenhouse gas (Figure 5.23). The potential for CO_2 sequestration in depleted oil and gas reservoirs is enormous. The Department of Energy has documented enough potential sites to sequester more than 152 billion tons of CO_2 in the U.S. and Canada. Currently, about 48 million tons of CO_2 is injected annually for EOR operations in the United States.

An international case study for CO_2-EOR is underway in southeast Saskatchewan, Canada. Carbon dioxide has been sequestered from coal gasification in North Dakota, and transported via pipeline to Canada, where it was used to extract oil from a Saskatchewan oilfield. Thus far, this initiative has shown to increase oil production by as much as 28,000 barrels a day for certain regional oil fields. Overall, an additional 130 million barrels of oil will be produced from this initiative, which will extend the production lifetime of the oil field by 25 years. It is estimated that 60% of the original oil deposit will be extracted by 2035, after which 30 million tons of CO_2 is projected to be stored—the equivalent of removing about 9 million cars off of roads for a year.

Petroleum experts have predicted that oil production will peak and then decline. Once we tap the easy sources, we are left with unconventional oil found in more difficult locations. It is found thousands of feet below seawater, extractable only with deep-water drilling rigs. It is found in the tarry oil sands of Canada. And it is found in the oil shales of Utah, Colorado, and Wyoming. The amounts involved—perhaps 3 trillion barrels—are impressive. Nonetheless, had this oil been easily recoverable, it would not today still lie embedded in the oil sands and shales.

5.11 | Natural Gas: A "Clean" Fossil Fuel?

Natural gas, which is primarily composed of methane, CH_4, provides heat for more than half of the homes in the United States. Heat from natural gas is furnished either directly at the home or via electricity produced by burning natural gas at a power plant. Among the fossil fuels, natural gas is relatively clean. When burned, it releases essentially no sulfur dioxide and relatively little particulate matter, carbon monoxide, and nitrogen oxides. No ash residue containing toxic metals remains after combustion. Although burning natural gas does produce CO_2, a greenhouse gas, the amount is less per unit of energy released than for the other fossil fuels.

Your Turn 5.23 Skill Building Coal Versus Natural Gas

The combustion of one gram of natural gas releases 50.1 kJ of heat.

a. Calculate the mass of CO_2 released when natural gas is burned to produce 1500 kJ of heat. Assume that natural gas is pure methane, CH_4.

b. Select one of the grades of coal from **Table 5.2**. Compare the mass of CO_2 produced when enough of this coal is burned to produce the same 1500 kJ of heat.
Hint: Assume coal is $C_{135}H_{96}O_9NS$. You calculated its molar mass in Your Turn 5.19.

Oil reservoirs are typically accompanied by pockets of natural gas. The controversial process of hydraulic fracturing, or **fracking**, is used for natural gas extraction. Although first attempted in the 1940s, only more recently has fracking been carried out on a large scale. In 2004, the first well was drilled into the Marcellus Shale underlying Pennsylvania, West Virginia, and parts of nearby states. In 2010, production of natural gas from the Marcellus Shale reached a billion cubic feet per day.

Did You Know? Since the late 1970s, hydraulic fracturing has been used to increase the yield of drinking water from selected wells in a number of countries, such as the United States, South Africa, and Australia.

Fracking involves drilling down into the gas- or oil-bearing rocks that lie 1–3 miles beneath Earth's surface. This technique is required to obtain natural gas or petroleum from hard rock formations (*e.g.*, shales) that have low permeabilities. However, fracking is also used to increase the rate at which fossil fuels are recovered from porous formations (*e.g.*, sandstones, limestones).

This technique is highly controversial because it involves high water consumption, and may impact the environment through water contamination, air emissions, and climate change, while also posing a potential health risk to humans. Furthermore, some reports link seismic events such as earthquakes and tremors to increasing fracking activities. Consequently, varying federal and regional regulations have been instituted across the globe, with some regions opting to completely ban this activity, such as France and the U.S. states of Vermont and New York.

A fracking fluid, consisting of water that contains a cocktail of substances, is injected under pressure in order to create cracks into which natural gas and petroleum can flow (Figure 5.24). The water also carries fine sand that props open these cracks. The next activity gives you the opportunity to pin down some of the details of fracking.

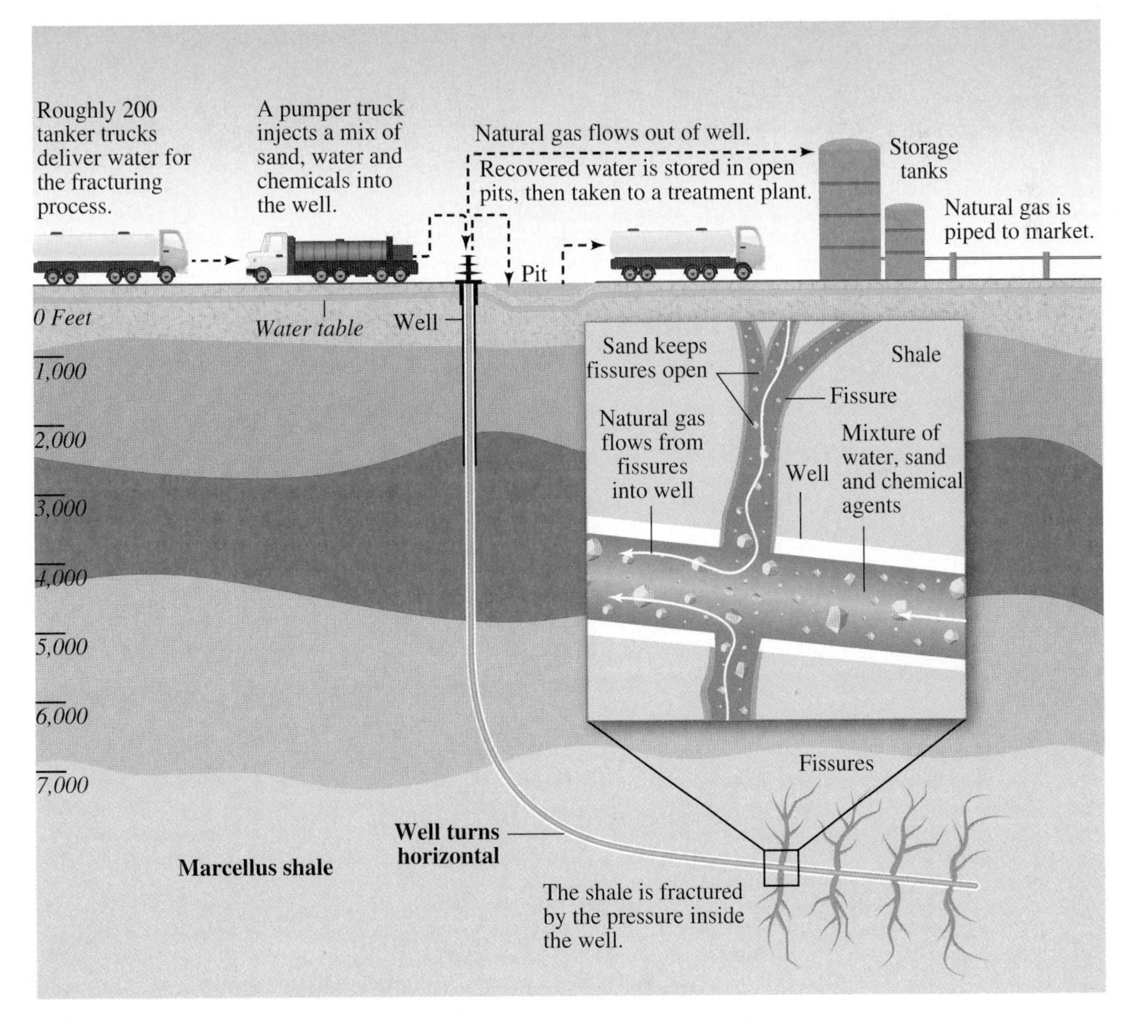

Figure 5.24

Schematic of fracking, used to extract natural gas from underground reservoirs.

Your Turn 5.24 Scientific Practices Fracking!

a. During the extraction of natural gas, the fracturing of shale is done hydraulically rather than with dynamite. What does the term *hydraulic* mean? Suggest reasons why dynamite wouldn't work.

b. How much water is typically injected in a well? What are the ingredients of the "cocktail" that this water contains? How difficult was this information to find? Provide an assessment of the likely reliability of your sources.

c. Some of the water returns to the surface as waste water. What are some of the options for handling this waste water?

5.12 Cracking the Whip: How Do We Obtain Useful Petroleum Products from Crude Oil?

Petroleum is a mixture of several thousand different compounds, with the great majority being hydrocarbons composed of 5–12 carbon atoms per molecule. Many of the hydrocarbons in petroleum are **alkanes**, hydrocarbons with only single bonds between carbon atoms.

Table 5.3 Selected Alkanes (Gases and Liquids)

Name and Chemical Formula	Boiling Point (physical state at 25 °C)	Structural Formula	Condensed Structural Formula
Methane CH_4	−161 °C (gas)	H H—C—H H	CH_4
Ethane C_2H_6	−89 °C (gas)	H H H—C—C—H H H	CH_3CH_3
Propane C_3H_8	−42 °C (gas)	H H H H—C—C—C—H H H H	$CH_3CH_2CH_3$
n-Butane C_4H_{10}	−0.5 °C (gas)	H H H H H—C—C—C—C—H H H H H	$CH_3CH_2CH_2CH_3$
n-Pentane C_5H_{12}	36 °C (liquid)	H H H H H H—C—C—C—C—C—H H H H H H	$CH_3CH_2CH_2CH_2CH_3$
n-Hexane C_6H_{14}	69 °C (liquid)	H H H H H H H—C—C—C—C—C—C—H H H H H H H	$CH_3CH_2CH_2CH_2CH_2CH_3$
n-Heptane C_7H_{16}	98 °C (liquid)	H H H H H H H H—C—C—C—C—C—C—C—H H H H H H H H	$CH_3CH_2CH_2CH_2CH_2CH_2CH_3$
n-Octane C_8H_{18}	125 °C (liquid)	H H H H H H H H H—C—C—C—C—C—C—C—C—H H H H H H H H H	$CH_3CH_2CH_2CH_2CH_2CH_2CH_2CH_3$

Note: n-butane, *n*-pentane, *n*-hexane, *n*-heptane, and *n*-octane all have other structural forms, known as *isomers* (see Section 5.13). The *n* stands for normal, the straight-chain isomer.

Figure 5.25

Photo of a crude oil refinery showing the tall distillation towers. Small amounts of natural gas are flared, as evidenced by the flames.

How are gasoline and other hydrocarbons produced from petroleum? The process takes place at an oil refinery, the icon of the petroleum industry (Figure 5.25). During the initial step in the refining process, the crude oil is separated into fractions, including the gasoline fraction. Refineries work their magic on crude oil using several processes, including that of **distillation**, a separation process in which a solution is heated to its boiling point and the vapors are collected and condensed. But just what happens to a liquid at the molecular level when its boiling point is reached?

Did You Know? Gasoline was first produced in the mid-1800s; however, it didn't become valuable until the advent of the automobile in the early 20th century.

The **volatility** of a liquid refers to how easily it is transformed into its gaseous phase—a process known as **vaporization**. When a liquid is heated, the **intermolecular forces** that hold together individual molecules are broken. The attractive forces between hydrocarbon molecules are referred to as **London dispersion forces**, and are quite weak relative to the covalent bonds holding the atoms together within the molecules (Figure 5.26). Whereas C–H covalent bonds in hydrocarbons are on the order of 400 kJ/mole, the strength of London dispersion forces are only 10–20 kJ/mole. Molecules that contain larger and heavier atoms will exhibit stronger dispersion forces than smaller and lighter ones. Furthermore, molecules with similar structures but greater molar masses will also experience stronger London dispersion forces.

At any temperature greater than absolute zero (0 K or –273 °C), all liquids will experience some degree of vaporization. As more molecules are released from the liquid state into the gaseous state, the **vapor pressure** of the liquid increases (Figure 5.27). The **boiling point** of a liquid is the temperature at which the vapor pressure of the liquid equals the ambient pressure. At sea level, the atmospheric pressure is 1 atm (101 kPa or 760 Torr). So, water at sea level needs to be heated to its boiling point of 100 °C so

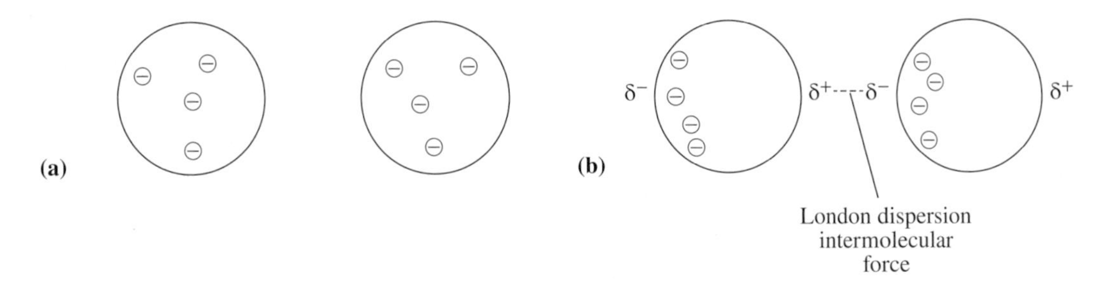

Figure 5.26

Illustration of London dispersion intermolecular forces. **(a)** Two separate atoms/molecules with a random distribution of electrons, which results in no measurable attraction between the species. The positively charged nucleus is not shown. **(b)** Since electrons are in constant motion, the approach of neighboring atoms/molecules may cause more electrons to be situated on one side of the species than the other. This creates partial negative/positive charges on the neighbors, generating a weak attraction. As the constituent atoms increase in size, the electrons are farther from the nucleus, making them easier to re-distribute when approaching their neighbors

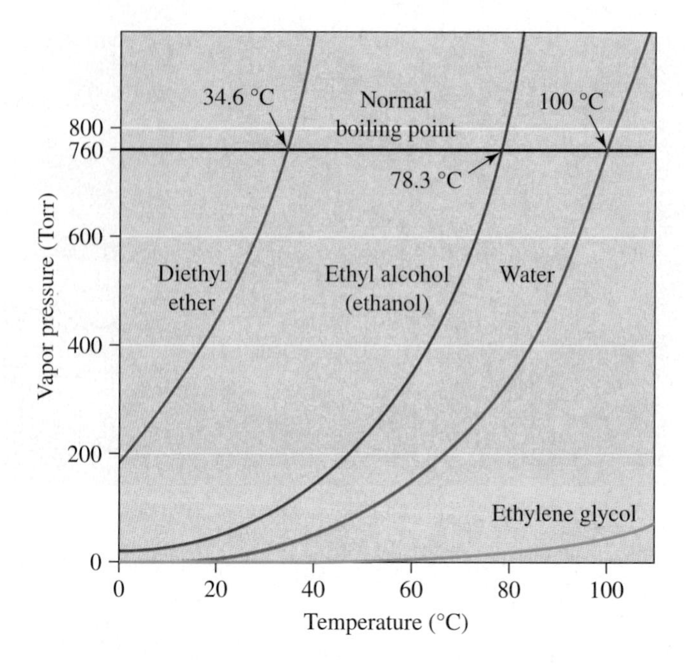

Figure 5.27

The relationship between the vapor pressure of certain liquids and temperature. The *normal boiling point* is defined as the vapor pressure of a liquid at an external pressure of 1 atmosphere (atm) or 760 Torr.

its vapor pressure will match that of the ambient atmosphere. The relationship between the boiling point of a liquid and its intermolecular forces (IMFs) is:

as the strength of IMFs ⇑ *the boiling point of the liquid will* ⇑

Your Turn 5.25 Skill Building Relative Boiling Points

Rank the following molecules in order of increasing boiling points, based on the relative strengths of their intermolecular forces.

a. I_2, Br_2, Cl_2, F_2
b. CH_4, CF_4, CBr_4, CI_4

To distill crude oil, it must first be pumped into a large vessel (the boiler in Figure 5.28) and heated. The distillation process then proceeds:

- As the temperature in the boiler increases, hydrocarbons with lower boiling points (lower molar masses) begin to vaporize.
- As the temperature further increases, compounds with higher boiling points (higher molar masses and/or more complex chemical structures) vaporize.
- Once vaporized, all compounds travel up the distillation tower. Compounds are condensed to liquids at different heights in the tower due to a decrease in temperature at certain heights.

Figure 5.28 illustrates a distillation tower, showing the possible mixtures of compounds that can be obtained. These include solids such as asphalt, liquids such as gasoline, and gases such as methane. The amounts of each depend on the type of crude oil. The "lightest" fraction collected is the *refinery gases*. These compounds have 1–4 carbon atoms per molecule and include methane, ethane, propane, and butane. Refinery gases are flammable and often used as fuel at the refineries. They also can be liquefied and sold for home use. Chemical manufacturers also may use refinery gases as a starting material for making new compounds.

Refinery gases include methane, the main component of natural gas. However, most natural gas is obtained directly from oil and gas wells, rather than from distillation at a refinery. The natural gas piped to your home is practically pure methane, but also includes ethane (2–6%) and other hydrocarbons of low molar mass. Natural gas may also contain small quantities of water vapor, carbon dioxide, hydrogen sulfide, and helium.

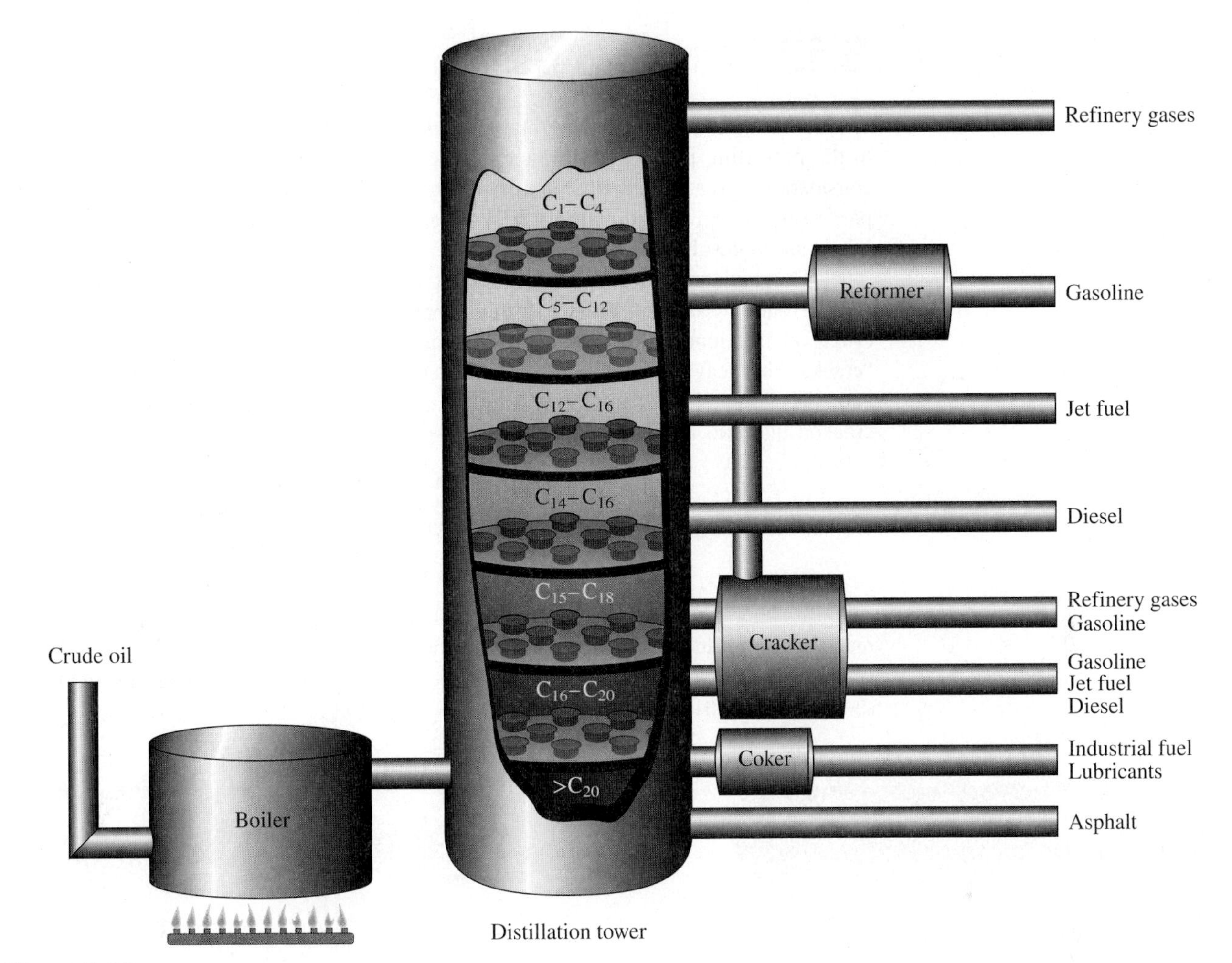

Figure 5.28

Illustration of a distillation tower for crude oil, showing the various fractions and typical applications for the products.

Source: Copyright © McGraw-Hill Education. Permission required for reproduction or display.

In addition to refinery gases, a wide range of compounds is produced at a refinery (Figure 5.29). From a barrel of crude (42 gallons), about 35 gallons are burned for heating and transportation. The remaining gallons are used primarily for non-fuel purposes, including the gallon or two that serve as "feedstocks" to produce plastics, pharmaceuticals, fabrics, and other carbon-based products. As these hydrocarbon feedstocks are nonrenewable resources, one can easily predict that one day petroleum products might become too valuable to burn.

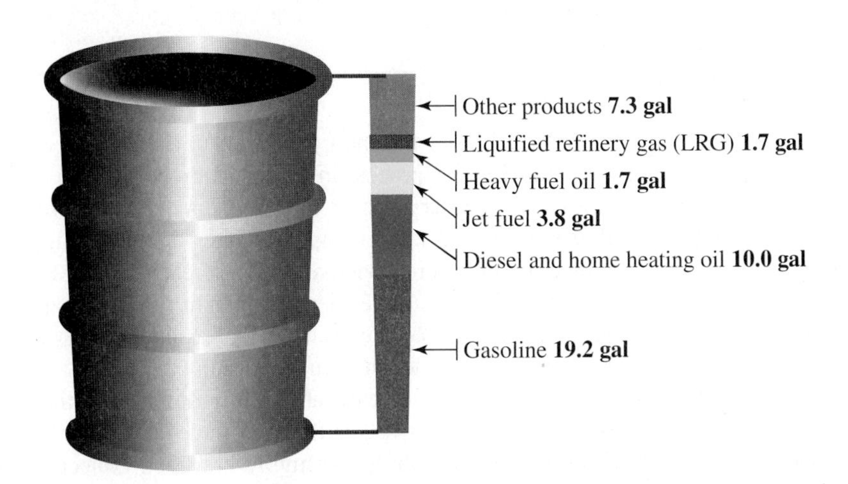

Figure 5.29

Products (in gallons) from the refining of a barrel of crude oil.

Source: U.S. Energy Information Administration, 2009

5.13 | What's in Gasoline?

The distribution of compounds obtained from distilling crude oil does not correspond to the prevailing pattern of commercial use. For example, the demand for gasoline is considerably greater than that for higher-boiling fractions. Chemists employ several processes, known as *cracking* or *reforming*, to change the natural distribution and to obtain more gasoline of higher quality.

Thermal cracking is a process that breaks large hydrocarbon molecules into smaller ones by heating them to a high temperature. In this procedure, the heaviest crude oil fractions are heated to a temperature between 400 and 450 °C. This heat "cracks" the heaviest tarry crude oil molecules into smaller ones useful for gasoline and diesel fuel. For example, at high temperature, one molecule of $C_{16}H_{34}$ can be cracked into two nearly identical molecules:

$$C_{16}H_{34} \xrightarrow{\text{heat}} C_8H_{18} + C_8H_{16} \qquad \textbf{[5.9a]}$$

Thermal cracking also can produce different-sized molecules:

$$C_{16}H_{34} \xrightarrow{\text{heat}} C_{11}H_{22} + C_5H_{12} \qquad \textbf{[5.9b]}$$

In either case, the total number of carbon and hydrogen atoms is unchanged from reactants to products. The larger reactant molecules simply have been fragmented into smaller, more economically important molecules. We can use space-filling models to show the size difference more clearly:

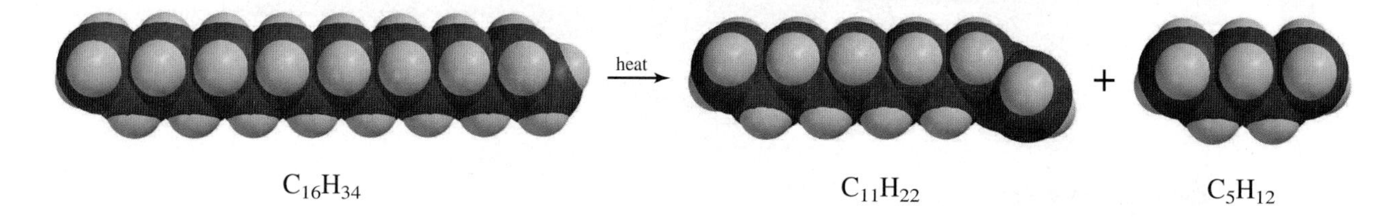

Your Turn 5.26 Skill Building More Practice with Cracking

a. Draw structural formulas for one pair of products formed when $C_{16}H_{34}$ is thermally cracked (see **Equations 5.9a, b**).
Hint: Draw the atoms of the product molecules in a linear chain and include one double bond (only in one product).

b. Revisit the alkanes shown in **Table 5.3**. Look closely to find a pattern for the number of H atoms per C atom. Use this pattern to write the generic chemical formula.

c. Write the generic chemical formula for a hydrocarbon with one C=C double bond.

d. Apply your knowledge of intermolecular forces to predict the relative boiling points of $C_{16}H_{34}$, $C_{11}H_{22}$, and C_5H_{12}. Look up the actual values and compare these to your predictions.

We discuss how catalysts affect the rates of chemical reactions in Section 5.14.

The problem with thermal cracking is the energy required to produce the required high temperatures. **Catalytic cracking** is a process in which catalysts are used to crack larger hydrocarbon molecules into smaller ones at relatively low temperatures, thus reducing energy use. Chemists at all major oil companies have developed important cracking catalysts and continue to find more selective and inexpensive processes.

Sometimes chemists want to combine molecules rather than split them apart. To produce more of the intermediate-sized molecules needed for gasoline, *catalytic combination* can be used. In this process, smaller molecules are joined.

Another important chemical process is **catalytic reforming**. Here, the atoms within a molecule are rearranged, usually starting with linear molecules and producing ones with more branches. As we will see, the more highly branched molecules burn more smoothly in automobile engines.

It turns out that molecules with the same molecular formula are not necessarily identical. For example, octane has the formula C_8H_{18}. Careful analysis reveals 18 different compounds with this formula. Molecules with the same molecular formula but with different chemical structures and different properties are called **isomers**. In *n*-octane (normal octane), the carbon atoms are all in a continuous chain (Figure 5.30a), whereas in iso-octane, the carbon chain has several branch points (Figure 5.30b). The chemical and physical properties of these two isomers are similar, but they are not identical. Molecules with linear structures will have stronger London dispersion forces than branched structures due to the stronger interaction of C–H groups along the entire chain length (Figure 5.31). For example, the boiling point of *n*-octane is 125 °C, compared with 99 °C for iso-octane.

Although the heats of combustion for *n*-octane and iso-octane are nearly identical, the two fuels burn differently in an auto engine. The more compact shape of the latter compound imparts a "smoother" burn. In a well-tuned car engine, gasoline vapor and air are drawn into a cylinder, compressed by a piston, and ignited by a spark. Normal combustion occurs when the spark plug ignites the fuel-air mixture and the flame front travels rapidly across the combustion chamber, consuming the fuel.

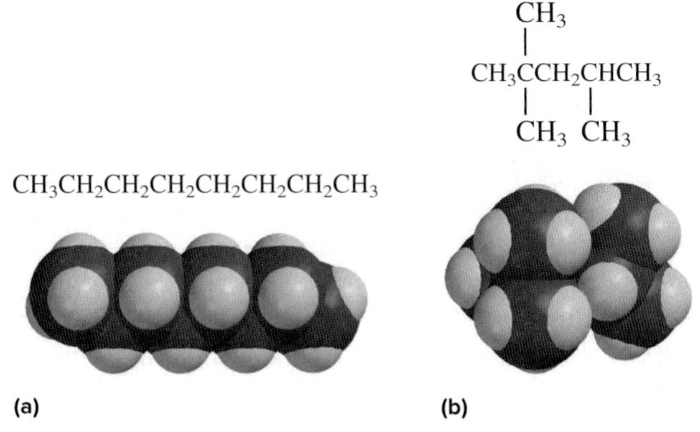

Figure 5.30

Structural formula and space-filling model for **(a)** *n*-octane, and **(b)** iso-octane.

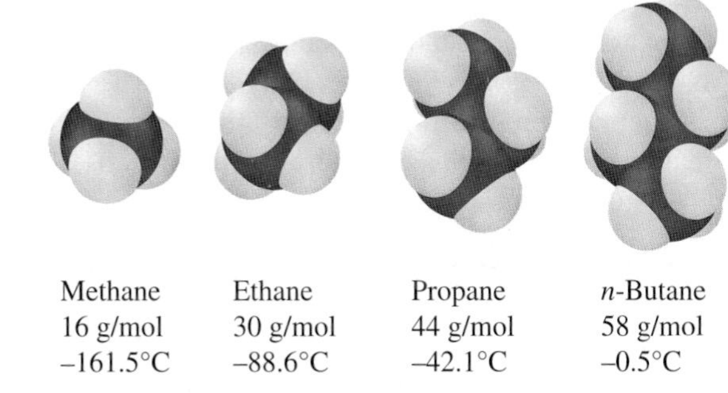

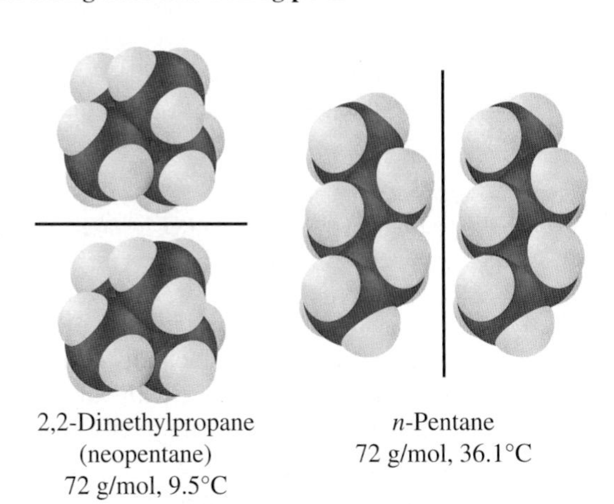

Figure 5.31

The influence of molecular structure on the boiling points of hydrocarbons. **(a)** As the molar mass of neighboring molecules increases, the strength of London dispersion intermolecular forces increases, causing an increasing boiling point. **(b)** As the surface area of neighboring molecular chains increases, the strength of London dispersion intermolecular forces increases, causing an increase in the boiling point.

Table 5.4 Octane Ratings of Several Compounds

Compound	Octane Rating
n-octane	–20
n-heptane	0
iso-octane	100
methanol (CH_3OH)	107
ethanol (CH_3CH_2OH)	108
methyl tertiary-butyl ether (MTBE, $CH_3OC(CH_3)_3$)	116

Sometimes, however, compression alone is enough to ignite the fuel before the spark occurs. This premature firing is called pre-ignition. It results in lower engine efficiency and higher fuel consumption because the piston is not in its optimal location when the burned gases expand. "Knocking," a violent and uncontrolled reaction, occurs after the spark ignites the fuel, causing the unburned mixture to burn at supersonic speed with an abnormal rise in pressure. Knocking produces an objectionable metallic sound, loss of power, overheating, and even engine damage when severe.

In the 1920s, knocking was shown to depend on the chemical composition of the gasoline. The "octane rating" was developed to designate a particular gasoline's resistance to knocking. Iso-octane performs exceptionally well in automobile engines and arbitrarily has been assigned an octane rating of 100. Like *n*-octane, *n*-heptane is a straight-chain hydrocarbon, but with one fewer $-CH_2$ group. It also has a high tendency to knock and is assigned an octane rating of 0 (Table 5.4). Hence, when you go to the gasoline pump and fill up with 87 octane, you are buying gasoline that has the same knocking characteristics as a mixture of 87% iso-octane and 13% *n*-heptane. Higher grades of gasoline also are available: 89 octane (regular plus) and 91 octane (premium) (Figure 5.32). These blends contain a higher percent of compounds with higher octane ratings.

Figure 5.32

Gasoline is available in a variety of octane levels.

© Justin Sullivan/Getty Images

Although *n*-octane has a poor octane rating, it is possible to catalytically re-form *n*-octane to iso-octane, thus greatly improving its performance. This rearrangement is accomplished by passing *n*-octane over a catalyst consisting of rare and expensive elements such as platinum (Pt), palladium (Pd), rhodium (Rh), and iridium (Ir). The reforming of isomers to improve their octane rating became important starting in the late 1970s because of the nationwide efforts to ban the use of tetraethyllead (TEL, $Pb(CH_2CH_3)_4$) as an antiknock additive.

Your Turn 5.27 You Decide Getting the Lead Out

Beginning in the 1920s, the octane-booster tetraethyllead (TEL) was mixed with gasoline to improve vehicle performance and fuel economy. The United States completed the ban on leaded gasoline in 1996 because of the hazards associated with lead exposure and its damaging effect on catalytic converters and spark plugs. However, other sources of lead still exist. Be a detective on the Internet to identify:

a. an occupational source of lead exposure.
b. a hobby that is a source of lead exposure.
c. a source of lead exposure that particularly affects children.

Elimination of TEL as an octane enhancer necessitated finding substitutes that were inexpensive, easy to produce, and environmentally benign. Several were evaluated including methanol, ethanol, and MTBE, each with an octane rating greater than 100 (Table 5.4). Fuels containing these additives are referred to as **oxygenated gasolines**. Since they contain oxygen, these gasoline blends burn more cleanly and produce less carbon monoxide than their non-oxygenated counterparts.

As pointed out earlier in Chapter 3, the volatile organic compounds (VOCs) in conventional gasoline play a role in tropospheric ozone pollution, especially in high-traffic areas. Since 1995, about 90 cities and metropolitan areas with the highest ground-level ozone levels have adopted the Year-Round Reformulated Gasoline Program mandated by the Clean Air Act Amendments of 1990. This program requires the use of **reformulated gasolines** (RFGs)—oxygenated gasolines that contain a lower percentage of volatile hydrocarbons found in non-oxygenated conventional gasoline. RFGs cannot contain more than 1% benzene (C_6H_6), a cancer-causing hydrocarbon, and must be at least 2% oxygen. Due to their composition, reformulated gasolines evaporate less readily than conventional gasolines and produce less carbon monoxide emissions. When RFGs were introduced in the 1990s, MTBE was the oxygenate of choice. However, concerns over its toxicity and its ability to leach from gasoline storage tanks into the groundwater have led many states to ban MTBE and switch to ethanol.

As an additive and a fuel in its own right, ethanol is described more fully in Section 5.15.

5.14 | New Uses for an Old Fuel

World supplies of coal are predicted to last for hundreds of years, much longer than current estimates of remaining available oil reserves. Unfortunately, the fact that coal is a solid makes it inconvenient for many applications, especially as a fuel for vehicles. Therefore, research and development projects are underway aimed at converting coal into fuels that possess characteristics similar to petroleum products.

Before large supplies of natural gas were discovered, cities were lighted with *water gas*, a mixture of carbon monoxide and hydrogen. Water gas is formed by blowing steam over hot coke, the impure carbon that remains after volatile components have been distilled from coal.

$$\underset{\text{coke}}{C(s)} + H_2O(g) \longrightarrow \underbrace{CO(g) + H_2(g)}_{\text{water gas}} \quad [5.10]$$

This same reaction is the starting point for the **Fischer–Tropsch process** for producing synthetic gasoline from coal. German chemists Emil Fischer (1852–1919) and Hans Tropsch (1889–1935) developed the process during the early 20th century. At that time, Germany had abundant coal reserves, but little petroleum.

The Fischer–Tropsch process can be described by this general equation:

$$n\,CO(g) + (2n + 1)\,H_2(g) \xrightarrow{\text{catalyst}} C_nH_{2n+2}(g, l) + n\,H_2O(g) \quad [5.11]$$

The hydrocarbon products can range from small molecules like methane, to the medium-sized molecules (n = 5–8) typically found in gasoline. This chemical reaction proceeds when the carbon monoxide and hydrogen are passed over a catalyst containing iron or cobalt.

To better understand the role of the catalyst, consider a typical exothermic reaction, as shown in Figure 5.33. The potential energy of the reactants (left side) is higher

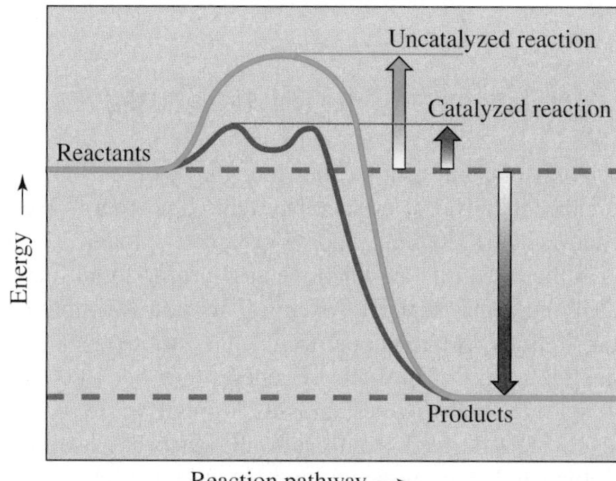

Figure 5.33

Energy-reaction pathway for the same reaction with (**blue** line) and without (**green** line) a catalyst. The green and blue arrows represent the activation energies. The red arrow represents the overall energy change for either pathway.

than the potential energy of the products (right side) because it is an exothermic reaction. Now examine the pathways that connect the reactants and products. The green line indicates the energy changes during a reaction in the absence of a catalyst. Overall, this reaction gives off energy, but the energy initially *increases* because some bonds break (or start to break) first. The energy necessary to initiate a chemical reaction is called its **activation energy** and is indicated by the green arrow for the uncatalyzed reaction. Although energy must be expended to get the reaction started, energy is given off as the process proceeds to a lower potential energy state. Generally, reactions that occur rapidly have low activation energies; slower reactions have higher activation energies. However, there is no direct relationship between the height of the activation barrier and the net energy change in the reaction. In other words, a highly exothermic reaction can have a large or a small activation energy.

Increasing the temperature often results in increased reaction rates; when molecules have extra energy, a greater fraction of collisions can overcome the required activation energy. Sometimes, however, increasing the temperature isn't a practical solution. The blue line in Figure 5.33 shows how a catalyst can provide an alternative reaction pathway and thus a lower activation energy (represented by the blue arrow), without raising the temperature. However, the net energy change is the same with and without the addition of a catalyst.

In the Fischer–Tropsch process, strong C≡O triple bonds must be broken for the reaction to proceed. Breaking this bond corresponds to an activation energy so large that the reaction simply does not proceed without the help of a metal catalyst. This is the point at which the metal catalyst enters the reaction. Molecules of CO can form bonds with the metal surface, and when this happens, the C≡O bonds weaken. The hydrogen molecules also attach to the metal surface, completely breaking the H–H single bonds. The rest of the reaction proceeds quickly, producing hydrocarbons with a higher molar mass.

The advantage of a catalyst is that it is not consumed and thus only small amounts of it are needed. Green chemists value catalytic reactions not only because small amounts of catalysts are employed, but also because the reaction often can be carried out at lower temperatures.

Historically, commercialization of Fischer–Tropsch technology has been limited. South Africa, a coal-rich and oil-poor nation, is the only country that synthesizes a majority of its gasoline and diesel fuel from coal. Any spike in oil prices, coupled with a plentiful domestic coal supply, may spark increased use of the Fischer–Tropsch process in other energy-hungry countries. As of 2008, China is constructing a coal-to-liquid fuel plant in Inner Mongolia. In the United States, an Australian energy corporation announced plans to build a $7 billion coal-to-liquids plant in Big Horn County, Montana, home to the Apsáalooke people (Crow Nation). The deposits there are estimated to contain almost 9 billion tons of coal.

Coal, whether solid or converted to liquid fuels, still burns to produce CO_2. Recent work by the National Renewable Energy Laboratory indicates that greenhouse gas emissions over the entire fuel cycle for producing coal-based liquid fuels are nearly twice as high as their petroleum-based equivalent. Thus, we need to continue the search for fuels to replace coal.

5.15 | From Brewery to Fuel Tank: Ethanol

"A sustainable society is one that is far-seeing enough, flexible enough, and wise enough not to undermine either its physical or social systems of support." We repeat the words from Donella Meadows (1941–2001), a biophysicist and the founder of the Sustainability Institute, noting that the rapid rate at which we are burning fossil fuels is indeed undermining our physical and social support systems. This rate is simply *not* sustainable.

What are our options? Some people think that a more sustainable energy future will require the increased use of **biofuels**, a generic term for a renewable fuel derived from a biological source, such as trees, grasses, animal waste, or agricultural crops. Biofuels can replace fuels derived from crude oil, such as gasoline and diesel fuel. Although most biofuels today are not being produced in a sustainable manner, research continues to evaluate how they could be in the future.

Like fossil fuels, all biofuels release CO_2 when burned. However, biofuels should release a lower net amount of CO_2 into the atmosphere than fossil fuels, which is definitely a plus. Why? The plants used to make today's biofuels originally absorbed CO_2 from the atmosphere while they were growing. Based on the principles of the First Law of Thermodynamics, whether burned as a fuel or not, these plants release this same amount of CO_2 back into the biosphere after they die. In contrast, fossil fuels have kept their carbon "locked up" underground if not extracted and burned as fuels. Therefore, when burning fossil fuels there is a net increase in CO_2 emissions into the atmosphere. The assertion that the net amount of CO_2 released from biofuels is smaller assumes that the energy used to produce and transport the biofuel does not cancel out this benefit. As you will see in the next section, this assertion needs to be questioned and challenged.

Wood, the most common biofuel, has been used throughout human history for cooking and heating. Have you ever wondered why wood burns? Wood contains **cellulose**, a naturally occurring compound composed of C, H, and O that provides structural rigidity in plants, shrubs, and trees. Similar to hydrocarbons, cellulose is made up of carbon and hydrogen. Unlike these, however, cellulose also contains oxygen, which lowers its energy content as a fuel. In fact, all of the biofuels we will describe in this section contain some oxygen. As their oxygen content increases, biofuels release proportionately less energy per mass than hydrocarbons during their combustion (revisit Figure 5.6).

Cellulose is a carbohydrate composed of a chain of thousands of glucose molecules linked together. This is why we equated burning wood to burning glucose in Figure 5.6. You may recognize glucose, $C_6H_{12}O_6$, as a sugar. Equation 5.12 lists the chemical equation for burning wood, which is the same equation as the one we provided for respiration ("burning glucose") in your body.

Natural and synthetic molecules composed of repeating sub-units, known as "polymers" will be discussed in Chapter 9.

$$\underset{\text{glucose}}{C_6H_{12}O_6} + 6\ O_2 \longrightarrow 6\ CO_2 + 6\ H_2O + \text{energy} \qquad [5.12]$$

Even though widely available in many parts of the world, the supply of wood is insufficient to meet our energy demands. Cutting down trees for fuel also destroys vegetation that effectively absorb CO_2 from our atmosphere. So instead of relying on wood, people in all sectors are eyeing liquid biofuels such as ethanol.

Ethanol is the same alcohol found in wine, beer, and spirits. It is a clear, colorless, and flammable liquid. Since ancient times, people have known how to ferment grain in order to produce ethanol. Admittedly, their purpose was to brew alcoholic beverages rather than to fuel automobile engines.

Recall that the "eth-" prefix implies two carbons in the molecular structure.

Which sugars and grains can be fermented? Almost any will, although the latter may require a catalyst to nudge the process along. The choice depends on both availability and politics. Today in the United States, most of the ethanol is produced by fermenting the sugars and starches found in corn (Figure 5.34). However, in early human history, corn

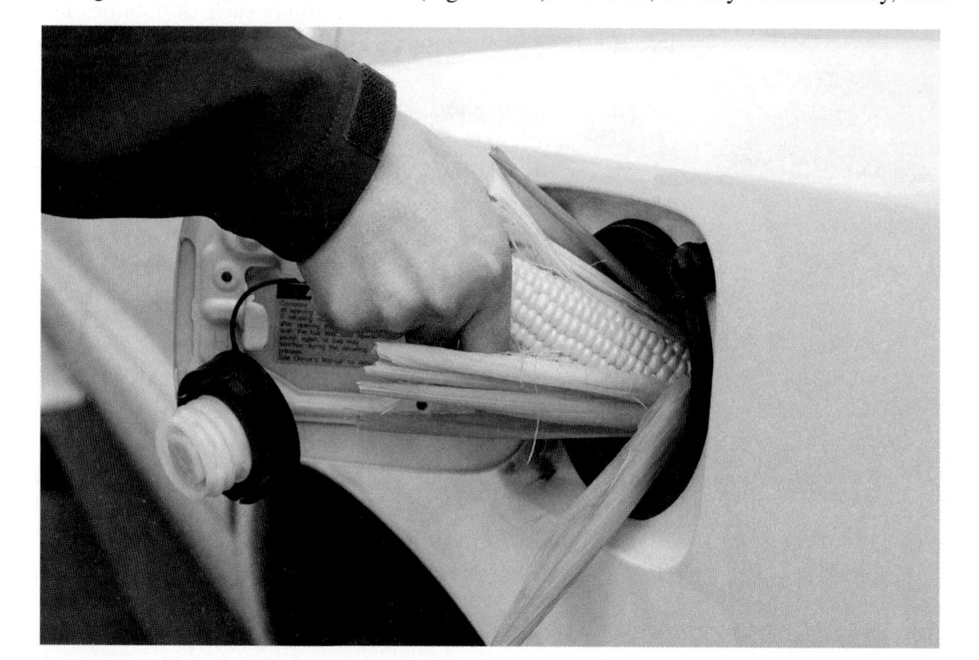

Figure 5.34

An advertisement for ethanol, a renewable fuel that can be produced from many different grains, including corn.

was not as widely available as it is today. As you'll learn in Chapter 13, the people of the New World bred corn from a wild strain and made a crop that readily grew. Those living elsewhere used other grains such as rice and barley to brew alcoholic beverages. This is why ethanol also goes by the name of grain alcohol.

In the context of oxygenated gasoline, we introduced ethanol in Section 5.13. For your convenience, we provide its Lewis structure:

```
     H   H
     |   |       ..
H — C — C — O — H
     |   |       ..
     H   H
```

ethanol

Ethanol is an example of an **alcohol**, a hydrocarbon substituted with one or more –OH groups (hydroxyl groups) bonded to its carbon atoms. Just like hydrocarbons, alcohols are flammable and burn to release energy. During complete exothermic combustion, the products are CO_2 and H_2O:

$$C_2H_5OH(l) + 3\ O_2(g) \longrightarrow 2\ CO_2(g) + 3\ H_2O(g) + 1240\ \text{kJ} \qquad \textbf{[5.13]}$$

Did You Know? The simplest alcohol, methanol (CH_3OH), is significantly more toxic to consume than ethanol. As little as 10 mL of pure methanol can cause permanent blindness, and 30 mL is potentially fatal.

Because alcohols contain one or more –OH groups, their properties differ from those of hydrocarbons. One difference is that humans can safely consume small amounts of ethanol in wine, beer, and other alcoholic drinks. In contrast, hydrocarbons have no appeal as a beverage, and many are *carcinogenic* (cancer-causing) upon skin contact or inhalation. Another difference is their solubility. For example, ethanol dissolves readily in water, whereas hydrocarbons are insoluble in water.

Alcohols are hydrocarbons that contain a **functional group**; that is, a distinctive arrangement of a group of atoms that imparts characteristic properties to the host molecule. To emphasize the hydroxyl group (–OH) present in any alcohol, chemists typically write C_2H_5OH for ethanol rather than C_2H_6O. Another example of a functional group is C=C, the double bond, mentioned in Your Turn 5.26. In the next section on biofuels, we offer a sneak peak at another functional group, the ester.

Enzymes are biological catalysts. The structures and functions of enzymes will be detailed in Chapters 11, 12, and 13.

Several steps are required to obtain ethanol from corn. The first involves making a "soup" of corn kernels and water. The second step uses **enzymes** to catalyze the breakdown of the starch molecules found in these kernels to make glucose. Like cellulose, **starch** is a carbohydrate found in many grains, such as corn and wheat. It is a natural polymer of glucose; like cellulose, starch molecules are composed of a chain of thousands of glucose molecules linked together. But unlike cellulose molecules, the links in starch are formed differently such that the enzymes in our bodies can break them. Thus, we can digest the starch in foods such as potatoes and rice. However, we cannot digest the cellulose found in paper and leafy foods such as lettuce, and thus refer to it as the "roughage" in our diets.

Figure 5.35

The Archer Daniels Midland ethanol plant in Peoria, Illinois.

The third step, fermentation, converts glucose to ethanol. Yeast cells take over by releasing different enzymes that catalyze this conversion:

$$\underset{\text{glucose}}{C_6H_{12}O_6} \xrightarrow{\text{yeast enzymes}} 2\ \underset{\text{ethanol}}{C_2H_5OH} + 2\ CO_2 \qquad \textbf{[5.14]}$$

The result is an alcoholic brew, and not a very tasty one, with an alcohol content of about 10%. To separate the ethanol, the final step is to distill the mixture. Recall how the components of crude oil can be separated by their boiling points at an oil refinery. The same principles apply here. Ethanol and water differ in their boiling points and can be separated by distillation. Check out Figure 5.35 to see the tall distillation towers at an ethanol plant. There are currently 195 ethanol plants operating in the continental United States.

Your Turn 5.28 You Decide A Picture Is Worth . . . ?

This data table shows the annual ethanol production (millions of gallons) in the United States.

Year	Ethanol	Year	Ethanol	Year	Ethanol
1980	175	1991	950	2002	2130
1981	215	1992	1,100	2003	2800
1982	350	1993	1,200	2004	3400
1983	375	1994	1,350	2005	3904
1984	430	1995	1,400	2006	4855
1985	610	1996	1,100	2007	6500
1986	710	1997	1,300	2008	9000
1987	830	1998	1,400	2009	10,600
1988	845	1999	1,470	2010	13,230
1989	870	2000	1,630	2011	13,900
1990	900	2001	1,770	2012	13,300

Source: Renewable Fuels Association

a. Find entries for more recent years to update this table.
b. Present this information to a public audience of your choice by some other visual means and explain why this source of fuel may be beneficial for meeting our future energy needs.

With more than 14 billion gallons of ethanol produced annually in 2014, the United States is the world's largest ethanol producer. Brazil is second, with about 8 billion gallons of ethanol. Together, these two countries account for about 85% of the world's production. Corn is the raw material in the United States. In contrast, Brazil derives almost all of its ethanol by fermenting sugarcane. Why the difference?

To answer this question, we again note that practically any sugar or grain can be fermented to produce ethanol. The substance fermented depends on its availability, economics, and politics. Sugarcane plants are rich in sucrose, also called "table sugar." Just as the glucose produced from the starch of corn kernels can be fermented to produce ethanol, so can sucrose. In Brazil, sugarcane is grown in areas that once were tropical rain forests. In the United States, corn is grown in the Midwest in areas that were once prairies and forests. In both cases, the use of land to produce biofuel remains a controversial topic.

Worldwide, another possible and less controversial source of ethanol is cellulose, a compound we mentioned earlier that gives support to plants, shrubs, and trees. **Cellulosic ethanol** is ethanol produced from any plant containing cellulose, typically cornstalks, switchgrass, wood chips, and other materials that are inedible by humans. Although you may not recognize the name switchgrass, if you live in the Plains regions of the United States, Canada, or Mexico, you most likely have seen it. Switchgrass is a native plant; Figure 5.36 shows one of its varieties.

Figure 5.36

Switchgrass, a perennial plant that is native to North America.

Source: Warren Gretz/NREL/U.S. Dept. of Energy

Derived from inedible plants like switchgrass, cellulosic ethanol has a widespread appeal. For years, chemists have been successful at producing small amounts of cellulosic ethanol in the laboratory. But carrying out the process in batches large enough to obtain millions of gallons of ethanol has turned out to be another matter entirely. Like starch, cellulose does not ferment and therefore must first be broken down into sugars. As of 2015, the enzymes that catalyze the breakdown of cellulose were expensive and their rate of reaction was slow.

Your Turn 5.29 Scientific Practices Biofuel from Nonfoods

List three desirable characteristics of sources of cellulosic ethanol, such as wood chips and switchgrass.

Figure 5.37

Gasoline is often blended with ethanol to make E10 ("gasohol") that is 90% gasoline, and 10% ethanol.

(left): © Jeffrey Sauger/Bloomberg via Getty Images; (right): © Ashley Cooper pics/Alamy Stock Photo

Regardless of its source, ethanol doesn't go straight into your gas tank, because most of our automobiles are not engineered to burn it alone. However, auto engines can run on "gasohol," a blend of gasoline with ethanol. Currently, more than 30 countries have mandated that gasoline contain up to 10% ethanol (Figure 5.37). This blend and other "oxygenated" fuels have the added benefit of higher octane ratings and reducing vehicle emissions that produce ground-level ozone.

Beginning with legislation launched in 2007, the United States sought to reduce its dependence on oil imports and increase its use of renewable fuels. As a result, a shift to E15 was proposed. As you might suspect, E15 is 15% ethanol, 85% gasoline. What you may not be anticipating, however, is the controversies that ensued given the relatively smooth transition to E10. Some constituencies were confident that the transition to a new fuel blend could be easily orchestrated. Others cited the need to carefully check whether increased concentrations of ethanol would corrode existing fuel tanks or adversely affect livestock feed prices. In addition, lawn mowers, boats, and snowmobiles currently cannot run on E15. As of 2015, the U.S. Environmental Protection Agency has approved the use of E15, but has not mandated it.

Whether using E10 or E15, don't confuse its octane rating with gas mileage. The octane rating relates to how smoothly the fuel burns in the engine, rather than to the energy content of the fuel. The octane rating of a gasoline increases as more ethanol is added. But with more ethanol, the gas mileage decreases slightly.

Why? Recall that ethanol releases a lower amount of energy per amount burned than do the hydrocarbons found in gasoline. Using *n*-octane as a representative hydrocarbon, here are the two combustion equations:

$$C_2H_5OH(l) + 3\ O_2(g) \longrightarrow 2\ CO_2(g) + 3\ H_2O(g) + 1240\ \text{kJ} \qquad [5.15]$$

$$C_8H_{18}(l) + \tfrac{25}{2}\ O_2(g) \longrightarrow 8\ CO_2(g) + 9\ H_2O(g) + 5060\ \text{kJ} \qquad [5.16]$$

Per gram, the values are 26.8 kJ/g for C_2H_5OH and 44.4 kJ/g for C_8H_{18} (Figure 5.6). Ethanol releases less energy because it contains oxygen. As a fuel, ethanol already is partially oxidized, or "burned."

As we pointed out at the start of this section, ethanol is not the only biofuel in town. The next section is devoted to *biodiesel*, another renewable fuel.

5.16 | From Deep Fryer to Fuel Tank: Biofuels

The production of biodiesel has grown dramatically in recent years. Its synthesis is so straightforward that you may have carried it out in your chemistry lab. Biodiesel is unique among transportation fuels in that it can be produced economically in small batches by individual consumers, including students. As we will discover later in this section, it is produced commercially as well.

Although biodiesel is made primarily from vegetable oils, animal fats work equally well. As you may already know, oils and fats are part of your diet and help fuel your body. When you dip your bread in olive oil or spread a roll with butter, you are preparing to consume an oil or a fat. Although biodiesel could be synthesized from olive oil or butter, both are too expensive (and too tasty) to use as a starting material. Instead, biodiesel is made from soy, rapeseed, or palm oil. It also can be produced from waste cooking oil, such as that used to make french fries. Figure 5.38 shows a carton of oil that is destined for a restaurant fryer, together with the waste cooking oil.

To understand why fats and oils can serve as starting materials for biodiesel, we need to know more about **triglycerides**, a class of compounds that includes both fats and oils. **Fats** such as butter and lard are triglycerides that are solids at room temperature. In contrast, **oils** such as olive oil and soybean oil are triglycerides that are liquids. Either as liquids or solids, triglycerides are the starting material for biodiesel. They occur naturally in both plants and animals.

Here is the structural formula for glyceryl tristearate, a triglyceride found in animal fat:

$$
\begin{array}{c}
H_3C-(CH_2)_{16}-C(=O)-O-CH \\
| \\
H_3C-(CH_2)_{16}-C(=O)-O-CH_2 \quad CH_2-O-C(=O)-(CH_2)_{16}-CH_3
\end{array}
$$

Actually, any triglyceride would serve our purposes because they all share common structural features. We picked this particular one because you will encounter it later in the context of nutrition (Chapter 11).

Glyceryl tristearate is a complex molecule. Even so, this structure contains three hydrocarbon chains, illustrated above in red. Can you see their resemblance

(a)

(b)

Figure 5.38

Restaurants purchase oil for frying in large amounts. Shown here is **(a)** a 35-pound box, and **(b)** hot oil out of the fryer. Depending on the cook, the cooking oil is changed frequently (shown here) or infrequently. In the latter case, the oil darkens with waste products.

(a–b): © Cathy Middlecamp

to hydrocarbon fuels? Each of these hydrocarbon chains, if snipped off, could serve as diesel fuel, a mixture of hydrocarbons with 14 to 16 carbon atoms (revisit Figure 5.28). When you eat foods containing triglycerides, each "diesel-like" hydrocarbon chain is metabolized slowly in your body to release energy and produce CO_2 and H_2O. Although diesel fuel burns in an engine much more rapidly and at a higher temperature, the net result is the same. Energy is released, and CO_2 and H_2O are produced.

Although soybean oil and other triglycerides will burn, they should not go straight into your gas tank. Before triglycerides can be utilized as a fuel, they need to be snipped into smaller pieces that are closer in size (for ease of evaporation) to the molecules in diesel fuel. One way to do this is to react them with an alcohol such as methanol (CH_3OH) and a catalytic amount of sodium hydroxide (NaOH). Equation 5.17 lists the chemical reaction, using glyceryl tristearate (an animal fat) as the starting material.

$$\underset{\text{glyceryl tristearate (a triglyceride)}}{C_{57}H_{110}O_6} + 3\ CH_3OH \xrightarrow{\text{NaOH}} 3\ \underset{\text{methyl stearate (a biodiesel molecule)}}{CH_3(CH_2)_{16}\overset{\overset{\displaystyle O}{\|}}{C}OCH_3} + \underset{\text{glycerol}}{C_3H_8O_3} \qquad [5.17]$$

One glyceryl stearate molecule produces three biodiesel molecules that each contain a long chain of carbon atoms. Depending on the fat or oil used as a starting material, other biodiesel molecules are possible, such as methyl lineolate:

$$CH_3CH_2CH_2CH_2CH_2CH{=}CHCH_2CH{=}CHCH_2CH_2CH_2CH_2CH_2CH_2CH_2\overset{\overset{\displaystyle O}{\|}}{C}OCH_3$$

In general:

- Biodiesel molecules contain a hydrocarbon chain, typically with 16–20 carbon atoms.
- The hydrocarbon chains usually contain one or more C=C bonds, especially if the triglyceride used as a starting material is an oil.
- In addition to the hydrocarbon chain, each biodiesel molecule also contains oxygen. The two O atoms form part of the *ester* functional group, which we will describe later in the context of polyesters in Chapter 9.
- Triglycerides (fats and oils) typically produce a mixture of different biodiesel molecules, unlike the triglyceride glyceryl stearate that produced just one product.

Also notice the three molecules of methanol in Equation 5.17. Methanol provides the $-OCH_3$ group to "cap" each carbon chain at the point it was snipped from the larger triglyceride molecule. Other alcohols, including ethanol, can work equally well. No matter which alcohol is used, the net result is three molecules of biodiesel. The next activity is designed to refresh your memory about alcohols such as ethanol and methanol. It also sets the stage for glycerol, the alcohol that is a product of the biodiesel synthesis.

Your Turn 5.30 Skill Building More About Alcohols

In this chapter, you already have encountered two alcohols: ethanol and methanol.

a. Draw the structural formula for each.

b. Here is a new alcohol: $CH_3CH_2CH_2OH$. Suggest a name.

c. There is another alcohol with the same chemical formula as $CH_3CH_2CH_2OH$; that is, an isomer of this alcohol. Draw its structural formula.

If you worked through the previous activity, it should be evident that many different alcohols can be derived from hydrocarbons. All you need to do is substitute an –OH group for one of the H atoms. Furthermore, alcohols can contain more than one

–OH group. The synthesis of biodiesel produces one such multifunctional alcohol, **glycerol.** In Equation 5.17, we gave only its chemical formula, $C_3H_8O_3$. From its structural formula, you can see that glycerol is a "triple" alcohol:

$$\begin{array}{ccccccc} & \text{H} & & \text{H} & & \text{H} & \\ & | & & | & & | & \\ \text{H}- & \text{C} & - & \text{C} & - & \text{C} & -\text{H} \\ & | & & | & & | & \\ & \text{OH} & & \text{OH} & & \text{OH} & \end{array}$$

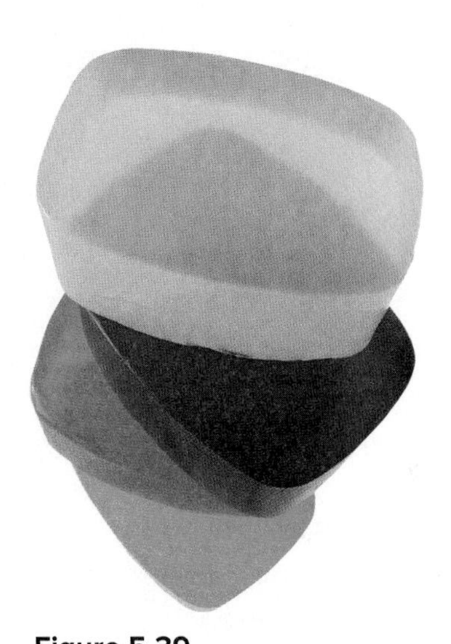

Figure 5.39

Glycerin (glycerol) is one of the ingredients in translucent soaps.

Glycerol, a by-product of the synthesis of biodiesel, is used in many different consumer products. You may know it by the name glycerin, a common ingredient in soaps and cosmetics (Figure 5.39). However, every 9 pounds of biodiesel nets 1 pound of glycerol, which has resulted in a glut of glycerol on the market. In 2006, Galen Suppes and coworkers at the University of Missouri earned a Presidential Green Chemistry Challenge Award for a process to convert glycerol to a different alcohol, propylene glycol:

$$\underset{\text{glycerol}}{\begin{array}{ccccccc} & \text{H} & & \text{H} & & \text{H} & \\ & | & & | & & | & \\ \text{H}- & \text{C} & - & \text{C} & - & \text{C} & -\text{H} \\ & | & & | & & | & \\ & \text{OH} & & \text{OH} & & \text{OH} & \end{array}} + H_2(g) \xrightarrow{\text{copper catalyst}} \underset{\text{propylene glycol}}{\begin{array}{ccccccc} & \text{H} & & \text{H} & & \text{H} & \\ & | & & | & & | & \\ \text{H}- & \text{C} & - & \text{C} & - & \text{C} & -\text{H} \\ & | & & | & & | & \\ & \text{OH} & & \text{OH} & & \text{H} & \end{array}} + H_2O(l) \quad [5.18]$$

The U.S. Food and Drug Administration (FDA) lists propylene glycol as "generally recognized as safe" and has approved it for use as a food additive. The compound also finds uses as a moisturizer in cosmetics and as a solvent for some drugs not soluble in water. As an antifreeze in vehicles and as a de-icer at airports, it is far less toxic than ethylene glycol, another compound used as an antifreeze. Propylene glycol produced in this manner is from a renewable resource, not requiring petroleum as a feedstock. Conversion of the glycerol into value-added products lowers the cost of biodiesel production, making it more competitive with petroleum-derived diesel fuel.

Your Turn 5.31 Scientific Practices Heat of Combustion for Biodiesel

Revisit the structural formula for a biodiesel molecule.

a. Would you predict that the heat released per gram of biodiesel burned is higher or lower than that of octane? Explain your reasoning.
Hint: Revisit Figure 5.6 to see the values for different fuels.

b. If instead the comparison were made on the basis of 1 mole of biodiesel versus 1 mole of octane, which would release more heat if burned?

c. It is more useful to make comparisons per gram of fuel rather than per mole of fuel. Explain.

Biodiesel is blended with petroleum-based diesel fuel, just like ethanol is blended with gasoline. For example, B20 is 20% biodiesel and 80% diesel fuel (Figure 5.40a). Blends up to 20% are fully compatible with any diesel engine, including those in medium- and heavy-weight trucks. As of 2016, biodiesel blends were available at more than 1000 retail locations in the United States, which can easily be located via the Internet (Figure 5.40b).

In this section and the previous one, we have discussed two biofuels: ethanol and biodiesel. In the process, we have hinted at the complexities and the controversies involved. In the final section of this chapter, we take a critical look not only at biofuels, but also at the larger energy picture. We need to find a sustainable way forward. But are biofuels really as sustainable as proponents claim?

(a)

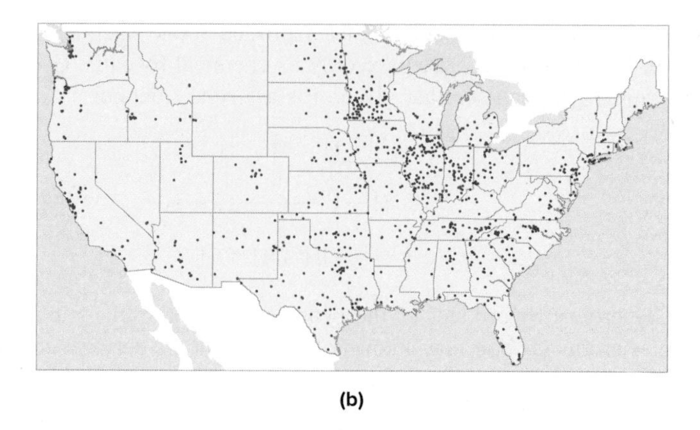

(b)

Figure 5.40

(a) B20 biodiesel is a mixture of 80% petroleum diesel and 20% biodiesel. **(b)** Locations of the biodiesel stations in the U.S. (2016).

5.17 | Are Biofuels Really Sustainable?

"Renewable" fuels are produced from resources that are naturally replenished on a human timescale. Examples include biodiesel and hydrogen obtained from a biological process.

Is growing our fuel a step on the path to a sustainable future? In this final section, we devote our full attention to this question. Underlying it are the economic, environmental, and societal costs of biofuels. Many have pointed out that we need to proceed along this path deliberately and with due process.

Scientists aren't the only ones responsible for a sustainable planet. Recall in Section 1.10 that we mentioned that three pillars of sustainability: the environmental pillar, the societal pillar, and the economic pillar. In the world of business, the bottom line has always included turning a profit, preferably a large one. Today, however, the bottom line includes more than this. For example, corporations are judged to be successful when they are fair and beneficial to workers and to the larger society. Another measure of their success is how well they protect the health of the environment, including the quality of the air, water, and land.

Taken together, this three-way measure of the success of a business based on its benefits to the economy, to society, and to the environment has become known as the **Triple Bottom Line**. One way to represent the Triple Bottom Line is with the overlapping circles shown in Figure 5.41. The economy must be healthy—that is, the annual reports need to show a profit. But no economy exists in isolation. It connects to a community whose members also need to be healthy. In turn, communities connect to ecosystems that need to be healthy. Hence, the figure includes not one, but three connecting circles. At the intersection of these circles lies the "Green Zone." This represents the conditions under which the Triple Bottom Line is met.

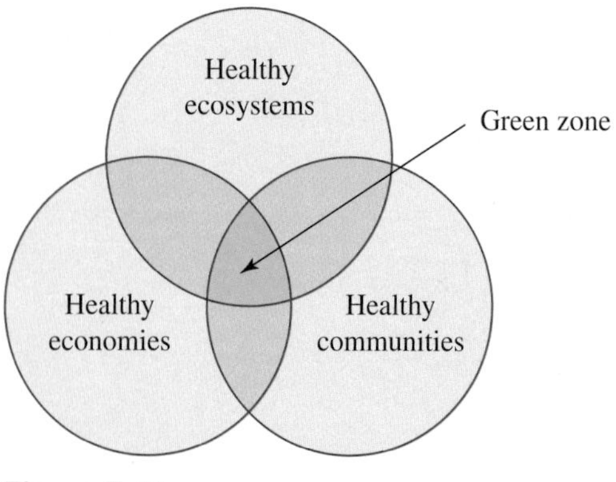

Figure 5.41

A representation of the Triple Bottom Line. The "Green Zone" (where the Triple Bottom Line is achieved) lies at the intersection of these three circles.

Ethanol and biodiesel currently claim only a small share of global final energy consumption—that is, the energy delivered to consumers for all uses. According to a recent report issued on the status of biofuels (Figure 5.42), renewable fuels represented just less than 20% of total energy use, with biofuels a mere 0.8%. Nonetheless, biofuel production has increased in recent years, a trend that is expected to continue. Furthermore, many nations are setting political and economic wheels in motion to encourage the further use of biofuels. Figure 5.42 also reveals that biofuels are only one of several renewable fuel options. Others include wind, water, solar, and geothermal. Here, we focus on biofuels because of their connections to gasoline and diesel fuel; as liquids, they can be pumped to fuel vehicles and aircraft. In the next chapter, we will discuss a variety of alternative energy sources.

In the 1990s, when excitement over biofuels was still in full swing, important problems with their large-scale production began to emerge.

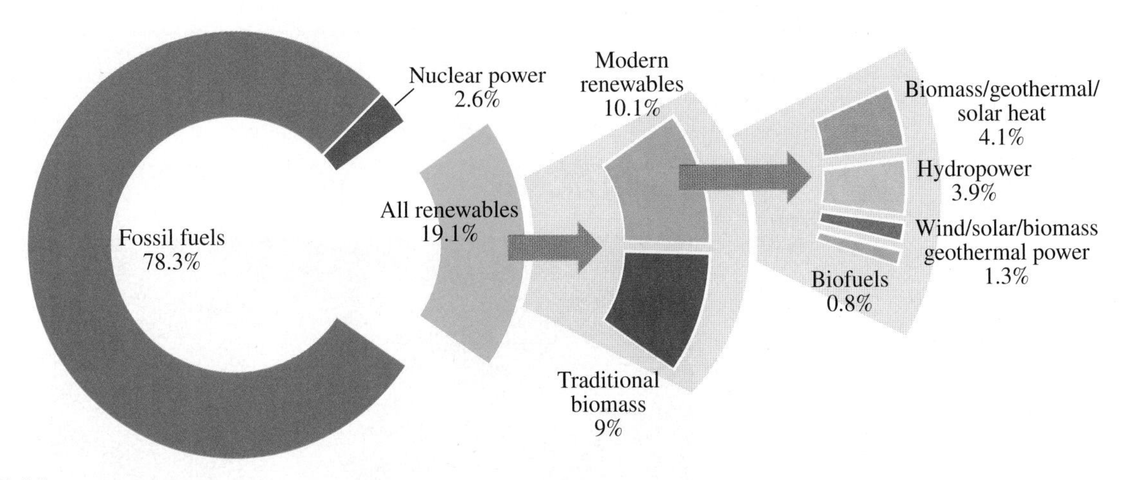

Figure 5.42

Renewable energy share of global final energy consumption (2015). Traditional biomass includes wood, agricultural waste, and animal dung.

Source: Renewable Energy Policy Network for the 21st Century (2015), Renewables 2015: Global Status Report, *Paris, REN21 Secretariat*

The claim that biofuels produce significantly lower greenhouse gas emissions compared with fossil fuels was contested. Some argued that biofuels would result in severe environmental consequences, including pollution and the loss of biodiversity—for example, through the destruction of rain forest to enable large-scale production of biofuels. Concerns were also raised over the competition that biofuels pose to food production, and their consequent effects on food security and food prices. Moreover, many worried about infringements of the rights of farmers, farm workers, and land holders—particularly in vulnerable populations in the developing world.

Concerns were also expressed related to the moral values of using biofuels: human rights, solidarity, sustainability, stewardship, and justice. From these values, the Nuffield Council report constructed ethical principles that could guide discussion, policy, and action (Table 5.5). While all of these principles warrant attention, two stand out as directly relevant to our discussion of the chemistry of biofuels: (1) a net reduction of greenhouse gas emissions, and (2) the larger issues of environmental sustainability. We will address each in turn.

Table 5.5	Ethical Principles to Apply to Current and Future Use of Biofuels
1.	Biofuels development should not be at the expense of peoples' essential rights (including access to sufficient food and water, health rights, work rights, and land entitlements).
2.	Biofuels should be environmentally sustainable.
3.	Biofuels should contribute to a net reduction of total greenhouse gas emissions and not exacerbate global climate change.
4.	Biofuels should develop in accordance with trade principles that are fair and recognize the rights of people to just reward (including labor rights and intellectual property rights).
5.	Costs and benefits of biofuels should be distributed in an equitable way.
6.	If the first five principles are respected and if biofuels can play a crucial role in mitigating dangerous climate change, then depending on additional key considerations, there is a duty to develop such biofuels. These additional key considerations are: absolute cost, alternative energy sources, opportunity costs, the existing degree of uncertainty, irreversibility, degree of participation, and the notion of proportionate governance.

Source: Nuffield Council on Bioethics, Biofuels: Ethical Issues, *2011, 84*

Do biofuels contribute to a net reduction of greenhouse gas emissions? To answer this question, remember that ethanol and biodiesel, like fossil fuels, contain carbon and thus produce carbon dioxide when burned. Try your hand at writing the combustion reactions in the next activity. Although the context is culinary, these chemical reactions also occur in automobile and truck engines.

Your Turn 5.32 Skill Building Cherries and Bananas

Have you ever been served cherries jubilee or bananas flambé? A chef first douses the fruit in a liquor and then lights it with a match. The ethanol in the liquor burns with a dramatic blue-tinged flame.

a. Write the chemical equation for the complete combustion of ethanol—that is, burning it with enough oxygen so that only carbon dioxide and water are produced.

b. Although it wouldn't be very tasty, you also could pour cooking oil or biodiesel on fruit and set it afire. Write the balanced chemical equation for the complete combustion of methyl stearate, $C_{19}H_{38}O_2$, the biodiesel product in **Equation 5.17**.

From "field-to-tank," do ethanol and biodiesel contribute to a net reduction of the greenhouse gas CO_2? Here, the basis of comparison is with gasoline and diesel fuel derived from crude oil. These fossil fuels are not **carbon neutral**; that is, the CO_2 produced by their combustion is *not* offset by some natural process such as photosynthesis or by some human system of offsets. Burning fossil fuels leads to a net increase of CO_2 in the atmosphere.

In contrast, biofuels are potentially more carbon neutral because they are derived from modern-day crops, grasses, and trees. The carbon released on combustion of biofuels is at least partially offset by the carbon these plants once absorbed via photosynthesis. So where do biofuels stand in regard to reducing greenhouse gas emissions? The answer depends on the particular biofuel and how much energy was required to produce it, which includes the energy required to plant and harvest the crop, produce the fertilizers, and water the crops. This is a moving target because, at least in some cases, the technologies have been improving with time.

Measuring the net reduction in CO_2 emissions for biofuels is challenging and controversial. To understand why, we again cite information from the **Nuffield Council report** on the ethics of biofuels:

- **Direct change in the use of land**
  This concept refers to converting natural land to cropland, such as deforestation or draining wetlands. Destroying existing natural lands to produce biofuels degrades the existing soil and vegetation. This process also removes an existing habitat that effectively sequesters a large amount of carbon. The challenge comes in determining both how much carbon is sequestered, and by which land types.
- **Indirect change in the use of land**
  This refers to converting existing pastures or croplands to crops for biofuels. This switch can involve using more fertilizer, more herbicides, and more water, all accompanied by energy use and additional greenhouse gas emissions. The challenge comes in measuring these over the life cycle of a biofuel crop.

- **Waste products from biofuel production**
  This refers to agricultural and industrial wastes that have no value for food or fuel. The challenge comes not only in measuring the greenhouse emissions, but also in assigning the emissions correctly to their source.

In spite of the inherent challenges, several constituencies have proposed values for the CO_2 emissions. In 2011, the biofuel industry estimated 10–15% less CO_2 emissions for corn-based ethanol and 40–45% less CO_2 from soybean-based biodiesel. Both values were in comparison to petroleum-based gasoline. In contrast, other groups propose that biofuels result in net increases of CO_2 compared to petroleum-based gasoline, arguing that the costs of land use and waste products needed to be more carefully accounted for. The debates are likely to continue, given the complexities in correctly assigning the CO_2 emissions to produce a fuel.

Are ethanol and biodiesel sustainable? Like the previous question of greenhouse gas emissions, this one cannot be answered in the abstract. Rather, the answers depend on where and how each particular biofuel is produced.

Let's begin with the tally for biodiesel because it is somewhat more straightforward than that of ethanol. Recall that biodiesel has several inherent advantages over ethanol. Its synthesis from oils is relatively easy and can be done either in small batches or on a large scale. Biodiesel blends well with existing diesel fuel and can be distributed via the same infrastructure. Like ethanol, and because it contains oxygen, it burns more cleanly than diesel fuel releasing lower amounts of particulate matter, carbon monoxide, and volatile organic compounds. In the balance, it seems to be a winner for improving public health, assuming that ethical principles (Table 5.5) have been followed for the local communities that produce biofuel. But can we assume this? Figure 5.43 shows a plantation and factory in Malaysia, one part of the world where palm oil is produced. The next activity offers the opportunity to further explore palm oil production.

Your Turn 5.33 You Decide Palm Oil, Biodiesel, and Ethics

Are there ethical issues with biodiesel that need attention? In 2008, an Oxfam report noted: "The big losers from the rich countries' biofuel boom are poor people, at risk from spiraling food prices, and a 'scramble to supply' that places their land rights, labor rights, and human rights under threat." Use palm oil as a case study. It is produced in many parts of the world, including Indonesia and Malaysia. Prepare a one-page briefing that identifies the key issues for your classmates.

(a)

(b)

Figure 5.43

(a) A plantation of young palm trees in Malaysia; **(b)** a palm oil factory in Malaysia.

a: © Romeo Gacad/AFP/Getty Images; b: © Universal Images Group via Getty Images

The sustainability of ethanol production is far more difficult to assess. As noted earlier, ethanol originates from several feedstocks, including corn and sugarcane. Understandably, the environmental tally for each one must be assessed separately. No matter which feedstock, scientists and citizens alike are questioning the sustainability of ethanol production. Recall the Triple Bottom Line (Figure 5.41): healthy economies, healthy communities, and healthy ecosystems.

To give you a better sense of how sustainable the production of ethanol might be, let's examine the details. We begin with the economic bottom line. As of 2016, with its precipitous drop in oil prices, it was still more expensive to produce a gallon of ethanol than a gallon of gasoline. So why the booming corn ethanol market? The answer varies by locale. In recent years, the U.S. government provided tax credits to ethanol producers. Although subsidies have encouraged the use of this fuel, they were controversial and ended in 2011 for ethanol produced from corn.

In terms of energy costs, the good news is that the Sun freely provides the energy for plants to grow. The bad news, however, is that growing corn requires additional energy inputs. Planting, cultivating, and harvesting all require energy. The same is true for watering the crops, producing and applying the fertilizers, manufacturing and maintaining the necessary farm equipment, and distilling the alcohol from the fermented grains. Currently, this energy is supplied by burning fossil fuels at a significant monetary cost and with significant carbon dioxide emissions. The overall energy cost for corn ethanol is difficult to quantify. Some studies estimate that for every joule put into ethanol production, 1.2 J is recovered. Others contend that the combined energy inputs outweigh the energy content of the ethanol produced.

Next, we consider the social bottom line. Many rural communities in the Midwest have benefited greatly from booming ethanol demand. Construction of a distillery provides not only jobs to local workers, but also a buyer for locally grown corn. Several communities, hit hard by the depressed viability of family farms, have experienced a resurgence thanks to demand for ethanol. There are certain drawbacks as well. Increased demand for corn leads to increased prices on many other products (especially foods); prices that everyone in that community (and others around the world) must pay.

Last, we turn to the environmental bottom line. Growing corn relies on the heavy use of fertilizers, herbicides, and insecticides. The manufacturing and transport of these chemicals requires the burning of fossil fuels, which in turn releases carbon dioxide. Furthermore, these chemicals, once introduced to the field, degrade the soil and water quality. Although corn growers can and do follow responsible practices, they certainly face challenges when called on to produce more corn.

Where to go from here? Clearly, the choices are not easy. We have burned our way through many of the easy and conventional sources of oil. Our future choices are complex and involve trade-offs.

We end this chapter by returning to words from the Nuffield Council report. They call to mind the **precautionary principle** alluded to in earlier chapters, reminding us that both to act and not to act carry risks. We hope these words not only will inspire you but also lead you to further investigate the issues and take action with appropriate caution.

Exhorting people to make lifestyle changes will, of course, continue to be one approach to help achieve an overall reduction of greenhouse gas emissions. However, this and other non-fuel-based power sources such as wind, wave, and solar energy will not be sufficient to reduce global dependence on fossil fuels for the foreseeable future.

We will need new sources of liquid fuels and new ways of producing current biofuels more efficiently. Advanced biotechnology, including genetic modification, could be an important part of the tool kit to help deliver on these needs. Precautionary safeguards have already been built into the development of advanced biotechnologies to ensure that this will not present any new hazards. Indeed, it is important that precautionary approaches are implemented in a balanced and equitable way—we should be as precautionary about the risks of doing nothing as we are about the risks of developing new technologies.

Your Turn 5.34 You Decide A Sustainable Future

In 2002, then Secretary General of the United Nations, Kofi Annan, called sustainability "... an exceptional opportunity—economically to build markets, socially to bring people in from the margins, and politically to reduce tensions over resources that could give every man and woman a voice and a choice in deciding their own future." Expand on the Secretary General's remarks, giving some specifics for each area mentioned.

Conclusions

Fire! To early humans, fire was a source of security. It warded off animals, brought the ability to cook and preserve food, and minimized the spread of some diseases. Fires were an important social vehicle, a place to gather and share stories. Fire also allowed people to venture into colder regions of the planet.

Today, combustion still remains central to our human community. We use it daily to cook, to heat or cool our dwellings, to produce goods and crops, and to travel the roads, rails, waterways, and skies of our planet. Few chemical reactions have as far-reaching consequences for our health, well-being, and productivity as the ability to burn fuels.

As we have seen in this chapter, the process of combustion converts *energy* into less useful forms. For example, when we burn a mixture of hydrocarbons such as gasoline, we dissipate some of the potential (stored) energy it contains in the form of heat. Although the *forms* of energy change, the total amount of energy before and after any transformation remains the same, according to the First Law of Thermodynamics.

We have also seen that the process of combustion converts *matter* into less useful and sometimes even undesirable forms. For example, the products of complete combustion—carbon dioxide and water—are not currently usable as fuels. Furthermore, as seen in the previous chapter, CO_2 is a greenhouse gas linked to global climate change. Products of incomplete combustion, such as carbon monoxide and soot, are undesirable because of their effects on human health. The same is true for NO_x air pollutants that are formed in the high temperatures of flames.

Today, fossil fuels power the planet. So do renewable fuels, but to a much smaller extent. What will future generations use for energy sources? Renewable biofuels such as ethanol and biodiesel will compose part of our energy future. Like all fuels, they need to align with our values, including stewardship and sustainability. However, there are other possible ways of satisfying our ever-increasing appetite for energy, such as nuclear or solar energies, which will be discussed in the next chapter.

Learning Outcomes

The numbers in parentheses indicate the sections within the chapter where these outcomes were discussed.

Having studied this chapter, you should now be able to:

- identify the chemical characteristics of a fuel and describe how they make a fuel useful (5.1)
- describe combustion as an energy-transfer process (5.2)
- compare and contrast the various forms of energy (5.3)
- draw and interpret energy diagrams for chemical reactions (5.3)
- distinguish between endothermic and exothermic processes (5.4)
- identify and compare units of energy (5.4)
- describe how energy transformations are measured (5.4)
- calculate and compare heats of combustions among fuels (5.4)
- illustrate how energy is released or absorbed when bonds break or form (5.5)
- predict the amount of carbon released from combustion of a fuel (5.5)
- explain how the type of fuel determines the type of waste products (5.5)
- identify potential consequences of the waste products generated from combustion of fuels (5.5)
- illustrate energy transfer in a combustion-based power plant, including points of inefficiency (5.6, 5.7)
- compare the efficiencies of energy transfer from the combustion of various fuels (5.7)
- identify the chemical composition of coal (5.8)
- calculate how much coal is needed to provide energy for a particular activity (5.8)
- calculate how much CO_2 is released by a certain amount of coal combustion (5.8)
- describe the advantages and disadvantages of burning coal and oil-derived fuels (5.8, 5.9, 5.11)
- describe how fuel industries extract oil from natural rock formations (5.10)

- describe the role of distillation to refine fuels (5.10)
- define catalysts and understand their role in chemical reactions (5.14)
- define hydrocarbons (5.12)
- name and provide the chemical formulas of simple hydrocarbons (5.12)
- define volatility (5.12)
- predict the relative boiling points of hydrocarbons, based on their relative intermolecular forces (5.12)
- describe how the octane rating is calculated and how it relates to combustion efficiency (5.12)
- describe the benefits and issues with adding ethanol to gasoline (5.13, 5.14)
- characterize renewable and nonrenewable fuels (5.14)
- compare and contrast biofuels with fossil fuels (5.15, 5.16)
- describe why biofuels are considered renewable (5.15, 5.16)
- evaluate the environmental, economic, and social impacts of both conventional fuels and biofuels (5.17)

Questions

Emphasizing Essentials

1. a. List five fuels. Name at least two properties that these fuels share.
   b. Of the fuels you listed, which are fossil fuels or those derived from them?
   c. Of the fuels you listed, which are renewable?
2. The combustion of coal releases several substances into the air.
   a. Of these substances, one is a gas that is produced in large amounts. Give its chemical formula and name.
   b. In contrast, the amount of SO_2 (sulfur dioxide) released is relatively small. Even so, this SO_2 is of concern. Explain why.
   c. Another gas produced in small amounts is NO (nitrogen monoxide). However, coal contains very little nitrogen. What is the origin of the nitrogen in NO?
   d. When coal burns, fine particles of soot may be released. What are the health concerns with $PM_{2.5}$, the smallest of these particles?
3. Revisit Figure 5.10, which depicts the components of an electric power plant.
   a. For a coal-fired power plant, where would the coal appear in the figure?
   b. Water is part of two separate loops. One loop connects the boiler and turbine and is usually under pressure. Explain why.
   c. Another loop brings in (and out) water from a lake or river. Explain why a large body of water is needed.
4. Energy exists in different forms in our natural world. In Figure 5.10, identify where:
   a. Potential (stored) energy of the fuel is converted to heat.
   b. Kinetic energy of water molecules is converted to mechanical energy.
   c. Mechanical energy is converted to electrical energy.
   d. Electrical energy is converted into forms such as heat and light.
5. A coal-burning power plant generates electrical power at a rate of 500 megawatts (MW), or 5.00×10^8 J/s. The plant has an overall net efficiency of 37.5% (0.375) for the conversion of heat to electricity.
   a. Calculate the electrical energy (in joules) generated in one year of operation and the heat energy used for that purpose.
   b. Assuming the power plant burns coal that releases 30 kJ/g, calculate the mass of coal (in grams and metric tons) that is burned in one year of operation. **Hint:** 1 metric ton = 1.3×10^3 kg = 1.3×10^6 g.
6. The energy of sunlight can be converted into the potential energy of glucose and oxygen.
   a. Name the process by which this conversion occurs.
   b. Name three fuels whose energy originates in sunlight.
7. Describe how grades of coal differ. What is the significance of these differences?
8. Although coal is an important fuel for producing electricity, it also has drawbacks. Name three of these.
9. Mercury (Hg) is present in trace amounts in coal, ranging from 50–200 ppb. Consider the amount of coal burned by the power plant in Your Turn 5.19. Calculate tons of mercury in the coal based on the lower (50 ppb) and higher (200 ppb) concentrations.
10. Name two ways in which all hydrocarbons are alike. Then, name two ways in which they differ.
11. Here are the condensed structural formulas for two alkanes: CH_3CH_3 and $CH_3(CH_2)_2CH_3$.
    a. What are the names for these compounds?
    b. For each one, give the chemical formula and draw a structural formula that shows all bonds and atoms.
    c. Comment on the relative advantages of chemical formulas, condensed structural formulas, and structural formulas in terms of convenience and information provided.

12. The structural formulas of straight-chain ("normal") alkanes containing one to eight carbon atoms are given in Table 5.3.
    a. Draw a structural formula for *n*-decane, $C_{10}H_{22}$.
    b. Predict the chemical formula for *n*-nonane (9 C atoms) and for *n*-dodecane (12 C atoms).
13. Here is a ball-and-stick representation for one isomer of butane (C_4H_{10}):

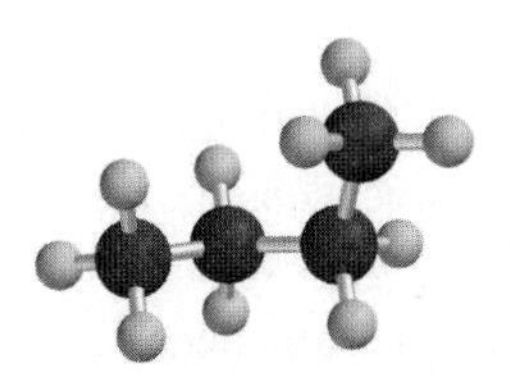

    a. Draw a structural formula for this isomer.
    b. Draw structural formulas for all other isomers. Watch out for duplicate structures.
14. Consider these three hydrocarbons:

Compound, Formula	Melting Point (°C)	Boiling Point (°C)
pentane, C_5H_{12}	−130.5	35.9
triacontane, $C_{30}H_{62}$	65.8	449.7
propane, C_3H_8	−187.7	−42.2

At room temperature (25 °C), categorize each one as a solid, liquid, or gas.

15. During petroleum distillation, kerosene and hydrocarbons with 12–18 carbons used for diesel fuel condense at position C marked on this diagram.

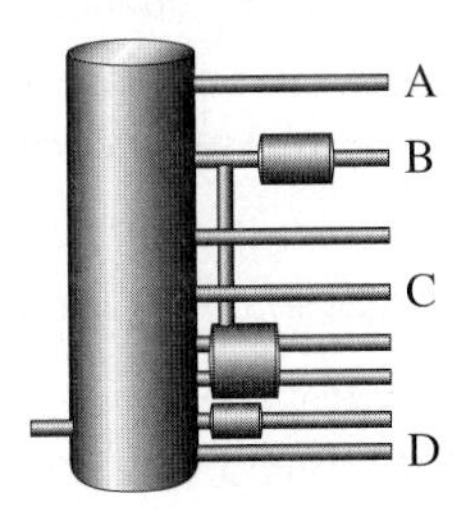

    a. Separating hydrocarbons by distillation depends on differences in a specific physical property. Which one?
    b. How does the number of carbon atoms in the hydrocarbon molecules separated at A, B, and D compare with those separated at position C? Explain your reasoning.
    c. How do the uses of the hydrocarbons separated at A, B, and D differ from those separated at position C? Explain your reasoning.
16. The complete combustion of methane is given in Equation 5.4.
    a. By analogy, write the balanced chemical equation for the combustion of ethane, C_2H_6.
    b. Rewrite this equation using Lewis structures.
    c. The heat of combustion for ethane, C_2H_6, is 52 kJ/g. How much heat is produced if 1.0 mol of ethane undergoes complete combustion?
17. a. Write the balanced chemical equation for the complete combustion of *n*-heptane, C_7H_{16}.
    b. The heat of combustion for *n*-heptane is 4817 kJ/mol. How much heat is released if 250 kg of *n*-heptane burns completely to produce CO_2 and H_2O?
18. A single-serving bag of potato chips has 70 Cal (70 kcal). Assuming that all of the energy from eating these chips goes toward keeping your heart beating, how long can these chips sustain a heartbeat of 80 beats per minute?
    Hint: 1 kcal = 4.184 kJ, and each human heartbeat requires approximately 1 J of energy.
19. A 12-oz serving of a soft drink has an energy equivalent of 92 kcal. In kilojoules, what is the energy released when metabolizing this beverage?
20. State whether these processes are endothermic or exothermic.
    a. Charcoal burns in an outdoor grill.
    b. Water evaporates from your skin.
    c. Glucose is synthesized in the leaves of a plant by photosynthesis.
21. Use the bond energies in Table 5.1 to calculate the energy changes associated with each of these reactions. Label each reaction as endothermic or exothermic.
    a. $N_2(g) + 3\ H_2(g) \longrightarrow 2\ NH_3(g)$
    b. $H_2(g) + Cl_2(g) \longrightarrow 2\ HCl_3(g)$
    Hint: Draw Lewis structures of the reactants and products to determine the number and kinds of bonds.
22. Use the bond energies in Table 5.1 to calculate the energy changes associated with each of these reactions. Then, label each reaction as endothermic or exothermic.
    a. $2\ H_2(g) + CO(g) \longrightarrow CH_3OH(g)$
    b. $H_2(g) + O_2(g) \longrightarrow H_2O_2(g)$
    c. $2\ BrCl(g) \longrightarrow Br_2(g) + Cl_2(g)$

23. Ethanol can be produced by fermentation. Another way to produce ethanol is the reaction of water vapor with ethene (ethylene), a hydrocarbon containing a C=C double bond.

$$CH_2CH_2(g) + H_2O(g) \longrightarrow CH_3CH_2OH(l)$$

a. Rewrite this equation using Lewis structures.

b. Use the bond energies in Table 5.1 to calculate the energy change for this reaction. Is the reaction endothermic or exothermic?

24. Here are structural formulas for ethane, ethene (ethylene), and ethanol.

```
   H  H          H       H          H  H
   |  |           \     /           |  |
H—C—C—H          C=C          H—C—C—O—H
   |  |           /     \           |  |
   H  H          H       H          H  H

  ethane          ethene            ethanol
```

a. Is ethane an isomer of ethene? Of ethanol? Explain.

b. Are any other isomers possible for ethene? Explain.

c. Are any other isomers possible for ethanol? Explain.

25. These three compounds all have the same chemical formula of C_8H_{18}. The hydrogen atoms and C–H bonds have been omitted for simplicity.

```
C
|
C
|           C              C
C           |              |
|       C—C—C            C—C
C           |              |
|           C              C
C           |              |
|           C—C        C—C—C
C           |              |
|           C              C
C
|
C
```

a. For each compound, draw structural formulas that show the missing H atoms. All should have 18 H atoms.

b. Which (if any) of these structural formulas are identical?

c. Draw the structural formulas for any two additional isomers of C_8H_{18}.

26. Catalysts speed up cracking reactions in oil refining and allow them to be carried out at lower temperatures. Describe two other examples of catalysts given in previous chapters of this text.

27. Explain why cracking is a necessary part of the refinement of crude oil.

28. Consider this equation representing the process of cracking.

$$C_{16}H_{34} \longrightarrow C_5H_{12} + C_{11}H_{22}$$

a. Which bonds are broken and which bonds are formed in this reaction? Use Lewis structures to help answer this question.

b. Use the information from part a and Table 5.1 to calculate the energy change during this cracking reaction.

29. What is a biofuel? Give three examples.

30. Consider these three alcohols: methanol, ethanol, and *n*-propanol (the straight-chain isomer).

a. These compounds all contain a common functional group. Name it.

b. These compounds all are flammable. Give names and chemical formulas for the products.

c. Predict which one of these compounds has the lowest boiling point. Explain your reasoning.

d. The chemical structure of one of these compounds is somewhat similar to glycerol. Which one and why?

31. Cellulose and starch both can be fermented to produce ethanol.

a. In terms of chemical structure, how are starch and cellulose similar?

b. In terms of a food source for humans, how are starch and cellulose different?

32. When glucose, $C_6H_{12}O_6$, is "burned" (metabolized) in your body, the products are carbon dioxide and water.

a. Write the balanced chemical equation.

b. The chemical equation for burning wood is essentially the same as that of metabolizing glucose. Explain why.

33. As biofuels, biodiesel and ethanol warrant a close comparison. Use these parameters as the basis.

a. The source

b. The chemical reaction that produces the fuel

c. The combustion products

d. The solubility in water (more about this in Chapter 8)

34. Compare and contrast a molecule of biodiesel with a molecule of ethanol. Use these parameters as the basis of your comparison.

a. The types of atoms each contains and their approximate relative proportions

b. The number of atoms each contains

c. The functional groups each contains

35. Use Figure 5.6 to compare the energy released for the combustion of 1 gallon of ethanol and 1 gallon of gasoline. Assume gasoline is pure octane (C_8H_{18}). Explain the difference.
Hint: You will need to consider their densities.

36. Explain the terms *conventional* and *unconventional* with respect to crude oil as an energy source. Give an example of each.

Concentrating on Concepts

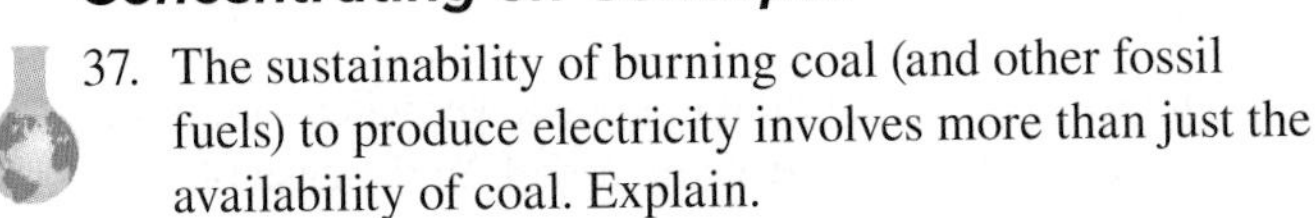

37. The sustainability of burning coal (and other fossil fuels) to produce electricity involves more than just the availability of coal. Explain.

38. In this chapter, we approximated the chemical formula of coal as $C_{135}H_{96}O_9NS$. However, we also noted that low-grade lignite (soft coal) has a chemical composition more similar to wood. Cellulose is a primary component in wood. Given this, predict an approximate chemical formula for lignite.

39. Use Figure 5.14 to compare the sources of U.S. energy consumption. Arrange the sources in order of decreasing percentage and comment on the relative rankings.

40. Compare the processes of combustion and photosynthesis in terms of energy released or absorbed, chemicals involved, and the ability to remove CO_2 from the atmosphere.

41. How might you explain the difference between temperature and heat to a friend? Use some practical, everyday examples.

42. Write a response to this statement: "Because of the First Law of Thermodynamics, there can never be an energy crisis."

43. The concept of entropy and probability is used in games like poker. Describe how the rank of hands (from a simple high card to a royal flush) is related to entropy and probability.

44. Bond energies such as those in Table 5.1 are sometimes found by "working backward" from heats of reaction. A reaction is carried out, and the heat absorbed or evolved is measured. From this value and known bond energies, other bond energies can be calculated. For example, the energy change associated with the combustion of formaldehyde (H_2CO) is −465 kJ/mol.

$$H_2CO(g) + O_2(g) \longrightarrow CO_2(g) + H_2O(g)$$

Use this information and the values found in Table 5.1 to calculate the energy of the C=O double bond in formaldehyde. Compare your answer with the C=O bond energy in CO_2 and speculate on why there is a difference.

45. Use the bond energies in Table 5.1 to explain why chlorofluorocarbons (CFCs) are so stable. Also explain why it takes less energy to release Cl atoms from CFCs than it does to release F atoms and connect this to HFCs as replacements for CFCs.

46. Halons are similar to CFCs but also contain bromine. Although halons are excellent materials for fighting fires, they more effectively deplete ozone than CFCs. Here is the Lewis structure for Halon-1211.

:Br:
|
:F—C—F:
|
:Cl:

a. Which bond in this compound is broken most easily? How is that related to the ability of this compound to deplete ozone?

b. In fire extinguishers, C_2HClF_4 is a possible replacement for halons. Draw its Lewis structure and identify the bond broken most easily.
Hint: The carbon atoms are the central atoms.

47. The energy content of fuels can be expressed in kilojoules per gram (kJ/g), as shown in Figure 5.6. From these values, how do fuels containing oxygen compare to those that do not? Now calculate the energy content for each of these fuels in kilojoules per mole (kJ/mol). What trend do you now observe?

48. A friend tells you that hydrocarbon fuels containing larger molecules liberate more heat per gram than those with smaller ones.

a. Use these data, together with appropriate calculations, to discuss the merits of this statement.

Hydrocarbon	Heat of Combustion
octane, C_8H_{18}	–5070 kJ/mol
butane, C_4H_{10}	–2658 kJ/mol

b. Based on your answer to part a, do you expect the heat of combustion per gram of candle wax, $C_{25}H_{52}$, to be more or less than that of octane? Do you expect the molar heat of combustion of candle wax to be more or less than that of octane? Justify your predictions.

49. The Fischer–Tropsch conversion of hydrogen and carbon monoxide into hydrocarbons and water was given in Equation 5.11:

$$n\,CO + (2n + 1)\,H_2 \longrightarrow C_nH_{2n+2} + n\,H_2O$$

a. Determine the heat evolved by this reaction when $n = 1$.

b. Without doing a calculation, do you think that more or less energy is given off per mole in the formation of larger hydrocarbons ($n > 1$)? Explain your reasoning.

50. Here is a ball-and-stick model of ethanol, C_2H_5OH or C_2H_6O.

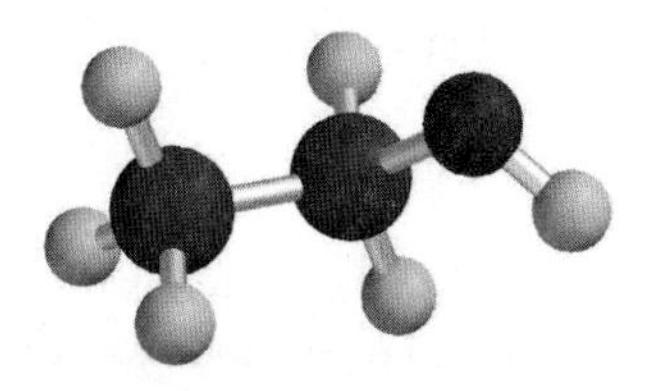

a. Dimethyl ether is an isomer of ethanol. Draw its Lewis structure.

b. People used to refer to "ether" as an anesthetic. What they meant was diethyl ether. Draw its structural formula.
Hint: *Diethyl* means there are a total of how many carbons?

c. The ether is a functional group not described in this chapter. Based on your answer to the two previous parts, what common structural feature do all ethers have?

51. Octane ratings of several substances are listed in Table 5.4.

a. What evidence can you give that the octane rating is not a measure of the energy content of a gasoline?

b. Octane ratings measure a fuel's ability to minimize or prevent engine knocking. Why is this important?

c. Why do higher octane blends cost more than lower octane ones?

d. A premium gasoline available at most stations has an octane rating of 91. What does this tell you about whether the fuel contains oxygenates?

52. Both *n*-octane and iso-octane have essentially the same heat of combustion. How is this possible given that they have different molecular structures?

53. All of these terms fit under the heading of fuels: renewable fuel, nonrenewable fuel, coal, petroleum, biodiesel, natural gas, and ethanol. Use a diagram to show the relationship among them. Also find a way to show where the terms *fossil fuel* and *biofuel* fit.

54. Use a diagram to show the relationship among these terms that relate to foods we eat: fat, lard, oil, triglyceride, butter, olive oil, and soybean oil. Although biodiesel is not a food, it still connects to these terms. Find a way to represent this connection.

55. On a timescale of a few years, the combustion of ethanol derived from biomass releases a *lower* net amount of CO_2 into the atmosphere than does burning gasoline derived from crude oil. People argue whether this statement is true or not. What is the point of contention?

56. Emissions of some pollutants are lower when biodiesel is used rather than petroleum diesel. In the case of biodiesel fuel, suggest a reason for lower emissions of the following:

a. sulfur dioxide, SO_2

b. carbon monoxide, CO

Exploring Extensions

57. Although coal contains only trace amounts of mercury, the amounts released into the environment by the burning of coal have significant consequences. Defend or refute this statement by gathering the appropriate evidence.

58. According to a statement once made by the U.S. EPA, driving a car is "a typical citizen's most polluting daily activity."

a. What pollutants do cars emit?

b. What assumptions does the truth of this statement depend on?

59. An article in *Scientific American* pointed out that replacing a 75-watt incandescent bulb with an 18-watt compact fluorescent bulb would save about 75% in the cost of electricity. Electricity is generally priced per kilowatt-hour (kWh). Using the price of electricity where you live, calculate how much money you would save over the life of one compact fluorescent bulb (about 10,000 h). *Note:* Standard incandescent bulbs last about 750 h.

60. C. P. Snow, a noted scientist and author, wrote an influential book called *The Two Cultures,* in which he stated: "The question, 'Do you know the Second Law of Thermodynamics?' is the cultural equivalent of 'Have you read a work by Shakespeare?'" How do you react to this comparison? Discuss his remark in light of your own educational experiences.

61. This chapter mentions several nonconventional sources of oil and gas, including drilling deep below seawater, hydraulic fracturing deep below the Earth, and extracting oil from shales and tarry oil sands. Pick one, describe it, and provide an analysis using the Triple Bottom Line: economic health, environmental health, and societal health.

62. Chemical explosions are *very* exothermic reactions. Describe the relative bond strengths in the reactants and products that would make for a good explosion.

63. The chapter pointed out that the FDA approved propylene glycol for use as a food additive. In which foods is it used and for what purposes?

64. Tetraethyllead (TEL) was first approved for use in gasoline in the United States in 1926. It wasn't banned until 1986. Construct a timeline that includes any four events in the 60 years of its use, including some that led to its ban.

65. Tetraethyllead (TEL) has an octane rating of 270. How does this compare with other gasoline additives? Examine a structural formula for TEL and propose a reason for the value of its octane rating in comparison to other additives.

66. Another type of catalyst used in the combustion of fossil fuels is the catalytic converter that was discussed in Chapter 2. One of the reactions that these catalysts speed up is the conversion of $NO(g)$ to $N_2(g)$ and $O_2(g)$.

    a. Draw a diagram of the energy of this reaction similar to the one shown in Figure 5.33.

    b. Why is this reaction important?

67. Figure 5.8 shows energy differences for the combustion of H_2, an exothermic chemical reaction. The combination of N_2 and O_2 to form NO (nitrogen monoxide) is an example of an endothermic reaction:

$$N_2(g) + O_2(g) \longrightarrow 2\ NO(g)$$

$$:\dot{N}=\ddot{O}:$$

The bond energy for N=O is 630 kJ/mol. Sketch an energy diagram for this reaction and calculate the overall energy change.

68. Because the United States has large natural gas reserves, there is significant interest in developing uses for this fuel. List two advantages and two disadvantages of using natural gas to fuel vehicles.

CHAPTER 6

Energy from Alternative Sources

REFLECTION

Where Does Your Energy Come From?

Visit the U.S. Energy Information Administration's website to find a state-by-state comparison of energy production and consumption.

a. Choose one state or territory and select it on the map. Scroll down to look at the information provided. From which sources does this state or territory get the energy that it consumes? What types of energy sources are used to produce energy in this state? How do these two sources differ?

b. Choose a second state and make a comparison with the state you chose in part **a**.

c. Of all the sources of energy you found in parts **a** and **b**, which are derived from fossil fuels? Which are not? This chapter will explore energy sources that are alternatives to fossil fuels.

The Big Picture

In this chapter, you will explore the following questions:

- How much energy is used in the world?
- What is solar energy?
- What is radioactivity, and what are some applications for radioactive elements in energy production?
- How do nuclear power plants produce electricity, and what are their environmental impacts relative to fossil-fuel power plants?
- What are other types of renewable energy sources, and how are they assessed in terms of environmental and economic impacts?

Introduction

People across the world use energy. As individuals, we use energy to warm and cool our living spaces, to cook food, to power electronic devices, to move ourselves around, and to provide light in the darkness. Together, the humans on the planet consume more than 500 exajoules of energy each year.

The metric prefix "exa" means 10^{18}, or one quintillion joules. First introduced in Section 5.4, the joule (J) is a unit of energy: $1 \text{ J} = 1 \text{ kg·m}^2/\text{s}^2$.

Currently, the majority of energy consumed across the globe comes from the fossil fuels introduced in Chapter 5, including coal, natural gas, and petroleum. The combustion of fossil fuels releases energy stored from sunlight that reached Earth hundreds of millions of years ago.

Your Turn 6.1 Scientific Practices Personal Energy

Review the past 24 hours, or even just the past few hours. What kinds of energy did you use? Where did the energy come from?

The energy we use may be provided in the form of natural gas piped into homes, gasoline to fuel cars, and electricity from wall outlets. Natural gas and gasoline are both fossil fuels, but where does the electricity provided by a wall outlet come from? The answer depends on where you live.

Source: NASA's Earth Observatory/NOAA/DoD

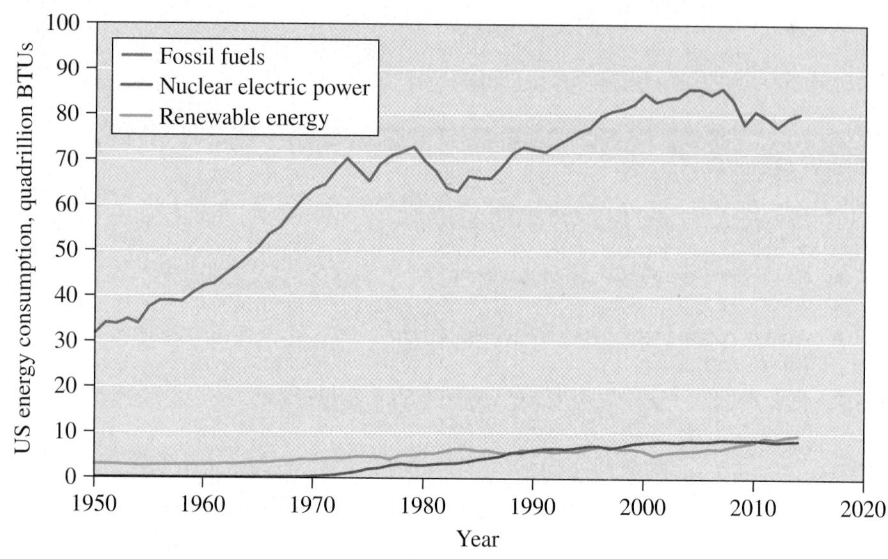

Figure 6.1

Annual U.S. energy consumption by source. During the same time period, the U.S. population approximately doubled. Renewable energy sources include hydroelectric, geothermal, solar photovoltaic, wind, and biomass.

Source: US Energy Administration, May 2015 Monthly Review.

Electricity generation in many regions of the world is driven by the availability of natural resources. The combustion of natural gas is the dominant source of electricity for those living in the Russian Federation, while electricity generated by hydroelectric dams powers much of the Pacific Northwest in the United States. Countries without abundant fossil fuel resources, such as France, may rely heavily on nuclear power.

Global energy use is increasing, and fossil fuels continue to provide the vast majority of energy consumed worldwide. Figure 6.1 shows that U.S. fossil fuel use has nearly tripled over the past 60 years.

In many parts of the world, people are turning more and more toward sources of energy that do not rely on the combustion of hydrocarbon-based fuels such as coal, oil, and natural gas to power their daily lives. Nuclear power plants, solar panels, wind turbines, and hydroelectric dams all generate electricity without consuming fossil fuels and thus produce zero carbon dioxide emissions.

Although these energy sources result in zero greenhouse gas (GHG) emissions when they generate electricity, certain processes used to build and fuel the plants are not as environmentally benign. There are GHG emissions associated with the construction and maintenance of the facility, as well as (for nuclear power) fuel processing and waste disposal. For instance, massive amounts of methane and other GHGs are produced during the flooding of land for a hydroelectric dam.

Nuclear energy is not considered to be a renewable energy source because it relies on the mining of uranium, a finite resource.

Solar, wind, hydroelectric, and geothermal energies are said to be **renewable** energy sources because they are continually and rapidly collected and replenished from natural resources. Hence, these energy sources are expected to continue well into the future.

This chapter will describe the chemistry that powers these alternative sources of energy, starting with nuclear power.

6.1 From Nuclear Energy to Bombs: The Splitting of Atomic Nuclei

The key to understanding the fundamentals of nuclear reactions is probably the most famous equation in all of the natural sciences, which summarizes the equivalence of energy, E, and matter, or mass, m:

$$E = mc^2 \qquad \textbf{[6.1]}$$

This equation dates from the early years of the 20th century, and is one of the many contributions of Albert Einstein (1879–1955). The symbol c represents the speed of light, 3.00×10^8 m/s, so c^2 is equal to 9.00×10^{16} m^2/s^2. The large value of c^2 means that it should be possible to obtain a tremendous amount of energy from a very small amount of matter traveling at the speed of light—whether in a power plant or in a weapon.

For more than 30 years, Einstein's equation was a curiosity. Scientists believed that it described the source of the Sun's energy, but as far as anyone knew, no one on Earth had ever observed a transformation of a substantial fraction of matter into energy. But in 1938, two German scientists, Otto Hahn (1879–1968) and Fritz Strassmann (1902–1980), discovered otherwise. When they bombarded U-238 with neutrons, they found what appeared to be the element barium (Ba-137) among the products. At first, the scientists were tempted to conclude that the element was radium (Ra, atomic number 88), a heavier member of the same group as barium in the periodic table. Up until this time, only elements with atomic numbers close to that of the bombarded element were observed. But for Hahn and Strassmann, the chemical evidence for barium was too compelling to ignore.

Figure 6.2

Lise Meitner is pictured shortly after her arrival in New York for a visit in January, 1946.

The German scientists were unsure how barium could have been formed from uranium, so they sent a copy of their results to their colleague, Lise Meitner (1878–1968), for her opinion (Figure 6.2). Dr. Meitner had collaborated with Hahn and Strassmann on related research, but was forced to flee Germany in March of 1938 because of the Nazi government. When she received their letter, she was living in Sweden. She discussed the strange results with her physicist nephew, Otto Frisch (1904–1979). In a flash of insight, she understood. Under the influence of the bombarding neutrons, the uranium atoms were splitting into smaller ones such as barium. The nuclei of the heavy atoms were dividing, like biological cells undergoing binary fission.

The word fission is applied to a physical phenomenon in the letter that Meitner and Frisch published on February 11, 1939, in the British journal *Nature*. In the letter, entitled "Disintegration of Uranium by Neutrons: A New Type of Nuclear Reaction," the authors stated the following:

> Hahn and Strassmann were forced to conclude that isotopes of barium are formed as a consequence of the bombardment of uranium with neutrons. At first sight, this result seems very hard to understand. . . . On the basis, however, of present ideas about the behavior of heavy nuclei, an entirely different and essentially classical picture of these new disintegration processes suggests itself. . . . It seems therefore possible that the uranium nucleus has only small stability of form, and may, after neutron capture, divide itself into two nuclei of roughly equal size. . . . The whole "fission" process can thus be described in an essentially classical way.

Although just over a page long, this letter was immediately recognized for its significance. Niels Bohr (1885–1962), an eminent Danish physicist, learned of the news directly from Frisch and brought a copy of the letter to the United States on an ocean liner several days before its publication. Within a few weeks of Meitner and Frisch's letter in *Nature*, scientists in a dozen laboratories in various countries confirmed that the energy released by the splitting of uranium atoms was that predicted by Einstein's equation. Lise Meitner's contributions to the discovery of nuclear fission were honored by naming element 109 meitnerium.

Did You Know? Although Meitner was nominated several times, Otto Hahn was the sole awardee of the 1944 Nobel Prize in chemistry for the discovery of nuclear fission. Hahn, Meitner, and Strassman were awarded the Fermi award (a U.S. honor) in 1966 for their work on nuclear fission. In 1997, meitnerium became the first element named after a non-mythological woman.

Nuclear fission is the splitting of a large nucleus into smaller ones with the release of energy. Energy is released because the total mass of the products is slightly less than the total mass of the reactants. Despite what you may have been taught, neither matter nor energy is individually conserved. Rather, they are conserved together. When matter "disappears," an equivalent quantity of energy "appears." Alternatively, one can view matter as a very concentrated form of energy; nowhere is it more concentrated than in an atomic nucleus. Remember that an atom is mostly empty space. If a hydrogen nucleus were the size of a baseball, then its electron would be found within a sphere half a mile in diameter. Since almost all the mass of an atom is associated with its nucleus, the nucleus is incredibly dense. Indeed, a pocket-sized matchbox full of atomic nuclei would weigh more than 2.5 billion tons! Given the energy–mass equivalence of Einstein's equation, the energy content of all nuclei is, relatively speaking, immense.

Only the nuclei of certain elements undergo fission, and these only under certain conditions. Three factors determine whether a particular nucleus will split: its size, the numbers of protons and neutrons it contains, and the energy of the neutrons that bombard the nucleus to initiate the fission. For example, relatively light and stable atoms such as oxygen, chlorine, and iron do not split, but extremely heavy nuclei may fission spontaneously. Some heavy nuclei, such as those of uranium and plutonium, can be made to split if hit hard enough with neutrons. Notably, one isotope of uranium fissions with neutrons of a more moderate speed, such as those employed in the reactor of a nuclear power plant.

Let's examine uranium more closely. All uranium atoms contain 92 protons. If these atoms are electrically neutral, these protons are accompanied by 92 electrons. In nature, uranium is found predominantly as two isotopes. The more abundant one (99.3% of all uranium atoms) contains 146 neutrons. The mass number of this isotope of uranium is 238; that is, 92 protons plus 146 neutrons. We represent this isotope as uranium-238, or more simply as U-238. The less abundant isotope (0.7%) contains 143 neutrons and 92 protons; namely, U-235.

The terms mass number and isotope were introduced in Section 1.4.

Your Turn 6.2 Skill Building Comparing Isotopes

A trace amount of a third isotope, U-234, is also found in nature. How do U-238 and U-234 compare in terms of the number of protons and the number of neutrons? Do their number of electrons differ?

More commonly, we specify an isotope with both its mass number and atomic number. The former is a superscript and the latter a subscript, both written to the left of the chemical symbol. Using this convention, uranium-238 becomes:

$$\begin{array}{r} \text{Mass number} = \text{number of protons} + \text{number of neutrons} \longrightarrow 238 \\ \text{Atomic number} = \text{number of protons} \longrightarrow 92 \end{array} \text{U}$$

Similarly, U-235 is written as $^{235}_{92}U$. Although $^{235}_{92}U$ and $^{238}_{92}U$ differ by a mere three neutrons, this difference translates to one key difference in nuclear properties. Under the conditions present in a nuclear reactor, $^{238}_{92}U$ does not undergo fission, yet $^{235}_{92}U$ does.

The process of nuclear fission is initiated by neutrons, but also can release neutrons, as seen by this example:

$$^{1}_{0}n + ^{235}_{92}U \longrightarrow [^{236}_{92}U] \longrightarrow ^{141}_{56}Ba + ^{92}_{36}Kr + 3\ ^{1}_{0}n \qquad \textbf{[6.2]}$$

Let's examine the components of Equation 6.2 from left to right. Initially, a neutron hits the nucleus of U-235. This neutron, $^{1}_{0}n$, has a subscript of 0, indicating no positive charges; the superscript is 1 because the mass number of a neutron is 1. The nucleus of $^{235}_{92}U$ captures the neutron, forming a heavier isotope of uranium, $^{236}_{92}U$. This isotope is written in square brackets indicating that it exists only momentarily. Uranium-236 immediately splits into two smaller atoms (Ba-141 and Kr-92) with the release of three more neutrons.

Nuclear equations are similar to, but not the same as, "regular" chemical equations. To balance a nuclear equation, you count the protons and neutrons rather than counting atoms as you would do in a chemical equation. A nuclear equation is balanced if the sum of the subscripts (and of the superscripts) on the left is equal to the sum of those on the right. Coefficients in nuclear equations, such as the 3 preceding the $^{1}_{0}n$ in Equation 6.2, are treated the same way as in chemical equations, multiplying the term that follows it. For example, examine the math to see why nuclear Equation 6.2 is balanced:

	Left	Right
Superscripts:	1 + 235 = 236	141 + 92 + (3 × 1) = 236
Subscripts:	0 + 92 = 92	56 + 36 + (3 × 0) = 92

When the nucleus of an atom of U-235 is struck with a neutron, many different fission products are formed. The activity that follows acquaints you with two other possibilities.

Your Turn 6.3 Skill Building Writing Nuclear Equations

With the help of a periodic table, write these two balanced nuclear equations. Both are initiated with a neutron.

a. U-235 fissions to form Ba-138, Kr-95, and neutrons.

b. U-235 fissions to form an element (atomic number 52, mass number 137), another element (atomic number 40, mass number 97), and neutrons.

Look again at nuclear Equation 6.2. Both sides contain neutrons, which might lead you to think that we should cancel them. Although you might do this in a mathematical expression, don't do it here. The neutrons on both sides of the equation are important! The one on the left initiates the fission reaction, whereas the ones on the right are produced by it. Each neutron that is produced can in turn strike another U-235 nucleus, causing it to split, which will release a few more neutrons. This is an example of a **chain reaction**, a term that generally refers to any reaction in which one of the products becomes a reactant, thus making it possible for the reaction to become self-sustaining. This particular rapidly branching nuclear chain reaction is self-sustaining and spreads in a fraction of a second (Figure 6.3). With this exact chain reaction, the first controlled nuclear fission took place at the University of Chicago in 1942.

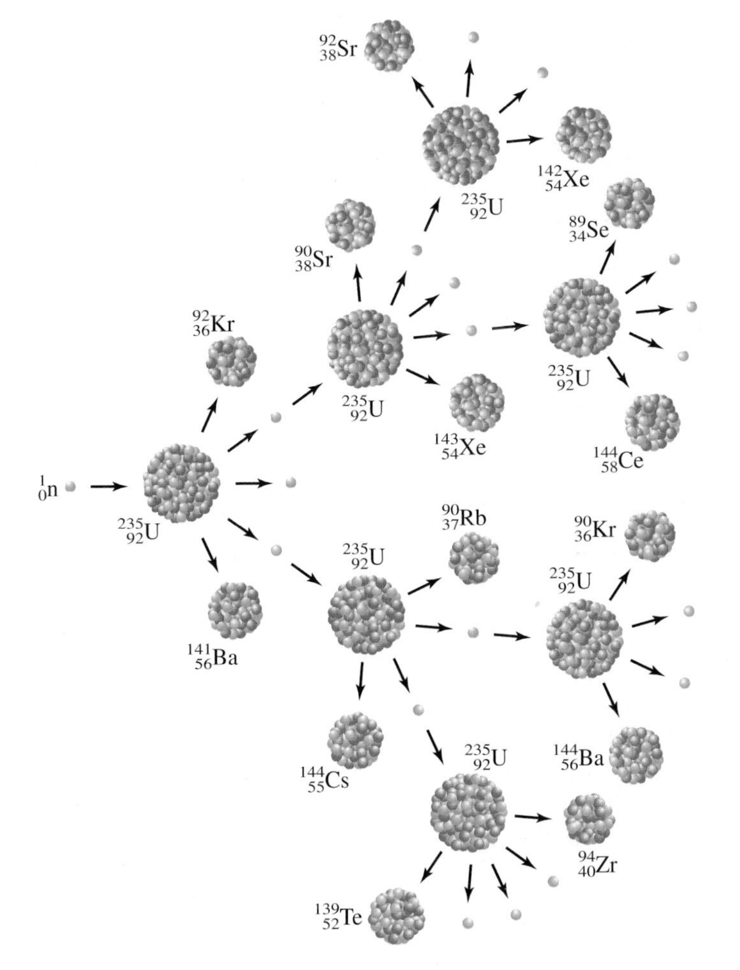

Figure 6.3

A neutron initiates the fission of uranium-235, starting a chain reaction. It should be noted that not every U-235 nucleus that is bombarded by a neutron results in the same products every time.

A **critical mass** is the amount of fissionable fuel required to sustain a chain reaction. For example, the critical mass of U-235 is about 15 kg (or 33 lb). If this mass of pure U-235 were brought together in one place with a source of neutrons, fission would occur spontaneously and would continue as long as the critical mass persists. Nuclear weapons work on this principle, although the energy released quickly blows the critical mass apart, stopping the fission reaction. But, as you will soon see, the uranium fuel used in a nuclear power plant is far from pure U-235 and is unable to explode like a nuclear bomb. There simply aren't enough neutrons around (and enough fissionable nuclei for these neutrons to hit) to produce an uncontrolled chain reaction characteristic of a nuclear explosion.

We mentioned earlier that energy is given off during fission because the mass of the products is slightly less than that of the reactants. However, from the nuclear equations we have just written, no mass loss is apparent because the sum of the mass numbers is the same on both sides. In fact, the actual mass does decrease slightly. To understand this, remember that the actual masses of the nuclei are not the mass numbers (the sum of the number of protons and neutrons); rather, they have measured values with many decimal places. For example, an atom of uranium-235 weighs 235.043924 u. Were you to keep all six decimal places and compare the masses on both sides of the nuclear equation for the fission of U-235, you would find that the mass of the products is less by about 0.1%, or 1/1000. This difference corresponds to the energy that is released.

Recall from Section 4.3 that a *unified atomic mass unit*, u, is 1/12 the mass of a C-12 atom, or 1.67×10^{-27} kg. This unit is convenient for expressing the mass of individual atoms.

How much energy would be released if all the nuclei in 1.0 kg (2.2 lb) of pure U-235 were to undergo fission? We can calculate an answer by using an equation closely related to $E = mc^2$; namely, $\Delta E = \Delta mc^2$. Here, the Greek letter delta (Δ) means "the change in," so now with a change in mass we can calculate a change in energy. Since 1/1000 of this mass is lost, the value for Δm, the change in mass, is 1/1000 of 1.00 kg, which is 1.00×10^{-3} kg. Now, we will substitute this value, and $c = 3.00 \times 10^8$ m/s, into Einstein's equation:

$$\Delta E = \Delta mc^2 = (1.00 \times 10^{-3}\ \text{kg}) \times (3.00 \times 10^8\ \text{m/s})^2 \qquad \textbf{[6.3]}$$

Completing the calculation gives an energy change in what may appear to be unusual units:

$$\Delta E = 9.00 \times 10^{13}\ \text{kg}\cdot\text{m}^2/\text{s}^2$$

As you saw in Chapter 5, the unit $\text{kg}\cdot\text{m}^2/\text{s}^2$ is identical to a joule (J). Therefore, the energy released from the fission of an entire kilogram of uranium-235 is a whopping 9.00×10^{13} J, or 9.00×10^{10} kJ.

To put things into perspective, 9.00×10^{13} J is the amount of energy released by the detonation of about 22 kilotons of the explosive trinitrotoluene, better known as TNT. By way of comparison, this is roughly twice the amount of energy released by the atomic bombs dropped on Hiroshima and Nagasaki in 1945! This energy originates from the fission of a single kilogram of U-235, in which a mass of approximately 1 gram (0.1% mass change) was transformed into energy.

Your Turn 6.4 Skill Building Comparing Nuclear to Coal

Select a grade of coal from **Table 5.2**. What mass of coal would be needed to produce the same amount of energy as the fission of 1 kg of U-235?

As it turns out, one cannot fission a kilogram or two of pure U-235 in one fell swoop. In an atomic weapon, for example, the energy released blasts the fissionable fuel apart in a fraction of a second, thus halting the chain reaction before all the nuclei can undergo fission. Nonetheless, the energy released is enormous—on the order of 10 kilotons of TNT for the atomic bomb dropped on the city of Hiroshima. Figure 6.4 shows an atomic explosion at the U.S. Nevada Test Site. Code named Priscilla, this test in 1957 had more than twice the explosive power of the bombs at Hiroshima and Nagasaki.

Figure 6.4

The nuclear test "Priscilla" was exploded on a dry lake bed northwest of Las Vegas, Nevada, on June 24, 1957.

Source: Photo courtesy of National Nuclear Security Administration/Nevada Site Office/U.S. Dept. of Energy

Fortunately, the energy of nuclear fission can be harnessed. This is exactly the objective of a nuclear power plant. Here, the energy is slowly and continually released under controlled conditions, as we shall see in the next section.

6.2 Harnessing a Nuclear Fission Reaction: How Nuclear Power Plants Produce Electricity

Section 5.6 described how a conventional power plant burns coal, oil, or some other fuel to produce heat. The heat is then used to boil water, converting it into high-pressure steam that turns the blades of a turbine. The shaft of the spinning turbine is connected to large wire coils that rotate within a magnetic field, thus generating electric energy. A nuclear power plant operates in much the same way except the energy released from the fission of atomic nuclei, such as U-235, is used to heat the water instead of combustion of a fossil fuel (Figure 6.5). Like any power plant, a nuclear facility is subject to the efficiency constraints imposed by the Second Law of Thermodynamics. The theoretical efficiency for converting heat energy to work depends on the maximum and minimum temperatures between which the plant operates. This net efficiency, typically 55–60%, is significantly reduced by other mechanical, thermal, and electric losses.

The Second Law of Thermodynamics has many versions. A form relevant to this section states that it is impossible to convert heat completely into work in a process that is a cycle. See Section 5.4.

The nuclear reactor is the heart of the power station. The reactor, together with one or more steam generators and the primary cooling system, is housed in a special steel vessel within a concrete dome-shaped containment building. The non-nuclear portion contains the turbines that run the electric generator, as well as the secondary cooling system. In addition, the non-nuclear portion must be connected to some means of removing excess heat from the coolants. Accordingly, a nuclear power station has one or more cooling towers, or is located near a sizeable body of water (or both). Look back at Figure 5.10, which shows a diagram of a fossil fuel power plant. This plant also requires a means of removing heat, as shown by the stream of cooling water.

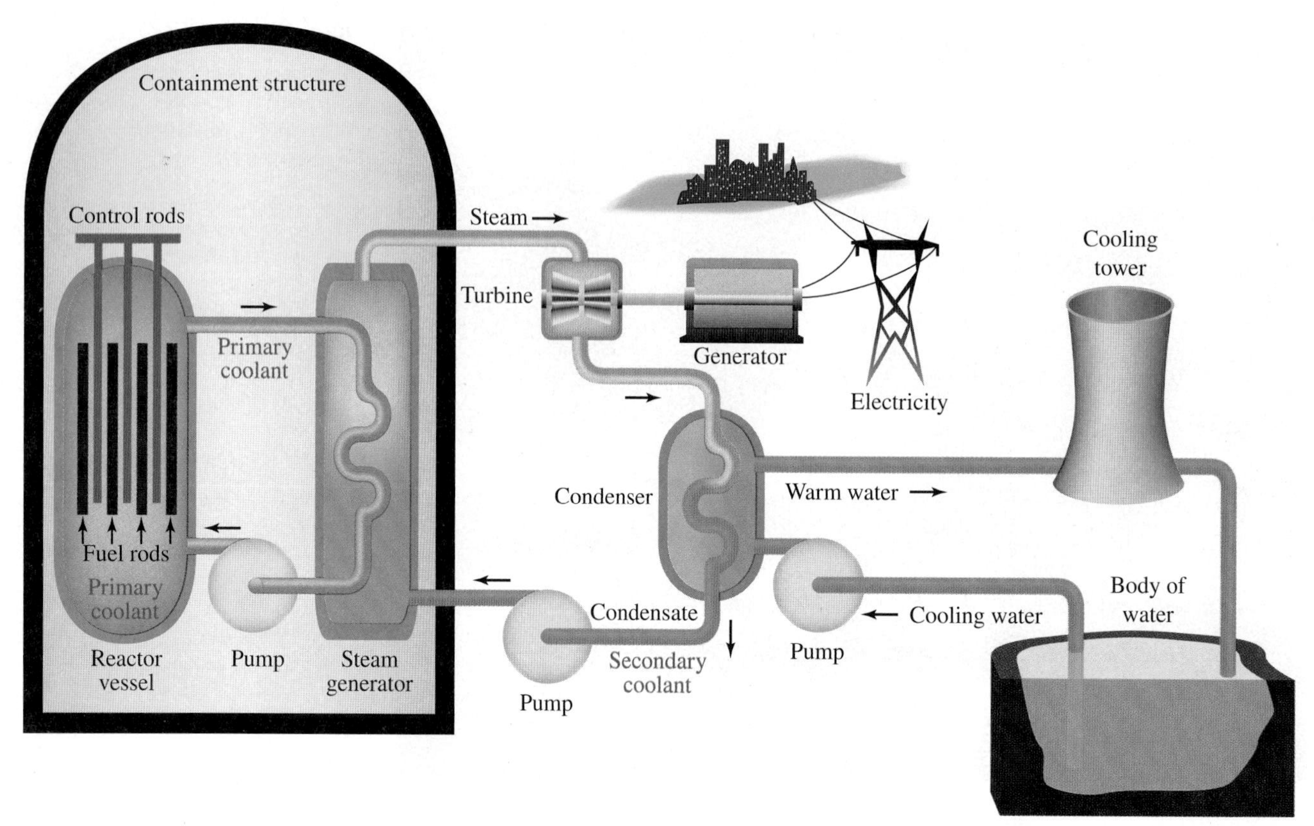

Figure 6.5

Diagram of a nuclear power plant. Components are not illustrated to scale.

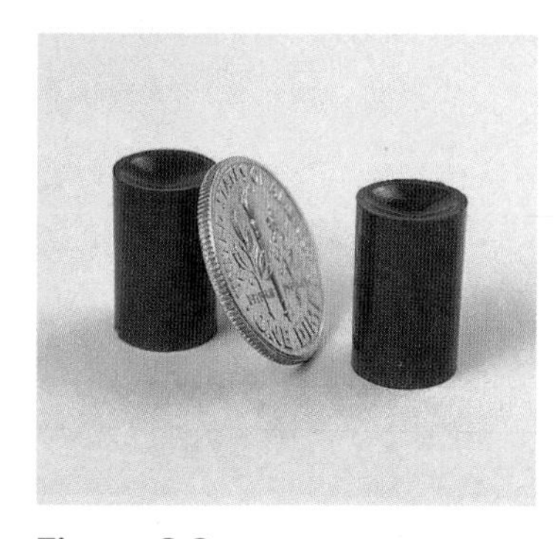

Figure 6.6

Comparative sizes of nuclear fuel pellets and a U.S. dime.

© McGraw-Hill Education. C.P. Hammond, photographer

Gamma rays were first introduced with the rest of the electromagnetic spectrum in Section 3.1

The uranium fuel in the reactor core is in the form of uranium(IV) oxide (UO_2), each comparable in height to the width of a dime, as shown in Figure 6.6. These pellets are placed end-to-end in tubes composed of an alloy of zirconium and other metals, which in turn are grouped into stainless steel–clad bundles (Figure 6.7).

Each rod contains at least 200 pellets. Once started, a fission reaction can sustain itself by a chain reaction. However, neutrons are needed to induce the process (Equation 6.2 and Figure 6.3). One means of generating neutrons is to use a combination of beryllium-9 and a heavier element such as plutonium. The heavier element releases alpha (α) particles, ${}^{4}_{2}\text{He}$:

$$ {}^{238}_{94}\text{Pu} \longrightarrow {}^{234}_{92}\text{U} + \underset{\text{alpha particle}}{{}^{4}_{2}\text{He}} \qquad [6.4] $$

These alpha particles in turn strike the beryllium, releasing neutrons, carbon-12, and gamma rays, ${}^{0}_{0}\gamma$:

$$ {}^{4}_{2}\text{He} + {}^{9}_{4}\text{Be} \longrightarrow {}^{12}_{6}\text{C} + {}^{1}_{0}\text{n} + \underset{\text{gamma ray}}{{}^{0}_{0}\gamma} \qquad [6.5] $$

The neutrons produced in this way can initiate the nuclear fission of uranium-235 in the reactor core.

Your Turn 6.5 Skill Building Poo–Bee and Am–Bee

A neutron source constructed with Pu and Be is a PuBe ("poo–bee") source. Similarly, the AmBe ("am–bee") source is constructed from americium and beryllium. Analogous to the PuBe source, write the set of reactions that produce neutrons from an AmBe source. Start with Am-241.

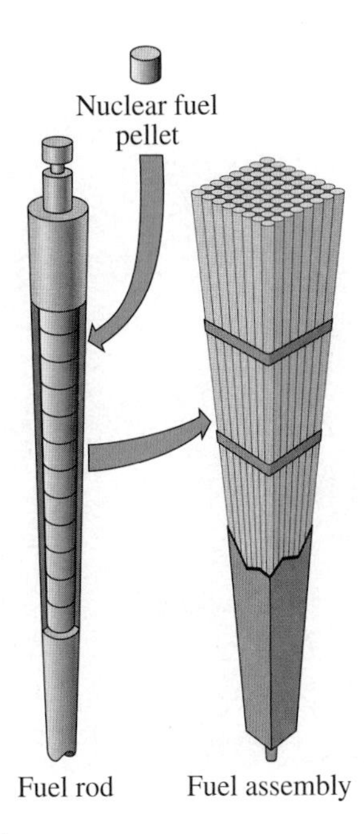

Figure 6.7

Schematic of fuel pellets, fuel rod, and fuel assembly making up the core of a nuclear reactor (*left*). The fuel assembly is submerged under water in an active reactor core (*right*).

Remember—one fission event produces two or three neutrons. The trick is to "sponge up" these extra neutrons, but still leave enough to sustain the fission reaction. A delicate balance must be maintained. With extra neutrons, the reactor runs at too high a temperature; with too few, the chain reaction halts and the reactor cools down. To achieve the needed balance, one neutron from each fission event should in turn cause another fission reaction.

With a natural abundance three times greater than uranium and the generation of less harmful long-term nuclear waste products during its use, thorium (Th) has been proposed as an attractive alternative to using uranium as a nuclear fuel.

Metal rods interspersed among the fuel elements serve as the neutron "sponges." These *control rods*, composed primarily of an excellent neutron absorber such as cadmium or boron, can be positioned to absorb fewer or more neutrons. With the rods fully inserted, the fission reaction is not self-sustaining. But as the rods are gradually withdrawn, the reactor can "go critical" and become self-sustaining, running at different rates depending on the exact position of the control rods. Over time, fission products that absorb neutrons build up in the fuel pellets. To compensate, the control rods can be withdrawn. Eventually, the reactor fuel bundles must be replaced.

Your Turn 6.6 You Decide Earthquake!

Look ahead to **Figure 6.16** to see that earthquakes may occur in the vicinity of nuclear reactors. Reactors near the epicenter should automatically shut down. Should the software be programmed to fully insert the control rods into the reactor core, or should they be pulled out? Explain.

The fuel bundles and control rods are bathed in the primary coolant, a liquid that comes in direct contact with them and carries away heat. In the Byron nuclear reactor (Figure 6.8) and in many others, the primary coolant is an aqueous solution of boric acid, H_3BO_3. The boron atoms absorb neutrons and thus control the rate of fission and temperature. Like the control rods, the solution serves as a moderator

Figure 6.8

The Byron nuclear power plant in Illinois. The two cooling towers (one shown with a cloud of condensed water vapor) are the most prominent features of this plant. The reactors, however, are located in the two cylindrical containment buildings with white roofs in the foreground.

© AP/Wide World Photos

for the reactor, slowing the neutrons, which makes them more effective in producing fission. Another major function of the primary coolant is to absorb the heat generated by the nuclear reaction. Because the primary coolant solution is at a pressure more than 150 times normal atmospheric pressure, it does not boil. It is heated far above its normal boiling point and circulates in a closed loop from the reaction vessel to the steam generators, and back again. This closed primary coolant loop thus forms the link between the nuclear reactor and the rest of the power plant (Figure 6.5).

Did You Know? Cooling towers are also used in some coal-fired plants, such as the one shown below near Tianjin, China.

© Bradley D. Fahlman

The heat from the primary coolant is transferred to what is sometimes referred to as the secondary coolant, the water in the steam generators that does not come in contact with the reactor. At the Byron nuclear plant (Figure 6.8), more than 30,000 gallons of water are converted to vapor each minute. The energy of this hot vapor turns the blades of turbines that are attached to an electric generator. To continue the heat-transfer cycle, the water vapor is then cooled, condensed back to a liquid, and returned to the steam generator. In many nuclear facilities, the cooling is done using large cooling towers that are commonly mistaken for the reactors. The reactor buildings are not as large.

Your Turn 6.7 Scientific Practices Clouds (Not Mushroom-Shaped)

Some days you can see a cloud coming out of the cooling tower of a nuclear power plant, as shown in **Figure 6.8**. What causes the cloud? Does it contain any nuclear products produced from the fission of U-235? Explain.

Nuclear power plants also use water from lakes, rivers, or the ocean to cool the condenser. For example, at the Seabrook nuclear power plant in New Hampshire, every minute about 400,000 gallons of ocean water flow through a huge tunnel (19 feet in diameter and 3 miles long) bored through rock 100 feet beneath the ocean floor.

A similar tunnel from the plant carries the water, now 22 °C warmer, back to the ocean. Special nozzles distribute the hot water so that the observed temperature increase in the immediate area of the discharge is only about 2 °C. The ocean water is in a separate loop from the fission reaction and its products. The primary coolant (water with boric acid) circulates through the reactor core inside the containment building. However, this boric acid solution is kept isolated in a closed circulating system, which makes the transfer of radioactivity to the secondary coolant water in the steam generator highly unlikely. Similarly, the ocean water does not come in direct contact with the secondary system, so the ocean water is well-protected from radioactive contamination. Clearly, the electricity generated by a nuclear power plant is identical to the electricity generated by a fossil fuel plant; the electricity is not radioactive, nor can it be.

While the electricity generated by nuclear and fossil fuel sources is identical, different sources of energy vary in how much electricity is generated per second. The rate of energy production is referred to as **power**, and a common unit of power is the joule per second (J/s), or **watt**. Single nuclear reactors typically have a maximum power capacity between 500 and 1300 megawatts (MW). In comparison, a typical coal-fired power plant generates electricity at a rate of 600 MW.

The metric prefix "mega" means one million.

Your Turn 6.8 Skill Building The Palo Verde Reactors

One of the most powerful nuclear plants in operation in the United States is the Palo Verde complex in Arizona. At maximum capacity, just one of its three reactors generates 1.2 billion joules of electric energy every second, or 1.2 GW of power. Calculate the total amount of electric energy produced by the three reactors per day and the mass of U-235 lost each day.

Hint: Start by calculating the quantity of energy generated not per second, but per day. Then, use the equation $\Delta E = \Delta mc^2$ and solve for the change in mass, Δm. Report the mass loss in grams.

The Palo Verde nuclear reactor

© *Royalty-Free/Corbis*

The topics we have been discussing—nuclear fission, uranium, nuclear fuel, nuclear weapons—all rest on an understanding of radioactivity. We now turn to this topic.

6.3 | What Is Radioactivity?

Our knowledge of radioactive substances is well over 100 years old. In 1896, the French physicist Antoine Henri Becquerel (1852–1908) discovered radioactivity. At the time, his research involved using photographic plates. Prior to use, these plates were sealed in black paper to keep them from being exposed to light. By accident, he left a mineral near one of these sealed plates and found that the plate's light-sensitive emulsion darkened. It was as though the plate had been exposed to light! Becquerel immediately recognized that the mineral emitted powerful rays that penetrated the lightproof paper.

Further investigation by the Polish scientist Marie Curie (1867–1934) (Figure 6.9) revealed that the rays were coming from a constituent of the mineral—the element uranium. In 1899, Curie applied the term **radioactivity** to the spontaneous emission of radiation by certain elements. Subsequent research by Ernest Rutherford (1871–1937) led to the identification of two major types of radiation. Rutherford named them after the first two letters of the Greek alphabet, alpha (α) and beta (β).

Figure 6.9

Marie Curie won two Nobel Prizes—one in chemistry, the other in physics—for her research on radioactive elements.

© *Hulton-Deutsch Collection/Corbis*

Alpha and beta radiation have strikingly different properties. A beta particle (β) is a high-speed electron emitted from the nucleus. It has a negative electric charge (1–) and only a tiny mass, about 1/2000 that of a proton or a neutron. If you are wondering

Table 6.1 Types of Nuclear Radiation

Name	Symbol	Composition	Charge	Change to the Parent Nucleus
alpha	${}^{4}_{2}He$ or α	2 protons 2 neutrons	2+	mass number decreases by 4 atomic number decreases by 2
beta	${}^{0}_{-1}e$ or β	1 electron	1−	mass number does not change atomic number increases by 1
gamma	${}^{0}_{0}\gamma$ or γ	photon	0	no change in either the mass number or the atomic number

how an electron (a beta particle) could possibly be emitted from a nucleus, stay tuned. We offer an explanation shortly.

In contrast, an alpha particle (α) is a positively charged particle emitted from the nucleus. It consists of two protons and two neutrons (the nucleus of a He atom) and has a 2+ charge since no electrons accompany the helium nucleus.

See Figure 3.4 for more information about the electromagnetic spectrum.

Gamma rays frequently accompany alpha or beta radiation. A gamma ray (γ) is emitted from the nucleus and has no charge or mass. It is a high-energy, short-wavelength photon. Just like infrared (IR), visible, and ultraviolet (UV) radiation, gamma rays are part of the electromagnetic spectrum and have more energy than X-rays. Table 6.1 summarizes these three types of nuclear radiation.

The term "radiation" tends to be confusing, because people don't always specify whether they mean electromagnetic radiation or nuclear radiation. As seen in Section 3.1, electromagnetic radiation refers to all the different types of light: radio, X-rays, visible, infrared, ultraviolet, microwave, and, of course, gamma rays. For example, it is perfectly correct to say "visible radiation" instead of "visible light." **Nuclear radiation**, however, refers to alpha, beta, or gamma radiation emitted from a nucleus. Watch out for one more source of confusion. Gamma rays are both a type of electromagnetic radiation and of nuclear radiation. When emitted from the nucleus of a radioactive substance, we refer to gamma rays as nuclear radiation. In contrast, when emitted from a galaxy far away, we call these gamma rays electromagnetic radiation.

Your Turn 6.9 You Decide "Radiation"

For each sentence, use the context to decipher whether the speaker is referring to nuclear or electromagnetic radiation.

a. "Name a type of radiation that has a shorter wavelength than visible light."
b. "The gamma radiation from cobalt-60 can destroy a tumor."
c. "Watch out for UV rays! If you have lightly pigmented skin, this radiation may cause a sunburn."
d. "Rutherford detected the radiation emitted by uranium."

When either an alpha or beta particle is emitted, a remarkable transformation occurs—the atom that emitted the particle changes its identity. For example, earlier with the PuBe neutron source (Equation 6.4), you saw that alpha emission resulted in the nucleus of plutonium becoming that of uranium. Similarly, when uranium emits an alpha particle, it becomes the element thorium. This nuclear equation shows the process for uranium-238:

$$ {}^{238}_{92}U \longrightarrow {}^{234}_{90}Th + {}^{4}_{2}He \qquad \textbf{[6.6]} $$

Notice that the sum of the mass numbers on both sides of the nuclear equation is equal: 238 = 234 + 4. The same is true for the atomic numbers: 92 = 90 + 2.

In some cases, the nucleus formed as the result of radioactive decay is still radioactive. For example, thorium-234, formed by the alpha decay of uranium-238, is radioactive. Thorium-234 undergoes subsequent beta decay to form protactinium (Pa):

$$^{234}_{90}\text{Th} \longrightarrow {}^{234}_{91}\text{Pa} + {}^{0}_{-1}\text{e} \qquad \textbf{[6.7]}$$

In contrast to alpha emission, with beta emission the atomic number increases by 1 and the mass number remains unchanged. Table 6.1 summarizes the changes that occur with both beta and alpha emission.

A concept that can help you make sense of this seemingly unusual set of changes is to regard a neutron as a combination of a proton and an electron. Beta emission can be thought of as breaking a neutron apart. Equation 6.8 shows this process, giving us an explanation of how an electron can be emitted from the nucleus:

$$^{1}_{0}\text{n} \longrightarrow {}^{1}_{1}\text{p} + {}^{0}_{-1}\text{e} \qquad \textbf{[6.8]}$$

During beta emission, the mass number (neutrons plus protons) in the nucleus remains constant because the loss of the neutron is balanced by the formation of a proton. For example, a neutron in thorium "became" a proton in protactinium. Because of this additional proton, the atomic number increases by 1. Again, this model can help you better visualize beta emission, but may not be exactly what is occurring.

Your Turn 6.10 Skill Building Alpha and Beta Decay

a. Write a nuclear equation for the beta decay of rubidium-86 (Rb-86), a radioisotope produced by the fission of U-235.

b. Plutonium-239, a toxic isotope that causes lung cancer, is an alpha emitter. Write the nuclear equation.

As we noted earlier, a nucleus may decay to produce another radioactive nucleus. In some cases, we can predict this, because *all* isotopes of *all* elements with atomic number 84 (polonium) and higher are radioactive. Thus, all the isotopes of uranium, plutonium, radium, and radon are radioactive because these elements all have atomic numbers greater than 83.

What about the lighter elements? Some of these are naturally radioactive, such as carbon-14, hydrogen-3 (tritium), and potassium-40. Whether an isotope is radioactive (referred to as a **radioisotope**) or stable depends on the ratio of neutrons to protons in its nucleus. With each emission of an alpha or a beta particle, this neutron-to-proton ratio changes. Eventually, a stable ratio is achieved, and the nucleus is no longer radioactive. Most of the atoms that make up our planet are *not* radioactive. They are here today, and you can count on their being here tomorrow—although possibly not located in the same spot you last saw them (such as the atoms that make up your keys).

In some cases, radioisotopes may decay many times before producing a stable isotope. For example, the radioactive decay of U-238 and Th-234 (Equations 6.6 and 6.7) are only the first two steps of a 14-step sequence! As shown in Figure 6.10, lead-206 is the end-product in this sequence. Similarly, lead-207 is also the end-product in a different sequence of 11 steps that begins with U-235. Each of these sequences is called a **radioactive decay series**; that is, a characteristic pathway of radioactive decay that begins with a radioisotope and progresses through a series of steps to eventually produce a stable isotope. Radon, a radioactive gas, is produced midway in both the U-238 and U-235 decay series. Thus, wherever uranium is present, so is radon.

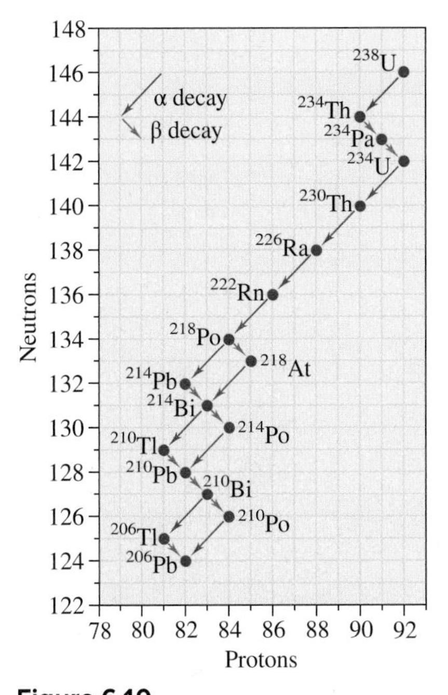

Figure 6.10
The naturally occurring radioactive decay series of uranium-238.

6.4 How Long Do Substances Remain Radioactive?

How long does a radioactive sample "last"? The answer depends on the radioisotope. Some radioisotopes decay quickly over a short period of time, whereas others undergo radioactive decay much more slowly. Each radioisotope has its own **half-life ($t_{1/2}$)**, the time required for the level of radioactivity to fall to one-half of its initial value. For example, plutonium-239, an alpha emitter formed in nuclear reactors fueled with uranium, has a half-life of about 24,110 years. Accordingly, it will take 24,110 years for half of a sample of Pu-239 to decay. After a second half-life (another 24,110 years, or 48,220 years total), the level of radioactivity will be one-fourth of the original amount. And in three half-lives (72,330 years total), the level will be one-eighth (Figure 6.11). From these times, you can see that it takes a very long time for the sample size of Pu-239 to decrease!

Other radioisotopes decay even more slowly. For example, the half-life of U-238 is 4.5 billion years. Coincidentally, this is approximately the age of the oldest rocks on Earth, a determination made by measuring their uranium content. The half-life for each particular isotope is a constant, and is independent of the physical or chemical form in which the element is found. Moreover, the rate of radioactive decay is essentially unaltered by changes in temperature and pressure. Table 6.2 shows that half-lives range from milliseconds to millennia.

From Table 6.2, you also can see that Pu-239 and Pu-231 have different half-lives. Other isotopes of plutonium have different half-lives as well. In 1999, Carola Laue, Darleane Hoffman, and a team at Lawrence Berkeley National Laboratory characterized plutonium-231. These researchers had to work fast because the half-life of Pu-231 is a matter of mere minutes! In general, each radioisotope has its own unique half-life, including isotopes of the same element.

We can use the half-life of a radioisotope to determine the percent of a sample that remains at some later point in time. For example, once Pu-231 is generated in a laboratory, what percent of the original sample remains after 25 minutes? To answer this, first recognize that 25 minutes is roughly three half-lives or 3×8.5 minutes. After one half-life, 50% of the sample has decayed and 50% remains. After two half-lives, 75% of the sample has decayed and 25% remains. And after three half-lives, 87.5% has decayed and 12.5% remains. These values are not exact because 25 minutes is not exactly three half-lives. Nonetheless, quick back-of-the-envelope calculations can be useful.

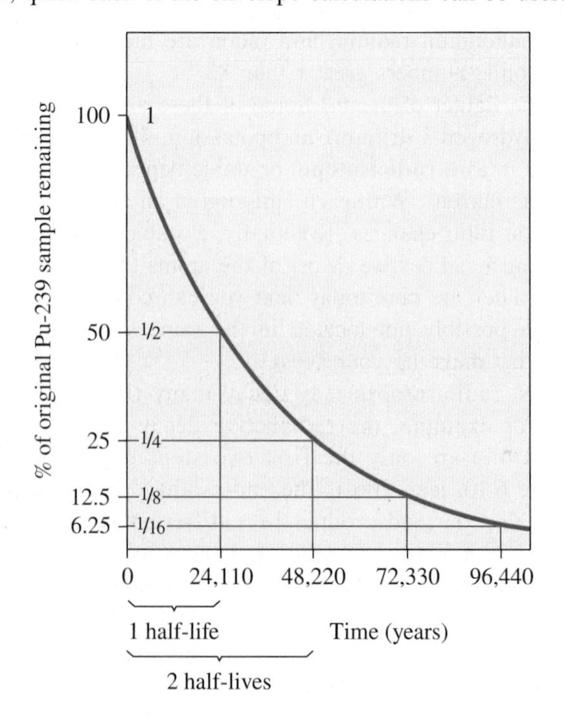

Figure 6.11

Decay of a sample of Pu-239 over time.

Table 6.2 Half-life of Selected Radioisotopes

Radioisotope	Half-life ($t_{1/2}$)	Found in the Used Fuel Rods of Nuclear Reactors?
uranium-238	4.5×10^9 years	Yes. Present originally in fuel pellet.
potassium-40	1.3×10^9 years	No
uranium-235	7.0×10^8 years	Yes. Present originally in fuel pellet.
plutonium-239	24,110 years	Yes. See Equation 6.4
carbon-14	5715 years	No.
cesium-137	30.2 years	Yes. Fission product.
strontium-90	29.1 years	Yes. Fission product.
thorium-234	24.1 days	Yes. Small amount generated in natural decay series of U-238.
iodine-131	8.04 days	Yes. Fission product.
radon-222	3.82 days	Yes. Small amount generated in natural decay series of U-238.
plutonium-231	8.5 minutes	No. Half-life is too short.
polonium-214	0.00016 seconds	No. Half-life is too short.

This question also could have been phrased in this way: "After 25.5 minutes, what percent of a sample of Pu-231 has *decayed*?" This question requires one more step. To find the amount decayed, simply subtract the percent that remains from 100%. If 12.5% remains, then 100% – 12.5% = 87.5% has decayed. Table 6.3 summarizes these changes for any radioisotope.

Your Turn 6.11 Skill Building Here Today ...

... and gone tomorrow? People sometimes use the value of 10 half-lives to indicate when a radioisotope will be gone; that is, when only a negligible amount of it will be present. What percent of the original sample remains after 10 half-lives? Add rows to **Table 6.3** so that it shows the mathematics of decay to 10 half-lives.

Let's do another back-of-the-envelope calculation with a different radioisotope. For example, if you had a sample of U-238, what percent would remain after 25 minutes? To answer this, recognize that minutes, days, or even months would be a mere instant in the span of a 4.5-billion-year half-life. Thus, essentially all of the uranium-238 would remain. The next two activities offer you more practice with half-life calculations.

Table 6.3 Half-life Calculations

# of Half-lives	% Decayed	% Remaining
0	0	100
1	50	50
2	75	25
3	87.5	12.5
4	93.75	6.25
5	97.88	3.12
6	98.44	1.56

Your Turn 6.12 Skill Building Tritium Calculation

Hydrogen-3 (tritium, H-3) is sometimes formed in the primary coolant water of a nuclear reactor. Tritium is a beta emitter with $t_{1/2}$ = 12.3 years. For a given sample containing tritium, after how many years will about 12% of the original sample remain?

Your Turn 6.13 Skill Building Radon Calculation

Radon-222 is a radioactive gas produced from the decay of radium, a radioisotope naturally present in many rocks.

a. What is the most likely origin for the radium present in rocks?
Hint: See Figure 6.10.

b. Radon activity is usually measured in picocuries (pCi). Suppose that the radioactivity from Rn-222 in your basement were measured at 16 pCi, a high value. If no additional radon entered the basement, how much time would pass before the level dropped to 0.50 pCi?
Hint: In dropping from 16 to 1 pCi, the radioactivity level halves four times: 16 to 8 to 4 to 2 to 1.

c. In regard to above, why is it incorrect to assume that no more radon will enter your basement?

The long half-lives of the isotopes produced from nuclear reactors mean that nuclear waste will emit radiation for thousands of generations to come. Today, most nuclear waste is stored at the site where it is generated, while countries try to develop plans for long-term underground storage facilities. These geological repositories can be designed to require minimal human intervention. In the U.S., the 1997 Nuclear Waste Policy Amendments Act designated Yucca Mountain (Figure 6.12) in Nevada as the sole site to be studied as an underground long-term nuclear waste repository.

(a)

(b)

Figure 6.12

(a) Map of Yucca Mountain and state of Nevada. **(b)** Yucca Mountain, looking south.

(b) Source: U.S. Dept. of Energy

In the years that followed, billions of dollars were spent to fund the development of the repository. As of 2016, however, it appears that the Yucca Mountain repository will not be completed due to ongoing scientific concerns.

One final difficulty with reactor waste is that the fission products, if released, may enter and accumulate in your body, with potentially fatal consequences. One culprit is strontium-90, a radioactive fission product that entered the biosphere in the 1950s from the atmospheric testing of nuclear weapons. Strontium ions are chemically similar to calcium ions; both elements are in Group 2 of the periodic table. Like Ca^{2+}, Sr^{2+} accumulates in milk and in bones. Thus, once ingested, radioactive strontium with its half-life of 29 years poses a lifelong threat. Along with I-131, Sr-90 was among the harmful fission products released in the vicinity of the Chornobyl reactor that exploded and caught on fire in Ukraine in 1986.

For more about the positive and negative effects of radiation on human health, see Chapter 12.

Your Turn 6.14 Skill Building Strontium-90

Sr-90 is one of the fission products of U-235 listed in Table 6.2. It forms in a reaction that produces three neutrons and another element. Write the nuclear equation.

Hint: Remember to include the neutron that induces the fission of U-235.

6.5 | What Are the Risks of Nuclear Power?

All nuclear plants use the process of fission to produce energy, and all produce radioactive fission products. Have these radioactive products posed a danger in the past? In this section, we consider the accidental release of radioisotopes into the environment. Although this is not the sole legacy of nuclear power, it nonetheless is a significant one.

On April 26, 1986, the engineers of the Chornobyl nuclear power plant in Ukraine (Figure 6.13), then part of the Soviet Union, were running a safety test when the reactor overheated. This plant had four reactors, two built in the 1970s and two more in the 1980s. Water from the nearby Pripyat River was used to cool the reactors. Although the surrounding region was not heavily populated, approximately 120,000 people lived within a 30-km radius, including the cities of Chornobyl (pop. 12,500) and Pripyat (pop. 40,000).

Chornobyl stands as the world's worst nuclear power plant accident. So what went wrong in Ukraine? During an electrical power safety test at the Chornobyl Unit 4 reactor, operators deliberately interrupted the flow of cooling water to the core. The temperature of the reactor rose rapidly. In addition, the operators had left an insufficient

Did You Know? The alternate spelling of "Chernobyl" is the transliteration of the Russian pronunciation. Чорнобиль (Chornobyl) is the Ukrainian word.

As shown in Figure 6.13, the black trefoil on a yellow background is the international radiation warning symbol. In the United States, magenta is used instead of black.

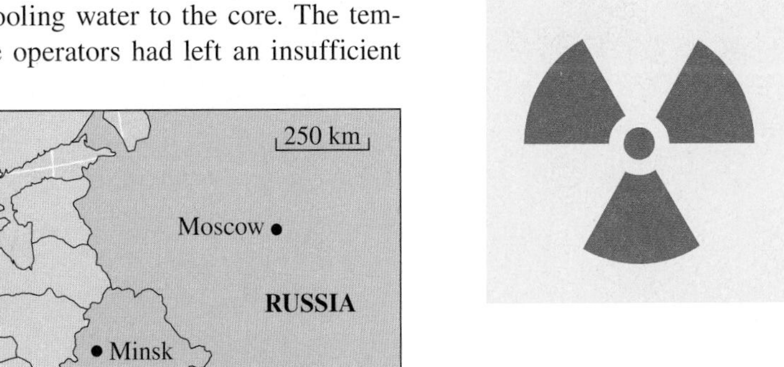

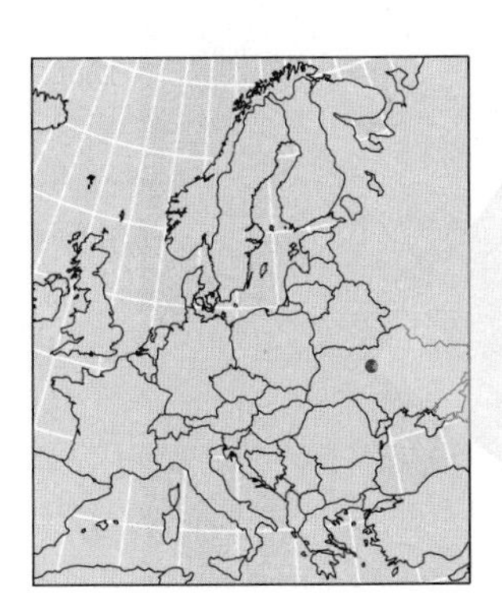

Figure 6.13

Chornobyl, Ukraine, in the former Soviet Union.

number of control rods in the reactor and other control rods couldn't be reinserted quickly enough. Furthermore, the steam pressure was too low to provide coolant, due to both operator error and faulty reactor design.

A chain of events quickly produced a disaster. An overwhelming power surge produced heat, rupturing the fuel elements and releasing hot reactor fuel particles. These, in turn, exploded on contact with the coolant water, and the reactor core was destroyed in seconds. The heat ignited the graphite used to slow neutrons in the reactor. When water was sprayed on the burning graphite, the water and graphite reacted to produce hydrogen gas:

$$2\ H_2O(l) + C(s,\ graphite) \longrightarrow 2\ H_2(g) + CO_2(g) \qquad \textbf{[6.9]}$$

In turn, the hydrogen exploded upon reaction with oxygen in the air:

$$2\ H_2(g) + O_2(g) \longrightarrow 2\ H_2O(g) \qquad \textbf{[6.10]}$$

The explosion blasted off the 4,000-ton steel plate covering the reactor (Figure 6.14).

Your Turn 6.15 Skill Building Hydrogen Explosion

Equation 6.10 represents the combustion of hydrogen; that is, the reaction of hydrogen and oxygen to produce water vapor. **Equation 5.7** provided the Lewis structures for this chemical reaction. Using the bond energies for those bonds broken and formed, the energy change for this reaction can be estimated. As shown in **Figure 5.8**, the value is −498 kJ for burning two moles of H_2.

a. Calculate the energy change per mole and per gram of H_2.

b. Of the fuels listed in **Figure 5.6**, methane releases the most heat per gram upon combustion. Burning hydrogen releases even more heat. Approximately how many times more?

Fires started in what remained of the building and burned for 10 days. Although a "nuclear" explosion was not possible, the fire and explosions of hydrogen blew vast quantities of radioactive material out of the reactor core and into the atmosphere. People living within 60 km of the power plant were permanently evacuated. The radioactive dust cut a swath across Ukraine, Belarus, and up into Scandinavia, affecting

Figure 6.14

An aerial view of the Chornobyl Unit 4 reactor taken shortly after the chemical explosion.

some who had not benefited from the power plant but nonetheless shared in its risks. The human toll was immediate. Several people working at the plant were killed outright, and another 31 firefighters died in the cleanup process.

One of the hazardous radioisotopes released was iodine-131, a beta emitter with an accompanying gamma ray:

$$^{131}_{53}\text{I} \longrightarrow {}^{131}_{54}\text{Xe} + {}^{\;0}_{-1}\text{e} + {}^{0}_{0}\gamma \qquad \textbf{[6.11]}$$

If ingested, I-131 can cause thyroid cancer. In the contaminated area near Chornobyl, the incidence of thyroid cancer increased sharply, especially for those younger than age 15. More than 6,000 cases of thyroid cancer have been reported among those who were children and adolescents living in Belarus, the Russian Federation, and Ukraine at the time of the accident. Fortunately, with treatment, the survival rate for thyroid cancer is high and most have survived. Apart from the dramatic increase in thyroid cancer incidence among those exposed at a young age, there is no clearly demonstrated increase in the incidence of solid cancers or leukemia due to radiation in the exposed populations.

Your Turn 6.16 Skill Building Iodine

When people speak of iodine, they may be referring to an iodine atom, an iodine molecule, or an iodide ion, depending on the context.

a. Draw Lewis structures to show the differences among these chemical forms of iodine.
b. Which one is the most chemically reactive and why?
c. Which chemical form of iodine-131 is implicated in thyroid cancer?

In 2012, construction of a steel arch, large enough to encompass a college football stadium with the Statue of Liberty at midfield, began at Chornobyl to encase the ruins of the plant (Figure 6.15). Recognizing that Chornobyl is a global problem, about 30 countries are contributing to the cost of the five-year project, estimated at \$1.5 billion. To this day, the area around the nuclear reactor has remained unsuitable for human habitation.

Figure 6.15

View of the containment structure being constructed at the site of the Chornobyl disaster.

Underlying the solemn facts of Chornobyl is the inevitable question: "Could it happen again?" The closest brush with nuclear disaster in the United States occurred in March of 1979, when the Three Mile Island power plant near Harrisburg, Pennsylvania, lost coolant and a partial meltdown occurred. Although some radioactive gases were released during the incident, no fatalities resulted. A 20-year follow-up study concluded in 2002 that the total cancer deaths among the exposed population were not higher than those of the general population. Nuclear engineers agree that no commercial nuclear reactors in the United States have the design defects that led to the Chornobyl catastrophe.

The world's most recent nuclear disaster occurred in 2011, when a pair of natural disasters, an earthquake and a tsunami, resulted in the meltdown of three units of the Fukushima Daiichi nuclear power plant in Japan (Figure 6.16).

The tsunami delivered a one-two punch. First, the flood waters knocked out the electrical generators necessary to pump the cooling water at the Fukushima power plant; as a result, the reactor cooling systems failed. The fuel inside reactors 1, 2, and 3 quickly heated, and the heat started a chemical reaction that generated hydrogen gas. Fearing a hydrogen gas explosion that would rival that of Chornobyl, plant workers vented the hydrogen. At the same time, this action released some of the radioactive fission products, including I-131, to the surrounding countryside. Despite the venting, explosive chemical reactions occurred at four of the six reactors, which released dangerous radioisotopes into the environment.

Over the following two weeks, the government instructed approximately 300,000 people living within 30 km of the power plant to leave. Tens of thousands of people continue to live in temporary housing while waiting to return to their homes. The International Atomic Energy Association recommended in 2013 that most evacuees be allowed to return to their homes, but a full lifting of the evacuation is not expected for several decades. Public health agencies estimate that more than 1,000 evacuees have died prematurely due to the physical and mental stress of the evacuation. However, the World Health Organization estimates that the level of exposure of the Japanese general public was so low that it does not expect to see any radiation-related long-term health effects. After the disaster, Japan's nuclear agency shut down all of the 48 nuclear power plants in the country. The first one restarted in the fall of 2014 under new safety rules.

Figure 6.16

Flooding from the tsunami that followed the 2011 Tohoku earthquake, a 9.0 on the Richter scale.

Your Turn 6.17 Scientific Practices Zirconium

At the Chornobyl nuclear power plant, hydrogen was generated by a reaction of water with the hot graphite, as described earlier in this section. At Fukushima, however, the hydrogen was generated by a reaction of water with the element zirconium found in the alloy composing the outer casing of the fuel rods.

a. Zirconium is the metal of choice for reactors due to several reasons, including that it does not absorb neutrons. Why is this a desirable property?

b. Zirconium, if heated to a high temperature (such as in an accident at a nuclear plant), has two undesirable properties: (1) it will swell and crack, and (2) it will react with water to produce hydrogen. Explain the danger that these present.

Today, nuclear plants and their past operations continue to be under intense scrutiny. Undoubtedly, nuclear energy will be part of our future; however, at present, it is not clear to what extent future generations will rely on nuclear energy.

6.6 | Is There a Future for Nuclear Power?

Should we build more nuclear power plants? The answer depends on both whom you ask and when you ask them. Some longtime opponents of nuclear energy are now in favor of it. Similarly, some who supported it are now questioning its societal costs, both to our current generation and those to come.

Let's say that you switch on your coffee pot. If you live in the United States, about one-fifth of the electricity is coming from a nuclear power plant. If you live in France, Belgium, or Sweden, the percentage is even higher. In either case, you can brew your coffee!

We have a high demand for electricity (and caffeine).

© McGraw-Hill Education. Photo by Eric Misko. Elite Images Photography.

Worldwide, nations differ in the extent to which they employ nuclear energy to generate electricity. For example, in the United States, 20% of commercial electric power is produced from 100 nuclear reactors, all licensed by the Nuclear Regulatory Commission. As of 2014, these reactors were operating at 62 sites in 31 states. As you can see in Figure 6.17, the electricity generated by these nuclear plants has increased over the years, despite the drop in number of operating reactors from its peak of 112 in 1990.

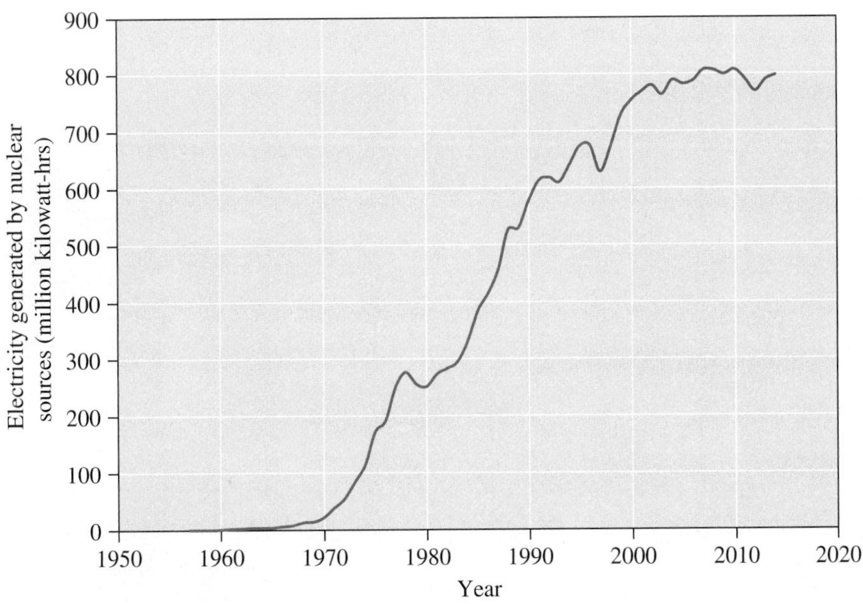

Figure 6.17

Nuclear power generation in the United States, 1957–2014. The power increase over time stems from both improved reactor efficiencies and upgrades to reactor components.

Source: Energy Information Administration

When you brew your coffee a decade from now, from where will the electricity originate? Nuclear power plants surely will be one source. Although no new plants have been built in the United States since 1978, several reactors are currently under construction at existing plants. As of 2015, these include two new units in Georgia at Plant Vogtle that eventually will make this the largest nuclear plant in the United States. Construction also continues in South Carolina at two additional units at the Virgil C. Summer nuclear generating station. In Tennessee, the Watts Bar unit 2 reactor began operation in May 2016.

Many nuclear plants like Vogtle (Figure 6.18) generate electricity using multiple reactors. In another example, the Palo Verde plant in Arizona (Your Turn 6.8) has three reactors.

Your Turn 6.18 Scientific Practices State-By-State

This map shows the 31 states (shaded in blue) that have nuclear power plants.

a. Select a state and prepare a summary of its nuclear power plants, their energy production, and any proposed changes.

b. As of 2014, Vermont was the state with the highest percent of nuclear energy (70%). Search the Internet to find at least two other states that use at least one-third nuclear energy.

c. From the map, select a state with no nuclear power plants. How is that state's electricity generated?

Construction sites for new nuclear power plants are enormous. They cover hundreds of acres and employ a workforce in the thousands, making them essentially cities themselves. The construction site at Plant Vogtle, pictured in Figure 6.18, even has its own railway!

The construction and continued operation of a commercial nuclear power plant is not only a matter of energy supply-and-demand, but also one of public acceptance. People have been lining up on one side or the other of the nuclear fence since nuclear power was first proposed back in the 1970s. Which side are you on?

Figure 6.18

Aerial view of the 550-acre construction site of the Vogtle nuclear power plant, units 3 and 4 (March 2012).

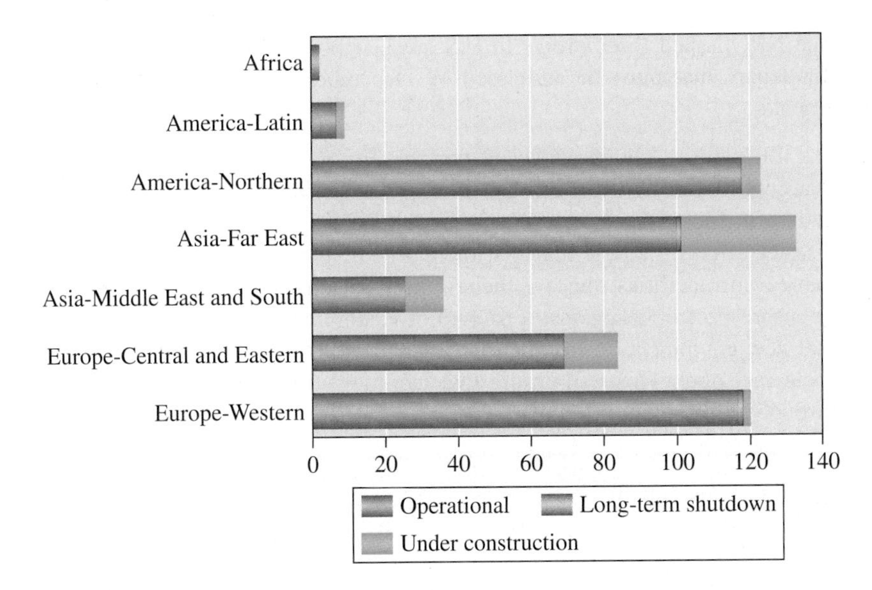

Figure 6.19

Worldwide distribution of nuclear power plants, including those under construction (June 2015).

Source: International Atomic Energy Agency

The larger picture of nuclear energy worldwide is one of change. In part, these changes stem from increased energy demand. Major commercial development of nuclear energy is clearly on the agenda of many nations. For example, although India only generated 3.5% of its electricity from 21 reactors in 2016, construction was underway on six new plants, with others planned or proposed. In 2016, China had 26 operational nuclear power reactors, with 23 under construction, and others planned or proposed.

However, as some nations move toward increased nuclear power, others are wary or even moving away from this controversial energy source. Most recently, this cautiousness is a result of the 2011 Fukushima incident.

Your Turn 6.19 Scientific Practices World-Wide Nuclear Use

a. Where are nuclear reactors located? Write a summary of the data in **Figure 6.19**.

b. Which countries are the top producers of uranium worldwide? How many of these countries have commercial nuclear reactors?

c. Suggest reasons why some countries, more than others, should develop nuclear energy.

People across the globe share the dream of clean and sustainable sources of energy for the future. Does this dream include nuclear energy? If so, should we build more nuclear power plants to achieve this dream? If you had asked this question in the United States back in the early 1960s, the answer would have been yes. At this time, the United States experienced a dramatic growth in the nuclear power industry, one that lasted until 1979 when the malfunction at Three Mile Island occurred. The fear that accompanied this incident certainly contributed to the end of the growth phase. More important at that time, however, were the economics of nuclear energy. With the retreat of fossil fuel prices and the added costs of nuclear safety and oversight imposed in the 1980s, it simply was not economically feasible for utilities to construct new nuclear plants.

What are the economic realities today? First, any new reactors will be built with improved designs, especially in light of the earthquake and tsunami that disabled reactors in Japan. Second, these designs will have a higher price tag.

One new design being developed is for smaller, modular nuclear reactors that can be factory-built. These reactors could be deployed with minimal construction on-site, speeding construction and making power plant expansion easier in order to meet increasing energy demands.

In terms of design, the near future of nuclear power, especially in the United States, is primarily focused on ensuring current nuclear power plants are prepared for extraordinary disasters such as what occurred at Fukushima. The United States Nuclear Regulatory Commission (NRC) stated in a report that "a sequence of events like the Fukushima, Japan, incident is unlikely to occur in the U.S.," but an "accident involving core damage and uncontrolled release of radioactivity to the environment, even one without significant health consequences, is inherently unacceptable."

The NRC issued three orders to U.S. nuclear power facilities, in reaction to the events in Japan, that must be addressed by December 2016. The orders include the requirements:

- that all facilities obtain sufficient portable safety equipment, such as devices that could burn off hydrogen generated in an accident, to support all reactors and spent fuel pools at a given site simultaneously. This is to ensure that if a disaster affects multiple reactors, there will be protection.
- that certain facilities improve their venting systems for boiling water reactors to ensure protection against a backup of steam and to control the temperature.
- that new equipment be installed in order to monitor water levels in each plant's spent fuel pool. This will ensure that facilities will know water levels throughout the plant.

Another recommendation of the NRC is to use more passive cooling systems in nuclear plants. Passive cooling systems use convection current rather than mechanical pumps to circulate coolant, making the system less susceptible to malfunctions.

Your Turn 6.20 You Decide Pros and Cons of Nuclear Power

Using the Internet, research some arguments both for and against using nuclear energy as a power source.

a. Explain your own thoughts on nuclear waste, mining, effects on climate change, cost, and human fear.
b. What countries, if any, have long-term plans for dealing with nuclear waste?
c. What does the cartoon (right) show about future energy concerns?
d. What do you feel is the future of nuclear power?

© Peter J. Welleman.

So where does that leave us? As you can see, there are no easy answers for the issue of nuclear power. Global demand for energy expands daily, as does the mass of radioactive waste from nuclear power plants with which we must cope. The era of climate change has dawned. Yet both real and perceived hazards associated with radioactivity, with mining and enriching uranium, and with nuclear weapons still remain. This presents a classic risk-benefit situation, and the final compromise has yet to be reached. For now, it is clear that nuclear power is not the cure-all for the world's energy woes. It is, however, the cause of some environmental and societal woes. Even so, it will remain a piece of the energy pie for years to come.

6.7 Solar Power: Electricity from the Sun

Given our increasing energy needs, it would surely make sense to take advantage of sunlight, a renewable energy source. The Sun's rays hit Earth every hour with enough energy to meet the world's energy demand for an entire year! Currently, however, less than 1% of the electric power generated in the United States comes directly from solar energy. So why does solar energy currently account for such a small part of our larger energy picture?

Although remarkable amounts of sunshine hit Earth daily, the rays do not strike any one site on the planet for 24 hours a day, 365 days a year. Furthermore, some parts of the planet receive too low an intensity of light to be practical for solar collecting. The disparity arises due to differences in geographical locations and local factors such as cloud cover, aerosols, smog, and haze. For example, examine the map shown in Figure 6.20. The data in the figure are reported in kilowatt-hours (kWh) per square meter per day for a flat-panel solar collector that is stationary. The next activity helps you explore the differences in daily solar energy during a calendar year.

1 kWh = 3,600,000 joules. Recall from Section 6.2 that 1 watt = 1 joule per second.

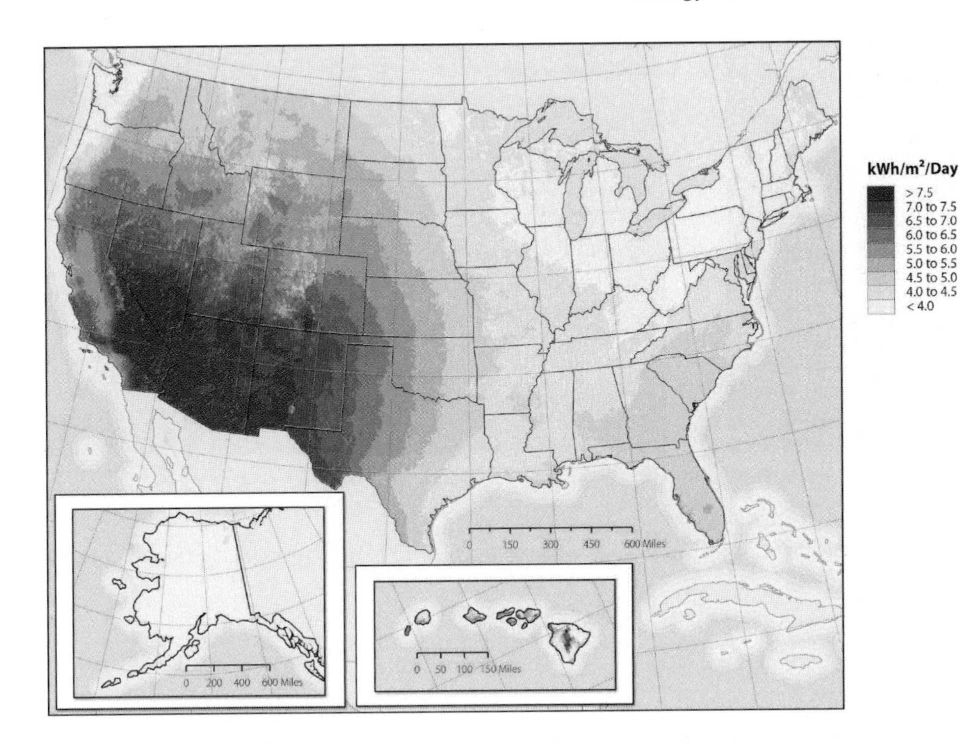

Figure 6.20

The average amount of daily solar energy received by a fixed photovoltaic panel oriented due south. *Note:* These energy levels would be higher if the panels tracked the path of the Sun, rather than being stationary.

National Renewable Energy laboratory (NREL) for the U.S. Department of Energy, 2012.

Your Turn 6.21 Scientific Practices Where Does the Sun Shine?

Thanks to a website provided by the U.S. National Renewable Energy Laboratory (NREL), you can view solar maps for different parts of the United States.

a. Select a state of your choice and view the data for each month of the year. What do you notice about how solar radiation varies throughout the year?

b. It should come as no surprise that California, Arizona, New Mexico, and Texas lead the United States in average annual solar radiation. Why do some parts of these states have higher values than others?

Your Turn 6.22 Skill Building Could Solar Energy Power a House in Your Neighborhood?

How feasible is solar energy for meeting your everyday energy demands?

a. Use **Figure 6.20**, or another resource from the Internet, to estimate how many kilowatt-hours (kWh) of solar energy fall on one square meter of land per day in your area.

b. A section of a homeowner's monthly electric bill is shown below. In your area, would this homeowner be able to power their household using only solar energy?

Your energy use

Meter # IN24775778

Schedule 07 (residential rate)

Service Period	**Meter Reading**
06/23/16	15335
05/21/16	15079
33 days of service	256 kWh

c. Assuming that this homeowner's monthly energy use is typical, how many households could be supported by a nuclear power plant support with a capacity of 750 MW? ***Hint:*** Start by calculating how many megawatt-hours (MWh) of energy the plant would generate in one day and how many kWh of energy the household uses in one day.

d. Does the answer you calculated in part **c** make sense? If so, explain. If not, describe the assumptions made that might lead to a larger or smaller number than expected.

(a)

(b)

Figure 6.21

(a) An aerial view of the Solar Millennium Andasol project in Spain, which has an approximate capacity of 150 MW of solar-thermal power.
(b) A close-up of a portion of the mirrored array.

Source: (a) © Langrock/Solar Millennium/SIPA/Newscom; (b) © Boris Roessler/Deutsche Presse-Agentur/Newscom

The challenge, then, is to locate the areas in which the incident average solar energy is high, and to collect this energy in sufficient quantities to produce electricity. There are two primary ways that energy from the Sun can be transformed into electricity to power our lives. Recall that electromagnetic radiation from the Sun is composed of a range of wavelengths with different energies. Thus, both the Sun's heat (longer wavelengths, lower energy) and its light (shorter wavelengths, higher energy) can be used to generate electricity.

Concentrating the Sun's energy in order to heat water is a solar-thermal process. When the Sun's heat is used to generate electricity, the process is known as *concentrating solar power* (CSP). CSP depends on solar collector devices such as the ones shown in Figure 6.21. These mirrored arrays concentrate sunlight in much the same way that a magnifying glass can focus light to burn a hole in a piece of paper. The heat from this concentrated light is then used to power a steam turbine and generate electricity as coal and nuclear power plants do. As of 2014, there were 17 large concentrated solar power plants in the world, with each generating between 100 and 400 MW of power.

This group of 17 plants taken together generates approximately 2800 MW of power, about the same as three nuclear power plants.

Your Turn 6.23 Scientific Practices Solar-Thermal Collectors

All solar collectors focus and concentrate the Sun's rays for the purpose of producing heat. However, they do so in different ways.

a. Searching the Internet, find and describe the designs for three different types of collectors.
b. How is each design matched to its end use? As part of your answer, include the scale of use—that is, for a single home, for a community, or for a business.
c. Name at least one limitation for each type of collector.

Figure 6.22

Photovoltaic (solar) cells are used to improve security, enhance safety, and direct pedestrians and vehicles.

Source: Warren Gretz/NREL/U.S. Dept. of Energy

A second way to tap into the Sun's energy is to use a **photovoltaic (PV) cell**—a device that converts light energy directly into electric energy, sometimes called a *solar cell.* It takes only a few PV cells to produce enough electricity to power your calculator or digital watch. Other common uses for photovoltaic cells include communication satellites, highway signs, security and safety lighting (Figure 6.22), automobile recharging stations, and navigational buoys. Cost savings can be substantial. For example, using solar cells rather than batteries in navigational buoys saves the U.S. Coast Guard several million dollars annually through reduced maintenance and repair.

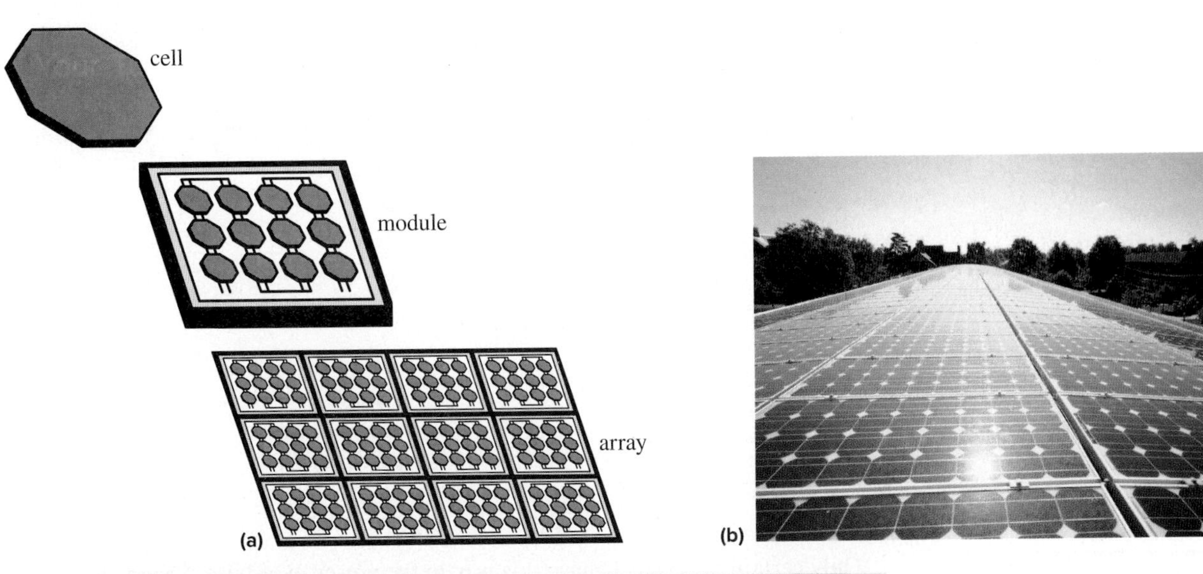

Figure 6.23

(a) Arrangement of photovoltaic cells used to make a module and an array. **(b)** A silicon solar array installed on a roof. **(c)** An aerial view of the Solarpark Gut Erlasee in Bavaria, Germany. At peak capacity, it can generate 12 MW. A typical nuclear power plant generates 1000 MW of electricity.

Source: (b): NREL/U.S. Dept. of Energy; (c): © Daniel Karmann/DPA/Corbis

If more power is required, PV cells can be combined into modules or arrays to make up solar panels, as shown in Figure 6.23. Many people today power their homes and businesses with solar PV systems—especially throughout Europe. Depending on the size of a home, it may use a dozen or more solar panels for power. These panels are usually mounted facing due south. Installing them on a system that rotates to track the Sun's path, and thus maximizing their exposure to sunlight, optimizes their efficiency but has a higher initial cost. For electric utility or industrial applications, hundreds of solar arrays are interconnected to form a large-scale PV system, such as the one shown in a field in Bavaria, Germany (Figure 6.23c).

6.8 Solar Energy: Electronic "Pinball" Inside a Crystal

How does a photovoltaic cell generate electricity? The answer lies in the behavior of the electrons in the cell material. When light shines onto a PV cell, it may pass through the cell, be reflected, or be absorbed. If absorbed, the energy may cause an excitation of the electrons in the atoms of the cell. These excited electrons escape from their normal positions in the cell material and become part of an electric current.

Did You Know? In 1839, A. E. Becquerel (1820–1891), a French physicist, discovered the process of using sunlight to produce electricity in a solid material.

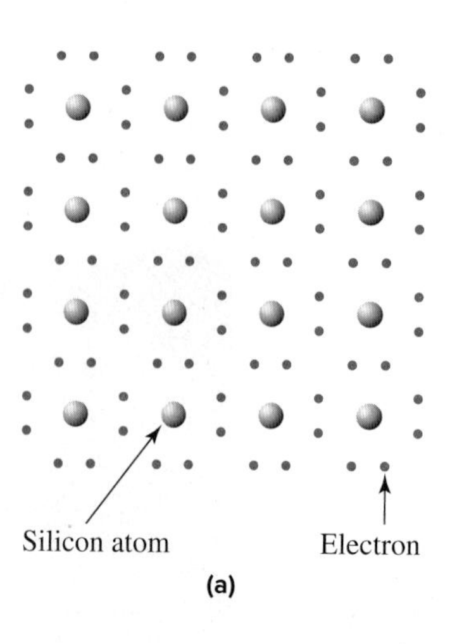

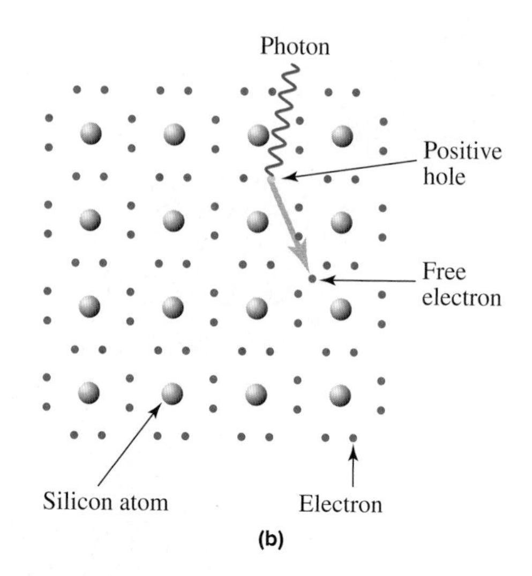

Figure 6.24

(a) Schematic of bonding in silicon. **(b)** Photon-induced release of a bonding electron in a silicon semiconductor. The release of a bonding electron creates a positively charged vacancy, referred to as a *hole*.

Conductivity and semiconductors (metalloids) were introduced in Section 1.5.

Did You Know? The element silicon was one of the first semiconducting materials developed for use in computers and in PV cells. In fact, many of the high-tech businesses that developed semiconductors were clustered in California's "Silicon Valley."

A crystalline structure has a regular repeating array of atoms or ions. One example is silicon dioxide (quartz), as shown in Figure 1.19.

Recall that Silicon is in Group 14 (IVA) and has four valence electrons. In comparison, gallium (Ga) is in Group 13 (IIIA) and has three valence electrons. Arsenic (As) is in Group 15 (VA) and has five valence electrons.

Only certain materials behave this way in the presence of light. Photovoltaic cells are made from a class of materials called **semiconductors**, materials that have a limited capacity of conducting an electric current. Most semiconductors are made from a crystalline form of silicon, a metalloid. A crystal of silicon consists of an array of silicon atoms, each bonded to four others (to satisfy the octet rule) by means of shared pairs of electrons (Figure 6.24a). These shared electrons are normally fixed in the bonds and unable to move through the crystal. Consequently, silicon is not a very good electrical conductor under ordinary circumstances. However, if a bonding electron absorbs sufficient energy, it can be excited and released from its bonding position (Figure 6.24b). Once freed, the electron can move throughout the crystal lattice, making the silicon an electrical conductor.

In reality, pure silicon semiconductors do not allow an electric current to flow unless they are doped. **Doping** is a process of intentionally adding small amounts of other elements, known as *dopants* (sometimes called impurities), to pure silicon. These dopants are chosen for their ability to facilitate the transfer of electrons. For example, about 1 ppm of gallium (Ga) or arsenic (As) is often introduced into the silicon. These two elements, and others from the same groups in the periodic table, are used because their atoms differ from silicon by only a single valence electron. Thus, when an atom of As is introduced in place of a Si atom in the lattice, an extra electron is added. This material with an excess of electrons is referred to as an **n-type semiconductor**. In contrast, the replacement of a Si atom with a Ga atom means that the crystal is now one electron "short." This material with a shortage of electrons (or an excess of "holes," the lack of an electron at a position in the crystal where one could/did exist) is referred to as a **p-type semiconductor**. Figure 6.25 illustrates doped n- and p-type semiconductors.

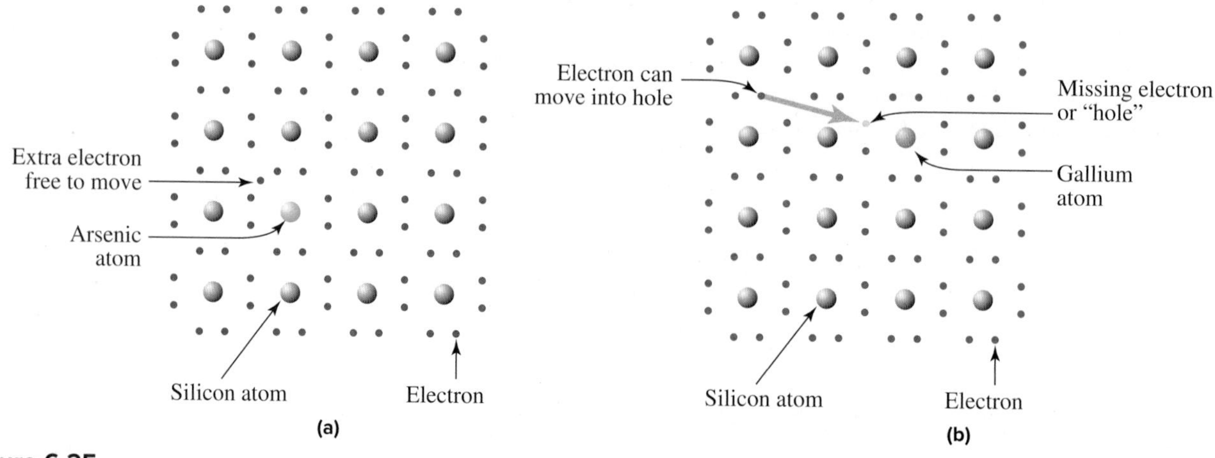

Figure 6.25

(a) An arsenic-doped n-type silicon semiconductor. **(b)** A gallium-doped p-type semiconductor.

Both types of doping increase the electrical conductivity of silicon because electrons can now move from an electron-rich to an electron-deficient environment.

Your Turn 6.24 Skill Building Doping Predictions

Some solar cell designs often use phosphorus and boron to dope silicon crystals.

a. Which will form an n-type semiconductor? Explain your reasoning.
b. Which will form a p-type semiconductor? Explain your reasoning.

To induce a voltage in a PV cell, two layers of n-type and p-type semiconducting materials are placed in direct contact. In order to generate an electric current, the light hitting the PV cell must have enough energy to set the electrons in motion from the n-type side to the p-type side through the electric circuit (Figure 6.26). The transfer of electrons generates a current of electricity that can be used to do all the things electricity does, including being stored in batteries for later use. As long as the cell is exposed to light, the current continues to flow, powered only by solar energy.

A photovoltaic cell is typically composed of multiple layers of doped n- and p-type semiconductors in close contact (Figure 6.26). The p–n junctions not only make possible the conduction of electricity, but also ensure that the current flows in a specific direction through the cell. Only photons with enough energy can knock electrons free from the doped material. These electrons then become part of the external electric circuit. For a PV cell to convert as much sunlight as possible into electricity, the semiconductors must be constructed in such a way to make the best use of the photons' energy. If not, the energy of the Sun is lost as heat or not trapped at all.

"Sandwiches" of n- and p-type semiconductors are used in transistors and other miniaturized electronic devices that have revolutionized communications and computing.

The fabrication of photovoltaic cells poses some significant challenges. The first is that although silicon is the second-most abundant element in Earth's crust, it is most frequently found combined with oxygen as silicon dioxide, SiO_2. You know this material by its common name, sand, or more correctly as quartz sand. The good news is that the starting material from which silicon is extracted is cheap and abundant. As seen in Chapter 1, the not-so-good news is that processes to extract and purify silicon are expensive. In order to be useful in solar devices, the silicon must be refined to a purity of at least 99.9999%.

Using silicon dioxide as a source of pure silicon was discussed in Section 1.8.

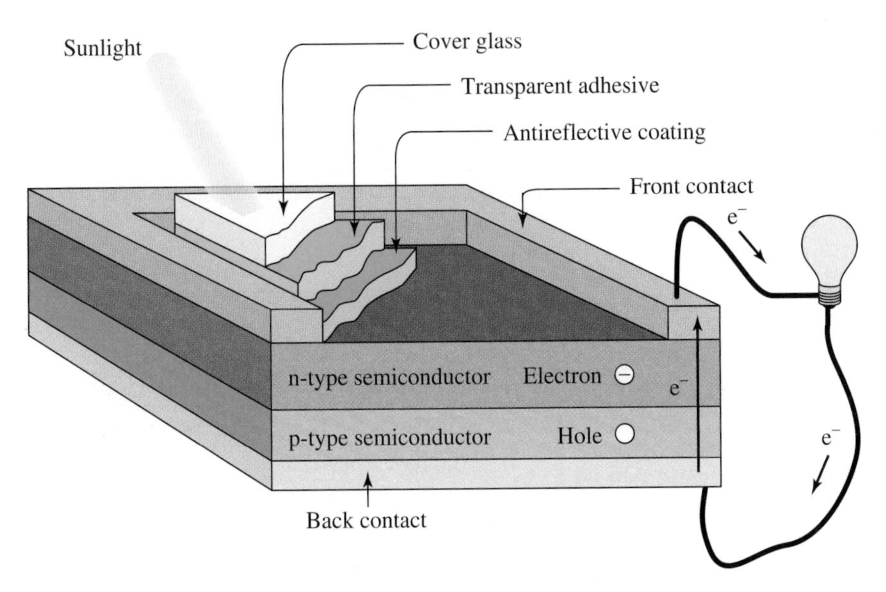

Figure 6.26

Schematic diagram of one layer of a solar cell showing the sandwiching of n-type and p-type semiconductors, and the "electron-hole pairs" generated by photons of sunlight. The movement of electrons and holes in opposite directions results in the flow of electrical current through the external circuit.

A second challenge is that the direct conversion of sunlight into electricity is not very efficient. A photovoltaic cell could, in principle, transform up to 31% of the radiant energy to which it is sensitive into electricity. However, some of the radiant energy is reflected by the cell or absorbed to produce heat instead of an electric current. As of 2015, a state-of-the-art commercial solar cell has an efficiency of only 21%, but even this is a significant increase over the first solar cells built in the 1950s, which had efficiencies of less than 4%. In Chapter 5, we lamented the 35–50% net efficiency of converting heat to work in a conventional power plant. It might seem that we should be even more distressed at the lower limits that can be achieved by photovoltaics. Remember, however, that the first use of solar cells was to provide electricity in NASA spacecraft. For that application, the intensity of radiation was so high that low efficiency was not a serious limitation, and costs were not of paramount concern. For commercial use on Earth, costs and efficiency are issues. Our Sun is an essentially unlimited energy source, and converting its energy to electricity, even inefficiently, is free from many of the environmental problems associated with burning fossil fuels or storing waste from nuclear fission. These considerations add impetus to the research and development of solar cells.

One approach to increasing commercial viability is to replace crystalline silicon with the noncrystalline form of the element, known as *amorphous* Si. Photons are more efficiently absorbed by less highly ordered Si atoms, a phenomenon that permits reducing the thickness of the silicon semiconductor to 1/60 or less of its former value. The cost of materials is thus significantly reduced.

Other researchers are developing multilayer solar cells. By alternating thin layers of p-type and n-type semiconductors, each electron has only a short distance to travel to reach the next p–n junction. This lowers the internal resistance within the cell and raises its efficiency. The maximum theoretically predicted efficiencies could improve to 50% for two junctions, to 56% for three junctions, and to 72% for 36 junctions. Furthermore, the use of other semiconductors in contact allows the device to absorb different regions of the electromagnetic spectrum, to cover the entire UV–IR range. In comparison, Si is most sensitive to the absorption of light in the blue region of the electromagnetic spectrum, which results in significant wavelength ranges not being effectively absorbed. As of 2015, the maximum efficiency actually demonstrated with a multjunction solar cell was 46%. Figure 6.27 gives a sense of just how thin these layers actually are.

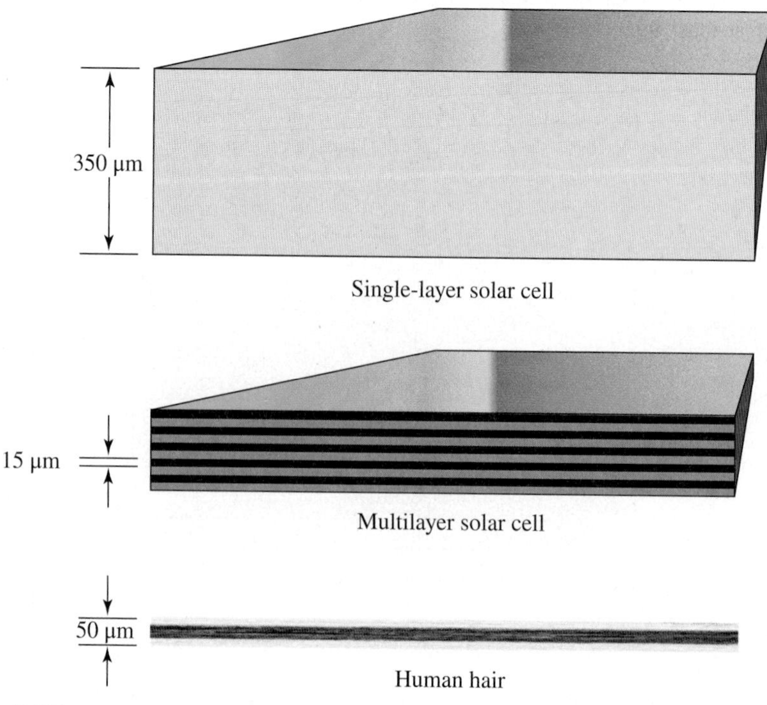

Micrometers (μm) were introduced in Chapter 1.

Figure 6.27

A comparison of the relative thickness of a solar cell layer, either in a single-layer or multilayer cell, to the diameter of an average human hair. *Note:* 1 μm = 10^{-6} m.

Thin-film solar cells are made from amorphous silicon or non-silicon materials such as cadmium telluride (CdTe) or a combination of copper, indium, gallium, and selenium (referred to as CIGS, $CuIn_xGa_{(1-x)}Se_2$, where $x = 0 - 1$). These thin films use layers of semiconductor materials only a few micrometers thick. For comparison, the width of a typical human hair is about 50 μm! Thin-film solar cells can even be incorporated into rooftop shingles and tiles, building facades, or the glazing for skylights because of their flexibility compared with more rigid traditional cells (Figure 6.28). Other solar cells are being made using various materials, such as solar inks employing conventional printing press technologies, solar dyes, nanoparticles known as "quantum dots," and conductive plastics. Solar modular units use plastic lenses or mirrors to concentrate sunlight onto small but very highly efficient PV materials. Utilities and industries experimenting with these solar lens materials find that, despite their higher initial cost, using a small amount of these more efficient materials is becoming more cost-effective.

Figure 6.28

Thin-film solar tiles on a roof.

Source: NREL/U.S. Dept. of Energy

Your Turn 6.25 Scientific Practices Solar PV Use

How are people today using solar photovoltaics? Search the Internet to answer this question for each group listed below.

a. farmers and ranchers
b. small–business owners
c. homeowners

Long-range prospects for photovoltaic solar energy are encouraging. Its cost is decreasing, while the cost of electricity generated from fossil fuels is increasing. Despite recent advances in technology, solar energy from either thermal or photovoltaic sources is still significantly more expensive than fossil fuels. And there is still the question of land use. At currently attainable levels of operating efficiency, the electricity needs of the United States have been estimated to require a photovoltaic generating station covering an area of 85 × 85 miles, roughly the size of New Jersey! Although photovoltaic power is steadily growing, it still represents a minute fraction of global power supplies.

Your Turn 6.26 You Decide Where to Site Solar Power?

Last we heard, New Jersey was not volunteering to be converted wholesale into a solar farm to power the rest of the United States.

a. Would New Jersey be a reasonable geographic location?
Hint: Revisit Figure 6.20.
b. Which locations in the United States show the most promise for solar energy collection?
c. Location isn't everything. Name two other factors that come into play in dedicating land to solar energy collection.

Because of the diffuse nature of sunlight, photovoltaic technology is well suited to distributed generation, as will be described for fuel cells in the next chapter. More than one-third of Earth's population is not hooked into an electric network. This is due

Figure 6.29

Photovoltaics can power water pumps in remote areas of the world where there is no access to electricity.

Source: NREL/U.S. Dept. of Energy

to the costs associated with constructing and maintaining equipment, and supplying the fuel to generate the electricity. Because PV installations are relatively maintenance-free, they are particularly attractive for electric generation in remote regions. For example, the highway traffic lights in certain parts of Alaska far from power lines operate on solar energy. A similar but more significant application of photovoltaic cells may be to bring electricity to isolated villages in economically disadvantaged countries. In recent years, more than 200,000 solar lighting units have been installed in residential units in Colombia, the Dominican Republic, Mexico, Sri Lanka, South Africa, China, and India. Photovoltaic cells are currently affecting the lives of millions of people across our planet (Figure 6.29).

Your Turn 6.27 Scientific Practices Local Solar Energy

Distributed generation! Use the resources of the Internet to learn more about how people are using solar energy locally, such as the Million Solar Roofs project. Then, propose a solar energy project in your class, or in the community of your choice. List at least five factors to consider before proceeding with the project.

Your Turn 6.28 Scientific Practices Solar–Thermal Versus Voltaic

The energy from the Sun can be converted into electricity by solar-thermal or photovoltaic routes.

a. Outline how sunlight is converted into electricity by each route.
b. Which approach is currently generating the most electricity worldwide?

Electricity generated by solar-thermal collectors and photovoltaic cells during the day must be stored using batteries for use at night. Nevertheless, the direct conversion of heat and sunlight to electricity has many advantages. In addition to relieving some of our dependence on fossil fuels, an economy based on solar electricity would reduce the environmental damage of extracting and transporting these fuels. Furthermore, it would help lower the levels of air pollutants such as sulfur oxides and nitrogen oxides. It would also help avert the dangers of climate change by decreasing the amount of carbon dioxide released into the atmosphere. Fossil fuels will certainly remain the preferred form of energy for certain applications. However, for the longer term, we can turn to a number of renewable energy sources, many of which are driven directly by the Sun or are a result of the solar heating of our atmosphere and water. The final section takes a brief look at how we can generate electricity from other sustainable renewable resources.

6.9 Beyond Solar: Electricity from Other Renewable (Sustainable) Sources

No single source of electricity can meet our global energy needs. We also know that no energy source comes without a cost, such as mining, pollution, greenhouse gases, or setting up distribution networks. Clearly, it is to our advantage to further develop and add a greater percentage of renewable sources than it is to continue to rely on fossil fuels and nuclear power. We discussed renewable sources such as biofuels and ethanol in Chapter 5. In this section, we turn to wind, water, and the heat given off by the core of our planet as renewable energy sources.

Wind

The Sun's heat ultimately drives the large-scale movements of the air on our planet that we know better as "wind." For centuries, humans relied on various forms of windmills that, in turn, spun wheels to grind grain or pump water. Wind turbines today make use of large blades, sometimes nicknamed by the locals as the "pin-wheels" that dot the landscape. These spin a shaft that turns a generator to produce electricity. Wind farms are located around the world in order to take advantage of prevailing winds. Such a farm is shown at Pakini Nui (South Point) on the Big Island of Hawaii (Figure 6.30).

Among the possible alternative sources of energy, wind power has seen the largest growth over the last decade. Figure 6.31 on the next page shows the amount of electricity generated by alternative energy sources since 1990. While hydroelectric power has generated the most electricity over this time period, wind power has dramatically increased since 2005. In addition to wholesale electrical generation for the

Did You Know? For hundreds of years, the Dutch have used windmills to pump water from their low-lying land, grind grain for bread, and saw wood for construction.

© Steve Allen/Brand X Pictures/Jupiter Images RF

Figure 6.30

Pakini Nui Wind Farm, completed in 2007, supplied 20.5 MW of power in 2013.

© Radius Images/Corbis RF

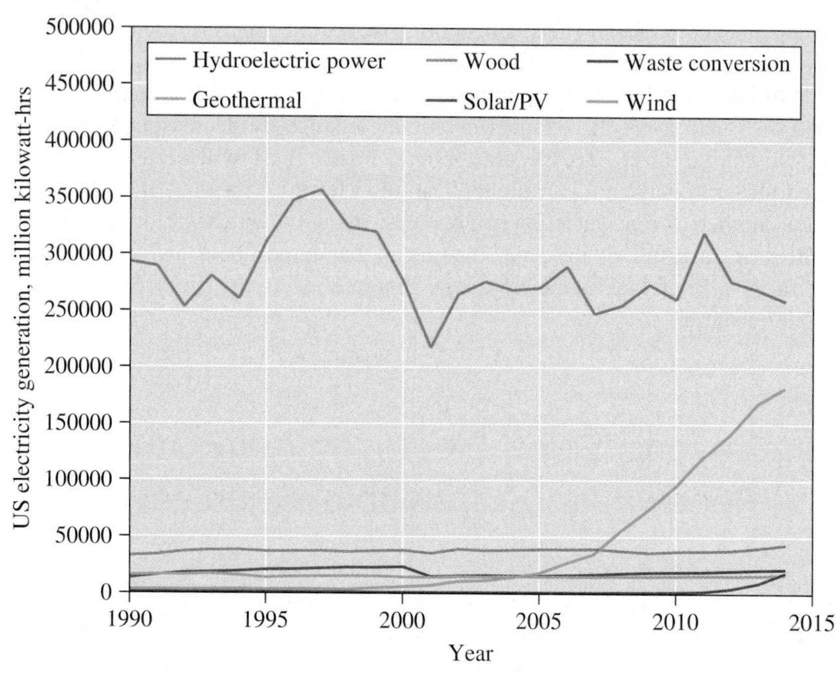

Figure 6.31

U.S. alternative energy electricity generation, 1990–2014. For comparison, in 2015, 2.7 trillion kWh of electricity was generated by the combustion of fossil fuels.

Source: Energy Information Administration

public, wind power is also used for individual homes and businesses, particularly in areas that are too distant to be on the power grid.

Did You Know? Each MW of power generated by a wind turbine requires up to 1 tonne of rare earth magnets. The challenges related to the global availability of the rare earths is discussed in Section 1.11.

The vast majority of electricity generation from wind is from wind farms, which feature hundreds of individual wind turbines. The average nuclear reactor in the U.S. generates about 1,000 MW of electricity, whereas coal plants average about 550 MW. In comparison, a single typical wind turbine generates only 1.5–2.5 MW of electrical power. However, these turbines, if arranged together in a wind-farm approach, can exceed the output of traditional power plants. For instance, the largest wind farm in the United States, the Alta Wind Energy Center in California, can generate 1,548 MW of electricity. The largest wind farm in the world, the Gansu wind farm in China, has an upper output of 5,160 MW.

In the United States, wind power contributes a significant portion of electricity generation in 39 states (Figure 6.32). The siting of a wind farm requires considerable planning and includes consideration for the maximum wind speed, the fraction of time that sufficient wind exists, land rights, and the impact on adjacent public and private land. Furthermore, there are various environmental impacts such as the effects on birds, bats, and other wildlife. A specific location may have excellent wind conditions with regular sustained winds, but would be a poor site because it is adjacent to a protected wildlife preserve with a large population of migrating birds.

The National Renewable Energy Laboratory (NREL) has surveyed the average windspeed at an elevation of 80 meters, the height of most turbines used by electric utility companies. As shown in the map in Figure 6.33, the U.S. region with the greatest wind speeds is the Great Plains states. At a height of 80 m, there is minimal blockage of wind by most trees, buildings, and other structures. Also, as the height above the ground increases, so does the average wind speed. Wind turbines require wind speeds of between 3.6–18 m/s (8–40 mph) to generate electricity. Slower speeds are inefficient in generating electricity, and may not even provide enough force to turn the blades. Faster speeds may cause damage to the mechanical and electrical components of the turbine. Wind turbines are turned off or on according to the wind speed to optimize the production of electricity.

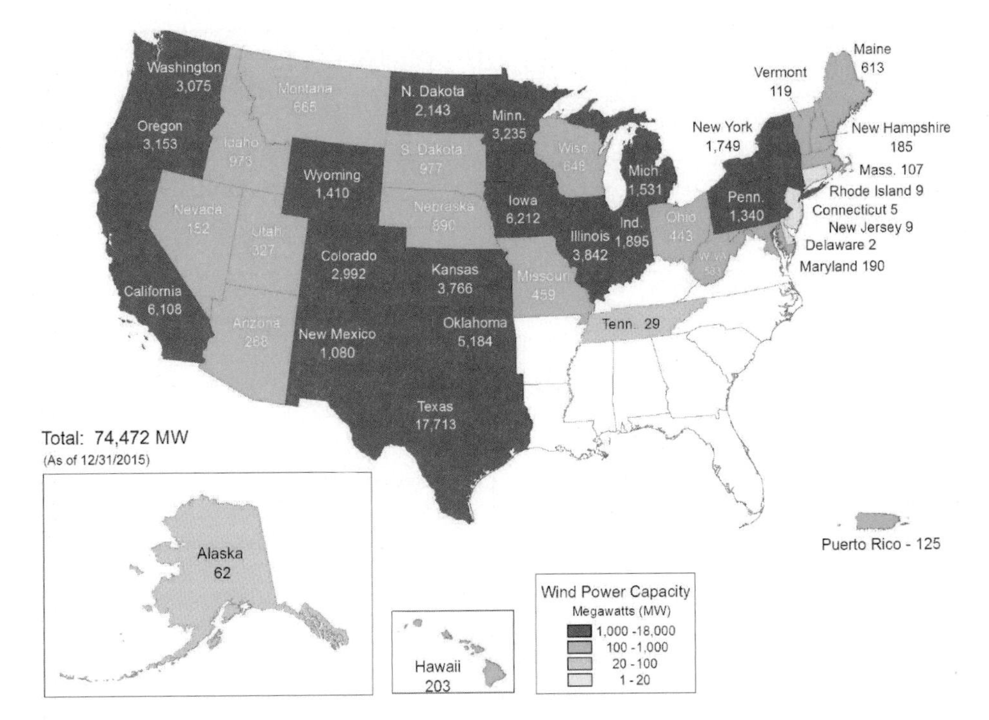

Figure 6.32

Wind-power capacity in the U.S. (2014).

Source: National Renewable Energy Laboratory (NREL) for the U.S. Department of Energy, 2014

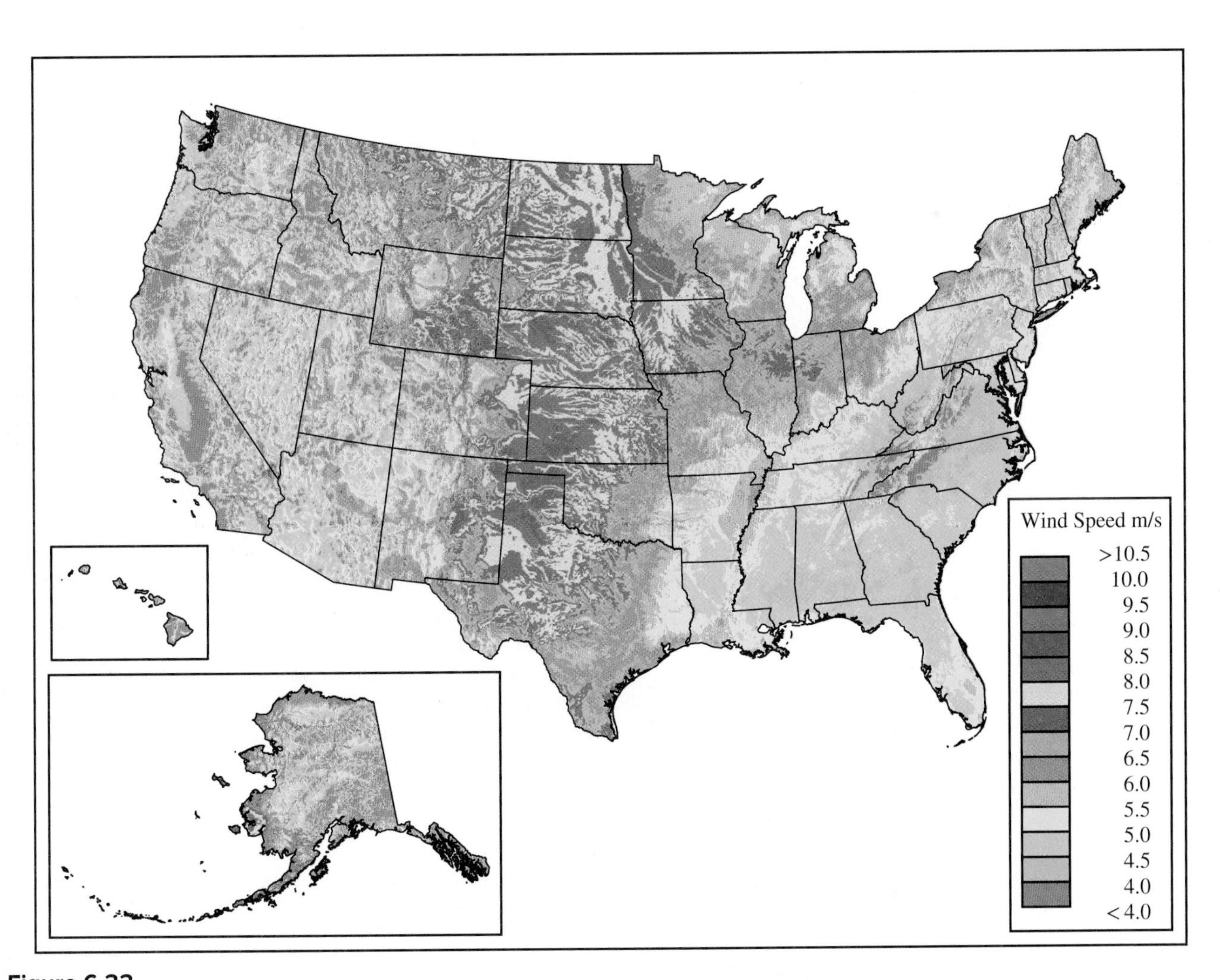

Figure 6.33

Annual average wind speed at 80 m height above ground in the U.S.

Source: National Renewable Energy Laboratory for the U.S. Department of Energy, 2014.

Your Turn 6.29 Scientific Practices Regional Variations in Wind

Consider the maps of the United States in **Figures 6.32** and **6.33**. Why do you think there is very little wind power in the southeastern U.S.?

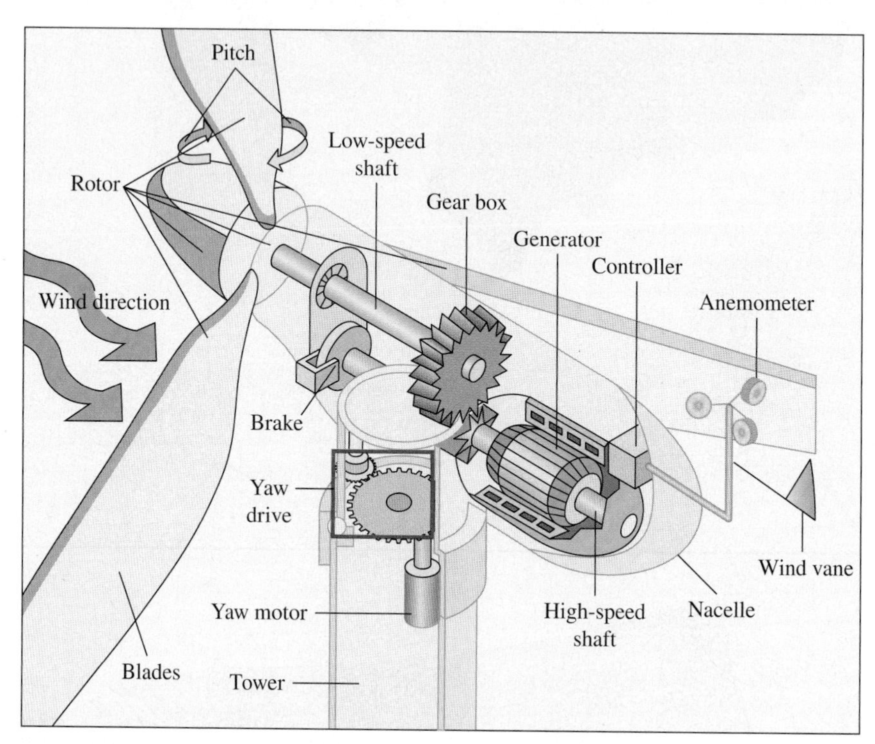

Figure 6.34

Schematic diagram of a wind turbine nacelle.

A wind turbine consists of a tower and blades. The tower elevates the turbine high enough to be exposed to sufficient wind speeds to produce electricity. The tower is mounted in massive concrete blocks of up to 102 m^3 (3,600 ft^3) and reinforced with 30 tons of steel. The body of the turbine, known as the *nacelle*, is mounted at the top of the tower. The nacelle contains the bulk of the mechanical and electrical operating pieces of the turbine, as seen in Figure 6.34. The blades are attached to a shaft that rotates an electromagnet within an electrical generator, similar to the generators found in traditional power plants. There is also a computer controller within the nacelle, which optimizes the operation of the turbine by controlling operations such as rotating the nacelle into the wind and changing the pitch of the blades.

In addition to ground-based wind turbines (also known as on-shore turbines), many utility companies are installing turbines that are located off the coast in oceans or large lakes. While there are currently no off-shore wind farms in the United States, there are an increasing number in Europe. For instance, about one-third of the United Kingdom's wind energy is supplied by off-shore wind farms.

Off-shore wind offers some advantages compared to on-shore wind farms. More than half of the population of the U.S. lives in coastal areas, so the electrical demand in these areas is high and the available land is limited. Off-shore winds also typically blow harder and more uniformly than on-shore winds, making off-shore wind farms more efficient than their on-shore cousins.

Water

For centuries, humans have harnessed the movement of water with devices such as water wheels. When water flows over a wheel, this wheel can turn other devices, including stones that can grind grain into flour. Similarly, small- and large-scale hydroelectric dams harness the movement of water. When water falls across turbine blades, the potential energy of water trapped in a reservoir behind the dam is converted into kinetic energy, which in turn is converted to electricity. Worldwide, only a few dams are still being constructed, as most large rivers are already in the service of hydroelectric projects.

The movement of ocean water in tides, currents, and waves can be harvested by a variety of principles to generate electricity. Some involve the turning of turbine blades, while others involve forcing compressed air through a turbine. All involve the kinetic energy of motion to turn a generator in order to produce electricity.

Did You Know? Two dams have recently been removed from rivers in the Pacific Northwest because the electricity generated was determined to be less than the cost to the environment.

Geothermal

Another renewable energy source is the heat given off from the core of our planet. Literally "earth heat," geothermal energy relies on drilling into underground reservoirs containing hot water or steam, and thus drawing heat from Earth. These heated sources of water can then be used to drive generators to produce electricity, or the hot water may be used directly to heat a home. Geothermal works well in locations known for volcanic activity that have "hot rock," such as Hawaii, which generates 25% of its energy from geothermal sources. Although the U.S. currently leads the world in total geothermal electric generating capacity, geothermal energy is showing tremendous growth in countries such as Kenya, Turkey, and Indonesia, among others (Figure 6.35).

The top five countries ranked in terms of geothermal electric generating capacity are: the U.S. (3.5 GW), the Philippines (1.9 GW), Indonesia (1.4 GW), Mexico (1.0 GW), and New Zealand (1.0 GW).

Your Turn 6.30 Scientific Practices Our Energetic Future

Renewable energy comes from the wind, the oceans, or geothermal sources, not just from the Sun or from biofuels. Pick one of these renewable energy sources and learn more about the technologies available to harness it.

a. Name the geographic restrictions (if any) to its use.
b. Prepare a list of the reasons to support this technology. Prepare a similar list for the "nay-sayers."
c. Predict how this technology will affect energy production output where you live.

This section offered a renewable energy sampler; no world view of energy resources would be complete without considering these and other sustainable sources of energy. The share that renewable sources occupy on the world's energy scene, as well as their economics, availability, and ease of use must be improved.

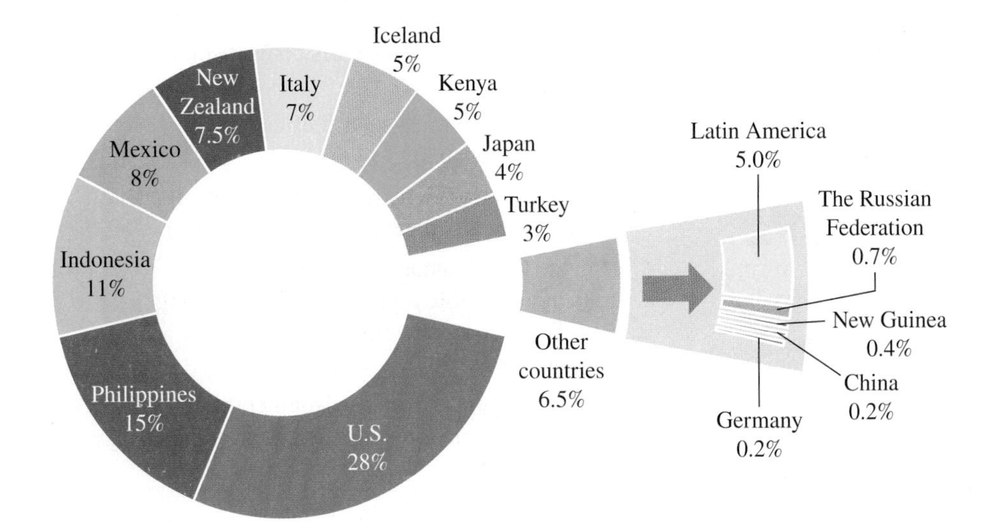

Figure 6.35

Geothermal power output by country (2014).

Source: Renewables 2015 Global Status Report

Conclusions

Almost 60 years have passed since the first commercial nuclear power plant began producing electricity in the United States. The glittering promise of boundless, unmetered electricity—drawn from the nuclei of uranium atoms—has proved illusory. But the needs of both our nation and the world for safe, abundant, and inexpensive energy are far greater today than they were in 1957. Therefore, scientists and engineers continue their atomic quest.

Where the search will lead is uncertain, but it is clear that people and politics will have a major say in ultimately making the decision. Reason, together with a regard for those who will inhabit our planet in both the near and far future, must govern our actions.

However, there are less controversial sources of electricity, such as photovoltaic cells that are used to tap the energy of the Sun. From photovoltaic roof shingles and integrated-building materials, to vehicles with exterior solar panels, the sky is literally the limit for the exciting future of solar power. Advances in research, together with changes in global economies, will continue to make this and other sustainable options such as wind, geothermal, and hydroelectric more fiscally and energetically feasible throughout the world.

Nuclear and renewable sources of energy provide electricity without releasing stored carbon into the atmosphere in the form of carbon dioxide. Wind and solar in particular are growing in popularity, taking over a larger portion of the energy market. However, solar and wind power are both variable; the Sun doesn't shine at a steady rate and the wind doesn't blow constantly. To make the most of these energy sources, they are paired with technologies that can store the energy for later use. The next chapter will explore the chemistry underlying these energy storage devices we know as batteries.

Learning Outcomes

The numbers in parentheses indicate the sections within the chapter where these outcomes were discussed.

Having studied this chapter, you should now be able to:

- calculate the energy released by a change in mass (6.1)
- write balanced nuclear equations (6.1)
- compare and contrast how nuclear transmutations differ from chemical reactions (6.1)
- describe how energy is produced from a nuclear power plant (6.2)
- diagram the components of a nuclear power plant (6.2)
- compare and contrast nuclear power plants with combustion power plants (6.2)
- compare energies derived from various types of fossil fuel sources with nuclear power (6.2)
- explain why some isotopes are radioactive (6.3)
- distinguish the term "radiation" in electromagnetic and nuclear contexts (6.3)
- compare and contrast the three fundamental types of nuclear radiation (6.3)
- identify the fundamental types of nuclear radiation (6.3)
- define half-life (6.4)
- describe how the amounts of radioisotopes change over time and the importance of half-life in nuclear power (6.4)
- describe the impact on long-term human health from nuclear waste and its disposal (6.5)
- outline the risk factors and benefits of using nuclear power (6.5)
- compare and contrast the nuclear power capacity of the U.S. with that of other countries (6.6)
- estimate the amount of energy that reaches Earth from the Sun and describe how we can use it (6.7)
- describe how solar radiation is used to generate electricity (6.7)
- illustrate how semiconductors convert solar radiation to electricity (6.8)
- compare and contrast solar-thermal and photovoltaic energy sources (6.8)
- describe how wind, hydroelectric, and geothermal energy sources can generate electricity (6.9)
- evaluate the advantages and disadvantages of using renewable energy sources, including how they incorporate green chemistry principles (6.9)
- evaluate the economic, environmental, human health, and societal costs of an alternative energy source (6.9)

Questions

Emphasizing Essentials

1. Name two ways in which one carbon atom can differ from another. Then, name three ways in which *all* carbon atoms differ from *all* uranium atoms.

2. The representations ^{14}N or ^{15}N give more information than simply the chemical symbol N. Explain.

3. a. How many protons are in the nucleus of this isotope of plutonium: Pu-239?
   b. The nuclei of all atoms of uranium contain 92 protons. Which elements have nuclei with 93 and 94 protons, respectively?
   c. How many protons do the nuclei of radon-222 contain?

4. Determine the number of protons and neutrons in each of these nuclei.
   a. ^{14}C, a naturally occurring radioisotope of carbon
   b. ^{12}C, a naturally occurring stable isotope of carbon
   c. ^{3}H, tritium, a naturally occurring radioisotope of hydrogen
   d. Tc-99, a radioisotope used in medicine

5. $E = mc^2$ is one of the most famous equations of the 20th century. Explain the meaning of each symbol in this equation.

6. Give an example of a nuclear equation and one of a chemical equation. In what ways are the two equations alike? Different?

7. What is an alpha particle and how is it represented? Answer these same questions for a beta particle and a gamma ray.

8. This nuclear equation represents a plutonium target being hit by an alpha particle. What do the superscript numbers mean? How about the subscripts? Show that the sum of the subscripts on the left is equal to the sum of the subscripts on the right. Then, do the same for the superscripts.

$$^{239}_{94}Pu + ^{4}_{2}He \longrightarrow [^{243}_{96}Cm] \longrightarrow ^{242}_{96}Cm + ^{1}_{0}n$$

9. For the nuclear equation shown in question 8:
   a. suggest the origin of the $^{4}_{2}He$ particle.
   b. $^{1}_{0}n$ is a product. What does this symbol represent?
   c. Curium-243 is written in square brackets. What does this notation convey?
   Hint: See Equation 6.1.

10. Californium, element number 98, was first synthesized by bombarding a target with alpha particles. The products were Californium-245 and a neutron. What was the target isotope used in this nuclear synthesis?

11. Explain the significance of neutrons in initiating and sustaining the process of nuclear fission. In your answer, define and use the term *chain reaction*.

12. Nuclear fission occurs through many different pathways. For the fission of U-235 induced by a neutron, write a nuclear equation to form:
    a. Bromine-87, Lanthanum-146, and more neutrons.
    b. a nucleus with 56 protons, a second with a total of 94 neutrons and protons, and a third with two additional neutrons.

13. This schematic diagram represents the reactor core of a nuclear power plant.

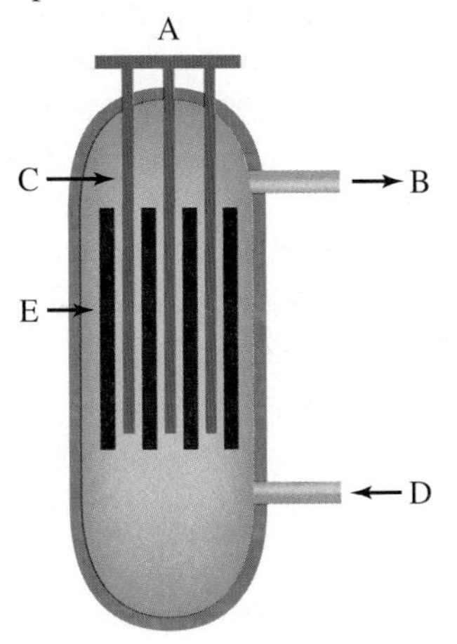

Match each letter in the figure with one of the following terms:

fuel rods
cooling water into the core
cooling water out of the core
control rod assembly
control rods

14. Identify the segments of the nuclear power plant diagrammed in Figure 6.5 that contain radioactive materials and those that do not.

15. Explain the difference between the primary coolant and the secondary coolant. The secondary coolant is not housed in the containment dome. Why not?

16. Boron can absorb neutrons and be used in control rods. Write the nuclear equation in which boron-10 absorbs a neutron to produce lithium-7 and an alpha particle.

17. Plutonium-239 decays by alpha emission (with no gamma ray), and iodine-131 decays by beta emission (and emits a gamma ray).
    a. Write the nuclear equation for each.
    b. Plutonium is most hazardous when inhaled in particulate form. Why is this?
    c. Iodine-131 can be hazardous if ingested. Where do all isotopes of iodine accumulate in the body?
    d. If 25 g of each sample were present, after three half-lives, how much would remain? Which substance would take the longest to decay to this amount? Explain.
    Hint: See Table 6.2.

18. Radioactive decay is accompanied by a change in the mass number, a change in the atomic number, a change in both, or a change in neither. For the following types of radioactive decay, which change(s) do you expect?
    a. alpha emission
    b. beta emission
    c. gamma emission

19. Figure 6.10 shows the radioactive decay series for U-238. Analogously, U-235 decays through a series of steps (α, β, α, β, α, α, α, β, α, β, α) to reach a stable isotope of lead. For practice, write nuclear reactions for the first six. Although some steps are accompanied by a gamma ray, you may omit this.
Hint: The result is an isotope of radon.

20. What percent of a radioactive isotope would remain after two half-lives, four half-lives, and six half-lives? What percent would have decayed after each period?

21. Estimate the half-life of radioisotope X from this graph.

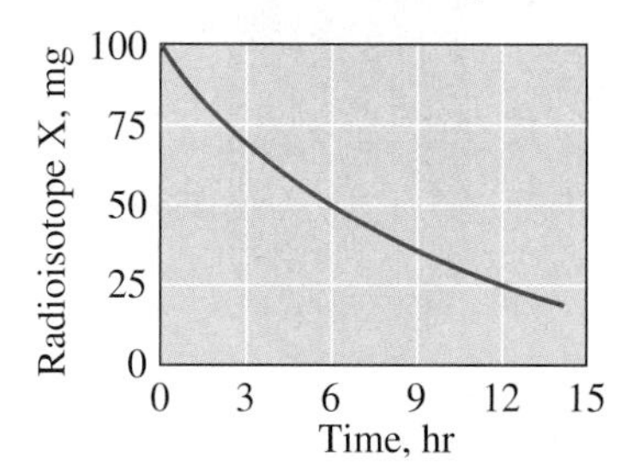

22. Every year, 5.6×10^{21} kJ of energy comes to Earth from the Sun. Why can't this energy be used to meet all of our energy needs?

23. The symbol • represents an electron and the symbol ● represents a silicon atom. The darker purple sphere in the center of the diagram represents either a gallium or an arsenic atom. Does this diagram represent a gallium-doped p-type silicon semiconductor or an arsenic-doped n-type silicon semiconductor? Explain your answer.

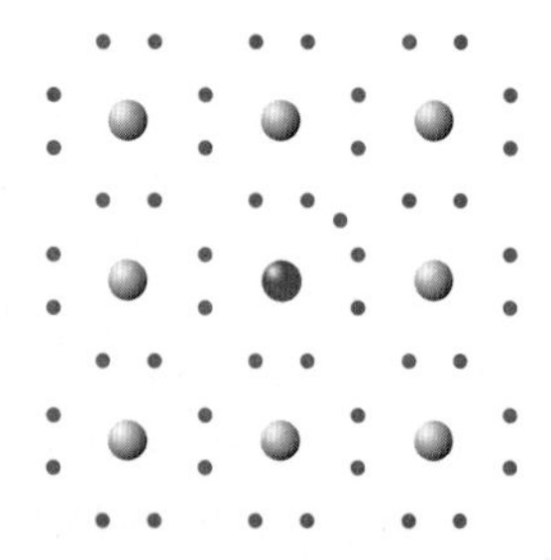

24. Describe the main reasons why solar cells have solar energy conversion efficiencies significantly less than the theoretical value of 31%.

Concentrating on Concepts

25. Alchemists in the Middle Ages dreamed of converting base metals, such as lead, into the precious metals gold and silver.
    a. Why could they never succeed?
    b. Today, can we convert lead or mercury into gold? Explain using specific nuclear reactions.
    c. Can nuclear reactions be used to synthesize consumer quantities of precious metals, such as gold and silver?

26. The isotopes U-235 and U-238 are alike in that they are both radioactive. However, these two isotopes have very different abundances in nature. Recall their natural abundances (found in Section 6.1) and explain the significance of this difference.

27. Consider the uranium fuel pellets used in commercial nuclear power plants.
    a. Describe one way in which U-235 and U-238 can be separated.
    b. Why is it necessary to enrich the uranium for use in the fuel pellets?
    c. Fuel pellets are enriched only to a few percent, rather than to 80–90%. Name three reasons why.
    d. Explain why it is not possible to separate U-235 and U-238 by chemical means.

28. a. Why must the fuel rods in a reactor be replaced every few years?
    b. What happens to the fuel rods after they are taken out of the reactor?

29. At full capacity, each reactor in the Palo Verde power plant uses only a few pounds of uranium to generate 1243 megawatts (MW) of power. To produce the same amount of energy would require about 2 million gallons of oil or about 10,000 tons of coal in a conventional power plant. How is energy produced in the Palo Verde plant, compared with conventional power plants?

30. One important distinction between the Chornobyl reactors and those in the United States is that those in Chornobyl used graphite as a moderator to slow neutrons, whereas U.S. reactors use water. In terms of safety, give two reasons why water is a better choice.

31. If you look at nuclear equations in sources other than this textbook, you may find that the subscripts have been omitted. For example, you may see an equation for a fission reaction written this way.

$$^{235}U + {}^{1}n \longrightarrow [^{236}U] \longrightarrow {}^{87}Br + {}^{146}La + 3\,{}^{1}n$$

    a. How do you know what the subscripts should be? Why can they be omitted?
    b. Why are the superscripts *not* omitted?

32. Coal can contain trace amounts of uranium. Explain why thorium must be found in coal as well.

33. Suppose somebody tells you that a radioisotope is gone after 10 half-lives. Critique this statement, explaining why it could be a reasonable assumption for a small sample, but might not be for a large one.

34. "Bananas are radioactive!" A vice president of nuclear services made this comment in a public lecture in the context of comparing the different sources of radiation to which people are exposed.

a. Why might he have made such an assertion?

b. Suggest a better way to have phrased this.

c. Should you stop eating bananas because they are radioactive? Explain.

35. Fossil fuels have been called the "Sun's ancient investment on Earth." Explain this statement to a friend who is not enrolled in your course.

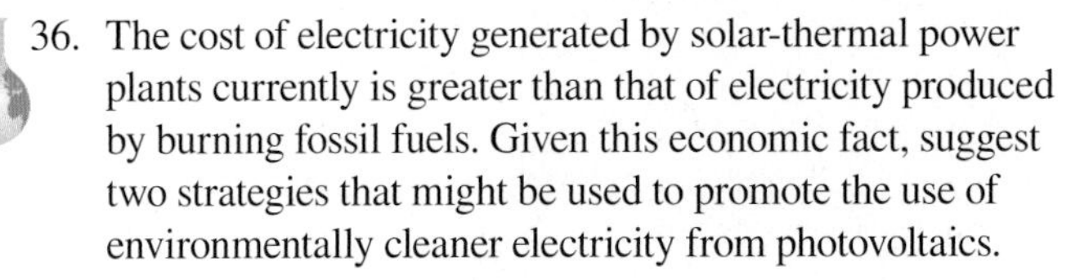

36. The cost of electricity generated by solar-thermal power plants currently is greater than that of electricity produced by burning fossil fuels. Given this economic fact, suggest two strategies that might be used to promote the use of environmentally cleaner electricity from photovoltaics.

37. Name two current applications of photovoltaic cells *other* than the production of electricity in remote areas.

Exploring Extensions

38. Explain the term *decommission*, as in "decommissioning a nuclear power plant." What technical challenges are involved? The resources of the Internet can help you.

39. Einstein's equation, $E = mc^2$, applies to chemical reactions as well as nuclear ones. An important chemical change studied in Chapter 5 was the combustion of methane, which releases 50.1 kJ of energy for each gram of methane burned.

a. Use Einstein's equation to determine what mass loss corresponds to the release of 50.1 kJ.

b. To produce the same amount of energy, what is the ratio of the mass of methane burned in a chemical reaction to the mass converted into energy?

c. Think about the mass conversion and combustion. Why does $E = mc^2$ not apply to a combustion reaction?

40. When 4.00 g of hydrogen nuclei undergoes fusion to form helium in the Sun, the change in mass is 0.0265 g and energy is released.

a. Which has more mass, the hydrogen or helium nuclei? Explain.

b. Use Einstein's equation, $E = mc^2$, to calculate the energy equivalent of this mass change.

41. Under conditions like those on the Sun, hydrogen can fuse with helium to form lithium, which in turn can form different isotopes of helium and of hydrogen. The mass of one mole of each isotope is given.

$$\underset{2.01345\text{ g}}{{}^{2}_{1}\text{H}} + \underset{3.01493\text{ g}}{{}^{3}_{2}\text{He}} \longrightarrow [{}^{5}_{3}\text{Li}] \longrightarrow \underset{4.00150\text{ g}}{{}^{4}_{2}\text{He}} + \underset{1.00728\text{ g}}{{}^{1}_{1}\text{H}}$$

a. In grams, what is the mass difference between the reactants and the products?

b. For one mole of reactants, how much energy (in joules) is released?

42. Lise Meitner and Marie Curie were both pioneers in developing an understanding of radioactive substances. You likely have heard of Marie Curie and her work, but may not have heard of Lise Meitner. How are these two women related in time and in their scientific work?

43. Advertisements for Swiss Army watches stress their use of tritium. One ad states that the "hands and numerals are illuminated by self-powered tritium gas, 10 times brighter than ordinary luminous dials." Another advertisement boasts that the "tritium hands and markers glow brightly making checking your time a breeze, even at night." Evaluate these statements and, after doing some Internet research, discuss the chemical form of tritium in these watches, and what its role is.

44. Deciding where to locate a nuclear power plant requires analysis of both risks and benefits associated with the plant. If you were to play the role of a CEO of a major electric utility considering whether to pursue permits for the construction of a nuclear power plant in your area, what risks and benefits would you cite?

45. Provide at least two similarities and two differences between a nuclear-fueled power plant (Figure 6.5) and a coal-fueled power plant (Figure 5.10).

46. At the cutting edge of technology, the line between science and science fiction often blurs. Investigate the "futuristic" idea of putting mirrors in orbit around Earth to focus and concentrate solar energy for use in generating electricity.

47. Building-integrated photovoltaic materials are becoming more popular, due to the relatively unsightly appearance of solar panels on rooflines of homes. Provide some examples of these materials. How do these building materials work and how durable are these materials with respect to extreme weather conditions (snow, sleet, hail, wind, etc.)?

48. Although silicon, used to make solar cells, is one of the most abundant elements in Earth's crust, extracting it from minerals is costly. The increased demand for solar cells has some companies worried about a "silicon shortage." Find out how silicon is purified and how the PV industry is coping with the rising prices.

49. Of the alternative forms of renewable energy presented in Section 6.9, which do you think is most promising? Do some Internet research to find some pros and cons to this energy source. On the basis of what you find, do you think it is a viable option for the future, or is more research necessary to implement it?

50. Figure 6.23c shows an array of photovoltaic cells installed at the Solarpark Gut Erlasee in Bavaria, Germany.

a. At present, where is the largest photovoltaic power plant located in your country?

b. Name two other locations of large-scale photovoltaic cell installations.

c. Name two factors that promote a centralized array rather than individual rooftop solar units.

CHAPTER 7 Energy Storage

Source: Shaddack via Wikimedia

REFLECTION

The Role of Batteries in Your Everyday Life

a. Our everyday life would not be the same without batteries. List all of your daily activities that involve the use of batteries and include the type of battery used for each of these activities.

b. Not all batteries are the same. This chapter will describe chemical reactions that occur inside various types of batteries. Among the batteries you listed in part **a** above, which are able to be recharged? For these rechargeable batteries, predict some factors that will influence their usable lifetime (*i.e.*, the number of possible charge/discharge cycles until they are no longer able to power a device).

The Big Picture

In this chapter, you will explore the following questions:

- What are the main types of batteries and how do they work?
- What are the differences between galvanic and electrolytic cells?
- How do the primary components of batteries store energy?
- How are batteries recycled?
- What are "hybrid" vehicles?
- What are the differences between supercapacitors and batteries?
- Are there benefits to using fuel cells instead of conventional gasoline-fueled vehicles?

Introduction

We rely on a flow of electrons—better known as electricity—to heat or cool our living and work spaces, to provide light to read by, and to power our electronic devices. For some applications, the electricity we use is generated at centralized power plants, such as those powered by fossil fuels (Chapter 5) or fissionable isotopes (Chapter 6). To a lesser extent, we also rely on wind, sun, and geothermal energy, as well as the potential energy of water trapped by dams, as sources to generate electric power.

However, for our mobile lifestyles, we are increasingly dependent on convenient-sized portable sources of electricity, better known as batteries. These long-lasting and reliable power sources fill a special energy niche. They power our cell phones, laptops, and other portable electronic devices, but their recent use in modes of transportation represents a modern renaissance. Although companies such as Tesla are looking to capture the bulk of the electric vehicle market, even established automakers have electric vehicles such as the Nissan Leaf and Chevrolet Volt. Most manufacturers now feature at least one hybrid model, which uses a combination of electrical and fuel sources to power the vehicle.

© Reed Richards/Alamy RF

Benjamin Franklin (1706–1790) first coined the term "battery" in 1749 to describe an apparatus used for his experiments. However, Alessandro Volta (1745–1827) is credited for fabricating the first functional battery, which was referred to as a *voltaic pile*. But does the first use of batteries actually date back much earlier—perhaps to ancient times? You decide.

Your Turn 7.1 You Decide The Baghdad Battery

Alessandro Volta is widely credited with the discovery of the battery in the early 1800s. However, ancient artifacts (**Figure 7.1**) have been found that could have been used as a battery as early as 242 AD. Using the Internet, describe the components and function of the so-called "Baghdad battery." Based on the critics' descriptions of this artifact, and recent experimental testing of this type of apparatus, do you believe this was used as a battery, or employed in some other application? Explain your reasoning.

Figure 7.1

A terracotta pot, rolled copper sheet, and an iron rod, which may date back to the Parthian period (between 250 BC and 224 AD).

Batteries have evolved a great deal since their early designs (Figure 7.2). Improvements to Volta's battery were made by John Daniell (1790–1845), Gaston Plante (1834–1889), Georges Leclanche (1839–1882), and Carl Gassner (1855–1942). The first rechargeable battery (Ni-Cd) was invented in 1899 by Waldmar Jungner (1869–1924); however, it was not widely available for consumer use until 1947. Current alkaline batteries, made popular by the Energizer and Duracell corporations, were developed by Lew Urry (1927–2004) at the Eveready Battery Company in 1949. Lithium–ion rechargeable batteries represent the most recent type of commercial battery, and were introduced in 1971. However, our thirst for longer battery life in portable electronics and modes of transportation continues to drive the further evolution of battery design.

In this chapter, we will describe the components, operating principles, and safety considerations for various types of batteries. We will also describe the environmental impacts of their production and end-of-use practices. Let's begin by peering through the casing of a battery to discover the chemical reactions that produce electrical energy.

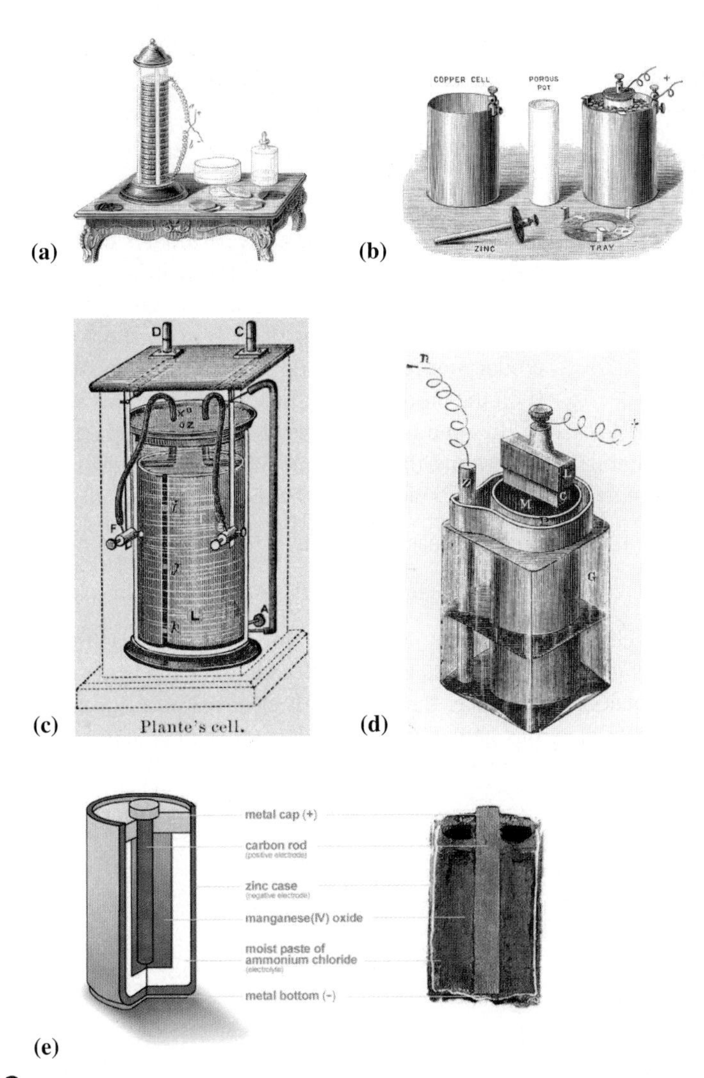

Figure 7.2

Illustrations of early battery designs. Shown are: **(a)** Volta's voltaic pile, composed of alternating zinc and copper discs separated by fabric pieces wetted by an acidic solution; **(b)** a Daniell cell, featuring a jar containing metallic copper immersed in an aqueous copper sulfate solution, and another connected jar with metallic zinc immersed in an aqueous zinc(II) sulfate solution; **(c)** a Plante cell (the lead-acid battery, which is used in vehicles to power the starter), consisting of rods of lead immersed in sulfuric acid solution; **(d)** a Leclanche cell, consisting of manganese(IV) oxide and zinc rods immersed in an ammonium chloride solution; **(e)** the first "dry cell" invented by Carl Gassner—a design that is still widely used for alkaline batteries today.

(a): © Hulton Archive/Getty Images; (b): © Mary Evans Picture Library/The Image Works; (c): © VintageMedStock/Alamy Stock Photo; (d): © Hulton Archive/Getty Images; (e): Source: Mcy jerry via Wikimedia Commons (https://commons.wikimedia.org/wiki/File:Zincbattery_%281%29.png)

Aqueous solutions, consisting of solutes dissolved in a water solvent, were described in Section 4.2.

7.1 | How Does a Battery Work?

Batteries represent a large and growing business worldwide due to consumer demand for portable electronics, which often starts at an early age (Figure 7.3). The workhorses of batteries are **galvanic cells**—compartments that are capable of converting the energy released from spontaneous chemical reactions into electrical energy. A collection of several galvanic cells wired together constitutes a true **battery**.

The branch of chemistry that deals with the transformation between chemical and electrical energies is known as **electrochemistry**.

All galvanic cells produce useful energy through the transfer of electrons from one substance to another. For this transfer process, you can write an overall chemical equation that may be divided into two parts. One is for **oxidation**, a process in which a chemical species loses electrons. The other is for **reduction**, a process in which a

Figure 7.3

A humorous, although realistic, view of the importance of batteries to many consumer products.

chemical species gains electrons. We refer to these two parts as **half-reactions** in the sense that each represents half of the overall process occurring in the galvanic cell. Each half-reaction is a chemical equation that shows the electrons either lost or gained by the reactants.

Half-reactions always occur in pairs and must include both ions and electrons. Even though electrons cannot be poured from a bottle into a flask, it is still helpful to show them in half-reactions so you can better understand what is taking place. Note that the electrons show up either on the products or reactants side of the half-reaction, but not on both. If on the products side, then the reactant has lost electrons; this is an *oxidation* half-reaction. In contrast, if the electrons are on the reactants side, then the reactant is gaining electrons, and this is a *reduction* half-reaction.

Half-reactions were first introduced in Section 1.7 to describe the isolation of metals from natural ores.

As an example, consider a simple type of battery consisting of zinc and copper:

$$\text{oxidation half-reaction:} \quad Zn \longrightarrow Zn^{2+} + 2\ e^- \qquad [7.1]$$

$$\text{reduction half-reaction:} \quad Cu^{2+} + 2\ e^- \longrightarrow Cu \qquad [7.2]$$

In this case, two electrons are lost by zinc in the oxidation half-reaction (Equation 7.1). But where do they go? These electrons were transferred to the ion being reduced. In order for the overall equation to balance, the number of electrons lost during oxidation must equal the number of electrons gained through reduction (Figure 7.4).

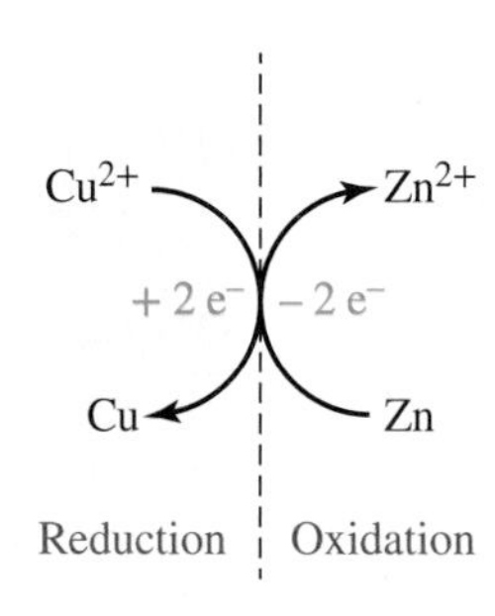

Figure 7.4

A scheme for redox reactions, showing the electrons released by an oxidation half-reaction being used for the reduction half-reaction.

We now can combine the two half-reactions to obtain the overall balanced equation:

$$Zn + Cu^{2+} + \cancel{2e^-} \longrightarrow Zn^{2+} + Cu + \cancel{2e^-} \qquad [7.3]$$

The electrons that appear on both sides of Equation 7.3 cancel, because the electrons *lost* by the metallic zinc are *gained* by the copper ions. So, we can rewrite the overall cell equation as:

$$Zn + Cu^{2+} \longrightarrow Zn^{2+} + Cu \qquad [7.4]$$

The oxidation state for ions is equivalent to their charge. For neutral atoms, the oxidation state is always zero.

The charge of the metal is called its **oxidation state**. Hence, in Equation 7.4, the oxidation state of zinc has increased from 0 to +2, whereas the oxidation state of copper has decreased from +2 to 0. As a general rule of redox reactions, *reduction* always results in a *decrease* in oxidation state; oxidation always results in an *increase* in oxidation state.

Your Turn 7.2 Skill Building Electrons in Half-Reactions

Categorize each as an oxidation half-reaction or a reduction half-reaction. Explain your reasoning.

a. $Al^{3+} + 3\ e^- \longrightarrow Al$
b. $Zn \longrightarrow Zn^{2+} + 2\ e^-$
c. $Mn^{7+} + 3\ e^- \longrightarrow Mn^{4+}$
d. $2\ H_2O \longrightarrow 4\ H^+ + O_2 + 4\ e^-$
e. $2\ H^+ + 2\ e^- \longrightarrow H_2$

The movement of electrons through an external circuit produces **electricity**, the flow of electrons from one region to another that is driven by a difference in potential energy. The electrochemical reaction provides the energy needed to operate a cell phone, a power tool, or countless other battery-operated devices. The chemical species oxidized and reduced in the cell must be connected in such a way to allow electrons released during the oxidation process to transfer to the reactant being reduced, while following an appropriate electrical path for use in the desired application.

Most galvanic cells convert chemical energy into electric energy with a net efficiency of about 90%. Compare this with the much lower efficiencies of 30–40% that characterize coal-fired power plants that generate electricity. Recognize, though, that electricity from these plants is used to recharge batteries. This is but one of many incentives to explore renewable energy sources.

Potential energy was first introduced in Section 5.3.

7.2 | Ohm, Sweet Ohm!

Almost everyone has inserted a battery into a flashlight, calculator, or digital camera. You may recognize the ones shown in Figure 7.5. One end of each of these batteries is marked with a + sign, while the other end displays a – sign. These markings point to the fact that electron transfer is at work. Figure 7.6 provides an illustration of an electrical circuit, showing the flow of electrons from the negative terminal of a battery through an electrically conductive wire to light a bulb. When the switch is open (as shown), the electrons are not able to return from the light bulb along the electrical pathway toward the positive terminal of the battery since the circuit has been interrupted. By flipping the switch, the electrical circuit is completed, and electrons may then flow to/from the battery through the light bulb. The cycle of electron flow will continue until all of the electrons are used up from the redox reactions occurring in the battery, or until the switch is opened, which again breaks the electrical circuit. As we will see shortly, a light bulb is called a **resistor** since it slows the flow of electrons through the external circuit.

Figure 7.5

Alkaline batteries from size AAA to D, all of which produce 1.5 V.

Alkaline cells each produce 1.5 V, but the larger cells can sustain a current through the external circuit for a longer time. The **current**, or rate of electron flow, is measured in amperes (amps, A) or, for smaller cells, milliamps (mA).

An amusing illustration of voltage, current, and resistance is shown in Figure 7.7. The relationship between these parameters is known as **Ohm's Law**, which is described by a simple equation:

$$V = I \cdot R \quad [7.5]$$

where: V = voltage (measured in volts, V)
I = current (measured in amps, A)
R = resistance (measured in ohms, Ω)

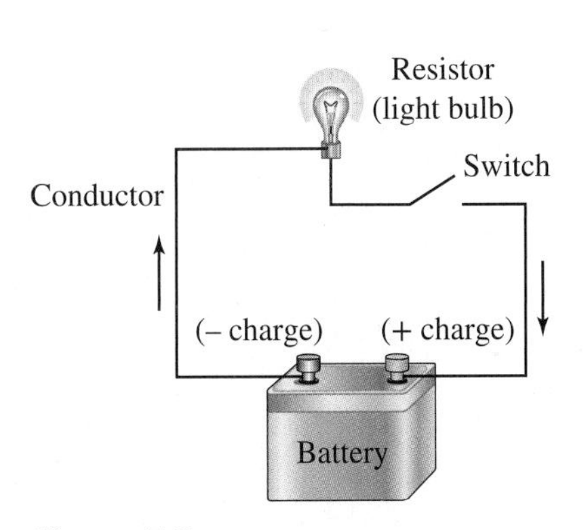

Figure 7.6

A simple schematic for an electrical circuit for a battery-powered light bulb.

Figure 7.7

An illustration of Ohm's Law. Voltage is related to the force of electron flow; current is related to the flow of electrons; resistance works to block the flow of electrons in an electrical circuit.

Your Turn 7.3 Skill Building Coulombs

The SI unit of electric charge is the coulomb, C, which is equivalent to the charge of 6.242×10^{18} protons. Express amps and ohms in terms of the base SI units of J, C, and s. Perform dimensional analysis for **Equation 7.5** to show that volts may be expressed as J/C.

An electrical circuit within a device uses electricity to perform a task such as power a lamp or run a handheld vacuum cleaner. Within this circuit, the flow of electrons is referred to as electrical **current**, whereas **voltage** is the difference in electrical energy (also known as electrical potential) between two points. As its name suggests, *resistance* is something that resists the flow of electrons. As seen in Equation 7.5, an increase in circuit resistance leads to a decrease in electrical current. Furthermore, as the voltage increases at a constant circuit resistance, so does the electrical current.

Your Turn 7.4 Scientific Practices Water Flow Analogy

The flow of water through pipes is commonly used as an analogy for the flow of electrons through an electrical circuit.

a. Construct a diagram for the flow of water through pipes, which shows the interrelation of voltage, current, and resistance.

b. Compare your illustration to those on the Internet. How accurate is this analogy, and what are some limitations for this description of electricity?

Anode = oxidation
Cathode = reduction

Some heavier Group 13 metals are most stable in the +1 oxidation state, such as indium (In) and thallium (Tl). Metals at the bottom of Group 14 such as tin (Sn) and lead (Pb) may be stable in either +2 or +4 oxidation states.

Revisit Section 4.1 to recall the naming scheme for transition metals.

Electron transfer in a battery takes place within its **electrodes**—electrical conductors within a cell that serve as sites for chemical reactions. At the **anode**, oxidation takes place and is the source of electrons flowing into the device's circuit. At the **cathode**, reduction takes place. The cathode receives the electrons sent from the anode through the external circuit to complete the reduction process.

Since battery electrodes serve as the location of oxidation and reduction reactions, they must be composed of substances that undergo facile electron loss or gain. For compounds containing ions from Group 1, 2, or 13 of the periodic table, the oxidation state of the metal is (almost) always +1, +2, or +3, respectively. As a consequence, the names of these ionic compounds such as "sodium chloride" do not need to include the charge of the metal, because it is predictable and unchanging. However, transition metals are stable at a variety of oxidation states (Figure 7.8), which makes them desirable for reduction/oxidation (also known as *redox*) reactions.

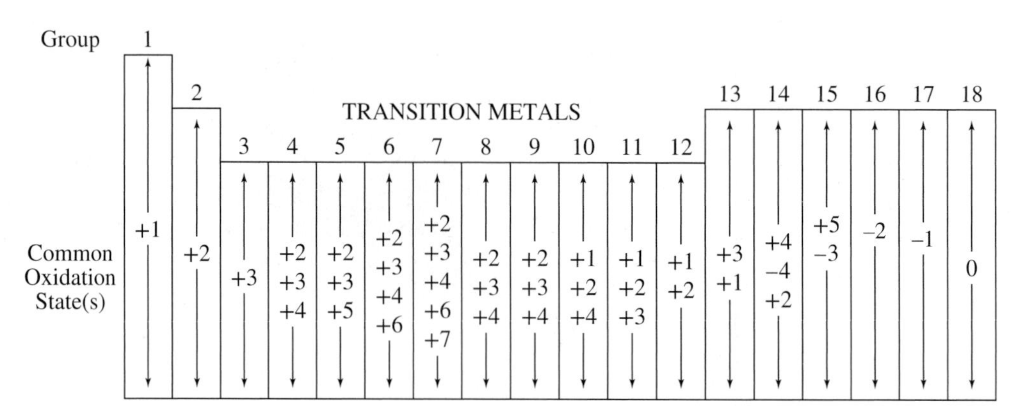

Figure 7.8

The common oxidation states of ions in various groups of the periodic table.

7.3 | Batteries, Batteries Everywhere!

Once the electrical circuit is complete, a voltage can be measured across the cell; that is, the difference in electrochemical potential between the two electrodes. Voltage is measured in units called volts, V. The greater the difference in potential between the two electrodes, the higher the voltage and the greater the energy associated with the electron transfer. For example, with a nickel–cadmium (Ni–Cd) cell, the maximum difference in electrochemical potential under the conditions specified is measured as 1.2 V. In contrast, alkaline cells deliver 1.5 V, and lithium-ion cells are capable of potentials in excess of 4 V! To produce the higher voltages needed to power larger devices (*e.g.*, power tools or automobile starter motors), several cells must be connected (Figure 7.9).

Did You Know? The unit "volt" honors the Italian physicist Alessandro Volta.

Figure 7.9

This 7.2-V Ryobi portable power drill comes with two Ni–Cd battery packs and a charging unit.

The voltage of a battery is primarily determined by its chemical composition, and is *not* related to the size of the battery. You can see from the examples listed in Table 7.1 that different voltages are produced using different chemical systems. Only a few volts are possible with a single galvanic cell. But, as we noted, higher voltages are possible by connecting cells. For example, in order to run a 19.2-V power drill, manufacturers sell a "battery pack" that contains multiple cells.

Table 7.1 Some Common Galvanic Cells

Type	Maximum Voltage (V)	Rechargeable?	Examples of Uses
nickel–cadmium (Ni–Cd)	1.25	yes	toys and portable electronic devices, including digital cameras, power tools
nickel–metal hydride (NiMH)	1.25	yes	replacing Ni–Cd for many uses in consumer devices; hybrid vehicles
alkaline	1.5	no	flashlights, small appliances, calculators, audio/video remote controls, toys
lithium (primary)	1.5–3.6	no	LED lighting, smoke alarms, watches, vehicle remotes and key FOBs
lead–acid	2.1	yes	automobiles (starting, lighting, and ignition)
lithium-ion, lithium-polymer	3.6	yes	laptop computers, cell phones, portable electronic devices, power tools

All alkaline batteries, from the tiny AAA size to the large D cells, produce the same voltage of 1.5 V. However, larger cells have a greater **capacity**—the ability to sustain the flow of electrons longer because they contain more material. The capacity of a battery is usually given in units of mAh; for instance, C, AA, and AAA alkaline batteries have capacities of 3800 mAh, 1100 mAh, and 540 mAh, respectively. Just as a gasoline-powered vehicle is able to travel farther on a larger tank of gas, an electric device is able to operate for longer periods of time on a larger-capacity battery pack.

The half-reactions for an alkaline cell are:

oxidation half-reaction (taking place at the anode):

$$Zn(s) + 2\ OH^-(aq) \longrightarrow Zn(OH)_2(s) + 2\ e^- \quad [7.6]$$

reduction half-reaction (taking place at the cathode):

$$2\ MnO_2(s) + H_2O(l) + 2\ e^- \longrightarrow Mn_2O_3(s) + 2\ OH^-(aq) \quad [7.7]$$

overall cell reaction (sum of the two half-reactions):

$$Zn(s) + 2\ MnO_2(s) + H_2O(l) \longrightarrow Zn(OH)_2(s) + Mn_2O_3(s) \quad [7.8]$$

Acids and bases will be described in more detail in Chapter 8. There you will learn that acids are generally defined as generating H^+ ions in solution, whereas basic solutions contain higher concentrations of OH^- ions.

For the compound MnO_2 found within alkaline batteries, there are two O^{2-} species that will give a total negative charge of 4–. Hence, to maintain an overall zero charge for the entire compound, the manganese must carry a charge of 4+. Hence, this compound is known as *manganese(IV) oxide*. Note that the cell is called "alkaline" because it operates in a basic, rather than acidic medium.

Compact, long-lasting cells may even find their way into your body. For example, the widespread use of cardiac pacemakers is largely due to the improvements made in the electrochemical cells rather than in the pacemakers themselves. Lithium–iodine cells are so reliable and long-lived that they are often the battery of choice for this application, lasting as long as 10 years before needing to be replaced.

7.4 (Almost) Endless Power-on-the-Go: Rechargeable Batteries

Imagine if the battery on your cell phone needed to be discarded and replaced every day after one use. Not only would that cost money and pollute the environment, but it would also be a time-consuming process for some portable electronic devices where the battery is seamlessly incorporated into its design. We take for granted the convenience of using a battery that can be recharged on demand in your office, on a plane or train, in your vehicle, or even outdoors by using a portable solar panel (Figure 7.10).

A primary (or disposable) battery, such as the alkaline batteries traditionally used in flashlights, cannot be recharged and must be thrown away after the charge is "used up."

Rechargeable batteries are called **secondary** batteries and use electrochemical reactions that can run in both directions. The transfer of electrons takes place both during the forward (discharging) and the reverse (recharging) processes.

As an example, let's consider the reversible reactions that occur inside a rechargeable Ni–Cd battery, used in many digital cameras. As the battery is discharged, atoms of cadmium become oxidized to Cd^{2+} at the anode. These ions, in turn, combine with OH^- to form cadmium(II) hydroxide, $Cd(OH)_2$ (Equation 7.9, left-right). Simultaneously, Ni^{3+}, present in the cathode as the hydrated form of nickel(III) oxide hydroxide, NiO(OH), is reduced to Ni^{2+} to form nickel(II) hydroxide, $Ni(OH)_2$ (Equation 7.10, left-right). These two chemical reactions happen because two electrons from the cadmium reaction are transferred to the nickel reaction, and this movement of electrons is where the power comes from.

$$Cd(s) + 2\ OH^-(aq) \underset{\text{recharging}}{\overset{\text{discharging}}{\rightleftharpoons}} Cd(OH)_2(s) + 2\ e^- \qquad [7.9]$$

$$2\ NiO(OH)(s) + 2\ H_2O(l) + 2\ e^- \underset{\text{recharging}}{\overset{\text{discharging}}{\rightleftharpoons}} 2\ Ni(OH)_2(s) + 2\ OH^-(aq) \qquad [7.10]$$

Figure 7.10

A portable solar-powered charging device for mobile devices.

Figure 7.11

A photo showing the consequences of a fire in a Li-ion laptop battery. The suspected cause of the fire was failure of the membrane separator, which resulted in a short circuit and thermal runaway.

© Kyodo/AP Images

All batteries, whether primary or secondary, require a **separator** that is placed between the anode and cathode to ensure that the electrodes do not come into physical contact. As you might expect, if the anode and cathode are allowed to touch in a battery, this would cause a short circuit and failure of the battery—often with dangerous consequences (Figure 7.11). Although early galvanic cells employed a salt bridge separator (Figure 7.12), typical separators in modern batteries are composed of a semi-permeable membrane. This membrane effectively separates the electrodes, while still allowing ions to pass through and thereby retain charge-neutrality during battery operation.

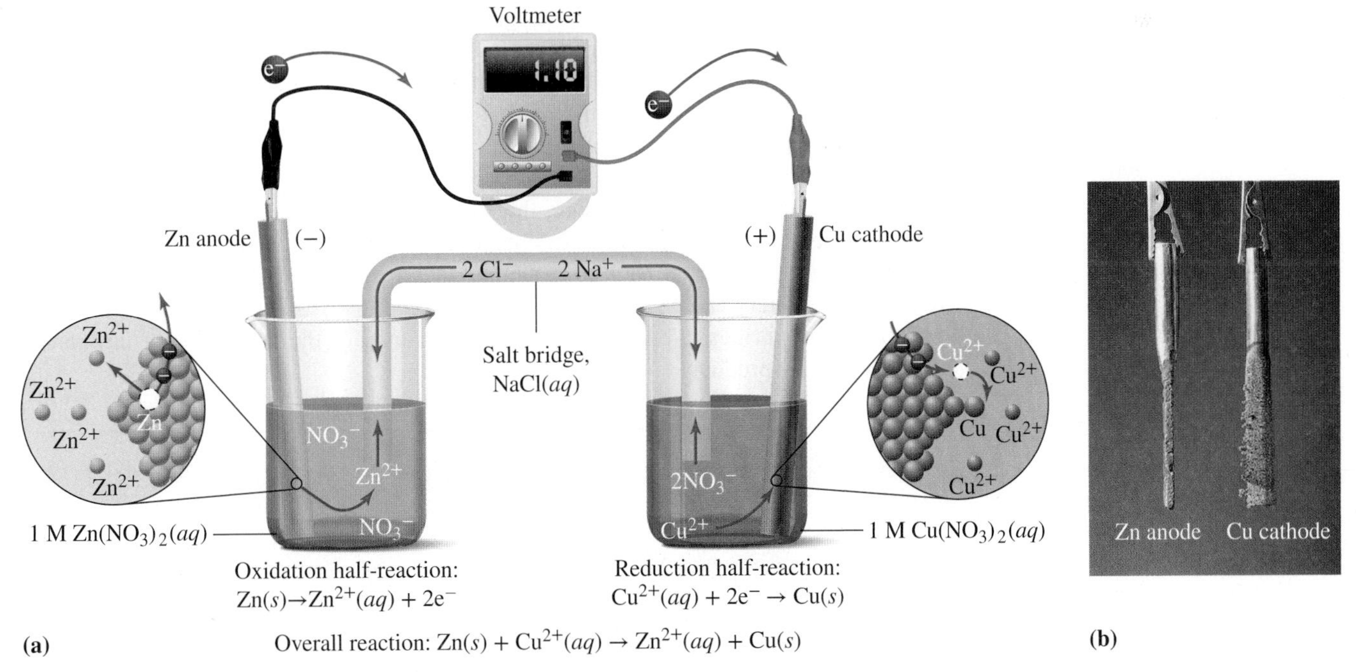

Figure 7.12

An illustration of a two-component Zn–Cu galvanic cell, containing a *salt bridge* separator (composed of aqueous sodium chloride, NaCl), and electrolytes composed of 1 M zinc(II) nitrate ($Zn(NO_3)_2$, anode compartment) and 1 M copper(II) sulfate ($CuSO_4$, cathode compartment). The image shown in **(b)** illustrates the loss of metallic Zn by oxidation and gain of Cu metal through reduction. *Note:* molarity (M) is a unit of concentration that will be explained in more detail in Chapter 8.

(b) © McGraw-Hill Education. Stephen Frisch, photographer

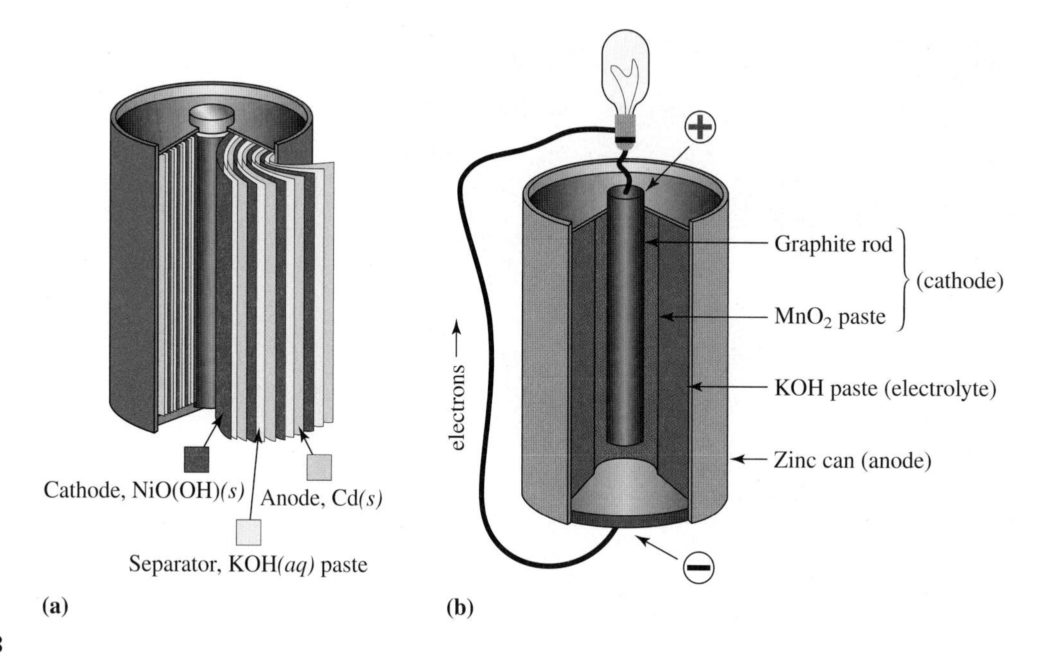

Figure 7.13

Representations of **(a)** a Ni–Cd rechargeable battery, showing how the components are layered to increase the surface area of the electrodes, and **(b)** a primary alkaline battery, in which the zinc container acts as the anode.

Electrolytes will be discussed in more detail in Chapter 8. For our discussion here, an electrolyte is an electrically conductive solution that contains the ions required to complete the chemical reactions occurring at the electrodes.

Did You Know? Potassium hydroxide, KOH, is also used in the production of drain cleaners, bleach, biodiesel, and soft soaps.

In order to facilitate ionic movement through the separator, an **electrolyte** must be used in all galvanic cells. With the exception of lead–acid batteries used for automobile starter motors, aqueous solutions are usually too hazardous to use in batteries because sooner or later they leak from the battery casing. For example, you may have seen the corrosive mess inside of a flashlight or child's toy from a leaking battery.

Most commercial batteries are known as **dry cells**, in which the electrolyte is immobilized as a paste, with only enough moisture available to allow ions to flow. Unlike the wet cells in early battery designs (Figure 7.2b and c), a dry cell can operate in any orientation without spilling, because it contains no free liquid. For rechargeable Ni–Cd (Figure 7.13a), primary alkaline (Figure 7.13b), or rechargeable NiMH (nickel–metal hydride) batteries, the electrolyte that fills the pores of the membrane separator is an aqueous paste of potassium hydroxide (KOH).

What features make a battery rechargeable? The key is that both the reactants and products are solids. Furthermore, the solid products cling to a stainless-steel grid within the battery rather than dispersing. If a voltage is applied to this grid, these products can be converted back to reactants, thus recharging the battery. Although a rechargeable battery can be discharged and recharged many times, eventually the accumulation of impurities, a breakdown of the separators, or the generation of unwanted side-reaction by-products ends its useful life.

Your Turn 7.5 Scientific Practices Lifetime of Rechargeable Batteries

Rechargeable batteries all have a lifespan, and will no longer function after many charging/discharging cycles. Using the Internet as a resource, investigate the following:

a. What is the "memory effect" found in some rechargeable batteries, and what operating conditions cause this phenomenon to occur?
b. Which batteries are most prone to this effect?
c. How can this effect be repaired?
d. Many believe that batteries will last longer if stored in a cold environment such as a refrigerator. Do you agree? Do all batteries (primary and secondary) benefit from being stored in a cold atmosphere?

Batteries come in many shapes and sizes, each one uniquely matched to its use. For example, in a hearing aid, the size and weight of the cell is of paramount importance. In contrast, an automobile battery must last for years and perform over a range of temperatures. To be successful in the eyes of today's consumers, batteries must be affordable, last a reasonable length of time, and be safe to use and recharge. Ultimately, to be successful in the years to come, batteries must also be designed so their materials can be recycled in a sustainable way.

Your Turn 7.6 Scientific Practices The Shift from Ni–Cd to NiMH

Look up the reactions occurring in nickel–metal hydride (NiMH) batteries, and compare them to those presented for Ni–Cd batteries (**Equations 7.9** and **7.10**). Why have NiMH batteries supplanted Ni–Cd batteries for consumer applications?

7.5 Lead–Acid: The World's Most Widely Used (and Heaviest!) Rechargeable Battery

Found under the front hood of most cars, the lead–acid battery is the workhorse of today's rechargeable batteries. In an automobile, it powers an electric motor that is used to start the car, a function that was once initiated by a hand crank. Although we take lead–acid batteries for granted, these represent one of the key technological developments that ushered in the rapid rise of the automobile, which has forever changed our world. The lead–acid battery comprises six electrochemical cells, each generating 2.0 V for a total of 12 V (Figure 7.14). Here is the overall chemical equation, the sum of the two half-reactions:

$$\underset{\text{lead}}{Pb(s)} + \underset{\text{lead(IV) oxide}}{PbO_2(s)} + \underset{\text{sulfuric acid}}{2\,H_2SO_4(aq)} \underset{\text{charging}}{\overset{\text{discharging}}{\rightleftharpoons}} \underset{\text{lead(II) sulfate}}{2\,PbSO_4(s)} + \underset{\text{water}}{2\,H_2O(l)} \qquad [7.11]$$

As the arrows in Equation 7.11 indicate, when the chemical reaction proceeds to the right, the battery is discharging. For example, using the battery to start the car or using the lights or radio with the engine off discharges the battery. However, once the engine is running, an alternator that is turned by the engine provides the current needed

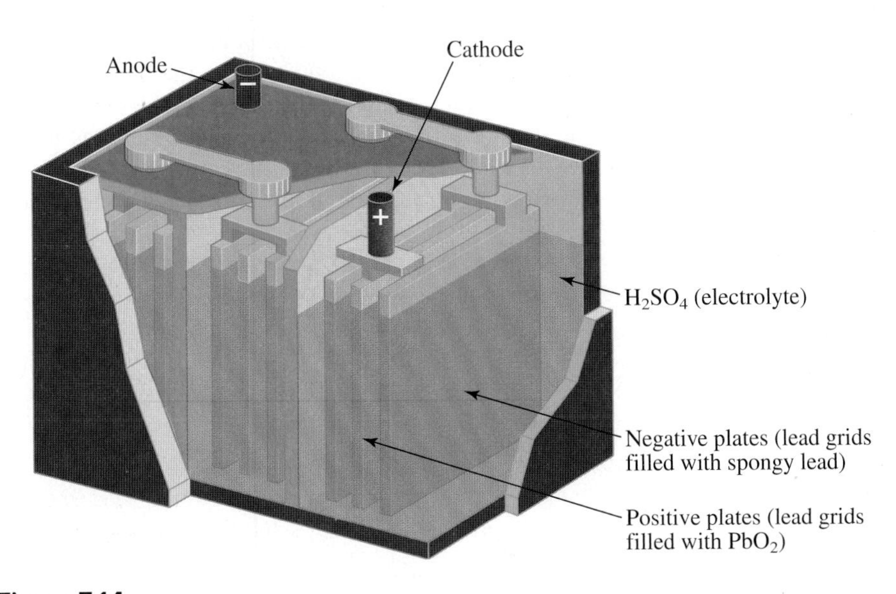

Figure 7.14

Cutaway view of a lead–acid battery.

to reverse the chemical reaction and recharge the battery. Fortunately, the battery can be discharged and recharged many times before it needs to be replaced. A high-quality battery can perform for five years or more.

Your Turn 7.7 Scientific Practices The Battery in Your Car

Let's take a closer look at the lead–acid storage battery, the one found in most cars (**Equation 7.11**).

a. Lead occurs in this equation as Pb, PbO_2, and $PbSO_4$, all solids. In which of these is lead in an ionic form? What are the charges (or oxidation states) of the ions? Which one of these symbols represents lead in its metallic form?

b. When lead is converted from its metallic form to an ionic form, are electrons lost or gained? Is this oxidation or reduction?

c. When the battery is discharging, is metallic lead oxidized or reduced?

Because lead–acid batteries have the advantage of being rechargeable and low cost, they may be used together with wind-turbine electric generators. The generator recharges the batteries during favorable wind conditions, whereas the batteries discharge during unfavorable winds. You can also find lead–acid batteries in environments where the emissions from internal combustion engines cannot be tolerated. The forklifts in warehouses, the passenger carts in airports, and the electric wheelchairs in supermarkets are typically powered by lead–acid batteries. Their weight may even be an advantage in stabilizing these vehicles.

In an automobile, however, the weight of the lead–acid battery is a disadvantage. Another disadvantage is the nature of chemical components in the battery. The anode (metallic lead), cathode (lead(IV) oxide), and the electrolyte (sulfuric acid solution) pose disposal challenges as toxic or corrosive chemicals. The next section discusses another type of rechargeable battery that is significantly lighter and more efficient than lead–acid, and thereby more attractive for electric vehicle applications.

7.6 | Vehicles Powered by Electricity

For batteries to be useful in modes of transportation, we wish to maximize the distance the vehicle can travel on a single charge, and minimize the time it takes to recharge the battery. Current electric vehicles (EVs) such as the Tesla Model S can travel up to 270 miles between charges, which is approaching the distance one can travel on a tank of gasoline before a fill-up. But, how much time does it take to charge the battery, and how does the cost of electricity required to charge the battery compare to the cost of filling up a fuel tank? Let's consider these questions in the following activity.

Your Turn 7.8 Scientific Practices Tesla Model S Battery Charging

At 270 miles, the Tesla Model S currently features the longest driving range per charge of all electric vehicles. In this activity, you will calculate how long it will take to charge the battery, as well as its associated electricity costs. *Power* that is released by an electric device is measured in units of *watts*, W. In order to calculate watts, you need to know the voltage, V, and current, I:

$$W = V \times I \quad [7.12]$$

a. Compare the power output for a standard 110 V/20 A electrical plug (a "level 1" charger) to that of a 240 V/40 A plug (a "level 2" charger), and that of a "dual charger" (240 V/80A).

b. Calculate the time required to charge a Tesla Model S using each charging scenario listed in **a**. Assume that the capacity of the battery is 85 kWh. Also assume a charging efficiency of 92%. ***Hint:*** Use dimensional analysis to ensure that the units properly cancel.

c. Using the Internet, find some locations of "supercharging stations" across the U.S. Using their maximum charging rates, how long would a single charge require?

d. Look up the current cost per kWh of electricity in your region, and calculate the costs per charge for each of the above charging scenarios.

e. Using the current price of gasoline, an average driving distance of 1100 miles per month, and the range of your favorite gasoline-powered vehicle versus the Tesla Model S, determine the cost difference in operating each vehicle per month and per year. What factors will influence the maximum driving range of the vehicle? How will these affect the operating costs for the electric car?

The charging efficiency is related to the efficiency of converting the alternating current, AC, power used by the charger into direct current, DC, power that is used by the battery.

As you calculated in the above activity, it takes a long time to charge the battery in an EV—much longer than it takes to fill up a gas tank on a traditional gasoline-powered vehicle. This shouldn't surprise you. It takes an hour or so to charge the battery on your laptop or cell phone, so the much larger battery packs in vehicles should take significantly longer to charge. Thankfully, batteries slowly discharge upon use, allowing us to drive many miles or surf the Web on our portable devices over extended periods of time. The reason for slow charging/discharging in batteries is due to the chemical nature of the battery itself. In previous sections, we described the chemical reactions that occur in batteries. These reactions require time to complete, and the speed of these reactions are affected by the operating conditions of the battery. For instance, using or storing a battery at elevated temperatures will cause the reactions to proceed faster, which results in faster charging times but shorter battery life. Consequently, using or storing a battery in cold environments will require longer charging times, but will generally result in a longer battery life.

In order to understand the rate (also known as **kinetics**) of these reactions, let's consider what goes on when a rechargeable battery is charged/discharged. The current battery-of-choice for portable electronics and vehicle applications is the Li-ion battery. As you survey the periodic table, it is no surprise why Li is so pervasive in portable battery applications. Lithium is the lightest metal in the periodic table. With a density of 0.535 g/cm^3, this corresponds to a density that is 2000% lighter than lead (11.34 g/cm^3), which is used in lead–acid batteries. For EV applications, it is paramount that the battery be small and lightweight, which maximizes the driving range of the vehicle. You can think of this as being equivalent to vehicles of today, composed of plastics and composites, being able to travel farther on a tank of gas than classic cars, which comprised heavyweight steel and chrome.

In addition to being lightweight, Li atoms and ions are much smaller than other metals. This means that more ions may be placed within the electrode during charging, which results in longer usage on a single charge. The **energy density** of a battery relates both the number of ions stored in the electrode material and the weight or volume of the battery (Equation 7.13). Figure 7.15 shows a comparison of the various types of rechargeable batteries, with those involving Li being the most preferable due to being lightweight (high gravimetric energy density) and smaller (high volumetric energy density).

Gravimetric energy density is commonly referred to as the specific energy density.

$$\text{Energy density} = \frac{\text{Voltage} \times \#\text{ of movable Li ions in electrodes}}{\text{Total battery weight (gravimetric) or volume (volumetric)}} \quad \textbf{[7.13]}$$

The reactions occurring in a Li-ion battery are much less complicated relative to alkaline or other rechargeable batteries discussed thus far. As shown in Figure 7.16, Li^+ ions simply shuttle back and forth between the two electrodes during

The term *intercalation* refers to the reversible insertion of a molecule or ion into compounds with a layered structure.

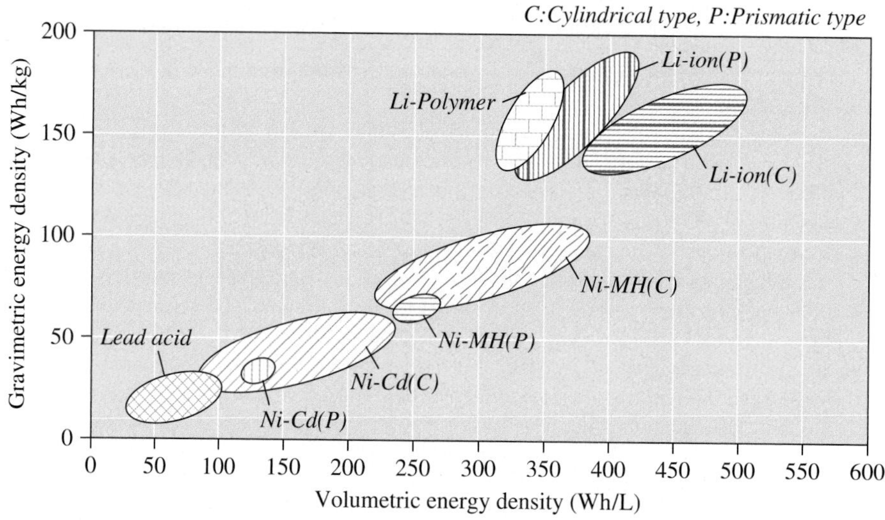

Figure 7.15

Comparison of the energy densities for various rechargeable batteries. Li-polymer batteries have similar electrochemical reactions as Li-ion varieties, but differ in their choice of electrolyte and packaging.

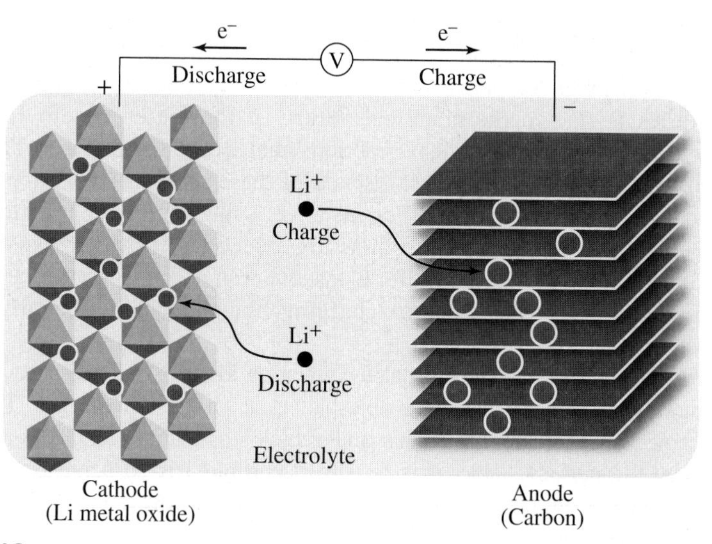

Figure 7.16

Schematic of the Li^+ migration during charging and discharging of a Li-ion battery.

charging and discharging. The specific half-reactions for a Li-ion battery are as follows:

$$\text{(Cathode)} \quad LiCoO_2(s) \underset{\text{discharging}}{\overset{\text{charging}}{\rightleftharpoons}} Li_{1-x}CoO_2(s) + x\,Li^+ + x\,e^- \quad [7.14]$$

$$\text{(Anode)} \quad x\,Li^+ + x\,e^- + 6\,C(s) \underset{\text{discharging}}{\overset{\text{charging}}{\rightleftharpoons}} Li_xC_6(s) \quad [7.15]$$

The battery can only function as long as there are available Li^+ ions and suitable electrode material to house the ions during its use and recharging cycles. As charging/discharging takes place, the structure and composition of the electrode surfaces will change, which will result in less effective storage of Li^+ in subsequent cycles. To improve the lifetime of a Li-ion battery, it is important that you don't over-discharge the battery. Allowing the battery to run down to <10% can cause some of the Li^+ ions to irreversibly react with the cathode material (Equation 7.16), which will lower its overall capacity because fewer ions will be available for intercalation.

$$Li^+ + LiCoO_2 \longrightarrow Li_2O + CoO \quad [7.16]$$

Figure 7.17

A Zotye M300 electric car catching fire in Hangzhou, China.

© Imaginechina via AP Images

You may have seen video or pictures of a Li-ion battery engulfed in flames (Figure 7.17). This is very rare, but could be caused by a number of issues such as a separator failure, leaking electrolyte from cracks in the battery pack, or failure of the charging control circuitry. If a Li-ion battery is overcharged, the cathode breaks apart and releases oxygen gas that causes an immediate exothermic reaction with the Li metal plated on the anode. This then causes further breakdown of the cathode, as well as products deposited on the anode, which generates sufficient heat to ignite the organic-based electrolyte. There are current efforts to incorporate fire-retardant chemicals within the electrolyte of Li-ion batteries in order to prevent such a thermal runaway event from occurring.

Did You Know? All Group 1 metals, including Li, are dangerously reactive toward oxygen, nitrogen, and water. As such, non-aqueous electrolytes must be used in Li-ion batteries, and battery packs must be hermetically sealed to prevent exposure to air.

7.7 | Storage Wars: Supercapacitors vs. Batteries

Since it takes time for Li^+ ions to migrate into the various layers of anode and cathode compartments, charging and discharging processes take time to complete. In general, the slower the charging rate, the more Li^+ ions are able to position themselves within the layered electrode structures. The same goes for all other types of rechargeable batteries—their charging (and discharging) rate is governed by the kinetics of the chemical reactions taking place. However, we are most interested in rapid charging, which will allow us to get back on the road in a shorter time.

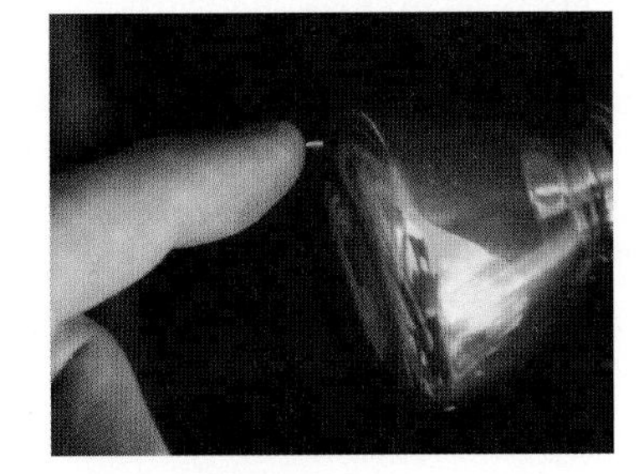

Figure 7.18

An image showing a shock from touching a doorknob through build up of static electricity.

© 2006 Richard Megna, Fundamental Photographs NYC

Enter supercapacitors. Similar to a battery, a **supercapacitor** consists of two charged electrodes, or plates, that are immersed in an electrolyte. However, instead of storing energy in the form of electrochemical reactions, supercapacitors store energy by means of a static charge. Think about the last time you walked across a carpeted room while dragging your feet. When you touched a conductive object such as a metal doorknob, the built-up static electricity was released through your finger (Figure 7.18). This build up of electrical charge is referred to as **capacitance**. As their name implies, supercapacitors (also referred to as *ultracapacitors*) are devices that store a significant amount of charge that may be released as electrical energy to power a device.

Capacitance is a key operating principle for mobile device touchscreens (Section 1.5).

As shown in Figure 7.19, the charges in a (super)capacitor are separated by a certain distance, d. The capacitance of a device will increase as the area of the plates, A, increases, or the distance between the plates decreases (Equation 7.17). The units of capacitance are given in farads, F, with supercapacitors exhibiting values in the 100–12,000 F range.

Note that a 1-farad capacitor can store 1 C of charge at 1 V.

$$C \propto \frac{A}{d} \quad [7.17]$$

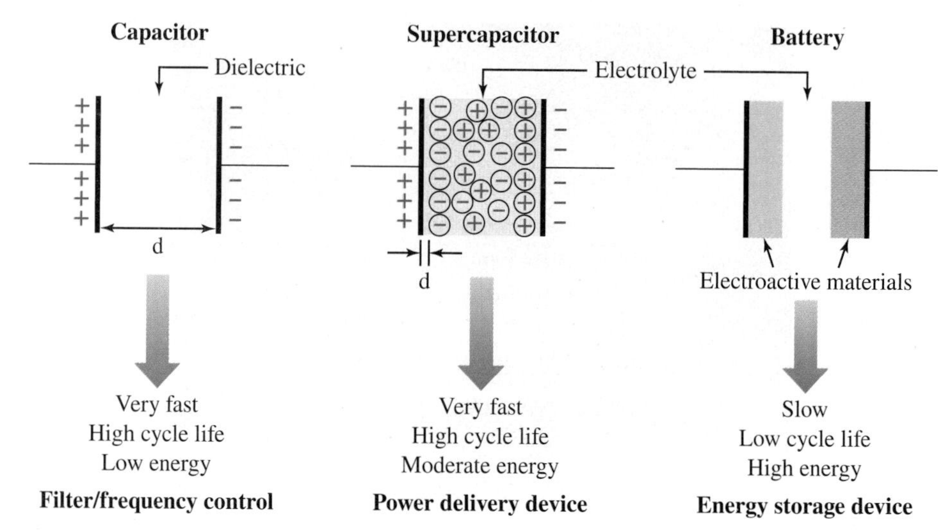

Figure 7.19

Schematics of a capacitor, supercapacitor, and battery.

Whereas all rechargeable batteries have a limited lifetime of available cycles, supercapacitors are able to be charged/discharged for millions of cycles. Imagine virtually unlimited use/recharge cycling of your cell phone without any loss in capacity! Furthermore, charging takes place in seconds, as compared to minutes or hours required to charge batteries. Additionally, supercapacitors are able to operate under a wider temperature range without premature failure.

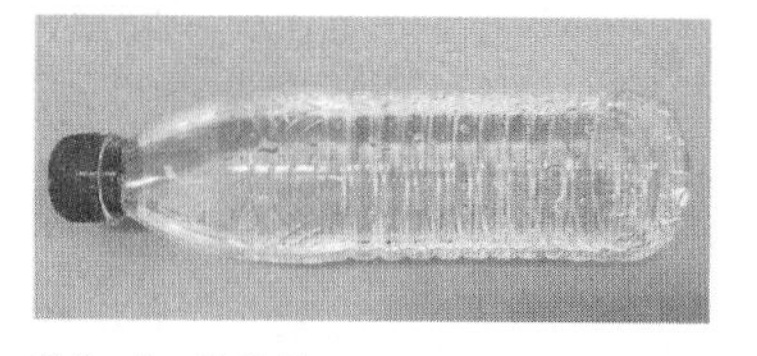

© Bradley D. Fahlman

Think of a simple water bottle as an analogy for energy and power densities. Whereas the size of the bottle represents the *energy* density, the size of its opening denotes the *power* density.

With so many desirable characteristics, why are batteries still preferred over supercapacitors for portable devices and vehicles? In fact, supercapacitors suffer from a few critical limitations. As shown in Table 7.2, supercapacitors have very low energy densities as compared to batteries. Whereas a battery stores its potential energy in chemical form, the potential energy in a supercapacitor is stored in an electric field, which corresponds to less energy per weight of material.

When a battery charges/discharges, it takes time to convert between chemical and electrical energies. From an application perspective, the longer it takes for a battery to deliver its power, the lower its **power density**. In contrast, supercapacitors have very high power densities and are able to charge/discharge almost instantaneously because they directly store electrical energy (Figure 7.20). However, it becomes difficult for a supercapacitor to maintain a constant voltage when being discharged. In contrast,

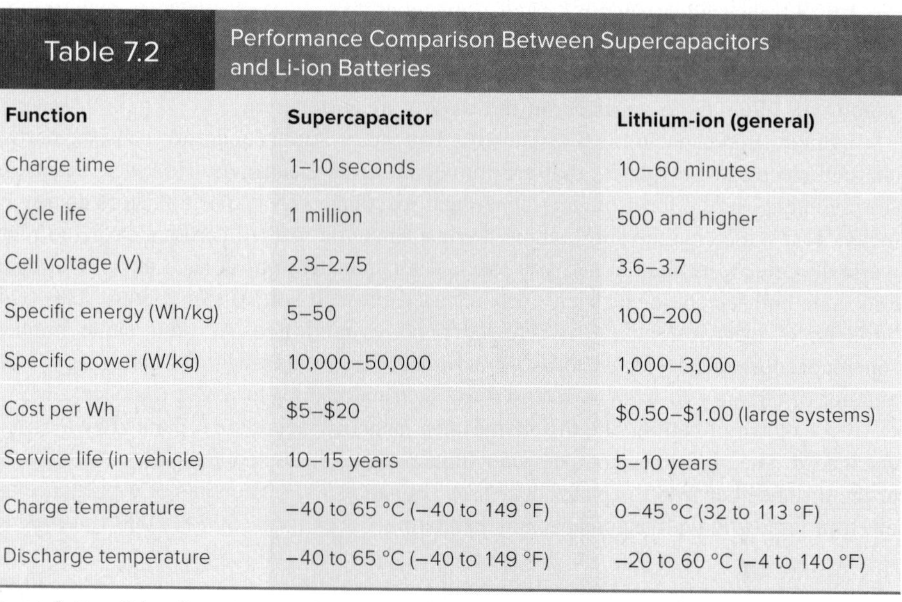

Table 7.2 Performance Comparison Between Supercapacitors and Li-ion Batteries

Function	Supercapacitor	Lithium-ion (general)
Charge time	1–10 seconds	10–60 minutes
Cycle life	1 million	500 and higher
Cell voltage (V)	2.3–2.75	3.6–3.7
Specific energy (Wh/kg)	5–50	100–200
Specific power (W/kg)	10,000–50,000	1,000–3,000
Cost per Wh	$5–$20	$0.50–$1.00 (large systems)
Service life (in vehicle)	10–15 years	5–10 years
Charge temperature	−40 to 65 °C (−40 to 149 °F)	0–45 °C (32 to 113 °F)
Discharge temperature	−40 to 65 °C (−40 to 149 °F)	−20 to 60 °C (−4 to 140 °F)

Source: Battery University

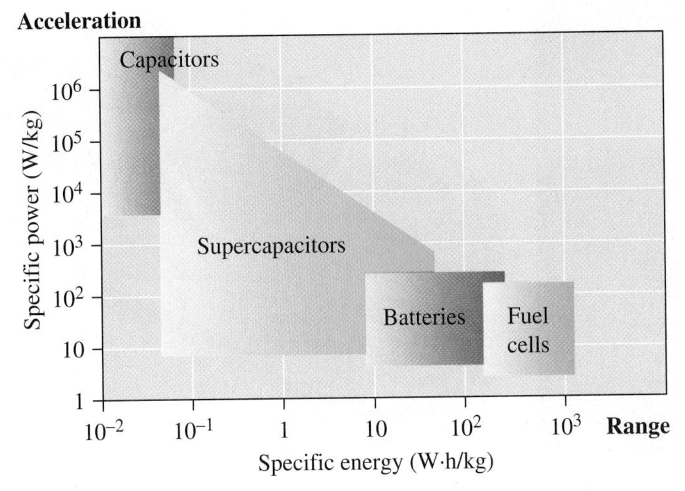

Figure 7.20

Comparison of various energy storage devices. Whereas specific power relates to acceleration of an electric vehicle, specific energy relates to its range. Fuel cells will be discussed later in the chapter.

batteries offer a constant voltage that can be turned on and off during discharging. Even when a supercapacitor is not connected to an external circuit, it tends to lose charge over time, a problem known as **self-discharge**. Some rechargeable batteries such as Ni–Cd also suffer from this problem; however, Li-ion batteries have an extremely low rate of self-discharge, retaining their charge over long periods of inactivity.

Supercapacitors are used in applications where a large amount of power is needed in a relatively short period of time. As such, supercapacitors have been used to deliver power for photographic flashes in digital cameras, and to provide backup power to memory cards in computers. Supercapacitors have also found applications in stabilizing the voltage of powerlines, which is especially useful for fluctuating wind or solar sources. Most recently, supercapacitors have been used alongside rechargeable batteries to improve the efficiency of hybrid electric vehicles by allowing faster acceleration from rest, while conserving battery capacity. In the next section, we will discuss these *hybrids*—vehicles that use a combination of electrical power and gasoline.

Your Turn 7.9 You Decide Supercapacitors Instead of Batteries?

This section discussed a variety of limitations of supercapacitors relative to batteries for automotive and portable electronics applications. Using a variety of sources, discuss the merit of the following statement: "New developments will bring supercapacitors into full competition with rechargeable batteries."

7.8 Higher MPGs with Less Emissions: Gasoline-Electric Hybrid Vehicles

As concerns grow about the cost and availability of gasoline and about the pollutants that gasoline-powered vehicles emit, more car shoppers are considering **hybrid electric vehicles (HEVs)**, better known as *hybrids*. These vehicles are propelled by a conventional gasoline engine in tandem with an electric motor powered by batteries. In 1999, the Honda Insight, a small two-seater, was the first sold in the United States. The Toyota Prius became available in Japan in 1997 and three years later in the United States. Today, most manufacturers, including luxury brands such as Porsche (Figure 7.21), produce at least one model of hybrid car, SUV, or truck.

Many hybrids, unlike conventional gasoline-powered cars, deliver better mileage in city driving than at highway speeds (Table 7.3 on pg. 289).

In Latin, *prius* means "to go before."

Figure 7.21

The 2016 Porsche Panamera S E-Hybrid and cutaway schematic showing the position of the drivetrain and batteries located in the trunk (**red**).

The Toyota Prius will change from nickel–metal hydride batteries to lithium-ion batteries with the 2016 model year. Most hybrid cars from other manufacturers already use lithium-ion batteries.

Kinetic energy was first introduced in Section 5.3.

In delivering about 50 miles per gallon, the Honda and Toyota hybrids burn about half the gasoline of a conventional car, thus emitting a half-equivalent of carbon dioxide. Hybrids feature gasoline engines that work together with nickel–metal hydride or lithium-ion batteries, an electric motor, and an electric generator. The electric motor draws power from the batteries to start the car moving and to power it at low speeds. Using a process called *regenerative braking*, the energy of the car's motion (kinetic energy) is transferred to the alternator, which in turn charges the batteries during deceleration and braking. The gasoline engine assists the electric motor during normal driving, with the batteries boosting power when extra acceleration is needed.

Given that the consumption of each gallon of gasoline releases about 18 pounds of CO_2 into the atmosphere, the average gasoline-powered vehicle emits from 6 to 9 tons of CO_2 each year. Unlike other vehicle emissions, such as NO_x and CO, pollution control technologies currently do not reduce CO_2 emissions. Rather, we must reduce the amount of carbon dioxide either by burning less fuel *or* by burning a fuel such as H_2 that does not contain, and therefore emit, carbon. Doing the math, each increase of 5 miles per gallon a year (for example, improving from 20 to 25 miles per gallon) can reduce CO_2 emissions by about 18 tons over a vehicle's lifetime. This calculation assumes a vehicle lifetime of 200,000 miles, which results in burning about 2,000 fewer gallons of gasoline.

Your Turn 7.10 You Decide Yes, Tons of CO_2!

Could an automobile *really* emit 7 tons of carbon dioxide in a year? Do a calculation to prove or disprove this. State all the assumptions that you make. Aside from buying a hybrid car, what are some other ways people can reduce transportation-related carbon dioxide emissions?

Automobile manufacturers now provide more than one option for fuel-efficient, low-emission vehicles, allowing customers to choose the option that best meets their transportation needs. Chevrolet introduced the first plug-in hybrid electric vehicle (PHEV) to the U.S. market in 2011. The Chevrolet Volt and other plug-in hybrid vehicles use rechargeable batteries for short daily commutes to run an electric motor and switch to a combustion engine to travel longer distances (Table 7.3). Furthermore, electrical energy provided by the battery decreases the

Table 7.3 Fuel Economy Leaders for the 2016 Model Year

Rank	Manufacturer/Model	Miles per Gallon (city/highway)
	Electric Vehicles (EVs)	
1	BMW i3 BEV	137/111
2	Chevrolet Spark EV	128/109
3	Volkswagen e-Golf	126/105
4	Nissan Leaf	126/101
5	Fiat 500e	121/103
	Plug-in Hybrid Electric Vehicles (PHEVs)	
1	BMW i3 REX	97/79
2	Chevrolet Volt	82/72
3	Hyundai Sonata	57/60
4	Toyota Prius Eco	58/53
5	Cadillac ELR	55/54
	Hybrid Electric Vehicles (HEVs)	
1	Toyota Prius	54/50
2	Volkswagen Jetta Hybrid	42/48
3	Ford Fusion Hybrid	44/41
4	Hyundai Sonata Hybrid SE	40/44
5	Lexus CT 200h	43/40

This list is taken from the top midsize cars, including plug-in hybrids and all-electric vehicles.

Source: http://www.fueleconomy.gov/feg/topten.jsp

direct emissions from the tailpipe and the amount of gasoline consumed, a major selling point.

The large lithium-ion batteries required for electric-only driving make plug-in hybrid vehicles more expensive than their gasoline-powered and gasoline-electric hybrid peers. However, nearly every major automotive manufacturer currently has research and development teams working on all-electric (EV) and PHEVs. Most of these companies agree that, in the future, plug-in electric cars could easily number in the millions on U.S. city streets, provided that battery technologies continue to improve and suitable economic incentives are developed. As of 2015, the U.S. offered a federal tax credit of $2,500 to $7,500 for the purchase of plug-in hybrids and all-electric vehicles, depending on the size of the battery. Even so, the numbers are daunting. The 170,000 PHEVs on the U.S. highways in 2015 were far outnumbered by the 250 million gasoline-powered cars already in use.

With such advantages offered by both HEVs and PHEVs, has the United States turned into a hybrid nation? The answer most certainly is no. Although the United States has the highest count of hybrid vehicles on its roads, HEVs and PHEVs represent only ~3% of all vehicles. In contrast, more than 20% of new car sales in Japan are HEVs. In the next two sections, we examine another way in which we might power our lifestyles: hydrogen fuel cells.

7.9 | Fuel Cells: The Basics

With fuel cells, we take another step on our journey to find fuels that release high amounts of energy and low amounts of pollutants. In Chapter 5, we compared the energy released, gram for gram, in the combustion of coal, hydrocarbons, and other combustible fuels. As we saw, methane was clearly the winner. Assuming the combustion products to be $CO_2(g)$ and $H_2O(g)$, the heats of combustion of coal (anthracite or bituminous) and *n*-octane (C_8H_{18}, a major component in gasoline) are 30 and 45 kJ per gram of fuel, respectively. In comparison, the heat of combustion of methane is 50 kJ/g.

However, when paired against methane, hydrogen easily wins the competition:

$$H_2(g) + \tfrac{1}{2}\,O_2(g) \longrightarrow H_2O(g) + 124.5\ \text{kJ/g} \qquad \textbf{[7.18]}$$

Therefore, hydrogen releases almost three times as much energy as methane per gram when burned! In addition to its superior energy production, using hydrogen raises another tantalizing prospect—the powering of motor vehicles with a fuel that would produce only water vapor as a product. Neither $CO(g)$ nor $CO_2(g)$ would be produced, although depending on the engine conditions and temperatures, some $NO_x(g)$ could conceivably form.

Your Turn 7.11 You Decide Hydrogen Versus Methane

Is hydrogen *really* that good of a fuel? Use the bond energy values from **Table 5.1** to find out. Clearly show how you performed your calculation, noting any assumptions you needed to make. Does the value you calculated match that of **Equation 7.18**?

Hint: You'll find most of the work done for you in Section 5.5, except the calculation there is done on a mole basis.

As with other flammable fuel sources such as methane or gasoline, when hydrogen is directly mixed with oxygen, a mere spark can set off an explosion. With its 7 million cubic feet of hydrogen gas, the Hindenburg was to airspace as the Titanic was to the high seas. When the airship caught fire in 1937 and plunged many of its passengers and crew to their deaths, hydrogen was indelibly stamped in our consciousness as an explosive fuel.

But suppose someone were to suggest a way to combine H_2 and O_2 to form H_2O without the hazards of combustion. Furthermore, let's suppose that this person also claimed that the reaction could be carried out with no direct contact between the hydrogen and the oxygen. The skeptical chemist in us might well dismiss such assertions as sheer nonsense—an outright impossibility. And yet, the operation of a fuel cell is a case in point. A **fuel cell** is an electrochemical cell that produces electricity by converting the chemical energy of a fuel directly into electricity without burning the fuel. William Grove (1811–1896), an English physicist, invented fuel cells in 1839. However, these cells remained a mere curiosity until the dawn of the Space Age. It was only in the 1980s, when the U.S. Space Shuttle *Apollo* carried three sets of 32 cells fueled with hydrogen, that fuel cells came into public view. The electricity generated by these cells powered the lights, motors, and computers on board the shuttle.

Unlike conventional batteries, such as those in flashlights, under the hood of a car, or powering your laptop computers, fuel cells operate on an external supply of fuel that is electrochemically oxidized inside the fuel cell. They also require an external supply of oxygen gas or other *oxidizing agent* to accept the electrons that are lost by the fuel. With the supply of fuel and oxidant continually being replenished, these so-called *flow batteries* produce electricity. They do not run down or need to be recharged in the same manner as conventional batteries do. Check out the location of

An *oxidizing agent* is a reactant that accepts electrons from a *reducing agent* during a reduction-oxidation ("redox") reaction. Therefore, an oxidizing agent is reduced (adds electrons), whereas a reducing agent is oxidized (releases electrons) during the reaction.

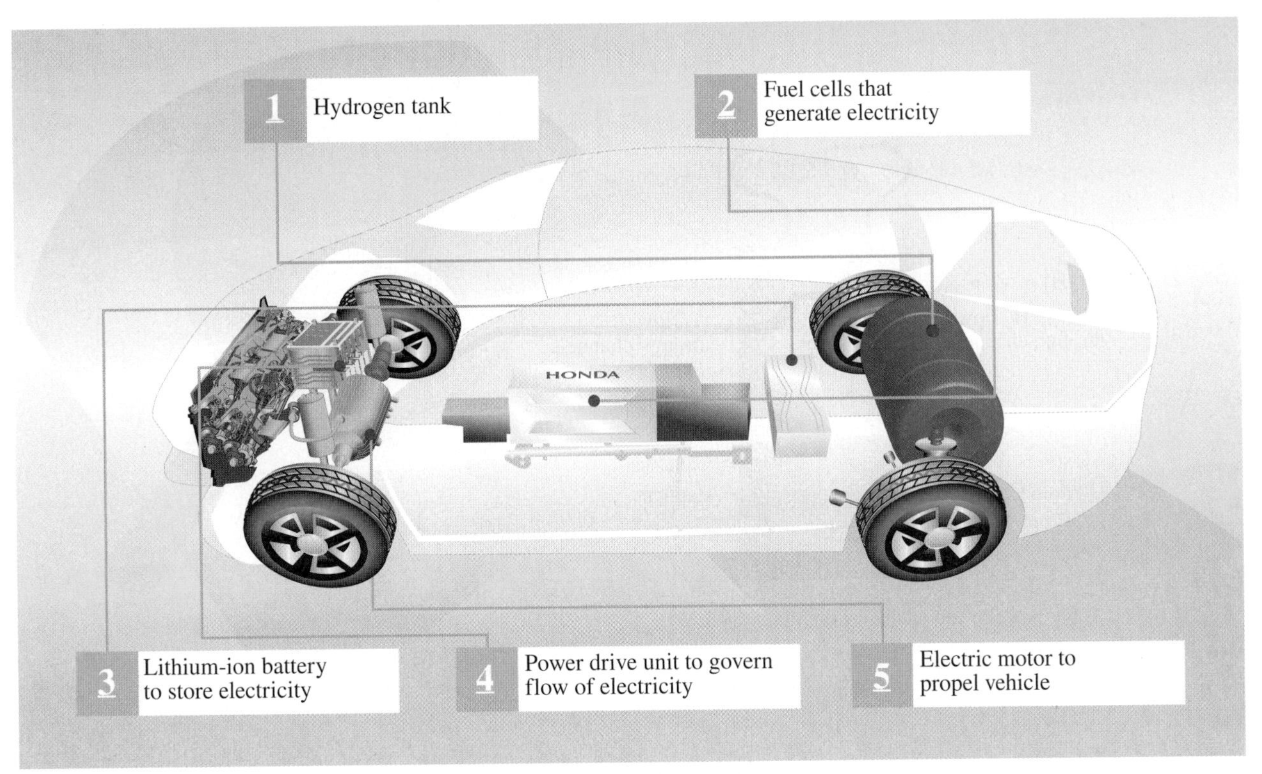

Figure 7.22

Schematic of a Honda FCX, powered by fuel cells, that shows the location of the fuel cell, hydrogen storage tank, and lithium-ion storage battery.

Source: American Honda Motor Company

the hydrogen fuel supply in Figure 7.22, a schematic drawing of a fuel cell vehicle (FCV). The hydrogen tanks are pressurized up to 5,000 pounds/inch^2 (psi), which is equivalent to 34 MPa.

What may surprise you about fuel cells is that the chemicals being oxidized and reduced are physically separated; that is, they do not come in direct contact with each other. Oxidation still occurs at the anode, and reduction at the cathode. However, instead of the anode *itself* being the source from which electrons are released, the anode is merely an electric conductor that provides a physical location in the cell at which the oxidation of the fuel takes place. Similarly, the cathode is an electric conductor where reduction of the oxygen takes place and does not enter into the reaction.

The electrolyte that separates the anode from the cathode serves the same purpose as in a traditional electrochemical cell; that is, to allow the flow of ions and hence the flow of charge. The earliest commercially available fuel cells used a corrosive acid (phosphoric acid, H_3PO_4) as an electrolyte. As a result, these fuel cells were closed systems that fully contained the liquid, not unlike the closed system of a conventional alkaline battery. Current designs of fuel cells are open systems that require a continued flow of fuel and oxidant, adding complexity and cost.

Today, fuel cells based on different electrolyte materials have been developed for a variety of applications. One type incorporates a solid polymer electrolyte separating the reactants. We use this type to explain the general operation of fuel cells. The polymer electrolyte membrane, also called a proton-exchange membrane (PEM), is permeable to H^+ ions and is coated on both sides with a platinum-based catalyst. These electrolytes operate at reasonably low temperatures, typically from 70–90 °C, and transfer electrons to rapidly provide electric power. As a result, PEM fuel cells are currently popular with automakers for new fuel cell prototype vehicles and for personal consumer applications. A typical design is shown in Figure 7.23.

H^+ is a hydrogen atom that has lost an electron, so it's just a proton and is often referred to as such.

In fuel cells, hydrogen is used as the fuel in conjunction with oxygen; the oxidation and reduction half-reactions are represented by Equations 7.19 and 7.20, respectively.

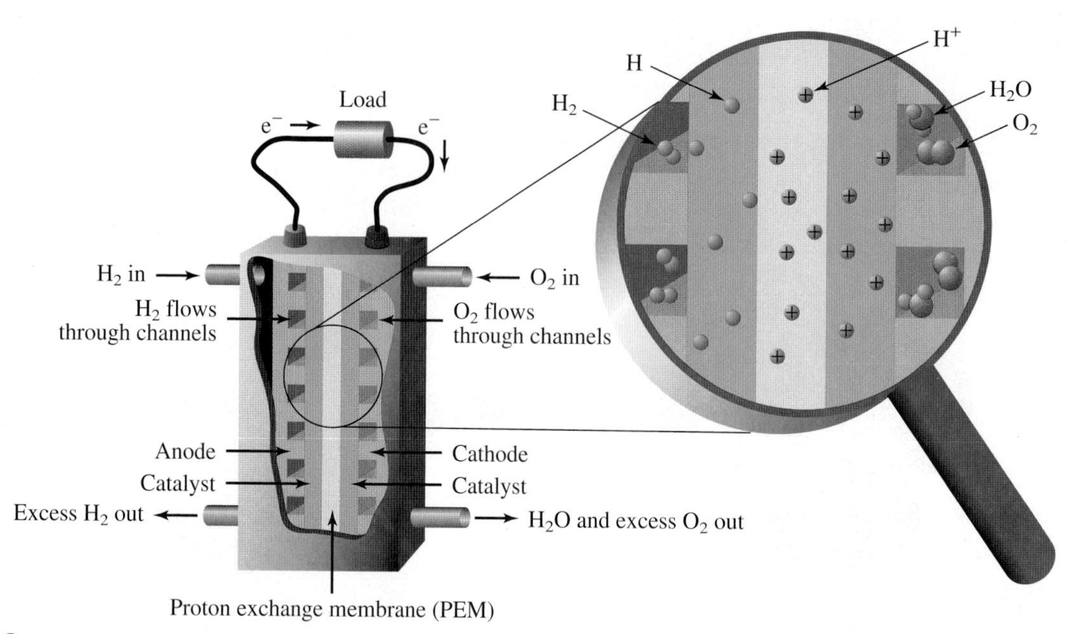

Figure 7.23

A PEM fuel cell in which H_2 and O_2 combine to form water without combustion.

As a molecule of hydrogen (H_2) passes through the membrane, it is oxidized and loses two electrons to form two hydrogen ions:

$$\text{oxidation half-reaction (anode)} \quad H_2(g) \longrightarrow 2\ H^+(aq) + 2\ e^- \qquad \textbf{[7.19]}$$

The hydrogen ions, H^+, flow through the proton exchange membrane and combine with oxygen (O_2). At the same time, they combine with two electrons to form water:

$$\text{reduction half-reaction (cathode)} \quad \tfrac{1}{2}\ O_2(g) + 2\ H^+(aq) + 2\ e^- \longrightarrow H_2O(g) \qquad \textbf{[7.20]}$$

As with galvanic cells, the overall cell equation is the sum of the two half-reactions:

$$H_2(g) + \tfrac{1}{2}\ O_2(g) + 2\ \cancel{H^+(aq)} + 2\ \cancel{e^-} \longrightarrow 2\ \cancel{H^+(aq)} + H_2O(g) + 2\ \cancel{e^-} \qquad \textbf{[7.21]}$$

In Section 5.5, we showed this chemical equation with whole-number coefficients as:

$$2\ H_2(g) + O_2(g) \longrightarrow 2\ H_2O(g)$$

The 2 e^- and 2 H^+ appearing on both sides of the arrow can be canceled:

$$H_2(g) + \tfrac{1}{2}\ O_2(g) \longrightarrow H_2O(g) \qquad \textbf{[7.22]}$$

The electrons flowing from the anode to the cathode of a fuel cell move through an external circuit to do work, which is the whole point of the device. Thus, in a fuel cell, a transfer of electrons occurs from H_2 to O_2. This occurs with no flame, with relatively little heat, and without producing any light. Because of these characteristics, the reaction is not classed as combustion. If only the energy-producing step is considered (admittedly omitting other parts of the energy picture), hydrogen fuel cells are considered a more environmentally friendly way to produce electricity than are coal-fired or nuclear power plants. No carbon-containing greenhouse gases are produced, no air pollutants are emitted, and no spent nuclear fuel needs to be disposed of. Water is the only chemical product if hydrogen is the fuel, an added benefit for the astronauts on the space shuttle, who relied on it as their source of water while in space.

The overall reaction (Equation 7.22) releases 249 kJ of energy per mole of water formed. But instead of liberating most of this energy in the form of heat, the fuel cell converts 45–55% of it to electric energy. This direct production of electricity eliminates the inefficiencies associated with using heat to do work to produce electricity. Internal combustion engines are only 20–30% efficient in deriving energy from fossil fuels. Table 7.4 shows a comparison of fuel combustion with fuel cell technology.

Table 7.4	Combustion versus Hydrogen Fuel Cell Technology			
Process	**Fuel**	**Oxidant**	**Products**	**Other Considerations**
combustion	hydrocarbons, alcohols H_2, wood, etc.	O_2 from air	H_2O, CO/CO_2, heat, light, sound	rapid process, flame present, low efficiency, useful for producing heat
hydrogen fuel cell	H_2	O_2 from air	H_2O, electricity, heat	slow process, no flame, quiet, efficient, useful for generating electricity

Your Turn 7.12 Scientific Practices Revisiting the PEM Fuel Cell

Spend time exploring the **Figures Alive!** animations of a PEM fuel cell on **Connect**.

a. How is a fuel cell different from other batteries described earlier in this chapter?
b. Can a PEM fuel cell be recharged? Explain.
c. Why is the combination of H_2 and O_2 in a fuel cell not classified as combustion? Explain.

Just as batteries, motors, and electric generators come in different sizes and types, so do fuel cells. Although the fuels and principles of operation are essentially the same, different electrolytes give each type of fuel cell unique characteristics that are appropriate for a given application. Many companies are experimenting with fuel cell vehicles. Like EVs, fuel cell vehicles (FCVs) are powered by electric motors. But they differ in that FCVs create their own electricity, whereas EVs draw electricity from an external source, storing that energy in an onboard battery.

An alternative to hydrogen gas is to use a hydrogen-rich fuel such as methanol or natural gas. These fuels must be converted into hydrogen gas by a reformer. This device uses heat, pressure, and a catalyst to run a chemical reaction that yields hydrogen as one of the products (Figure 7.24). Because they are liquids under standard conditions, methanol and ethanol could be pumped at conventional gas stations. However, the onboard reformers add cost and maintenance demands to the vehicle. Additionally, greenhouse gases and other air pollutants are generated in the reforming process.

As a source of electricity, fuel cells have a broad range of applications. Hospitals, airports, banks, police stations, and military installations all now make use of them for

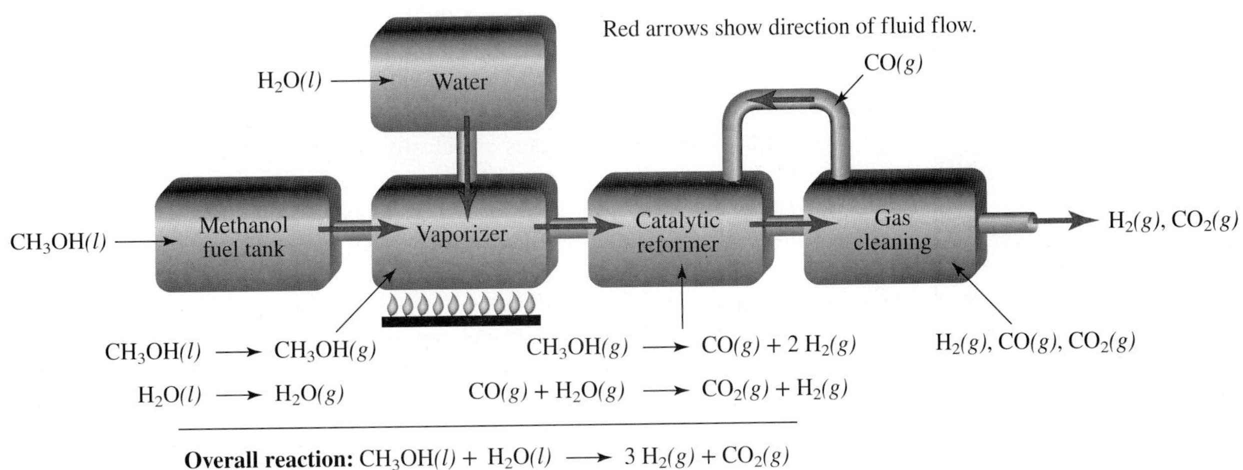

Figure 7.24
This schematic shows how hydrogen gas is obtained from methanol via a reforming process.

standby and backup-power applications. Fuel cells are a form of **distributed generation**; that is, they generate electricity on-site right where it is used, avoiding the losses of energy that occur over long electric transmission lines. As such, they serve as an alternative to central electric utility power plants. Also under development is the powering of portable electronic devices such as cell phones and laptop computers with miniature fuel cells. Such devices offer an advantage over batteries because they would not require time-consuming electric recharging, but rather could be refueled by simply swapping out or refilling a fuel cartridge.

Before our societies can fully benefit from hydrogen fuel cell technologies, scientists and engineers need to meet several technological challenges. The first is to store, transport, and eventually distribute hydrogen to the consumer. A second challenge is to produce enough hydrogen to meet the projected demand. The next section examines both of these challenges in more depth.

7.10 | Hydrogen for Fuel Cell Vehicles

Imagine needing to refuel your fuel cell vehicle with hydrogen at a "gas station." Currently, refueling stations are few and far between. As of 2015, the U.S. Department of Energy reported about 69 fueling stations that were approved for operation in the United States. A FCV can travel about 300 miles before refueling, which is certainly competitive with mileage achieved with a conventional gasoline-fueled engine.

Did You Know? By way of comparison, 12 L of gasoline has a mass of 9 kg.

The pressure, volume, and temperature of a gas are related through the *ideal gas law:* PV = nRT, where P = pressure (in atm), V = volume (in L), n = moles, R = gas constant (0.08206 L·atm/mol·K), T = temperature (in K)

Since hydrogen is a gas, it requires a different system for storage and transfer from that used for gasoline. As a gas, hydrogen also takes up a lot of space. For example, at sea level and room temperature, H_2 occupies a volume of about 11 L (almost 4 gal) per gram! In order to avoid having an enormous fuel tank, your vehicle must store hydrogen in a gas cylinder under pressure. To replenish the hydrogen in this cylinder, you must refuel with an airtight connection through a hose that can withstand high pressures, as shown in Figure 7.25. Although the refueling process requires a different system than a gasoline pump used for a gasoline-powered vehicle, the process is similar in that there is a nozzle and you squeeze a trigger to start the flow of hydrogen.

Compressing H_2 into metal cylinders results in a heavy and somewhat unwieldy tank. Hence, chemical engineers are investigating other methods for storing and transporting H_2 that could reduce space and the need for high-pressure gas compression. One promising technology is that some compounds, if subjected to hydrogen under high pressure, can absorb the hydrogen molecules like a sponge absorbs water. Then, by either decreasing the hydrogen pressure or increasing the temperature, the H_2 can be re-released on demand (Figure 7.26). For example, metal hydrides can perform in this way. Lithium hydride, LiH, is one example. A chemical formula of LiH may appear strange to you and for good reason. The problem is not the lithium ion (Li^+), as this should be an old friend by now. Rather, it is the hydride ion, H^-, a chemical species that differs markedly from the hydrogen ion, H^+. The hydride ion, with two electrons instead of one for neutral hydrogen, plays an important role in the reversible storage of hydrogen gas.

Figure 7.25

Refueling a Honda FCX Clarity, a hydrogen-powered vehicle.

© National Hydrogen Association (www.HydrogenAssociation.org)

Metal-hydride storage systems are ideally suited for PEM fuel cells that require high-purity hydrogen. Because metal hydrides are selective and absorb only hydrogen and not larger gas molecules such as CO, CO_2, or O_2, they act simultaneously as a storage material and a way of filtering out other gases. The use of these materials addresses the need for storage technologies that take up less vehicle space needed for people and cargo, while safely allowing more fuel on board for longer-range travel.

Another challenge is the projected demand for hydrogen as a fuel. Where is all the hydrogen going to come from? On one hand, things look promising because hydrogen is the most plentiful element in the universe. Over 93% of all atoms are hydrogen atoms! Although hydrogen is not nearly this abundant on Earth, there is still an immense supply of the element. On the other hand, essentially all of the hydrogen on our planet is in some form other than H_2. Hydrogen gas is too reactive to exist for long in this form, so it is primarily found in its oxidized form of H_2O, better known as

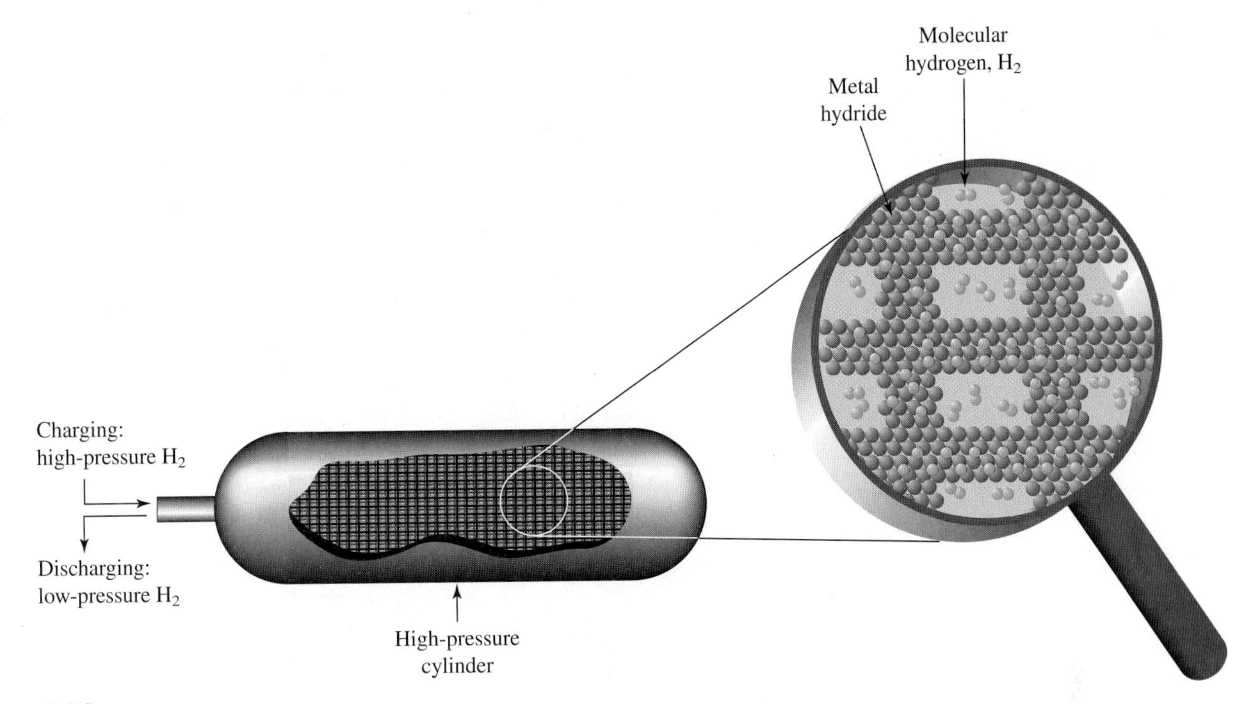

Figure 7.26

Illustration of absorption and release of hydrogen from a metal hydride.

water. Therefore, to obtain hydrogen for use as a fuel, we must form it from water or other hydrogen-containing compounds, a process that requires energy.

Fossil fuels, including natural gas and coal, are one possible source of hydrogen. In particular, methane, the major component of natural gas, currently is the chief source of hydrogen. Hydrogen can be produced from CH_4 via an endothermic reaction with steam:

$$165 \text{ kJ} + CH_4(g) + 2\ H_2O(g) \longrightarrow 4\ H_2(g) + CO_2(g) \qquad [7.23]$$

Another possible way of producing hydrogen from methane is via an endothermic reaction with carbon dioxide:

$$247 \text{ kJ} + CO_2(g) + CH_4(g) \longrightarrow 2\ H_2(g) + 2\ CO(g) \qquad [7.24]$$

You can see the downside of this reaction—it requires a significant energy input. However, companies such as the Hydrogen Energy Corporation now use a solar mirror array that can focus sunlight to heat the reactants, CO_2 and CH_4. Not only can this technology produce hydrogen, but it can also do so from a waste gas generated by a landfill.

Your Turn 7.13 Skill Building Back to Bond Energies

a. Use the average bond energy values in **Table 5.1** to check the energy required by the reactions in **Equations 7.23** and **7.24**. Show your work.

b. Did the values you calculated in part **a** match the values given in the equations? Explain.
Hint: Revisit Section 5.5.

Still, each of the reactions just described contains a major flaw; either carbon dioxide or carbon monoxide is produced. Is there another source of hydrogen around? In Jules Verne's 1874 novel, *Mysterious Island,* a shipwrecked engineer speculates about the energy resource that will be used when the world's coal supply has been used up. "Water," the engineer declares, "I believe that water will one day be employed as fuel, that hydrogen and oxygen which constitute it, used singly or together, will furnish an inexhaustible source of heat and light."

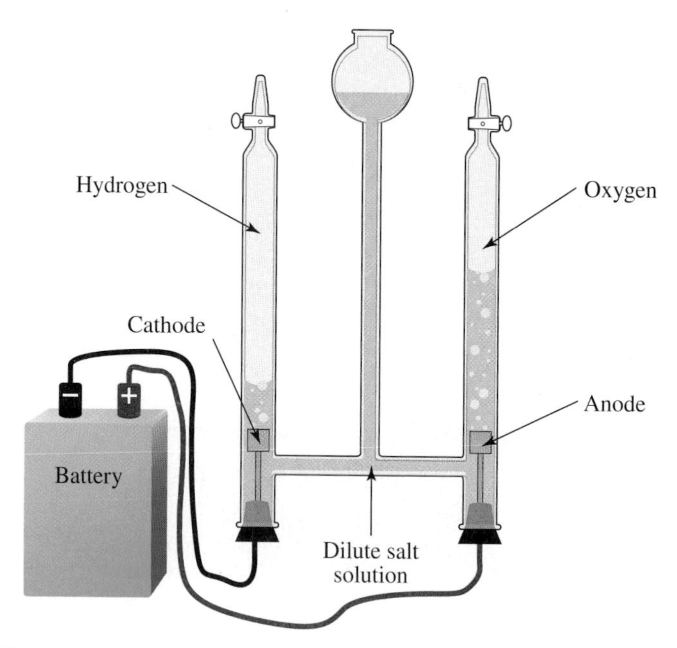

Figure 7.27

Experimental setup for the electrolysis of water.

Is this simply science fiction, or is it energetically and economically feasible to break water into its elemental components? To assess the credibility of the claim by Verne's engineer, we need to examine the energy requirements of this chemical reaction. In Section 7.9, we noted that the formation of 1 mol of water from hydrogen and oxygen releases 249 kJ of energy. An identical quantity of energy must be absorbed to reverse the reaction to produce hydrogen:

$$249 \text{ kJ} + H_2O(g) \longrightarrow H_2(g) + \tfrac{1}{2}\, O_2(g) \qquad \textbf{[7.25]}$$

The most convenient method of decomposing water into hydrogen and oxygen is by **electrolysis**, the process of passing a direct current of electricity of sufficient voltage through water to decompose it into H_2 and O_2 (Figure 7.27). This process takes place in an **electrolytic cell**, a type of electrochemical cell in which electrical energy is converted to chemical energy. An electrolytic cell is the opposite of a galvanic cell, where chemical energy is converted to electric energy. When water is electrolyzed in an electrolytic cell, the number of hydrogen molecules generated is twice that of the number of oxygen molecules, as shown in Equation 7.25. This suggests that a water molecule contains twice as many hydrogen atoms as oxygen atoms, testimony to the formula H_2O.

Water electrolysis produces no CO or CO_2 and requires about half the energy input per mole of H_2 than the methane reaction shown in Equation 7.23. From a thermodynamic point of view, it takes energy to split water into oxygen and hydrogen. Figure 7.28 shows the energy differences involved.

Of course, the question remains: How will the electricity be generated for large-scale electrolysis? Most electricity in the United States is produced by burning fossil fuels in conventional power plants. If we only had to contend with the First Law of Thermodynamics, the best we could possibly achieve would be to burn an amount of fossil fuel equal in energy content to the hydrogen produced in electrolysis. But we

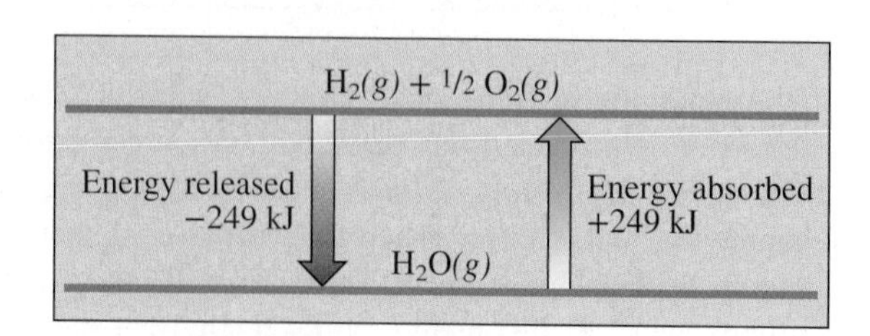

Figure 7.28

Energy differences in the hydrogen-oxygen-water system.

must also deal with the consequences of the Second Law of Thermodynamics. Because of the inherent and inescapable inefficiencies associated with transforming heat into work, the maximum possible efficiency of an electric power plant is 63%. When we add the additional energy losses caused by friction, incomplete heat transfer, and transmission over power lines, it would require at least twice as much energy to produce the hydrogen than we could obtain from its combustion. This is comparable to buying eggs for 10¢ each and selling them for 5¢, which is no way to do business.

Another way to produce hydrogen is to use heat energy to decompose water. Simply heating water to decompose it thermally into H_2 and O_2 is not commercially promising. To obtain reasonable yields of hydrogen and oxygen, temperatures of more than 5,000 °C are required. To attain such temperatures is not only extremely difficult, but also requires enormous amounts of energy—at least as much as is released when the hydrogen burns. Thus, we have again reached a point where we are investing a great deal of time, effort, money, and energy to generate a quantity of hydrogen that, at best, returns only as much energy as we invested. In practice, a good deal less energy is obtained.

Boiling water does not result in a chemical change, but rather a physical change. That is, heating water to boiling simply increases the distance among water molecules. However, when water thermally decomposes at very high temperatures, a chemical change takes place wherein water molecules break down into new molecules, H_2 and O_2.

Instead of burning fossil fuel to generate the enormous amount of heat needed to split water, another option is to use a sustainable source of energy, the radiant energy of the Sun. Photons of visible light have enough energy to split water. Unfortunately, water doesn't absorb light at these wavelengths (which is why water is colorless). New materials are being designed to use the power of the Sun to help drive the splitting of water. One type of photoelectrochemical cell, or a galvanic cell, contains a platinum (Pt) cathode and an anode covered with nanoparticles of titanium(IV) oxide (TiO_2) coated with dye molecules. The dye molecules are tuned to absorb light in the most intense part of the solar spectrum. When submerged in an aqueous electrolyte solution and exposed to light, some of the electrons in the dye are promoted to higher-energy states, high enough that they are transferred quickly to the TiO_2. Once there, the electrons can leave the electrode and move through an electric circuit (Figure 7.29).

The loss of electrons, as you have learned, corresponds to oxidation, and in this case the oxygen in water can be oxidized to O_2. After passing through the external circuit,

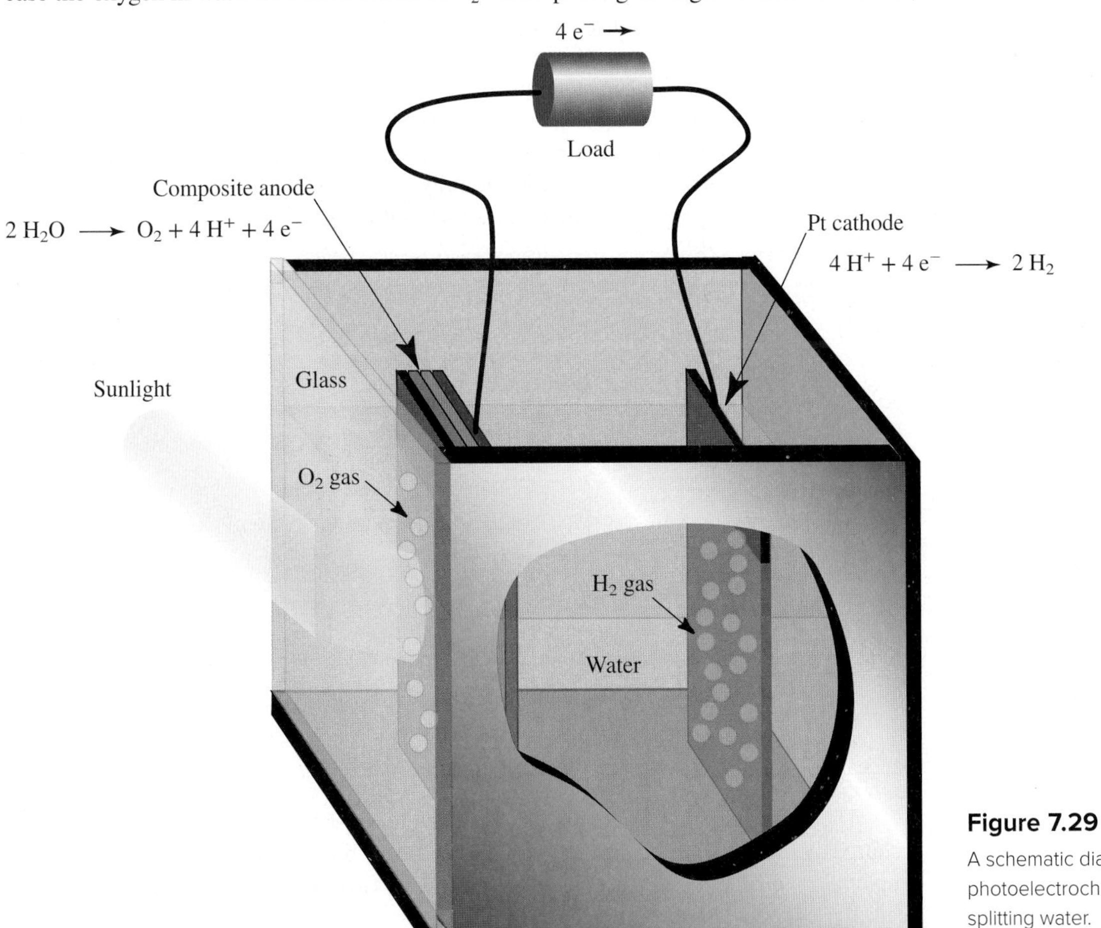

Figure 7.29

A schematic diagram of a photoelectrochemical cell for splitting water.

the electrons arrive at the platinum cathode where they reduce hydrogen ions to H_2. Efficiencies of modern devices are less than 10%, but are expected to increase.

Figure 7.30

Some strains of algae can produce hydrogen via photosynthesis.

© Michael Barnes, University of California

Your Turn 7.14 Scientific Practices Light that Splits Water

The energy required for **Equation 7.25** corresponds to a wavelength of 420 nm.

a. Which region of the electromagnetic spectrum does this fall in?
Hint: Refer to Figure 3.4.

b. It is advantageous to use light energy directly, as opposed to the heat energy of the Sun to split water. Explain.

In a vivid example of *green chemistry*, scientists are looking to biological organisms to produce hydrogen. Certain species of unicellular green algae produce hydrogen gas during photosynthesis (Figure 7.30). The advantage is that sunlight provides the energy, rather than fossil fuel combustion. At present, the efficiency of the process is far too low to be commercially viable. However, having new strains of algae that are more efficient in utilizing sunlight could tip the economic balance. A current area of research is to genetically engineer such types of algae—both a promising and a controversial area of inquiry.

7.11 | My Battery Died—Now What?

Batteries have made cell phones, tablets, laptops, and hand-held calculators so commonplace that we tend to take them for granted. Even developing countries have become increasingly reliant on battery-powered electronics. Yet the battery in your cell phone, car, or even in a solar installation costs more than what one pays for it in a store or online. There is also an environmental price tag; that is, an "external cost" that is borne by all. Part of this stems from the "ingredients" found in just about any battery, namely, one or more metals. These metals must be mined from Earth and refined from the ores in which they occur. As discussed in Chapter 1, the process of mining is extremely energy-intensive and produces mine tailings and other waste. The refining process also requires energy and produces pollutants. For example, metal refining often results in the release of sulfur dioxide because so many metals occur naturally as sulfides. The next activity gives you the opportunity to examine the details of a metal-refining process.

Your Turn 7.15 Skill Building Metal Refining (Smelting)

The nickel–metal hydride battery used in power tools and the lead–acid battery used in cars require nickel and lead, respectively. These metals are smelted from sulfur-containing ores such as NiS and PbS.

a. Name three attributes that distinguish a metal from a nonmetal.
Hint: Review Section 1.6.

b. To produce elemental nickel, oxygen gas is reacted with an ore of nickel and sulfur, represented as NiS:

$$NiS(s) + O_2(g) \longrightarrow Ni(s) + SO_2(g)$$

Is Ni in the ore oxidized or reduced?

c. Write the analogous chemical equation for lead, and again identify the species oxidized and reduced.

d. Why is the release of SO_2 a serious problem?
Hint: Revisit Chapter 2.

Smelting is the process of heating and chemically processing an ore. The smelting of sulfide ores was mentioned in the context of air quality (Section 2.13).

The environmental price tag also includes the disposal of "dead" batteries. Even rechargeable batteries eventually have to be replaced, because at some point the voltage drops below usable levels and electrons no longer flow. Although the battery may be dead, the chemicals are still hazardous. Thus, communities either must pay the cost of cleaning up the landfills where batteries are improperly disposed of, or they must pay to properly recycle the batteries.

One way to reduce battery waste is to think "cradle-to-cradle." The end of the life cycle of one item should dovetail with the beginning of the life cycle of another, so that everything is reused rather than added to the waste stream. If each battery served as the starting material for a new product, then the metals these batteries contain would not be lost to the landfill. This also is called "closed-loop recycling." The economics make sense, especially when it is cheaper to extract and reuse a metal (such as from a discarded battery) than to mine new ore and refine it. With a cradle-to-cradle approach, the item to be recycled is sent to a company that pulls out the desired metal and then returns it to another manufacturer. Unfortunately, far too few batteries are recycled in this manner worldwide.

The term cradle-to-cradle was introduced in Section 1.10.

Keeping toxic materials out of the environment also makes sense. For example, the components of an automobile battery—metallic lead, lead(IV) oxide (PbO_2), and sulfuric acid (H_2SO_4)—are toxic or corrosive. Other metals commonly used in batteries, including cadmium and mercury, are equally if not more toxic. Disposing of these batteries in landfills contaminates the land, the surface water, and ultimately the groundwater with these metals. In addition, some metals become lost to the manufacturing supply chain as they are too widely dispersed to effectively mine. The next activity explores a possible future scenario if we continue along this path.

Your Turn 7.16 You Decide Could Metals Become Extinct?

In 2015, the American Chemical Society published this "Periodic Table of Endangered Elements":

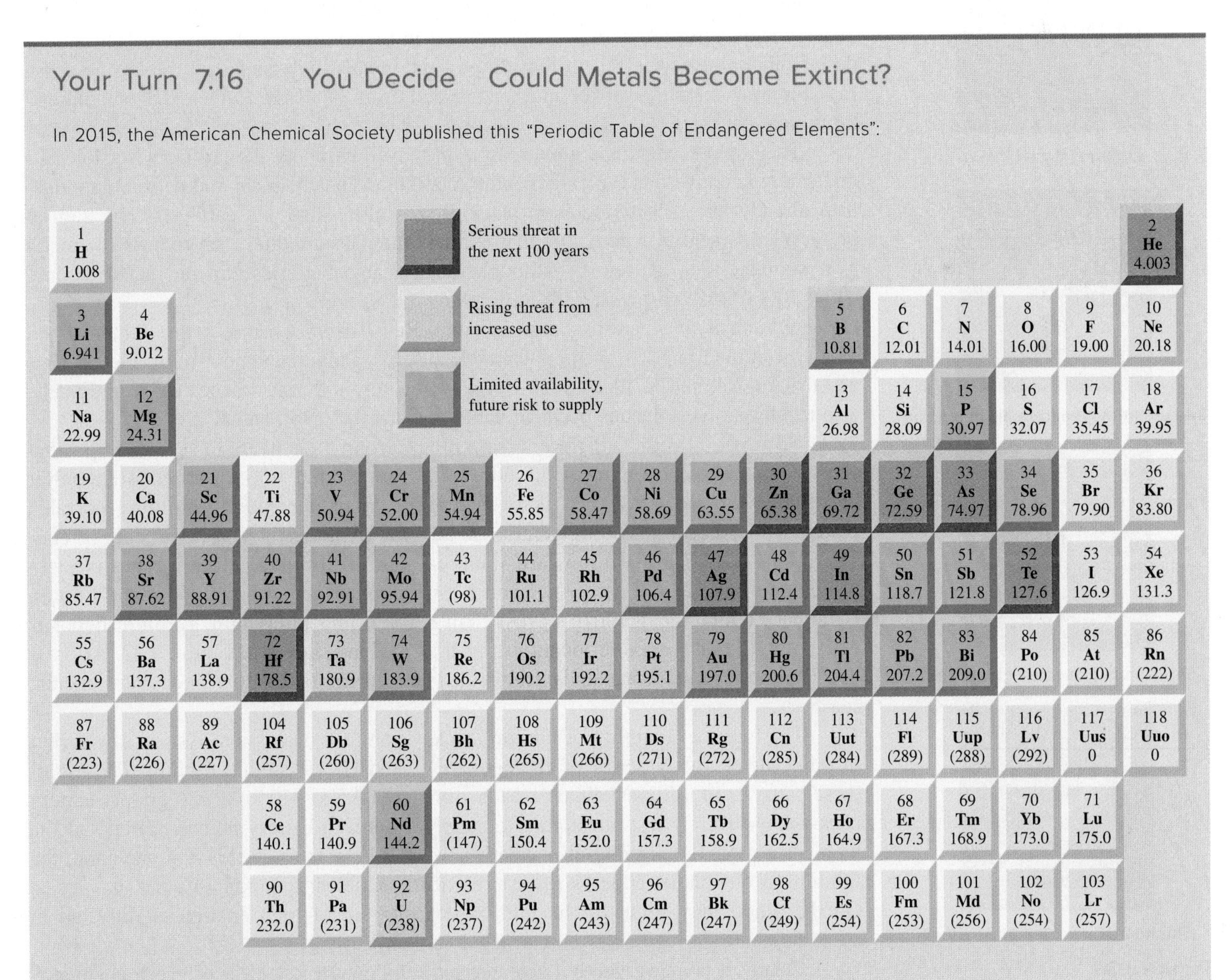

a. Is it really possible for a metal to become "extinct"?

b. Which of the highlighted elements above are currently in demand for energy storage devices (batteries, fuel cells, supercapacitors)?

c. Are there any non-highlighted elements that are currently being used, or tested for future use, in energy-storage applications?

The lead–acid battery represents one success story. Today, most state laws require retailers that sell lead–acid batteries to also collect them for recycling. The EPA reports that since 1988, more than 90% of lead–acid batteries have been recycled in the United States. The next activity enables you to learn more about battery recycling.

Your Turn 7.17 Skill Building Battery Recycling

What can you do to keep the metals used in batteries from being lost to a landfill? The answer depends on the battery type. Search the Internet to answer the following:

a. Which types of batteries are more commonly recycled: rechargeables (secondary), or non-rechargeables (primary)?
b. Why is recycling a Ni–Cd battery more critical than recycling an alkaline one?
c. List some reasons why household battery-recycling programs have not been as effective as those for recycling car batteries.

Lithium (stored in oil)

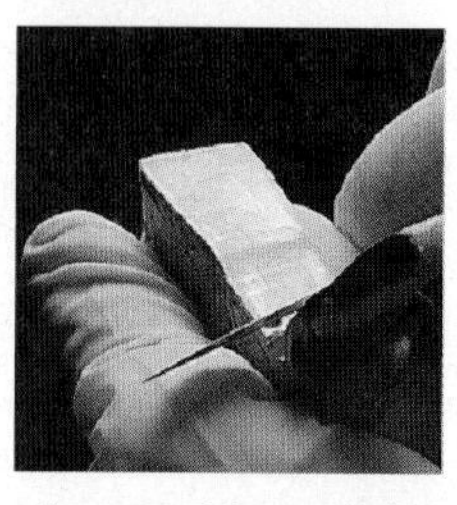
Sodium (removed from oil, being cut)

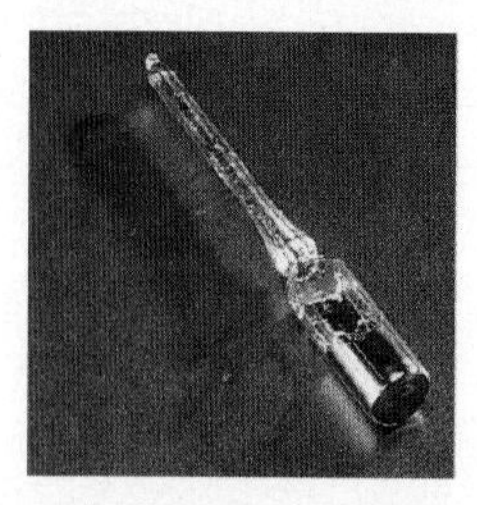
Potassium (in sealed glass tube)

Rubidium (in sealed glass tube)

Figure 7.31

Selected Group 1 elements.

In an article on the future of metals, Thomas Graedel, an industrial ecologist at Yale, points out that "metals have limits in the same way that crude oil and clean water do." If metals are to remain available for use in the future, we all need smarter battery designs that allow for efficient recycling of the metals they contain. It makes no sense for cadmium, mercury, nickel, and lead to end up in landfills. Rather, they should end up in new batteries.

There is precedent and good reason for recycling metals. Recall from Section 2.13 that catalytic converters typically contain platinum. The chemical and petroleum industries already have set protocols in place for recycling platinum catalysts. The item to be recycled is sent to a company that extracts the platinum and then returns the metal to a manufacturer for reuse. Although rechargeable batteries can be recycled, only about 10% of all cell phones are recycled.

Lithium is an interesting case in point. Recall from Chapter 1 that lithium is an alkali metal in Group 1 of the periodic table, just like sodium and potassium (Figure 7.31), and that Li, Na, and K all are highly reactive metals, with one valence electron. Finally, also recall that these metals occur in nature as ions: Li^+, Na^+, and K^+.

However, lithium is different from other Group 1 elements in several important ways. Compared to the atoms of sodium and potassium, lithium atoms are smaller and lighter. Having less mass is an advantage when it comes to building portable batteries. Furthermore, being smaller is also an advantage for specific energy considerations in that lithium ions are small enough to fit within certain types of electrode materials, in contrast to Na^+ and K^+, which may be too large. However, lithium is far less abundant in Earth's crust than sodium and potassium. Lithium deposits tend to be found in remote locations, such as the one shown in Figure 7.32. At present, lithium is mined primarily from salt lakes (brine lakes) that once were ancient sea beds.

The future availability of lithium is a key point for discussion. As you saw earlier in this chapter, the batteries of choice for EVs and HEVs contain lithium. The use of these batteries in millions of cars—each with about 4.5 kg (10 lb) of lithium per battery pack—may severely test our ability to supply lithium to battery manufacturers. At the moment, people are arguing over whether we will be able to do this. On one hand, the lithium on our planet appears to be present in sufficient quantity to meet our needs. On the other, only some of the lithium deposits are of high enough quality and in accessible enough regions to be extracted economically.

Clearly, it is in everybody's best interest to follow the key ideas of **green chemistry** as we manufacture batteries now and in the future. This means that we must minimize or avoid the use of toxic metals in batteries *and* use smart battery designs that enable metals in short supply to be efficiently recycled.

Figure 7.32

One of the world's largest lithium deposits, a salt lake in a desert area of Chile. Lithium is in the form of soluble chloride and carbonate salts, LiCl and Li_2CO_3.

Conclusions

We look to many different forms of electron transfer to meet our energy needs. Batteries can store chemical energy and convert it to a flow of electrons useful for many applications. Supercapacitors directly store electrical energy and quickly release it to a device on demand. Hybrid vehicles use new battery technologies combined with internal combustion engines to improve fuel efficiency. Fuel cells also represent efficient ways to produce electricity, and are becoming more popular for transportation and large-scale electricity production.

We hope that our discussions of energy in this book have provided you with enough background that you have gained a perspective on the complexity of energy issues that we face. We also hope you are in a position to take stock of the situation and look ahead to the future. A few facts seem beyond debate. The world's thirst for energy will not diminish; it most assuredly will continue to grow. Moreover, the ways in which we currently generate energy are not sustainable. We have been the beneficiaries of a bountiful resource base from Earth. In turn, we have an obligation to provide ample sources of energy and effective and low-cost means of energy storage for generations yet to come.

Learning Outcomes

The numbers in parentheses indicate the sections within the chapter where these outcomes were discussed.

Having studied this chapter, you should now be able to:

- define oxidation and reduction (7.1)
- define electrochemistry (7.1)
- identify and isolate half-reactions from an overall redox reaction (7.1)
- given an overall redox reaction, distinguish which chemical species is oxidized and which is reduced (7.1)
- diagram the components of a galvanic cell and outline the directions of electron and ion migration (7.1)
- define Ohm's law and use analogies to describe electron flow through an electrical circuit (7.2)
- describe the design, operation, applications, and advantages of several different types of batteries (7.3–7.5)
- relate a battery's voltage to the range of possible applications (7.6)
- outline the factors that could contribute to hazardous conditions in batteries (7.6)

- describe the function of a capacitor (7.7)
- compare and contrast supercapacitors and batteries in terms of total energy capacity and rate of energy delivery (7.7)
- identify some applications for supercapacitors and hybrid capacitors (7.7)
- classify the uses for the term "hybrid" related to energy storage (7.8)
- evaluate the principles, advantages, and challenges of producing and using hybrid vehicles (7.8)
- describe the design, operation, applications, and advantages of typical fuel cells (7.9)
- diagram the components of a proton-exchange fuel cell and outline the reactions taking place within the compartments (7.9)
- contrast galvanic and electrolytic cells (7.10)
- estimate the energy costs and gains of producing hydrogen and using it as a fuel (7.10)
- summarize the social, environmental, and economic costs of recycling energy storage devices (7.11)

Questions

Emphasizing Essentials

1. Define the terms *oxidation* and *reduction*. Why must these processes take place together?
2. Which of the following half-reactions represent oxidation and which reduction? Explain your reasoning.
   a. $Fe \longrightarrow Fe^{2+} + 2\,e^-$
   b. $Ni^{4+} + 2\,e^- \longrightarrow Ni^{2+}$
   c. $2\,Cl^- \longrightarrow Cl_2 + 2\,e^-$
3. Which chemical species gets oxidized and which gets reduced in the following overall chemical equation:

$$2\,Zn(s) + O_2(g) \longrightarrow 2\,ZnO(s)$$

4. What is the difference between a galvanic cell and a true battery? Give an example for each.
5. Two common units associated with electricity are the volt and the amp. What does each unit measure?
6. Consider the galvanic cell pictured. A coating of impure silver metal begins to appear on the surface of the silver electrode as the cell discharges.

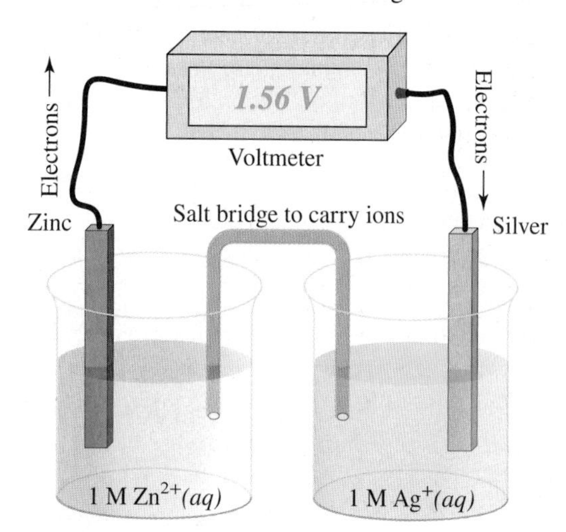

   a. Identify the anode and write the oxidation half-reaction.
   b. Identify the cathode and write the reduction half-reaction.

7. In the lithium-iodine cell, Li is oxidized to Li^+, and I_2 is reduced to $2\,I^-$.
   a. Write the oxidation half-reaction and the reduction half-reaction that take place in this cell.
   b. Write the overall reaction that occurs in the cell.
   c. Identify the half-reaction that occurs at the anode and at the cathode.
8. a. How does the voltage from a tiny AAA alkaline cell compare with that from a large D alkaline cell? Explain.
   b. Can both batteries sustain the flow of electrons for the same amount of time? Explain.
9. Identify the type of galvanic cell commonly used in each of these consumer electronic products. Assume none uses solar cells.
   a. laptop computer
   b. cell phone
   c. digital camera
   d. calculator
10. The mercury battery has been used extensively in medicine and industry. Its overall cell reaction can be represented by the following equation.

$$HgO(l) + Zn(s) \longrightarrow ZnO(s) + Hg(l)$$

   a. Write the oxidation half-reaction.
   b. Write the reduction half-reaction.
   c. Why is the mercury battery no longer in common use?
11. a. What is the function of the electrolyte in a galvanic cell?
   b. What is the electrolyte in an alkaline cell?
   c. What is the electrolyte in a lead–acid storage battery?

12. These two *incomplete* half-reactions in a lead–acid storage battery do not show the electrons lost or gained. The reactions are more complicated, but it is still possible to analyze the reactions that take place.

$$Pb(s) + SO_4^{2-}(aq) \longrightarrow PbSO_4(s)$$

$$PbO_2(s) + 4\,H^+(aq) + SO_4^{2-}(aq) \longrightarrow PbSO_4(s) + 2\,H_2O(l)$$

   a. Balance both equations with respect to charge by adding electrons as needed.
   b. Which half-reaction represents oxidation and which reduction?
   c. One of the electrodes is made of lead; the other is lead(IV) oxide. Which is the anode and which is the cathode?

13. During the conversion of $O_2(g)$ to $H_2O(l)$ in a fuel cell, the following half-reaction takes place.

$$\tfrac{1}{2}\,O_2(g) + 2\,H^+(aq) + 2\,e^- \longrightarrow H_2O(l)$$

   Does this half-reaction represent an example of oxidation or reduction? Explain.

14. How does the reaction between hydrogen and oxygen in a fuel cell differ from the combustion of hydrogen and oxygen?

15. This diagram represents the hydrogen fuel cell that was used in some of the earlier space missions.

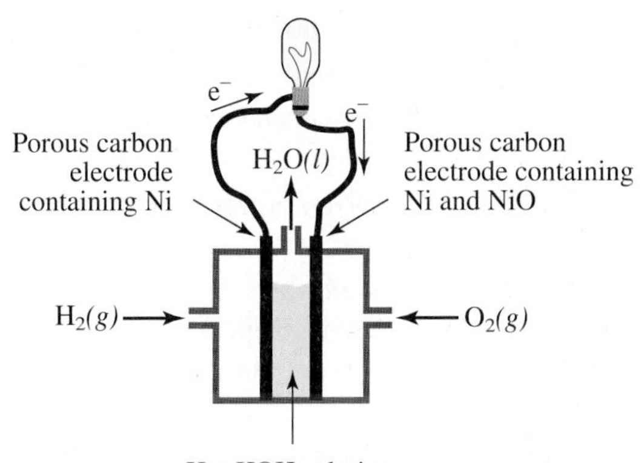

   The chemistry in the hydrogen–oxygen fuel cell can be represented by these half-reactions:

$$H_2(g) \longrightarrow 2\,H^+(aq) + 2\,e^-$$

$$\tfrac{1}{2}\,O_2(g) + 2\,H^+(aq) + 2\,e^- \longrightarrow H_2O(l)$$

   Which half-reaction takes place at the anode and which at the cathode? Explain.

16. What is a PEM fuel cell? How does it differ from the fuel cell represented in Question 15?

17. How do PEM fuel cells allow H_2 and O_2 to combine to form water without combustion?

18. In addition to hydrogen, methane also has been studied for use in PEM fuel cells. Balance the given oxidation and reduction half-reactions, and write the overall equation for a methane-based fuel cell.

   Oxidation half-reaction:

$$\_\,CH_4(g) + \_\,OH^-(aq) \longrightarrow \_\,CO_2(g) + \_\,H_2O(l) + \_\,e^-$$

   Reduction half-reaction:

$$\_\,O_2(g) + \_\,H_2O(l) + \_\,e^- \longrightarrow \_\,OH^-(aq)$$

19. Relative to a vehicle with an internal combustion engine, list two advantages offered by hydrogen FCVs.

20. Potassium and lithium both are reactive Group 1 metals. Both form hydrides, highly reactive compounds.
   a. Potassium reacts with H_2 to form potassium hydride, KH. Write the balanced chemical equation.
   b. KH reacts with water to produce H_2 and potassium hydroxide. Write the balanced chemical equation.
   c. Offer a reason why LiH (rather than KH) has been proposed as a means of storing H_2 for use in fuel cells.

21. What challenges keep hydrogen fuel cells from being a primary energy source for vehicles?

Concentrating on Concepts

22. Explain the concept of energy density of a battery and write out the energy density formula.

23. Describe how a normal AA battery stores and conducts electrons into useful energy.

24. List some differences between a rechargeable battery and one that must be discarded. Use a Ni–Cd battery and an alkaline battery as examples.

25. What is the difference between an electrolytic cell and a fuel cell? Explain, giving examples to support your answer.

26. Provide some differences between a lead–acid storage battery and a fuel cell.

27. Describe the importance of a separator in primary and secondary batteries. What would happen if the anode and cathode were allowed to touch inside a battery? Explain.

28. The company ZPower is promoting its silver–zinc batteries as replacements for lithium-ion batteries in laptops and cell phones.
   a. What advantages do silver–zinc batteries have over current Li-ion batteries?
   b. Write the oxidation and reduction half-reactions for ZPower's proposed battery using this overall cell equation as a guide. Indicate which reactant gets oxidized and which gets reduced.

$$Zn + Ag_2O \longrightarrow ZnO + 2\,Ag$$

29. The battery of a cell phone discharges when the phone is in use. A manufacturer, while testing a new "power boost" system, reported these data.

Time, min:sec	Voltage, V
0:00	6.56
1:00	6.31
2:00	6.24
3:00	6.18
4:00	6.12
5:00	6.07
6:35	6.03
8:35	6.00
11:05	5.90
13:50	5.80
16:00	5.70
16:50	5.60

a. Prepare a graph of these data.

b. The manufacturer's goal was to retain 90% of its initial voltage after 15 minutes of continuous use. Has that goal been achieved? Justify your answer using your graph.

30. Assuming that HEVs are available in your area, draw up a list of at least three questions you would ask the auto dealer before deciding to buy or lease one. Offer reasons for your choices.

31. Describe some advantages and disadvantages of HEVs. If you owned an HEV, how would this affect your lifestyle?

32. You never need to plug in Toyota's gasoline–battery hybrid car to recharge the batteries. Explain.

33. Occasionally, the power goes out. When there is no electricity for an extended period of time, are HEVs, PHEVs, and EVs affected any differently than gasoline-powered vehicles?

34. Hydrogen is considered an environmentally friendly fuel, producing only water when burned in oxygen. Name two positive effects that the widespread use of hydrogen would have on urban air quality.

35. Fuel cells were invented in 1839 but never developed into practical devices for producing electric energy until the U.S. space program in the 1960s. What advantages did fuel cells have over previous power sources?

36. Hydrogen and methane both can react with oxygen in a fuel cell. They also can be burned directly. Which has greater heat content when burned, 1.00 g of H_2 or 1.00 g of CH_4?
Hint: Write the balanced chemical equation for each reaction and use the bond energies in Table 5.1 to help answer this question.

37. Engineers have developed a prototype fuel cell that converts gasoline to hydrogen and carbon monoxide. The carbon monoxide, in contact with a catalyst, then reacts with steam to produce carbon dioxide and more hydrogen.

a. Write a set of reactions that describes this prototype fuel cell, using octane (C_8H_{18}) to represent the hydrocarbons in gasoline.

b. Speculate as to the future economic success of this prototype fuel cell.

38. Consider this representation of two water molecules in the liquid state.

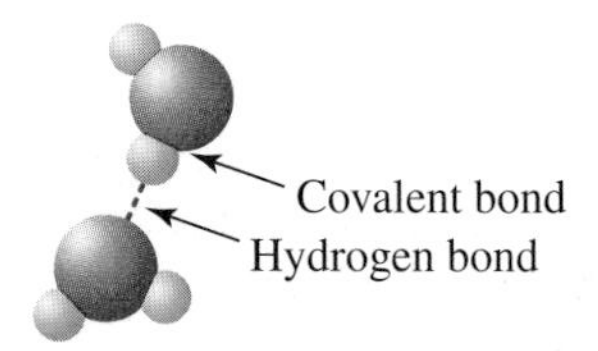

a. What happens when water boils? Does boiling break covalent bonds within molecules or does it disrupt hydrogen bonds between molecules?
Hint: Revisit Chapter 5.12.

b. What happens when water undergoes electrolysis? Does this break covalent bonds within molecules or does it disrupt hydrogen bonds between molecules?

39. Describe some similarities and differences between supercapacitors and traditional batteries.

40. After reading this chapter, which battery source do you think is the most efficient and most effective for your daily activities? Explain.

41. Why isn't the electrolysis of water the best method to produce hydrogen gas?

42. Describe three different ways scientists can extract hydrogen from different compounds found on Earth. Comment on the relative sustainability of each of these methods.

43. Small quantities of hydrogen gas can be prepared in the lab by reacting metallic sodium with water, as shown in this equation.

$$2\ Na(s) + 2\ H_2O(l) \longrightarrow H_2(g) + 2\ NaOH(aq)$$

a. Calculate the grams of sodium needed to produce 1.0 mol of hydrogen gas.

b. Calculate the grams of sodium needed to produce sufficient hydrogen to meet an American's daily energy requirement of 1.1×10^6 kJ.

c. If the price of sodium were $165/kg, what would be the cost of producing 1.0 mol of hydrogen? Assume the cost of water is negligible.

44. a. As a fuel, hydrogen has both advantages and disadvantages. Set up parallel lists for the advantages and disadvantages of using hydrogen as the fuel for transportation and for producing electricity.

b. Do you advocate the use of hydrogen as a fuel for transportation or for the production of electricity? Explain your position in a short article for your student newspaper.

45. Although Alessandro Volta is credited with the invention of the first electric battery in 1800, some feel this is a reinvention. Research the "Baghdad battery" to evaluate the merit of this claim.

Exploring Extensions

46. Oxidation and reduction also take place during combustion, the process of burning a fuel in oxygen. Because no metal electrodes are present during combustion, the electron transfer is harder to track. In this case, oxidation occurs when a chemical species loses H atoms or gains O atoms. Similarly, reduction occurs when a chemical either gains H atoms or loses O atoms.

a. Use these new definitions to determine which species is oxidized and which is reduced in the equation below, the combustion of hydrogen. Explain.

$$H_2(g) + \frac{1}{2}\, O_2(g) \longrightarrow H_2O(g)$$

b. Determine which species is oxidized and which is reduced in each of the following combustion reactions. Explain.

$$C + O_2 \longrightarrow CO_2$$

$$2\ C_8H_{18} + 17\ O_2 \longrightarrow 16\ CO + 18\ H_2O$$

47. "The Earth is a metal-rich rock. I can't see the human race running out of metals when it will be possible to mine in new places or recycle or simply reduce consumption. We probably won't be able to live on the planet due to global warming or other environmental problems before we run into a metal supply problem." These comments were offered by geologist Maurice A. Tivey of Woods Hole Oceanographic Institution in an article in *Chemical & Engineering News* published in June 2009.

a. Do you agree with the writer's sentiment about not running out of metals? Explain.

b. Name two challenges connected with increasing the recycling of batteries.

48. What is *the tragedy of the commons*? How does this concept apply to our practice of using metals such as mercury and cadmium in batteries?

49. How can the principles of green chemistry be applied during the development of new technologies for batteries and fuel cells? Give three specific examples.

50. If all of today's technology presently based on fossil fuel combustion were replaced by hydrogen fuel cells, significantly more H_2O would be released into the environment. Is this of concern? Research other consequences that might be anticipated from switching to an economy powered by hydrogen, a so-called hydrogen economy.

51. Consider these three sources of light: a candle, a battery-powered flashlight, and an electric light bulb. For each source, provide

a. the origin of the light.

b. the immediate source of the energy that appears as light.

c. the original source of the energy that appears as light.
Hint: Trace this back stepwise as far as possible.

d. the end-products and by-products produced from using each.

e. the environmental costs associated with each.

f. the advantages and disadvantages of each light source.

52. Iceland is taking bold steps to cut its ties to fossil fuels. Part of the plan is to demonstrate that the country can produce, store, and distribute hydrogen to power both public and private transportation.

a. Name three factors that motivate Iceland to cut its ties to fossil fuels.

b. What tangible outcomes have resulted to date?

c. Can lessons learned in Iceland be relevant where you live? Explain.

CHAPTER 8

Water Everywhere: A Most Precious Resource

© Tiago Fioreze

REFLECTION

Water Everywhere

Think about the water you drink and use on a daily basis. Where does this water come from, and where does the waste water eventually go? How do you think the water habits of a community can affect the natural water supply?

The Big Picture

In this chapter, you will explore the following questions:

- What are the unique properties of water?
- Where is the water we and other animals (and plants) use and drink?
- How does water interact with other chemicals?
- How do the properties of water change through its interaction with other components?
- How can we improve the quality of water?

© Maryia Bahutskaya/Getty Images

© Arnulf Husmo/Getty Images

Introduction

Calm and rough. Life and death. Thirsty and quenched. Plentiful and scarce. All of these terms can describe the most important resource for life on Earth—water. Water plays a role in nearly everything that takes place on our planet. We humans are in fact 60% water, and 71% of Earth is covered with water. Have you ever imagined a world without water? What if you were not able to take a drink of water?

Did You Know? Scientists look for water when they search for life on other planets.

Your Turn 8.1 You Decide Opposites Attract

Examine the pictures of water above. What do they tell you about water? How do they relate to the wide-ranging effects of water on our lives? After answering these questions, brainstorm a list of other opposites that can be represented by water and then answer what it would be like to have a world with no water.

Although oceans are home to a wealth of plant and animal life, they are not hospitable to the creatures that dwell on land. As Rachel Carson (1907–1964) noted in *Silent Spring,* "By far the greater part of the Earth's surface is covered by its enveloping seas, yet in the midst of this plenty we are in want. By a strange paradox, most of the Earth's abundant water is not usable for agriculture, industry, or human consumption because of its heavy load of sea salts." We who live on land need fresh water and must obtain it either through natural processes such as rain and snowfall, or though energy-intensive water purification technologies.

Scientist, conservationist, and author Rachel Carson helped launch the environmental movement in 1962 with the publication of her book *Silent Spring.*

Unfortunately, fresh water is not an unlimited resource on our planet. Furthermore, it is not renewable fast enough to meet the needs of our increasing world population. As a result, water has become a strategic resource. Its scarcity brews conflicts and raises questions of who has the right to access and use it.

Whether found in oceans, lakes, or rivers, water is a compound with unique properties. Some of these are important in understanding large-scale processes on our planet. For example, water is the only common substance that can exist as a solid, liquid, or vapor at average Earth temperatures. In its three states, water affects both the daily weather of a region and, over a longer time span, the climate.

Other properties of water help protect ecosystems. For instance, unlike most solids, ice is less dense than its liquid counterpart. Because ice floats on water, ecosystems in lakes and streams can survive beneath the ice during frigid winter days. Water also absorbs more heat per gram than most other substances, allowing bodies of water on Earth to serve as heat reservoirs. As a result, oceans and lakes change their temperatures slowly, acting to moderate temperature swings.

Still other properties of water are important for smaller-scale processes. For example, water dissolves many substances. It is the essential medium for the biochemical reactions in the cells of all living species, including humans. Your body can go weeks without food, but only days without water. If the water content in your body were reduced by 2%, you would get thirsty. With a 5% water loss, you would feel fatigue and have a headache. At a 10–15% loss, your muscles would become spastic and you would feel delirious; greater than 15% dehydration will kill you.

The amount of water you should drink daily depends on your size, age, health, and level of physical activity.

Your Turn 8.2 You Decide Do You Know Everything About Water?

Based on the previous few paragraphs, assess your knowledge of the properties of water. Which properties are you most familiar with, and which do you need to learn more about? What else would you like to learn about water?

Before we launch into the details, we ask that you first consider how water is part of your daily routine. You may sip from a tap, bottle, or can. You may steam some vegetables, wash laundry, or flush a toilet. Or you may sit by a river, casting for fish. The next activity gives you the opportunity to document how water plays a role in your life.

Your Turn 8.3 Scientific Practices Keep a Water Log

Pick a two-day period that represents typical activities for you. Record all of your activities that involve water by time and activity. Also capture:

a. The role that water played in your life. For example, are you consuming it? Are you using it in some process? Is it part of your outdoor experience?
b. The source of the water, the quantity involved, and where it went afterward.
c. The degree to which you got the water dirty.

Your Turn 8.4 You Decide Beyond Toilets?

Flushing a toilet is just one part of your daily water routine. Select a water-use calculator that is available on the Internet, and input your data from the diary you created in **Your Turn 8.3**.

a. What surprised you about your water use?
b. What actions would you consider taking based on the results of the calculator?

1 L = 1000 mL
1 gal = 3.8 L

According to the U.S. Geological Survey, more than 390 liters (~100 gallons) of water per day are required to support the lifestyle of the average U.S. citizen. As you undoubtedly discovered in your water log, we use water for many purposes.

You may remember the line "Water, water everywhere, nor any drop to drink" (from "The Rime of the Ancient Mariner"). What if we changed that to "Water, water, everywhere ... but maybe not where it used to be"? Water is an essential chemical to life. Without it, life simply does not exist on Earth. This chapter describes the unique properties of water, how chemicals behave in it, how this affects the quality of water, and how larger global issues are impacting water in lakes, streams, and oceans.

8.1 | Solids and Liquids and Gases, Oh My!

Clearly, water is essential to our lives. What may not be as apparent is that water has a number of unusual properties. In fact, these properties are quite peculiar, and we are *very* fortunate they are. If water were a more conventional compound, life as we know it could not exist.

Let's begin our investigation of water by taking a look at the three phases of matter that are present on Earth—namely solid, liquid, and gas. These all play a critical role in our daily lives, but are often taken for granted. We breathe in gases on a daily basis; the components in air were described in Chapter 2. We drink liquids regularly in the form of water, soda, coffee, etc., and also eat many solids such as candy, french fries, and potato chips. Therefore, we are inundated with these phases every day. But what are the defining principles of solids, liquids, or gases? Let's find out by examining their properties.

Imagine for a moment that you are looking at an open box just like the one in Figure 8.1. If you put matter in that container, it will either take the shape of the box or remain in its current shape. Solids will retain their shape—for instance, if you put a pencil in the box, the pencil will remain shaped as a pencil and not transform to look like the box. However, if you pour water from a glass into the box, the water will take the shape of the box, and will not remain in the same shape as the glass. This is also true for gases. Thus, solids do not take the shape of their container, whereas liquids and gases do.

Figure 8.1

An open box.

Continuing to look at the open box in Figure 8.1, some forms of matter can expand to fill the box. For instance, if we place a gold coin in the box, will it expand? Unfortunately, no—we would be rich if it did! The same goes for liquids. If we pour a glass of water in the box, it will also not expand to fill the box. If it did, we would never have a shortage of our favorite drinks, as one drop could expand to fill our glasses. However, if we place a gas in the box, it *will* expand. If you consider perfume opened at the front of a room, even the people in the back of the room will smell the perfume as it fills the container (the room). Thus, solids and liquids do not fill their containers, whereas gases do. The next activity will allow you to create a table that is useful when looking at the properties of solids, liquids, and gases.

Your Turn 8.5 Skill Building Macroscopic Properties of Solids, Liquids, and Gases

Answer the following questions for solids, liquids, and gases. Provide an example to support each of your answers.

a. Does it have a definite volume?
b. Does it have a definite shape?
c. Will it take the shape of its container?
d. Will it completely fill its container?

Based on your answers from Your Turn 8.5, we are able to create a table describing the macroscopic properties of solids, liquids, and gases:

Table 8.1 Macroscopic Properties of Solids, Liquids, and Gases

	Takes the shape of its container?	Completely fills its container?	Definite volume?	Definite shape?
Solid	No	No	Yes	Yes
Liquid	Yes	No	Yes	No
Gas	Yes	Yes	No	No

The behavior of substances can give us insight into the arrangement and attraction of their atoms and molecules. Based on Table 8.1, we can make predictions about the attractions and relative spacing among their constituent species. In general, the closer that atoms or molecules are to one another, the more attracted they are, which leads to a more rigid material, such as a solid. On average, the atoms or molecules of gases are much farther apart than in solids or liquids (Figure 8.2). In the next activity, you will use these depictions and the definitions you created for solids, liquids, and gases to create an even more descriptive model of these phases at the particulate, atomic, or molecular level.

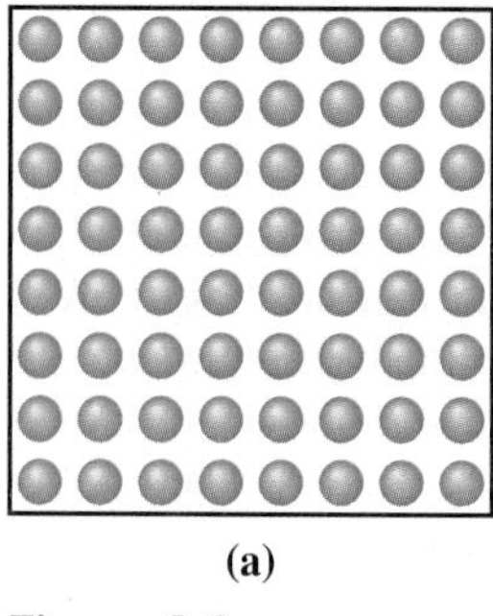
(a)

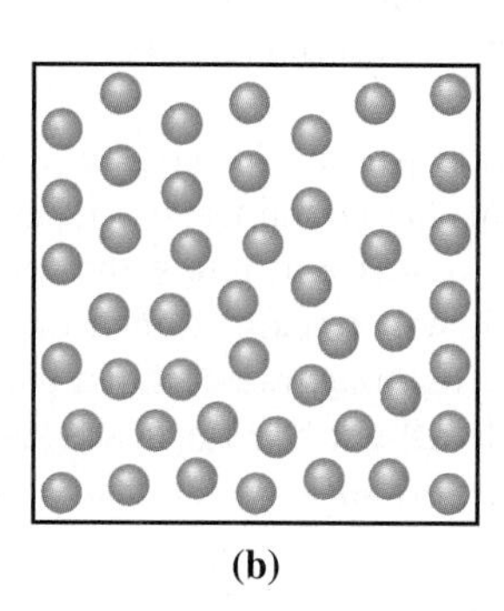
(b)

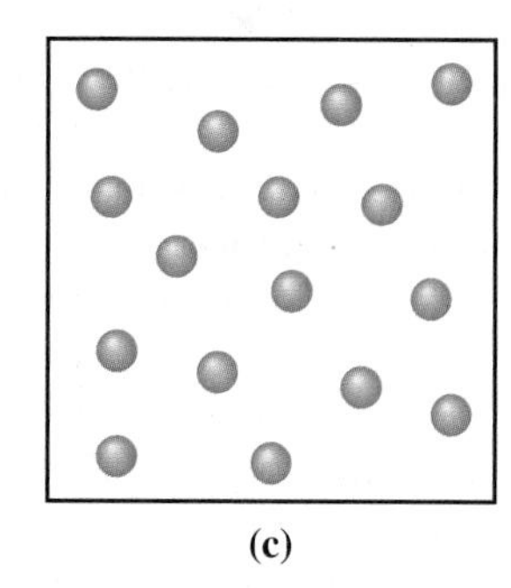
(c)

Figure 8.2

Atomic or molecular representations of **(a)** a solid, **(b)** a liquid, and **(c)** a gas.

Your Turn 8.6 Scientific Practices Creating a Model for Phases

What conclusions can you draw from **Figure 8.2**? Do these representations follow the properties outlined in **Table 8.1**? Create a representation for 15 particles of a solid, liquid, and a gas in a container of your choice.

Hint: Make sure you ensure your representation follows all the statements in Table 8.1. For example, your solid should not fill the container, nor take the shape of the container.

Scientists make models such as those you created in Your Turn 8.6 to describe phenomena. These models will be helpful as we look at the reasons why water may exist in one of three phases at varying ambient temperatures and pressures on Earth.

8.2 | The Unique Composition of Water

In the previous section, we looked at three states of matter from both a macroscopic and molecular perspective. We will now use that information to further investigate the unique and unusual properties of water. Water is a liquid under **standard temperature**

and pressure (STP); that is, a temperature of 25 °C and pressure of 1 atm. This is surprising because almost all other compounds with a similar molar mass to water are gases under these conditions. For instance, consider these three gaseous components found in air: N_2, O_2, and CO_2. Their molar masses are 28, 32, and 44 g/mol, respectively—all greater than that of water (18 g/mol).

For molecular compounds, generally as the molar mass increases, the boiling point also increases.

Water also has an anomalously high boiling point of 100 °C (212 °F); in contrast, liquids with similar molecular structures—such as hydrogen sulfide, H_2S—have much lower boiling points. Further, when water freezes, it exhibits another somewhat bizarre property—it expands. Most liquids contract when they solidify. These and other unusual properties of water are derived from its molecular structure and the interactions among the individual molecules.

First, recall the chemical formula of water, H_2O. This is probably the world's most widely known bit of chemical trivia. Next, recall that water is a covalently bonded molecule with a bent shape. Figure 8.3 shows the same representations of the water molecule that we used in Chapter 4:

In Section 4.7, you learned how to predict molecular shapes based on the Lewis structure of a molecule.

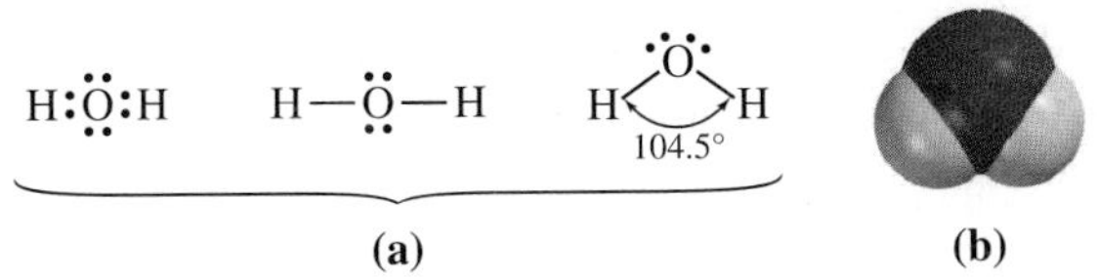

H:Ö:H   H—Ö—H   H–O–H 104.5°

(a)   (b)

Figure 8.3

Representations of water, H_2O. **(a)** Lewis structures and structural formula; **(b)** space-filling model.

Let's focus on how the electrons are shared in the O–H covalent bond. Despite drawing a line showing that the two atoms are connected, the electrons are not shared evenly. Experimental evidence indicates that the O atom attracts the shared electron pair more strongly than does the H atom. In chemical language, oxygen is said to have a higher electronegativity than hydrogen. **Electronegativity** is a measure of the attraction of an atom for an electron in a chemical bond. The values have no units and are relative measurements to one another, ranging to a maximum of 4.0 for fluorine (Table 8.2). The greater the electronegativity, the more an atom attracts the electrons in a chemical bond toward itself.

Electronegativity values were developed by the chemist, peace activist, and Nobel Prize winner, Linus Pauling (1901–1994).

Table 8.2 Electronegativity Values for Selected Elements

Group 1	2	13	14	15	16	17	18
H 2.1							He *
Li 1.0	Be 1.5	B 2.0	C 2.5	N 3.0	O 3.5	F 4.0	Ne *
Na 0.9	Mg 1.2	Al 1.5	Si 1.8	P 2.1	S 2.5	Cl 3.0	Ar *

*Noble gases rarely (if ever) bond to other elements, and therefore do not have EN values.

The greater the difference in electronegativity between two bonded atoms, the more *polar* the bond is. For example, the electronegativity difference between oxygen and hydrogen is 1.4. In contrast, this difference in a S–H bond is only 0.4—much less polar than an O–H bond. The electrons in an O–H bond are pulled closer to the more electronegative oxygen atom. This unequal sharing results in a partial negative charge (δ^-) on the O atom, and a partial positive charge (δ^+) on the H atom, as shown in Figure 8.4.

An arrow is used to indicate the direction in which the electron pair is displaced, which is often referred to as a **bond dipole**. The result is a **polar covalent bond**, a

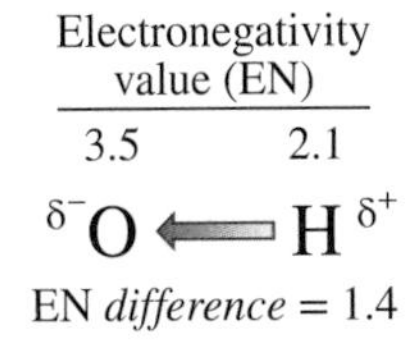

Electronegativity value (EN)

3.5   2.1

$^{\delta-}O \Longleftarrow H^{\delta+}$

EN *difference* = 1.4

Figure 8.4

Representation of the polar covalent bond between a hydrogen and oxygen atom. The electrons are pulled toward the more electronegative oxygen atom.

covalent bond in which the electrons are not equally shared but rather are closer to the more electronegative atom. In contrast, a **nonpolar covalent bond** is a covalent bond in which the electrons are shared equally or nearly equally between atoms. As an example of nonpolar covalent bonds, consider molecules existing in our atmosphere, such as N_2 and O_2. Because both atoms in the molecule are identical, there is no electronegativity difference between the atoms, which corresponds to equal sharing of their electrons.

Your Turn 8.7 Skill Building Polar Bonds

For each pair of bonds, which is more polar? In the bond you select, the electron pair is more strongly attracted to one of the atoms. Which one?

Hint: Use Table 8.2.

a. H–F or H–Cl **b.** N–H or O–H **c.** N–O or O–S **d.** H–H or Cl–C

We have made the case that bonds can be polar, some more than others. What about molecules as a whole? To help you predict if a molecule is polar, we offer two useful generalizations:

- A molecule that contains only nonpolar bonds *must be* nonpolar. For example, homonuclear diatomic elements such as O_2, N_2, Cl_2, and H_2 molecules are nonpolar.
- A molecule that contains polar covalent bonds *may or may not be* polar. The polarity depends on the shape of the molecule. For example, the water molecule contains two polar bonds, and the molecule is polar (Figure 8.5). Each H atom carries a partial positive charge (δ^+), and the oxygen atom carries a partial negative charge (δ^-). With these two polar bonds and a bent geometry, the water molecule is polar.

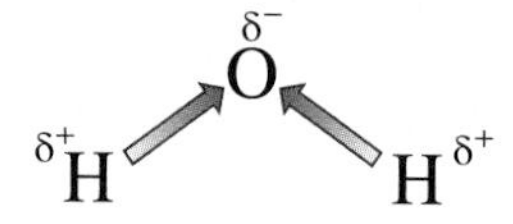

Figure 8.5

H_2O, a polar molecule with polar covalent bonds.

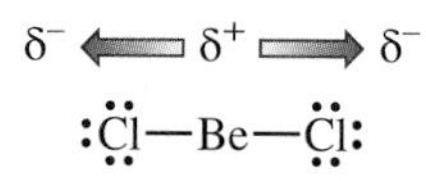

Figure 8.6

Beryllium dichloride, $BeCl_2$, a nonpolar molecule with polar covalent bonds.

You may think of the polar bonds as "tug-of-wars" between the two atoms. If the bond dipoles are not positioned 180° from one another and do not cancel, such as in water, the molecule is *polar*. However, let's now consider $BeCl_2$ (Figure 8.6). If you employ the strategy used in Chapter 4 to construct its Lewis structure and determine its shape, you will determine that the molecule is linear instead of bent. Since the "tug-of-war" between the bond dipoles cancel one another, $BeCl_2$ is a nonpolar molecule—even though it contains polar covalent bonds.

The central beryllium atom in $BeCl_2$ represents an exception to the octet rule, containing less than an octet of electrons.

Many of the unique properties of water are a consequence of its polarity. But before we continue the story of water, take a moment to complete this activity.

Your Turn 8.8 Skill Building Polarity of Molecules

Revisit the carbon dioxide molecule.

a. Draw the Lewis structure for carbon dioxide.
b. Are the covalent bonds in CO_2 polar or nonpolar? Use **Table 8.2**.
c. Similar to **Figures 8.5** and **8.6**, draw a representation for CO_2.
d. In contrast to the H_2O molecule, the CO_2 molecule is nonpolar. Explain.
e. Complete **a–c** for any molecular compound of your choice, and determine whether it's polar or nonpolar.

8.3 | The Key Role of Hydrogen Bonding

Consider what happens when two water molecules approach each other. Because opposite charges attract, a H atom (δ^+) on one of the water molecules is attracted to the O atom (δ^-) on the neighboring water molecule. This is an example of an **intermolecular force**; that is, a force that occurs *between* molecules.

With more than two water molecules, the story gets more complicated. Examine each H_2O molecule in Figure 8.7 and note the two H atoms and two nonbonding pairs of electrons on the O atom. These allow for multiple intermolecular attractions. This set of attractions among molecules is called "hydrogen bonding." A **hydrogen bond** is an electrostatic attraction between a H atom, which is bonded to a highly electronegative atom (O, N, or F), and a neighboring O, N, or F atom—either in another molecule, or in a different part of the same molecule. Do not confuse hydrogen "bonds" with covalent bonds. Typically, hydrogen bonds are only about one-tenth as strong as the covalent bonds connecting atoms *within* molecules. Also, the atoms involved in hydrogen bonding are farther apart than they are in covalent bonds. In liquid water, there may be up to four hydrogen bonds per water molecule, as shown in Figure 8.7.

Compare:

- *Intermolecular* forces are *between* molecules, whereas *intramolecular* forces are *within* a single molecule.
- *Intercollegiate* sports are played *between* colleges, whereas *intracollegiate* (or *intramural*) sports are played *within* colleges.

The relative strengths of interactions are (typically): ionic bonds > covalent bonds ≫ hydrogen bonds > London dispersion forces

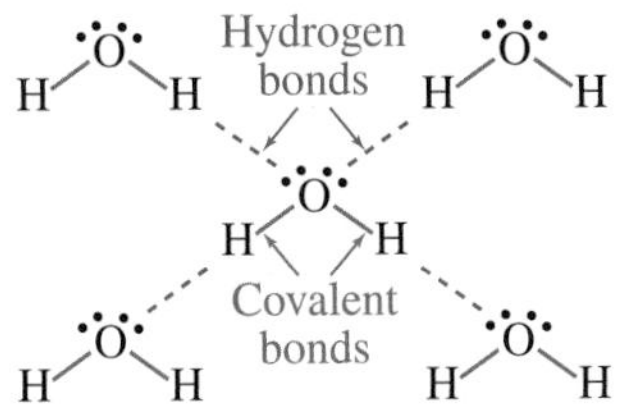

Figure 8.7

The inter- and intramolecular forces within and among water molecules (distances not to scale).

An interactive illustration of hydrogen bonding is found on **Figures Alive!** in **Connect**

Your Turn 8.9 Skill Building Hydrogen Bonding

a. Explain what the dashed lines between water molecules in **Figure 8.7** represent.

b. In the same figure, label the atoms on two adjacent water molecules with δ^+ or δ^-. How do these partial charges help explain the orientation of the molecules?

c. Illustrate hydrogen bonding in four molecules of NH_3.

Although hydrogen bonds are not as strong as covalent bonds, hydrogen bonds are still quite strong compared with other types of intermolecular forces. The boiling point of water gives us evidence for this assertion. For example, consider hydrogen sulfide, H_2S, a molecule that has the same shape as water but does not contain hydrogen bonds. Due to its relatively weak intermolecular forces, H_2S boils at about −60 °C and so is a gas at room temperature. In contrast, water boils at 100 °C. Because of hydrogen bonding, water is a liquid at room temperature, as well as at body temperature (about 37 °C). In fact, life's very existence on our planet depends on this fact!

Sulfur is less electronegative than oxygen and nitrogen. Although N–H or O–H groups can form hydrogen bonds with other molecules, S–H groups are unable to do so.

Your Turn 8.10 Scientific Practices Bonds Within and Among Water Molecules

Are covalent bonds broken when water boils? Explain with drawings.

Hint: Start with molecules of water in the liquid state, as shown in Figure 8.7. Make a second drawing to show water in the vapor phase.

Hydrogen bonding can also help you understand why ice cubes and icebergs float. Ice is composed of a regular array of water molecules in which every H_2O molecule is

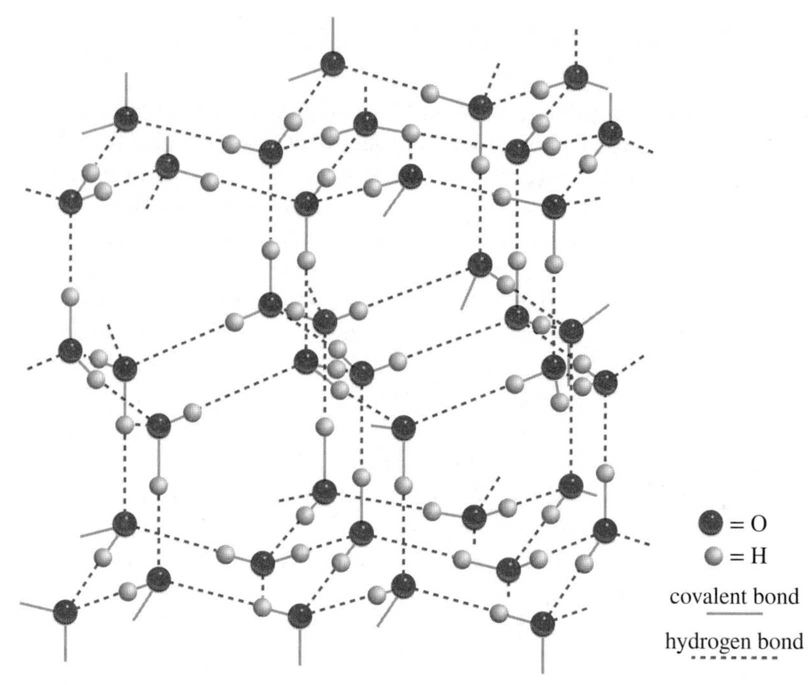

Figure 8.8

The hydrogen-bonded lattice structure of the common form of ice. Note the open channels between "layers" of water molecules that cause ice to be less dense than liquid water.

hydrogen bonded to four others (Figure 8.8). Note the empty space in the form of hexagonal channels (channels that look like a hexagon). When ice melts, the pattern is lost, and individual H_2O molecules can enter the open channels. As a result, the molecules in the liquid state are more closely packed than in the solid state. Thus, a volume of 1 cm^3 of liquid water contains more molecules than 1 cm^3 of ice. Consequently, liquid water has a greater mass per cubic centimeter than ice. This is simply another way of saying that the **density**, the mass per unit volume, of liquid water is greater than that of ice.

For any liquid at any temperature, one can assume that 1 cm^3 = 1 mL.

Did You Know? Water is most dense at 4 °C. At 0 °C, it is slightly less dense.

We usually express the mass of water in grams. Expressing its volume is a bit trickier. We use either cubic centimeters (cm^3), or milliliters (mL)—the two units are equivalent. The density of liquid water is 1.00 g/cm^3 at 4 °C, and varies only slightly with temperature. So, for convenience, we sometimes say that 1 cm^3 of water has a mass of 1 g. On the other hand, 1.00 cm^3 of ice has a mass of only 0.92 g, so its density is 0.92 g/cm^3. The bottom line? The ice cubes in your favorite beverage float rather than sink.

Unlike water, most substances are denser as solids (Figure 8.9). The fact that water shows the reverse behavior means that, in winter, ice floats on lakes rather than sinks. This topsy-turvy behavior also means that surface ice, often covered by snow, can act as an insulator and keep the lake water beneath from freezing. Aquatic plants and fish thus can live in a freshwater lake during cold winters. And when the ice melts in spring, the water that is formed sinks, which helps mix the nutrients in the freshwater ecosystem. Needless to say, water's unique behavior has implications both for the biological sciences and for life itself.

The phenomenon of hydrogen bonding is not restricted to water. It can occur in other molecules that contain covalent O–H or N–H bonds. The hydrogen bonds help stabilize the shape of large biological molecules, such as proteins and nucleic acids. For example, DNA molecules form hydrogen bonds between *different* strands of DNA. In contrast, proteins can form hydrogen bonds with different regions within the *same* molecule. Again, hydrogen bonding plays an essential role in the processes of life.

Look for more about the structures of proteins and DNA in Chapter 13.

We end this section by examining one last unusual property of water, its uncommonly high capacity to absorb and release heat. **Specific heat** is the quantity of heat energy that must be absorbed to increase the temperature of 1 gram of a substance by 1 °C. The specific heat of water is 4.18 J/g·°C. This means that 4.18 J of energy is needed to raise the temperature of 1 g of liquid water by 1 °C. Conversely, 4.18 J of

As discussed in Section 5.4, joules and calories are units of energy. The specific heat of water can also be expressed (using calories) as 1.00 cal/g·°C.

Figure 8.9

The solid phase of paraffin is denser than its liquid phase and thus sinks to the bottom of the container (left). In comparison, the solid phase of water (ice) is less dense than its liquid phase and floats to the top (right).

heat must be removed in order to cool 1 g of water by 1 °C. Water has one of the highest specific heats of any substance and is said to have a high heat capacity. Because of this, it is an exceptional coolant and can be used to carry away the excess heat in a car radiator, in a power plant, or in the human body.

Because of water's high specific heat, large bodies of water influence regional climate. Water evaporates from seas, rivers, and lakes because the bodies of water absorb heat. By absorbing vast quantities of heat, the oceans and the droplets of water in clouds help moderate global temperatures. Because water has a higher capacity to "store" heat than the ground does, when the weather turns cold, the ground cools more quickly. Water retains more heat and is able to provide more warmth for a longer time to the areas bordering it. Such properties should be familiar to anyone who has ever lived near a large body of water.

Your Turn 8.11 You Decide A Barefoot Excursion

Have you ever walked barefoot across a carpeted floor and then onto a tile or stone floor? If not, try it and see what you notice. Based on your observation, does carpet or tile have a higher heat capacity? Why?

Your Turn 8.12 Scientific Practices How Important is Water?

We have discussed several properties of water that make it unique. Create a data table to summarize these properties.

We have just examined some of the critical properties of water that influence life on our planet. Before we explore its ability to dissolve many different substances, we seek a broader picture of where water comes from, how we use it, and which issues are related to its use.

8.4 | Where, Oh Where Is All the Water?

Devastating cholera outbreaks can occur after major earthquakes such as one that happened after a 2010 earthquake in Port-au-Prince, Haiti.

Just as we need clean, unpolluted air to breathe, we also need **potable water**; that is, water safe for drinking and cooking. We also may bathe and wash dishes with potable water. In contrast, non-potable water contains contaminants that include particulates from dirt, toxic metals such as arsenic, or bacteria that cause cholera. Although not drinkable, non-potable water still has its uses. For example, water from rivers or lakes may be hauled in trucks (Figure 8.10a) and used to wash sidewalks, to reduce roadway dust, or to irrigate.

If first treated at a municipal water plant, non-potable water finds additional uses. This reclaimed water, sometimes called recycled water, is distributed to communities through "purple pipes," as shown in Figure 8.10b. It can be used to irrigate athletic fields, flush toilets, or fight fires. To keep water flowing, community water utilities match the type of water available with its best use.

(a)

(b)

Figure 8.10

(a) Water truck at the University of Alaska, Fairbanks, with a warning that the water is not fit to drink.
(b) Reclaimed water is pumped in purple pipes.

(a): © Cathy Middlecamp; (b): Reclaimed water booster pump station piping at City of Surprise, AZ. Courtesy of Malcolm Pirnie, © the Water Division of ARCADIS

Your Turn 8.13 You Decide to Function Matching Form

In some communities, reclaimed or recycled (non-potable) water is used to wash cars, water gardens, and flush toilets.

a. List three other activities for which non-potable water could be used.
b. What conditions might prompt a community to use non-potable water?
c. Does your community use reclaimed or recycled water? Find out for which purposes, if any.
d. Revisit your water diary from the beginning of the chapter. Identify areas of your personal water usage that could use non-potable water. How would this affect your water habits?

Where is fresh water found on our planet? The most convenient source for human activities is **surface water**, the fresh water found in lakes, rivers, and streams (Figure 8.11). Less convenient to access is **groundwater**, fresh water found in

Figure 8.11

Lakes and reservoirs provide much of our drinking water. This one, Hetch Hetchy, provides water to San Francisco, California.

© Cathy Middlecamp

underground reservoirs also known as *aquifers*. People worldwide pump groundwater from wells drilled deep into these underground reservoirs. Fresh water is also found in our atmosphere in the form of mists, fogs, and humidity.

Your Turn 8.14 Scientific Practices Sources of Water

a. Determine whether your drinking water is obtained through a surface-water or groundwater source. Provide a map showing where your water comes from.
b. Find an area (besides where you live) that uses surface water. Determine whether there have been any conflicts over that source.
c. Consult a map of aquifers in the United States. Which area of the country depends on groundwater the most to meet their needs? Do you think this factors into population density?

How much of the water on our planet is fresh water? Amazingly, only about 3%; the remainder is salt water. As shown in Figure 8.12, about two-thirds of this fresh water is locked up in glaciers, ice caps, and snowfields. Additionally, about 30% is found underground and must be pumped to the surface in order to use it. Lakes, rivers, and wetlands account for a mere 0.3% of the fresh water. Think of it this way. If all the water on our planet were represented by the contents of a 2-L bottle, only 60 mL of this would be fresh water. The water easily accessible to us in lakes and rivers would be about four drops!

Seawater is drinkable only if we remove its salt through a process called *desalination*, a process we will describe later.

Your Turn 8.15 You Decide A Drop to Drink

We just stated that four drops in 2 liters corresponds to the amount of fresh water available for our use. Is this accurate? Make a determination of your own.

Hint: Use the relationships in Figure 8.12, and assume 20 drops per milliliter.

Your Turn 8.16 Scientific Practices Modeling Water on Earth

We provided you with a diagram of water present on Earth in **Figure 8.12**. However, this diagram is not complete, because we did not provide a breakdown of the remaining 0.9% of fresh water into further fractions. Research the breakdown of fresh water on Earth, and create your own detailed diagram for the allocation of water on Earth.

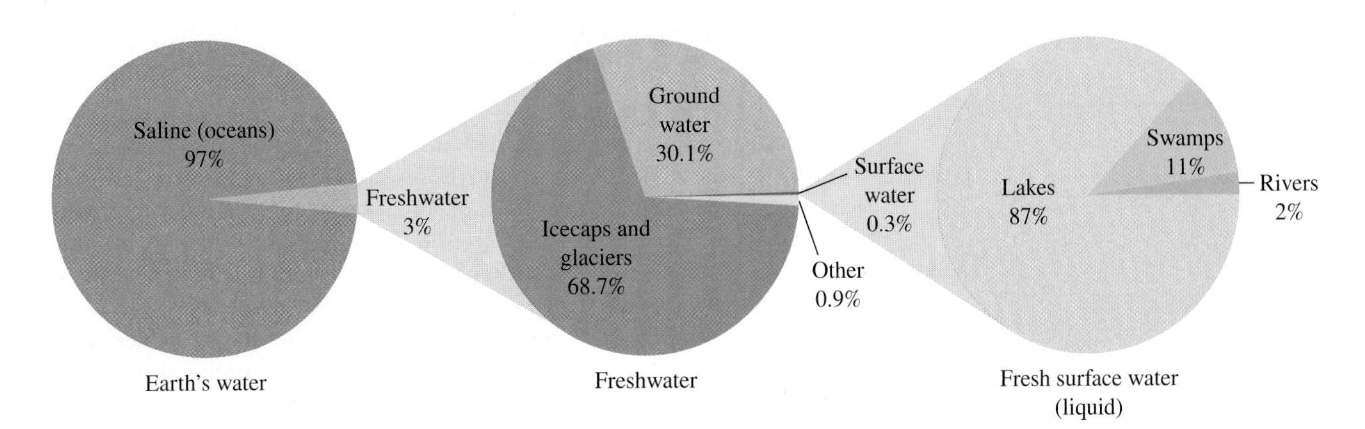

Figure 8.12
The distribution of fresh water on Earth.

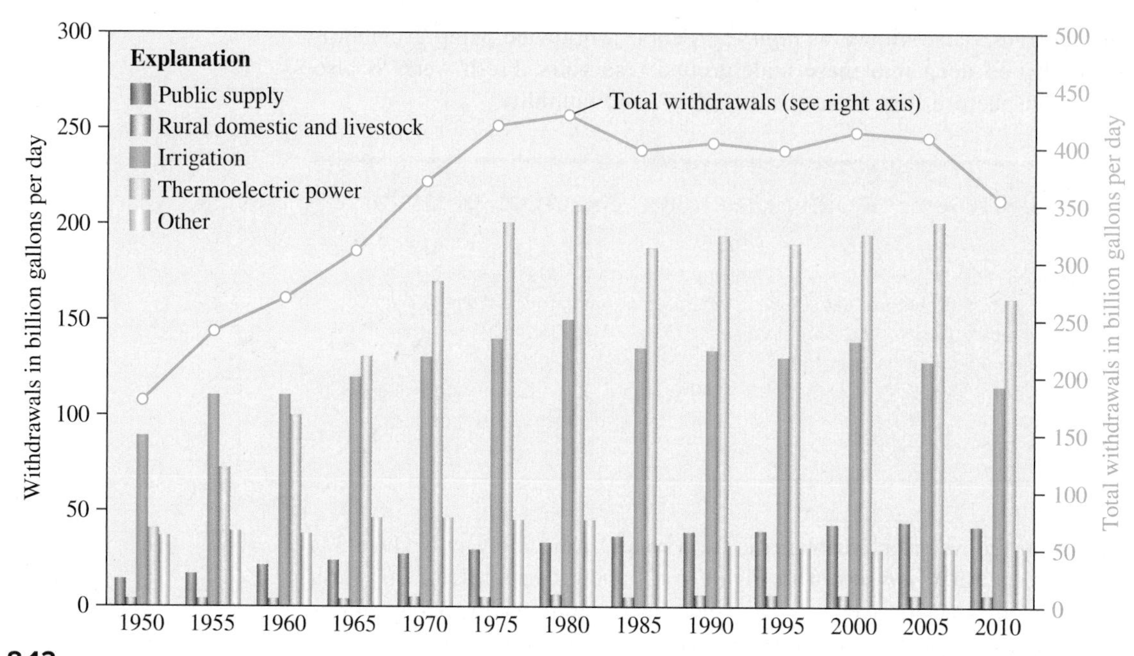

Figure 8.13

Total fresh and salt water withdrawals in the United States, 2010.

Source: Estimated Use of Water in the United States in 2010, *USGS*

How do we use water? Predictably, the answer depends on where you live. In the United States, the U.S. Geological Survey (USGS) estimates that of the 355 billion gallons of water withdrawn daily, 86% comes from fresh water and 14% from salt water. Figure 8.13 shows four primary activities responsible for this water use, with the production of electricity being the largest. About 160 billion gallons of water daily, or 45% of the total water withdrawn, is used as a coolant in electric power plants—coal, natural gas, and nuclear. The next-largest uses are for crop irrigation and for homes, schools, and businesses, accounting for another 32% and 12%, respectively.

Your Turn 8.17 Scientific Practices Water in Your Area

The United States Geological Survey provides data for individual states. Research your state on the Internet and determine how your state compares with the rest of the United States. Also, the data we have presented is from 2010. Search for more current data, and comment on some reasons for a decline in total water use from 2005–2010. Use the Internet to find out some "other" uses for water, which accounted for 50 billion gallons per day in 2010 (**Figure 8.13**).

Worldwide, agriculture accounts for about 30% of the global water consumption. Crops such as wheat, rice, corn, and soybeans are grown by farmers across the globe, each requiring several thousand liters of water, on average, in order to produce one kilogram of food. These values, reported in Table 8.3, are examples of **water footprints**; that is, estimates of the volume of fresh water used to produce particular goods or to provide services. The values in Table 8.3 are global averages. The actual value for a water footprint depends both on the country and on the particular region within the country in which the crop is grown. For example, according to the Water Footprint Network, corn grown in the United States has an average water footprint of 760 L. In comparison, the values in China and India are 1,160 L and 2,540 L, respectively. Over time, footprint values change if there are changes in rainfall or in agricultural practices.

Table 8.3 Water Footprints for Meats and Grains

Food (1 kg)	Water footprint (L, global average)
corn (maize)	1,200
wheat	1,800
soybeans	2,100
rice	2,500
chicken	4,300
pork	6000
sheep	8,700
beef	15,400

Source: Water Footprint Network, 2012

Your Turn 8.18 You Decide Differences in Water Footprints

Based on the data in **Table 8.3**, how do crops compare to meat, in terms of water usage. What are some reasons for this?

Water footprints also can be estimated for other products as well. For example, consider a 250-mL glass of cow's milk. On average, the volume of water used to produce this is 255 L—almost a thousand times the volume of one glass of milk! This includes the water needed to care for the cow and the water used to grow the food that it eats. It also includes the water used at a dairy farm to collect the milk and clean the equipment. You can check out the water needed to produce other beverages, foods, and consumer goods in Table 8.4.

Your Turn 8.19 You Decide Where Did All the Water Come From?

Choose two items from those listed in **Table 8.4** and brainstorm all the areas where water is used in the production of those items. Are there ways that the water footprint could be lowered for those items? Describe how this might be done.

Table 8.4 Water Footprints for Various Products

Product	Water footprint (L, global average)
1 cup of coffee (250 mL)	260
1 cup of tea (250 mL)	27
1 banana (200 g)	160
1 orange (150 g)	80
1 glass of orange juice (200 mL)	200
1 egg (60 g)	200
1 chocolate bar (100 g)	1,700
1 cotton T-shirt (250 g)	2,500

Source: Water Footprint Network, 2012

Water footprint values are inexact and, as a result, controversial. Our intent in providing them is not to label items as good or bad. Rather, these values are meant to remind you that water is used to produce goods, and to provide you with a more inclusive picture of water use. Large water footprints can encourage us to irrigate more efficiently and to design industrial practices that conserve water, as we'll see in the next section.

8.5 | Help! There Is Something in My Water

Figure 8.14

Some people (but not all) can take the safety of, and access to, drinking water for granted.

© Jaimie Duplass/Shutterstock.com

In some nations, water is truly a bargain right out of the faucet at home. For example, the average price for 1,000 gallons (3,800 L) of tap water in the United States is about two dollars. So inexpensive, this tap water is supplied free in drinking fountains along streets, in parks, or in public buildings (Figure 8.14).

However, what if you could not turn on a tap or buy bottled water? Some people inhabit regions where they must walk miles to reach a water source, fill a container, and carry it home (Figure 8.15a). Others, because an emergency has interrupted their usual water supply, must depend on a water truck or bottled water donations to supply their needs (Figure 8.15b).

Still others need engineers to design megastructures that move water from one region of the country to where they live. For example, aqueducts in the United States move water from the Colorado River to cities in the Southwest. Major diversions of water are often accompanied by unintended consequences, as we will discuss in a later section. Unfortunately, the water found on our planet does not always match where people need to use it. Several issues further complicate the availability of water, such as global climate change, overconsumption and inefficient use of water, and contamination. We now discuss each of these in turn.

Global Climate Change

Just as carbon cycles from place to place on our planet, so does water. For example, rain or snow falls on land and becomes part of lakes and rivers. Some of this water seeps through the soil into **aquifers**. Other water finds its way to the ocean, or is trapped for a time in snow or glaciers. Still other water evaporates and becomes the water vapor in our atmosphere. Natural processes continually recycle water on our planet.

(a)

(b)

Figure 8.15

(a) Young girls walking home with water vessels. **(b)** A member of the National Guard helps distribute bottled water to residents of Flint, Michigan, due to high levels of lead found in the drinking water supply.

(a): © Noah Seelam/AFP/Getty Images; (b): © Sarah Rice/Getty Images

(a)

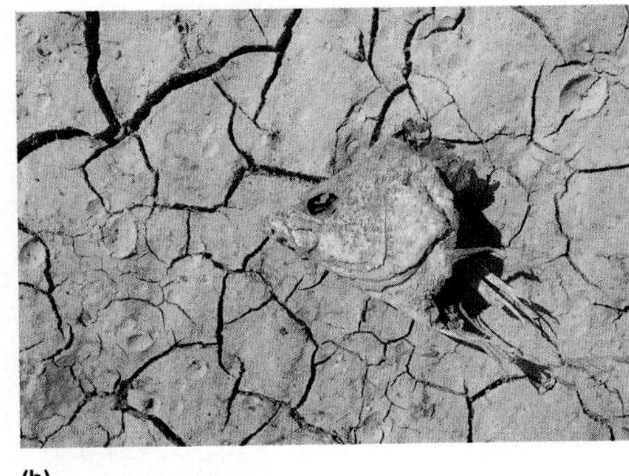
(b)

Figure 8.16

(a) Parched soils feed a dust storm approaching a town in southeastern Australia. **(b)** Shrinking waters at a dam during the Big Dry (a huge drought in Australia) left thousands of fish stranded.

(a): © AP Photo/Denis Couch; (b): © Jack Atley/Bloomberg via Getty Images

Climate plays an important role in the timing of the water cycle, and therefore the distribution of water on the planet. For example, glaciers accumulate snowpack during winter months and then release a regular stream of water during summer months. The great glaciers of the Himalayas feed seven of the largest rivers in Asia, ensuring a reliable water supply for 2 billion people—more than 25% of the world's population.

Violent storms and floods bring water in ferocious abundance, as witnessed by periodic flooding across the globe. At the other extreme, drought creates crippling water shortages (Figure 8.16). The timing of the water cycle also affects events in Earth's ecosystems. As another example, insects, birds, and plants need to appear in the right order so that the birds can feed, the insects can pollinate, and the plants can grow. If birds migrate earlier in the spring, they may arrive before enough insects have hatched for food. Conversely, if too many insects hatch before the birds are present to eat them, the insects may devastate crops. Either way, water is a key variable supporting ecosystems in which these creatures live.

Your Turn 8.20 You Decide Weather and Water

Identify a recent drought or flood that caused hardship for people and/or for an ecosystem. For an audience of your choice, write a paragraph that describes the hardship, who or what was impacted, and how some of the challenges were met.

Overconsumption and Inefficient Use

In many places, water is being pumped out of the ground faster than it is replenished by the natural water cycle. For example, much of the bountiful grain harvest from the central United States stems from using water from the High Plains Aquifer. This vast aquifer trapped water from the last Ice Age, and runs from South Dakota to Texas (Figure 8.17). It is an unsustainable practice to pump water from aquifers faster than they recharge. Continuous pumping can bring harmful outcomes as well. For example, if water is removed from a geologically unstable area near the coast, salt water may intrude into a freshwater aquifer.

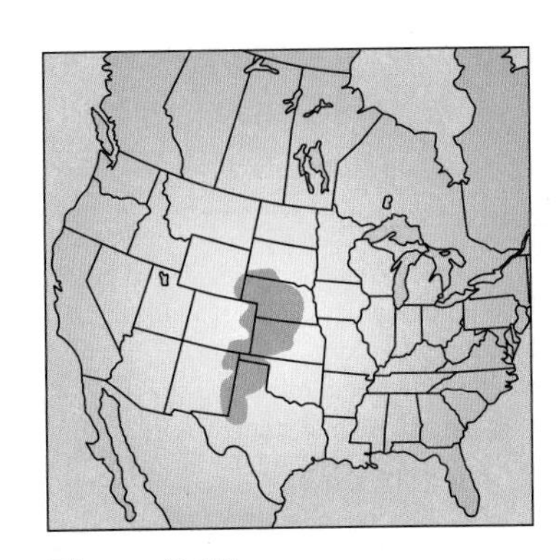

Figure 8.17

One of the world's largest aquifers, the High Plains Aquifer, is shown in dark blue on this map.

Overdrawing reserves of surface water creates problems as well. For example, consider Kazakhstan and Uzbekistan, countries that border the Aral Sea. Until recently,

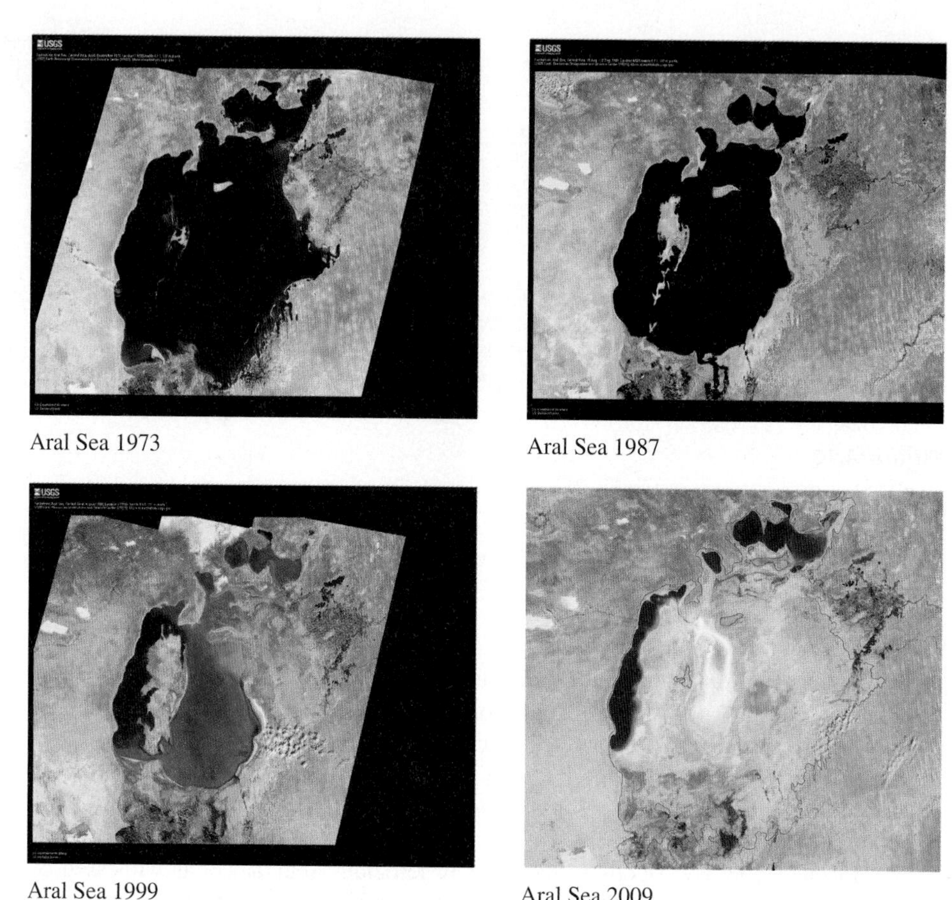

Figure 8.18

The Aral Sea has lost more than 80% of its water over a period of 30 years. The rivers that fed it were diverted to irrigate crops.

(1973, 1987, 1999) *Source: U.S. Geological Survey; (2009): Source: NASA image created by Jesse Allen*

this sea was the world's fourth-largest inland body of fresh water. In the 1960s, workers in the former Soviet Union built a network of canals that diverted this water from the rivers that fed the Aral Sea in order to grow cotton in the arid climate. Not only were the rivers feeding the Aral Sea diverted, but also the water was used inefficiently. For example, the water for cotton irrigation was transported in open canals, resulting in loss through evaporation.

Consequently, the Aral Sea dried up, as shown in Figure 8.18. Although the ecosystem once was rich as a fishery, today only a few salty pools of water remain. The United Nations has called this the greatest environmental disaster of the 20th century. Dust that is laden with toxins, pesticides, and salt now blows in the region, causing health problems and contributing to poverty.

Practices that can conserve water include using efficient ways to irrigate fields, replacing grass lawns with native vegetation, and repairing leaky pipes in aging water distribution systems.

Such water diversion stories present us with examples of the **tragedy of the commons** (revisit Section 2.14). The water from the aquifers and surface water is the resource used in common, yet no one in particular is responsible for its use. If water is overdrawn for agriculture or some other purpose, this act can be to the detriment of all who depend on this common and necessary resource.

Your Turn 8.21 You Decide Water Misuse Around You

Perform a search for evidence of water misusage in your area. How was the misuse handled, and were guidelines set to prevent misuse in the future?

Contamination

We expect access to water that is safe; that is, devoid of harmful chemicals and microbes. However, a 2010 joint report released by the World Health Organization/United Nations Children's Fund (WHO/UNICEF) indicated that almost a billion people, primarily in developing nations, lacked safe drinking water. Each day, more than 3,000 infants and children die because of contaminated water, sometimes indicated by its appearance (Figure 8.19), but other times not.

What makes water safe to drink? "Safe" water doesn't mean it is pure—it will have substances dissolved in it. Many of these substances are part of the world's natural systems. For example, beneficial minerals found in groundwater contribute calcium and magnesium ions to the water. However, other natural substances can be harmful. The U.S. Environmental Protection Agency (EPA) defines a water contaminant as anything physical, chemical, biological, or radioactive that is harmful to human health or degrades the taste or color of the water. The EPA regulates more than 90 substances known to contaminate drinking water.

Figure 8.19

People kayak in the Animas River near Durango, Colorado, in water colored from a mine waste spill in July 2015. Although this water source is obviously contaminated, clear water may actually be just as harmful to human health and the environment.

© Jerry McBride/The Durango Herald via AP

The recent water crisis in Flint, Michigan, has renewed our focus on water contamination, teaching us to be more cautious of the water sources we take for granted. Whereas the EPA guideline for lead in drinking water is 15 ppm, several water samples taken from homes in the Flint area contained Pb levels higher than 100 ppm. Unfortunately, this contamination may create lifelong problems for children exposed to high levels of lead in drinking water, including low scholastic performance. Such a preventable situation shows us the importance of regular water monitoring, and ensuring that testing is done in an ethical and transparent manner.

Your Turn 8.22 Scientific Practices Water Contamination

Unfortunately, contaminated water is not an infrequent event. Research an incident where water was contaminated and led to health concerns. Write a paragraph reporting on the incident and what has been done to remedy the situation.

Not all contaminants found in water are monitored or regulated. For example, personal care products such as cosmetics, lotions, and fragrances contribute thousands of chemicals to wastewater. In addition, trace amounts of pharmaceutical drugs end up in our wastewater stream and quite possibly in our drinking water as well. At present, we are in only the early stages of understanding the effects of these substances in our water. The next activity will help you assess your use of personal care products.

Your Turn 8.23 You Decide "To Clean" or "To Dirty"?

We use personal care products with a goal in mind. For example, we use shampoo to clean hair, pat on shaving lotion to refresh the skin, or apply hand lotion to soften it. However, after the product has done its job, what happens to it?

a. List several personal care products that you use daily.

b. Suggest several routes by which these personal care products could end up in water.

c. Revisit your list from part **a.** How might you apply green chemistry key ideas in your daily use of personal care products? For example, would using less shampoo still be effective in cleaning your hair?

d. Revisit your water diary. In which cases do personal care products affect your water use?

We hope this section has increased your awareness of how water is used, misused, and sometimes picks up contaminants—either of natural or human origin. This last point warrants our closer attention. What is it about water that allows contamination to happen so easily? In the next section, we will turn to topics that help us better understand why water is able to dissolve and mix with so many substances.

Your Turn 8.24 Scientific Practices Water Issues and Your Diary

Revisit your water diary from the beginning of this chapter. How would climate change issues, overconsumption, inefficient use, and contamination affect how you currently use water? Point out the areas that would most be affected by these factors.

8.6 How Much Is OK? Quantifying Water Quality

Water dissolves a remarkable variety of substances. As we will see, some of them, including salt, sugar, ethanol, and the air pollutant SO_2, are *very* soluble in water. In comparison, limestone rock, oxygen, and carbon dioxide dissolve only in tiny amounts. To build your understanding about water quality, you need to know *what* dissolves in water, *why* it dissolves, and *how* to specify the concentration of the resulting solution. This section tackles solution concentrations; the section that follows addresses solubility.

Let's begin with some useful chemical terminology. Water is a **solvent**—a substance, often a liquid, that is capable of dissolving one or more pure substances. The solid, liquid, or gas that dissolves in a solvent is called the **solute**. The result is called a **solution**—a homogeneous (of uniform composition) mixture of a solvent and one or more solutes. In this section, we are particularly interested in **aqueous solutions**, solutions in which water is the solvent.

Because water is such a good solvent, it practically never is "100% pure." Rather, it contains impurities. For example, when water flows over the rocks and minerals of our planet, it dissolves tiny amounts of the substances that they contain. Although this usually causes no harm to our drinking water, occasionally the ions dissolved in water are toxic. The water on our planet also comes in contact with air. When it does, it dissolves tiny amounts of the gases in the air, most notably oxygen and carbon dioxide. Some air pollutants are *very* soluble in water. So when it rains, the water actually cleans some of the pollutants out of the air, including SO_2 and NO_x, resulting in *acid rain*. As we will see later in this chapter, the acidic solutions that form can have serious consequences for the environment.

The reactions responsible for acid rain were introduced in Chapter 2.

Humans also contribute to the number of substances dissolved in water. When we wash clothes, we add not only the spent detergent, but also whatever made our clothing dirty in the first place. When we flush a toilet, we add liquid and solid wastes. Our urban streets add solutes to rainwater during the process of storm run-off. And our agricultural practices add fertilizers and other soluble compounds to water.

What does water's property of being a good solvent mean for our drinking water? In order to assess water quality, you need to know several things. One is a way to specify *how much* of a substance has dissolved, so that you can compare the value with a known standard. In other words, you need to understand the concept of concentration. This was first introduced in Chapter 2 in relation to the composition of air. For example, O_2 and N_2 are about 21% and 78% of dry air, respectively. Now we examine this concept in terms of substances dissolved in water. As we will see, percent and parts per million are valid ways of expressing concentrations for aqueous solutions as well.

To get started with solution concentrations, let's use a familiar analogy—sweetening a cup of tea (Figure 8.20). If 1 teaspoon of sugar is dissolved in a cup of tea, the resulting solution has a concentration of 1 teaspoon per cup. Note that you would have this same concentration if you were to dissolve 3 teaspoons of sugar in 3 cups of tea, or half a teaspoon in half a cup of tea. If your recipe is tripled or halved, the sugar and tea are adjusted proportionally. Therefore, the **concentration**—the ratio of the amount of solute to the amount of solution—is the same in each case.

Figure 8.20

Sweetening a cup of tea.

© Neil Rutledge/Alamy Stock Photo

Solute concentrations in aqueous solutions follow the same pattern but are expressed with different units—percent (%), parts per million (ppm), parts per billion (ppb), and molarity (M). Three of these should already be familiar to you. The fourth, molarity, uses the concept of moles introduced in Section 4.4.

Percent (%) means parts per hundred. For example, an aqueous solution containing 0.9 g of sodium chloride (NaCl) in 100 g of solution is a 0.9% solution by mass. This concentration of sodium chloride is referred to as "normal saline" in medical settings when given intravenously. You may find the antiseptic isopropyl alcohol in your medicine cabinet as a 70% aqueous solution by volume. It contains 70 mL of isopropyl alcohol in every 100 mL of aqueous solution. Percent is used to express the concentration of a wide range of solutions.

For solutions with low concentrations, the mass of the solution is approximately the mass of the solvent.

But when the concentration is very low, as is the case for many substances dissolved in drinking water, **parts per million (ppm)** is more commonly used. For example, water that contains 1 ppm of calcium ions contains the equivalent of 1 gram of calcium (in the form of the calcium ion) dissolved in 1 million grams of water. The water we drink contains substances naturally present in the parts per million range. For example, the acceptable limit for nitrate ions, NO_3^-, found in well water in some agricultural areas, is 10 ppm; the limit for fluoride ions, F^-, is 4 ppm.

Although parts per million is a useful concentration unit, measuring 1 million grams of water is not very convenient. We can do things more easily by switching to the unit of a liter. One ppm of any substance in water is equivalent to 1 mg of that substance dissolved in a liter of solution. Here is the math:

$$1 \text{ ppm} = \frac{1 \cancel{\text{ g solute}}}{1 \times 10^6 \cancel{\text{ g water}}} \times \frac{1000 \text{ mg solute}}{1 \cancel{\text{ g solute}}} \times \frac{1000 \cancel{\text{ g water}}}{1 \text{ L water}} = \frac{1 \text{ mg solute}}{1 \text{ L water}}$$

1000 grams (1×10^3 g) of H_2O can be taken to have a volume of 1 L. However, strictly speaking, this is true only at 4 °C.

Municipal water utilities may use the unit mg/L to report the minerals and other substances dissolved in tap water. For example, Table 8.5 shows a tap water analysis from an aquifer that supplies a Midwestern community in the United States.

Some contaminants are of concern at concentrations much lower than parts per million, and are reported as **parts per billion (ppb)**. In aqueous solutions, 1 ppb = 1 μg/L, whereas 1 ppm = 1 mg/L. Another way to think about ppm and ppb is to assume that 1 ppm corresponds to 1 second in nearly 12 days. Then, 1 ppb corresponds to 1 second in 33 years. Yet another way of looking at these units is that one part per billion corresponds to a few centimeters on the circumference of Earth—very small indeed!

Table 8.5 Tap Water Mineral Report

Cation	mg/L	Anion	mg/L
Calcium ion (Ca^{2+})	97	Sulfate ion (SO_4^{2-})	45
Magnesium ion (Mg^{2+})	51	Chloride ion (Cl^-)	75
Sodium ion (Na^+)	27	Nitrate ion (NO_3^-)	4
		Fluoride ion (F^-)	1

Your Turn 8.25 Scientific Practices Concentration Analogies

We have presented a couple of analogies for ppm and ppb. Devise two analogies to describe percent, parts per million, and parts per billion. For each analogy, choose an audience that you could share the analogy with in order to describe concentration.

Mercury in water is present in a soluble form (Hg^{2+}) rather than as elemental Hg ("quicksilver").

One contaminant found in the range of parts per billion is mercury. For humans, the primary source of exposure to mercury is food, mainly fish and fish products. Even so, the concentration of mercury in water needs to be monitored. One part per billion of mercury (Hg) in water is equivalent to 1 gram of Hg dissolved in 1 billion grams of water. In more convenient terms, this means 1 microgram (1 μg or 1×10^{-6} g) of Hg dissolved in 1 liter of water. The U.S. acceptable limit for mercury in drinking water is 2 ppb:

$$2 \text{ ppb Hg} = \frac{2 \text{ g Hg}}{1 \times 10^9 \text{ g H}_2\text{O}} \times \frac{1 \times 10^6 \text{ μg Hg}}{1 \text{ g Hg}} \times \frac{1000 \text{ g H}_2\text{O}}{1 \text{ L H}_2\text{O}} = \frac{2 \text{ μg Hg}}{1 \text{ L H}_2\text{O}}$$

Confirm that the units cancel.

Your Turn 8.26 Skill Building Mercury Ion Concentrations

a. A 5-L sample of water contains 80 μg of dissolved mercury ion. Express the mercury ion concentration of the solution in ppm and ppb.

b. Would your answer in part **a** be in compliance with the U.S. acceptable limit for drinking water? Explain.

Your Turn 8.27 Scientific Practices Concentration Data

Search the Internet for the current acceptable metal ion concentrations for water in the United States. Compare this with data of acceptable levels for the same metal ions in your area. Are the limits the same? Why or why not?

Molarity (M), another useful concentration unit, is defined as a unit of concentration represented by the number of moles of solute present in 1 liter of solution:

$$\text{Molarity (M)} = \frac{\text{moles of solute}}{\text{liter of solution}}$$

The great advantage of molarity is that solutions of the same molarity contain exactly the same number of moles of solute, and hence the same number of molecules (ions or atoms) of solute. The mass of a solute varies depending on its identity. For example, 1 mole of sugar has a different mass than 1 mole of sodium chloride. But if you take the same volume, all 1 M solutions (read as "one molar") contain the same number of solute molecules.

The molar mass of NaCl (58.5 g/mol) is calculated by adding the molar mass of sodium (23.0 g/mol) plus the molar mass of chlorine (35.5 g/mol).

As an example, consider a solution of NaCl in water. The molar mass of NaCl is 58.5 g/mol; therefore, 1 mol of NaCl has a mass of 58.5 g. By dissolving 58.5 g of NaCl in some water and then adding enough water to make exactly 1.00 L of solution, we would have a 1.00 M NaCl aqueous solution.

(aq) is short for aqueous, indicating that the solvent is water.

Figure 8.21 shows the preparation of a 1.00 M solution of sodium chloride. Note the use of a **volumetric flask**, a type of glassware that contains a precise volume of solution when filled to the mark on its neck. But because concentrations are simply ratios of solute to solvent, there are many ways to make a 1.00 M NaCl*(aq)* solution.

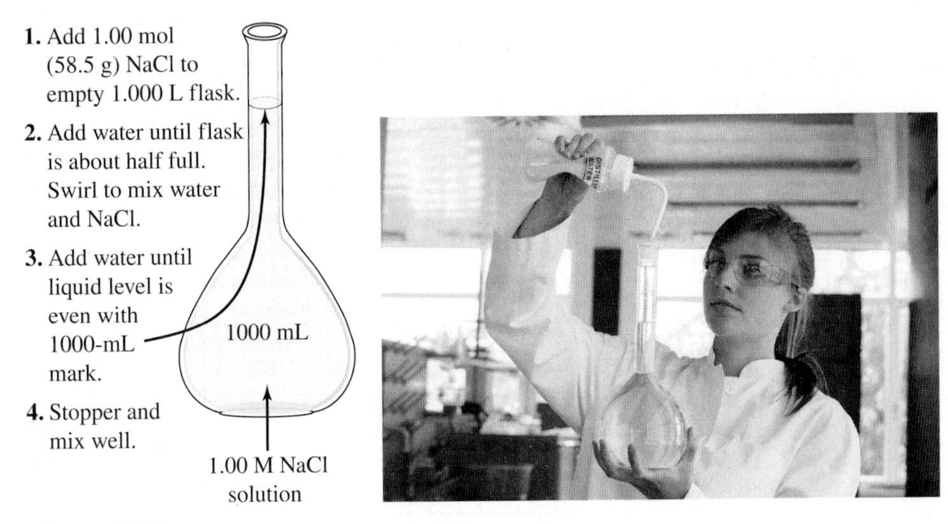

1. Add 1.00 mol (58.5 g) NaCl to empty 1.000 L flask.
2. Add water until flask is about half full. Swirl to mix water and NaCl.
3. Add water until liquid level is even with 1000-mL mark.
4. Stopper and mix well.
1000 mL
1.00 M NaCl solution

Figure 8.21

Preparing a 1.00 M NaCl aqueous solution.

Another possibility is to use 0.500 mol NaCl (29.2 g) in 0.500 L of solution. This requires the use of a 500-mL volumetric flask, rather than the 1-L flask shown in Figure 8.21.

$$1 \text{ M NaCl}(aq) = \frac{1 \text{ mol NaCl}}{1 \text{ L solution}} \quad \text{or} \quad \frac{0.500 \text{ mol NaCl}}{0.500 \text{ L solution}}, \text{ etc.}$$

Let's say you have a water sample with 150 ppm of dissolved mercury, Hg^{2+}. What is this concentration expressed in molarity? You might do the calculation this way:

$$150 \text{ ppm Hg}^{2+} = \frac{150 \cancel{\text{mg Hg}^{2+}}}{1 \text{ L H}_2\text{O}} \times \frac{1 \cancel{\text{g Hg}^{2+}}}{1000 \cancel{\text{mg Hg}^{2+}}} \times \frac{1 \text{ mol Hg}^{2+}}{200.6 \cancel{\text{g Hg}^{2+}}} = \frac{7.5 \times 10^{-4} \text{ mol Hg}^{2+}}{1 \text{ L H}_2\text{O}}$$

Remember that 1 ppm = 1 mg/L, and that the molar mass of Hg is 200.6 g/mol.

Thus, a sample of water containing 150 ppm of mercury also can be expressed as 7.5×10^{-4} M Hg^{2+}.

Your Turn 8.28 Skill Building **Moles and Molarity**

a. Express a concentration of 16 ppb Hg^{2+} in units of molarity.

b. For 1.5 M and 0.15 M NaCl, how many moles of solute are present in 500 mL of each?

c. A solution is prepared by dissolving 0.50 mol NaCl in enough water to form 250 mL of solution. A second solution is prepared by dissolving 0.60 mol NaCl to form 200 mL of solution. Which solution is more concentrated? Explain.

d. A student was asked to prepare 1.0 L of a 2.0 M $CuSO_4$ solution. The student placed 40.0 g of $CuSO_4$ crystals in a volumetric flask and filled it with water to the 1,000-mL mark. Was the resulting solution 2.0 M? Explain.

In this section, we made the case that water is an excellent solvent for a wide variety of substances, and that we can express the concentration of these substances numerically. As promised, the next section helps you build an understanding of how and why substances dissolve in water.

8.7 | A Deeper Look at Solutes

Salt and sugar both dissolve in water. However, one of these compounds is ionic and the other is molecular. What are the differences between ionic and molecular compounds? Refresh your memory by completing the following activity.

Your Turn 8.29 Skill Building Ionic Versus Molecular

Throughout our study of chemistry, we have discussed ionic and molecular compounds in a variety of contexts such as fuels, global climate change, and portable electronic devices. Review your current knowledge of the differences in these compounds in the chart below. You may add more rows with new information and other information you learn in this chapter such as "behavior in water."

	Ionic	Molecular
Definition		
Structure		
Macroscopic properties		
Intermolecular forces		
Bonding model		
Bond strength		
Nomenclature (naming rules)		
Other		

As we pointed out earlier, about 97% of the water on our planet is found in the salt water of the oceans. This source of water contains much more than simple table salt (NaCl) dissolved in water. You are now in a position to understand why so many other ionic compounds can be found dissolved in our oceans.

Recall from Section 8.2 that water molecules are polar. When you take salt crystals and dissolve them in water, the polar H_2O molecules are attracted to the Na^+ and Cl^- ions contained in these crystals. The partial negative charge (δ^-) on the O atom of a water molecule is attracted to the positively charged Na^+ cations of the salt crystal. At the same time, the H atoms in H_2O, with their partial positive charges (δ^+), are attracted to the negatively charged Cl^- anions. Over time, the ions comprising the salt are separated and then surrounded by water molecules. Equation 8.1 and Figure 8.22 represent the process of forming an aqueous sodium chloride solution.

$$NaCl(s) \xrightarrow{H_2O} Na^+(aq) + Cl^-(aq) \quad [8.1]$$

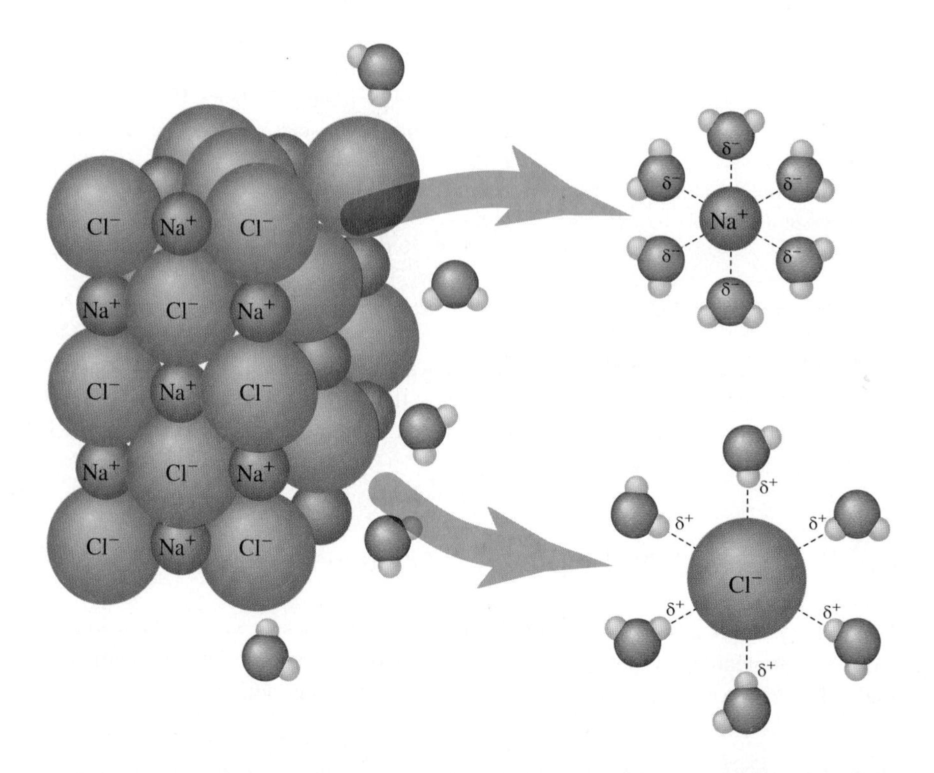

Figure 8.22

Sodium chloride dissolving in water.

The process is similar for forming solutions of compounds containing polyatomic ions. For example, when solid sodium sulfate dissolves in water, the sodium ions and sulfate ions simply separate. Note, however, that the sulfate ion stays together as a unit:

$$Na_2SO_4(s) \xrightarrow{H_2O} 2\,Na^+(aq) + SO_4^{2-}(aq) \quad [8.2]$$

Many ionic compounds dissolve in this manner. This explains why almost all naturally occurring water samples contain various amounts of ions. The same is also true for our bodily fluids, as these also contain significant concentrations of ionic solutes, referred to as **electrolytes**. When a strong electrolyte such as NaCl or another ionic compound is placed into a polar solvent such as water, the compound completely dissociates into positively charged ions (known as cations) and negatively charged ions (known as anions). As seen in Chapter 7, if an electrical potential (voltage) is applied to such a solution, the cations and anions are drawn to oppositely charged electrodes, giving rise to electrical conductivity. Hence, whereas pure water is not electrically conductive, salt solutions are conductive.

Your Turn 8.30 You Decide Electricity and Water Don't Mix

Small electric appliances such as hair dryers and curling irons carry prominent warning labels advising the consumer not to use the appliance near water. Why is water a problem, since pure water does not conduct electricity? What is the best course of action if a plugged-in hair dryer accidentally falls into a sink full of water?

If the solubility principles we just described applied to *all* ionic compounds, our planet would be in trouble. When it rained, ionic compounds such as calcium carbonate (limestone) would dissolve and end up in the ocean! Fortunately, many ionic compounds are only slightly soluble, or have extremely low solubilities. The differences arise because of the sizes and charges of the ions, how strongly they attract one another, and how strongly the ions are attracted to water molecules.

Table 8.6 is your guide to solubility. For example, calcium nitrate, $Ca(NO_3)_2$, is soluble in water, as are all compounds containing the nitrate ion. Calcium carbonate, $CaCO_3$, is insoluble, as are most carbonates. By similar reasoning, copper(II) hydroxide, $Cu(OH)_2$, is insoluble, but copper(II) sulfate, $CuSO_4$, is soluble. Table 8.7 summarizes some environmental consequences of solubility, as they pertain to the dissolution of minerals.

The composition of minerals and rocks was introduced in Section 1.6.

Your Turn 8.31 Skill Building Water Solubility of Ionic Compounds

Which of the compounds below are soluble in water? Use **Table 8.6** as your guide.

a. Ammonium nitrate, NH_4NO_3, a component of fertilizers.
b. Sodium sulfate, Na_2SO_4, an additive in laundry detergents.
c. Mercury(II) sulfide, HgS, known as the mineral cinnabar.
d. Aluminum hydroxide, $Al(OH)_3$, used in water purification processes.

Table 8.6 Water Solubility of Ionic Compounds

Ions	Solubility of Compounds	Solubility Exceptions	Examples
Group 1 metals, NH_4^+	all soluble	none	$NaNO_3$ and KBr. Both are soluble.
nitrates	all soluble	none	$LiNO_3$ and $Mg(NO_3)_2$. Both are soluble.
chlorides	most soluble	silver, mercury(I), lead(II)	$MgCl_2$ is soluble. AgCl is insoluble.
sulfates	most soluble	strontium, barium, lead(II), silver(I)	K_2SO_4 is soluble. $BaSO_4$ is insoluble.
carbonates	mostly insoluble*	Group 1 metals, NH_4^+	Na_2CO_3 is soluble. $CaCO_3$ is insoluble.
hydroxides, sulfides	mostly insoluble*	Group 1 metals, NH_4^+	KOH is soluble. $Sr(OH)_2$ is insoluble.

* Insoluble means that the compounds have extremely low solubilities in water (less than 0.01 M). All compounds have at least a very small solubility in water.

Table 8.7 Environmental Consequences of Solubility

Source	Ions	Solubility and Consequences
salt deposits	sodium and potassium halides*	These salts are soluble. Over time, they dissolve and wash into the sea. Thus, oceans are salty and seawater cannot be used for drinking without expensive purification.
agricultural fertilizers	nitrates	All nitrates are soluble. The runoff from fertilized fields carries nitrates into surface and groundwater. Nitrates can be toxic, especially for infants.
metal ores	sulfides and oxides	Most sulfides and oxides are insoluble. Minerals containing iron, copper, and zinc are often sulfides and oxides. If these minerals had been soluble in water, they would have washed out to sea long ago.
mining waste	mercury(I), lead(II)	Most mercury and lead compounds are insoluble. However, they may leach slowly from mining waste piles and contaminate water supplies.

* Halides, such as Cl^- and I^-, are anions of the atoms in Group 17, the halogens.

Now that we have looked at how ionic compounds dissolve in water, let's look at molecular compounds. From the previous discussion, you might have gotten the impression that only ionic compounds dissolve in water. But remember that sugar dissolves in water as well. The white granules of "table sugar" that you use to sweeten your coffee or tea are *sucrose*, a polar molecular compound with the chemical formula $C_{12}H_{22}O_{11}$.

Figure 8.23

Structural formula of sucrose. The covalently bound –OH groups are shown in red.

When sucrose dissolves in water, the sucrose molecules disperse uniformly among the H_2O molecules. However, unlike ionic compounds, the sucrose molecules remain intact and do *not* separate into ions. Evidence for this includes the fact that aqueous sucrose solutions do not conduct electricity (Figure 8.24). However, even though the sugar molecules act as non-electrolytes, they still interact with water molecules, since they are both polar and are attracted to one another. Furthermore, the sucrose molecule contains eight –OH groups and three additional O atoms that can participate in hydrogen bonding with water (Figure 8.23). Solubility is always promoted

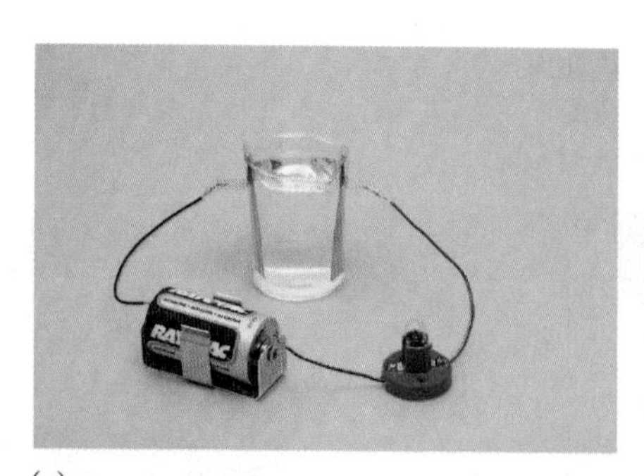

(a)

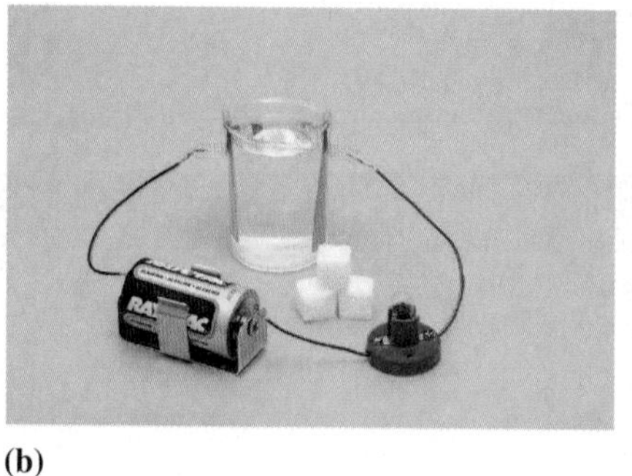

(b)

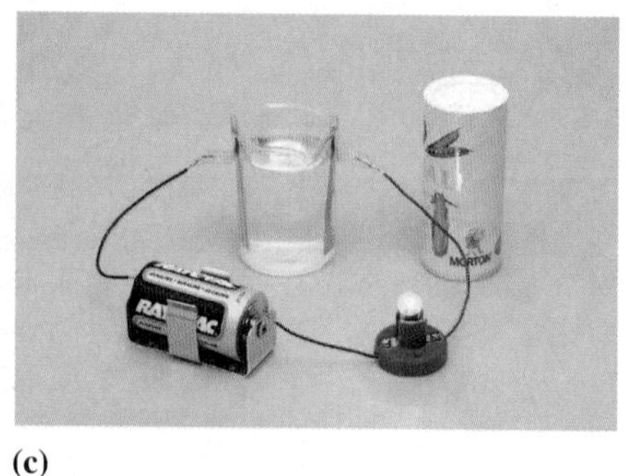

(c)

Figure 8.24

Conductivity experiments. A conductivity meter such as this apparatus shown here indicates whether electricity is being conducted. The light bulb will only glow if the electrical circuit is completed, which is only possible if the electrodes are immersed in an electrically conductive solution. Shown are: **(a)** Distilled water (non-conducting). **(b)** Sugar dissolved in distilled water (non-conducting). **(c)** Salt dissolved in distilled water (conducting).

(a–c): © GIPhotoStock/Science Source

```
    H   H                          H   H
    |   |                          |   |
H — C — C — Ö — H          H — Ö — C — C — Ö — H
    |   |                          |   |
    H   H                          H   H
```

ethanol   ethylene glycol

Figure 8.25

Lewis structures of ethanol and ethylene glycol. The –OH groups are shown in red.

when an attraction exists between the solvent molecules and the solute molecules or ions. This suggests a general solubility rule: *Like dissolves like.*

Let's also consider two other familiar polar molecular compounds, both of which are highly soluble in water. One is ethylene glycol, the main ingredient in antifreeze; and the other is ethanol, or ethyl alcohol, found in beer and wine. These molecules both contain the polar –OH group and are classified as alcohols (Figure 8.25).

As shown in Figure 8.26, the H in the –OH group of an ethanol molecule can hydrogen bond, just as was the case for water:

— covalent bond

---- hydrogen bond

Figure 8.26

Hydrogen bonding between an ethanol molecule and three water molecules.

This is why water and ethanol have a great affinity for each other. Any bartender can tell you that alcohol and water form solutions in all proportions. Again, both molecules are polar, and *like dissolves like.*

Ethylene glycol is another example of an alcohol, sometimes called a "glycol." Ethylene glycol is added to water, such as the water in the radiator of your car, to keep it from freezing. As an antifreeze additive, it is also one of the **volatile organic compounds** (VOCs) that some water-based paints emit when drying. Examine its structural formula in Figure 8.25 to see that it has two –OH groups available for hydrogen bonding. These intermolecular attractions give appreciable water solubility to ethylene glycol, a necessary property for any antifreeze.

It has often been observed that "oil and water don't mix." Water molecules are polar, and the hydrocarbon molecules in oil are nonpolar. When in contact, water molecules tend to attract to other water molecules; in contrast, hydrocarbon molecules stick with their own. Since oil is less dense than water, oil slicks float on top of water (Figure 8.27).

Figure 8.27

Oil and water are not miscible with each other.

Your Turn 8.32 Skill Building More About Hydrocarbons

Hydrocarbon molecules such as pentane and hexane contain C–H and C–C bonds. Using the electronegativity values in **Table 8.2**, predict whether these bonds are polar or nonpolar. Why are these molecular compounds nonpolar?

Hint: Consider their molecular geometries alongside the bond dipoles.

Since water is a poor solvent for grease and oil, we cannot use water to wash these off. Instead, we wash our hands (and clothes) with the aid of soaps and detergents. These compounds are **surfactants**, compounds that help polar and nonpolar compounds mix, sometimes called "wetting agents." The molecules of surfactants contain both polar and nonpolar groups. The polar groups allow the surfactant to dissolve in water, while the nonpolar ones are able to dissolve the grease.

Your Turn 8.33 Scientific Practices Surfactants to the Rescue!

Knowing that surfactants have both polar and nonpolar ends, sketch what you think a surfactant molecule looks like. Then create a diagram showing a mixture of surfactant molecules, nonpolar molecules, and water molecules. Check the accuracy of your depictions with those appearing on the Internet or in a textbook.

Another way to dissolve nonpolar molecules is to use nonpolar solvents. Like dissolves like! Nonpolar solvents (sometimes called "organic solvents") are widely used in the production of drugs, plastics, paints, cosmetics, and cleaning agents. As another example, dry cleaning solvents are typically composed of chlorinated hydrocarbons. One example, "perc," is a cousin of ethene. Take ethene (sometimes called ethylene), a compound with a C=C double bond, and replace all the H atoms with Cl atoms. The result is tetrachloroethylene—also called perchloroethylene, or "perc" for short.

```
H       H          Cl       Cl
 \     /             \     /
  C=C                 C=C
 /     \             /     \
H       H          Cl       Cl
```

ethylene tetrachloroethylene (''perc'')

Perc and other chlorinated hydrocarbons like it are carcinogens or suspected carcinogens. They have serious health consequences, whether we are exposed to them in the workplace or as contaminants of our air, water, or soil.

Green chemists aim to redesign processes so that they don't require solvents. But if this is not possible, they try to replace harmful solvents like perc with ones that are friendly to the environment. One possibility is liquid carbon dioxide. Under conditions of high pressure, the gas you know as CO_2 can condense to form a liquid. Compared with organic solvents, $CO_2(l)$ offers many advantages. It is nontoxic, nonflammable, chemically benign, non-ozone-depleting, and it does not contribute to the formation of smog. Although you may be concerned with the fact that it is a greenhouse gas, carbon dioxide that is used as a solvent is a recovered waste product from industrial processes and it is generally recycled.

Adapting liquid CO_2 to dry cleaning posed a challenge, as it is not very good at dissolving oils, waxes, and greases found in soiled fabrics. To make carbon dioxide a better solvent, Joe DeSimone (b. 1964), a chemist and chemical engineer at the University of North Carolina, Chapel Hill, developed a surfactant to use with $CO_2(l)$. For his work, DeSimone received a 1997 Presidential Green Chemistry Challenge Award. His breakthrough process paves the way for designing environmentally benign, inexpensive, and easily recyclable replacements for conventional organic and water solvents currently in use. DeSimone was instrumental in the beginnings of Hangers Cleaners, a dry cleaning chain that uses the process he developed.

Your Turn 8.34 You Decide Liquid CO_2 as a Solvent

a. How does using liquid carbon dioxide as a solvent compare to organic solvents?
b. Comment on this statement: "Using carbon dioxide as a replacement for organic solvents simply replaces one set of environmental problems with another."
c. If a local dry cleaning business switched from "perc" to carbon dioxide, how might this business report a different triple bottom line?

The tendency of nonpolar compounds to dissolve in other nonpolar substances explains how fish and animals accumulate nonpolar substances such as PCBs (polychlorinated biphenyls) or the pesticide DDT (dichlorodiphenyltrichloroethane) in their fatty tissues. When fish ingest these, the molecules are stored in body fat (nonpolar) rather than in the blood (polar). PCBs can interfere with the normal growth and development of a variety of animals, including humans, in some cases at concentrations of less than 1 ppb.

The higher you go in the food chain, the greater concentrations of harmful nonpolar compounds like DDT you will find. This is called **biomagnification**, the increase in concentration of certain persistent chemicals in successively higher levels of a food chain. Figure 8.28 shows a biomagnification process that was studied extensively in the 1960s. At that time, DDT was shown to interfere with the reproduction of peregrine falcons and other predatory birds at the top of their food chain. In 1962, Rachel Carson's publication of *Silent Spring* also linked a decline in the song bird populations with their exposure to pesticides.

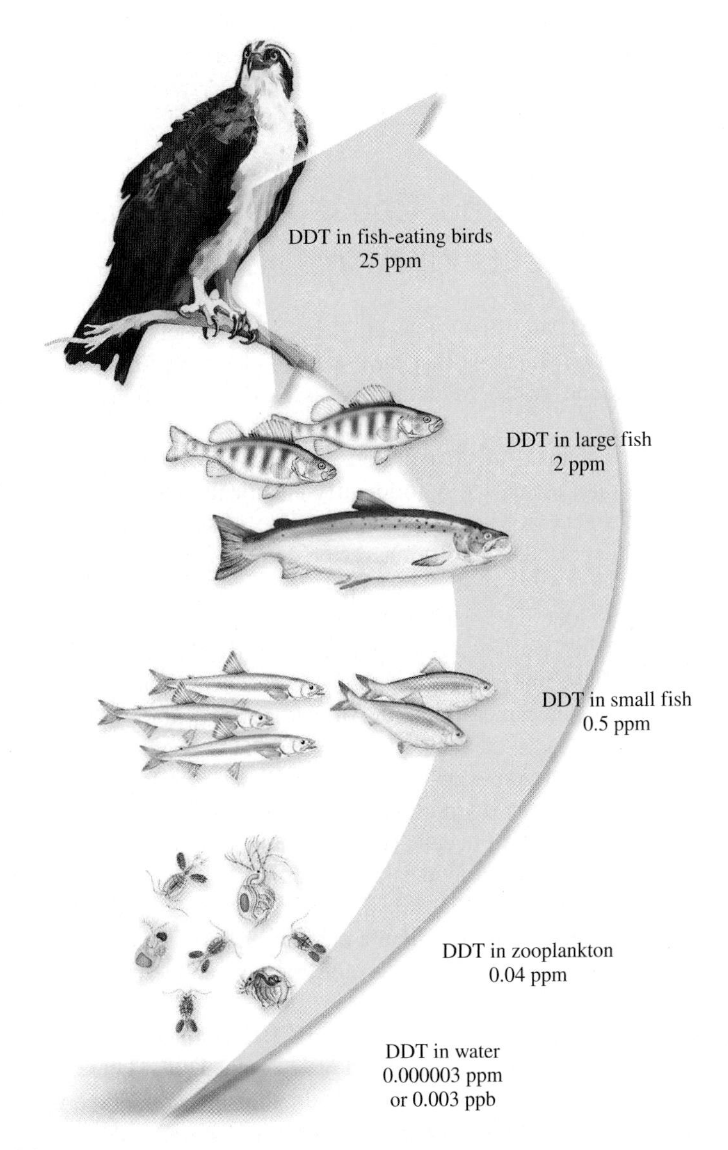

Figure 8.28

Organisms in the water take up and store DDT. They are eaten by larger creatures that in turn are eaten by still larger ones. Consequently, creatures highest on the food chain will accumulate the highest concentration of DDT.

Source: From William and Mary Ann Cunningham. Environmental Science: A Global Concern, *10th ed., 2008. Reprinted with permission of the McGraw-Hill Education.*

Before we move on to our investigation of water as a solvent for acids and bases, let's review the solvation processes for ionic and molecular compounds and relate them to our water diaries in the following two activities.

Your Turn 8.35 Scientific Practices Summarizing Solvation

Create a comparison table for the aqueous solvation of ionic versus molecular compounds. At a minimum, you should include a description of how each dissolves, properties of the solution, environmental impacts, and applications for the solution.

Your Turn 8.36 You Decide Solutes in Your Water Diary

Examine your water diary. In which areas does your water use depend upon dissolved ionic and molecular compounds? Do any of your water uses add solutes to the water?

8.8 Corrosive and Caustic: The Properties and Impacts of Acids and Bases

The role of acids and bases in food chemistry will be explored in Chapter 10.

Did You Know? The plant dye *litmus* changes from blue to pink in acid. Interestingly, the term *litmus test* has also come to refer to something that quickly reveals a politician's point of view.

A video in **Connect** demonstrates that increasing the acidity of water can cause the dissolution of an eggshell through a chemical reaction.

Water is known as the "universal solvent," because it dissolves a variety of solutes. Let's now consider two solutions that have a large effect on Earth and in your daily lives, namely acids and bases. Acids and bases are both solutions, and many utilize water as the solvent.

Historically, chemists identified acids by properties such as their sour taste. Although tasting is not a smart way to identify chemicals, you undoubtedly know the sour taste of acetic acid in vinegar. The sour taste of lemons comes from acids as well (Figure 8.29). Acids also show a characteristic color change with indicators such as litmus.

Another way to identify an acid is by its chemical properties. For example, under certain conditions, acids can react with and dissolve marble, eggshell, or the shells of marine creatures. These materials all contain the carbonate ion (CO_3^{2-}), either as calcium carbonate or magnesium carbonate. An acid reacts with a carbonate to produce carbon dioxide. This gas is the "burp" when carbonate-containing stomach antacid tablets react with acids in your stomach. As we will see in a later section, this chemical reaction also explains the dissolution of the skeletons of carbonate-based sea creatures such as coral in acidified oceans (Figure 8.30).

At the molecular level, an **acid** is a compound that releases hydrogen ions, H^+, in aqueous solution. Remember that a hydrogen atom is electrically neutral and consists of one electron and one proton. If the electron is lost, the atom becomes a positively charged ion, H^+. Because only a proton remains, sometimes H^+ is referred to as a **proton**.

Figure 8.29

Citrus fruits contain both citric acid and ascorbic acid.

© Nancy R. Cohen/Getty Images RF

For example, consider hydrogen chloride (HCl), a compound that is a gas at room temperature. Hydrogen chloride is composed of HCl molecules. These dissolve readily in water to produce a solution that we name hydrochloric acid. As the polar HCl molecules dissolve, they become surrounded by polar water molecules. Once dissolved, these molecules break apart into two ions: $H^+(aq)$ and $Cl^-(aq)$. This equation represents the two steps of the reaction:

$$HCl(g) \xrightarrow{H_2O} HCl(aq) \longrightarrow H^+(aq) + Cl^-(aq) \qquad [8.3]$$

We also could say that HCl *dissociates* into H^+ and Cl^-. No HCl molecules remain in solution because they dissociate completely in water, which is referred to as a **strong acid**.

Figure 8.30

Example of a coral reef.

There is a slight complication with the definition of acids as substances that release H^+ (protons) in aqueous solutions. By themselves, H^+ are much too reactive to exist as such. Rather, they attach to something else, such as water molecules. When dissolved in water, each HCl molecule donates a proton (H^+) to an H_2O molecule, forming H_3O^+, a **hydronium ion**. Here is a representation of the overall reaction:

$$HCl(aq) + H_2O(l) \longrightarrow H_3O^+(aq) + Cl^-(aq) \qquad \textbf{[8.4]}$$

The solution represented on the product side in *both* Equations 8.3 and 8.4 is called *hydrochloric acid*. It has the characteristic properties of an acid because of the presence of H_3O^+. Chemists often simply write H^+ when referring to acids (*e.g.*, in Equation 8.3), but understand this to mean H_3O^+ (hydronium ion) in aqueous solutions. Figure 8.31 shows the Lewis structure of the hydronium ion.

In addition to forming H_3O^+, H^+ ions can combine with more molecules of water to form $H_5O_2^+$, $H_7O_3^+$, etc. in aqueous solutions.

To name acids acids that do not contain oxygen in the anion, the prefix "hydro–" and the suffix "–ic" are added to the root name of the anion (*e.g.*, fluoride anion: "hydrofluoric acid")

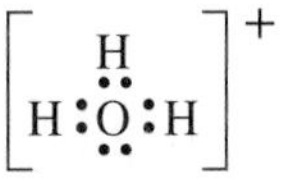

$$\left[\begin{array}{c} \text{H} \\ \text{H}\!:\!\ddot{\text{O}}\!:\!\text{H} \end{array}\right]^+$$

Figure 8.31

Lewis structure for the hydronium ion. Notice that oxygen follows the octet rule in this structure.

Your Turn 8.37 Skill Building Acidic Solutions

For each of the strong acids shown below, write a balanced chemical equation that shows the release of a proton, H^+ when dissolved in water. Also provide an equation that shows the formation of a hydronium ion.

Hint: Remember to include the charges on the ions. The net charge on both sides of the equation should be the same.

a. HI*(aq)*, hydroiodic acid **b.** HNO_3*(aq)*, nitric acid **c.** H_2SO_4*(aq)*, sulfuric acid

Your Turn 8.38 You Decide Are All Acids Harmful?

Although the word *acid* may conjure up all sorts of pictures in your mind, every day you eat or drink various acids. Check the labels of foods or beverages and make a list of the acids you find. Speculate on the purpose of each acid.

Hydrogen chloride is but one of several gases that dissolves in water to produce an acidic solution. Sulfur dioxide and nitrogen dioxide are two others. These two gases are emitted during the combustion of certain fuels (particularly coal) to produce heat and electricity. SO_2 and NO_2 both dissolve in rain and mist. When they do so, they form acids that in turn fall back to Earth's surface in rain and snow.

But before delving into the acidity in rain caused by nitrogen oxides and sulfur dioxide, let's focus on carbon dioxide. With an atmospheric concentration of about 400 ppm in 2015, carbon dioxide is at a far higher concentration than either sulfur dioxide or nitrogen dioxide. Just as solids vary in their solubility in water, so do gases. Compared with more polar compounds such as SO_2 and NO_2, carbon dioxide is far less soluble in water. Even so, it dissolves to produce a weakly acidic solution.

Given that an acid is defined as a substance that releases hydrogen ions in water, how can carbon dioxide act as an acid? There are no hydrogen atoms in carbon dioxide! The explanation is that when CO_2 dissolves in water, it produces carbonic acid, $H_2CO_3(aq)$. Here are some ways to represent the process:

$$CO_2(g) \xrightarrow{H_2O} CO_2(aq) \quad \textbf{[8.5a]}$$

$$CO_2(aq) + H_2O(l) \longrightarrow H_2CO_3(aq) \quad \textbf{[8.5b]}$$

The carbonic acid dissolves to produce H^+ and the hydrogen carbonate ion, also known as the bicarbonate ion:

$$H_2CO_3(aq) \rightleftharpoons H^+(aq) + HCO_3^-(aq) \quad \textbf{[8.5c]}$$

Strong and weak acids are analogous to strong and weak electrolytes—both are defined by their relative degrees of dissociation into ionic species.

As indicated by a double-arrow symbol (described in more detail below), this reaction occurs only to a limited extent, producing only tiny amounts of H^+ and HCO_3^-. Accordingly, we say that carbonic acid is a **weak acid**; that is, an acid that dissociates only to a small extent in aqueous solution.

Carbon dioxide is only slightly soluble in water; accordingly, only a tiny amount of the dissolved carbonic acid dissociates to produce H^+. However, these reactions are happening on a large scale across the planet. The carbon dioxide can dissolve in water in the troposphere, resulting in acidic rain, or in the planet's oceans, lakes, and streams.

Did You Know? Dilute basic solutions have a soapy feel because bases can react with the oils of your skin to produce a tiny bit of soap.

No discussion of acids would be complete without discussing their chemical counterparts, bases. For our purposes, a **base** is a compound that releases hydroxide ions (OH^-) in aqueous solution. Aqueous solutions of bases have their own characteristic properties attributable to the presence of $OH^-(aq)$. Unlike acids, bases generally taste bitter and do not lend an appealing flavor to foods. Aqueous solutions of bases have a slippery, soapy feel. Common examples of bases include household ammonia (an aqueous solution of NH_3) and NaOH (sometimes called lye). The cautions on oven cleaners (Figure 8.32) warn that lye can cause severe damage to eyes, skin, and clothing.

Water is composed of H^+ and OH^-, so it is an example of an *amphoteric* substance, one that can act as *either* an acid or a base.

Many common bases are compounds containing the hydroxide ion. For example, sodium hydroxide (NaOH), a water-soluble ionic compound, dissolves in water to produce sodium ions (Na^+) and hydroxide ions (OH^-):

$$NaOH(s) \xrightarrow{H_2O} Na^+(aq) + OH^-(aq) \quad \textbf{[8.6]}$$

Although sodium hydroxide is very soluble in water, most compounds containing the hydroxide ion are not, according to the solubility rules of ionic compounds (Table 8.6). As you might expect, bases that dissociate completely in water, such as NaOH, are called **strong bases**.

Your Turn 8.39 Skill Building Basic Solutions

For each of the bases shown below, write a chemical equation that shows the release of a hydroxide ion(s), OH^-, when dissolved in water.

a. KOH(s), potassium hydroxide.
b. LiOH(s), lithium hydroxide.
c. $Ca(OH)_2(s)$, calcium hydroxide.

Note: Whereas acids with more than one proton (known as *polyprotic acids*) lose one H^+ at a time, bases lose all OH^- groups at once.

Figure 8.32

Oven cleaning products may contain NaOH, commonly called lye.

© McGraw-Hill Education. Photo by Eric Misko, Elite Images Photography

Some bases, however, do not contain the hydroxide ion, OH^-, but rather react with water to form it. One example is ammonia, a gas with a distinctive sharp odor. Unlike carbon dioxide, ammonia is very soluble in water. It rapidly dissolves in water to form an aqueous solution:

In some industrial applications, ammonia (rather than HCFCs) is used as a refrigerant gas. Great care needs to be taken to prevent the exposure of workers to ammonia, because the gas can react with moist lung tissue, resulting in injury or death.

$$NH_3(g) \xrightarrow{H_2O} NH_3(aq) \qquad \textbf{[8.7a]}$$

On a supermarket shelf, you may see a 5% (by mass) aqueous solution of ammonia called "household ammonia." This cleaning agent has an unpleasant odor; if it gets on your skin, you should wash it off with plenty of water.

The chemical behavior of aqueous ammonia is difficult to simplify, but we will do our best to represent it for you with a chemical equation. When an ammonia molecule reacts with a water molecule, the water molecule transfers H^+ to the NH_3 molecule. An ammonium ion, $NH_4^+(aq)$, and a hydroxide ion, $OH^-(aq)$, are formed. However, this reaction only occurs to a small extent; that is, only a tiny amount of $OH^-(aq)$ is produced.

The ammonium ion, NH_4^+, is analogous to the hydronium ion, H_3O^+, in that each was formed by the addition of a proton (H^+) to a neutral compound.

$$\underset{\text{Weak Base}}{NH_3(aq)} + \underset{\text{Acid}}{H_2O(l)} \rightleftharpoons \underset{\text{Conjugate Acid}}{NH_4^+(aq)} + \underset{\text{Conjugate Base}}{OH^-(aq)} \qquad \textbf{[8.7b]}$$

When a reversible reaction reaches equilibrium, both forward and reverse reactions still continue to occur, but the rates of these reactions are constant.

As shown in Equation 8.7b, an acid will donate a proton, H^+, to a base. In this case, water acts as the acid, which donates a proton to NH_3. The double-arrow of this reaction indicates that this is an **equilibrium reaction**, one that proceeds in both directions to continually form both products and reactants. In the reverse (right–left) direction, the NH_4^+ ion donates a proton to OH^-. Hence, NH_4^+ is referred to as a **conjugate acid** of the base NH_3, and OH^- is the **conjugate base** of the acid H_2O. To indicate more clearly that aqueous ammonia is a base, some people use the representation $NH_4OH(aq)$. If you add up the atoms (and their charges), you will see that $NH_4OH(aq)$ is equivalent to the left-hand side of Equation 8.7b. It is unlikely, however, that this species exists intact within an aqueous solution of ammonia.

NH_4^+/NH_3 and H_2O/OH^- are often referred to as conjugate acid-base pairs, with each pair differing by a single proton, or H^+.

The source of the hydroxide ion in household ammonia now should be apparent. When ammonia dissolves in water, it releases small amounts of the hydroxide ion and the ammonium ion. Aqueous ammonia is an example of a **weak base**; a base that dissociates only to a small extent in aqueous solution. Accordingly, for the equilibrium reaction shown in Equation 8.7b, the concentration of reactants will be much greater than products. In contrast, the analogous reaction with a strong base such as NaOH would be highly product-favored. Instead of writing the reaction as a reversible

The $H_2O(l)$ solvent can be removed from both sides of Equation 8.8 to yield Equation 8.6, the simple dissociation of NaOH in an aqueous solution.

equilibrium, reactions involving strong acids and bases are best represented by a traditional arrow:

$$\underset{\text{Strong Base}}{NaOH(aq)} + \underset{\text{Acid}}{H_2O(l)} \xrightarrow[\text{non-reversible}]{100\%} Na^+(aq) + OH^-(aq) + H_2O(l) \quad [8.8]$$

Your Turn 8.40 Scientific Practices Acids and Bases in Your Water Diary

Revisit your water diary from the beginning of the chapter. Where do acids and bases show up in the diary? Do you need to make adjustments to include acids and bases? Do any of your actions add acidic and basic components to water? Explain.

8.9 Heartburn? Tums® to the Rescue: Acid/Base Neutralization!

As seen in the previous section, acids and bases react with each other—often very rapidly. This happens not only in laboratory test tubes, but also in your home and in almost every ecological niche of our planet. For example, if you put lemon juice on fish, an acid–base reaction occurs. The acids found in lemons neutralize the ammonia-like compounds that produce the "fishy smell." Similarly, if the ammonia fertilizer on a corn field comes in contact with the acidic emissions of a power plant nearby, an acid–base reaction occurs.

Let us first examine the acid–base reaction of solutions of hydrochloric acid and sodium hydroxide. When the two are mixed, the products are sodium chloride and water:

$$\underset{\text{acid}}{HCl(aq)} + \underset{\text{base}}{NaOH(aq)} \longrightarrow NaCl(aq) + H_2O(l) \quad [8.9]$$

This is an example of a **neutralization reaction**, a chemical reaction in which the protons from an acid combine with the hydroxide ions from a base to form water molecules. The formation of water can be represented like this:

$$H^+(aq) + OH^-(aq) \longrightarrow H_2O(l) \quad [8.10]$$

What about the sodium and chloride ions? Recall from Equations 8.3 and 8.6 that the $HCl(g)$ and $NaOH(s)$, when dissolved in water, completely dissociate into ions. We can rewrite Equation 8.9 to show this, which is often referred to as a *total ionic equation*:

$$H^+(aq) + Cl^-(aq) + Na^+(aq) + OH^-(aq) \longrightarrow Na^+(aq) + Cl^-(aq) + H_2O(l) \quad [8.11]$$

Neither $Na^+(aq)$ nor $Cl^-(aq)$ take part in the neutralization reaction; they remain unchanged. Canceling these ions, referred to as *spectator ions*, from both sides again gives us Equation 8.10, which summarizes the chemical changes taking place in an acid–base neutralization reaction. This form of the neutralization equation that omits the spectator ions is often designated as the *net ionic equation*.

Your Turn 8.41 Skill Building Neutralization Reactions

For each acid–base pair, write a balanced neutralization reaction. Then, rewrite the equation in total ionic and net ionic forms. What is the relevance of the final simplified step in each case?

a. $HNO_3(aq)$ and $KOH(aq)$ **b.** $HCl(aq)$ and $NH_4OH(aq)$ **c.** $HBr(aq)$ and $Ba(OH)_2(aq)$

A **neutral solution** is neither acidic nor basic; that is, it has equal concentrations of H^+ and OH^-. Pure water is a neutral solution. Some salt solutions are also neutral, such as the one formed by dissolving solid NaCl in water. In contrast, acidic solutions contain a higher concentration of H^+ than OH^-, and basic solutions contain a higher concentration of OH^- than H^+.

It may seem strange that acidic and basic solutions contain both hydroxide ions *and* protons. But when water is involved, it is not possible to have H^+ without OH^- (or vice versa). A simple, useful, and very important relationship exists between the concentration of protons and hydroxide ions in any aqueous solution:

$$[H^+][OH^-] = 1 \times 10^{-14} \quad \textbf{[8.12]}$$

The square brackets indicate that the ion concentrations are expressed in molarity, M, and $[H^+]$ is read as "the hydrogen ion (or proton) concentration." When $[H^+]$ and $[OH^-]$ are multiplied, the product is a constant with a value of 1×10^{-14}, as shown in Equation 8.12. This shows that the concentrations of H^+ and OH^- depend on each other. When $[H^+]$ increases, $[OH^-]$ decreases, and when $[H^+]$ decreases, $[OH^-]$ increases. However, both ions are always present in aqueous solutions.

Acidic solution: $[H^+] > [OH^-]$
Neutral solution: $[H^+] = [OH^-]$
Basic solution: $[H^+] < [OH^-]$

Knowing the concentration of H^+, we can use Equation 8.12 to calculate the concentration of OH^- (or vice versa). For example, if rainwater has a H^+ concentration of 1×10^{-5} M, we can calculate the OH^- concentration by substituting 1×10^{-5} M for $[H^+]$:

$$(1 \times 10^{-5}\ \text{M}) \times [OH^-] = 1 \times 10^{-14}$$

$$[OH^-] = \frac{1 \times 10^{-14}}{1 \times 10^{-5}}$$

$$[OH^-] = 1 \times 10^{-9}\ \text{M}$$

Since the hydroxide ion concentration (1×10^{-9} M) is smaller than the hydrogen ion concentration (1×10^{-5} M), the solution is acidic.

In pure water or in a neutral solution, the concentrations of the hydrogen and hydroxide ions both equal: 1×10^{-7} M. Applying Equation 8.12, we can see that $[H^+][OH^-] = (1 \times 10^{-7}\ \text{M})(1 \times 10^{-7}\ \text{M}) = 1 \times 10^{-14}$.

By definition, the product of the two concentrations is unitless.

Your Turn 8.42 Skill Building Acidic and Basic Solutions

For parts **a** and **c** below, calculate $[OH^-]$. For **b**, calculate $[H^+]$. Then, classify each solution as either acidic, neutral, or basic.

a. $[H^+] = 1 \times 10^{-4}$ M **b.** $[OH^-] = 1 \times 10^{-6}$ M **c.** $[H^+] = 1 \times 10^{-10}$ M

Your Turn 8.43 Skill Building Ions in Acidic and Basic Solutions

These solutions represent either strong acids or strong bases. Classify each as acidic or basic. Then, list all of the ions present in order of decreasing relative amounts in each solution.

a. $KOH(aq)$ **b.** $HNO_3(aq)$ **c.** $H_2SO_4(aq)$ **d.** $Ca(OH)_2(aq)$

How can we know if the acidity of seawater, rain, or another solution is cause for concern? To make a judgment, we need a convenient way of reporting how acidic or basic a solution is. The pH scale is such a tool because it relates the acidity of a solution to its H^+ concentration.

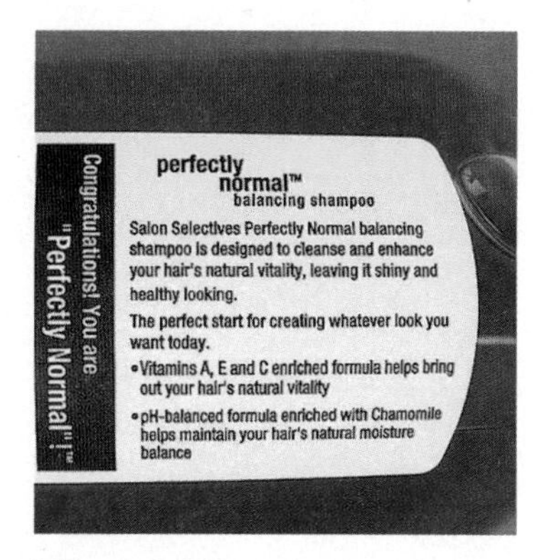

Figure 8.33

This shampoo claims to be "pH-balanced"; that is, adjusted to be closer to neutral. Soaps tend to be basic, which can be irritating to skin.

8.10 Quantifying Acidity/Basicity: The pH Scale

The term "pH" may already be familiar to you. For example, test kits for soils and for the water in aquariums and swimming pools report the acidity in terms of pH. Deodorants and shampoos claim to be pH-balanced (Figure 8.33). And, of course, articles about acid rain make reference to pH. The notation pH is always written with a small p and a capital H and stands for "power of hydrogen." In the simplest terms, pH is a number, usually between 0 and 14, that indicates the acidity (or basicity) of a solution.

At the midpoint on the scale, pH 7 separates acidic from basic solutions. Solutions with a pH less than 7 are acidic, and those with a pH greater than 7 are basic (alkaline). Solutions of pH 7 (such as pure water) have equal concentrations of H^+ and OH^- and are said to be neutral.

The pH values of common substances are displayed in Figure 8.34. You may be surprised that you eat and drink so many acids. Acids naturally occur in foods and contribute distinctive tastes. For example, the tangy taste of McIntosh apples comes from malic acid. Yogurt gets its sour taste from lactic acid, and cola soft drinks contain several acids, including phosphoric acid. Tomatoes are well known for their acidity, but with a pH of about 4.5, they are in fact less acidic than many other fruits.

Universal indicator paper is a quick way to determine the pH of a solution. However, for more accurate results, pH meters are used.

Did You Know? For highly acidic or basic solutions, the pH may lie outside the 0-to-14 range.

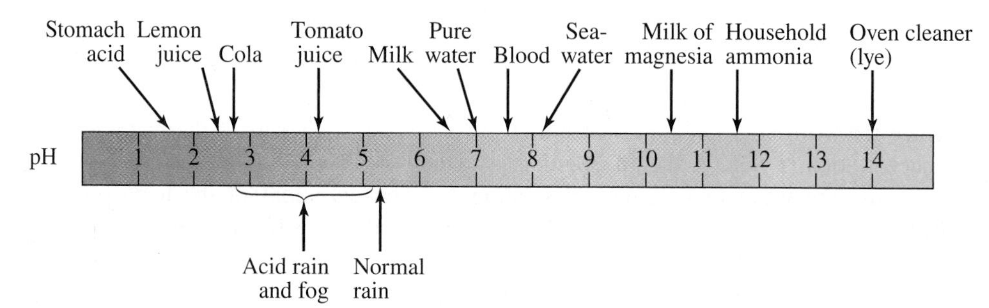

Figure 8.34

Common substances and their pH values.

Your Turn 8.44 Skill Building pH Simulations

Work through the simulations is found on **Figures Alive!** in **Connect** to practice pH. Comment on the models presented.

Your Turn 8.45 Skill Building Acidity of Foods

a. Rank tomato juice, lemon juice, milk, cola, and pure water in order of increasing acidity. Check your order against **Figure 8.34**.

b. Pick any other five foods and make a similar ranking. Search the Internet to find their actual pH values.

c. Is there anything in your water diary that has a pH value? What are the values?

Is the water on the planet acidic, basic, or neutral? Water would be expected to have a pH of 7.0, but Figure 8.34 shows that the pH of water depends on where it is found. "Normal" rain is slightly acidic, with a pH value between 5 and 6. Even though the acid formed by dissolved carbon dioxide is a weak acid, enough H^+ is produced to lower the pH of rain. In contrast, seawater is slightly basic, with a pH of approximately 8.2.

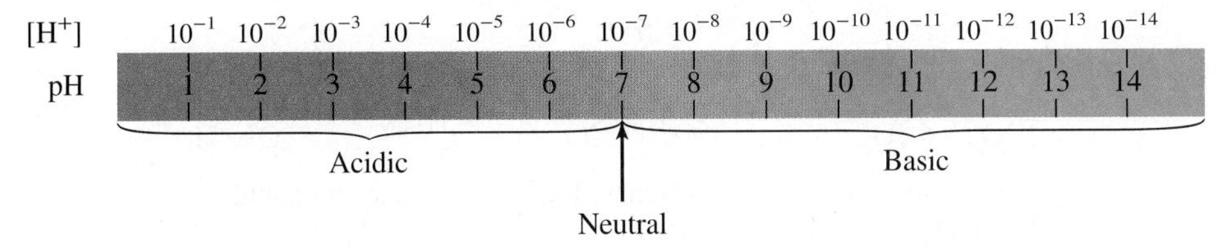

Figure 8.35

The relationship between pH and the concentration of H^+ in moles per liter (M). As pH increases, $[H^+]$ decreases (and $[OH^-]$ increases]).

As you might have guessed, pH values are related to the hydrogen ion concentration. If $[H^+] = 1 \times 10^{-3}$ M, then the pH is 3. Similarly, if $[H^+] = 1 \times 10^{-9}$ M, the pH is 9. Equation 8.12 shows that the hydrogen ion concentration multiplied by the hydroxide ion concentration is a constant, 1×10^{-14}. When the concentration of H^+ is high (and the pH is low), the concentration of OH^- is low. Likewise, as pH values rise above 7.0, the concentration of hydrogen ions decreases and the concentration of hydroxide ions increases. As the pH value *decreases,* the acidity *increases.* For example, a sample of water with a pH of 5.0 is 1/10 the acidity than one with a pH of 4.0. This is because a pH of 4 means that the $[H^+]$ is 0.0001 M. By contrast, a solution with a pH of 5 is more dilute, with a $[H^+] = 0.00001$ M. This second solution is *less* acidic with only 1/10 the hydrogen ion concentration of a solution of pH 4. Figure 8.35 shows the relationship between pH and the hydrogen ion concentration.

The equation used to calculate pH from $[H^+]$ is: $pH = -\log [H^+]$. See Appendix 3 for more details regarding logarithmic functions.

Your Turn 8.46 Skill Building Small Changes, Big Effects

Compare the pairs of samples below. For each, which one is more acidic? Include the relative difference in hydrogen ion concentration between the two pH values.

a. Rainwater, pH = 5.0; lake water, pH = 4.0.
b. Ocean water, pH = 8.3; tap water, pH = 5.3.
c. Tomato juice, pH = 4.5; milk, pH = 6.5.

Your Turn 8.47 You Decide On the Record

A legislator from the Midwest is on record with an impassioned speech in which he argued that the environmental policy of the state should be to bring the pH of rain all the way down to zero. Assume that you are an aide to this legislator. Draft a tactful memo to your boss to save him from additional public embarrassment.

Your Turn 8.48 Scientific Practices pH Range of Your Water Diary

Review your water diary. Did you include all items that you drank? What was the range of pH values and $[H^+]$ concentrations for the items you consumed?

8.11 | Acid's Effect on Water

Changes in pH can affect both the balance in our oceans and in our other sources of water. In this section, we will look at how this can affect life on this planet. How can seawater be basic, when rain is naturally acidic? Indeed, this is the case, as shown in Figure 8.34.

A solution that contains ionic species that help maintain a constant pH is referred to as a *buffer solution.* Buffers will be discussed in more detail in Section 12.2.

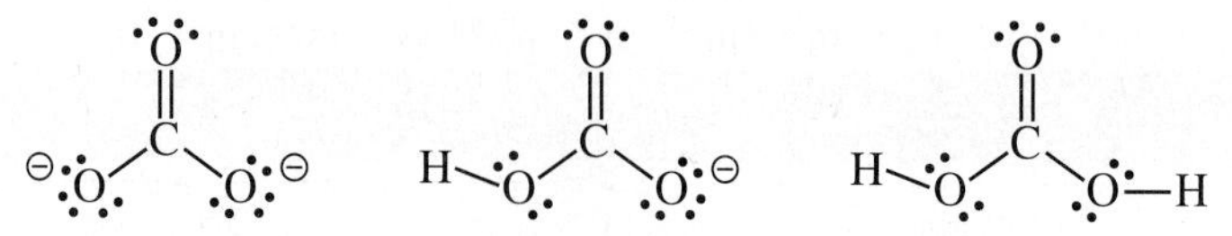

Figure 8.36

Lewis structures for the carbonate and bicarbonate ions, as well as carbonic acid.

To name acids that contain oxygen atoms in their anions, follow this simple rule: If the anion name ends with the "–ate" suffix, change it to "–ic" for the acid (*e.g.*, carbon<u>ate</u> anion: "carbonic acid"). If the anion name ends with the "–ite" suffix, change it to "–ous" for the acid (*e.g.*, nitr<u>ite</u> anion: "nitrous acid").

Ocean water contains small amounts of three chemical species that arise from dissolved carbon dioxide, and play a role in maintaining the ocean pH at approximately 8.2. These three species—the carbonate ion, the bicarbonate ion, and carbonic acid (Figure 8.36) interact with each other as well. These species also help maintain your blood at a pH of about 7.4.

Many organisms, such as mollusks, sea urchins, and coral have connections to this ocean chemistry because they build their shells out of calcium carbonate, $CaCO_3$. Changing the amount of one chemical species in the ocean (such as carbonic acid) can affect the concentration of the others, in turn affecting marine life. The amount of carbon dioxide released into the atmosphere over the past 200 years has increased (Section 4.9). As a result, more carbon dioxide is dissolving into the oceans and reacting to form carbonic acid. In turn, the pH of seawater has dropped by roughly 0.1 pH unit since the early 1800s. This may sound like a small number; however, remember that each full unit of pH represents a 10-fold difference in the concentration of H^+. A decrease of 0.1 pH unit corresponds to a 26% increase in the amount of H^+ in seawater. The lowering of the ocean pH due to increased atmospheric carbon dioxide is called **ocean acidification**.

Ocean pH can vary by ± 0.3 pH units, depending on latitude and region.

How can such a seemingly small change in pH pose a danger to marine organisms? Part of the answer lies in the chemical interactions between $CO_3^{2-}(aq)$, $HCO_3^-(aq)$, and $H_2CO_3(aq)$. The H^+ produced by the dissociation of carbonic acid reacts with carbonate ion in seawater to form the bicarbonate ion:

$$H^+(aq) + CO_3^{2-}(aq) \longrightarrow HCO_3^-(aq) \qquad \textbf{[8.13]}$$

The net effect is to reduce the concentration of carbonate ions in seawater. The calcium carbonate in the shells of sea creatures then begins to dissolve in response to the decreased concentration of carbonate ions in seawater:

$$CaCO_3(s) \xrightarrow{H_2O} Ca^{2+}(aq) + CO_3^{2-}(aq) \qquad \textbf{[8.14]}$$

The interactions of carbonic acid, bicarbonate ions, and carbonate ions are summarized in Figure 8.37. As carbon dioxide dissolves in ocean water, it forms carbonic acid. This in turn dissociates to produce "extra" acidity in the form of H^+. The H^+ ions react with carbonate ions, thereby depleting it and producing more bicarbonate ions. Calcium carbonate then dissolves to replace the carbonate that was depleted.

Ocean scientists predict that within the next 40 years, the carbonate ion concentration will reach a low enough level that the shells of sea creatures near the ocean surface will begin to dissolve. In fact, one study has shown that the Great Barrier Reef off the coast of Australia is already growing at slower and slower rates. However, other factors could be to blame. For example, ocean warming also contributes to the poor health of coral reefs. One can examine growth rings in a slice of coral, much as one can view tree rings (Figure 8.38).

To date, only a small number of researchers have focused on the effects of thinning shells on sea creatures. However, negative effects on whole ecosystems have been projected. For example, weaker (or missing) coral reefs could fail to protect coastlines from harsh ocean waves. Coral reefs also provide fish species with their habitat, and

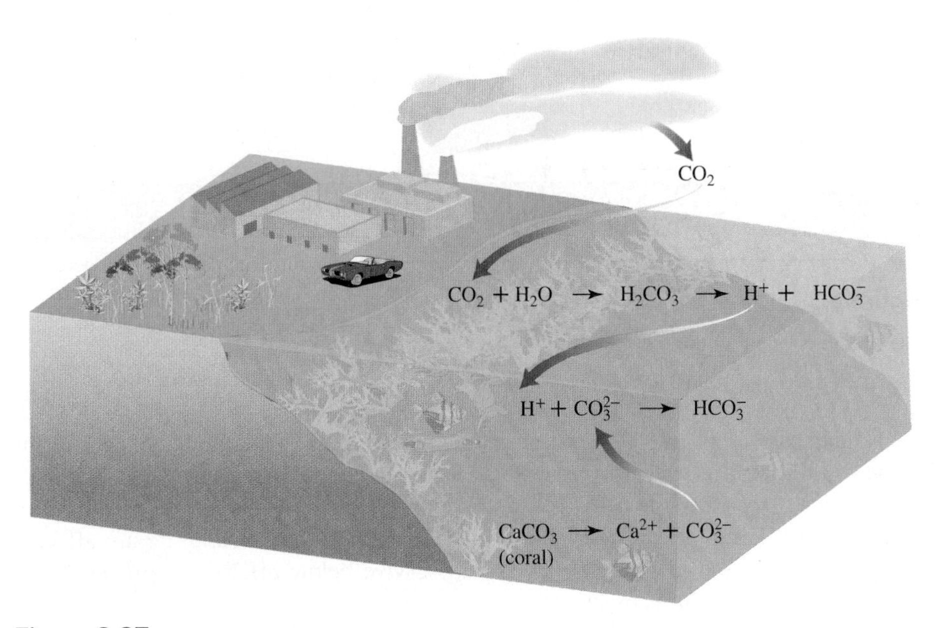

Figure 8.37

Chemistry of CO_2 in the ocean.

damage to the reefs would translate into losses of marine life. Finally, a weakening of the reefs would make them more susceptible to further damage from storms and predators.

Can the ocean heal itself? Although we don't know the answer for sure, nonetheless we can speculate from what we know of past events. When changes in ocean pH have occurred over a very long period of time, the ocean has been able to compensate. This happens because large collections of sediment at the bottom of the ocean contain massive amounts of calcium carbonate, mostly from the shells of long-deceased marine creatures. Over long time periods, these sediments dissolve to replenish the carbonate lost to reaction with excess H^+. But today's changes in ocean pH have happened rapidly on the geologic time scale. In just 200 years, the pH of the ocean has dropped to a level not seen in the past 400 million years. Because the acidification is occurring over a relatively short time and in water close to the surface, the sediment reserve has not had time to dissolve and counteract the effects of the added acidity.

Even if the amount of carbon dioxide in the atmosphere were to immediately level off, the oceans would take thousands of years to return to the pH measured in pre-industrial times. Coral reefs would take even longer to regenerate, and any species lost to extinction, of course, would not return.

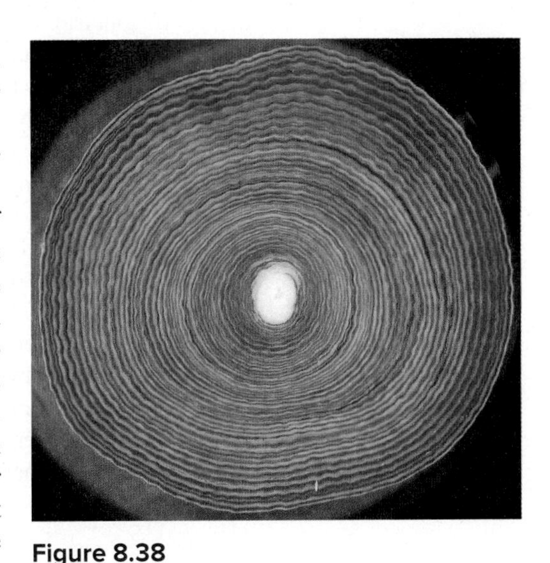

Figure 8.38

A thin slice of coral. Special lighting reveals annual growth rings. A recent study has shown that some corals have seen a dramatic decrease in their growth rate over the past 20 years.

Your Turn 8.49 You Decide International Response to Ocean Acidification

In 2008, a group of scientists met in Monaco to raise awareness about ocean acidification. They issued the Monaco Declaration, calling on the countries of the world to reverse carbon dioxide emissions trends by 2020. Have more recent gatherings of scientists and negotiators created a worldwide policy to address ocean acidification? Do research of your own and summarize your findings.

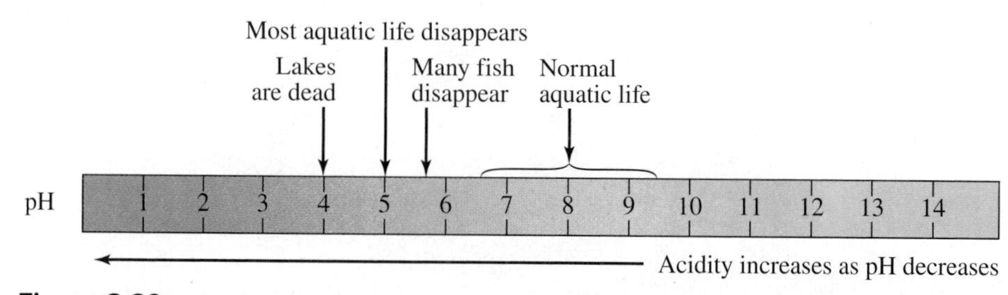

Figure 8.39

Aquatic life and pH.

Humans are not the only creatures bearing the costs of acidification. Organisms in the world's surface waters experience a change in environment when acid rain (also called *acidic precipitation*) fills lakes and streams. Healthy lakes have a pH of 6.5 or slightly above. If the pH is lowered below 6.0, fish and other aquatic life are affected (Figure 8.39). Only a few hardy species can survive below pH 5.0. At pH 4.0, lakes become essentially dead ecosystems.

Numerous studies have reported the progressive acidification of lakes and rivers in certain geographic regions, along with reductions in fish populations. In southern Norway and Sweden, where the problem was first observed, one-fifth of the lakes no longer contain any fish, and half of the rivers have no brown trout. In Southeastern Ontario, the average pH of lakes is now 5.0, well below the pH of 6.5 required for a healthy lake. In Virginia, more than one-third of the trout streams are episodically acidic, or at risk of becoming so.

Many areas of the Midwestern United States have no problem with acidification of lakes or streams, even though the Midwest is a major source of acidic precipitation. This apparent paradox can be explained quite simply. When acid rain falls on or runs off into a lake, the pH of the lake drops (becomes more acidic) unless the acid is neutralized, or somehow used by the surrounding vegetation. In some regions, the surrounding soils may contain bases that can neutralize the acid. The capacity of a lake or other body of water to resist a decrease in pH is called its **acid-neutralizing capacity**. The surface geology of much of the Midwest is limestone, $CaCO_3$. As a result, lakes in the Midwest have a high acid-neutralizing capacity because limestone slowly reacts with acid rain. Perhaps most importantly, the lakes and streams also have a relatively high concentration of calcium and bicarbonate ions. This occurs as a result of the reaction of limestone with carbon dioxide and water:

$$CaCO_3(s) + CO_2(g) + H_2O(l) \longrightarrow \underset{\text{calcium ion}}{Ca^{2+}(aq)} + \underset{\text{bicarbonate ion}}{2\ HCO_3^-(aq)} \qquad \textbf{[8.15]}$$

Because acid is consumed by the carbonate and bicarbonate ions, the pH of the lake remains more or less constant.

Your Turn 8.50 Skill Building The Bicarbonate Ion

A bicarbonate ion produced in **Equation 8.15** can also accept a hydrogen ion, H^+.

a. Write the balanced chemical equation.
b. Is the bicarbonate ion functioning as an acid or a base?

In contrast to the Midwest, many lakes in New England and northern New York (as well as in Norway and Sweden) are surrounded by granite, a hard, impervious, and much less-reactive rock. Unless other local processes are at work, these lakes have very little acid-neutralizing capacity. Consequently, many show a gradual acidification.

As it turns out, understanding the acidification of lakes is a good deal more complicated than simply measuring pH and acid-neutralizing capacities. One level of complexity is added by annual variations. Some years, for example, heavy winter snowfalls persist into the spring and then melt suddenly. As a result, the runoff may be more acidic than usual, because it contains all the acidic deposits locked away in the winter snows. A surge of acidity may enter the waterways at just the time when fish are spawning or hatching and are more vulnerable. In the Adirondack Mountains of northern New York, about 70% of the sensitive lakes are at risk for episodic acidification, in comparison with a far smaller percent that are chronically affected (19%). In the Appalachians, the number of episodically affected lakes (30%) is seven times those chronically affected.

As you can see, pH differences in water play a huge role in biodiversity, habitats, and the overall environment. In the final sections of this chapter, we are going to revisit providing clean freshwater for people to drink.

8.12 | Treating Our Water

This section explores both what takes place to make water clean (at a local water treatment plant), and what happens after we make it dirty (at a sewage treatment plant). Let us begin with what takes place at a local drinking water treatment plant. We assume that the plant gets water from an aquifer or lake. For example, if you live in San Antonio, water is pumped from the Edwards Aquifer. Or, if you live in San Francisco, the water comes from a reservoir in the Hetch Hetchy valley, more than 100 miles away.

In a typical water treatment plant (Figure 8.40), the first step is to pass the water through a screen that physically removes large impurities such as weeds, sticks, and beverage bottles. The next step is to add aluminum sulfate ($Al_2(SO_4)_3$) and calcium hydroxide ($Ca(OH)_2$). Take a moment to review these two chemicals.

Your Turn 8.51 Skill Building Water Treatment Chemicals

a. Write chemical formulas for these ions: sulfate, hydroxide, calcium, and aluminum.

b. What are some compounds that can be formed from these four ions? Write their chemical formulas.

c. The hypochlorite ion (ClO^-) plays a role in water purification. Write chemical formulas for sodium hypochlorite and calcium hypochlorite.

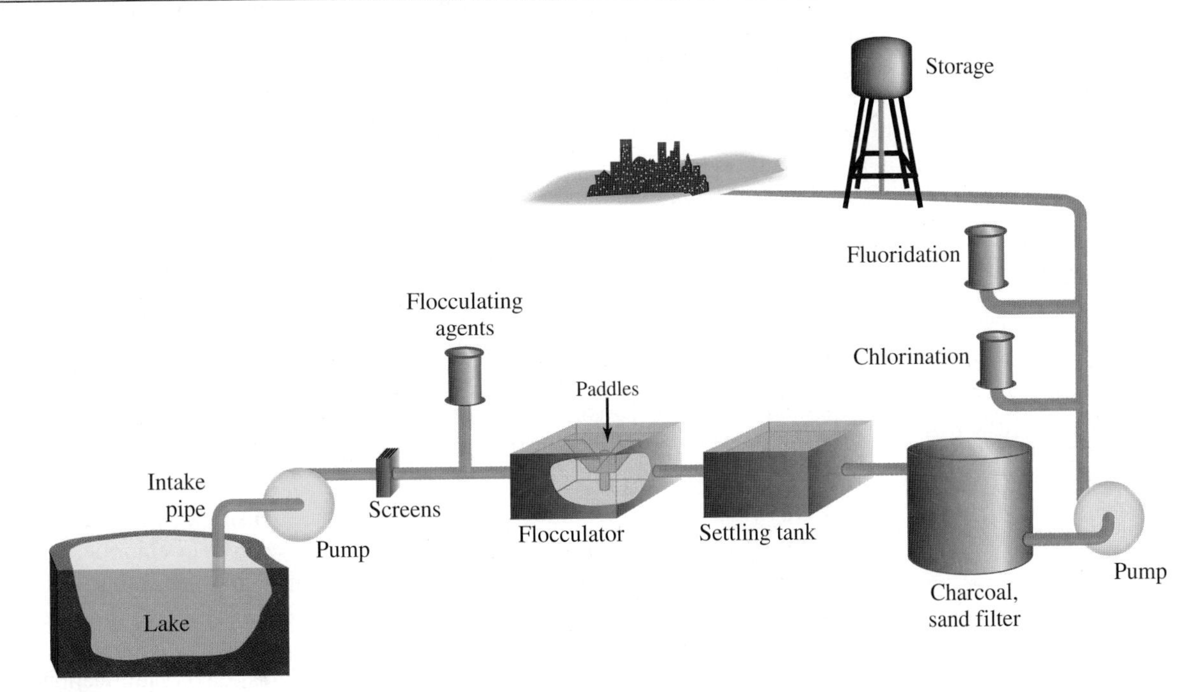

Figure 8.40

A typical municipal water treatment facility. Components are not illustrated to scale.

Did You Know? Tamarind seed kernel powder is an inexpensive, natural, and biodegradable flocculating agent.

Did You Know? Chlorine can only kill the microorganisms with which it comes in contact. Chlorine does not kill bacteria or viruses that are enclosed inside particles of silt or clay. This is one reason why particles need to be removed before the chlorination step.

NaClO is present in Clorox® and other brands of laundry bleach. $Ca(ClO)_2$ is commonly used to disinfect swimming pools.

HClO is sometimes written as HOCl to show the order in which the atoms are bonded.

Aluminum sulfate and calcium hydroxide are *flocculating agents*; that is, they react in water to form a sticky floc (gel) of aluminum hydroxide, $Al(OH)_3$ (Equation 8.16). This gel collects suspended clay and dirt particles on its surface. As the $Al(OH)_3$ gel slowly settles, it carries particles with it that were suspended in the water. Any remaining particles are removed as the water is filtered through charcoal or gravel and then sand.

$$Al_2(SO_4)_3(aq) + 3\ Ca(OH)_2(s) \longrightarrow 2\ Al(OH)_3(s) + 3\ CaSO_4(aq) \quad [8.16]$$

The crucial step comes next—disinfecting the water to kill disease-causing microbes. In the United States, this is most commonly done with chlorine-containing compounds. Chlorination is accomplished by adding chlorine gas (Cl_2), sodium hypochlorite (NaClO), or calcium hypochlorite ($Ca(ClO)_2$). All of these compounds generate the antibacterial agent hypochlorous acid, HClO. A very low concentration of HClO, 0.075 to 0.600 ppm, remains to protect the water against further bacterial contamination as it passes through pipes to the user. **Residual chlorine** refers to the chlorine-containing chemicals that remain in the water after the chlorination step. These include hypochlorous acid (HClO), hypochlorite ions (ClO^-), and dissolved elemental chlorine (Cl_2).

Before chlorination, thousands died in epidemics spread via polluted water. In a classic study, John Snow (1813–1858), an English physician, was able to trace a mid-1800s cholera epidemic in London to water contaminated with the excrement of cholera victims. Another example occurred in 2007 in war-torn Iraq. After extremists put chlorine tanks on suicide truck bombs earlier that year, authorities kept tight controls on chlorine. The chlorine killed two dozen people in several attacks, sending up noxious clouds that left hundreds of people panicked and gasping for breath. At one point, a shipment of 100,000 tons of chlorine was held up for a week at the Jordanian border amid fears for its safe passage through Iraq. With the water infrastructure disrupted and the quality of water and sanitation poor, levels of fecal coliform bacteria increased dramatically, resulting in thousands of Iraqis contracting cholera.

Even in peacetime when the transportation of chlorine is relatively safe, chlorination has its drawbacks. The taste and odor of residual chlorine can be objectionable, and is commonly cited as a reason why people drink bottled water or use filters to remove residual chlorine. A more serious drawback is the reaction of residual chlorine with other substances in the water to form by-products in drinking water at concentrations that may be toxic. The most widely publicized, **trihalomethanes (THMs)**, are compounds such as $CHCl_3$ (chloroform), $CHBr_3$ (bromoform), $CHBrCl_2$ (bromodichloromethane), and $CHBr_2Cl$ (dibromochloromethane) that form from the reaction of chlorine or bromine with organic matter in drinking water. Like HClO, hypobromous acid (HBrO) used to disinfect spa tubs can generate trihalomethanes.

Your Turn 8.52 Skill Building THMs at a Glance

a. Draw Lewis structures for any two THM molecules.
b. THMs differ from CFCs in their chemical composition. How?
c. THMs differ from CFCs in their physical properties. How?

The toxicity of ozone, O_3, in the lower atmosphere (troposphere), as well as its benefits within the upper atmosphere (stratosphere), were described in Sections 2.8 and 3.5, respectively.

Did You Know? Both ozone and chlorine kill bacteria by attacking their cell walls.

Many European and a few U.S. cities use ozone to disinfect their water supplies. One advantage is that a lower concentration of ozone relative to chlorine is required to kill bacteria. Furthermore, ozone is more effective than chlorine against water-borne viruses. But ozonation also comes with disadvantages. One is cost. Ozonation only becomes economical for large water-treatment plants. Another is that ozone decomposes quickly, and hence does not protect water from possible contamination as it is piped through the municipal distribution system. Consequently, a low dose of chlorine must be added to ozonated water as it leaves the treatment plant.

Disinfecting water using ultraviolet (UV) light is gaining in popularity. By UV, we mean UVC, the high-energy UV radiation that can break down DNA in microorganisms, including bacteria. Disinfection with UVC is fast, leaves no residual by-products, and is economical for small installations, including rural homes with unsafe well water. Like ozone, however, UVC does not protect the water after it leaves the treatment site. Again, a low dose of chlorine must be added. Depending on local needs, one or more additional purification steps may be taken after disinfection at the water treatment facility. Sometimes the water is sprayed into the air to remove volatile chemicals that create objectionable odors and taste. If little natural fluoride is present in the water supply, some municipalities add fluoride ions (~1 ppm NaF) to protect against tooth decay. Learn more about fluoridation in the next activity.

Your Turn 8.53 You Decide Keep Your Teeth!

Until recently, losing your teeth was common as you grew older. The culprit was dental caries, a disease in which bacteria attack enamel and cause infections.

a. Community water fluoridation is cited as one of 10 greatest public health achievements of the 20th century by the U.S. Centers for Disease Control and Prevention. Explain why.
b. Although important in all communities, water fluoridation is especially important for low-income communities. Explain.
c. In some communities, water fluoridation is highly controversial. What are the arguments against adding fluoride to drinking water?

We just described how water is treated before it is ready to drink out of the tap. But once we turn on the tap, we start the process of getting the water dirty again. We add waste to the water each time it leaves our bathrooms in a toilet flush, runs down the drain after a soapy shower, or goes down the sink after we wash the dishes. Clearly, it makes sense to use as little water as possible because if we dirty it, it has to be cleaned again before being released back to the environment. Remember green chemistry! It is better to prevent waste than to treat or clean up waste after it is formed.

How do we remove waste from water? If the drains in your home are connected to a municipal sewage system, then the wastewater flows to a sewage treatment plant. Once there, it undergoes similar cleaning processes to those for water treatment, with the exception of end-stage chlorination, before it is released back to the environment.

Cleaning sewage is more complicated, though, because it contains waste in the form of organic compounds and nitrate ions. To many aquatic organisms, this waste is a source of food! As these organisms feed, they deplete oxygen from surface waters. **Biological oxygen demand (BOD)** is a measure of the amount of dissolved oxygen that microorganisms use up as they decompose organic waste found in water. A low BOD is one indicator of good water quality.

Nitrates and phosphates both contribute to BOD, because these ions are important nutrients for aquatic life. An overabundance of either can disrupt the normal flow of nutrients and lead to algal blooms (Figure 8.41) that clog waterways and deplete oxygen from the water. In turn, this reduced oxygen can lead to massive fish kills. The problem of reduced oxygen in water is compounded by the fact that the solubility of oxygen in water is so very low in the first place.

Some treatment plants are using wetland areas to capture nutrients such as nitrates and phosphates before the water is returned to the surface water or recharges the groundwater. Plants and soil microorganisms in these wetland areas (marshes and bogs) facilitate nutrient recycling, thus reducing the nutrient load in the water. If the water produced from treated sewage is clean enough, why not just use it as a source of drinking water? Singapore's growing population relies on several potable water sources. One of these, NEWater, is purified wastewater. The next activity gives you the opportunity to explore this controversial use of reclaimed water, and the final section will look at ways scientists and communities are using chemistry to allow more people to receive potable water.

Figure 8.41

A pond with algal bloom in Brookmill Park, Great Britain.

> **Your Turn 8.54 You Decide Toilet to Tap?**
>
> Communities are considering using reclaimed water as a source of drinking water. If the quality of the water produced from the sewage treatment process matched the quality of the water in our current drinking water system, would you accept treated sewage water as drinking water? Comment either way.

> **Your Turn 8.55 Scientific Practices and Your Diary Water Treatment**
>
> Revisit your water diary. How was the water you used over the diary period treated? Research the treatment processes in your community. Are there ways for you to conserve the amount of water that needs to be treated?

8.13 | Water Solutions for Global Challenges

According to the United Nations website, "Water is crucial for sustainable development, including the preservation of our natural environment and the alleviation of poverty and hunger. Water is indispensable for human health and well-being."

In this final section, we showcase efforts that demonstrate the sustainable use of water. The first relates to the production of fresh water from salt water. The second describes how individuals in developing nations can purify their own drinking water.

Fresh Water from Salt Water

"Water, water everywhere, nor any drop to drink." These words from *The Rime of the Ancient Mariner* are as true today as they were in 1798 when written by Samuel Coleridge. The high salt content (3.5%) of seawater makes it unfit for human consumption.

Figure 8.42

A desalination plant at Jebel Ali in the United Arab Emirates.

While some creatures can live in salt water, neither the ancient mariner nor we can subsist on drinking it.

Today, we are able to tap the sea as a source of water for both agriculture and drinking. **Desalination** is any process that removes sodium chloride and other minerals from salty water, thus producing potable water. In 2013, the International Desalination Association reported that more than 17,000 desalination plants worldwide produced more than 80 billion liters of water daily. With demand for fresh water ever increasing, we now are witnessing the construction of many new desalination facilities, currently in 150 countries including Spain, the United States, China, and Australia, as well as the Middle East and North Africa. One of the world's largest in the United Arab Emirates is shown in Figure 8.42.

One means of desalination is **distillation**, a separation process in which a liquid solution is heated and the vapors are condensed and collected. Impure water is heated; as the water vaporizes, it leaves behind most of its dissolved impurities. However, distillation requires energy! Figure 8.43 shows this energy being provided by a Bunsen burner in one case, and by the Sun in the other. Recall that water has a high specific heat and requires an unusually large amount of energy to convert to a vapor. Both properties result from the extensive hydrogen bonding among water molecules.

Distillation was discussed in Section 5.12, for the fractionation of crude oil into various fuels.

Large-scale distillation operations employ new technologies with impressive names such as *multistage flash evaporation*. Although these technologies have increased energy efficiency over the basic distillation process shown in Figure 8.43a, their energy requirement is still high and is usually provided by burning fossil fuels. An alternative is to purify water using smaller solar distillation units, as shown in Figure 8.43b.

Other desalination options exist. For example, **osmosis** is the passage of water through a semipermeable membrane from a solution that is less concentrated to a solution that is more concentrated. The water diffuses through the membrane, and the solute does not. This is why the membrane is called "semipermeable." However, with an input of energy, osmosis can be reversed. **Reverse osmosis** uses pressure to force the movement of water through a semipermeable membrane from a solution that is more concentrated to a solution that is less concentrated. To use this process to purify water, pressure is applied to the saltwater side, forcing water through the membrane to leave the salt and other impurities behind (Figure 8.44).

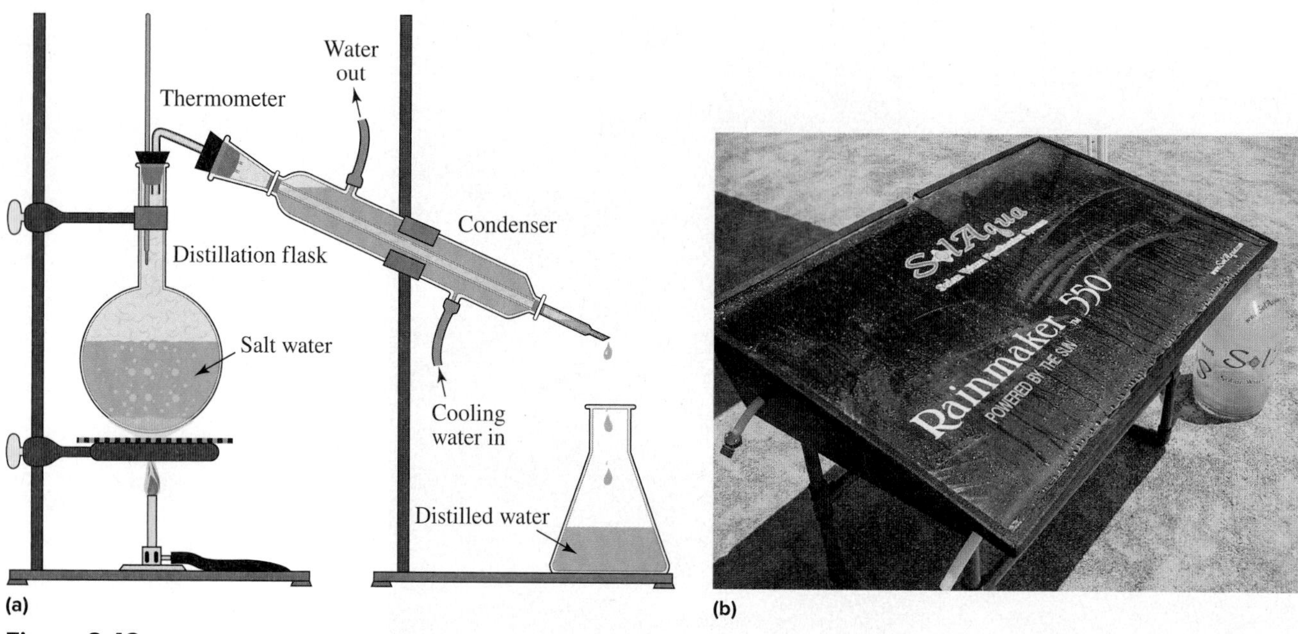

(a)

(b)

Figure 8.43

(a) Laboratory distillation apparatus. **(b)** Tabletop solar still.

(b): © and courtesy of SolAqua

As might be expected, producing the required pressure in reverse osmosis systems is energy intensive. Reverse-osmosis technology can be used to produce some bottled water, as well as ultra-pure water used in the microelectronic and pharmaceutical industries. Portable units are also suitable for use on sailboats (Figure 8.45).

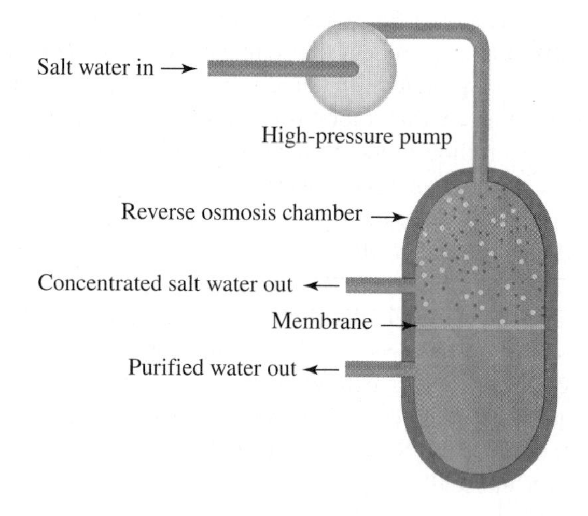

Figure 8.44

Water purification by reverse osmosis.

Your Turn 8.56 You Decide At What Cost?

An Internet blogger proclaimed, "Desalination will make it possible for us to get clean water. This will solve our water shortages." Revisit the green chemistry key ideas to help you refute these claims.

Point-of-Use Straws

With advances in sanitation and management of water-borne diseases over the past century, many in developed countries have access to high-quality drinking water that meets certain standards. However, worldwide, more than one billion people are sickened

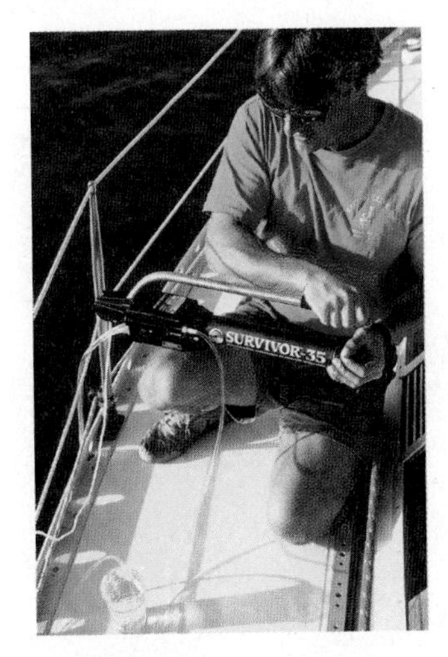

Figure 8.45

A small reverse-osmosis apparatus for converting seawater to potable water.

© *Courtesy of Katadyn*

Figure 8.46

Children using personal LifeStraws to drink.

© *Vestergaard Frandsen*

or die each year due to cholera, typhoid, and other diseases caused by microbes in untreated water. A European company, Vestergaard Frandsen, developed the LifeStraw, which removes virtually all bacteria and protozoan parasites from water. LifeStraws are used in many parts of the globe, including in time of need following a natural disaster.

Aptly named, the personal LifeStraw is a type of pipe filter through which to consume water, as shown in Figure 8.46. This unit can be used to drink from a stream, river, or lake. Lasting about a year, it can purify about 1,000 liters of water. The larger LifeStraw family unit contains a different filter that removes bacteria and further improves the quality of water. The family unit filters up to 18,000 liters of water for about three years.

However, the personal LifeStraw has limitations. It is not a long-term solution to the lack of potable water. In addition, it doesn't remove metals such as arsenic or mercury, or the viral microbes responsible for diarrhea. Both types of LifeStraws provide an interim solution in regions where fresh water is contaminated with microbes.

Your Turn 8.57 You Decide Periodic Error

The company that produces LifeStraws has a set of FAQs on the Internet. One reads: "Does LifeStraw filter heavy metals like arsenic, iron, and fluoride?" What might the "informed chemist" say about the phrasing of this answer: "No, the present version does not filter any of the heavy metals."

Your Turn 8.58 Scientific Practices The Future of Water

a. Brainstorm two ideas that might help us keep water clean.

b. Identify an important global water issue. Suggest two factors that make it important. Name two ways in which people currently are addressing this issue.

Your Turn 8.59 Scientific Practices Final Analysis of the Water Diary

Revisit your water diary. After studying this chapter, what are your thoughts on your current usage of water? Suggest a different method for tracking your water usage data.

Conclusions

Like the air we breathe, water is essential to our lives. It bathes our cells, transports nutrients through our bodies, provides most of our body mass, and cools us when it evaporates. Water is also central to our way of life. We drink it, cook with it, clean things in it, use it to irrigate our crops, and manufacture goods with it. However, as we do these things, we add waste to the water. Although fresh water purifies itself through a cycle of evaporation and condensation, we humans are dirtying water faster than nature can regenerate clean water.

Remember: *It is better to prevent waste than to treat or clean up waste after it is formed.* So catch the rainwater and use it on a garden, rather than letting it run off and join the streams of runoff that pick up pollutants. Instead of using the garbage disposal to grind up food wastes, put the scraps in a compost pile and save the tap water. Turn off the faucet when you are brushing your teeth, limit your time in the shower, and fix that dripping faucet and running toilet!

You may feel like your efforts are a mere drop in a much larger bucket. Indeed, they are. But remember that like the raindrop shown in a winning Earth Day poster (Figure 8.47), your efforts are part of the bigger water picture on this planet.

Although fresh water is a renewable resource, the demands of population growth, rising affluence, and other global issues are amplifying shortages of this essential commodity. If we are to achieve sustainability, we must think water! That's right, "Think water!" Your life and the lives of other creatures depend on it.

One little raindrop
Falling from the darkened sky
Landing in the lake

Figure 8.47

An Earth Day Haiku Poster Winner, 2008.

Artwork © Robert Schill. Photograph by Sally Mitchell.

Learning Outcomes

The numbers in parentheses indicate the sections within the chapter where these outcomes were discussed.

Having studied this chapter, you should now be able to:

- classify and characterize the states of matter present on Earth (8.1)
- describe the conditions that allow water to be present in all three physical states on Earth (8.1)
- draw and describe the composition, shape, and polarity of water molecules (8.2)
- explain how the composition, shape, and polarity lead to its unique properties and interactions (8.3)
- define and illustrate hydrogen bonding among polar molecules (8.3)
- identify the properties of water that make it essential to life (8.3)
- describe and illustrate water's unique molecular composition as a solid and a liquid (8.4)
- identify sources and locations of water on Earth (8.4)
- describe the magnitudes and fractions of water that are actually fresh water (8.4)
- compare and contrast the properties of various sources of water (8.4)
- define potable water (8.5)
- analyze data to evaluate water use, consumption, and contamination (8.6)
- describe water as the "universal solvent" and why it can "carry" so many other chemicals (8.7)
- recall differences between ionic and molecular compounds (8.7)
- model solvation of ionic and molecular compounds in water (8.7)
- define molarity as a unit of concentration (8.8)
- calculate various concentrations of substances in water and use difference units of concentrationc (8.8)
- explain how various solutes can change the properties of water (8.9)
- define acids, bases, and pH (8.10)
- relate the pH scale to common household chemicals (8.10)
- describe how small pH changes indicate large changes in concentration (8.11)
- explain how changes in water properties alter its ability to support life, ecosystems, and biodiversity (8.12)
- describe how water can be treated (filtration, distillation) on small and large scales (8.12)
- outline and illustrate the desalination process and explain how it makes water usable (8.13)
- explain how people in different parts of the world obtain and make water usable (8.13)

Questions

Emphasizing Essentials

1. In any language, water is the most abundant compound on the surface of the Earth.
   a. Explain the term *compound* and also why water is *not* an element.
   b. Draw the Lewis structure for water and explain why its shape is bent.
2. Today we are creating dirty water faster than nature can clean it for us.
   a. Name five daily activities that dirty the water.
   b. Name two ways in which polluting substances naturally are removed from water.
   c. Name five steps you could take to keep water cleaner in the first place.
3. Life on our planet depends on water. Explain each of these.
   a. Bodies of water act as heat reservoirs, moderating climate.
   b. Ice protects ecosystems in lakes because it floats rather than sinks.
4. Why might a water pipe break if left full of water during extended frigid weather?
5. The following are four pairs of atoms. Consult Table 8.2 to answer these questions.

   N and C   S and O
   N and H   S and F

   a. What is the electronegativity difference between the atoms?
   b. Assume that a single covalent bond forms between each pair of atoms. Which atom attracts the electron pair in the bond more strongly?
   c. Arrange the bonds in order of increasing polarity.
6. Consider a molecule of ammonia, NH_3.
   a. Draw its Lewis structure.
   b. Does the NH_3 molecule contain polar bonds? Explain.
   c. Is the NH_3 molecule polar? **Hint:** Consider its geometry.
   d. Would you predict NH_3 to be soluble in water? Explain.
7. In some cases, the boiling point of a substance increases with its molar mass.
   a. Does this hold true for hydrocarbons? Explain with examples.
   b. Based on the molar masses of H_2O, N_2, O_2, and CO_2, which would you expect to have the lowest boiling point?
   c. Unlike N_2, O_2, and CO_2, water is a liquid at room temperature. Explain.
8. Both methane (CH_4) and water are compounds of hydrogen and another nonmetal.
   a. Give four examples of nonmetals. In general, how do the electronegativity values of nonmetals compare with those of metals?
   b. How do the electronegativity values of carbon, oxygen, and hydrogen compare?
   c. Which bond is more polar, the C–H bond or the O–H bond? Justify your answer.
   d. Methane is a gas at room temperature, but water is a liquid. Explain.
9. This diagram represents two water molecules in a liquid state. What kind of bonding force does the arrow indicate?

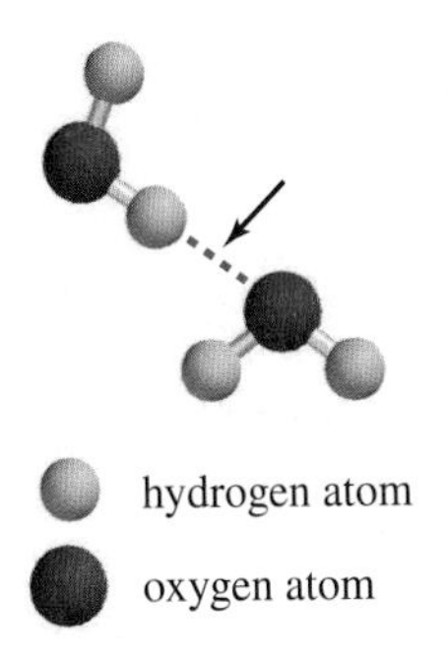

10. For each of these atoms, draw a Lewis structure. Also draw the Lewis structure for the corresponding ion.
    a. Cl   b. Ba
    c. S   d. Li
    e. Ne
11. a. Draw the Lewis structure for the water molecule.
    b. Draw Lewis structures for the hydrogen ion and the hydroxide ion.
    c. Write a chemical reaction that relates all three structures from parts a and b.
12. The density of water at 0 °C is 0.9987 g/cm^3; the density of ice at this same temperature is 0.917 g/cm^3.
    a. Calculate the volume occupied at 0 °C by 100.0 g of liquid water and by 100.0 g of ice.
    b. Calculate the percentage increase in volume when 100.0 g of water freezes at 0 °C.

Concentrating on Concepts

13. Consider these liquids:

Liquid	Density, g/mL
dishwashing detergent	1.03
maple syrup	1.37
vegetable oil	0.91

a. If you pour equal volumes of these three liquids into a 250-mL graduated cylinder, in what order should you add the liquids to create three separate layers? Explain.

b. Predict what would happen if a volume of water equal to the other liquids were poured into the cylinder in part **a** and the contents then were mixed vigorously.

14. Let's say the water in a 500-L drum represents the world's total supply. How many liters would be suitable for drinking?

15. Based on your experience, how soluble is each of these substances in water? Use terms such as *very soluble, partially soluble,* or *not soluble.* Cite supporting evidence.

a. orange juice concentrate

b. household ammonia

c. chicken fat

d. liquid laundry detergent

e. chicken broth

16. a. Bottled water consumption was reported to be 29 gallons per person in the United States in 2011. The 2010 U.S. census reported the population as 3.1×10^8 people. Given this, estimate the total bottled water consumption.

b. Convert your answer in part **a** to liters.

17. NaCl is an ionic compound, but $SiCl_4$ is a molecular compound.

a. Use Table 8.2 to determine the electronegativity difference between chlorine and sodium, and between chlorine and silicon.

b. What correlations can be drawn about the difference in electronegativity between bonded atoms and their tendency to form ionic or covalent bonds?

c. How can you explain, on the molecular level, the conclusion reached in part **b**?

18. The maximum contaminant level (MCL) for mercury in drinking water is 0.002 mg/L.

a. Does this correspond to 2 ppm or 2 ppb mercury?

b. Is this mercury in the form of elemental mercury ("quicksilver") or the mercury ion (Hg^{2+})?

19. The acceptable limit for nitrate, often found in well water in agricultural areas, is 10 ppm. If a water sample is found to contain 350 mg/L, does it meet the acceptable limit?

20. A student weighs out 5.85 g of NaCl to make a 0.10 M solution. What size volumetric flask does he or she need?

21. Solutions can be tested for conductivity using this type of apparatus.

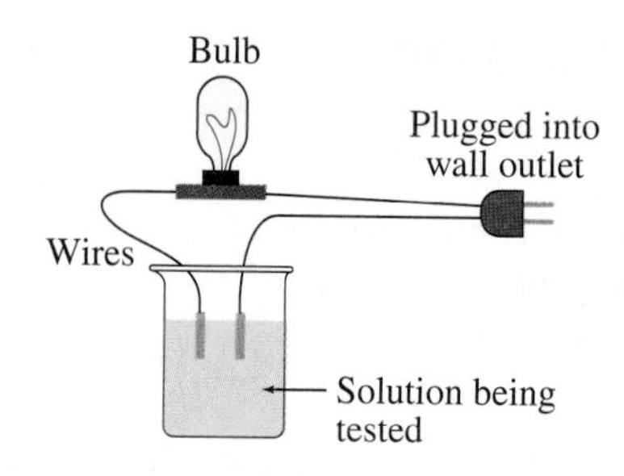

Predict what will happen when each of these dilute solutions is tested for conductivity. Explain your predictions briefly.

a. $CaCl_2(aq)$

b. $C_2H_5OH(aq)$

c. $H_2SO_4(aq)$

22. An aqueous solution of KCl conducts electricity, but an aqueous solution of sucrose does not. Explain.

23. Based on the generalizations in Table 8.6, which compounds are likely to be water-soluble?

a. $KC_2H_3O_2$

b. LiOH

c. $Ca(NO_3)_2$

d. Na_2SO_4

24. For a 2.5 M solution of $Mg(NO_3)_2$, what is the concentration of each ion present?

25. Calcium carbonate is a salt. Write its chemical formula. Would you expect calcium carbonate to be soluble or insoluble in water?

26. Write a chemical equation that shows the release of one hydrogen ion from a molecule of each of these acids.

a. $HBr(aq)$, hydrobromic acid

b. $H_2SO_3(aq)$, sulfurous acid

c. $HC_2H_3O_2(aq)$, acetic acid

27. Classify the following aqueous solutions as acidic, neutral, or basic.

a. $HI(aq)$

b. $NaCl(aq)$

c. $NH_4OH(aq)$

d. $[H^+] = 1 \times 10^{-8}$ M

e. $[OH^-] = 1 \times 10^{-2}$ M

f. $[H^+] = 5 \times 10^{-7}$ M

g. $[OH^-] = 1 \times 10^{-12}$ M

28. For parts **d** and **f** of question 27, calculate the $[OH^-]$ that corresponds to the given $[H^+]$. Similarly, for parts **e** and **g**, calculate the $[H^+]$.

29. In each pair below, the $[H^+]$ is different. By what factor of 10 is it different?
    a. pH = 6 and pH = 8
    b. pH = 5.5 and pH = 6.5
    c. $[H^+] = 1 \times 10^{-8}$ M and $[H^+] = 1 \times 10^{-6}$ M
    d. $[OH^-] = 1 \times 10^{-2}$ M and $[OH^-] = 1 \times 10^{-3}$ M

30. Which of these has the *lowest* concentration of hydrogen ions: 0.1 M HCl, 0.1 M NaOH, 0.1 M H_2SO_4, or pure water? Explain your answer.

31. Consider these ions: nitrate, sulfate, carbonate, and ammonium.
    a. Give the chemical formula for each.
    b. Write a chemical equation in which the ion (in aqueous form) appears as a product.

32. Write a chemical equation that shows the release of hydroxide ions as each of these bases dissolves in water.
    a. KOH(*s*), potassium hydroxide
    b. $Ba(OH)_2$(*s*), barium hydroxide

33. Explain how you would prepare these solutions using powdered reagents and any necessary glassware.
    a. Two liters of 1.50 M KOH
    b. One liter of 0.050 M NaBr
    c. 0.10 L of 1.2 M $Mg(OH)_2$

34. a. A 5-minute shower requires about 90 L of water. How much water would you save for each minute that you shorten your shower?
    b. Running the water while you brush your teeth can consume another liter. How much water can you save in a week by turning it off?

35. Which gas is dissolved in water to produce each of the following acids?
    a. carbonic acid, H_2CO_3
    b. sulfurous acid, H_2SO_3

36. Write a balanced chemical equation for each acid–base reaction.
    a. Potassium hydroxide is neutralized by nitric acid.
    b. Hydrochloric acid is neutralized by barium hydroxide.
    c. Sulfuric acid is neutralized by ammonium hydroxide.

37. Give names and chemical formulas for five bases of your choice. Name three observable properties generally associated with bases.

38. Give names and chemical formulas for five acids of your choice. Name three observable properties generally associated with acids.

39. Use the Internet to determine which has the higher water footprint, a 100-gram chocolate bar or a 16-ounce glass of beer. Explain the difference.

40. Explain why water is often called the *universal solvent*.

41. Is there any such thing as "pure" drinking water? Discuss what is implied by this term, and how the meaning of this term might change in different parts of the world.

42. Some vitamins are water-soluble, whereas others are fat-soluble. Would you expect either or both to be polar molecules? Explain.

43. At the edge of a favorite fishing hole, a new sign is posted that reads "Caution: Fish from this lake may contain over 1.5 ppb Hg." Explain to a fishing buddy what this unit of concentration means, and why the caution sign should be heeded.

44. This periodic table contains four elements identified by numbers.

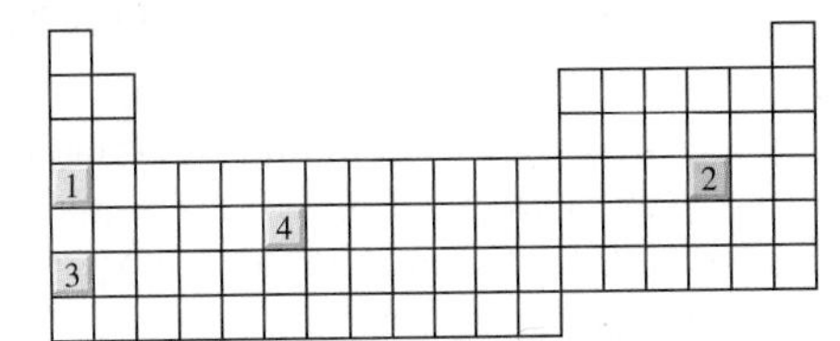

    a. Based on the trends you observe from Table 8.2, which of the four elements would you expect to have the highest electronegativity value? Explain.
    b. Based on trends within the periodic table, rank the other three elements in order of decreasing electronegativity values. Explain your ranking.

45. A diatomic molecule XY that contains a polar bond *must* be a polar molecule. However, a triatomic molecule XY_2 that contains a polar bond *does not necessarily* form a polar molecule. Use some examples of real molecules to help explain this difference.

46. Imagine you are at the molecular level, watching water vapor condense.

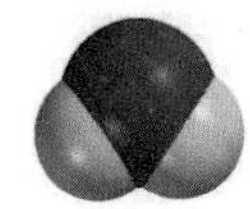

    a. Sketch four water molecules using a space-filling representation similar to this one. Sketch them in the gaseous state and then in the liquid state. How does the collection of molecules change when water vapor condenses to a liquid?
    b. What happens at the molecular level when water changes from a liquid to a solid?

47. Propose an explanation for the fact that NH_3, like H_2O, has an unexpectedly high specific heat.

48. a. What type of bond holds together the two hydrogen atoms in the hydrogen molecule, H_2?
    b. Explain why the term *hydrogen bonding* does *not* apply to the bond within H_2.
    c. Explain why the term hydrogen bonding does *not* apply to the bond within H_2O, but *does* apply to a sample of water.

49. Consider ethanol, an alcohol with the chemical formula of C_2H_5OH.
    a. Draw the Lewis structure for ethanol.
    b. A cube of solid ethanol sinks rather than floats in liquid ethanol. Explain this behavior.

Exploring Extensions

50. The unusually high specific heat of water helps keep our body temperature within a normal range despite age, activity, and environmental factors. Consider some of the ways the body produces and loses heat. How would these differ if water had a low specific heat?

51. Health goals for contaminants in drinking water are expressed as MCLG, or maximum contaminant level goals. Legal limits are given as MCL, or maximum contaminant levels. How are MCLG and MCL related for a given contaminant?

52. Some areas have a higher than normal amount of THMs (trihalomethanes) in the drinking water. Suppose you are considering moving to such an area. Write a letter to the local water district asking relevant questions about the drinking water.

53. Infants are highly susceptible to elevated nitrate levels because bacteria in their digestive tract convert nitrate ion into nitrite ion, a much more toxic substance.
    a. Give chemical formulas for both the nitrate ion and nitrite ion.
    b. Nitrite ion can interfere with the ability of blood to carry oxygen. Explain the role of oxygen in respiration.
    c. Boiling nitrate-containing water will not remove nitrate ion. Explain.

54. In 2016, testing indicated that the drinking water supply in Flint, Michigan, had concentrations of dissolved lead that far exceeded those established by the Safe Drinking Water Act.
    a. What was the major source of lead in the drinking water?
    b. What are possible health effects of the elevated levels of lead?
    c. What is being done to improve water quality in this community?
    d. What other cities or states have recently reported high levels of lead?

55. Explain why desalination techniques, despite proven technological effectiveness, are not used more widely to produce potable drinking water.

56. In 2005, the Great Lakes–St. Lawrence River Basin Sustainable Water Resources Agreement set the stage to coordinate water management and protect water from use by those outside the region.
    a. List states and provinces involved with this unique transboundary agreement.
    b. What was the impetus behind protecting these waters?

57. Liquid CO_2 has been used successfully for many years to decaffeinate coffee. Explain how and why this works.

58. How can you purify your water when you are hiking? Name two or three possibilities. Compare these methods in terms of cost and effectiveness. Are any of these methods similar to those used to purify municipal water supplies? Explain.

59. Hydrogen bonds vary in strength from about 4 to 40 kJ/mol. Given that the hydrogen bonds between water molecules are at the high end of this range, how does the strength of a hydrogen bond *between* water molecules compare with the strength of a H–O covalent bond *within* a water molecule? Do your values bear out the assertion that hydrogen bonds are about one tenth as strong as covalent bonds?

60. Levels of naturally occurring mercury in surface water are usually less than 0.5 mg/L.
    a. Name three human activities that add Hg^{2+} ("inorganic mercury") to water.
    b. What is "organic mercury"? This chemical form of mercury tends to accumulate in the fatty tissues of fish. Explain why.

61. We all have the amino acid glycine in our bodies. Here is the structural formula.

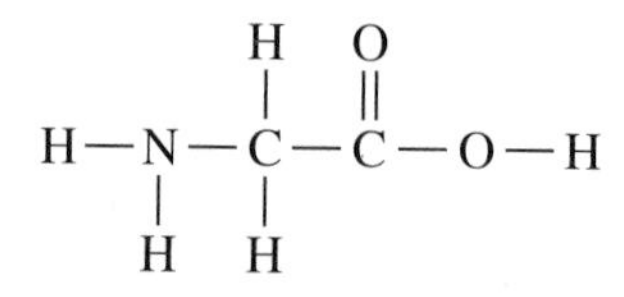

    a. Is glycine a polar or nonpolar molecule? Explain.
    b. Can glycine exhibit hydrogen bonding? Explain.
    c. Is glycine soluble in water? Explain.

62. Hard water may contain Mg^{2+} and Ca^{2+} ions. The process of water softening removes these ions.
    a. How hard is the water in your local area? One way to answer this question is to determine the number of water-softening companies in your area. Use the Internet to find out if your area is targeted for the marketing of water-softening devices.
    b. If you chose to treat your hard water, what are the options?

63. Suppose you are in charge of regulating an industry in your area that manufactures agricultural pesticides. How will you decide if this plant is obeying necessary environmental controls? Which criteria affect the success of this plant?

64. Before the U.S. EPA banned their manufacture in 1979, PCBs were regarded as useful chemicals. What properties made them desirable? Besides being persistent in the environment, they bioaccumulate in the fatty tissues of animals. Use the electronegativity concept to show why PCB molecules are nonpolar and thus fat-soluble.

65. The PUR "Purifier of Water" is a point-of-use system.
   a. How does this system work?
   b. Compare it to the personal LifeStraw by listing benefits offered by each system.

66. In the United States, the EPA has set SMCLs (secondary maximum contaminant levels) for substances in drinking water that are not health threatening. Visit the EPA website to learn more about one of these substances and prepare a summary of your findings.

67. The EPA uses an extensive process to add contaminants to its list of regulated substances. Search the Internet for information on the Unregulated Contaminant Monitoring (UCM) program.
   a. What is the UCM, and when does it occur?
   b. What is the importance of its Contaminant Candidate List (CCL), and how does it relate to the precautionary principle?
   c. List some general categories of substances included in the CCL. Include one specific substance from the most current list.

68. List a recent theme for World Water Day. Prepare a short presentation of this theme in a format of your choice.

69. Carbon dioxide is a gas found in our atmosphere.
   a. What is the approximate concentration?
   b. Why is its concentration in the atmosphere increasing?
   c. Draw the Lewis structure for the CO_2 molecule.
   d. Would you expect carbon dioxide to be highly soluble in seawater? Explain.

CHAPTER 9

The World of Polymers and Plastics

© Bignai/Shutterstock.com

REFLECTION

Recycling Plastics

A video in **Connect** shows how a soda bottle may be recycled into clothing. After watching that video, look at the plastic items around you and answer these questions.

a. Find three different items made from plastic with three different recycling symbols (a number within a triangle).
b. What do these symbols mean?

As we move through this chapter, you will learn how various plastics are made and how their chemical structures differ.

The Big Picture

In this chapter, you will explore the following questions:

- What are polymers?
- Where can you find polymers in your everyday life?
- How are polymers synthesized?
- What kinds of polymers can be recycled?
- How are polymers recycled?
- What are some unique applications for polymers?

The EVOLUTION of CELL PHONES

Source: https://netherregioniii.files.wordpress.com/2012/06/evolutionofcellphone.jpg

Introduction

Imagine carrying a brick around in your pocket wherever you go. A rough, hard, heavy brick that was your connection to the digital world. It would barely fit in your pocket, if at all. Rubbing your hand or face across it would scratch your skin. Dropping it would cause it to shatter, potentially hurting your feet. Without polymers, this would be what your cell phone would look and feel like. Not a convenient tool to carry around to connect with family and friends, but an inefficient, heavy anchor that would likely remain unused and might be easily damaged.

Throughout history, humans have continually improved the tools around them to better conform to their needs and desires by making materials stronger, lighter, and more durable. One class of materials that has transformed our world is known as *polymers*. By definition, a **polymer** is a chemical compound that is in the form of long, repeating chains. Man-made, or synthetic polymers, are amazingly tailorable and can be used to create materials with properties ideal for different applications. Need a lightweight but strong material that resists scratches? There's a polymer for that. Need a smooth surface that won't irritate skin or clothing? How about a material that is slightly bendable and doesn't break when dropped? All of these properties can be engineered into a synthetic polymer that can become part of a better cell phone.

Did You Know? Some polymers are natural, such as spider silk; others are man-made, or synthetic, such as Kevlar®. In either case, polymers are large molecules made from many small starting materials.

Whereas synthetic polymers were first developed in the early 20th century, natural polymers have been around since life itself began. Cellulose, starch, and other complex carbohydrates are examples of natural polymers. Natural rubber is a polymer obtained from rubber trees. Even the code for life itself, DNA, is a natural polymer.

In this chapter, you will learn about what makes up a polymer, how the structure of the polymer can give rise to a variety of properties, how these amazing materials are created, and how products made of polymers can be recycled into other materials for further use.

9.1 | Polymers Here, There, and Everywhere

Did You Know? Some artificial turfs are made from recycled plastic. In some climates, artificial turf lowers water use as well. This can help in locations that experience severe droughts, such as the Southwestern United States.

Polymers have revolutionized the world around us. For example, in the world of sports, football is often played on artificial turf by players wearing plastic helmets. Tennis balls are made from synthetic polymers. Carbon fibers embedded in plastic resins provide the strength, flexibility, and lightweight construction required in bicycles, fishing rods, and sailboat hulls. Hockey players skate on rinks of Teflon™ or high-density polyethylene when natural ice is not available. Although wooden canoes still have their appeal, most canoes today are made of polymers.

All plastics are polymers, but not all polymers are plastics!

In the previous paragraph, notice that we referred to both *polymers* and *plastics*. These two terms are related, and are oftentimes used interchangeably. Whereas the word *polymer* encompasses both natural and synthetic polymers, the term *plastic* is only appropriate for some synthetic varieties. Look for some examples in the following activity.

Your Turn 9.1 Scientific Practices Tennis Anyone?

Examine this photo of tennis players. Choose three applications for polymers from the photo. Describe the polymer properties that make each one well-suited for its intended use.

© andresr/Getty Images RF

© CMCD/Getty Images RF

Although polymers are everywhere, you may need to train your eye to recognize them. Not all have the look and feel of a yellow rubber duck! Some are transparent, such as the clear plastic wrap on food, whereas others are opaque, such as the container that holds liquid laundry detergent. Some are rigid, such as nylon automotive parts, while others are more flexible, such as a plastic spatula. Some polymers are drawn into fibers to weave clothing and carpet, whereas others are molded into different shapes.

In principle, synthetic polymers can be made from many different starting materials. In practice, most come from a single raw material: crude oil. As you learned in Section 5.10, oil is no longer as easy to obtain on our planet as it used to be. Today, crude oil is the starting material for many plastics, pharmaceuticals, fabrics, and other carbon-based products. However, plastics can also come from renewable materials, as we will discuss toward the end of this chapter.

Dmitri Mendeleev (1834–1907), the great Russian chemist who organized the periodic table as we know it today, remarked that burning petroleum as a fuel "would be akin to firing up a kitchen stove with bank notes."

Both the origin and fate of polymers are of interest to us. For instance, modern cell phones and electronics contain many plastic components. However, in 2013, only 9.2% of the total plastic discarded in the United States was recycled. To understand the complexities surrounding the sources and ultimate fate of polymers, you first need to know something about their chemical structures and how they are made—the topic of the next section.

9.2 | Polymers: Long, Long Chains

Polymers are often referred to as *macromolecules*. For convenience, it is common to use the unit Dalton, Da, to express the molar mass of macromolecules such as polymers. For instance, a molar mass of 10,000 g/mol would be equivalent to 10 kDa.

Rayon, nylon, and polyurethane. Teflon™, Lycra®, Styrofoam™, and Formica®. These seemingly different materials are all synthetic polymers. What they have in common is most easily evident at the molecular level. This consists of long chain(s) of atoms covalently bonded to one another. A polymer can easily contain thousands of atoms, and have a molar mass of more than 1,000,000 g/mol!

Monomers (*mono* meaning "one"; *meros* meaning "unit") are the small molecules used to synthesize polymers. Each monomer is analogous to a link in a chain. Polymers (*poly* means "many") can be formed from one monomer, or from a

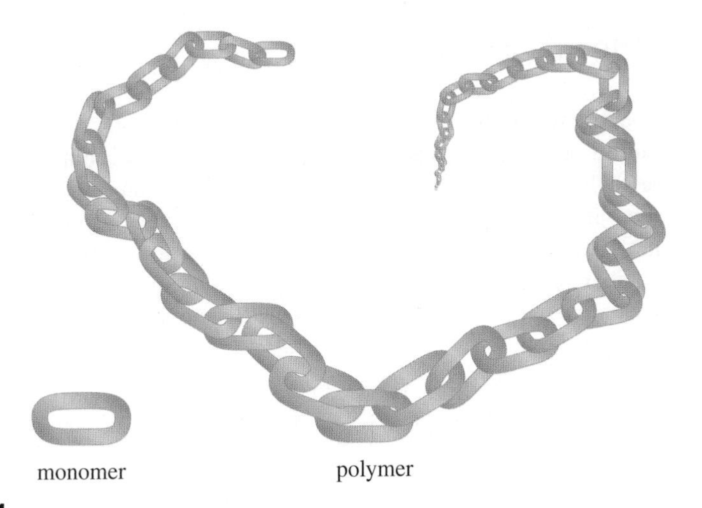

Figure 9.1

Representations of a monomer (*single link*) and a polymer (*long chain*) made from one type of monomer.

© Everday Objects OS06/Getty Images RF

combination of two or more different monomers. The long chain shown in Figure 9.1 may help you to imagine a polymer made from identical monomers; that is, identical links in a chain.

Keep in mind that chemists did not invent polymers. For example, the natural polymers of glucose, cellulose, and starch were described earlier in the context of biofuels (Chapter 5). Other natural polymers include wool, cotton, silk, natural rubber, skin, and hair. Like synthetic polymers, natural ones exhibit a stunning variety of properties. They give strength to an oak tree, delicacy to a spider's web, softness to goose down, and flexibility to a blade of grass (Figure 9.2).

Some early synthetic polymers were developed as substitutes for expensive or rare natural polymers such as silk and rubber. Others were developed to deliver comparable strength at a lower mass. For example, contrast the density of steel, about 8 g/cm^3, with that of plastics, 1–2 g/cm^3. As a result, an automobile body constructed with plastics weighs less than its steel counterpart, thus requiring less fuel to operate. Similarly, plastic packaging reduces weight and helps save fuel during shipping.

The concept of *density* was introduced in Section 1.9.

Figure 9.2

Oak logs and grass both contain the natural polymer cellulose. Glucose is the monomer.

(both): © Cathy Middlecamp

9.3 | Adding Up the Monomers

How do monomers combine to make a polymer? In the previous section, we used a chain to represent a polymer but made no mention of how the chain was formed. In this section, we will provide details of how covalent bonds connect the monomers.

Polyethylene is our first example. As the name indicates, polyethylene is a polymer of ethylene, $H_2C{=}CH_2$ (Equation 9.1). Ethylene is a common name for ethene, the smallest member in the family of hydrocarbons containing a C=C double bond. In the polymerization reaction, n molecules of the ethylene monomer combine to form polyethylene:

$$n\, H_2C{=}CH_2 \xrightarrow{R\cdot} \left[\!-\mathrm{CH_2}-\mathrm{CH_2}-\!\right]_n \quad [9.1]$$

Polyethylene is also called polyethene in the United Kingdom, reflecting the fact that ethene (not ethylene) is the systematic name. "Eth" indicates two carbon atoms, and "-ene" indicates a C=C double bond.

The coefficient n in front of the ethylene monomer specifies the number of molecules reacting. In turn, this determines the molar mass of the polymer, typically between 10,000 and 100,000 g/mol, but can run into the millions. On the right side, the n appears as a subscript, indicating that each monomer has become part of the long chain. The large square brackets enclose the repeating unit of the polymer.

Polyethylene is the sole product. The monomers add to one another to form a long chain of n units. As a result, we call this **addition polymerization**, a type of polymerization in which the monomers add to the growing chain in such a way that the polymer contains all the atoms of the monomer.

Notice the R· over the arrow in Equation 9.1. So that you can better appreciate its significance, we will tell you a bit more about ethylene, the monomer. Produced at oil refineries, ethylene is a flammable, colorless gas with a faint gasoline-like odor. These properties are quite unlike the odorless solid, polyethylene (Figure 9.3a). Although not classified as an air pollutant, ethylene nonetheless is a VOC (volatile organic compound). As you learned in Chapter 2, VOCs in the atmosphere are precursors to the buildup of photochemical smog. Accordingly, safety precautions are needed when transporting ethylene from refineries to sites at which polyethylene is produced. To conserve space, the ethylene gas is pressurized and refrigerated to liquefy it. In this form, it is transported in tank cars that bear labels like the one shown in Figure 9.3b.

(a)

(b)

Figure 9.3

(a) Bottles made from polyethylene. **(b)** A sign posted on a railway tank car that transports liquefied ethylene. The 1038 identifies it as ethylene, the red diamond indicates high flammability, and the 2 indicates moderate reactivity.

(a): © Cathy Middlecamp

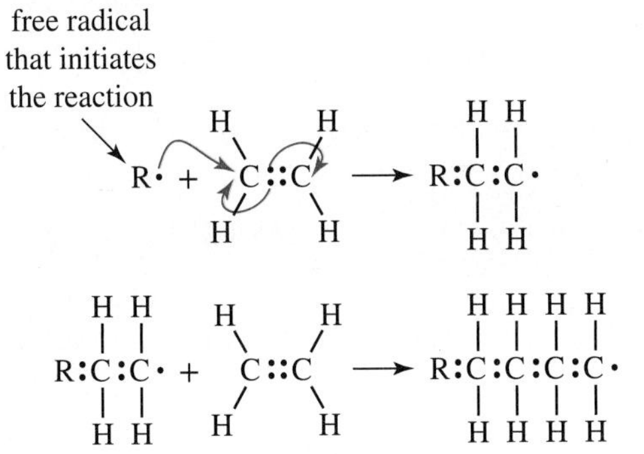

free radical
that initiates
the reaction

$$\mathrm{R\cdot} + \begin{array}{c}\mathrm{H}\quad\ \ \mathrm{H}\\ \diagdown\ \ \diagup\\ \mathrm{C{::}C}\\ \diagup\ \ \diagdown\\ \mathrm{H}\quad\ \ \mathrm{H}\end{array} \longrightarrow \begin{array}{c}\mathrm{H\ \ H}\\ |\ \ \ |\\ \mathrm{R{:}C{:}C\cdot}\\ |\ \ \ |\\ \mathrm{H\ \ H}\end{array}$$

$$\begin{array}{c}\mathrm{H\ \ H}\\ |\ \ \ |\\ \mathrm{R{:}C{:}C\cdot}\\ |\ \ \ |\\ \mathrm{H\ \ H}\end{array} + \begin{array}{c}\mathrm{H}\quad\ \ \mathrm{H}\\ \diagdown\ \ \diagup\\ \mathrm{C{::}C}\\ \diagup\ \ \diagdown\\ \mathrm{H}\quad\ \ \mathrm{H}\end{array} \longrightarrow \begin{array}{c}\mathrm{H\ \ H\ \ H\ \ H}\\ |\ \ \ |\ \ \ |\ \ \ |\\ \mathrm{R{:}C{:}C{:}C{:}C\cdot}\\ |\ \ \ |\ \ \ |\ \ \ |\\ \mathrm{H\ \ H\ \ H\ \ H}\end{array}$$

Figure 9.4

Scheme showing the mechanism for the polymerization of ethylene.

Does liquid ethylene polymerize in the tank car? Fortunately, no. Clearly the end user of the ethylene would be distressed to receive a tank car full of solid polyethylene! In order to initiate the polymerization reaction, a free radical (R·) is required, as shown over the arrow in Equation 9.1. This free radical represents one of a variety of chemical species, all with an unpaired electron.

Recall that the hydroxyl free radical, ·OH, was described earlier in Section 2.13.

To initiate the process of forming the polymer chain, R· attaches to $H_2C{=}CH_2$ (Figure 9.4). To understand what happens next, recall that the double bond in ethylene contains *four* electrons. After an ethylene molecule reacts with R·, only *two* of these electrons remain in a C–C single bond. The other *two* electrons move (shown by the red arrows) to form two new bonds—one to R·, and the other as an unpaired electron at the end of the molecule, thus providing a site at which another monomer can add.

Your Turn 9.2 Scientific Practices Polymer Reactions

An animated version of **Figure 9.4** is found on **Figures Alive!** in **Connect**. Once you watch the animation, how do you think the process of polymerization could be stopped?

As each ethylene monomer adds, a new C–C bond forms and the chain grows. This process repeats many times. Occasionally, the ends of two polymer chains join and stop the chain growth. The process stops when the supply of monomers is exhausted. The result of all this chemistry is that gaseous ethylene is converted to solid polyethylene.

Industrial chemists use several synthetic routes to produce polyethylene. The most common uses a metal catalyst and mild temperatures.

Although we placed R· over the arrow in Equation 9.1, we also could have represented the reaction in this way:

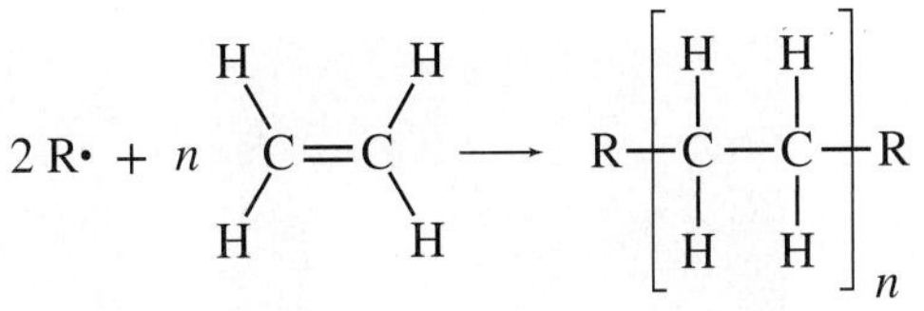

$$2\,\mathrm{R\cdot} + n\ \begin{array}{c}\mathrm{H}\quad\ \ \mathrm{H}\\ \diagdown\ \ \diagup\\ \mathrm{C{=}C}\\ \diagup\ \ \diagdown\\ \mathrm{H}\quad\ \ \mathrm{H}\end{array} \longrightarrow \mathrm{R}\left[\begin{array}{c}\mathrm{H\quad H}\\ |\quad\ \ |\\ \mathrm{C{-}C}\\ |\quad\ \ |\\ \mathrm{H\quad H}\end{array}\right]_n\mathrm{R} \qquad [9.2]$$

Because the R group that "caps" each end of the molecule is such a small part of the much longer chain, we will continue our practice of omitting R· as a reactant, as we did in Equation 9.1.

The numerical value of *n*, and hence the length of the chain, can vary. During the manufacturing process, *n* will be adjusted in order to create specific properties for the polymer. Moreover, within a single sample the individual polymer molecules can have varying lengths. In every case, however, the molecules contain a chain of carbon atoms. In essence, the molecules in polyethylene resemble those in a hydrocarbon such as octane, except that they are much, much longer.

Your Turn 9.3 Skill Building Polymerization of Ethylene

In **Equation 9.1** for the polymerization of ethylene, assume that $n = 4$.

a. Rewrite **Equation 9.1** to indicate this change.

b. Draw the structural formula of the product without using brackets. Remember to put an R group at each end of the chain.

c. In terms of its molecular structure, how does the product differ from octane?

9.4 | Got Polyethylene?

Polyethylene is found in many packaging materials, including plastic bags, milk jugs, detergent containers, and "bubble wrap" (Figure 9.5). Yet, as we have seen in the previous section, all polyethylene is made from the monomer ethylene. How can these materials with seemingly different properties all be made of the same substance?

Your Turn 9.4 Scientific Practices Polyethylene Hunt

As we will describe in this section, polyethylene containers, bags, and packaging materials are marked either as low-density (LDPE) or as high-density (HDPE). Using the recycling code as your guide, locate several items made from each. Do LDPE and HDPE differ in flexibility? Is one more translucent? Is one more often colored with a pigment than the other? Summarize your findings in a brief report.

4 LDPE 2 HDPE

If the actual width of a polyethylene molecule was that of a piece of spaghetti, its length would be almost a half-mile long! In fact, the actual width of a polyethylene molecule is on the order of 0.5 nm—*i.e.*, 20,000 times smaller than an individual hair fiber!

Another type of intermolecular force, hydrogen bonding, was discussed in Section 8.3 and is responsible for the physical properties of water, such as its anomalous boiling point and surface tension.

The different properties of polyethylene stem largely from differences in their long molecular chains. Relatively speaking, these molecules are very long indeed. Imagine a polyethylene molecule to be as wide as a piece of spaghetti. Now imagine that the polyethylene used to make plastic bags contains molecular chains arranged somewhat like cooked spaghetti on a plate. The strands are not very well aligned, although in some regions the molecular chains run parallel to one another. Moreover, the polyethylene chains, like the spaghetti strands, are not covalently bonded to one another.

Recall that in Section 5.12 we used **London dispersion forces** to describe the attractive forces holding hydrocarbons together, which give rise to their varying boiling points. These **intermolecular forces** differ from the covalent bonds that exist *within* molecules arising from shared pairs of electrons. Consequently, chemical changes are governed by the strengths of covalent bonds, whereas physical changes involve intermolecular forces.

Figure 9.5

Packing material and containers made from polyethylene.

All: © McGraw-Hill Education. Jill Braaten, photographer

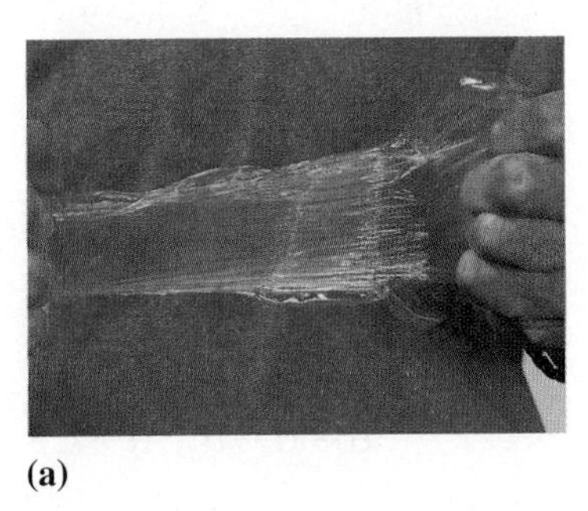
(a)

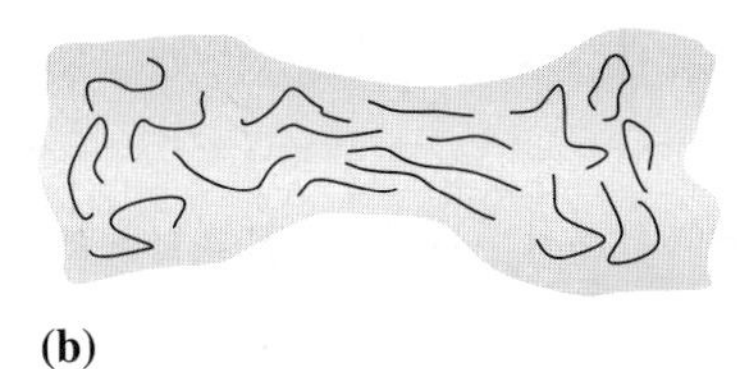
(b)

Figure 9.6

(a) A plastic bag stretched until it "necks." **(b)** A representation of "necking" at the molecular level.

(a): © Bill Aron/PhotoEdit, Inc.

Dispersion forces are significant in large molecules, such as polymers.

Dispersion forces arise in a polymer because each atom in the long polymeric chain contains electrons. These electrons are attracted to the atoms on neighboring chains; the degree of attraction between strands of polyethylene results from the large number of atoms involved. The attraction is a bit like that between the two halves of Velcro™. The larger the surface area of one Velcro® strip, the better it will hold to the other. Individual intermolecular forces between atoms are very small, but polymers have significant intermolecular forces because of the large number of these interactions.

Evidence of the molecular arrangement of polyethylene can be obtained by doing a short experiment. Cut a strip from a heavy-duty polyethylene bag, grab the two ends of the strip, and pull. A fairly strong pull is required to start the plastic stretching, but once it begins, less force is needed to keep the stretch going. The length of the plastic strip increases dramatically as the width and thickness decrease (Figure 9.6a). A small shoulder forms on the wider part of the strip and a narrow neck almost seems to flow from it in a process called "necking." Unlike the stretching of a rubber band, the necking effect is not reversible. Eventually the plastic thins to the point that it tears.

Figure 9.6b represents the necking of polyethylene from a molecular point of view. As the strip narrows, the molecular chains shift, slide, and align parallel to one another in the direction of pull. In some plastics, such stretching (sometimes called "cold drawing") is carried out as part of the manufacturing process to alter the three-dimensional arrangement of the chains in the solid. As the force and stretching continue, the polymer eventually reaches a point at which the strands can no longer realign, and the plastic breaks. Paper, a natural polymer, tears when pulled because the strands (fibers) are rigidly held in place and are not free to slip like the long molecules in polyethylene.

Your Turn 9.5 Scientific Practices "Necking" Polyethylene

Necking permanently changes the properties of a piece of polyethylene.

a. Does necking affect the number of monomer units, *n*, in the average polymer?
b. Does necking affect the bonding between the monomer units within the polymer chain?

Differences in the physical properties of polymers can also arise as a result of the extent of branching within the polymer chain. This is the case with high-density polyethylene (HDPE) and low-density polyethylene (LDPE), as shown in Figure 9.7. As you may have discovered in Your Turn 9.4, the plastic bags dispensed in the produce aisles of supermarkets are usually LDPE. These bags are stretchy, transparent, and not very strong. Their molecules consist of about 500 monomeric units and the central polymeric chain has numerous branches, like limbs radiating from a central tree trunk (Figure 9.7b).

```
                                                       |
                                                     H—C—H
                                                       |
                                                     H—C—H
  H  H  H  H  H  H  H  H  H                     H  H   |   H  H  H  H  H
  |  |  |  |  |  |  |  |  |                     |  |   |   |  |  |  |  |
—C—C—C—C—C—C—C—C—C—                           —C—C—C—C—C—C—C—C—
  |  |  |  |  |  |  |  |  |                     |  |   |   |  |  |  |  |
  H  H  H  H  H  H  H  H  H                     H  H   |   H  H  |  H  H
                                                     H—C—H  H—C—H
                                                       |       |
                                                            H—C—H
                                                               |
                                                            H—C—H
                                                               |
```

(a)

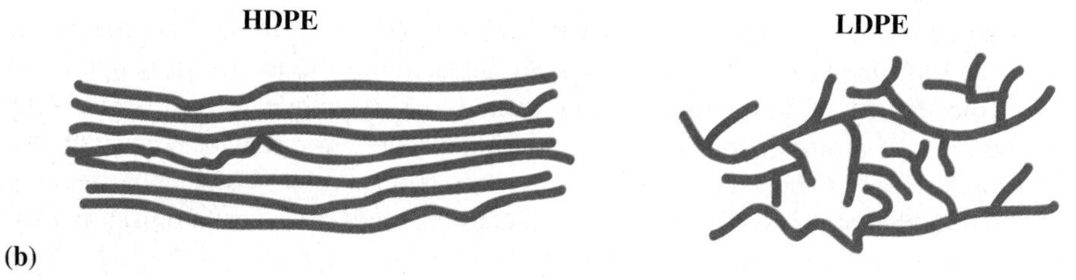

(b)

Figure 9.7

High-density (linear) polyethylene and low-density (branched) polyethylene: **(a)** structural formulas, and **(b)** schematic representations.

Your Turn 9.6 Scientific Practices Ban the Bag or Bag the Ban?

Many cities, a few states, and even whole countries have enacted legislation to ban single-use plastic bags. There are a large number of supporters and opponents to these bag bans. Search the Internet and compile a list of pros and cons for bans of single-use plastic bags.

This low-density form was the first type of polyethylene to be manufactured. About 20 years after its discovery, chemists were able to adjust reaction conditions to prevent branching and thus HDPE was born. In their Nobel Prize-winning research, Karl Ziegler (1898–1973) and Giulio Natta (1903–1979) developed catalysts that enabled them to make linear (unbranched) polyethylene chains of about 10,000 monomer units. With no side branches, these long chains arranged in parallel, unlike the irregular tangle of the polymer chains in LDPE (Figure 9.7). HDPE exhibits a highly ordered molecular structure and has a slightly higher density, greater rigidity, more strength, and a higher melting point than LDPE.

Catalysts were discussed in Section 5.14.

Your Turn 9.7 Scientific Practices HDPE and LDPE

The densities of HDPE and LDPE are 0.96 g/cm^3 and 0.93 g/cm^3, respectively. Use **Figure 9.7** to rationalize the slight difference in densities.

As you might expect, HDPE and LDPE have different uses. High-density polyethylene is used to make many different types of plastic bottles, toys, stiff or "crinkly" plastic bags, and heavy-duty pipes. A newer use of HDPE was spurred by surgery patients with blood-borne diseases, such as HIV/AIDS. Without suitable protection, surgeons would run the risk of being infected. AlliedSignal, Inc. produced a linear

In 1999, AlliedSignal and Honeywell merged to form Honeywell International.

polyethylene fiber called Spectra® that could be fabricated into liners for surgical gloves. These gloves are reported to have 15 times more resistance to cuts than medium-weight leather work gloves, but are still thin enough to allow a keen sense of touch. A sharp scalpel can be drawn across the glove with no damage to the fabric. Such strength is in marked contrast to the properties of everyday plastic gloves used by health professionals.

Your Turn 9.8 Scientific Practices Shopping for Polymers

The *Macrogalleria*, a "Cyberwonderland of Polymer Fun," was created with the support of many sponsors, including the American Chemical Society.

a. Search for *The Macrogalleria* on the Internet and find its virtual shopping mall. Visit stores to locate at least six different items made from LDPE or HDPE. List your findings.

b. Why do you think this site was named the *Macrogalleria*?
Hint: Refer to Chapter 11 to learn about macronutrients and micronutrients.

It would be a mistake to conclude that polyethylene is restricted to the extremes represented by highly branched or strictly linear forms. By modifying the extent and location of branching in LDPE, its properties can be varied from the soft and wax-like coatings on milk cartons to stretchy plastic food wrap. HDPE is rigid enough to be used for plastic milk bottles. The hot water of a dishwasher will melt neither HDPE nor LDPE, but either may melt if left near a hot frying pan or heating element.

Polyethylene has one more property of interest—namely, that it is a good electrical insulator. During World War II, polyethylene was used by the Allied Forces to coat electrical cables in aircraft radar installations. Sir Robert Watt (1892–1973), who discovered radar, described polyethylene's critical importance. "The availability of polythene [polyethylene] transformed the design, production, installation, and maintenance problems of airborne radar from the almost insoluble to the comfortably manageable. A whole range of aerial and feeder designs otherwise unattainable was made possible, a whole crop of intolerable air maintenance problems was removed. And so polythene [polyethylene] played an indispensable part in the long series of victories in the air, on the sea, and on land, which were made possible by radar."[1]

Your Turn 9.9 Scientific Practices Other Types of Polyethylene

In addition to LDPE and HDPE, polyethylene is manufactured as "MDPE" and "LLDPE." Use the Internet to find out about these and other types of polyethylene. How do their properties differ?

9.5 | The "Big Six": Theme and Variations

Today, more than 60,000 synthetic polymers are known. Although polymers were developed for many specialized uses, six types account for roughly 75% of those used in both Europe and the United States. We refer to these everyday polymers as the "Big Six," and you can find them in Table 9.1: polyethylene (low- and high-density), polyvinyl chloride, polystyrene, polypropylene, and polyethylene terephthalate.

Terephthalate is pronounced "ter-eh-THAL-ate." The "ph" is silent.

[1] Quoted by J. C. Swallow in "The History of Polythene" from *Polythene—The Technology and Uses of Ethylene Polymers,* 2nd ed., edited by A. Renfrew. London: Iliffe and Sons, 1960.

Table 9.1 The Big Six

Polymer Recycle Symbol	Monomer	Properties of Polymer	Uses of Polymer
Polyethylene 4 LDPE	Ethylene $H_2C{=}CH_2$	Translucent if not pigmented. Soft and flexible. Unreactive to acids and bases. Strong and tough.	Bags, films, sheets, bubble wrap, toys, wire insulation.
Polyethylene 2 HDPE	Ethylene $H_2C{=}CH_2$	Similar to LDPE. More rigid, tougher, slightly more dense.	Opaque milk, juice, detergent, and shampoo bottles. Buckets, crates, and fencing.
Polyvinyl chloride 3 PVC, or V	Vinyl chloride $H_2C{=}CHCl$	Variable. Rigid if not softened with a plasticizer. Clear and shiny, but often pigmented. Resistant to most chemicals, including oils, acids, and bases.	Rigid: Plumbing pipe, house siding, charge cards, hotel room keys. Softened: Garden hoses, waterproof boots, shower curtains, IV tubing.
Polystyrene 6 PS	Styrene $H_2C{=}CH(C_6H_5)$	Variable. "Crystal" form transparent, sparkling, somewhat brittle. "Expandable" form lightweight foam. Both forms rigid and degraded in many organic solvents.	"Crystal" form: Food wrap, CD cases, transparent cups. "Expandable" form: Foam cups, insulated containers, food packaging trays, egg cartons, packaging peanuts.
Polypropylene 5 PP	Propylene $H_2C{=}CH(CH_3)$	Opaque, very tough, good weatherability. High melting point. Resistant to oils.	Bottle caps. Yogurt, cream, and margarine containers. Carpeting, casual furniture, luggage.
Polyethylene terephthalate 1 PETE, or PET	Ethylene glycol $HO{-}CH_2CH_2{-}OH$; Terephthalic acid $HOOC{-}C_6H_4{-}COOH$	Transparent, strong, shatter-resistant. Impervious to acids and atmospheric gases. Most costly of the six.	Soft-drink bottles, clear food containers, beverage glasses, fleece fabrics, carpet yarns, fiber-fill insulation.

Note: The structures of the first five monomers differ only by the atoms show in blue.

In contrast to thermoplastics, some plastics are *thermosetting*. These solidify or "set" irreversibly with heat. Examples include rubber-soled footwear and antique Bakelite ovenware.

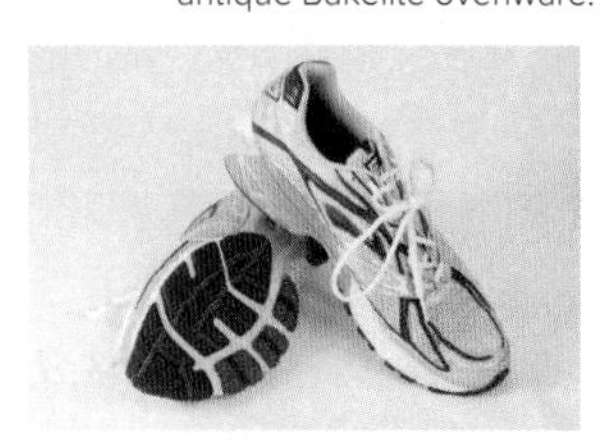

© McGraw-Hill Education

Table 9.1 also lists properties of these six polymers. All are solids that can be colored with pigments. All are also insoluble in water, although some degrade or soften in the presence of hydrocarbons, fats, and oils. All are **thermoplastic polymers**, meaning that with heat, they can be melted and reshaped over and over again. However, they exhibit a range of melting points depending on the route by which they were manufactured. Of the Big Six, polyethylene has the lowest melting point, with LDPE and HDPE melting at about 120 °C and 130 °C, respectively. In contrast, polypropylene (PP) melts at 160–170 °C.

Depending on the arrangement of their molecules, polymers have varying degrees of strength. At the microscopic level, the molecules in some parts of the polymer may have an orderly repeating pattern, such as one would find in a crystalline solid (Figure 9.8). In these **crystalline regions**, the long polymer molecules

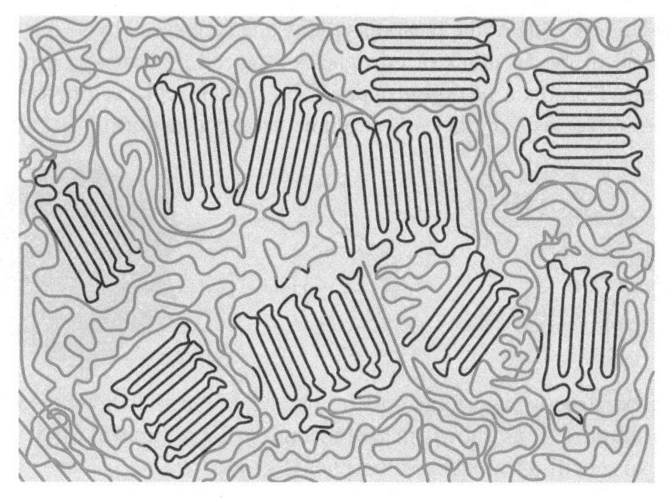

Figure 9.8

A semicrystalline polymer with crystalline regions **(red)** and amorphous regions **(green)**.

are arranged neatly and tightly in a regular pattern. In other parts of the same polymer, you can find **amorphous regions**. Here, the long polymer molecules are found in a random, disordered arrangement and are packed more loosely. Because of their structural regularity, the crystalline regions impart strength and abrasion resistance, such as in HDPE and PP. Although some polymers are highly crystalline, most still include amorphous regions. These regions impart flexibility. For example, the amorphous regions in PP give it the ability to be bent without breaking. The range of properties among polymers means that they are differently suited for specific applications. The next exercise provides an opportunity to match polymers with their uses.

Your Turn 9.10 Scientific Practices Uses of the Big Six

Use **Table 9.1** and other information provided about the Big Six to answer these questions.

a. Which polymer would not be suitable for margarine tubs because it softens with oil?
b. Which polymers are transparent? Which one is used in clear soft-drink bottles?
c. Which one is tough and used for bottle caps? Name another application in which toughness is important.
d. Which ones are listed as unreactive to acids and can serve as containers for acidic beverages, such as orange juice?

From Table 9.1, you can see that five monomers are used to make six different polymers. Here, we focus on three monomers closely related to ethylene: vinyl chloride, propylene, and styrene.

$$H_2C{=}CH_2 \qquad H_2C{=}CHCl \qquad H_2C{=}CH(CH_3) \qquad H_2C{=}CH(C_6H_5)$$

ethylene — vinyl chloride — propylene — styrene

Your Turn 9.11 Scientific Practices Identifying Monomers

An interactive learning resource regarding the monomers employed in PVC, Saran®, and Teflon™ may be found on **Figures Alive!** in **Connect**.

In vinyl chloride, one of the H atoms of ethylene is replaced by a Cl atom. Similarly, in propylene, one of the H atoms of ethylene is replaced by a *methyl* group ($-CH_3$). In contrast, styrene, features a phenyl group, $-C_6H_5$, which replaces one of the H atoms. The *phenyl* group consists of six carbon atoms arranged to form a hexagon:

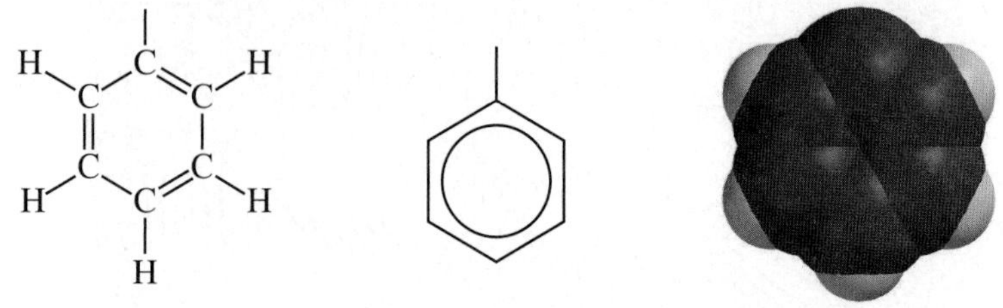

Because the first structural formula for the phenyl group is tedious to draw, the ring sometimes is simplified, as shown in the second structure above. Shown in the third structure is a space-filling model for the phenyl group.

Your Turn 9.12 Skill Building Benzene and Phenyl

The difference between a phenyl group, $—C_6H_5$, and the compound benzene, C_6H_6, is simply one H atom.

a. Both the phenyl group and benzene have two *resonance* structures. Draw them. ***Hint:*** Resonance was introduced in Section 3.7.

b. Given these resonance structures, why is the shorthand symbol of a circle within a hexagon a particularly good representation of both benzene and the phenyl group?

As you might suspect, vinyl chloride, propylene, and styrene undergo addition polymerization, just like ethylene. But the results are somewhat different. To see why, let's look at what happens when n molecules of vinyl chloride polymerize to form polyvinyl chloride, PVC.

In Equation 9.3, the Cl atom could be drawn in any of the four positions that attach to the C atoms in the vinyl chloride monomer. They are all equivalent due to the symmetry of the molecule.

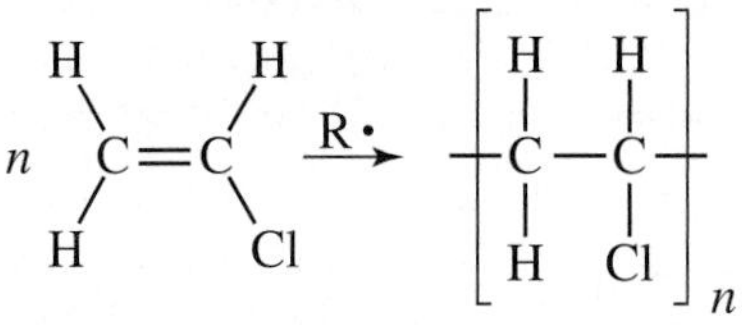

$$n\ \mathrm{CH_2{=}CHCl} \xrightarrow{R\cdot} \left[\!-\mathrm{CH_2{-}CHCl}-\!\right]_n \qquad [9.3]$$

tail head

The Cl atom creates an asymmetry in the monomer. Arbitrarily, think of the carbon atom bearing two H atoms as the "tail," and the carbon with the Cl atom as the "head."

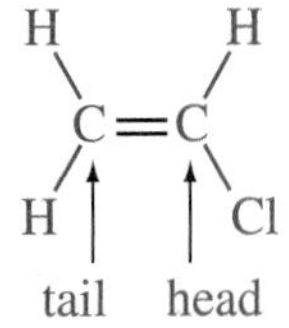

When vinyl chloride monomers add to form polyvinyl chloride, they orient in one of three ways, as shown in Figure 9.9:

- head-to-tail, with the Cl atoms on every other C atom
- alternating head-to-head/tail-to-tail, with the Cl atoms next to each other
- a random mix of the previous two arrangements

The head-to-tail arrangement is the usual product for polyvinyl chloride.

The arrangement of monomers in the chain is one factor that affects the flexibility of the polymer. Thus, each arrangement of PVC has somewhat different properties, with the most regular one—repeating head-to-tail—being the stiffest because the molecules pack more easily together to form crystalline regions. The stiffer PVC finds use in drain and sewer pipes, credit cards, house siding, furniture, and various automobile parts. The random arrangement is still stiff, but somewhat less so.

PVC can be further softened with **plasticizers**, compounds that are added in small amounts to polymers to make them softer and more pliable. Plasticizers work by fitting in between the large polymer molecules, thus disrupting the regular packing of the molecules. Flexible PVC that contains plasticizers is familiar in shower curtains, "rubber" boots, garden hoses, clear IV bags for blood transfusions, artificial leather

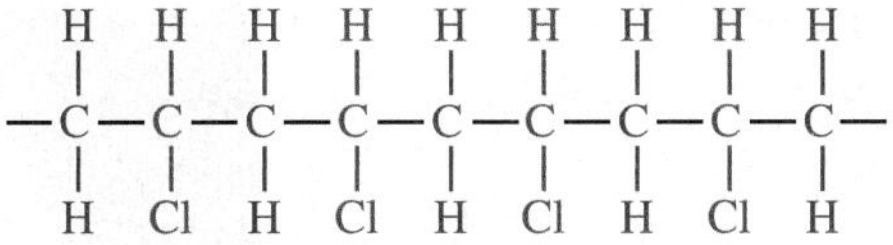

```
   H   H   H   H   H   H   H   H   H
   |   |   |   |   |   |   |   |   |
 —C —C —C —C —C —C —C —C —C—
   |   |   |   |   |   |   |   |   |
   H  Cl   H  Cl   H  Cl   H  Cl   H
```

Head-to-tail, head-to-tail

```
   H   H   H   H   H   H   H   H   H   H
   |   |   |   |   |   |   |   |   |   |
 —C —C —C —C —C —C —C —C —C —C—
   |   |   |   |   |   |   |   |   |   |
  Cl   H   H  Cl  Cl   H   H  Cl  Cl   H
```

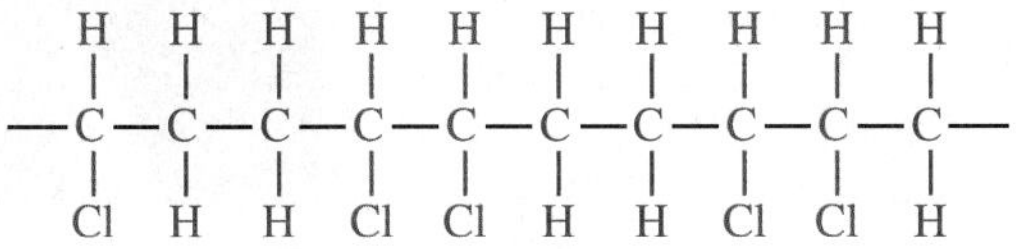

Tail-to-tail, head-to-head

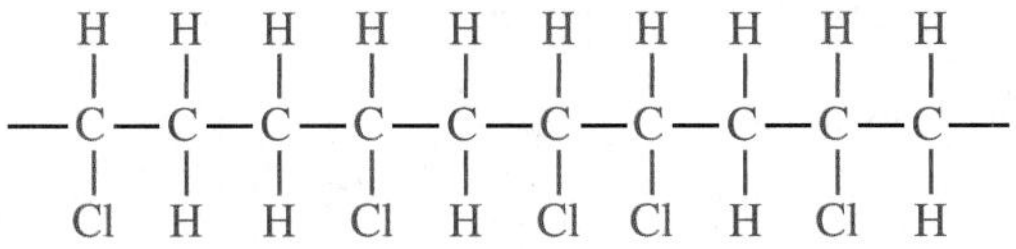

```
   H   H   H   H   H   H   H   H   H   H
   |   |   |   |   |   |   |   |   |   |
 —C —C —C —C —C —C —C —C —C —C—
   |   |   |   |   |   |   |   |   |   |
  Cl   H   H  Cl   H  Cl  Cl   H  Cl   H
```

Random

Figure 9.9

Three possible arrangements of the monomers in PVC.

("patent leather"), and flexible insulation coatings on electrical wires. Flame retardants are also commonly added to polymer mixtures to prevent fires in offices, homes, and small enclosed spaces such as boats and airplanes. Additives to plastics are controversial for several reasons, as we'll see in Section 9.11.

Next, let us consider the polymerization of propylene to form polypropylene, PP. Again, several arrangements are possible because of the asymmetry of the monomer. A particularly useful form of polypropylene is the repeating head-to-tail—head-to-tail arrangement. This regularity imparts a high degree of crystallinity and makes the polymer strong, tough, and able to withstand higher temperatures. These properties are reflected in its uses. For example, indoor–outdoor carpeting is often made using the strong fibers of polypropylene.

Just as ethylene is also called ethene, propylene is also called propene.

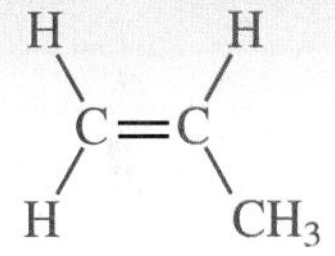

```
 H       H
  \     /
   C = C
  /     \
 H       CH3
```

Your Turn 9.13 Scientific Practices "The Tough One"

Polypropylene may not be as familiar to you as polyethylene or PET, in part because many polypropylene items don't carry a recycling symbol.

a. As just mentioned, polypropylene can be drawn into fibers such as those used in indoor–outdoor carpeting. Suggest two other uses for polypropylene fiber where toughness is desired.

b. Although HDPE is used in many food containers, polypropylene is used for margarine containers. Toughness is not the issue. So what is the issue?

Finally, let us examine the polymerization of *n* molecules of styrene to form polystyrene (PS), an inexpensive and widely used plastic. Here is a representation of the addition polymerization scheme:

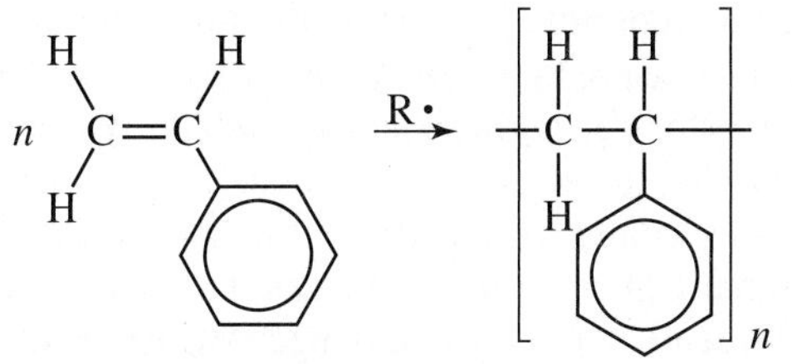

```
    H       H                 [ H   H    ]
     \     /        R·        [ |   |    ]
 n    C = C       ——→       —[—C — C—  ]—
     /     \                  [ |   |    ]
    H     (C6H5)              [ H  (C6H5)]n
```

[9.4]

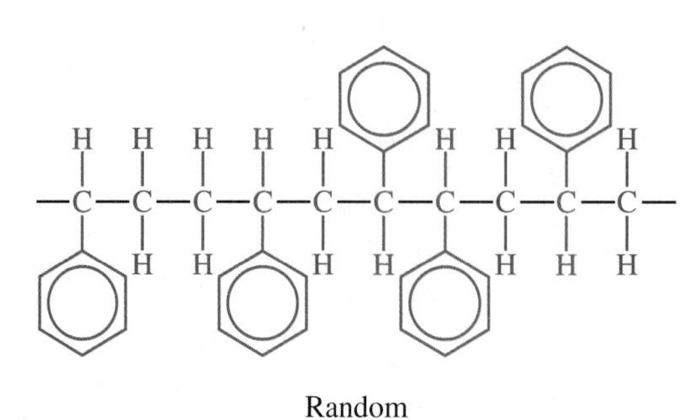

(a)

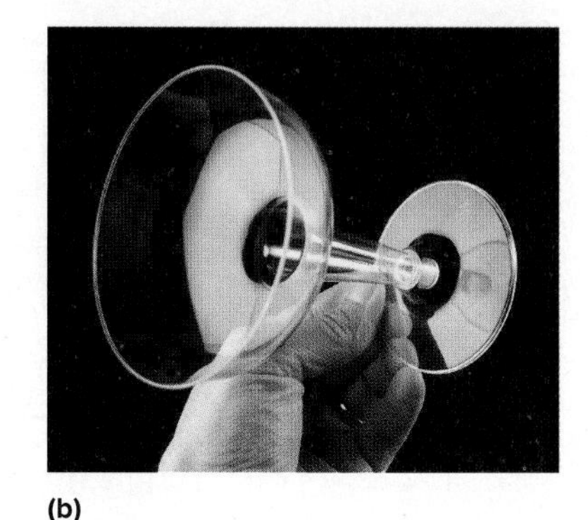

(b)

Figure 9.10

(a) The random arrangement of polystyrene. **(b)** Partyware made from "crystal" (general purpose) polystyrene.

(b): © McGraw-Hill Education. Jill Braaten, photographer

Polystyrene is a hard plastic with little flexibility. Like the other Big Six, it melts when heated (thermoplastic) and casts well into molds. Transparent cases for DVDs and clear-plastic party glasses and plates also are made from polystyrene. So are hard exteriors of many laptop computers and cell phones.

Most commercial polystyrene has the random arrangement of the monomers shown in Figure 9.10a. In this form, sometimes referred to as general purpose or "crystal" polystyrene, the polymer is hard and brittle. Have you ever squeezed too hard on a clear-plastic party glass causing it to split? It probably was polystyrene (Figure 9.10b).

Your Turn 9.14 Skill Building Polystyrene Possibilities

Show the arrangement of atoms in a polystyrene chain in the repeating head-to-tail arrangement. Why do you think this arrangement is favored rather than the head-to-head arrangement?

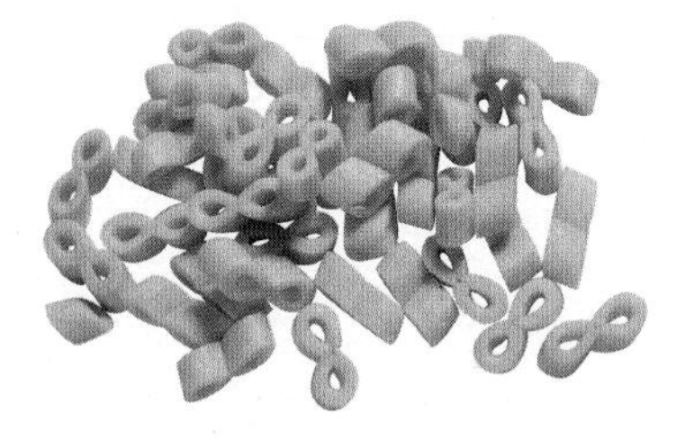

© Image Club RF

Using CO_2 in place of CFCs as a blowing agent illustrates a key principle of green chemistry introduced in Section 2.16: the fabrication and use of substances that are not toxic.

Did You Know? Styrofoam™ is a brand name of polystyrene foam insulation produced by the Dow Chemical Company.

Hot beverage cups, egg cartons, and packing peanuts are also made from polystyrene, sometimes called expandable polystyrene (EPS). These items are made from small hard "expandable" polystyrene beads. These beads contain 4–7% of a **blowing agent**; that is, either a gas or a substance capable of producing a gas to manufacture a foamed plastic. For PS, the blowing agent is typically a low-boiling liquid such as pentane, C_5H_{12}. If the beads are placed in a mold and heated with steam or hot air, the pentane vaporizes. In turn, the expanding gas expands the polymer. The expanded particles are fused together into the shape determined by the mold. Because it contains so many bubbles, this plastic foam is not only light but is also an excellent thermal insulator.

Chlorofluorocarbons, better known as CFCs, were once on the list of compounds used as blowing agents. Because CFCs destroy stratospheric ozone (Chapter 3), their use was phased out in 1990. Pentane in its vapor form and carbon dioxide were two possible replacements. For example, the Dow Chemical Company developed a process that uses pure carbon dioxide as a blowing agent to produce Styrofoam™ for packaging material, which eliminated the use of CFC-12. Furthermore, the CO_2 used is a by-product from existing commercial and natural sources, such as cement production and natural gas wells. Thus, it does not contribute additional CO_2, a greenhouse gas, to the atmosphere. Dow received a 1996 Presidential Green Chemistry Challenge Award for developing these alternative reaction conditions.

9.6 | Cross-Linking Monomers

Monomers make the polymer! As you saw in the previous section, changes in the monomer lead to changes in the properties of the polymer. To understand different monomers, we need to revisit the concept of **functional groups**; that is, distinctive arrangements of groups of atoms that impart characteristic chemical properties to the molecules that contain them (Table 9.2). For example, the *hydroxyl* functional group (–OH) was introduced earlier in the context of ethanol, a biofuel (Section 5.15). This group is present in all compounds classified as alcohols, including one that is of interest in this chapter, ethylene glycol.

Chapters 5 and 12 mention functional groups in the context of biofuels and drug molecules, respectively.

Table 9.2 Selected Functional Groups

Name	Chemical Formula	Structural Formula
hydroxyl (in alcohols)	–OH	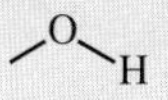
carboxylic acid	–COOH	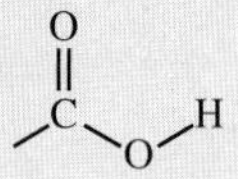
ester	–COOC–	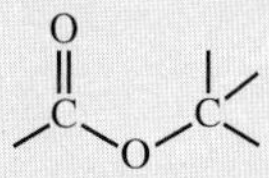
amine	$-NH_2$	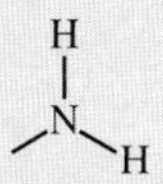
amide	$-CONH_2$	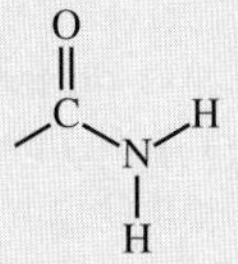

We will now introduce several new functional groups, starting with the *carboxylic acid* group:

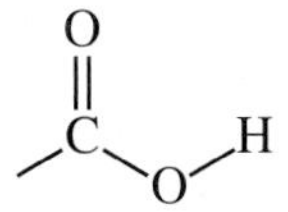

Although the carboxylic acid group contains an –OH group, it is *not* an alcohol. Rather, think of –COOH as a single unit.

Table 9.3 shows that carboxylic acid groups naturally occur in foods such as vinegar and cheese. If you examine the entries in the table closely, you will see that a molecule can contain more than one carboxylic acid functional group; for example, terephthalic acid and adipic acid. A molecule can also have two different functional groups, such as lactic acid.

Carboxylic acids are closely related to another new functional group, the *ester*. An ester can be represented by this structural formula:

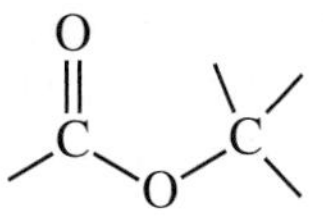

Armed with the ability to recognize alcohols, carboxylic acids, and esters, you are now ready to explore *polyesters*, the topic of this section.

Table 9.3 Selected Carboxylic Acids

Name	Chemical Formula	Information
ethanoic acid	$CH_3-C(=O)-O-H$	Naturally occurring in vinegar. Also called acetic acid.
propanoic acid	$CH_3CH_2-C(=O)-O-H$	Naturally occurring in some cheeses, providing a "sharp" taste. Also called propionic acid.
benzoic acid	$C_6H_5-C(=O)-O-H$	Another naturally occurring carboxylic acid. Used as a food preservative.
terephthalic acid	$H-O-C(=O)-C_6H_4-C(=O)-O-H$	One of the monomers used to produce PET.
adipic acid	$H-O-C(=O)-(CH_2)_4-C(=O)-O-H$	One of the monomers used to produce a type of nylon.
lactic acid	$CH_3CH(OH)-C(=O)-O-H$	The monomer for polylactic acid (PLA), a bio-based polymer.

Because polyethylene terephthalate contains no polyethylene, it is sometimes written as poly(ethylene terephthalate). This reduces the confusion, at least when the name is read. Spoken, the two names still sound the same.

Did You Know? PET is commonly used as the material in screen protectors for cell phones.

The star on the polyester stage is PET, also known as PETE. Both abbreviations stand for polyethylene terephthalate ester. Because PET is semi-rigid, clear, and reasonably gas-tight (Figure 9.11), its most common use is in beverage bottles. Polyesters can also be drawn into sheets and fibers. For example, Mylar™ is a trade name for thin plastic sheets of PET such as those used to make shiny festive balloons. When filled with helium, these balloons remain aloft for many hours or days because polyester is so impervious to gases. Eventually, however, the helium atoms escape slowly over time and the balloons deflate.

Figure 9.11

Two-liter soft-drink bottles made from PET.

© McGraw-Hill Education. Jill Braaten, photographer

In contrast to polyethylene, PET is *not* formed by addition polymerization. Rather, it is formed by **condensation polymerization**, a process in which the monomers join by releasing (eliminating) a small molecule, usually water. Thus, condensation polymerization always has a second product in addition to the polymer itself. Many natural polymers are formed by condensation reactions, including cellulose, starch, wool, silk from spiders, and proteins. Synthetic polymers include Dacron®, Kevlar®, and different types of nylon.

Also in contrast to polyethylene, PET is *not* formed from a single monomer. Rather, PET is a **copolymer**, a polymer formed by the combination of two or more different monomers. One monomer, ethylene glycol, is an alcohol that contains two hydroxyl groups, one on each carbon atom. The other monomer, terephthalic acid, contains two carboxylic acid groups, one on each side of the benzene ring. In essence, each monomer is ready to react with two functional groups. Revisit Table 9.1 to see structural formulas for each of the two monomers.

To understand how copolymers form, let's work with only one molecule of each monomer. Here is how these monomers can join:

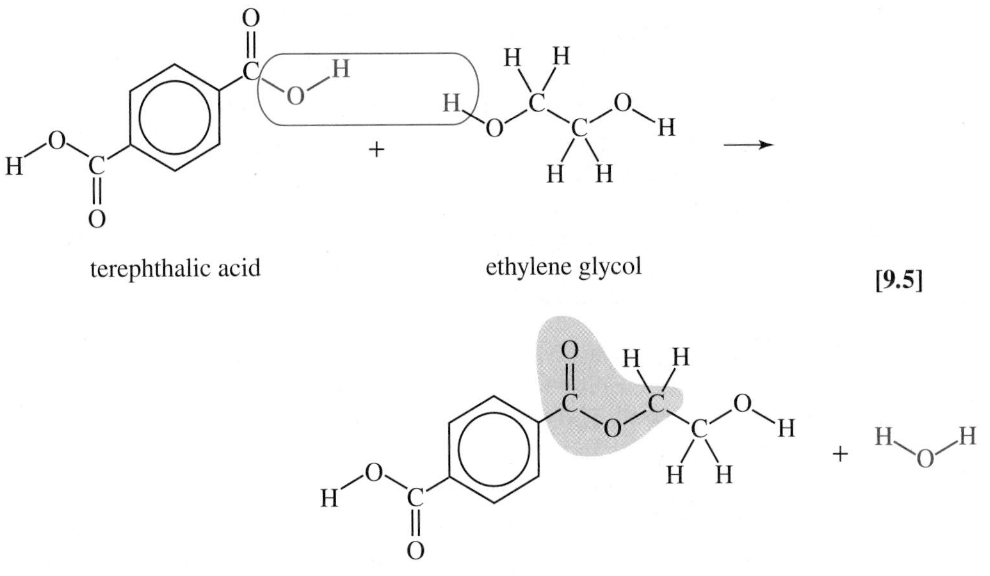

terephthalic acid + ethylene glycol → [9.5]

$+ H_2O$

Written and circled in red, the –OH in the carboxylic acid group and the H atom in the hydroxyl group react to produce water, HOH. The remaining portions of the alcohol and the carboxylic acid connect to form the ester functional group, highlighted in blue.

Notice that the product has functional groups that are sites for additional chain growth: –COOH on the left end, and –OH on the right. The former can react with the –OH of another ethylene glycol molecule; the latter can react with the –COOH of another terephthalic acid molecule. Each time, a molecule of water is released and an ester group is formed. This process, represented in Figure 9.12, occurs multiple times to yield polyethylene terephthalate. The result is a *polyester*, so named because an ester connects the monomers.

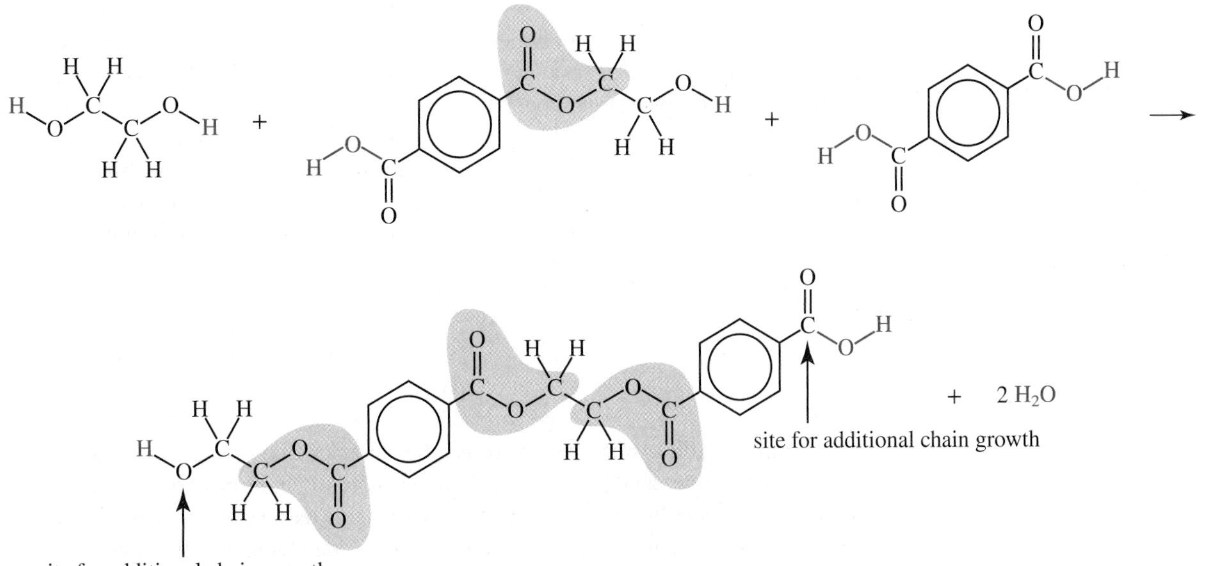

$+ 2\,H_2O$

site for additional chain growth

site for additional chain growth

Figure 9.12

Two different monomers are used to build PET, a polyester. The ester functional groups are highlighted in **blue**.

Your Turn 9.15 Skill Building Esters and Polyesters

You have seen that terephthalic acid and ethylene glycol can react. Now consider ethanoic acid (acetic acid) and ethanol (ethyl alcohol):

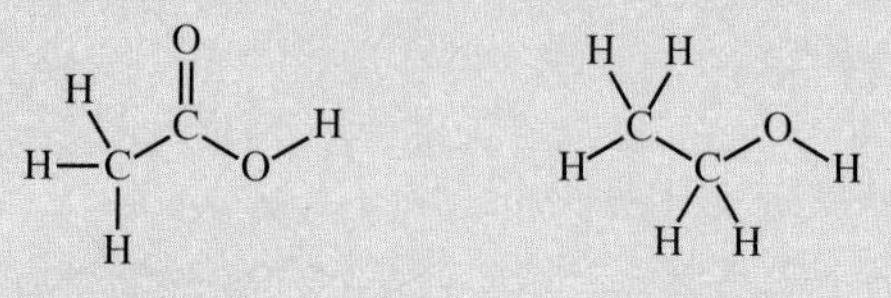

ethanoic acid ethanol

a. Show how this carboxylic acid and alcohol can react to form an ester.
Hint: Remember a water molecule is formed as a product.

b. Could ethanoic acid and ethanol react to form a polyester? Explain your reasoning.

PET is not the only polyester in town! By varying the number and type of carbon atoms in the monomers, chemists have synthesized other polyesters with trade names such as Dacron®, Polartec®, Fortrel®, and Polarguard®. Polyester spins readily into fibers that are easy to wash and quick to dry. Polyester also blends well with other fibers, such as cotton or wool. The next activity describes polyethylene naphthalate (PEN), a polyester that has better temperature resistance than PET.

Your Turn 9.16 Skill Building From PET to PEN

In both PET and PEN, the alcohol monomer is ethylene glycol; however, the organic acid monomers differ slightly. Here is the organic acid monomer in PEN, naphthalic acid:

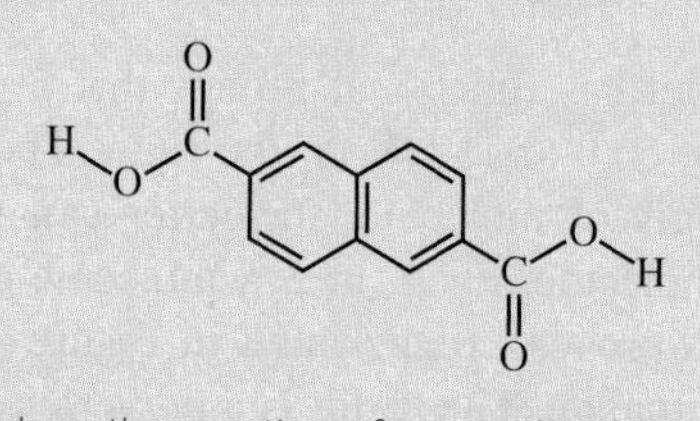

Use structural formulas to show the reaction of two molecules of naphthalic acid with two molecules of ethylene glycol.

In our discussion of polymers so far, all have been based on chains primarily composed of carbons. These may be considered *organic* polymers. Recall that in Section 2.7 we defined organic compounds as those that always contain carbon, usually hydrogen, and sometimes other elements such as oxygen and nitrogen. There are a number of *inorganic* polymers that are based principally on elements other than carbon. The most widely used inorganic polymers are silicones. Rather than carbon, these polymers rely upon a backbone of alternating silicon and oxygen atoms:

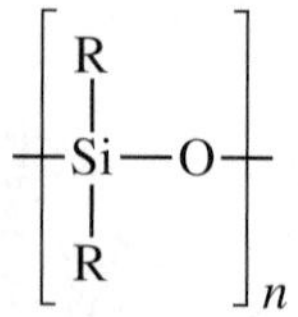

$$\left[-\mathrm{Si(R)_2}-\mathrm{O}- \right]_n$$

Similar to PET, silicones are formed by a condensation reaction. Sometimes water is a by-product of the condensation reaction, but more often an alcohol such as methanol or ethanol is produced instead of water. The side-chains designated by R in the structure are typically carbon chains of varying lengths. These side-chains greatly affect the properties of the silicone, giving them just as wide a range of properties as their organic polymer cousins. Indeed, due to these tailorable properties, silicones have found applications in a variety of products from lubricating oils and paints, to cooking utensils and caulks.

9.7 | From Proteins to Stockings: Polyamides

No discussion of condensation polymerization can be complete without examining two specific types of polymers. The first is proteins, which are natural polymers such as those found in our muscles, fingernails, and hair; the second is nylons, which are synthetic substitutes that brilliantly duplicate some of the properties of silk, a naturally occurring protein. In 2011, according to the Chemical Heritage Foundation, manufacturers worldwide produced around 8 million pounds of nylon, roughly 12% of all synthetic fibers.

Amino acids are the monomers from which our body builds proteins. Each amino acid molecule contains two functional groups: an amine ($-NH_2$) and a carboxylic acid ($-COOH$). Twenty different amino acids occur naturally, each differing in one of the groups bonded to the central carbon atom. This side-chain is represented with an R, as shown below in the general structural formula for an amino acid. In some amino acids, R consists of only carbon and hydrogen atoms; in others, R may include oxygen, nitrogen, or even sulfur atoms. Some R groups have acidic properties, others are basic.

See Sections 11.6 and 12.6 for more information about amino acids and proteins.

Chemists use R as a placeholder in a molecule. With amino acids, R represents one of 20 side-chains. Earlier in this chapter, R• was used to represent a free radical such as Cl• or •OH.

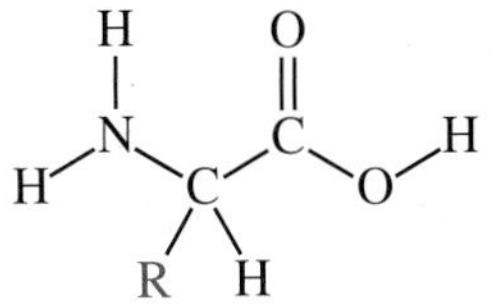

As monomers, amino acids join to form a long chain via condensation polymerization. However, keep in mind three key differences between a condensation polymer such as PET and any given protein:

- PET is a polyester. In contrast, proteins are **polyamides**; that is, condensation polymers that contain the amide functional group.
- PET is built from two monomers, ethylene glycol and terephthalic acid, that are in a 1:1 ratio. In contrast, proteins can contain up to 20 different monomers (amino acids) in any ratio.
- In proteins, each amino acid has two *different* functional groups, $-NH_2$ and $-COOH$; in PET, the two monomers have two *identical* functional groups, either $-OH$ or $-COOH$.

The amide functional group is shown in Table 9.2.

To see how these differences play out, examine this reaction between two amino acids. One has the side chain R; the side chain on the other amino acid is labeled as R′:

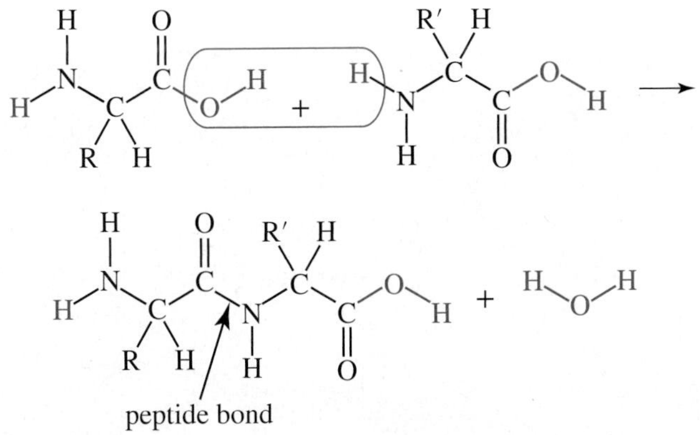

[9.6]

peptide bond

In this reaction, an amide is formed and a molecule of water is eliminated. This amide contains a C–N bond, referred to as a **peptide bond**, the covalent bond that forms when the $-COOH$ group of one amino acid reacts with the $-NH_2$ group of another, thus joining the two amino acids. In the sophisticated chemical factories of the cells of any organism, this condensation reaction between different amino acids is repeated many times to form the long polymeric chains that we call proteins. Given that 20 different amino acids exist in nature, a great variety of proteins can be synthesized. Some contain hundreds of amino acids, others only a few.

Only in the context of proteins is a C–N bond referred to as a peptide bond.

Figure 9.13

Wallace Carothers, the inventor of nylon.

Chemists sometimes attempt to replicate the chemistry of nature. For example, a brilliant chemist working for the DuPont Company, Wallace Carothers (1896–1937) (Figure 9.13), was studying many polymerization reactions, including the formation of peptide bonds (Equation 9.6, for example). Instead of using amino acids, Carothers tried combining adipic acid and hexamethylenediamine:

$$\mathrm{HO{-}C({=}O){-}(CH_2)_4{-}C({=}O){-}OH} \qquad \mathrm{H_2N{-}(CH_2)_6{-}NH_2}$$

adipic acid hexamethylenediamine

Note that adipic acid has a carboxylic acid at each end of the molecule and hexamethylenediamine has an amine group on each end. As in protein synthesis, the acid and amine groups react to form an amide and release water. But unlike protein synthesis, the resulting polymer, better known as nylon, is formed from only two monomers. Here is how the monomers join:

$$\underset{\text{adipic acid}}{\mathrm{HO{-}C({=}O){-}(CH_2)_4{-}C({=}O){-}OH}} + \underset{\text{hexamethylenediamine}}{\mathrm{H_2N{-}(CH_2)_6{-}NH_2}} \longrightarrow \mathrm{HO{-}C({=}O){-}(CH_2)_4{-}C({=}O){-}NH{-}(CH_2)_6{-}NH_2} + \mathrm{H_2O} \quad [9.7]$$

(The OH of adipic acid and an H of the amine are circled; arrows label the terminal HO–C and the terminal N as "site for additional chain growth.")

DuPont executives decided that nylon had promise, especially after company scientists learned to draw it into thin filaments. These filaments were strong, smooth, and very much like the protein spun by silkworms. Therefore, nylon was first introduced to the world as a substitute for silk. Nylon was one of the first **biomimetic** materials—components for use in human applications that are developed using inspiration from nature. The world greeted the release of nylon with bare legs and open pocketbooks! Four million pairs of nylon stockings were sold in New York City on May 15, 1940, the first day they became available (Figure 9.14). But, in spite of consumer passion for "nylons," the civilian supply soon dried up as the polymer was diverted from hosiery to parachutes, ropes, clothing, and hundreds of other wartime

Figure 9.14

Customers eagerly lined up to buy nylon stockings in 1940, when they were first commercially available.

uses. By the end of World War II in 1945, nylon had repeatedly demonstrated that it was superior to silk in strength, stability, and resistance to rot. Today, this polymer, with its many modifications, continues to find wide applications in carpets, sportswear, camping equipment, the kitchen, and the laboratory.

Kevlar® ends our tales of condensation polymers. We remind you that it is a polyamide, just like the silk spun by silkworms and spiders.

Your Turn 9.17 Skill Building Kevlar®

Kevlar® is a polyamide used in bulletproof vests and the cases of some cell phones. Like PET, one of the monomers is terephthalic acid. The other monomer, phenylenediamine, contains two amine functional groups:

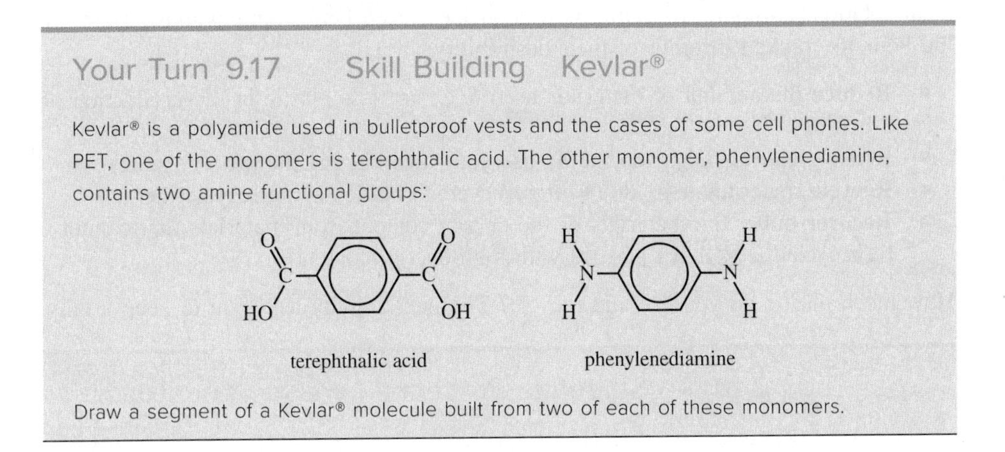

Draw a segment of a Kevlar® molecule built from two of each of these monomers.

9.8 Dealing with Our Solid Waste: The Four Rs

Speaking of spider webs, this natural polymer has many useful properties, including strength, the ability to stretch, and enough stickiness to ensnare prey. Orb spiders, like the one shown in Figure 9.15, are notoriously picky builders and spin new webs each day. This daily web construction could easily exhaust the resources available to the spider. So, how does an orb spider manage to spin so much silk and still survive? Most simply, it recycles! Orb spiders have the ability to ingest old spider silk and recover the raw materials from which they are built. While the actual chemical processes are not fully understood, up to two-thirds of the existing web goes into making a new one. Humans need to mimic this spider, lest we, too, run out of resources and create an overwhelming amount of waste. As a 2010 report from multiple European plastics industries points out, "Plastic is simply too valuable to throw away."

Did You Know? The silk of an orb spider ranks among the toughest biological materials ever studied, an order of magnitude stronger than a similar piece of Kevlar®.

Figure 9.15

Golden orb spider and web.

© Edwin Remsberg/Alamy Stock Photo

Indeed, we humans have produced a lot of plastic! In 1950, the value was just under 2 million metric tons worldwide. Over the years, the amount of plastic produced has increased steadily, reaching 311 million metric tons worldwide in 2014. Without question, we need sustainable answers to the question of how to deal with plastic waste.

© Ingram Publishing/Fotosearch RF

Most likely, you have taken the garbage to the curb for pickup. You also may have watched the waste being driven away in a truck, never to be seen again (by you at least). Plastics are part of this waste, and sending plastic to a landfill is far from an ideal solution. Although recycling is a good idea, even better options exist. Here are the Four Rs, ranked in order of their desirability:

- **Reduce** the amount of materials used (*e.g.*, use less plastic in the production of a bottle)
- **Reuse** materials (*e.g.*, repeatedly use your own bags at the grocery store)
- **Recycle** materials (*e.g.*, don't throw beverage bottles away, recycle them)
- **Recover** either the materials or the energy content from materials that cannot be recycled (*e.g.*, burn plastics with high-energy content)

How much plastic do you use and recycle? The next activity asks you to keep a tally.

Your Turn 9.18 Scientific Practices Plastic You Toss: Part I

Keep a journal of all the plastic you either throw away or recycle in one week. Include plastic packaging from food and other products that you purchase.

a. Estimate the mass of this plastic. Is it a few grams, a kilogram, or more?
b. Which is greater, the mass of the plastic you throw away or the mass you recycle?

Keep your journal handy because you will be asked to revisit it.

Let's now examine each of the Four Rs as options for dealing with plastics.

Chapter 11 will offer a perspective on why it makes sense to reduce your consumption of sugared beverages. In turn, this might reduce your use of plastics.

Reduce! Source reduction is always the option of choice. This means using less material and generating less waste later on. Source reduction conserves resources, reduces pollution, and minimizes toxic materials in the waste stream. As an example, consider beverage bottles. Through an improved design, a 2-L soda bottle now uses about a third less plastic than when it was introduced in 1970; similarly, a 1-gallon milk jug now weighs less than it did a few decades ago.

Corporations are recognizing that reduced packaging offers economic incentives, such as lower costs for shipping and lower landfill costs for waste. For example, Edward Humes' book "Force of Nature: The Unlikely Story of Wal-Mart's Green Revolution" (2011) described Walmart's goal of reducing packaging by 5% for the 329,000 items on its shelves by 2013, with 2008 as the baseline year. Corporate leaders realized that sustainability wasn't just a way of being cleaner and more efficient, but it also seemed to be driving innovation.

Speaking of packaging, keep an eye out for innovations. **Sustainable packaging** is the design and use of packaging materials to reduce their environmental impact and improve the sustainability of all practices. Criteria established in 2011 by the Sustainable Packing Coalition include that such packaging:

- is beneficial, safe, and healthy for individuals and communities throughout its life cycle
- meets market criteria for both performance and cost
- optimizes the use of renewable or recycled source materials
- is physically designed to optimize materials and energy
- is manufactured using clean production technologies and best practices
- is effectively recovered and utilized

As you might expect, polymer chemists and chemical engineers are key players in this endeavor.

Reuse! Reusing something means not disposing of it after a single use. In the checkout line, supermarket clerks once gave their customers only two choices, "Paper or plastic?" Today, however, they may ask if you brought your own bag. The next activity expands on the idea of reusing bags.

Your Turn 9.19 You Decide Paper, Plastic . . . Neither?

Grocery stores are not the only place in which people could rethink their use of plastic and paper bags. List three other possibilities. For each, tell whether or not you would be willing to change your "bag habits" and reuse your own bag.

As another example, consider how polystyrene foam packing "peanuts" can be reused. While only a tiny part of the waste stream, these peanuts are a huge nuisance once they escape their intended use. They end up just about everywhere, including waterways, roads, and fields. Because they are only about 5% polystyrene by weight, they have little recycling value. Reusing these peanuts definitely is the option of choice. Actually, the same is true for all polystyrene foam packing materials. If you have worked at a retail store or shipping desk, chances are you have seen some type of "in-house" reuse or recycling.

"Expanded" polystyrene that is used in packaging peanuts was described in Section 9.5.

Recycle! You probably see recycling containers just about everywhere—in campus buildings, sports centers, airports, and hotels. Reasons for recycling include:

- reduction of waste at landfills and incinerators
- prevention of air, water, and soil pollution during the manufacturing process
- decrease in emissions of greenhouse gases during manufacturing
- conservation of natural resources such as petroleum, timber, water, and minerals

How well are we doing? First the good news. In 2013, the Environmental Protection Agency reported that, on average, each person in the United States recycled 1.5 pounds of their individual waste generation, which is about 4.4 pounds per day! However, the percentage of waste recycled is increasing. Items that people deposit in bins or at the curbside include aluminum cans, office paper, cardboard, glass, and plastic containers. In addition, about 0.4 pounds per person of waste (such as grass clippings and food scraps) is composted, and another 0.5 pounds of waste per person is incinerated to produce energy daily. Given these reductions, the amount of waste sent to the landfill is now averaging 2.4 pounds per person per day.

And now the bad news. As you will see in the next section, roughly 12% of what we discard is plastic. Depending on the type of plastic, our recycling efficiency varies, as the next activity will reveal.

Your Turn 9.20 Scientific Practices Plastics Recycling Scorecard

According to the EPA, here is the U.S. recycling scorecard for 2013. Durable goods include items such as luggage, plastic furniture, and garden hoses. Nondurable goods include plastic pens and safety razors.

Use of Plastic	Weight Generated (millions of tons)	Weight Recovered (millions of tons)
Durable goods	12.07	0.83
Nondurable goods	6.47	0.13
Containers/Packaging	13.98	2.04

a. For each type of plastic, calculate the plastic recovered as a percent of the waste generated.

b. Nondurable and durable goods tend to have low recycling rates. List three examples of each, and suggest reasons why.

If you did the calculations, you saw that we recycle plastics at a surprisingly low rate. This may seem at odds with all the milk jugs and plastic bottles you see being recycled in your own community. Certain plastics are recycled more consistently than others. For example, in the United States, polyethylene milk containers are recycled at a rate of 28% and clear PET soft drink bottles at 31%. While these numbers may seem high, nonetheless around 70% of these containers are still being tossed out.

Your Turn 9.21 You Decide Plastic You Toss: Part II

Earlier, in **Your Turn 9.18**, you kept a journal of all the plastic you discarded in a week. Revisit what you wrote. Do the yearly recycling figures just cited ring true for you? That is, do you throw out far more plastic than you recycle? Briefly report on how your own plastic use stacks up against the national averages. Remember that your journal may not reveal your use of durable and nondurable goods over a longer period of time.

Incineration and landfills create relatively few jobs in comparison to recycling programs. A commitment to increasing recycling can benefit a local economy.

Recover! What about incineration; that is, recovering the energy in plastics by burning them as fuels? Because the Big Six and most other polymers closely resemble hydrocarbon fuels, incineration would seem to be an excellent way to dispose of them, reducing demand on landfills. The chief products of combustion are carbon dioxide, water, and a good deal of energy. In fact, pound for pound, plastics have a higher energy content than coal. In the United States, plastics accounted for only about 13% of the weight of municipal solid waste in 2013; however, they represent approximately 30% of its energy content.

But incineration of plastics has drawbacks. The gases produced by combustion may be "out of sight," but they best not be "out of mind." Because their chemical compositions are similar to fossil fuels, burning plastics produces CO_2, a greenhouse gas. Of special concern in incineration are chlorine-containing polymers such as polyvinyl chloride that release hydrogen chloride (HCl) during combustion. Because HCl dissolves in water to form hydrochloric acid, such smokestack exhaust could make a serious contribution to acid rain. Burning chlorine-containing plastics can produce other toxic gases. So in terms of the overall benefit, including the energy involved, recycling is always preferable to incineration.

Your Turn 9.22 Skill Building Burning a Plastic

Under conditions of complete combustion, polypropylene burns to produce carbon dioxide and water.

a. Write a balanced chemical equation. Assume an average chain length of 2500 monomers.

b. If the combustion is incomplete, other products form. Name two possibilities.

c. Using the Internet as a resource, list some additives that are commonly added to polymers such as polypropylene. Could these also be released as harmful environmental pollutants during combustion? Explain your answer.

The plastic that we do not reuse, recycle, or recover eventually ends up in a landfill (the "out of sight, out of mind" approach), or as trash in the environment. Both are problematic. Although landfill space is still available, landfills have drawbacks. They take up space in congested areas, they have costs associated with their construction and upkeep, they leak and attract vermin, and they emit methane, a greenhouse gas.

The majority of plastics do not biodegrade in the landfill (or anywhere else). Most bacteria and fungi lack the enzymes necessary to break down synthetic polymers. Some microbes, however, possess the enzymes to break down naturally occurring polymers, such as cellulose. For example, in Chapter 4 you read about the release of methane by cattle. Actually, the methane is produced when bacteria obtain energy by decomposing cellulose in the cow's rumen. In the same chapter, you also learned that methane is generated by natural decomposition of organic materials in landfills, another result of bacterial activity.

Figure 9.16

Some buried wastes can remain intact for a long time. This newspaper from 1952 was excavated 37 years later.

Even natural polymers do not decompose completely in landfills. Modern waste disposal facilities are covered and lined to deter leaching of waste and waste by-products into the surrounding ground. These landfill linings and coverings create anaerobic (oxygen-free) conditions that impede the breakdown of these wastes. As a result, many supposedly biodegradable substances decompose slowly, or not at all. Excavation of old landfills has unearthed old newspapers that are still readable (Figure 9.16) and 5-year-old hot dogs that, while hardly edible, are at least recognizable.

Ideally, landfill liners last forever. However, over time, liners break down or rupture.

Your Turn 9.23 Scientific Practices Landfill Liners

Landfill liners include natural clay and human-made plastics. For example, thick sheets of high-density polyethylene may be employed. Even the best HDPE liners, however, can crack and degrade.

a. From **Table 9.1**, which types of chemicals soften HDPE?

b. Name five substances sent to the landfill that could degrade a HDPE liner over time.

Given the problems associated with landfill disposal and incineration of natural and synthetic polymers, *recycling* has an important role to play. However, in contrast to landfilling, recycling polymers requires an input of energy. Furthermore, if the waste plastic is dirty or of low quality, more energy may be needed to recycle it than to manufacture it from new plastic. Nonetheless, recycling is one of several ways to divert plastic from landfills and incinerators. In the next section, we examine the bigger picture of garbage.

9.9 | Recycling Plastics: The Bigger Picture

Together with other waste, the plastic that you discard is part of a bigger picture. In the United States, the EPA has been keeping statistics about municipal solid waste—better known as garbage—for more than 30 years. **Municipal solid waste** (MSW) includes everything you discard or throw into your trash, including food scraps, grass clippings, and old appliances. MSW does not include all sources, such as waste from industry,

Worldwide, it is estimated that 1.3 billion tons of MSW is generated each year, which may double by the year 2025.

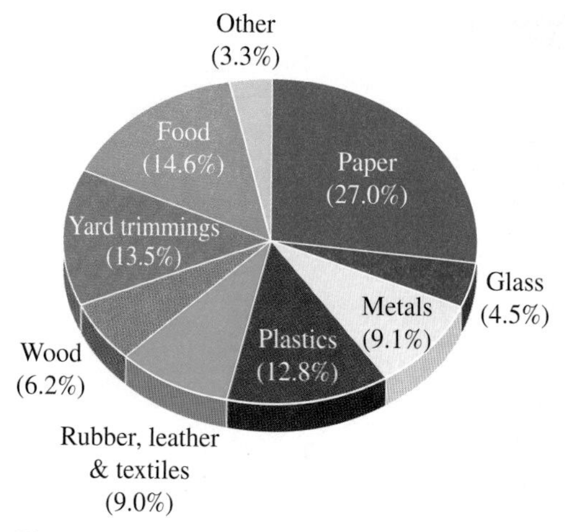

Figure 9.17

What's in your garbage? Composition by weight of municipal solid waste before recycling (254 million tons, 2013).

Source: U.S. Environmental Protection Agency, Advancing Sustainable Materials Management: 2013 Fact Sheet, June 2015

agriculture, mining, or construction sites. In the United States, municipal solid waste has been averaging about 254 million tons per year. What is the largest single item? Paper, as you can see from Figure 9.17. Materials of biological origin such as paper, wood, food scraps, and yard trimmings make up the majority of the materials classified as municipal solid waste. All of these can be dealt with by one of the Four Rs (reduce, reuse, recycle, recover).

How much of this MSW is plastic? Consult Figure 9.17 to see that plastic is roughly 13% of what U.S. citizens discard, the equivalent of 32.5 million tons in 2013. The U.S. EPA reports data for three types of plastics:

- durable items, such as plastic furniture, bowls, and garden hoses
- nondurable items, such as disposable plastic cups, plates, trash bags, pens, and safety razors
- packaging, such as beverage bottles and food containers

How much of this plastic do we recycle? Figure 9.18 shows the amount in millions of tons over the years for *total* MSW recycling in the United States, not just for plastics. The overall rate of recycling of MSW in recent years has reached 34%.

By way of comparison, in 2013 only 3.5% of the plastic in the United States was recycled. Why so little? The devil is in the details! Some items can be easily recycled; others present nothing short of a logistical nightmare. Furthermore, some types of plastics have a ready market, while others do not. Table 9.4 reveals the relatively high recycling rate for plastic bottles in comparison to the overall 3.5% plastic recycling rate. Figure 9.19 further illustrates how plastics recycling compares with other products. As discussed in Chapter 7, lead–acid batteries continue to represent one of the most widely recycled products.

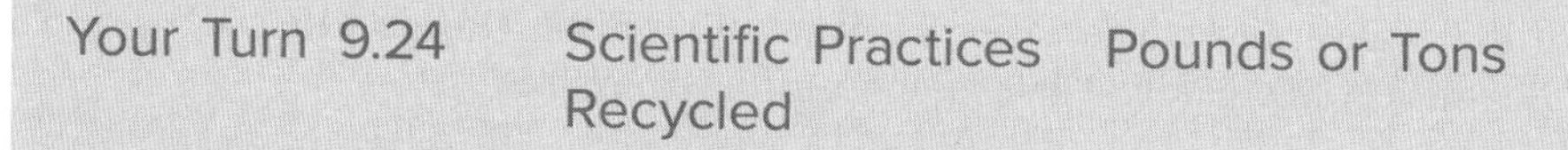

Your Turn 9.24 Scientific Practices Pounds or Tons Recycled

Examine the values in **Table 9.4** from the American Chemistry Council (ACC).

a. Are these values comparable to those quoted by the EPA in **Your Turn 9.20**? Assume that the tons quoted were short tons; that is, 2,000 pounds per ton.

b. The EPA reported the amounts recycled using one unit (million tons), and the ACC reported in another (million pounds). Several explanations are possible. Propose one.

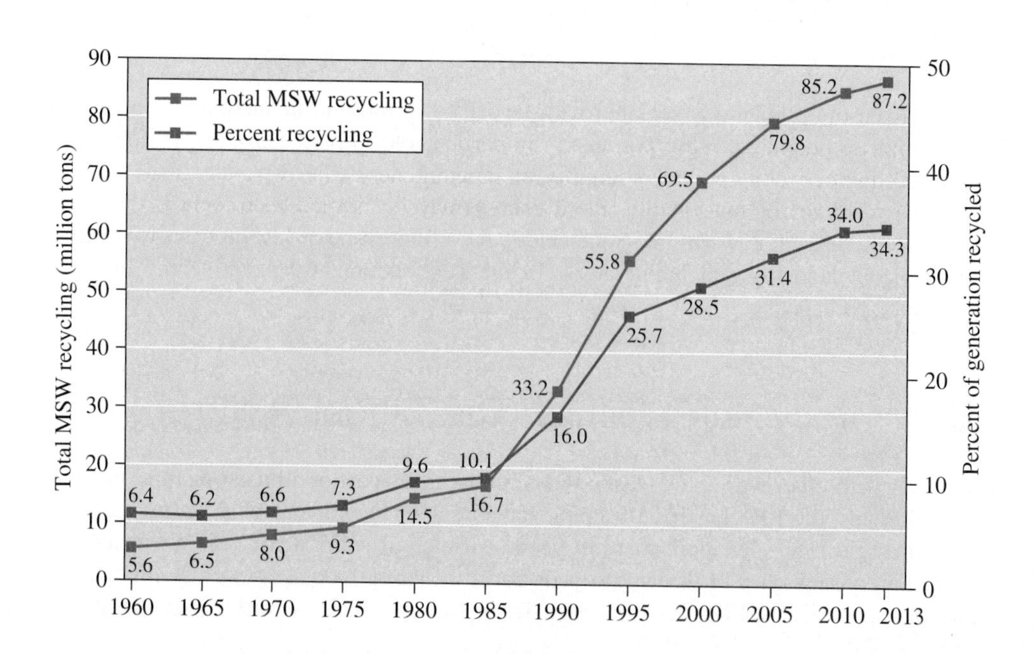

Figure 9.18

Amount and percent of municipal solid waste recycled in million tons, 1960–2013.

Source: U.S. Environmental Protection Agency, Advancing Sustainable Materials Management: 2013 Fact Sheet, June 2015

Table 9.4 Recycled Plastic Bottles in 2013

Plastic	Amount Recycled in 2013 (million pounds)	Recycling Rate
PET	1798	31.2%
HDPE	1045	31.6%
PVC	0.4	0.5%
PP	62.0	31.8%
LDPE	0.3	0.4%

Source: American Chemistry Council, 2013, National Post-Consumer Plastic Bottle Recycling Report

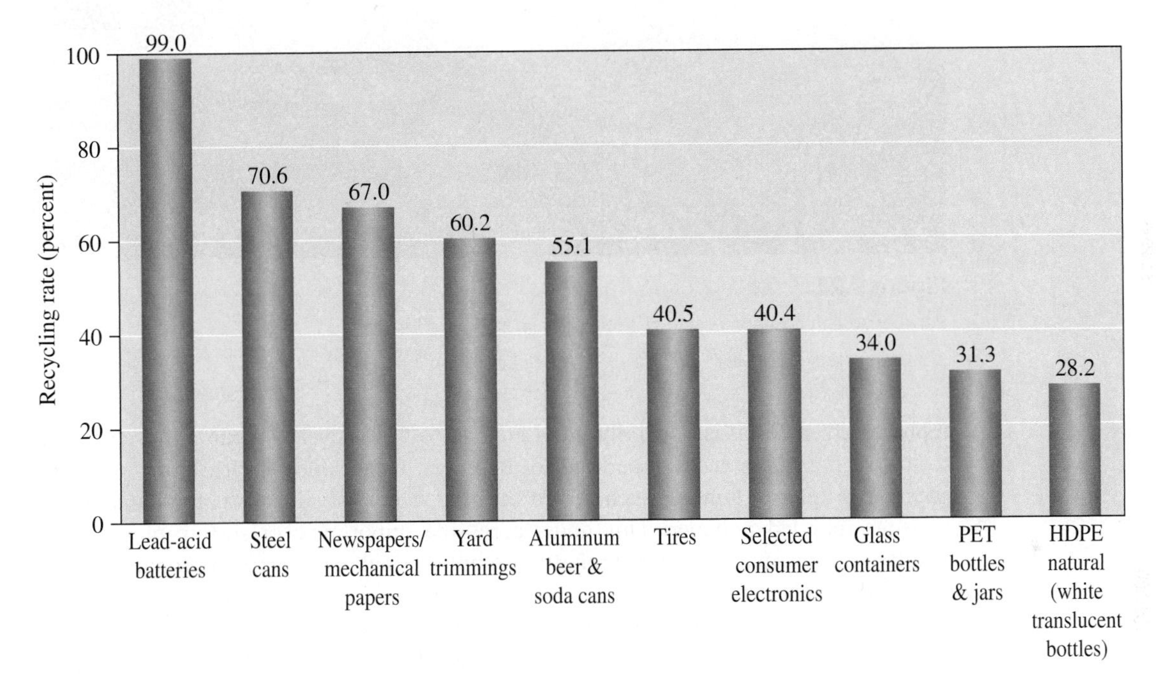

Figure 9.19

Recycling rates of selected products, 2013. The reported rates do not include combustion with energy recovery.

Source: U.S. Environmental Protection Agency, Advancing Sustainable Materials Management: 2013 Fact Sheet, June 2015

For recycling to be successful and self-sustaining, a number of factors must be coordinated. These involve not only science and technology, but also economics and sometimes politics—especially at the local level. The best recycling involves a closed loop (Figure 9.20) in which plastics are collected, sorted, and then converted into products that consumers buy, use, and later recycle.

In order to recycle, it is first necessary to collect the plastic. Several options are available: collecting at curbside, at local drop-off centers, and through bottle bill programs involving a deposit and refund. For recycling to be successful, a dependable supply of used plastic must be consistently available at designated locations.

Once collected, the plastic needs to be transported to a facility where it can be sorted (Figure 9.21) and prepared for some marketable commodity, such as the manufacturing of outdoor furniture, toys, or even clothing. The symbols that appear on plastic objects (revisit Table 9.1) help facilitate the sorting process. Because of the large volume of material, automated sorting methods have been developed. Once sorted, the polymer is melted, which can then be used directly in the manufacturing of new products. Alternatively, it can be solidified, pelletized, and stored for future use.

If a mixture of various polymers is melted, the product tends to be darkly colored and has different properties depending on the nature of the mixture. This type of

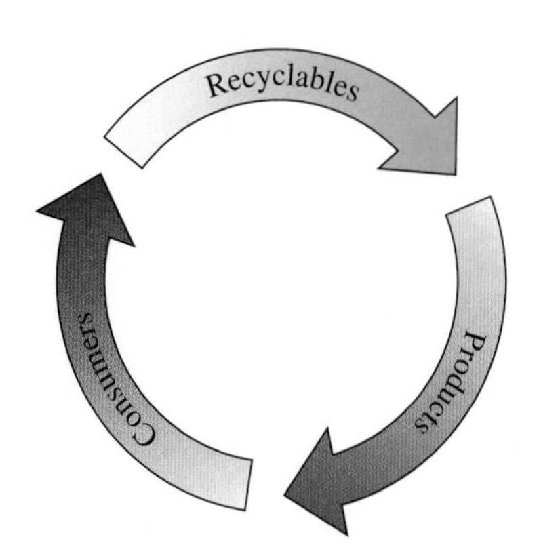

Figure 9.20

Recycling ideally is a never-ending loop.

Source: National Association for PET Container Resources.

Figure 9.21

Polymer sorting and processing at a recycling facility.

reprocessed material is generally good enough to "downcycle," meaning to convert it to lower-grade uses such as parking lot bumpers, disposable plastic flower pots, and inexpensive plastic lumber. Such mixed material is not as valuable as the pure, homogeneous recycled polymer. This underscores the importance of sorting plastics. For similar reasons, manufacturers prefer to use only a single polymer in a product to avoid the need to separate.

Given a supply of plastic (ideally clean and sorted), the manufacturers can get to work. The items produced are composed of varying percentages and types of recycled materials. The terminology is confusing. **Recycled-content products** are those made with materials that otherwise would have been in the waste stream. These include items manufactured from discarded plastic, as well as rebuilt items such as plastic toner cartridges that are refilled. Trash bags, laundry detergent bottles, and carpeting are common plastic items that may qualify as recycled-content products. Some playground equipment and park benches also are made from discarded plastic.

Recycled products are now beginning to provide the origin of the recycled material. **Post-consumer content** is material that previously was used individually that otherwise would have been discarded as waste. Recycling this waste—office paper, foam packing, and beverage bottles—is one way to keep it out of the landfill. **Pre-consumer content** is waste left over from the manufacturing process itself, such as scraps and clippings. Pre-consumer fabrics, such as polyester fabric scraps from the clothing industry, can be recycled rather than discarded.

A product that is designated as *recyclable* simply means that it *can* be recycled. The term may be misleading because a recycling pathway may not exist. Recyclable products do not necessarily contain any recycled materials.

Your Turn 9.25 Scientific Practices Recyclable and Recycled

Give three examples of items that you might purchase and recycle. Also give three examples of recycled-content products. Can an item fall into both categories?

To complete the cycle shown in Figure 9.20, the recycled items are marketed and (ideally) purchased by consumers. Without a product and buyers, recycling programs are doomed to fail. In fact, recycling laws in a number of cities have not been implemented and enforced because one of the links in this polymeric chain of supply, collecting, sorting, processing, manufacturing, and marketing was missing.

Consult Table 9.4 to see that as consumers, we are moderately adept in dropping PET beverage bottles into recycling bins (Figure 9.22). In fact, more than 1 billion pounds of PET is recycled in the United States! Since PET is more successfully recycled than most plastics, it warrants a closer look. PET soft drink bottles need special handling before they can be melted and reused. The bottles are usually sorted to remove other types of plastic, such as PVC. If left in the batch, PVC can weaken the final product. Any labels, bottle caps, or food that adhered to the plastic also must be separated or scrubbed off. Bottle caps, for example, are usually made out of the tougher polypropylene. The next exercise shows how PET can be separated from other polymers by density. This is helpful in the case of PET mixed with PVC because these can look alike.

Figure 9.22
PET beverage bottles are widely recycled.

© *Thinkstock/SuperStock RF*

Your Turn 9.26 Scientific Practices Float or Sink?

Here are density values for PET and for three other plastics likely to be found with it in a recycling bin.

Plastic	Density (g/cm^3)
PET	1.38–1.39
HDPE	0.95–0.97
PP	0.90–0.91
PVC	1.30–1.34

When dropped into a liquid, a plastic will float or sink depending on the density of the liquid. Here are the densities for several liquids that do not degrade the four plastics above.

Liquid	Density (g/cm^3)
methanol	0.79
42% ethanol/water mixture	0.92
38% ethanol/water mixture	0.94
water	1.00
saturated solution of $MgCl_2$	1.34
saturated solution of $ZnCl_2$	2.01

Given a PET sample contaminated with HDPE, PP, and PVC, propose a way to separate the PET from the other three plastics. Assume that all density values were measured at the same temperature.

When you recycle a PET beverage bottle, it may come back to life as part of another beverage bottle. More likely, though, the polyester is downwardly recycled (downcycled) to produce items of lower purity. For example, the PET may be melted and spun into polyester carpeting, clothing (Figure 9.23), "fleece" bed sheets, and the fabric uppers in jogging shoes. Five recycled 2-L bottles can be converted into a shirt or the insulation for a ski jacket; it takes about 450 such bottles to make a 9- × 12-foot polyester carpet.

A reasonable question to ask at this point is what happens to all of these products made from recycled PET. Shaw Industries won a 2003 Presidential Green Chemistry Challenge Award for providing an answer with its development of Eco-Worx broadloom carpet and carpet tiles. By removing the PVC from the backing of these carpets, the product became 100% recyclable into other products. Additional environmental benefits of this carpeting include lower VOC emissions and lower transportation costs, because the carpet tiles are lighter in weight. As of 2012, this

(a)

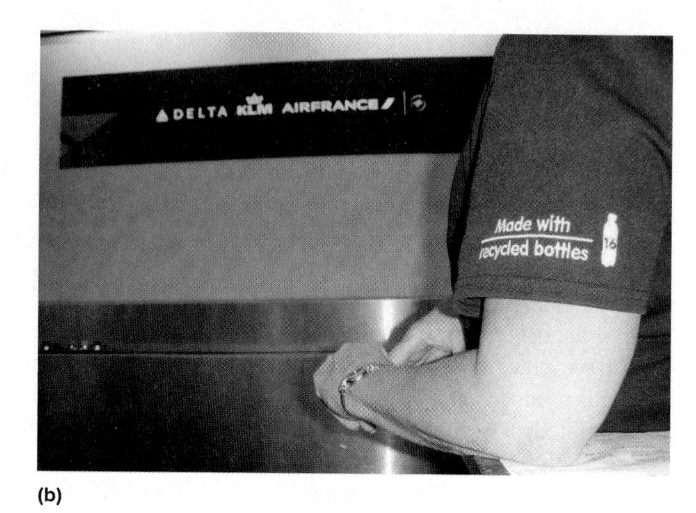

(b)

Figure 9.23

(a) Activewear made from recycled PET. **(b)** Airline uniform made from recycled PET bottles.

(a): © Vanderlei Almeida/AFP/Getty Images; (b): © Cathy Middlecamp

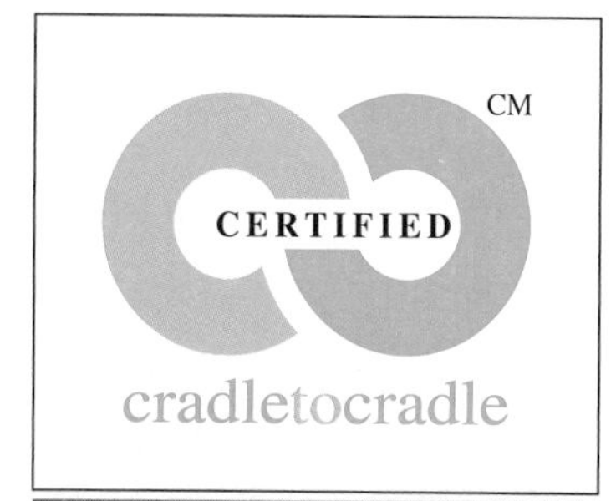

SILVER

Architect William McDonough and chemist Michael Braungart are co-authors of the book *Cradle-to-Cradle*.

carpeting contained 40% recycled content and was labeled "We want it back." As more of this carpet is used, recycled, and again recycled, this percent is expected to grow. This is another example of the concept of **cradle-to-cradle**; in fact, this line of carpet was awarded a silver cradle-to-cradle award from MBDC (McDonough Braungart Design Chemistry).

This section explored the complexities of recycling. But remember that recycling is not the only game in town. There is no single, best solution to the problems of plastic waste, or more generally, *all* solid waste. Incineration, reuse, recycling, and source reduction all provide benefits, and all have costs. Therefore, it is likely that the most effective response will be an integrated waste management system that employs multiple strategies. Ultimately, such a system would optimize efficiency, conserve energy and material, and minimize cost and environmental damage.

Your Turn 9.27 Scientific Practices In a Store Near You

Unless people buy products made from recycled plastics, manufacturers will have little financial incentive to produce them. Find five recycled-content plastic items available for sale.

a. Identify the polymer(s) in each, and the % recycled content, if provided.
b. Comment on the consumer appeal of the item, including whether or not you would purchase it.

Your Turn 9.28 Scientific Practices More Plastics than Fish?!

The World Economic Forum recently published a report entitled "The New Plastics Economy: Rethinking the Future of Plastics," in which it was asserted that by 2050 there will be more plastics than fish (by weight) in oceans. Download the report from the Internet and describe the assumptions that contribute to this dire forecast. What steps can be taken to prevent this scenario, and what are the biggest threats to realizing this so-called "new plastics economy"?

9.10 | From Plants to Plastics

As we mentioned earlier, most synthetic polymers are produced from petroleum, a nonrenewable resource. Even so, some of these polymers find their origin in renewable materials such as wood, cotton fibers, straw, starch, and sugar. What makes these plant-based polymers different from their petroleum-based cousins?

- They are **compostable**; that is, under the conditions of either a home composter or an industrial composter, they are able to undergo biological decomposition to form a material (compost) that contains no materials that are toxic to plant growth.
- Some polymers can be "unzipped" and converted back into monomers, and remade into virgin polymer.
- Their synthesis generally requires fewer resources, results in less waste, and uses less energy than petroleum-based polymers.
- They do not contain chlorine or fluorine.

Because of these characteristics, such polymers are termed "eco-friendly." However, employ this term with caution. As you will see, the composting of biopolymers is not as straightforward as it may sound. In addition, in order to compare different polymers, the waste and energy costs from manufacture, use, and the end of a product's life all need to be considered. Again, the best solution is to *reduce* what you consume rather than to switch to any particular type of plastic.

While polylactic acid (PLA) is not the only plastic produced from plants, it serves as the poster child for eco-friendly plastics. Like the Big Six, PLA is a thermoplastic polymer that softens with heating and can be molded. Because PLA is a polyester that has a similar look and feel to PET, it is used to produce some of the same items as PET, including clear shiny bottles, transparent food packaging, fibers for clothing, and plasticware (Figure 9.24). PLA also is used as a coating on paper cups and plates to make them water-resistant.

Unlike PET with a melting point >250 °C, PLA softens around 140 °C (60 °F). As a result, if you leave an item made from PLA in a car on a sunny day, you may return to find it has melted. So unless blended with other resins to improve its temperature stability, PLA is limited to uses at lower temperatures.

(a)

(b)

(c)

Figure 9.24

(a) PLA cups can be colorless, transparent, and water-resistant, just like PET. **(b)** PLA can be pigmented, again like PET. **(c)** Paper cups coated with PLA.

(a): © GIPhotoStock X/Alamy Stock Photo; (b): © Voinakh/Shutterstock.com; (c): © Tim Gainey/Alamy Stock Photo

Did You Know? Lactic acid is naturally present in the biosphere. It gives sour milk its taste, and is partly responsible for making your muscles ache after vigorous exercise.

As its name suggests, polylactic acid is a polymer of lactic acid. This monomer has two functional groups: a carboxylic acid group and a hydroxyl group. Here is its structural formula:

$$HO-\underset{\underset{CH_3}{|}}{CH}-\overset{\overset{O}{\parallel}}{C}-OH$$

Like PET, polylactic acid is a condensation polymer, and releases a molecule of water each time a covalent bond forms in the polymer chain.

Your Turn 9.29 Skill Building The Chemistry of PLA

We don't show the chemical reaction for the formation of PLA from lactic acid because it does not proceed in a single step and is complicated. Even so, you should be able to write a chemical formula for PLA.

a. Circle and label the functional groups in the monomer, lactic acid.
b. When lactic acid polymerizes, explain how you know it is a condensation reaction.
c. What is the repeating unit in PLA?

PLA, marketed by the trade name Ingeo™, is primarily manufactured in the U.S. by NatureWorks, a subsidiary of Cargill. It is also manufactured in the Netherlands by PURAC Biomaterials, and by several manufacturers in Japan and China.

As an eco-friendly polymer, PLA has its share of controversies. One set arises from its synthesis from corn. Like corn used to produce ethanol (Chapter 5), corn that is used to produce PLA competes with its use as animal food. Furthermore, the runoff from cornfields may produce nutrient-rich waterways (Chapter 11), and the corn may be genetically engineered to resist pests (Chapter 13).

PLA can also be produced from carbohydrate-containing plants other than corn. For example, a Dutch company uses sugarcane and a Japanese manufacturer uses tapioca root. In the United States, PLA is largely manufactured from the starch of corn kernels leading to the nickname of "corn plastic."

In the biowaste stream (as part of municipal solid waste), PLA makes up only about 1%. Yard trimmings and food scraps are the major items.

Unlike petroleum-based polymers, PLA is compostable, but the process goes slowly without the heat supplied by an industrial composter. How slowly? In a backyard compost heap, the process takes up to a year. In contrast, industrial composters do the job in 3–6 months. However, many communities do not have access to such composters, at least not at present. Note that there is no ecological benefit to tossing PLA in a landfill; the actual breakdown of anything in a landfill is slow.

Can PLA be recycled? In theory, yes, but at present no. In fact, having PLA in the recycling stream is of concern to those who recycle PET, because, as one recycler quipped, "the two mix as well as oil and water." Recyclers currently collect and bale PET bottles and then process the plastic, eventually making it into new shirts, containers, fiberfill, or carpeting. PLA, if present in more than small amounts, needs to be separated from the PET recycling stream.

Your Turn 9.30 Scientific Practices Detective Work on Your Campus

Meals are served on most college and school campuses, up to thousands each day. Most likely, professionals in your food service department have given serious thought to which cups, plates, forks, spoons, chopsticks, and napkins to use. Be a detective and learn about the sustainable practices on your campus. What happens to plates, cups, and utensils? Are they washed and reused? Are they discarded? If so, are any made from PLA? What are the controversies? Prepare a one-page briefing on a particular item, for example, hot beverage cups.

9.11 | A New "Normal"?

One of the key components of sustainability is the concept of **shifting baselines**; that is, the idea that what people expect as "normal" on our planet has changed over time. Our use of plastics is a good example. Many people are still alive who remember "how it used to be" before the advent of plastics. Today, they are likely to have gray hair and were children back perhaps as early as the 1930s. Even the baby boomers of the 1950s remember collecting glass bottles such as that shown in Figure 9.25a to reclaim the two-cent deposit.

Think about plastic beverage bottles, for instance. It wasn't until 1970 that the first 2-L (64-oz) bottles appeared on supermarket shelves. The early models were made from PET and fitted with an opaque base cup for added strength (Figure 9.25b). PepsiCo was the first to sell soft drinks in 2-L bottles, and other beverage companies quickly followed suit.

Before the advent of plastic, most beverages were bottled in glass. Even as late as the 1970s, milk was brought to homes in glass bottles, with the empty bottles collected at the time of delivery. Portion sizes were smaller as well. For example, Coca-Cola bottles once held 10 ounces (Figure 9.25a); in contrast, today's aluminum cans hold 12 ounces and plastic bottles are larger still. Rather than dispensing cans, vending machines of the past dispensed bottles, and wooden racks stood nearby to receive the empties. However, glass bottling is not necessarily "greener" or more sustainable than bottling in plastic. The next activity invites you to explore the two options.

(a) (b)

Figure 9.25

(a) A 1960s 10-ounce glass Coke bottle; returnable. **(b)** A 1970s 2-L bottle, with plastic cup at the base for additional strength.

(a): © jvphoto/Alamy Stock Photo; (b): © Todd Franklin/Neato Cool Creative, LLC

Did You Know? Vending machines that dispensed aluminum cans were invented around 1965.

Your Turn 9.31 Scientific Practices Glass or Plastic?

a. Even though selling milk in glass bottles may be coming back in style, plastic jugs or plastic-coated cartons are still the norm in most places. List two advantages and disadvantages of using glass bottles. Do the same for using plastic bottles.

b. Today, if not sold in aluminum cans, soft drinks are sold in plastic bottles and beer is sold in ones made of glass. Research and report on at least two reasons for the difference.

© McGraw-Hill Education. Mark A. Dierker, photographer.

Plastic debris! Not only has our use of plastic become the norm, but it has also become the norm to find plastic debris everywhere—streets, backyards, streams, beaches, and even wilderness areas. The trouble lies in the durability of plastic. Once a piece of plastic finds its way into the local environment, it does not dissolve, break down in sunlight, or decompose—at least not at any appreciable rate. Rather, it tends to break into smaller and smaller pieces that widely disperse. The very properties that make plastics so useful in the first place mean that the pieces of plastic persist for years and years. Does this sound familiar? See if the next activity helps jog your memory.

Your Turn 9.32 Scientific Practices Lessons from Refrigerators Past

Chlorofluorocarbons, better known as CFCs, were once widely used in refrigerators, aerosol sprays, foams, and medical inhalers.

a. Why were CFCs phased out?

b. Some CFCs remain in the atmosphere for 100 years or more. Explain how this property of CFCs is connected to the fact that they have been phased out.

c. Name some properties that polymers such as HDPE, LDPE, PVC, and PS share with CFCs.

d. Unlike CFCs, it is highly unlikely that plastics will be phased out. Offer some reasons why.

e. Some believe that we cannot sustain our current use of plastics. Give evidence that either supports or contradicts this statement.

However, there is more to the story than the plastic bottles and wrappers that are seen around you that litter the landscape. Also ubiquitous in nature—including in our bodies—are the invisible substances that leach out of plastics. Do the environmental math. What is added to a polymer is slowly subtracted with the passage of time. Why? Plasticizers are not chemically bonded to the plastic. Rather, they are mixed in to make plastics softer and more pliable. Over time, they slowly leach out into the biosphere. The next activity introduces you to di-2-ethylhexyl phthalate (DEHP), a controversial plasticizer.

Your Turn 9.33 Skill Building Meet DEHP

DEHP belongs to a common class of plasticizers called *phthalates* (THAL-ates). Phthalates are esters of phthalic (THAL-ic) acid, an isomer of terephthalic acid, one of the monomers used to synthesize PET.

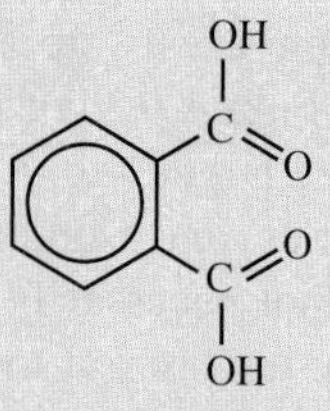

phthalic acid

a. Explain the meaning of the term *ester*.

b. Below is the structural formula for DEHP. Circle the two ester groups in this molecule.

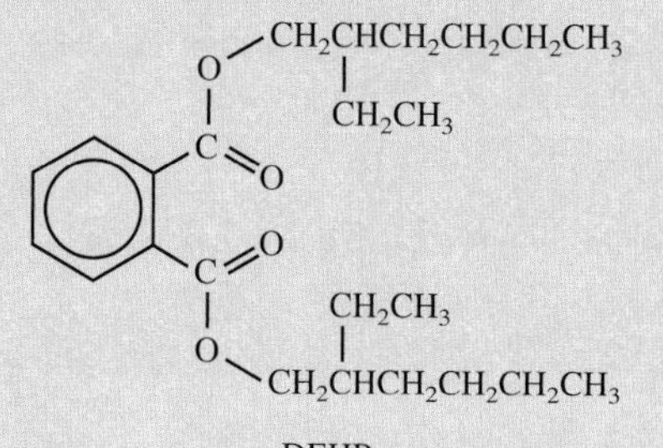

$CH_2CHCH_2CH_2CH_2CH_3$

CH_2CH_3

CH_2CH_3

$CH_2CHCH_2CH_2CH_2CH_3$

DEHP

c. Draw a structural formula for the alcohol that reacted with terephthalic acid to form this ester.

As you saw in the previous activity, the DEHP molecule has two long "wavy" side chains attached to a benzene ring. Imagine what happens when DEHP, perhaps as much as 30% by weight, is mixed in with a repeating head-to-tail arrangement of PVC. This arrangement of PVC tends to be stiff because its molecules pack well together and form crystalline regions. However, with the addition of DEHP, the regular packing of the PVC polymer chains is disrupted and the polymer becomes much more flexible.

Why is there a controversy? DEHP, like other phthalates, is a suspected **endocrine disrupter**, a compound that affects the human hormone system, including hormones for reproduction and sexual development. Estrogen is one such hormone, and unfortunately DEHP seems to have biological activity similar to that of estrogen. DEHP is also a suspected human carcinogen.

The chemical structure and biological activity of estrogen will be discussed in Chapter 12.

And why is it difficult to resolve the controversy? Although the evidence against DEHP has been mounting for decades, the research dots have been difficult to connect. Part of the difficulty lies in the low concentrations involved—parts per billion. Even so, in 2011, the U.S. Food and Drug Administration set the allowable limit for DEHP in bottled water at 0.006 mg/liter or 6 ppb. The very fact that DEHP might be present in bottled water may come as a surprise to you! But remember what we stated earlier: Compounds that originate in plastics have made their way almost everywhere in the environment, including our bodies.

Another reason why it is difficult to resolve the controversy is because not one, but many endocrine disrupters are present in our environment. Whereas some are naturally present, others have been added by humans. As an example of the latter, you may have heard of bisphenol A (BPA), another compound that mimics estrogen. BPA is transferred to the environment from several sources, including some plastic bottles.

Did You Know? Bisphenol A has been known to mimic the effects of estrogen since the 1930s.

A third difficulty lies in the fact that it is unethical to test compounds like BPA on humans. Although such research quickly could resolve the arguments, it is neither possible nor desirable. One way to get around this is to study those who have been inadvertently exposed to BPA.

Many plastic products on the market now boast that they are "BPA-free."

In spite of the difficulties, in some cases, potentially harmful substances have been banned by law. For example, it made sense to ban DEHP in infant pacifiers because babies receive repeated exposure, and because research on animals showed that DEHP affected male sexual development. Similarly, DEHP and other related plasticizers have been banned in children's toys. These bans are an example of the **precautionary principle**. This principle stresses the wisdom of acting, even in the absence of complete scientific data, before the adverse effects on human health or the environment become significant or irrevocable.

In most other cases, though, the choices are still being debated. The extremes range from banning the chemicals entirely to allowing their indiscriminate use. Neither extreme currently is in practice. So now it is a matter of reaching consensus on allowable uses. A report about BPA in *Chemical & Engineering News*, the weekly news magazine of the American Chemical Society, assessed the difficulties that you and all citizens face:

> *As this debate has unfolded, the public has been bombarded with a steady flow of studies, reports, claims, counter claims, conflicts of interest, lawsuits, and congressional inquiries regarding BPA. Both sides of the debate have been active in promoting their views to the media and the public. And both sides accuse each other of using spin tactics to create uncertainty about BPA, not unlike the socioscientific debates that have unfolded over cigarette smoking and climate change.* (June 6, 2011, p. 13)

We end this chapter with a quote: *"Nature doesn't have a design problem. People do."*[2] We have designed marvelous plastics that serve us in ways that a century ago we couldn't even dream of. At the same time, we have failed to design systems that carry these materials smoothly, safely, and economically from cradle to cradle.

Conclusions

Synthetic polymers are at the very center of modern living, yet their existence depends on a precious resource that we are consuming—crude oil. We have come not only to depend on synthetic polymers, but also in many cases to take them for granted to the point of being wasteful. Once more, we encounter a chemical topic that has the potential to inspire us to revisit the issue of our lifestyle and its sustainability.

Over time, chemists have created an amazing array of polymers and plastics—new materials that have made our lives more comfortable and more convenient. In many cases, these plastics represent a significant improvement over the natural polymers they replace. Furthermore, products that we use today would be impossible without synthetic polymers: cell phones, DVDs, breathable contact lenses, fleece clothing, kidney dialysis equipment, and artificial hearts. We have become dependent on polymers, and it verges on the impossible to abandon their use.

The chemical industry has responded to consumers. Together with those who work in the corporate world, we must learn to cope with plastic waste, while at the same time save raw materials and energy for tomorrow. To create a new world of plastics and polymers will require the intelligence and efforts of policy planners, legislators, economists, manufacturers, consumers, and, of course, chemists. This chapter showed that efforts at reducing, reusing, recycling, and recovering are well under way.

As we've seen in previous chapters, everything is interconnected. In this chapter, we looked at the connections of polymers to their raw materials—petroleum or plants—as well as at their connections to waste (or compost) in the environment. The chapter ended with an unexpected connection, that of additives to plastics that leach into the environment and have drug-like properties similar to estrogens. We use many plastic containers and wraps to store and sometimes cook food, so we should keep this in mind as we move into the kitchen laboratory—the topic of the next chapter.

[2] M. Braungart "Cradle to Cradle," North PointPress: New York, 2002.

Learning Outcomes

The numbers in parentheses indicate the sections within the chapter where these outcomes were discussed.

Having studied this chapter, you should now be able to:

- describe some of the properties of polymers (9.1–9.7)
- list some applications and uses for polymers around you (9.1–9.7)
- predict how the properties of polymers are useful for different applications (9.1–9.7)
- distinguish between polymers and plastics (9.1)
- classify polymers as either natural or synthetic (9.2)
- list various types of natural polymers (e.g., spider silk, paper (cellulose)) (9.2, 9.10)
- illustrate the macro- and molecular-scale structure of polymers (9.3–9.7)
- differentiate between addition and condensation polymerization reactions (9.3–9.7)
- illustrate the difference between crystalline and amorphous regions of polymers (9.5)
- identify different types of addition polymers (9.3–9.5)
- identify different types of condensation polymers (9.5–9.7)
- predict the type of polymer that can be produced from a monomer (9.3–9.7)
- explain how the structure of polymers affects their overall properties (9.3–9.7)
- list the main types of polymers and their applications (9.5)
- describe how polymers degrade (9.8)
- evaluate what types of polymers can be recycled (9.9)
- describe the recycling process for polymers (9.9)
- explain the benefits, types, and applications of biodegradable polymers (9.10)
- describe different additives that can be used with polymers and how each changes the property of the polymer (9.5, 9.11)

Questions

Emphasizing Essentials

1. Give two examples of natural polymers and two of synthetic polymers.
2. Think about your intended profession or career path. How can you contribute in a meaningful way to reducing our solid waste? Suggest three ways.
   Hint: Thinking about the Four Rs may be of help.
3. Equation 9.1 contains an n on both sides of the equation. The one on the left is a coefficient; the one on the right is a subscript. Explain.
4. In Equation 9.1, explain the function of the R· over the arrow.
5. Describe how each of these strategies would be expected to affect the properties of polyethylene. Also provide an explanation at the molecular level for each effect.
   a. increasing the length of the polymer chain
   b. aligning the polymer chains with one another
   c. increasing the degree of branching in the polymer chain
6. Figure 9.3a shows two bottles made from polyethylene. How do the two bottles differ at the molecular level?
7. Ethylene (ethene) is a hydrocarbon. Give the names and structural formulas of two other hydrocarbons that, like ethylene, can serve as monomers.
8. Why is a repeating head-to-tail arrangement not possible for ethylene?
9. Determine the approximate number of $H_2C{=}CH_2$ monomeric units, n, in one molecule of polyethylene with a molar mass of 40,000 g/mol. How many carbon atoms are in this molecule?
10. A structural formula for styrene is given in Table 9.1.
    a. Redraw it to show all of the atoms present.
    b. Give the chemical formula for styrene.
    c. Calculate the molar mass of a polystyrene molecule consisting of 5,000 monomers.
11. Vinyl chloride polymerizes to form PVC in several different arrangements, as shown in Figure 9.9. Which example is shown here?

```
 Cl  H   H   Cl  Cl  H
 |   |   |   |   |   |
—C——C——C——C——C——C—
 |   |   |   |   |   |
 H   H   H   H   H   H
```

12. Here are two segments of a larger PVC molecule. Do these two structures represent the same arrangement? Explain your answer by identifying the orientation in each arrangement.
    Hint: See Figure 9.9.

```
 Cl  H   Cl  H   Cl  H
 |   |   |   |   |   |
—C——C——C——C——C——C—
 |   |   |   |   |   |
 H   H   H   H   H   H
```

and

```
 H   H   H   H   H   H
 |   |   |   |   |   |
—C——C——C——C——C——C—
 |   |   |   |   |   |
 Cl  H   H   Cl  Cl  H
```

13. Butadiene, $H_2C_5CH–HC_5CH_2$, can be polymerized to make a synthetic rubber. Would this be by addition or condensation polymerization?
14. Which of the "Big Six" most likely would be used for these applications?
    a. clear soda bottles
    b. opaque laundry detergent bottles
    c. clear, shiny shower curtains
    d. tough indoor–outdoor carpet
    e. plastic baggies for food
    f. packaging "peanuts"
    g. containers for milk
15. a. Analogous to Equation 9.3, write the polymerization reaction of n monomers of propylene to form polypropylene.
    b. Analogous to Figure 9.9, show a random arrangement of the monomers in a segment of polypropylene.
16. Many containers are made from plastic. Check the recycling code on 10 containers of your choice (see Table 9.1). In your sample, which polymer did you most frequently encounter?
17. Name the functional group(s) in each of these monomers.
    a. styrene
    b. ethylene glycol
    c. terephthalic acid
    d. the amino acid in which R = H
    e. hexamethylenediamine
    f. adipic acid
18. Circle and identify all the functional groups in this molecule:
19. Kevlar® is a type of nylon called an *aramid*. It contains rings similar to that of benzene. Because of its great mechanical strength, Kevlar® is used in radial tires and in bulletproof vests. Your Turn 9.17 gives the structures for the two monomers: terephthalic acid and phenylenediamine. Name the functional groups in both the monomers and in the polymer.
20. Table 9.3 gives structural formulas for ethanoic acid and propanoic acid. From these two names, you should be able to determine the naming pattern.
    a. How would a carboxylic acid containing five carbon atoms be named?
    b. Methanoic acid is the smallest carboxylic acid. Also known as formic acid, it is one of the components in the sting of an ant bite. Draw the structural formula for methanoic acid.
    c. Butanoic acid, like propanoic acid, has a sharp smell. Draw the structural formula for butanoic acid.
21. Silk is an example of a natural polymer. Name three properties that make silk desirable. Which synthetic polymer has a chemical structure modeled after silk?

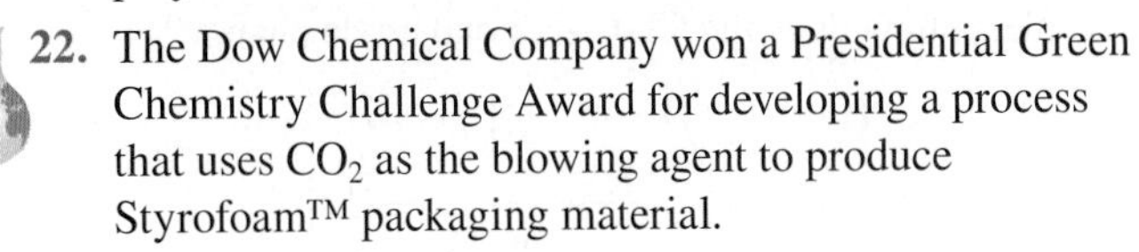

22. The Dow Chemical Company won a Presidential Green Chemistry Challenge Award for developing a process that uses CO_2 as the blowing agent to produce Styrofoam™ packaging material.
    a. What is a blowing agent?
    b. What compound does CO_2 likely replace in the process, and why is this substitution environmentally beneficial?
23. Suggestions for reducing your waste include: (1) buying in bulk and/or economy sizes and (2) avoiding individually packaged servings. Let's say you followed this practice for these cases. Which plastic would you use less of? Would you use more of something else?
    a. For use in your refrigerator, buying a half-gallon plastic jug of milk rather than 2 quarts.
    b. For guests at a reception, purchasing 2-liter bottles of lemonade rather than individual bottles.
    c. Buying more concentrated laundry detergent in a smaller plastic bottle.
24. Recycled products now are beginning to provide the origin of the recycled material.
    a. Give examples of post-consumer content and of pre-consumer content.
    b. Do recyclable products contain recycled materials?

Concentrating on Concepts

25. Draw a diagram to show the relationships among these terms: *natural, synthetic, polymer, nylon, protein.* Add other terms as needed.
26. Currently, many 2-liter beverage bottles are made of PET with polypropylene caps. Why is polypropylene a good choice for a bottle cap? What difficulty does using polypropylene present in the recycling of PET bottles?
27. Glucose from corn is the source of some new bio-based polymer materials. Glucose also is the monomer in cellulose. Earlier in this text, you encountered glucose in the chemical reaction of photosynthesis. What is photosynthesis and from what compounds is glucose produced?
28. The properties of a polymer depend, in part, on which chemical elements it contains. Name three additional things that influence the properties of a particular polymer.
29. Many monomers contain a C=C double bond. Select such a monomer and draw its structural formula together with the corresponding polymer. Describe the similarities and differences between the monomer and the polymer.

30. What structural features must a monomer possess to undergo addition polymerization? Explain, giving an example. Do the same for condensation polymerization.

31. This equation represents the polymerization of vinyl chloride. At the molecular level as the reaction takes place, how does the Cl–C–H bond angle change?

$$n\ \mathrm{H_2C{=}CHCl} \xrightarrow{R\cdot} \left[-\overset{\mathrm{H}}{\underset{\mathrm{H}}{\mathrm{C}}}-\overset{\mathrm{H}}{\underset{\mathrm{Cl}}{\mathrm{C}}}- \right]_n$$

32. Polyacrylonitrile is a polymer made from the monomer acrylonitrile, CH_2CHCN.

   a. Draw the Lewis structure for this monomer. **Hint:** The N atom is attached via a triple bond.

   b. Polyacrylonitrile is used in making Acrilan™ fibers used widely in rugs and upholstery fabric. If ignited, this fiber can release a poisonous gas. In the case of a fire, what danger might rugs and upholstery made of this polymer present?

33. Roy Plunkett, a DuPont chemist, discovered Teflon™ while experimenting with gaseous tetrafluoroethylene. Here is the monomer.

$$\mathrm{F_2C{=}CF_2}$$

   a. Analogous to Equation 9.1, write the chemical reaction for the polymerization of *n* molecules of tetrafluoroethylene to form Teflon™.

   b. Why is a repeating head-to-tail arrangement not possible for this polymer?

   c. Teflon™ is a solid and CFC-12 (CCl_2F_2) is a gas. Nonetheless, they both contain C–F bonds. What other characteristics do Teflon™ and CFC-12 have in common?

34. Equation 9.1 shows the polymerization of ethylene. From the bond energies of Table 5.1, is this reaction endothermic or exothermic?

35. Would your answer from question 34 differ if tetrafluoroethylene were used as the monomer? See question 33 for the monomer.

36. Do you expect the heat of combustion of polyethylene, as reported in kilojoules per gram (kJ/g), to be more similar to that of hydrogen, coal, or octane, C_8H_{18}? Explain your prediction.

37. Recycling is not the same as waste prevention. Explain.

38. Here is a recycling symbol that is more colorful than the standard ones used on many plastic containers.

   a. What is PLA?

   b. Why is corn depicted in the center of the symbol?

   c. This symbol is printed in green ink, presumably to convey that this polymer is "green." Give two reasons why PLA is considered an eco-friendly polymer.

   d. For each of your reasons in the previous part, provide information counter to your argument.

39. Consider the polymerization of 1000 ethylene molecules to form a large segment of polyethylene.

$$1000\ \mathrm{CH_2 = CH_2} \xrightarrow{R\cdot} \mathrm{-\!\!\left(CH_2CH_2\right)\!\!-}_{1000}$$

   a. Calculate the energy change for this reaction. **Hint:** Remember that polystyrene foam is made with blowing agents.

   b. To carry out this reaction, must heat be supplied or removed from the polymerization vessel? Explain.

40. Here is the structural formula for Dacron™, a condensation polyester.

$$\left[-\mathrm{O-CH_2-CH_2-O-\overset{O}{\overset{\|}{C}}-C_6H_4-\overset{O}{\overset{\|}{C}}}- \right]_n$$

Dacron™ is formed from two monomers, one with two hydroxyl groups (–OH) and the other with two carboxylic acids (–COOH). Draw a structural formula for each monomer.

41. When you try to stretch a piece of plastic bag, the length of the piece of plastic being pulled increases dramatically and the thickness decreases. Does the same thing happen when you pull on a piece of paper? Why or why not? Explain on a molecular level.

42. Consider Spectra®, AlliedSignal Inc.'s HDPE fiber, used as liners for surgical gloves. Interestingly, Spectra is linear HDPE, which is usually associated with being rigid and not very flexible.

   a. Suggest a reason why LDPE cannot be used in this application.

   b. Name two other possible uses of a fabric made of Spectra®.

43. The Four Rs are reduce, recycle, reuse, and recover.

   a. Give an example of each, naming the plastic involved.

   b. A possible fifth R is "rethink." For example, plastic waste can be rethought in terms of benefits to public health. Give an example of a connection between waste reduction and public health.

44. All the Big Six polymers are insoluble in water, but some dissolve or at least soften in hydrocarbons (see Table 9.1). Use your knowledge of molecular structure and solubility to explain this behavior.

45. Polystyrene foam packing peanuts are degraded when immersed in acetone (a solvent in some nail-polish removers). If the acetone is allowed to evaporate, a solid remains. What is this solid? Explain what happened. **Hint:** Remember that polystyrene foam is made with blowing agents.

46. Today, some packing peanuts are made from plant-based materials rather than polystyrene foam.

a. Starch is one of the options. What is starch and what is its source?

b. Name two advantages and two disadvantages of starch packing peanuts.

c. Name an option for disposing of starch packing peanuts.

47. Explain the concept of shifting baselines. Then, give two examples each in regard to:

a. plastic items used in packaging.

b. plastic item contamination in waterways.

48. DEHP is a plasticizer that is an example of a phthalate, an ester of phthalic acid.

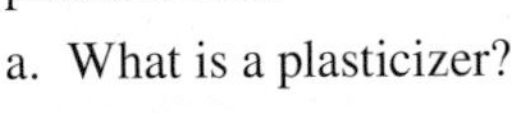

a. What is a plasticizer?

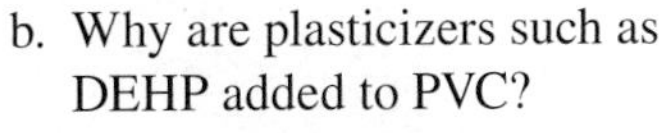

b. Why are plasticizers such as DEHP added to PVC?

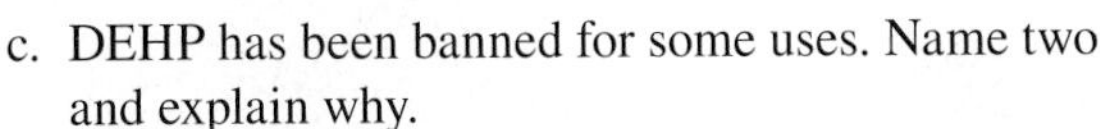

c. DEHP has been banned for some uses. Name two and explain why.

Exploring Extensions

49. a. Name two functional groups not discussed in this chapter. Give an example of a molecule containing each one.
Hint: Look ahead to Chapter 10.

b. Find the structural formula for the acetone molecule mentioned in question 45. What functional group does it contain?

50. Cotton, rubber, silk, and wool are natural polymers. Consult other sources to identify the monomer in each of these polymers. Which are addition polymers and which are condensation polymers?

51. The Great Pacific Garbage Patch ("Plastic Trash Vortex") supposedly consists of plastic broken into small fragments that lie below the surface of the ocean and wreak havoc on marine life and on those that eat it, including humans. In 2011, a high-ranking person in the plastics industry informally offered an opinion to an author of this textbook. "Personally," he said, "I think it's a hoax." Is he correct or is he misinformed about the facts? Use well-credentialed sources to make your case.

52. A Teflon™ ear bone, fallopian tube, or heart valve? A Gore-Tex® implant for the face or to repair a hernia? Some polymers are biocompatible and are now used to replace or repair body parts.

a. List four properties desirable for polymers used *within* the human body.

b. Other polymers are used *outside* your body but in close contact with it, such as those in contact lenses. What are contact lenses made of? What properties are desirable?

53. PVC, also known as "vinyl," is a controversial plastic. Comment on the controversies, either from the standpoint of a consumer or a worker in the vinyl industry.

54. Learn the story of the discovery of Kevlar®. This polymer was originally sought for use in radial tires but found other applications as well. Write a short report citing your sources.

55. Isoprene polymerizes to form polyisoprene, a natural rubber. Here is the structural formula of isoprene, with its carbons numbered.

$$\underset{1}{CH_2}=\underset{\underset{CH_3}{|}}{\overset{2}{C}}-\underset{3}{CH}=\underset{4}{CH_2}$$

When isoprene monomers add, polyisoprene has a C=C double bond between carbon atoms 2 and 3. How does this double bond form?
Hint: Each C=C bond contains four electrons. Each new C–C bond that forms to link two monomers only needs two electrons, one from each of the monomers that joined to form it.

56. Synthetic rubber is usually formed through addition polymerization. An important exception is silicone rubber, which is made by the condensation polymerization of dimethylsilanediol. Here is a representation of the reaction.

$$n\ HO-\underset{CH_3}{\overset{CH_3}{Si}}-OH \longrightarrow \left[-O-\underset{CH_3}{\overset{CH_3}{Si}}-O-\right]_n + n\ H_2O$$

a. Predict two properties for this polymer. Explain the basis for your predictions.

b. Silly Putty™ is a popular form of silicone rubber. Name two of its properties.

c. Name two other household uses for silicone rubber.

57. Given the number of personal computers in use today, there is good reason to keep keyboards, monitors, and "mice" out of the landfill.

a. Which polymers do your computer and its accessories contain?

b. What are the options for recycling the plastics in computers?

58. Some regions in the United States have bottle bills that require a deposit on some or all containers. Some grocers, beverage companies, and bottle associations stand strongly against bottle bills. In contrast, some consumer groups and environmental groups argue strongly for them. Draft a one-page position statement that speaks either for or against bottle bills.

59. Cargill won a 2007 Presidential Green Chemistry Challenge Award for using soybeans instead of petroleum to produce polyols. What is a polyol? How are polyols used to produce "soybean plastics"?

CHAPTER 10 Brewing and Chewing

© Rawpixel.com/Shutterstock.com

Health Disclaimer: This chapter is meant to discuss the science of different ways we prepare food and beverages for consumption. Although we do touch on general ideas about food and our health, we do not discuss any details of benefits or detriment to one's health in these preparations. Although the next chapter will provide information about food, nutrition, and health, you may also want to conduct your own research with trained medical and nutritional health specialists about any health impacts of the foods and beverages we discuss.

REFLECTION

Flavor Beads?

Connect provides a video that introduces this chapter. "Flavor beads" were used in a drink order. Using the Internet as a resource, what are these beads composed of, and what reaction(s) are taking place to give the visible color change?

The Big Picture

In this chapter, you will explore the following questions:

- What's in a mouthful?
- Why is the kitchen like a laboratory?
- What is significant about heat transfer in the kitchen?
- What are some chemical processes we use to prepare our food?

Introduction

Accompanying this paragraph are words of wisdom from that fictional, though masterful, chocolatier and candy maker, Mr. Willy Wonka. From its wonderful smell, smooth texture, soothing taste, and lingering finish, chocolate represents a remarkable culinary and scientific creation. Ironically, it all starts from raw cacao beans that have a repulsive taste and appeal. However, through a laborious multistep set of chemical and physical processes, the beans become the wonderful chocolate many of us crave and enjoy.

"Invention, my dear friends, is 93% perspiration, 6% electricity, 4% evaporation, and 2% butterscotch ripple."

—Willy Wonka

We eat and drink for a variety of reasons. Fundamentally, foods and beverages are necessary for our survival, providing the nutrients and energy we need. However, many of us also eat food because it tastes good, makes us feel good, and is part of our customs and traditions. Regardless of whether we are herbivores, carnivores, or omnivores, our food types, the ways in which we prepare food, and our consumption practices are often tied to our principles, lifestyles, families, cultures, and histories. These traditions involve stories, rituals, choices, and even religious observations. This chapter surveys aspects of the chemistry of food and drink.

Your Turn 10.1 You Decide What's in a Mouthful?

Traditionally, regions of the tongue have been identified that are consistent with a particular type of taste. Gather a sample of lemon juice, sugar, and salt. With a toothpick, apply samples of each to the tip, sides, and back of the tongue. Which part of your mouth responds to each food?

Your Turn 10.2 You Decide The Truth Behind the Tongue Map

Tongue map—true or false? In the previous activity, you were asked to map different types of taste to regions of your tongue. Were different regions of your tongue able to detect sweet or sour tastes? Do a bit of research on the Internet to find out the origin and current thoughts regarding the so-called "tongue map."

10.1 What's in a Mouthful? The Science of Taste

The bottom-line answer to the question posed in this section title is: "a whole bunch of chemical interactions!" Our tongue is a remarkable organ and an essential nerve center that also helps move food around to aid in our chewing. Our tongue also detects taste—or the more technical term, *gustatory*—sensations (Figure 10.1). Most scientists, chefs, and other experts agree that there are six basic tastes our tongue detects: sweet, salty, bitter, sour, savory (umami), and fatty. From those six basic tastes, countless flavor combinations can be made and experienced. The mouth helps determine a food's chewiness, oiliness, texture, and viscosity.

The molecules in food interact with the macromolecular sensors on the surface of our tongue. Like other places in our bodies, our tongue has nerve cells that have both chemical receptors and ion channels. We perceive taste because a particular molecule binds to these receptors, which triggers a particular signal to the brain. Alternatively, brain signals may also be generated by a change in the number of ions passing through an ion channel. For example, as you saw in Chapter 8, sodium chloride completely dissociates into ions when dissolved in water (the solvent of our saliva):

$$NaCl(s) \xrightarrow{H_2O(l)} Na^+(aq) + Cl^-(aq) \quad [10.1]$$

The sodium-ion channels of our tongue detect the increased concentration of sodium ions from these food sources, and send signals to our brain that trigger the detection of saltiness. Similarly, proton-ion channels detect acidity:

$$\underset{\text{vinegar (acetic acid)}}{CH_3COOH} \rightleftharpoons \underset{\text{acetate ion}}{CH_3COO^-} + \underset{\text{proton}}{H^+} \quad [10.2]$$

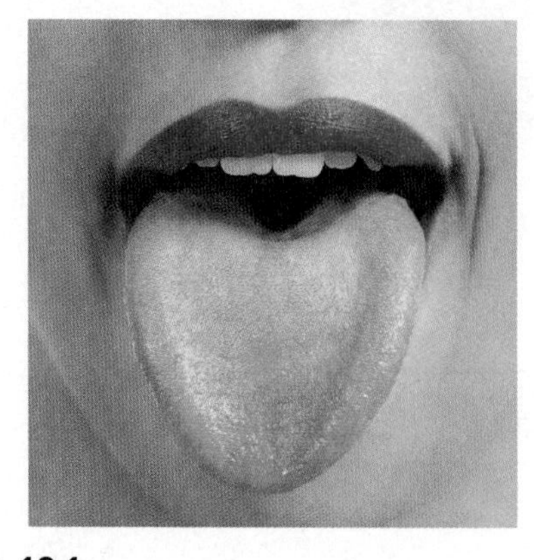

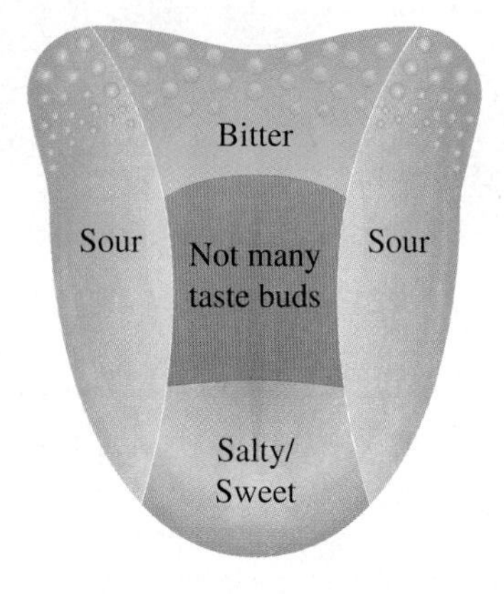

Figure 10.1

Picture of a tongue and a traditional diagram of taste buds on the tongue. However, recent research has indicated that the entire tongue can sense all six tastes more or less equally.

These experiences do not stop at the tongue. The chemicals from food can also interact in many other locations in our body that affect our mood and health. There have been significant research efforts to understand receptors and chemical reactions occurring on the tongue (Figure 10.1). However, many more studies are needed to understand all of the processes that help us enjoy those first contacts with food on our tongue.

10.2 | How Does Smell Affect Taste?

The smell of food is actually where our experience starts. Those wonderful aromatic compounds enter our nostrils and bind with odor receptors in the nasal cavity. Messages are then sent to our brain centers, setting up what we expect to taste in the food. For many, the smells of food may elicit powerful memories—grandma's fresh bread, a summer barbeque, hot spiced apple cider. The aroma of chocolate may be no different—its beckoning in hot chocolate after a winter's snowball fight, or its welcoming in a dessert cake at the end of a family meal. In fact, more than 600 chemicals come together to give chocolate its aroma and taste. It is no wonder that such a product can produce such a pleasing response.

Your Turn 10.3 You Decide How Does Aroma Affect Taste?

You will need samples of apple and potato chopped into small pieces. Close your eyes and have someone mix up the samples. Pinch your nose closed and taste both of the samples, taking a drink of water between the two different foods. Can you tell the difference between the apple and potato? Explain what you experienced. Try this with other foods—for example, flavor extracts or jelly beans.

(top): © www.nick-moore.com/Moment Open/Getty Images; (bottom): © Andreas Altenburger/Shutterstock.com

Let's return to chocolate. The Aztecs called it Theobroma—food of the gods. Food researchers have studied this food, and have discovered a number of healthy attributes for our bodies. Its smell originates soothing feelings, and once chocolate touches the tongue, the cascade of reactions to our taste brain centers are initiated. No matter how you eat chocolate, its distinctive taste is likely to put one in a pleasing or calm mood, even if just for a moment.

There are chemical reasons for the relaxing feeling we have after eating chocolate. One is that chocolate contains *tryptophan* (Figure 10.2), a chemical precursor for the body's synthesis of *serotonin*, a neurotransmitter responsible for making us feel happy. Chocolate has also been found to contain *anandamide*, another neurotransmitter that targets the same brain receptors as *tetrahyrdocannibol* (THC), the active component of marijuana.

Mood and health are certainly connected, and chocolate also contains health-beneficial chemicals. Among them are *flavonols* (Figure 10.2), or phytochemicals, that

Recent research has suggested that flavonols may even play a role in cancer prevention.

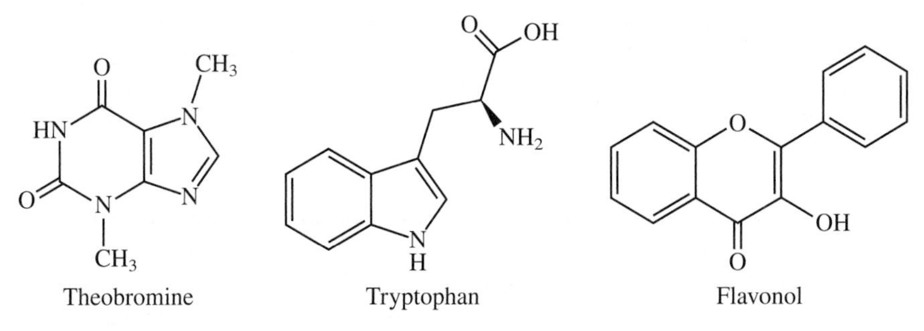

Figure 10.2

Chemical structures for theobromine, tryptophan, and flavonol—key chemicals found in chocolate.

are also present in teas, red wine, and fruits and berries. These compounds have antioxidant properties and support a healthy heart. Furthermore, recent research has indicated that these compounds may even have a role in cancer prevention. Chocolate also contains magnesium, phosphorus, and potassium, essential elements for the body. Dark chocolate tends to be healthier than milk or white chocolate, because it has less saturated fat. Although one could argue whether or not chocolate is a "perfect" food, it certainly is among those that provide the total package—great taste, good feelings, and good health.

Cocoa Nib Hot Chocolate
(Yield: four 6-oz servings)

Chemicals
1 ounce cocoa nibs
16 ounces whole milk
6 ounces 65% to 70% bittersweet chocolate, finely chopped
3 ounces sugar
2 ounces water
¼ teaspoon kosher salt

Procedure
- Pulse the cocoa nibs in a spice grinder 3 to 4 times until the nibs are coarsely chopped. Place the nibs in a 1-quart microwave-safe measuring cup and add the milk. Microwave on high for 3 to 4 minutes or until the milk reaches 160 °F. Steep at room temperature for 30 minutes.
- Meanwhile, combine the chocolate, sugar, water and salt in the carafe of a 1-liter French press. Set aside.
- After steeping, return the nib-milk mixture to the microwave and heat on high for 2 minutes until it simmers or reaches 185 °F. Strain the hot nib-milk mixture through a fine-mesh strainer into the French press carafe. Set aside for 1 minute, and then stir to combine the chocolate and milk. Pump the plunger of the French press 10 to 15 times to froth and aerate. Serve immediately.

Recipe credited to Alton Brown, 2011.

Your Turn 10.4 You Decide A Chocolate Taste Test!

Find three different bars of plain chocolate. They could be different brands, or dark, milk, and white chocolate. Put a small sample of each in your mouth, one at a time. Try tasting it in two different ways: (1) By putting it in your mouth and chewing it. How does the sample taste? What do you hypothesize to be different about each type of chocolate? (2) Now, repeat using the same three types of chocolate, but this time place an ample amount of each in your mouth, let it melt, and think about the flavors. What is the texture of each sample? How does the sample taste? What do you hypothesize to be different about each type of chocolate?

It has been suggested that letting chocolate melt and spread across your tongue is the best way to enjoy its full flavors. This warms the chocolate and allows the various components, including flavorful oils, to "bloom" and release their full flavor. Melting it in this way also allows the chocolate to spread across your tongue and contact more of those vital flavor receptors.

So, what's in a mouthful? Nothing should be taken for granted. What we consume directly affects our longevity, energy, and mood. In order to eat and experience these wonderful sensations though, we must first prepare the food.

10.3 | The Kitchen Laboratory

We transform the food we grow, cultivate, and raise. Oftentimes, we do this to make it safe for eating, such as reaching a temperature that will kill any unwanted organisms that might make us sick. At other times, we transform our food to make it taste better, or differentiate it from traditional ways of preparing it. As a result, the kitchen, in so many ways, becomes a place to experiment, explore, and refine the possibilities for enhancing the experiences, and even the nourishment, we gain from food and drink.

Maybe you had this kind of aunt or grandmother. The one who made perfect bread or biscuits every time she tried, but who never lifted a measuring cup or spoon. She cooked by feel, and by sight. What researchers have come to find is that your aunt or grandmother was not just "winging it." She was instead following a practiced and carefully acquired regiment that had amazing consistency. Those hand- and sight-line "measurements" made by grandma actually do have remarkable precision. Otherwise, those biscuits would not have tasted the same or held the same texture and fluffiness each time. Although there are certainly wonderful moments and opportunities to be spontaneous in the kitchen, good cooking does require—like good scientific research—a careful set of protocols and replication methods.

© Mat Hayward/Shutterstock.com

Just like a different pathway can be used to synthesize many chemicals in the laboratory, there are many ways to produce a food product. For instance, examine Table 10.1 compiled from a randomly selected set of five recipes from an Internet search for chocolate chip cookie recipes.

Table 10.1 Selected Ingredient Measurements Per Dozen Chocolate Chip Cookies

Recipe	Flour (cups)	Total Sugar (cups)	Butter (cups)	Baking Soda (teaspoons)	Brief Description of Resulting Cookies
A	0.45	0.30	0.20	0.20	"Perfect blend of textures"
B	0.65	0.45	0.27	0.20	"Wonderfully combined textures"
C	0.50	0.27	0.20	0.15	"Crisp bottom, soft top"
D	0.60	0.40	0.20	0.15	"Crisp and crunchy"
E	0.45	0.30	0.20	0.10	"Soft and chewy"
Average	0.53	0.34	0.21	0.16	
Standard Deviation	0.09	0.08	0.03	0.04	

You might at first be tempted to state that we have just contradicted ourselves. That is, there is much variation in these recipes for chocolate chip cookies. And you would be correct, until you examine the details of the properties of the cookies made from each recipe. Each variation leads to a different version of the product. Some are soft and chewy, while others are crisp and crunchy. So, yes, they all produce chocolate chip cookies, but not all cookies are the same!

Your Turn 10.5 You Decide The "Best" Chocolate Chip Cookie

Maybe you have an image of the ideal chocolate chip cookie. What change in ingredient(s) do you predict causes a crunchy versus a chewy cookie? Can you confirm your predictions by making changes to these ingredients? What can you conclude after conducting your experiments?

10.4 | The Science of Recipes

Amounts matter. More specifically, amounts in proportion matter. A recipe is really nothing more than a good *stoichiometry* problem. That is, if you change the amounts of one ingredient, the amount of the other ingredient(s) needs to change in proportion. One-dozen chocolate chips cookies may be nice, but two-dozen may be even better. The cookies that are produced in either amount, though, still need to taste the way you want. Otherwise, why make more?

The term *stoichiometry* is related to moles (Chapter 1) and the Law of Conservation of Mass (Chapter 2), which states that the total mass of reactants must equal the total mass of products.

Your Turn 10.6 Skill Building Cookie Recipes

You need to make 100 cookies for your school bake sale. One cookie recipe makes 25 cookies, and uses the following ingredients:

Ingredients for Chocolate Chip Cookies

½ cup butter
1 cup chocolate chips
½ cup brown sugar
½ cup white sugar
1 egg
½ teaspoon vanilla
1 ¼ cups flour
¾ teaspoon baking soda
¼ teaspoon salt

How much of each ingredient do you need?

The term **limiting reagent** is often used in chemistry to describe the reactant that is totally consumed during a chemical reaction, and thereby limits the amount of product that may be formed.

Your Turn 10.7 Skill Building "Limiting Reagents" in the Kitchen

The "perfect quesadilla" can be made with two large flour tortillas (200 g per tortilla) and 1 cup of cheese (50 g). You open the refrigerator and find 350 g of cheese that is about to expire. If you use all of the cheese, how many quesadillas can you make? If you have a total of eight tortillas, which ingredient will be completely consumed, and which will be left over?

Recipes are carefully designed chemical reactions that produce the product(s) we want. We can change amounts, but can only do so in proportion with the other ingredients. Too much flour in the gravy—no good. Too much salt in the cake—no good. Too much sugar in the dressing—no good. Too much butter—well, maybe that's good (though perhaps not for your health, as we will detail in Chapter 11)! Once again, we change the food we harvest, and not haphazardly. We carefully follow steps so that the product is something we can make, eat, and enjoy again and again!

If you happen to find a recipe from a different country, you might find that their system of measuring ingredients is different from the one you use. For example, a country from Europe might use milliliters (mL) to measure the volume of a liquid, while in the United States the recipe might use ounces (oz) or teaspoons (tsp). Again, emphasizing the importance of protocols, as in good scientific experimentation, you need to use the correct amounts of ingredients for the recipe. Therefore, it is important to know how to convert between the various systems of measurements.

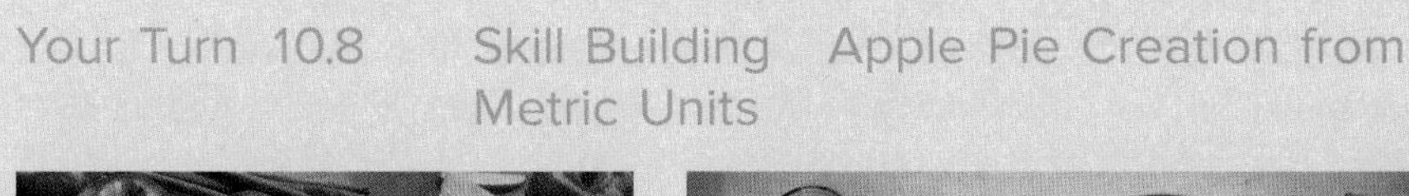

Your Turn 10.8 Skill Building Apple Pie Creation from Metric Units

(left): © Mitch Hrdlicka/Getty Images RF; (right): © Jiratthitikaln Maurice/Shutterstock.com

Below is an apple pie recipe using the metric system to describe the amounts of ingredients to use. Conduct an Internet search to translate between the metric system and the English system—used primarily in the United States. Convert the units below into units of cups, teaspoons, and tablespoons.

Ingredients for Apple Pie in Metric Units

1.5 kilograms of apples
150 grams sugar
25 milliliters of cornstarch
4 milliliters of cinnamon
0.75 milliliters of salt
0.75 milliliters of nutmeg
40 grams butter

In a stir-fry or soup recipe, one can usually get away with a little extra dash of this or that. In baking, though, precision and accuracy are critical for creating the particular tastes and textures referred to in many of the examples in the preceding text. Although there is an ongoing debate regarding which is better to use—volumes or masses for ingredients—many professionals use masses for their recipes (Table 10.2).

Think for a minute about flour. Flours are fine powders, and their particles can pack differently based just on how you measure the flour from the bag or jar. Scooping the measuring cup into the bag versus using a spoon to transfer the flour to the measuring cup actually can produce different amounts of flour (by mass) that ends up

Table 10.2 Conversions Between Volumes and Masses for Common Baking Ingredients

Ingredient	Volume (cups)	Mass (grams)
Butter (Salted or Unsalted)	1/2 cup 1 cup	113 grams 226 grams
Flour (All Purpose or Plain)	1 cup	130 grams
Flour (Cake)	1 cup	120 grams
Flour (Whole Wheat)	1 cup	130 grams
Potato Flour	1 tablespoon 1/2 cup	12 grams 80 grams
Cornstarch (Corn Flour)	1 tablespoon	10 grams
Ground Almonds (Almond Meal or Flour)	1 cup	90 grams
Cornmeal	1 cup	120 grams
Sugar (Granulated White Sugar)	1 cup	200 grams
Sugar (Brown) (lightly packed)	1 cup	210 grams
Confectioners' Sugar (Powdered or Icing)	1 cup	120 grams
Chocolate Chips	1 cup	170 grams
Cocoa Powder (Regular Unsweetened or Dutch Processed) (can vary by brand)	1 tablespoon 1 cup	6 grams 100 grams
Graham Cracker Bread Crumbs	1 cup	100 grams
Old Fashioned Rolled Oats	1 cup	95 grams

in the recipe. Also consider brown sugar. Many recipes call for the cook to pack the brown sugar into a measuring cup. If we pack the sugar differently, do you think that different amounts (again by mass) would be put into the recipe? At the end of the day, maybe these small variations between using volumes and masses may not be completely noticeable in the everyday kitchen. However, in many restaurants, where it is critical to serve a repeatable and reliable product with almost no variation, these discrepancies in measurements may create unwanted and noticeable differences in their food.

Your Turn 10.9 Skill Building Density Calculations

Based on the metric conversions you found in **Your Turn 10.8**, determine the density (in units of g/mL) for a few of the cooking ingredients from **Table 10.2**. Which flour has the highest density?

10.5 Kitchen Instrumentation: Flames, Pans, and Water

After adhering to the careful measurements of recipe ingredients, the next step is to assemble and transform the ingredients. Let's start with a process that is perceivably rather simple in the kitchen: boiling pasta. Here is one set of instructions:

- Step 1: Make sure there is at least enough water to cover the pasta. Too little water will all get absorbed by the pasta before it has finished cooking. Too much is not a problem; it will just take longer for the water to reach a boil. Add some salt and a swirl of oil, if desired.

- Step 2: Add uncooked pasta to boiling water.
- Step 3: Return water to boil and set timer to a cook time for *al dente*. This is an Italian phrase for "to the tooth," which means the pasta has a good firm, but cooked, texture.
- Step 4: Remove pasta from hot water and drain. Serve with your favorite *gravy* (the true Italian phrase for pasta sauce!).

Contrary to popular belief, the addition of a few pinches of salt does not alter the cooking time or temperature of the boiling water. In fact, it would take 114 g of salt per 1 L of water to cause only a 1 °C increase in its boiling point, certainly more salt than one would care to have in their food!

Your Turn 10.10 You Decide Varying Steps in Cooking Instructions

You just read a set of instructions that will give a particular product of cooked pasta. Now, imagine changing aspects of those steps. What would happen if you do not add salt? What happens if you cook your pasta longer than the *al dente* cook time? What if you are cooking pasta at higher-than-sea-level altitudes—should the cooking time remain the same?

Using hot water for cooking may seem very simplistic; however, this technique actually employs a remarkable amount of scientific principles. First and foremost, boiling water cooks our food because of the Laws of Thermodynamics. Heat is transferred from objects at a higher temperature to objects at a lower temperature. Our pasta or vegetables cook when placed into boiling water because heat is transferred from the hot water to the foods initially held at room or refrigerator temperatures.

Water is a good solvent to use for cooking because of its high **latent heat**. In other words, its ability to absorb significant amounts of heat before undergoing a phase change from liquid to gas. This is partly accounted for because of water's chemical composition (H_2O) and the ways in which the water molecules interact with one another through their intermolecular forces (Figure 10.3 and Section 8.3). This is why water can absorb much heat before changing state; overcoming those types of interactions requires a significant amount of energy. At its boiling point, the water molecules themselves have more energy than at room temperature. Accordingly, some of this energy is transferred to the food to cook it.

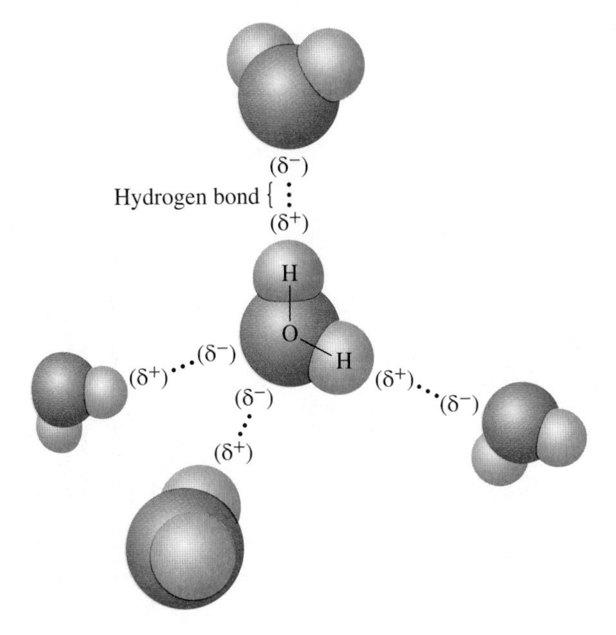

Figure 10.3

Intermolecular forces, known as hydrogen bonds, among water molecules.

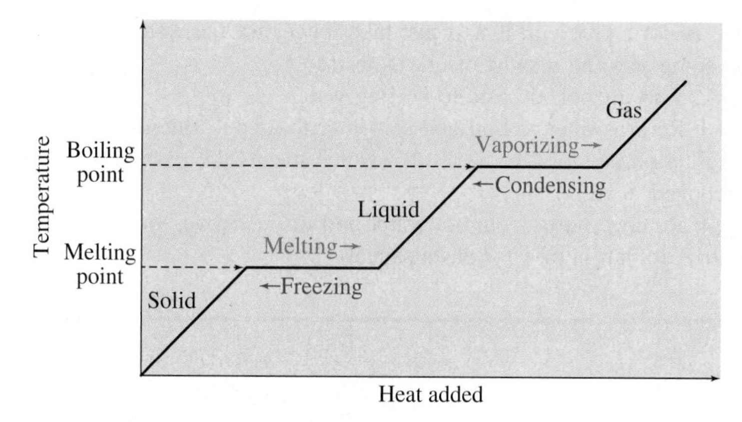

Figure 10.4

Heating curve for all substances that change between solid, liquid, and gaseous states.

Why do different altitudes require different cooking times when using boiling water? To answer this, we need to return to the concept of boiling. Boiling water is not just about achieving a particular temperature. Water boils at a certain temperature *and* pressure. Recall from Chapter 5 that we define the **boiling point** as the point when the vapor above the liquid has at least matched the ambient pressure acting on the liquid. Additionally, when a substance is undergoing a phase change, its temperature does not change until the phase change is complete (Figure 10.4). This is because the energy being transferred during a phase change is used to either break or form intermolecular forces, and not used to change the kinetic energy of the molecules themselves. Hence, when we cook something with boiling water, we can no longer increase the temperature of the water molecules until they have all changed to steam.

Weaker gravitational forces at higher elevations results in a lower availability of gas molecules in the upper atmosphere. Consequently, there are fewer molecular collisions, which corresponds to a lower atmospheric pressure.

Since higher altitudes have lower atmospheric pressures, it takes less energy to have the vapor pressure of the liquid reach and exceed the external air pressure. As a result, water will reach its boiling point at a lower temperature in higher altitudes (Figure 10.5), which necessitates longer cooking times.

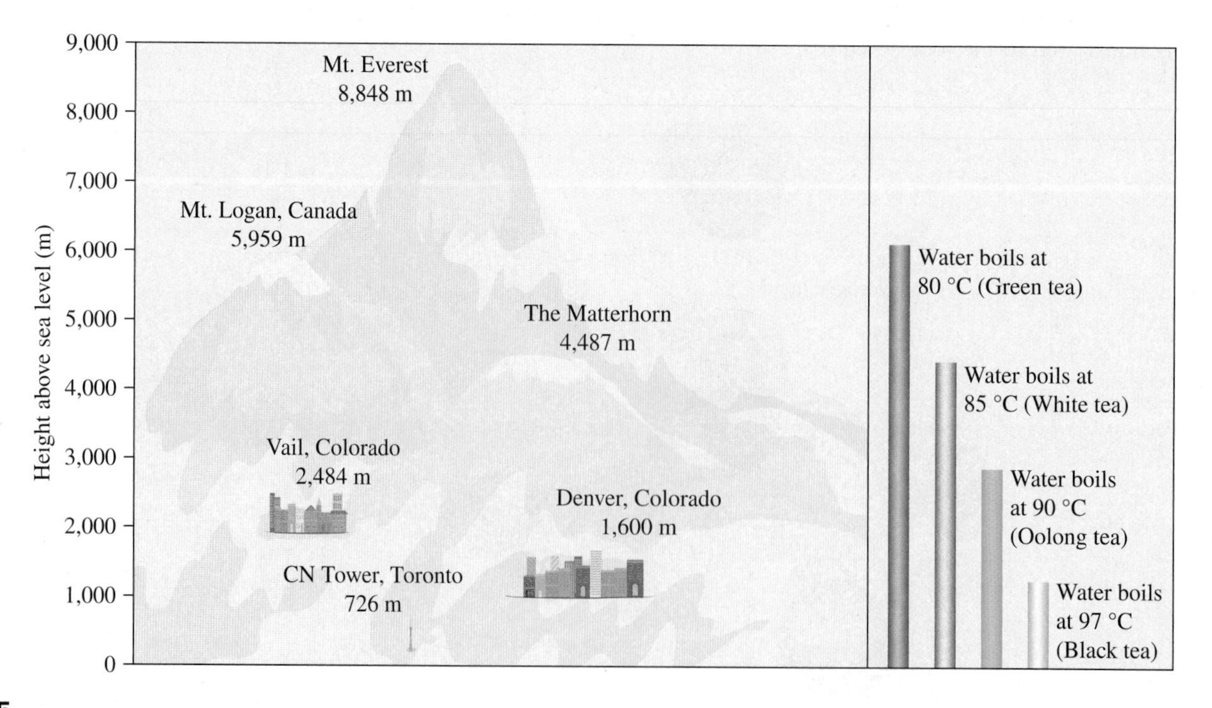

Figure 10.5

The boiling point of water and ideal steeping temperatures of tea varieties with changes in altitude. Interestingly, the flavors from green tea leaves are suitably extracted from lower-temperature boiling water. In contrast, black tea leaves require higher temperatures (and therefore lower altitudes) to release the desired flavors.

Table 10.3 Altitude Adjustment of Pressure Cookers

Altitude (ft.)	15 psi Pressure Cooker (psi)	14.5 psi Pressure Cooker (psi)	13 psi Pressure Cooker (psi)	11.5 psi Pressure Cooker (psi)
10,000	10.4	9.9	8.4	6.9
9,000	10.8	10.3	8.8	7.3
8,000	11.2	10.7	9.2	7.7
7,000	10.8	10.3	8.8	7.3
6,000	12.1	11.6	10.1	8.6
5,000	12.5	12	10.5	9.0
4,000	**13**	12.5	11	9.5
3,000	**13.5**	**13**	11.5	10
2,000	**14**	**13.5**	12	10.5
1,000	**14.5**	**14**	12.5	11
0	**15**	**14.5**	**13**	11.5

*Red indicates the pressures normally found in standard pressure cookers

The pressure unit psi refers to *pounds per square inch*.
1 psi = 0.068 atm.

Your Turn 10.11 Scientific Principles Pasta Cooking Times

Predict the difference in boiling time for spaghetti noodles in Denver, Colorado (the "Mile High City"), versus the city of Los Angeles (approximately at sea level).

Pressure cookers also take advantage of the relationship between boiling temperature and pressure. Because a pressure cooker is sealed, once the water is turned to steam it is trapped in the pot. The pressure above the water also increases so the boiling point of water increases, raising the temperature of the water (and food). However, one must also take into account a change in altitude when cooking with pressure cookers. The higher the altitude, the lower the actual pressure in the cooker (Table 10.3).

Regardless of the temperature used in cooking, we use heat to transform our foods both physically (*e.g.,* melting ice) and chemically. For instance, let's consider the breakfast staple food of poultry eggs, which may be boiled, fried, poached, or baked. Each of these preparations requires heat transfer to both physically and chemically change the raw egg to the version we wish to eat. When cooking on a natural gas stove, you might ignite a burner around medium-high heat. Then, you would select a small pan and possibly add some oil or butter. Once the butter or oil sizzles slightly, you would crack the egg into the pan. The translucent portion of the egg first begins to turn white, and the yolk remains a thick yellow liquid. After a few minutes, you might gently take a spatula under the egg and turn it over (the bolder cook may lift the pan by its handle and flip the egg) taking care not to "break" the yolk. You let it go a minute or two more to complete cooking of the egg white, but not the yolk because you want the "over easy" product. If done properly, you will have accomplished a slightly brown and crisp edge to your egg, which is then slid onto a plate and enjoyed with a piece of toast.

The difference between physical and chemical changes is discussed in Chapter 1.

© *Bradley D. Fahlman*

Now, let's take a closer look at the physical and chemical changes that have occurred during the routine practice of cooking an egg. This symphony that produced such a wonderful food was conducted under the Laws of Thermodynamics. In chemical or physical processes, energy is neither created nor destroyed, but can be transferred and/or transformed. Additionally, within these laws, we also know that when energy is in the form of heat, it can only be transferred from a hotter object to a less hot (or cold) object. When you turn on the stove, a flame emerges or a coil warms. The energy of the stovetop is transferred to the pan. The energy of the pan is then transferred to the oil or butter,

"Cold" is not actually a scientific term. Scientists usually just refer to the amount of thermal energy a substance has, typically indicated by its temperature. "Cold" substances just have less thermal energy than the surrounding room or our skin—noticeable during a touch.

causing it to melt, thus creating a physical change. The final transfer of thermal energy to the raw egg is sufficient to break intermolecular forces present in the proteins of the egg white, which results in a chemical change (Figure 10.6). These alterations to the molecules result in changes to the egg's properties, and the clear white liquid is transformed into an opaque white solid. Since the composition of the egg yolk is different from the white, it does not respond to the same amount of heat in an identical manner. The amount of heat transferred during the same time duration is not quite sufficient to transform it from a golden yellow gel to a light yellow solid. However, leaving the egg in the heated pan will eventually accomplish this product if it is desired.

The edge of the white has a further opportunity to turn crispy brown. The transferred heat has initiated yet another chemical reaction called the **Maillard reaction**. This reaction occurs at high temperatures and features a chemical reaction between the functional groups present in sugars (the primary component of carbohydrates) and amino acids (the building blocks of proteins) within foods (Figure 10.7). The products of this reaction are the browned crust that forms on cooked eggs, meats, breads, cakes, and cookies, which adds a delightful texture and slightly roasted sweetness to the flavor.

Safety Disclaimer: The health benefits (or detriments) of grilling and browning our foods are controversial. Certainly, cooking our meats and proteins to a particular temperature kills potentially harmful microorganisms. However, some by-products of the Maillard reaction, as well as carbon-dense compounds that result from charring our food, are considered possibly carcinogenic.

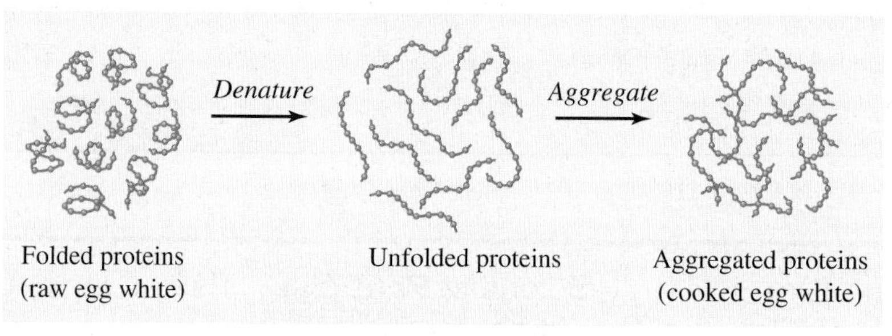

Figure 10.6

Schematic of the denaturation process of egg whites.

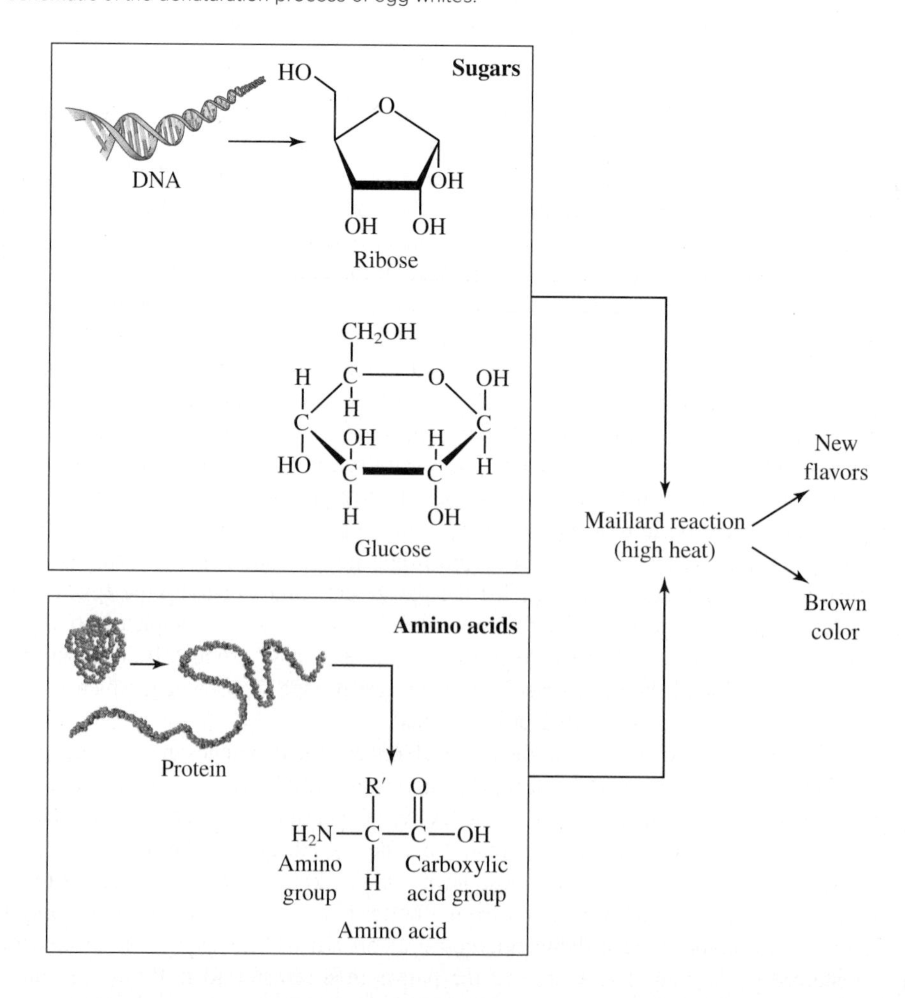

Figure 10.7

Schematic describing the browning and flavoring of foods via the Maillard reaction. More details regarding the structure and function of DNA and proteins will be provided in later chapters.

10.6 Cooking in a Vacuum: Not Just for Astronauts!

Another means of cooking food is called *sous vide*. This method started as a French technique and literally means "under vacuum." This method carefully controls the temperature at which food is cooked, which is usually the same temperature at which it is to be eaten. *Sous vide* still uses water as the medium for cooking; however, unlike boiling or steaming, the food never actually contacts the water. *Sous vide* works by using a temperature-controlled water bath appliance, much like might be used in a chemistry laboratory. In this appliance, circulated water is brought to a certain temperature, say 49 °C (120 °F) for a salmon fillet (Figure 10.8). The food is seasoned as desired, and placed in a plastic bag that is vacuum sealed. Upon placing the bag into the water bath, temperature control is maintained and the food begins to cook. *Sous vide* is usually a longer process for cooking; for example, several meat recipes require 48–72 hours of cooking. Despite the long cook time, the food will not overcook as long as the temperature of the water bath is maintained. However, because the entire piece of meat is at the same temperature, char or browning is not present when the food comes out of the bag. Many recipes suggest a quick turn on a hot grill or skillet to accomplish the Maillard reaction, if desired!

So, why cook without air? The answer is really of a practical nature. Unpackaged foods get messy in a water bath, and bags with air bubbles in them float in water, resulting in uneven heating of the food. But, there are scientific reasons for doing so as well. Consistent temperature control is critical to most recipes; however, air is a poor conductor of heat and can create uneven heating. Additionally, water, the important reason food stays juicy, will evaporate if there is available space. If you remove the air (and extra volume), water does not evaporate from the food. Hence, some benefits of *sous vide* cooking are a better consistency of temperature and texture of the food product.

Figure 10.8

Sous vide cooking.

Fragrant *Sous Vide* Salmon
(Yield: 4 servings)

Part I: Chemicals for Fish Spice Mix
50 g hazelnuts
44 g sesame seeds
12 g coriander seeds
10 g poppy seeds
4 g dried, ground ginger
4 g table salt
2.5 g dried chamomile blossoms

Procedure for Fish Spice Mix
- Preheat oven to 350 °F.
- Roast the hazelnute until the skins turn dark brown, 10-12 minutes.
- Once cooled, rub the nuts with a cloth to remove the skins. Discard the skins.
- Chop the nuts.
- Toast the sesame seeds in a dry frying pan over medium-heat, stirring constantly, until they begin to pop, about 3 minutes.
- Toast the coriander seeds in a dry frying pan over medium-high heat, stirring constantly, until they become golden brown and fragrant, about 3 minutes.
- Crush in a coffee grinder or with a mortar and pestle.
- Combine all ingredients in a coffee grinder and grind to a coarse powder. Work in batches if necessary.

Part II: Chemicals for Fragrant Salmon
1 kg water
50 g table salt
40 g sugar
600 g of salmon fillets (about four fillets)
120 g of either olive oil or melted butter
80 g unsalted butter
15 g fish spice mix

Procedure for Fragrant Salmon
- Stir together water, salt, and sugar until completely dissolved to make a brine. Submerge the salmon in the brine in a zip-top bag and refrigerate for 3-5 hours.
- Preheat a water bath to 115 °F. Remove salmon from brine and place each filler in its own zip-top bag with 30 g of oil or butter. Remove as much air as possible from the bags, and seal them.
- Cook *sous vide* to a core temperature of 113 °F, or about 25 minutes for fillets that are about 2.5 cm/1 in thick.
- Transfer cooked fillets gently from bags to a plate.
- Melt the butter in a nonstick frying pan over medium-low heat.
- Add the fish spice and increase the heat until the butter just starts to bubble.
- Add the fillets, and cook while basting with the hot butter for about 30 seconds per side. Serve immediately.

Recipe from Myhrvold, N.; Bilet, M. "**Modernist Cuisine at Home**," The Cooking Lab, 2012.

Your Turn 10.12 Scientific Principles The Sustainability of Cooking Techniques

Think about the different ways of cooking that we have talked about in this chapter so far. Make a list of pros and cons about each method. Based on your list, which method is more environmentally sustainable? Explain.

Your Turn 10.13 You Decide Dangers from Chemical Leaching?

Modernist Cuisine, one of the defining volumes for *sous vide* cooking, claims that bags made expressly for cooking are safe.

a. What type(s) of polymers are used in such specialized *sous vide* bags?
b. Using the Internet as a resource, are these specialized cooking bags immune to the leaching of unwanted chemicals into our food? Consider what happens to the polymer(s) at the temperatures commonly used for *sous vide* cooking.
c. Would the threat of chemical leaching into food be greater if other inexpensive bags (*e.g.*, Ziploc™) were used instead of the specialized *sous vide* variety? Explain.

Some common types of polymers and their properties were described in Chapter 9.

10.7 | Microwave Cooking: Fast and Easy

As our work schedules get busier and work days get longer, many of us do not always have sufficient time to prepare a complex meal from scratch. In our increasingly busy society, microwave cooking has risen to become the most popular form of cooking because of its speed and simplicity. However, food quality is not nearly as desirable as that cooked over a stove or in an oven, where one has more control over the chemical reactions and resulting development of flavors.

"What do you mean, you're not hungry? - Your mother spent two minutes cooking that in the microwave just for you!"

Original Artist., Reproduction rights obtained from www.cartoonstock.com

Your Turn 10.14 Skill Building The Electromagnetic Spectrum Revisited

In Chapter 3, we discussed the various regions of the electromagnetic (EM) spectrum.

a. Describe the relative energies and wavelengths of the UV, IR, and microwave regions, and diagram how each of these energies would affect a water molecule (*i.e.*, bond breaking, bond stretching/vibration, or molecular rotation).
b. A company claims to have a new type of cooking apparatus using radio waves to cook food. Their claim is that the food is cooked more uniformly and that food quality is better than using microwaves. Do you believe their claims? Explain your answer.

Due to its lower energy and longer wavelength than UV, visible, or IR radiation, microwave radiation is not sufficient to cause rupturing of individual chemical bonds. Instead, the microwave radiation is absorbed by the water, fat, and sugar molecules in food, which causes these molecules to rotate (Figure 10.9). Since the molecules rotate some 2.5 million times per second, they can easily bump into and rub against

Did You Know? If you look closely at the front window of your microwave oven, you will notice a metal mesh. This is designed to be transparent to visible light, but opaque to microwave radiation. That is, the holes in this mesh are smaller than the wavelengths of the microwaves, but large enough so visible light may still pass through.

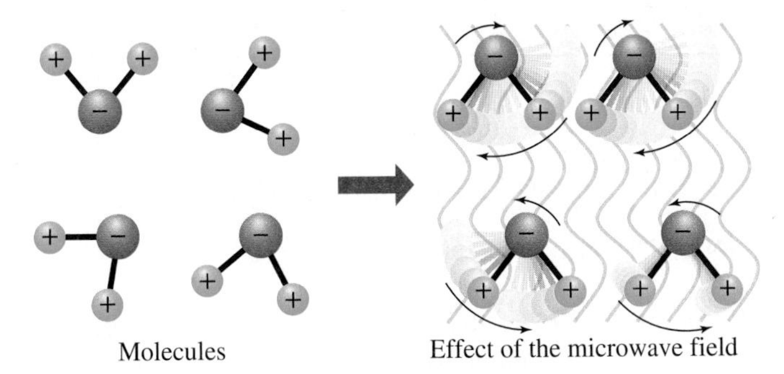

Figure 10.9
Illustration of the rotation of polar molecules in response to an applied external microwave field.

one another, resulting in the production of heat due to frictional forces. Microwaves penetrate the food item to a depth of approximately 1–1.5 inches. Accordingly, in thicker pieces of food, the microwaves don't effectively reach the center. Unlike a conventional oven in which food is heated by hot air, food cooked in a microwave oven normally does not become brown and crispy because the air inside the oven is at room temperature.

Your Turn 10.15 Skill Building Microwave Cooking Versus Conventional Ovens

Sketch and compare the differences in heat transfer occurring as food is cooked in a microwave oven versus a conventional oven. Some claim that microwave ovens cook food from the inside out. Is this accurate?

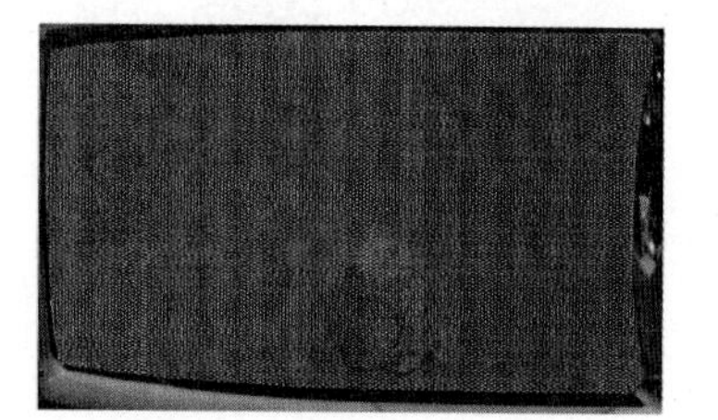

© McGraw-Hill Education. Mark A. Dierker, photographer

New, smooth aluminum foil should only be used; wrinkled foil can cause increased reflection of microwaves. Additionally, the foil should be placed no closer than one inch from the oven walls. If the oven has metal shelves or a metal turntable, foods should not be placed within foil containers or metal pans, and the foil used for food shielding should not touch the metal shelves/turntable.

Metals do not absorb microwave radiation, and instead reflect these wavelengths. Accordingly, a metal such as aluminum is used on the sides of the oven to prevent the microwaves from escaping and irradiating objects outside the oven, such as yourself!

So, if metal is used inside the microwave oven, why is it so dangerous to place a metal object inside the oven? As you discovered in Chapter 1, metals are great conductors of electricity. Hence, when microwaves irradiate a metal, electrons on its surface move rapidly to the side, which prevents the microwaves from being absorbed by the food item. The radiation is reflected, which forms an arc (visible sparks) between the metal object and the metal walls of the oven. This can cause failure of the microwave source, known as a *magnetron*, and can often damage the walls of the oven.

Contrary to popular belief, it can actually be safe to place small amounts of a metal, such as aluminum foil, into a microwave oven. However, it should never be used to completely cover a food item, because the microwaves would not be absorbed and would result in the dangerous reflection/arcing situation described above. However, small pieces of *non-wrinkled* aluminum foil may be used to cover certain areas of foods, such as poultry drumsticks, to prevent overcooking.

Your Turn 10.16 You Decide Sustainability of Cooking Methods

Calculate how much electricity is consumed, and the amount of associated greenhouse gas (GHG) is produced, to boil water using a microwave versus using a standard electrical cooktop stove. Compare these values for the GHG emissions that would result from boiling water using a gas stove. Considering the average time spent per year on cooking, would there be a significant difference in the overall sustainability of each of these cooking methods?

10.8 Cooking with Chemistry: No-Heat Food Preparation

Just like us, the harmful microbes that may be present in food need water to survive. Accordingly, we can preserve food items by simply removing water. Drying and curing have long been a part of our human practices for helping to sustain our food supplies—especially during harsh environmental conditions when food items could not grow or

Figure 10.10

Dried fruit and cured meats.

(left): © Pakhnyushchy/Shutterstock.com; (right): © Pixtal/age fotostock RF

In smoked meats, the pink color does not indicate that the meat is undercooked!

readily reproduce. A slow drying process simply removes the water present in fruits, vegetables, and meats, thus preserving them (Figure 10.10).

As an application of osmosis, sodium chloride is used to remove water from a food item in a process known as *salting*. Salting that includes nitrites (NO_2^-) and nitrates (NO_3^-) also kills bacteria due to their antioxidative properties, and actually adds flavor to meat. The nitrites and nitrates break down further in the meat into nitric oxide (NO), which binds to iron in hemoglobin and prevents further oxidation of the meat.

Smoking meats, though technically still a heat-transfer application, is done over long periods of time. Smoke, usually formed from some type of wood burning, interacts with the meat over many hours. The specific protein that is key in the smoking process is *collagen*, found mostly in the connective tissue of the animal. Over suitable times and temperatures, the collagen converts to gelatin (Figure 10.11). Gelatin is more water-soluble than collagen, and gives way to a desirable tender meat product. Additionally, the much preferred *smoke ring*, the pink discoloration just under the meat crust, forms because nitrogen dioxide (NO_2) from the smoke interacts with meat compounds and forms a more acidic environment that changes the meat to a pink color.

Smoking is not just for meats. Potatoes, corn, eggplant, zucchini, and tomatoes are among the other favorites for smoking, and create a variety of tantalizing flavors that do not resemble their raw counterparts. Nuts, too, can be smoked and eaten as-is, or even transformed into unique varieties of nut-based, non-dairy cheeses.

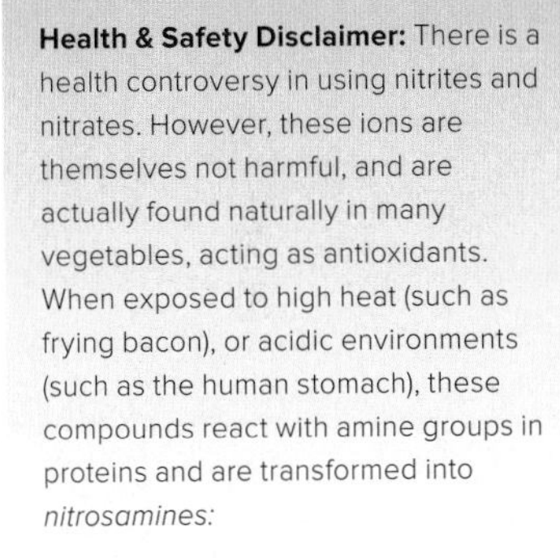

Health & Safety Disclaimer: There is a health controversy in using nitrites and nitrates. However, these ions are themselves not harmful, and are actually found naturally in many vegetables, acting as antioxidants. When exposed to high heat (such as frying bacon), or acidic environments (such as the human stomach), these compounds react with amine groups in proteins and are transformed into *nitrosamines:*

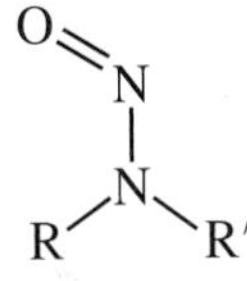

Nitrosamine compounds are also found in latex products and tobacco, and have been identified as carcinogens.

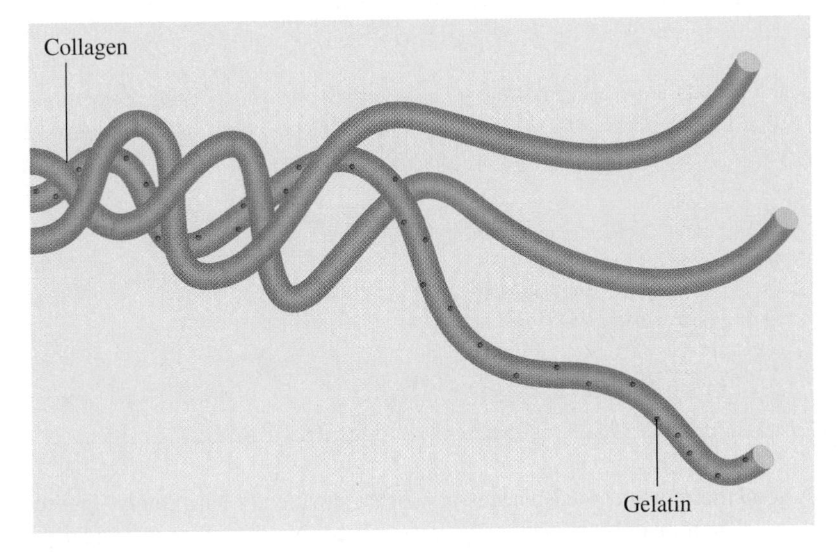

Figure 10.11

Illustration of the thermal unfolding of the triple-helix structure of collagen to form gelatin strands.

Did You Know? Gelatin is the main chemical in Jell-O™.

The chemistry of acids and bases was described in Chapter 8.

Another way to cause chemical changes to our food is through the use of acidic marinades. Upon extended contact with fish fillets, citrus juices will impart a wonderful flavor and also serve to denature proteins, much like the heat of a skillet or grill. There are a variety of styles for *ceviche* derived from different cultures, but all involve some type of acid as a primary ingredient. The higher-acidic environment can also kill many harmful microbes.

Pickling is a wet-curing process that uses both salts (and/or nitrates) and acids. Unlike preparing ceviche, pickling usually involves much longer periods of time exposure—weeks or months—to alter the food product. For example, consider sauerkraut or kimchi, two food products derived from different cultural traditions. At the simplest level, both of these dishes involve exposing raw cabbage to vinegar in an *anaerobic* (oxygen-free) environment over time in order to change the vegetable into a new food product.

Your Turn 10.17 Scientific Practices Identifying the Preparation of Your Food

Keep a journal of the foods you eat over three days. From the foods you consumed, can you identify any of the food preparation techniques you just learned about (drying and curing, smoking, acid marination, or pickling)?

10.9 How Can I Tell When My Food Is Ready?

Health & Safety Disclaimer: You may have noticed on restaurant menus the following message: "Consuming raw or undercooked meats, poultry, seafood, shellfish, or eggs may increase your risk of foodborne illness." This recognizes that cooking times for flavor may not necessarily be safe for consumption. Foods not cooked to certain temperatures may still contain harmful bacteria that can make us sick. The U.S. Department of Agriculture (USDA) food codes require that you be informed of this when choosing to have food prepared below these suggested cooking temperatures.

Because of the replication and refinement of recipes over time, most that involve cooking food provide some type of temperature for heat transfer (*e.g.*, "set oven at 425 °F") and time for maintaining the transfer. These protocols provide a reproducible way for the food to reach the appropriate levels of flavor and texture.

Various types of thermometers are used to help detect the optimal temperature for heat transfer to produce the desired food product. In addition to checking the internal temperature, reliable thermometers are also necessary for frying foods in oils. For example, in order to fry chicken, several recipes state that the oil should be 149–163 °C (350–375 °F), and then maintained between 177–191 °C (300–325 °F) while the chicken cooks. At lower temperatures, one risks producing unhealthy, undercooked, and soggy chicken.

Your Turn 10.18 Scientific Practices Cooking Temperatures

Using the Internet or your favorite culinary book, list the recommended temperatures required to cook the meats below. Which meat preparations are below the minimum temperatures suggested by U.S. Department of Health & Human Service?

a. Beef and lamb steaks, rare
b. Beef and lamb steaks, medium-rare
c. Beef and lamb steaks, medium
d. Beef and lamb steaks, medium-well
e. Beef and lamb steaks, well
f. Chicken
g. Turkey
h. Ham

A good thermometer is also necessary for the creation of hard candy (Figure 10.12). The type of hard candy is generally determined by the stage of sugar cooking (Table 10.4). Hence, a good candy thermometer is necessary to help decipher among these critical stages and the desired candy outcome.

Table 10.4 Stages and Temperatures Involved in Candy Making

Stage	Temperature Range (°C)	Sugar Concentration (%)
Thread (*e.g.*, syrup)	110–112	80
Soft ball (*e.g.*, fudge)	112–116	85
Firm ball (*e.g.*, soft caramel candy)	118–120	87
Hard ball (*e.g.*, nougat)	121–130	90
Soft crack (*e.g.*, salt water taffy)	132–143	95
Hard crack (*e.g.*, toffee)	146–154	99
Clear liquid	160	100
Brown liquid (*e.g.*, liquid caramel)	170	100
Burnt sugar	177	100

*All temperatures correspond to 1 atm, the atmospheric pressure at sea level.

In addition to the **quantitative** information provided by thermometers, changes in textures and firmness can also indicate the relative cooking completion in a **qualitative** manner. Recognize once again that all of these property changes that we have described in reaching cooking outcomes result because the chemicals in the food are reacting and being transformed into new chemicals with new properties. Grill masters often use the "fist test" to determine whether meat is fully cooked. This consists of making a fist and tapping the flesh between your thumb and index finger, and then comparing the touch to the cooking meat. For example, a loosely clinched fist and firmness indicates a "rare" temperature, while a fully clinched fist indicates a "well done" temperature.

In Table 10.4, the first column could be interpreted as texture assessments. Now, one does not verify these different stages by touching the sugar at the temperature listed. Instead, these are determined by taking a small sample of the sugar at the measured temperature and dropping it in cold water. Once cool enough, the pinch test can help one determine the stage. Give it a try. Find the proper equipment, including a good pot, a candy thermometer, sugar, recipes, cold water, some "pure imagination," and go make some candy!

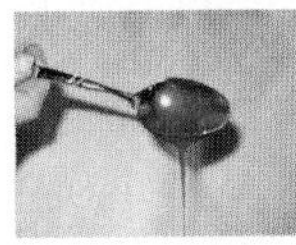
Thread stage

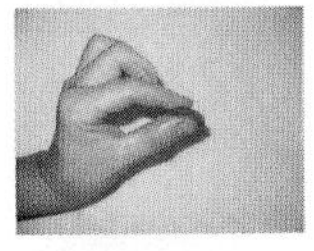
Soft ball stage

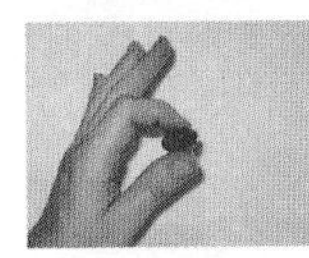
Firm ball stage

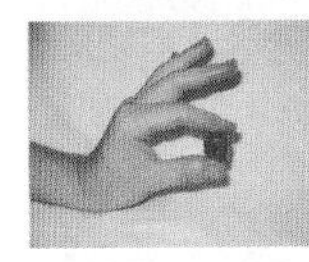
Hard ball stage

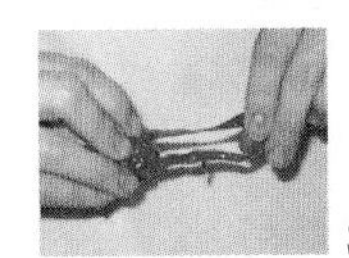
Soft crack stage

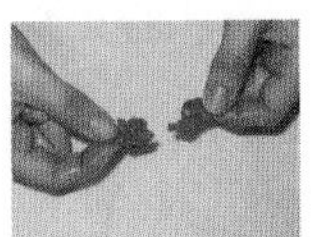
Hard crack stage

Caramel stage

Figure 10.12

The seven stages of candy making.

(a–g): © Elizabeth LaBau, www.sugarhero.com

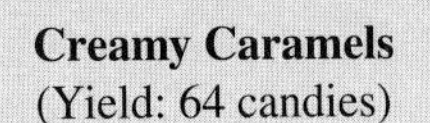

Creamy Caramels
(Yield: 64 candies)

Chemicals

½ cup finely chopped pecans
2 cups sugar
2 cups heavy whipping cream
¾ cup light corn syrup
½ cup margarine or butter

Procedure

- Butter a square pan, 8 × 8 × 2 or 9 × 9 × 2 inches. Spread pecans in pan.
- Heat remaining ingredients to boiling in a 3-quart saucepan over medium heat, stirring constantly.
- Cook mixture, stirring frequently, to 245 °F on a candy thermometer, or until a small amount of mixture dropped into very cold water forms a firm ball that holds its shape until pressed.
- Spread prepared mixture over nuts in pan and allow to cool.
- Cut into 1-inch (2.54 cm) squares.
- If desired, wrap squares individually in plastic wrap or waxed paper.

Recipe from "**Betty Crocker's Cookbook**," General Mills, Inc., Prentice Hall: New York, 1991.

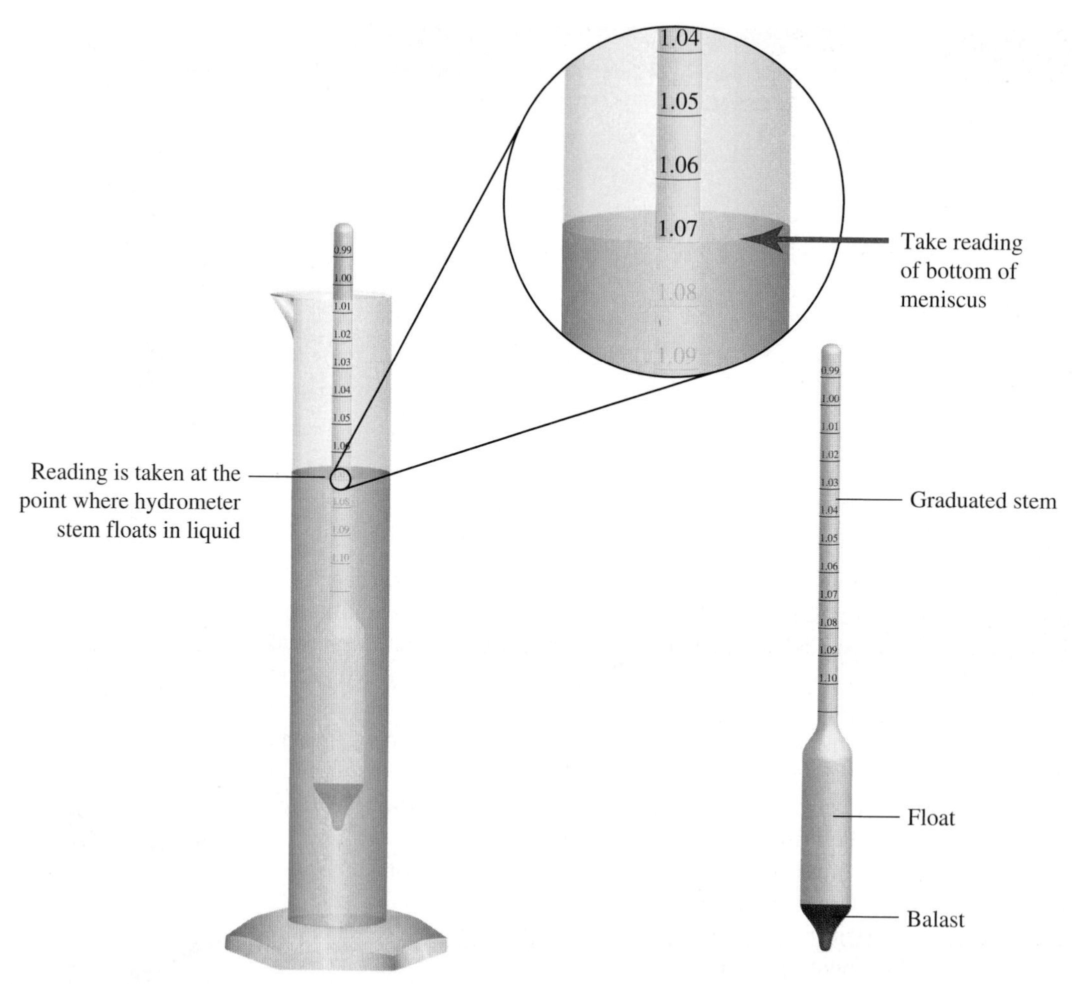

Figure 10.13

A simple hydrometer. The difference in the specific gravity readings of a liquid before (original gravity) and after the fermentation process (final gravity) is used to calculate its alcohol by volume (ABV).

Veruca Salt, poor girl, was a bad egg!

Density, a relatively simple property—the ratio of mass to volume in a substance or mixture—can reveal vital information about the suitability for consumption of a cooking product. If you have seen the first "Willy Wonka" movie (1971), you may recall that Mr. Wonka used this idea to detect the good from bad golden eggs laid by his geese. He checked the mass of eggs of the same volume; bad eggs float in tap good eggs float in some concentrations of salt water. Why is this? Egg shells are actually a bit porous, and over time bacteria can creep into the egg. As the bacteria grow inside, they produce hydrogen sulfide (H_2S), a gas that produces that foul "rotten egg" smell. Sure, if you crack open a bad egg you will certainly detect the malodor instantly. However, within the intact egg, the gas builds up and makes the egg less dense than a good egg, and thus you have the reason a bad egg floats.

The process of fermentation for beer and wine making will be discussed later in this chapter.

Vintners (wine makers) also take advantage of changes in density caused by changes in sugar content during fermentation. A **hydrometer** is an instrument used to measure the changes in density in liquids (Figure 10.13). As the yeast consumes the sugar, the density of the wine mixture decreases, and the hydrometer sinks to a different level. The vintner can then decide when to stop the fermentation process based on the desired sugar level.

A description of how light propagates through a material is referred to as its **refractive index**.

Refractometry represents another method that may be used to determine the sugar content of a solution. All solutions absorb light and cause light to bend (*i.e.*, to be *refracted*); a refractometer measures the extent of this refraction. Since solutions represent a different medium than air for the passage of light beams, light travels at different speeds through solutions compared to air. The higher the density of a solution, the more the light will bend while passing through it. For all solutions, there is a linear relationship between its refractive index and **specific gravity**—the ratio of the density of solution to the density of the pure solvent, usually water (Figure 10.14).

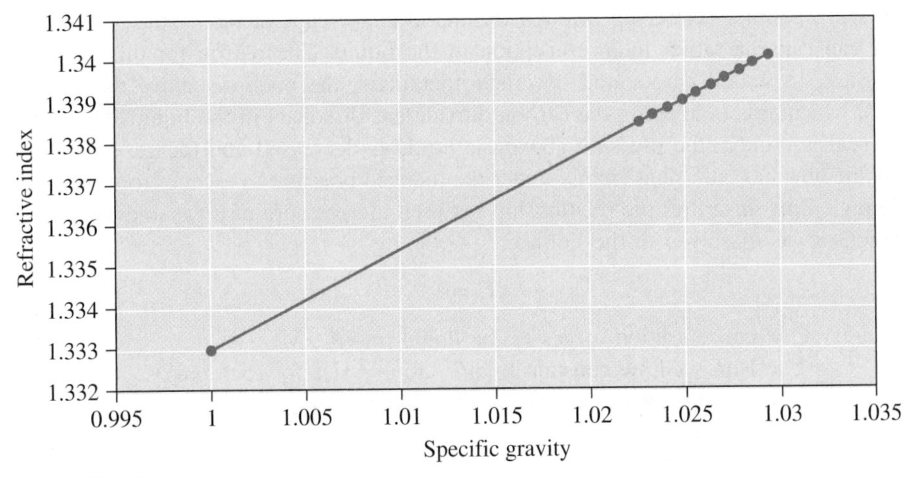

Figure 10.14

Linear relationship between the refractive index and specific gravity of a seawater solution (at 20 °C).

The **Brix scale** (Figure 10.15) is used to quantitatively express the sugar content of an aqueous solution, in which one degree Brix (°Bx) is equal to 1 g of sucrose per 100 g of solution (*i.e.*, 1%(w/w), also referred to as a Brix%). Because the temperature of a solution will affect its density, the refractive index or specific gravity measurements should be determined at a constant temperature, typically 20 °C (68 °F).

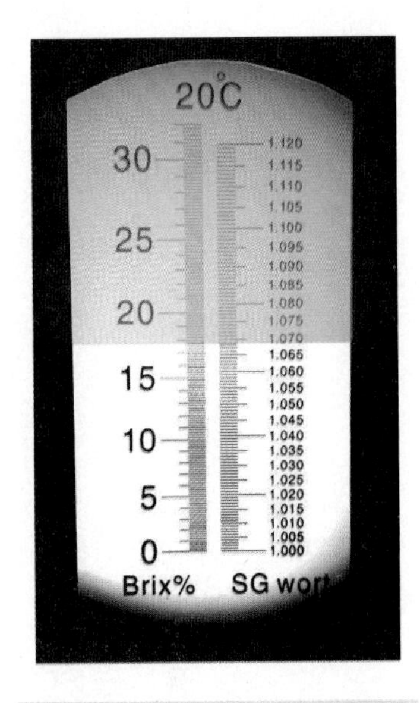

Figure 10.15

A refractometer and Brix scale for determining the sugar content of aqueous solutions. A few drops of liquid are placed onto a prism and light is allowed to pass through. The values of the specific gravity (SG) and °Bx (or %(w/w)) sugar content of the solution are simply read from the scale where the blue and white regions intersect.

(both): © McGraw-Hill Education. Mark A. Dierker, photographer

Your Turn 10.19 Skill Building The Brix Scale

For the refractometer reading shown in **Figure 10.15**, what is the concentration of sugar in the aqueous solution? Express the concentration as both %(w/w) and as molarity (M).
Hint: The density of the solution is 1.06 g/mL.

Your Turn 10.20 You Decide Fickle Soda Cans

Maybe you have been to a party or outdoor barbeque where they placed a variety of regular and diet sodas in a cooler or tubs with ice water. Are all the cans floating or at the bottom? Or, do some float and some sink? Think about what might be causing this difference, in terms of the compositions of the various types of sodas. Conduct some research to determine why this occurs.

10.10 Exploiting the Three States of Matter in Our Kitchen

The combination of chemicals in our foods, and variety of preparation techniques involve a variety of physical states. The wonderful aromas described earlier come from aromatic chemicals, which frequently form gases at slightly elevated temperatures. These chemicals then entice us—for example, the aroma of chicken soup—or unfortunately repulse us (remember those rotten eggs)! Water vapor (steam) is used to cook food because of its increased ability to transfer heat to food. However, liquids and solids also show up in countless ways during food preparation. What can be more unique, though, are the ways in which these states can interact to form a variety of food products.

Carbonated beverages are rather common to us. Soft drinks, beer, and champagne all take advantage of dissolved carbon dioxide. If you view a sealed plastic soda bottle, you probably do not notice any bubbles. However, the minute you open the bottle (break

Popping the cork on a champagne bottle can be a dangerous event. Use caution and aim away from other people or breakable items!

the seal), bubbles scatter up through the liquid and escape at the surface. Sometimes this can cause a rather intense eruption in the liquid. The reason for this is simple: pressure. In sealed carbonated beverage containers, the pressure above the liquid is high, which aids in keeping the carbon dioxide gas dissolved in the liquid. The minute the seal is broken, the pressure above the liquid is decreased and the gaseous carbon dioxide now has sufficient energy under the reduced pressure to escape from the liquid. **Henry's Law** describes the relationship between the pressure of a gas and the concentration of gas dissolved in the liquid:

$$C = kP \qquad [10.3]$$

Where: C = concentration of gas in the liquid (mol/L, M)
k = Henry's Law constant (mol/L · atm; varies for each gas)
P = pressure of the gas above the solution (atm)

Your Turn 10.21 Skill Building Flat Cola

Using Henry's Law, compare the concentration of dissolved carbon dioxide in a sealed 500 mL bottle of Coca-Cola (pCO_2 = 1.25 atm), to that after the bottle is opened at 25 °C. The Henry's Law constant for carbon dioxide is 0.031 mol/L·atm. How much CO_2 (in grams) escapes after the bottle has been opened? ***Hint:*** Assume the ambient atmospheric pressure is 1.00 atm.

The trapping of a gas in a liquid or solid may result in a frothy mixture known as a *foam*. Maybe you are familiar with foams already—whipped cream, shaving cream, or foam soap. In the culinary world, these can be used to create subtle nuances of flavor and texture to a dish. How does something like *apple pie with vanilla and cinnamon cream foam* sound? Interesting? Delicious? Foam foods can be made by starting with a thick flavored liquid, and thickened with starch, gelatin, eggs, or agar. Then, using a device such as a siphon, a gas such as nitrous oxide (N_2O) is injected under high pressure into the liquid. *Et voilà*, when the nozzle is triggered, a foam is produced. Tiny gas bubbles get injected into, and trapped in, the thick liquid, creating an edible foam. Although you may have heard that nitrous oxide can cause a "high" when inhaled, when eaten it is a safe non-buzz ingredient!

Strawberry Juice with Apple Foam
(Yield: four 6-oz servings)

Chemicals
24 ounces of strawberry juice (buy, or make your own with juicer!)
12 ounces fresh green apple juice (again, use a juicer, if desired)

Procedure
- Evenly distribute the strawberry juice to 4 chilled wine or tumbler glasses.
- Add apple juice to a whipping siphon and tighten the lid. Do not overfill it.
- Charge the siphon with only one cartridge of nitrous oxide. Shake vigorously for 5-10 seconds (resting is unnecessary – the gas dissolves quickly).
- Turn the siphon upside down and press the lever to dispense the apple foam gently on top of the strawberry juice.
- Enjoy!

Myhrvold, N. and Bilet, M. (2012) **Modernist Cuisine at Home.** The Cooking Lab, LLC, pp. 18 and 161.

Of course, foams can also be formed with a little elbow grease by whisking air into the liquid. This is certainly a more laborious process, but can also create wonderful results that you may be familiar with—whipped cream and meringues are just two.

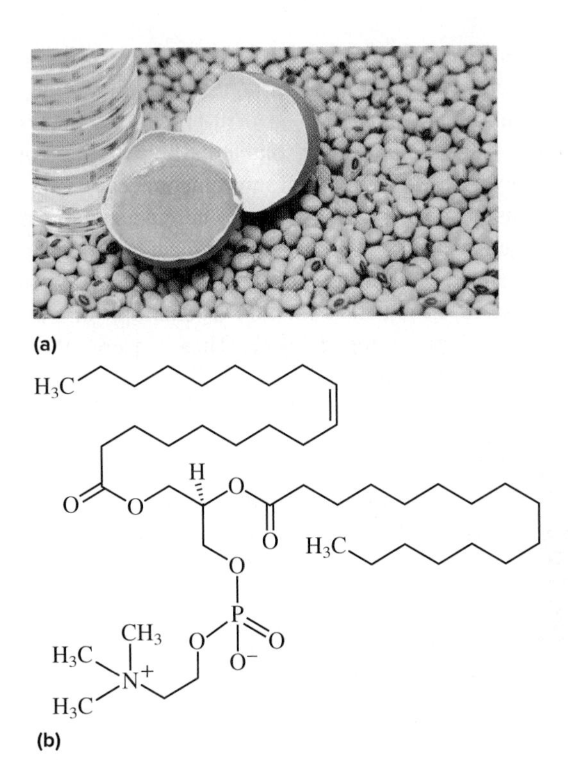

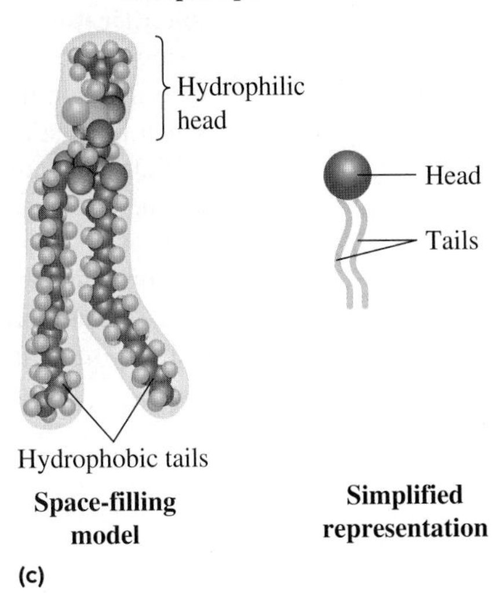

Figure 10.16

Three perspectives of lecithin. Shown are: **(a)** the macroscopic view of egg yolks, and **(b)** structural and **(c)** symbolic views of a phospholipid molecule that comprises lecithin.

(a): © PeoGep/Shutterstock.com

If you have ever made pasta and added a drizzle or two of oil to the water, you likely noticed that the oil floats on the surface. As you discovered in Chapter 8, oil and water do not mix. Whereas oil is nonpolar, water is polar. With some assistance, though, the two can be held together. This is the basis for mayonnaise, which is a type of mixture called an **emulsion**. A considerable amount of shaking of the oil and water mixture can produce a short-lasting emulsion. However, if let to set, the two substances separate once again.

A discussion of recognizing nonpolar and polar molecules, as well as intermolecular forces, is presented in Chapter 8.

A way to create a more permanent emulsion, like mayonnaise, is to add an *emulsifying agent* such as egg yolks. Among other chemicals, the egg yolks contain *lecithin*. Lecithin contains phospholipid molecules that are **amphiphilic**, which means they are attracted to both nonpolar and polar molecules (Figure 10.16). In forming mayonnaise, the polar portion of the lecithin attracts water molecules, whereas the nonpolar portion of the lecithin attracts the oil. These two incompatible chemicals are then held together by this molecular hinge.

Garlic & Red Pepper Aioli
(Yield: 1¼ cups)

Chemicals

6 large garlic cloves, peeled
¾ teaspoon kosher salt
2 egg yolks
1 teaspoon freshly squeezed lemon juice
½ cup roasted red peppers (jarred or homemade), roughly chopped
¾ cup virgin olive oil

Procedure

- Place the garlic in a blender or food processor and pulse until finely chopped.
- Add the salt, egg yolks, lemon juice, and roasted red peppers. Process until well combined.
- While the machine is still running, add the oil in a thin, steady stream until it is completely incorporated and the mixture is thickened.
- Taste and adjust seasoning as necessary.

Recipe from Emerill Lagasse, downloaded from http://emerils.com/127770/roasted-red-pepper-a%C3%AFoli, February 29, 2016.

Another application of intermolecular forces in the kitchen is the process of **spherification** (Figure 10.17). Much like a latex balloon holds a gas, a gel-like, edible membrane can be used to contain small "bites" of flavorful liquid. One version of spherification uses sodium alginate, which is a water-soluble carbohydrate compound. When in solution, the long carbohydrate chains float around unconnected; however, if calcium ions are added, the chains become interconnected. The calcium ions act like fasteners, snapping the chains together and forming a larger molecular network in the form of an impermeable gel around the remaining water solution. Adding a flavored sodium alginate solution drop-wise to an aqueous solution of calcium ions produces little solid spheres with the flavored liquid trapped inside. These spheres then "pop" and release their liquid when you bite into them.

The structure and properties of carbohydrates will be discussed in Chapter 11.

(a)

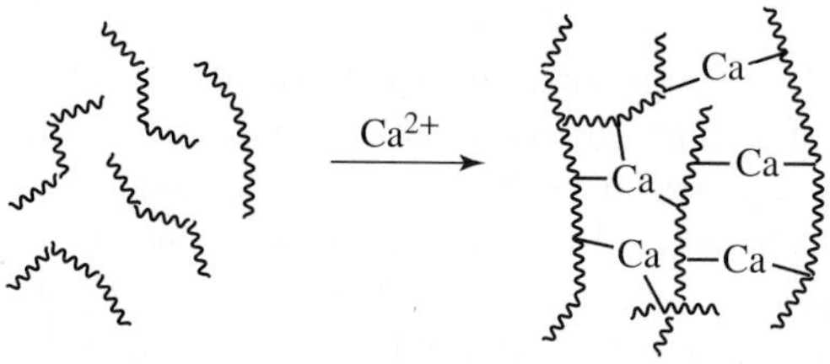

(b)

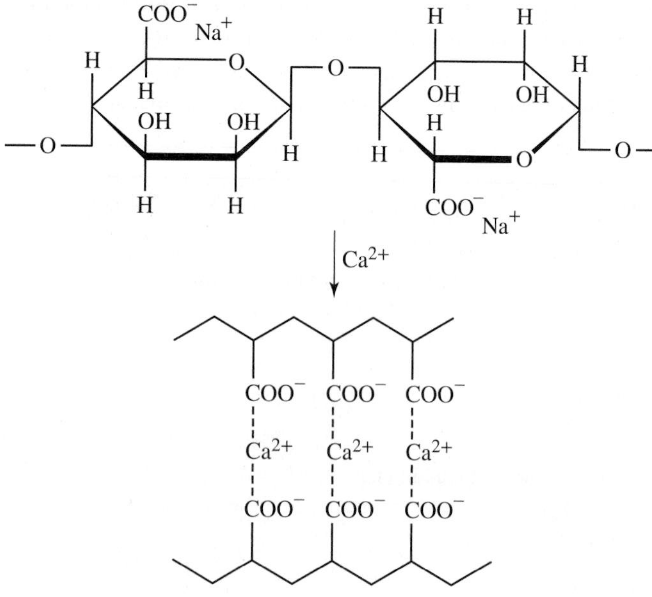

(c)

Figure 10.17

Three views of the process of spherification. Shown are: **(a)** the macroscopic view of the product; **(b)** a schematic view, showing the aggregation of individual alginate strands caused by the presence of calcium (Ca^{2+}) ions; and **(c)** the symbolic view, showing the chemical equation of alginate molecules bound together by replacing sodium (Na^{+}) with calcium (Ca^{2+}) ions.

10.11 The Baker's and Brewer's Friend: Fermentation

Fermentation is a natural process that occurs via the anaerobic metabolism of sugar in the presence of micro-organisms such as yeast. These organisms take simple sugars like glucose or fructose and convert them into molecules that provide the organism with energy. As seen in Figure 10.18, the by-products of this process are ethanol (C_2H_5OH) and carbon dioxide (CO_2).

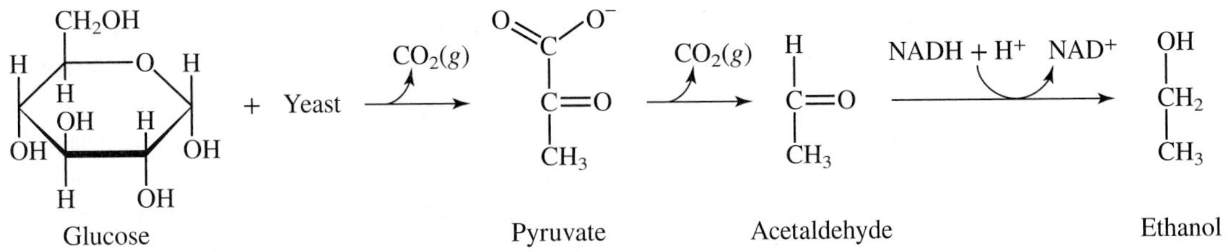

Glucose + Yeast $\xrightarrow{CO_2(g)}$ Pyruvate (CH_3COCOO^-) $\xrightarrow{CO_2(g)}$ Acetaldehyde (CH_3CHO) $\xrightarrow{NADH + H^+ \;\; NAD^+}$ Ethanol (CH_3CH_2OH)

Figure 10.18

The chemical reactions involved during fermentation. Glucose combines with yeast to form pyruvate. An enzyme known as *nicotinamide adenine dinucleotide* (*NAD*) assists in the transformation of acetaldehyde into ethanol.

Yeast that is used in bread making provides the leavening agent, carbon dioxide. Within bread dough, yeast ferments and produces the ethanol and carbon dioxide by-products. The carbon dioxide forms bubbles in the dough, which are trapped in a semi-solid matrix formed by the long-chain molecules—usually proteins or polysaccharides. Holes are left in the bread as it sets, while the ethanol merely evaporates during the baking process. The process results in bread with a wonderful, airy texture.

The structure of proteins and polysaccharides will be discussed in Chapter 11.

In the production of beer, wine, and spirits, the same process is used, except that the ethanol is not allowed to escape. Because the ethanol is contained, a variety of products may be formed, depending on the starting mixture. The primary agents used in alcoholic beverage production are fruits (wines and ciders), grains (beer, whiskey, and vodka), rice (sake), or honey, sugarcane, and molasses (mead and rums). Alcohol itself has a rather undesirable, bitter taste. However, the alcohol derived from fermentation processes results in flavors that are unique and distinctive.

Alcohol is both fat and water soluble. The dual-solubility property of ethanol is due to the very short hydrocarbon component and the –OH group. This allows ethanol to travel just about anywhere in the human body, and it does. About 10–20% of the consumed amount is readily absorbed by the stomach, with the remaining amount taken in by the small intestines. Its high solubility allows for passage into and through cell membranes, flooding nearly all of our organs. The effects can include relaxation, lifted inhibitions, possible euphoria, false confidence, silliness, slurred speech, slower response rate and motor skills, loss of orientation, an impaired sense of equilibrium, and finally, loss of consciousness. What a remarkable array of effects from a small molecule composed of merely nine atoms, with a molar mass of only 46 g/mol (Figure 10.19).

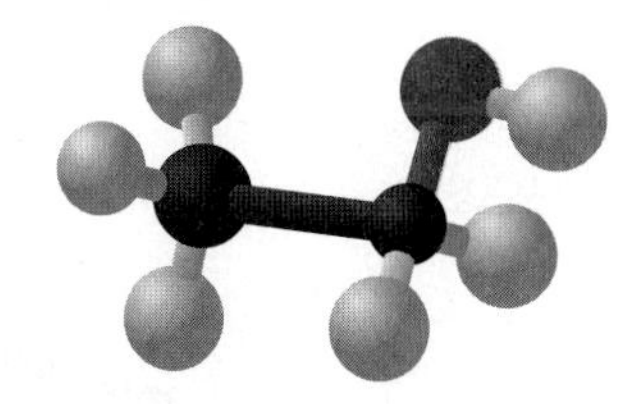

Figure 10.19

A ball-and-stick molecular model for ethanol.

10.12 From Moonshine to Sophisticated Liqueurs: Distillation

As seen in Chapter 5, distillation is a separation and purification technique that is used in industrial chemistry laboratories throughout the world. However, this process is also used in the production of many alcoholic beverages. Beer, wine, and ciders, though fermented, are not distilled, but spirits such as brandies and vodka are. How does this process work?

Did You Know? The distillation of wine results in brandy.

The products of fermentation form a liquid mixture. The liquid components of these mixtures, including the drinkable alcohol, have different boiling points. And so, if continuous heat is added to the mixture, the substance with the lowest boiling point will vaporize first. As you can see in Figure 10.20, the distillation apparatus is sealed in such a way that the evaporated liquid is captured and moves through water-cooled tubing, known as the **condenser**. This process of purification of the fermented mixture creates refinements and nuances in the spirit. Depending on what primary agents (*e.g.*, fruits or grains) were used in the fermentation process, the stages of distillation can produce different refinements, purities, and flavors.

Most states require a license for distillation, since without proper equipment, one can inadvertently produce **methanol** (CH_3OH) as a by-product, which may lead to adverse health effects or death. Even though methanol is chemically very similar to ethanol, the way the body metabolizes these alcohols is quite different. As little as 10 mL of methanol can break down in the body to make formic acid, which attacks the optic nerve and causes blindness. As little as 30 mL of pure methanol—the equivalent of a single shot—can be fatal. Commercial distillers carefully control the fermentation and distillation processes to reduce methanol production, which is confirmed by stringent quality-control testing in federally accredited laboratories.

A beverage only begins to be called a *spirit* at 20% alcohol by volume or greater, and it can contain no added sugars or flavorings. Spirits with added flavorings and/or sugars are called *liqueurs*.

One of the results of distillation is to increase the concentration of alcohol in the spirit. In the United States, the proof level of the spirit is displayed on the label to identify how much alcohol the beverage contains. As shown in Table 10.5, the U.S. proof number is two times the alcohol percent, by volume.

When alcohol is consumed, it enters the bloodstream and causes the pituitary gland in the brain to block the creation of the hormone *vasopressin*. As a result, the kidneys send water directly to the bladder instead of reabsorbing it in the body. According to studies, drinking about 250 mL (8.5 oz) of an alcoholic beverage causes the body to expel 800–1,000 mL of water—three to four times as much liquid is lost as ingested! This so-called diuretic effect decreases as the alcohol in the bloodstream decreases; however, this process is the leading cause of hangover symptoms because the body loses so much water. Electrolytes such as sodium, potassium, and magnesium are lost through urination. They are necessary for proper cell, nerve, and muscle functions, and as the electrolyte concentrations diminish, headache, nausea, and fatigue set in.

Did You Know? The so-called hangover is referred to as *veisalgia* in medical terminology, which is derived from the Norwegian word *kveis* for indisposition brought on by intemperance, and the Greek word *algia* for pain.

The metabolism of ethanol in the body generates the compound *acetaldehyde* (Figure 10.18), which is highly unstable and readily forms free radicals in the body. Long-term exposure to this toxic chemical can damage organs—especially the liver and kidneys. Alcohol also converts glycogen that is stored in the liver into glucose, which is excreted in urine.

A drug called *Antabuse* was designed to fight alcoholism. It blocks the enzyme, *acetaldehyde hydrogenase*, which breaks down acetaldehyde. Symptoms such as severe vomiting and headaches result from a high residual concentration of acetaldehyde, which makes one wary of their next drink.

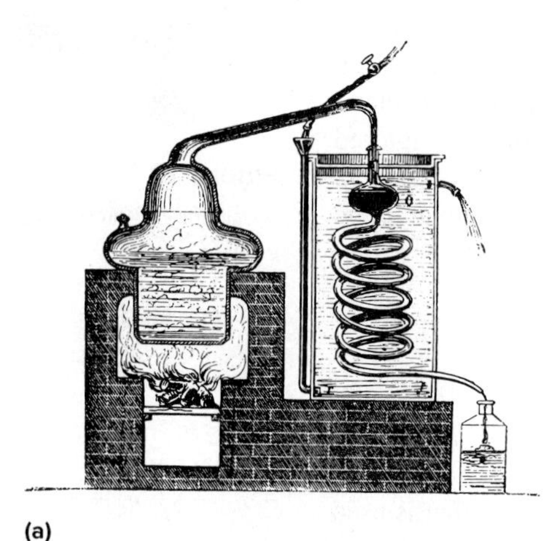

(a)

(b)

Figure 10.20

(a) Schematic of a distillation apparatus used for whiskey production. **(b)** Photo for the production of brandy from the distillation of wine.

(a): © Morphart Creation/Shutterstock.com; (b) © Carl Court/Getty Images

Table 10.5	Alcohol Content in Typical Alcoholic Beverages	
Beverage	**Typical Proof**	**Percent Alcohol by Volume (%ABV)**
Beer	<20	<10
Wine	25	12.5
Fortified wine	40–60	20–30
Vodka and brandy	80	40
Whiskey	100	50
Tsikoudia (a grape-distilled spirit from Eastern Europe)	120	60
Absinthe	140	70
Neutral grain alcohol	180–200	90–100

Once the body metabolizes the alcohol, the body sends a message to replenish its water deficiency, usually in the form of a dry mouth. Additionally, the water-deprived organs in the body steal water from the brain, which pulls on the membranes that connect the brain to the skull. This results in a headache.

10.13 | Extraction: Coffees and Teas

Coffees and teas have an important role in many cultures and traditions throughout the world, dating back to the earliest civilizations. In brewing coffees and teas, we are interested in extracting compounds from a solid: leaves or beans. Similar to how different chocolates can be made, so too can a variety of teas and coffees be produced. The variations are endless, from the different starting materials used, to different drying and roasting techniques, to the inclusion of different additives for flavoring.

To create drinkable coffee, the starting point is a far cry from that dark liquid in your mug. Raw, green coffee beans virtually have no flavor. Pouring hot water over them will simply produce an astringent and bitter tasting brew. However, roasting the beans initiates a host of reactions that generate a large number of products, all of which add to the varied nuance of coffee flavors. Some tea leaves are also roasted, which results in the various flavors comprising green and black teas.

Brewing coffee or tea is itself another application of a laboratory technique: extraction. The beans and leaves of coffees and teas are solid mixtures. Thus, to assist with the extraction from coffee beans, we grind them to increase their surface area. Smaller tea leaves and powders also have more surface area. Then, we expose these materials to a hot extraction medium (water). Fortunately, the desirable components of the coffee or tea tend to be the first to percolate out. The undesired components take longer to extract, and so we don't keep the hot water in contact with the solids for too long.

Many of us are drawn to coffees and teas because they contain caffeine (Figure 10.21). Appropriate doses of caffeine affect the brain and increase mental alertness. And again, like chocolate, teas and coffees contain antioxidants and flavonoids, the chemicals we described earlier that have been shown to improve heart health and interfere with cancer growth.

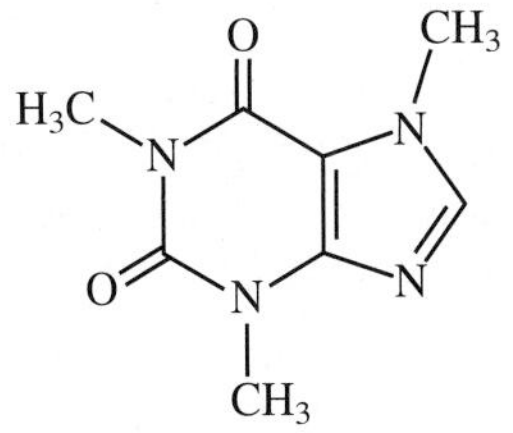

Figure 10.21
The caffeine molecule.

Conclusions

The food and drink we prepare and ingest provides us with nutrients, taste, and even satisfaction and pleasure. We transform food by changing its molecular compositions and structures to yield new substances with different properties, such as flavors. Or, we merely change it physically to a different state or separate it from other components in a mixture, so as to extract or purify it. Most of the time, we cook food by transferring heat in some way. This, however, occurs through a variety of methods (*e.g.,* steak cooked on the grill or *sous vide*) and leads to an equal variety of outcomes (char and cool center, or equal temperature throughout). The recipes, techniques, and utensils we use can often mimic those found in a research lab—for example, glassware, water baths, and digital balances, just to name a few. At the end of the day, good cooking relies on good chemistry. So happy cooking and bon appetit!

Learning Outcomes

The numbers in parentheses indicate the sections within the chapter where these outcomes were discussed.

Having studied this chapter, you should now be able to:

- describe how our tongue and nose detect flavors (10.1)
- explain how changing temperature changes the flavors of food (10.2–10.3)
- describe the importance of quantities and measurement in cooking (10.2)
- describe some neurotransmitter responses in the brain generated from eating food (10.3)
- calculate and assess metric-based units versus "kitchen" units (cups, teaspoons, etc.) (10.4)
- explain why eggs turn white when you cook them (10.4)
- explain why you need to adjust recipes when cooking at higher altitudes (10.5)
- provide examples of how changing conditions and/or ingredients change the product (10.4–10.9)
- predict the chemical processes that can occur in the kitchen (10.4–10.9)
- identify the types and chemical characteristics of food (10.4–10.9)
- explain how a microwave cooks our food (10.7)
- describe how leavening agents work (10.8)
- describe heat-free cooking (10.8)
- explain how curing meat preserves it (10.8)
- illustrate some techniques that may be used to determine the sugar content of solutions (10.9)
- calculate the concentration of a dissolved gas based on its partial pressure above a liquid (10.10)
- explain the processes of fermentation (10.11)
- explain how distillation is used for the preparation of alcoholic beverages (10.12)
- describe the effects of drinking alcoholic beverages on the body (10.12)
- define extraction and its use for coffee and tea preparation (10.13)

Questions

Emphasizing Essentials

1. The amount of salt typically used in cooking is 5 g/dm^3. Convert this into the following units: g/mL, mg/L, mg/m^3.
2. The addition of 1 mole of any solute to 1 kg of water raises the boiling point of water by 0.52 °C. Calculate the boiling point of water that contains the concentration of salt provided in question 1. How much salt would need to be added to 1 L of water in order for its boiling point to reach 101 °C?
3. What is the density of pure water (in mg/L) and how is this affected by dissolving solutes such as salt or sugar?
4. It has been suggested that some people can taste salt to levels of 0.5 g/dm^3, but this varies significantly between testers. How could you experimentally determine your own threshold of salt tasting?
5. Cooking involves a series of chemical reactions. The rate of these reactions roughly double for every 10 °C rise in temperature. If the pressure of a cooker is increased from normal atmospheric pressure (1 atm) to 2 atm, how many times faster will food be cooked?
6. Convert the following cooking units into their appropriate SI units provided in parentheses):
   a. 2 cups of water (L)
   b. 2 teaspoons of salt, NaCl (kg)
   c. 3 hours at 300 °F (seconds and °C)
7. Flavor-causing molecules must be volatile in order to reach the nose and be detected. What structural features of a molecule affect its volatility?
8. Describe the process of browning of fruit, such as apples and bananas, once they have been cut or bruised. Is this the same process as that resulting from carmelization or the Maillard reaction? Explain.
9. How does the addition of vitamin C work to slow down the browning of fruit? What other methods can be used to prevent this browning reaction?
10. Describe why cakes may "fall."

Concentrating on Concepts

11. Research some techniques used to produce sea salt and comment on which method is likely to yield a higher purity of salt.

12. Provide a molecular description of why the color and texture of green vegetables are altered during overcooking?
13. Can sea salt be labeled as "pure" but still contain trace elements such as magnesium and calcium? Explain.
14. Hikers often complain that it is difficult to make a decent cup of tea at high altitudes. Suggest why this problem may be encountered.
15. Is it possible to properly cook a boiled egg at the top of Mt. Everest? Explain.
16. Using balanced chemical equations, provide an explanation of why baking powder is an ingredient in many baked goods.
17. Bread rises due to fermentation. Research approximately how much alcohol remains behind in bread. What factors affect the alcohol content of bread?
18. The terms "strong" and "weak" are typically used to describe acids. How are these terms different than the terms "concentrated" and "dilute"?
19. Suggest which of the following molecules will respond to microwave radiation in the same manner as water: carbon tetrachloride (CCl_4), ammonia (NH_3), and carbon dioxide (CO_2). Explain your choices.
20. What is the name of the substance responsible for the hot flavor of chili peppers, and what is its molecular structure? Use the Internet to find your answers. Also, when chilies are placed in your mouth, why is their burning sensation not greatly affected by drinking water? What drinks would be more effective in reducing this "burning"?
21. Describe why silver cookware tarnishes, especially in the presence of sulfur-containing foods such as eggs.

Exploring Extensions

22. Pick a couple of your favorite flavors—perhaps those such as banana, peppermint, or buttered popcorn. Research the identity of the key chemicals in these flavors and draw their structures. Do they have anything in common with other flavor molecules you looked up or saw in this chapter? What are those features? How are these molecules different from one another?
23. There is a fun movement happening in the culinary world called "flavor tripping" (Mr. Wonka would have loved it!). This technique involves eating something called a "miracle berry" and consuming other foods that you may not normally eat in their raw or isolated forms. For example, moments after eating a miracle berry, an ordinary lemon when eaten tastes like sweet lemonade. Given what you have learned about taste in this chapter, and after conducting some of your own research, can you provide a chemical explanation for why this berry changes the way things taste? What other foods might be interesting to try in this way?
24. The food we eat does not just affect our bodies. How food is grown or raised, and then produced and processed, can also impact the environment. Current modern agricultural practices are actually a leading contributor to climate change. Over the next couple of days, keep a food journal of what and how much you eat. After you complete your journal:
    a. Select three foods you think ranked highest in promoting your health (state the criteria on which you base your ranking).
    b. Select the three foods you think ranked highest in promoting the health of the environment (again, state the criteria on which you base your ranking).
25. What are "food miles"? Select three of your favorite foods and conduct a food mile analysis. What are the ways we can reduce food miles? Do you think it is important to do so? Why or why not?
26. Select your favorite baked good; for example, chocolate chip cookies, blueberry muffins, rye bread, or something else. Conduct an Internet search to find different recipes for this same item.
    a. What do these recipes have in common?
    b. How are these recipes different?
    c. Can you predict how any variations among the recipes will lead to different outcomes in the baked good?
27. One section in this chapter described the Maillard reaction, a key reaction that brings distinctive flavors to browned foods. Food that is boiled in water or cooked in a microwave does not go through this reaction. Why not?
28. The various types of cooking oils have a range of properties such as taste, color, and their physical state at room temperature. Three commonly used oils—canola, olive, and coconut—provide one set of examples of these property differences. Conduct some research to discover the chemical nature of these oils and how this accounts for their varying properties in cooking and taste.
29. In the coffee industry, equipment ranging from simple French presses to elaborate machines costing thousands of U.S. dollars are used to "brew" coffee. Conduct some research about different coffee-making equipment and describe the scientific principles they rely on to make a unique cup of "joe."
30. Teflon™-coated cookware is widely used in kitchens throughout the world.
    a. What is the molecular structure of Teflon™ and how does this coating prevent food from sticking to the pan?
    b. If Teflon acts as a non-stick coating, how does Teflon itself stick to the pan?
31. If fresh pineapple, kiwi, or papaya fruit is added to Jell-O®, will it prevent the Jell-O® from setting? Explain.
32. Using your knowledge of intermolecular forces and solubility, describe the chemical and physical processes involved in making ice cream.
33. The "five-second rule" is often used to forgive our clumsiness and consume food that has been dropped on the floor. Using the Internet as a resource, describe the validity of this popular deadline.
34. Describe the origin of "asparagus pee"—the odd odor in one's urine after eating asparagus.

CHAPTER 11

Nutrition

© Stock Footage, Inc./Getty Images

REFLECTION

Food Choices

Every day, you are presented with a variety of food choices. So let's reflect on your diet. What did you eat yesterday?

a. Make a list, starting with your first cup of coffee (or however you chose to start your day), through your last evening snack.

b. From your list, select three food items you believe ranked highest in promoting your health. Name the criteria on which you based your ranking.

c. Select from your list the three food items you believe ranked highest in terms of promoting the health of the land, air, and water on our planet. Again, name your criteria.

The Big Picture

In this chapter, you will explore the following questions:

- What are the components of food, and what are their effects on our health?
- What practical information can we glean from food labels?
- How safe is our food supply?
- What are the environmental implications of agricultural practices?
- How do we feed our hungry and ever-expanding world population?

© Valentyn Volkov/Shutterstock.com

Introduction

Imagine never eating another hamburger. Put the thought out of your mind of picking up a piece of fried chicken. And your eggs definitely will come *without* bacon from now on. Would this be your worst nightmare? To some, being a vegetarian is akin to being deprived of those foods they most crave; well, with the exception of coffee and chocolate. All too often, though, we set up our food selection in terms of all or nothing. No ice cream, because it has too many Calories. No red meat, because it is unhealthy. No soft drinks, because they are loaded with sugar. And no diet soft drinks either, because they contain artificial sweeteners. No, no, NO!

The term Calorie was first introduced in Chapter 5 as a unit of energy. One Calorie is equivalent to 1000 calories (1 kcal), or 4184 joules (4.184 kJ).

Could your choices possibly be more nuanced? Unless dictated by an allergy, a specific health concern, or a deeply held belief, what you eat need not be an all-or-nothing proposition. For instance, if you were to eat no meat one day a week, you could improve the odds that, as you age, you would have a higher quality of life. Why exactly is this? As a vegetarian, you are likely to get more of the whole grains, fruits, and vegetables your body needs. At the same time, you are likely to get less saturated fat. Do the math—one day a week *does* matter. One day in seven is just short of a 15% change in your food intake. Make that two days a week and you have changed your diet by more than 25%.

In this chapter, we will make the case that both what you dine on and what you skip not only has an effect on your health, but also affects the health of our planet. If you are one of the people on the planet fortunate enough to have adequate food to eat, some very good reasons exist to eat simply and to vary your diet. In fact, both *what* you eat and *how much* you eat may be two of the most important decisions you make over the course of your life.

11.1 | You Are What You Eat

Whether you sit down to a gourmet meal or gobble junk food on the run, you eat because you need the energy and micronutrients that food provides. We need food as an energy source to power muscles, to send nerve impulses, and to transport molecules and ions in our bodies. In addition, food serves as the raw material for bodies, including new bone, blood cells, enzymes, and hair. Food also supplies nutrients essential for **metabolism**—the complex set of chemical processes that are essential in maintaining life.

The foods you eat also provide an important source of water, which plays an essential role in our everyday health. This compound serves both as a reactant and a product in metabolic reactions, as a coolant and thermal regulator, and as a solvent for the countless substances that are essential for life. In fact, our human bodies are approximately 60% water.

Eating properly means more than filling your stomach. It is possible to eat, even to the point of being overweight, and still be malnourished. **Malnutrition** is caused by a diet lacking in proper nutrients, even though the energy content of the food may be adequate. Contrast malnutrition with **undernourishment**, a condition in which a person's daily caloric intake is insufficient to meet metabolic needs. Today, people worldwide are malnourished and undernourished; yet increasingly, others are more overweight than ever before. In 2014, the Centers for Disease Control and Prevention reported that 69% of all adults in the United States are classified as overweight, with more than half of that population classified as obese. This epidemic of obesity is caused by several factors, including eating the wrong foods, eating too much food, and lacking physical activity.

Your Turn 11.1 You Decide A Lifetime of Food

During your lifetime, it has been claimed that you will eat about 700 times your adult body weight. Is this statement in the ballpark? Do a calculation to find out. State your assumptions clearly.

Hint: You might assume a life span of 78 years and that your present weight is your adult weight. Estimate the weight of food eaten daily at present, and use these data to project your lifetime food consumption.

Think about the foods you ate yesterday. Did they come to you with minimal or no processing, such as an apple, a baked potato, or a juicy pork chop? Or, did you obtain these same foods as apple sauce, a bag of frozen french fries, or sliced smoked bacon? The latter are **processed foods**—foods that have been altered from their natural state by techniques such as canning, cooking, freezing, and adding chemicals such as thickeners or preservatives. The typical diet in many countries contains numerous processed foods.

Processed foods in the United States must list nutritional information on their labels. These labels, such as the one shown in Figure 11.1, include the **macronutrients**—fats, carbohydrates, and proteins that provide essentially all of the energy and most of the raw material for body repair and synthesis. Sodium and potassium ions are present in much lower concentrations, but these ions are essential for the proper electrolyte balance in the body. Several other minerals, and an alphabet soup of vitamins, are listed in terms of the percent of recommended daily requirements supplied by a single serving of the product. All of these substances, whether naturally occurring or added during processing, are chemicals. In fact, all food is inescapably and intrinsically chemical, even food that claims to be organic or "natural."

Table 11.1 indicates the mass percentages (grams of component per 100 g of food item) of water, fat, carbohydrate, and protein in several familiar foods. For this particular selection of foods, the variation in composition is considerable. But in every case, these four components account for almost all of the mass present. Water ranges from a high of 89% in 2% milk to a low of 1% in peanut butter. Peanut butter is comparable to steak and fish in terms of protein; in this table, it leads in fat content. Chocolate chip cookies are the highest in carbohydrates because of their high sugar and refined flour content.

Compare this table with similar data for the human body (Figure 11.2). You are what you eat, but only to a certain extent. You are more like steak than chocolate

Nutrition Facts

8 servings per container

Serving size	**2/3 cup (55g)**
Amount per serving	
Calories	**230**
	% Daily Value*
Total Fat 8g	**10%**
Saturated Fat 1g	**5%**
Trans Fat 0g	
Cholesterol 0mg	**0%**
Sodium 160mg	**7%**
Total Carbohydrate 37g	**13%**
Dietary Fiber 4g	**14%**
Total Sugars 12g	
Includes 10g Added Sugars	**20%**
Protein 3g	
Vitamin D 2mcg	10%
Calcium 260mg	20%
Iron 8mg	45%
Potassium 235mg	6%

* The % Daily Value (DV) tells you how much a nutrient in a serving of food contributes to a daily diet. 2,000 calories a day is used for general nutrition advice.

Figure 11.1

Nutrition information reported on a food label.

Table 11.1 Percent Water, Fat, Carbohydrate, and Protein in Selected Foods

Food	Water	Fat	Carbohydrate	Protein
white bread	37	4	48	8
2% milk	89	2	5	3
chocolate chip cookies	3	23	69	4
peanut butter	1	50	19	25
sirloin steak	57	15	0	28
tuna fish (canned)	63	2	0	30
black beans (cooked)	66	<1	23	9

Source: U.S. Department of Agriculture, Agriculutural Research Service, Home and Garden Bulletin, *72, 2002.*

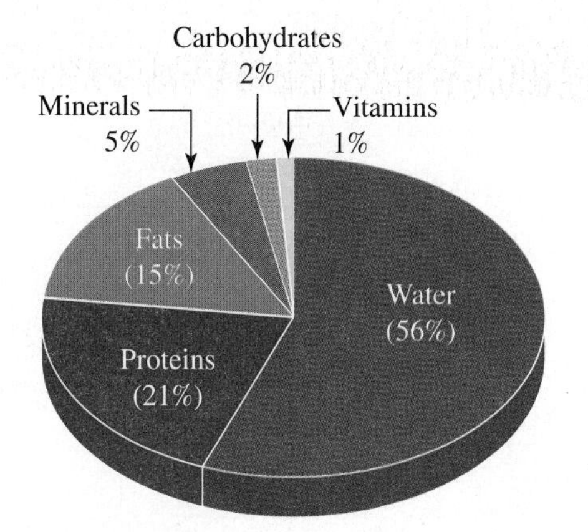

Figure 11.2

Composition of the human body, in percent by mass.

chip cookies. You contain more water and fat than bread, and contain more protein than milk. From these data, you can determine that a 150-pound (68-kg) person consists of about 90 pounds (41 kg) of water and about 30 pounds (14 kg) of fat. The remaining 30 pounds is almost all protein, carbohydrate, and calcium and phosphorus in the bones. The other minerals and vitamins weigh less than 1 pound (0.5 kg), indicating that a little bit of each goes a long way, a point that will be discussed in Section 11.7.

In the following sections, we take a look at each food component in turn—fats, carbohydrates, proteins, minerals, and vitamins. As you will see, each one is unique in several regards.

11.2 From Buttery Popcorn to Cheesecake: Lipids

As discussed in Sections 8.2 and 8.3, the principle of "like dissolves like" is used to determine whether a solute is soluble in a solvent. Since fats/oils are nonpolar solutes, they are insoluble in polar solvents such as water or short-chain alcohols (*e.g.*, methanol—CH_3OH, ethanol—C_2H_5OH, etc.)

From your experiences with ice cream, butter, and cheese, you probably know that fats can help impart a desirable flavor and texture to food. More generally, fats are greasy, slippery, and low-melting solids that are not soluble in water. When melted, fats float on the top of broth or soup since their densities are less than the underlying aqueous liquid. Some of the most delectable foods such as cakes, frosting, sour cream, and most pastries are loaded with fats (and Calories, as we will see later).

You may also know about oils, such as those obtained from corn and soybeans. Perhaps you have noticed that peanut oil often forms a layer on top of your peanut butter. Maybe you like eating bread dipped in olive oil. Or maybe in the past you have prepared a loaf of nut bread using canola oil as the shortening. Many oils are of plant origin, and share many of the properties of animal-based fats. However, unlike fats, they are liquids at room temperature.

This group is characteristic of an ester:

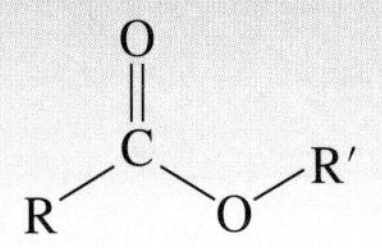

$$R-C(=O)-O-R'$$

where R and R' are abbreviations for any group in which a hydrocarbon group is attached to the rest of the molecule (*e.g.*, $-CH_3$, $-CH_2CH_3$, etc.)

As we pointed out earlier when describing biodiesel in Section 5.16, the molecules that make up fats and oils share a common structural feature. They are both **triglycerides**; that is, molecules that contain three ester functional groups. They are formed from a chemical reaction between three fatty acids and the alcohol *glycerol*. **Fats** are triglycerides that are solids at room temperature, whereas **oils** are triglycerides that are liquids at room temperature. In turn, all triglycerides are **lipids**, a class of compounds that includes not only triglycerides, but also related compounds such as cholesterol and other steroids. Figure 11.3 shows the lipid family tree.

We introduced several new terms in the previous paragraph, and we will now describe each one by one. First up is *fatty acids*; examine Figure 11.4 to see an example, known as stearic acid. Like all fatty acids, the stearic acid molecule has two important characteristics. One is a long hydrocarbon chain with an even number of carbon atoms, typically 12 to 24. This hydrocarbon chain gives fats and oils their

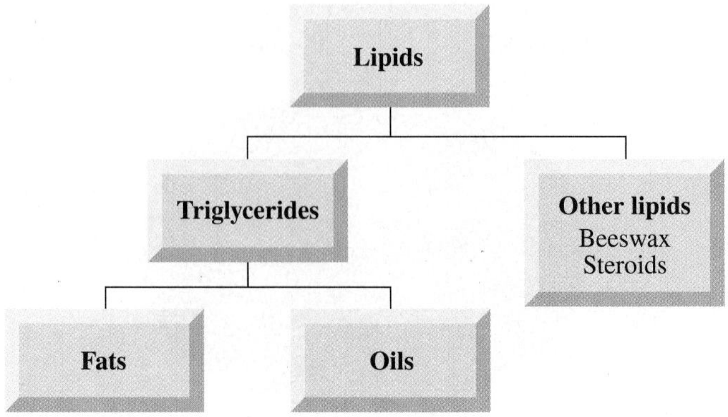

Lipids
Triglycerides
Fats
Oils
Other lipids
Beeswax
Steroids

Figure 11.3

Types of lipids.

$CH_3(CH_2)_{16}COOH$

condensed structural formula

$CH_3CH_2CH_2CH_2CH_2CH_2CH_2CH_2CH_2CH_2CH_2CH_2CH_2CH_2CH_2CH_2CH_2-C(=O)OH$

semicondensed structural formula

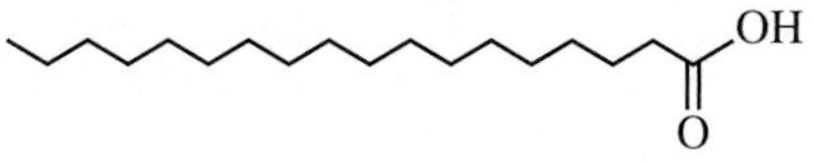

line-angle drawing

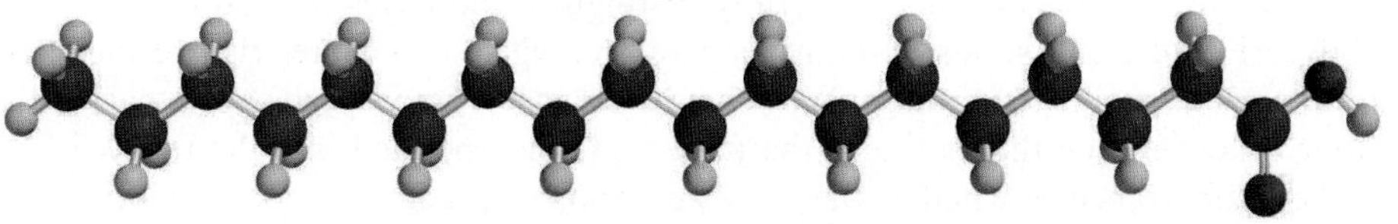

ball-and-stick model

Figure 11.4

Molecular representations of stearic acid, $C_{17}H_{35}COOH$, an example of a fatty acid.

Chemists routinely use a **line-angle drawing** to represent the structure of a molecule (Figure 11.4). One carbon atom is assumed to occupy each vertex position. Any line extending from the backbone signifies another carbon atom (actually a $-CH_3$ group), unless the symbol for another element is given. Hydrogen atoms are not indicated in the line-angle drawing, but are implied as required by the octet rule.

$CH_2(OH)CH(OH)CH_2OH$

condensed structural formula

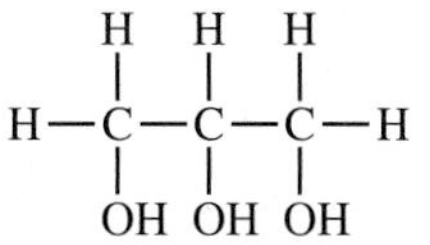

structural formula

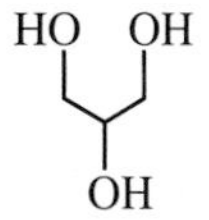

line-angle drawing

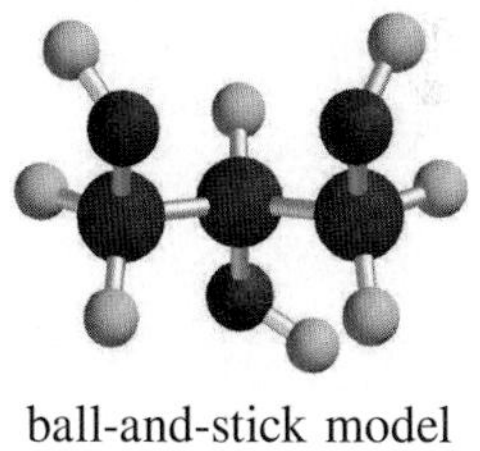

ball-and-stick model

Figure 11.5

Molecular representations of glycerol, an alcohol.

characteristic greasiness. The other is the carboxylic acid group, –COOH, at the end of the hydrocarbon chain. The carboxylic acid group accounts for the "acid" in the names of these compounds.

Next, we turn to the term **glycerol**, an alcohol that we briefly mentioned in Chapter 5 when describing biodiesel. Glycerol is a sticky, syrupy liquid that is sometimes added to soaps and hand lotions. The molecular representations of a molecule of glycerol (Figure 11.5) show that it is an alcohol with three –OH groups.

Of interest to us here is that each –OH group of the glycerol molecule can form an ester with a fatty acid molecule. The result is a triglyceride. For example, three stearic acid molecules can combine with a glycerol molecule to form glyceryl tristearate, a triglyceride. The three ester functional groups in glyceryl tristearate are highlighted in red in Equation 11.2.

3 fatty acid molecules + 1 glycerol molecule ⟶ 1 triglyceride molecule + 3 water molecules **[11.1]**

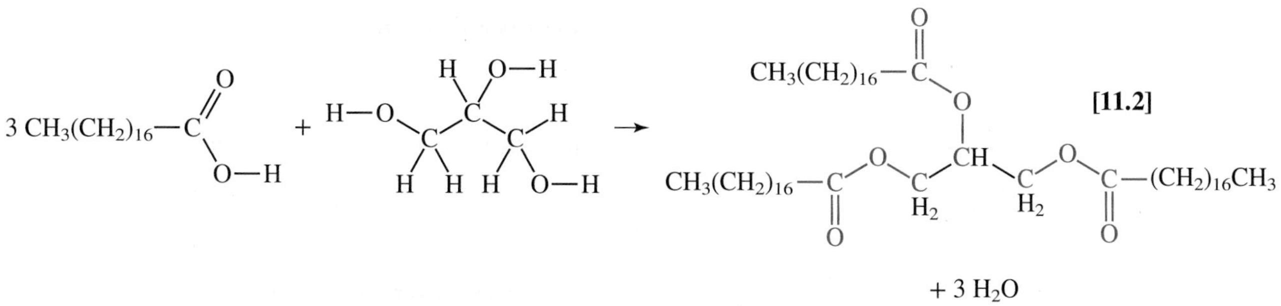

$$3\,CH_3(CH_2)_{16}-C(=O)-O-H + \text{glycerol} \longrightarrow \text{glyceryl tristearate} + 3\,H_2O \qquad [11.2]$$

The process shown in Equation 11.2 is the basis for forming most animal fats and vegetable oils. In most cases, enzymes catalyze this process. Almost all of the fatty acids in our bodies are transported and stored in the form of triglycerides.

Your Turn 11.2 Skill Building Triglyceride Formation

Like stearic acid, palmitic acid is a component of animal fats.

a. Construct a line-angle drawing for palmitic acid, $CH_3(CH_2)_{14}COOH$.
b. Name the functional group responsible for the acidic properties of fatty acids.

See **Figures Alive!** on **Connect** for more representations of triglycerides.

Finally, we need to more carefully distinguish the terms *fat* and *oil*. Fats are triglycerides that are solids at room temperature, whereas oils are triglycerides that are liquids at room temperature. Why the difference? The properties of a particular fat or oil depend on the nature of the fatty acids incorporated into the triglyceride. Of key importance is whether the fatty acid molecule contains one or more C=C double bonds.

A fatty acid is **saturated** if the hydrocarbon chain contains only single bonds between the carbon atoms. In a saturated hydrocarbon chain, the C atoms contain the maximum number of H atoms that can be accommodated and is therefore saturated in hydrogen. This is the case with stearic acid. In contrast, fatty acids are **unsaturated** if they contain one or more C=C double bonds because there is room for H_2 to react and become saturated.

Unsaturated fatty acids are either monounsaturated or polyunsaturated. For example, oleic acid, with only one double bond between carbon atoms per molecule, is classified as **monounsaturated**. In contrast, linoleic acid (two C=C double bonds per molecule), and linolenic acid (three C=C double bonds per molecule) are both examples of polyunsaturated fatty acids. A **polyunsaturated** fatty acid contains more than one double bond between carbon atoms. In Figure 11.6, each of the unsaturated fatty acids contains 18 carbon atoms, but differs in the number and placement of the C=C double bonds. Your Turn 11.3 gives you a chance to work with different unsaturated fatty acids.

$CH_3(CH_2)_7CH{=}CH(CH_2)_7COOH$

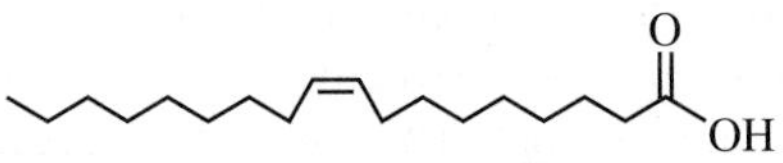

O
OH

oleic acid, a **monounsaturated** fatty acid

$CH_3(CH_2)_4CH{=}CHCH_2CH{=}CH(CH_2)_7COOH$

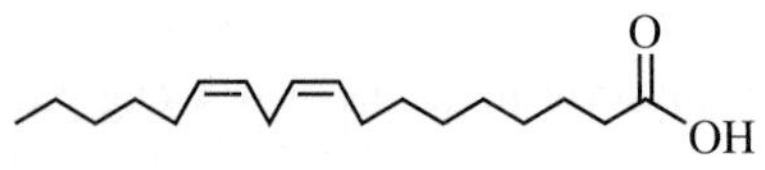

O
OH

linoleic acid, a **polyunsaturated** fatty acid

$CH_3CH_2CH{=}CHCH_2CH{=}CHCH_2CH{=}CH(CH_2)_7COOH$

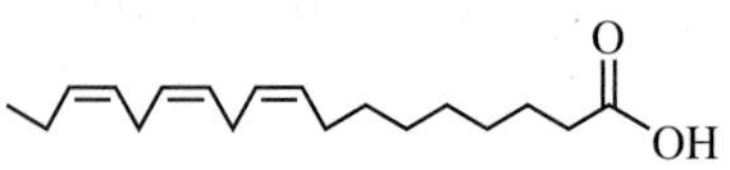

O
OH

linolenic acid, a **polyunsaturated** fatty acid

Figure 11.6

Examples of unsaturated fats.

Your Turn 11.3 Skill Building Unsaturated Fatty Acids

a. What structural feature identifies oleic, linoleic, and linolenic acids as unsaturated fatty acids?

b. Lauric acid, $CH_3(CH_2)_{10}COOH$, is a component of palm oil. Draw the line-angle representation for lauric acid, and classify it as either saturated or unsaturated.

The three fatty acids that form a triglyceride molecule can be identical, two can be the same, or all three can be different. Moreover, these fatty acids can be saturated or unsaturated. They also can be sequenced differently in the molecule. All of these factors contribute to the variety of fats and oils that we find in animals and plants. Solid or semisolid animal fats, such as lard and beef tallow, tend to be high in saturated fats. In contrast, olive, safflower, and other plant oils consist mostly of unsaturated fats.

Tallow is a processed form of beef or mutton fat. In addition to being used for shortening, tallow also appears in soap and biodiesel production.

Table 11.2 indicates some trends within a given family of fatty acids. For example, in saturated fatty acids, the melting points increase as the number of carbon atoms per molecule (and the molecular mass) increases. On the other hand, in a series of fatty acids with a similar number of carbon atoms, increasing the number of C=C double bonds decreases the melting point. Thus, when the melting points of the 18-carbon fatty acids are compared, saturated stearic acid (no C=C double bonds) is found to melt at 70 °C, oleic acid (one C=C double bond per molecule) melts at 16 °C, and linoleic acid (two C=C double bonds per molecule) melts at –5 °C. These trends carry over to the triglycerides containing the fatty acids and explain why fats rich in saturated fatty acids are solids at room and body temperatures, whereas ones with a high degree of unsaturation are liquids.

Normal body temperature is 37 °C, whereas room temperature is approximately 20–25 °C.

Your Turn 11.4 Scientific Practices Melting Points

Draw the molecular structures for some saturated fatty acids and those with C=C double bonds within their framework. Provide a rationale for decreasing melting points of the fatty acids with increasing degrees of unsaturation.

Hint: Recall the concept of intermolecular forces described in Chapter 5.

Table 11.2 Comparing Fatty Acids

Name	Number of C Atoms per Molecule	Number of C=C Double Bonds per Molecule	Melting Point (°C)
Saturated Fatty Acids			
capric acid	10	0	32
lauric acid	12	0	44
myristic acid	14	0	54
palmitic acid	16	0	63
stearic acid	18	0	70
Unsaturated Fatty Acids			
oleic acid	18	1	16
linoleic acid	18	2	−5
linolenic acid	18	3	−11

Your Turn 11.5 Scientific Practices Hydrocarbons, Triglycerides, and Biodiesel

a. How does a molecule of octane compare with one of a fat or oil? List two differences.
b. How does a biodiesel molecule compare with one of a fat or oil?
Hint: See Chapter 5.

As you might guess, fats and oils not only differ in their physical properties, but they also differ in how they affect your health. We turn to this topic in the next section.

11.3 Fats and Oils: Not Necessarily a Bad Thing!

Did You Know? Our brains are rich in lipids, which are important for normal brain function.

People tend to be preoccupied with dietary fat, because fats pack more Calories than any other nutrient. But fats are far more than just a fuel. Fats enhance our enjoyment of food, improve "mouth feel," and intensify certain flavors. Almost every dessert tastes better with a bit of whipped cream! Fats are also essential for life. They provide insulation that retains body heat, which helps to cushion internal organs. Moreover, triglycerides and other lipids, including cholesterol, are the primary components of cell membranes and nerve sheaths.

Fortunately, our bodies can synthesize almost all fatty acids from the foods we eat. The exceptions are linoleic and linolenic acids (Figure 11.6). These two fatty acids must be present in our diet because our bodies cannot produce them. Generally, this does not pose a problem because many foods, including plant oils, fish, and leafy vegetables, contain linoleic and linolenic acid.

Figure 11.7 reveals some surprising differences in the composition of fats and oils we consume. For example, flaxseed oil is particularly rich in alpha-linolenic acid

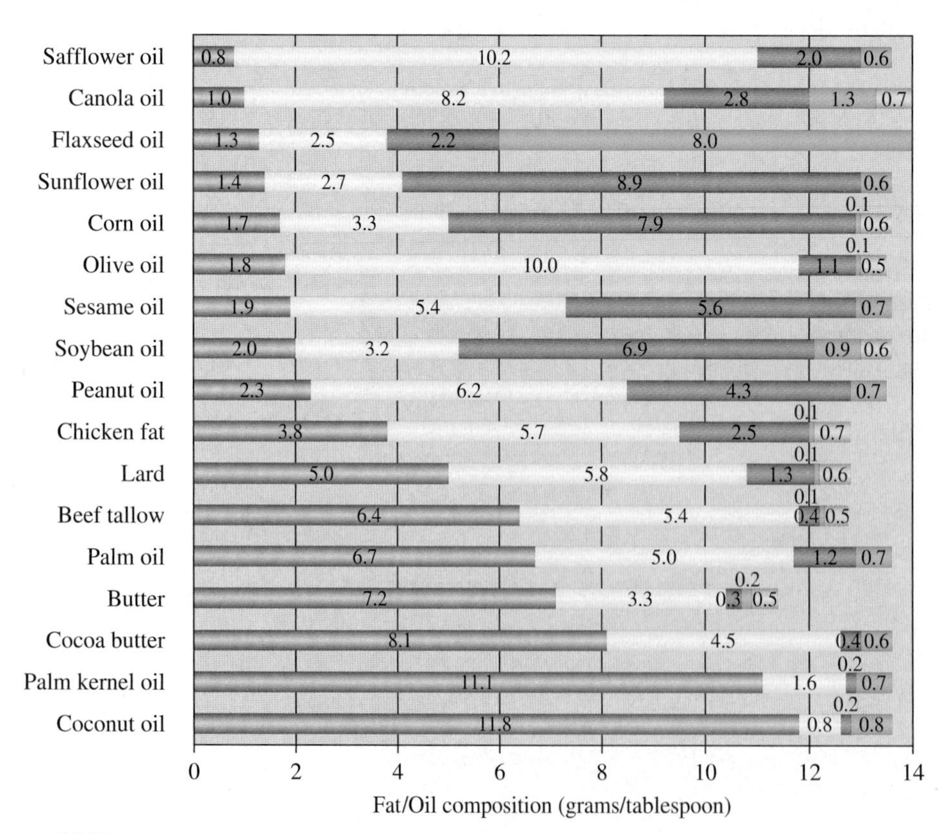

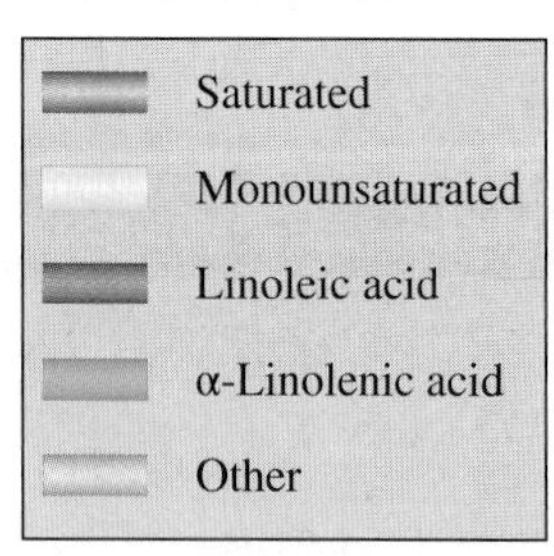

Figure 11.7

Comparative compositions of various saturated and unsaturated oils and fats.

See **Figures Alive!** on **Connect** to identify some fatty acids in food products.

(α-linolenic acid, or ALA), a polyunsaturated fatty acid that is being studied for its health benefits. Palm kernel and coconut oils contain much more saturated fat than corn and canola oil. Ironically, the coconut oil used in some nondairy creamers contains about 87% saturated fat, far more than the percentage found in the cream it replaces. In fact, coconut oil contains more saturated fat than pure butterfat. Concern over the high degree of saturation in coconut and palm oil accounts for the statement sometimes printed on food labels: "Contains no tropical oils."

The solid form of coconut oil is called *coconut butter*. It melts to form an oil at around room temperature.

Your Turn 11.6 Scientific Practices The Chemistry of Cooking Oil

a. Consider this label from a popular brand of cooking oil. Is the major component likely to be safflower oil, canola oil, or soybean oil? Explain.

b. This brand of cooking oil has one unusual ingredient: vitamin E. Do you think this is a part of the oil itself, or an added component? We will provide more details about vitamins later in Section 11.7.

Nutrition Facts
Serving Size 1 Tbsp (15 mL)
Servings Per Container about 63

Amount Per Serving	
Calories 120	Cal. from fat 120
	% Daily Value*
Total Fat 14g	21%
Saturated Fat 1g	6%
Trans Fat 0g	
Polyunsaturated 11g	
Monounsaturated 2g	
Cholesterol 0g	0%
Sodium 0g	0%
Total Carbohydrate 0g	
Protein 0g	
Vitamin E 20%	

Not a significant source of dietary fiber, sugars, vitamin A, vitamin C, calcium, and iron
*Percent Daily Values are based on a 2,000 Calorie diet.

However, the higher degree of unsaturation in oils comes with a drawback. You may have noticed the slight rancid odor that oils acquire over time. The reason for this is because C=C double bonds are more susceptible to reaction with the oxygen in the air than are C–C single bonds. The "off-flavor" that you may detect in an oil is most likely a result of such reactions with oxygen. As a result, oils are sometimes treated to increase their saturation, which improves the shelf life of the food containing the oil.

One way to more fully saturate an oil or a fat is by **hydrogenation**, a process in which hydrogen gas, in the presence of a metal catalyst, adds to a C=C double bond and converts it to a C–C single bond:

See **Figures Alive!** on **Connect** for details regarding the reaction of H_2 with unsaturated fats.

$$-\underset{|}{\overset{\text{H}}{\overset{|}{\text{C}}}}=\underset{|}{\overset{\text{H}}{\overset{|}{\text{C}}}}- + H_2 \xrightarrow{\text{metal catalyst}} -\underset{\text{H}}{\underset{|}{\overset{\text{H}}{\overset{|}{\text{C}}}}}-\underset{\text{H}}{\underset{|}{\overset{\text{H}}{\overset{|}{\text{C}}}}}- \quad [11.3a]$$

When oils are hydrogenated, some or all of their C=C double bonds are converted to C–C single bonds, increasing the degree of saturation and raising the melting point. As a result, the oil becomes more margarine-like; that is, semisolid and spreadable. Converting all C=C double bonds to C–C single bonds would create an undesirable hard-to-spread solid. By carefully selecting the temperature and pressure, the extent of hydrogenation can be controlled in order to yield products with the desired melting point, softness, and spreadability. Equation 11.3b shows this reaction with linoleic acid, one of the fatty acids in the triglycerides of peanut oil.

[11.3b] (skeletal structure of linoleic acid, C(=O)OH) $+ H_2 \longrightarrow$ (skeletal structure with one C=C, C(=O)OH)

Note that only one of the double bonds in linoleic acid was hydrogenated. The resulting customized fats and oils are used in margarines, cookies, and candy bars.

Although shelf life and spreadability are important considerations, they are not the only ones. It turns out that some fats and oils are healthier for your heart than others. To understand why, we need to look more closely at the geometry of the hydrogen atoms attached to the C=C double bonds of triglycerides. In most natural unsaturated fatty acids, the hydrogen atoms attached to the carbon atoms are on the *same* side of the C=C double bond. We call this bonding arrangement *cis*:

In a *cis* isomer, the H atoms are on the *same* side of the double bond:

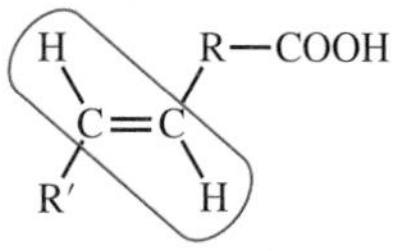

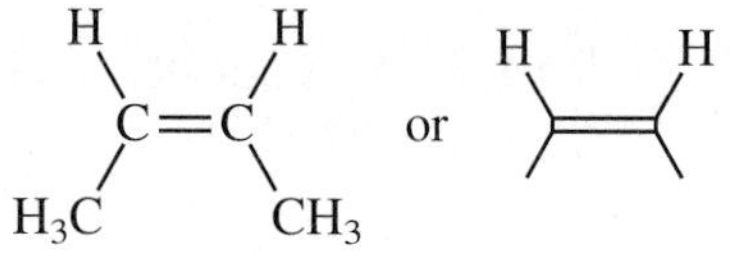

cis-2-butene

Alternatively, the hydrogen atoms can be diagonally across from each other on the C=C double bond. We call this bonding arrangement *trans*:

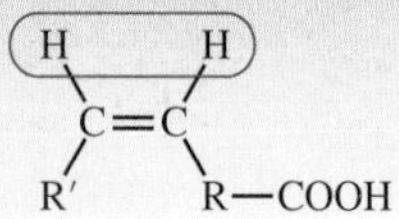

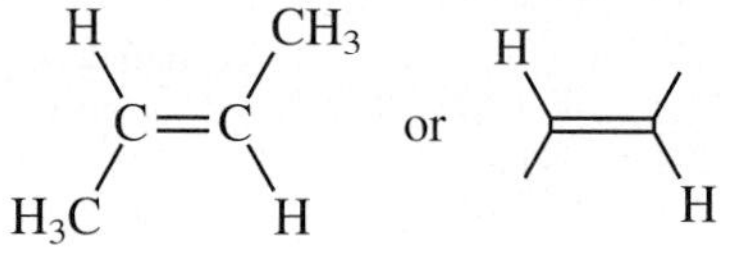

In a *trans* isomer, the H atoms are on the *opposite* side of the double bond:

trans-2-butene

For example, oleic acid and elaidic acid are monounsaturated fatty acids that both have the same chemical formula, $C_{18}H_{34}O_2$, so they are **isomers** of one another. However, their properties, uses, and health effects are different. Oleic acid, a *cis* fatty acid, is a major component of the triglycerides in olive oil. In contrast, elaidic acid, a *trans* fatty acid, is found in some soft margarines made via hydrogenation. Compare the structures shown in Figure 11.8.

Low-density lipoprotein (LDL) cholesterol is referred to as "bad" because of its tendency to build up on artery walls, causing heart attacks and strokes.

***Trans* fats** are triglycerides that are composed of one or more *trans* fatty acids. Scientific studies show that *trans* fats raise the level of triglycerides and "bad" cholesterol in the blood. This finding came as somewhat of a surprise because partially hydrogenated fats still contain some C=C double bonds, and unsaturation is definitely a plus in a healthy diet. However, *trans* fats are similar in properties to saturated fats. With their long "straight" hydrocarbon chains, saturated fat molecules tend to pack well together, one reason why they are solids at room temperature. With their *cis* geometries, the molecules of naturally occurring unsaturated edible oils have "bends" that do not pack as well, which is one reason they are liquids at room temperature.

Revisit Equation 11.3b to see that hydrogenation is not an all-or-nothing proposition. When partial hydrogenation takes place, some C=C bonds in the oil or fat remain. We might expect these molecules to be *cis* because C=C bonds naturally occur this way in fats and oils. However, the process of hydrogenation converts some of these *cis* C=C bonds to the *trans* configuration (Equation 11.3c). The fats containing these *trans*

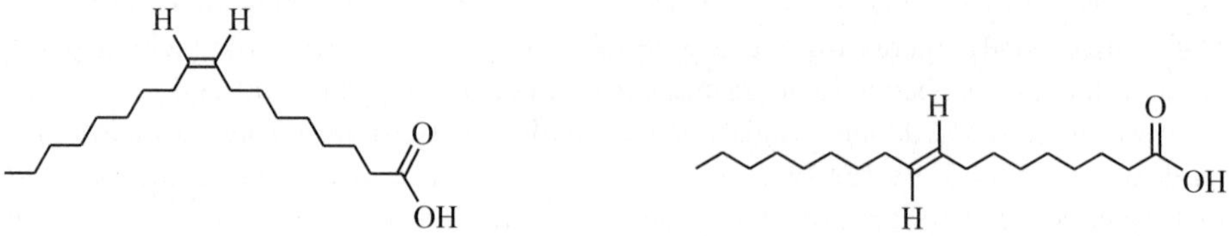

oleic acid, a *cis* fatty acid

elaidic acid, a *trans* fatty acid

Figure 11.8

Molecular structures of oleic acid (left) and elaidic acid (right).

 See **Figures Alive!** on **Connect** for interactive representations of fatty acids.

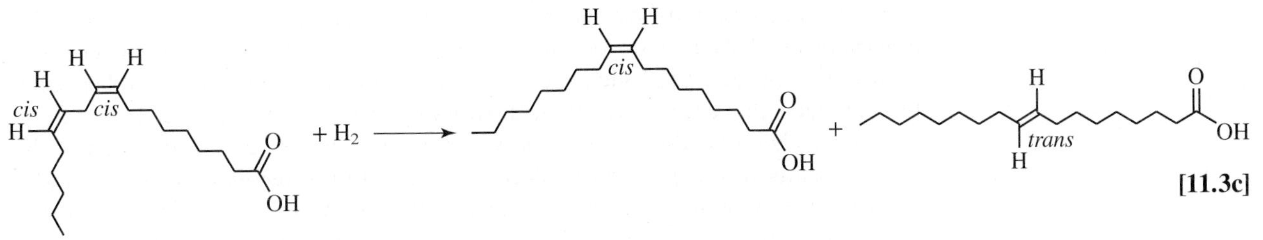

[11.3c]

fatty acids more closely resemble the shape of saturated fats, and therefore behave in a similar manner within the body. What types of fats and oils are in your margarine? Check out the next activity.

Your Turn 11.7 Scientific Practices Margarines and Fat Content

The following table lists the fat content for butter and three margarines, each corresponding to a serving size of 1 tbsp (14 g).

	Butter	Land O' Lakes (stick)	I Can't Believe It's Not Butter (tub)	Benecol Spread (tub)
Total fat	11 g	11 g	9 g	8 g
Saturated fat	7 g	2 g	2 g	1 g
Trans fat	0 g	2.5 g	0 g	0 g
Polyunsaturated fats	1 g	3.5 g	3.5 g	2 g
Monounsaturated fats	3 g	2.5 g	2 g	4.5 g

a. Which margarine has the highest percent of saturated fat? How does it compare with butter?

b. In butter, what percent of the total fat is polyunsaturated?

c. Conduct a mini-survey. List the fat content for three different butters or margarines. Among those listed in the table above, and those in your survey, which would be the most "healthy" for your diet? Explain your rationale.

d. The rise in obesity has long been blamed on fat consumption in the diet. However, the consumption of Calories from fat has decreased since the 1980s, while the rate of obesity continues to grow. Investigate reports about the current medical recommendations for avoiding obesity to find other factors that are likely responsible.

The Harvard-based Nurses' Health Study and the Health Professionals Follow-Up Study tracked 90,000 women and 50,000 men over several decades. This study revealed that a participant's risk of heart disease was strongly influenced by the type of dietary fat consumed. Eating *trans* fat increased the risk substantially; saturated fat increased it slightly. In contrast, unsaturated fat decreased the risk. Therefore, the total fat intake alone is not the cause of heart disease.

In March 2003, Denmark became the first country to strictly regulate foods containing *trans* fats. Canada followed suit in 2004. Starting in January 2006, the U.S. Food and Drug Administration (FDA) required that food labels include values for *trans* fat. In 2007, New York City adopted a regulation that forced restaurants to eliminate the use of partially hydrogenated vegetable oils and spreads, the main sources of *trans* fats in most diets. As a result of the ban, many restaurant chains have since chosen to eliminate their use of *trans* fats worldwide.

Although most medical professionals now recommend that consumers avoid products with *trans* fats, this may not be easy because of the way foods are labeled. In the United States, a label may show "zero grams *trans* fat" as long as this food contains less than 0.5 grams *trans* fat per serving. Thus, with multiple servings of what is not truly 0 grams, the amount of *trans* fat can add up.

Manufacturers are responding by looking for substitutes for *trans* fats. Using tropical oils such as palm or coconut oil would not be acceptable to consumers because of their high percentage of saturated fats. Some manufacturers are adding other oils to their products that are polyunsaturated, such as sunflower oil or flaxseed oil.

Food chemists have also been busy discovering alternatives to hydrogenation that produce semisolids, but do not produce *trans* fats. **Interesterification** is any process in which the fatty acids on two or more triglycerides are scrambled to produce a mixture of different triglycerides (Figure 11.9). If you perform this process with a low-melting triglyceride (an oil) and a high-melting triglyceride (a fat), the result is a mixture of triglycerides with an intermediate melting point, a semisolid fat.

One way of carrying out an interesterification reaction uses a base as a catalyst. The use of basic solutions raises concerns for worker safety, results in significant loss of the oils, requires large amounts of water, and produces waste that is high in biological oxygen demand, BOD.

Biological oxygen demand was discussed in Section 8.12.

Enzymes can also catalyze this reaction. However, using enzymes came with a high price tag until Novozymes and the Archer Daniels Midland Company teamed up

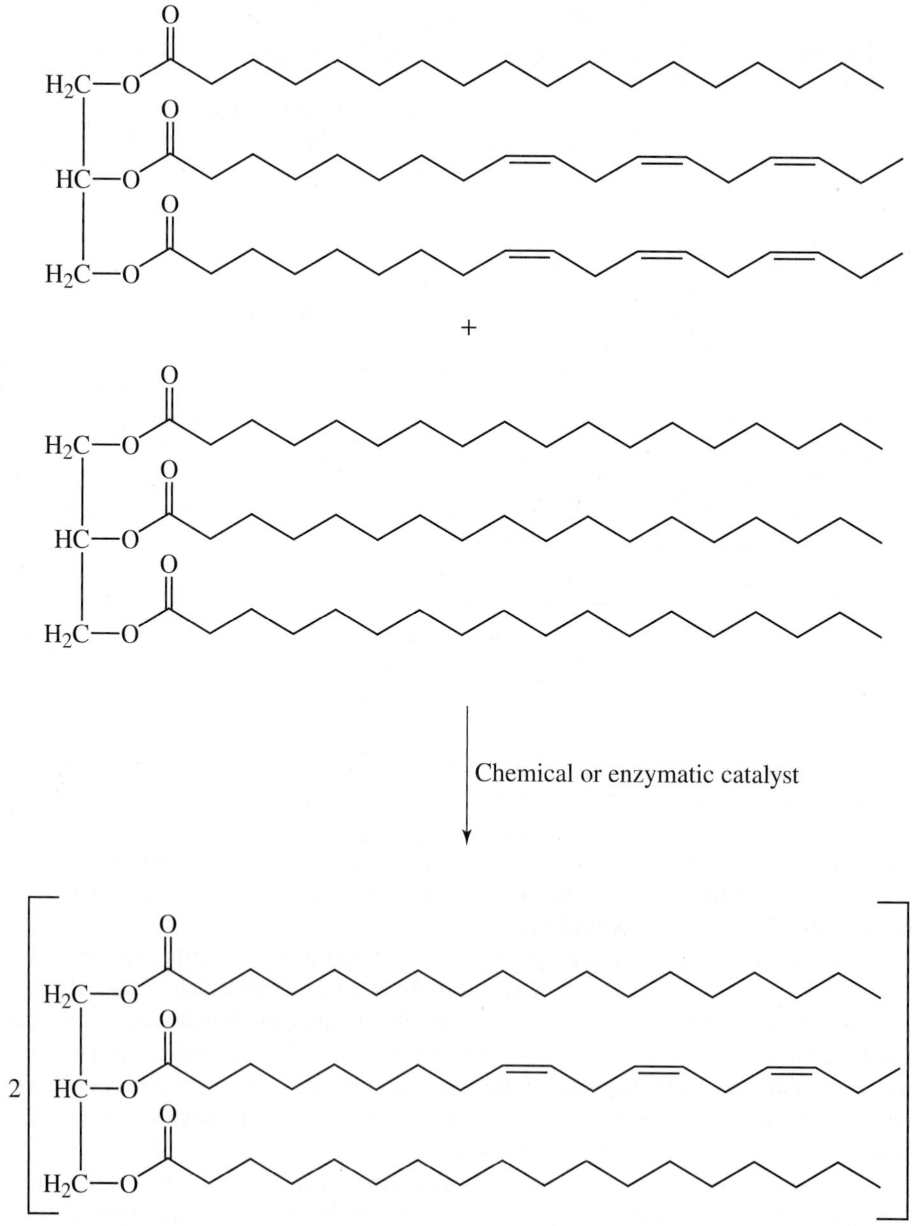

$H_2C—O$ O
$HC—O$ O
$H_2C—O$ O

+

$H_2C—O$ O
$HC—O$ O
$H_2C—O$ O

Chemical or enzymatic catalyst

2 [$H_2C—O$ O
$HC—O$ O
$H_2C—O$ O]

Figure 11.9

An example of interesterification. A triglyceride with two linolenic acid side chains (top) and a fully saturated triglyceride (middle) are combined to produce two molecules with an intermediate degree of chain saturation (bottom).

to refine the reaction. For their efforts, they shared a Presidential Green Chemistry Challenge Award in 2005. Not only is their method more cost-effective, but it also has environmental benefits. These include a large reduction in both water usage and in the BOD of the aqueous waste streams. Furthermore, there is less decomposition of oils during the process, and the need for a base catalyst is eliminated.

Your Turn 11.8 Scientific Practices Next-Generation Interesterification

a. Revisit the key ideas of green chemistry discussed in **Chapter 2**. Which of the key ideas are met by the use of enzymes for interesterification?

b. What are some new enzymatic and non-enzymatic processes that are being used for interesterification?

With this, we end our discussion of fats and oils. The next section tells the sweet tale of sugars, and their not-as-sweet relatives, starches.

11.4 | Carbohydrates: The Sweet and Starchy

Sugars are the sweet-tasting members of the carbohydrate family. Examples that you may recognize include glucose and fructose, both naturally occurring in fruits, vegetables, and honey. In addition, glucose and fructose are components of high-fructose corn syrup (Figure 11.10).

Starch, a polymer of glucose introduced in Chapter 5, is another carbohydrate. It is found in nearly all types of grains, potatoes, and rice. Although pleasing to our taste buds, starch lacks a sweet taste and takes a bit longer to digest than sugars. Whether sweet or starchy, carbohydrates have the job of providing energy to the cells in our bodies.

Figure 11.10

High-fructose corn syrup is a mixture of fructose, glucose, and water. In pure form, sugars are white crystalline solids.

Carbohydrates are also used commercially to produce ethanol, an energy source for vehicles. As discussed in Chapter 5, the starch found in corn kernels is currently fermented to produce millions of gallons of ethanol each year. But corn starch is not unique. The sugar or starch of almost any plant can be fermented to produce ethanol, including that found in alcoholic beverages.

Carbohydrates are compounds that contain carbon, hydrogen, and oxygen, with H and O atoms found in the same 2:1 ratio as in H_2O. This composition gives rise to the name *carbohydrate,* which implies "carbon plus water." However, the H and O atoms in carbohydrates are not in the form of H_2O molecules. Rather, these atoms are part of a larger molecule, typically a ring (or a long chain of rings, in the case of starch and cellulose). Verify this for yourself by examining the structural formulas of these sugars in Figure 11.11.

As seen in the figure below, glucose and fructose both have the same chemical formula, $C_6H_{12}O_6$, but different structures. These *isomers* are relatively easy to tell apart: glucose has a six-membered ring and fructose a five-membered ring. In contrast,

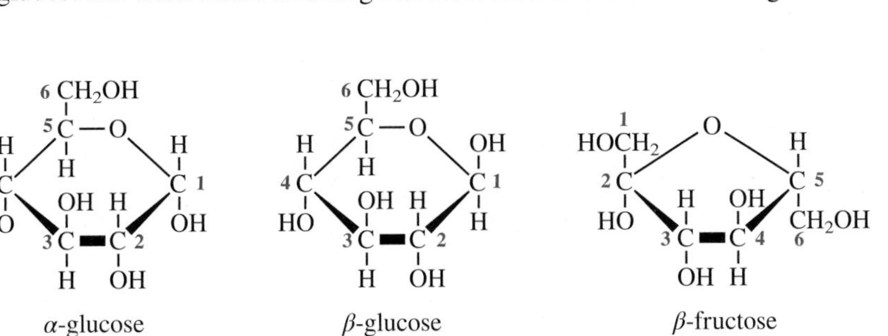

Figure 11.11

Molecular structures of glucose and fructose.

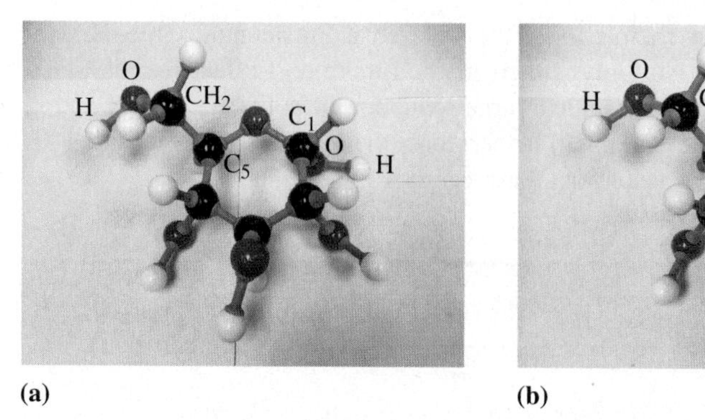

Figure 11.12

Molecular models illustrating the structural differences between **(a)** α- and **(b)** β-glucose. Notice that in α-glucose, the –OH group on carbon 1 is on the opposite side of the ring from the $-CH_2OH$ group attached to carbon 5. In β-glucose, the –OH and $-CH_2OH$ groups are on the same side of the ring.

the alpha (α) and beta (β) isomers of glucose are hard to distinguish without a model (Figure 11.12).

Fructose is an example of a **monosaccharide**; that is, a single sugar. So is glucose. However, sucrose (table sugar) is a **disaccharide**, a "double sugar" formed by joining two monosaccharide units. In forming a sucrose molecule, an α-glucose and a β-fructose unit are connected by a C–O–C linkage created when an H atom and an –OH group are split off to form a water molecule. Does this sound familiar? It is a condensation reaction analogous to those in Chapter 9 that formed polyesters. Figure 11.13 shows the chemical reaction that forms the sucrose molecule (a disaccharide) and releases a water molecule.

Your Turn 11.9 Scientific Practices The Sweet Story

Table 11.3 below compares different sugars. See **Table 11.4** for quantitative sweetness values.

Table 11.3

Sugar	Also Known As	Sweetness	Calories/gram	Chemical Formula
Glucose	Blood sugar	Least sweet	3.87	$C_6H_{12}O_6$
	Grape sugar			
	Corn sugar			
Fructose	Fruit sugar	Sweetest	3.87	$C_6H_{12}O_6$
Sucrose	Table sugar	Intermediate	4.01	$C_{12}H_{22}O_{11}$
	Powdered sugar			
	Granulated sugar			

a. Propose a reason why these three sugars differ in their degree of sweetness.
b. Explain why these sugars are almost identical in their Calories per gram.

α-glucose + β-fructose $\longrightarrow$ sucrose + H_2O

Figure 11.13

Formation of sucrose, a disaccharide.

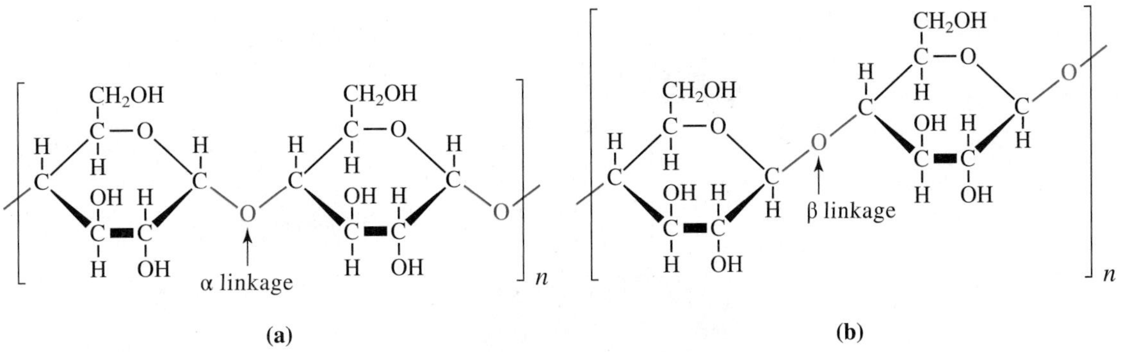

(a) (b)

Figure 11.14

The bonding between glucose units in **(a)** starch and **(b)** cellulose.

As you can see from the previous activity, the sugars we consume are remarkably similar in their chemical composition, in their Calories, and even in their sweet taste. Our taste buds detect a more intense sweetness from fructose than they do glucose or sucrose. So does it matter which sugars you eat? Actually, there is a better question to ask. How *much* sugar are you eating? The quick answer may be "too much."

Monosaccharides can also combine to form much bigger molecules. **Polysaccharides** are condensation polymers made up of thousands of monosaccharide units. As the name implies, these macromolecules consist of "many sugar units." Analogous to the formation of sucrose (Figure 11.13), the formation of a polysaccharide releases a water molecule each time a monosaccharide is incorporated into the polymer chain. Familiar examples of polysaccharides include starch and cellulose (Figure 11.14), which lack the sweetness of simple sugars.

Our bodies can digest starch by breaking it down into glucose; in contrast, we cannot digest cellulose. Consequently, we depend on starchy foods such as potatoes or pasta rather than devouring paper or toothpicks. The difference in digestibility stems from a subtle difference in how the glucose monomers are connected. In Figure 11.14, compare the alpha linkage between the glucose units in starch with the beta linkage between glucose units in cellulose. The enzymes of many mammals—including humans—are unable to catalyze the breaking of beta linkages in cellulose. Consequently, we can't dine on grass or trees.

In contrast, cows, goats, and sheep manage to break down cellulose with a little help. Their digestive tracts contain bacteria that decompose cellulose into glucose. The animals' own metabolic systems then take over. Termites also contain cellulose-hungry bacteria, which is why these insects can damage wooden structures.

When we have excess glucose in our bodies, it is polymerized to glycogen with the help of insulin and stored in our muscles and liver. When our glucose levels slip below normal, the glycogen is converted back into glucose. Glycogen has a molecular structure similar to that of starch, except that its chains of glucose units are longer and more branched. Glycogen is vitally important because it stores energy for use in our bodies. It accumulates in muscles and especially in the liver, where it is available as a quick source of internal energy.

Are "carbs" the culprit in rising obesity rates? Diet books about carbohydrate intake have spanned the spectrum from banning all carbs to taking a "good carb versus bad carb" approach. The latter claims that "bad" carbohydrates cause a quick rise in blood sugar, followed by a spike in blood insulin level. As shown in Figure 11.15, **insulin** is a hormone secreted by the pancreas that allows the cells in your body to absorb and store sugar that is in the blood. The sugar that is not immediately broken down for energy is converted to fat and stored in your cells. Conversely, glucagon is another hormone secreted by the pancreas that essentially has the opposite effect of insulin: It promotes the use of stored glucose in cells. The release of glucagon decreases after a "glucose spike," such as results from eating "bad carbs," but it is known to increase after consumption of proteins. Therefore, some "low-carb" diets promote

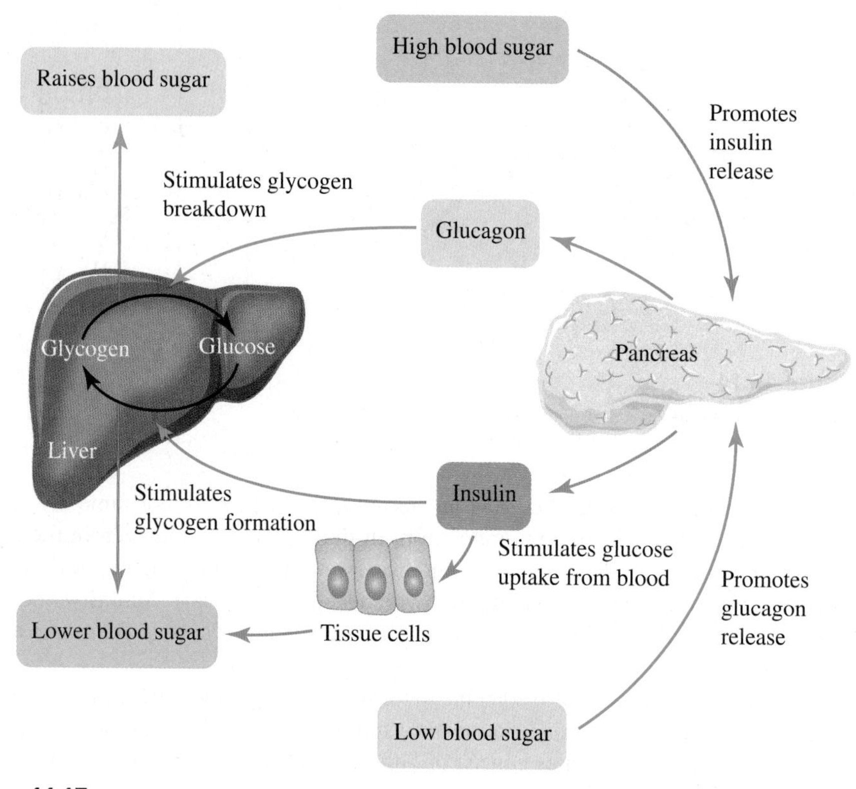

Figure 11.15

The role of insulin and glucagon in controlling the blood sugar levels in our body.

eating more protein as a way to use up stored Calories. The long-term health effects of this approach are yet to be seen. Clearly, both nutrition and dieting involve a complex bit of chemistry!

As you may know, a healthful diet derives more of its carbohydrates from polysaccharides than from simple (and sweet) sugars. In the next section, we delve into topics related to sweetness.

11.5 How Sweet It Is: Sugars and Sugar Substitutes

Got a sweet tooth? We seem to be born with a preference for sweets, and for most of us this lasts throughout our lives. Sweeteners come as syrups, small crystals in packages, and as cubes or tablets that you can drop into a cup of coffee. Some of these sweeteners are natural, whereas others are artificial (synthetic). Some are very sweet, while others less so. Some have been available since antiquity, while others are relatively new on the market.

Honey is primarily composed of fructose and glucose. Both cane sugar and beet sugar are primarily sucrose.

How sweet is sweet? Table 11.4 indicates the relative sweetness of some common natural sweeteners, all compared with sucrose, which is assigned a value of 100. From this table, one would have to use less fructose to equal the sweetness of a teaspoon of sucrose (table sugar). In contrast, lactose (milk sugar) would require more than 6 teaspoons.

Table 11.4 Approximate Relative Sweetness Values

NATURAL SWEETENERS					
Lactose	**Maltose**	**Glucose**	**Honey**	**Sucrose**	**Fructose**
16	32.5	74.3	97	100	173

Source: International Food Information Council

Does it matter which sugar you consume? Yes and no. As we pointed out earlier, for most people the issue is not which sugar. Rather, it is too much of all of them combined. Again, a healthful diet derives more of its carbohydrates from polysaccharides than from simple sugars. If you overindulge in sugar, you increase your risk both for becoming obese and for the diseases that accompany obesity, such as diabetes and high blood pressure.

Let's begin by getting a handle on how much sugar you consume. The next activity allows you to explore one possible source of sugar intake.

Your Turn 11.10 You Decide Your Favorite Cola or UnCola

Soda has been referred to as "liquid candy." Is this a fair characterization? Make an argument *pro* or *con*. In either case, cite the grams of sugar involved.

One reason that you consume sugars is because foods naturally contain them. All the sugars listed in Table 11.4 occur naturally. For example, fructose occurs in many fruits, and lactose occurs in milk. Another reason that you consume sugar is because you add it either during cooking or at the table, or because food companies add it during processing. For instance, manufacturers of processed foods add sugar to peanut butter, spaghetti sauce, and bread. These and many other products have small amounts of sugar added to improve the taste, the texture, or the shelf life. According to data released in the 2005–2010 National Health and Nutrition Examination Survey, 13% of the total Calories consumed by people in the United States are from added sugars. This translates to about 17–22 teaspoons of added sugar daily. At about 4 grams of sugar per teaspoon and 4 Calories per gram, this corresponds to about 270–350 Calories daily from added sugar!

In order to more clearly identify the added sugar content of food (as well as *trans* fats and other minerals), the U.S. FDA has recently proposed new labels (Figure 11.16b), which may soon appear on food products in a supermarket near you.[1]

Nutrition Facts

Serving Size 2/3 cup (55g)
Servings Per Container About 8

Amount Per Serving	
Calories 230	Calories from Fat 72
	% Daily Value*
Total Fat 8g	12%
Saturated Fat 1g	5%
Trans Fat 0g	
Cholesterol 0mg	0%
Sodium 160mg	7%
Total Carbohydrate 37g	12%
Dietary Fiber 4g	16%
Sugars 12g	
Protein 3g	
Vitamin A	10%
Vitamin C	8%
Calcium	20%
Iron	45%

* Percent Daily Values are based on a 2,000 calorie diet. Your daily value may be higher or lower depending on your calorie needs.

	Calories:	2,000	2,500
Total Fat	Less than	65g	80g
Sat Fat	Less than	20g	25g
Cholesterol	Less than	300mg	300mg
Sodium	Less than	2,400mg	2,400mg
Total Carbohydrate		300g	375g
Dietary Fiber		25g	30g

(a)

Nutrition Facts

8 servings per container
Serving size **2/3 cup (55g)**

Amount per serving	
Calories	**230**
	% Daily Value*
Total Fat 8g	**10%**
Saturated Fat 1g	**5%**
Trans Fat 0g	
Cholesterol 0mg	**0%**
Sodium 160mg	**7%**
Total Carbohydrate 37g	**13%**
Dietary Fiber 4g	**14%**
Total Sugars 12g	
Includes 10g Added Sugars	**20%**
Protein 3g	
Vitamin D 2mcg	10%
Calcium 260mg	20%
Iron 8mg	45%
Potassium 235mg	6%

* The % Daily Value (DV) tells you how much a nutrient in a serving of food contributes to a daily diet. 2,000 calories a day is used for general nutrition advice.

(b)

Figure 11.16

The current food label format **(a)** and the proposed new label format for food products **(b)**.

Source: U.S. Food and Drug Administration

[1] http://www.fda.gov/Food/GuidanceRegulation/GuidanceDocumentsRegulatoryInformation/LabelingNutrition/ucm385663.htm

Your Turn 11.11 You Decide Sugar Consumption in the U.S.

Using the Internet, review the latest sugar consumption data from the U.S. National Health and Nutrition Examination Survey.

a. Describe the trends in sugar consumption with gender, age, race and ethnicity, and income. Are any other factors responsible for differing levels of sugar consumption among these groups?
b. How do the trends in sugar consumption between women, men, and children compare to the dietary sugar suggestions provided by the American Medical Association, American Heart Association, and the World Health Organization?
c. How does the sugar consumption (and country-specific dietary sugar recommendations, if present) in other countries compare to the U.S.? Can you think of some reasons for these discrepancies?

High-fructose corn syrup (*HFCS*) is used to sweeten many beverages and foods. Depending on where you live, HFCS goes by different names. In Europe, it is called *iso-glucose*, and in Canada it goes by *glucose-fructose*. In this textbook, we refer to it as HFCS. Corn syrup primarily contains glucose. But if you treat corn syrup with enzymes, you can convert the glucose to fructose, which is sweeter. Several different "blends" of HFCS exist, depending on the end use. For example, a typical blend used in soft drinks is about 55% fructose, with the remainder being glucose.

Why is HFCS added to foods? Actually, many reasons exist. Food manufacturers claim that the free monosaccharides in high-fructose corn syrup provide better flavor enhancement, stability, freshness, color, texture, "pourability," and consistency in foods, relative to sucrose. Although opponents of HFCS have suggested that this corn sweetener is metabolized differently from sucrose, this does not appear to be true. The American Medical Association (AMA) has concluded that "it appears unlikely that HFCS contributes more to obesity or other conditions than sucrose does."[2] Metabolically, HFCS appears to be similar to sucrose in our bodies.

We end this section by turning to artificial (synthetic) sweeteners, also called sugar substitutes. As you can see from Table 11.5, these are far sweeter than natural sugars. The Calorie content per gram of aspartame is about the same as sucrose. However, since it is about 200 times sweeter than sucrose, 1/200 of a teaspoon of aspartame is used as the equivalent of a teaspoon of sucrose. Saccharin, aspartame, sucralose, neotame, and acesulfame potassium are the five artificial sweeteners approved in the United States. Other countries have a slightly different list.

Did You Know? Although artificial sweeteners have 0 Calories, they are often combined with other sugars such as dextrose and/or maltodextrin to provide bulk. Hence, on a per-mass basis, most commercial sweetener formulations only reduce the number of Calories by 10–15% as compared to sugar. However, artificial sweeteners are much sweeter than sugar, so smaller quantities relative to sugar are often added to food.

So are artificial sweeteners the way to go? These compounds certainly do offer the advantage of fewer Calories. For example, the sugar in a 12-ounce can of soda accounts for about 140 Calories. In terms of a 2,000-Calorie diet, this translates to about 7%. By contrast, the same beverage sweetened with a pure synthetic sweetener would have 0 Calories. Although people have been concerned about the health effects of using artificial sweeteners, studies indicate that the sweeteners currently on the market are safe for most people. However, a small percentage of the population must avoid aspartame, which we will describe in the next section.

Table 11.5 Approximate Relative Sweetness Values*

SYNTHETIC SWEETENERS				
Acesulfame potassium	**Aspartame**	**Neotame**	**Saccharin**	**Sucralose**
200	200	7,000–13,000	300	600

*As a reference for artifical sweeteners, sucrose has a sweetness value of 1.
Source: International Food Information Council.

[2] *Journal of American College of Nutrition*, **2009**, *28(6)*, 619–626 (DOI: 10.1080/07315724.2009.10719794).

Your Turn 11.12 You Decide Chemicals!

A well-known U.S. medical clinic posted this statement online: "Artificial sweeteners are chemicals or natural compounds that offer the sweetness of sugar without as many Calories."

a. What is the problem with the categories "chemicals or natural compounds"?
b. How might you rewrite this language?

11.6 | Proteins: First among Equals

The word *protein* is derived from *protos,* Greek for "first." However, the name is misleading. Life depends on the interaction of thousands of chemicals, and to assign primary importance to any single compound or class of compounds is too simplistic. Nevertheless, proteins are an essential part of every living cell. They are major components in hair, skin, and muscle. They also transport oxygen, nutrients, and minerals through the bloodstream. Many of the hormones that act as chemical messengers are proteins, as are most of the enzymes that catalyze the chemistry of life.

A **protein** is a polyamide or polypeptide; that is, a polymer built from amino acid monomers. The great majority of proteins are made from various combinations of 20 different naturally occurring amino acids. Molecules of amino acids share a common structure. As shown below, four chemical species are attached to a carbon atom: a carboxylic acid group (green), an amine group (yellow), a hydrogen atom, and a side chain designated R (red):

Did You Know? There are approximately 300 different types of amino acids in nature; however, only 20 of these serve as building blocks of proteins, which are known as *proteinogenic* amino acids.

$$H_2N{-}CH(R){-}C({=}O){-}OH$$

Variations in the R side chain differentiate individual amino acids. As seen in Figure 11.17, the structure of the R group governs the chemical properties of an amino acid. Some feature polar side groups, which are hydrophilic ("water-loving") and can form hydrogen bonds with water, whereas others contain nonpolar groups and are thereby repelled by water—referred to as hydrophobic. Still other amino acids contain positively or negatively charged groups that lead to pH-dependent properties.

The topics of polarity, hydrogen bonding, and pH were introduced in Chapter 8.

Polar side chains can lead to more types of interactions: ionic or hydrogen bonds. Side chains that contain acidic or basic groups (such as carboxylic acids or amines) often become charged ions and attract their opposites to form ionic bonds. If you examine Figure 11.17 closely, you should see potentially favorable interactions between amino acids such as lysine and aspartate through the side-chain functional groups: an amine and a carboxylic acid. Polar side chains often contain hydroxyl groups or amides. Many of these side chains will form hydrogen bonds, a high-energy and specific interaction. Hydrogen bonds form between electronegative atoms, N and O being the most important in biochemistry, and a hydrogen atom bound to another electronegative atom. Once again, considering the examples in Figure 11.17, serine is a great example of an amino acid that can form H-bonds through its alcohol group in the side chain.

Nonpolar side chains tend to be hydrophobic, meaning they specifically associate with each other to avoid interaction with water. Hydrophobic groups also form vital energetic interactions, called **London dispersion** forces. If you examine the options in Figure 11.17, you will see that nonpolar amino acids are as numerous and varied as polar amino acids. A side chain with a planar phenyl group (phenylalanine) will create a different protein shape relative to an amino acid with a longer carbon chain (such as leucine).

Dispersion forces were discussed in the context of hydrocarbon boiling points in Section 5.12.

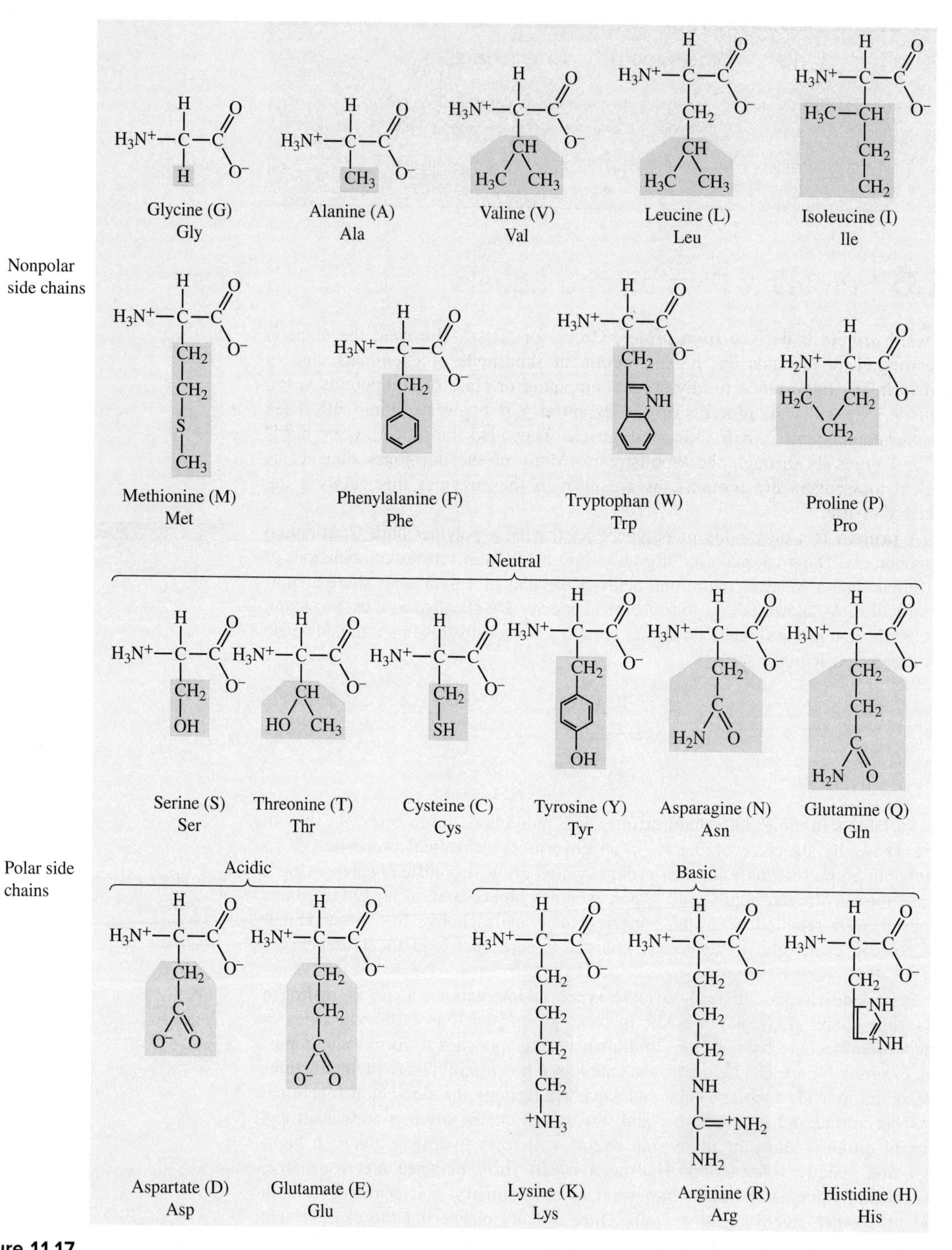

Figure 11.17

A comparison of the R-group side chains (highlighted) of the 20 naturally occurring amino acids.

Two amino acids can combine by a condensation reaction between the amine group on one amino acid, and a carboxylic acid group on the other. For example, glycine can react with alanine, as shown in Equation 11.4a:

This is the same type of cross-linking polymerization that was discussed in Sections 9.6–9.7, which is used to fabricate polymers such as nylon, PET, Kevlar™, and many others.

glycine + alanine → dipeptide + H_2O (water) **[11.4a]**

Note that the acidic –COOH group of the glycine molecule reacts with the $-NH_2$ group of the alanine molecule. In the process, the two amino acids link through the C–N peptide bond shown in the blue shaded area. In addition, a H_2O molecule is produced. Once incorporated into the peptide chain, the amino acids are called **amino acid residues**.

Equation 11.4a labels the product as a **dipeptide**, a compound formed from two amino acids. Glycine and alanine can form two different dipeptides. Equation 11.4b shows the other possibility:

alanine + glycine → dipeptide + H_2O (water) **[11.4b]**

This time in the condensation reaction, alanine provided the –COOH group and glycine the $-NH_2$ group.

Look closely and you will see that the two dipeptides in Equations 11.4a and b are different. In the first dipeptide, the unreacted amine group is on the glycine residue and the unreacted acid group is on the alanine residue; in the second dipeptide, the $-NH_2$ is on the alanine residue and the –COOH is on the glycine residue.

The point of all this is that the order of amino acid residues in a peptide makes a difference. A particular protein structure depends not only on which amino acids are present, but also on their sequence in the protein chain. Assembling the correct amino acid sequence to make a particular protein is like putting letters in a word: If they are in a different order, a completely new meaning results. Thus, a tripeptide consisting of three different amino acids is like a three-letter word containing the letters *a, e,* and *t*. There are six possible combinations of these letters. Three of them—*ate, eat,* and *tea*—form recognizable English words; the other three—*aet, eta,* and *tae*—do not. Similarly, some sequences of amino acids may be biological nonsense.

Still restricting ourselves to three-letter words and only the letters *a, e,* and *t,* but allowing the duplication of letters, we can make perfectly good words such as *tee* and *tat,* and lots of meaningless combinations such as *aaa* and *tte*. There are, in fact, a total of 27 possibilities, including the six identified earlier. Just as words can use letters more than once, most proteins contain specific amino acids more than once.

Your Turn 11.13 Skill Building Making Tripeptides

The equations in this section show that glycine (Gly) and alanine (Ala) can combine to form two dipeptides: GlyAla and AlaGly. If these two amino acids can be used more than once, two other dipeptides are possible: GlyGly and AlaAla. Thus, four different dipeptides can be made from two different amino acids. Eight different tripeptides can be made from supplies of two different amino acids, assuming that each amino acid can be used once, twice, three times, or not at all. Use the symbols Gly and Ala to write down representations of the amino acid sequence in all eight of these tripeptides.

Hint: Start with GlyGlyGly.

Normally, the body does not store a reserve supply of protein, so foods containing protein must be eaten regularly. As the principal source of nitrogen for the body, proteins are constantly being broken down and reconstructed. A healthy adult on a balanced diet is in nitrogen balance, excreting as much nitrogen (primarily as urea in the urine) as she or he ingests. Growing children, pregnant women, and persons recovering from long-term debilitating illness or burns have a positive nitrogen balance. This means that they consume more nitrogen than they excrete because they are using the element to synthesize additional protein. A negative nitrogen balance exists when more protein is being decomposed than is being made. This occurs in starvation, when the energy needs of the body are unmet from the diet, and muscle is metabolized to maintain physiological functions. In effect, the body feeds on itself.

Another cause of a negative nitrogen balance may be a diet that does not include sufficient quantities of **essential amino acids**. These components are those required for protein synthesis, but are not synthesized by the human body. Of the 20 natural amino acids that make up our proteins, we can synthesize only 11 in our bodies from simpler molecules. Consequently, we must obtain the other nine amino acids from the foods we eat. If your diet is missing any of the nine essential amino acids identified in Table 11.6, the result can be severe malnutrition.

Good nutrition thus requires protein in sufficient quantity and suitable quality. Beef, fish, and poultry contain all the essential amino acids in approximately the same proportions found in the human body. Therefore, all of these are "complete" proteins. However, most people of the world depend on grains and other vegetable crops rather than on meat or fish. If such a diet is not sufficiently diversified, some essential amino acids may be lacking. For example, Mexican and Latin American diets tend to be rich in corn and corn products, a protein source that is *incomplete* because corn is low in tryptophan, an essential amino acid. A person may eat enough corn to meet the total protein requirement but still be malnourished because of insufficient tryptophan.

Fortunately for millions of vegetarians, a reliance on vegetable protein does not doom them to malnutrition. The trick is to apply a principle that nutritionists call **protein complementarity**. This consists of combining foods that complement essential amino acid content so that the total diet provides a complete supply of amino acids used for protein synthesis. Although some may worry that vegetarians need to strictly adhere to this principle, most are likely to do it automatically. Say, for example, you eat a peanut butter sandwich. Bread is deficient in lysine and

Table 11.6 The Essential Amino Acids

histidine	lysine	threonine
isoleucine	methionine	tryptophan
leucine	phenylalanine	valine

isoleucine, but peanut butter supplies these amino acids. On the other hand, peanut butter is low in methionine, but it is provided by the bread. The traditional diets in many countries also tend to meet protein requirements. For example, in Latin America, beans are used to complement corn tortillas, and soy foods are eaten with rice in parts of Southeast Asia and Japan. People in the Middle East combine bulgur wheat with chickpeas or eat hummus—a paste made from sesame seeds and chickpeas—with pita bread. In India, lentils and yogurt are eaten with unleavened bread. Thus, it is likely that if you follow a balanced vegetarian diet, you will ingest sufficient quantities of essential amino acids, assuming that you are eating an adequate number of Calories.

We end this section by revisiting sweetness, the topic of the previous section. It may surprise you to learn that aspartame, a sugar substitute, is a dipeptide! Aspartame is composed primarily of the amino acids aspartic acid and phenylalanine (Figure 11.18). It is one of the most highly studied food additives, and for the vast majority of consumers, aspartame is a safe alternative to sugar. One group of people, however, definitely should not use aspartame. The warning on packets of artificial sweeteners and products containing aspartame is explicit: "Phenylketonurics: Contains Phenylalanine."

This is a case where one person's treat is another person's poison. Phenylalanine is an essential amino acid converted in the body to tyrosine, a different amino acid. Individuals with phenylketonuria, a genetically transmitted disease, lack the enzyme (phenylalanine hydroxylase) that catalyzes this transformation. Consequently, the conversion of dietary phenylalanine to tyrosine is blocked and the phenylalanine concentration in blood and tissues rises (Figure 11.19). To compensate for the elevated

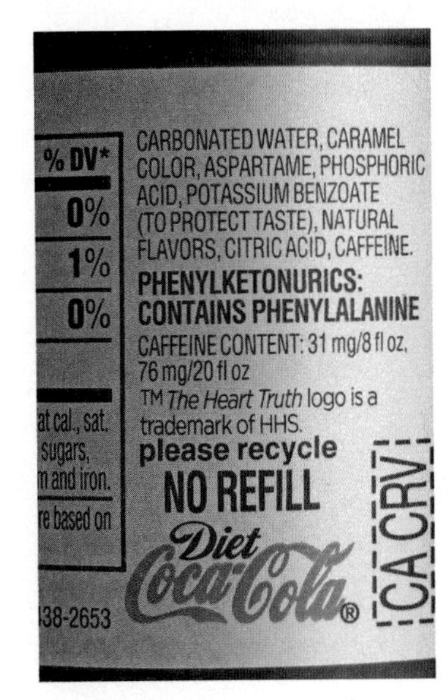

© Jill Braaten

Figure 11.18

The structural formula of aspartame. The blue shading illustrates the peptide bond linking the amino acids aspartic acid and phenylalanine.

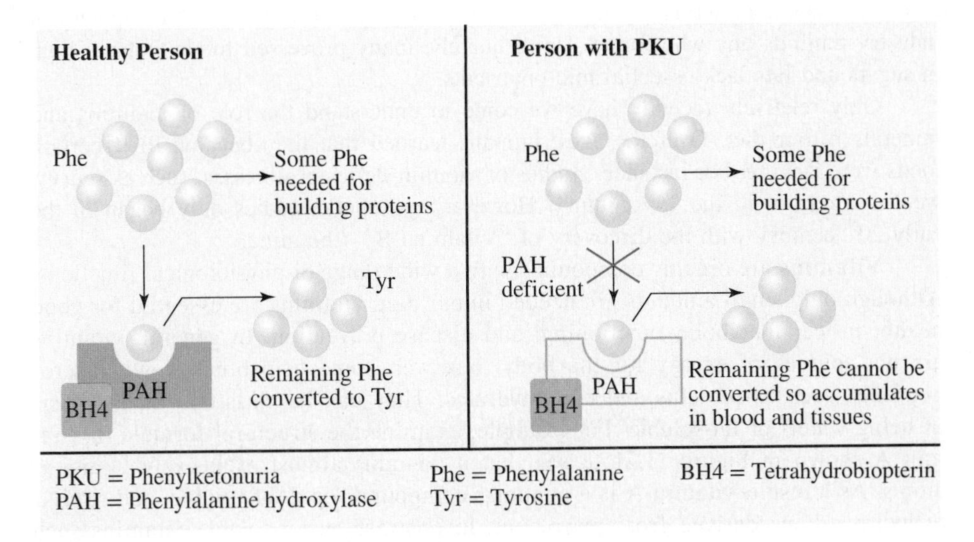

Figure 11.19

A schematic illustration of phenylketonuria, PKU.

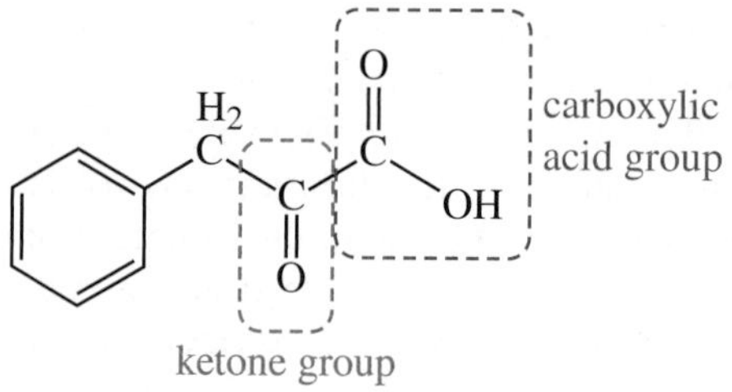

carboxylic acid group

ketone group

Figure 11.20

The molecular structure of phenylpyruvic acid. Because the structure contains both a ketone and carboxylic acid group, it is referred to as a *keto acid*.

Note: A ketone functional group contains a central carbonyl (C=O) connected to two R groups.

phenylalanine, the body converts it to phenylpyruvic acid, excreting large quantities of this acid in the urine. Phenylpyruvic acid is termed a *keto acid* because of its molecular structure (Figure 11.20); hence, the disease is known as phenyl*keto*nuria or PKU. People with the disease are called phenylketonurics.

Excess phenylpyruvic acid causes severe mental retardation. Therefore, the urine of newborn babies is tested for this compound using special test paper placed in the diaper. Infants diagnosed with PKU must be put on a diet severely limited in phenylalanine. This means avoiding excess phenylalanine from milk, meats, and other sources rich in protein. Because phenylalanine is an essential amino acid, a minimum amount of it must still be available, even in phenylketonurics. Supplemental tyrosine may also be needed to compensate for the absence of the normal conversion of phenylalanine to tyrosine. A phenylalanine-restricted diet is recommended for phenylketonurics, at least through adolescence. Adult phenylketonurics also must limit their phenylalanine intake, and hence curtail their use of aspartame.

This last discussion has demonstrated that even small quantities of a substance can make a difference in your diet. The next section looks at other substances that are found in your diet in small quantities.

11.7 Vitamins and Minerals: The Other Essentials

Vitamins and minerals are **micronutrients**, substances that are needed only in miniscule amounts but still are essential to life. Nearly everyone in the United States knows that vitamins and minerals are important, but a thriving multimillion-dollar supplement industry reminds any who forget. Unfortunately, many processed foods that are high in sugars and fats lack essential micronutrients.

Did You Know? The particular designation of "B_1," was the label on the test tube in which the sample was first collected. In 1912, Polish biochemist Casimir Funk (1884–1967) first coined the name *vitamine* (short for *vital amine*), to describe a class of compounds he was studying that were vital for life and contained the amine ($-NH_2$) group. The final "e" in its spelling disappeared in 1920 after it was suspected that not all vitamins contained amines.

Only relatively recently have we come to understand the role of vitamins and minerals in our diet. Over the ages, humans learned that they became ill if certain foods were lacking. For instance, studies of vitamin-deficient illnesses, such as scurvy, were carried out in the 18th century. However, systematic studies only began in the early 20th century with the discovery of "Vitamine B_1" (thiamine).

Vitamins are organic compounds with a wide range of physiological functions. Although only small amounts are needed in our diet, vitamins are essential for good health, proper metabolic functioning, and disease prevention. In general, vitamins are not sources of energy for the body; however, some help break down macronutrients, such as proteins and carbohydrates. They can be classified on the basis of being water- or fat-soluble. For example, examine the structural formula of vitamin A shown in Figure 11.21 to see that it contains almost exclusively C and H atoms. As a result, vitamin A is a nonpolar compound, lipid/fat-soluble, and similar to hydrocarbons derived from petroleum. In contrast, water-soluble vitamins such as vitamin C (Figure 11.21) often contain several –OH groups that can hydrogen bond with water molecules.

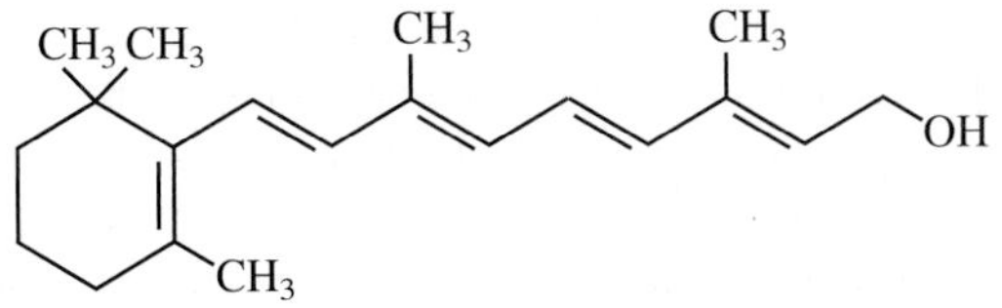

vitamin A, a lipid-soluble vitamin

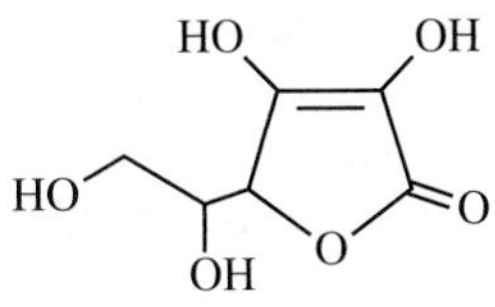

vitamin C, a water-soluble vitamin

Figure 11.21

Examples of lipid-soluble and water-soluble vitamins.

Your Turn 11.14 You Decide Classifying Vitamins

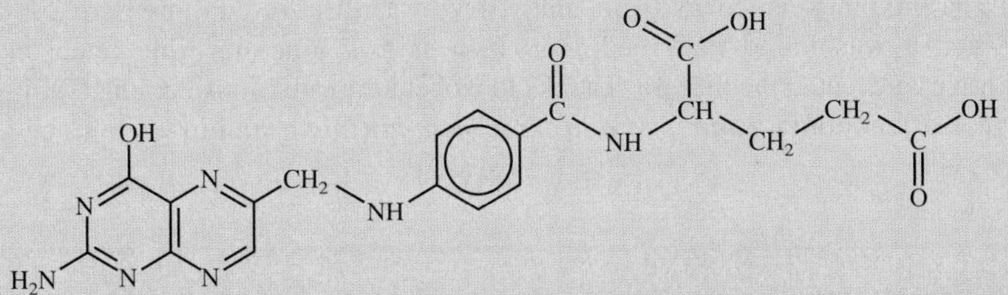

Folic acid (above) helps prevent certain types of anemia and aids in nucleic acid synthesis. This vitamin is particularly important for pregnant women. Do you expect that it would be soluble in fat tissue (lipids) or in the bloodstream and cell tissue (water)? Explain your reasoning.

The solubility of vitamins has significant implications for health. Because of their fat-solubility, vitamins A, D, E, and K are stored in cells rich in lipids, where they are available on biological demand. If swallowed in excess, fat-soluble vitamins can build up to a toxic level. For example, high doses of vitamin A can result in troublesome symptoms such as fatigue and headache, and more serious ones such as blurred vision and liver damage. Although the toxic level of vitamin D is not known, similar illness can result if too much vitamin D is ingested, including heart and kidney damage. High levels of these vitamins are not reached via diet; rather, they are a result of excessive use of vitamin supplements.

In contrast, water-soluble vitamins are excreted in the urine rather than stored in the body. As a result, you need to eat foods containing these vitamins frequently. Unfortunately, even water-soluble vitamins can accumulate at toxic levels when taken in large doses, although such cases are rare. For most people, a balanced diet provides the necessary vitamins and minerals, making vitamin supplements unnecessary. The one exception seems to be vitamin D, which is synthesized in the skin by using the energy of sunlight, rather than ingested. Recent research on vitamin D has led more physicians to check vitamin D blood levels as part of an annual physical exam, and to use the results to determine whether taking a supplement is necessary.

Your Turn 11.15 You Decide Can You Turn Orange from Eating Carrots?

Using the Internet, find actual cases where the consumption of large amounts of carrots or oranges caused skin pigmentation to turn yellow/orange. Do you think the observed skin discoloration is due to water-soluble or fat-soluble components in these foods? Explain.

Common names for the vitamin B family are:
B_1 — thiamine
B_2 — riboflavin
B_3 — niacin
B_5 — pantothenic acid
B_6 — pyridoxine
B_7 — biotin
B_9 — folic acid
B_{12} — cobalamins

Many of the water-soluble vitamins serve as **coenzymes**, molecules that work in conjunction with enzymes to enhance their activity. Members of the vitamin B family are particularly adept in acting as coenzymes. Niacin plays an essential role in energy transfer during glucose and fat metabolism. The synthesis of niacin in the body requires the essential amino acid tryptophan. Thus, a diet deficient in tryptophan may lead to niacin deficiency. Such a deficiency causes pellagra, a serious condition that is characterized by "the 4Ds" of diarrhea, dermatitis, dementia, and death. This disease is still common today in parts of the world, including several African nations.

Some vitamins were discovered when observers correlated diseases with the lack of specific foods. For example, vitamin C (ascorbic acid) must be supplied in the diet, typically via citrus fruits and green vegetables. An insufficient supply of the vitamin leads to scurvy, a disease in which collagen, an important structural protein, is broken down.

This practice also led to British sailors being called "limeys."

© *SunnyS/Shutterstock.com*

The link between citrus fruits and scurvy was discovered more than 200 years ago when it was found that feeding British sailors limes or lime juice on long sea voyages prevented the disease. Thanks to Nobel Laureate Linus Pauling (1901–1994) who in 1970 authored *Vitamin C and the Common Cold*, vitamin C continues to be in the public eye.

Your Turn 11.16 You Decide Megadoses of Vitamin C

Decades ago, Linus Pauling claimed that large doses of vitamin C were therapeutic in preventing the common cold.

a. What range of vitamin C daily constitutes a "megadose"?
b. Find evidence to either support or refute the claim of preventing the common cold. Cite your sources.
c. Interview three people of different age groups, including a nurse or physician if you are able. Ask if they take vitamin C, and if so, why.

We would be remiss not to mention vitamin E, which actually consists of several closely related fat-soluble vitamins rather than a single compound. Vitamin E is only synthesized by plants, and in varying amounts. Vegetable oils and nuts are good sources of it. Nonetheless, it is so widely distributed in foods that it is difficult to create a diet deficient in vitamin E. Since the 1990s, this vitamin has been in the news as part of the antioxidant system that protects the body from chemically active and damaging free radicals. Although at one time taking vitamin E supplements was recommended, this is no longer the case. Skin preparations are another matter, though. Many products contain vitamin E, and claim that it prevents or helps heal skin damage. Investigate these claims for yourself in the next activity.

Your Turn 11.17 You Decide Vitamin E and Your Skin

Check the advertisements and you will see that many hand lotions and beauty creams contain vitamin E.

a. Identify three skin products that contain vitamin E.
b. How is vitamin E thought to help protect your skin?
c. Although it might seem logical that vitamin E would be good for the skin, it is difficult to find evidence. Investigate this topic to see for yourself. Use the resources on the Internet to assist you.

Minerals are ions or ionic compounds that, like vitamins, have a wide range of physiological functions. You may be familiar with minerals such as sodium and calcium, but actually the list is much longer. Depending on how much of them you need, minerals are classified as either macro, micro, or trace:

- **Macrominerals**: Ca, P, Cl, K, S, Na, and Mg. As ions, these elements are necessary for life, but not nearly as abundant in our bodies as O, C, H, and N. You need to ingest macrominerals daily, typically in the range of 1–2 g.
- **Microminerals**: Fe, Cu, and Zn. The body requires lesser amounts of these. You may recognize iron as a component of hemoglobin, a protein in the blood that carries oxygen.
- **Trace minerals**: I, F, Se, V, Cr, Mn, Co, Ni, Mo, B, Si, and Sn. These usually are measured in microgram (μg) quantities. Although the total amount of trace elements in the body is tiny, their small quantity is required for good health.

The periodic table in Figure 11.22 displays the essential dietary minerals. The metals exist in the body as cations; for example, Ca^{2+} (calcium ion), Mg^{2+} (magnesium ion), K^+ (potassium ion), and Na^+ (sodium ion). The nonmetals are typically present as anions. For example, chlorine is found as Cl^- (chloride ion), and phosphorus appears as PO_4^{3-} (phosphate ion).

The physiological functions of minerals are widely diverse. Calcium is the most abundant mineral in the body. Along with phosphorus and smaller amounts of fluorine, it is a major constituent of bones and teeth. Blood clotting, muscle contraction, and transmission of nerve impulses all require the calcium ion, Ca^{2+}. Sodium is also essential for life, but not in the excessive amounts supplied by the diets of many people today. The salt we eat does not necessarily come from the salt shaker. Rather, you perhaps unknowingly add salt to your diet from sauces, snack foods, fast foods, and even canned soup! Labels are required to list the "sodium" content, meaning the number of milligrams of Na^+ per serving. For example, different brands of tomato soup may have between 700–1,260 mg of Na^+ per serving. Compare this with the recommended daily value of no more than 2,400 mg (2.4 grams) of Na^+. The major concern with excess dietary sodium is its correlation with high blood pressure for some individuals. They may be advised by their doctors to limit their sodium intake.

Did You Know? The Latin word for salt is *sal*. Salt was so highly valued in Roman times that soldiers were paid in *sal*, thereby forming the root for the modern word *salary*.

Your Turn 11.18 You Decide Sodium in Your Diet

Compare the sodium content for foods in the same category, such as different brands of pretzels, bread, frozen pizza, salad dressing, or even tomato soup. Have your findings surprised you, or influenced your future choices?

Hint: 1 g = 1,000 mg.

Oranges, bananas, tomatoes, and potatoes all help supply the recommended daily requirement of 2 grams of potassium (in the form of K^+), another essential mineral. You may have heard both K^+ and Na^+ referred to as the "electrolytes" of sports drinks.

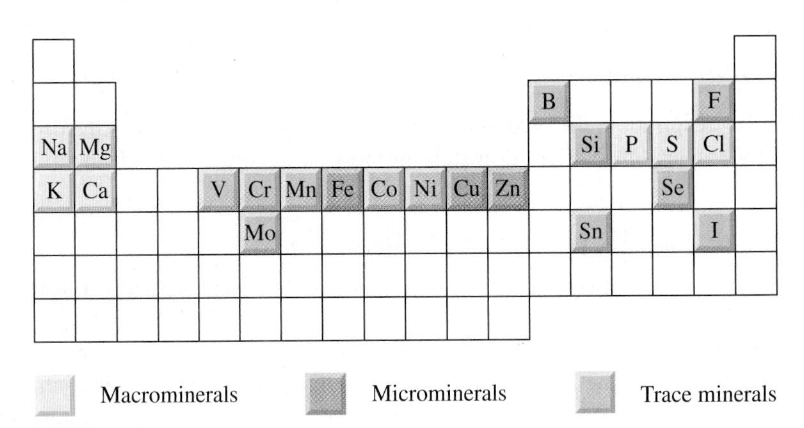

Figure 11.22

Periodic table indicating the dietary minerals necessary for human health.

Sodium and potassium ions are close chemical cousins, as they are both in Group 1 of the periodic table. They have similar chemical properties and physiological functions. Within cells, the concentration of K^+ is considerably greater than that of Na^+. The reverse situation holds true in the lymph and blood serum outside the cells, in which the concentration of K^+ is low, and that of Na^+ is high. The relative concentrations of K^+ and Na^+ are especially important for the rhythmic beating of the heart. Individuals who take diuretics to control high blood pressure may also take potassium supplements to replace K^+ excreted in the urine. However, such supplements should be taken only under the advice of a physician, because they also can dramatically alter the potassium–sodium balance in the body and lead to cardiac complications.

Seafood is one rich source of iodine. Another is iodized salt (sodium chloride) to which 0.02% of potassium iodide has been added.

In most instances, microminerals and trace elements have very specific biological functions and are incorporated in relatively few biomolecules. Iodine is an example. Most of the body's iodine is found in the thyroid gland incorporated into thyroxine, a hormone that regulates metabolism. Excess thyroxine production is associated with hyperthyroidism (Graves' disease), in which basal metabolism is accelerated to an unhealthy level, rather like a racing engine. In contrast, a thyroxine deficiency, sometimes caused by a lack of dietary iodine, slows metabolism and results in tiredness and listlessness. The tendency of the thyroid gland to concentrate iodine makes possible the use of radioactive I-131 in treating thyroid disorders, and in imaging the thyroid gland for diagnostic purposes. The next activity gives you the opportunity to combine what you have learned about radioactive I-131 and iodine as a trace mineral.

Your Turn 11.19 Scientific Practices Radioactive Iodine

Hyperthyroidism (Graves' disease) is also referred to as having an overactive thyroid gland.

a. I–131 is used to treat hyperthyroidism. Explain how ingestion of this radioisotope can lead to a reduction in the function of the thyroid gland.

b. I–131 treatment has both risks and benefits for a patient. List two of each.

c. Patients treated with I–131 temporarily carry a source of radioactivity in their bodies. After 10 half-lives, a radioisotope can be said to be "gone." How much time is this for a patient treated with I–131?

Hint: See Table 6.2.

11.8 | Food for Energy

The energy needed to keep our bodies warm and to run our complex chemical, mechanical, and electrical systems comes from the food we eat: fats, carbohydrates, and proteins. As noted in several earlier chapters, this energy initially arrives on Earth in the form of sunlight, and then is absorbed by green plants. In the process of photosynthesis, CO_2 and H_2O are combined to form $C_6H_{12}O_6$ (Equation 11.5). Hence the Sun's energy is stored in chemical bonds of the monosaccharide we know as glucose ($C_6H_{12}O_6$).

$$\text{energy (from sunshine)} + 6\,CO_2 + 6\,H_2O \xrightarrow{\text{chlorophyll}} C_6H_{12}O_6 + 6\,O_2 \quad [11.5]$$

During respiration, the outcome from photosynthesis is reversed. Glucose is converted into simpler substances (ultimately in most cases to CO_2 and H_2O), and energy is released:

$$C_6H_{12}O_6 + 6\,O_2 \longrightarrow 6\,CO_2 + 6\,H_2O + \text{energy} \quad [11.6]$$

The energy balance between Equations 11.5 and 11.6 is schematically illustrated in Figure 11.23.

In addition to having a supply of sufficient energy, our bodies must have some way of regulating the rate at which the energy is released. Without such control, your body temperature would wildly fluctuate. The automobile provides an analogy. Dropping a lighted match into the fuel tank would burn all the gasoline at once, and likely the entire car as well! Under normal operating conditions, just enough fuel is delivered

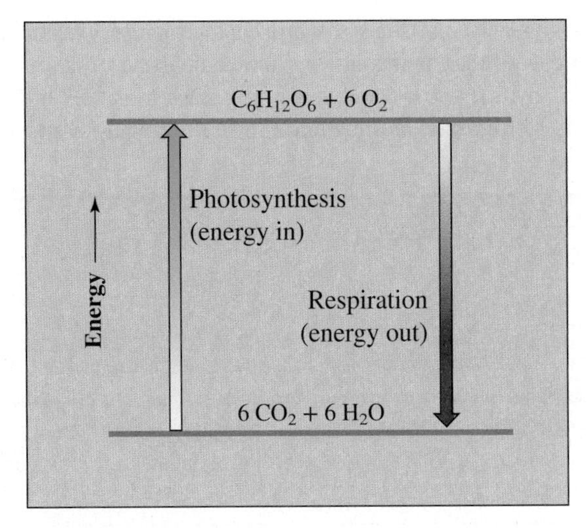

Figure 11.23

Glucose energy balance (photosynthesis and respiration).

to the ignition system to supply the automobile with the energy it needs, without raising the temperature of the car and its occupants beyond reason. By releasing a little energy at a time, the efficiency of the process is enhanced. So it is with our body. The conversion of foods into carbon dioxide and water occurs over many small steps, each one involving enzymes, enzyme regulators, and hormones. As a result, energy is released gradually, as needed, and body temperature is maintained within normal limits.

The energy in Calories associated with the metabolism of a gram of fats, carbohydrates, and proteins is given in Table 11.7. On a Calorie-per-gram basis, fats provide about 2.5 times as much energy as do proteins and carbohydrates. This observation makes it easy to understand the popularity of low-fat diets for losing weight. Although proteins, like carbohydrates, yield about 4 Cal/g if metabolized, proteins are not used in the body primarily as an energy source. Rather, proteins are used for building skin, muscles, tendons, ligaments, blood, and enzymes.

1 dietary calorie = 1 Cal
1 kcal
1,000 calories

Table 11.7 Average Energy Content of Macronutrients

Macronutrient	Energy Content (Cal/g)
fats	9
carbohydrates	4
proteins	4

The reason for the difference in energy content between fats and carbohydrates is evident from their chemical composition. Compare the chemical formula of a fatty acid, lauric acid ($C_{12}H_{24}O_2$) with that of sucrose (table sugar, $C_{12}H_{22}O_{11}$). Both compounds have the same number of carbon atoms per molecule and very nearly the same number of hydrogen atoms. When molecules such as these "burn" as fuel in your body, the C and H atoms that they contain combine with oxygen to form CO_2 and H_2O, respectively. But more oxygen is required to burn a gram of lauric acid, $C_{12}H_{24}O_2$, than one gram of sucrose, $C_{12}H_{22}O_{11}$. Examine these two reactions:

More information about lauric acid is revealed by writing its condensed structural formula, $CH_3(CH_2)_{10}COOH$.

$$\underset{\text{lauric acid}}{C_{12}H_{24}O_2} + 17\ O_2 \longrightarrow 12\ CO_2 + 12\ H_2O + 8.8\ \text{Cal/g} \quad [11.7]$$

$$\underset{\text{sucrose}}{C_{12}H_{22}O_{11}} + 12\ O_2 \longrightarrow 12\ CO_2 + 11\ H_2O + 3.8\ \text{Cal/g} \quad [11.8]$$

In the language of chemistry, the sugar is already more "oxygenated" or more "oxidized" than the fatty acid. Weaker C–H bonds (416 kJ/mol) have already been replaced by stronger O–H bonds (467 kJ/mol) in the sucrose. The result is that even though fewer

Oxygenated fuels were discussed in Section 5.13.

O=O double bonds (498 kJ/mol) must be broken for sucrose to combine with O_2, less energy is released overall than is the case for the combustion of lauric acid.

Given how many tasty foods contain fat, it is easy to get an unhealthy percentage of our daily Calories from fats, as illustrated in the following activity.

Your Turn 11.20 You Decide "Low-Fat" Cheese

A popular brand of low-fat shredded cheddar cheese advertises that it provides 1.5 g of fat per serving. Of this 1.5 g of fat, 1.0 g is saturated fat. In addition, a serving of this cheese is 28 grams (or 1/4 cup) and has 50 Calories, with 15 of these coming from fat. Is this a "low-fat" cheese? Support your answer with some numbers. Note that the dietary recommendations provided by the U.S. Department of Agriculture (USDA) and the U.S. Department of Health and Human Services (HHS) are that 20–35% of Calories in adults should come from fat.

The guidelines also recommend that less than 10% of Calories come from saturated fatty acids and that *trans* fat consumption be kept as low as possible.

So, how many Calories does a person need? The answer is, "It depends." The number of Calories your diet should supply each day is a function of your level of activity, the state of your health, your gender, age, body size, and a few other factors. Table 11.8 summarizes the daily food energy intakes that have been recommended for people in the United States. The estimated Calorie requirements are presented by gender and age groups at three different activity levels. Growing children (not included in the table) need a larger energy intake, both to fuel their high level of activity and to provide raw material for building muscle and bone. Children are particularly susceptible to undernourishment and malnutrition. Indeed, mortality rates among infants and young children are disproportionately high in famine-stricken countries.

Your Turn 11.21 You Decide Calories by Gender and Age

Consider the information in **Table 11.8** and the *Dietary Guidelines for Americans*, which can be found on the Internet.

a. Do males and females of the same age require the same number of Calories for the same level of activity? Explain.
b. As an active male or female grows older, how does the estimated Calorie requirement change?
c. Do other countries have similar dietary guidelines? If so, how do these differ from those established in the U.S.?

Did You Know? Each heartbeat requires about 1 J (4.18 cal) of energy.

Where does all this food energy go? The first call on the Calories you consume is to keep your heart beating, your lungs pumping air, your brain active, all major organs working, and your body temperature at about 37 °C. These requirements define the **basal metabolic rate (BMR)**, the minimum amount of energy required daily to support basic body functions. This corresponds to approximately 1 Calorie per kilogram (2.2 pounds) of body mass per hour, although it varies with size and age.

To put this on a personal basis, consider a 20-year-old female weighing 55 kg (121 pounds). If her body has a minimum requirement of 1 Cal/(kg·h), her daily basal metabolic rate will be 1 Cal/(kg·h) × 55 kg × 24 h/day, or about 1,300 Cal/day. According to Table 11.8, the recommended daily energy intake for a woman of this age and weight is a maximum of 2,200 Cal if she is moderately active. This means that 59% of the energy derived from this food goes just to keep her body systems going.

$$\frac{1{,}300\ \text{Cal}}{2{,}200\ \text{Cal}} \times 100\% = 59\%$$

Table 11.8 Estimated Calorie Requirements (United States)

	ACTIVITY LEVEL		
Age (yr)	**Sedentary***	**Moderately Active†**	**Active‡**
Females			
14–18	1,800	2,000	2,400
19–30	2,000	2,000–2,200	2,400
31–50	1,800	2,000	2,200
51+	1,600	1,800	2,000–2,200
Males			
14–18	2,200	2,400–2,800	2,800–3,200
19–30	2,400	2,600–2,800	3,000
31–50	2,200	2,400–2,600	2,800–3,000
51+	2,200	2,200–2,400	2,400–2,800

**Sedentary* means a lifestyle that includes only the light physical activity associated with typical day-to-day life.

†*Moderately active* means a lifestyle that includes physical activity equivalent to walking about 1–3 miles per day at 3–4 miles per hour, in addition to the light physical activity associated with typical day-to-day life.

‡*Active* means a lifestyle that includes physical activity equivalent to walking more than 3 miles per day at 3–4 miles per hour, in addition to the light physical activity associated with typical day-to-day life.

Source: Dietary Guidelines for Americans USDA 2005

Where do the rest of her Calories go? The First Law of Thermodynamics decrees that the energy must go somewhere. If she "burns off" the extra Calories through exercise, none will be stored as added fat and glycogen. But, if the excess energy is not expended, it will accumulate in chemical form.

How hard and how long we have to work (or play) to burn dietary Calories is reported in Table 11.9. In Table 11.10, exercise is related in readily recognizable units such as hamburgers, potato chips, and cookies. Of course, by combining the information in this section with the information in earlier parts of this chapter about the types of nutrients in food, it should be clear that a healthful diet cannot be achieved simply by consuming the correct number of Calories. A 2,000-Calorie diet of only potato chips and cookies would leave a person malnourished. Proper nutrition is not simply a matter of how much, but also of what kind of food a person consumes.

Table 11.9 Energy Expenditure for Common Physical Activities*

Moderate Physical Activity	**Cal/hr**	**Vigorous Physical Activity**	**Cal/hr**
hiking	370	running (10 mph)	1,050
light gardening/yard work	245	heavy yard work (chopping wood)	440
dancing (ballroom, fast)	315	swimming (freestyle laps)	510
golf (walking, carrying clubs)	245	aerobics	480
bicycling (<10 mph)	279	bicycling (12–14 mph)	559
walking (3.5 mph)	196	jogging (5 mph)	490
weight lifting (light workout)	140	weight lifting (vigorous workout)	350
stretching	105	basketball (competitive game)	490

*Values taken from CalorieLab.com, and include both resting metabolic rate and activity expenditure for a 70-kg (154-pound) person. Calories burned per hour are higher for persons heavier than 70 kg and lower for persons who weigh less.

Table 11.10 How Much Must I Exercise if I Eat This Food?*

Food	Calories	Walking at 3.5 mph (min)	Jogging at 5 mph (min)
apple	125	27	13
beer, 8 ounces	100	21	10
chocolate chip cookie	85	11	5
hamburger	350	75	35
ice cream, 4 ounces	175	38	18
pizza, cheese, 1 slice	180	39	18
potato chips, 1 ounce	108	23	11

*Values include both resting metabolic rate and activity expenditure for a 70-kg (154-pound) person.

Your Turn 11.22 Skill Building Basketball and Calories

A 70-kg person consumes a meal consisting of two hamburgers, 3 oz of potato chips, 8 oz of ice cream, and an 8-oz beer. Calculate the number of Calories in the meal, and the number of minutes the person would have to vigorously play basketball in order to "work off" the meal.

Your Turn 11.23 You Decide Dietary Advice

Which foods should you eat less of, and which ones more? One day the experts say one thing; the next day they seem to say the opposite. With so much information available, you may be confused or feel overwhelmed.

a. Using the Internet, search for the dietary advice illustrated by "the Basic Four" (1958), "the Food Pyramid" (1991), "MyPyramid" (2005), and "MyPlate" (2011). What do all of these have in common? What are key differences among these suggestions?

b. Cite research studies that support the above dietary suggestions. What are the merits and weaknesses of MyPlate versus those of MyPyramid? Are there additional components that should be included in these plans?

11.9 | Food Safety: What Else Is in Our Food?

Another source of diet-related illness is from food allergies, which only affect certain people, and are not able to be prevented by legislation and food inspection protocols.

It is estimated that 400,000–500,000 deaths occur every year in the U.S. due to unhealthy eating and inactivity. The dietary suggestions outlined in the previous section are designed to prevent diet-related diseases such as hypertension (high blood pressure), heart disease, liver disease, cancer, stroke, diabetes, and obesity. Although much less frequent, there are more than 3,000 deaths and 150,000 cases of food poisoning in the U.S. annually due to the presence of **foodborne illnesses**—the presence of bacteria, viruses, parasites, and chemical toxins that can go undetected in our foods. Of course, this begs the question: How safe is our food supply?

Did You Know? The first food legislation was instituted in the UK with the Baking Laws of 1155. This law banned the use of sand in bread that was used to make it heavier.

To ensure that food is safe to consume, the processes of food production, processing, transportation, and handling/storage must all be carried out in such a way to prevent contamination with harmful substances. No food producer or restaurateur would purposely wish to harm their consumers; however, sometimes shortcuts in food preparation may be taken in order to increase profits or save time. Consequently, all developed countries have devised some type of food legislation, and international food laws have been developed to protect the quality of foods that are shipped between countries.

In order for food laws to be successful, there has to be some means of policing to check that the food industry is in compliance with the legislation. There are three common policing methods used to determine whether levels of contaminants exceed the limits established in the law(s). Depending on the offense, court action could be taken.

- **Food surveillance**: Random samples of food and produce are taken from farms or grocery stores, and analyzed for residues of chemicals and/or microbes. Staple foods such as bread, milk, and potatoes are analyzed most frequently because any contamination would affect large populations of consumers.
- **Total diet surveys**: Samples of consumers' meals are analyzed for contaminants. Typically, a group of consumers are recruited and asked to prepare duplicate meals for a prescribed period of time. They eat one meal, and the other is sent to a laboratory for analysis.
- **Enforcement sampling**: Samples are taken when/where there is concern that a food safety problem may exist, and this might lead to a food recall. It should be noted that food recalls are mostly due to the voluntary reporting of a food safety issue by a producer, without the need for external policing. For instance, in October 2015, Hormel Foods found metal shavings in some cans of Skippy peanut butter due to malfunctioning equipment in their packaging plant. A product recall was announced by the company after a pallet of the contaminated product was inadvertently released before their quality assurance laboratory discovered the contaminants.

Of course, legislation and monitoring are significantly more complex in our globalized society. Notwithstanding the benefits of eating locally described later, we are accustomed to consuming bananas from Costa Rica, coffee from Colombia, wine and cheese from France, and olive oil from Italy. But, when we import food from another country, should we assume the food item was properly checked by their national surveillance protocols? Not necessarily, thus requiring rigorous testing facilities at major ports of entry. The Food Safety Modernization Act of 2010 has also added requirements to increase the number of routine inspections of foreign food facilities that export to the United States, under the jurisdiction of the U.S. FDA. This is designed to help identify potential safety issues before products arrive in the United States, and determine the compliance status of foreign facilities to the FDA's safety standards.

Your Turn 11.24 You Decide Food Safety

Using the Internet, find a recent example of a food recall.

a. How was this food safety issue discovered, and how long did it take for it to be detected?
b. Who was to blame for the food recall? Were there any legal actions taken against this party?
c. Do you have concerns about the overall safety of our foods? What other methods could be used to detect food safety issues?

Your Turn 11.25 You Decide Food Additives

Although legislation and monitoring practices are quite effective in preventing widespread cases of foodborne illnesses, there are more than 1000 food additives that are currently unregulated by the FDA.

a. Using the Internet as a resource, list five of these unregulated food additives. What types of processed foods contain these ingredients, and what is the purpose for them?
b. Describe the potential health impacts of these additives. Evaluate the reliability of the claimed health implications based on the presence, or lack, of scientific evidence.

Armed with your knowledge of the ingredients contained in food items, let's now focus on the bigger picture of producing food on our planet.

11.10 | The Real Costs of Food Production

Throughout history, food has played a pivotal role in human health and well-being. Millions have starved from a lack of food, while others have died from eating too much of it. Wars have been fought over food, and countries destabilized by the lack of it. In previous chapters, we discussed how the air we breathe and the water we drink are connected to regional and global issues. We now do the same for food. For instance, producing food connects to several water-related issues, including:

- Depletion of aquifers and drying up of rivers due to irrigation
- Contamination of groundwater by insecticides and herbicides
- Increase of nutrient pollution due to fertilizer runoff

Keep these water-related issues in mind as you do the next activity.

The United Nations General Assembly designated the first World Water Day on March 22, 1993. This event focuses on a different global water-related issue each year.

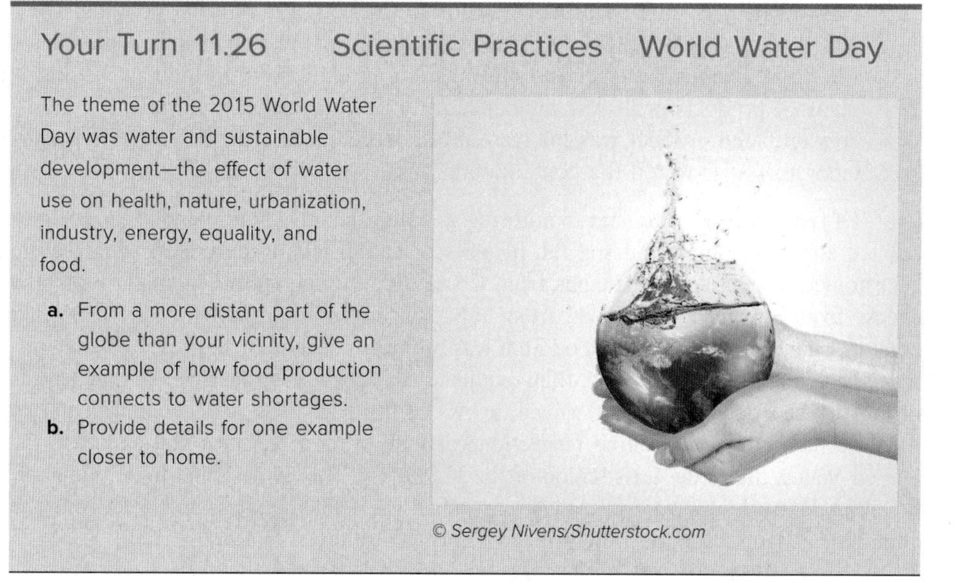

Your Turn 11.26 Scientific Practices World Water Day

The theme of the 2015 World Water Day was water and sustainable development—the effect of water use on health, nature, urbanization, industry, energy, equality, and food.

a. From a more distant part of the globe than your vicinity, give an example of how food production connects to water shortages.
b. Provide details for one example closer to home.

© Sergey Nivens/Shutterstock.com

UNESCO, or the United Nations Educational, Scientific, and Cultural Organization, was created in 1945 to help establish long-term peace between nations on the basis of humanity's moral and intellectual solidarity.

Every day, you consume water. However, our water consumption involves more than what we obtain directly from bottles, fountains, or our favorite cuisine. Water is also indirectly consumed during the *production* of food. A 2014 UNESCO report indicated that it requires about 15,000 liters of water to bring one kilogram of beef protein to the table—enough to make only nine quarter-pounder hamburgers! It's not that cattle drink this much water. Rather, this value reflects how much water is used to produce the grain that feeds the cattle. Let's now examine the connections between food and land use.

We use land both for raising crops and grazing animals. In turn, these activities connect to many issues, including:

- Loss of forest ecosystems to create land for agriculture
- Erosion of topsoil from overplanting and overgrazing
- Loss of biodiversity from the repeated planting of single crops

As you can see from the bar graph of Figure 11.24a, foods differ in the amount of land required to produce them. One reason for the differences is that grain may be needed to feed the animals, as shown in Figure 11.24b. However, as estimates, these values do not necessarily apply to all meat and dairy products. For example, in some regions, animals are grass-fed and consume little or no grain.

Estimates, such as those shown in Figure 11.24, are based on a set of assumptions. As you will see shortly, the values may be higher or lower depending on which assumptions are used. For example, this particular set of values is somewhat high

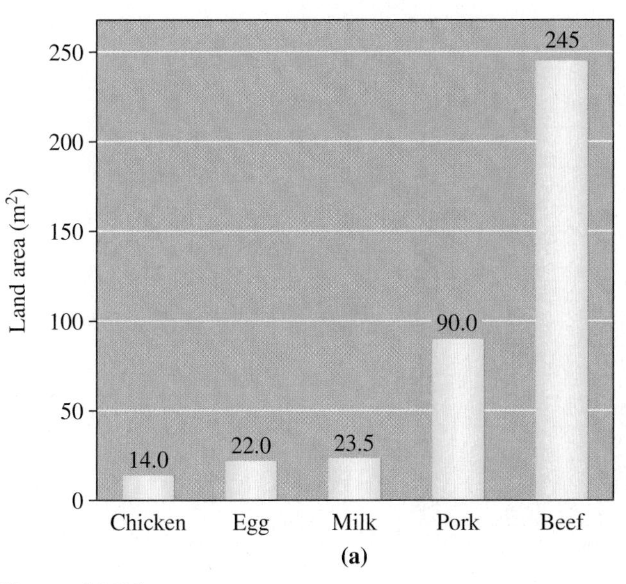

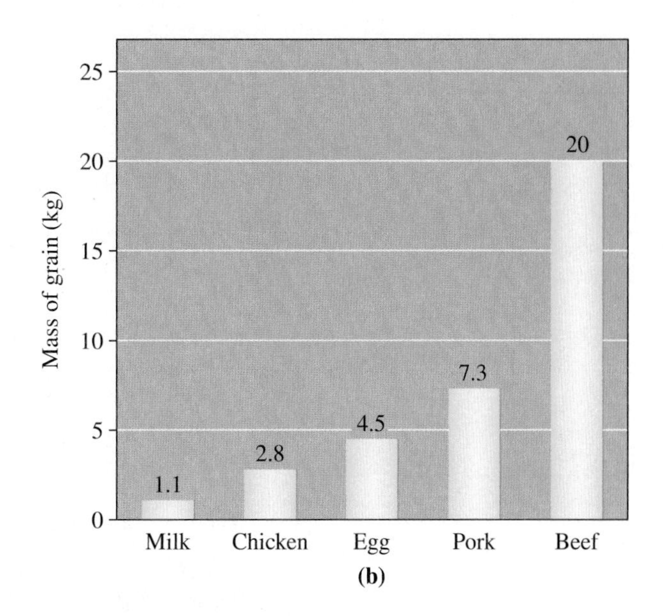

Figure 11.24

Land and grain estimates for food production. **(a)** Square meters of land required to bring 1 kg of each food to the table. **(b)** Kilograms of grain required to bring 1 kg of each food to the table.

Source: Feeding the World: A Challenge for the Twenty-First Century, *Smil, V., MIT Press: 2001.*

because the food (in kg) brought to the table includes only the edible parts of the animal, not the whole animal. No matter which assumptions are used, the trend is clear: Grain-fed livestock require more land and grain than other varieties of animals. The next activity helps you to be more discerning as you evaluate the values shown in Figure 11.24.

Your Turn 11.27 Scientific Practices Checking the Assumptions

Below are some of the factors that affect an estimate of the amount of land or feed grain that is needed to bring one kilogram of beef to the table. Examine each factor. Does it raise, lower, or have no effect on the estimate?

a. The yield in grain (corn or soy) per acre.
b. The portion of the animal's lifespan included in the estimate.
c. The breed of livestock (e.g., Angus, Hereford, Jersey, Texas Longhorn, etc.).

In general, meat consumption and production are increasing worldwide (Figures 11.25 and 11.26). Fueled in large part by the rising affluence of people in developing countries, this trend is expected to continue through 2030. However, can Earth continue to feed its people a few decades from now if we continue on this path of increased meat consumption? To answer this question, let's do some math. From Figure 11.25, we can obtain the approximate values for the projected annual meat consumption (in kg/person/year) in 2030. We then can pair each type of meat with its corresponding grain requirement from Figure 11.24b:

$$10 \text{ kg beef/person} \times 20 \text{ kg grain/kg beef} = 200 \text{ kg grain/person}$$

$$15 \text{ kg pork/person} \times 7.3 \text{ kg grain/kg pork} = 110 \text{ kg grain/person}$$

$$17 \text{ kg chicken/person} \times 2.8 \text{ kg grain/kg chicken} = 48 \text{ kg grain/person}$$

Doing the math, this gives us 358 kg of grain per person in 2030. If we assume a world population of 8 billion people (a low estimate) for 2030, the world grain production would need to be about 2.9 trillion kilograms to produce this much meat.

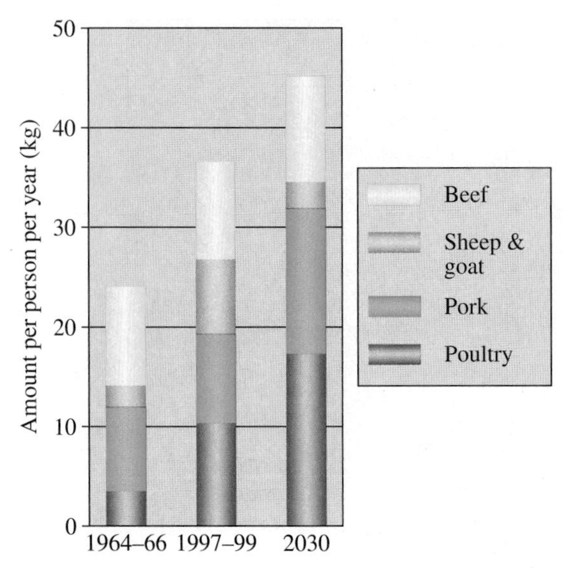

Figure 11.25

World average meat consumption.

Source: Agriculture, FAO, 2012

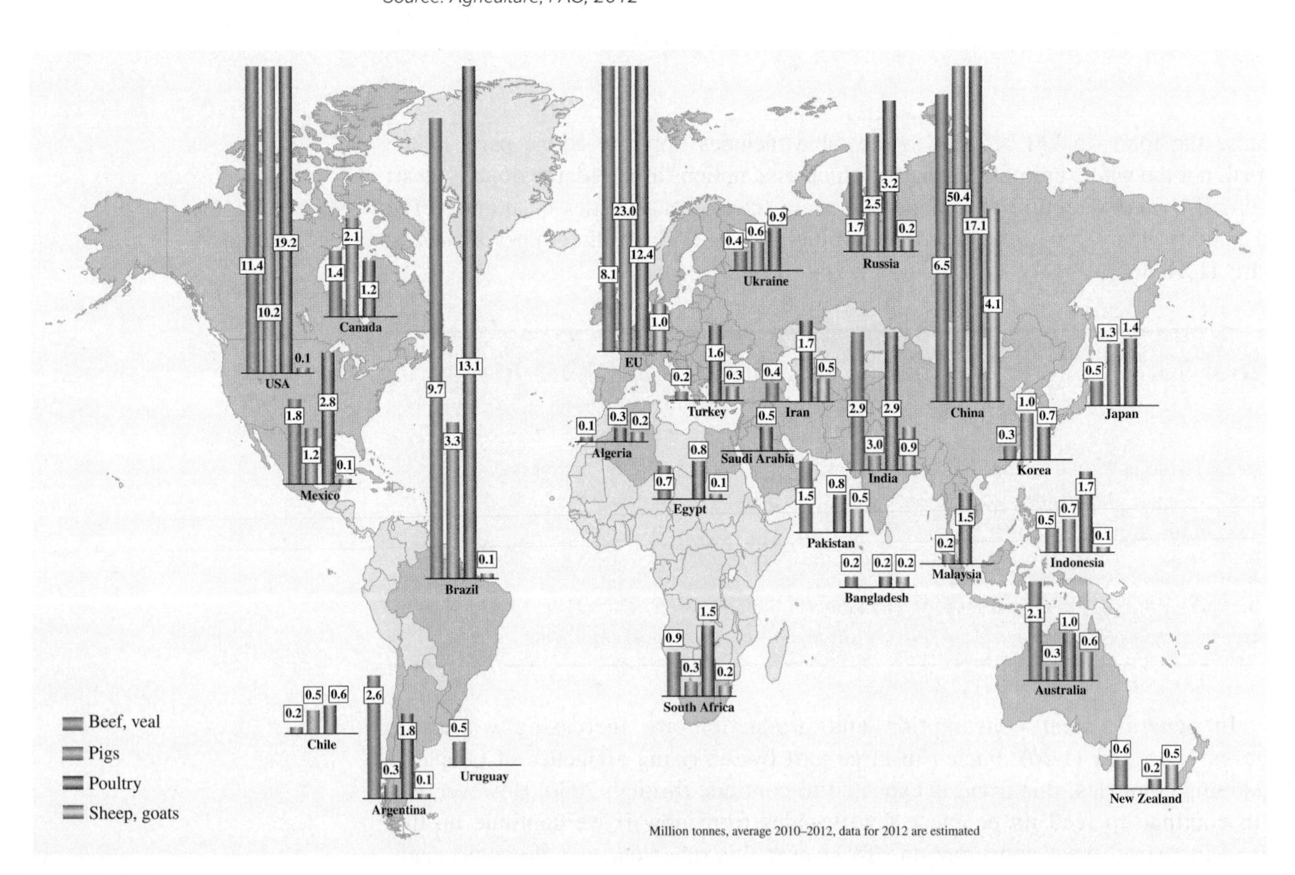

Figure 11.26

Worldwide meat production.

Source: Meat Atlas, Heinrich Boll Foundation, 2014

By way of comparison, in 2009 the world grain production *for all uses* was approximately 2 trillion kilograms (2 billion metric tons). Even if a lower set of values for estimated grain requirements were used than those supplied in Figure 11.24b, there still would not be enough grain available. Although the shortfall possibly could be made up by increasing crop yields, this is unlikely, as we will see in the final section of this chapter.

11.11 From Field to Fork I: The Carbon Footprint of Foods

In the next two sections, we will describe the connections among food production, energy use, and global climate change. We launch our discussion with two closely related topics: eating locally, and monitoring the "food miles" of grocery items.

In regard to the first, people desire to eat local for many reasons. Picking a tomato from the vine or visiting a local farmer's market truly can be a pleasure. Table 11.11 lists five other reasons people choose to eat local. In the activity that follows, you will have the opportunity to examine the validity of these claims.

"Food miles" approximate the distance that a food item travels from the location it was grown to the location where it was consumed; that is, farm to plate. They are one measure of food sustainability because they reflect energy use.

Table 11.11 Reasons to "Eat Local"

1. Eating local means more for the local economy. A dollar spent locally generates twice as much income for the local economy. When businesses are not owned locally, money leaves the community with every transaction.
2. Locally grown produce is fresher and tastes better. Produce at your local farmer's market often has been picked within 24 hours of your purchase. This freshness not only affects the taste of your food but also the nutritional value. Have you ever tried a tomato that was picked within 24 hours?
3. Locally grown fruits and vegetables have longer to ripen. Because the produce is handled less, locally grown fruit does not have to stand up to the rigors of shipping. You will get peaches so ripe that they fall apart as you eat them, and melons that were allowed to ripen on the vine until the last possible minute.
4. Buying local food keeps us in touch with the seasons. By eating with the seasons, our foods are at their peak taste and less expensive.
5. Supporting local producers supports responsible land development. When you buy locally, you give those with local open space—farms and pastures—an economic reason to resist further development.

© Steve Bower/ Shutterstock.com

Source: Adapted from "10 Reasons to Eat local Food," Jennifer Maizer, EatlocalChallenge.com

Your Turn 11.28 You Decide Eat Local

© Arina P. Habich/Shutterstock.com

Critically examine each statement in **Table 11.11**.

a. For a statement of your choice, provide a supporting example. For a second statement, provide a counterexample.
b. Because this list is not complete, suggest two other entries to add to the list.
c. Suggest an item on this list that should be removed or revised. Explain.

Table 11.12 Energy Use and Emissions for Different Modes of Transportation

	Rail	Water	Road	Air
Primary energy consumption (kJ/tonne-km)	210	160	2,400	6,900
Specific total emissions (Tg CO_2 Equivalent)				
Carbon dioxide	40	39	1,400	150
Methane	0.1	0.0	1.6	0.0
Dinitrogen monoxide	0.3	0.7	15	1.4
HFCs	2.6	0.0	58	0.0

Note: 1 metric ton (Tonne) = 1,000 kg or 2,205 lb. 1 Tg (teragram) = 1 million metric tons

Sources: Transportation Energy Data Book, September 2015, Oak Ridge National Laboratory; U.S. Transportation Sector Greenhouse Gas Emissions, 1990–2013, Environmental Protection Agency

In regard to the second topic, people question food miles; that is, how far food travels to reach their table. Depending on the food, it could travel a few yards from your garden or several thousand miles from another country. Most of our foods must be transported, which results in a varying degree of energy use and atmospheric emissions (Table 11.12). Would it help if all food were produced locally? Not necessarily. For example, it may be less energy-efficient to grow tomatoes locally in a greenhouse than to import them from a warmer climate. Transportation, as well as other energy costs, must be balanced against the food production costs.

If your primary reason for eating locally is to reduce fossil fuel consumption, you may be interested in a 2008 study from Carnegie Mellon University. The researchers reported that transportation represents only 11% of life-cycle greenhouse gas emissions, while final delivery from producer to retail contributes only 4%. So, if the majority of emissions and energy cost with food are not transportation-related, where do they originate?

Most carbon footprints use a time frame of one year.

To answer this question, examine Figure 11.27, which shows carbon footprint data for food reaching a grocery chain in Great Britain. From Chapter 4, recall that carbon footprints estimate carbon dioxide emissions in a given time frame. Determining a "carbon footprint" for a particular food requires a set of assumptions; consequently, you will see different values for the same food product.

Cows produce methane in their digestive tracts in the range of 200 pounds per animal per year.

As seen in the figure below, the majority of the carbon footprint (67%) comes from food production on the farm. For example, operating farm machinery produces carbon dioxide, and using fertilizer stimulates microbes in soils to produce nitrous oxide. Farm animals also produce methane. In addition to the production

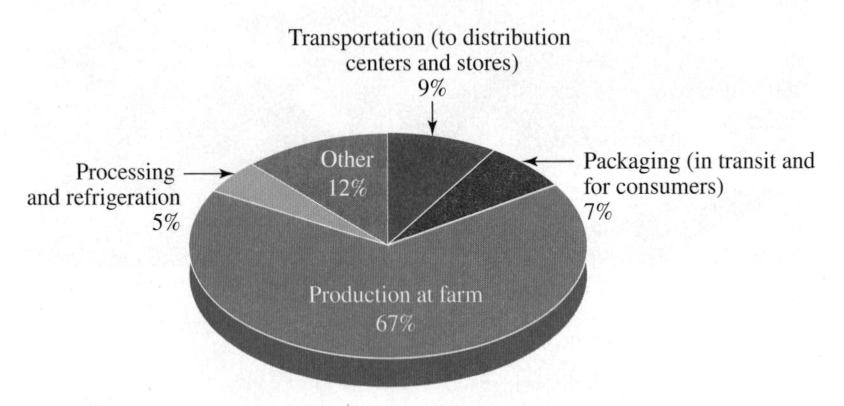

Figure 11.27

Carbon footprint of food before it reaches a grocery chain in Great Britain.

Source: Berners-Lee, M. *How Bad Are Bananas?* The Carbon Footprint of Everything, Greystone Books: Vancouver, BC, 2011.

of food, the figure points to other areas in which food contributes to the carbon footprint:

- **Transportation (9%)**: Relatively low, unless air freight
- **Packaging (7%)**: Primarily from disposal of the packaging
- **Processing and refrigeration (5%)**

Many small factors, including the leakage of refrigerant gases (all greenhouse gases), contribute as well. Although not included in Figure 11.27, food waste and packaging also increase the carbon footprint, which in some countries is up to 25% of all purchased food.

Your Turn 11.29 You Decide Local vs. Organic

Organic foods are defined as being grown without synthetic fertilizers or pesticides, and are not processed using solvents, irradiation, or synthetic food additives. Additionally, the agricultural practices involved in organic food production use methods and materials that minimize negative impacts on the environment. Considering food cost, quality, and nutritional value, as well as the energy consumption and overall environmental impacts of food production (carbon footprint, food miles, etc.), provide pros and cons for each of the following sources of food:

a. local and organic
b. non-local and organic
c. local and non-organic
d. non-local and non-organic

Some foods on your dinner table have higher carbon footprints than others. Looking at the graph in Figure 11.28, you probably notice that steak and cheese are the highest of the items included. Why might this be the case? Revisit Figure 11.24 to see that the production of beef, unless grass-fed, requires grain production. In turn, both of these are associated with greenhouse gas emissions. You also may have noticed the low values for tomatoes and carrots. However, if these items had been transported instead of grown in your backyard, these values would increase. Similarly, had the tomatoes been grown in a greenhouse, the carbon footprint value would be higher.

Before we bring this section to a close, we return to the theme mentioned at the start of this chapter: eating vegetarian. According to the Carnegie Mellon food-miles study mentioned earlier, replacing a single meat-based meal with a vegetarian option saved the equivalent of driving about 1,200 fewer miles annually. In contrast, an all-local diet *every day of the week* saved the equivalent of driving 1,000 fewer miles annually. In the next activity, you can connect these numbers to carbon dioxide emissions.

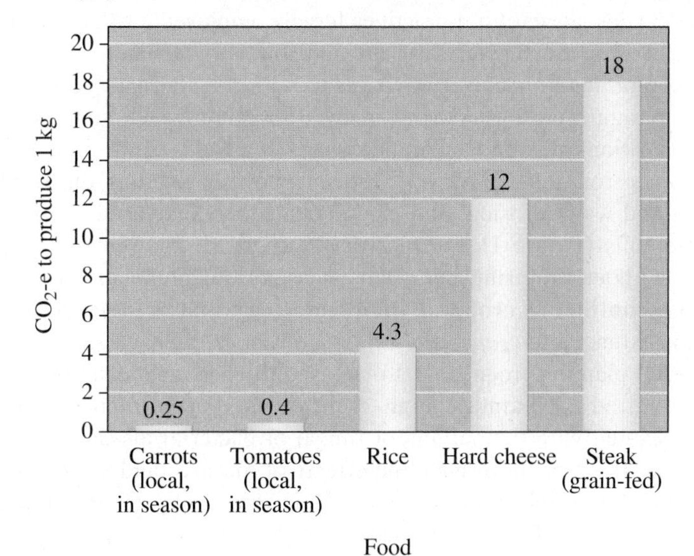

Figure 11.28

Carbon dioxide emissions (CO_2-e) from food production.

Source: Berners-Lee, M. *How Bad Are Bananas?* The Carbon Footprint of Everything, Greystone Books: Vancouver, BC, 2011.

The term carbon dioxide equivalent, CO_2-e, includes all greenhouse gases, not just CO_2. For example, in its ability to warm the atmosphere, 1 kg of methane is equivalent to 21 kg of carbon dioxide.

Your Turn 11.30 Skill Building Local Produce and CO_2 Emissions

In **Your Turn 4.15**, we noted that a clean-burning automobile engine emits about 5 pounds of carbon (~2,300 grams) in the form of CO_2 for every gallon of gasoline burned.

a. How many pounds of CO_2 is this?
b. If you saved the equivalent of 1,000 fewer miles each year by eating locally, approximately how many pounds of CO_2 is this? List any assumptions you make.

Your Turn 11.31 You Decide Lowering the Carbon Footprint of Your Foods

In this section, we have described the concept of food miles and the benefits of eating local.

a. What actions make the most sense to lower the carbon footprint of the foods we eat?
b. What are some assumptions used in determining the carbon footprint of food items?
c. List five general suggestions that are likely to help you lower the carbon footprint of the foods you consume.

In this section, we have examined ways to decrease the carbon footprint of our food. However, in the next section, you will discover that agricultural practices are also affecting where nitrogen compounds are found, and how they move through the air, water, and land of our planet.

11.12 From Field to Fork II: The Nitrogen Footprint of Foods

In previous decades, agricultural practices largely employed a tillage rotation, which was thought to improve the fertility and moisture of the land. This consisted of growing a crop on land once every two years, with the alternate years used for soil recovery by maintaining the land in a plowed (or "fallowed") state. The so-called *crop-fallow* rotations resulted in higher yields for fields sown into plowed land, versus *crop-crop* rotations. However, frequent plowing exposes the soil to greater loss by wind and water erosion, and may contribute to increased salinity (salt content) in some soils (Figure 11.29). In order to decrease land erosion, and increase farm profits without acquiring more land, agricultural practices have now largely shifted toward **continuous cropping** (minimum/zero tillage), in which the frequency of cropping is extended to *crop-crop-fallow*, or even *crop-crop-crop* rotations. However, when such planting frequency is increased, higher rates of nitrogen-based fertilizer are required to maintain high crop yields. Furthermore, there are often problems associated with the buildup of fungal or bacterial diseases, and herbicide-resistant weeds. In order to mitigate the effects of diseases and weeds, most continuous-croppers rotate the type of crop from one year to the next, such as corn-soybean-corn, etc. The overall yield is much higher for crops grown with some degree of rotation, relative to those in which the same type of crop is continually planted on the same land (Figure 11.30).

Figure 11.29
Photo of the Dust Bowl of the 1930s—a vivid reminder of the effects of unsustainable agricultural practices.

Source: Photo by George E. Marsh, NOAA, Department of Commerce

A **fertilizer** is any natural or synthetic material that is applied to soils or directly to plants in order to supply plant nutrients and promote plant growth. The elemental

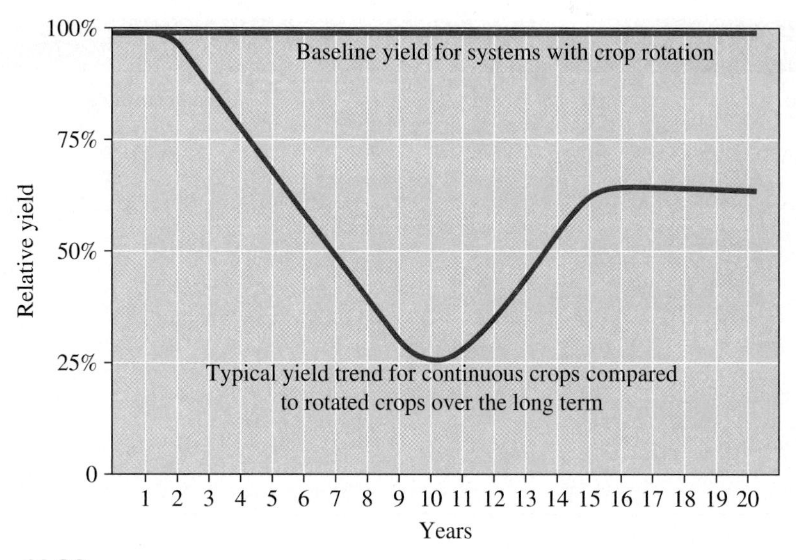

Figure 11.30

A general comparison between the yield of continuous crops and those with crop rotation. It may take about a decade to develop sufficient quantities of soil microbes that can reverse the effects of soil-borne diseases.

composition of fertilizers typically consists of nitrogen, phosphorus, potassium, sulfur, carbon, and hydrogen. Although many of these elements are readily available in soils for uptake by plants, usable forms of nitrogen are scarce; hence, we need to supply this in the form of fertilizers.

You might be wondering how nitrogen levels can be so low in soils when N_2 makes up so much of our atmosphere. Although abundant, the nitrogen molecule is *not* in a chemical form that most plants can use. As you may recall, N_2 is far less reactive than O_2.

Your Turn 11.32 Skill Building Unreactive Nitrogen

How does the bond energy of the triple bond in N_2 compare with other bond energies, such as the O–H bond in water, or O=O bond in O_2?

Hint: Refer to Chapter 5.

In order to grow, plants need access to a form of nitrogen that reacts more easily than N_2, such as the ammonium ion (NH_4), ammonia (NH_3), or the nitrate ion (NO_3^-). These and other forms are called **reactive nitrogen**, the compounds of nitrogen that cycle through the biosphere and interconvert with each other. Some reactive forms of nitrogen are listed in Table 11.13. As you might imagine, the air pollutants NO and NO_2 that were discussed in Chapter 2 are included in the list. These forms of nitrogen all occur naturally, and are present on our planet in relatively small amounts. Other forms of reactive nitrogen also exist, such as the amine functional group ($–NH_2$).

Your Turn 11.33 Skill Building Reactive Nitrogen

Select one of the compounds in **Table 11.13**. Give evidence, either in the form of an observation or a chemical equation, that this compound is reactive.

Hint: Revisit Chapter 2. Also draw on your personal knowledge.

Table 11.13 Some Reactive Forms of Nitrogen*

Name	Chemical Formula
nitrogen monoxide (nitric oxide)	NO
nitrogen dioxide	NO_2
dinitrogen monoxide (nitrous oxide)	N_2O
nitrate ion	NO_3^-
nitrite ion	NO_2^-
nitric acid	HNO_3
ammonia	NH_3
ammonium ion	NH_4^+

*These forms of nitrogen all are naturally occurring.

Figure 11.31

Nodules on the root of a soya plant that contain nitrogen-fixing bacteria.

This icon represents the bacteria responsible for the interconversion of nitrogen-containing species.

Although we categorized N_2 as generally unreactive, one reaction involving the nitrogen molecule is of utmost importance: biological nitrogen fixation. Plants such as alfalfa, beans, and peas "fix" (remove) N_2 from the atmosphere (Figure 11.31). To be more accurate, it is not the plants themselves, but rather the bacteria living on or near the roots of these plants that fix the nitrogen. As part of their metabolism, **nitrogen-fixing bacteria** remove nitrogen gas from the air and convert it to ammonia. When the ammonia dissolves in water, it produces the ammonium ion, NH_4^+ (revisit Chapter 8, Equations 8.7a and 8.7b). This polyatomic ion is one of two forms of reactive nitrogen that most plants can absorb. Recall from Chapter 8 that compounds containing the ammonium ion tend to be water-soluble. Here is the pathway:

$$N_2 \xrightarrow[\text{nitrogen fixation}]{} NH_3 \xrightarrow{H_2O} NH_4^+$$

The other form of reactive nitrogen that plants can absorb is the nitrate ion, NO_3^-. Again, compounds containing the nitrate ion are always soluble in water. **Nitrification** is the process of converting ammonia in the soil to the nitrate ion. Bacteria are involved in this two-step process:

$$NH_4^+ \xrightarrow[\text{bacteria in the soil}]{} NO_2^- \xrightarrow[\text{bacteria in the soil}]{} NO_3^-$$

Finally, to come full-circle, bacteria help with **denitrification**, the process of converting nitrate to nitrogen gas. In so doing, these bacteria harness the energy released when the stable N_2 molecule forms. Recall from Chapter 5 the large amount of energy released in forming the triple bond found in the nitrogen molecule, a very stable molecule indeed!

Depending on the soil conditions, the pathway may occur in steps that include NO and N_2O. Thus, these reactive forms of nitrogen can also be converted to N_2 and released from the soil:

$$NO_3^- \xrightarrow[\text{bacteria in the soil}]{} NO \xrightarrow[\text{bacteria in the soil}]{} N_2O \xrightarrow[\text{bacteria in the soil}]{} N_2$$

All of these pathways are part of the **nitrogen cycle**, a set of chemical pathways whereby nitrogen moves through the biosphere. Figure 11.32 assembles the separate pathways shown above into a simplified version of the nitrogen cycle. In this cycle, all species are forms of reactive nitrogen except for N_2.

Your Turn 11.34 Scientific Practices The Nitrogen Cycle

a. Ammonia (NH_3) is applied to soil in the form of anhydrous ammonia; that is, ammonia without water. Ammonia is very soluble in water. Explain why.
Hint: Revisit Chapter 8.

b. When ammonia is mixed with water, the resulting solution is basic. Explain why, including the ammonium ion (NH_4^+) as part of your answer.
Hint: Revisit Chapter 8.

c. According to **Figure 11.32**, which chemical species of nitrogen is used (assimilated) by plants?

d. Microbes in the soil interconvert the chemical species of nitrogen. Before the ammonium ion can be assimilated by plants, another chemical species is formed. What is it?

Remember that reactive forms of nitrogen are needed for plant growth. Because bacteria in the soil cannot supply ammonia, ammonium ions, or nitrate ions in the amounts needed for optimal plant growth, farmers use fertilizers. A few centuries ago, fertilizers were obtained by mining deposits of saltpeter (ammonium nitrate) from the deserts of Chile or by collecting guano, a nitrogen-rich deposit from bird and bat droppings in Peru. Neither source, however, was sufficient to meet the demand of the growing world population. An additional drain on the supply of nitrates was their use in the production of gunpowder and other explosives such as TNT. By the early 1900s, the search was on for a way to synthesize reactive nitrogen compounds from the abundant N_2 in the air.

How are fertilizers obtained in the large quantities needed for present-day agriculture? The answer lies in a second important reaction of N_2, one that literally captures it out of the air to synthesize ammonia:

$$N_2(g) + 3\,H_2(g) \longrightarrow 2\,NH_3(g) \qquad [11.9]$$

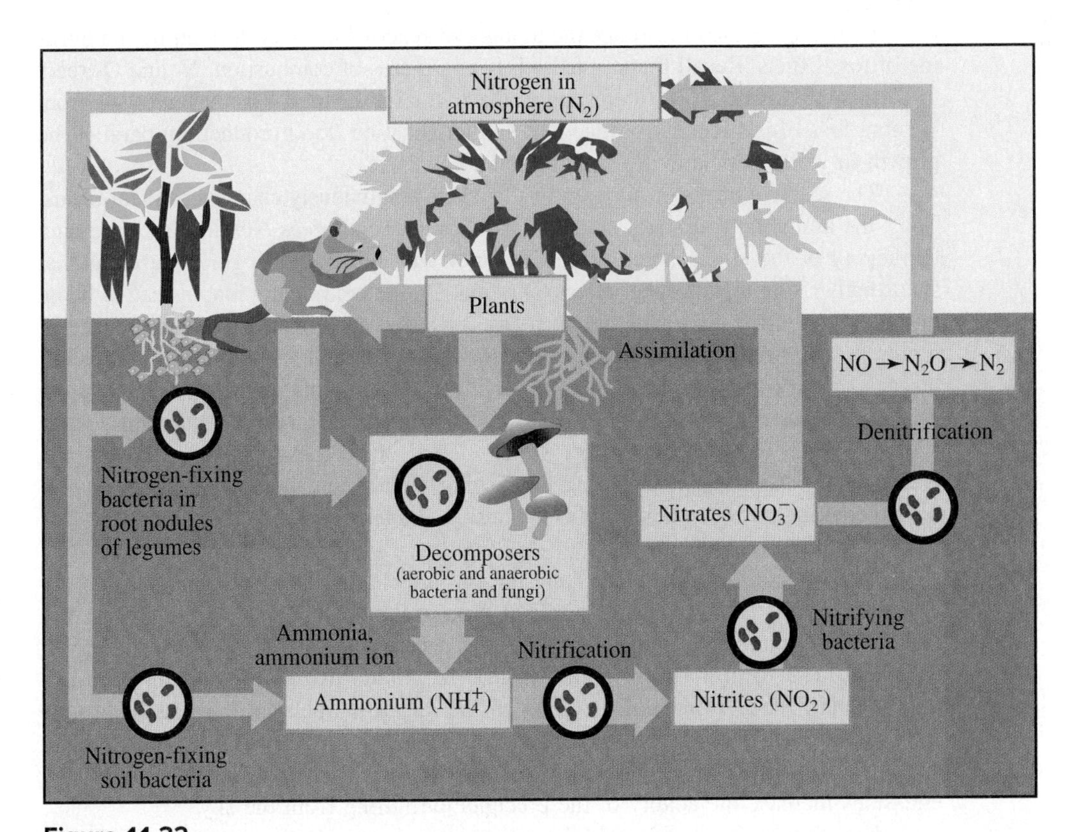

Figure 11.32
The nitrogen cycle (simplified), a set of chemical pathways whereby nitrogen moves through the biosphere.

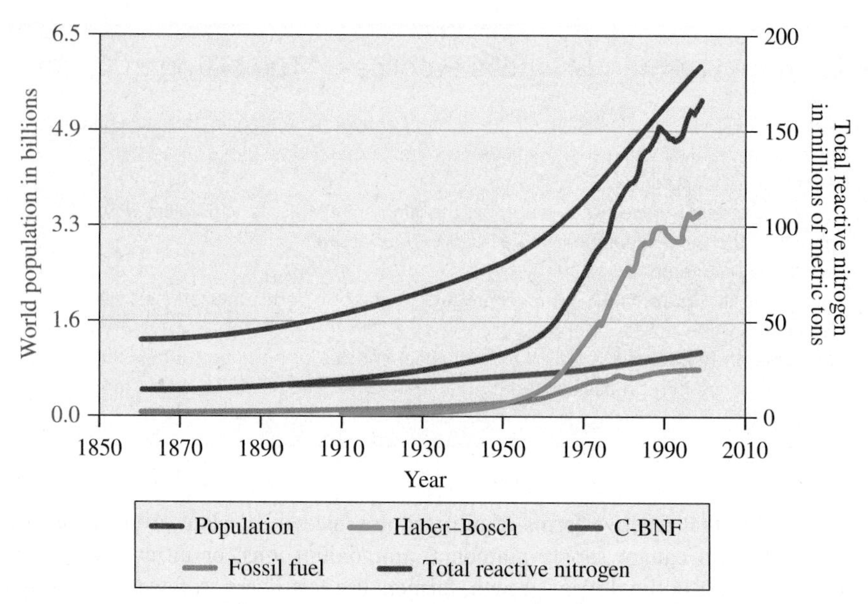

Figure 11.33

Global changes in reactive nitrogen produced by various sources (million metric tons, scale on the right). The red line is the world's population (billions, scale on the left).

Note: C-BNF is the reactive nitrogen created from the cultivation of legumes, rice, and sugarcane.

Source: BioScience by American Institute of Biological Sciences, April ***2003****, Vol. 53, No. 4, p. 342. DOI: 10.1525/bio:201262.4.6. Copyright © 2003 by University of California Press–Journals. Reproduced with permission of University of California Press–Journals via Copyright Clearance Center.*

In 1918, Fritz Haber (1868–1934) received the Nobel Prize in chemistry for synthesizing NH_3 from N_2 and H_2. In 1931, Carl Bosch (1874–1940) received the Nobel Prize for using this synthesis commercially.

Equation 11.9 is known as the **Haber-Bosch process**. This procedure allows for the economical production of ammonia, which in turn has enabled the large-scale production of fertilizers and nitrogen-based explosives (Figure 11.33). As a fertilizer, ammonia can be directly applied to the soil, or can be applied as ammonium nitrate or ammonium phosphate.

Figure 11.33 also illustrates the amount of reactive nitrogen formed by the burning of fossil fuels. Recall that at the high temperatures of combustion, N_2 and O_2 react to form nitric oxide, NO. Interestingly, the increases in total reactive nitrogen from burning fossil fuels (energy production) and fertilization (food production) parallel the growth in world population (people production).

The reactive forms of nitrogen in this cycle continuously change chemical forms. Thus, the ammonia that starts out as a fertilizer may end up as NO, in turn increasing the acidity of the atmosphere. Or, the NO may end up as N_2O, a greenhouse gas that is currently rising in atmospheric concentration. Or, the ammonium ion, instead of being tightly bound to the soil, may end up being converted and leached out as the nitrite or nitrate ion, in turn contaminating a water supply. For example, algal blooms in the Gulf of Mexico are caused by the runoff of fertilizers from farms in the watershed of the Mississippi River in states from as far away as Montana, Minnesota, and Pennsylvania (Figure 11.34). These blooms affect the livelihood of all those who use the common waters of the Mississippi, including the fishing industry in the Gulf. Although unintended consequences can be minimized by using less fertilizer and timing its application more carefully, we have not yet been successful in avoiding such harm.

Now we can begin to understand the full effects that NO emissions have on our environment. First, the oxides of nitrogen form ground-level ozone in the presence of sunlight, contributing to photochemical smog, as we saw in Chapter 2. Second, NO_x emissions are a form of reactive nitrogen, just like the fertilizers used for food production. NO is formed from unreactive N_2 in the air when fuels are burned. The more fuels are burned, the more N_2 is changed into a reactive form. Both NO_x and fertilizer use are affecting the balances within the nitrogen cycle on our planet. Finally, NO_x emissions increase the acidity of the precipitation falling from the sky.

Now that we have addressed the carbon and nitrogen footprints of food products, we are still faced with the problem of ensuring enough food is available, and safe to

Figure 11.34

Brown, nutrient-rich water from the Mississippi River meets the Gulf of Mexico, creating a "dead zone." The zone is approximately 6,000–7,000 square miles, varying in size seasonally.

© Nancy Rabalais, Louisiana Universities Marine Consortium

consume, for an ever-growing population. We address this critical issue in the final section of the chapter.

Your Turn 11.35 Scientific Practices Putting It All Together

Using your knowledge of the carbon cycle (**Chapter 4**) and the nitrogen cycle (**Figure 11.32**), perform a life-cycle analysis (LCA) on one of your favorite foods. Ensure that you deal with the energies and environmental emissions involved during the fabrication, transportation, and end-of-use for the food packaging, in addition to the lifetime of the food item itself. Compare your results to the LCA of similar food products that are available on the Internet.

11.13 | Food Security: Feeding a Hungry World

Our ancient ancestors were hunter–gatherers, spending most of each day searching for their next meal. About 10,000 years ago, humans learned to grow crops and domesticate animals, thus launching the agricultural revolution. At that time, the population of the entire Earth was estimated to be 4 million people, or roughly that of Los Angeles today.

Over the next 8,000 years, the global population grew to 170 million people, or about half the size of the current U.S. population. By 1000 AD, the population had risen to 310 million, and 200 years ago the population finally topped 1 billion people. Today, the world population is over 7 billion and within the next 40 years will likely rise to 9 billion (Figure 11.35). Clearly, we have many mouths to feed.

The concept of *food security*, according to the World Health Organization, deals with three primary issues:

- **Food availability**: having sufficient quantities of food available on a consistent basis.
- **Food access**: having sufficient resources to obtain appropriate foods for a nutritious diet.
- **Food use**: appropriate use of food based on knowledge of basic nutrition and care, as well as adequate water and sanitation.

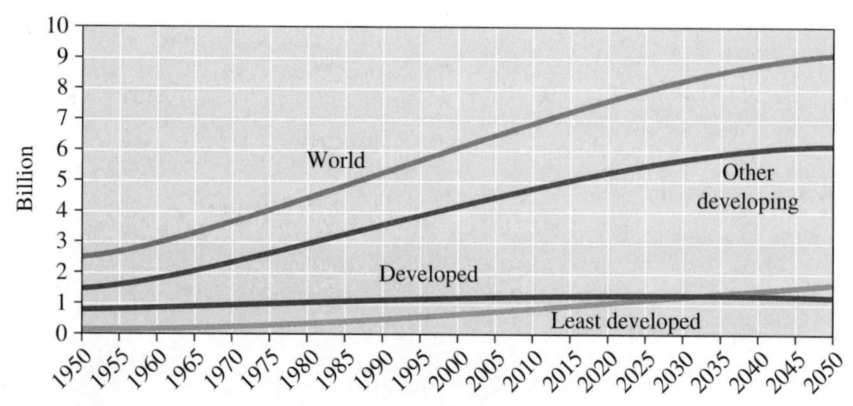

Figure 11.35

World population growth, 1950–2050 (projected).

Source: Food and Agriculture Organization of the United Nations

The term *undernourishment* refers to the number of people who are unable to consume enough food to conduct an active and healthy lifestyle. Although < 5% of the population is undernourished in developed regions such as North and South America, Europe, Australia, and New Zealand, developing regions such as Africa, Asia, Latin America, and the Caribbean have a current undernourishment rate of 12.9%.

Thomas Malthus (1766–1834), and more recently, entomologist Paul Ehrlich (b. 1932) predicted that the human population of Earth would outstrip food production. During the first 100 years after Malthus published his essay, the population of Earth grew by 60% to 1.6 billion. During the next 100 years (the 20th century), the population exploded with an additional 4.4 billion people. The Food and Agriculture Organization (FAO) of the United Nations estimated that we have managed to increase the world food supply to currently meet the needs of over 89% of the people on Earth. This represents a significant improvement from 25 years ago, when the worldwide undernourishment level was almost 20%, and much higher in certain parts of Africa, Asia, and the Caribbean (Figure 11.36).

Given that the population continues to grow, so must food production. This is particularly important considering that global food consumption per person is projected to steadily increase into the foreseeable future (Figure 11.37). Two methods have largely been responsible for food increases of the past: planting crops on more land, and increasing crop yields. However, can this increase to our food supply continue? Neither method has much room for growth. Almost all of the world's biologically productive land is now in use, and arable land is only likely to be increased by 5% from current levels as of 2050, according to FAO estimates (Figure 11.38). Furthermore, in many areas, cropland is actually shrinking due to factors such as desertification, soil nutrient depletion, erosion, and urban development.

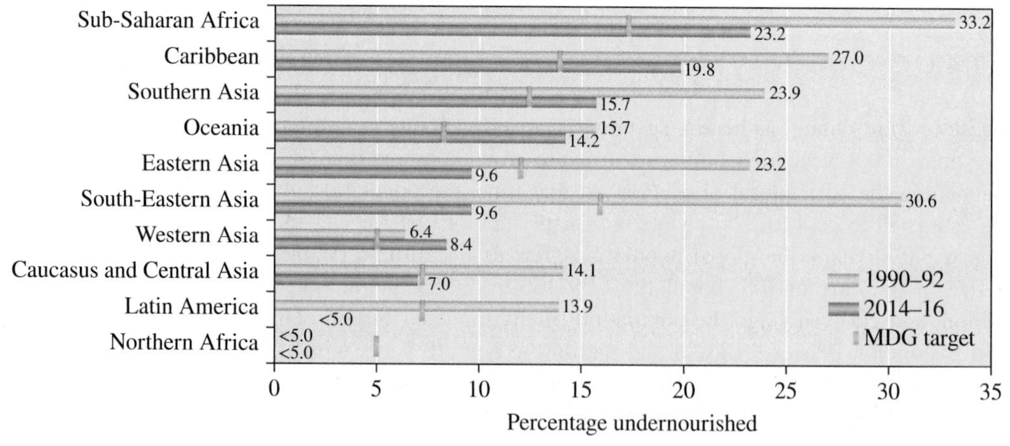

Figure 11.36

Comparison of undernourishment trends across various developing regions, from 1990–92 and 2014–16. The MDG target corresponds to the Millennium Development Goal of "...reducing the number of undernourished people to half their present level no later than 2015...," established at the World Food Summit at Rome, Italy, in 1996.

Note: The Oceania region represents regions such as Melanesia, Micronesia, and Polynesia.

Source: The State of Food Insecurity in the World, *Food and Agriculture Organization of the United Nations, 2015*

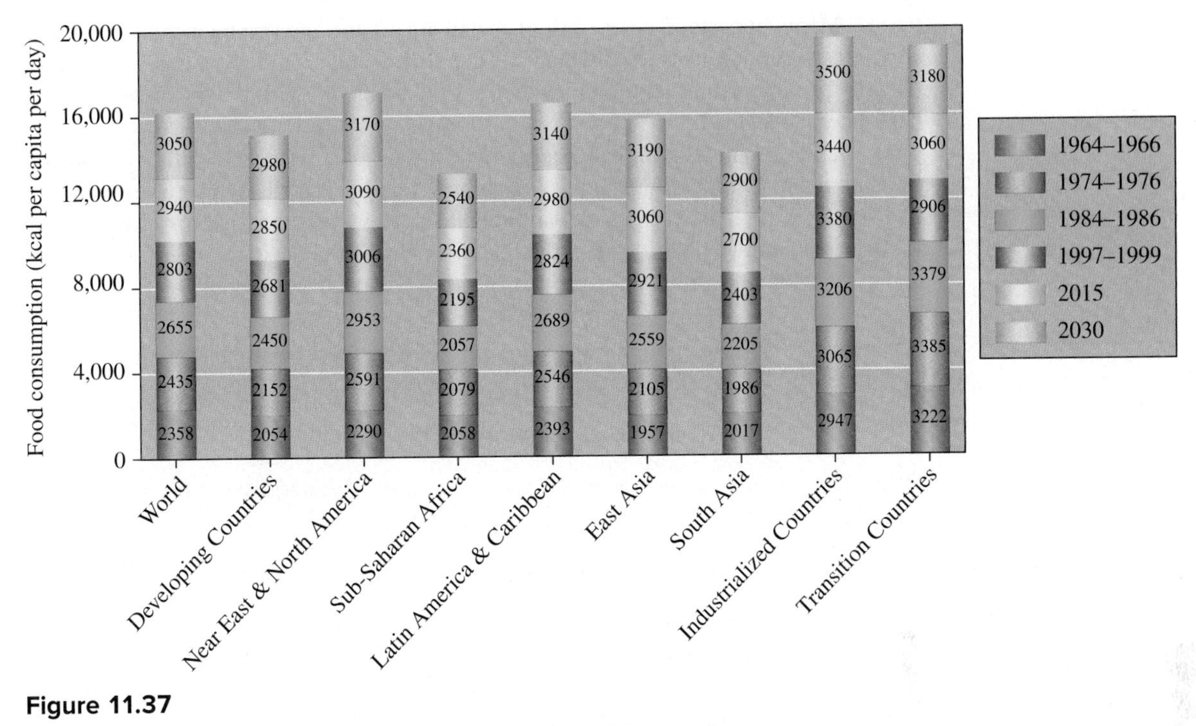

Figure 11.37

Global and regional per capita food consumption, from 1964–2030 (projected).

Source: Food and Agriculture Organization of the United Nations, 2015

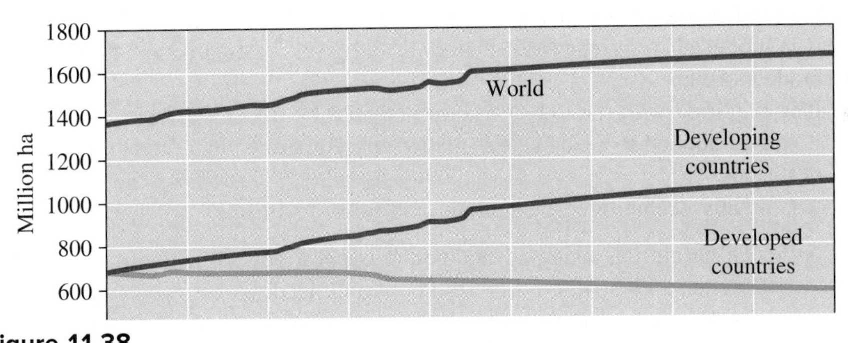

Figure 11.38

Trends in arable land use, from 1961–2050 (projected).

Source: Food and Agriculture Organization of the United Nations, 2015

The 1940s saw the beginning of the *Green Revolution*. In the decades that followed, agricultural productivity per acre of corn, rice, and wheat more than doubled. Many factors were responsible, including the use of fertilizers and pesticides, irrigation, mechanization, double-cropping, and most importantly, the advent of high-yielding crop varieties. Billions of people across the world benefited.

In spite of its successes, the Green Revolution also resulted in economic, environmental, and societal costs. As discussed in previous sections, producing crops requires water and energy, and results in the emission of greenhouse gases. Furthermore, modern agricultural practices require the use of supplemental nitrogen fertilizers such as ammonia, urea, or nitrates, all of which are forms of "reactive nitrogen."

The use of synthetic pesticides, though integral in raising crop yields, represents yet another environmental cost of the Green Revolution. Like nitrogen fertilizers, pesticides impact our health and the health of the planet. The use of insecticides saves millions of pounds of crops from being devoured by pests in the field, but at the same time these chemicals can kill beneficial insects along with the intended target. Many persistent pesticides also end up in the environment as a chemical cocktail, the consequences of which we are still deciphering. More environmentally friendly pesticides that target only specific organisms or elicit a plant's own defenses have been developed.

Methyl bromide is an extremely effective pesticide used in sterilizing soils and as a fumigant. It kills a wide variety of insects, and is particularly useful in preparing soil for crops such as strawberries and tomatoes. However, given its toxicity and ability to deplete stratospheric ozone, the use of methyl bromide was phased out in 2005 under the Montreal Protocol, except for some specialized uses.

As of 2016, allowable specialized uses ("critical exceptions") of methyl bromide include its use in California for growing strawberries, and the prevention of certain insect infestations in dry cured pork products, both for which no good alternative exists.

Your Turn 11.36 Skill Building Methyl Bromide and the Ozone Hole

Draw the structural formula for methyl bromide. Why would you expect it to deplete the ozone layer?

Hint: Revisit Chapter 3.

© *PhotoAlto/PunchStock RF*

The search for pesticides that are less harmful to our health and to the health of the planet presents a challenge to green chemists. Harpin, a naturally occurring protein, is a replacement for some uses of methyl bromide. EDEN Bioscience Corporation won a Presidential Green Chemistry Challenge Award in 2001 for its discovery that applying harpin to the stems and leaves of plants triggered a plant's natural defense mechanisms to diseases caused by bacteria, fungi, and nematodes. Some advantages of harpin include the following:

- It does not directly kill, so the pest is unlikely to acquire resistance to it.
- It is produced from the fermentation of a genetically engineered laboratory strain of *E. coli* that is benign.
- It is produced from renewable materials and gives waste products that are biodegradable.
- It is classified with the lowest hazard potential by the U.S. EPA.
- It can be applied to fields using smaller concentrations than many other pesticides.
- It is rapidly decomposed by sunlight and microorganisms.

Although harpin has many advantages, it is not a perfect solution. Some plants respond to it better than others. Furthermore, harpin needs to be reapplied at several-week intervals. Because it is relatively new, over time some unintended consequences may be discovered.

Your Turn 11.37 You Decide Food (In)Security?

Even with sustainable agricultural practices, there are many other factors that could influence the availability of food for the growing population of Earth. List five factors that are unpredictable, but could cause the price of food to rise, causing widespread food insecurity.

Before we end our discussion of food production, let's briefly revisit the topic of biofuels from Chapter 5. The data presented in Your Turn 5.28 showed that ethanol production increased sharply in the United States from 2000–2010. Furthermore, this ethanol was produced almost exclusively from feed corn. In recent years (2007–2011), according to the USDA, corn production has reached 12 billion to 13 billion bushels. Assuming 56 pounds of shelled corn per bushel, this translates to about 700 billion pounds or 350 million tons of corn annually. From the following activity, you can see the urgency of developing processes that produce ethanol from nonedible biomass (cellulose).

Your Turn 11.38 You Decide Food or Fuel?

a. Using the data just provided, together with the values in **Your Turn 5.28**, what percentage of the U.S. corn crop has been used in recent years to produce ethanol? Assume that 100 gallons of ethanol can be produced from 1 ton (2000 lb) of corn.
b. In 2007, the U.S. Energy Independence and Security Act set a mandate of producing 36 billion gallons of biofuels by 2022. If this goal were met by producing ethanol solely from corn, how many million tons of corn would be required?

We end this final section by returning to the question raised earlier of whether eating less meat translates to more available food. Indeed, eating less meat does have benefits in terms of energy use, land use, and water use. Eating the saturated fats of beef in moderation (together with eating more fruits, vegetables, and whole grains) is also part of a recommended diet. But those who study the issues have asserted that cutting meat consumption represents only a small contribution toward improving global food security. Even so, the small contributions of many people add up to larger benefits. What actions should each of us take? This final activity offers personal advice in simple terms.

Your Turn 11.39 You Decide Food Choices Revisited

In the opening activity of the chapter, you created a list of your food choices. Review your ranking of those foods you believed were highest in promoting your health, as well as the health of the planet.

a. Based on the topics discussed in this chapter, how would you re-rank your food choices?
b. What changes (if any) are you planning to make to your diet, based on your own health, and the health of the planet?
c. Use the Internet to find out how much food is lost (during production or processing) or wasted (discarded by retail markets and consumers) each year. List some practices that could reduce food waste.

Conclusions

Even though our individual tastes vary, our biological needs are much the same. We need carbohydrates and fats as our energy sources; fats for cell membranes, synthesis, and lubrication; proteins to build muscle and create the enzymes that catalyze the intricate chemistry of life; and vitamins and minerals to help make that chemistry happen.

What and how much we eat affect not only our own health but also the health of the planet. In this chapter, we saw some of the human health and environmental consequences of our food choices. Some foods, especially beef, disproportionately use water, grain, fuel, and land to produce. Some crops, including corn, can stress not only the land of the farmer but also the ecosystems many miles downstream.

Meeting the dietary needs of all on our planet is one of the greatest challenges of our time. A little knowledge of chemistry allows us to both raise and answer good questions. But chemical knowledge alone cannot lead us to a more peaceful, more prosperous, and healthier world. Individual and community choices, which in turn are determined by wisdom from economic, social, religious, and political communities, also will help lead the way forward. In the next chapter, we will build on your knowledge of food chemistry and nutrition, as we delve into the role of chemistry in the treatment of illnesses.

Learning Outcomes

The numbers in parentheses indicate the sections within the chapter where these outcomes were discussed.

Having studied this chapter, you should now be able to:

- define metabolism (11.1)
- identify the macromolecules present in food (proteins, carbohydrates, lipids) (11.1–11.7)
- define lipids (11.2)
- illustrate the general structure of a lipid (11.2)
- contrast macro- and micronutrients (11.2)
- distinguish between saturated and unsaturated fats (11.3)

- discriminate between *trans* and *cis* fats (11.3)
- describe how we get energy from lipids (11.3)
- summarize the process of hydrogenation and analyze its effect on food (11.3)
- define carbohydrates (11.4)
- illustrate the general structure of a carbohydrate (11.4)
- describe how we get energy from carbohydrates (11.4)
- describe the polymerization process for carbohydrates (11.4)
- compare and contrast the chemical properties of different sweeteners (11.5)
- define amino acids (11.6)
- define proteins (11.6)
- summarize how amino acids connect to form proteins (11.6)
- explain why a varied diet is important for protein nutrition (11.6)
- describe using diagrams why some people should not consume artificial sweeteners containing aspartame (11.6)
- distinguish between vitamins and minerals (11.7)
- relate the function of vitamins to their water- or fat-soluble characteristics (11.7)
- characterize carbohydrates, lipids, and proteins in terms of energy content (11.8)
- calculate the calories required to sustain life (11.8)
- evaluate the energy content and nutritional value of foods based on information in food labels (11.8)
- describe the worldwide protocols that are in place to maintain high levels of food safety (11.9)
- map the water use for growing fruits and vegetables or raising livestock (11.10)
- compare the food supply to the amount of food needed on Earth (11.10)
- evaluate "organic foods" compared to non-organic varieties (11.11)
- analyze the effects of pesticides and herbicides on nutrition and the environment (11.12)
- describe the role of fertilizers in soil (11.12)
- illustrate and explain the nitrogen cycle (11.12)
- track the life cycle of fertilizer (11.12)
- determine how the current agricultural practices affect our oceans and drinking water (11.12)
- outline a strategy to combat food shortage (11.13)
- describe the changing use of pesticides and herbicides in agricultural practices (11.13)

Questions

Emphasizing Essentials

1. Eating properly involves more than filling your stomach. Explain the difference between malnutrition and undernourishment.
2. What is a processed food? Give five examples of processed foods, including ones that you eat.
3. Macronutrients provide a source of energy and raw materials for your body.
   a. Name the three different types of macronutrients.
   b. How do macronutrients differ in energy content? **Hint:** See Table 11.7.
4. Although water is not considered a macronutrient, it clearly is essential to maintaining health. Name three roles that water plays in our bodies. **Hint:** Refer back to Chapter 5.
5. Consider this pie chart.

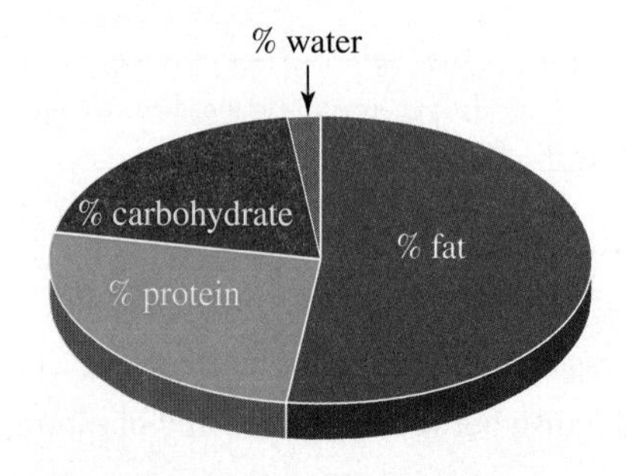

Based on the relative percentages of protein, carbohydrate, water, and fat, is this pie chart more likely to represent steak, peanut butter, or a chocolate chip cookie? Explain your choice.

6. Of the foods listed in Table 11.1,
   a. identify the best sources of carbohydrates and arrange them in decreasing order.
   b. identify the best sources of protein and arrange them in decreasing order.
   c. identify two foods that should be avoided if you are controlling your intake of fat.
7. At a local restaurant, an 18-ounce steak is the manager's special. Use Table 11.1 to calculate the ounces of protein, fat, and water in a portion of this size.
8. Although fats and fatty acids are related, they differ in terms of the size of their molecules, their functional groups, and their role in your diet. Elaborate on each of these differences.
9. Here is the condensed structural formula for lactic acid: $CH_3CH(OH)COOH$.
   a. Draw a structural formula for lactic acid that shows all the bonds and atoms.
   b. If considered as a fatty acid, would lactic acid be saturated or unsaturated?
   c. Is lactic acid a fatty acid? Explain.

10. Both unsaturated fats and saturated fats are triglycerides. Explain how they differ in terms of their observable properties, chemical structures, and role in your diet.

11. Using Figure 11.7, identify the fat or oil that contains the highest number of grams per tablespoon of:

   a. polyunsaturated fat.
   b. total unsaturated fat.
   c. monounsaturated fat.
   d. saturated fat.

12. Compare and contrast a *trans* fat to natural unsaturated and saturated fats in terms of:

   a. chemical structure.
   b. physical properties.
   c. effects on your health.

13. Name foods in which you would be likely to find these carbohydrates.

   a. lactose
   b. sucrose
   c. fructose
   d. starch

14. Explain each term and give an example.

   a. monosaccharide
   b. disaccharide
   c. polysaccharide

15. Starch and cellulose are both polysaccharides. How are these two compounds similar in terms of their chemical structures? How are they different in terms of our ability to digest them?

16. Fructose, $C_6H_{12}O_6$, is an example of a carbohydrate.

   a. Rewrite the chemical formula of fructose to show that a *carbohydrate* can be thought of as "carbon plus water."
   b. Draw a structural formula for any isomer of fructose.
   c. Do you expect different isomers of fructose to have the same sweetness? Explain.

17. Fructose and glucose both have the chemical formula of $C_6H_{12}O_6$. How do their structural formulas differ?

18. How many grams of sucralose is required to match the sweetness of 1 g of sucrose?
Hint: See Tables 11.4 and 11.5.

19. Chemical names, especially for organic compounds, can give information about the structure of the molecules that these compounds contain. What does the term *amino acid* suggest about its molecular structure?

20. Proteins are polymers, sometimes referred to as polyamides. Similarly, nylons also are polymers and polyamides.

   a. What is the amide functional group?
   b. Compare and contrast proteins and nylon in terms of
      i. the functional groups present on the monomer(s).
      ii. the variety of different proteins as compared to that of different nylons.

21. Analogous to Equation 11.4a, show how glycine and phenylalanine react to form a dipeptide.

22. Some amino acids are called "essential amino acids." Explain why.

23. Why are people with phenylketonuria able to drink beverages sweetened with sucralose but should avoid those sweetened with aspartame?

24. Explain the nutritional significance of the elements shaded on this periodic table.

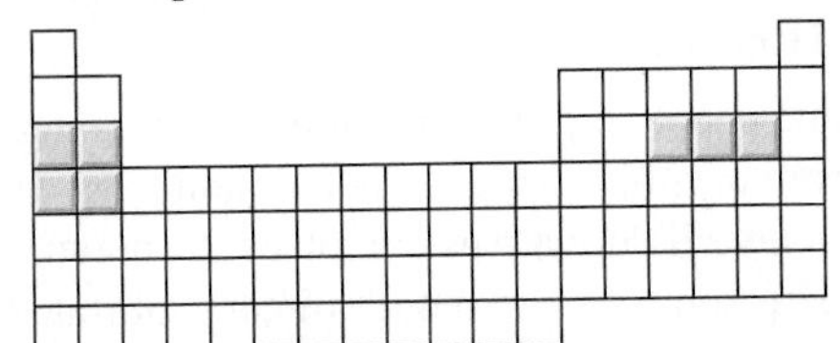

25. In recent decades, which two methods primarily have been used to increase food production?

26. One theme in this chapter is that what you eat affects not only your health but also the health of the planet. Provide two examples that illustrate this theme.

27. Suggest at least two connections between food production and water quality. Do the same for water use.

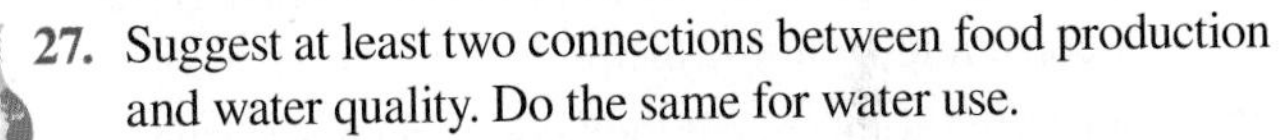

28. In general, it requires more water and land to produce beef, chicken, and pork than it does to produce grains such as corn and soybeans. Give two reasons why.

29. Plants require nitrogen in a "reactive" form.

   a. Explain the meaning of the term *reactive nitrogen.*
   b. Give an example of a form of nitrogen that is nonreactive.
   c. Describe one natural and one unnatural source of reactive nitrogen.

Concentrating on Concepts

30. The old adage "you are what you eat" is everywhere, including in this chapter. List your food consumption over the past 24 hours and compare it to Figure 11.2. Does your diet contradict or confirm the adage?

31. Explain to a friend why it is impossible to go on a highly advertised "all organic, chemical-free" diet.

32. Indicate whether each statement is true, true only in some circumstances, or false. Explain your reasoning.

   a. Plant oils are lower in saturated fat than animal fats.
   b. Given time and exposure to the air, fats and oils become rancid.
   c. Fats are not necessary in our diet because our bodies can manufacture them from other substances we eat.

33. The label of a popular brand of soft margarine lists "partially hydrogenated soybean oil" as an ingredient. Explain the term *partially hydrogenated.* Why must the label report partially hydrogenated soybean oil rather than simply soybean oil?

34. A well-known medical clinic used the phrase "double trouble" to refer to *trans* fatty acids. Explain the logic behind this.

35. Explain why the process of interesterification is a useful alternative to hydrogenation.

36. Although interesterification is a useful alternative to hydrogenation, the process had several drawbacks until a green chemistry solution was developed.
    a. What were the drawbacks?
    b. What solution was developed?
    c. Which of the green chemistry key ideas did this solution address?
    Hint: Revisit Chapter 2.

37. Some people prefer to use nondairy creamer rather than cream or milk. Some but not all nondairy creamers use coconut oil derivatives to replace the butterfat in cream. Is a person trying to reduce dietary saturated fats wise to use nondairy creamers such as these? Explain.

38. Low-Calorie and zero-Calorie substitutes have been developed for fat ("fake fats" such as Olean) and sugar (sucralose and aspartame). Why aren't comparable zero-Calorie substitutes for protein being developed? Even so, some protein substitutes exist. Which groups of people might choose them?

39. Your friend wants to cut food costs and has learned that peanut butter is a good protein source. What additional information should your friend consider before eating peanut butter as a major dietary protein source?
    Hint: See Table 11.1.

40. Here is the composition of a fast-food meal. Do calculations to determine whether the meal meets the guideline that 8–10% of total Calories should come from saturated fats.

	Cheeseburger	French Fries	Shake
Calories	330	540	360
Calories from fat	130	230	80
total fat (g)	14	26	9
saturated fat (g)	6	4.5	6
cholesterol (mg)	45	0	40
sodium (mg)	830	350	250
carbohydrates (g)	38	68	60
sugars (g)	7	0	54
proteins (g)	15	8	11

41. American diets depend heavily on bread and other wheat products. A slice of whole wheat bread (36 g) contains approximately 1.5 g of fat (with 0 g saturated fat), 17 g of carbohydrate (with about 1 g of sugar), and 3 g of protein.
    a. Calculate the total Calorie content in a slice of this bread.
    b. Calculate the percent of Calories from fat.
    c. Do you consider bread a highly nutritious food? Explain your reasoning.

42. Describe three ways in which agriculture is connected to the use of fossil fuels. Which aspects will eating locally alter?

43. Ethanol is an example of a biofuel.
    a. From which macronutrient does it originate: fats, carbohydrates, or proteins?
    b. Name two foods now used to produce ethanol for vehicles.
    c. By what process is the ethanol produced from these foods?
    d. Describe one of the current controversies in producing ethanol.

44. Biodiesel is another example of a biofuel. Answer the same questions for biodiesel that were asked in Question 43 about ethanol.

Exploring Extensions

45. Considering your diet over the past 24 hours, use the ideas in this chapter to describe two ways to improve your diet for your health and two ways to improve your diet for the heath of the planet.

46. Nature holds some surprises! An avocado is a tropical fruit; in fact, it is a fruit with a high fat content. However, this fat differs from the type found in coconut or palm oil. What is the primary type of fat found in avocados? How does this fat compare to tropical oils (coconut and palm) in terms of its effects on your health?

47. Estimate your average yearly intake in grams of sugar from soft drinks and fruit juices, listing the assumptions you made in arriving at this estimate. By what amount (in grams) would this estimate increase if you included the sugar you added to beverages such as coffee and tea?

48. Here is information about the sugar content of different foods.

Food Product	Sugar	Calories	Serving Size
Altoids, peppermint	2 g	10	3 pieces (2 g)
Ginger snaps	9 g	120	4 cookies (28 g)
Critic's Choice Tomato Ketchup	3 g	15	1 tbsp (13 g)
Del Monte Pineapple Cup	13 g	50	Individual cup (113 g)
Dr Pepper soft drink	40 g	150	1.5 cups
French Vanilla Coffee-Mate	5 g	40	1 tbsp (15 mL)
Hostess Twinkies	14 g	150	1.5 ounces
LifeSavers, Wint O Green	15 g	60	4 mints (16 g)
Tropicana HomeStyle Orange Juice	22 g	110	8 ounces (1 cup)
Snickers bar	29 g	200	2.1 ounces
Sunkist orange soda	52 g	190	1.5 cups
Wheatables crackers	4 g	130	13 crackers (29 g)

a. Examine this list. Which item has the highest ratio of grams of sugar to the number of Calories (g sugar/Cal) in one serving?

b. The sugar content of some of these foods may surprise you. If so, which ones?

c. Do you predict that the type(s) of sugar found in Dr Pepper would be the same as those found in Sunkist orange soda? In the orange juice or in the pineapple cup? Explain.

d. The complete label for Wint O Green LifeSavers shows 16 g of total carbohydrates per serving, 15 g of which is sugars. What might account for the other 1 g of carbohydrates?

49. A yellow packet of Splenda sugar substitute contains the compound sucralose. Use the resources of the Internet to answer these questions.

a. How many Calories does a packet of Splenda contain?

b. Splenda's slogan is "Made from Sugar, So It Tastes Like Sugar." Does it appear to you to be a helpful statement or misleading advertising? Explain your point of view.

50. Consider this structural formula for one of the forms of vitamin K.

a. Do you expect it to be water-soluble or fat-soluble? Explain.

b. What role does vitamin K play in your body?

c. People rarely, if ever, experience vitamin K deficiencies. Propose a reason why.

51. Compare these two pie charts for the percentage of macronutrients in soybeans and wheat.

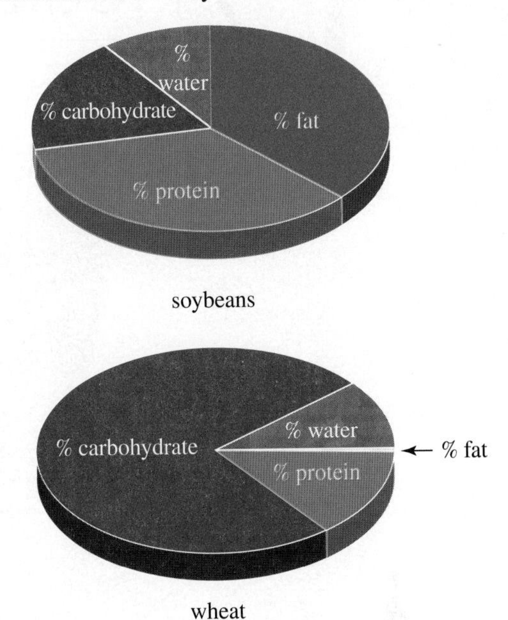

a. Explain why the World Health Organization has helped develop soy- rather than wheat-based food products for distribution in parts of the world where protein deficiency is a problem.

b. Suggest cultural reasons why soy might be preferable to wheat in some areas of the world.

52. Some view climate change and food security as "the twin grand challenges" that we face today. Name two ways in which these challenges connect to each other.

53. This chapter (together with Section 5.15) provided data for ethanol production through 2012. Use the resources of the Internet to update this information, particularly in regard to ethanol feedstocks, and the energy required to produce ethanol. Use this information to explain how bioethanol fuel meets, or does not meet, the goals of environmental sustainability and food security.

CHAPTER 12 Health & Medicine

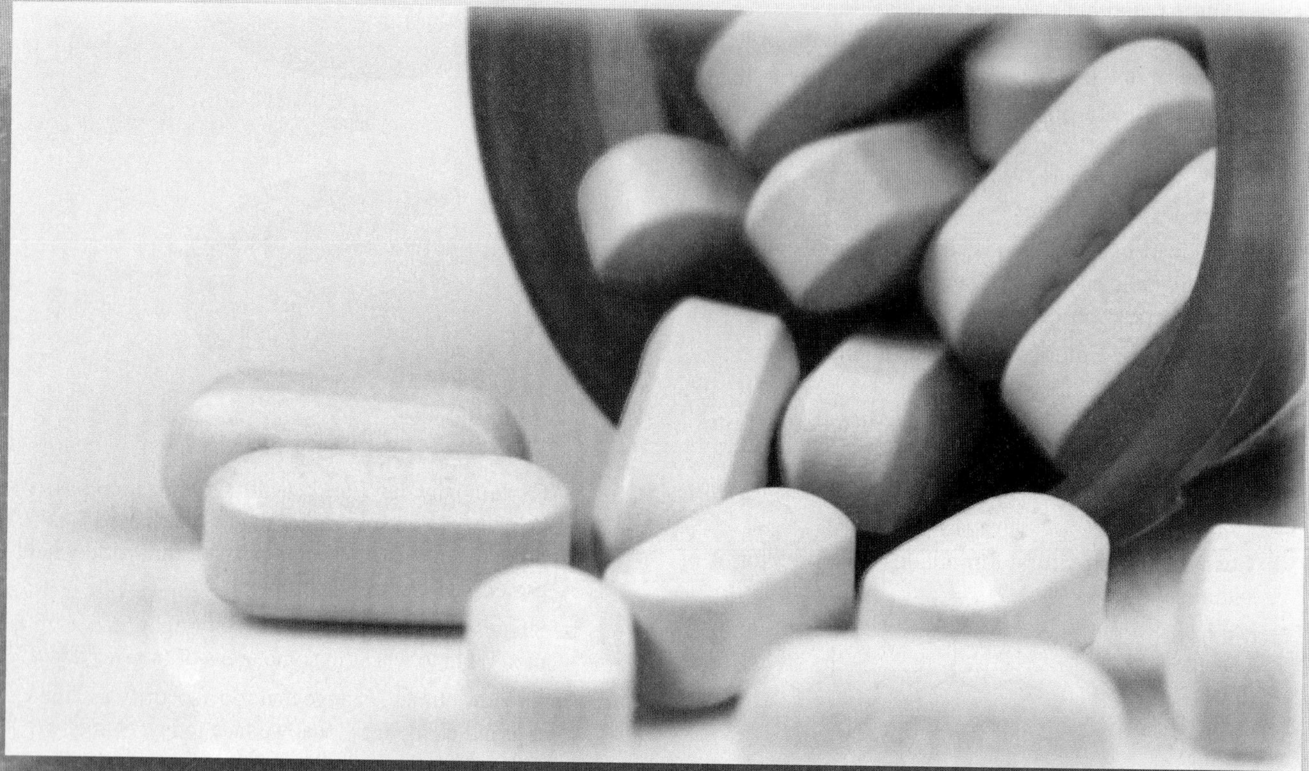

REFLECTION

Delving into Medicinal Chemistry

Reflect upon your own interactions with doctors. How often do you consider the chemistry behind your health, their diagnostic tools, or their advice? List five common medical treatment paths. For each, decide whether it cured a problem, treated a symptom, or prevented a future problem.

The Big Picture

In this chapter, you will explore the following questions:

- How does the healthy body work?
- How do different types of medications work?
- How is nuclear chemistry used as a medical tool?
- How are new pharmaceuticals developed?

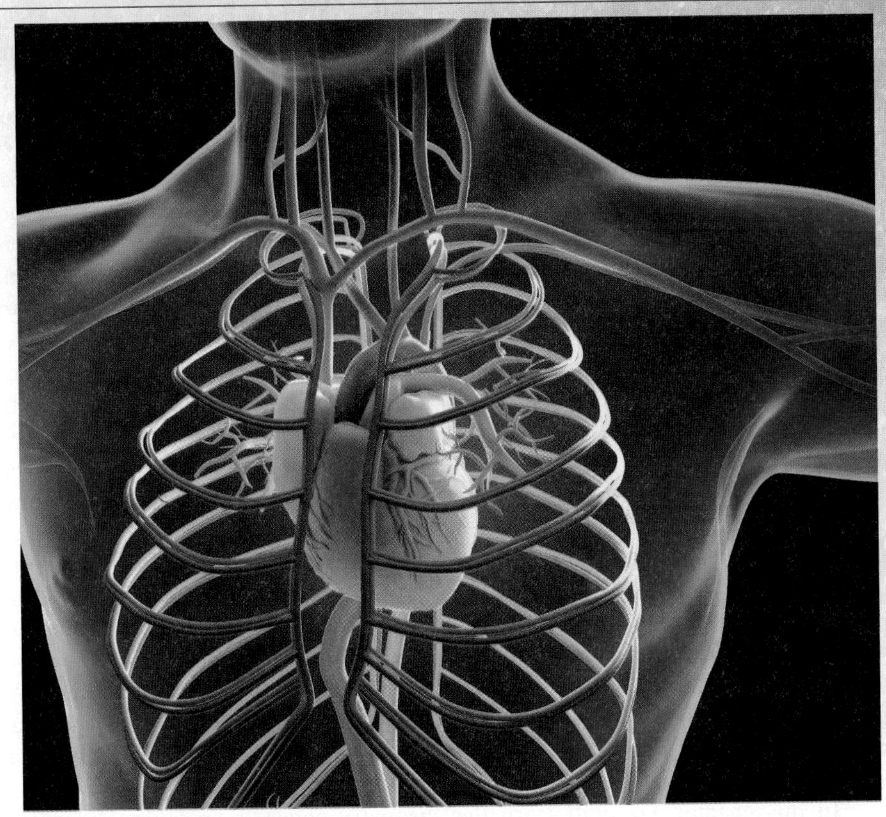

© Andrzej Wojcicki/Science Source

Introduction

Grasp a pencil. This momentary (and possibly pointless) act required one conscious choice, but a series of stepwise, though rapid, events. You had to scan for a pencil, move your hand, confirm by feel and sight that you were applying pressure, then tighten your hand around it. Each of these individual steps required the rapid firing of neurons; some signals traveled as far as your brain and back to your fingertips. Even this conscious activity requires a multitude of actions that require no thought. All the while, the heme in your blood cells pulls oxygen from your lungs to transport it toward tissues in need. Your digestive system is processing your most recent meal and, perhaps, signaling for a switch in whether your organs store or use sugars. Neurons are firing electrochemical codes to create new links that will record this information and maybe associate it with the music track in the air or the feel of that pencil in your hand. The majority of these actions simply happen in an automated sort of way, but on the molecular level these processes are very complex.

Your Turn 12.1 Scientific Practices Follow the Hormone

Using the Internet as a resource, identify a hormone produced by a healthy thyroid and then find two roles it plays in the body. How does the hormone travel from the thyroid to those organs?

The human body is like a factory full of many interconnected assembly lines keeping the whole system operational. Chemists, often within the specialty of biochemists, study the mechanisms by which each line carries out their specific role. While each line can be considered in isolation, the connections among them is what drives the complexity that defines you. What happens when one or many assembly lines malfunction; when the proper functioning of a bodily mechanism goes haywire? From your perspective, you see the malfunctions as a disease or set of symptoms.

Biologists, biochemists, and chemists study the functioning, as well as malfunctioning, of the human body to understand illness and design treatments. From your perspective, you miss this cause-and-effect. When you go to your local urgent care center with severe abdominal pain, the physician is trying to quickly determine the specific pathology of your ailment. In the case of the severe abdominal pain, the doctor will aim to narrow down a list of possible diagnoses with the aim of curing the ailment or, at a minimum, alleviating the symptoms. During the springtime, many people develop the seasonal allergy symptoms of watery eyes, runny nose, and sneezing. Again, people flock to their primary care physician to get prescriptions for nasal steroid sprays and other anti-allergy medicines to reduce these disruptive symptoms. Most of us don't even notice how our bodies work until something goes wrong. While we are content to apply modern solutions to return us to comfort and a functioning physiology, we do not see the chemistry and biochemistry behind the symptoms and treatment.

Did You Know? The father of medicine, Hippocrates (460–370 BC), considered tuberculosis to be the most deadly disease, but modern medicine can now treat tuberculosis with antibiotics.

Pharmaceuticals are therapeutic substances intended to prevent, moderate, or cure illnesses. **Medicinal chemistry** is the science that deals with the discovery or design of new therapeutic chemicals and their development into useful medicines. We are very fortunate that modern medicine has developed drugs, which can readily halt bothersome symptoms and cure pathologies. Therapeutic drugs have made a radical improvement in the longevity and quality of our lives.

In this chapter, you will learn the chemistry concepts necessary to explain how our bodies operate in times of sickness and health. You will also explore how advances in chemistry are directly involved with the treatment and management of diseases.

12.1 | A Life Spent Fighting Against Equilibrium

Admittedly, this textbook has generally implied that reactions proceed in one direction—from reactants to products. While this statement can be true, it is less frequent than suggested. Reactions are often reversible, and reactions in biochemistry are almost always so. They proceed forward only until a balance of the product and reactants is reached. In this state, at **equilibrium**, the system on a macroscopic level will appear static with a constant concentration of products balanced by a constant concentration of reactants.

See **Connect** for a video demonstration involving these two containers.

To create a visual analogy, imagine that you have two identical containers: one filled with water and one empty. Once you connect the two containers, the water moves from the filled container into the empty one until the level in each is at equal height. The relative height of water in each container in the final state is determined by the amount of water used. To consider this analogy on a microscopic level, individual molecules of water travel back and forth between the containers, but the overall amount in each is unchanged. Reactions will behave the same way; macroscopically, they will halt when the system balances at a specific ratio of products and reactants. On the molecular level, the reactants and products will continue exchanging. The ideal, stable ratio of product to reactant concentrations is defined by the specific reaction and the relation to its environment. This value is so important that we provide it with a special term: the **equilibrium constant**, K_{eq}, which is unitless.

To understand the equilibrium constant generally, let's start with a simple conversion of A to B (Equation 12.1). In this system, K_{eq} is simply the ratio of the concentration of your product, B, over the concentration of your reactant, A (Equation 12.2).

$$A \rightleftharpoons B \qquad [12.1]$$

$$K_{eq} = \frac{[B]}{[A]} \qquad [12.2]$$

Once you look up or calculate the equilibrium constant, you can immediately evaluate it for whether the reactant or product is favored. By "favored," we mean which has a greater concentration after equilibrium has been established. If $K_{eq} > 1$, then the concentration of the product, [B], must be favored. In contrast, if $K_{eq} < 1$, then the concentration of the reactant, [A], must be favored (Figure 12.1). However, what if $K_{eq} = 1$? This simply indicates that both reactant and product concentrations are equivalent, and neither is favored.

By convention, a reactant or product enclosed in brackets indicates its concentration—typically, in units of moles/L, or M. For instance, $[Na^+]$ indicates the concentration of sodium ions.

Your Turn 12.2 Skill Building Finding Equilibrium

Glucose and fructose are monosaccharides that are isomers of each other; they have the same numbers and types of atoms ($C_6H_{12}O_6$), but in a different structural arrangement. Unlike some isomers, glucose and fructose can be reversibly interconverted.

a. Write the balanced chemical reaction for the conversion of glucose to fructose.
b. Write the formula for the equilibrium constant, K_{eq}.
c. In a sample at equilibrium, [glucose] = 6.02 mM and [fructose] = 4.45 mM. Calculate the equilibrium constant.

Return to Chapters 10 and 11 to review the role of sugars in your food.

We can expand this formula to any reaction. For instance, take the arbitrary conversion of two reactants into two products (Equation 12.3). The concentrations of each are taken into account, as well as the number of moles of each required in the balanced chemical reaction. Specifically, the lowercase numbers in this equation represent the number of moles of each in the balanced chemical reaction.

$$a\,A + b\,B \rightleftharpoons c\,C + d\,D \qquad K_{eq} = \frac{[C]^c[D]^d}{[A]^a[B]^b} \qquad [12.3]$$

By comparing the equilibrium constant of a reaction to the current concentrations of reactants and products, we can make predictive statements about the relative changes

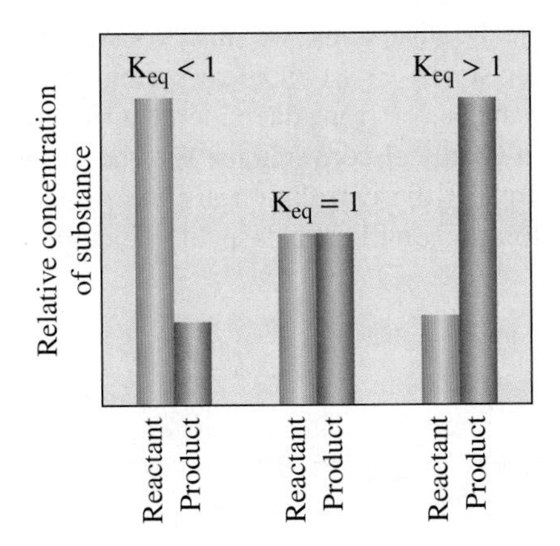

Figure 12.1
An illustration of the magnitude of the equilibrium constant, relative to the concentrations of products (**blue**) and reactants (**green**).

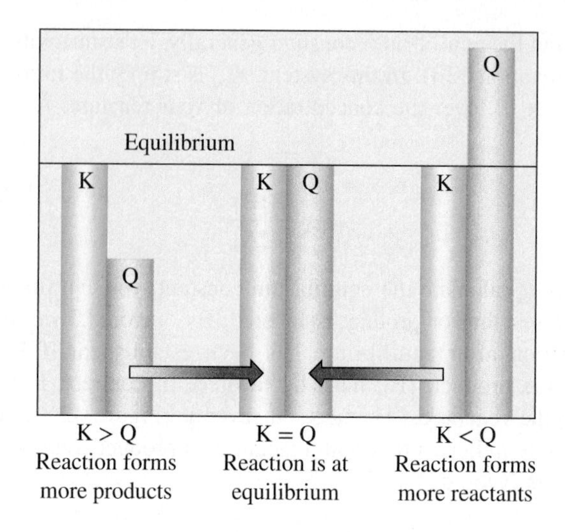

Figure 12.2

The relationship between the reaction quotient, Q, and equilibrium constant, K.

http://chemwiki.ucdavis.edu/Physical_Chemistry/Equilibria/Chemical_Equilibria/The_Equilibrium_Constant/Calculating_An_Equilibrium_Concentration_From_An_Equilibrium_Constant/Writing_Equilibrium_Constant_Expressions_Involving_Gases/Gas_Equilibrium_Constants%3A_Kc_And_Kp

in concentrations that will occur *en route* toward establishing equilibrium. The current ratio of [products]/[reactants] is known as the **reaction quotient**, Q, and it may be greater than, less than, or equal to the equilibrium constant, K_{eq}.

Think again at those two containers of water connected by a tube. Let's designate the filled container as the reactants (*i.e.,* initial state) and the empty container as the products (*i.e.,* final state). Initially, the concentration of water in the two containers is highly unbalanced, and is clearly not in equilibrium. Because the second water container is initially empty, the [products] are equal to zero; therefore, 0 M/[reactants] = 0, so $Q < K_{eq}$. In order to reach equilibrium, the water level in the full container must drop, and the water level in the empty container must increase. In other words, the concentration of products must increase at the expense of the reactants. In contrast, a system may exhibit conditions where $Q > K_{eq}$. In this case, the system would prefer to regenerate some of the reactants by consuming products. Lastly, if $Q = K_{eq}$, then the system is at equilibrium, and there will be no net increase in the concentrations of products or reactants. To summarize, a comparison of product versus reactant concentrations at any stage during a reversible reaction can tell us whether a system is stable, or must move in the forward or reverse direction in order to establish equilibrium (Figure 12.2).

To study biological systems, we need to expand our concept of reactions. Instead of simply chemical conversions, we must consider the possibility of moving between two positions or states. In a biological system, the act of a chemical moving from free to bound states, or from one side of a membrane to another, can be far more important than chemical conversions. For the purposes of the equilibrium constant, all other aspects of the calculation are the same. Let's walk through the equilibrium constant for the binding/release of molecule A to/from molecule B (Figure 12.3).

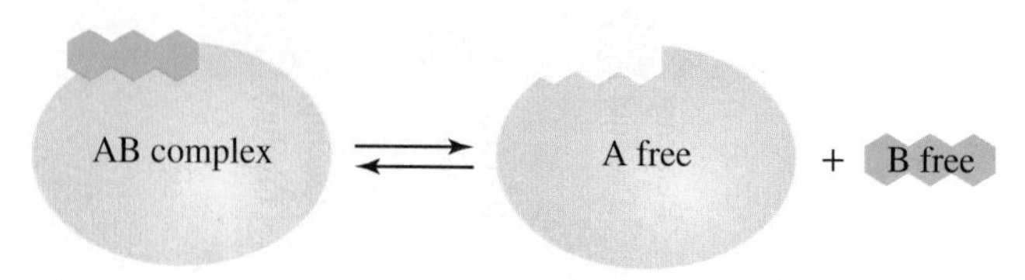

Figure 12.3

The reversible dissociation of molecules A and B shown from the complex or associated state to their free states.

Your Turn 12.3 Skill Building Finding Equilibrium in Binding

Epinephrine (better known as adrenaline) is a small organic molecule that binds to your cells at a specific site called the β-adrengenic receptor, and which gives you that familiar heightened energy level. The release of epinephrine from the receptor has a K value equal to 5×10^{-6}. For a solution of epinephrine with the β-adrengenic receptor, does the epinephrine prefer to stay bound or free?

In biochemistry, we so often discuss the equilibrium constant for the dissociation reaction that we give it a special constant, K_d.

On its own, the equilibrium constant does not tell us how quickly we can go between states, only if the system prefers one state to another. Imagine pushing a stone from one valley to another with a hill or barrier between (Figure 12.4). Whether the barrier is low or high, the stone will be more stable on lower ground. The higher the barrier in between, the more difficult it will be to push the stone between the valleys. Chemically, the study of the relative energy of the two end states (A and B) is **thermodynamics**, while the study of the rate of conversion between those chemical states is **kinetics**. While the equilibrium constant can provide us with everything we need to know about the thermodynamics of a system, its kinetics requires a larger set of information. Luckily, much of biochemistry operates quite well in the realm of thermodynamics alone.

Your Turn 12.4 Scientific Practices Equilibrium in Your Blood

Your blood cells are rich with an iron-carrying molecule called hemoglobin. One primary responsibility of hemoglobin is to pick up oxygen in our lungs and deposit it in our tissues by transferring the oxygen to a second molecule called myoglobin. Do you expect myoglobin or hemoglobin to have a higher equilibrium constant for this exchange to be favorable? Explain your reasoning.

In order to survive, all organisms concentrate the specific chemicals necessary for function (such as nutrients) and pump out or block any material that would be harmful. In short, life requires pushing away from equilibrium. Instead, biochemical systems will reach a steady-state within their environment. From an outside perspective, this means that the overall concentrations of reactants and products will remain unchanged, but staying there requires a constant input of energy and materials. To continue with our previous analogy, consider a set of water-filled containers and one has a leak. While the water levels can be kept constant, work will be necessary to add new water into the system. To connect to the concept of equilibrium, this condition is often called "dynamic equilibrium" in addition to steady-state.

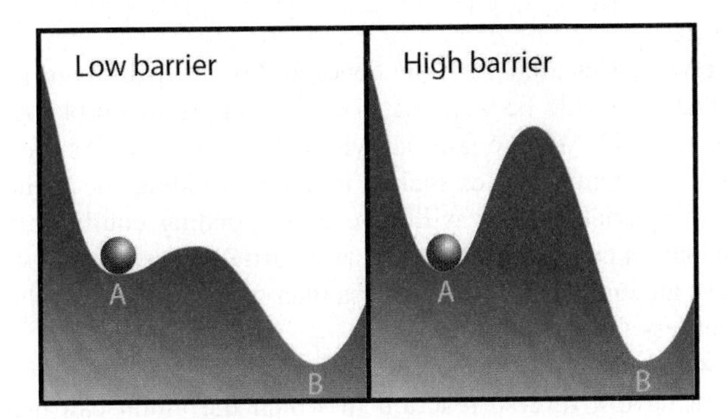

Figure 12.4

Visualizing barriers to a reaction progress.

12.2 Keeping Our Bodies in Equilibrium

You likely recognize the symptoms of too much acid in your stomach: a gurgling, often unpleasant sensation followed by potentially harmful reflux. Your stomach is out of balance and, while long-term and frequent acid reflux may indicate a physiological imbalance worthy of consulting your physician, the most immediate cause is likely something acidic you ate. Lucky for you, this set of symptoms does not extend to other areas of your body. Your tissues, blood, even the interior compartments of your cells have a defined and well-regulated pH. These areas are resilient to the addition of acidic or basic chemicals because they employ buffers.

A **buffer** is a system that responds only gradually or slightly to an external influence. The label of buffer is quite apt here because its behavior corresponds well to the familiar verb. A good wool coat buffers you against cold gusts of wind. Analogously, buffers—composed of acids and bases—shield our bodies from large shifts in pH.

Your Turn 12.5 Skill Building Reviewing Acids and Bases

List at least three acids and three bases that you know from previous chapters. Describe the context of each.

Return to Section 8.8 to review the identity and properties of acids and bases.

Acids and bases are specific chemicals that effectively change the concentration of hydrogen and hydroxide ions in solution. An acid in water separates into an anion and a hydronium ion. Take for example, the behavior of hydrochloric acid in water:

$$HCl(aq) + H_2O(l) \rightleftharpoons H_3O^+(aq) + Cl^-(aq) \quad [12.4]$$

The concentrations of pure solids or liquids are never included in the expression for K_{eq}. These components are present in large excess, so their concentrations remain constant throughout the reaction.

This reaction can be written with or without explicitly showing water. This type of dissociation would not be favored were water not present and so unique in its chemistry. This reaction, and others like it, can be described with an equilibrium constant. While interaction with water is a necessary part of the reaction, the concentration of any liquid is considered unchanged. The equilibrium constant will be independent of the amount of any liquid present:

$$K_{eq} = \frac{[H_3O^+][Cl^-]}{[HCl]} \quad [12.5]$$

Theoretically, this reaction could be reversed and the two ions would associate again. However, the equilibrium constant of HCl dissociation is extremely high (1×10^6), meaning the products are strongly favored. Accordingly, as detailed in Section 8.8, the reaction may be best summarized as the following, which uses a single-headed arrow—an indication of a very low likelihood of reversibility:

$$HCl(aq) + H_2O(l) \longrightarrow H_3O^+(aq) + Cl^-(aq) \quad [12.6]$$

Remember, a low pH indicates a high concentration of hydrogen ions, while a high pH is a low concentration of hydrogen ions. Pure water has equal concentrations of H^+ and OH^- and has a pH = 7.

In order to reverse Equation 12.4, the concentration of hydronium ions would have to be so high that we would be well out of the pH ranges found biologically.

Likely, the acids you are familiar with are strong acids. As seen with HCl above, **strong acids** will completely dissociate in water, yielding the highest value possible of free ions; numerically, they will have corresponding equilibrium constants much greater than one. In contrast, **weak acids** will partially dissociate in water; numerically, they will have an equilibrium constant less than one, showing that the dissociated state is present but less favorable.

If the dissociation of an acid to a free hydrogen ion and anion is reversible, we must then have a reverse reaction in which the anion can pick up a hydrogen ion. In this direction, the anion is acting as a base. Specifically, the base created when an acid loses its hydrogen ion is called the *conjugate base*. Conjugate bases are only truly important for weak acids, as the large equilibrium constant for strong acids make the reverse reaction (the association of a conjugate base and hydrogen cation) too unlikely.

Conjugate acids and bases are discussed in Section 8.8.

Your Turn 12.6 Skill Building Acids and Their Conjugate Bases

For each of the acidic compounds below, write the reaction for the dissociation of the acid, and then put a box around the conjugate base.

a. CH_3COOH (acetic acid)
Hint: Only the hydrogen on the right of the chemical formula can dissociate as an ion.
b. H_2CO_3 (carbonic acid)
Hint: For acids with multiple hydrogens that can dissociate, there will be multiple conjugate bases.
c. H_3PO_4 (phosphoric acid)

A solution of a weak acid and its conjugate base has the tendency to buffer a solution because it resists a change in the overall concentration of hydrogen ions. If an acid is added to pure water, there will be a significant change in the concentration of hydrogen ions and therefore pH. For instance, consider the addition of 0.05 mole of HCl to 1.0 L of pure water. Whereas pure water had an initial pH of 7.0, the pH instantly drops to 1.3 after the addition of acid!

$pH = -\log [H^+]$
$[H^+] = 10^{-pH}$

Let's now consider what happens if the same concentration of acid is added to 1.0 L of a buffer consisting of 0.1 M HF (a weak acid) and 0.1 M NaF (the conjugate base of the weak acid, HF).

$$HF(aq) + H_2O(l) \rightleftharpoons F^-(aq) + H_3O^+(aq) \quad [12.7]$$

$$K_{eq} = \frac{[F^-][H_3O^+]}{[HF]} \quad [12.8]$$

The equilibrium constant for weak acids is often designated as K_a.

In order to determine how the pH of a buffer solution is affected by adding an acid or a base, the Henderson-Hasselbalch equation may be employed:

$$pH = pK_a + \log \frac{[\text{conjugate base}]}{[\text{weak acid}]} \quad [12.9]$$

Just as $pH = -\log[H^+]$, the pK_a is the $-\log[K_a]$. For HF, the pK_a is 3.14; accordingly, the pH of the above HF/NaF buffer solution would be 3.14 (*i.e.*, $3.14 + \log\left(\frac{0.1\text{ M}}{0.1\text{ M}}\right)$). When 0.05 mole of acid is added to 1.0 L of this buffer, the $H^+(aq)$ ions will react with the conjugate base $F^-(aq)$ ions of the buffer, resulting in the formation of more $HF(aq)$:

Appendix 3 provides more details regarding the use of logarithmic functions for pH calculations.

$$H^+(aq) + F^-(aq) \longrightarrow HF(aq) \quad [12.10]$$

Therefore, the $[F^-]$ will decrease by 0.05 M and the [HF] will increase by 0.05 M. This leads to a slightly lower final pH of 2.66 (*i.e.*, $3.14 + \log \frac{(0.1 - 0.05)}{(0.1 + 0.05)}$). Whereas adding acid to pure water caused a decrease of 5.7 pH units, the addition of acid to a buffer only resulted in a drop of 0.48 pH unit.

A buffer is most useful when the solution's pH is equal to the pK_a value. This is achieved by mixing equal moles of the acid and conjugate base. Table 12.1 lists the K_a and pK_a values for a variety of weak acids. A fair amount of acid can be added to a buffer solution before the conjugate base in the system is depleted and the pH begins to decrease (becoming more acidic) in proportion to the acid added. In the opposing direction, we can also add a base to react with the conjugate acid. Once the conjugate acid is depleted, the pH begins to increase (becoming more basic). In fact, a buffer will resist against additions of acids and bases up to ±1 pH unit around the value of pK_a (Figure 12.5).

Table 12.1 Dissociation of Some Weak Acids with K_a and pK_a Values

Acid (name)	Conjugate Base (name)	pK_a	K_a
HCOOH (formic acid)	$HCOO^-$ (formate ion)	3.8	1.78×10^{-4}
CH_3COOH (acetic acid)	CH_3COO^- (acetate ion)	4.8	1.74×10^{-5}
H_3PO_4 (phosphoric acid)	$H_2PO_4^-$ (dihydrogen phosphate ion)	2.1	7.24×10^{-3}
$H_2PO_4^-$ (dihydrogen phosphate ion)	HPO_4^{2-} (hydrogen phosphate ion)	6.9	1.39×10^{-7}
HPO_4^{2-} (hydrogen phosphate ion)	PO_4^{3-} (phosphate ion)	12.4	3.98×10^{-4}
H_2CO_3 (carbonic acid)	HCO_3^- (bicarbonate ion)	6.3	5.1×10^{-7}
HCO_3^- (bicarbonate ion)	CO_3^{2-} (carbonate ion)	10.3	5.62×10^{-11}
NH_4^+ (ammonium)	NH_3 (ammonia)	9.3	5.62×10^{-10}

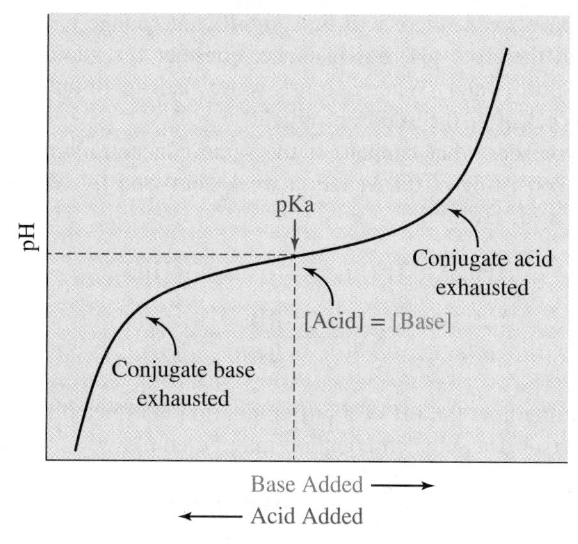

Figure 12.5
Useful range of buffering against acid or base additions around the pK_a.

Your Turn 12.7 Scientific Practices A Buffer Choice

A practicing scientist must judge a potential buffer for its utility in mimicking different cellular environments. Consider the list of target pH values and match them to the best possible chemical mixture for creating that pH, see **Table 12.1** for a list.

a. pH 2.5
b. pH 7.4
c. pH 6.8
d. pH 5.4

Biological systems require these traits of weak acid and base mixtures in order to absorb any local changes in hydrogen ion concentrations. The pH of your tissues will be resilient to most of your activities, even if reactions taking place within your cells produce or use up hydrogen ions. Importantly, not all of your systems use the same buffer or the same pH. Your tissues will have preferential pH levels. Even within your cells, you will find compartments buffered at different pH levels. These systems are all ideally set at pH values that optimize the reactions and biological machinery that operates within them.

Thus far, we have described biology in terms of the chemical balance required for a healthy system. Still, to completely understand the chemical transformations and the development of medicines, we need to gather some tools to understand the chemical framework of biology, built upon the versatile carbon atom.

12.3 Carbon: The Essential Building Block of Life

Carbon is the basis of all life-forms on our planet. This element is so ubiquitous in nature that a sub-discipline of chemistry, **organic chemistry**, is devoted to the study of carbon compounds. The name *organic* is historical and suggests a biological origin, but this is not necessarily true. In practice, most organic chemists investigate compounds, of biological origin or human design, in which carbon is combined with a relatively small number of other elements: hydrogen, oxygen, nitrogen, sulfur, chlorine, phosphorus, and bromine.

Your Turn 12.8 Skill Building Compound Categories

Using the elements listed above, list five organic and five inorganic compounds. For each, describe the context in which you learned about the compound.

To specify an organic compound from the myriad of possibilities, you must be able to name it correctly. An international committee called the International Union of Pure and Applied Chemistry (IUPAC) established and periodically updates a formal set of nomenclature rules so each of the known compounds can be uniquely named. However, many of these compounds have been known for a long time by common names such as alcohol, sugar, or morphine. When a headache strikes, even chemists do not call out for 2-acetyloxybenzoic acid; they simply say, "Give me some aspirin!" We test our blood glucose levels, not our blood (2R,3S,4R,5R)-2,3,4,5,6-pentahydroxyhexanal levels. A mouthful like this is the cause of great merriment to those who like to satirize chemists. Nonetheless, chemical names are important and unambiguous to those who know the system. You can rest easy because in this chapter, we will use common names in almost all cases.

We remind you that a few basic rules for bonding in organic molecules can help you find order in the chaos of millions of organic compounds. We introduced one of these in Section 4.7, the octet rule. When bonded, each carbon atom has a share in eight electrons, an octet. Eight electrons can be arranged to form four bonds, with a pair of shared electrons in each covalent bond. Carbon almost always forms four bonds, which may include four single bonds, or some combination of single, double, and triple bonds. Some possibilities are illustrated in Figure 12.6.

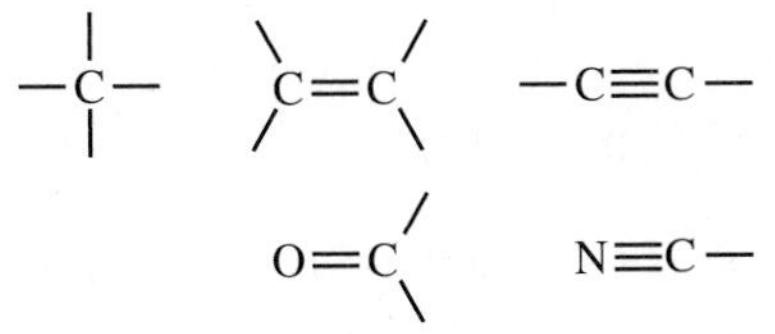

—C— C=C —C≡C—

O=C N≡C—

Figure 12.6

Some common bonding arrangements for carbon.

Your Turn 12.9 Skill Building Checking on Carbon

a. Examine the carbon atoms illustrated in **Figure 12.6**. Does each one follow the octet rule?

b. Can you recall a molecule in which carbon does not form four bonds?
Hint: This molecule is one of the air pollutants discussed in Chapter 2.

Other elements exhibit different bonding behaviors in organic compounds. A hydrogen atom is always attached to another atom by a single covalent bond. An oxygen atom typically attaches either with two single bonds (to two different atoms) or one double bond (to a single atom). A nitrogen atom commonly forms three single bonds (to three different atoms), but also can form either a triple bond (to one other atom), or a single and a double bond (to two different atoms). Both oxygen and nitrogen satisfy the octet rule, because in addition to these chemical bonds, they have at least one pair of nonbonding electrons.

The same number and kinds of atoms can be arranged in different ways, helping to explain why there are so many different organic compounds. **Isomers** are molecules with the same chemical formula (same number and kinds of atoms), but with different structures and properties. In Chapter 5, you encountered two of the isomers of C_8H_{18}: *n*-octane (straight chain) and iso-octane (branched).

In this chapter, we revisit the concept of isomers, this time using C_4H_{10}. Analogous to C_8H_{18}, we can draw both a straight chain and a branched isomer. Here are the structural formulas; note that *n*-butane is represented by a more realistic zigzag form:

```
                                   H  H
                                    \/
     H  H  H  H                  H—C  H
      \/    \/                       \/
  H   C      C                H      C      H
   \ / \    / \  H             \   /  \   /
    C    C       /              C       C
   / \  / \                    / \     / \
  H  H  H  H                  H  H    H   H

     n-butane                     iso-butane
```

Convince yourself that although these compounds both have the same chemical formula, the way in which the atoms are connected is different.

Both the linear isomer and the more complex iso-butane can be represented with a condensed structural formula. Here we show the options for iso-butane:

$$\begin{array}{c}CH_3\\|\\CH_3—CH—CH_3\end{array} \quad \text{or} \quad CH_3CH(CH_3)CH_3 \quad \text{or} \quad CH_3CH(CH_3)_2$$

Although a $-CH_3$ group at the left end of a chain could be drawn as H_3C-, for ease we often reverse the order to CH_3- with the understanding that the bond is to the C atom, not to the H atoms.

These condensed structural formulas are trickier to interpret. The parentheses around the $-CH_3$ groups indicate that they are attached to the C atom to their left. Note that with three $-CH_3$ groups attached to the central C atom, a "branch" has been introduced into the molecule.

Figure 12.7 shows three representations of *n*-butane and iso-butane. The first column shows the simple structural formula, the second a ball-and-stick model. The third column shows a space-filling model that presents a more realistic view of the molecular shape.

Only two isomers of C_4H_{10} exist. As the number of atoms in a hydrocarbon increases, so does the number of possible isomers. In addition to *n*-octane and iso-octane, you can draw 16 other isomers for C_8H_{18}. And for $C_{10}H_{22}$, you could draw 75 isomers if you had the patience! Given a chemical formula, calculations to determine the number of isomers can be complex.

Your Turn 12.10 Skill Building Switching Representations

Expand each of these condensed structural formulas into a complete structural formula.

a. $CH_3CH(CH_3)CH_2CH_3$

b. $CH_3CH(CH_2CH_3)CH_3$

c. $CH_3CH_2C(CH_3)_3$

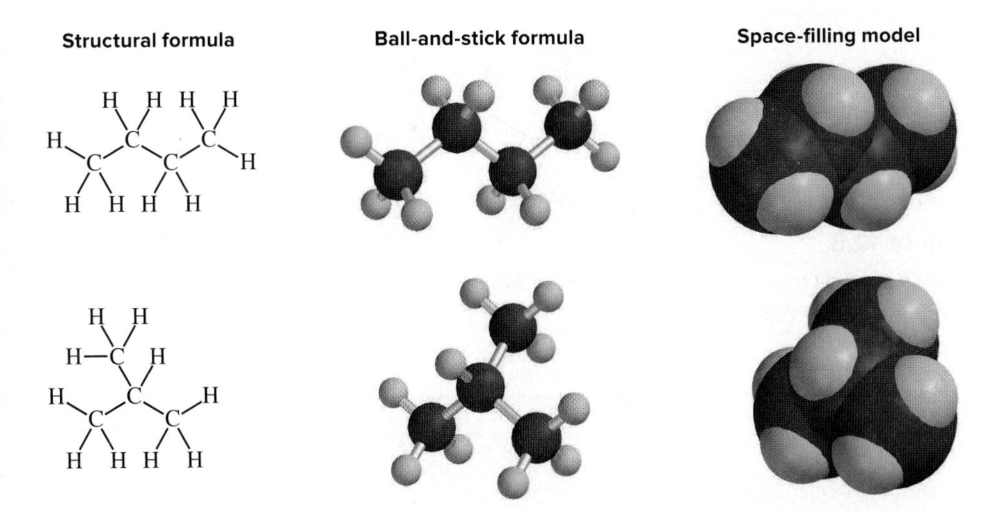

Figure 12.7

Three representations of the isomers of *n*-butane (top) and iso-butane (bottom).

As seen in Chapter 11, line-angle drawings are often used to represent the structural formulas for large molecules. In these illustrations, the hydrogen atoms are omitted for clarity; hence, it is important to remember that each carbon atom has four bonds, sharing a total of eight electrons. Table 12.2 shows various ways to represent *n*-butane, iso-butane, and two other simple molecules.

Table 12.2 Molecular Representations

Compound	Chemical Formula	Structural Formula	Line-Angle Drawing
n-butane	C_4H_{10}		
iso-butane	C_4H_{10}		
n-hexane	C_6H_{14}		
cyclohexane	C_6H_{12}		

Your Turn 12.11 Skill Building Practice with Line-Angle Drawings

Revisit the compounds in the **Your Turn 12.10**. For all three, draw line-angle representations.

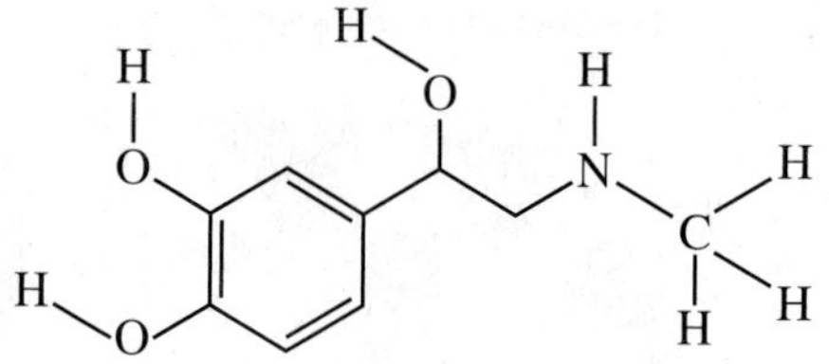

Figure 12.8

The structure of epinephrine, more commonly known as adrenaline, a vital chemical signal between neurons as well as a hormone that regulates metabolism.

Your Turn 12.12 Skill Building Practice with Isomers

a. Are *n*-butane and iso-butane isomers? Explain.

b. Are *n*-hexane and cyclohexane isomers? Explain.

c. Three isomers have the formula C_5H_{12}. For each, draw a structural formula, a condensed structural formula, and a line-angle drawing.

Many molecules, including thyroxine, dopamine, aspirin, and glucose, have carbon atoms arranged in a ring. For example, examine the structure of cyclohexane, C_6H_{12}, in Table 12.2. The ring in cyclohexane has six carbons. Rings most commonly contain five or six carbon atoms because this number seems to confer some stability—not too floppy and not too constricted. In epinephrine (Figure 12.8), however, the six-membered rings are based on benzene, C_6H_6, (Figures 12.9 and 12.10) rather than on cyclohexane.

Resonance forms were first discussed in Section 3.7.

A C–C single bond is 0.154 nm in length, while a C=C double bond is 0.134 nm. In benzene, the C–C bond lengths are all 0.139 nm.

Figure 12.9(b) shows a line-angle drawing for benzene, which illustrates alternating single and double bonds. However, this is only one possibility; there are two equivalent structures for benzene, known as **resonance forms**. The structure of benzene is a hybrid of the two resonance structures, which is represented by the dashed structure in Figure 12.10 or the circle shown in Figure 12.9(c). Since each of the C–C bond lengths are experimentally determined to be equivalent, each C–C bond in benzene may be thought of as a hybrid between a C–C single bond and a C=C double bond. This same hexagonal structure is found in the $-C_6H_5$ phenyl group that is part of many molecules, including polymers such as polystyrene (Section 9.5).

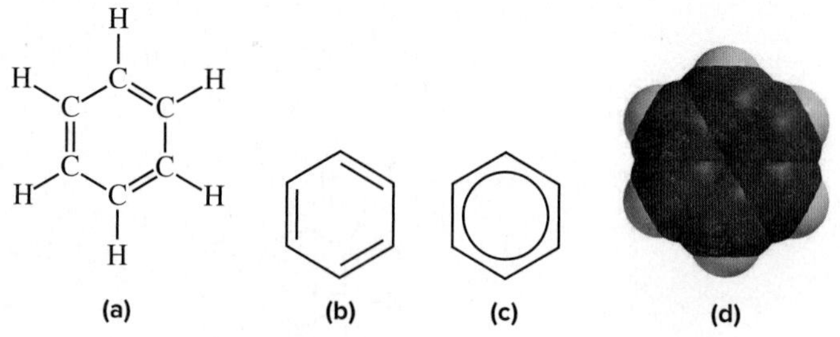

Figure 12.9

Representations of benzene, C_6H_6.

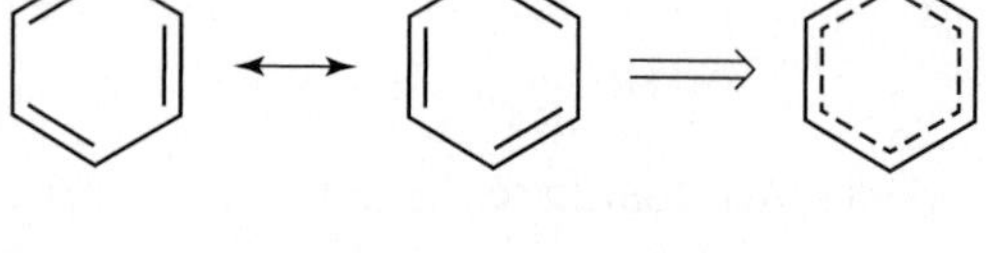

Figure 12.10

Resonance structures of benzene (hydrogen atoms are omitted for clarity).

12.4 | Functional Groups

Central to the study of biological interactions are functional groups. **Functional groups** are distinctive arrangements of groups of atoms that impart characteristic physical and chemical properties to the molecules that contain them. Indeed, these groups are so important that we often show them in structural formulas and represent the remainder of the molecule with an R. The R is generally assumed to include at least one carbon or hydrogen atom. You already encountered some functional groups in Chapter 9. For instance, the generic formula for an alcohol is ROH, as in methanol, CH_3OH (an alcohol derived from degradation of wood), and ethanol, CH_3CH_2OH (alcohol derived from fermentation of grains and sugar). The presence of the –OH group attached to a carbon makes the compound an alcohol.

An alcohol has an –OH group *covalently* bonded to the rest of the molecule. This is different from the hydroxide ion, OH^-, which is *ionically* bonded to a cation.

Similarly, a carboxylic acid group, commonly written as $-\mathrm{C}(=\mathrm{O})-\mathrm{O}-\mathrm{H}$, –COOH, or $-CO_2H$ confers acidic properties to the molecule. In aqueous solutions, a H^+ ion (a proton) is transferred from the –COOH group to a H_2O molecule to form a hydronium ion, H_3O^+. This leaves behind an anion that is commonly written as $-COO^-$.

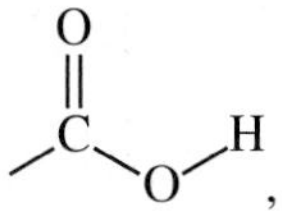

We represent an organic acid with the general formula RCOOH. In acetic acid (CH_3COOH), the acidic component in vinegar, the R group is $-CH_3$, the methyl group. The carboxylic acid, in fact, is what gives acetic acid its characteristic sharp-sour smell and taste. Table 12.3 lists eight functional groups found in drugs and other organic compounds, and Figure 12.11 displays the thyroxin molecule, which contains a variety of functional groups. Each functional group is characteristic of an important class of compounds.

Your Turn 12.13 Skill Building Line-Angle Drawings

For each of these condensed structural formulas, construct a line-angle drawing. Name the functional group in each one.

a. $CH_3CH_2CH_2COCH_3$
b. $CH_3CH_2CH(CH_3)CH_2OH$
c. $CH_3CH(NH_2)CH_2CH_3$
d. $CH_3COOCH_2CH_3$
e. CH_3CH_2CHO

Your Turn 12.14 Skill Building Functional Groups in Dopamine

Draw the structure of dopamine ($C_8H_{11}NO_2$) in both structural and line-angle drawings. Using the Internet, how does this structure differ from epinephrine? Draw a box around the ring(s) and a circle around the various functional groups. What are the names of these functional groups?

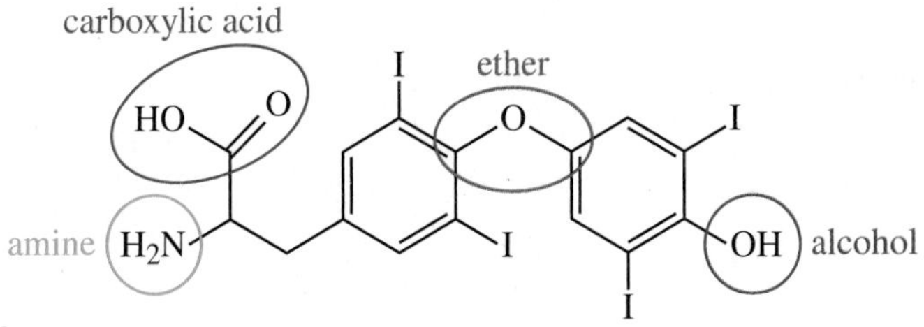

carboxylic acid · ether · amine · alcohol

Figure 12.11

Structural formula of thyroxin, a vital hormone precursor molecule, with selected functional groups (an amine, an ether, an alcohol, and an acid) circled.

Table 12.3 Some Important Organic Functional Groups

Functional Group	Generic Formula	SPECIFIC EXAMPLES		
		Name*	**Structural Formula**	**Condensed Structural Formula**
alcohol	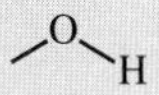 $-O-H$	ethanol (ethyl alcohol)	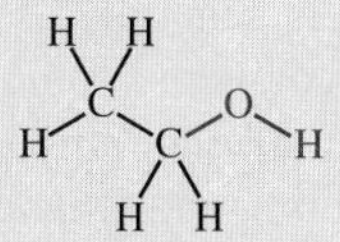 H_3C-CH_2-O-H	CH_3CH_2OH
ether	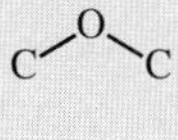 $C-O-C$	dimethyl ether	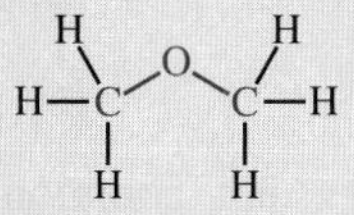 $H_3C-O-CH_3$	CH_3-O-CH_3 or CH_3OCH_3
aldehyde	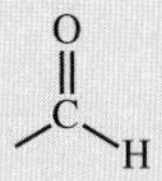 $-C(=O)-H$	propanal	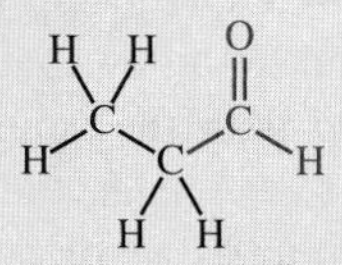 $H_3C-CH_2-C(=O)-H$	$CH_3CH_2-C(=O)-H$ or CH_3CH_2CHO
ketone	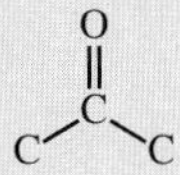 $C-C(=O)-C$	2-propanone (dimethyl ketone, acetone)	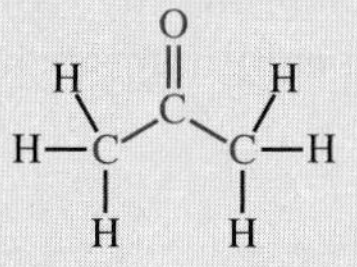 $H_3C-C(=O)-CH_3$	$CH_3-C(=O)-CH_3$ or CH_3COCH_3
carboxylic acid	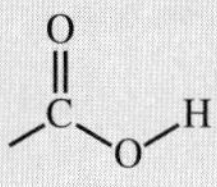 $-C(=O)-O-H$	ethanoic acid (acetic acid)	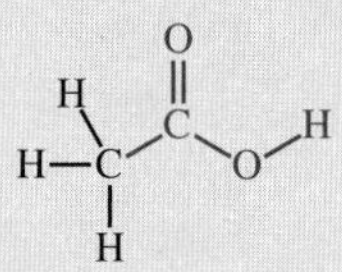 $H_3C-C(=O)-O-H$	$CH_3-C(=O)-OH$ or CH_3CO_2H or CH_3COOH
ester	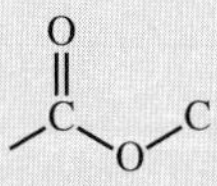 $-C(=O)-O-C$	methyl ethanoate (methyl acetate)	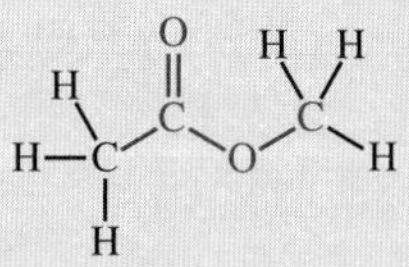 $H_3C-C(=O)-O-CH_3$	$CH_3-C(=O)-OCH_3$ or CH_3COOCH_3
amine	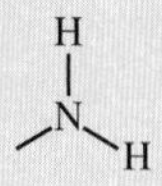 $-N(H)-H$	ethylamine	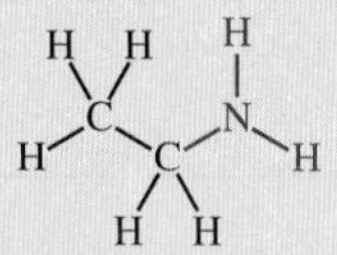 $H_3C-CH_2-N(H)-H$	$CH_3CH_2NH_2$
amide	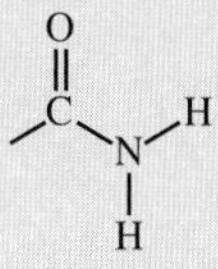 $-C(=O)-N(H)-H$	propanamide	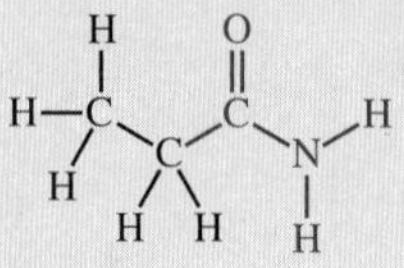 $H_3C-CH_2-C(=O)-N(H)-H$	$CH_3CH_2-C(=O)-NH_2$ or $CH_3CH_2CONH_2$

*IUPAC names, common names in parentheses

Functional groups in a biomolecule behave as the magnets of biological interactions. On a first pass, they distinguish sections of a molecule as polar or nonpolar; biochemistry can get more specific though. Within the subcategory of polar, we have specific functional groups that can be charged, with a tendency to form ionic bonds, or uncharged, with the talent to form hydrogen bonds. Nonpolar groups also build complexity. We can have a flexible chain of CH_2 groups, a more confined six-carbon ring, or a planar phenyl ring. These types of interactions and their exact position on a molecule provides the basis of the lock-and-key model of biological binding.

Revisit Chapter 8 for more on polar versus nonpolar molecules.

Models of biological binding will be expanded upon in Section 12.7.

Functional groups can play a role in the solubility of a compound, an important consideration in the targeting, regulation, uptake, and rate of reaction for any chemicals in the body. The general solubility rule "like dissolves like" that was introduced in Chapter 8, applies in the body as well as in a beaker. Functional groups containing oxygen and nitrogen atoms (for example, –OH, –COOH, and $–NH_2$) usually increase the polarity of a molecule. This in turn enhances its solubility in other polar substances such as water.

By contrast, hydrocarbons that do not contain such functional groups are typically nonpolar and will not dissolve in polar solvents. For example, *n*-octane (C_8H_{18}) is nonpolar and insoluble in water. However, it does dissolve in nonpolar solvents such as hexane (C_6H_{14}) and dichloromethane (CH_2Cl_2). For the same reasons, biomolecules with significant nonpolar character tend to accumulate in cell membranes and fatty tissues that are largely hydrocarbon and nonpolar.

Your Turn 12.15 Scientific Practices Accumulation of Vitamins

The solubility of vitamins has dramatic and important consequences for your recommended daily dose. If a vitamin is fat soluble, your body will collect and store it. If a vitamin is water soluble, your body will tend to flush any excess from your system. Look up the structures for vitamins A, C, D, and K. For which vitamins should you watch your dose? Explain your answers.

12.5 | Give These Molecules a Hand!

Biochemistry is further complicated because many of its interactions involve a common but subtle phenomenon called optical isomerism, or chirality. **Chiral or optical isomers**, have the same chemical formula, but differ in their three-dimensional molecular structure and their interaction with polarized light. Chirality most frequently arises when four different atoms or groups of atoms are attached to a carbon atom. A compound having such a carbon atom can exist in two different molecular forms that are non-superimposable mirror images of each other. One optical isomer will rotate polarized light in a clockwise manner, which is called the dextro (D) isomer. The other isomer is called the levo (L) isomer, and it rotates polarized light in a counterclockwise manner.

Polarized light waves move in a single plane; nonpolarized light waves move in many planes.

Non-superimposable mirror images should be familiar to you. You carry two of them around with you all the time—your hands. If you hold them palms up, you can recognize them as mirror images. For example, the thumb is on the left side of the left hand and on the right side of the right hand. Your left hand looks like the reflection of your right hand in a mirror. But your two hands are not identical. Figure 12.12 illustrates this relationship for both hands and molecules.

The molecule in Figure 12.12 could be drawn as a wedge-dash drawing: Cl, H, C, Br, F. The Cl and the H atoms are in the plane of the page, whereas the dashed line to the Br indicates that the Br atom is behind the page, extending away from the viewer. The solid wedge going to the F indicates that the F atom is in front of the page, oriented toward the viewer.

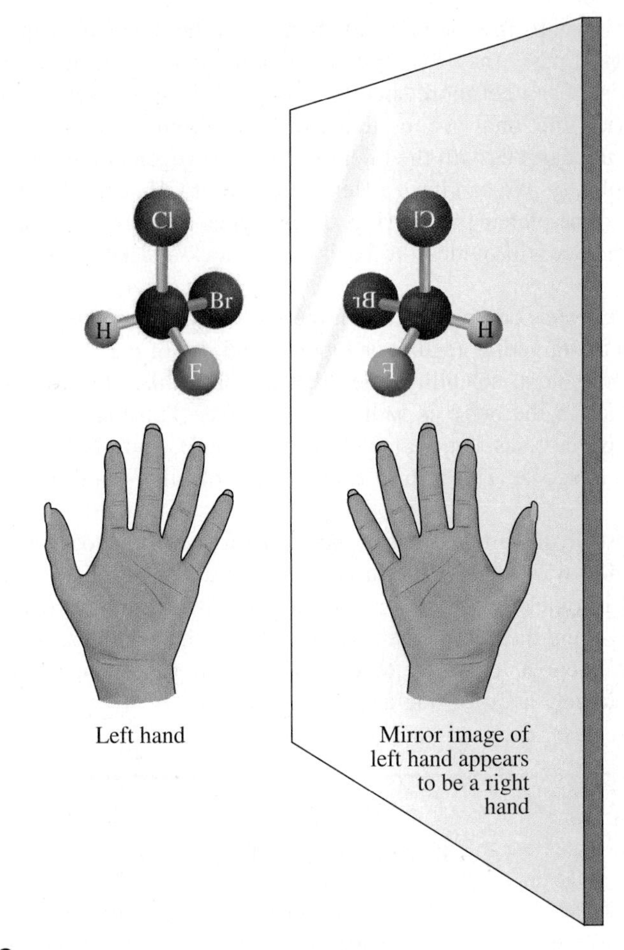

Figure 12.12

Mirror image of a molecular model and a hand. The molecule CHBrClF is chiral and its shape is tetrahedral.

As shown in Figure 12.13, four atoms or groups of atoms bonded to a central carbon atom are in a tetrahedral arrangement. The positions of these four atoms correspond to the corners of a three-dimensional tetrahedron with equal triangular faces. The “handedness” of these molecules gives rise to the term *chiral,* from the Greek word for hand.

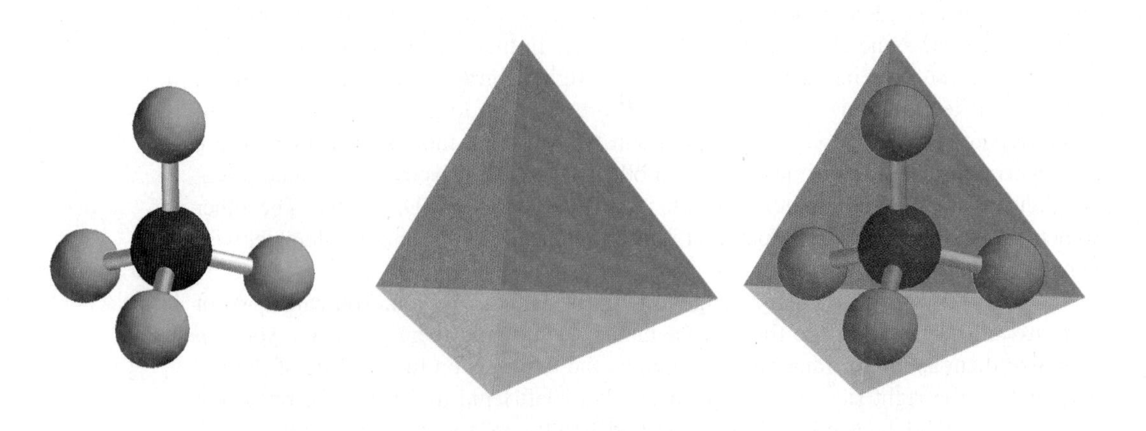

Tetrahedral molecule Tetrahedron Tetrahedral molecule inside a tetrahedron

Figure 12.13

A tetrahedron has four triangular faces.

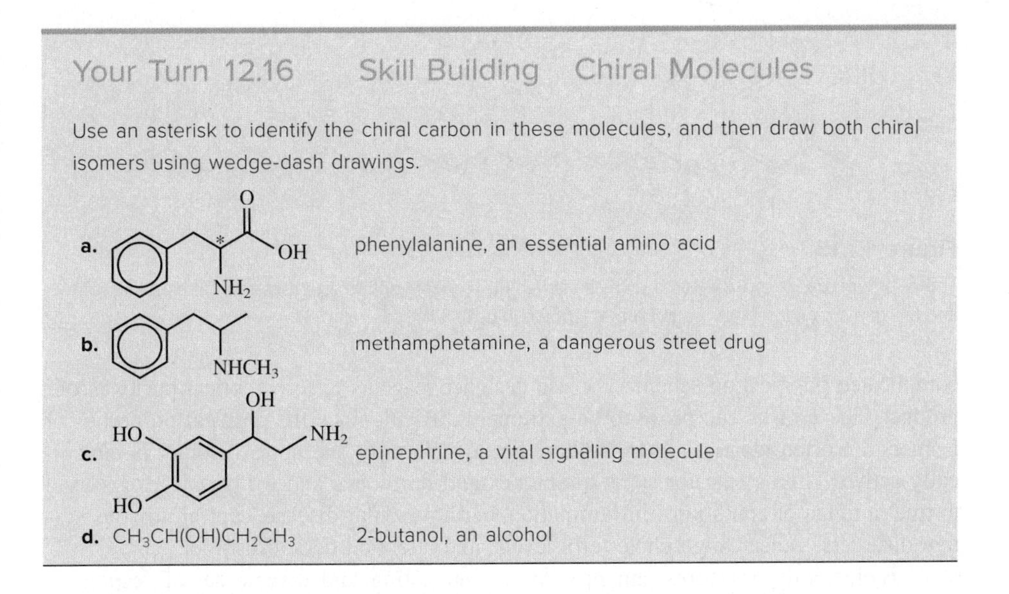

Many biologically important molecules, including sugars and amino acids, exhibit chirality. This is significant because, although most chemical and physical properties of a pair of optical isomers are very nearly identical, their biological behavior can differ markedly. Generally, the explanation for this difference is related to the necessity of a good molecular fit between a molecule and its receptor site. Maybe Lewis Carroll's Alice had some inkling of this when, in *Through the Looking Glass,* she remarked to her cat, "Perhaps looking-glass milk isn't good to drink."

Sugars and amino acids were discussed in Chapter 11 in the context of food.

You can illustrate the relationship between chirality and biological activity by taking things into your own hands. Your right hand fits only a right-handed glove, not a left-handed one. Similarly, a right-handed molecule fits only a site that complements and accommodates it. Any molecule containing a carbon atom with four different atoms or groups attached to it will exist as chiral isomers, only one of which usually fits into a particular asymmetrical receptor site (Figure 12.14).

The extreme molecular specificity created by chirality complicates the study of biochemistry as well as the medicinal chemist's task of synthesizing drugs, but dramatically improves the sensitivity and nuance of biological interactions. To interact, a molecule must include the appropriate functional groups in the correct three-dimensional configuration to give the molecule its desired biological activity.

In many chemical reactions, the "right" and "left" optical isomers are produced simultaneously. Such a situation results in a **racemic mixture**, consisting of equal amounts of each optical isomer. But in biological systems, the machines that synthesize new molecules all have inherent chirality themselves. To create synthetic mimics of biologically active molecules, we must use expensive chiral starting materials, go through

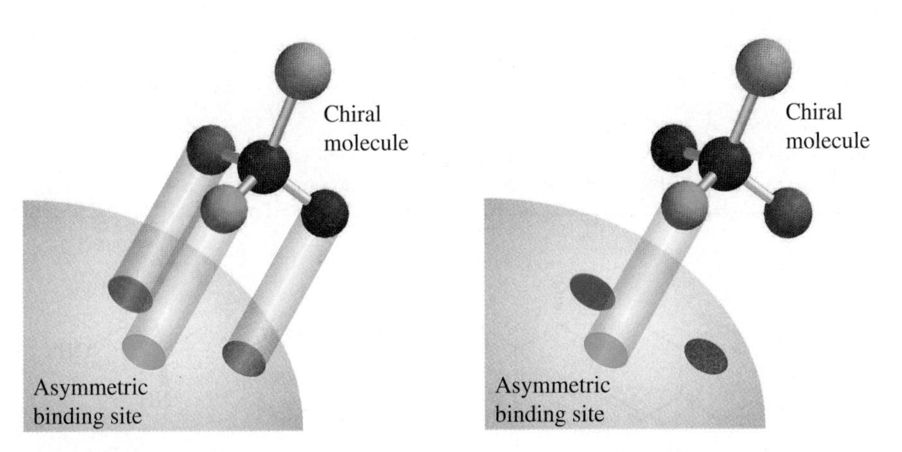

Figure 12.14

A chiral molecule binding to an asymmetrical site (left); its mirror image cannot.

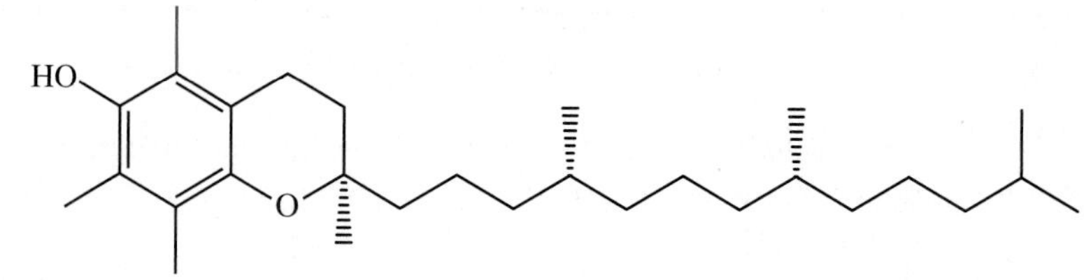

HO O

Figure 12.15

The α-form of tocopherol (better known as vitamin E). The only difference between the alpha (α) form and the gamma (γ) form is a methyl group (shown in red).

complicated purification schemes, or use designer reagents to force optical purity in our product. These tasks can prove to be extremely difficult since the physical properties of isomers are often identical. Why bother? Frequently, only one optical isomer is biologically active. This is true for some antibiotics and hormones and for certain drugs used to treat a range of conditions: inflammation, cardiovascular disease, central nervous system disorders, cancer, high cholesterol levels, and attention deficit disorder.

Only plants produce vitamin E. Animals, such as you, must find a good source to collect this valuable antioxidant.

Biologically, mixtures can provide an interesting and diverse set of responses. Vitamin E supplements are considered necessary for antioxidant properties, including potent anti-cancer properties as well as a vital role in fetal development. However, vitamin E is not one compound. The number of chiral centers (as well as a few biosynthetic precursors) means that vitamin E is a complex mixture of compounds. Biologically, the D-isomer of the α-form is particularly noted in fetal development, while the D-isomer γ-form is the most potent against some cancers (Figure 12.15). Your systems carefully select between optical and structural isomers to absorb and concentrate specific forms of the vitamin E family into appropriate tissues. The synthetic form of vitamin E is a racemic mixture, which has only half of the activity of the racemically pure form. Simply, the body's import pathways (like much of your body) are chiral and refuse to import or use some of the mirror-image compounds. However, any excess is likely flushed from your system as waste and does you no harm.

A racemic mixture can also be damaging if the optical isomers are both active in your system. Naproxen (Figure 12.16), a common pain reliever, is one example of many in which chiral purity is not only preferred, it is required. One form of naproxen relieves pain; the other causes liver damage.

One last example involves a treatment for Parkinson's disease. L-DOPA (Figure 12.16) is a synthetic precursor of the neurotransmitter, dopamine. The initial use of racemic DOPA for treatment of this disease brought on adverse effects such as anorexia, nausea, and vomiting. However, the use of the single isomer greatly reduced these side effects.

Your Turn 12.17 Skill Building Examining Naproxen and L-DOPA

Carefully examine the structural formulas for naproxen and L-DOPA given in **Figure 12.16**.

a. For each drug, which is the chiral carbon atom?

b. Identify all of the functional groups present in both drugs.

c. Draw the structural formulas for the mirror images of both molecules.

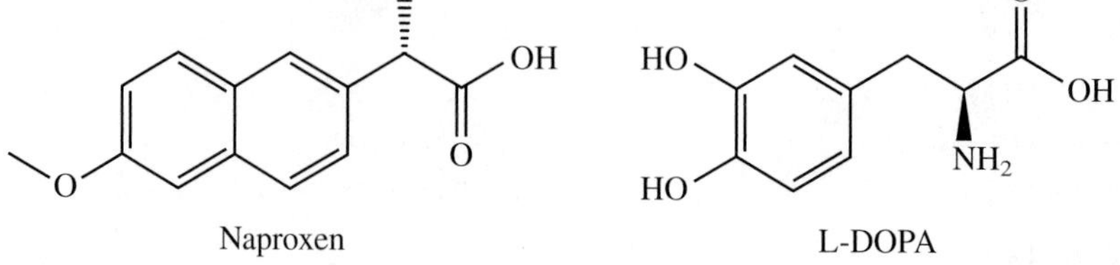

OH O O Naproxen

HO HO O OH NH_2 L-DOPA

Figure 12.16

Biologically active forms of naproxen and L-DOPA.

12.6 | Life via Protein Function

Small molecules do not act in isolation. They elicit your physiological response through their interactions with the comparatively massive macromolecules in your systems. You may already be familiar with some of these large molecules. If your neurons were firing, you will remember grabbing that pencil at the beginning of this chapter. In an automated sort of way, your body underwent numerous actions—only some of which were as a direct result of your conscious action. At some point, every action required a change in the behavior of a macromolecule.

Biological macromolecules can have molar masses of thousands to millions of grams per mole.

Did You Know? Your blood type is largely determined by the presence of specific oligosaccharides (polymers composed of 2–10 simple sugar units) on the outside of cells.

Life depends on four major classes of macromolecules: lipids, carbohydrates, nucleic acids, and proteins (Figure 12.17). As described in Chapter 11, lipids provide

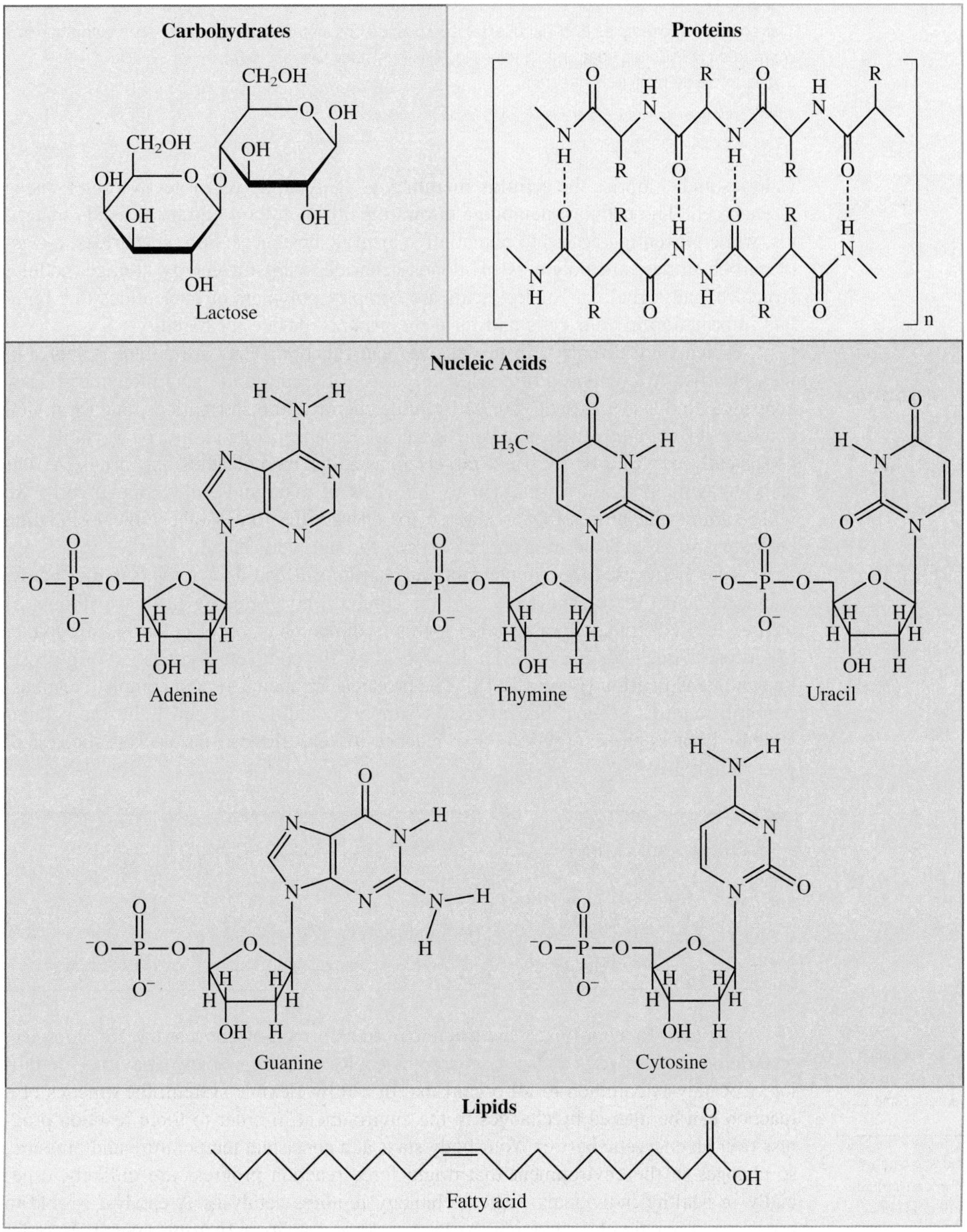

Figure 12.17

Macromolecule classes: carbohydrates, proteins, nucleic acids, and lipids.

(a)

(b)

Figure 12.18

Two contrasting examples of sugar use. **(a)** Combustion of simple sugars is cleanly seen in the burning of wood that produces heat, light, and chaos. This is in stark contrast to **(b)**, a more efficient use of sugars to create muscle contraction.

Discovery of many previously overlooked functions of nucleic acids is a rapidly growing research field. Look to Chapter 13 for fascinating new information about nucleic acids.

In Chapter 9, we discussed proteins as polymers. In Chapter 11, we discussed them as nutrients.

Revisit Chapter 5 to explore the combustion of fuels.

Calories and comprise the **cellular membrane**—a dynamic yet protective outer casing of each cell. The cellular membrane allows organisms to concentrate valuable materials, while preventing access to potentially harmful substances. Polysaccharides, a class of carbohydrates, are polymers of monosaccharides vital for energy storage, cellular structure, and signaling. Nucleic acids are complex polymers of nucleotides that regulate information in your cells and form the primary device for heredity.

Proteins are diverse both in structure and function. They are polymers, polyamides specifically, composed of a selection of 20 different amino acid monomers. They contribute to the structure of your cells, guide chemical transformations, and form vital conduits for signaling between your systems. Because proteins are so versatile and consistently prove to be the final targets for smaller biomolecules and drugs in your cells, we will focus our discussion to this class of macromolecules. Specifically, we will examine the processing of glucose for energy, the routes of chemical signaling between organs, and the transport of oxygen through your blood.

Your body uses the simple monosaccharide glucose as a universal fuel across your cells and tissues. You have seen the combustion of glucose in this textbook, as well as in vivid living color in a fire—this is the combustion of cellulose, a polymer of glucose. Your body uses a different process for the complete combustion of glucose, known as **respiration** (Figure 12.18). The products are identical, and complete combustion still requires oxygen, but instead of simply providing heat and light, the cellular systems harness the energy stored in glucose to keep the cell (as well as the rest of your body) alive.

Your Turn 12.18 Skill Building Glucose Reaction

The chemical formula for glucose is $C_6H_{12}O_6$.

a. Write the balanced reaction for the complete chemical combustion of glucose.
b. Determine how much oxygen is required to completely combust 10 grams of glucose.

Review chemical catalysis and activation energy in less biological contexts in Chapters 2 and 5.

It is important to note that a catalyst does not change the equilibrium constant of the products versus reactants.

In order to coordinate the chemical transformations necessary for complete respiration, biology uses a class of macromolecules called enzymes. **Enzymes** are biological catalysts common to all organisms. In a fully flexible system, the kinetics of a reaction can be altered by changes in the environment in order to force reaction progress over an energetic barrier. Your body stays at a consistent temperature and pressure, so changes in the environment that might force reaction progress are unlikely, especially in healthy individuals. Instead, biology requires catalysis. A catalyst is able to accelerate a reaction by changing the barrier, known as its **activation energy**. In order to accelerate and control reactions, enzymes contain an **active site**, or catalytic region.

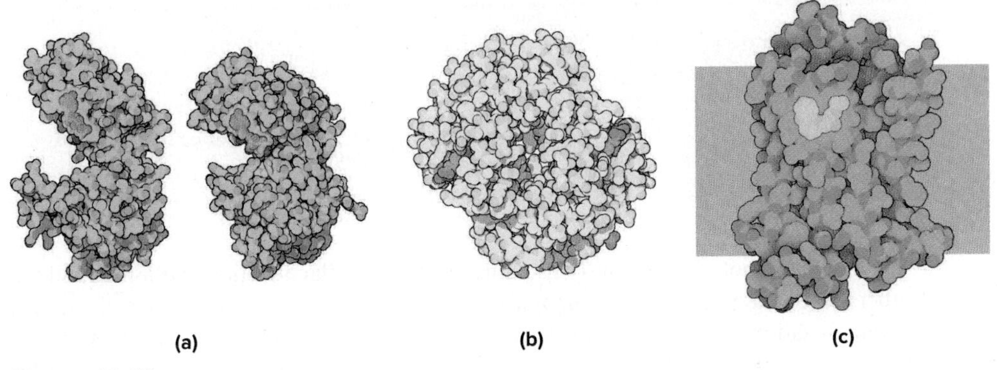

Figure 12.19

Proteins, a vital class of macromolecules, in 3-D. **(a)** The enzyme, phosphoglycerate kinase, that performs the seventh step of many necessary to thoroughly convert glucose to useful cellular energy.[1] **(b)** Hemoglobin holds oxygen on the iron in the center of the organic molecule heme (**red**).[2] **(c)** A hormone receptor on a cellular membrane (grey) bound to its hormone (**orange**).[3]

In the active site, the enzyme selectively binds only specific reactant(s) to promote the desired reaction (Figure 12.19).

To break down glucose in the body, your system requires something more complex than a lit match. Cellular respiration in more complex organisms requires three different stages involving the carefully regulated chemical coordination of at least 25 enzymes located in three distinct cellular compartments.

Stage one of cellular respiration is the most universal. **Glycolysis** converts intracellular glucose into higher-energy three-carbon sugars, while collecting some useful chemical energy. When our muscles require energy immediately and do not have time to wait for slower aerobic processes, we are able to depend on glycolysis alone as well. Glycolysis operates even in the absence of oxygen, conditions referred to as *anaerobic respiration*—required for many lower-order microorganisms (*e.g.,* yeasts, bacteria, internal parasites). This first stage of respiration can be effectively reversed to convert smaller sugars, as well as the products of oxidation of fatty acids or proteins, back to glucose for storage as the polysaccharide, glycogen.

Did You Know? The term *glycolysis* is derived from the Greek terms for the process of breaking (*lysis*) something sweet (*glycol*).

Glycolysis in the absence of further respiration produces *lactic acid*—believed to be the chemical source responsible for the familiar, and painful, ache after rapid exercise.

The remaining stages of respiration require oxygen and produce carbon dioxide, while gathering far more energy per molecule of glucose. Specifically, through a complex series of reactions, each molecule of glucose can be converted into ~32 molecules of adenosine triphosphate (ATP, Figure 12.20). Why bother converting one chemical fuel into another? ATP is the chemical fuel of choice for almost all costly transformations and changes in the cell. The molecule ATP provides a nicely sized universal currency for the regular work of the cell. Other fuels, such as fat and protein, follow similar paths to feed into the same processes as carbohydrates.

The process of manipulating energetic molecules, either breaking them down to produce ATP or building them up to store energy is called **metabolism**.

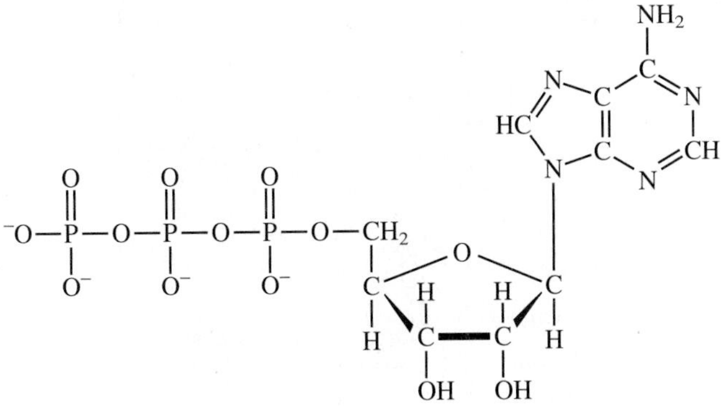

Figure 12.20

Adenosine triphosphate, ATP.

[1] http://pdb101.rcsb.org/motm/50

[2] http://pdb101.rcsb.org/motm/41

[3] http://pdb101.rcsb.org/motm/100

Your Turn 12.19 You Decide ATP Formation

The molecular structure of glucose is comprised of only 24 atoms, but is converted into 32 molecules of ATP. Wouldn't this conversion violate the Law of Conservation of Mass? Explain.

Cells do not spontaneously use glucose; this cellular function and a multitude of others is coordinated across your body and triggered. We normally think of internal communication as consisting of electrical impulses traveling along a network of nerves. This is true for the system that triggers movement, breathing, heartbeats, and reflex actions. However, most of the body's messages are conveyed not by electrical impulses, but through chemical processes. In fact, it is much more efficient to release chemical messengers into the bloodstream where they can be circulated to appropriate body cells, than to "hardwire" each individual cell with nerve endings.

The chemical signals produced by your endocrine glands to regulate other systems are called **hormones** (Figure 12.21). These messengers encompass a wide range of functions and a similarly wide range of chemical compositions and structures. Thyroxine, a hormone secreted by the thyroid gland, is essential for regulating metabolism. The ability of the body to carry a stable quantity of glucose through the blood depends on insulin. This hormone, a small protein built from 51 polymerized amino acids, is secreted by the pancreas. Persons who suffer from diabetes are often required to take daily injections of insulin. Yet another well-known hormone is adrenaline (epinephrine), a small molecule that prepares the body for "fight or flight" in the face of danger.

Chapter 13 presents the miracle of modern medicine that allows for the life-saving production of insulin synthetically.

As most hormones remain outside of the cell, they require receptors on the cells they affect. A **receptor** is a specific class of biomolecule generally embedded in the cellular membrane. In order to respond to extracellular hormones, the receptor must span the membrane so it can sense and respond to the presence of the hormone, while also promoting activity within the cell. On the outside of the cell, a membrane receptor is specifically engineered to respond perfectly and selectively to one type of hormone. Through the cellular membrane, the receptor changes slightly to convey the message "epinephrine here," like a doorbell for the cell. This results in a cascade of

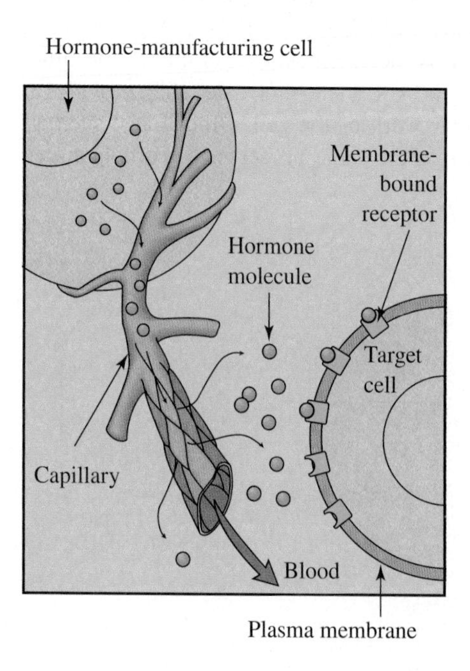

Figure 12.21

Chemical communication in the body. Hormone molecules travel through the bloodstream from the cell where they are positioned to the target cell containing the correct receptor.

events tuned exactly to the purpose of the hormone. The receptor changes in shape, allowing the hormone to dock but not pass, while still permitting information to be transferred through the membrane. Cells in different organs will have different receptors to tailor their response to specific hormones according to the big-picture role of the organ in your physiology.

As previously mentioned, the final stages of complete respiration of glucose require oxygen. Because this is a universal fuel, the oxygen you inhale must be spread universally throughout your tissues. Your blood cells carry oxygen molecules from your lungs to your tissues, but they cannot do it without a particular biomolecule. Your red blood cells carry oxygen through the action of **hemoglobin**, a class of biomolecules specifically designed to transport oxygen through binding to an iron-containing organic molecule called heme.

Learn more about hemoglobin and some specific complications in the disease sickle cell anemia in Chapter 13.

Typically, the exposure of iron to oxygen gas will result in the formation of iron(III) oxide, Fe_2O_3, known as rust. Why, then, doesn't the iron in your blood oxidize to form rust? The hemoglobin carefully transports the oxygen bound to the iron while preventing the reaction of oxygen and iron. In addition, hemoglobin has tuned its binding strength according to the amount of oxygen available. In the lungs where oxygen concentration is high, the binding strength is very high so hemoglobin picks up four molecules of oxygen—its maximum capacity. In the tissues where the oxygen concentration is lower, the binding strength falls, so hemoglobin releases all of the oxygen at once. The message to release oxygen is reinforced by changes in pH and the presence of small molecules that are produced by tissues in need of oxygen.

Your Turn 12.20 Scientific Practices Extra Protein Structures

The structures of proteins, such as the ones shown in **Figure 12.19**, have been gathered in a repository online, the Protein Data Bank. Explore the Molecule of the Month archive on the site to find a protein not discussed here. Summarize your discovery. What role does the protein play in biology? What features of the protein are most interesting?

12.7 | Life Driven by Noncovalent Interactions

Each scenario described in the previous section required controlled and specific interactions between small molecules and proteins. How can this class of polyamides provide the sensitivity necessary for all of the functions described (as well as the numerous functions we have not even discussed)? Proteins have the same polyamide backbone that you may find familiar from nylons, but they also have a diverse variety of functional groups that extend from this backbone structure (Figure 12.22).

Nylons and other polyamide-based polymers were discussed in Section 9.7.

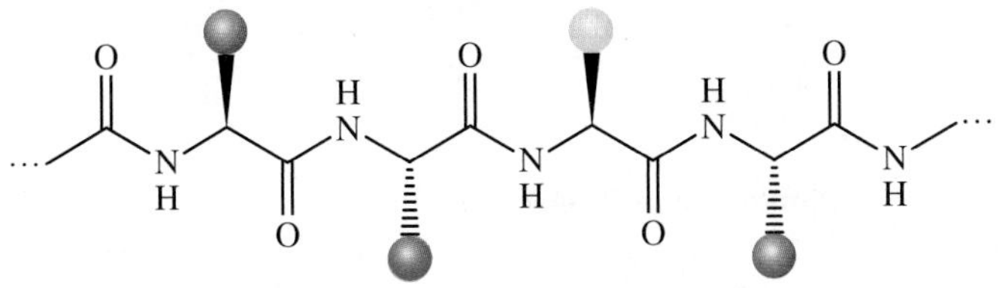

Figure 12.22

A symbolic view of diversity in proteins. A variety of functional groups (colored spheres) extend from the polyamide backbone of the protein segment.

Your Turn 12.21 Scientific Practices Some Context

Reconsider your work in past chapters. List three examples of polyamides that you have explored previously and the context of each.

Recall that proteins are polymers composed of amino acids monomers. Biology commonly uses 20 different versions of these monomers to generate a large variety of proteins. Each amino acid has a carboxylic acid (–COOH), an amine (–NH_2), and a side chain group. The side chain plays no part in the process of condensation polymerization, but simply sticks off of the polymer and interacts with other side chains, as well as the molecules around the protein. Reexamine Figure 11.17 to explore the variety of functional groups and preferred intermolecular forces.

Your Turn 12.22 Skill Building Amide Practice

Choose two amino acids of your choice and draw their combination to form a dipeptide. Is your dipeptide polar, nonpolar, or mixed?

Hint: Remember that amines and carboxylic acids react to form the amide functional group.

The amino acids interact with the structural facets of their relevant small-molecule targets. The three-dimensional structures of proteins have a limited range of motion because their overall architecture is relatively well-defined. The result is that small molecules must present a stable and well-defined set of functional groups in order to bind with high affinity and specificity to a receptor protein, or within an enzyme active site. This is the basis of discriminating between closely related molecules, even those with subtle differences such as optical isomers.

Although the structures of proteins have relatively well-defined shapes, they do have some room for motion. On most proteins, this literal wiggle-room changes the overall shape rather imperceptibly from an outside perspective, but the implications for binding to smaller molecules can be quite significant. Three models describe the breadth of possible binding modes, given this limited variable of protein movement. In the *lock-and-key* model (Figure 12.23(a)), the the exact shapes of both molecules are set before the association occurs. Selectivity occurs because the small molecule has the correct functional groups positioned perfectly to interact with the well-defined positions of amino acids in the protein target.

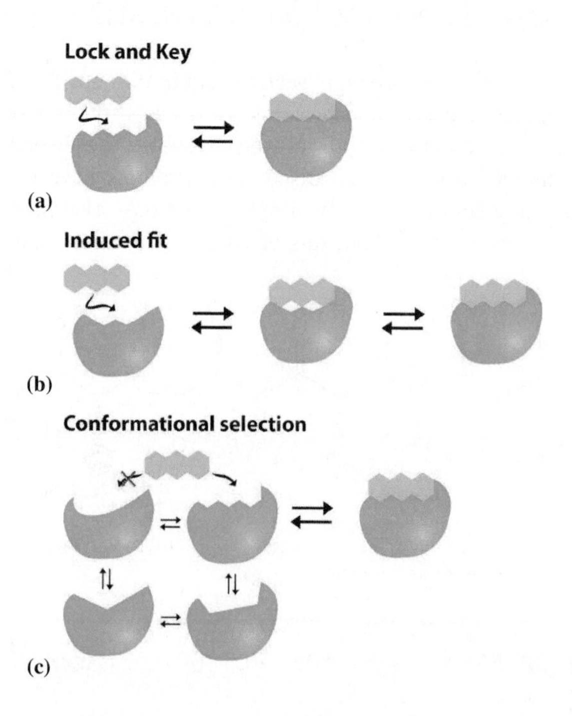

Figure 12.23

A comparison of three binding models in biological interactions.

The *induced fit* and *conformational selection* models (Figure 12.23(b) and (c)) include the concept of motion in either the protein or its interacting partner. Induced fit suggests that the molecules shift into the correct binding position only upon binding. Specificity in this model comes from the fact that only the correct molecule will be able to reach that perfect binding interaction. Conformational selection suggests that one of the interacting partners switches between multiple three-dimensional shapes before the binding event. The binding partner picks out the specific shape necessary to correctly associate. Again, specificity in this model comes from the fact that only the correct molecule will be able to reach the correct conformation necessary for the perfect binding interaction.

Your Turn 12.23 Scientific Practices Exploring Interactions

Reexamine the structure of epinephrine in **Figure 12.10**. Which category of amino acids shown in **Figure 11.17** (nonpolar or acidic, basic, or neutral polar) should associate with each of the functional groups in order to bind with and respond to adrenaline?

12.8 Steroids: Essential Regulators for Life (and Performance Manipulators!)

As a key family of compounds, **steroids**, are a class of naturally occurring or synthetic fat-soluble organic compounds that share a common carbon skeleton arranged in four rings. Biologically, they serve as the best illustration of the relationship between form and function. The naturally occurring members of this ubiquitous group of substances include structural cell components, metabolic regulators, and the hormones responsible for secondary sexual characteristics and reproduction. Table 12.4 lists some of the diverse functions of steroids.

Secondary sexual characteristics are physical characteristics that first appear during puberty and do not have a direct reproductive function.

Table 12.4 Steroid Functions

Function	Example Molecules
Regulation of secondary sexual characteristics	estradiol (an estrogen), testosterone (an androgen)
Regulation of the female reproductive cycle	progesterone
Regulation of metabolism	cortisol
Digestion of fat	cholic acid
Component of cellular membranes	cholesterol
Stimulation of muscle and bone growth	gestrinone, trenbolone

Did You Know? This class of molecules has also been used for artificial manipulation of our biology as synthetic drugs for birth control, abortion, bodybuilding, and sports performance enhancement.

In spite of their tremendous range of physiological functions, all steroids are built on the same molecular skeleton. While this may seem like a risky recipe for loss of specificity, these compounds actually provide a marvelous example of the chemical economy in living systems. The common characteristic of steroids is a molecular framework consisting of 17 carbon atoms arranged in four rings, illustrated below:

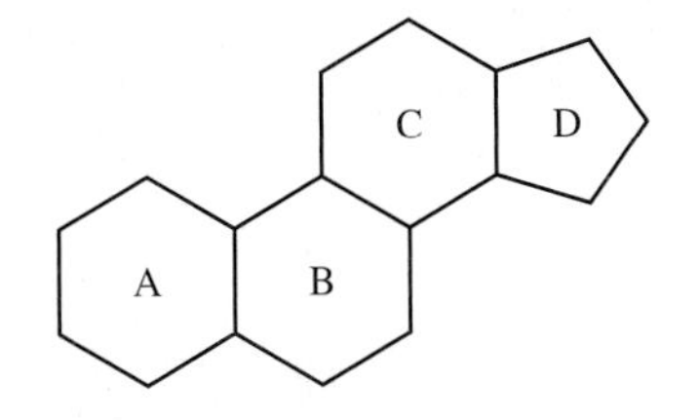

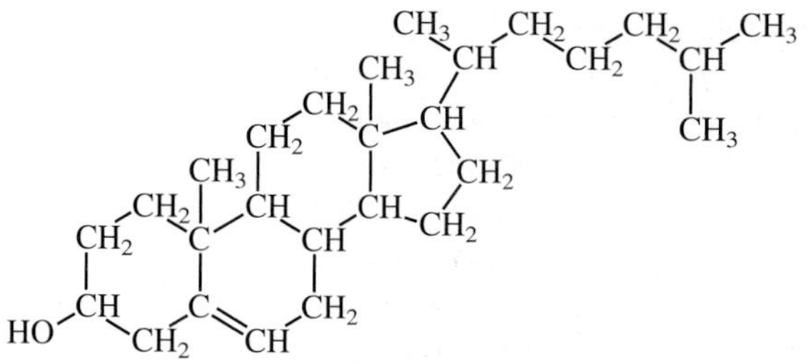

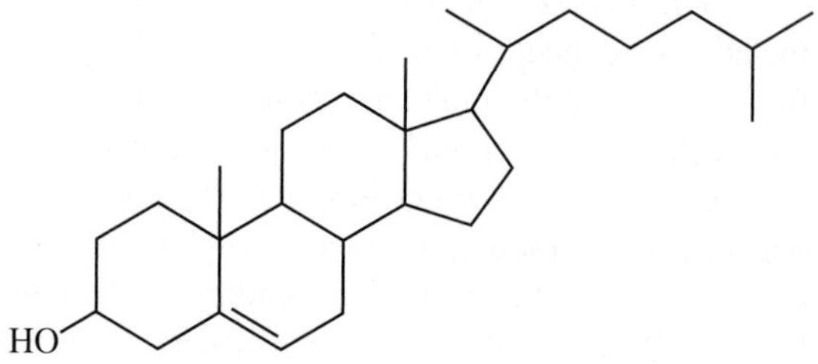

Figure 12.24

Structural representations of cholesterol.

Recall that in such a representation (a line-angle drawing), carbon atoms are assumed to occupy the vertices of the rings but are not explicitly drawn. The three six-membered carbon rings of the steroid skeleton are designated A, B, and C, and the five-membered ring is designated D. Although the steroid framework appears flat as drawn, it actually is three-dimensional in shape. The dozens of natural and synthetic steroids are all variations on this theme. Some differ only slightly in structural detail, but have radically different physiological functions. Extra carbon atoms or functional groups at critical positions on the rings are responsible for this variation.

The steroid *cholesterol* is a major component of cell membranes and is shown in Figure 12.24. The figure on the left includes all the atoms in the molecule; the one on the right gives the skeletal representation using a line-angle drawing.

Careful examination of Figure 12.25 illustrates how subtle molecular differences can result in profoundly altered physiological properties. The difference between a molecule of estradiol and one of testosterone lies only in one of the rings. Each of these forms a family of molecules with even more subtle chemical changes.

Your Turn 12.24 You Decide Men and Women

Are the only differences between men and women due to a carbon atom and a few hydrogen atoms? Elaborate, based on the molecular structures and properties of estradiol and testosterone, as well as the role of other hormones found in males and females.

Your Turn 12.25 Skill Building Structural Similarities of Steroids

Using the Internet, evaluate the pairs of steroids listed below. Write their chemical formulas and identify some structural similarities in each pair.

a. estradiol and progesterone
b. corticosterone and cortisone
c. cholic acid and cholesterol

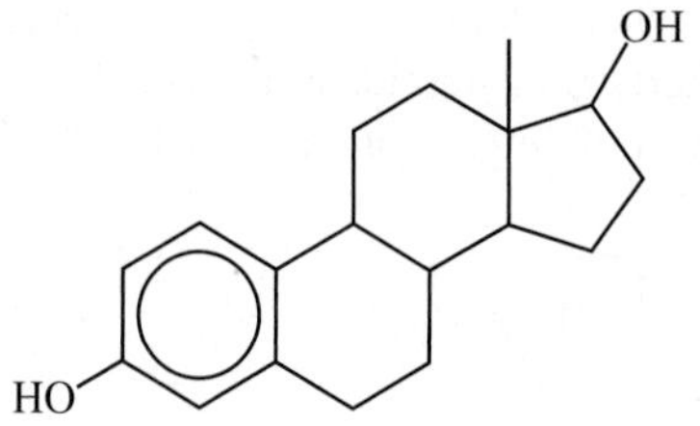

estradiol

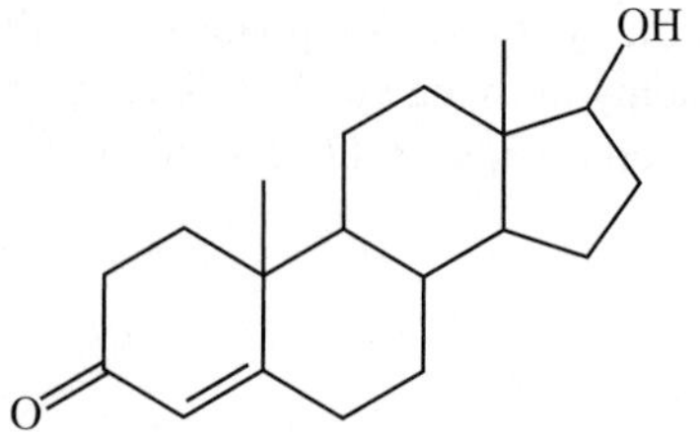

testosterone

Figure 12.25

Estradiol and testosterone.

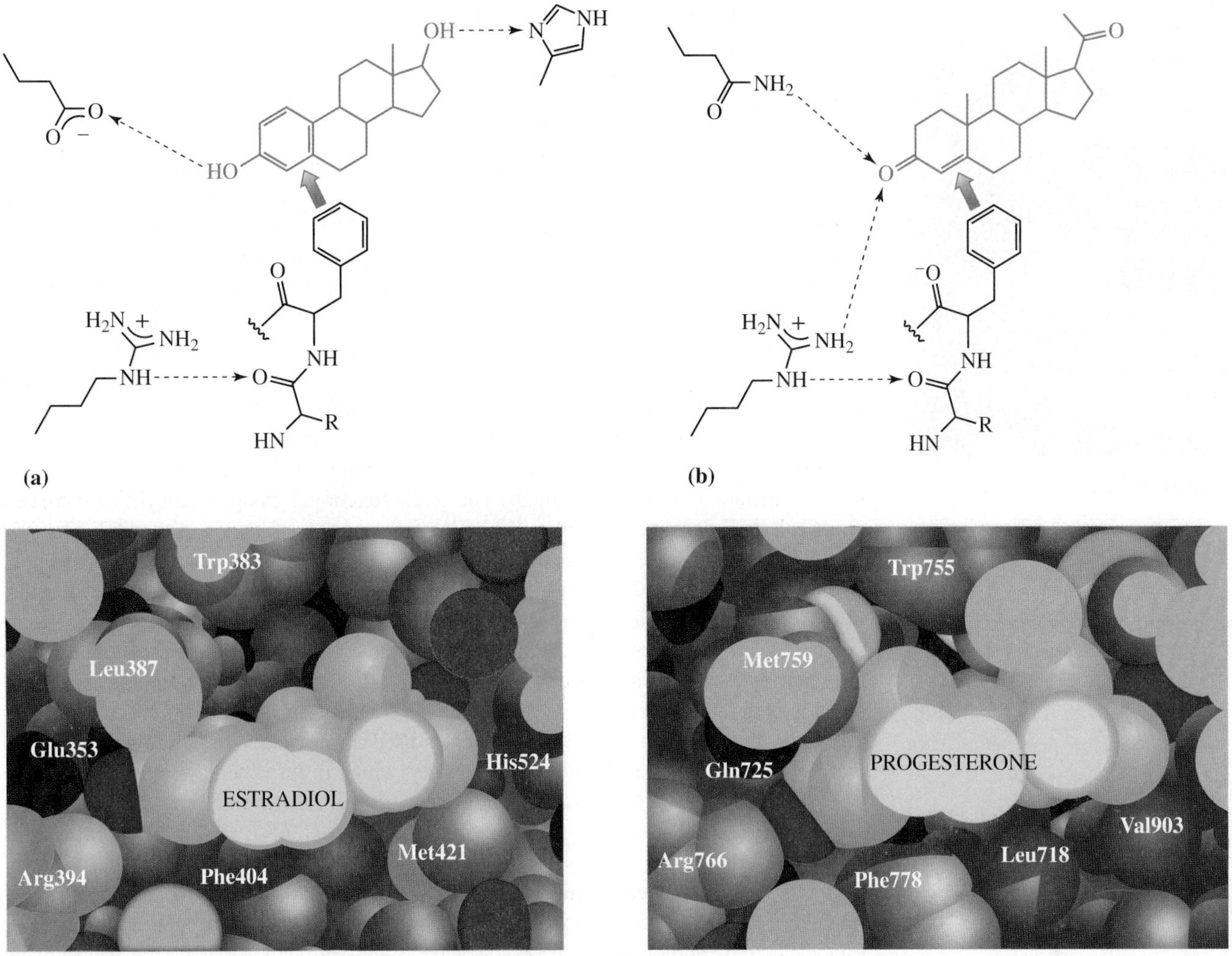

Figure 12.26

Selective receptor binding pockets for estradiol (an estrogen) versus progesterone (an androgen).

Source: (a–b) From David M. Tanenbaum, et al., "Crystallographic comparison of the estrogen and progesterone receptor's ligand binding domains," *Proceedings of the National Academy of Sciences*, **1998**, Vol. 95, No. 11, Figure A3. Copyright (1998) National Academy of Sciences, U.S.A.

To complete our study of steroids, let's examine the proteins that bind them. Figure 12.26 illustrates the difference in binding with target intracellular receptor proteins between an estrogen and an androgen. Differing amino acid residues in the active site result in differences in intermolecular forces, which leads to preferential binding of one molecule relative to another.

12.9 | Modern Drug Discovery

Modern pharmacology has its origins in folklore. The use of herbs, roots, berries, and barks for relief from illnesses can be traced to antiquity, as illustrated in documents recorded by ancient Chinese, Indian, and Near East civilizations. More recently, chemists have designed, synthesized, and characterized a vast array of prescription and over-the-counter drugs. Today, drugs help patients regulate their blood sugar, blood pressure, cholesterol, and allergies. The course of the HIV/AIDS epidemic has brought about new treatment options, which now includes preventative drugs. Effective anticancer drugs and powerful analgesics now exist. Other drugs can even manage mental disorders that once were thought untreatable. How is it that modern chemists continue to refine our medical options?

The evolution of willow bark tea to aspirin and further modifications to this painkiller's structure is a fantastic case study in both historical and modern drug design. In the 4th century BC, Hippocrates (460–370 BC) described a tea made by

Figure 12.27

The white willow tree, *Salix alba*, the original source of an inspirational drug.

© Terry Wild Stock

boiling willow bark (Figure 12.27) in water. The concoction, common to many different cultures, was said to be effective against fevers. Over the centuries, the folk remedy ultimately led to the synthesis of a true wonder drug, *aspirin*, a drug with continued potential to aid millions of people.

From the 4th century BC to the 18th century AD, the possible pharmaceutical benefits of willow bark were left unexplored. Edmund Stone (1702–1768), an English clergyman, set the stage for modern explorations with a report to the Royal Society in 1763 on the success of powdered willow bark as a treatment. Chemists were subsequently able to isolate small amounts of yellow, needle-shaped crystals of a substance from the willow bark extract. Because the tree species was *Salix alba*, this new substance was named salicin. Experiments showed that salicin could be chemically separated into two compounds, only one of which reduced fevers and inflammation, salicyl alcohol. Once within the body, metabolism converts an alcohol in this active compound to a carboxylic acid and generates the true active ingredient in willow bark tea, salicylic acid (Figure 12.28).

After its discovery, salicylic acid was used to treat pain, fever, and inflammation. Unfortunately, it not only had a very unpleasant taste, but its acidity also led to acute stomach irritation in some individuals. The acidic functional group of salicylic acid created a difficult drug design conundrum. While it is necessary for function in the body, the side effects are serious. The most intuitive solution is a rather simple neutralization reaction. The acid can be neutralized with a base, either sodium hydroxide or calcium hydroxide, to form a salt of the acid. The resulting salts have fewer side effects than the parent compound. Generating the salt of a promising, but acidic or basic compound continues to be an important strategy in drug development. The salt form of common organic acids and bases is preferable because it is less reactive, has less of an odor, and is more water soluble. An estimated half of all drug molecules used in medicine are administered as salts that improve their water solubility and stability, which in turn increase their shelf life.

Acid–base neutralization reactions were discussed in Chapter 8.

In the history of the development of aspirin, neutralization alone was not enough. To further the potential of the drug, Felix Hoffmann (1868–1946) and his colleagues at IG Farbenindustrie, a major German chemical firm now incorporated in Bayer, explored a series of covalent modifications of salicylic acid. Rather than simply neutralizing the carboxylic acid, the chemists most successfully modified the structure of salicylic acid by adding more carboxylic acids! Specifically, they reacted those carboxylic acids with the alcohol on salicylic acid. Remember that carboxylic acids and alcohols react to form a novel functional group, the ester.

Did You Know? Legend has it that Hoffmann's motivation was more than just scientific curiosity or assigned task. His father regularly took salicylic acid as treatment for arthritis and suffered greatly from its side effects.

Your Turn 12.26 Skill Building Ester Formation

Draw structural formulas for the esters that form when these alcohol and acid pairs react.

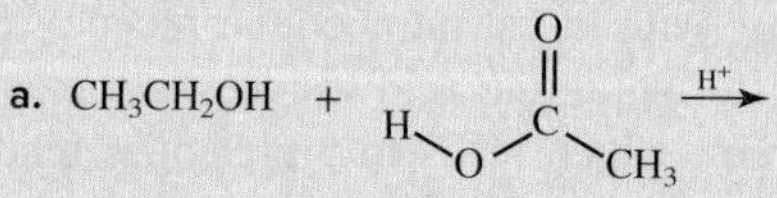

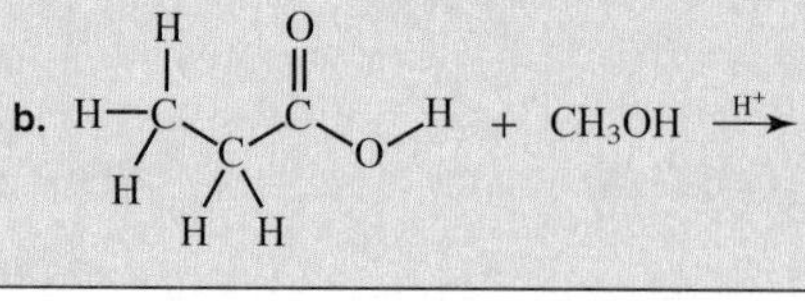

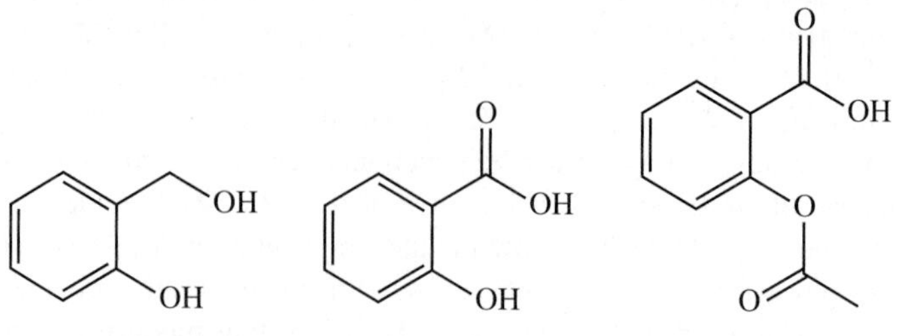

a. $CH_3CH_2OH + H{-}O{-}C({=}O){-}CH_3 \xrightarrow{H^+}$

b. $H{-}C(H)(H){-}C(H)(H){-}C({=}O){-}O{-}H + CH_3OH \xrightarrow{H^+}$

Salicyl alcohol   Salicylic acid   acetylsalicylic acid

Figure 12.28

Chemical variations to a wonder drug: salicyl alcohol, salicylic acid, and acetylsalicylic acid, a compound more commonly known as aspirin.

While Hoffman and his colleagues generated a variety of new esters with different carbon lengths and functional groups, the best hit was one of the simplest. The reaction of salicylic acid with acetic acid produced acetylsalicylic acid (Equation 12.11). Although the carboxylic acid is untouched, the ester modification still changes the traits of the compound enough to greatly reduce nausea and other adverse side effects.

$$\text{salicylic acid} + \text{acetic acid} \xrightarrow{H^+} \text{acetylsalicylic acid} + \text{water} \qquad [12.11]$$

Because aspirin retains the –COOH group of the original salicylic acid, it still has some of the undesirable acidic properties of the parent compound. However, the ester group (yellow area in Equation 12.11) makes the compound more palatable and less irritating to the stomach lining. Once aspirin reaches the bloodstream, Equation 12.11 is reversed. The ester splits into acetic acid and the active salicylic acid, and the latter compound exerts its antipyretic (fever-reducing) and analgesic (pain-reducing) properties.

Your Turn 12.27 Scientific Practices Common Drugs, Common Structural Features

Approximately 40 alternatives to aspirin have been produced—with ibuprofen and acetaminophen being the most familiar. The compounds are extraordinarily similar as compounds with similar molecular structures that often share useful physiological properties.

a. Look up the structures for ibuprofen and acetaminophen and identify two shared and two different structural features with aspirin.

b. Look up the symptoms that each medication is advertised to treat. Do you find any similarities or differences?

c. Look up the top side effects for each medication. Do you find any similarities or differences?

Your Turn 12.28 You Decide Supersize My Aspirin

A friend who suffers from heart disease has been told by the doctor to take one aspirin tablet a day. To save money, your friend often buys the large 300-tablet bottle of aspirin. You, on the other hand, rarely take aspirin, but cannot pass up a good bargain. You also buy the large bottle.

a. Why is the "giant economy size" bottle of aspirin not as good a deal for you as it is for your friend?

b. What chemical evidence supports your opinion?

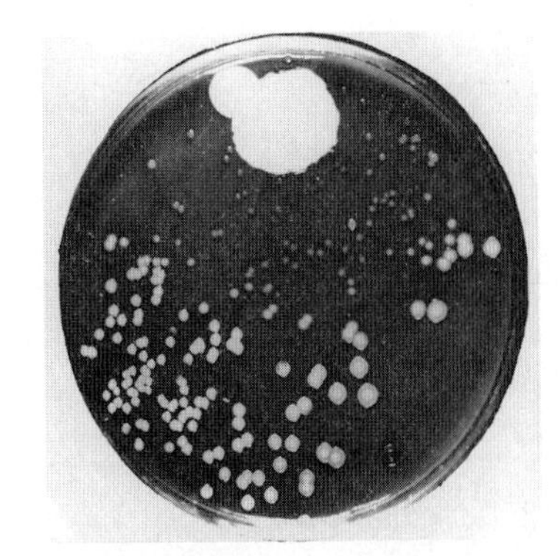

Figure 12.29
Photograph of Sir Alexander Fleming's original culture plate including the fungus *Penicillium notatum* at 12 o'clock; the smaller white spots are areas of bacterial growth.
© *Biophoto Associates/Science Source*

Penicillin is another example of a miracle drug whose origin lies in natural sources. Molds had been used for treating infections for 2500 years, although their effects were unpredictable and sometimes toxic. The penicillin story includes an accidental discovery in 1928 by the British bacteriologist Alexander Fleming (1881–1955). Fleming's curiosity was aroused by the chance observation that in a container of bacterial colonies, the area contaminated by the mold *Penicillium notatum* was largely free of bacteria (Figure 12.29). Spores from the mold, part of an experiment in a nearby

Figure 12.30

A sign posted at the entrance of a new facility for penicillin production during World War II.

Source: Library of Congress, Prints & Photographs Division [LC-USZC4-1986]

lab, drifted into Fleming's laboratory and accidentally contaminated some Petri dishes containing *Staphylococcus* (bacteria) growing on a nutrient medium. He correctly concluded that the mold produced a substance that inhibited bacterial growth, and he named this biologically active material *penicillin.*

Fleming's experience allowed him to interpret the chance phenomenon, recognizing that some unknown substance produced by the *Penicillium* was a potential antibacterial agent. "The story of penicillin," Fleming wrote, "has a certain romance in it and helps to illustrate the amount of chance, of fortune, of fate, or destiny, call it what you will, in anybody's career." But, of course, the discovery would not have happened without Fleming's powers of observation and insight. This episode illustrates the often misquoted maxim of the great French scientist, Louis Pasteur (1822–1895): "In the fields of observation, chance favors only the prepared mind." Most versions of this famous aphorism neglect the "only." It was *only* because Fleming's mind was prepared that he was able to capitalize on this chain of unlikely events.

The process of taking penicillin from the Petri dish to the pharmacy was not much different from what is done today. The first step was a systematic effort to isolate the active agent produced by *Penicillium notatum.* Once identified, the substance had to be purified and concentrated by analytical and chemical techniques. Finally, the efficacy of penicillin in treating humans had to be demonstrated. World War II gave increased impetus to this research and to the development of new methods for preparing large quantities of penicillin. Because the scientists were successful in doing so, thousands of lives were saved during the war (Figure 12.30), and millions since then.

Treatments of infections once considered incurable—pneumonia, scarlet fever, tetanus, gangrene, and syphilis—were revolutionized by Fleming's discovery. The discovery of penicillin may have been serendipitous, but the development of the next several generations of antibiotics in this class involved systematic and careful research. Small structural changes to a drug have resulted in more than a dozen different penicillins currently in clinical use, including: penicillin G (the original discovered by Fleming, and the form that causes an allergic reaction in about 20% of the population), ampicillin, oxacillin, cloxacillin, penicillin O, and amoxicillin (the pink, bubble-gum-flavored concoction you might have been given as a child). Amoxicillin is still available in capsule form; it is commonly prescribed for being effective against a broad spectrum of bacteria and is usually well tolerated.

Your Turn 12.29 Scientific Practices Drugs by Chance

Modern methods of drug discovery involve systematic studies of compounds with only small variations in structure and computer modeling, among other techniques. Sometimes side effects of a drug may open the door for its usefulness in treating other illnesses. There are many examples where a new drug was discovered by "chance." Find an example of a drug that was discovered by unusual circumstances and describe its discovery.

Did You Know? A new class of antibiotics was discovered in early 2015. One member of this new class, known as teixobactin, targets lipid molecules that bacteria use to build their cell walls. Since it's difficult for bacteria to modify these molecules, it is expected to take much longer for resistance to develop.

The effectiveness of penicillin has unfortunately led to extreme overuse. As a result, cunning bacterial bugs have developed mechanisms for rendering penicillin (along with other antibiotics) useless. We are now witnessing strains of resistant bacteria or "superbugs," a phenomenon Fleming predicted back in 1945. Bacteria develop resistance to penicillin by secreting an enzyme that attacks the penicillin molecule before it can act. Some of the newer antibiotics differ in their effectiveness at killing certain bacteria and their susceptibility to the enzymes the organisms produce. Closely related to the penicillins are the cephalosporins (cephalexin, or Keflex™) that are particularly effective against some resistant strains of bacteria. Careful research on structural modifications has led to other important medicines like cyclosporine, a drug that prevents tissue rejection. Its development made possible the revolutionary success of organ transplant surgery.

Although these examples of aspirin and penicillin have been steeped in history, natural sources and medicinal lore continue to be explored with the aim of discovery. Potentially, each new organic molecule may have a distinct or interesting medicinal purpose.

12.10 | New Drugs, New Methods

Modern drug discovery is a larger, broader, and more complex search than the stories in the previous section suggest. Our targets change. Some targets are organisms themselves—like bacteria and viruses—and evolve resistance to our methods of interference. Some recent targets, like cancer, require extreme precision. The biology of cancerous tissue shares many similarities with non-cancerous tissue. Successfully eradicating cancer cells without any deleterious effects to non-cancerous healthy tissue is challenging, if not impossible. Increasingly, we are less impressed with new compounds that include a textbook's worth of possible consequences. We have a lower tolerance for side effects and require drugs to sensitively and specifically alter one, and only one, system. To both aid and complicate our search, we have more biological and biochemical information to define our targets and understand our physiology. This section explores the methods and challenges associated with the development of new pharmaceuticals.

Let's start with the complications of target biology. Although drugs vary in their versatility, many of them act only against particular diseases or infections. This specificity is consistent with the relationship that exists between the chemical structure of a drug and its therapeutic properties. Drugs can be broadly classified into two groups: those that produce a physiological response in the body and those that inhibit the growth of substances that cause infections. Aspirin falls in the first group; it acts against a class of enzymes involved in inflammation response. So do synthetic hormones and psychologically active drugs. These compounds typically initiate or block a chemical action that generates a cellular response, such as a nerve impulse or the synthesis of a protein. Antibiotics exemplify drugs that prevent the reproduction of foreign invaders. They do so by inhibiting an essential chemical process in the infecting organism. Thus, they are particularly effective against bacteria. All act by making a specific interaction with a target macromolecule, often a protein, and preventing the normal function of that macromolecule.

Your Turn 12.30 You Decide Friend or Foe?

Make two lists of drugs for each of the two broadly classified groups: those that bring about a desired physiological response and those that kill foreign invaders. Propose three drugs for each list, using examples not given in this section.

Why is targeting a foreign invader so much simpler than altering our physiology? Remember from our discussion of steroids that biology is synthetically economical. If our systems have two very similar jobs to do, our cells will reuse the plans for the protein that is best evolved to do that job. In addition, our physiological pathways are remarkably interconnected. Successfully designing a molecule to target only one of two or more closely related macromolecules requires designing very nuanced and specific structures, which are able to satisfy the perfect set of noncovalent interaction points. Foreign organisms, like bacteria or viruses, use dramatically different basic plans for their proteins. The challenge with foreign organisms lies in the discovery of new compounds to skirt around evolved resistance to currently used drugs.

How do we accomplish the basic work of finding and then optimizing a new potential pharmaceutical? We need a starting-point chemical with the potential to work against our target disease, either a specific protein in a physiological path or whole organism. Our modern understanding of biology and biochemistry allows us to generate complex models of disease. We can synthesize and isolate target proteins to examine their three-dimensional structure, as well as explicitly measure their binding affinity with different compounds. We have numerous model cells, including both cancerous and non-cancerous mammalian cells. We even have the ability to modify these cells to display known traits of specific problematic diseases. Even further, we have generated whole organisms, from simple worms to zebrafish and mice, again modified to emulate the contrast between health and disease. Through these, we often study cancer and neurodegenerative disorders, such as Huntington's and Alzheimers, in the context of a more complicated set of physiological pathways. Against these model diseases, we can test potential compounds for even slightly favorable structural features and begin to sense potential harmful side effects. Early tests can even examine for specific activity against one target, such as a particular receptor or class of cancer, versus another.

Model mammalian cell lines are vital for research, but ethically challenging. The HeLa cancer model cells, a model cell line used in initial testing of the polio vaccine, came from the cancer of a woman named Henrietta Lacks.

The number and type of potential compounds that we examine depends greatly on our biochemical knowledge of the target. For these tests, pharmaceutical companies and national research institutes house large libraries of chemicals containing small framework molecules with an interesting variety of functional groups, as well as known current drugs and any unique natural products. The more we know about our target system, the smaller and less diverse our library test can be. If, as an example, we already know that a molecule has a favorable natural substrate, we can start our library with small chemical variations of that compound. In the absence of any structural knowledge, pharmaceutical hunts are more likely to begin with a broader set of small framework molecules.

From our library search, any favorable interaction can be built upon. The molecules are examined for common structural features with promising traits combined into one molecule. The ideal search yields functional groups of the proper polarity in the right places without any excess. Therefore, one important strategy in designing a drug is to determine its **pharmacophore**, the three-dimensional arrangement of atoms or groups of atoms responsible for the biological activity of a drug molecule.

Medicinal chemists then synthesize a molecule having that specific active portion, but with a much simpler, non-active remainder. The researchers then custom design the molecule to meet the requirements of the receptor site. With each round of chemical design, chemists can perform further tests of activity against model proteins, cells, and organisms (Figure 12.31). The process of systematically changing the structure of a drug molecule with assessment of the resulting changes in activity is known as a **structure–activity relationship (SAR) study**. A good medicinal chemist concurrently aims for a compound that best fits the target site, while minimizing any interaction with other systems. Specifically, we must evaluate the drugs for an inability to bypass our inherent biological barriers against foreign compounds, as well as the possibility that our own metabolic pathways will act on the drug to generate unexpected and damaging side products from metabolism in the body.

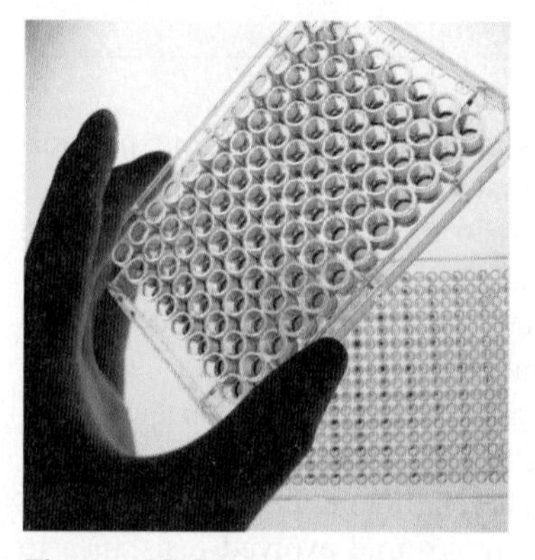

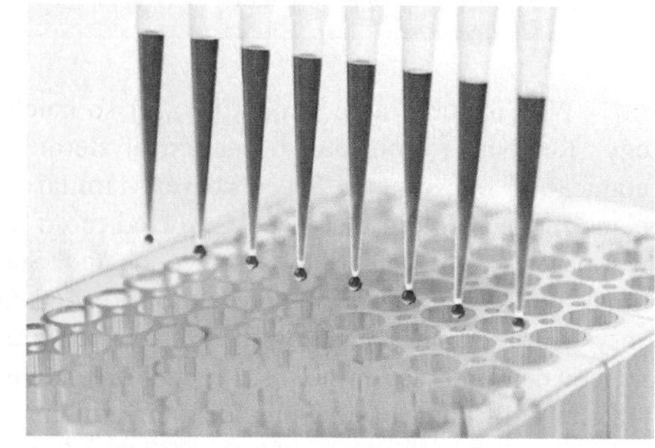

Figure 12.31

High-throughput screening of new drug candidates.

Your Turn 12.31 You Decide What Should a Drug Be Like?

Make a list of the properties you think a drug should have. Then, compare your list with those of a classmate. Note similarities and differences.

a. Describe the motivation for the items on your list.
b. Are any items missing from your list that you should now include?
c. Are any items present on your list that you should now delete?

An outstanding example of this approach is provided by opiate drugs such as morphine. Morphine, a complex molecule, is difficult to synthesize. However, the pharmacophore responsible for opiate activity has been identified and is highlighted in Figure 12.32. The flat benzene ring fits into a corresponding flat area of the receptor, and the nitrogen atom binds the drug molecule to the site. Incorporating this particular portion into other less complex molecules, such as meperidine (more commonly known by its brand name, Demerol®), creates opiate activity. Meperidine is much less addictive than morphine, but also less potent.

The discovery that only certain functional groups are responsible for the therapeutic properties of pharmaceutical molecules was an important breakthrough in drug design. Sophisticated computer graphics are now used to model potential drugs and receptor sites. Thanks to these 3-D representations, medicinal chemists can "see" how drugs interact with a receptor site. Computers can then be used to search for compounds that have structures similar to that of an active drug. Chemists can also modify structures in computer models and visualize how the new compounds will function.

Your Turn 12.32 Scientific Practices 3-D Drugs

See for yourself the shapes of drug molecules by visiting 3Dchem.com. On screen, you can rotate three-dimensional representations of best-selling drugs such as amoxicillin, Lipitor®, and ibuprofen.

a. Select several drugs and examine their three-dimensional structure. How do these computer representations differ from the structural formulas of drugs shown in this chapter?
b. What advantages do the computer representations have over two-dimensional drawings? What are their limitations compared with "real" molecules? Are there any disadvantages?

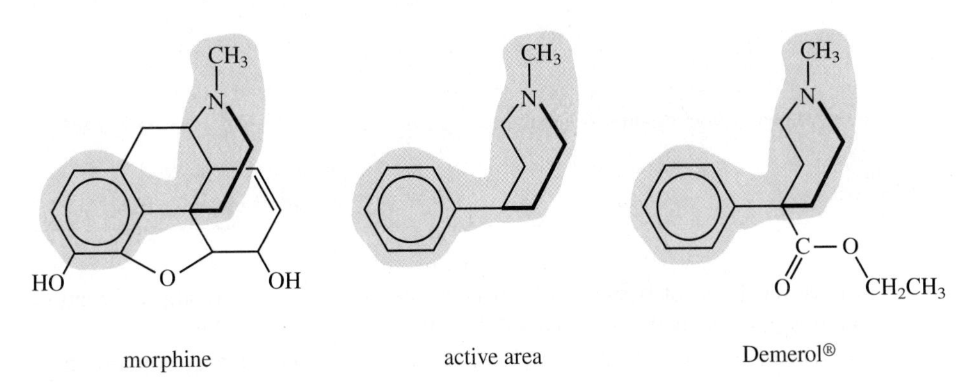

Figure 12.32

Molecular structures of morphine and meperidine (Demerol®). The highlighted "active areas," or pharmacophores, are the portions of the molecule that interact with the receptor. The darker lines indicate that these bonds are in front of the others, or are coming out of the plane of the page.

Outside of classic drug development, medicinal chemistry has contributed to dramatic improvements in medical imaging. While the use of chemotherapy is vital in the treatment of cancer, the best and most successful course of treatment is still surgical. We require high-resolution methods to image the detailed inner workings of our systems and detect the exact position of problem areas, ideally before we go under a knife.

Doctors regularly use radioactive iodide to image and later treat thyroid cancer. This process works due to a few key dependable chemical processes. First, the thyroid must collect and concentrate iodide for use in the synthesis of the hormone thyroxin. We can depend on our thyroid to respond to ingestion of iodide by pulling it from the blood into the gland. We, fortunately here, cannot depend on the thyroid to differentiate between isotopes of iodide due to their very similar molar masses. Second, the radioactive iodide dependably and predictably decays on a known timescale measured by its half-life. Upon decay, the release of high-energy particles will be detectable through the tissues and skin. These traits in combination result in fantastic images of problematic thyroid glands.

Your Turn 12.33 Skill Building Isotopes and Ionizing Radiation

Review the classes of radiation you have seen in previous chapters or the Internet: alpha, beta, gamma, or positron emission. Which would be best for detection through tissues and skin? Which the worst? Explain your choices.

Since their discovery, radioactive elements have been both fascinating and frightening. Their potential for both miracles and terror in the arena of power and bombs has already been discussed in Chapter 6.

Our era of medicine is unique in the use of nuclear isotopes for both treatment and imaging, and the difference between detection and damage is rather fine. The idea itself is not novel. Pierre and Marie Curie hypothesized that the ionizing rays of radiation would be medically beneficial; sadly, both Marie and their daughter Irene passed from diseases likely caused by their unprotected handling of these risky materials. The same radiation that allowed us to detect the thyroid, and the boundaries of a tumor within it, can instead ionize compounds within the tissues. These ionization events create reactive free radicals within biological systems and may irreversibly alter the cells. As a natural process, damaged cells should undergo *apoptosis*, or programmed cell death. However, too much damage can disrupt the apoptosis pathways and lead instead to uncontrolled cancerous growth. Under controlled circumstances, selective and targeted use of ionizing radiation is useful to promote the death of cancerous cells.

Your Turn 12.34 Scientific Practices Treatment After Terror

Imagine that a train carrying waste from a nearby nuclear plant recently crashed in your city or town. You know from **Chapter 6** that the area is likely contaminated with radioactive I-131.

a. What simple salt can you take immediately to prevent damage to your thyroid?
b. What symptoms in the population should medical professionals be watchful for?

Unfortunately, most tissues will not conveniently and automatically concentrate their own imaging agents the way that iodide collects in the thyroid. Instead, traditional medicinal chemistry tools must be applied. **Radiopharmaceuticals** are organic molecules carrying radioactive isotopes that are optimized to create contrast between tissue areas. Through the development of selective and targeted pharmaceuticals, we can bring the sensitivity of thyroid imaging, as well as treatment, to a far broader range of medical concerns.

Your Turn 12.35 You Decide The "Big Question" Revisited

Consider the introductory video on **Connect** again, then answer these questions based on your current knowledge.

a. Considering the principles of equilibrium, evaluate the potential consequences of low [I^-].

b. Describe the role of the thyroid and the consequences of its malfunction for Megan.

c. Based on your understanding of drug design, what variables might you consider for Megan's hormone replacement therapy.

Conclusions

Medicinal chemistry as a field has become rather complicated. Gone are the days of "magic bullets" against foreign invaders. Our medicinal targets now require a nuanced understanding of you—your body, your physiology, and every detailed tangled function of your cells, proteins, and biomolecules. While our understanding of tissue biochemistry, molecular function, and genetic diversity has seen dramatic and sudden improvements in recent history, new treatment options will only appear through integrated, interdisciplinary teamwork. Eric Lander, a leader of the human genome project and presidential advisor on science and technology, gave the following advice regarding the recent avalanche of data on human systems, "It's just a parts list. And if you have a parts list for a 747, that doesn't mean you can fly the plane. It doesn't mean you can build the plane. It doesn't even mean you know what the parts do. It just means you have a list of all the parts." While he was only discussing the limitations of the groundbreaking Human Genome Project, this advice is nonetheless relevant across fields that are working to bring about a bright new future in medicine. We have a wonderful, promising, though possibly overwhelming parts list, and only through interdisciplinary teamwork across fields can we hope to positively influence and change the future of health and medicine.

Learning Outcomes

The numbers in parentheses indicate the sections within the chapter where these outcomes were discussed.

Having studied this chapter, you should now be able to:

- define equilibrium (12.1)
- evaluate a system based upon its position relative to equilibrium (12.1)
- compare and contrast the meaning of equilibrium with steady-state (12.1)
- apply the concept of equilibrium to acids and bases (12.2)
- describe the nature of a buffered system (12.2)
- appraise weak acids and bases to predict their buffering capacity (12.2)
- recognize that biological conditions can vary by tissues and cell compartments (12.2)
- label molecules as organic, inorganic, and biological (12.3)
- convert representations of compounds to line-angle drawings (12.3)
- classify groups of related compounds as isomers (12.3)
- recognize and name select functional groups (12.4)
- relate functional group traits to potential interactions (12.4)
- examine structure and functional groups to solubility (12.4)
- define and use the term *chirality* (12.5)
- relate chirality to specificity in biology (12.5)
- connect macromolecule classes to biological functions (12.6)
- define and relate the terms enzyme, substrate, and active site (12.6)
- give examples of catalysis inside and outside of the body (12.6)
- relate the function of enzymes and substrates to metabolism (12.6)
- understand the role of metabolism of food in the production of ATP (12.6)
- define the role of hormones in regulating biological systems (12.6)
- explain the role of membrane proteins in cellular communication and transport (12.6)
- categorize amino acids by type of noncovalent interactions (12.7)
- apply the concepts of amino acids and proteins from nutrition to biological activity (12.7)
- examine the interaction of small molecules and proteins (12.7)
- relate structure to function in the category of steroids (12.7)
- describe the key stages in development of aspirin and penicillin (12.8)
- classify medication types by purpose (12.9)
- contrast the risk of evolved resistance in bacteria, viruses, and cancers with other illnesses (12.9)
- summarize the routes of drug development (12.9)
- evaluate testing and optimization of new drugs (12.9)
- explain how structure-activity relationships (SARs) are used for the development of new drugs (12.10)
- describe how radiopharmaceuticals are used to diagnose and treat illness (12.10)

Questions

Emphasizing Essentials

1. The field of chemistry has many sub-disciplines.
   a. What do biochemists study?
   b. What do organic chemists study?
2. As seen in Section 11.5, high-fructose corn syrup (HFCS) is produced from the partial conversion of glucose to fructose. The equilibrium constant of this reaction is

$$K = \frac{[\text{fructose}]}{[\text{glucose}]} = 0.74$$

   a. Is glucose or fructose favored in this reaction?
   b. In a system at equilibrium, if [glucose] = 0.22 mM, what must the [fructose] be?
3. Assume that you have a simple reaction of $A \rightleftharpoons B$ at equilibrium. Use your understanding of the reaction quotient, Q, and equilibrium constant, K, to explain the result of each of the following.
   a. The amount of A decreases by half.
   b. The amounts of both A and B are doubled.
4. Write the equilibrium constant for the dissociation shown in Figure 12.3.
5. Nitrous acid (HNO_2) has a K_a value of 4.0×10^{-4}, which corresponds to a pK_a of 3.39.
   a. What salt would you use to prepare a good buffer solution with nitrous acid?
   b. If NaOH were added to the buffer solution, what would it react with?
   c. Compare how the pH would change if 0.05 moles of NaOH were added to the buffer solution compared to a container with just water.
6. Use the Henderson-Hasselbalch equation and Table 12.1 to calculate the pH of the following solutions:
   a. 0.05 M formic acid and 0.1 M sodium formate.
   b. 0.2 M ammonium chloride and 0.1 M aqueous ammonia.
   c. 0.1 M acetic acid and 0.1 M sodium acetate.
7. Write the structural formula and line-angle drawings for each different isomer of C_6H_{14}.
   Hint: Watch out for duplicate structures.
8. Consider the isomers of C_4H_{10}. How many different isomers could be formed by replacing a single hydrogen atom with an –OH group? Draw the structural formula for each.
9. For each compound, identify the functional group present.
   a. 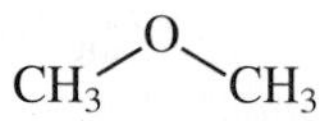 CH_3–O–CH_3
   b. 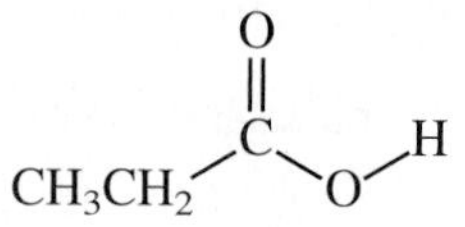 CH_3CH_2–C(=O)–O–H
   c. CH_3CH_2–C(=O)–CH_3
   d. 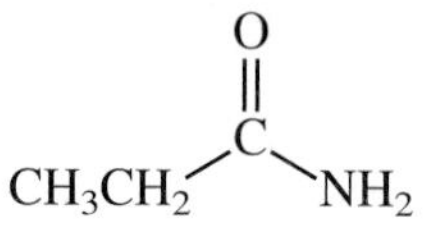 CH_3CH_2–C(=O)–NH_2
   e. 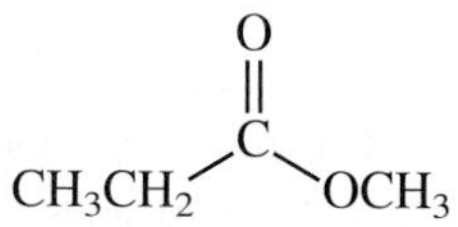 CH_3CH_2–C(=O)–OCH_3
10. Draw the simplest compound that can contain each of these functional groups. In some cases, only one carbon atom is required; in other cases two.
    a. an alcohol
    b. an ether
    c. an ester
    d. a carboxylic acid
    e. an aldehyde
    f. a ketone
11. For each of these, identify the functional group. Then, draw an isomer that contains a different functional group.
    a. CH_3CH_2—OH
    b. 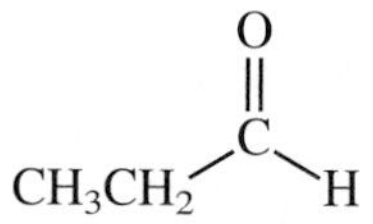 CH_3CH_2–C(=O)–H
    c. CH_3CH_2–C(=O)–OCH_3
12. Histamine is a vital, and often annoying, part of your body's immune response. It causes runny noses, red eyes, and other symptoms. Here is its structural formula. 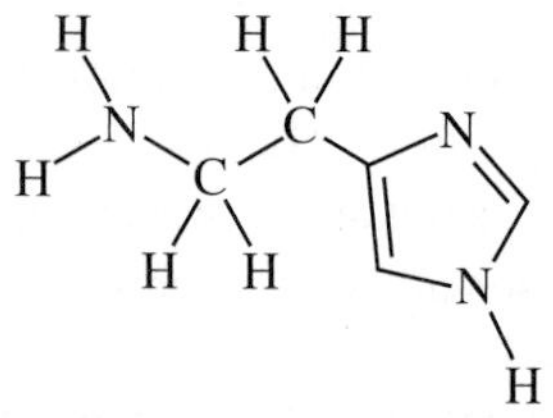
    a. Give the chemical formula for this compound.
    b. Circle the amine functional groups in histamine.
    c. Which part (or parts) of the molecule make the compound water-soluble?
13. Estradiol is relatively insoluble in water but readily soluble in most organic solvents. Explain this solubility behavior based on its structural formula.
    Hint: See Figure 12.25.

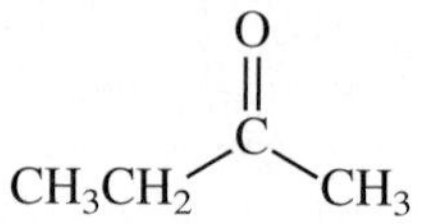

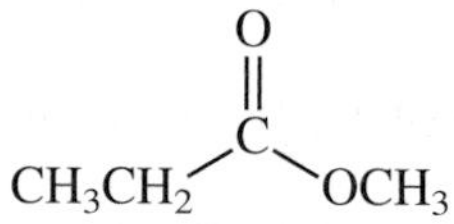

14. Usually, carbon forms four covalent bonds, nitrogen three, oxygen two, and hydrogen only one bond. Use this information to draw structural formulas for:

    a. A compound that contains one carbon atom, one nitrogen atom, and as many hydrogen atoms as needed.

    b. A compound that contains one carbon atom, one oxygen atom, and as many hydrogen atoms as needed.

15. Which of these molecules has chiral forms?

    a. NH_2 (top), $CH_3—C—CH_3$, OH (bottom)

    b. OH (top), $H—C—CO_2H$, CH_3 (bottom)

    c. NH_2 (top), $CH_3—C—CO_2H$, C≡N (bottom)

    d. OH (top), $CH_3—C—CO_2H$, CH_3 (bottom)

16. Which of these molecules has chiral forms?

    a. NH_2 (top), $CH_3—C—CH_2CH_3$, OH (bottom)

    b. OH (top), $H—C—C_2H_5$, H (bottom)

    c. NH_2 (top), $CH_3—C—CO_2H$, CH_2OH (bottom)

    d. OH (top), $CH_3—C—CO_2H$, CH_2SH (bottom)

17. Define and relate the two terms: hormone and receptor. Can one function without the other?

18. Refer to Figure 11.17. Select two examples of amino acids with side chains that are categorized as nonpolar. Describe what characteristics make them nonpolar.

19. Refer to Figure 11.17. Select two examples of amino acids with side chains that are categorized as acidic. Describe what characteristics make them acidic.

20. Molecules as diverse as cholesterol, sex hormones, and cortisone contain common structural elements. Use a line-angle drawing to show the structure they share.

21. Ester formation reactions were vital in the discovery of aspirin. Draw structural formulas for the esters formed when acetic acid reacts with these alcohols.

    a. *n*-propanol, $CH_3CH_2CH_2OH$

    b. iso-propanol, $(CH_3)_2CHOH$

    c. *tert*-butanol, $(CH_3)_3COH$

22. Identify the functional groups in morphine and meperidine. Can these molecules be assigned to a particular class of compound (*i.e.*, an alcohol, ketone, or amine)? Explain. **Hint:** See Table 12.3 for structural formulas.

23. What is meant by the term pharmacophore?

24. Sulfanilamide is the simplest sulfa drug, a type of antibiotic. It appears to act against bacteria by replacing para-aminobenzoic acid, an essential nutrient for bacteria, with sulfanilamide. Use these structural formulas to explain why this substitution is likely to occur.

    H—N(H)—C_6H_4—S(=O)(=O)—N(H)—H

    sulfanilamide

    H—N(H)—C_6H_4—C(=O)—O—H

    para-aminobenzoic acid

Concentrating on Concepts

25. Explain why an equilibrium constant cannot tell you how quickly a reaction will turn into a product.

26. Use the information in Table 12.1 to redraw Figure 12.5 for the following acids and bases.

    a. acetic acid

    b. ammonia

    c. carbonic acid

    Hint: Carbonic acid can be deprotonated twice!

27. Draw structural formulas for each of these molecules and determine the number and type of bonds (single, double, or triple) for each carbon atom.

    a. H_3CCN (acetonitrile, used to make a type of plastic)

    b. $H_2NC(O)NH_2$ (urea, an important fertilizer)

    c. C_6H_5COOH (benzoic acid, a food preservative)

28. In Your Turn 12.12, you were asked to draw structural formulas for the three isomers of C_5H_{12}. One student submitted this set, with a note saying that six isomers had been found. Help this student see why some of the answers are incorrect.
Note: The hydrogen atoms have been omitted for clarity.

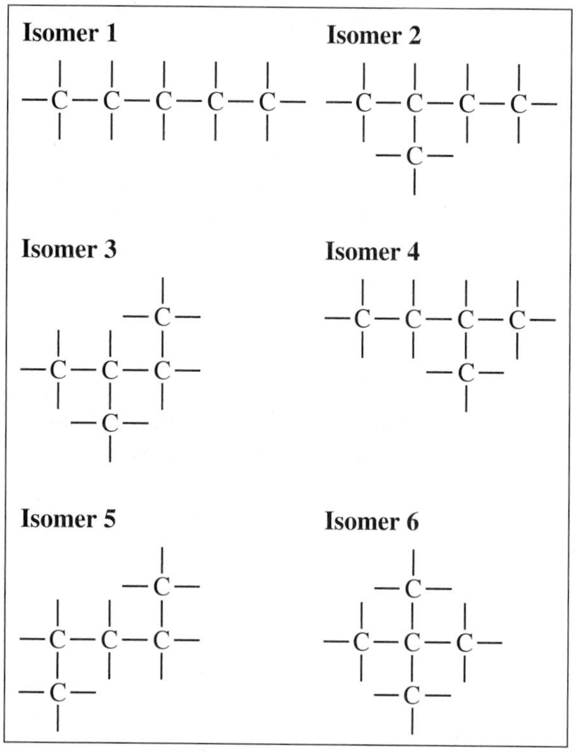

Isomer 1 **Isomer 2**

Isomer 3 **Isomer 4**

Isomer 5 **Isomer 6**

29. Styrene, $C_6H_5CH_5CH_2$, the monomer for polystyrene described in Section 9.5, contains the phenyl group, $-C_6H_5$. Draw structural formulas to show that this molecule, like benzene, has resonance structures.

30. Antihistamines are widely used drugs for treating symptoms of allergies caused by reactions to histamine compounds. This class of drug competes with histamine, occupying receptor sites on cells normally occupied by histamine. Here is the structure for a particular antihistamine.

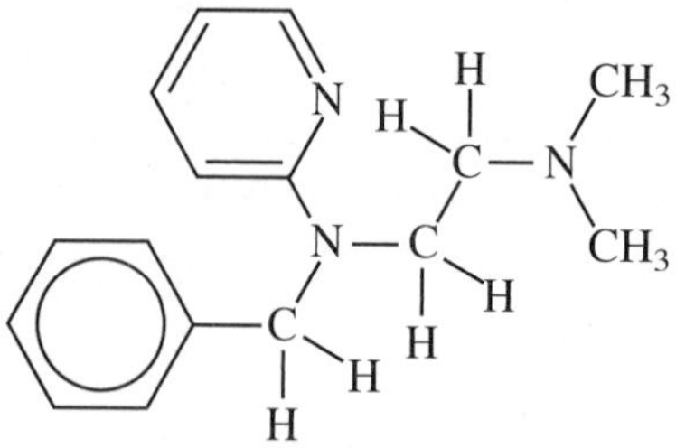

a. Give the chemical formula for this compound.

b. What similarities do you see between this structure and that of histamine (shown in Question 12) that would allow the antihistamine to bind to the same receptor as histamine?

31. Explain why type of functional groups on a substrate can be important for specific binding to enzymes.

32. Explain why specific positions of functional groups on a substrate can be important for specific binding to enzymes.
Hint: Revisit Figure 12.14.

33. Pair the three categories of macronutrients from Chapter 11 with the following roles in biology:

a. Cellular membrane

b. Energy storage as glycogen

c. Enzymes in metabolism

34. Redraw Figure 12.4 showing the relative barrier for a process with and without an enzyme.

35. Figure 12.21 represents chemical communication within the body. Write a paragraph explaining what this figure means to you in helping to explain chemical communication.

36. Describe the lock-and-key analogy for the interaction between drugs and receptor sites. Use the analogy in a discussion as if you were explaining this to a friend.

37. The lock-and-key analogy for binding clearly references a common everyday object. Create a similar visual analogy for the induced-fit model of binding.

38. Consider this statement: "Drugs can be broadly classed into two groups: those that produce a physiological response in the body and those that inhibit the growth of substances that cause infections." Into which class does each of the following drugs fall?

a. aspirin
b. estrogen
c. (Keflex™) antibiotic
d. penicillin
e. morphine

39. The text states that some racemic mixtures contain chemicals that are effective against disease, others are ineffective but harmless, and still others are potentially harmful. What methods in the text might help determine into which of these three categories a recently discovered substance fits?

40. Consider the structure of morphine in Figure 12.32. Codeine, another strong analgesic with narcotic action, has a very similar structure in which the –OH group attached to the benzene ring is replaced by an $-OCH_3$ group.

a. Draw the structural formula for codeine and label its functional groups.

b. The analgesic action of codeine is only about 20% as effective as morphine. However, codeine is less addictive than morphine. Is this enough evidence to conclude that replacement of –OH groups with $-OCH_3$ groups in this class of drugs will always change the properties in this way? Explain.

41. Dopamine is found naturally in the brain. The drug L-DOPA is found to be effective against the tremors and muscular rigidity associated with Parkinson's disease. Identify the chiral carbon in L-DOPA, and comment on why L-DOPA is effective, whereas D-DOPA is not.
Hint: Revisit Figure 12.16 for the structure of L-DOPA.

Exploring Extensions

42. Return to Figure 12.17 to examine the side chains of the amino acids lysine and aspartic acid. Propose how a pH buffer would affect the behavior of these side chains.
43. Before the cyclic structure of benzene was determined (see Figure 12.9), there was a great deal of controversy about how the atoms in this compound were arranged.
    a. Count valence electrons for C and H in C_6H_6. Then, draw the structural formula for a possible linear isomer.
    b. Give the condensed structural formula for your answer in part a.
    c. Compare your structure with those drawn by classmates. Are they all the same? Why or why not?
44. Thalidomide was first marketed in Europe in the late 1950s. It was used as a sleeping pill and to treat morning sickness during pregnancy. At that time, it was not known to cause any adverse effects. By the late 1960s, however, the drug was banned. Use the Internet to gather information to write a short paper that describes the optical isomers of thalidomide, why the drug was banned, and why the FDA did not approve thalidomide for use in the United States until recently. For what purpose has the FDA recently approved the use of thalidomide?
45. Use the lock-and-key model discussed in Section 12.7 to offer a possible explanation as to why individuals who suffer from lactose intolerance can digest sugars such as sucrose and maltose, but not lactose. Use the resources of the Internet to find the structure of lactose.
46. This chapter presented example proteins in important biological roles, including the metabolic enzyme, phosphoglycerate kinase, in Section 12.6. Explore the Molecule of the Month archive of the Protein Data Bank to find another enzyme involved in metabolism. Summarize your discovery. What role does the protein play in biology? What features of the protein are most interesting?
47. Merck and Codexis dramatically improved their synthetic route to sitagliptin, a treatment for type 2 diabetes, through the use of enzymes. The work won a 2010 Presidential Green Chemistry Challenge Award. How does this process differ from the earlier one for manufacturing sitagliptin? Do the changes match your understanding of enzymes? Write a brief report on your research, citing your sources.
48. The steroids in our body require carrier molecules to transport them through our blood to different organs. Examine the structures of steroid molecules in Section 12.8 to propose a reason that steroids require assistance to travel in blood.
49. Danco Laboratories, the U.S. company that produces RU-486 (mifeprex, the "abortion pill"), makes claims about the safety of this steroidal drug, including a comparison to aspirin. Do some research on RU-486 and write a short report on the drug. Include its structure, mode of action, and safety record.
50. One avenue for successful drug discovery is to use the initial drug as a prototype for the development of other similar compounds called analogs. Cyclosporine™, a major anti-rejection drug used in organ transplant surgery, is considered an example of a drug discovered in this way. Research the discovery of this drug to verify this statement. Write a brief report describing your findings, citing your sources.
51. Dorothy Crowfoot Hodgkin first determined the structure of a naturally occurring penicillin compound. What in her background prepared her to make this discovery? Write a short report on the results of your findings, citing your sources.
52. The antibiotic ciprofloxacin hydrochloride (Cipro™) treats bacterial infections in many different parts of the body. This drug made headlines in 2001 for use in patients who had been exposed to the inhaled form of anthrax. Use the Internet or another source to obtain the structure of Cipro™. Draw its structure and identify the functional groups.

CHAPTER 13

Genes and Life

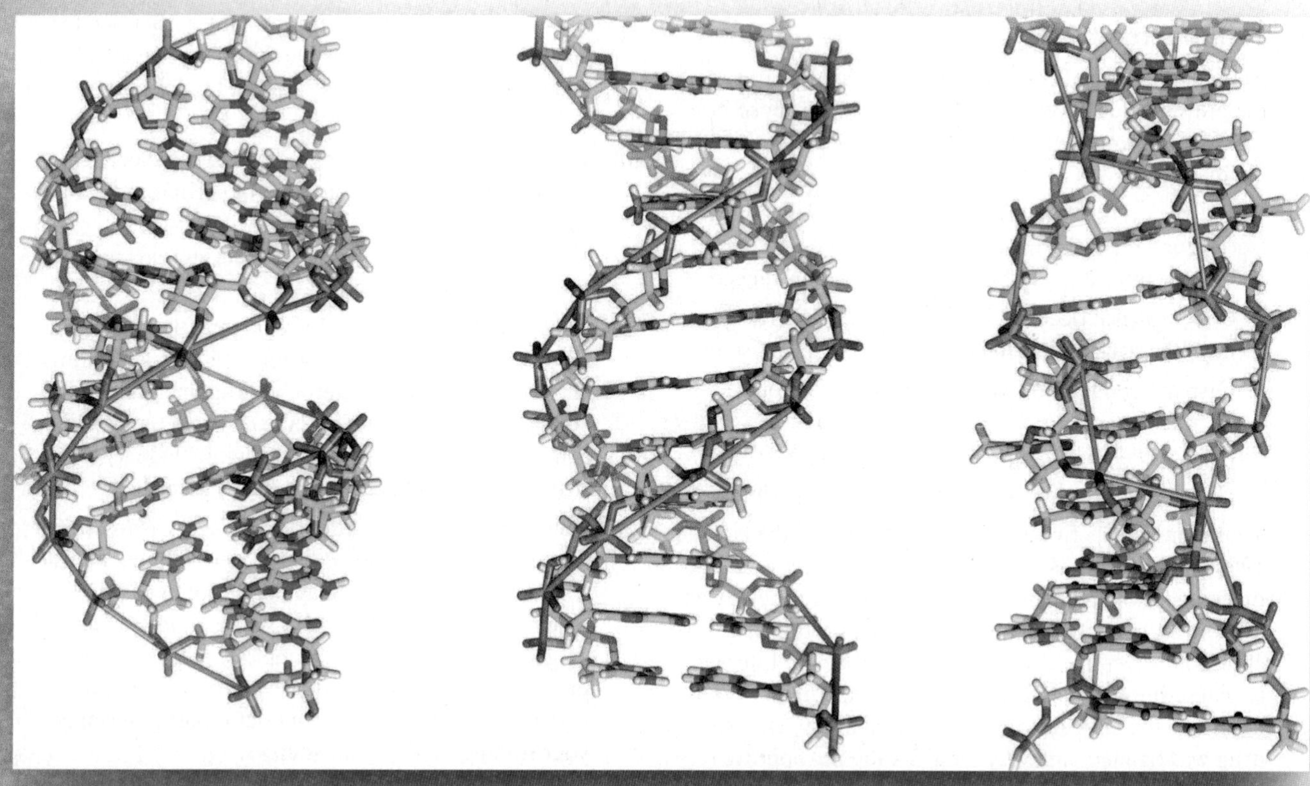

© Richard Wheeler

REFLECTION

Genetic Traits

Make a list of five different physical traits that you believe are genetically inherited. Survey at least 10 people in your family or class to determine which of those traits these individuals have. What similarities do you find? What differences do you find? For those people related to you, are all of the traits the same? For those people not related to you, are all of the traits different?

The Big Picture

In this chapter, you will explore the following questions:

- How was genetic engineering used to develop synthetic insulin?
- What large challenges are genetic modifications trying to solve?
- What role does DNA play in life?
- What does it mean for something to be "genetically modified"?
- What are the risks associated with genetically modified organisms?

A DNA sculpture becomes an intersection of art, science, and play at the Lawrence Hall of Science in Berkeley, California.

"No branch of science has created more acute or more subtle and interesting ethical dilemmas than genetics. . . . it is genetics that makes us recall, not simply our responsibilities to the world and to one another, but our responsibilities for how people will be in the future. For the first time we can begin to determine not simply who will live and who will die, but what all those in the future will be like."

Source of quote: Justine Burley and John Harris Eds. Companion to Genethics, *2002.*

Photos: © Michael Halberstadt/Siliconvalleystock.com

Introduction

Do you know someone who has diabetes? It is one of the oldest-known diseases, with historical evidence of its occurrence dating back to 1500 BC. People with diabetes have an imbalance in a hormone called **insulin** that is produced in the pancreas. As a result, their bodies have difficulty regulating blood sugar. Some people are born with diabetes (called type 1), while others develop it at an older age (type 2). In either case, diabetes can be a deadly disease if not treated and controlled.

Type 2 diabetes accounts for approximately 90–95% of all diabetes cases.

Did You Know? People have known for thousands of years that diabetics excrete sugar into their urine—a side effect of overwhelming the kidneys with too much blood glucose. In fact, in one of the early tests for diabetes, doctors poured urine on the ground to see whether it attracted insects. If insects crowded around the puddle, it meant they were attracted to sugar, a dead giveaway for diabetes.

Only recently has an effective treatment for diabetes been developed. Well into the 1900s, the only available treatment doctors could offer children diagnosed with type 1 diabetes was a starvation diet, which usually did not help to extend the life expectancy of the children beyond their teenage years. However, in the early 1920s, Canadian researchers were able to confirm the role of the insulin hormone in regulating blood sugar by carrying out research on dogs whose pancreases had been removed. Within a few months, the first children in a diabetes ward at Toronto General Hospital were injected with insulin extracted from the pancreases of calf fetuses. When the

Insulin from fetal sources did not need to be separated from digestive enzymes found in adult pancreases.

bovine insulin was sufficiently purified and given to afflicted children in appropriate doses, the results were impressive. Comatose children on the brink of death regained consciousness; with careful monitoring and ongoing treatment, the children were able to live long lives.

The first human to be injected with insulin was Leonard Thompson, a 14-year-old boy who suffered from late stages of type I diabetes in 1922. After insulin treatments, Thompson lived another 13 years before dying from pneumonia at the age of 27.

LEONARD THOMPSON
First patient to receive insulin in Toronto.

Source: Thomas Fisher Rare Book Library, University of Toronto

Soon, the pharmaceutical company Eli Lilly and Company purchased the patent from the University of Toronto, and took over production of insulin on a mass scale. The proximity of the Indianapolis-based company to the slaughterhouses in Chicago allowed access to an abundance of cow and pig pancreases, and soon insulin was on sale in the United States. However, about 8000 pounds of animal pancreas glands were required to produce one pound of insulin. Great care had to be taken in purifying the insulin from the animal sources, because impurities could lead to severe allergic reactions in the patients. Though similar to human insulin, the hormone derived from cow and pig pancreases is not identical, and some diabetes patients suffered from inflammation around the injection site.

Today, millions of people worldwide rely on daily doses of insulin to control their diabetes and allow them to be active and healthy. But the vast majority of all insulin today is no longer derived from animal sources. Instead, bacteria have been *genetically engineered* to produce human insulin. This chapter is about the chemistry of life, through a discussion of the fundamental building blocks of genetics—the system of molecules that control the traits of living organisms.

13.1 | A Route to Synthetic Insulin

For more than 50 years, animal-derived insulin was the standard treatment for diabetes patients. Then, in 1978, scientists at the biotechnology company Genentech and the City of Hope National Medical Center in California developed a process to synthesize mass quantities of human insulin with utmost purity. In a process that remains the standard today, the genetic code for human insulin is inserted into the existing code of a bacterial cell. The bacterial cell then produces insulin. The resulting synthetic insulin is chemically identical to the insulin produced in the human pancreas. Genentech's synthetic insulin, marketed under the trade name Humulin™, was approved for sale in 1982. Today, millions of people worldwide rely on it to control their diabetes.

Bacteria are single-celled organisms.

How can we get the bacterium to produce human insulin? We do this by "teaching" it to make some new chemicals. Inside each cell is the complete set of instructions, a guidebook if you like, on how to grow and reproduce. The guidebook passes from one generation to the next, often completely unchanged. This guidebook, termed the **genome**, is the primary route for inheriting the biological information required to build and maintain an organism.

The most common strain of corn, B73, contains more than 32,000 genes in its genome, more genes than are in human DNA. Researchers spent four years cataloging the massive amount of information in the corn so that we may understand the genes behind beneficial traits, such as higher yields and disease or drought resistance.

The genome is divided into short sections of instructions to produce specific reactions known as chemicals, or events, in the cell. These specific pieces are the basic units of heredity, **genes**—short pieces of the genome that code for the production of proteins. A change within a gene changes an inheritable trait. For a corn plant, a change in the gene for color may switch the corn kernels from light yellow to white. But small changes within a gene are not enough to make a cell produce a new chemical, such as insulin. We need a more dramatic change.

Did You Know? Although some *E. coli* strains such as O104 or O157 cause severe illness and deaths, there are hundreds of other varieties that are harmless to humans. Several of these forms are now used in the production of antibiotics, vaccines, cancer-targeting agents, and even biofuels such as butanol.

What we really need to do is to insert a whole new set of instructions (that is, a gene) into the genome of the bacterium. By inserting the genetic code for human insulin into the bacterium, we create bacteria that can produce the essential insulin hormone. The bacterium of choice is a strain of *Eschericia coli.* All of us have *E. coli* strains in our intestines. However, the particular strain chosen for the synthesis of insulin is designed to be unable to survive in living organisms, so there is no risk that a person could be host to a colony of insulin-producing bacteria.

Bacteria and other living organisms are more complex than you might think. What exactly are we modifying when we genetically alter something? We turn to this topic in the next section.

13.2 | DNA: A Chemical that Codes Life

With each passing second, a single cell is host to millions of chemical reactions. Some of these reactions decompose compounds, and others synthesize them. Some reactions transfer chemical signals, while others process them. Some reactions release energy, whereas others utilize it. One very special chemical lies at the heart of this dazzling chemical complexity.

We require a lot from this very special chemical. As cells grow and multiply, this chemical must replicate itself without error. It must remain largely unharmed and unchanged by its environment. This one chemical must organize and securely store a lot of information. This information is context-sensitive, as some reactions are ongoing while others start and stop depending on specific signals. In short, we need a highly advanced database in chemical form.

The chemical we have just described is **deoxyribonucleic acid (DNA)**—the biological polymer that carries genetic information in all species. DNA is the template of life, containing all of the biochemical information to make a full corn plant, for example. DNA can replicate easily, transfer information, and respond to feedback within the cell.

Like other plants and animals, you have a special template of life written on a tightly coiled thread of DNA. Unraveled, the DNA in *each* of your cells is about 2 meters (roughly 2 yards) long. If all of the DNA in all 100 trillion of your cells were placed end to end, the resulting ribbon would stretch from here to the Sun and back, more than 600 times! But as you will soon discover, this astronomical figure is far from the most astounding feature of this amazing molecule.

The estimates for the number of cells in your body range from 50–100 trillion. The bacterial cells outnumber the human cells by an estimated factor of 10.

Any strand of DNA—long or short—consists of three fundamental chemical units: nitrogen-containing bases, deoxyribose sugars, and phosphate groups. All are illustrated in Figure 13.1.

Revisit Section 8.8 to review a simple nitrogen-containing base, ammonia.

DNA contains four nitrogen-containing bases, each with different sizes and shapes. The larger bases, adenine (A) and guanine (G), have a six-membered ring and a five-membered ring of atoms fused together. The smaller bases, cytosine (C) and thymine (T), have only a six-membered ring. Notice that all of these compounds have nitrogen atoms embedded in their rings, leading to the name "nitrogen-containing bases." These bases also contain oxygen atoms that can participate in hydrogen bonding.

Your Turn 13.1 Skill Building Small, but Important Differences

The size or number of rings is not the only difference between the nitrogen-containing bases. Let's examine each of the four bases more closely.

a. Draw Lewis structures for each of the four bases. Be sure to show the lone pairs of electrons on the nitrogen and oxygen atoms.
b. Identify the H atoms that could form hydrogen bonds with water or other nucleic acids.
c. Now identify the other (non-hydrogen) atoms that could participate in hydrogen bonding.
Hint: Revisit Section 8.3 for information on hydrogen bonds.

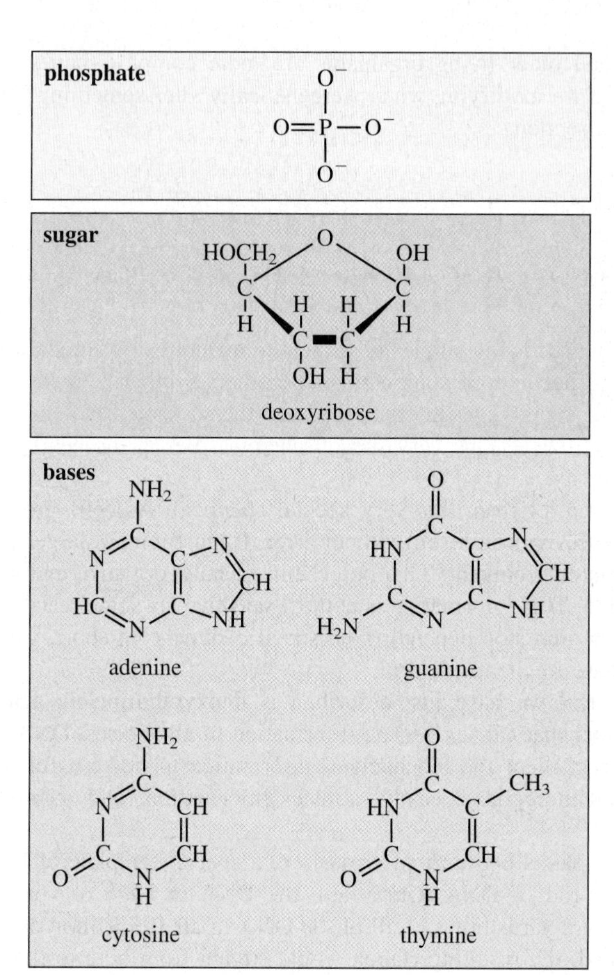

Figure 13.1

The components of deoxyribonucleic acid, DNA. See **Figures Alive!** on **Connect** for more details regarding the composition of DNA.

Monosaccharides were introduced in Section 11.4.

DNA molecules are built from sugar molecules. Unlike the nitrogen-containing bases in DNA, only one sugar is present, deoxyribose. Deoxyribose is a monosaccharide, a "single sugar," with chemical formula $C_5H_{10}O_4$ (Figure 13.1). The next activity gives you the opportunity to learn more about this sugar.

Your Turn 13.2 Skill Building Chemical Cousins Ribose and Deoxyribose

Ribose is a close molecular cousin to deoxyribose. It is the monosaccharide found in ribonucleic acid (RNA). Compare the structural formula of deoxyribose (**Figure 13.1**) with that of ribose:

a. Give the chemical formula for each sugar.

b. How do the structural formulas of these two sugars differ?

c. Carbohydrates are compounds that typically have the chemical formula $C_nH_{2n}O_n$, as noted in **Section 11.4**. Do both ribose and deoxyribose fit this pattern?

d. In these two sugar molecules, which atoms would form hydrogen bonds with water or other nucleic acids?

In addition to a nitrogen-containing base and a sugar, DNA molecules also contain the phosphate group; in essence, a phosphate ion that has become attached. However, depending on the pH, phosphate may be in the form of PO_4^{3-}, HPO_4^{2-} or $H_2PO_4^-$ (Figure 13.2). If three of the oxygen atoms in the phosphate are paired with H^+, the chemical form is H_3PO_4, or phosphoric acid. These hydrogen ions make nucleic acids acidic.

The phosphate ion was introduced in Section 4.1. Here is one of its resonance structures:

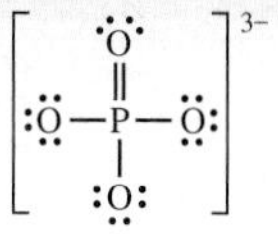

$$\left[\ :\ddot{O}-P(=\ddot{O})(-\ddot{O}:)-\ddot{O}: \ \right]^{3-}$$

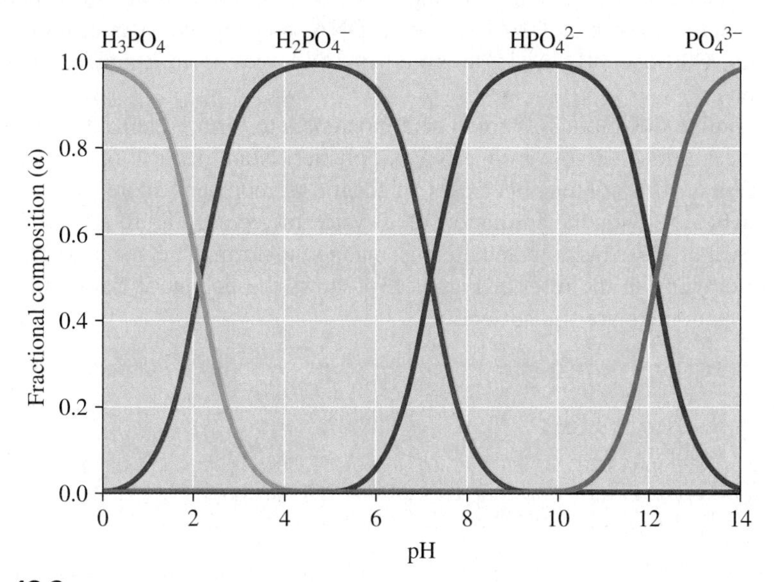

Figure 13.2

The composition of phosphate species present in aqueous solutions at varying pH ranges.

The data presented in Figure 13.2 is related to the following acid/base equilibria:

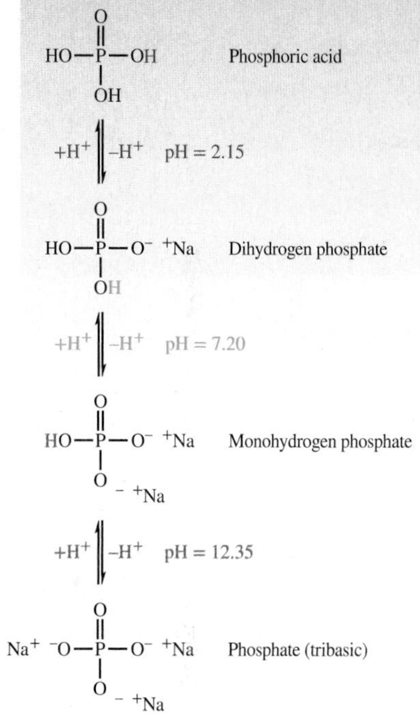

HO—P(=O)(OH)—OH Phosphoric acid

$+H^+$ ⇅ $-H^+$ pH = 2.15

HO—P(=O)(OH)—O^- ^+Na Dihydrogen phosphate

$+H^+$ ⇅ $-H^+$ pH = 7.20

HO—P(=O)(O^- ^+Na)—O^- ^+Na Monohydrogen phosphate

$+H^+$ ⇅ $-H^+$ pH = 12.35

Na^+ ^-O—P(=O)(O^- ^+Na)—O^- ^+Na Phosphate (tribasic)

Your Turn 13.3 Skill Building Speciation of Phosphate Ions

By interpreting the data represented in **Figure 13.2**, describe the fractional (or percent) composition of the phosphate ions present in aqueous solutions at the following pH values:

a. pH = 1 (acidic)

b. pH = 10 (basic)

c. pH = 5.5 (acidic)

d. pH = 13.5 (basic)

e. pH = 7 (neutral solution)

All three of the fundamental chemical units—the nitrogen-containing base, the sugar, and the phosphate group, have a valuable role to play in the structure of DNA. Joined together, these three pieces make up one monomer that in turn polymerizes to form DNA. Each monomer is called a **nucleotide**; that is, a covalently bonded combination of a base, a deoxyribose molecule, and a phosphate group. For example, Figure 13.3 shows

Recall from Chapter 9 that polymers are large molecules made up of many smaller monomers.

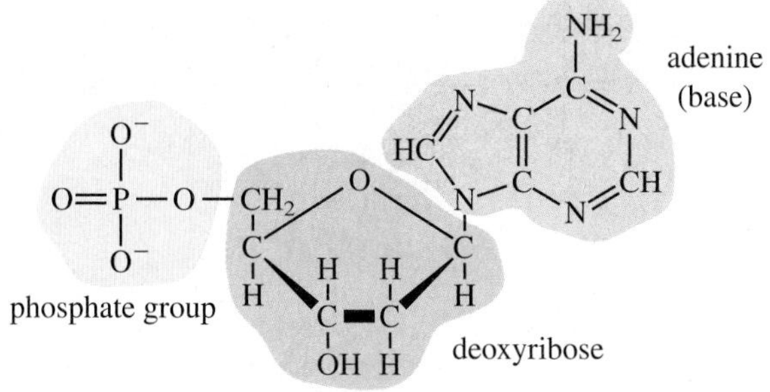

phosphate group deoxyribose adenine (base)

Figure 13.3

A nucleotide monomer built from a phosphate group, deoxyribose, and the base adenine.

the nucleotide named *adenine phosphate*. You can see that the sugar is bonded both to the phosphate group and to the base adenine. Similar nucleotides can be formed using the other three nitrogen-containing bases of DNA: guanine, cytosine, and thymine.

Note in Figure 13.3 that one –OH on the deoxyribose ring remains available to react. It does so with the phosphate group of another nucleotide. A condensation reaction occurs, thereby connecting the two nucleotides. If this happens repeatedly between nucleotides, the result is a long chain with an alternating sugar–phosphate backbone, better known as DNA. A typical DNA molecule consists of thousands of nucleotides. Consequently, a single strand of DNA may have a molecular mass in the millions.

See Chapters 9 and 11 for other examples of condensation polymerization.

Not unlike the joining of amino acid monomers to form proteins, the assemblage of nucleotides (monomers) to form DNA (a polymer) is an example of a condensation polymerization. The polymer increases in length as more and more nucleotides are joined, each time with the formation of a water molecule. Figure 13.4 shows four nucleotides that have been linked in this manner to form a segment of DNA. The schematic drawing in the inset of Figure 13.4 shows the polymeric nature of DNA in which the monomers are the nucleotides.

Your Turn 13.4 Skill Building Another Nucleotide

Analogous to **Figure 13.3**, draw the structural formula for the nucleotide containing cytosine.

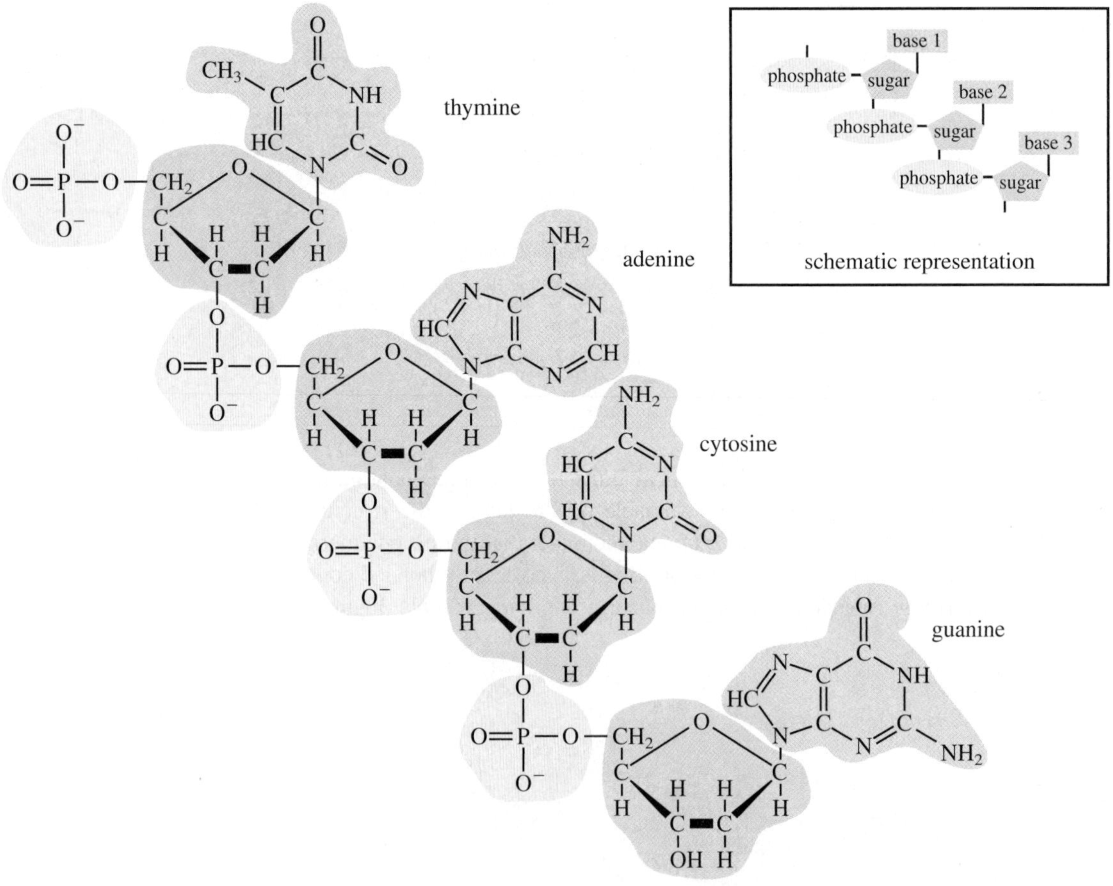

thymine

adenine

cytosine

guanine

base 1
phosphate – sugar
base 2
phosphate – sugar
base 3
phosphate – sugar

schematic representation

Figure 13.4

A segment of DNA represented chemically and schematically (insert). The phosphate group connects one deoxyribose to an adjacent one. Each of the four bases—thymine (T), adenine (A), cytosine (C), or guanine (G)—is attached to a deoxyribose sugar.

See **Figures Alive!** on **Connect** to explore how the building blocks of nucleotides combine to form DNA strands.

13.3 | The Double Helix Structure of DNA

DNA is a gorgeous molecule. The opening photo of this chapter displays a sculptor's rendition of DNA with two silvery strands curving in a gentle spiral, both elegant and simple. Hidden within its structural simplicity is a powerful chemical code for information. The structure—both how the nucleotides are covalently bonded and how the strands pack together—contributes to the function of DNA. Understanding how DNA performs its many functions first required solving the puzzle of the DNA structure.

To see the shape and submicroscopic details of DNA, scientists turned to the technique of X-ray diffraction. This technique has revolutionized our understanding of molecular structures and chemistry by helping us visualize chemical shapes. **X-ray diffraction** is an analytical technique in which a crystal is hit by a beam of X-rays to generate a pattern that reveals the positions of the atoms in the crystal. The X-ray photons interact with the electrons of the atoms in the crystal and are diffracted, or scattered (Figure 13.5). The crucial point is that the X-rays are only scattered at certain angles that are related to the distance between atoms, and that information can be used to determine the structures of a variety of crystalline materials. The X-ray diffraction pattern of a DNA fiber was obtained in late 1952 by the British crystallographer Rosalind Franklin (1920–1958) (Figure 13.6).

Return to Section 3.1 to find the X-ray region in the electromagnetic spectrum.

James Watson (b. 1928) and Francis Crick (1916–2004) combined Franklin's X-ray diffraction data with earlier chemical and biological analyses to create a model of the structure of DNA. The pattern in Franklin's diffraction photograph was consistent with a repeating helical arrangement of atoms, similar to a loosely coiled spring. Moreover, the X-ray photographs contained evidence of a repeated pattern separated by 0.34 nm within a DNA molecule. The Watson–Crick model explained this repetition by twisting the strands of DNA into a **double helix**, a spiral consisting of two strands that coil around a central axis (Figure 13.6). The base pairs are parallel to each other, perpendicular to the axis of the DNA molecule, and separated by 0.34 nm, the same distance calculated from the diffraction pattern. In addition, Franklin's results also suggested a second repetition pattern separated by 3.4 nm. Watson and Crick took this to be the length of a complete helical turn consisting of 10 base pairs.

James Watson, Francis Crick, and Maurice Wilkins (1916–2004) shared the 1962 Nobel Prize in physiology or medicine for their contributions to the structural understanding of DNA. Although crystallography data from Rosalind Franklin was vital, she died in 1958 and was not eligible for the Nobel Prize in 1962.

The letters in our English words have directionality. For example, the words *ward* and *draw* have the same letters in the same order, but the meaning is different because the direction is reversed. The same is true of DNA, with the bases within its backbone structure being analogous to letters, and the structure of the DNA backbone defining the

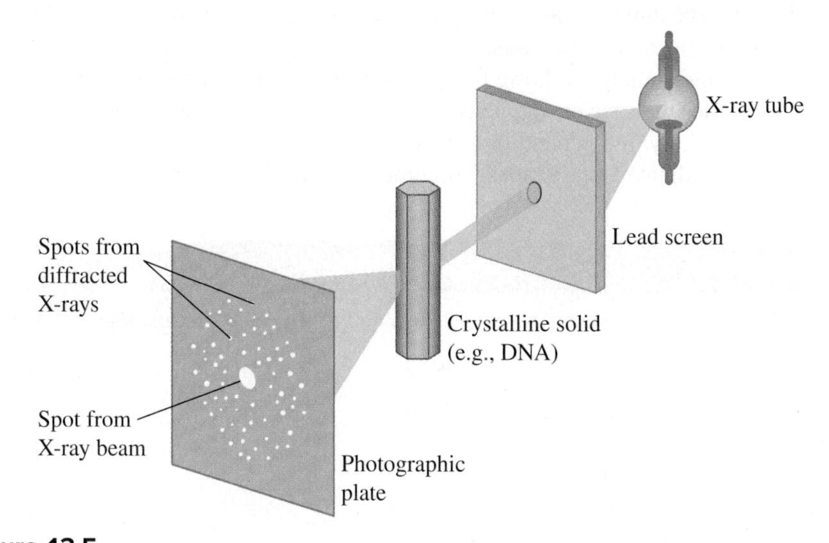

Figure 13.5

An illustration of X-ray diffraction. A crystalline solid is placed in the beam of X-ray radiation, and is rotated while being impinged with X-rays. Information regarding the molecular structure of the solid is obtained from the symmetry of the spots resulting from diffraction of the incident X-ray beam.

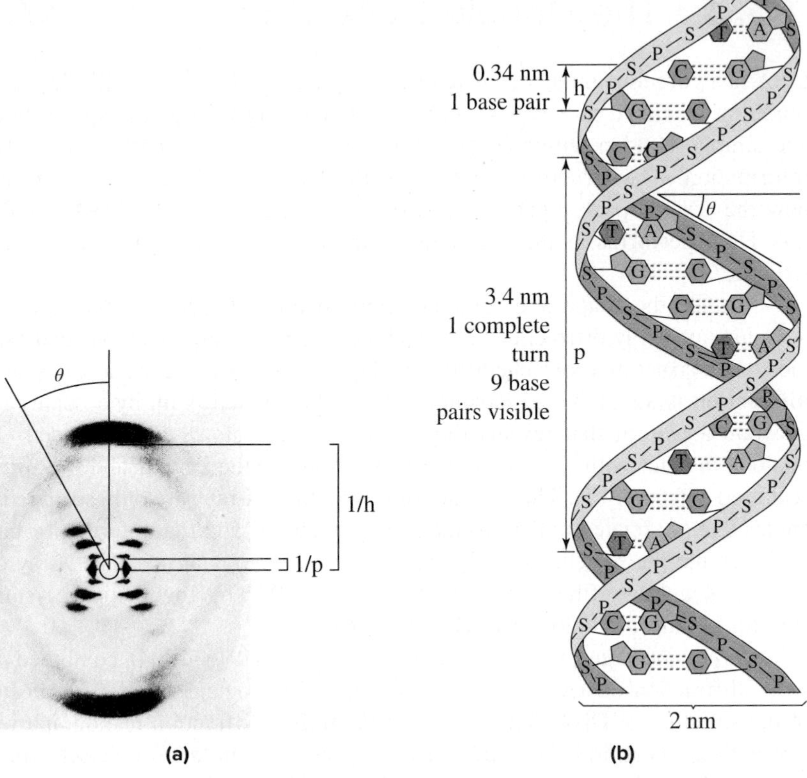

Figure 13.6

(a) Rosalind Franklin's X-ray diffraction photo for a hydrated DNA fiber. The cross in the center is indicative of a helical structure, and the darkened arcs at the top and bottom are due to the stack of base pairs. **(b)** A model of DNA with P = phosphate group; S = the sugar, deoxyribose; and the bases A = adenine; T = thymine; C = cytosine; G = guanine. The sugar and phosphate groups alternate on the backbone, and the four bases attach to this backbone.

directionality. For example, the base string *T-A-C* does not have the same meaning as *C-A-T*. Look carefully at the alternating phosphate and deoxyribose groups in the DNA backbone (Figure 13.4) to see how the deoxyribose ring connects directly to the phosphate below it, and the one above it links through another carbon. The different types of chemical bonds make one direction different from the other. When the two strands of the DNA double helix come together, one strand must run in the opposite direction from the other.

The base-pairing rules are termed *Chargaff's rules* after the discoverer, Austrian chemist Erwin Chargaff (1905–2002).

Early chemical analyses showed that the nitrogen-containing bases in DNA come in pairs. No matter the species, the percent of A almost exactly equals the percent of T (Table 13.1). Similarly, the percent of G is identical to the percent of C. The structural model of DNA validated these rules. Adenine (A) and thymine (T) bases fit almost

Table 13.1 Base Compositions of DNA for Various Species

Specific Name	Common Name	Adenine	Thymine	Guanine	Cytosine
Homo sapiens	human	31.0	31.5	19.1	18.4
Drosophila melanogaster	fruit fly	27.3	27.6	22.5	22.5
Zea mays	corn	25.6	25.3	24.5	24.6
Neurospora crassa	mold	23.0	23.3	27.1	26.6
Escherichia coli	bacterium	24.6	24.3	25.5	25.6
Bacillus subtilis	bacterium	28.4	29.0	21.0	21.6

Source: From I. Edward Alcamo, DNA Technology: The Awesome Skill, *2E © 2000 McGraw-Hill Education.*

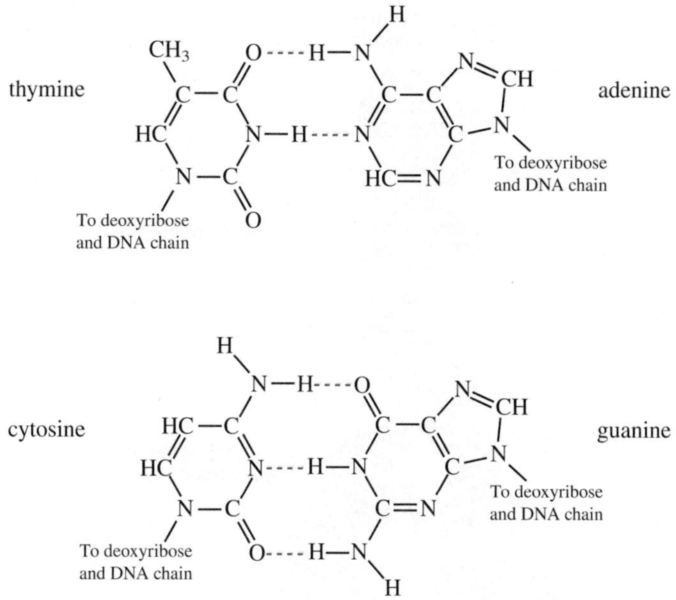

thymine
adenine
To deoxyribose and DNA chain
To deoxyribose and DNA chain

cytosine
guanine
To deoxyribose and DNA chain
To deoxyribose and DNA chain

Figure 13.7

Base-pairing of adenine with thymine, and cytosine with guanine in DNA. Chemical bonds are solid black lines, and the hydrogen bonds are dashed **red** lines.

See **Figures Alive!** on **Connect** to check your knowledge of base-pairing in DNA.

perfectly together, like pieces in a jigsaw puzzle. A closer look shows these two bases linked by two hydrogen bonds (Figure 13.7). Similarly, cytosine (C) and guanine (G) are linked by three hydrogen bonds. This base-pairing phenomenon is the molecular basis underlying both the structure and much of the function of DNA. To repeat: A pairs with T, and G pairs with C.

Your Turn 13.5 Skill Building Complementary Base Sequences

Adenine and thymine are said to be *complementary bases*. So are cytosine and guanine. In both cases, the bases form hydrogen bonds when they pair. Using one-letter codes, write out the base sequences that are complementary to each of these codes.

a. ATACCTGC **b.** GATCCTA

The structure of DNA, featuring the puzzle-piece pairing of its nucleotides, inspired another vital discovery. One side of the DNA strand contains all the information required to generate its partner strand! Thus, a single strand of DNA can guide the generation of its complement. **Replication** is the process of cell reproduction in which the cell must copy and transmit its genetic information to its progeny. The process is well understood, and is diagrammed in Figure 13.8.

Before a cell divides, the double helix rapidly, but only partially, unwinds. This results in a region of separated strands of DNA, as pictured in the middle portion of Figure 13.8. Free nucleotides in the cell are selectively hydrogen-bonded to these two single strands that serve as templates for a new DNA molecule: A to T, T to A, C to G, and G to C. Held in these positions, the nucleotides bond together by the action of an enzyme, a biological catalyst. By this mechanism, each strand of the original DNA generates a complementary copy of itself. The original strand, and its newly synthesized complement, coil to form a new molecule identical to the first. Similarly, the other separated strand of the original molecule twines around its new partner, forming another duplicate molecule. Thus, where there was originally one double helix, now there are two identical copies.

Return to Section 12.6 to review enzymes.

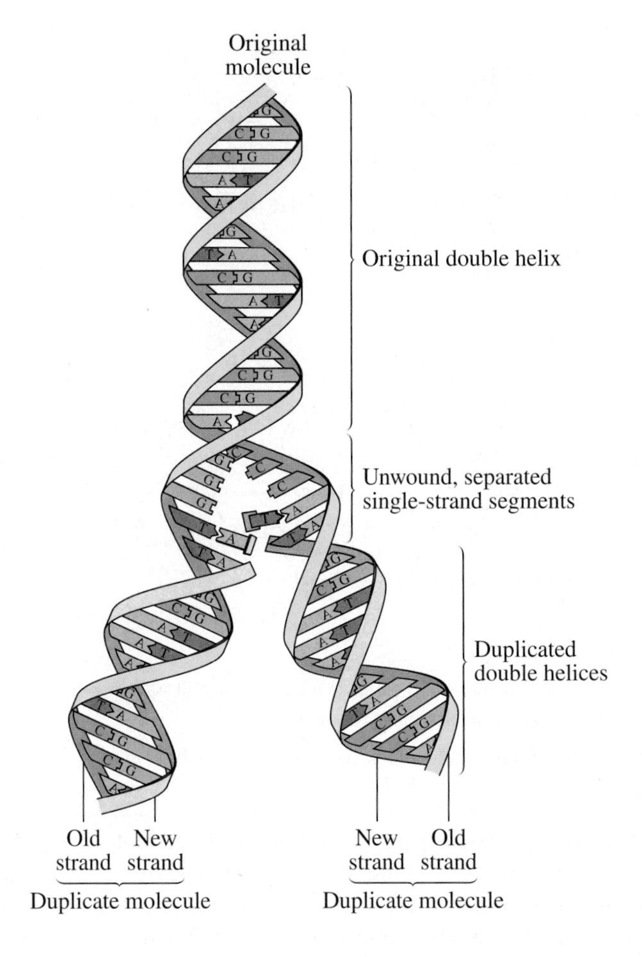

Figure 13.8

Diagram of DNA replication. The original DNA double helix (top portion of figure) partially unwinds, and the two complimentary portions separate (middle). Each of the strands serves as a template for the synthesis of a complimentary strand (bottom). The result is two complete and identical DNA molecules.

 See **Figures Alive!** on **Connect** for an illustration of this process.

Your Turn 13.6 Scientific Practices One Gorgeous Molecule!

Revisit the photograph that opens the chapter. It shows a sculpture of DNA that people can climb on. As a piece of art, it represents some parts of the DNA molecule better than others.

a. List three disadvantages of this DNA representation. What chemical details are omitted? What information is lost?
b. Now list three advantages. What information is highlighted? What is gained?
c. Find another artistic rendition of DNA, possibly on your campus or on the Internet, and repeat parts **a** and **b**. Cite your sources.

Your Turn 13.7 Skill Building DNA Sequence Repair

DNA must be copied with utmost perfection. After replication, enzymes scan over strands of DNA to identify and correct errors in base pairing.

ATGCCATGAA
TACGGTATTT

a. Find the error in base pairing for this set of DNA strands and circle it.
b. Do you expect the mismatched strands to be more or less stable than a correct pair? Explain your reasoning.

In most organisms, the newly copied DNA does not remain extended as a double helix, but becomes coiled even further. This not only saves space but it also organizes and protects the genetic information. The coiling is carefully regulated so that small portions of DNA can be accessed when specific stored information is needed. This complete set of genetic information is packaged into **chromosomes**, rod-shaped, compact coils of DNA and specialized proteins packed in the nucleus of cells.

Your Turn 13.8 Skill Building Is Your DNA Doin' the Twist?

The distance between base pairs in the double helix structure of DNA is 0.34 nm.

a. Calculate the length in centimeters (cm) of human chromosome 11 when extended in a double helix. Chromosome 11 consists of 135,000,000 base pairs.

b. Chromosomes can be visualized best immediately before cell division. In this compact state, the longest axis of chromosome 11 is approximately 4 μm. By what factor has the DNA been further condensed?
Hint: 1 μm is 1×10^{-6} m.

c. Suggest a reason why this level of compaction is necessary.
Hint: A typical human cell is only 10 μm in diameter.

Every time a cell splits to reproduce, the complete set of chromosomes must be uncoiled and replicated perfectly so that each new cell contains an identical set. Some cell types, such as those in the skin and in a cancerous tumor, divide more rapidly than others. These cells are more susceptible to collecting and passing on DNA damaged by ionizing radiation, free radicals, or chemical agents.

Return to Section 12.10 to review ionizing radiation, and Section 2.13 to review free radicals.

13.4 | Cracking the Chemical Code

It can be overwhelming thinking about all the complex chemistry happening in our cells every minute, chiefly because DNA molecules organize a *lot* of information! The billions of base pairs repeated in every corn cell provide the blueprint for producing one corn plant. These base pairs are ordered into specific sequences and grouped, sometimes into genes, to code for the production of proteins. Other information is present in the DNA, too, but our understanding of how it is used is still in its infancy.

Although the information is carried in DNA, it is expressed in other (smaller) molecules. The best understood are proteins. Proteins are found throughout our bodies in skin, muscle, hair, blood, and the thousands of enzymes that regulate the chemistry of life. By directing the synthesis of proteins, DNA can dictate many of the characteristics of an organism.

Section 11.6 defined proteins in the context of the foods we eat. Section 9.7 described proteins as polymers (polypeptides or polyamines).

Proteins are large molecules formed by the linking of amino acids. Recall that the 20 amino acids that commonly occur in proteins can be represented by this general structural formula, which is reproduced from Chapter 11:

Refer back to Figure 11.16 for more details regarding the general structure of an amino acid.

$$\mathrm{H_2N{-}CH(R){-}C({=}O){-}OH}$$

The amine group is $-NH_2$, the carboxylic acid group is $-COOH$, and R represents a side chain that is different for each of the 20 amino acids. In a condensation reaction, the –COOH group of one amino acid reacts with the $–NH_2$ group of another. In this process, a peptide bond is formed and a molecule of H_2O is formed. When many amino acids are connected, the result is a protein; that is, a polymer built from amino acid monomers. We can also describe a protein as a long chain of **amino acid residues**, which implies that the amino acids have been incorporated into the peptide chain.

The discovery of the molecular code for genetic information is arguably history's most amazing example of *cryptography*, the science of writing in secret code.

The information in a sequence of DNA nucleotides translates via a code into a sequence of specific amino acids in a protein. The code cannot be a simple one-to-one correlation between bases and amino acids because there are differing numbers of each. DNA has only four bases. If each base corresponded to an individual amino acid, DNA could encode for only four amino acids. But 20 amino acids appear in our proteins. Therefore, the DNA code must consist of at least 20 distinct code "words," with each word representing a different amino acid. Furthermore, the "words" must be selected from a pool of only four letters—A, T, C, and G—or, more accurately, the bases corresponding to those letters.

Some simple statistics can help us determine the minimum length of these code words. To find out how many words of a given length can be made from an alphabet of known size, raise the available number of letters to the power n, corresponding to the number of letters per word:

$$\text{words} = (\text{letters})^n$$

For example, using four letters to make two-letter words generates 4^2, or 16 different words. Thus, DNA bases taken in pairs (akin to two letters per word) could only code for 16 amino acids, which is insufficient to provide a unique representation for each of the 20 amino acids. So, we repeat the calculation and assume that the code is based on three sequential bases or, if you prefer, three-letter words. Now, the number of different triplet-base combinations is 4^3, or $4 \times 4 \times 4 = 64$. This system provides more than enough capacity to do the job.

Your Turn 13.9 Skill Building Quadruplet–Base Code

Suppose that the DNA code used four sequential base pairs instead of a triplet-base code. How many different four-base sequences would result?

Protein synthesis occurs in the ribosome, which is found in all living cells.

The three-letter groupings of nucleotides are the basis of the information transfer from DNA to proteins. Each grouping, or **codon**, is a sequence of three adjacent nucleotides that either guides the insertion of a specific amino acid, or signals the start or end of protein synthesis. If you were to use the letters A, T, C, and G in a game of Scrabble©, you could generate 64 different three-letter combinations. A few, CAT, TAG, and ACT, for example, make sense. However, most combinations such as AGC, TCT, and GGG are meaningless—at least in English. Nature does far better than that; 61 of the 64 possible triplet codons specify amino acids. Thus, the codon sequence CAC in a DNA molecule signals that a molecule of the amino acid histidine should be incorporated into the protein, TTC codes for phenylalanine, and CCG stands for proline. The three-base sequences that do not correspond to amino acids are signals to stop the synthesis of the protein chain. Figure 13.9 illustrates a simplified example of a nine-base nucleic acid segment, and how it codes for three amino acids.

Our calculations showed that three-letter groupings are the minimum necessary to cover the 20 amino acids, but did not show us how all 64 codons are used. The code has redundancy. Many amino acids have more than one codon. For example, leucine, serine, and arginine have six codons each. Also, three different codons tell protein synthesis to "stop." On the other hand, two amino acids (tryptophan and methionine), and the signal to start protein synthesis, are represented by only a single codon.

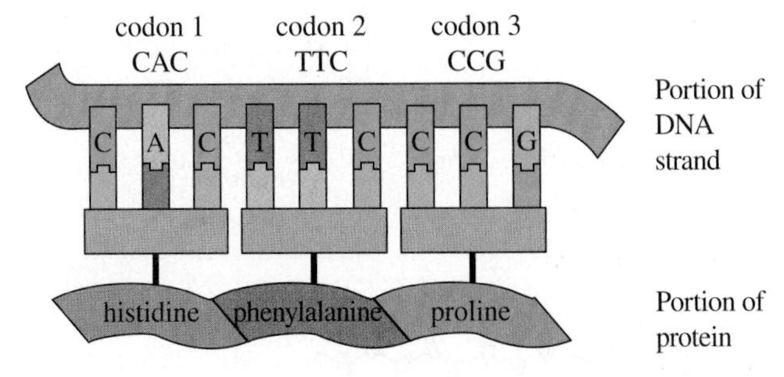

Figure 13.9

A nine-base nucleic acid sequence showing three codons.

> **Your Turn 13.10 Scientific Practices Duplicate Codons**
>
> Suggest some advantages of a genetic code in which several codons represent the same amino acid.

The genetic code is identical in all living things. With only a handful of exceptions, the instructions to make people, bacteria, and trees are written in the same molecular language of those 64 codons. In the genetic code, we have a "Rosetta Stone" to translate any genetic sequence from any organism. The significance of this statement will become apparent later in this chapter, when we discuss an example of genetically modified corn.

Did You Know? The markings on the Rosetta Stone were in several languages and helped to decipher Egyptian hieroglyphs.

13.5 | Proteins: Form to Function

Proteins are polymers. Admittedly they don't much resemble the clear polymer PET that may hold your favorite soft drink. Similarly, they don't seem to have much in common with the tough polypropylene used to make carpets. Nonetheless, proteins are big molecules built from little ones.

More specifically, proteins are polyamides. Like nylon (Section 9.7), they are built from the chemical reaction of carboxylic acids and amines. Unlike nylon, though, a protein is built from 20 different amino acid monomers. Comparing proteins to nylon may leave you with the image of proteins as long strands, rather than complex three-dimensional molecules. However, this image is far too simplistic. Proteins exist in a complicated environment—the cell. Just like necklaces in a messy jewelry box, proteins refuse to stay extended and neat. Unlike jewelry, each protein collapses into a unique, and often very specific, three-dimensional form. The final shape may look a mess, but that exact shape is necessary for the function performed in the organism.

> **Your Turn 13.11 Scientific Practices How Is Hamburger Like Nylon?**
>
> Take a moment to refresh your knowledge about two polyamides: the nylon from a sports jersey, and the protein found in hamburger.
>
> **a.** What functional group do nylon and meat proteins have in common?
>
> **b.** Nylon is usually synthesized from two types of monomers. In contrast, proteins are synthesized from one type of monomer. What functional group(s) do the monomers for each contain?
>
> **c.** Are nylon and protein addition polymers or condensation polymers?

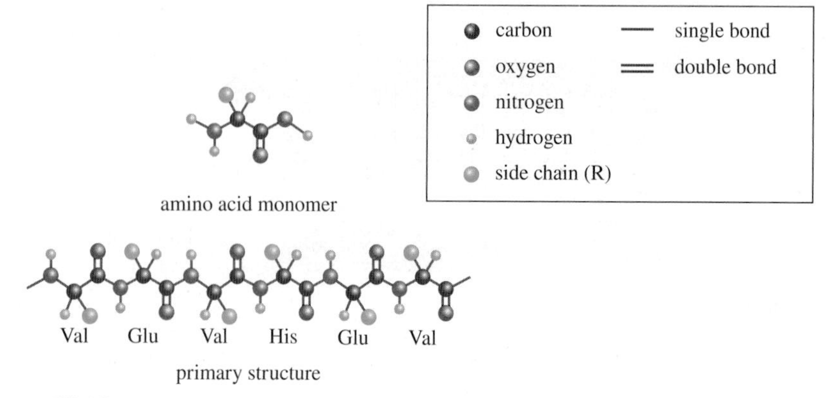

Figure 13.10

Representation of the primary structure of a protein.

We begin our discussion of protein shapes with the **primary structure** of a protein; that is, the unique sequence of the amino acids that make up each protein (Figure 13.10). Primary structure is the first and most basic identifier of a protein—the list of amino acids read over the length of the polymer. Knowing that a short protein contains three valines (Val), two glutamic acids (Glu), and one histidine (His) may tell you the size and a few other details, but it is not sufficient to specify a protein and explore its shape and function. The order and sequence of the amino acids matters. For example, Val-Glu-Val-His-Glu-Val is a different protein from Val-Val-Val-His-Glu-Glu. These short peptides behave differently as well!

Figure 11.17 categorized the 20 amino acids based on the properties of their side chains.

Section 9.4 discussed attractive dispersion forces between nonpolar groups in the context of plastics.

Revisit Section 8.2 for more about polar and nonpolar molecules.

Recall that each amino acid has a side-chain group. These side chains interact with one another, or with the molecules around and inside the protein. Side chains can attract and "lock" together, and in doing so, they hold a protein in a particular overall shape. The order and identity of the amino acids defines how and where those side-chain links can form. Each amino acid plays a role; changing just one can change the shape and, as a result, the function of a protein.

We will group the side chains into two categories: polar (either charged or neutral) or nonpolar. Like oil and water, nonpolar and polar side chains tend to separate. In the typical environment for a protein (water), the polar side chains can stay in the water. In contrast, nonpolar side chains group inside the protein, avoiding unfavorable interactions with water in favor of attractive dispersion forces between the side chains.

Polar side chains can lead to more types of interactions: ionic attractions or hydrogen bonds. Side chains that contain acidic or basic groups (such as carboxylic acids or amines) often become charged ions, and attract their opposites. Uncharged, yet polar, side chains often contain hydroxyl groups or amides. Many of these side chains attract one another by forming hydrogen bonds.

One very special amino acid contains a thiol group (–SH) in its side chain. In proteins, thiol groups perform an important and highly specialized function; namely, two thiol groups can react to form disulfide (S–S) bonds between the adjacent sulfur atoms. These strong bonds covalently link two different regions of a protein together. All of these tendencies—the nonpolar side chains grouping together, the polar groups making ionic or hydrogen bonds, and the thiol groups forming disulfide bonds—serve to give each sequence of amino acids a unique structure.

We continue our discussion by examining the **secondary structure** of a protein; that is, the folding pattern within a segment of the protein chain. Many, but not all, protein chains form regular, repeating structures from the particular bond angles and attractions between neighboring amino acids. The two most common are the α-helix, a spiraling strand, and the β-pleated sheet, extended strands stretching alongside each other with a slight zigzag.

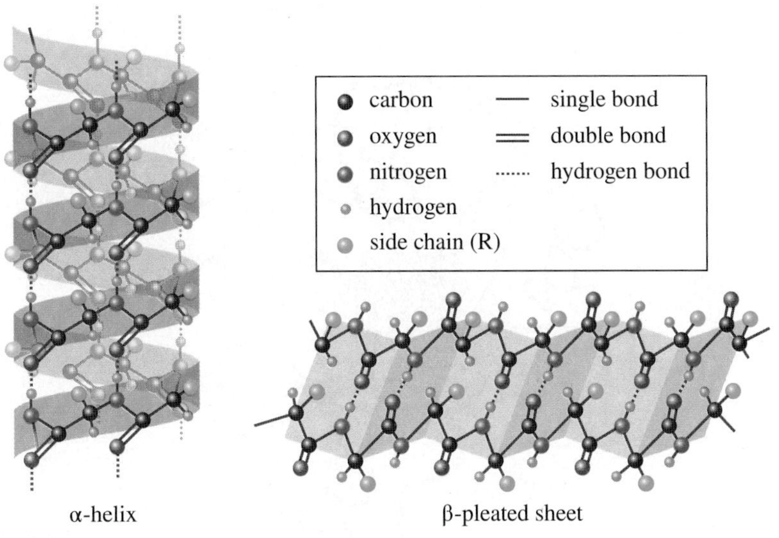

Figure 13.11

Representations of secondary structures of a protein. The two major types of secondary structures are the α-helix and the β-pleated sheet.

Both forms of secondary structure depend on the tendency of the protein backbone to form intramolecular hydrogen bonds. Figure 13.11 shows the hydrogen bonds between the backbone O and the N–H of the amide group as dotted lines. The number and regular spacing of these hydrogen bonds can pull together and align a protein strand, stabilizing the secondary structure. The choice of secondary structure, or even the complete lack of it, can be loosely predicted from the primary structure. Some side chains tend to pack well into β-pleated sheets, others tend toward the α-helix, and others even predispose the chain to disorder.

Your Turn 13.12 Skill Building Intra- Versus Intermolecular Hydrogen Bonding

Using the Internet as a resource, illustrate three examples of polymers that exhibit intermolecular hydrogen bonding, and three different examples of polymers with intramolecular hydrogen bonding.

Proteins are large, three-dimensional molecules, but both primary and secondary structures are relatively flat. We need a more "global" description of their shape, which is designated as its **tertiary structure**. A tertiary structure represents the overall molecular shape of the protein that is defined by the interactions between amino acids far apart in sequence, but close in space. If you imagine a protein as a jumbled charm bracelet, the secondary structure might be the kinks in the chain itself, but the tertiary structure forms from the contact between the Eiffel Tower charm near the clasp, and the Statue of Liberty charm near the middle. The overall fold results in a net increase in stability, maintained through hydrogen bonds, ionic bonds, disulfide bonds, and interactions (or avoidance of interactions) between the side chains and water.

From only 20 amino acids, proteins make a variety of shapes and serve a large array of functions. Their three-dimensional shapes dictate specific functionalities. For instance, the shape of enzymes must create an **active site** in an enzyme that binds only specific reactants and accelerates the desired reaction (Figure 13.12). Enzymes are the

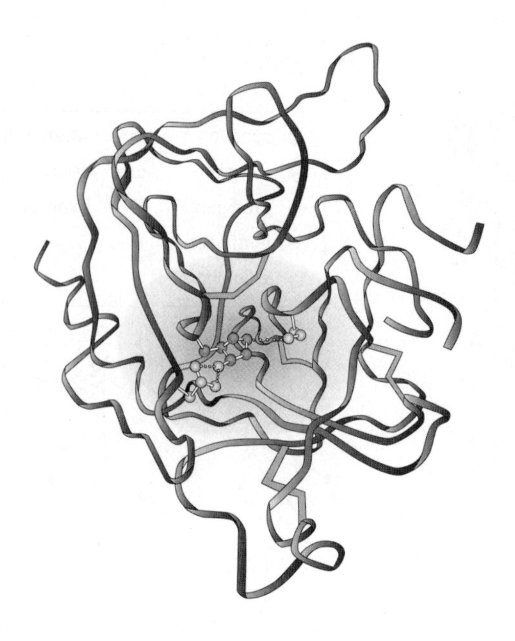

Figure 13.12

Tertiary structure of the enzyme chymotrypsin showing its active site. The "ribbon" portion (gray) represents the amino acid chain; the central colored portion is the active site at which the enzymatic chemistry takes place.

most commonly discussed type of protein, but a number of other examples exist. Some proteins bind DNA either to protect it or send a signal. Again, form suits function. When these proteins fold, they display positively charged side chains to attract and bind the negatively charged DNA. Another type of protein funnels material through the membrane of the cell. This is accomplished by forming channels that shuttle a specific chemical across the outer layers of the cell, while keeping the cell impermeable to undesirable chemicals.

A subtle change in the primary structure of a protein can have a profound effect on its properties. Notice the word *can* in the previous statement. Sometimes, a change in an amino acid leaves both the protein's shape and function unchanged. For instance, a nonpolar leucine can be switched to a valine, also nonpolar, by just removing a $-CH_3$ group. The protein may be a little less stable, but on the whole the same. Change the wrong glutamic acid (in which the side chain is often negatively charged) to a nonpolar valine, however, and the disease sickle cell anemia occurs.

Did You Know? Sickle cell disease affects a sizeable population, more than 70,000 individuals in the United States alone, and lowers life expectancy from almost 79 years to an average of 55 years.

Hemoglobin is the blood protein that transports oxygen. The single alteration of a particular glutamic acid to valine in the primary structure of hemoglobin creates a variant called hemoglobin S, and the condition called sickle cell disease. This substitution causes hemoglobin to convert to an abnormal shape at low oxygen concentrations, forcing red blood cells to distort into rigid sickle or crescent shapes (Figure 13.13). Because these cells lose their normal flexibility, they cannot pass

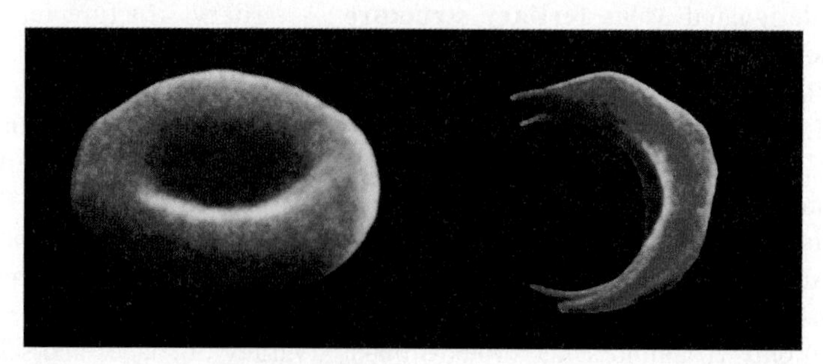

Figure 13.13

An image from a scanning electron microscope showing both normal red blood cells (left) and those distorted into a sickle or crescent shape (right).

through the tiny openings of the capillaries in the spleen and other organs. Some of the sickle cells are destroyed, and anemia results. Others clog organs so badly that the blood supply to these organs is reduced.

Capillaries are the smallest vessels in the blood circulatory system, and are only about one red blood cell wide.

Your Turn 13.13 Scientific Practices Function Follows Form

In sickle cell anemia, a glutamic acid residue in the sequence of hemoglobin is replaced with a valine residue.

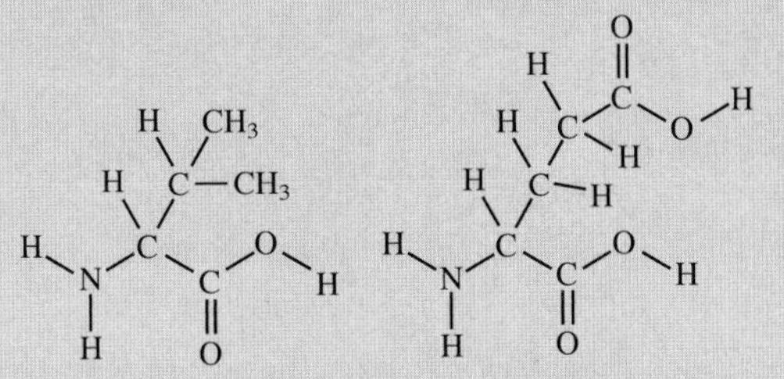

valine glutamic acid

a. Describe the structural difference between these two amino acids: valine and glutamic acid.
b. Predict the solubility for each of these amino acids in water.
c. Explain how these differences could give rise to the deformed cells typified by sickle cell anemia.

13.6 | The Process of Genetic Engineering

We started this chapter with thoughts of using bacterial cells to generate the vast amounts of human insulin required to maintain the health of the world's 382 million people who struggle with diabetes. We noted that this challenge is addressable with genetic engineering, and posed the question: "What are we modifying when we genetically modify something?"

By now, we hope you know the answer. We modify the DNA in the cell. If we change the genes, we change the proteins synthesized by these genes. Ultimately, we change the chemistry of the cell.

Throughout history, humans have manipulated genes. This may surprise you, as you might think that our ability to modify genes has come about only recently. But consider, for example, how we have cultivated plants. We tended to grow ones that carried specific traits, such as a better taste or appearance. The others we rejected. To produce these strains with new and unique traits, we crossbred different strains. The process took many years, but eventually we "domesticated" plants and created the crops that feed us today. Our crops are so far removed from their wild forebearers that we would hardly recognize them.

Corn is an excellent example, being indigenous to the Americas. The region's native people manipulated the genes of the teosinte plant (one that bore seeds on the end of its stalks rather than on the body of the plant) to get the growth pattern seen in today's corncob (Figure 13.14). Domesticating the teosinte plant led to a food that was both more nutritious and more abundant.

Domesticating a plant is a process of genetic modification. Even without understanding the chemistry, we selected plants with certain sequences of DNA and rejected ones with other sequences. Slowly over time, we encouraged changes

Figure 13.14

Corn's early much smaller ancestor, teosinte, is shown here next to a modern larger ear of corn.

in the DNA, allowing one sample of DNA to carry on, spread, and survive. Traits that conferred fitness that we could not see, taste, smell, or feel were often lost. Fast-growing yet scrawny plants were discarded. So were deep, persistent roots that were impossible to till. Similarly, disease resistance that was not immediately required was lost. Modern-day crops are now unable to survive without humans to tend to them.

Nature carries out genetic modification as well. For example, all fields contain bacteria, and plants are susceptible to the different strains of bacteria to greater or lesser degrees. Let's suppose that a new, highly virulent bacterium appears in the field. Perhaps this bacterium arose through a mutation, a small random change in its genome. Over time, most of the wild grain plants in the field succumb to the new pathogenic bacterium strain. But one particular plant contains the cell chemistry to resist the new onslaught. Suddenly, a gene that may not have mattered before is the key to survival. That plant is able to resist the bacterium, while others cannot. The plant spreads, and the gene survives.

The stress on the plant may take other forms as well, such as a three-year drought or a more aggressive weed. But the process is the same. Nature generally selects for plants that are more self-sufficient; humans tend to select for plants that look and taste better. In either case, the result is changes in the genome of the plant. The process of selective breeding, either in the wild or in human agriculture, is a long, slow, and somewhat random route toward modifying the gene pool. Neither natural nor artificial selection is genetic engineering. **Genetic engineering**, as we know it, is the direct manipulation of the DNA in an organism.

While the terms are often interchanged, we use *genetic engineering* to describe the technical process, and *genetic modification* to specifically discuss food products.

The easiest organisms to manipulate are small single-celled bacteria. These bacteria contain **plasmids**, or rings of DNA, in addition to their chromosomes. Scientists can now routinely remove, change, and replace the plasmids to create new chemistry in the bacteria. They use special enzymes to cut open the plasmid at specific sites. Scientists then copy the DNA containing a promising gene from another organism, and that DNA is inserted into the plasmid ring from the bacterium. The result is a new interspecies DNA plasmid, or **vector**—a modified plasmid used to carry DNA back into the bacterial host (Figure 13.15). Once inside the cell, the chemistry of the bacterium takes over. The bacterium grows rapidly and produces new cells. Soon, the scientist has millions of copies of the "guest" gene and its protein product.

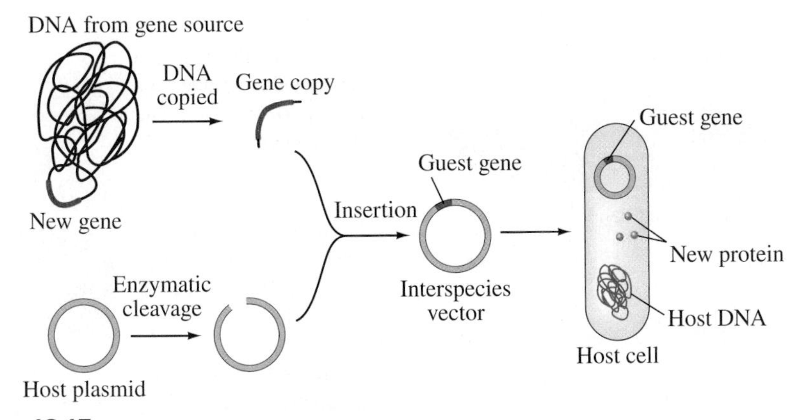

Figure 13.15

A general representation of the process of genetic engineering.

Inserting foreign DNA becomes a bit trickier as one climbs the evolutionary ladder from bacteria to plants. Higher organisms are better at protecting themselves against foreign DNA. For example, plants have thick cell walls. For another, many organisms have chemical mechanisms designed to detect and destroy foreign DNA. Even so, scientists have found ways to get around such defenses. One is to hijack a special soil bacterium that has the ability to infect plants. This bacterium creates a bridge into the plant cell and, in the process, transfers its own DNA into the genome of the plant. The bacterial genes induce points of large abnormal growth (Figure 13.16). You may have seen these growths (tumors) on a variety of plants and trees, including apple trees, rose bushes, and some vegetable plants.

Agrobacterium tumefaciens is commonly used to transform plants, particularly in the commercial seed industry.

Scientists are adept at disabling the bacterial gene to use the bacterium to insert a different gene of interest. If the gene of interest is the specific sequence from the soil bacterium, *Bacillus thuringiensis* (or *Bt*), that produces a *Bt* toxin, then the plant acquires the instructions to make the toxin itself (Figure 13.17). Not only is there *Bt* corn, but also *Bt* cotton, *Bt* potatoes, *Bt* tomatoes, and *Bt* rice, all producing toxins against crop-destroying pests!

The *Bt* toxin is but one example of genetically engineered possibilities. Not all examples of genetically engineered crops are **transgenic**, in which a transfer of genes occur across species. Genetic engineering is used to achieve the same aims as selective

Figure 13.16

A crown gall tumor (*Agrobacterium radiobacter*) on a chrysanthemum plant.

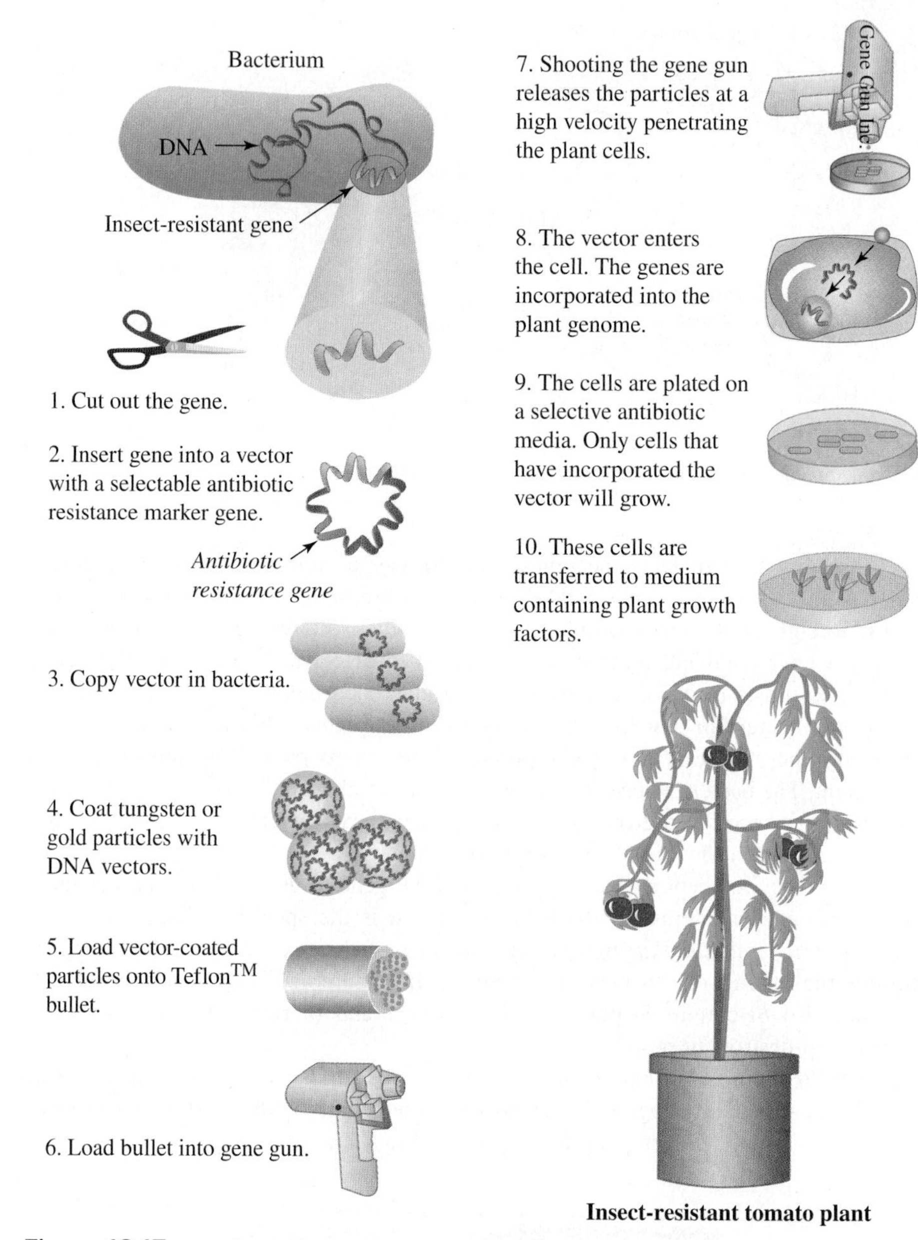

Bacterium
DNA
Insect-resistant gene
1. Cut out the gene.
2. Insert gene into a vector with a selectable antibiotic resistance marker gene.
Antibiotic resistance gene
3. Copy vector in bacteria.
4. Coat tungsten or gold particles with DNA vectors.
5. Load vector-coated particles onto Teflon™ bullet.
6. Load bullet into gene gun.
7. Shooting the gene gun releases the particles at a high velocity penetrating the plant cells.
Gene Gun Inc.
8. The vector enters the cell. The genes are incorporated into the plant genome.
9. The cells are plated on a selective antibiotic media. Only cells that have incorporated the vector will grow.
10. These cells are transferred to medium containing plant growth factors.
Insect-resistant tomato plant

Figure 13.17

An illustration of how an insect-resistant tomato plant is produced through genetic engineering.

breeding, but with far more speed and control. Imagine you have a rice crop that grows well, tastes great, and cooks with a perfect level of stickiness, but it is being destroyed by a local plant virus. At the same time, a wild relative of your crop grows just beside it, which is resistant to the virus. Instead of securing the resistance trait by selective breeding, you can find and copy just the resistance gene from the wild plant, then splice it into your weak but otherwise perfect crop. Technically, the genetic engineering process is identical, but the result is a plant with only rice genes.

Transgenic rice plants have been developed for use in sub-Saharan Africa where the yellow mottle virus destroys much of the rice crop each year (Figure 13.18). At first, the gene that offered resistance to the rice plants was taken from a surprising source: the virus itself! The result was a transgenic plant with a small amount of viral code. The gene source is perhaps a concern, particularly to those who worry that consuming the viral proteins may result in allergic reactions. Hence, a preferable gene source is currently in development. Scientists are finding new potential genes by exploring the reason for viral resistance found in some rice varieties.

Figure 13.18

Virus-resistant transgenic rice.

© Dung Vo Trung/Corbis

Perhaps you have a trait in mind, but have no source for the gene. In such a case, both traditional selective breeding and whole-gene transfer won't work. Neither of these can function if a trait does not already exist. Although in theory you could wait for a trait to evolve through natural, random mutations, most likely you would run low on patience. To speed up the process of evolution, scientists use either chemicals or ionizing radiation to produce random mutations in a batch of seeds. Upon planting, some seeds do not grow at all, others grow but with no apparent changes, and still others show unique traits. Once a useful trait appears, scientists can use either selective breeding to refine a new crop with the positive trait, or genetic engineering to isolate the gene responsible for the trait and transfer it into a different plant.

Your Turn 13.14 Scientific Practices Ionizing Radiation

Ionizing radiation is a term commonly used by both scientists and health professionals. Using the information presented in previous chapters, answer the following questions.

a. What is an ion? Give two examples.
b. Some species have unpaired electrons. Give two examples.
c. How does radiation produce ions with unpaired electrons?
d. How do these free radicals lead to random mutations in DNA?

As humans, we have a responsibility to act wisely, both for current and future generations. Our needs are great and our societal problems are pressing. But can our *inaction* result in more harm than good? The sheer power of genetic engineering can frighten us; you may be automatically concerned at even the idea of transgenic organisms. In the next section, we describe the benefits of genetic engineering to the chemical industry, both current and projected. In the section after that, we delve into other benefits and potential risks of genetic engineering.

13.7 Better Chemistry Through Genetic Engineering

Pest and herbicide resistance are not the only reasons for genetic modification. Scientists have also modified the genes of corn, soybeans, and wheat, with the aims of making them more resistant to disease, more tolerant of stresses such as salt, heat, or drought, and more nutritious. Although farmers have benefited, many others have as well. Scientists have designed plants to absorb toxic metals from contaminated soil. Some are developing crops such as soybeans that produce high yields of biofuel per acre. Others are engineering bacteria to detect and remediate radioactive contamination. Of interest to us in this section is that genetic engineering can incorporate the key ideas of green chemistry into large-scale chemical production.

A synthetic route to a desired chemical may require toxic chemicals, large amounts of solvents, and high temperatures. Although such processes yield many useful chemicals, they also produce a staggering amount of waste—up to 100 times the weight of the compound! One way to reduce the ecological footprint is to use enzymes, the biological catalysts described earlier. These "biological machines" perform reactions just as you would conventionally within flasks or beakers, but faster and safer with fewer toxic reagents, at lower temperatures, and with less waste. Another plus is that enzymes can be used over and over again. Minimal waste, reusable reagents, and low toxicity are the hallmarks of green chemistry.

Growing bacteria with natural or artificial genes is now a step in manufacturing small-molecule drugs such as insulin (Figure 13.19). As discussed earlier, this application of genetic engineering is not only one of the most rapidly growing uses of the technology but also the oldest, dating back to the development of synthetic insulin in the late 1970s. The process is particularly easy when the gene, or something similar, already exists. Sometimes, however, special tricks are required, as was the case in the synthesis of the drug atorvastatin (Figure 13.20). Enzymes for specific reaction steps

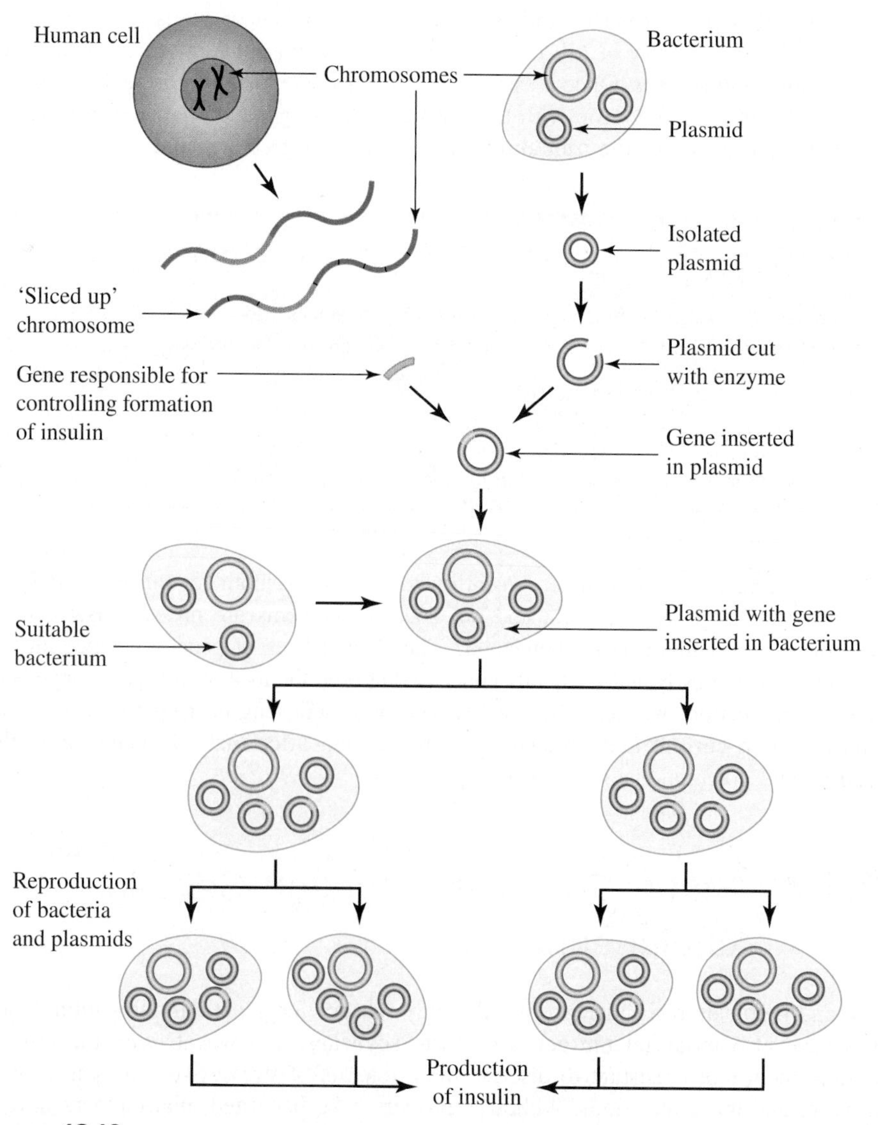

Figure 13.19

An illustration of how human insulin is produced through the use of bacteria.

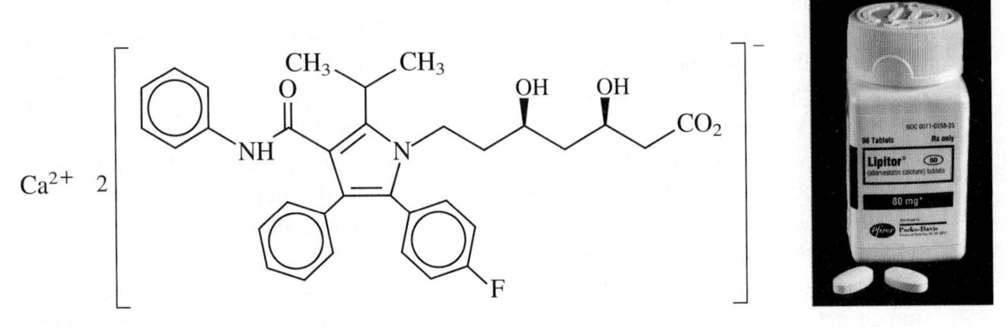

CH3
CH3
O
OH
OH
NH
N
CO2
Ca2+ 2
F
Lipitor
80 mg

Figure 13.20

Atorvastatin, the active ingredient in Lipitor®, requires enzyme-generated building blocks. The cholesterol-lowering drug produced by Pfizer had annual sales exceeding $10 billion before going off patent in December 2011.

could not be found, so they had to be evolved. Scientists mimic natural selection by creating an environment where the bacteria must evolve a new trait in order to survive—a process known as **directed evolution**. Typically, the scientists start with a random variety of DNA sequences transferred into a population of bacteria. A new enzyme emerges over multiple generations from growth under certain conditions, such as providing only certain chemicals as food.

Directed evolution can also be performed without using bacteria by selecting, mutating, and multiplying DNA with enzymes.

Engineered organisms can even produce plastics. Most synthetic polymers are fabricated in large chemical plants using a process that consumes large amounts of chemical reagents and energy. Furthermore, these chemical reagents are often derived from petroleum. Scientists at Metabolix® have engineered organisms to improve the sustainability of traditional polymer syntheses. These organisms produce monomers from renewable materials such as corn, sugarcane, and vegetable oil, and then catalyze the polymerization reaction. One example of a resulting polymer is polyhydroxybutyrate (PHB). This bioplastic is used to produce items such as plastic utensils and coatings for cups, much like polypropylene (PP). However, unlike PP, PHB is biodegradable. This process uses materials of low toxicity, is extremely efficient, and lowers greenhouse gas emissions.

Return to Chapter 9 for a review of plastic production, uses, and consequences.

Scientists and engineers at Metabolix® earned a Presidential Green Chemistry Challenge Award in 2005 for their innovative and sustainable bioplastic technology.

Imagine using our farms to create not just polymers and drugs, but also vaccines. This is no longer a utopian dream, because our transgenic plants can now produce both. For example, vaccines against infectious diseases of the intestinal tract have been produced in potatoes and bananas (Figure 13.21). Anticancer antibodies have been expressed

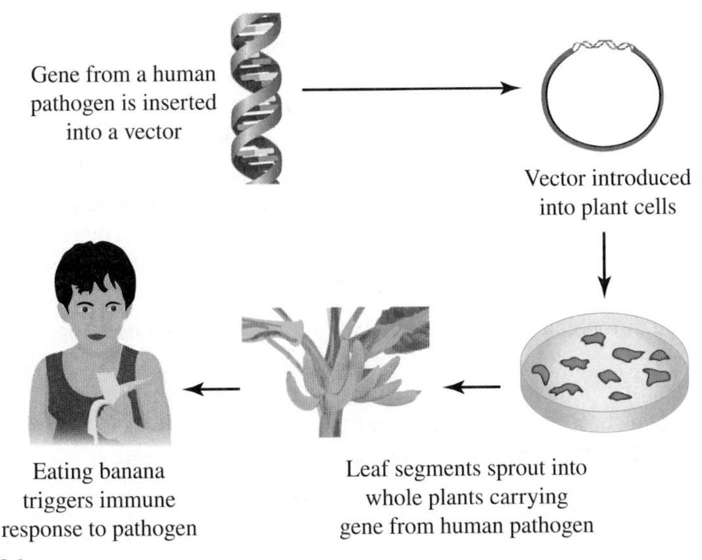

Gene from a human pathogen is inserted into a vector
Vector introduced into plant cells
Eating banana triggers immune response to pathogen
Leaf segments sprout into whole plants carrying gene from human pathogen

Figure 13.21

The fabrication of transgenic plants with edible vaccines using a recombinant DNA approach.

and introduced into wheat. Peptide drugs against HIV/AIDS have been produced from tobacco fields. Also, most vaccines require refrigeration or other special handling, together with trained professionals to administer them. In some countries, health professionals cannot afford even the needles to inoculate people, leading to possible infections from reused needles. Vaccines produced within edible products may be difficult to correctly dose, but would be easy to administer and transfer. In addition to yielding crops for food, these fields could provide the hope of low-cost, readily available vaccines. Thus, fields of transgenic plants can go hand-in-hand with a good public health policy.

The technology is still developing, but the bonuses are already both apparent and immense. Compared with traditional methods, a genetically engineered enzyme, microorganism, or crop can yield high-purity products while reducing waste and by-products. Genetically engineered routes increase the yield with fewer steps and often eliminate labor, energy, and resource-intensive purification processes. They avoid the use of toxic and corrosive chemicals, which is better for the environment and increases overall worker safety.

This all being said, the promise of developing technology and society's desperate needs can lead to rash choices. In light of this, the risks of genetically engineered cures may be large, perhaps unacceptably so. Let's now delve into the potential risks of genetic engineering.

13.8 | The Great GMO Debate

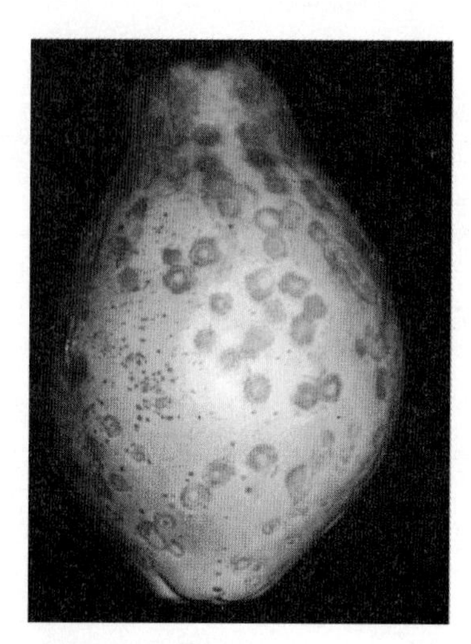

Figure 13.22

Photo of the papaya ringspot virus.

Source: Courtesy S. Ferreiria

In this chapter, we have highlighted a few genetically modified organisms (GMOs) that were created for beneficial purposes, such as to produce useful chemicals or reduce the use of other chemicals. Another case study to illustrate the utility of genetic engineering is related to papaya production in Hawaii. In the 1990s, an outbreak of the papaya ringspot virus took a severe toll on fruit production (Figure 13.22). Researchers at the University of Hawaii developed a type of papaya that incorporated DNA from the ringspot virus, making papayas resistant to the disease. Seeds for the genetically engineered papaya were then distributed to Hawaiian farmers for free, and now more than 80% of the papaya crop in Hawaii consists of this virus-tolerant variety.

Genetically engineered crops are also being used in the fight against malnutrition. Researchers at the Swiss Federal Institute of Technology and the University of Freiberg developed a strain of rice that provides a significant amount of vitamin A (or beta carotene), a vitamin critical in the development and function of eyesight. An estimated 667,000 children worldwide die every year from vitamin A deficiency. The regions of the world with the highest risk for this deficiency are in Southeast Asia, Africa, and South America. This strain of rice, called "golden rice" because of the yellow color of the beta carotene in the grain (Figure 13.23), provides the daily recommended allowance of vitamin A in less than half a cup of rice. The developers of the rice are providing the rice royalty-free to farmers making less than $10,000 annually, which represents 99% of all farmers in the regions most affected by vitamin A deficiency.

Although there are clearly a multitude of advantageous uses for genetic modification, it is not a technology without its detractors. Indeed, the topic of GMOs has sparked protests, labeling campaigns, and environmental vandalism (Figure 13.24). The arguments between those against GMOs and those that support their use are often heated. We try to explore these arguments in the following paragraphs.

Because GMOs are currently a hot topic among scientists, we've provided you with peer-reviewed references for more information. The footnote references are formatted as: *Journal title* **year**, *volume*, page #. The digital object identifier (DOI) is also provided for each reference.

Genetic engineering is not natural.

Many people feel the insertion of DNA from a different species into a plant, also called *transgenetic manipulation*, is not a natural process. People argue that the transfer of genes in this way is beyond the role of humans.

Figure 13.23

A comparison of genetically modified golden rice with traditional white rice.

© International Rice Research Institute (IRRI)

Figure 13.24

Protesters in a field of genetically modified crops near Ghent, Belgium, in 2011.

© Orlando Kissner/AFP/Getty Image

- **Counterpoint**. The introduction of foreign DNA into a plant's genome is actually a natural process. In fact, recent research has shown that bacteria similar to those that genetic engineers use to modify corn with the *Bt* gene, inserted its own bacterial DNA into sweet potatoes in South America at least 8,000 years ago. What used to be a normal plant root became swollen with starch, and was integrated into the diet of early Native Americans. Several varieties of sweet potato now exist throughout the world, and it is the world's seventh most important staple crop, being the primary source of Calories for millions of people. The exchange of DNA across species is not a rare event and most typically occurs in single-celled organisms such as bacteria, but the transfer of genes has also occurred between millet and rice.

Research reference: *PLOS Biology* **2006**, *4*, e5 (DOI: 10.1371/journal.pbio.0040005)

Research reference: *Proceedings of the National Academy of Sciences* **2015**, *112*, 5844 (DOI: 10.1073/pnas.1419685112)

One paper published in 2012 in *Food and Chemical Toxicology* claimed that genetically modified maize caused serious disease in rats. However, their studies used faulty test methods (including the use of rats that were predisposed to tumors!) and inadequate data to support its conclusions. As a result, the paper was retracted. Source: *Nature* (News), November 2013 (DOI: 10.1038/nature.2013.14268)

GMOs cause health problems such as allergies and cancer.
There is some concern that genetic engineering would introduce dangerous compounds such as allergens or toxins into the food chain. Some of these compounds may be carcinogenic, and lead to tumors in those who eat the GMO.

- **Counterpoint**. While it is possible for a new gene to express a protein that is an allergen or toxin, research scientists work with food regulators and rigorously test their creations, because they would be legally liable for any toxicity. For example, in 1996, a company that was trying to insert a beneficial Brazil nut gene into soybeans halted the project when it was discovered that the resulting soybeans also carried a nut allergen. While the majority of testing data is treated as proprietary information by biotech companies and not released to the public, thousands of published safety reports on GMOs have shown no evidence that the foods are dangerous.

Research references: *New England Journal of Medicine* **1996**, *334*, 688 (DOI: 10.1056/NEJM199603143341103); *Nature Biotechnology* **2008**, *26*, 73; (DOI: 10.1038/nbt1343) *Allergy* **2005**, *60*, 559 (DOI: 10.1111/j.1398-9995.2005.00704.x)

GMOs cause farmers to over-use pesticides and herbicides.
GMO opponents often point out that farmers growing genetically engineered crops have increased the amount of pesticides or herbicides used in protecting their crops. One type of corn, for example, is genetically engineered to be resistant to a type of herbicide called glyphosate. The use of glyphosate has dramatically increased since the introduction of this type of corn.

- **Counterpoint**. Some types of crops have been developed that actually *reduce* reliance on pesticides. A case in point is the *Bt* corn described earlier in this chapter. The plant itself grows its own pesticide—one that is specific to insects, and does not interact with the biology of animals—in levels much lower than that required to treat the outside surface of the plant. As for glyphosate, its increased use has led to the decrease in use of other herbicides that are more toxic. The application of glyphosate typically calls for 360 mL/acre, or about a soda can worth of herbicide for about 43,500 ft^2. To put this in perspective, an American football field (excluding the end zones) is 48,000 ft^2.

Did You Know? Glyphosate has 1/25 the toxicity of caffeine.

Research reference: *Environ. Sci. Europe* **2012**, *24*, 24 (DOI: 10.1186/2190-4715-24-24)

GMOs create "super insects" and "super weeds."
In the ecosystem surrounding a genetically modified crop, the targeted weeds, bugs, or bacteria must move, die, or evolve. If the species evolves, the GM crop may have no resistance to it. Historically, we solved the problem of the evolution of a pest by finding a natural trait within the existing crop varieties, and then using selective breeding to create a new, stronger hybrid. As we come to depend on a narrower set of crop varieties, we no longer have the genetic variety to utilize.

Research reference: *Nature* **2013**, *497*, 24 (DOI: 10.1038/497024a)

- **Counterpoint**. Although new resistance traits easily evolve in insects in laboratories, only a few resistant insects have been reported in fields. Regulators in both the United States and the European Union require the cultivation of traditional crops, without engineered toxins or resistance traits, alongside GM varieties. The mix is intended to prevent evolution resistance in pests because resistance to the crop will not be a survival requirement. As the engineered crops become more widespread, scientists may need to further test how much of the non-engineered crop is enough to minimize the risk of newly resistant pests.

Research reference: *Turk.J. Agric. For* **2015**, *39, 531* (DOI: 10.3906/tar-1408-69)

Farmers can't replant genetically modified seeds.
While a myth exists that some seed companies use what is called a "terminator gene" to prevent the seeds from GMO crops from being able to be replanted, there have been no crops developed that incorporate this technology. Seed companies often do, however, require agreements that prevent farmers from replanting seeds gathered from a GMO crop, ensuring future purchases of new seed every year.

http://www.monsanto.com/sitecollectiondocuments/technology-use-guide.pdf

- **Counterpoint**. Before the advent of genetic engineering, most of the corn grown in the United States and the European Union were commercially developed hybrids that, when replanted, led to a mixture of inferior corn variants with lower yields. Beginning in the 1930s, seed companies began developing high-yield hybrid corn, and by 1965, 95% of the corn crops planted in the U.S. were hybrids, requiring planting of newly purchased seed to keep yields high. Although this trend started with corn hybrids, farmers growing other crops also began to repurchase new seeds rather than replant saved seeds. Today, even those farmers not using GMOs prefer to purchase new seed every year to ensure a higher yield in their harvests.

As you can see, there have been many arguments back and forth over the use and safety of genetically modified organisms. The first commercially approved GMO, a transgenic tomato that was more resistant to rotting after being picked, went to market in 1994. In the twenty-plus years since then, many other crops have been approved and commercialized. Recent developments include a type of apple that doesn't brown when cut, and a potato that produces significantly less acrylamide (a possible carcinogen) when fried.

Arctic apples are engineered to not produce a compound called polyphenol oxidase, which causes apples to brown when cut or bruised.

You may be surprised to know that you almost certainly have eaten a genetically modified food item. As of 2015, the U.S. Department of Agriculture estimated that 92% of the corn planted in the United States was genetically modified. The vast majority (94% in the U.S.) of soybeans and cotton were also genetically modified. Considering the variety of products that include high-fructose corn syrup or soy-based vegetable oil, you likely have already consumed a genetically engineered product.

Your Turn 13.15 You Decide GMO Labeling

Use the Internet to find out the current status of mandatory GMO food labeling in the U.S. List some pros and cons of the proposed legislation. Do you think these food labels would be useful for consumers? Justify your answer based on the arguments for/against GMOs provided in this section.

Conclusions

More than 30 years after the development of Humulin™, and more than 20 years after the commercialization of the first genetically modified crop, there is still controversy surrounding genetic engineering. There are many parallels between the cases of global warming and genetically modified organisms. In both, the majority of scientists in the field are convinced of the soundness of the science. In both, an incomplete understanding of the science has led to the spread of misinformation and the growth of fanatical opposition to using the knowledge gained from scientific research. It is somewhat ironic that the vocal opponents to GMOs oftentimes are fervent supporters of the need to address global warming, and vice versa.

A common accusation of those against GMOs is that the science supporting the use of genetically modified organisms is funded by the large agricultural industry. While seed companies do fund research in this area, there is still significant research in non-corporate labs that is funded by nonprofit and public funding agencies. For instance, the development of ringspot-resistant papaya was carried out at the University of Hawaii and Cornell University. Research on golden rice was supported in part by grants from the Rockefeller Foundation and the Bill and Melinda Gates Foundation.

Regardless of the funding source of research, scientists have a fundamental duty to approach their work in an unbiased manner. While a researcher will often be expecting a certain result from their experiments, a poorly designed experiment biased toward these results will lead to false conclusions, wasting time and money in the best-case scenario, and potentially resulting in the loss of lives in the worst-case scenario. The checks and balance of peer-reviewed work and scientific consensus ensure that scientific research—especially on controversial topics—is based on facts and logical reasoning.

Your Turn 13.16 You Decide Always Consider the Source!

Earlier in this chapter, we discussed the retraction of a scientific study based on inconclusive evidence.

a. Using the Internet, find two other instances of academic dishonesty, which resulted in the retraction of the scientific article from a peer-reviewed journal. What were the issues surrounding the study?

b. Read the article "Allergenicity Assessment of Genetically Modified Crops—What Makes Sense?" (*Nature Biotechnology*, **2008**, *26*, 73; DOI: 10.1038/nbt1343), which suggests that the studies to determine the allergenicity of GM crops have not always been carried out using sound scientific basis. Describe the problems cited by these authors, and (using the Internet) determine whether allergy testing of GM foods has addressed these issues in recent years.

c. Find additional scientific studies, from peer-reviewed journals, that refute the common arguments against GMOs. Find some opponent views to GMOs from the Internet, magazines, or newspapers. Do these opponents cite the results of any scientific studies? If so, where have these studies been published (*e.g.*, online, newspapers, etc.), and have the results been given the opportunity to be reviewed by other experts in the field (*i.e.*, the "peer-review" process of publishing scientific data)?

d. Evaluate the report published in 2016 by the National Academies of Science (NAS), which may be accessed at http://nas-sites.org/ge-crops/. Do you agree with all of their findings? If not, explain your viewpoint, citing sources that contradict the conclusions of the NAS.

Learning Outcomes

The numbers in parentheses indicate the sections within the chapter where these outcomes were discussed.

Having studied this chapter, you should now be able to:

- discuss the complications of corn farming as an example of inspiration for genetic engineering (13.1)
- understand that cells function via a complex series of chemical reactions (13.2)
- discuss deoxyribonucleic acid (DNA) as a storage device of information to run chemical reactions in cells (13.2)
- describe the chemical composition of DNA (13.2)
- interpret evidence for the double-helical structure of DNA and its base pairing (13.3)
- understand the structural basis of DNA replication (13.3)
- explain how the genetic code is written in groupings of three DNA bases called codons (13.4)
- understand how the codons relate to amino acids across organisms (13.4)
- discuss the primary, secondary, and tertiary structure of proteins (13.5)
- recognize the general properties of amino acid side chains (13.5)
- relate the properties of amino acids to the interactions formed in protein structure (13.5)
- discuss, with examples, how small changes in a protein sequence may cause disease (13.5)
- understand the essential steps in carrying out recombinant DNA techniques (13.6)
- describe what is meant by transgenic organisms and give examples (13.6)
- give brief examples of natural selection, selective breeding, and genetic engineering (13.6)
- discuss, with examples, how genetic engineering has changed the chemical industry (13.7)
- analyze controversial issues associated with transgenic organisms and GM food (13.8)
- debate issues associated with the prudent and ethical applications of genetic engineering (13.8)

Questions

Emphasizing Essentials

1. The theme of this chapter is that DNA guides the chemistry of every living organism on the planet. Name three traits you possess that are dictated by your DNA.
2. List two industries that have changed and propose one that you expect to change with the rise of genetic engineering.
3. What is the difference between a genome and a gene?
4. Consider the structural formulas in Figure 13.1.
   a. What functional group(s) are found in the adenine molecule?
   b. What functional group(s) are found in the deoxyribose molecule?
   c. From what you learned in Section 11.4, why is deoxyribose a sugar and adenine is not?
5. a. What three units must be present in a nucleotide?
   b. What type of bonding holds these three units together?
6. Consider the acid–base equilibria of the phosphate ion in Figure 13.2. Describe the fractional composition of the phosphate ions present inside the mitochondria where the pH is ~8.
7. DNA contains the four bases: adenine, cytosine, guanine, and thymine. Name two similarities among the four bases. Highlight one feature unique to each.
8. Circle and name the functional groups in this nucleotide. Also label the sugar, the base, and the phosphate group.

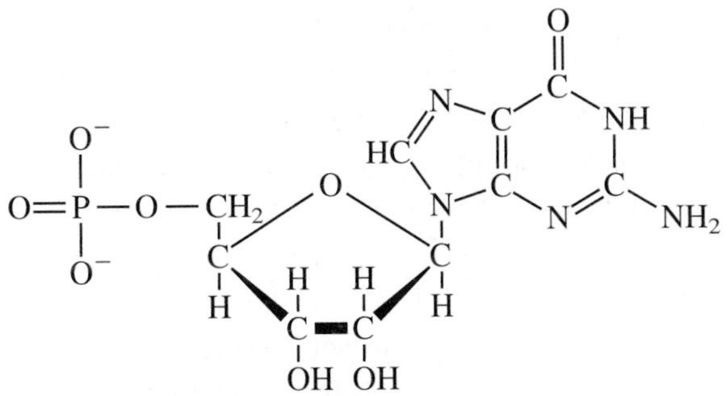

9. Compare the DNA segment in Figure 13.4 with the nucleotide shown in Question 8.
   a. Circle the two functional groups that react to form a polymer similar to DNA.
   b. The sugar in this nucleotide is ribose rather than deoxyribose. Suggest an appropriate name for a polymer of this nucleotide.
10. a. What does each letter in DNA stand for?
    b. Examine Figure 13.4. Which aspects of the DNA molecule does the name DNA highlight?
    c. Which aspects of the DNA molecule are not part of the name?
11. Here is the structural formula for the base thymine, as attached to the DNA chain.

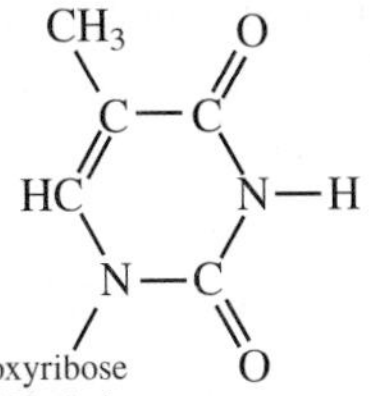

    To deoxyribose and DNA chain

    a. Label the H atoms that can hydrogen bond with a water molecule.
    b. Use electronegativity differences to explain why only these H atoms can form hydrogen bonds.
    c. Which atoms would form hydrogen bonds with water or other nucleic acids?
12. Explain why the base sequence ATG is different from the base sequence GTA.
13. Given a short sequence of DNA: TATCTAG
    a. Write and align a DNA code that complements the sequence given.
    b. Draw connecting lines between the sequences to represent the number of hydrogen bonds between each base pair.
14. Amino acids are the monomers used to build proteins.
    a. Draw the *general* structural formula for an amino acid.
    b. Name the functional groups in the structural formula you just drew.
15. Define a codon and its role in the genetic code.
16. Polar amino acids can be classified as acidic, basic, or neutral.
    a. Draw an example of a possible amino acid for each of the three types of polar amino acid.
    b. For each example, determine if the side chain can make hydrogen bonds, ionic bonds, or both.
    c. Describe each category in more detail. What functional groups would you expect in the side chains?
17. Describe what is meant by the primary, secondary, and tertiary structure of a protein.
18. Explain how an error in the primary structure of a protein in hemoglobin causes sickle cell anemia.
19. Explain one similarity and one difference between selective breeding and the genetic engineering of plants.
20. Describe three benefits of using designed enzymes in organic synthesis.

Concentrating on Concepts

21. Diagram the steps to produce insulin from a cow or pig versus synthetic insulin from bacteria.
22. Figure 13.4 represents a segment of DNA.
    a. What part of each nucleotide becomes the backbone of the polymer?
    b. What part hangs off the backbone?
    c. How is the backbone represented in sculpture at the beginning of the chapter and Figure 13.6?
23. Use Figure 13.7 to explain why adenine-thymine base pairs are less stable than cytosine-guanine base pairs.
24. a. What are X-rays?
    b. How is X-ray diffraction used to determine a crystal structure?
    c. The first X-ray diffraction patterns were of simple salts, such as sodium chloride. The X-ray diffraction studies of nucleic acids and proteins did not come until much later. Suggest two reasons why.
25. Explain why enzymes, or biological catalysts, can use one strand of DNA as a template to create a perfect complement strand.
26. Ionizing radiation (or the free radicals it generates in the cell) can break covalent bonds to disrupt one or more strands of DNA. A cell can more easily repair damage to a single strand in a DNA double helix than repair a double-strand break. Why is this?
27. Many compounds damage DNA and very effectively kill bacterial cells. Why are these compounds not typically used as antibacterial medicines?
28. Errors in DNA duplication can alter the base sequence of a strand permanently. But not all of these errors result in the incorporation of an incorrect amino acid in a protein for which the DNA codes. Explain why a single base change may not change the amino acid.
29. Almost all organisms use the same four bases and the same codons. Explain how the discussion of genetically engineering bacteria to produce insulin demonstrated this fact.
30. Insulin production through the genetic engineering of *E. coli* has not received the same amount of public concern as genetic engineering in plants. Propose two reasons insulin production has not received the same amount of public concern as genetically engineered crops.
31. Consider the idea of mixing genes as an improvement on nature.
    a. Describe what is meant by the term *transgenic organisms.*
    b. Consider the use of selective breeding, genetic engineering, and ionizing radiation to alter the genetic makeup of plants. List an advantage and disadvantage of each.

Exploring Extensions

32. A number of companies were involved in the early production of insulin. Find one recent press release each from Lilly and Genentech. In one to two paragraphs, summarize the press releases, and then compare this recent work to the company's role in the development of insulin.
33. Of the major players in the discovery of the structure of DNA, Rosalind Franklin was not included in the 1962 Nobel Prize given for solving the structure of DNA. What was her background and experience that enabled her to make significant contributions? Explore the reasons why she did not receive adequate credit and recognition for her work. Summarize your findings, citing your sources.
34. The genetic traits leading to sickle cell disease are more common in people of African, African-American, or Mediterranean heritage. Using the Internet, explain a proposed reason that the sickle cell trait has persisted rather than being discarded through evolution.
35. Finding the structure of proteins can be challenging. Explore the program Foldit, a game and protein structure simulator created and supported by the University of Washington.
    a. In one paragraph, describe how the game combines human ingenuity and computer simulations.
    b. Explore one of the results from the program and write a two-paragraph report.
36. List two advantages and two disadvantages of issuing patents for genetically modified plants and seeds.
37. To clone or to breed? Grieving owners may find the opportunity to clone (genetically replicate) a beloved dog too tantalizing to resist. Even so, many professional dog breeders and their organizations advise against cloning. Phil Buckley, a spokesperson for the Kennel Club in England, argues, "Canine cloning runs contrary to the Kennel Club's objective to promote in every way the general improvement of dogs." Explain why cloning cannot improve dog breeds.

38. Transgenic plants have not been widely accepted in all countries.

a. For the European Union, find two examples of transgenic plants that have been banned. Compare these to two that were allowed. Discuss the differences between those allowed and those rejected.

b. Create a timeline of five events that were key to the rapid increase or subsequent leveling (or both) of the adoption of transgenic crops in the United States. Briefly discuss each of your choices.

39. One reason why science fiction is successful is that it starts with a known scientific principle and extends, elaborates, and sometimes embroiders it. "Jurassic Park" began with the known scientific principle of copying and manipulating DNA and expanded on it to focus on the production of prehistoric creatures. Now it is your turn. Choose any scientific principle from this text. Then write a one- or two-page outline for a story based on that principle. Be sure to identify the chemical concepts and any pseudoscience you employ.

40. Recently developed techniques have dramatically changed our ability to alter DNA in human cells. Use the Internet to gather information on this rapidly advancing medical tool. Write a one- to two-page report on gene therapy, including specific examples of how diseases are being treated and how patients are faring.

41. Find a transgenic organism not discussed in the text. Describe the motivation for engineering this organism, its gene source, and a general description for the genetic modification.

42. You are the head of a government facing another year of a long drought and a serious risk of famine. Another nation has offered you a supply of genetically modified rice to feed your people. List two advantages and two disadvantages of accepting the aid. Decide whether you would accept it and explain why.

CHAPTER 14

Who Killed Dr. Thompson? A Forensic Mystery

REFLECTION

Forensic Evidence Collection

A number of techniques are required to analyze evidence in a forensics investigation. However, the most crucial component of the investigation is related to the crime scene itself. With emotions running high, those among the first to respond to a crime must take steps to ensure that the crime scene is not contaminated. Furthermore, forensic investigators must collect and record all relevant evidence in an appropriate manner. For each of the following personnel, list three important operating protocols that you think would help protect the integrity of the crime scene:

a. Initial responder (police officer, firefighter, or emergency medical technician (EMT))
b. Detectives
c. Forensic scientists

The Big Picture

In this chapter, the chemical principles discussed in earlier chapters will be used to solve a fictitious mystery. Some topics from each chapter that will be woven into the storyline include:

- The periodic table, mixtures (Ch. 1)
- Air quality (Ch. 2)
- The electromagnetic spectrum (Ch. 3)
- Isotopes (Ch. 4)
- Combustion reactions (Ch. 5)
- Solar power (Ch. 6)
- Electric vehicles (Ch. 7)
- Polarity and intermolecular forces (Ch. 8)
- The composition of plastics (Ch. 9)
- Enzymes (Ch. 10, 11)
- Drug design (Ch. 12)
- DNA fingerprinting (Ch. 13)

Friday, Aug. 1—7:08 pm: A Relaxing Evening Interrupted

After another long and productive work week, Professor David Thompson relaxed with a glass of wine, contemplating the future with joyful anticipation. In just two weeks, he would deliver a landmark presentation at the American Chemical Society meeting in Boston—marking the culmination of his 25-year research career.

For the last 20 years, his research has focused on the development of "smart" cancer treatment drugs. His most promising molecule, called Zeta-12, features a targeting agent that attaches itself to individual cancer cells, and then releases the drug to kill the cells with high specificity and efficiency. He has received continual funding from the National Institutes of Health (NIH), which helped fund the basic studies and extensive clinical trials for the drug. His presentation at the upcoming meeting would detail the successful completion of Phase III clinical testing, leading to FDA approval, and a possible trillion-dollar market. Next week, he plans to partner with a major pharmaceutical company that would supply Zeta-12 to cancer treatment centers across the globe.

Your Turn 14.1 Scientific Practices Cancer Treatment Drugs

Using the Internet, find examples of "smart" cancer treatment drugs that are currently in development.

a. What are the possible benefits of these drugs, relative to those currently used for chemotherapy?

b. In drug discovery, what are the differences between Phase I, Phase II, and Phase III clinical testing?

Thompson's phone rang, interrupting his train of thought. *I'll just let it ring, since it's probably Julie complaining about her alimony*, he thought. It had been six months since their heated divorce—an all-too-common ending for marriages with a workaholic spouse. Convinced that even his ex-wife couldn't rob him of his joy, he walked across the room to his desk and answered the call anyway, before it went to voicemail.

"Dr. Thompson?" the baritone voice asked, the tone urgent.

"Yes," Dr. Thompson replied, concerned.

"There's been an accident at your laboratory, and we need you to tell us—"

"What kind of an accident?" Dr. Thompson interrupted.

"A raging fire. The hazmat group is on site right now, and we need to know what chemicals you have so we can safely extinguish the fire."

"Are you kidding? We have a few solvents in the lab that contain sodium metal."

"Which solvents?"

Upset by the seemingly irrelevant question, Dr. Thompson stammered. "Toluene, … THF, ether, … acetonitrile, hexanes, and, um, … dichloromethane."

"Was there anyone working in your lab tonight?"

"No, I work alone."

"Thanks, Dr. Thompson. We'll do what we can to save your lab."

"I'm on my way!"

Solvent Stills: An Effective but Dangerous Way to Purify Solvents

In Section 5.12, we discussed the process of **distillation** for the separation of various fuel products from crude oil. A **solvent still** (Figure 14.1) employs a similar process to remove moisture and oxygen from organic solvents. The resultant "dry" and oxygen-free

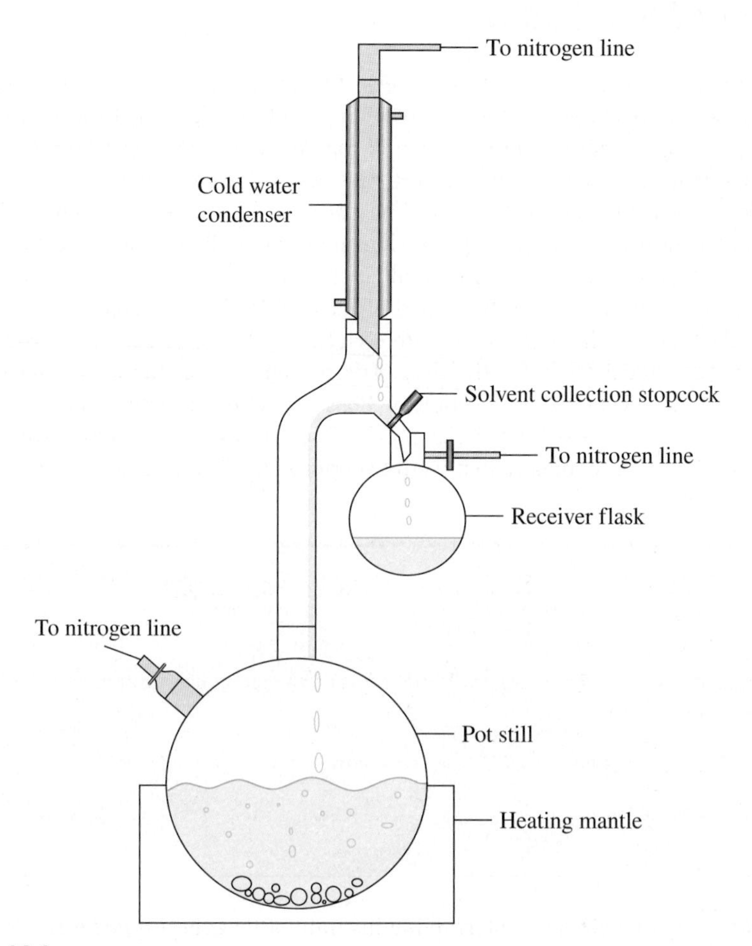

Figure 14.1

Schematic of a solvent still.

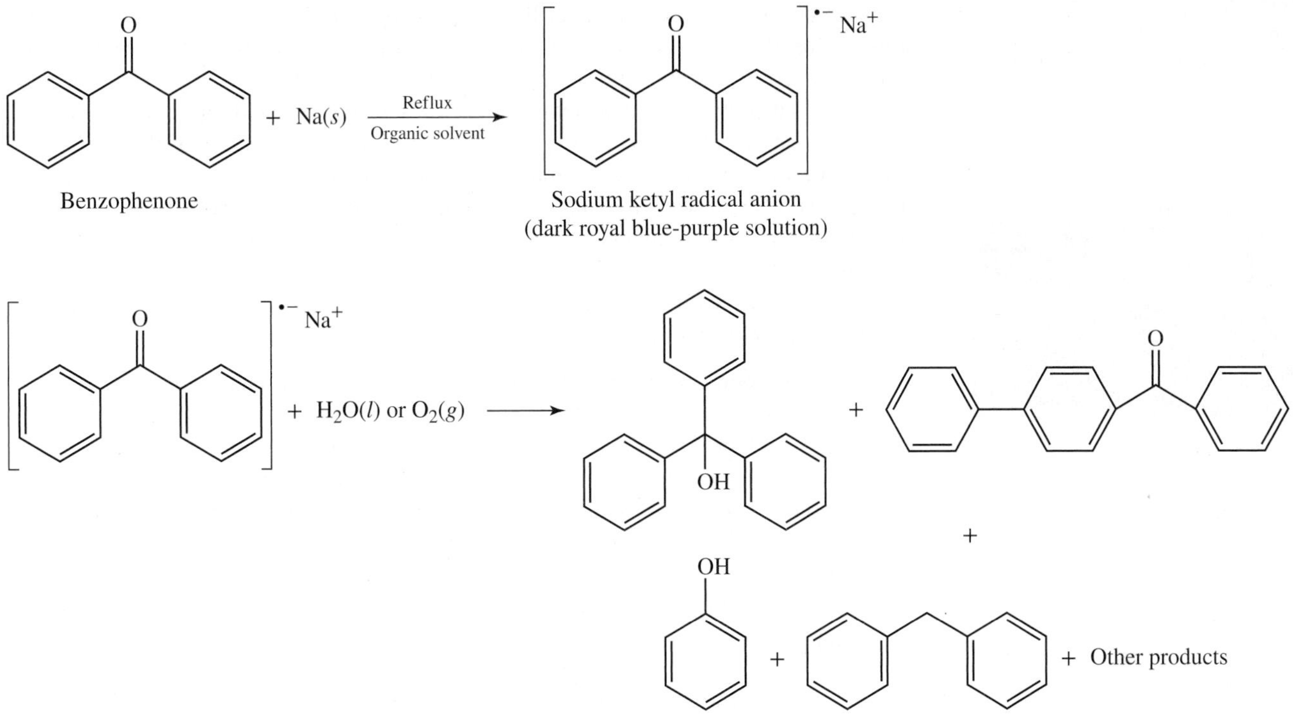

Benzophenone + Na(s) —Reflux / Organic solvent→ Sodium ketyl radical anion (dark royal blue-purple solution)

+ $H_2O(l)$ or $O_2(g)$ → OH + OH + + Other products

Figure 14.2

The reactions of sodium and benzophenone with oxygen and water within a solvent still.

solvents are used for many chemical reactions, in which traces of water or oxygen would result in contamination and unsuccessful syntheses. As the organic solvent is heated to boiling, its vapor rises up the glassware and is condensed back to its liquid state using a water-filled condenser. The repeated process of heating a solvent to boiling and condensing its vapor is known as **reflux**. The moisture/air-free solvent is collected in the receiver flask, which may be removed for use via syringe. A nitrogen bubbler system is used to flush the solvent still with an inert gas to prevent moisture and oxygen from dissolving in the solvent, and prevent a build-up of pressure that would result in a dangerous explosion.

Metallic sodium serves as a drying agent, reacting with both water and oxygen dissolved in the solvent, thereby removing these impurities. Non-chlorinated organic solvents, such as benzene, toluene, pentane, hexanes, etc., are purified by adding benzophenone and sodium metal, and then heating the mixture to reflux under pure nitrogen. Recall from Chapter 1 that metals such as sodium are strong **reducing agents**. As shown in Figure 14.2, each sodium atom donates an electron to a benzophenone molecule, resulting in a sodium ion (Na^+) and a radical anion, evidenced by a characteristic blue/purple color. Within a few hours of adding sodium to the colorless organic solvent, the solution becomes a royal blue-purple color, which indicates that a ketyl radical anion has been formed. The radical anion then reacts with trace amounts of water and oxygen present in the solvent to form a number of reaction products (Figure 14.2). Hence, a persistent blue-purple color indicates that the solvent is free (<10 ppm) from both moisture and oxygen.

Some safer alternatives for solvent purification consist of stainless steel columns packed with silica (SiO_2) or alumina (Al_2O_3) powders with high surface areas (>200 m^2/g) and small particle sizes (~5 nm), molecular sieves (aluminosilicate minerals with an average pore size of 4 Å), and a copper catalyst powder—used to react with/remove oxygen.

Although Dr. Thompson often heard of explosions and fires from solvent stills, he meticulously maintained his stills and provided rigorous training for his students to ensure this would not happen on his watch. As he rushed to put on his shoes and fumbled with his car keys, he recalled a stainless steel solvent purification system he

A tetrahydrofuran (THF) solvent still in Dr. Thompson's synthetic laboratory, showing the characteristic dark royal blue-purple color for solvent that is both air- and moisture-free.

For a description of safer alternatives for solvent purification, see: Grubbs, R. H. *et al.*, "Safe and Convenient Procedure for Solvent Purification" *Organometallics* **1996**, *15*, 1518 (DOI: 10.1021/om9503712)

© Bradley D. Fahlman

saw months earlier at a conference. *Why didn't I upgrade to this safer alternative?* he thought. Speeding through the neighborhood and taking the on-ramp to the freeway, his mind raced to inventory the items that could be lost in the fire. How could he ever replace the lab notebooks that detailed his life's work? What if the notebooks were needed for the patent prosecution process? As he screeched to a halt at the side of the building, he was relieved to see so many fire trucks and personnel already on the scene. His feelings of panic waned as he remembered that his lab notebooks were in a separate lab. *Surely, the firefighters would be able to stop the blaze before it reached them*, he thought.

The hazmat team was well aware of the five classes of fires, and the types of fire extinguishers required:

- Class A (ordinary combustible materials such as wood, paper, plastics): use water only.
- Class B (flammable liquids such as organic solvents, gasoline, petroleum oil, paint, as well as flammable gases such as methane, propane, butane): use carbon dioxide or dry-chemical extinguishers (*e.g.*, ammonium phosphate, sodium bicarbonate, potassium chloride).
- Class C (electrical fires): use the same extinguishers as Class B fires.
- Class D (combustible metals such as Na, K, Mg): use dry-powder extinguishers (graphite or sodium chloride powders pressurized with nitrogen gas).
- Class K (cooking oils and greases): use wet-chemical extinguishers (*e.g.*, aqueous potassium carbonate, containing detergents to improve the wettability of the nonpolar grease).

Nonpolar and polar molecules were discussed in Chapters 5 and 8, respectively.

Your Turn 14.2 Scientific Practices Tricky Firefighting

a. What should be used by the hazmat crew to extinguish the fire? What would happen if they simply used water?

b. Speculate why it would not be wise to use sodium metal and benzophenone to purify chlorinated solvents, such as dichloromethane.

Dr. Thompson's lab contained wooden cabinets and desks; however, the fuel sources were primarily organic solvents (Class B) and combustible sodium metal (Class D). Accordingly, the fire crew entered the burning lab, with one team focused on containing the fire around the solvent stills using dry-powder extinguishers. The second crew sprayed the rest of the lab with water to extinguish the burning of paper, wood, and plastic materials.

Your Turn 14.3 Scientific Practices The Fire Triangle Revisited

Explain how dry-powder, wet-chemical, and dry-chemical fire extinguishers work, by referring back to the fire triangle discussed in Chapter 5.

Under the sounds of breaking glass, crackling flames, and pounding metal, Dr. Thompson stood helpless, watching the firefighters scramble to take control of the fire. He scurried back and forth along the perimeter of the scene, asking crew members to divulge details about his lab, to no avail. At long last, firefighters began exiting the building and congregated outside the building, a sign that the fire was finally extinguished. Thompson approached the fire commander, who was speaking with witnesses.

"Can I have a look at the damage to my lab?"

"Not yet. You won't be allowed in for at least a few days until we finish our investigation. We'll call you if we have any questions. Please go home and try to get some sleep."

"Will you let me know if the firefighters were able to save my lab notebooks?"

"Sure. Where were they located?"

"In 305B, the workroom connected to my synthetic lab."

"Okay, will do. We'll be in touch."

Dr. Thompson headed to his car, taking one last look at the biochemistry building, unsure when he would be able to evaluate the extent of the damages.

Many labs have now transitioned to a secure electronic laboratory notebook in the cloud. This seems especially common in the pharmaceutical industry, while not so prevalent in academia.

Friday, Aug. 1—10:13 pm: The Aftermath

Luckily, no one was hurt or injured by the devastating fire. Two graduate students from another research lab had reported the fire after hearing a loud explosion. Their office was located along the same hallway as Dr. Thompson's lab. One of the students thought he heard the door of Dr. Thompson's laboratory shut about 30 minutes before the explosion, which was typically very quiet during the evening hours. Dr. Thompson normally didn't work at this time in the evening, and the janitors passed through much later, so it was strange that someone would be in his lab.

Figure 14.3

The extensive fire damage to Dr. Thompson's laboratory.

© *Mark Cabot/Alamy Stock Photo*

City fire department investigator Jim Williams was called to the scene to determine whether the fire was natural, accidental, or deliberate in nature. As was his ordinary protocol for fire investigations, Investigator Williams suited up into disposable coveralls that covered his hair, gloves, and shoes. The smell of smoke still saturated the stairwell, which became more pronounced as he climbed the flights of stairs toward the third-floor laboratory. Although he had investigated many lab fires across the state, what he saw as he stepped into Dr. Thompson's laboratory was still surprising (Figure 14.3).

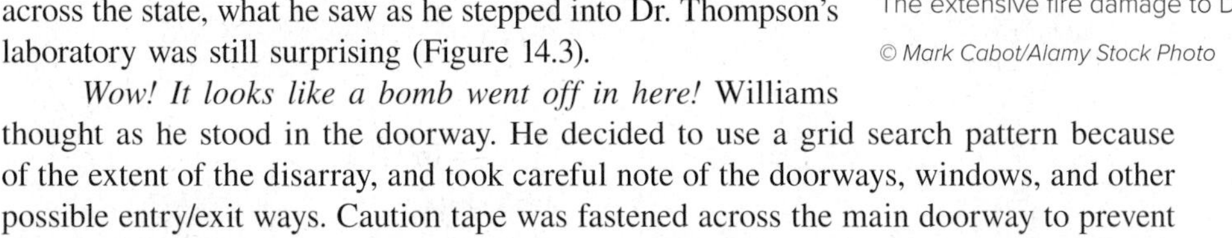

Wow! It looks like a bomb went off in here! Williams thought as he stood in the doorway. He decided to use a grid search pattern because of the extent of the disarray, and took careful note of the doorways, windows, and other possible entry/exit ways. Caution tape was fastened across the main doorway to prevent unauthorized people from entering the lab until his investigation was completed.

The laboratory space featured two adjoining rooms (Figure 14.4). The first was a large (15 m × 15 m) synthetic lab with numerous organic and inorganic chemicals, solvents, and associated storage areas. A workroom was connected through an open doorway, which contained desks, filing cabinets, and a bookshelf that held Dr. Thompson's treasured lab notebooks. Sadly, both labs were totally decimated by the impacts of fire, soot, and smoke. Although the light switch in the lab was in the on position, darkness filled the room; the electricity was turned off by the firefighters. Upon inspection of the room with a flashlight, Williams noticed that the lightbulbs had become deformed and cracked due to the fire's extreme heat.

The extent of damage to the workroom was somewhat suspicious to Williams. Even though both labs were connected by an open door, he was skeptical that the fire could spread so efficiently to the bookshelf that was tucked around the corner in the adjoining room. In comparison, there was much less fire and smoke damage to the desks and filing cabinet on the opposite side of the workroom. Perhaps most concerning was the presence of ruptured solvent containers near the stills, and empty metal acetone canisters outside of the flammable cabinet, near the doorway that connected the two rooms. Lastly, as reported by witnesses, Williams was also suspicious about the possibility of someone present in Dr. Thompson's laboratory just minutes before the fire was reported. He drew detailed sketches and took numerous photographs of both labs, compiling notes about fire damage to the walls and floors. Williams soon realized this case might be out of his expertise: so many chemicals, unfamiliar equipment, and complicated methods. He decided to call the state arson forensic expert to help with the investigation.

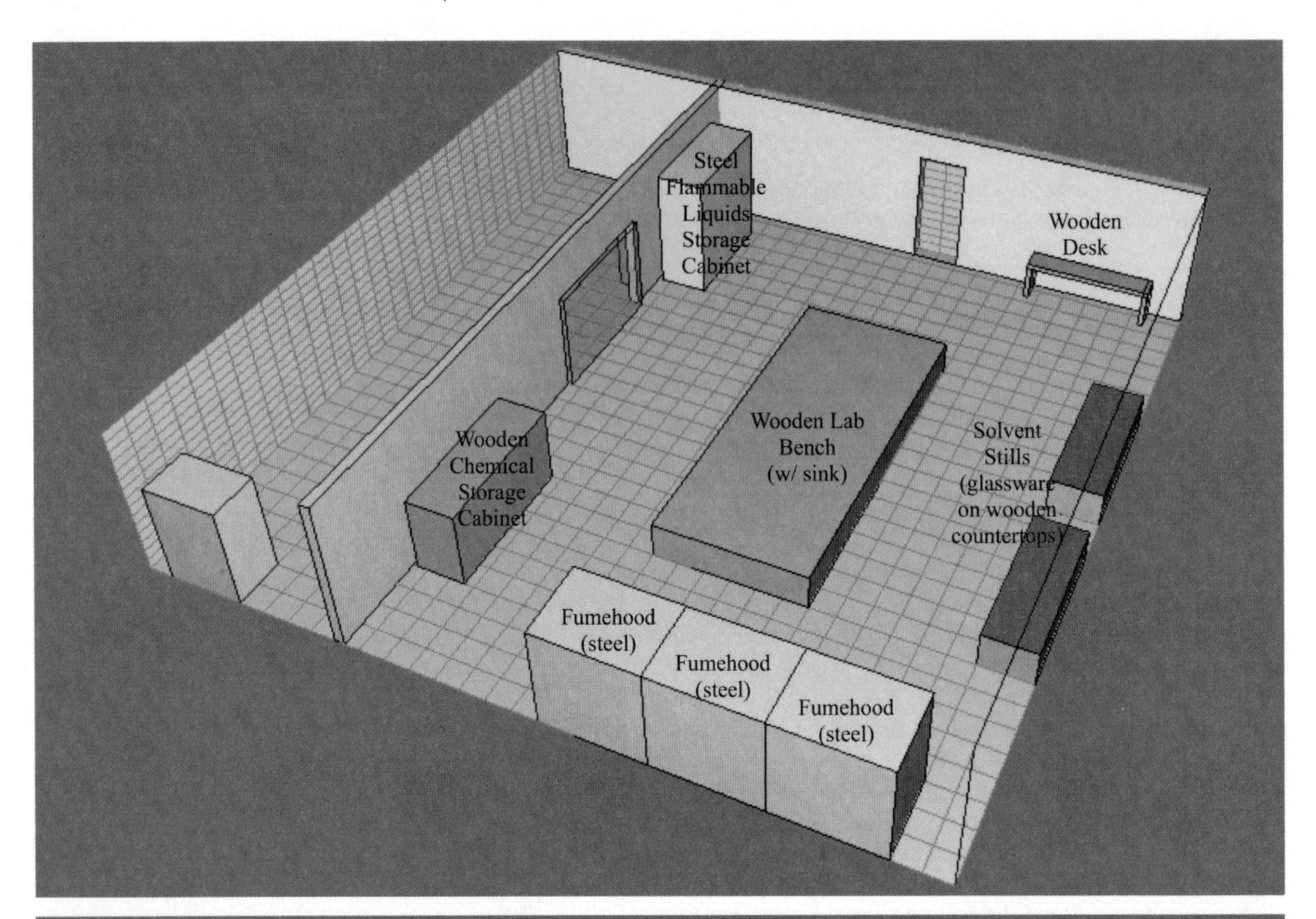

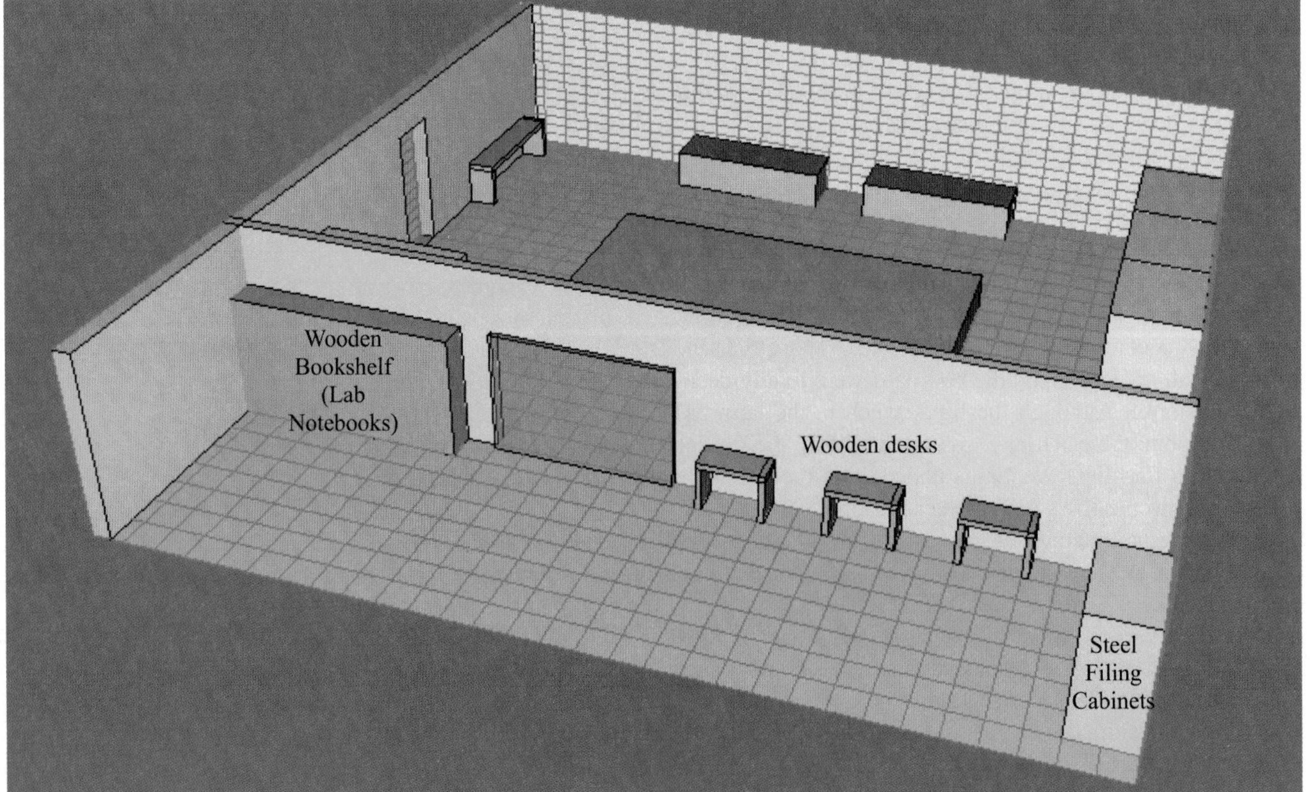

Figure 14.4

Schematic of Dr. Thompson's laboratory. Shown is the main synthetic laboratory (top), which housed the solvent stills believed to be the primary source of the fire. The adjoining room (bottom) housed student desks, filing cabinets, and a bookshelf with laboratory notebooks that documented the many years of research by Dr. Thompson and his employees.

Saturday, Aug. 2—8:05 am: Accidental or Deliberate?

A photographer and an arson investigator arrived on the scene, to follow up on Williams' questions and determine whether the fire was started deliberately. Williams discussed his findings with the forensic arson investigator, Dr. Keisha White, who asked additional questions about the conditions of the lab, Dr. Thompson's whereabouts at the time of the fire, and whether there were any witnesses or if anyone was seen entering or leaving the building at the time of the fire.

Dr. White was the lead forensic arson investigator for the state and was skilled at determining probable causes of fires at both residences and businesses. Fitted with protective coverings, her forensic team processed the scene (taking photographs, compiling detailed notes about the scene conditions, and collecting evidence).

As expected, the most extensive fire damage was observed near the flammable solvent stills. Two groups of solvent stills were housed in the laboratory. On the left were toluene, hexanes, and tetrahydrofuran (THF); on the right were diethyl ether, dichloromethane, and acetonitrile. Because these solvents were thought to be the origin of the fire, the arson investigation team needed to consider the relative flammabilities of the solvents present in Dr. Thompson's laboratory.

CH_3

Toluene

O

Diethyl ether

O

Tetrahydrofuran (THF)

n-hexane (+ other isomers)

H H Cl C Cl

Dichloromethane (also known as methylene chloride)

$H_3C—C{\equiv}N$

Acetonitrile

Your Turn 14.4 Scientific Practices Isomers Revisited

One of the solvents is listed as "hexanes." Draw the structural formulas and line-angle diagrams for five possible isomers of hexane. ***Hint:*** Review Section 5.13.

The flammability of a liquid is related to the concentration of its vapor in air. As you may recall from Chapter 5, the vapor pressure of a liquid is directly dependent on its temperature. That is, as its temperature is increased, the intermolecular forces among the liquid molecules are broken, which results in higher concentrations of gaseous molecules. The lowest concentration of solvent vapor that can ignite in air is known as the **lower flammability limit (LFL)**. The **flash point** of a solvent occurs when the vapor pressure of a solvent equals its LFL, and combustion is therefore possible in the presence of an ignition source. The flash point of a fuel will allow combustion, but will not sustain it. In contrast, the **autoignition temperature** corresponds to the minimum temperature at which the vapor spontaneously ignites, even in the absence of an ignition source.

In order to properly assess the dangers associated with the storage and use of flammable and combustible liquids, a classification system has been devised by the National Fire Protection Association (NFPA). This scheme ranks the relative fire hazards of liquids based on their flash and boiling points (Table 14.1).

Table 14.1 Relative Fire Hazards of Liquids

Class	Flash Point (°C)	Boiling Point (°C)	Examples
IA	<22.8	<37.8	ethylene oxide, methyl chloride, pentane
IB	<22.8	≥ 37.8	acetone, benzene, ethanol, gasoline, iso-propanol
IC	≥ 22.8	<37.8	butanol, diethyl glycol, styrene, turpentine
II	≥ 37.8	<60	camphor oil, diesel fuel, pine tar, Stoddard solvent
IIIA	≥ 60	<93	creosote oil, formaldehyde, formic acid, fuel oil #1
IIIB	≥ 93	>93	castor oil, coconut oil, fish oil, olive oil

Your Turn 14.5 Scientific Practices Flash Points

a. At the flash point, is a chemical or physical change taking place? How about at the boiling point?
b. What are some differences and similarities between "flammable" and "combustible" liquids?
c. The NFPA classification scheme presented above is based on an atmospheric pressure of 1 atm (sea level). Would the flash points of the liquids be higher or lower at higher altitudes? What would this mean with respect to storing or handling these liquids at a higher altitude location, such as Denver, Colorado?

Your Turn 14.6 You Decide Fire Hazard Ranking

Using the NFPA classification scheme, provide the Class ranking for each of the solvents present in Dr. Thompson's laboratory.

Vapor pressure, volatility, heat of combustion, and activation energy are discussed in Chapter 5.

Table 14.2 lists some thermal parameters for solvents found in Dr. Thompson's laboratory. The boiling point of a solvent occurs when the vapor pressure of the solvent equals the external atmospheric pressure. However, the flash point often occurs at temperatures much less than its boiling point—as long as the LFL is exceeded, combustion is possible. In assessing the overall flammability of a solvent, the flash point is more diagnostic than its boiling point. For instance, toluene has a relatively high boiling point (low volatility), but a relatively low flash point (high flammability). In contrast, dichloromethane has a lower boiling point (more volatile), but a much higher flash point (less flammable). This difference is related to the relative reactivity of a particular solvent with oxygen during combustion. The heat of combustion for toluene is 3900 kJ/mol, whereas the heat of combustion for dichloromethane is only 605 kJ/mol. The more heat that is generated by the reaction of a fuel and an oxidizer, the easier it is to exceed the **activation energy** of the reactants, resulting in ignition of the fuel/oxidizer mixture. Therefore, dichloromethane is less flammable than toluene since its combustion is less exothermic, or energetically preferred.

Your Turn 14.7 You Decide Spontaneous Combustion?

a. If an ignition source (*e.g.*, a spark from a match) is present, which solvent vapors will catch fire at room temperature (20–25 °C)?
b. Based on the autoignition temperatures of the solvents found in Dr. Thompson's laboratory, could any of these solvents combust without the presence of an outside source of ignition?

Table 14.2 Thermal Properties for the Solvents Found in Dr. Thompson's Lab

Solvent	Boiling Point (°C)	Flash Point (°C)	Autoignition Temp. (°C)
Diethyl ether	35	−45	160
Dichloromethane	40	100	600
Acetone	56	−20	465
Tetrahydrofuran	66	−14	321
Hexanes	69	−26	223
Ethanol	78	17	365
Acetonitrile	82	2	523
Toluene	111	6	530

Your Turn 14.8 Scientific Practices Combustion Reactions and Air Quality

a. Write balanced equations for the combustion of three solvents of your choosing found in Dr. Thompson's lab.

b. What other products, such as CO and particulate matter, will likely be generated from the combustion of solvents and common building materials that may be present in the lab? Which of these products will likely exceed the air quality standards discussed in **Chapter 2**?

c. Atmospheric monitoring discovered a mercury concentration of 3000 ng/m^3—10 times higher than the EPA safe standard of 300 ng/m^3. List some possible sources of mercury resulting from the lab fire. It should be noted that Dr. Thompson's laboratory did not contain any mercury-based thermometers, thermostats, manometers, or bottles of elemental mercury.

Broken glassware from the solvent stills was scattered throughout the lab. The wall immediately behind the solvent stills, and the flooring in front and along the sides of the stills, were severely damaged. In addition, the wooden desks in the lab were heavily burned, charred, and covered with a thick layer of soot. Due to the presence of water-reactive metals, the lab did not have an automatic sprinkler system, which would have limited the extent of damage to wooden desks, tables, shelves, and cabinets.

The arson investigation team is trained to pay particular attention to the following suspicious items that may be present in the aftermath of a fire:

- Suspicious burn patterns on floors and/or walls, such as fire trails or multiple points of origin
- Degree of charring of materials
- Flaking of plaster or concrete
- Distortion of materials softened or weakened by heat
- Soot and smoke damage
- Empty containers of fuel at the scene, especially those found without a cap fastened
- Presence of an accelerant, such as open containers or pools of unburned ignitable liquids
- Evidence of tampering with a sprinkler system, if present

For the combustion of a fuel located near a wall, the resultant burn pattern will adopt a characteristic shape at early stages of a fire (Figure 14.5(a)). As the flames become more intense and travel upward to intersect the ceiling, the wall burn pattern will become more columnar, with the appearance of circular burn patterns on the ceiling (Figure 14.5(b)). A more complex burn pattern may be present if an accelerant has been poured or sprayed onto the wall by an arsonist. Dr. White and her team will also look for what *may not be* present, such as indications of an accidental or natural cause for the fire.

Diagnostic burn patterns can also be found on floors of an arson scene that match the pool and flow from pouring a liquid accelerant (a so-called fire trail). In addition, the floor will often exhibit an intermixed array of light, medium, and heavy burn patterns (*e.g.,* Figure 14.6(a)), depending on the type of accelerant used.

However, some solvents will result in little or no scorching of the floor (Figure 14.6(b)–(c)). Why does a solvent such as gasoline cause so much damage to the underlying floor, whereas the combustion of acetone or ethanol is hardly noticeable? During the combustion of an organic solvent, oxygen must combine with the gaseous solvent molecules in an appropriate concentration. As the oxygen content of the fuel increases (*e.g.,* ethanol/acetone contains more oxygen than gasoline), the fuel burns more completely, leaving behind little or no residue.

It should be noted that burn patterns on a floor may also be a result from molten substances that have fallen from the ceiling, rather than the presence of a liquid accelerant. In fact, there are documented cases of fires erroneously classified as arson due to floor burn patterns, a problem derived from drawing a conclusion based on only one piece of evidence.

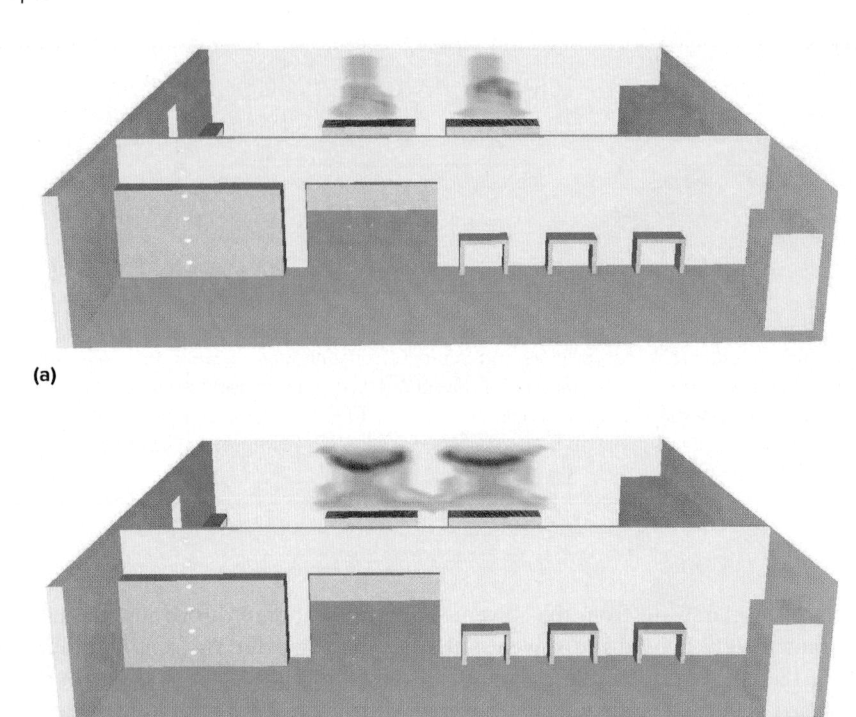

(a)

(b)

Figure 14.5

Simulated wall burn patterns from solvent still fires in Dr. Thompson's laboratory. Shown are **(a)** early, and **(b)** later, stages of the fire. Simulations were performed using PyroSim software (Thunderhead Engineering).

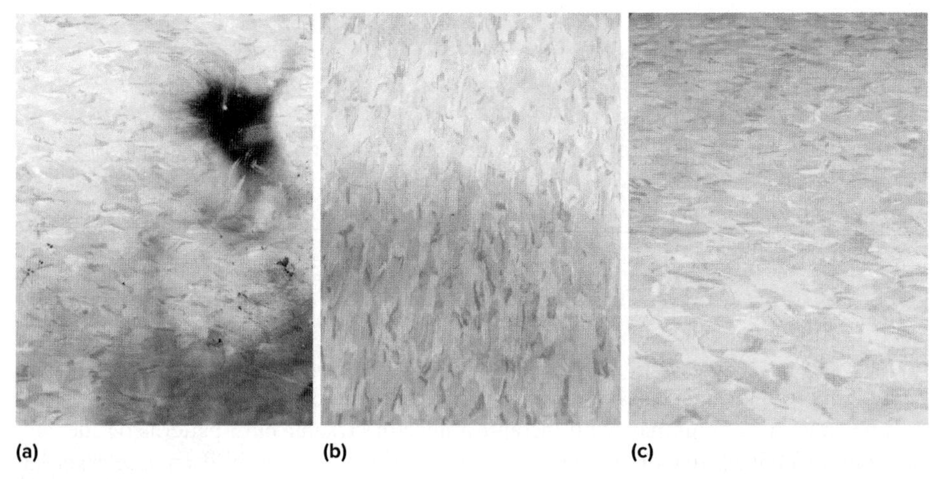

(a) (b) (c)

Figure 14.6

Burn patterns in commercial vinyl flooring from **(a)** gasoline, **(b)** acetone, and **(c)** ethanol.

Furthermore, gasoline releases a tremendous amount of heat and burns more slowly than oxygenated fuels such as acetone or ethanol. Whereas the combustion of gasoline releases 47 kJ of heat per gram of fuel, ethanol and acetone release less heat—29.7 kJ/g and 30.8 kJ/g, respectively. In order for damage to occur to underlying flooring as a result of fuel combustion, sufficient heat must be transferred from the burning fuel to the floor material (vinyl tile, etc.). Accordingly, it takes more acetone or ethanol to generate enough heat to cause fire damage to surrounding materials, relative to gasoline. This is especially the case with volatile solvents, since the liquid fuel tends to protect the flooring via evaporation rather than causing damage from combustion.

As noted the previous evening by first responders, Dr. White confirmed the presence of ruptured canisters of toluene and hexanes near the solvent stills, and empty cans of acetone located near the doorway that connected the workroom to the main lab. Not knowing whether it was Dr. Thompson's policy to store solvent near the stills instead of in his flammable solvents cabinet, she placed a call to his home.

"Hello, Dr. Thompson?"

"Yes ..."

"This is Dr. Keisha White, an arson investigator with the State Forensic Laboratory working the investigation into the fire at your lab."

"Okay … Is there a problem?"

"No, just a routine investigation. I do have some questions for you, if you have a minute."

"Sure."

"Do you keep your solvent canisters in the flammable solvents cabinet?"

"Yes, of course."

"Is there a possibility that you left some solvent containers near the stills? Maybe when you last refilled the flasks, for instance?"

"No, not a chance. I'm cautious when it comes to safety and I always ensure that solvents are put back in the cabinet after I use them."

"Who else has access to your lab?"

Dr. Thompson considered that question for a moment. "Just me, but I had a postdoc working with me a year and a half ago. He no longer works at the university."

"Does he still have a key to your lab?"

"No. He turned it in to the department office before he left."

"Okay. Is your lab always locked?"

"Yes. My lab is secured at all times."

"Is it possible that someone made a copy of your lab key before turning it back in? I can tell you that there was no sign of forced entry, or tampering with the lock."

"Hmmm. The keys have 'do not copy' on them, but I guess it's possible."

"We found some ruptured solvent canisters near the stills in your lab. We also found some empty cans of acetone near the door connecting your workroom. Does this seem suspicious to you?"

"Yes. I know for a fact that these bottles were always stored in the flammable cabinet."

"Thanks. That's all I wanted to know. Do you have any enemies?"

"No. I'm at a loss."

"Thanks again. We'll continue our investigation and will let you know more details as soon as possible."

"Thanks. Please call me anytime if you have questions."

As he hung up the phone, Dr. Thompson immediately thought of his colleague, Dr. John Littleton, who spent three years as a postdoctoral researcher in his lab. Although many students had contributed to Dr. Thompson's research program through the years, no one else had a greater impact on his project. In fact, his soon-to-be announced breakthrough in cancer treatment would not have been possible without Dr. Littleton's contributions. As was his policy, Dr. Thompson had all students sign a nondisclosure agreement prior to beginning their research with him. In addition, he had all students sign another form that waived their rights to being named as a co-inventor on any patents and any financial gain that would result from such patents.

During his last year of work, Dr. Littleton began to realize the full impact that his project would represent for cancer treatment throughout the world. He asked Dr. Thompson numerous times to include him on patents but was always met with opposition. This created increasing conflict between the two scientists, which ended less than two years ago with Dr. Littleton leaving the university to start his own research group at a nearby college. Since that time, Dr. Thompson has refused to work with any students due to the proprietary nature of his work, and his concern about information security. After the phone call from the forensic investigator, Dr. Thompson began to wonder if his former postdoc had enacted revenge by setting fire to his lab, including the lab notebooks that contained the important data from his research project.

Suspicions regarding the role of arson continued to rise when a member of Dr. White's investigative team discovered some matchsticks—two used, and one unused—behind the metal flammable solvent storage cabinet. A photograph was taken to mark the location of the matchsticks, and Dr. White was called over to view the evidence herself. The matches were carefully placed into plastic evidence bags and labeled for later analyses in the crime lab.

In addition, the investigative team had to determine whether acetone was poured on the floor or lab notebooks to ignite the fire in the workroom. If solvent cans were usually stored in a storage cabinet, their placement on the floor of the lab indicates that the solvent may have been used to start a fire in the workroom. The investigators postulated that the arsonist might have poured acetone on the lab notebooks in the workroom, perhaps creating a trail toward the solvent stills in the adjacent lab. As the fire reached the solvent canisters in front of the solvent stills, it would heat the canisters to the point of rupture. The ruptured cans could then continue to spray fuel into the fire for a long time, with additional fuel being added explosively as the solvent stills caught fire. With suspicions of arson heightened, the canisters of solvent found in the lab and workroom were packed to test for the presence of fingerprints at the crime lab.

When an accelerant is used to start a fire, trace residues are often found at the source of the fire, which are still measurable within 24–72 hours after the fire has been extinguished. This is especially the case when an accelerant such as gasoline is poured onto a combustible material, such as newspaper or books. Because air may not completely reach the interior of the material, a small amount of fuel remains trapped during combustion. To determine the presence of trace solvent, pieces of vinyl flooring were cut out from locations in front of the solvent stills, between the lab and workroom, and in front of the bookshelf in the workroom. In addition, the charred remains of the lab notebooks were collected and sealed for transport to the arson lab for analysis. In order to prevent evaporation and the loss of volatile liquids, the samples were placed into new, air-tight paint containers for transport to the crime lab. Final photographs were taken of both labs, and the rooms were sealed off until further analyses could be completed at the crime lab.

Your Turn 14.9 Scientific Practices Evidence Collection

Using the Internet, provide details about how samples suspected to contain residues of fire accelerants are best collected and transported to a forensics lab for testing.

Fire Modeling

Using detailed dimensions and drawings of Dr. Thompson's laboratory, Dr. White performed a computer simulation of the fire, assuming that the solvent stills were the primary sources of fuel. Simulations, such as those using the computer program PyroSim, are powerful techniques to provide information about the potential origin point(s) of a fire. In addition, the simulated burn and smoke patterns can be compared to those actually observed in the lab—an important tool to either implicate or rule out arson as a probable cause of a fire.

The simulated ambient temperature of the lab resulting from a fire originating at the solvent stills is shown in Figure 14.7. Whereas the wall behind the solvent stills likely reached temperatures as high as 800–1000 °C, the rest of the lab exhibited temperatures in the range of 400–650 °C. From the autoignition temperatures of common materials (Table 14.3), it is no surprise that the experimental lab was mostly consumed by the fire.

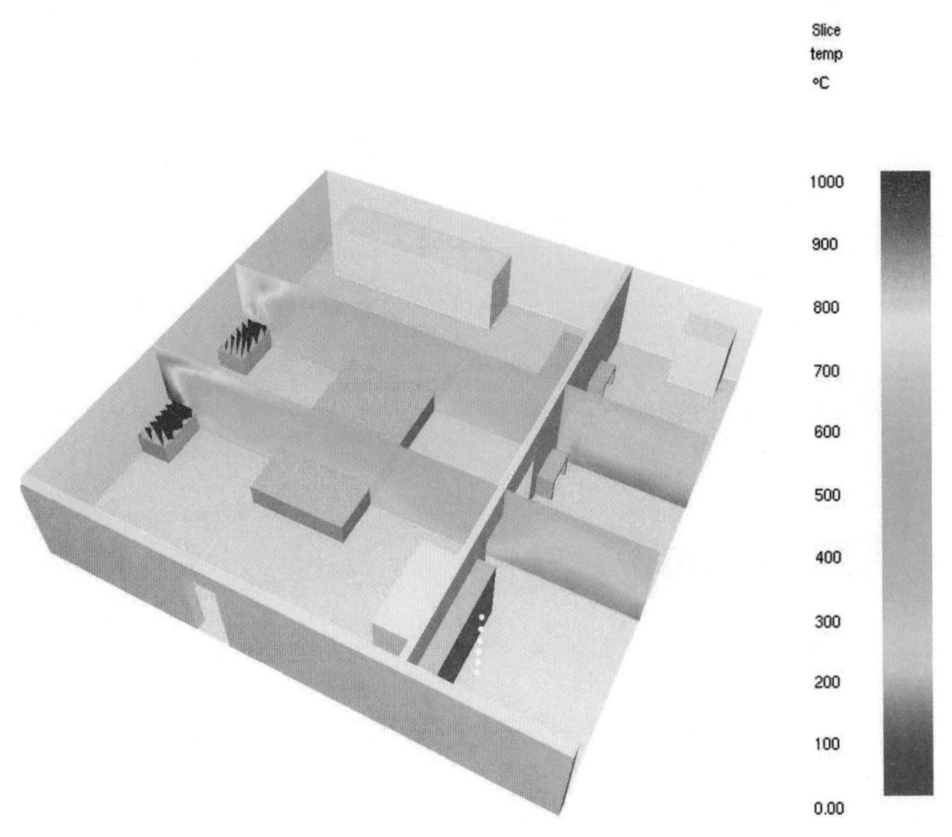

Figure 14.7

A simulated cross-section temperature profile of the synthetic lab, for late stages of a fire originating from both solvent stills. The white dots in front of the bookshelf in the workroom show the location of temperature sensors used in the modeling calculations. Calculations performed using PyroSim software (Thunderhead Engineering).

Table 14.3 Autoignition Temperatures for Common Materials

Material	Autoignition Temp. (°C)
Paper	218–246
Leather	200–212
Cotton	267
Nylon	289–377
Polycarbonate (PC)	478
Polyethylene (PE)	226
Polyethylene tetraphthalate (PET)	460
Polypropylene (PP)	201
Polystyrene (PS)	226
Poly(vinylchloride) (PVC)	455
Natural rubber	191–331
Wood	300–482

The calculated temperature profile for the workroom is shown in Figure 14.8, which shows temperatures in the range of 160–230 °C for most of the room, with higher temperatures (290–370 °C) near the ceiling of the connecting doorway. For modeling purposes, temperature sensors were added in front of the bookshelf in the workroom to determine

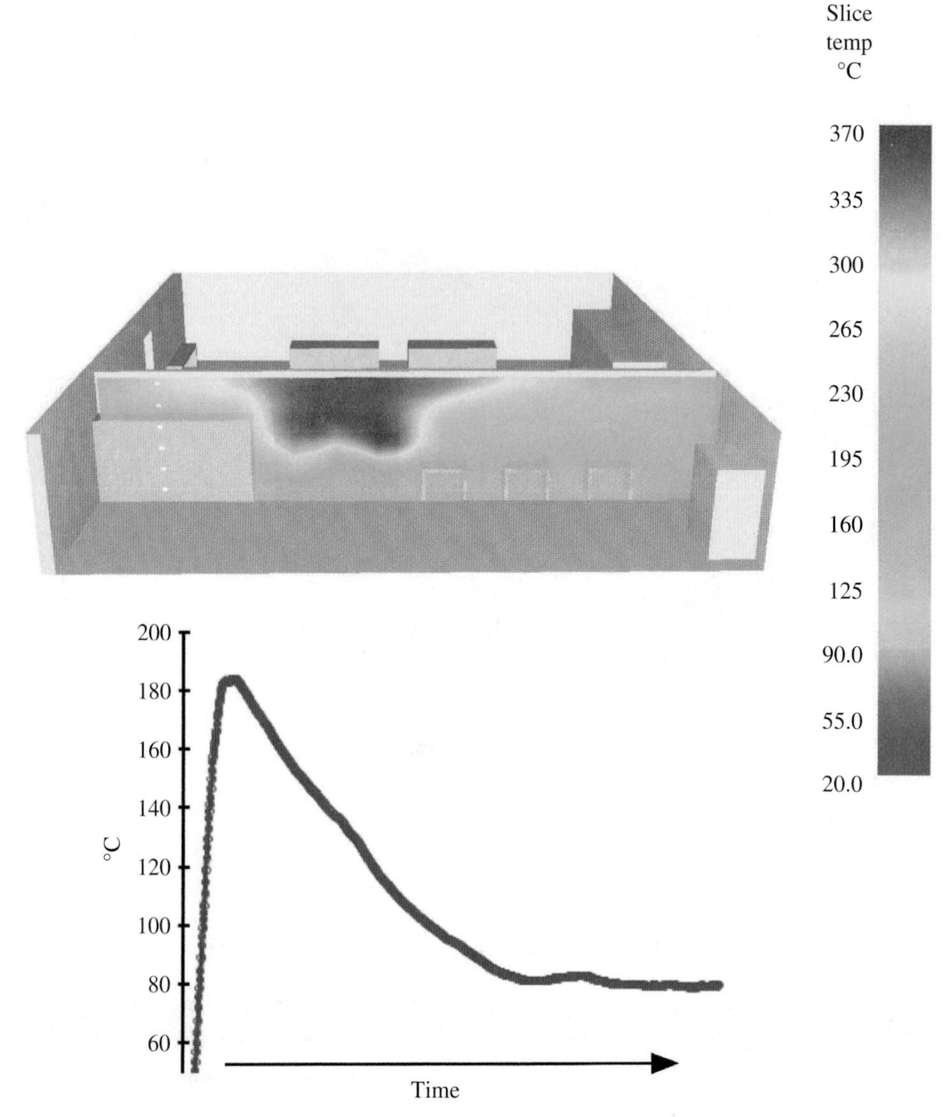

Figure 14.8

(Top) A simulated cross-section temperature profile of the workroom, for late stages of a fire originating from both solvent stills. The white dots in front of the bookshelf in the workroom show the location of temperature sensors used in the modeling calculations. (Bottom) The calculated temperature profile from sensors in front of the bookshelf. Calculations performed using PyroSim software (Thunderhead Engineering).

the extent of heat likely experienced by Dr. Thompson's lab notebooks. Calculations showed that the highest temperature of these sensors (Figure 14.8) was only *ca.* 190 °C.

Your Turn 14.10 You Decide Workroom Fire Damage

The key question posed by Williams' initial investigation of the lab concerned whether sufficient heat could be produced by the solvent stills to ignite the lab notebooks in the adjacent workroom. Using the simulated temperature profiles for the workroom shown in **Figure 14.8**, and the simulated temperature range of the sensors in front of the bookshelf, do you think that sufficient heat could be produced *in the absence of an accelerant* to ignite and burn the lab notebooks? Note that the covers of the lab notebooks were made of imitation leather—cotton fibers that are coated with the polymer polyvinyl chloride, PVC.

It should be noted that multiple points of origin in a fire is not necessarily a sign of arson. Heat from a fire may melt and ignite the asphalt from a collapsing ceiling, or the polystyrene light diffusers used in ceiling light fixtures. These liquids can produce a number of seemingly unconnected points of origin, which might be mistaken as pools of an accelerant.

On the basis of computer models, the temperatures found in the workroom were probably not high enough to ignite the vinyl-bound lab notebooks. However, the main experimental laboratory likely experienced temperatures above the autoignition point of wood, paper, and most plastics found in the lab. This would add to the heat evolved from the main laboratory, and would likely result in flashover conditions and significantly higher temperatures in the workroom than predicted by the calculations. Even so, the presence of empty and damaged solvent canisters and an unburned matchstick near the bookshelf strongly indicate arson as a probable cause of the lab fire.

The term **flashover** is used when the majority of the exposed surfaces in a room are heated to their autoignition temperatures, and emit flammable gases, which provide additional fuel to the fire.

Behind-the-Scenes at the Crime Lab

Perhaps the most straightforward and diagnostic analysis to identify possible suspects in a crime is fingerprint analysis. This information is extremely useful because no two sets of fingerprints are considered to be the same, including those of identical twins.

Latent fingerprints must be made visible before they can be analyzed. One common method of fingerprinting consists of distributing a powder onto a surface, which adheres to the residue deposited from the finger's touch. To avoid smudging the print, sometimes a magnetic powder is used in which the powder is poured onto a surface and then spread evenly using a magnet, instead of dusting the powder with a brush. To allow the best visibility, powders of contrasting or fluorescent colors may be used, depending on the color of the surface to be treated. Once the fingerprint is visible, photographs are taken. The powdered impression is then removed via fingerprint lifting tape and packaged for analysis and identification at the crime lab.

Patent prints are left when someone inadvertently uses a substance such as grease, paint, blood, or ink to leave a visible print on a surface. In contrast, **latent prints** are not visible to the naked eye and are left by the natural oils and residues from one's fingers.

As part of their ongoing investigation, police officials collected fingerprints from Dr. Thompson and were able to obtain fingerprints for some of his former graduate students who still resided in the county. It was also university policy to fingerprint all janitorial staff prior to their employment, because they have direct contact with students under the age of 21. Most of the postdoctoral researchers who had worked with Dr. Thompson had since left the state or country.

Fingerprint analysis was carried out on lifted prints from the doorknobs, cabinets, and fume hood sashes by the State Forensic Laboratory at Dr. Thompson's request. However, in a fire scenario, the presence of fingerprints is complicated by the presence of soot. In these areas, the prints were visualized using alternate light sources (ALSs) to illuminate the prints at varying wavelengths. The residue that makes up a latent print will fluoresce and, using special filters and goggles, the prints can be seen and photographed without being touched. The full and partial fingerprints taken from various surfaces in the lab were mostly Dr. Thompson's and janitorial crews'. The other unidentified latent prints were entered into the Integrated Automated Fingerprint Identification System, or IAFIS, a forensic database of known and crime scene prints. No match was found in the database for the unidentified latent prints taken from the laboratory.

Next, the forensics investigators analyzed the empty solvent canisters found at the crime scene. For this analysis, the solvent containers were placed into an airtight tank known as a fuming chamber (Figure 14.9). A few drops of liquid cyanoacrylate (**I**), the main ingredient of superglue, was then added to an open container and heated to a temperature of 49–65 °C to vaporize the liquid.

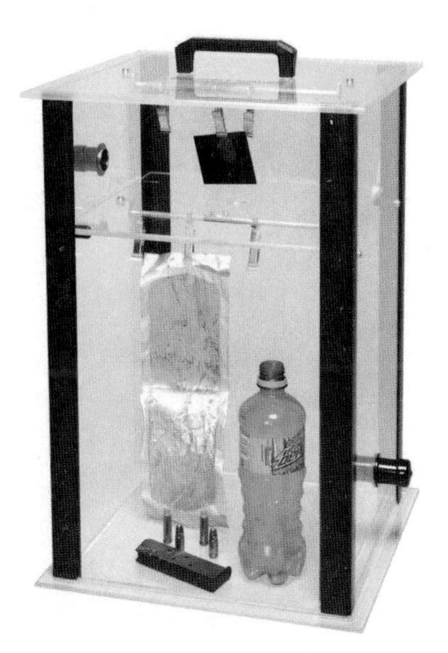

Figure 14.9

A portable superglue fuming chamber used for fingerprint analysis.

$$H_2C=C(C\equiv N)-C(=O)-O-CH_3$$

(**I**)

The gaseous cyanoacrylate reacts with the amino acids, fatty acids, and proteins in the exposed latent fingerprint. Upon further contact with residual moisture in the fuming chamber, a white, sticky film is generated that adheres to the ridges of the fingerprint.

Did You Know? Attempts at obscuring fingerprints by criminals are always unsuccessful. It is impossible to obliterate all of the fingerprint ridge characteristics on one's hands.

A number of fingerprints were found on the solvent canisters resulting from the fuming technique. Not surprisingly, most of the prints matched those from

Dr. Thompson and from his former graduate students. However, once again, there were thumb and index-finger prints that were not identified. Dr. Thompson would be contacted in the following days to obtain the names of those who might have had access to his laboratory, and whose fingerprints were not yet on file with the State Police.

With fingerprinting of the lab completed, Dr. White and her team of forensic scientists turned their attention to detecting the presence of accelerants in the flooring samples taken from Dr. Thompson's laboratory. A small hole was punched into the top of the air-tight container and the hole covered with a silicone septum. Upon heating the container to 60 °C for 30 minutes, any volatile residue present in the debris is evolved and trapped in the airspace of the container, referred to as its *headspace*. A few microliters of the vapor were removed by a syringe and analyzed using a technique known as **gas chromatography (GC)**.

Other types of chromatography are frequently used, based on the choice of mobile phase or type of separation being investigated. For example, liquid chromatography (LC) uses a liquid mobile phase; the use of a high-pressure liquid is known as high-performance liquid chromatography (HPLC).

A chromatographic separation employs the principle of "Like Retains Like"; this is analogous to the "Like Dissolves Like" principle discussed in Section 8.7.

Gas chromatography is a simple and rapid analytical technique commonly used to identify separate components in mixtures of liquids or gases. It consists of a liquid or gas sample being injected onto the beginning of a column. The sample's components are moved through the column by a flow of an inert gas, known as the *mobile phase*. The column contains a material—known as the *stationary phase*—that binds the components of the mixture to a varying degree, based on similar polarities. For instance, if a nonpolar stationary phase is used, it will bind the nonpolar components of the sample more strongly, which allows the more polar substances to pass through the column more quickly (Figure 14.10). As the various components reach the detector, a peak in the detector signal is recorded by a plotter or viewed on the screen of a connected computer. The time it takes for a component to flow through a column and reach the detector is known as the *retention time* of the substance. The longer a retention time for a component, the more closely its polarity matches that of the column material.

Your Turn 14.11 You Decide Peak Identification Using Gas Chromatography

The chromatogram shown in **Figure 14.10** at the top of the next page is from a mixture of caffeine, toluene, *n*-heneicosane, pyridine, and *n*-octyl acetate injected onto a column containing a nonpolar stationary phase. Using the Internet as a resource, draw the molecular structures for each compound, and rank them in order of increasing polarity. Identify the five peaks shown in the chromatogram below.

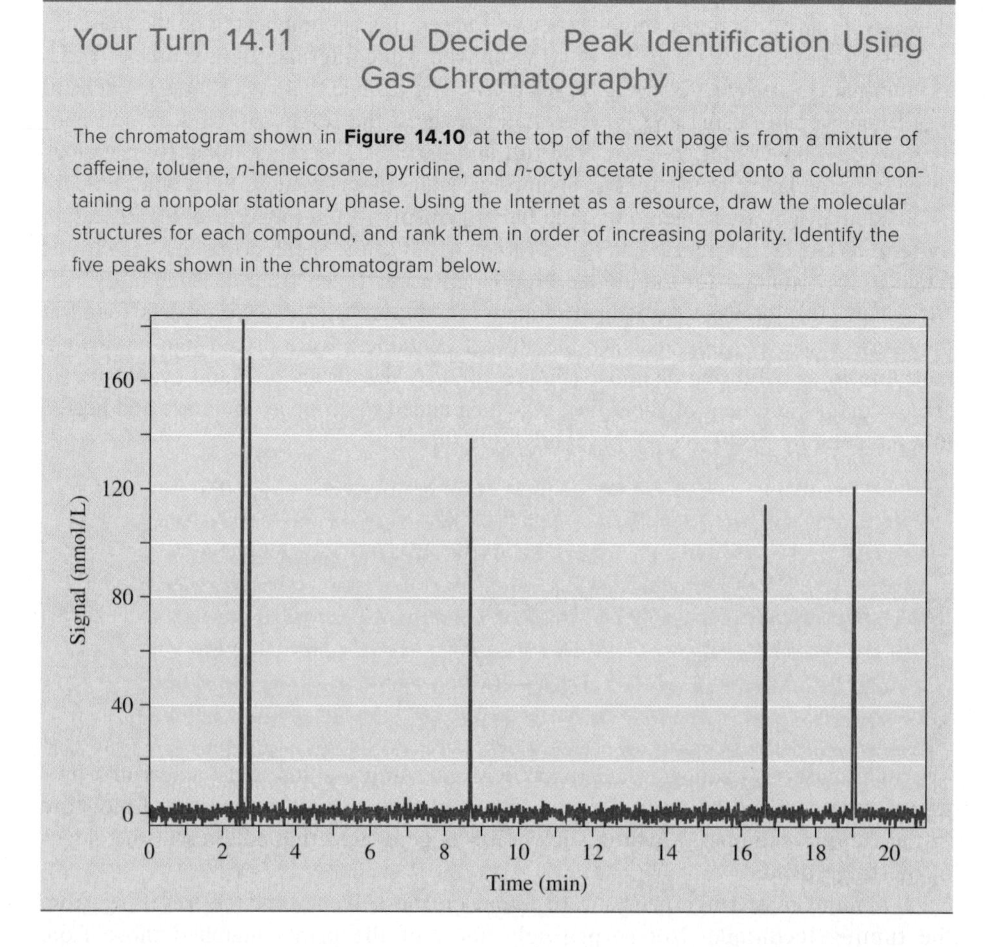

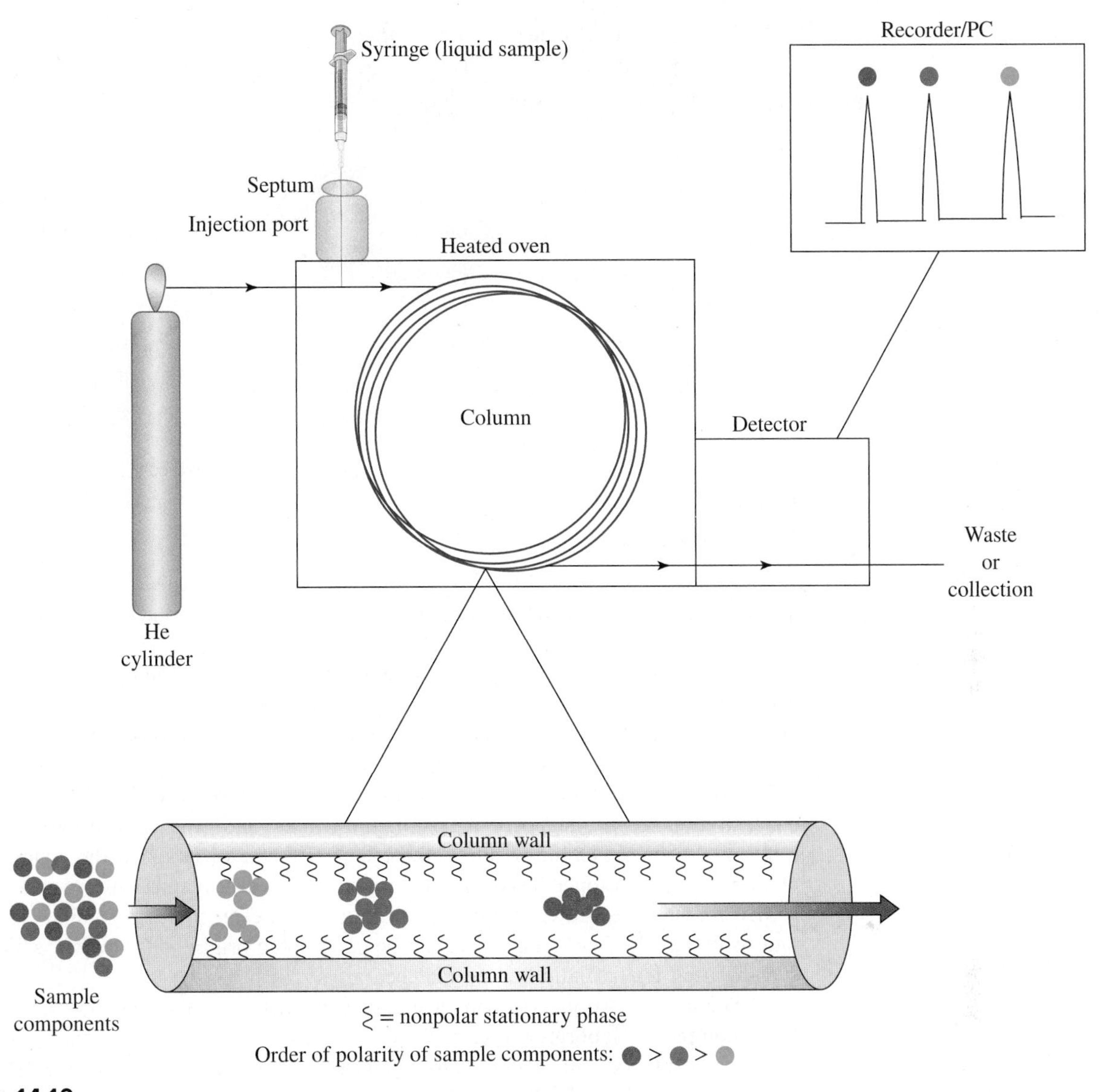

Figure 14.10

Schematic of the components of a gas chromatograph, showing separation of three sample components based on their relative polarities. The most nonpolar component is retained the most strongly by the nonpolar stationary phase.

The forensic scientists performed GC analysis on the floor samples taken from Dr. Thompson's laboratory. In the samples taken near the solvent stills, they found trace residues of benzene, toluene, dichloromethane, and a variety of compounds likely resulting from the burning of vinyl flooring and wood cabinets. No residues of an accelerant such as acetone were obtained from GC analysis of fragments of unburned lab notebooks, or floor samples taken from either lab. Although this may indicate that arson is not the cause of the fire, the lack of accelerant residues may be due to the fire extinguishing procedure and high volatility of the suspected accelerants. Empty containers of acetone were found in the doorway between the two rooms; however, acetone is miscible with water. When firefighters sprayed water onto the burning bookshelf and workroom, any unconsumed acetone would have been washed away, leaving no residue behind. Near the solvent stills where firefighters used dry chemical to extinguish the fire, the remaining acetone could have evaporated overnight, due to its extremely high volatility, before the forensics team arrived to collect samples. This is why fire investigations are so complicated, not only at the scene but also in the laboratory analysis.

The last pieces of evidence taken from the lab were the matchsticks, found in the workroom near the bookshelf. Dr. White had recalled reading about an arson case in the United Kingdom where **stable isotope analysis** was successfully used to match

The use of stable isotopes to understand the extent of climate change was discussed in Section 4.9. Stable isotopes also find many applications as markers in chemical reactions to understand reaction mechanisms, and to determine information about the ages and origins of rock, air, or water bodies.

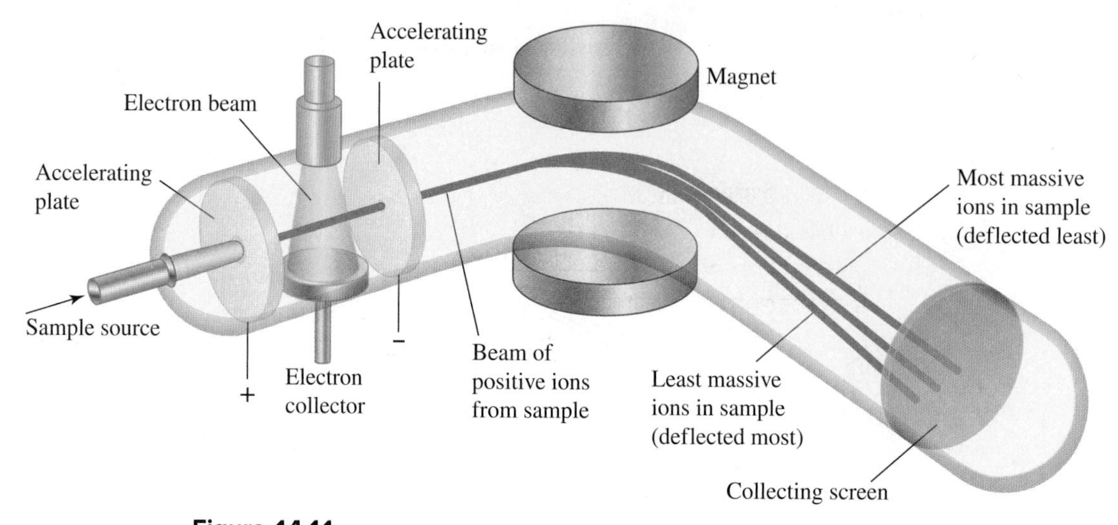

Figure 14.11

The operating principle of a mass spectrometer.

the wood from a matchstick found at the crime scene to the wood contained in matchsticks found at the perpetrator's apartment.

Depending on the growing conditions of trees, such as temperature, humidity, and nutrient supply, there will be observable differences in the ^{13}C isotopic fraction among different samples of wood. In addition, the ^{18}O concentrations of wood will vary depending on the relative uptake of oxygen atoms within trees from atmospheric CO_2 and O_2, as well as the water taken up through the root system. The ^{2}H isotopic concentration is also diagnostic regarding the climate and geographical location where a tree was grown. Not surprisingly, there is significant variation in the concentration of each of these isotopes between trees grown in different plantations, as well as variability among trees grown in the same location. However, the combination ^{2}H:^{18}O:^{13}C ratio is extremely diagnostic regarding the geographic origin of the wood, which may even be used for partially burned matches.

A mass spectrometer is often placed at the column outlet of a gas chromatograph, and provides details about the atomic and isotopic compositions of a sample. Solid samples are typically vaporized via *flash combustion*, an extremely rapid and complete combustion process taking place at very high temperatures (>1500 °C) and high oxygen concentrations.

Sections of wood 2–3 mm in length were cut from the matchsticks found at Dr. Thompson's laboratory. The wood samples were frozen, ground into small pieces, and placed into a desiccator containing diphosphorus pentoxide (P_2O_5) to remove traces of residual water from the samples. The most common analysis to determine isotopic ratios is **mass spectrometry**. As shown in Figure 14.11, vapor from a sample is ionized by bombarding it with electrons. The ions are then separated according to their mass-to-charge ratio by accelerating them in a vacuum (pressures of 10^{-6}–10^{-8} Torr) and passing them through a magnetic field. Ions of the same mass:charge ratio will experience the same degree of deflection and reach the detector at the same time.

The standard typically used for ^{18}O and ^{2}H is standard mean ocean water (SMOW), whereas ^{13}C employs standard forms of graphite, cellulose, or other carbon-containing compounds such as limestone from the Pee Dee Formation in South Carolina.

The variations in the natural abundance of stable isotopes are expressed using the delta (δ) notation as shown in Equations 14.1 and 14.2:

$$\text{Ratio(R)} = \frac{\text{Abundance of the heavy isotope}}{\text{Abundance of the light isotope}} \quad [14.1]$$

$$\delta = \left(\frac{R_{sample} - R_{standard}}{R_{standard}}\right) \quad [14.2]$$

The ‰ symbol is often designated as permil, from the Latin *per mille*, by analogy with per centum, for which percent (%) is derived. This is a preferred unit for isotopic ratio analyses since the differences between natural samples often occur at and beyond the third significant figure of the isotope ratio.

The δ values are typically multiplied by 1000 to yield units per thousand (‰). Hence, a negative ‰ value indicates a lower isotopic concentration (depleted in a particular isotope) relative to the standard. Triplet analyses of each wood piece resulted in an average ^{2}H:^{18}O:^{13}C ratio of −112 ± 1.3‰ : −5.35 ± 0.8‰ : −27.3 ± 2.1‰. These results were entered into the main case file to provide a matching profile, should matchsticks be found on the person or premises of a future suspect.

The match heads were analyzed by **scanning electron microscopy (SEM)**. Instead of using light to observe sample features via a traditional microscope, an

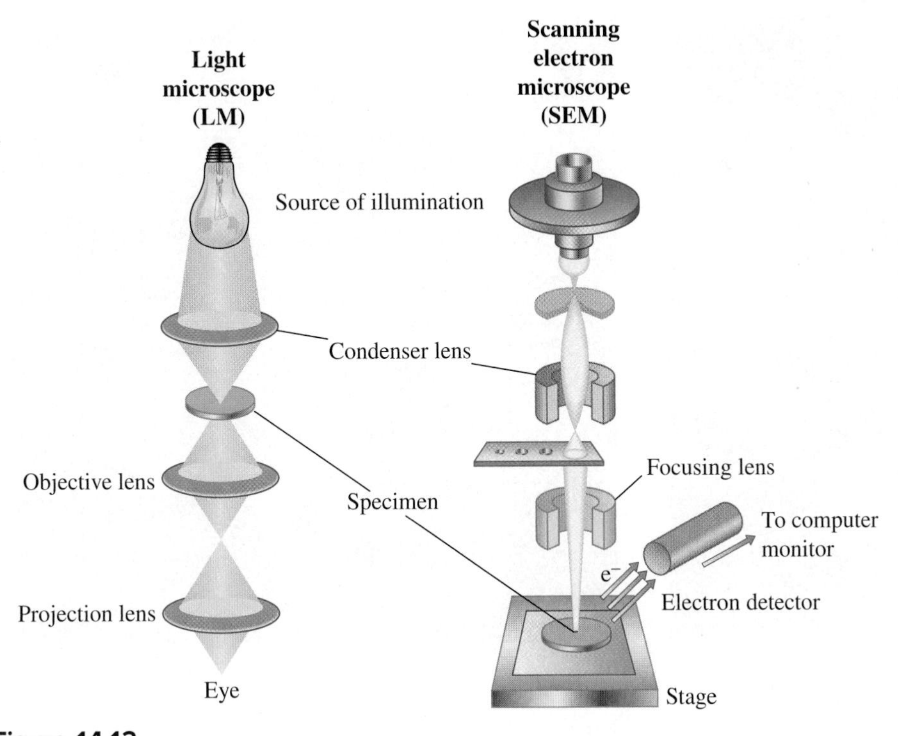

Figure 14.12

Comparison between a light microscope and scanning electron microscope.

Whereas common light microscopes can provide images of surface features with dimensions of 300-500 nm, a scanning electron microscope typically has a resolution of <20 nm.

electron microscope uses a high-energy beam of electrons to glean information about a sample's surface topography and composition (Figure 14.12). When an electron beam contacts the surface of a sample, energy is transferred to the sample atoms, which releases electrons (known as *secondary electrons*). The observed contrast in an SEM image is due to the emission of secondary electrons from different regions of the surface, which provides details about its topography. A *backscattered electron* image may also be obtained, which is related to scattering of the incident electron beam due to interactions with the atomic nuclei present within the sample. Furthermore, X-rays are also emitted from the atoms of the sample; their observed wavelengths are characteristic of the type and concentration of atoms that are present on the surface.

The SEM image of a match head shown in Figure 14.13(a) is a safety match, the type Dr. Thompson normally used in his research laboratory. Safety matches require you to strike the match on the surface of the matchbox. The tips of safety matches contain potassium chlorate ($KClO_3$) and glass (SiO_2). The striking surface also contains glass and an allotrope of phosphorus called red phosphorus. Red phosphorous is reasonably stable at room temperature, but the heat and friction generated by striking the match head against the striker surface converts the red phosphorus to the much more reactive white phosphorus. White phosphorus readily reacts with oxygen producing a large amount light and heat energy, igniting the rest of the match. The glass material in both the match head and striking surface provide rough surfaces that easily generate the friction required to ignite the phosphorus. Potassium chlorate is added to matches as a source of oxygen. Chlorates decompose under heat to form a salt and oxygen gas. The increased amount of oxygen speeds the combustion of the phosphorus and the wood or cardboard base of the match.

"Strike anywhere" matches, in contrast to safety matches, contain phosphorus in the match head itself. Thus, the match can be struck on any rough surface, generating friction and heat to ignite the phosphorus of the match. "Strike anywhere" matches (shown in Figure 14.13(c)) are less commonly found than the safety match variety as they are at higher risk of accidentally igniting.

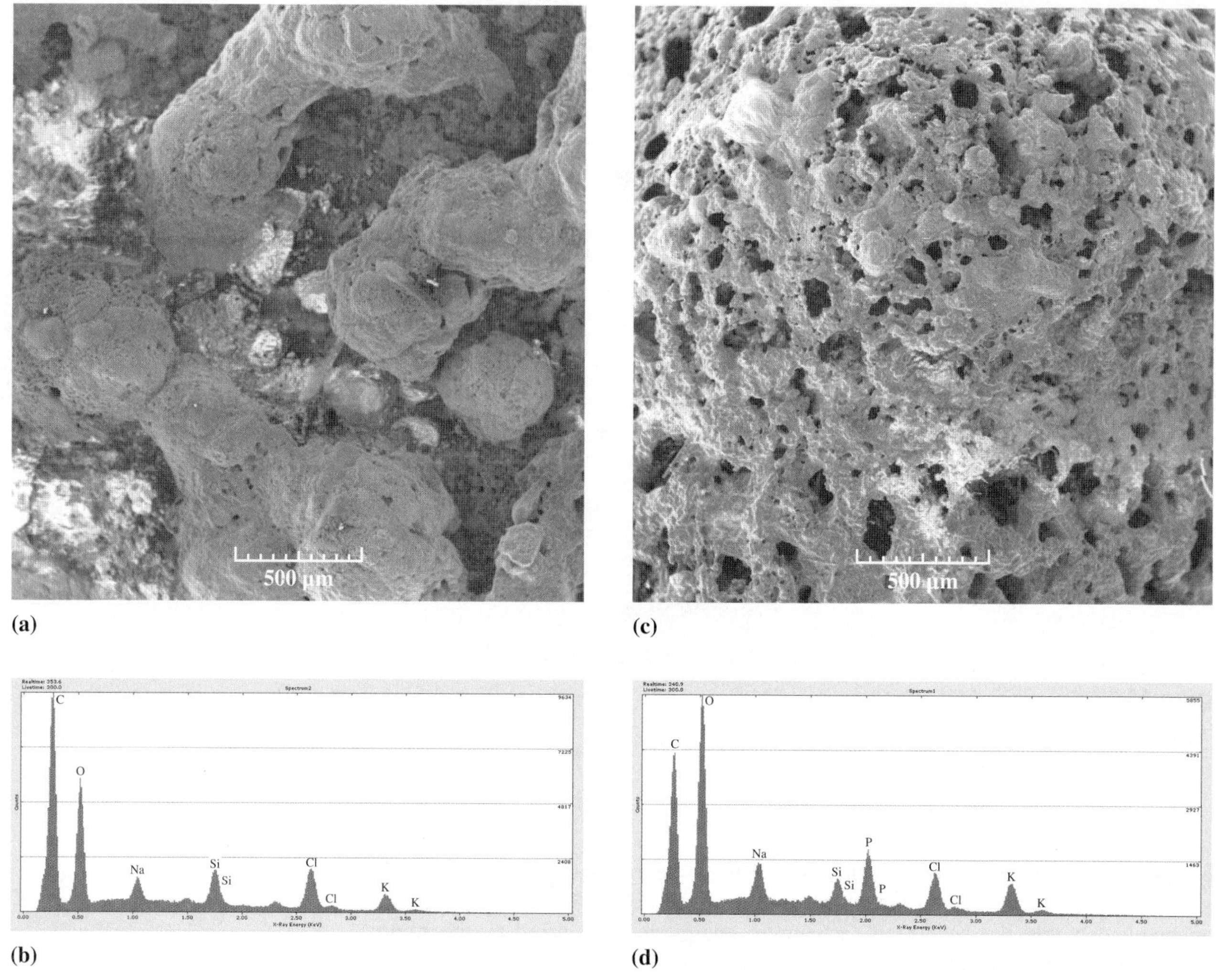

Figure 14.13

SEM images and X-ray elemental analyses of struck match heads from the lab's regular safety matches (**a** and **b**, respectively) and the match found near the bookshelf (**c** and **d**, respectively).

Figures 14.13(b) and 14.13(d) compare the X-ray spectrum of a match normally stocked in the lab to that of a match found near the bookshelf after the fire. X-ray analysis of the matches found in Dr. Thompson's lab revealed phosphorous as a significant component of the match head. This indicated that the matches were of the "Strike anywhere" variety, and were brought in from outside the lab, possibly being used as the ignition source for the fire.

Wednesday, Aug. 13—1:03 pm: Access to the Lab Restored

Almost two weeks had passed since the lab fire, so Dr. White called Dr. Thompson to update him on the status of the investigation.

"Hello, Dr. Thompson?"

"Yes."

"This is Dr. White, from the State Forensic Laboratory investigating your lab fire."

"Oh, good! Have you completed your investigation?"

"Yes, we have. We have now cleared your lab and you can access it to begin the cleanup."

"Great! So … was it really arson?"

"I'm afraid I can't say, it's still an open investigation. I can tell you we didn't find any residue of accelerants in either lab or on your notebooks. However, as we discussed a week or so ago, the empty acetone containers were suspicious. We also found some matchsticks, which should not have been there … correct?"

"Correct. I once had a postdoc who smoked. But it's a smoke-free campus. My goodness, I hope he never smoked in the lab!"

"I really would like to know your thoughts on other people who might have access to your lab. Since we last spoke, were you able to think of anyone? We found some fingerprints that didn't match you or your former graduate students. Did anyone else work with you who might have gained entry? Again, the door was locked and there was no sign of tampering or forced entry, so it would have to be someone with a key. Is there *anyone* you can think of who might wish you harm?"

"Professionally, no." Dr. Thompson barked a short laugh. "My ex-wife Julie told me a couple of months ago that she would destroy my life if she got the chance."

"Really? Would she have a key to your lab?"

"Actually, now that you mention it, she might. Julie had an affair with a former postdoc from my lab, who has since left the country. However, I don't remember if he turned in his keys when he left the campus …"

"Okay. We will definitely get her in for fingerprinting. Anyone else have access now or previously?"

"Yes. A couple of former postdocs: Dr. John Littleton and Dr. Avery Smith."

"Did you have a good relationship with them?"

"Yeah … it was fine … I can't think of them having a reason to do this."

"Okay, thanks. We'll speak with them as well to see if they can come down for fingerprinting. Of course, even if their prints match those that were unidentified from your lab, that doesn't put them in your lab during the night of the fire."

"Yes, that's true. However, if Julie's prints are there, she is definitely guilty!"

"Well, let's not get too ahead of ourselves. We will let you know. Thanks again for your time, and please let me know if you think of anyone else, or notice anything else out of the ordinary when you get back to your lab."

"Okay, will do. Thanks a lot. I'm heading to my lab now."

Dr. Thompson was not forthcoming to the police regarding his former postdoc's motive. With his important presentation now days away, he couldn't admit that he had used someone else's data without giving proper acknowledgement. If Dr. Littleton was responsible for the fire, he vowed to find out on his own. As Dr. Thompson drove to the lab, he placed a call to his former postdoc's office.

"Hello, you have reached the voicemail of Dr. John Littleton. Please leave your name and number, and I will get back to you. Thanks."

"Hi John, it's Dave. I need to speak with you about my lab. It's totally gone and I think you know why. Call me back."

As he hung up the phone, he noticed a voicemail left by his ex-wife earlier that morning.

"Hi Dave. I heard what happened to your lab. Have fun cleaning that up! Karma sucks, doesn't it?"

Infuriated by the voicemail, Dr. Thompson almost turned the car around to confront Julie. *How could she do something like this? I'm going to the cops tomorrow to have them listen to that voicemail. She's going to jail for this one!*

Dr. Thompson didn't recall the drive to his lab taking this long. As he flew down side streets to get there faster, he thought about how long it would take to get his lab back up and running. Maybe he would have to hire a student to help him set up again. He pulled into the faculty parking lot on campus, choosing a spot nearest the sidewalk to shorten his walk to the building. Now, finally in the building, he bounded up the stairs two at a time. *One more floor. There, made it.*

As Dr. Thompson entered the lab, his knees buckled. He became nauseated and fell to the floor. Power had been restored to the lab, but the lighting fixtures still weren't working. Investigators had left five large portable work lights at different parts of both rooms, which provided sufficient light to observe the full extent of the damages. There was no question that his entire lab would have to be rebuilt. Floors, ceilings, and furnishings were all severely burned. Soot covered the walls and surfaces of both rooms, completely masking the light blue color he proudly chose last year for a refreshing paint job. Whether it was an accident or deliberate didn't matter right now. Dr. Thompson was present at the University of California, Irvine, on July 23, 2001, and had witnessed the infamous accidental lab fire in Bill Evans' laboratory. He couldn't believe that the same had now happened at his own facility.

The accidental explosion of a benzene still in Dr. Evans' laboratory resulted in an injury to a Ph.D. student and a fire that caused approximately $3.5 million in damage, including repair/refurbishment costs.

Dr. Thompson sat expressionless on the soot-covered student desk in the workroom, poking through a pile of lab notebooks from his past years of research. Some pages were still partially legible, although none of the books would be useful anymore. As he reminisced on the excitement that had come with each discovery proudly detailed in the notebook, the hours passed. Back during the early investigations that were responsible for his breakthrough, he had never envisioned the research getting to the point of commercialization.

He still had his data, manuscripts, and analyses saved on his laptop, but the original discoveries recorded in his lab notebooks were now gone forever, reduced to a pile of soot. Throughout the day and early evening, a constant line of students and colleagues came into the lab to express their sympathies. Filled with disbelief about the damage, many offered to help with the cleanup, but Dr. Thompson, though greatly appreciative, seldom looked up from his beloved notebooks.

Wednesday, Aug. 13—9:57 pm: What Now?

With a heavy heart, Dr. Thompson put aside the remains of his notebooks and stood up. He shuffled through the lab one last time, closing the door behind him. He walked slowly to his car, thinking about the presentation he would deliver in a few days. All he wanted to do was go home and sit on the patio to tweak his presentation notes. The car started immediately as it always had, but then stalled. Repeatedly, he turned the ignition, without any success. Even holding the accelerator to the floor as the ignition was turned did nothing to revive his car. *Great! Now what?*

He reached for his phone to call roadside assistance, but the battery on his cell phone was dead. *Damn it! Why can't this phone make it through an entire day? Can anything else go wrong today?* Dr. Thompson often admitted his disgust with the current state of battery technology to his coworkers and friends. Although he thought of upgrading his cell phone to a new model with a supposedly longer battery life, he was skeptical of manufacturer claims and felt his phone worked just fine for his meager needs. Dr. Thompson was close friends with a researcher at Argonne National Laboratory, and joked that he could solve the battery dilemma in two years if given the resources.

Dr. Thompson got out of his car and began walking toward the biochemistry building to call roadside assistance. He always carried his black leather laptop case, which contained important data, manuscripts, and presentations. It was quiet on campus, as no one was around this time of night. As he took a step off the sidewalk to cross the street, a car sped toward him without warning. Unable to dodge the speeding vehicle, Dr. Thompson was hit dead-on, flipping him over the roof, and killing him instantly.

Under the cloak of darkness from the moonless night, the driver stopped the car, careful not to screech to a halt and leave identifying skid marks. With considerable effort, the murderer pulled Dr. Thompson's body off the road to delay its discovery until morning.

The driver quickly surveyed the damage to the car, retrieving a large piece of plastic from the front bumper, and making sure to collect Dr. Thompson's briefcase *en route* back to his vehicle. The car then sped away into the night.

Thursday, Aug. 14—5:42 am: A Gruesome Discovery

Sheila Jackson, a junior undergraduate student, was in the midst of training for a full marathon, and her early morning run often took her through campus. Today she was scheduled for 15 miles, and she was ready for the challenge. At mile three, she passed by the faculty parking lot, and noticed something that grabbed her attention. She slowed to a stop, and discovered a body that laid motionless on the side of the road. She knew immediately this was serious, as she saw blood coming from his head, and a blood trail leading from his body to the center of the roadway. She called 9-1-1, and waited on the other side of the street for police to arrive.

Within minutes, campus police arrived on the scene, with the city police showing up a few minutes later. The police questioned Sheila about her discovery, and then drove her home. The emergency medical technicians (EMTs) called a local medical doctor to obtain an official pronouncement of death. At 6:58 am, Dr. Thompson was pronounced dead at the scene by Marlene Jacobs, MD. His internal body temperature was 83 °F, indicating his death had occurred many hours prior.

Did You Know? The time of death in crime scene investigations is notoriously difficult to determine. However, as a general rule, a deceased body originally at 98.6 °F will lose heat at a rate of 1.5–2 °F per hour until the body reaches the ambient temperature.

Investigators closed all roads around the hit-and-run accident, and began their survey of the crime scene (Figure 14.14). The streets were dry and the ambient temperature was 58 °F. A few small fragments of clear plastic were found on the roadside near the victim's body, which might have originated from the perpetrator's vehicle. The lack of any visible skid marks on the street near the location of the body indicated that this could be an intentional murder. If someone had accidentally hit a person crossing an intersection, they would have braked hard or swerved to avoid a collision. Either of these preventative actions would have left skid marks on the road. However, another possibility is that the perpetrator had been distracted by texting or talking and didn't notice a person crossing in front of his/her vehicle.

The headlights of modern vehicles are composed of polycarbonate because of its light weight and transparency.

There was a visible pattern of blood leading from the middle of the road, where the victim's body probably first fell, to the side of the road where the body was found. Perhaps the victim had mustered just enough strength to crawl off the roadway by himself. However, it might also be possible that the perpetrator had dragged Dr. Thompson's lifeless body off the roadway after the collision. The astute police investigator examined Dr. Thompson's wrists, shoes, ankles, and neck with a forensic light source (UV light with a wavelength of 365 nm). A variety of fingerprint and palm prints were noticeable on Dr. Thompson's wrists. These prints could have been deposited by the perpetrator as they dragged the body off the roadway.

Other portable devices featuring a green laser (532 nm wavelength) may also be used for the observance of forensic trace evidence.

The police searched the surrounding area for additional pieces of evidence. Other bits of glass or plastic from the vehicle might have fallen off after the collision. The sides and middle of the roadway were carefully searched in a 200-yard radius from the scene of the crime. Although no additional fragments from a vehicle were discovered, a collection of five cigarette butts and used matchsticks were found in a temporary parking/pickup area about 100 yards from the crime scene. Were these from the perpetrator, who might have patiently waited for Dr. Thompson to cross the street? The butts and matchsticks were collected with plastic forceps and placed in a labeled plastic bag for analyses at the crime lab. Additionally, a black cell phone with a cracked screen and no remaining battery power was found on the side of the road, about 5 meters from where Dr. Thompson's body was found. This also was carefully packaged for transport to the forensics crime lab.

In order to re-create the events of this tragic evening, investigators had to determine whether Dr. Thompson was walking toward, or returning from, his car that evening. Car keys were found in the right front pocket of his trousers, and were successfully used to unlock Dr. Thompson's vehicle, a 2013 white Ford Focus that was parked a short distance from the hit-and-run scene. Investigators dusted the door handles, steering wheel, and gear selector for fingerprints using black powder and forensic lifting tape. After thorough investigation of the vehicle for signs of suspicious activity, the lead investigator attempted unsuccessfully to start Dr. Thompson's car. The battery and starter were deemed to be functional, so there must have been another reason for the vehicle failure. A tow truck was called to deliver

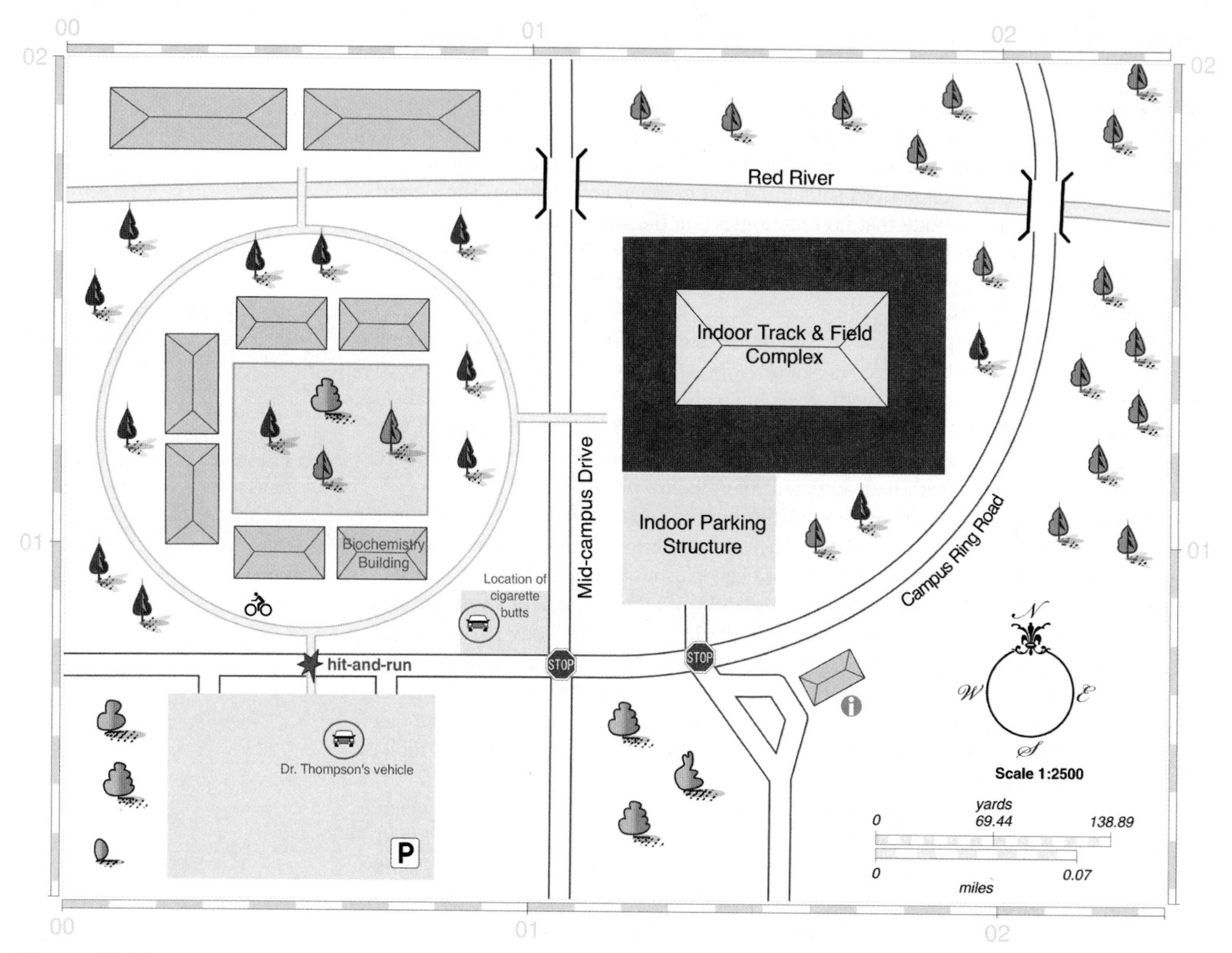

Figure 14.14

Map of the area surrounding the hit-and-run crime scene.

Dr. Thompson's car to the crime lab for a more thorough investigation. The investigator surmised that Dr. Thompson might have experienced difficulties in starting the car, and was returning to his office to call for help when he was struck down by a vehicle. But if that were the case, why didn't he simply use his cell phone to call roadside assistance from his car? The black cell phone, believed to belong to Dr. Thompson, was too badly damaged to surmise whether the battery was charged at the time immediately before the accident.

Behind-the-Scenes at the Crime Lab

Dr. White was informed of the fatal accident and contacted the police to offer her services to assist with the investigation. Although she focused on cases involving arson and gun-related homicides, it was quite possible that the hit-and-run accident was related to the fire in Dr. Thompson's laboratory. Her interest in this vehicular homicide peaked when she heard of matchsticks being found a short distance away.

The isotopic ratios of the wood from the matchsticks found near the crime scene were determined using the same procedure as before. In order to match samples from two different locations, extreme attention was given to details regarding sample preparation and analysis. The ^{2}H:^{18}O:^{13}C ratio of samples taken near the hit-and-run accident were $-114 \pm 2.8‰ : -5.51 \pm 1.3‰ : -26.9 \pm 2.1‰$.

Your Turn 14.12 You Decide Matchstick Analysis

Based on the isotopic ratios of the wood from the matchsticks found in Dr. Thompson's laboratory, and those found at the hit-and-run crime scene, do you think there is a match? If so, does this prove that the same person was involved in both crimes? Explain your reasoning.

The fingerprints on Dr. Thompson's lifeless wrists were analyzed at the crime lab. Magnetic Jet Black powder was applied to the surfaces of his wrists using a magnetic brush. The fingerprint (Figure 14.15) was carefully transferred onto lifting tape, for entry into the system for fingerprint matching. Fortunately for investigators, the area with the highest density of latent prints was in the hairless region just below the victim's palm, giving rise to high-quality impressions. In addition, the perpetrator likely exerted significant pressure in depositing their fingerprints onto the victim's wrists while moving the body. From other published reports of this type of forensic analyses, the overall quality of prints was found to be strongly dependent on the pressure used while depositing the marks.

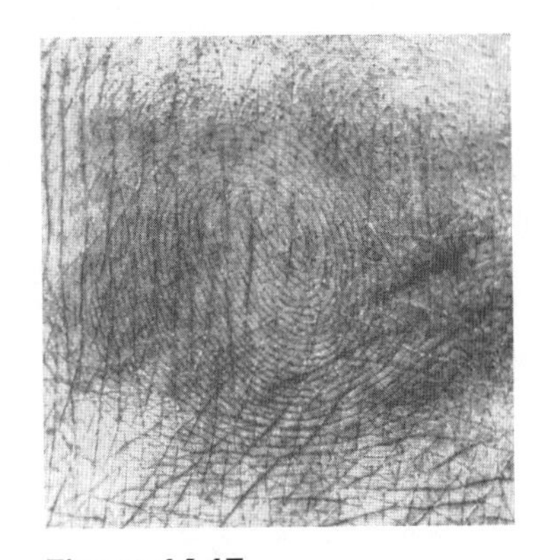

Figure 14.15
Image of a fingerprint developed from the application of a magnetic black powder.

© Matej Trapecar

After four hours of waiting, the fingerprint database did not provide a match for those found on Dr. Thompson's wrists. However, it was noted that the prints found in this crime matched some of those found on the solvent bottles in Dr. Thompson's laboratory. Furthermore, the partial and full prints lifted from various portions of Dr. Thompson's wrists and arms were from only one person, and that person was somehow involved.

An autopsy was performed on Dr. Thompson. Not surprisingly, the cause of death was determined to be multiple blunt force traumas to the head and torso. The injuries were consistent with being struck by a vehicle. Toxicology results did not show any observable concentrations of alcohol in his system. His stomach was nearly empty, indicating a long period (>6 h) since he had eaten his last meal, with the presence of starch grains and caffeine. His medical records were obtained to determine whether suicide was a factor. Perhaps Dr. Thompson was depressed about the lab fire, and/or had been diagnosed with a terminal illness. After viewing his medical records, it was determined that Dr. Thompson did not suffer from a terminal illness, and was in good physical condition.

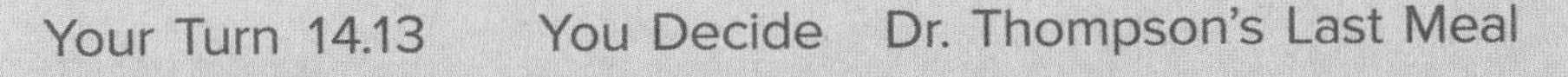

Your Turn 14.13 You Decide Dr. Thompson's Last Meal

Based on the contents of Dr. Thompson's stomach, as revealed by the autopsy, what are some possible foods he might have consumed at his last meal? If he had drank a glass of red wine at his last meal, approximately how long would it take for traces of ethanol to be detected in his blood? What factors influence the rate of alcohol metabolism in the body?

Your Turn 14.14 You Decide Healthy or Not?

During his annual health evaluation performed one month earlier, a blood test for Dr. Thompson revealed the following levels:

Total cholesterol: 229 mg/dL
HDL cholesterol: 42 mg/dL
LDL cholesterol: 187 mg/dL
Triglycerides: 158 mg/dL

Using the Internet and information provided in **Chapter 11** as resources, determine whether Dr. Thompson was in good health (relative risk of heart disease).

Episode 15 in the 2004 season of "MythBusters" evaluated the popular myth of adding sugar to a gas tank to prevent a car from working.

Investigators also examined Dr. Thompson's car to determine the cause of its mechanical issue and potential relevance to the murder case. No blockage in the tailpipe was noted, but the gasoline contained large amounts of water. Although placing sugar in a gas tank is widely thought to incapacitate a vehicle, this has been shown to be a myth.

Water is much more dangerous to engines than sugar because of the concept of "like dissolves like" that was discussed in Chapter 8, and density that was described in Chapter 1. Figure 14.16 illustrates the molecular structures of gasoline, sugar, and water. Whereas the components of gasoline are nonpolar, both sugar and water molecules are quite polar. Therefore, sugar will not dissolve and water will not be miscible in gasoline. So, why is one of these components "better" than the other, as far as causing damage to internal combustion engines? The answer is related to the relative densities of these substances. The density of gasoline is in the range of 0.71–0.77 kg/L, as compared to 1.587 kg/L and 1.0 kg/L for sucrose and water, respectively. Therefore, gasoline will float on top of water, and sugar will drop to the bottom of the gas tank. If a few cups of water are poured into the gas tank, the fuel pump will fill the fuel lines with water instead of gasoline, which would result in major engine damage. Because the gas cap on Dr. Thompson's car was not fastened tightly, it is clear someone wanted to purposely disable his car.

The addition of sugar into a gas tank would probably enact the same response as dumping in sand—a nasty mixture to clean up, with the solids dropping to the bottom of the tank instead of being transported into the fuel lines.

The picture was beginning to make more sense now to investigators. The murderer must have poured water into Dr. Thompson's gas tank, and then waited for him to return to his vehicle. Once he was unable to start his car, Dr. Thompson returned to his office to call for assistance, most likely due to a dead battery in his cell phone. He would have to cross the street, which was always desolate at that time of the night. But how did the perpetrator know Dr. Thompson's cell phone would be non-functional? Perhaps it was an educated guess, based on the known inability of his phone to last through an entire day. However, it is conceivable that there would have been a backup plan if Dr. Thompson would have been able to use his cell phone to call for assistance that night.

Traces of dark blue paint were found on Dr. Thompson's clothing, which likely originated from the perpetrator's vehicle. The location of the paint was carefully documented, and the victim's clothing was then scraped over clean, white craft paper. The debris collected from scraping was collected and examined using a variety of methods, starting with a stereomicroscope to determine the presence of fibers, flakes, and other microparticulates.

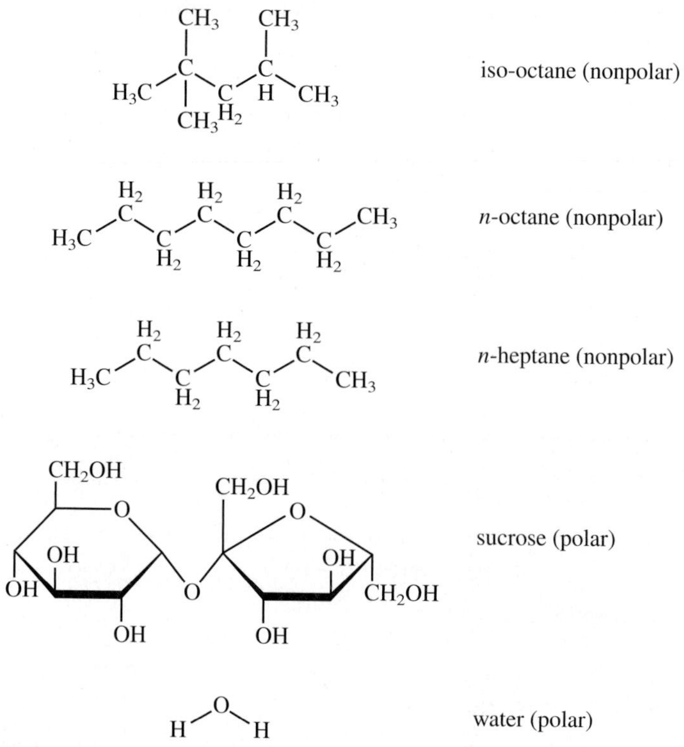

iso-octane (nonpolar)

n-octane (nonpolar)

n-heptane (nonpolar)

sucrose (polar)

water (polar)

Figure 14.16

Molecular structures of some major components of gasoline (iso-octane, *n*-octane, and *n*-heptane), table sugar (sucrose), and water.

Infrared (IR) spectroscopy (Figure 14.17) was used to determine the functional groups present in the paint formulation, which was then entered into the Paint Data Query (PDQ) database. The paint matched that of a 2014 metallic-blue Tesla Model S. The use of an electric car for the murder was clearly an insidious choice. Whereas Dr. Thompson could have easily heard the approach of a gasoline-powered vehicle, the much quieter operation of an electric car would have provided no warning to the victim, and no opportunity to jump to safety.

IR spectroscopy was described in Section 4.8.

The PDQ database, hosted by the Royal Canadian Mounted Police in Canada, contains nearly 20,000 samples of paint systems, representing more than 74,000 individual paint layers used on most domestic and foreign vehicles sold in North America and Australasia.

Your Turn 14.15 Scientific Practices Peak Matching

Using the Internet as a resource, assign functional groups that correspond to five major peaks that appear in the IR spectrum in **Figure 14.17**.

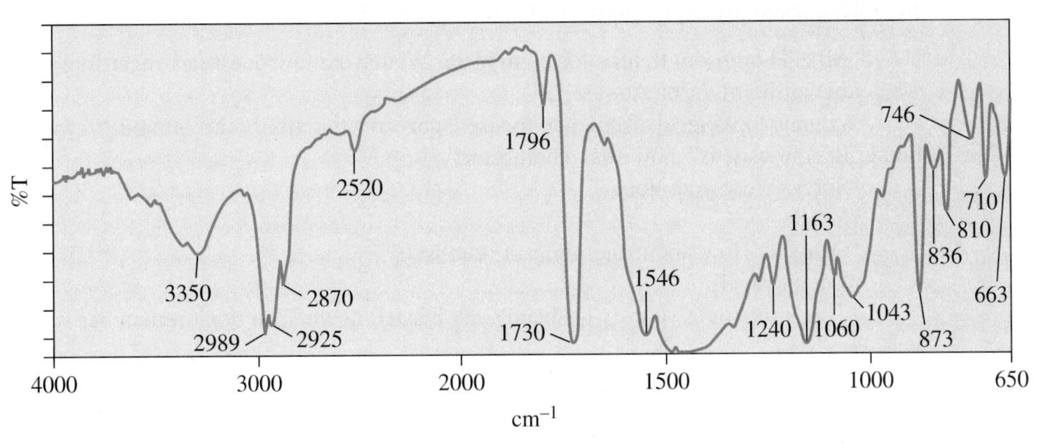

Figure 14.17

Infrared spectrum of the paint fragment recovered from the victim's clothing.

Your Turn 14.16 Scientific Practices IR Spectrum of Polycarbonate

The image below shows the IR spectrum for the polycarbonate fragments found at the crime scene that had likely come from the headlight lens of the perpetrator's vehicle. Assign functional groups to the regions labeled 1–6.

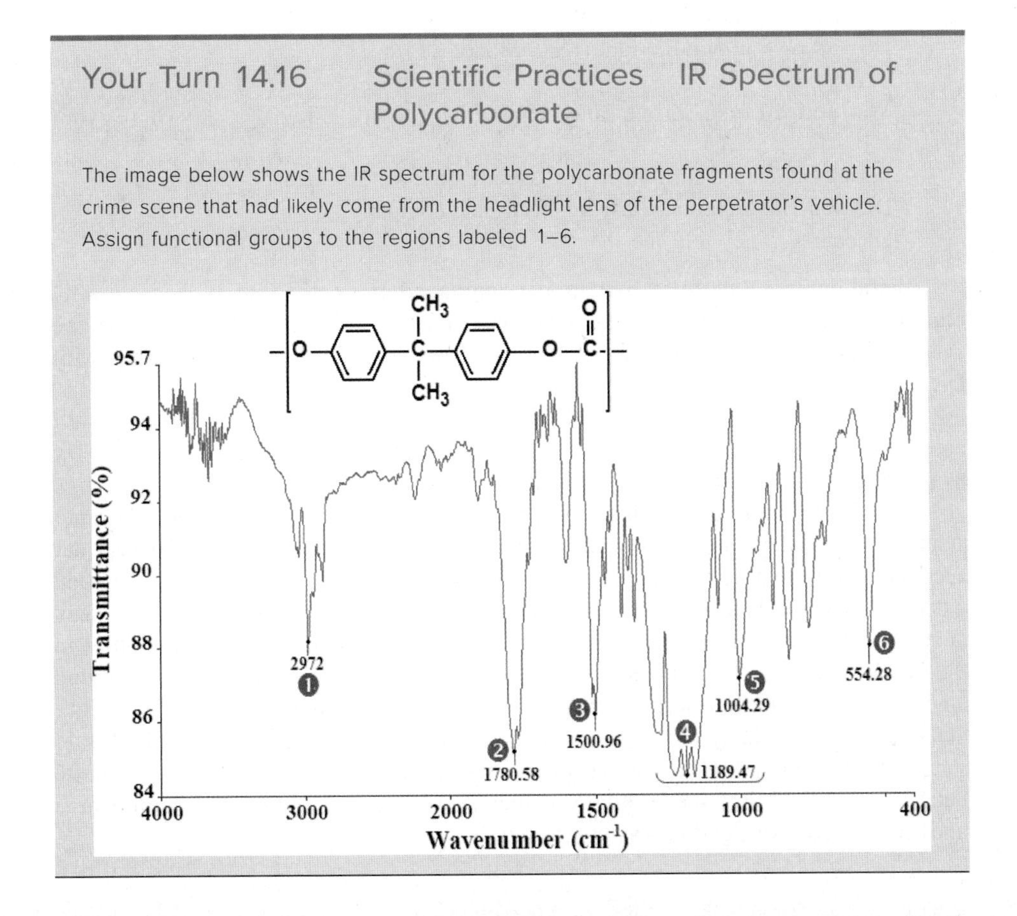

Investigators were pleased that a match was made to a vehicle with relatively low production numbers. In 2014, 31,655 Tesla Model S vehicles were sold worldwide, with 16,550 sold in the U.S. The State Police department searched for registered owners of a 2014 Tesla S within the state, which identified 231 households. However, only 35 owners took delivery of a Tesla S in metallic blue. As Dr. White examined the list of owners sent from the State Police, her eyes immediately honed in on Julie Thompson—the victim's ex-wife. The police called her and requested she visit headquarters to answer some questions.

Friday, Aug. 22—9:03 am: The Questioning of Julie Thompson

Julie Thompson sat expressionless in Interrogation Room B.

"Mrs. Thompson, thanks for coming in. We have a few questions regarding your whereabouts the night of August 13."

Without looking up from her focused gaze on the floor, she uttered "I was at home. By the way, it's now Ms. Thompson."

"Sorry. Were you alone?"

"Yes."

"Okay ... How about the night of August 1?"

"I don't recall."

"That's the night your ex-husband's lab burned down. You don't remember where you were?"

"No. I think I was shopping at the time."

"Ms. Thompson, you should realize that you are a suspect in both crimes. If I were you, I would take these questions more seriously."

"I am. If I can't recall, I can't recall ..."

"Do you have keys to your ex-husband's lab?"

"No. Why would I?"

"Do you own a metallic-blue Tesla Model S?"

"Yes."

"We noticed you drove another vehicle today. May I ask where your Tesla is?"

"It's in the shop. I hit a deer just outside of town last week."

"Did you file a claim with your insurance company?"

"Of course. They told me they should have it repaired sometime next week."

Establishing that Julie had in fact damaged the front end of her metallic blue Tesla S, police continued to question her about her whereabouts during the evenings of the hit-and-run and lab fire incidents, as well as her relationship with Dr. Thompson.

Her bitterness and anger boiled over at various times during the interview—"I hate him right now, but I wouldn't do something stupid like burn down his lab or try to kill him!"

The investigative team contacted Julie's insurance company, which confirmed her story of an accident involving wildlife. The tow truck driver also corroborated her account, stating that he was called to a wooded area about five miles from town, and a deer was clearly the cause of her accident. Julie's fingerprints were taken by the lead detective. Her fingerprints were not a match for those found on the acetone bottles in the lab. In fact, her fingerprints did not match any of the unmatched prints found anywhere in the lab. Furthermore, her prints did not match those found on the wrists of Dr. Thompson's body at the scene of the hit-and-run incident. Although Julie said she never smoked, a swab of her saliva was taken for DNA matching on the cigarette butts found at the crime scene. Julie was then sent home and told not to leave the area until the police investigation was completed.

Monday, Aug. 25—8:31 am: The Questioning of Dr. Littleton

With Julie released as a primary suspect, the attention of the police shifted to other possible suspects. The phone records for Dr. Thompson's cell phone indicated that he called a number from a neighboring town just hours before he was murdered. The number was tracked to Dr. Littleton in the biochemistry department at a nearby college. Police investigators placed calls to his office and home to ask him to come in for questioning. The next day, Dr. Littleton arrived at police headquarters.

"Thanks for coming in so quickly, Dr. Littleton. I am Detective Bentley and this is my colleague Detective Stevens."

"My pleasure. Anything to help out."

"Please describe your relationship with Dr. Thompson."

"I was a postdoc in his lab about a year and a half ago. I'd worked with him a total of three years, and then moved on to start my independent career."

"Did you have a good working relationship with him?"

"Yes. No problems."

"Was there a reason why he called you the night of his murder?"

"I'm not sure. I've been trying to understand that myself. He said that his lab was destroyed, but I already heard about that from a colleague in his department."

"From the accounts of others in Dr. Thompson's department, it seems you two didn't part on the best of terms. Is that correct?"

"I'm really not sure what that means. I landed a great position, so he must have written me a good letter of recommendation."

"Did you know Dr. Thompson was scheduled to speak at the American Chemical Society meeting in Boston a few weeks ago?"

"No, I wasn't aware of it. He hasn't attended meetings for a while ..."

"Where were you on the evenings of August 1 and August 13?"

"I was out of town at a conference between August 9 and the 14th, and I was at an event hosted by the Dean of our college on August 1."

"Where was your conference, and when was your presentation?"

"In Atlanta, and my talk was in the afternoon on August 13."

"So that's about two hours away from your former employer?"

"More like three-plus with traffic!"

"Is there a way to confirm your attendance at either of these events?"

"Well, I have confirmation of my conference registration, and you could ask my colleagues if they remember seeing me at the fundraiser."

"Okay, thanks. Please send me a copy of your conference registration, and we'll contact a few of your colleagues."

"Do you know of anyone who owns a Tesla Model S?"

"No. I wish! Those are nice cars! Why a Tesla?"

"We have reason to believe the vehicle that struck Dr. Thompson was a Tesla."

"Oh ... A shame to damage such a nice car!"

"Well, let's not forget about the person it hit!"

"Yeah, you're right ..."

The questions continued for another two hours, with attempts to unveil more details about Dr. Littleton's relationship with Dr. Thompson. Based on all accounts from acquaintances, Dr. Littleton was extremely angry with his former advisor, and wanted appropriate credit for his part of the work. Dr. Littleton denied these claims, and said they had worked out their differences before he left. Without any evidence to link him to either crime, the police let him go.

Dr. White strongly believed that Dr. Littleton wasn't telling investigators the whole truth. The search for registered Teslas in the state of Georgia revealed two rental agencies in the Atlanta area. Could Dr. Littleton have rented a Tesla and driven back to kill his former advisor?

Your Turn 14.17 You Decide Battery Range

The distance from the scene of the hit-and-run accident to the convention center in Atlanta, Georgia, is 140 miles each way. Based on the reported battery range of the Tesla Model S (assume a 90-kWh battery), would the driver have to stop for recharging between these locations, or could the car make a round trip on one full charge? What factors would affect the battery range of this vehicle?

Tuesday, Aug. 26—2:05 pm: Road Trip to Atlanta

Police detectives Bentley and Stevens drove to Atlanta to follow up on the suspicions of Dr. White. Exotic Rentals was visited first, and a warrant was secured to search the rental records. No mention of Dr. Littleton was found in the log book, and the lone Tesla Model S was parked in front of the parking lot, with no visible front-end damage. The Tesla was being charged with a solar-powered charger.

Your Turn 14.18 Scientific Practices Solar Charging Station

Using the Internet, research the cost and configuration of three different solar charger kits. How long would it take for these units to fully charge a Tesla Model S battery (assume 90 kWh)? Compare the costs/kW for the different solar charging kits.

The agents visited the second rental agency, Dream Cars, which was located about 25 minutes away. As one investigator flipped through the pages of the log book, he asked the shop manager:

"Is your Tesla currently being rented? I don't see it out front."

"No. The last guy who rented it hit a deer. That thing did a lot of damage. It'll be a while before we can get it back on the road."

As the investigator inspected the log book one last time, the name *Johnathon Littleton* jumped off the page. They had their suspect. The agents called back to headquarters to issue an arrest warrant for Dr. Littleton, and a search warrant for his home and office.

Back in the Crime Lab

Dr. Littleton was taken into custody on the evening of August 27 and remained in jail overnight. Fingerprints were taken, as well as a cheek swab to retrieve a sample of his DNA. DNA analysis was performed on the cigarette butts found near the hit-and-run scene and was compared with that of Dr. Littleton.

As discussed in Chapter 13, DNA is located in the nucleus of cells throughout the body. However, DNA must first be extracted from other cellular material, as well as debris such as clothing or cigarette butts. Commonly, a mixture with equal parts water, phenol (C_6H_6OH), and chloroform ($CHCl_3$) is used to extract DNA from its host matrix.

Your Turn 14.19 Scientific Practices Solvent Extraction

Look up the densities of water, phenol, and chloroform. ***Hint:*** Recall that "like dissolves like."

a. Once this DNA-extraction mixture is shaken vigorously, only two layers form, one having twice as much volume as the other. What are the compositions of each of the layers?

b. Based on your knowledge of DNA structure presented in **Chapter 13**, in which layer will the DNA be found?

Concentrating on Concepts

12. What solvents in Table 14.1 would ignite with a match inside a cold room maintained at –20 °C?

13. X-ray spectrometers often measure the energy of X-ray photons in units of kiloelectron volts (keV). This is the amount of energy gained by an electron after being accelerated by 1000 V of electricity. One keV is equivalent to 1.6×10^{-16} J. What is the wavelength of an X-ray from a phosphorous atom with an energy of 2.0 keV?

14. Infrared spectra are typically measured as a function of wavenumbers instead of wavelengths. Wavenumbers are simply the inverse wavelength ($1/\lambda$). A typical infrared spectrum will range from 4000 cm^{-1} to 400 cm^{-1}. What is this range in nm?

15. It has been said that "all evidence found at a crime scene has been transferred from one location, object, or person to another." Explain this concept, which is referred to as the *exchange principle* and provide three examples of transfer evidence commonly found at a crime scene.

16. Polymerase chain reaction (PCR) is typically carried out in modern equipment called a *thermocycler.* Thermocyclers heat and cool a mixture of DNA, enzymes, single nucleotides, and other reagents to optimize the conditions for enzyme function. For each cycle in temperature, the number of DNA strands doubles. So, for n cycles, the number of DNA strands is 2^n. How many cycles would be required to create a million copies from just one DNA strand?

17. Is it possible to obtain latent fingerprints from fabric surfaces? Explain.

18. Inspect the IR spectrum below and list the functional groups that are likely present in the sample. Can you indicate the concentrations of functional groups based on their relative peak heights? Explain.

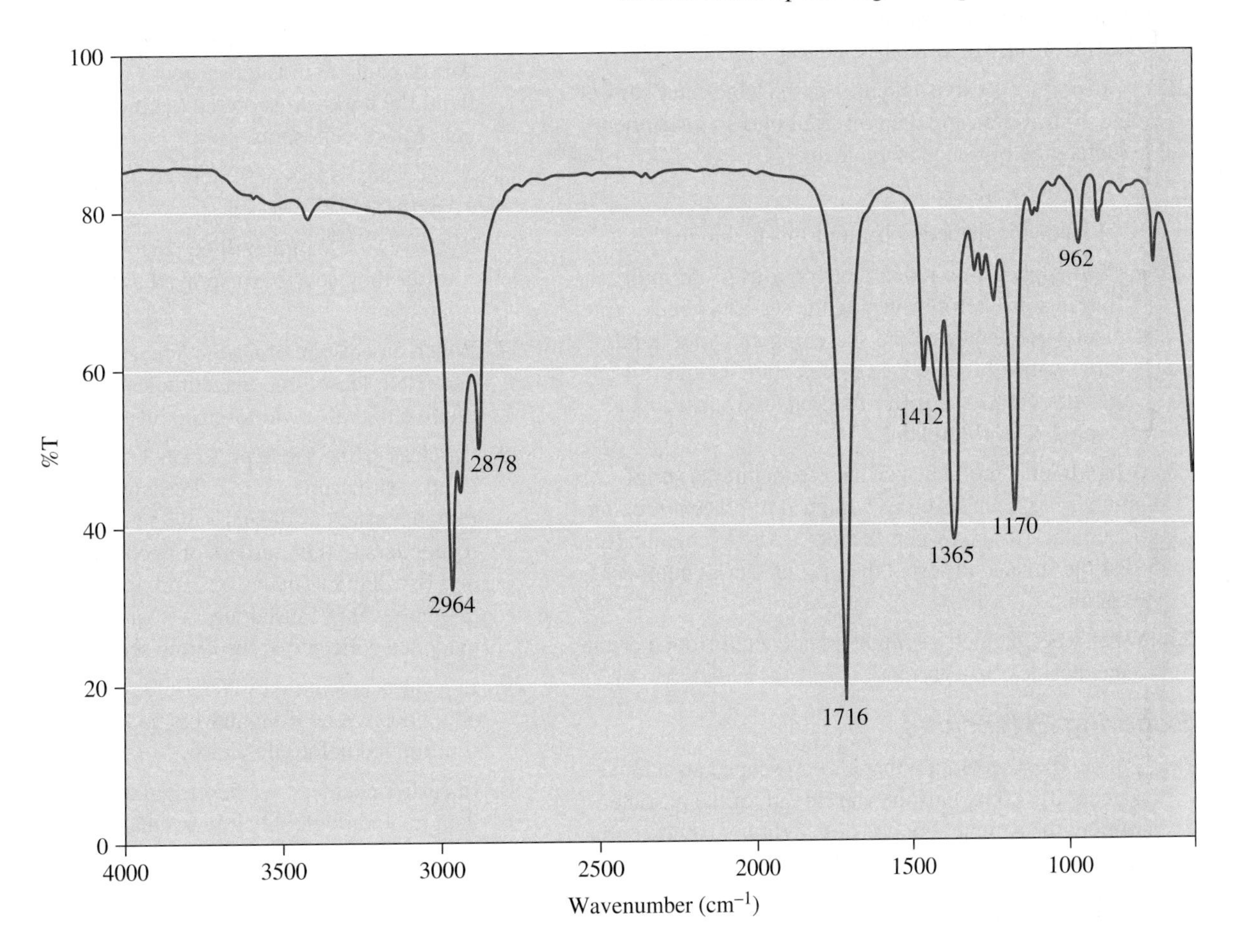

19. Could you discriminate between the three compounds below using IR spectroscopy? What other techniques could be used to determine the identity of each compound?

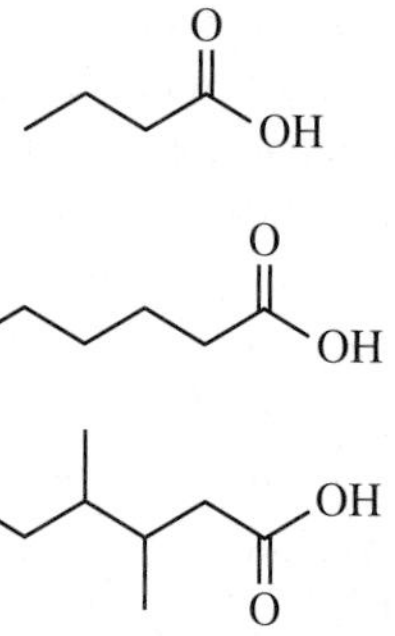

20. A sample containing heptane, 2,3-dimethyl octane (an isomer of decane), octane, pentane, and methyl cyclohexane is injected onto a GC column with a nonpolar stationary phase. What will be the order of elution for these compounds?
Hint: Consider their relative boiling points.

21. At airports, a swab is taken from the hands and bags of random travelers and then inserted into an instrument to detect the presence of explosives.
    a. How does this work?
    b. What is the detection limit of this technique?
    c. You apply a skin lotion that contains glycerin or nitrates just before entering the security check. You are selected for random screening and your hands are swabbed for explosives detection. Could this result in a false-positive reading (and subsequent long delays!)? Explain.

22. A handgun is found at a crime scene, but the serial number has been severely scratched to prevent it being traced back to the owner. Is there a way to chemically treat the surface to reveal the original serial number? Explain.

23. What other biometrics might be collected from a crime scene besides fingerprints?

Exploring Extensions

24. The smallest resolution that a microscope can achieve is generally about half the wavelength of the particle probing the sample. For a visible light microscope, this means the smallest features that can be see are about 300 nm in size. The electrons in an electron microscope behave similar to light in a light microscope, having their own wavelength that can be calculated using the equation: $\lambda = (6.63 \times 10^{-34}$ J•s)/(m × v), where m is the mass of the electron (in kg) and v is the velocity (in m/s). What is the best resolution of an electron microscope using electrons traveling at 6.0×10^{7} m/s?

25. Unlike what is commonly shown on TV, bloodstains do not illuminate as bright spots under UV light without first applying a chemical compound. In order to observe fluorescent spots, a solution of Luminol and an oxidizing agent must first be sprayed on the blood spots.
    a. Draw the structure of Luminol and name the various functional groups present in its chemical structure.
    b. What component(s) of blood catalyze the fluorescence of Luminol?
    c. What is the purpose of the oxidizing agent in the Luminol solution?

26. A forensic scientist sprayed a Luminol solution on a pair of pants found at a suspect's home. Under UV light, the dark stains on the denim appeared as bright blue fluorescent spots.
    a. Are these spots definitely composed of blood? Explain.
    b. Can DNA testing still be performed on bloodstains after they have been sprayed with a luminol solution?

27. Watch an episode of *Crime Scene Investigation* (*CSI*) and list at least three inaccuracies of their investigation relative to real-world forensic investigations.

28. Careless crime scene protocols will affect the outcome of an entire criminal case. Perhaps one of the most famous instances of such negligence was found for responders to the murders of Nicole Brown Simpson and Ronald Goldman. Research this case and list three anomalies that caused jurors to question the validity of evidence collected at the crime scene.

29. In a car accident investigation, describe a technique that could determine if the drivers had their headlights illuminated before the crash.

30. In earlier chapters, we described the investigation of data related to global climate change. Can climatology be classified as forensic science? Explain.

Appendix 1

Measure for Measure

Metric Prefixes, Conversion Factors, and Constants

Metric Prefixes

Prefix	Symbol	Value	Scientific Notation
pico	p	$1/10^{12}$ or 0.000000000001	10^{-12}
nano	n	$1/10^{9}$ or 0.000000001	10^{-9}
micro	µ	$1/10^{6}$ or 0.000001	10^{-6}
milli	m	1/1000 or 0.001	10^{-3}
centi	c	1/100 or 0.01	10^{-2}
deci	d	1/10 or 0.1	10^{-1}
deka/deca	da	10	10^{1}
hecto	h	100	10^{2}
kilo	k	1000	10^{3}
mega	M	1,000,000	10^{6}
giga	G	1,000,000,000	10^{9}
tera	T	1,000,000,000,000	10^{12}

Conversion Factors

Length

1 centimeter (cm) = 0.394 inches (in.)

1 meter (m) = 39.4 inches (in.) = 3.28 feet (ft)
= 1.08 yards (yd)

1 kilometer (km) = 0.621 miles (mi)

1 inch (in.) = 2.54 centimeters (cm) = 0.0833 feet (ft)

1 foot (ft) = 30.5 centimeters (cm) = 0.305 meters (m)
= 12 inches (in.)

1 yard (yd) = 91.44 centimeters (cm) = 0.9144 meters (m)
= 3 feet (ft) = 36 inches (in.)

1 mile (mi) = 1.61 kilometers (km)

Volume

1 cubic centimeter (cm^3) = 1 milliliter (mL)

1 liter (L) = 1000 milliliters (mL)
= 1000 cubic centimeters (cm^3)
= 1.057 quarts (qt)

1 quart (qt) = 0.946 liter (L)

1 gallon (gal) = 4 quarts (qt) = 3.78 liters (L)

Mass

1 gram (g) = 0.0352 ounces (oz) = 0.00220 pounds (lb)

1 kilogram (kg) = 1000 grams (g) = 2.20 pounds (lb)

1 pound (lb) = 454 grams (g) = 0.454 kilograms (kg)

1 tonne or metric ton (t) = 1000 kilograms (kg)
= 2200 pounds (lb)
= 1 long ton (t) = 1.10 tons (T)

1 ton (T) = 909 kilograms (kg)
= 2000 pounds (lb)
= 1 short ton (T)
= 0.909 tonnes (t)

Time

1 year (yr or y) = 365.24 days (d)

1 day (d) = 24 hours (hr or h)

1 hour (hr or h) = 60 minutes (min)

1 minute (min) = 60 seconds (s)

Energy

1 joule (J) = 0.239 calories (cal)

1 calorie (cal) = 4.184 joules (J)

1 exajoule (EJ) = 10^{18} joules (J)

1 kilocalorie (kcal) = 1 dietary Calorie (Cal)
= 4184 joules (J)
= 4.184 kilojoules (kJ)

1 kilowatt-hour (kWh) = 3,600,000 joules (J)
= 3.60×10^{6} J

Constants

Speed of light (*c*) = 3.00×10^{8} meters per second (m/s)

Planck's constant (*h*) = 6.63×10^{-34} joule-seconds (J · s)

Avogadro's number = 6.02×10^{23} objects per mole (objects/mol)

Unified atomic mass unit (u) = 1 Dalton (Da)
= 1.67×10^{-27} kg

Appendix 2

The Power of Exponents

Scientific (or exponential) notation provides a compact and convenient way of writing very large and very small numbers. The idea is to use positive and negative powers of 10. Positive exponents are used to represent large numbers. The exponent, which is written as a superscript, indicates how many times 10 is multiplied by itself. For example,

$$10^1 = 10$$
$$10^2 = 10 \times 10 = 100$$
$$10^3 = 10 \times 10 \times 10 = 1000$$

Note that the positive exponent is equal to the number of zeros between the 1 and the decimal point. Thus, 10^6 corresponds to 1 followed by six zeros, or 1,000,000. This same rule applies to 10^0, which equals 1. One billion, 1,000,000,000, can be written as 10^9.

When 10 is raised to a negative exponent, the number being represented is always less than 1. This is because a negative exponent implies a reciprocal, that is, 1 over 10 raised to the corresponding positive exponent. For example,

$$10^{-1} = 1/10^1 = 1/10 = 0.1$$
$$10^{-2} = 1/10^2 = 1/100 = 0.01$$
$$10^{-3} = 1/10^3 = 1/1000 = 0.001$$

It follows that the larger the negative exponent, the smaller the number. The negative exponent is always one more than the number of zeros between the decimal point and the 1. Thus, 1×10^{-4} is equal to 0.0001. Conversely, 0.000001 in scientific notation is 1×10^{-6}.

Of course, most of the quantities and constants used in chemistry are not simple whole-number powers of 10. For example, Avogadro's number is 6.02×10^{23}, or 6.02 multiplied by a number equal to 1 followed by 23 zeros. Written out, this corresponds to $6.02 \times$ 100,000,000,000,000,000,000,000, or 602,000,000,000,000,000,000,000. Switching to very small numbers, a wavelength at which carbon dioxide absorbs infrared radiation is 4.257×10^{-6} m. This number is the same as 4.257×0.000001, or 0.000004257 m.

Your Turn — Skill Building: Scientific Notation

Express these numbers in scientific notation.

a. 10,000 **b.** 430
c. 9876.54 **d.** 0.000001
e. 0.007 **f.** 0.05339

Answers

a. 1×10^4 **b.** 4.3×10^2
c. 9.87654×10^3 **d.** 1×10^{-6}
e. 7×10^{-3} **f.** 5.339×10^{-2}

Your Turn — Skill Building: Decimal Notation

Express these numbers in conventional decimal notation.

a. 1×10^6 **b.** 3.123×10^6
c. 2.5×10^4 **d.** 1×10^{-5}
e. 6.023×10^{-7} **f.** 1.723×10^{-16}

Answers

a. 1,000,000 **b.** 3,123,000
c. 25,000 **d.** 0.00001
e. 0.0000006023 **f.** 0.0000000000000001723

Appendix 3

Clearing the Logjam

You may have encountered logarithms in mathematics courses but wondered if you would ever use them. In fact, logarithms (or "logs" for short) are extremely useful in many areas of science. The essential idea is that they make it much easier to deal with very large *ranges* of numbers, for example, moving by powers of 10 from 0.0001 to 1,000,000.

It is likely that you have met logarithmic scales without necessarily knowing it. The Richter scale for expressing magnitudes of earthquakes is one example. On this scale, an earthquake of magnitude 6 is 10 times more powerful than one of magnitude 5. An earthquake of magnitude 8 would be 100 times more powerful than one of magnitude 6. Another example is the decibel (dB) scale. Each increase of 10 units represents a 10-fold increase in sound level. Therefore, a normal conversation between two people 1 m apart (60 dB) is 10 times louder than quiet music (50 dB) at the same distance. Loud music (70 dB) and extremely loud music (80 dB) are 10 times and 100 times louder than a normal conversation, respectively.

A simple exercise using a pocket calculator can be a good way to learn about logs. You will need a calculator that "does" logs and preferably has a "scientific notation" option. Start by finding the logarithm of 10. Simply enter 10 and press the "log" button. The answer should be 1. Next, find the log of 100 and then the log of 1000. Write down the answers. What pattern do you see? This pattern may be more obvious if you recall that 100 can be written as 10^2 and 1000 is the same as 10^3. Predict the log of 10,000 and then check it out. Then, try the log of 0.1, or 10^{-1}, and the log of 0.01 or 10^{-2}. Predict the log of 0.0001 and check it out.

So far so good, but we have been considering only whole-number powers of 10. It would be helpful to be able to obtain the logarithm of *any* number. Once again, your handy little calculator comes to the rescue. Try calculating the logs of 20 and 200, then 50, or 5×10^1, and 500, or 5×10^2. Predict the log of 5×10^3, or 5000. Now for something slightly trickier: the log of 0.05. Finally, try the log of 2473 and the log of 0.000404. In each of the three cases, does the answer seem to be in the right ballpark? Remember that your calculator will happily provide you with many more digits than have any meaning, so you will need to do some reasonable rounding.

In Section 8.10, the concept of pH is introduced as a quantitative way to describe the acidity of a substance. A pH value is simply a special case of a logarithmic relationship. It is defined as the negative of the logarithm of the H^+ concentration, expressed in units of molarity (M). Square brackets are used to indicate molar concentrations. The mathematical relationship is given by the equation $pH = -\log [H^+]$. The negative sign indicates an inverse relationship; as the H^+ concentration decreases, the pH increases. Let us apply the equation by using it to calculate the pH of a beverage with a hydrogen ion concentration, or $[H^+]$, of 0.000546 M. We first set up the mathematical equation and substitute the hydrogen ion concentration into it.

$$pH = -\log [H^+] = -\log (5.46 \times 10^{-4} M)$$

Next, we take the negative logarithm of the H^+ concentration by entering it into a calculator and pressing the log button, then the "plus/minus" button to change the sign. It may display 3.262807357 if you have not preset the number of digits, but common sense prompts you to round the displayed value. That is, the number of *significant figures* (revisit Section 1.6) in the concentration will define the number of *decimal places* in the pH value, and vice versa. For instance, the pH of a solution with $[H^+] = 5.46 \times 10^{-4}$ M (3 sig figs) should be reported as 3.263 (3 figures after the decimal place). Apply the same procedure to calculate the pH of milk with a hydrogen ion concentration of 2.2×10^{-7} M (pH = 6.66).

If we can convert hydrogen ion concentration into pH, how do we go in the reverse direction, that is, how to convert pH into a hydrogen ion concentration? Your calculator can do this for you if it has a button labeled "10^x." Alternatively, it may use two buttons: first "Inv" and then "log." To demonstrate the procedure, suppose you wish to find the hydrogen ion concentration of human blood with a pH of 7.40. Proceed as follows: Enter 7.40, use the "plus/minus" button to change the sign to negative, and then hit 10^x (or follow whatever steps are appropriate for your calculator). The display should give the hydrogen ion concentration as 4.0×10^{-8} M, using the appropriate number of significant figures. Now apply the same procedure to calculate the H^+ concentration of an acid rain sample with a pH of 3.60.

Your Turn — Skill Building: pH

Find the pH concentration in each sample.

a. tap water, $[H^+] = 1.0 \times 10^{-6}$ M

b. milk of magnesia, $[H^+] = 3.2 \times 10^{-11}$ M

c. lemon juice, $[H^+] = 5.0 \times 10^{-3}$ M

d. saliva, $[H^+] = 2.0 \times 10^{-7}$ M

Answers

a. 6.00 **b.** 10.49

c. 2.30 **d.** 6.70

Your Turn — Skill Building: H^+ Concentration

Find the H^+ concentration in each sample.

a. tomato juice, pH = 4.5.

b. acid fog, pH = 3.3.

c. vinegar, pH = 2.5.

d. blood, pH = 7.6.

Answers

a. 3.2×10^{-5} M **b.** 5.0×10^{-4} M

c. 3.2×10^{-3} M **d.** 2.5×10^{-8} M

Appendix 4

Answers to Your Turn Questions

Chapter 1

REFLECTION

a. Answers will vary; some possibilities are lightweight, thin, fast, small dimensions, etc.

b. Again, answers will vary. Lithium found in batteries is located in Chile, Australia, and other countries in a variety of lithium-containing minerals. Plastics are fabricated in most countries, and are derived from petroleum-based starting materials (see Ch. 9). Glass is also fabricated in most countries and is derived from sand.

c. Answers may vary; a possibility is Si and O_2 to form SiO_2, the composition of glass.

d. The average life of a smartphone in the U.S. is currently estimated to be 21 mo.

1.1 **a.** A hair fiber can vary in size but is commonly referenced as being 10 μm or 0.1 mm in diameter.

b. Dividing the length of your phone in mm by .1 mm will give you the number of hairs it would take to equal the length of a cell phone. For example, it would take 1,380 hairs to span a phone that was 138 mm long.

1.2 Most of the listed materials would not elicit a response from a touchscreen. However, objects that conduct electricity such as a battery will elicit a response.

1.3 Elements with measurable abundances that are found in a cell phone include: H, Li, Be, C, N, O, S, Mg, Al, Ti, V, Mn, Cd, Fe, Co, Si, Cu, Zn, As, Nb, Mo, Ag, Sn, Sb, Ba, Ta, W, Au, Pb, and Ni.

1.4 **a.** Some common macroscopic objects with dimensions on the order of …

(i) millimeters: the length of an ant, the width of a notebook

(ii) centimeters: the length of a pencil, the diameter of a pizza

(iii) meters: the length of a car, the width of a football field

b. A cell phone that is 138 mm × 67 mm × 7 mm is 13.8 cm or .138 m long, 6.7 cm or .067 m wide, and .07 cm or .007 m thick.

1.5 **a.** Heterogenous mixture

b. Element (atom)

c. Homogenous mixture (although many wines have sediments suspended in them)

d. Element (molecule)

e. Homogenous mixture

f. Ionic compound

g. Molecular compound

1.6 **a.** Taking the mass of one Ti atom over the total mass of Ti and O atoms equals 47.88/79.88 = 59.93% Ti and taking the mass of two O atoms over the total mass of Ti and O atoms equals 32.00/79.88 = 40.06% O

b. 63.19% Mn; 36.81% O

c. 79.89% Cu; 20.11% O

1.7 You can eliminate alumina because there is no hydrogen in that compound. To determine whether it is boehmite or gibbsite, you could determine the atomic percentages of aluminum and oxygen in each compound and match those to the atomic percentages you determined from the unknown compound.

1.8 **a.** 31 protons and 31 electrons

b. 50 protons and 50 electrons

c. 82 protons and 82 electrons

d. 26 protons and 26 electrons

a. 2 protons, 2 electrons, 0 neutrons

b. 24 protons, 24 electrons, 28 neutrons

c. 13 protons, 13 electrons, 14 neutrons

d. 33 protons, 33 electrons, 42 neutrons

1.9 One way to design a stylus would be to wrap aluminum foil around a Q-tip® that has been dampened with water. The electrically conductive foil and water allow the current to flow and the touchscreen to respond.

1.10 Scandium and yttrium are found in naturally occurring minerals with other elements. Originally, these rare earth metals were found in minerals in Scandinavian countries, but now they are found in minerals all over the world. The mineral known to contain the most scandium is pretulite ($ScPO_4$), which is found in Austria. The mineral with the most yttrium is iimoriite ($Y_2(SiO_4)(CO_3)$), which is found in Japan.

1.11 **a.** Four significant figures, trailing zeroes that follow a non-zero digit and decimal point are significant.

b. Three significant figures, all zeroes embedded between non-zero digits are significant.

c. One significant figure, leading zeroes placed ahead of non-zero digits are not significant.

d. Four significant figures, all zeroes embedded between non-zero digits are significant.

1.12 **a.** 5.0 g (two significant figures) ÷ 0.031 mL (two significant figures) = 16 g/mL. For division, answer should be reported to the least amount of significant figures used (two).

b. 15.0 m (three significant figures) × 0.003 m (one significant figure) = .05 m^2. For multiplication, answer should be reported to the least amount of significant figures used (one).

c. 1.003 g (three decimal places) + 0.01 g (two decimal places) = 1.01 g. For addition, answer should be reported to the least number of decimal places used (two), not the least number of significant figures used (one).

d. 1.000 mL (three decimal places) – 0.1 mL (one decimal place) = 0.9 mL. For subtraction, answer should be reported to the least number of decimal places used (one), not the least number of significant figures used (one), although in this case the answer also only has one significant figure.

1.13 **a.** Zn^{2+}, O^{2-}

b. Cu^{2+}, Cl^{-}

c. Ca^{2+}, S^{2-}

d. Ti^{4+}, O^{2-}

1.14 **a.** Desirable properties for a cell phone might include a battery with a longer life or a screen that is indestructible.

b. Incorporating elements like lithium could help make a phone with a longer battery life. Using aluminum and oxygen in the form of sapphire glass could help make a more durable screen.

1.15. **a.** P^{5+} O^{2-}

b. Cl^{0}

c. Zn^{0}

d. C^{4+} O^{2-}

e. S^{6+} F^{-}

1.16 **a.** U.S. National debt = 1.94×10^{13} dollars (as of July 2016)
World population = 7.4×10^{9} people

b. Length: 1.38×10^{2} mm, 1.38×10^{1} cm, or 1.38×10^{-1} m
Width: 6.7×10^{1} mm, 6.7×10^{0} cm, or 6.7×10^{-2} m
Thickness: 7×10^{0} mm, 7×10^{-1} cm, or 7×10^{-3} m

1.17 **a.** Transmitted

b. Reflected

c. Reflected

d. Reflected

e. Absorbed

f. Reflected

g. Absorbed

1.18 Aluminum-based frames are less dense than iron/steel-based frames, and that translates to a more fuel-efficient vehicle and reduces the amount of carbon emissions. Additionally, aluminum is infinitely recyclable, meaning aluminum used in cars now can be recycled many times without any loss of functionality.

1.19 Other than charging, your smartphone uses energy every time you search the Internet, watch a video, or use social media. Students should also describe the implications of increased Internet traffic (data centers, large supercomputing facilities, servers, etc.).

1.20 While modern cell phones require a lot more energy to produce than older versions, it still takes more energy to produce PCs, with many more complex computer chips and interconnected parts. Further, it takes significantly more energy to operate PCs relative to cell phones. If we replace "energy hogs" of PCs, TVs, and gaming consoles with smartphones, less energy will be spent for fabrication as well as use over their lifetimes. This, of course, assumes that cell phones will be used as replacements, and consumers won't simply keep adding to their collections of electronic "toys."

1.21 The Aluminum Association (www.aluminum.org) has published several statistics regarding the cost and energy that is saved by recycling aluminum as opposed to aluminum mining from ore. For example, recycling aluminum saves more than 90% of the energy that would be needed to create a comparable amount of the metal from raw materials.

Chapter 2

REFLECTION

a. Possible answers are indoor: paint, perfumes, deodorants, cooking, incense; outdoor: flowers, decaying leaves, plastics from a hot dish in the summertime, cooking (*e.g.*, bbq)

b. Most of these chemicals are harmless to human health; however, some people may have allergies toward certain chemicals that are quite serious in some instances.

2.1 Answers will vary, depending on the answers to the previous activity. At this stage, students will likely indicate that they don't yet understand the effects of the chemicals emitted into the air. However, after the chapter is completed, they will have a better idea regarding their effects to health and the environment. Basically, each of those indicated above will not have an appreciable effect on air quality, because they are emitted in such small concentrations. However, if pollutants are selected that were mentioned in the introductory video such as ozone, then it should be indicated that they will have a negative effect on air quality.

2.2 $\frac{0.5\text{ L}}{\text{breath}} \times \frac{12\text{ breaths}}{\text{min}} \times \frac{60\text{ min}}{\text{hr}} \times \frac{24\text{ hr}}{\text{day}} = \times \frac{7{,}000\text{ L}}{\text{day}}.$

The estimate I used was from a doctor's website, which estimated the average person in a resting state inhales .5 L in each breath. I estimated that I take about 12 breaths in one minute. Activities such as exercise, anxiety, or sleep may cause the quantity to change.

2.3 We inhale a mixture of nitrogen, oxygen, argon, carbon dioxide, water, and other gases in trace amounts. We exhale a mixture of the same chemicals although the relative amounts of each chemical change; most notably the oxygen quantity decreases and the carbon dioxide quantity increases.

2.4 **a.** Other chemicals you might smell in the air include cleaning products, rotting food, a burning candle, flowers blooming, freshly mowed grass, or the smell after a rain storm.

b. Smells that might alert us to a hazard include smoke, which is a mixture of gases released when something is burned; mercaptan, an additive to natural gas that gives it its "rotten egg" smell to alert you to a possible gas leak in your home; and acrolein, a toxic chemical produced when fats and cooking oils are heated for prolonged periods of time at high temperatures.

2.5 With an increased amount of oxygen in the atmosphere, corrosion would occur more quickly and combustion reactions would occur more readily and burn more efficiently. While this may not seem detrimental in all aspects (think better gas mileage!), the global ramifications of this change would be enormous; things that previously only got hot or smoked, like burnt toast, would now readily catch on fire.

2.6
- one second in nearly 12 days (1,036,800 seconds)
- one step in a 475 mile journey

$$\frac{1\text{ step}}{2.5\text{ ft}} \times \frac{5280\text{ ft}}{1\text{ mile}} \times 568\text{ miles} = 1{,}199{,}616\text{ steps}$$

- four drops of ink in a 55-gallon barrel of water

$$55\text{ gal} \times \frac{3785\text{ mL}}{1\text{ gal}} \times \frac{20\text{ drops}}{1\text{ mL}}$$
$$= 4{,}163{,}500\text{ drops of water}$$

So all are pretty fair estimations.

2.7 **a.** 9 ppm is equal to 0.0009%.

b. 78% nitrogen is equal to 780,000 ppm.

2.8 **a.** NO, NO_2, N_2O, N_2O_4

b. SO_2 is sulfur dioxide and SO_3 is sulfur trioxide.

2.9 **a.** The *eth-* in ethanol indicates 2 C atoms in the carbon chain.

b. The *meth-* in methylene indicates 1 C atom in the carbon chain.

c. The *prop-* in propane indicates 3 C atoms in the carbon chain.

2.10 For ideas, return to this chapter's opening video. Exhale. What are you breathing out? Due to the production of portable electronics (discussed in Chapter 1) and other consumer products, our air is not as clean as it once was. How do you think the pollutants created from the portable electronics production process could impact your health? When we breathe we are taking in both the expected substances that make up air (nitrogen, oxygen, argon, carbon dioxide, and water) as well as any other gases and fine particulates that may be present.

2.11 **a.** O_3. This is a hard call, as no common exposure period exists on which to base the comparison. Clearly, CO is not the most toxic, as all its standards are higher. It is not NO_2, because SO_2 has a lower 1-hr average standard. Between SO_2 and O_3, ozone has the stricter standard because of the lower 8-hr average in comparison to the 3-hr average standard for sulfur dioxide.

b. Claim is supported because the levels of exposure for $PM_{2.5}$ are lower than for PM_{10}.

2.12 **a.** She would not exceed the 1-hr limit of 210 $\mu g/m^3$. 44 μg per hour/0.625 m^3 of air per hour only equals an exposure of 70. $\mu g/m^3$.

b. At the same inhalation rate, she would not surpass the 1-hr or 3-hr rates of 210 mg/m^3 or 1300 mg/m^3. If the smelter is releasing 44 μg/hour, her rate of inhalation is still 70 $\mu g/m^3$ in each hour.

2.13 For the most part, the WHO has stricter standards for pollutants. In particular, the WHO says a SO_2 concentration of 500 $\mu g/m^3$ should not be exceeded over average periods of 10 minutes while the EPA says concentrations of 1300 $\mu g/m^3$ should not be exceeded over average periods of 3 hours. Source: http://www.who.int/mediacentre/factsheets/fs313/en/

2.14 Revisit Your Turn 2.1 on "air prints." Examples of preventing air pollution include (1) not burning leaves (produces smoke and particulate matter), but rather letting them compost or otherwise decompose, (2) using renewable energy sources such as geothermal or solar power to heat your home; also, burning low-sulfur coal rather than high-sulfur coal, using a scrubber to remove SO_2 if burning high-sulfur coal, or conserving so as to burn less coal of any type, (3) choosing methods of transportation, such as bicycling or walking, that do not release air pollutants.

2.15 Answers will vary, depending on location and weather.

2.16 **a.** $2\ H_2 + O_2 \longrightarrow 2\ H_2O$

b. $N_2 + 2\ O_2 \longrightarrow 2\ NO_2$

2.17 Equation 2.6 contains 16 C, 36 H, and 50 O on each side. Equation 2.7 contains 16 C, 36 H, and 34 O on each side.

2.18 **a.** For x = 1, nitrogen monoxide (NO) is the emission of interest. For x = 2, nitrogen dioxide (NO_2) is the emission of interest.

b. Nitrogen monoxide is produced when nitrogen and oxygen atoms from air react in the high pressure and temperature conditions of an engine.

c. In the year that this graph was produced, CO_2 was not classified as an air pollutant in the United States. Accordingly, it has no green line indicating an acceptable range. While there are many efforts currently underway to reduce greenhouse gas emissions from motor vehicles, there are currently no standards in place.

2.19 Your list should include: O_2, N_2, CO_2, CO, H_2O, NO, soot (particulate matter), and VOCs. The exhaust also contains tiny amounts of Ar and even tinier amounts of He, but we usually omit these gases as they are inert and low in concentration.

2.20 **a.** $Ag_2S(s) + O_2(g) \longrightarrow 2\ Ag(s) + SO_2(g)$

b. $CuS(s) + O_2(g) \longrightarrow Cu(s) + SO_2(g)$

2.21 **a.** Other gasoline-powered machines or vehicles include some lawn mowers, leaf blowers, forklifts, chain saws, snow blowers, and electrical generators.

b. One example is lawn and garden equipment, such as mowers and blowers. In 2008, the U.S. EPA issued more stringent exhaust standards and established new evaporative emission standards for the fuel tanks and fuel lines used in engines of this type. The new regulations were fully implemented for all lawn and garden equipment beginning in 2012.

2.22 Driving practices that conserve fuel are driving slower (maximizing the fuel efficiency of your vehicle), driving on the highway, rolling the windows down instead of using air conditioning, and turning the car off while parked. Conversely, driving practices that expend more fuel than necessary include idling stationary for long periods of time, speeding, running the air conditioning, and driving in congested city traffic.

2.23 **a.** The colors indicate the relative safety of the air in those regions. Green is good, yellow is moderate, orange is unhealthy for sensitive groups, and red is unhealthy.

b. Some groups at the highest risk from particle pollution are children, the elderly, people who work outdoors, and those with chronic diseases such as asthma, emphysema, heart disease, and diabetes.

c. Answers will vary based on location.

2.24 **a.** The air is hazardous to one or more groups in red and orange areas. Cities closest to these areas are Los Angeles and Sacramento.

b. The ozone level peaks around 5:00 pm.

c. Yes. Once the sun sets, the ozone levels drop. Sunlight and heat are necessary for ozone to result from VOCs and NO_x. Some ozone may linger, but peak levels are experienced when the sun is shining.

2.25 An example of this might be: The process of ozone formation begins with inefficient combustion in a car engine. Two products of this inefficient combustion are NO and VOCs. Over time, NO is converted to NO_2 through reaction with the produced VOCs and ·OH. NO_2 reacts with sunlight to form NO and free O atoms, which can then react with O_2 in the atmosphere to make O_3.

2.26 **a.** No, at this level the ozone concentration will pose no risk to a healthy person who wishes to exercise outside.

b. Answers will vary depending on location.

2.27 **a.** A concentration of 1943 μg of particulate matter per cubic meter of air exceeds the National Ambient Air Quality Standards for both PM_{10} and for $PM_{2.5}$.

b. Breathing fine particles at this level is hazardous for everybody. The primary danger is to the cardiovascular system, as the particles, when inhaled, pass into the bloodstream and cause or further aggravate heart disease.

2.28 Indoor activities that generate pollutants include burning incense, painting or varnishing (except if using low-VOC paint), cigarette or cigar smoking, frying foods (especially when something burns), using some cleaning products such as ammonia or spray oven cleaner, using aerosol hair sprays and some hair-coloring products, using some furniture polishes, or using spray insecticides.

Chapter 3

REFLECTION

Possible answers are protective clothing, sunscreens, sunblocks, umbrella, shade from a tree, building, etc. The most effective will be those that don't allow any sun rays to contact skin (clothing); less effective will be sunscreens (effectiveness depends on how much the user puts on and if he/she covers all exposed skin areas).

3.1 Answers will vary based on student data.

3.2 **a.** The sun emits energy in the form of ultraviolet, infrared, and visible light radiation.

b. Being out in the sun too long causes painful burns and blisters on the surface of the skin. It can only be assumed that damage is also occurring under the skin that we cannot see. The UV light has shorter wavelengths than visible or IR rays. Our skin and eyes are sensitive to these photons and can become damaged by this form of light energy.

c. Lasting effects of sun damage are evidenced by wrinkled, leathery skin, and the presence of solar lentigines (sun spots) on the skin. Also, cataracts and other eye problems can develop from prolonged UV exposure.

d. The sun is necessary for the skin's production of vitamin D, a nutrient that is necessary for your body to be able to absorb calcium.

3.3 **a.** $525 \text{ nm} \times \dfrac{1 \times 10^{-9} \text{ m}}{1 \text{ nm}} = 5.25 \times 10^{-7} \text{ m};$

$$\frac{3.00 \times 10^{8} \text{ m}\cdot\text{s}^{-1}}{5.25 \times 10^{-7} \text{ m}} = 5.71 \times 10^{14} \text{ s}^{-1}$$
$$= 5.71 \times 10^{14} \text{ Hz}$$

b. $5.71 \times 10^{14} \text{ s}^{-1} \times \dfrac{60 \text{ s}}{\text{min}} = 3.43 \times 10^{16} \text{ waves/min};$

$$\frac{3.43 \times 10^{16} \text{ waves}}{\text{min}} \times \frac{60 \text{ min}}{\text{hr}}$$
$$= 2.06 \times 10^{18} \text{ waves/hr}$$

c. Amplitude is the height of the wave. It has nothing to do with the wavelength, frequency, nor speed of the wave; it has to do with the intensity of the wave.

3.4 **a.** Red light has the longest wavelength at 700 nm (and therefore the lowest frequency) meaning violet light has the highest frequency with a wavelength of 400 nm.

b. 500 nm is equal to 500×10^{-9} m, which in proper scientific notation is 5×10^{-7} m.

3.5 **a.** In order of increasing wavelength: ultraviolet radiation < visible radiation < infrared radiation < microwave radiation.

b. A radio wave is on the order of 10^{1} m, while an X-ray is on the order of 10^{-10} m. This is a difference of 12 orders of magnitude, which means X-rays are 10^{12} times as energetic as radio waves!

3.6 **a.** The greatest portion of energy from the sun reaches Earth as infrared radiation.

b. The most intense radiation emitted by the sun is visible light with a wavelength around 500 nm (green visible light).

3.7 Using Figure 3.4, red light emits around 700 nm and blue light emits around 475 nm. (Although any value between 620–750 nm for red light and between 450–495 nm for blue light is correct.)

Using Equation 3.3:

$$E_{\text{red light}} = \frac{(6.626 \times 10^{-34} \text{ J}\cdot\text{s})(3.00 \times 10^{8} \text{ m/s})}{700 \times 10^{-9} \text{ m}}$$
$$= 2.84 \times 10^{-19} \text{ J}$$

$$E_{\text{blue light}} = \frac{(6.626 \times 10^{-34} \text{ J}\cdot\text{s})(3.00 \times 10^{8} \text{ m/s})}{475 \times 10^{-9} \text{ m}}$$
$$= 4.18 \times 10^{-19} \text{ J}$$

While these values are both incredibly small, compared to one another, blue light emits 150% more energy than red light emits.

No, blue light could not have been used. Blue light is too energetic and would overexpose the images being developed. The low-energy red light allows for more control of developing the images.

3.8 **a.** In order of increasing wavelength UVC < UVB < UVA

b. No, the numerator in Equation 3.3 is a constant, making energy inversely proportional to wavelength.

c. No, UVC radiation is absorbed by O_2 and O_3 in the stratosphere prior to reaching Earth's surface. Sunscreen needs to protect against UVA and UVB radiation that reaches Earth's surface. In fact, UVC is used to sterilize medical equipment by killing all bacteria on the surface.

3.9 320 nm/242 nm = 1.32

A 242-nm photon has roughly 130% more energy than a 320-nm photon!

OR

Using Equation 3.3:

$$E_{\text{242-nm photon}} = \frac{(6.626 \times 10^{-34} \text{ J}\cdot\text{s})(3.00 \times 10^{8} \text{ m/s})}{242 \times 10^{-9} \text{ m}}$$
$$= 8.21 \times 10^{-19} \text{ J}$$

$$E_{\text{320-nm photon}} = \frac{(6.626 \times 10^{-34} \text{ J}\cdot\text{s})(3.00 \times 10^{8} \text{ m/s})}{320 \times 10^{-9} \text{ m}}$$
$$= 6.21 \times 10^{-19} \text{ J}$$

$$\frac{8.21 \times 10^{-19} \text{ J}}{6.21 \times 10^{-19} \text{ J}} = 1.32$$

Using these calculations, a 242-nm photon still has roughly 130% more energy than a 320-nm photon!

3.10 **a.** The National Cancer Institute's Surveillance, Epidemiology, and End Results Program provides statistics for cancer incidence and mortality-rate trends in the U.S. For melanoma, the incidence rate has more than doubled since 1975, while the mortality rate has remained essentially constant. While incidence rates of other cancers such as breast and prostate cancer are also higher than they were 40 years ago, incidence rates of colorectal cancer have declined dramatically over the past 20 years. Additionally, mortality rates of all three cancers have dropped in the past 40 years and five-year survival rates have increased.

b. Lighter skin has less melanin, which provides a natural protection against the harmful effects of ultraviolet radiation.

c. You should think not only about protecting yourself for short-term UV exposure such as for a day at the beach, but also for UV exposure that builds up over time. For example, according to the Skin Cancer Foundation, UV exposure we receive driving in the car or shining through office windows can lead to significant skin damage over time. To avoid this low but constant exposure, it's advisable to have a UV protection on car or office windows, or to avoid direct sun exposure, even when indoors.

3.11 A well-known example of something that is required for good health but dangerous in high quantities is water. Water is necessary to regulate body temperature and to maintain a number of important processes within your body. However, drinking too much water too quickly can lead to serious complications and death.

3.12 **a.** In the United States, the UV index is higher in summer than it is in winter. This is due to the fact that the Sun is at a steeper angle and there are more hours of daylight.

b. During the summer months, the UV index is higher closer to the equator in states like Hawaii and gets lower as you get farther away from the equator. This is because the Sun's rays are most intense closest to the equator.

3.13 The National Conference of State Legislatures website (www.ncsl.org) provides a state-by-state comparison of current legislation for minors using tanning beds.

3.14 **a.** Perhaps in the discussion of climate change students have heard of ozone depletion or the hole in the ozone layer. In the 1990s, the ozone hole was a topic covered heavily in the media, whereas in recent times the phrases "global warming" and "climate change" are more of the focus.

b. The ozone layer is located within the stratosphere and is made up of both O_2 and O_3 molecules.

3.15 **a.** The approximate altitude of maximum ozone concentration is 23 km (14 miles) above sea level.

b. The highest concentration of ozone in the stratosphere is 12,000 ozone molecules per billion molecules and atoms of all types, which is 12,000 ppb.

c. In ambient air, ozone levels can be between 20 and 100 ppb, or more. The EPA suggests limiting your ozone exposure to 75 ppb in an 8-hr period.

3.16 **a.** The average area of the ozone hole over Antarctica in 2015 was 25.6 million km^2. This is the highest recorded value for the area of the "hole" since 2006.

b. The mean lowest reading observed for ozone was 116.5 DU. This is the lowest recorded value since 2011.

3.17 **a.** 1 H atom × 1 valence electron per atom = 1 valence electron

1 Br atom × 7 valence electrons per atom = 7 valence electrons

Total = 8 valence electrons

Here is the Lewis structure:

H—B̤̈r:

b. 2 Br atom × 14 valence electrons per atom = 14 valence electrons

Here is the Lewis structure:

:B̤̈r—B̤̈r:

3.18 **a.** H—S̤̈—H

b. :C̤̈l—C(:F̤̈:)(:F̤̈:)—C̤̈l: or :F̤̈—C(:F̤̈:)(:C̤̈l:)—C̤̈l:

3.19 **a.** :C≡O:

b. Ö=S̈—Ö̤: or :Ö̤—S̈=Ö

c. :Ö̤—S(=Ö:)—Ö̤: ↕ :Ö=S(—Ö̤:)—Ö̤: ⟷ :Ö̤—S(=Ö)—Ö̤:

3.20 Ozone concentration will vary based on the relative amount of sunlight a region receives. Because UV radiation is necessary for ozone production, it would be expected that higher concentrations of ozone would be found near the equator or in regions experiencing summer. Conversely, lower concentrations of ozone would be found at the poles or in regions experiencing winter.

3.21 **a.** The map is oriented over the Antarctic pole. This is because the depletion of the ozone happened most in this region of the atmosphere, so scientists have the most accurate data for this region.

b. 220 DU is the boundary between normal fluctuation of ozone levels and catalytic loss of ozone resulting in a "hole."

c. The ozone hole begins to appear in August, grows to its maximum in September, and slowly disappears throughout the month of December.

d. Regardless of the month selected (August–December), the ozone hole grows in size and the level of ozone decreases dramatically from 1979 until the present.

3.22 Since the late 2000s, ozone levels have (on average) rebounded since reaching all-time lows in the mid- to late 1990s. While these values are much lower than those observed in the late 1980s, it seems efforts to curb ozone destruction are making a difference. Given the recent trends, it could be predicted that ozone levels should continue to rise over the next three years. This prediction can be monitored using the Ozone Hole Watch website used in Your Turn 3.21.

3.23 Radical species are starred.

$$^{*}\cdot\ddot{\text{O}}\text{—H} + \text{H—CH}_2\text{—H} \longrightarrow \text{H—}\ddot{\text{O}}\text{—H} + {}^{*}\cdot\text{CH}_3$$

$$^{*}\cdot\text{CH}_3 + \ddot{\text{O}}\text{=}\ddot{\text{O}} \longrightarrow \cdot\ddot{\text{O}}\text{—}{}^{*}\ddot{\text{O}}\text{—CH}_3$$

$$\cdot\ddot{\text{O}}\text{—}{}^{*}\ddot{\text{O}}\text{—CH}_3 + {}^{*}\dot{\text{N}}\text{=}\ddot{\text{O}} \longrightarrow {}^{*}\cdot\ddot{\text{O}}\text{—CH}_3 + :\ddot{\text{O}}\text{—}{}^{*}\dot{\text{N}}\text{=}\ddot{\text{O}}$$

3.24 Having free radicals in the atmosphere is not only beneficial to the healthy functioning of the atmosphere but is necessary for the steady state reactions that occur in the atmosphere to continue.

3.25 From the figure, stratospheric ozone and stratospheric chlorine levels are inversely related. As chlorine levels in the stratosphere rise, ozone levels deplete by an equal amount. Conversely, when stratospheric chlorine is removed from the stratosphere, ozone levels increase proportionately. As shown in Equations 3.7 and 3.8, chlorine radicals react with ozone to create more chlorine radicals capable of destroying more ozone.

3.26 The message of this cartoon is satirical in nature because a youth uses something known to be harmful to the ozone layer—an aerosol can of paint—to spread a conservation message about the ozone layer. However, on a much deeper level, this cartoon highlights the importance that each of the more than 7 billion people inhabiting Earth should consider regarding the global implications of their actions.

3.27 While ozone levels have rebounded slightly in recent years, there is no data to suggest that the ozone hole is being repaired. The only evidence demonstrated in these figures is that ozone depletion was not observed until concentrations of reactive halogens in the atmosphere increased. Global initiatives to reduce the amounts of reactive halogens in the atmosphere are underway, and until reactive halogen concentrations in the stratosphere are drastically reduced, it is unknown if the ozone hole can be repaired.

3.28 Comparing the structures of the three molecules presented in this section, there are several surface features that account for their properties. Between HCFC-22 and HFC-32, both have a central carbon atom and two fluorine atoms. HCFC-22 has one hydrogen atom and one chlorine atom, while HFC-32 has two hydrogen atoms. The C–H and C–F single bonds in both atoms should account for their similar desirable qualities such as being more easily broken down in the lower atmosphere, a shorter atmospheric lifetime, and not depleting stratospheric ozone. The two C–H single bonds in HFC-32 as opposed to the one C–H bond and one C–Cl bond in HCFC-22 account for their differences in behavior—HCFC-22 contributes to ozone depletion, while HFC-32 contributes to global warming. Comparing these molecules to HFO-1234yf, the olefin-containing molecule has four C–F single bonds, one C=C double bond, and two C–H single bonds. The C–H and C–F single bonds again account for the similar desirable behavior as the others. However, the reactive C=C double bond shortens the atmospheric lifetime of the molecule even further.

3.29 One response to your friend to help demonstrate the influence we have over the environment could center on the idea that while there is some fluctuation to the ozone layer based on location and time of year, ozone levels over Antarctica dropped off significantly in the early and mid-1980s. Research into this observation in the late 1980s showed elevated levels of chlorine-containing oxides in the ozone hole and depleted ozone concentrations. Since reducing use and production of chlorine-containing compounds, the ozone levels have increased, but still aren't back to where the levels were in the 1970s.

3.30 **a.** To be classified as $PM_{2.5}$, the particle diameter must be no larger than 2.5 μm in diameter, so a dust particle that is 6 μm in diameter would be classified as PM_{10}.

b. Since 10^3 nm = 1 μm: $6\ \mu\text{m} \times \dfrac{1 \times 10^3\ \text{nm}}{1\ \mu\text{m}}$

$= 6 \times 10^3$ nm or 6,000 nm.

3.31 **a.** The UV index on a sunny summer day would probably fall in the high or very high category, depending on your location. Unprotected skin could burn in as few as 10–20 minutes on a day like this and even less if you have sensitive skin.

b. Since SPF refers to the ratio of time a person can stay in the sun without burning both with and without sunscreen, it can be calculated by multiplying the SPF by the number of minutes it would take you to get a sunburn. For example, if you typically develop a sunburn after 20 minutes of unprotected sun exposure, applying an SPF 70 sunscreen should give you 20 × 70 or 1400 minutes (23 hours) of protection. But this is assuming you don't sweat or lose the layer of protection in other ways.

c. While an SPF 70 sunscreen should in theory provide all-day protection from UV radiation, many guidelines are often overlooked. For example, SPF ratings are based on applying a much larger "dose" than most people apply. It is estimated that the average person is only getting 20–50% of the protection advertised on the label by not applying enough of the sunscreen. Additionally, it is recommended to reapply sunscreen every two hours to ensure the best protection.

3.32 **a.** The difference between SPF 15 and SPF 30 sunscreen is the amount of time one could theoretically remain in the sun without getting a sunburn. An SPF 30 sunscreen should provide twice the amount of time of protection than an SPF 15 sunscreen, and this is achieved by blocking more of the harmful UV rays emitted by the sun.

b. Choosing a higher SPF sunscreen allows for more time in the sun before developing a sunburn. The chances of getting a sunburn are lower with a higher SPF simply because you are more likely to reapply the sunscreen before the time of protection has "expired." Because the sunscreen shields your skin from more UV rays, you can avoid getting sunburned for a longer period of time.

3.33 **a.** The new FDA regulations required companies to provide more information on sunscreen bottles so that consumers can make a more informed decision on the best product to purchase. Examples of these changes include: specifically labeling whether a product protects against UVA or UVB, including warnings on sunscreens with an SPF between 2 and 14, and replacing the use of the words "waterproof" or "sweatproof" with "water resistant" and the specific number of minutes a person can expect to get the declared level of SPF protection while swimming or sweating.

b. The most common active ingredients in chemical-based sunscreens are oxybenzone, avobenzone, octisalate, octocrylene, homosalate, and oxtinoxate. The most common active ingredients in mineral-based sunscreens are zinc oxide and titanium dioxide.

3.34 The bigger the particle, the less protection because there are more holes between them for the UV radiation to travel. Just as sand grains are able to pack more tightly than marbles, smaller particles are more densely packed. The more dense the particles are packed, the less space between them, and the more protection they provide.

Chapter 4

REFLECTION

Based on the opinion of individual students. Students should be able to determine that CO_2 is from combustion of hydrocarbon fuels (gasoline, coal, etc.). It would be interesting for instructors to engage students with a discussion of "climate change" vs. "global warming."

4.1 It can be hypothesized that the condition of Earth as described in scenario **i** is the future of the planet if we do not take action to protect Earth and our atmosphere. By drastically reducing and eliminating harmful pollution, it is possible to begin to reverse the negative effects chemicals have had on the atmosphere and change the trajectory of Earth from scenario **i** to scenario **ii**.

4.2 In a superficial sense, students may only know that carbon is important to life in that "all living things have carbon." While this is true, it is important to emphasize that it is because of the properties carbon exhibits that it is so important. In addition to being the fourth most abundant element on the planet, carbon's unique structure allows it to bond in many ways and with many other elements. Carbon is the fundamental component for the macromolecules that make up proteins, nucleic acids (RNA and DNA), carbohydrates, and lipids in living organisms, as well as playing an important role in the structure of the food we use to sustain ourselves.

4.3 **a.** CaS, calcium sulfide

b. KF, potassium fluoride

c. MnO and MnO_2, manganese(II) oxide and manganese(IV) oxide (or manganese dioxide)

d. $AlCl_3$, aluminum chloride

e. $CoBr_2$ and $CoBr_3$, cobalt(II) bromide and cobalt(III) bromide

Note: for all ionic compounds, the metal (cation) comes first in the chemical formula.

4.4 **a.** Na_2SO_4

b. $Mg(OH)_2$

c. $Al(C_2H_3O_2)_3$

d. K_2CO_3

4.5 **a.** potassium nitrate

b. ammonium sulfate

c. sodium bicarbonate (or sodium hydrogen carbonate)

d. iron(II) sulfate

e. magnesium phosphate

4.6 **a.** NaClO

b. $MgCO_3$

c. NH_4NO_3

d. $Ca(OH)_2$

4.7 **a.** Burning fossil fuels and deforestation add CO_2 to the atmosphere. Other processes such as respiration and evaporation also add CO_2 to the atmosphere but are balanced out by photosynthesis and condensation.

b. While carbon dioxide is removed from the atmosphere through reforestation, photosynthesis, and precipitation, these processes do not result in a net loss of CO_2 from the atmosphere.

c. The largest carbon reservoirs are carbonate minerals in rocks and fossil fuels.

d. Growing populations have contributed to an increase in burning fossil fuels and deforestation. These actions directly increase atmospheric CO_2 levels, as well as deplete these important carbon reservoirs.

e. As discussed in Your Turn 4.2, the unique properties of carbon allow it to move to and from different reservoirs based on how it is bonded to other atoms and its location on Earth. For example, a carbon atom exhaled as a CO_2 molecule may be converted to starch ($C_6H_{12}O_6$) and stored in a plant through photosynthesis. The plant may be eaten and used as fuel by another organism or it may decompose and become part of the soil.

4.8 A 2013 *Science* publication estimated that during the years 2000–2012, 2.3 million km^2 of forest were lost globally, with only 800,000 km^2 of forest created. Tropical regions accounted for 32% of this loss and demonstrated a significant trend in forest loss which increased by 2101 square kilometers per year. While Brazil has drastically reduced rainforest destruction, this gain was offset by losses in Indonesia, Malaysia, Paraguay, Bolivia, Zambia, and Angola. Based on these trends, it can be assumed that without intervention, the rainforest will continue to be destroyed at an alarming rate. Source: High-resolution global maps of 21st century forest change cover (http://science.sciencemag.org/content/342/6160/850.full).

4.9 **a.** The atomic number of nitrogen is 7 and the atomic mass of nitrogen is 14.01 u.

b. A neutral atom of N-14 has 7 protons, 7 neutrons, and 7 electrons.

c. A neutral atom of N-15 has 7 protons, 8 neutrons, and 7 electrons.

d. An atomic mass of 14.01 means N-14 has the greatest natural abundance (>99%).

4.10 To verify these claims, let's start with marshmallows. The surface area of the United States is roughly 3.8×10^6 miles^2 and the dimensions of a single marshmallow are roughly 1 inch long × 1 inch wide (we will ignore height for now).

First, calculate how many marshmallows are in each square mile of the United States:

$$\frac{6.02 \times 10^{23} \text{ marshmallows}}{3.08 \times 10^6 \text{ mi}^2} = \frac{1.95 \times 10^{17} \text{ marshmallows}}{\text{mi}^2}.$$

Next, calculate the number of marshmallows that can fit in a single layer in a single square mile.

$$1 \text{ mi}^2 \times \frac{27{,}878{,}400 \text{ ft}^2}{1 \text{ mi}^2} \times \frac{144 \text{ in}^2}{1 \text{ ft}^2} = 4.01 \times 10^9 \text{ in}^2.$$

Since a single marshmallow occupies a surface area of 1 in^2, there are 4.01×10^9 marshmallows/mi^2.

Finally, using the two values calculated above, you can determine how many marshmallows thick each layer would be and convert that to miles.

1.95×10^{17} marshmallows/mi^2/4.01×10^9 marshmallows/mi^2 = 4.9×10^7 layers, in 1-inch tall marshmallows, which equates to 4.9×10^7 inches × 1 mile/63,360 inches. Converting to miles gives about 760 miles per marshmallow layer—that's still a pretty tall tower of marshmallows.

For the second analogy, determine how many pennies the entire population could spend in a lifetime (assuming each person started spending on their first day of life and lives to be 100 years old).

1 million dollars = 100,000,000 pennies.

$$\frac{1 \times 10^8 \text{ pennies}}{1 \text{ h}} \times \frac{24 \text{ h}}{1 \text{ day}} \times \frac{365 \text{ days}}{1 \text{ y}} \times \frac{100 \text{ y}}{1 \text{ lifetime}}$$

$\times 7 \times 10^9$ lifetimes = 6.13×10^{23} pennies spent. However, because it is an unreasonable assumption that every person on Earth would spend for 100 full years, it is not unreasonable that half of the pennies would be left over.

Ideas for additional analogies that may help demonstrate this concept are people lined up around the globe or stacks of objects to reach the moon.

4.11 **a.** 12 g/6.02×10^{23} C atoms = 1.99×10^{-23} g/C atom.

b. The mass of 5×10^{12} carbon atoms × 1.99×10^{-23} g/C atom = 9.95×10^{-11} g = 1×10^{-10} g

c. The mass of 6×10^{15} C atoms × 1.99×10^{-23} g/C atom = 1×10^{-7} g

While many students may have predicted these numbers to all be small, others may have thought that a large number of atoms, such as 5 trillion, might actually be enough to measure. In reality, you would need almost 5 trillion billion (10^{20}) atoms in order to weigh out .01 g of carbon atoms.

4.12 **a.** 1 mol O_3 = 3 mol O

= 3 mol O × 16.0 g O/1 mol O

= 48.0 g O_3

b. 1 mol N_2O = 44.0 g N_2O

c. 1 mol CCl_3F = 137.4g CCl_3F

4.13 **a.** The mass ratio is found by comparing the molar mass of S with the molar mass of SO_2.

32.1 g S/64.1 g SO_2 = 0.501 S/SO_2

b. To find the mass percent of S in SO_2, multiply the mass ratio by 100.

0.501 S × 100 = 50.1% S in SO_2

c. 1.0 mol N_2O = 44.0 g N_2O (2 mol N/1 mol N_2O) (28.0 g N/44.0 g N_2O) = 0.636 N/N_2O.
0.636 N × 100 = 63.6% N in N_2O.

4.14 **a.** $1.9 \times 10^7 \text{ t } SO_2 \times \frac{3.21 \times 10^7 \text{ t S}}{6.41 \times 10^7 \text{ t } SO_2} = 9.5 \times 10^6 \text{ t S}$

–or–

$1.9 \times 10^7 \text{ t } SO_2 \times .501$ (from Your Turn 4.13) = $9.5 \times 10^6 \text{ t S}$

b. $1.42 \times 10^8 \text{ t } SO_2 \times .501 = 7.11 \times 10^7 \text{ t S}$

4.15 According to www.autos.com, the weight of cars and trucks can vary from 1.5 tons (compact cars) to 6 tons (full-size pickups and SUVs). Gas mileage varies proportionately with the size of the vehicle. Assuming a car with an average weight of 2 tons (2000 pounds) and an average gas mileage of 25 miles per gallon, the carbon emissions can be calculated as follows:

12,000 miles/25 miles per gallon = 480 gallons of gas used per year

480 gallons of gas used per year × 5 pounds of C atoms per gallon = 2400 pounds of C atoms released per year.

Using this data, the statement should be amended to say that the average American car releases *more* than its weight in carbon into the atmosphere each year. However, this estimation is based on the assumption that the car engine in question is clean-burning. In reality, many vehicles on the road do not burn fuel as efficiently and they emit much higher quantities of CO_2 than this assumption accounts for.

4.16 **a.** In order of increasing wavelength: ultraviolet < visible < infrared

b. In order of increasing energy: infrared < visible < ultraviolet

4.17 A model that could be used to help describe this process could be to use money. Imagine the Sun gives Earth a daily allowance of $100 that Earth spends and returns to space. Earth gives $23 to the atmosphere, $46 to the surface, and returns $31 immediately. Over the course of the day, the surface of Earth only spends $9 of its $46 but gives the remaining 80% ($37) to the atmosphere to spend. Once the atmosphere spends its $60, the cycle is complete.

4.18 Arrhenius was referring to the burning of fossil fuels in which carbon is burned in oxygen to produce carbon dioxide. CO_2 is a gas, but it's formed via a chemical change (burning), not a physical change (evaporating).

4.19 There are naturally occurring events such as volcanic eruptions, which add CO_2 to the atmosphere, and limestone formation that removes CO_2 from the atmosphere.

4.20 Model or diagram will vary. The atmosphere acts as the greenhouse windows, trapping the air inside. The sun shines on Earth, just as it does on a greenhouse, to trap the gases inside the enclosure.

4.21 The warming of Earth with greenhouse gases is not only good, it is necessary for human survival. However, an increase in greenhouse gases causes a breakdown in the natural energy cycle and an increase in unwanted effects such as more prevalent and harder-to-control wildfires, as well as the melting of the polar ice caps.

4.22 **a.** 90° (at right angles). The two H atoms across from each other would be at 180°.

b. The tetrahedral shape is the most stable arrangement because it allows the C–H bonds (made up of electrons) to maximize their distance from one another with bond angles of 109.5°.

c. The C atom would occupy the central space where the three leg "bonds" and vertical shaft "bond" meet. The H atoms would be attached to the end of each of the 4 "bonds."

4.23 **a.** CCl_4 : 32 valence e^-

Lewis **Molecular**

~109.5° tetrahedral

b. CCl_2F_2 : 32 valence e^-

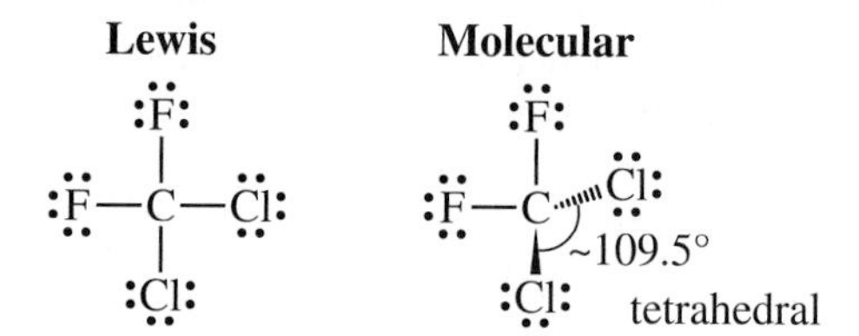

c. H_2S : 8 valence e^-

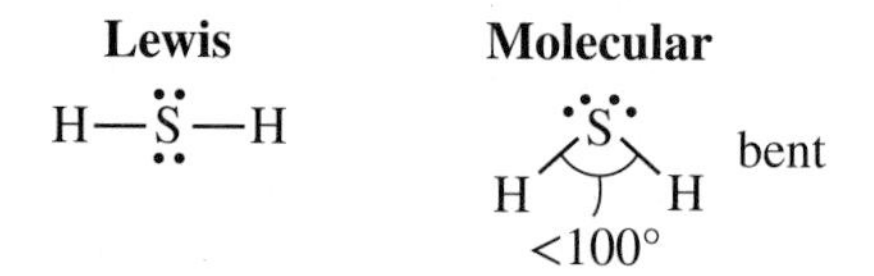

4.24 SO_2 : 18 valence e^-

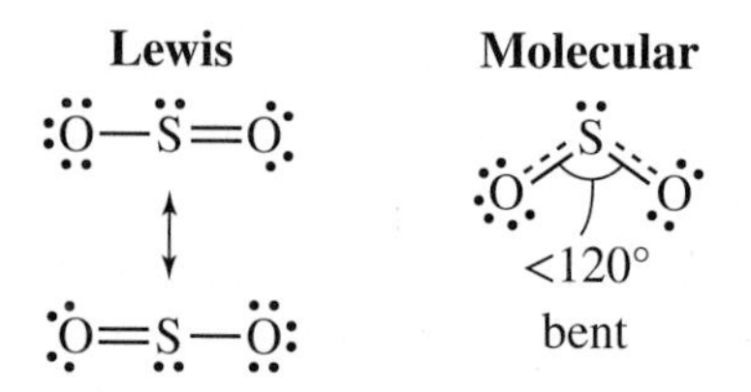

SO_3 : 24 valence e^-

Lewis **Molecular**

trigonal planar

120°

4.25 a. NO_2 : 17 valence e^-

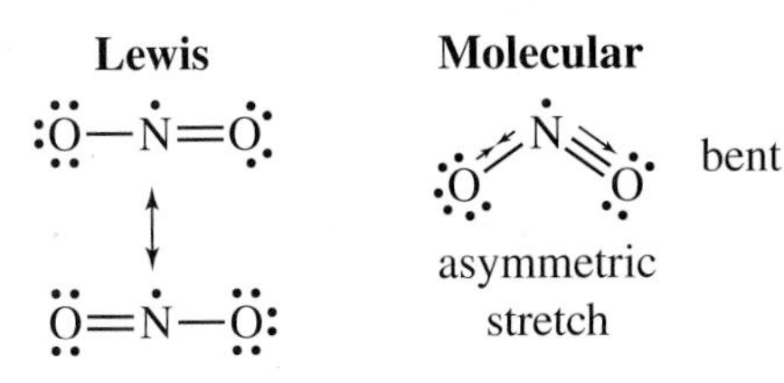

b. CH_4 : 8 valence e^-

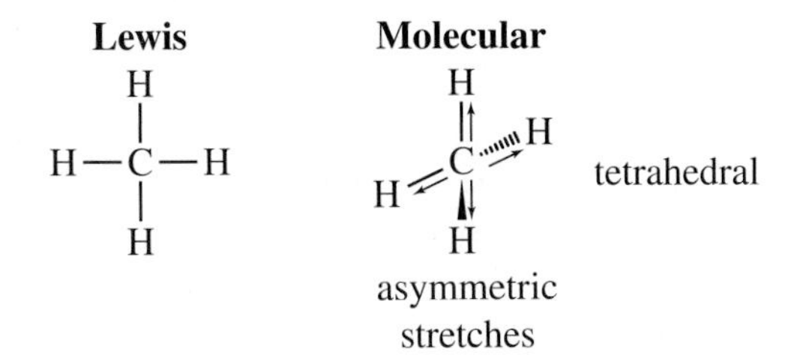

c. O_3 : 18 valence e^-

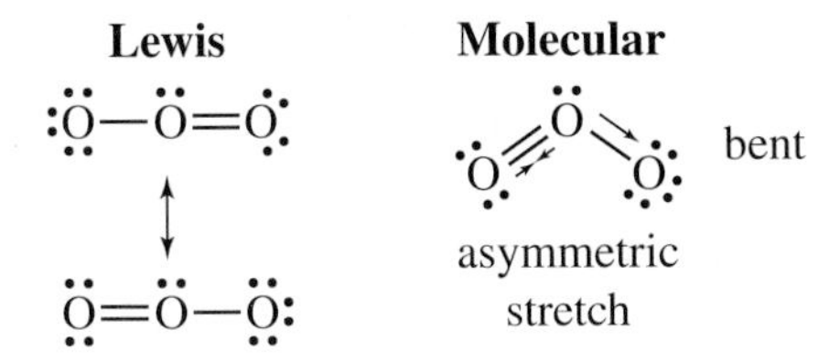

d. NH_3 : 8 valence e^-

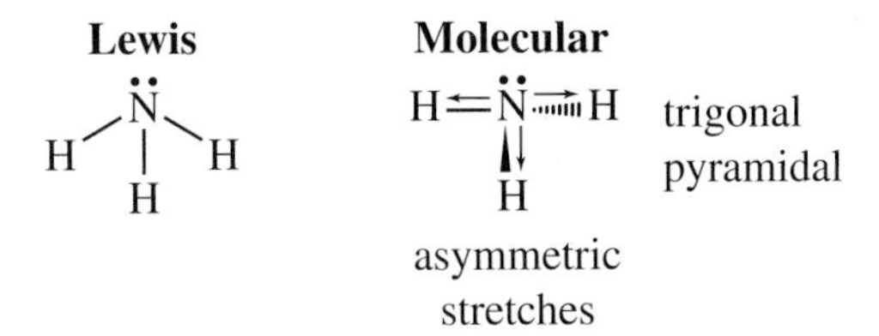

4.26 a. The scale in Figure 4.17 is in wavenumbers. Hence, students will need to convert to wavelengths. For instance, 1500 cm^{-1} = 6.7×10^{-4} cm; 3800 cm^{-1} = 2.6×10^{-4} cm. In terms of microns, the strongest IR absorbances for water vapor occur around 2.63 μm and 6.67 μm.

b. Because less energy is needed to bend a molecule than to stretch it, the longer wavelength vibrations at 6.67 μm should represent bending and the shorter wavelength vibrations at 2.63 μm should represent stretching.

4.27 The IR spectra of other greenhouse gases show similar stretching and bending frequencies as carbon dioxide and water vapor. Although, the energy required to cause these vibrations and the intensity of each absorption varies slightly by molecule. Non-greenhouse gases do not exhibit the same vibrations as greenhouse gases because they do not have the same molecular asymmetries. Additionally, the vibrations that greenhouse gases and non-greenhouse gases have in common are not the vibrations that account for greenhouse gas behavior.

4.28 a. Atmospheric warming has had many negative effects on Earth, such as warming ocean waters, melting arctic sea ice, and an increase in extreme weather events.

b. The greenhouse effect has undoubtedly caused climate change because the average global temperatures reached record-setting highs in 2015.

As illustrated above, climate change is occurring. If your students aren't convinced, have them check out the NASA global climate change website to see the evidence (http://www.climate.nasa.gov/evidence).

4.29 Weather describes the conditions of a region over a short period of time, while climate describes the conditions over decades. Weather refers to daily conditions such as high and low temperature, wind, rain, snow, sun, etc., while climate refers to average values taken over long periods. Weather can impact farming and wildlife on a short timescale such as what happens in the northern United States when an unexpected frost damages a specific crop. Climate has a much more important role in determining the types of plants and animals that are indigenous to a region or are able to survive and propagate in a region. Global warming is a description of a climate phenomenon that has the potential to destroy ecosystems across the globe.

4.30 This increase is a direct result of modern industrial practices and emissions from motor vehicles. Deforestation is another contributing factor to rising carbon dioxide levels.

4.31 The concentration of greenhouse gases such as N_2O and CH_4 have increased while the concentration of CFCs such as CFC-11, CFC-12, and CFC-13 have decreased. Despite decreased concentrations of CFCs, greenhouse gas levels continue to rise.

4.32 a. The atmospheric CO_2 concentration in May 1966 was 324 ppm; in May 2016 it was about 407 ppm. The percent increase is: 407 ppm – 324 ppm/ 324 ppm × 100 = 25.6%

b. Within a given year, the atmospheric CO_2 concentration varies by 5–7 ppm.

c. Photosynthesis removes CO_2 from the atmosphere. Spring begins in the northern latitudes in April; October is the start of spring in the southern latitudes. Both the landmasses (and number of green plants) are greater in the northern hemisphere, so the season in the northern hemisphere control the fluctuations.

4.33 a. Using the figure, we can conservatively estimate the concentration of CO_2 in 1860 to be about 290 ppm and the concentration of CO_2 in 2016 to be about 400 ppm. This relates to a percent change of about 38%.

b. Again using the figure, we can conservatively estimate the concentration of CO_2 in 1957 to be about 310 ppm. This relates to a percent change of about 30%. This means that almost 80% of the increase in atmospheric CO_2 concentrations has occurred in only the last 60 years.

4.34 Based on the figures, these conclusions are accurate. Given that current atmospheric carbon dioxide levels are above 400 ppm, it is clear that this is well above any previous recorded maximum. Additionally, the nature of the previous cycles shows the ebb and flow occurring over thousands of years as opposed to the rapid rise in concentration to record-setting levels that has been seen in only the last 50–60 years.

4.35 According to NASA (www.climate.nasa.gov), as of 2013 the global temperature has risen 1.4 °F since 1880. However, using the latest measurement from January of 2015, that value could actually be as high as 1.8 °F.

4.36 There are many examples in which there is correlation but no evidence of causation. The most popular example of this linked autism spectrum disorders with vaccinations. Other examples include a positive correlation between family involvement and student academic success or more simply an athlete who always seems to win when he wears his "lucky" shoes.

4.37 Depending on location, a significant temperature change may not be reported for over the last 100 years and regional temperature changes can vary quite significantly. For the continental United States, the average regional temperature from January to May has changed anywhere from a slight decrease to 5.09 °F in the last 5 years (www.ncdc.noaa.gov). Globally, North America, Africa, and Asia have experienced similar trends in increasing temperature and the polar regions have experienced the most noticeable increase in temperature over the last 100 years.

4.38 Based on the things we've learned throughout this chapter, the claim that the Sun is responsible for global warming is not substantiated. There is no evidence of Earth receiving increased amounts of radiation from the sun and data has actually shown that solar irradiance and global average temperature are trending in opposite directions.

4.39 Light is emitted from the sun in the form of ultraviolet, visible, and infrared radiation. The largest percentage of sunlight is made up of visible light.

4.40 **a.** A completely white surface will reflect 100% of incident radiation. The theory behind painting roofs white would be to increase Earth's albedo and reduce further warming.

b. "Green roofs" as a form of "reforestation" could also reduce atmospheric CO_2 levels in addition to increasing Earth's albedo. However, these types of gardens are susceptible to changing seasons and require more energy and resources to maintain.

4.41 **a.** Aerosols from volcanic eruptions and solar irradiance

b. Increases in greenhouse gas concentration and changes in Earth's albedo are additional forcings required to accurately re-create the temperature data for the 20th century.

4.42 Most of these conclusions are substantiated by evidence collected by NASA's global climate change project, the EPA, or the NOAA (to name a few).

4.43 In the Maldives, the tourism industry is an example of the tragedy of the commons. As the Maldives is forced to spend a significant portion of its annual budget on energy, the booming tourism industry has become an important source of revenue to the islands' inhabitants. However, rising sea levels threaten these popular vacation spots as beach erosion, saltwater contamination, and flooding are already common occurrences. Additionally, as popularity grows, many of the island resources are becoming depleted. The tourism industry is vital to building up the infrastructure of the islands, but the delicate ecosystems of the islands are becoming damaged in the process. One possible solution is to increase the renewable energy capacity on the islands to free up monetary resources for other important causes.

4.44 Plankton is a crucial food source to many aquatic creatures and it produces about half of the oxygen used to sustain life on Earth. Ocean acidification reduces the amount of plankton and would result in global food shortages because the marine food chain would completely break down.

4.45 **a.** Using an Internet search to find "carbon footprint calculator" will lead students to many options. Commonly requested information might include: number of people in your household, geographical location, size of your home, types of vehicles and miles driven per year, and types of appliances used in your home. Other questions relate to efforts made in your daily life to reduce your footprint, such as recycling, composting, turning off and unplugging appliances when not in use, carpooling, and growing your own food.

b. The information requested by each site does vary slightly, but all require information related to individual usage of the largest sources of carbon emissions such as motor vehicle and home energy use.

4.46 **a.** According to the United States Census Bureau the world population is estimated at roughly 7.3 billion people.

b. Using these figures there is roughly 4 acres of land available for each person in the world.

4.47 **a.** According to the United States Census Bureau the United States population is estimated at roughly 320 million people.

b. 3.2×10^6 people $\times$ 17 acres/per person = 5.4×10^9 acres

c. 5.4×10^9 acres needed for U.S. population/ 3.0×10^{10} acres available $\times 100 = 18\%$ of global biologically productive space needed for U.S. population.

4.48 **a.** Countries with large populations and industrial practices like the United States, India, China, Bangladesh, Mexico, and Brazil would be predicted to have large amounts of CO_2 emissions.

b. Per capita emissions are greater in countries with large industrial practices but with smaller populations such as in the Middle East.

4.49 **a.** $19 \text{ tons } CO_2 \times \frac{2000 \text{ lb } CO_2}{1 \text{ ton } CO_2} \times \frac{1 \text{ tree}}{25 \text{ lb } CO_2}$

= 1,520 trees. If 50 lb per tree is used, the answer is 760 trees.

b. Estimating global CO_2 emissions at 6 Gt (13.2 trillion pounds) and assuming each tree absorbs 50 lb CO_2:

1.2×10^{10} trees $\times$ (50 lb CO_2/1 tree) = 6×10^{11} pounds CO_2 absorbed

(6×10^{11} pounds CO_2/1.32×10^{13} pounds) $\times$ 100 = 4.5% of global emissions absorbed. If 25 lb per tree is used, the answer is 2.3%.

4.50 According to data supplied by the Department of Energy and Climate Change, Britain was able to reduce greenhouse gas emissions by about 24% by 2010 and has since further reduced emissions another 12–13%. Several countries such as Italy, Finland, Germany, Poland, and Sweden also reduced greenhouse gas emissions by 20% or more, but others such as New Zealand, Malta, and Canada actually saw greenhouse gas emissions increase by more than 35%.

4.51 It is unreasonable to think that greenhouse gas emissions can be completely eliminated. Therefore, knowing that greenhouse gas emissions are a certainty, lawmakers should focus on how to limit the total amount of emissions that reach the atmosphere.

4.52 While the United States, China, and India occupy the three top spots for total carbon dioxide emissions, the United States also ranks 13th for per capita carbon dioxide emissions. China and India are ranked 50th and 147th in per capita emissions, suggesting that the United States is heavily contributing to global climate change. The United States has a responsibility to reduce its per capita greenhouse emissions and could set a global precedence in doing so.

4.53 Revisit Your Turn 4.7, 4.21, and 4.28 for suggestions on how these can be answered.

Chapter 5

REFLECTION

a. $4460 \text{ km} \times \frac{0.6214 \text{ mi}}{1 \text{ km}} \times \frac{1 \text{ gal}}{30 \text{ mi}} = 92 \text{ gal}$

b. 139 gal

c. $\frac{26.1 \text{ lb corn}}{1 \text{ gal}} \times 139 \text{ gal} \times \frac{1 \text{ acre}}{7110 \text{ lb corn}} = 0.51 \text{ acre}$

5.1 **a.** Desirable fuels ignite easily at low temperatures and produce a large amount of heat when they combust. Desirable fuels are also inexpensive to produce and can be stored and transported safely.

b. A good match between a fuel and its intended use would be one that does not require excessive amounts of fuel to get the job done and leaves behind the least amount of residue.

c. Solid fuels like wood and coal have several disadvantages. In addition to both producing harmful gases, leaving solid residues, and being nonrenewable resources, coal dust has been proven to cause significant health problems. Liquid and gaseous fuels like petroleum and natural gas burn more smoothly than solid fuels, although both still produce gases that are known to be harmful to the environment.

d. A better alternative to petroleum would burn cleanly in addition to being inexpensive to produce, could be stored and transported safely, and would produce a large amount of heat when it combusts.

5.2 There are two reasons you might see steam rising from a compost pile. The simple explanation for a new compost pile could be escaping water vapor. For a pile that has actually started to "compost," the bacteria that eat the raw material in the heap produce heat as a byproduct of their metabolism. This heat can raise the temperature of the pile to as high as 70 °C (160 °F)!

5.3 **a.** China produces the most coal (2012). The United States, Russia, and Saudi Arabia produce the most petroleum (2014). The United States and Russia produce the most natural gas (2014).

b. The U.S. has the largest reserve of coal (2011). Canada, Saudi Arabia, and Venezuela have the largest reserves of petroleum (2015). Qatar has the largest reserves of natural gas (2015).

c. Natural resources are often purchased from countries with large reserves by countries with large demand. While the United States ranks only 11th in petroleum reserves, being the #1 producer reduces domestic dependence on foreign oil and prevents countries with large petroleum reserves from driving up oil prices.

d. The trends have not changed dramatically over the last 10 years, although the United States became the largest producer of petroleum and increased their reserves of natural gas in that time. China has remained the largest producer of coal for the past 30 years, but has almost tripled their production in the last 10 years. A driving force for the change in U.S. production and reserve capacity of fossil fuels is most likely linked to attempts to rebuild the struggling economy.

5.4 According to the U.S. Energy Information Administration, given the current U.S. production and U.S. reserve estimates, there is enough coal to last about 256 years, enough natural gas to last about 84 years, and enough petroleum to last about 34 years (compared to the year 2014). However, these values do not take into account global production or reserve values, changes in fossil fuel consumption, or importantly, the research being done to improve current energy technology.

5.5 **a.** $C_6H_{12}O_6 + 6\ O_2 \longrightarrow 6\ CO_2 + 6\ H_2O$

b. $CH_4 + 2\ O_2 \longrightarrow CO_2 + 2\ H_2O$

c. $2\ C_4H_{10} + 13\ O_2 \longrightarrow 8\ CO_2 + 10\ H_2O$

5.6 $C_3H_8 + 5\ O_2 + N_2 \longrightarrow 3\ CO_2 + 4\ H_2O + N_2$

Because the nitrogen does not react, we typically omit it from the balanced chemical equation. Notice N_2 is, on both the reactants and products side, unchanged.

5.7 **a.** Reactants

b. Products

c. Products

d. Products

5.8 **a.** $217\ \text{kcal} \times \frac{1000\ \text{cal}}{1\ \text{kcal}} \times \frac{4.184\ \text{J}}{1\ \text{cal}} \times \frac{1\ \text{kJ}}{1000\ \text{J}} = 908\ \text{kJ}$

b. 1 J is equivalent to lifting 100 g a distance of 1 meter (acceleration due to gravity = 10 m/s^2). Lifting 1 kg a distance of 2 m requires 20 times more energy, or 20 J. To burn 908 kJ (or 9.08×10^5 J) you would have to lift 45,400 textbooks!

5.9 **a.** The previous problem assumed that all of the Calories in the pizza could be metabolized into energy available to do work. That is not a reasonable assumption.

b. In reality, the amount of books this piece of pizza would allow you to lift would be much lower, as not all of the energy released from metabolizing food is transformed into usable energy.

5.10 As a general rule, hydrocarbons have high heats of combustion and make the best fuels. Coal makes a good fuel because it is made of carbon; however, the lack of hydrogen atoms makes it less viable. Ethanol is made primarily of carbon and hydrogen, but the presence of an oxygen atom makes it less viable as a fuel source. Because of the bonds that are broken and formed during combustion, the two substances have similar heats of combustion. They are made of different substances, and different bonds are broken and formed, however the net energy result is similar.

5.11 **a.** A hot pack is the result of an exothermic reaction and a cold pack is the result of an endothermic reaction.

d. The reaction between calcium chloride and water is exothermic, while the reactions of sodium chloride, ammonium chloride, potassium chloride, and sodium bicarbonate with water are endothermic.

e. Calcium chloride would be an effective salt for a hot pack, while ammonium chloride would be effective for a cold pack. Students should consider those salts with the largest temperature changes to be the most effective.

5.12 The bond energy for O_3 is intermediate between the O–O single bond (146 kJ/mol) and O=O double bond (498 kJ/mol) values; that is, it is less than the bond energy for the O=O double bond in O_2. Energy is inversely proportional to wavelength, so the higher bond energy of O_2 requires radiation of shorter wavelength to break its bonds.

5.13 $2\ C_2H_2 + 5\ O_2 \longrightarrow 4\ CO_2 + 2\ H_2O$

Bonds to break:		Bonds to make	
2 C≡C	2 × (+813 kJ/mol)	8 C=O	8 × (−803 kJ/mol)
4 H–C	4 × (+416 kJ/mol)	4 H–O	4 × (−467 kJ/mol)
5 O=O	5 × (+498 kJ/mol)		
	+5780 kJ/mol		−8292 kJ/mol

Heat = −2512 kJ/2 mol C_2H_2

Heat of combustion = −1256 kJ/mol C_2H_2, or

$$\frac{-1256\ \text{kJ}}{\text{mol}} \times \frac{1\ \text{mol}}{26\ \text{g}} = -48.3\ \text{kJ/g}\ C_2H_2$$

5.14 Examples include engines that convert chemical energy to mechanical energy (potential energy stored in bonds is released upon combustion to power parts of a machine), light bulbs that convert electrical energy to heat and light, or a wood stove that converts chemical energy into heat.

5.15 **a.** $\frac{5 \times 10^{12}\ \text{J}}{\text{x}} = .38;\ \text{x} = \frac{1.32 \times 10^{13}\ \text{J}}{\text{day}}$

$= \frac{1.32 \times 10^{10}\ \text{kJ}}{\text{day}}$

$\frac{1.32 \times 10^{10}\ \text{kJ}}{\text{day}} \times \frac{1\ \text{g}}{30\ \text{kJ}} \times \frac{1\ \text{kg}}{1000\ \text{g}} \times \frac{\$30}{1000\ \text{kg}}$

= $13,000/day

Coal costs for Plant A = $13,000/day. Coal costs for Plant B = $11,000/day.

b. When coal combusts, 12 g of carbon produces 44 g of carbon dioxide. Plant A burns

$1.3 \times 10^{10}\ \text{kJ} \times \frac{1\ \text{g}}{30\ \text{kJ}} = 4.3 \times 10^8$ g of coal per day, creating 1.6×10^9 g CO_2. Plant B burns 3.6×10^8 g of coal per day, creating 1.3×10^9 g of CO_2. Therefore, Plant B emits 3.0×10^8 (30 million) fewer grams of CO_2 each day than Plant A.

5.16 **a.** Examples of some energy losses that occur while driving include the heat and gas emissions produced in the engine during fuel combustion.

b. If it only takes 15% of the energy from fuel combustion to move an average size car (~2000 pounds), it would take an additional 2–4% of energy to move the passengers (assuming a passenger load of 300–500 pounds).

5.17 **a.** As coal (a solid) is combusted, carbon dioxide (a gas) and water vapor (a gas) are produced. Carbon dioxide as a by-product of coal combustion is released to the environment and increases the entropy of the universe.

b. This decrease in entropy must have been accompanied by an input of energy (by a human) and by an increase of entropy elsewhere in the universe (from the food burned as fuel by the human).

5.18 **a.** While coal has steadily remained a widely used fuel source since the 1860s, petroleum and natural gas demands have risen since the 1900s to make up the highest percent of fuel consumption in the United States.

b. Based on Figure 5.14, coal makes up about 20% of the total energy that is produced.

c. The global trends are similar to trends in the United States, except Asia where coal consumption far exceeds other sources of fuel. One suggestion for this difference is that China is the world leader in coal production.

5.19 **a.** Calculate the approximate molar mass of coal.

$$135 \cancel{\text{mol C}} \times \frac{12.0 \text{ g C}}{1 \cancel{\text{mol C}}} = 1620 \text{ g C}$$

$$96 \cancel{\text{mol H}} \times \frac{1.0 \text{ g H}}{1 \cancel{\text{mol H}}} = 96 \text{ g H}$$

$$9 \cancel{\text{mol O}} \times \frac{16.0 \text{ g O}}{1 \cancel{\text{mol O}}} = 144 \text{ g O}$$

$$1 \cancel{\text{mol N}} \times \frac{14.0 \text{ g N}}{1 \cancel{\text{mol N}}} = 14.0 \text{ g N}$$

$$1 \cancel{\text{mol S}} \times \frac{32.1 \text{ g S}}{1 \cancel{\text{mol S}}} = 32.1 \text{ g S}$$

The sum of these elemental contributions for $C_{135}H_{96}O_9NS$ is 1906 g/mol. Therefore, every 1906 g of coal contains 1620 g of carbon. Similarly, 1906 tons of coal contain 1620 tons of carbon.

$$\text{Mass of carbon} = 1.5 \times 10^6 \text{ tons } C_{135}H_{96}O_9NS \times \frac{1620 \text{ tons C}}{1906 \text{ tons } C_{135}H_{96}O_9NS} = 1.3 \times 10^6 \text{ tons C}$$

b. $1.5 \times 10^6 \text{ tons} \times \frac{2000 \text{ lb}}{1 \text{ ton}} \times \frac{454 \text{ g}}{1 \text{ lb}} \times \frac{30 \text{ kJ}}{\text{g}} = 4.1 \times 10^{13} \text{ kJ}$

c. 4.7 million tons

5.20 **a.** Coal contains small amounts of sulfur that combine with oxygen during combustion to produce SO_2. Volcanoes are a natural source of SO_2 in the atmosphere.

b. Although coal does contain small amounts of nitrogen that combine with oxygen during combustion, the major portion of NO is formed from the reaction of N_2 and O_2 from the air at the high temperatures produced during the combustion process. Other sources of NO include engine exhaust, lightning, and grain silos.

5.21 **a.** Coal fires produce particulates such as soot (damages infrastructure and respiratory systems) and fly ash (contains concentrated amounts of toxic elements like mercury, cadmium, and lead). It also produces harmful gas pollution in the form of nitrogen and sulfur oxides that contribute to acid rain and global warming.

b. As evidence that clean coal could be a viable option in the future, according to the Global CCS (Carbon Capture and Storage) Institute there are 15 CCS projects in operation with another seven in development (compared to only eight in operation in 2011). These 22 projects have the capacity to capture 40 million tons of CO_2 yearly, and there are another 18 CCS projects in various stages of development.

5.22 Conversion factors:

7.33 barrels = 1 metric ton

1 barrel = 42 gallons; 1 gallon = 3.8 L

According to Figure 5.21, worldwide consumption of petroleum is currently forecasted at about 93 million barrels/day which is equivalent to 3.9 billion gallons/day (9.3×10^7 barrels (42 gal/barrel)), 15 billion liters/day (3.9 billion gal (3.8 L gal)), or 13 million metric tons/day (93 million barrels (1 metric ton/7.33 barrels)). Worldwide petroleum production is forecasted at roughly 4.5 million barrels/day which is equivalent to about 190 million gallons/day, 720 million liters/day, or 620,000 metric tons/day.

5.23 **a.** $1500 \text{ kJ} \times \frac{1 \text{ g } CH_4}{50.1 \text{ kJ}} \times \frac{1 \text{ mol } CH_4}{16 \text{ g } CH_4} \times \frac{1 \text{ mol } CO_2}{1 \text{ mol } CH_4} \times \frac{44 \text{ g } CO_2}{1 \text{ mol } CO_2} = 82 \text{ g } CO_2$

b. For bituminous coal from Maryland, which releases 30.7 kJ/g on average, 150 g of CO_2 is released when 1500 kJ of heat is produced.

5.24 **a.** "Hydraulic" refers to work done by the pressure created by forcing water, oil, or another liquid through a comparatively narrow pipe or orifice. In fracking, water is pumped down a well where the pressure builds and fractures the shale. Dynamite would not work for this purpose because natural gas is flammable and the use of dynamite would result in dangerous explosions due to combustion reactions with the gas that is trying to be extracted.

b. According to the U.S. Geological Survey, it may take as few as 2 million gallons of water to frack a well or as much as 16 million gallons of water. Additionally, a well may be fracked multiple times. According to FracFocus, the national hydraulic fracturing chemical registry, there is an extensive list of chemicals that are commonly used as additives in fracking fluid. Fracking fluid also uses small particulates such as sand to help keep the fissures open.

c. According to the EPA, the wastewater from fracking can be disposed of into deep injection wells, treated and disposed of to surface water bodies, or recycled for use in future hydraulic fracturing operations.

5.25 **a.** Boiling point is predicted to be higher for molecules with stronger intermolecular forces and for molecules with a greater number of electrons (for those with the same type of intermolecular forces). Therefore, in order of increasing boiling point, $F_2 < Cl_2 < Br_2 < I_2$.

b. Following the same rules as in a., $CH_4 < CF_4 < CBr_4 < CI_4$.

5.26 **a.** One possible pair of products is C_8H_{18} and C_8H_{16}. Their structural formulas are:

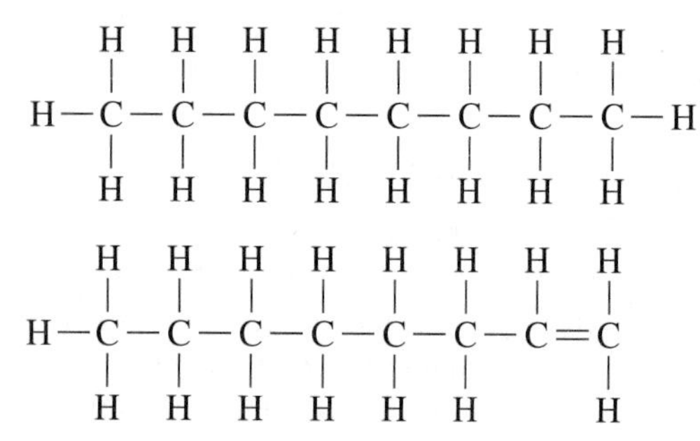

b. C_nH_{2n+2} (where n is an integer)

c. The generic chemical formula is C_nH_{2n} (where n is an integer).

d. Based on the rules used in Skill Building 5.25, relative boiling points should be (and are) C_5H_{12} (36.1 °C) < $C_{11}H_{22}$ (192.7 °C) < $C_{16}H_{34}$ (287 °C).

5.27 According to the United States Department of Labor Occupational Safety and Health Administration, lead exposure is most common to:

a. Industrial workers involved in the production, use, maintenance, recycling, and disposal of lead materials and products such as plumbing fixtures, solder, rechargeable batteries, lead bullets, leaded glass, brass or bronze objects, and radiators. Construction workers involved in the removal, renovation, or demolition of structures painted with lead pigments; or the installation, maintenance, and demolition of lead pipes and fittings, lead linings in tanks and radiation protection, leaded glass, soldering, or other work involving lead metal or lead alloys

b. Hobbyists involved in the use of objects listed above, such as firing ranges, radiator repair, or lead–acid battery recycling

c. Children living in places with high levels of lead in water, food, or old deteriorating paint.

5.28 **a.** 2013: 13,310 million gallons

2014: 14,340 million gallons

2015: 14,810 million gallons

b. A bar graph would be a useful visualization of these values to explain how production of ethanol has increased dramatically in recent years. Ethanol can be produced by fermenting a variety of different sugars and grains, which are crops that many farmers can easily produce.

5.29 The most desirable characteristic of using switchgrass and wood chips is that they are abundant, relatively renewable, and unlike corn and sugarcane, are not feedstocks.

5.30 **a.** Structural formulas for methanol and ethanol

```
    H                 H   H
    |                 |   |
H—C—O—H         H—C—C—O—H
    |                 |   |
    H                 H   H
 methanol          ethanol
```

b. Propanol, because (like propane) it has three carbon atoms. More properly, this compound is *n*-propanol or 1-propanol.

c.

```
CH3—CH—CH3
     |
     OH
```

This is iso-propanol (2-propanol), better known as "rubbing alcohol."

5.31 **a.** Because of the oxygen present, you could expect there would be less heat released. And because of fewer carbons, you would also expect less heat to be released. So overall, biodiesel releases less energy.

b. You would have to know the molar mass of the biodiesel to find its energy content per mole.

c. Depending on how you are measuring the two substances to be compared, you could argue either value is useful. Particle per particle, the per mole measurement is more meaningful, but if you are measuring your fuels by mass, then per gram is more useful.

5.32 **a.** $C_2H_5OH + 3\ O_2 \longrightarrow 2\ CO_2 + 3\ H_2O$

b. $2\ C_{19}H_{38}O_2 + 55\ O_2 \longrightarrow 38\ CO_2 + 38\ H_2O$
(*Hint:* Balance the C and H atoms first.)

5.33 Key points include:

- Palm oil is already the most widely consumed vegetable oil in the world, and is used in a variety of food products as well as soaps, detergents, and shampoos.
- The use of palm oil as a biofuel increases the demand for palm oil and creates problems for workers in countries that do not have labor laws in place. Often, these workers are forced to work longer hours for little pay to meet the demand.
- Additionally, this demand drives up the price of products that use this oil so that the cost of living is driven up and those already in tough financial situations are hurt even further.

5.34 Moving toward sustainability creates domestic job opportunities, reduces the cost of living for everyone, and reduces the amount of control big businesses have in government policies.

Chapter 6

REFLECTION

Answers will vary depending on the chosen states. Sources such as natural gas or coal are derived from fossil fuels, whereas photovoltaic, wind, or hydroelectric are not.

6.1 Examples of types of energy a student may use in a typical day include: mechanical energy in the form of driving a car, walking to class, or doing exercise; chemical energy in the form of using battery-powered electronics; and electrical energy in the form of using lights and other appliances that use electricity.

6.2 U-234 has 92 protons and 142 neutrons, while U-238 has 92 protons and 146 neutrons. Since isotopes differ only in the number of neutrons, both U-234 and U-238 have 92 electrons.

6.3 **a.** ${}^{1}_{0}n + {}^{235}_{92}U \longrightarrow {}^{138}_{56}Ba + {}^{95}_{36}Kr + 3\,{}^{1}_{0}n$

b. ${}^{1}_{0}n + {}^{235}_{92}U \longrightarrow {}^{137}_{52}Te + {}^{97}_{40}Zr + 2\,{}^{1}_{0}n$

6.4 For anthracite coal:

$$9.0 \times 10^{10}\text{ kJ} \times \frac{1.0\text{ g anthracite coal}}{30.5\text{ kJ}} \times \frac{1\text{ kg}}{1000\text{ g}}$$

$= 3.0 \times 10^{6}$ kg anthracite coal

The equivalent masses for the other grades are calculated in the same way: bituminous coal, 2.9×10^{6} kg; subbituminous coal, 3.8×10^{6} kg; lignite (brown coal), 5.6×10^{6} kg; peat, 6.9×10^{6} kg.

6.5 ${}^{241}_{95}Am \longrightarrow {}^{237}_{93}Np + {}^{4}_{2}He$

${}^{9}_{4}Be + {}^{4}_{2}He \longrightarrow {}^{12}_{6}C + {}^{1}_{0}n + {}^{0}_{0}\gamma$

6.6 In an emergency such as an earthquake, the fuel rods should be inserted in preparation for shutdown to slow the nuclear fission reaction in the reactor core.

6.7 The cloud is tiny droplets of condensed water vapor (some people call this "steam," however steam is invisible until it condenses). The cloud does not contain any nuclear fission products.

6.8 $$1.2 \times 10^{9}\text{ J/s} \times \frac{60\text{ s}}{1\text{ min}} \times \frac{60\text{ min}}{1\text{ hr}} \times \frac{24\text{ hr}}{1\text{ day}} \times 3\text{ reactors}$$

$= 3.1 \times 10^{14}$ J of electric energy are generated in a single day when the Palo Verde complex is operating at maximum capacity.

Rearranging $\Delta E = \Delta mc^2$ to solve for the mass of U-235 lost gives:

$\Delta m = \Delta E/c^2 = 3.1 \times 10^{14}\text{ J}/(3.00 \times 10^{8}\text{ m/s})^2 =$ .0034 kg or 3.4 g of U-235 lost each day.

6.9 **a.** Electromagnetic radiation

b. nuclear radiation

c. electromagnetic radiation

d. nuclear radiation

6.10 **a.** ${}^{86}_{37}Rb \longrightarrow {}^{86}_{38}Sr + {}^{0}_{-1}e$

b. ${}^{239}_{94}Pu \longrightarrow {}^{235}_{92}U + {}^{4}_{2}He$

6.11 After 10 half-lives, 0.0975% of the original sample remains

# of half-lives	% decayed	% remaining
7	99.22	0.78
8	99.61	0.39
9	99.805	0.195
10	99.9025	0.0975

6.12 Using Table 6.3 to verify, after 36.9 years (3 half-lives), 12.5% of the original tritium will remain.

6.13 **a.** Radium is a product in the natural decay series of uranium-238, the isotope of uranium with highest natural abundance. Therefore, if uranium is present in the rocks and soils, radium will be present as well.

b. If 0.5 pCi remain, that means 0.5 pCi/16 pCi = 3.12% remain, so five half-lives (5×3.8 days = 19 days) are required for the level of radioactivity to drop.

c. Radon will continue to enter your basement because the uranium in the soils and rocks underneath your home continuously produce it.

6.14 ${}^{1}_{0}n + {}^{235}_{92}U \longrightarrow {}^{143}_{54}Xe + {}^{90}_{38}Sr + 3\,{}^{1}_{0}n$

6.15 **a.** -498 kJ/2 mol $H_2 = -249$ kJ/mol

-249 kJ/mol (1 mol H_2/2 g H_2) $= -125$ kJ/g

b. From Figure 5.6, methane releases -50.1 kJ/g during combustion. -125 kJ/g/-50.1 kJ/g = 2.5. Hydrogen releases about two and half times more energy!

6.16 **a.** $:\ddot{\underset{..}{I}}\cdot$ $:\ddot{\underset{..}{I}}:\ddot{\underset{..}{I}}:$ $[:\ddot{\underset{..}{I}}:]^-$

atom   molecule   ion

b. The iodine atom is the most reactive because it has one unpaired electron, making it a free radical.

c. The element iodine (including its radioisotope, I-131) is taken up by the thyroid gland in the chemical form of I^-, the iodide ion.

6.17 **a.** Low neutron absorption is an important property for any structural material used in a nuclear reactor because the neutrons that are generated need to interact with the nuclear fuel in order to sustain the chain reaction taking place in the reactor's core.

b. In the event of an accident at a nuclear power plant, it is important that the fuel rods can be cooled until the reaction stops. In the event that the cooling mechanism fails, the fuel rods will get too hot (thousands of °C) and the zirconium metal becomes capable of oxidizing residual water in the core. This reaction "rusts" the protective zirconium casing of the fuel rods and produces highly flammable hydrogen gas. The hydrogen gas is capable of producing an explosion if it is not vented quickly enough. If hot enough, the zirconium alloy, the nuclear fuel, and any machinery remaining in the core can melt and form a dangerous radioactive substance called corium that is capable of eating through the concrete walls of a containment building.

6.18 **a.** The Nuclear Energy Institute (www.nei.org) provides helpful information about nuclear power usage in the United States and provides fact sheets for each state summarizing their current energy usage.

b. Connecticut (47.1%), New Jersey (46.7%), South Carolina (54%), and Illinois (48.4%) use the highest percentages of nuclear energy.

c. Depending on the state, energy sources could be coal, natural gas, hydroelectric, oil, or another renewable energy source.

6.19 **a.** Currently, nuclear reactors are located on four continents, with the largest number of operational reactors in Europe, North America, and Asia.

b. According to the World Nuclear Association, Kazakhstan produced 40% of the world's supply of uranium in 2015. Canada produced 22% of the world's supply and Australia produced 9%. Canada has 19 nuclear reactors that produce about 15% of the nation's energy needs. Australia and Russia do not currently have operational nuclear reactors.

c. According to the International Atomic Energy Agency, developing countries would benefit the most from developing nuclear energy. Nuclear technology can be used in the fight against cancer, cardiovascular and other non-communicable diseases, as well as to help combat child malnutrition.

6.20 **a.** The most important aspect in forming an opinion on any of the listed topics is to make an informed decision using reputable resources. Nuclear power has demonstrated its viability as an alternative energy source; however, it is still a developing technology that will continue to be improved upon.

b. Currently no long-term plan for dealing with nuclear waste exists in any country, although an international used-nuclear-fuel-waste facility has been proposed in Southern Australia.

c. The cartoon depicts a future in which there are renewable energy alternatives available to power household appliances.

d. One school of thought is that nuclear power is a viable alternative to fossil fuels and is a solution to the global energy crisis, while others feel that although nuclear power may be a solution to dwindling fossil fuel reserves, the radioactive waste generated from this process has the potential to put public safety at risk if it is not handled correctly.

6.21 **a.** Solar radiance increases gradually across the United States from winter to late summer before gradually declining again.

b. Solar radiance depends on local factors such as cloud cover, aerosols, smog, and haze.

6.22 **a.** http://www.efficientenergysaving.co.uk/solar-irradiance-calculator.html provides online solar irradiance estimates by location.

b. Using 256 kWh in 33 days means, on average, this homeowner would need to produce 7.76 kWh of solar energy per day in order to power their home. Compared to the data in Figure 6.20, this should be enough energy to power most homes in the U.S.

c. 7.50×10^8 W/7.76×10^3 W = 9.7×10^4 houses that could be powered by this nuclear plant.

d. Some things to keep in mind would be that nuclear power plants do not continually run at full capacity and that continually supporting 10,000 homes would mean that the nuclear reactor would continually be depleted. Therefore, the amount of homes that could be supported by this nuclear power plant would be much lower.

6.23 From The NEED project (www.NEED.org), examples of solar thermal collectors include: parabolic troughs, solar power towers, and dish/engine systems. Parabolic troughs use long reflecting troughs that focus the sunlight onto a pipe. The fluid circulating inside the pipe collects the energy and produces steam by transferring the heat to a heat exchanger. Depending on size, this type of solar collector could be used to power large communities. A solar power tower uses a large field of rotating mirrors to track the sun and focus the sunlight onto a thermal receiver on the top of a tall tower. The fluid in the receiver collects the heat and uses it to generate electricity or store it for later use. This type of collector can also be large enough to provide power to large communities of people. Dish systems concentrate sunlight and use an engine located at the focal point to create electricity. Due to the small size of these units, they are best used to power a single home or small business. All three types of collectors require a continuous supply of strong sunlight, such as that found in desert regions, to function at their full potential.

6.24 **a.** Doping with phosphorous forms an n-type semiconductor. Phosphorous is in Group 15 and has more electrons per atom than a silicon atom does.

b. Doping with boron forms a p-type semiconductor. Boron is in Group 13 and has one fewer electron per atom than a silicon atom.

6.25 **a.** Uses include pumping water for livestock and lighting in areas without electricity. Even on farms and ranches with electricity, solar PVs can reduce electric utility bills.

b. Uses include providing electricity to power lighting, heating, and other electrical needs in a building or in a process used by the business. For example, PVs have been used to generate enough electricity to power the electrical needs of a small microbrewery.

c. Uses include supplying electricity to part or all of a home. At present, some homes that have a PV system also can be used for electrical back-up locations in places where storms occur that may disrupt the power.

6.26 **a.** The climate of New Jersey is not continually hot and dry, and turning New Jersey into a solar farm would not be an efficient use of that land.

b. The desert regions of the southwestern United States would be the most effective place to construct a solar farm.

c. Large solar farms can damage the local desert ecosystem if not managed carefully. It would be important to protect other vital resources in the process of building and operating a solar farm.

6.27 Factors to consider before implementing a solar energy initiative could include: the financial burden placed on the public and how that cost would be offset by the initiative, what type of solar collector the initiative will use, how much power the solar collector would generate, how much energy would be needed to power the community, and how new construction in the community would be supported by the initiative.

6.28 **a.** Solar thermal energy is turned into electricity by creating high-pressure steam that is used to turn a turbine. The turbine spins a shaft that spins coils of copper wire inside a ring of magnets, which creates an electric field that produces electricity. A photovoltaic cell relies upon the energy of the Sun to energize electrons to free them from their atoms and create a flow of electrons through a conducting wire.

b. In recent years, photovoltaic systems have become a better and more viable option for solar energy than solar-thermal collectors. Photovoltaic systems have the ability to be installed almost everywhere that a solar-thermal system could be installed, while solar-thermal systems can't typically replace photovoltaic ones. Photovoltaic systems also require less irradiance and do not face the technological challenges that solar-thermal systems face.

6.29 The largest challenge to wind power in the southeastern U.S. is the Appalachian mountain range, which blocks the wind and creates uneven territory.

6.30 The National Energy Education Development Project's Energy Infobook (www.NEED.org) provides an excellent summary of the many types of renewable energy sources available, which includes not only how each type of energy works, but also the advantages and limitations of each type.

Chapter 7

REFLECTION

a. Answers will vary. Some possibilities include: flashlights (alkaline), cellphones (Li-ion), car starters (lead–acid).

b. Li-ion and lead–acid are rechargeable. Some factors include operating temperature, battery age, and extent of discharge.

7.1 According to translations of a paper published in a 1938 issue of the German journal *Forshungen and Fortschrittte*, a "battery" was found at a dig site of Khujut Rabu, near Baghdad, in 1936. The find was composed of a small flat-bottomed clay jar (5.5 inches tall, 3-inch mouth diameter) with an iron rod (3 inches long) inside that was surrounded by a thin sheet of copper (3.8 inch length by 1 inch width). The iron rod had an asphalt plug that fit into the opening of the copper cylinder, and the iron rod extended beyond the plug. Although the neck of the jar was broken, there was evidence that the jar was once sealed with asphalt. The author of the paper, William Konig, suggested that this device could have been used as a battery for electroplating, but more research was necessary. Recent reproductions of the jar have shown that when filled with an electrolyte such as grape juice, lemon juice, or vinegar, the device is capable of producing small amounts of electric potential (0.5 to 2 V). However, critics point out several problems with current theories of how these batteries were used. For example, the jar needs to be filled (and re-filled) with an electrolyte to function as a battery, and the sealed jar would make that extremely difficult. Additionally, the jar did not have terminals. While the iron rod extended beyond the jar, the copper cylinder did not, meaning that no wires could be attached to complete a circuit. Decayed papyrus has been found in other similar jars, suggesting that these jars were used to store scrolls to protect them from the elements.

7.2 **a.** Reduction. The aluminum ion gained three electrons.

b. Oxidation. Zinc metal lost two electrons.

c. Reduction. The manganese ion gained three electrons.

d. Oxidation. Two water molecules were split into four hydrogen ions, one molecule of oxygen, and four electrons.

e. Reduction. Two hydrogen ions gained two electrons and formed one molecule of hydrogen.

7.3 $1 \text{ amp} = \frac{\text{C}}{\text{s}}$ and $1\ \Omega = \frac{\text{J·s}}{\text{C}^2}$. Substituting these values into Equation 7.5 gives $\text{V} = \left(\frac{\cancel{\text{C}}}{\cancel{\text{s}}}\right)\left(\frac{\text{J·}\cancel{\text{s}}}{\cancel{\text{C}^2}}\right)$ which simplifies to $\text{V} = \frac{\text{J}}{\text{C}}$.

7.4 A great interactive example of this can be found at the HyperPhysics website hosted by the Georgia State University Department of Physics and Astronomy (http://hyperphysics.phy-astr.gsu.edu/hbase/electric/watcir.html#c2). This example shows water being pumped through a closed system of pipes. The pump is analogous to a battery in that it takes in water of low pressure and ejects water of high pressure (voltage); the water is analogous to current as it flows through the pipes, and the pipes themselves are analogous to the resistance found in wires. Pipes with small diameters have higher resistance than pipes with large diameters, and the pipes may develop mineral build-up over time, which increases resistance.

7.5 **a.** The memory effect occurs when rechargeable batteries are not fully charged and discharged between charging cycles. The battery "remembers" the shortened lifetime and the capacity is reduced. This term is more accurately referred to as "voltage depression."

b. Nickel–cadmium (Ni–Cd) batteries are most prone to suffering from voltage depression.

c. This effect can often be repaired by fully charging and discharging a battery.

d. Batteries should be stored in a cool, dry environment and should not be stored in hot or humid conditions. The ideal temperature to store batteries is 15 °C (~60 °F). So depending on the climate where you live, storing batteries in the refrigerator might be a good strategy, although a pack of silica gel should be placed with the batteries to absorb moisture.

7.6 Overall, the chemistry of both types of batteries is very similar. While the NiMH battery costs more and does not last as long as a Ni–Cd battery, it has a larger capacity, gives more power, and is less prone to the memory effect; it only requires a full discharge once every 30 cycles.

7.7 **a.** Lead exists as Pb^{+4} in PbO_2 and as Pb^{+2} in $PbSO_4$. The symbol Pb represents lead in its metallic form.

b. Electrons are lost from Pb to form cations. This is oxidation.

c. Metallic lead is oxidized during the discharge process, as two electrons are lost to form the Pb^{+2} ion.

7.8 **a.** Using W = V × I; Level 1: (110 V)(20 A) = 2200 W
Level 2: (240 V)(40 A) = 9600 W
Dual: (240 V)(80 A) = 19,200 W

b. Level 1: $85\text{ kWh} \times \frac{1000\text{ W}}{1\text{ kW}} \times \frac{1}{(.92)(2200\text{ W})}$
$= 42\text{ h}$

Level 2: $85\text{ kWh} \times \frac{1000\text{ W}}{1\text{ kW}} \times \frac{1}{(.92)(9600\text{ W})}$
$= 9.6\text{ h}$

Dual: $85\text{ kWh} \times \frac{1000\text{ W}}{1\text{ kW}} \times \frac{1}{(.92)(19{,}200\text{ W})}$
$= 4.8\text{ h}$

c. There are currently 680 supercharging stations located across the globe. These charging stations are said to deliver as much as 120 kWh while charging. This relates to a charge time of 85 kWh/120 kWh = .71 h or 42.5 min. However, the charge rate is 40 minutes to reach 80% of battery capacity and 75 minutes to reach 100% of battery capacity.

d. Using an example rate of 13¢ per kWh,

Level 1: 2.2 kW × 42 h × $0.13/kWh = $12.01
Level 2: 9.6 kW × 9.6 h × $0.13/kWh = $11.98
Dual: 19.2 kW × 4.8 h × $0.13/kWh = $11.98
Supercharger: 120 kW × .71 h × $0.13 = $11.08

e. Starting with the Tesla Model S, which is said to get 270 miles per charge:

1100 miles per month/270 miles per charge = 4.07 charges per month
Level 1: $12.01 per charge × 4.07 charges per month = $48.88 per month
Level 2: $11.98 per charge × 4.07 charges per month = $48.76 per month
Dual: $11.98 per charge × 4.07 charges per month = $48.76 per month
Supercharger: $11.08 per charge × 4.07 charges per month = $45.10 per month

Using an average car with a 13-gallon gas tank that gets 25.1 miles per gallon and using the national average of $2.21 per gallon (summer 2016):
13 gallons per tank × 25.1 miles per gallon = 326.3 miles per tank.
1100 miles per month/326.3 miles per tank = 3.37 tanks per month
13 gallons/tank × $2.21/gallon = $28.73 per tank
$28.73 per tank × 3.37 tanks per month = $96.82 per month

A Tesla Model S is the more economical choice. However, as mentioned above, the charging time of the Tesla is actually longer than estimated because the charger is designed to reduce charging power as the battery reaches capacity. In most cases, the cost per month to drive a Tesla is underestimated. However, the usage would have to be double the estimate in order to come close to matching the cost to driving an average vehicle. Another factor to consider is the cost of the car itself. A Tesla Model S has a price tag of $70,000, before customizations.

7.9 A 2014 report by the IDTechEx Company details the performance achievements and objectives of the 80 current manufacturers of supercapacitors over the next 10 years. This report highlights the current efforts underway to improve supercapacitor technology given the incredible need for energy alternatives.

7.10 The first example below shows how a student would most likely do this using the following assumptions and example data: 1 gallon of gasoline burned = 18 pounds of CO_2 released, an average 25.1 mpg, and an average monthly mileage of 1100 miles.
1100 miles × 12 months = 13,200 miles per year
13,200 miles per year/25.1 miles per gallon = 525.9 gallons of gas used per year
525.9 gallons per year consumed × 18 pounds of CO_2 released per gallon = 9,500 pounds (~5 tons) of CO_2 released.

The next example uses a back calculation to determine the mpg of a car that would emit 7 tons of CO_2 in an average year. Again, using the value that one gallon of gasoline burned = 18 pounds of CO_2 emitted and using an average yearly driving distance of 13,200 miles:
7 tons CO_2 = 14,000 pounds CO_2
14,000 pounds CO_2/18 pounds CO_2 per gallon = 777.78 gallons of gas used in a year
13,200 miles per year/777.78 gallons per year = 17 mpg. A car that gets 17 mpg (such as a small pickup truck) will emit 7 tons of carbon dioxide in a year. A larger vehicle such as a standard pickup truck only gets 15 mpg and will emit almost 8 tons of CO_2 in a year!

7.11 In Equation 7.19, 1 mol of H–H bonds and 1/2 mol of O=O bonds are broken.
= 1 mol (436 kJ/mol) + 1/2 mol (498 kJ/mol)
= 436 kJ + 249 kJ
= 685 kJ

In equation 7.19, 2 mol of O–H bonds are formed.
= 2 mol (467 kJ/mol)
= 934 kJ

For the overall reaction, (+685 kJ) + (−934 kJ) = −249 kJ/mol or −124.5 kJ/g

The negative sign indicates the reaction is exothermic, meaning 124.5 kJ of energy is released per gram of H_2 combusted. In contrast, methane produces 50 kJ/g.

7.12 **a.** A battery makes electricity from stored energy, while a fuel cell makes electricity in a fuel tank. A battery has a finite amount of energy it can store, while a fuel cell can continually make more energy.

b. A PEM fuel cell does not need to be recharged because it relies on an external fuel source, such as a hydrogen fuel tank. When the tank runs out, you can exchange it for a full one.

c. The chemistry that occurs within a fuel cell is not combustion because the fuel is not burned.

7.13 **a.** In Equation 7.24, 4 mol of C–H bonds and 4 mol of O–H bonds are broken.
= 4 mol (416 kJ/mol) + 4 mol (467 kJ/mol)
= 1664 kJ + 1868 kJ
= 3532 kJ

In Equation 7.24, 4 mol of H–H bonds and 2 mol of C=O bonds are formed.
= 4 mol (436 kJ/mol) + 2 mol (803 kJ/mol)
= 1744 kJ + 1606 kJ
= 3350 kJ

For the overall reaction, (+3532 kJ) + (−3350 kJ) = +182 kJ

In Equation 7.25, 2 mol of C=O bonds and 4 mol of C–H bonds are broken.
= 2 mol (803 kJ/mol) + 4 mol (416 kJ/mol)
= 1606 kJ + 1664 kJ
= 3270 kJ

In Equation 7.25, 2 mol of H–H bonds and 2 mol of C≡O bonds are formed.
= 2 mol (436 kJ/mol) + 2 mol (1073 kJ/mol)
= 872 kJ + 2146 kJ
= 3018 kJ

For the overall reaction, (+3270 kJ) + (−3018 kJ) = +252 kJ

b. Although there is general agreement (+182 kJ vs. +165 kJ), remember that Table 5.1 gives *average* bond energies, not specific energies associated with the bonds in these compounds (with the exception of CO_2). Also, states are not taken into account for the data in Table 5.1.

7.14 **a.** Energy with a wavelength of 420 nm falls in the visible region (violet).

b. Heating water to thermally decompose it into H_2 and O_2 has not yet proven to be effective; temperatures higher than 5,000 °C are required.

7.15 **a.** Metals are elements that are good conductors of electricity and heat. They have a shiny appearance. In contrast, nonmetals that are poor conductors of electricity and heat are not shiny. Additionally, metals have low electronegativity values while nonmetals do not. This helps explain why metals tend to lose electrons to form cations, while nonmetals tend to gain electrons to form anions.

b. NiS contains the Ni^{2+} ion, which is reduced in the reaction; the ion gains two electrons to form Ni metal.

c. $PbS(s) + O_2(g) \longrightarrow Pb(s) + SO_2(g)$. In this reaction, Pb^{+2} is reduced because it gains two electrons to form Pb metal. Additionally, sulfur is oxidized because it loses six electrons to go from S^{2-} to S^{4+}. And oxygen is also reduced because each oxygen atom gains two electrons to form O^{2-}.

d. Sulfur dioxide released from smelting-containing ores contributes to air pollution. The EPA monitors the level of SO_2 as part of ambient air quality standards, due to the serious nature of this air pollutant as a respiratory irritant.

7.16 **a.** No, a metal cannot become extinct, but it can become completely unavailable for commercial applications.

b. Some examples currently used in energy-storage applications include silver, lithium, cadmium, lead, mercury, platinum, nickel, and tin.

c. A 2011 *JACS* publication demonstrated the first metal-free electrocatalyst for fuel cell applications that utilized carbon nanotubes.

7.17 a. There are some statistics that almost 90% of all rechargeable lead–acid batteries are recycled in the United States, while less than 2% of consumer disposable batteries are recycled.

b. If a Ni–Cd battery is not disposed of properly and ends up in a landfill, the metal cylinder will corrode and the cadmium will pollute the water supply. This represents a serious environmental concern.

c. Household batteries are often referred to as "disposable batteries." This name implies that the battery should be disposed of after use. The majority of people who throw away batteries are unaware of the toxic metals housed within the battery and the implications this has for the environment. On the other hand, car batteries are often replaced by auto mechanics who know how to properly recycle the battery.

Chapter 8

REFLECTION

Answers will vary. Some sources of water are municipal water treatment plants or wells. The water-use habits of a community such as overuse or pollution from the use of pesticides or herbicides can affect the availability of clean water for others.

8.1 Student answers will vary. One answer could be that water can be moved easily by wind or it takes the shape of its environment. The image of a calm lake shows how abundant water is and how it is part of a larger ecosystem. The image of rough seas shows how water can be damaging to humans. The list of opposites will vary. If we had a world without water, life would cease to exist. The world would become dusty and dry.

8.2 Student answers will vary here depending on their prior knowledge of water.

8.3 a. Student answers will vary here. However, students may discuss showering/bathing as a cleaning process. They might discuss drinking water to stay hydrated, using water to wash clothing/pets/cars.

b. Sources of water will vary depending on the location of the student, well, municipal water, etc. The student could take it even further and say where the municipal water comes from.

c. The degree to which they got the water dirty will also vary given the task.

8.4 a. Student answers will vary in their level of surprise in their water usage.

b. The actions they could take vary to taking shorter showers, not leaving the water running while brushing their teeth, or collecting rainwater for outdoor tasks such as watering plants or washing the car.

8.5 a. Solids: Yes; Liquids: Yes; Gases: No

b. Solids: Yes; Liquids: No; Gases: No

c. Solids: No; Liquids: Yes; Gases: Yes

d. Solids: No; Liquids: No; Gases: Yes

8.6 It can be concluded that these boxes are not containers but zoomed in areas. The boxes do not follow the properties shown in the table. For example, the solid appears to take the shape of the container. This representation shows how the atoms are positioned with respect to each other.

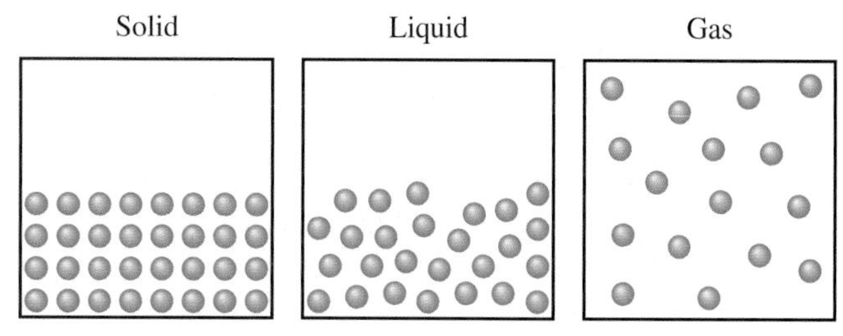

8.7 a. H–F; The electron pair is more attracted to the F atom.

b. O–H; The electron pair is more attracted to the O atom.

c. N–O; The electron pair is more attracted to the O atom.

d. Cl–C; The electron pair is more attracted to the Cl atom.

8.8 a. $\ddot{\underset{\cdot\cdot}{O}}=C=\ddot{\underset{\cdot\cdot}{O}}$

b. The covalent bonds in CO_2 are polar.

$$\overset{\delta^-}{O} \Leftarrow \overset{\delta^+}{C} \Rightarrow \overset{\delta^-}{O}$$

c.

$$\delta^- \Leftarrow \delta^+ \Rightarrow \delta^-$$
$$\ddot{\underset{\cdot\cdot}{O}}=C=\ddot{\underset{\cdot\cdot}{O}}$$

d. The CO_2 molecule is linear and the charges are symmetrical. There isn't a region of unequal sharing as seen in H_2O's bent molecule.

e. Answers will vary depending on the molecule that the student chooses.

8.9 a. The dashed lines represent hydrogen bonds that are intermolecular forces between the hydrogen on one water molecule and the oxygen of another water molecule.

b. The partial charges show how the partial positive of the hydrogen line up with the partial negative of the oxygen in the water molecules.

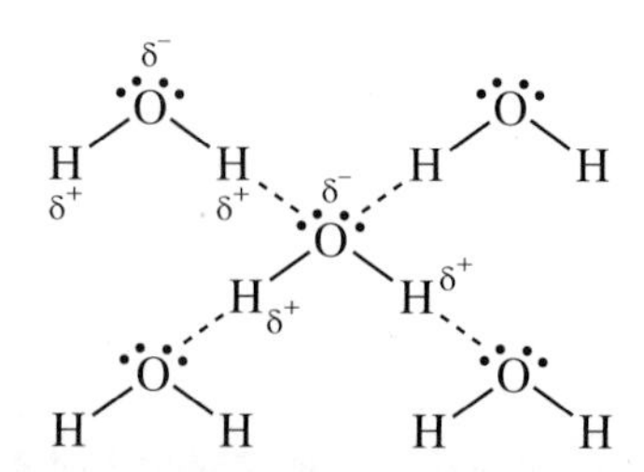

c.

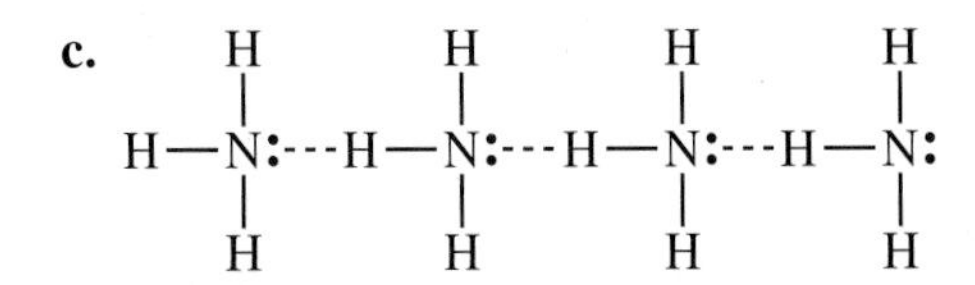

8.10 No, covalent bonds are not broken when water boils. Rather, the hydrogen bonds between the molecules are broken. The drawing shows that the molecules of water are still composed of H_2O before and after boiling.

Water liquid

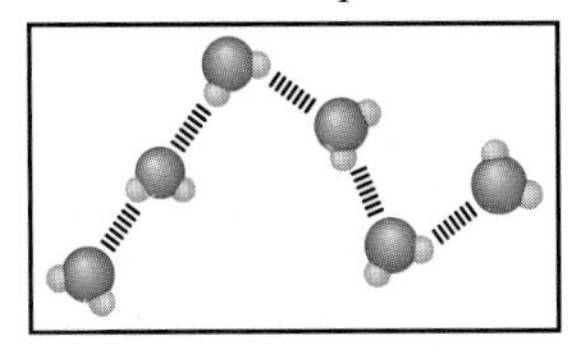

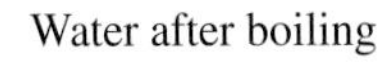

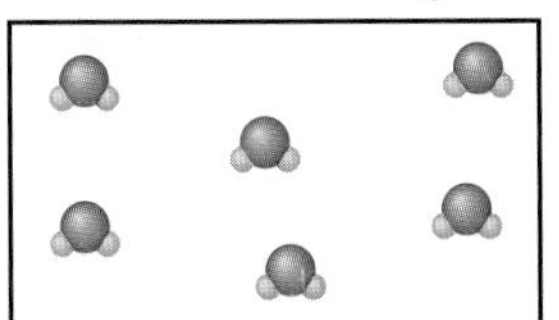

8.11 The carpet feels warm and the stone or tile floor feels cool to your bare feet. The stone floor has a lower heat capacity than the carpet. It is denser than the carpet and a better conductor of heat. It takes less energy to make a 1 °C temp change for a substance with a lower heat capacity, so you notice a more drastic temperature change on the tile because heat is being more rapidly removed from your feet.

8.12 Student answers will vary here, but should include: high specific heat, its solid is less dense than its liquid form, its density, its bent shape, and its polarity.

8.13 **a.** Non-potable water could be used to wash clothing, to put out fires, and for decoration purposes in fountains.

b. Student answers may vary, but an answer could be: A community might use non-potable water if fresh drinking water is scarce.

c. Student answers will vary.

d. Student answers will vary.

8.14 **a.** Student answers will vary.

b. Student answers will vary. An example is California uses surface water for irrigation. If surface water is collected and used, there can be conflict because then groundwater supplies are not replenished.

c. Student answers will vary. Rural areas depend on groundwater. Texas and Florida also depend largely on groundwater. Given the large populations of Texas and Florida, it doesn't factor in population density in that case. However, in the case of well water, population density is a factor.

8.15 Student answers will vary. Sample math is shown.

$$2000 \text{ mL} = \text{Earth's water}$$
$$2000 \text{ mL} \times 0.03 = 60 \text{ mL} = \text{Freshwater}$$
$$60 \text{ mL} \times 0.003 = 0.18 \text{ mL} = \text{Surface water}$$
$$0.18 \text{ mL} \times 89\% = 0.1602 \text{ mL} = \text{Lakes + Rivers}$$
$$0.1602 \text{ mL} \times \frac{1 \text{ drop}}{0.05 \text{ mL}} = 3.204 \text{ drops of water}$$

8.16 Student answers and diagrams will vary. A good resource for them to use is http://water.usgs.gov/edu/earthwherewater.html.

8.17 Student answers will vary depending on the state in which they live. "Other" uses for water can include mining, commercial, and aquaculture.

8.18 Crops use far less water than meat. This is because the animals used for meat need water to stay hydrated, and the food they eat includes the crop water usage as well.

8.19 Student answers will vary. Suggestions for reducing water footprints could be to use local ingredients, less processing of foods, better irrigation practices, etc.

8.20 Student answers will vary depending on the event they choose and the ecosystem/area that was affected.

8.21 Student answers will vary depending on the area in which they live. Answers may include pollution or overdrawing reserves of surface water.

8.22 Student answers will vary depending on the event chosen. Answers may include discussions of oil spills or chemical leaks.

8.23 **a.** Student answers may vary. They may include toothpaste, shampoo, conditioner, hand soap, body wash, face wash, makeup, moisturizer, etc.

b. These products enter the water supply as you wash them off your skin. Wastewater treatment plants do not filter out all chemicals and contaminants. For example, microbeads from products can pass through the filters into the environment.

c. Student answers will vary. One could use products that have less harmful chemicals to the water supply. One could wash their hair less often and use less shampoo. Many people overuse to get more suds and it is not necessary.

d. Student answers will vary, but showering and brushing teeth will likely be on the list. When you wash your hair, you keep running water until all of the suds are out of your hair.

8.24 Student answers will vary in how their personal habits would be affected by these issues. Contamination would likely affect a student the most in the short term as the water would not be potable.

8.25 Student answers will vary but must include the proper relationship between percent, parts per million and parts per billion.

8.26 **a.** $5\cancel{\text{L}} \times \frac{1000 \cancel{\text{mL}}}{1 \cancel{\text{L}}} \times \frac{1 \text{ g}}{1 \cancel{\text{mL}}} = 5000 \text{ g } H_2O$

$$80 \text{ µg mercury ion} \times \frac{1 \text{ g}}{1{,}000{,}000 \text{ µg}} = 0.00008 \text{ g}$$

$$\frac{0.00008 \text{ g mercury ion}}{5000 \text{ g } H_2O} = 0.000000016 \times 10^6 = 0.016 \text{ ppm}$$

$$\times 10^9 = 16 \text{ ppb}$$

b. The EPA cutoff for Mercury in water is 2 ppb (0.002 ppm). This would not be an acceptable limit.

8.27 Student answers will vary depending on where they live. States should follow national regulations, at the very least, but may have even stricter guidelines.

8.28 **a.** $16\text{ ppb} = \frac{16\ \cancel{\mu g}}{1\text{ L}} \times \frac{1\text{ g}}{1{,}000{,}000\ \cancel{\mu g}} = \frac{16\text{ g}}{1{,}000{,}000\text{ L}}$

$= 0.000016\frac{\text{g}}{\text{L}}$

$\frac{0.0000\ 16\ \cancel{\text{g Hg}^{2+}}}{1\text{ L}} \times \frac{1\text{ mol}}{200.59\ \cancel{\text{g Hg}^{2+}}} = 7.98 \times 10^{-8}\text{ M}$

b. $\frac{1.5\text{ moles NaCl}}{1\ \cancel{\text{L}}} \times 0.5\ \cancel{\text{L}} = 0.75\text{ moles NaCl}$

$\frac{0.15\text{ moles NaCl}}{1\ \cancel{\text{L}}} \times 0.5\cancel{\text{L}} = 0.075\text{ moles NaCl}$

c. 1st Solution:

$\frac{0.50\text{ moles NaCl}}{0.250\text{ L}} = 2\text{ M}$

2nd Solution:

$\frac{0.60\text{ moles NaCl}}{0.200\text{ L}} = 3\text{ M}$

The second solution is more concentrated.

d. $40\text{ g CuSO}_4 \times \frac{1\text{ mol}}{159{,}609\text{ g}} = 0.251\text{ moles CuSO}_4$

$\frac{0.251\text{ moles}}{1\text{ L}} = 0.251\text{ M}$

The student made a solution that was 0.251 M and not 2.0 M. The student would have needed to add 2 moles or 319.28 g of copper(II) sulfate to 1000 mL to get this molarity.

$2\text{ M} = \frac{\text{x moles CuSO}_4}{1\text{ L}}$

$\text{x} = 2\text{ moles CuSO}_4 \times \frac{159.609\text{ g}}{1\text{ mol CuSO}_4} = 319\text{ g CuSO}_4$

8.29

	Ionic	**Molecular**
Definition in terms of electrons	Transferred electrons	Shared electrons
Structure	Formula units made by attractive forces between anions and cations	Molecules
Macroscopic properties	Form crystals; higher melting point; mostly soluble in water; high boiling point	Can be solids, liquids, or gases; low melting point; solubility in water depends on polarity of molecule; low boiling point
Atomic properties	Metals and nonmetals transferring electrons	Nonmetals sharing electrons
Diagram of a bond	Student diagrams will vary	Student diagrams will vary
Bond strength	The charges pull the cations and anions together, resulting in a generally stronger bond.	Not as strong as ionic for the most part. Triple bonds are stronger than double bonds are stronger than single bonds.
Nomenclature	Name the cation and then the anion, only use Roman numerals for metals with multiple possible charges.	Elements are put in increasing group number; number of each atom type gets a prefix; last element has -ide on end.
Other	Student answers will vary	Student answers will vary

8.30 Water contains dissolved ions that conduct electricity. The best course of action is to unplug the hair dryer from the wall without touching the hair dryer or the water. Then, remove the hairdryer from the water.

8.31 **a.** (ammonium nitrate) and **b.** (sodium sulfate) are soluble in water. Because of the sulfide and hydroxide anions in the compounds **c.** and **d.** are insoluble in water.

8.32 Using the electronegativity values, one would expect the C–H bond and the C–C bond to be nonpolar. These molecules are nonpolar because the electronegativity differences between them are very small.

8.33 Student drawings will vary. A sample answer is:

Polar molecules   Polar head   Nonpolar tail   Nonpolar molecules

8.34 **a.** Liquid carbon dioxide is a nonpolar solvent and has a greater solvent power than some organic solvents.

b. Student answers will vary. While all applications will have some environmental impact, the environmental impact of carbon dioxide as a solvent is less than that of organic solvents. It isn't toxic or flammable.

c. The social aspect will increase with less toxic chemicals being used around their employees. The environment aspect will be increased with less toxic chemicals affecting the environment. The economic aspect will decrease as the cost of the carbon dioxide method is higher than the perc method. Note: The triple bottom line was introduced in Section 5.17.

8.35 Student tables will vary. Soluble ionic compounds will dissociate into cations and anions. Each will be surrounded by water molecules with the partial positive side toward the anion and the partial negative side toward the cation. Ionic solutions conduct electricity. Ionic compounds that are dissolved in water can be consumed by plants and animals having a negative impact. Ionic solutions can be used to transport necessary ions for sustaining life.

Molecular compounds can be polar (*e.g.*, H_2O or sugar) or nonpolar (*e.g.*, I_2, N_2). Polar molecular compounds will be surrounded by water molecules with the partial positive side of water toward the partial negative portion of the polar molecule. The partial negative side of the water molecule will orient toward the partial positive portion of the polar molecule. Solutions of dissolved molecular compounds do not conduct electricity. These molecules can result in bioaccumulation in an ecosystem. Polar solutions can be used to transport molecules from one medium to another. Often times, one chemical will evaporate. This is how nail polish remover works.

8.36 Student answers will vary. Adding salt to boiling water to make pasta would be an example of ionic compounds. Adding Kool-aid to water would be an example of molecular compounds. Washing a car would be an example of adding solutes to the water.

8.37 **a.** $HI(aq) + H_2O(l) \longrightarrow I^-(aq) + H_3O^+(aq)$

b. $HNO_3(aq) + H_2O(l) \longrightarrow NO_3^-(aq) + H_3O^+(aq)$

c. $H_2SO_4(aq) + H_2O(l) \longrightarrow HSO_4^-(aq) + H_3O^+(aq)$

$HSO_4^-(aq) + H_2O(l) \longrightarrow SO_4^{2-}(aq) + H_3O^+(aq)$

8.38 Student answers will vary. Examples could be lemon juice or vinegar as acids. They could be added for taste.

8.39 **a.** $KOH(s) \xrightarrow{H_2O(l)} K^+(aq) + OH^-(aq)$

b. $LiOH(s) \xrightarrow{H_2O(l)} Li^+(aq) + OH^-(aq)$

c. $Ca(OH)_2(s) \xrightarrow{H_2O(l)} Ca^{2+}(aq) + 2\ OH^-(aq)$

8.40 Student answers will vary depending on their experiences.

8.41 Neutralization Reactions:

a. $HNO_3(aq) + KOH(aq) \longrightarrow KNO_3(aq) + H_2O(l)$

b. $HCl(aq) + NH_4OH(aq) \longrightarrow NH_4Cl(aq) + H_2O(l)$

c. $2\ HBr(aq) + Ba(OH)_2 \longrightarrow BaBr_2(aq) + 2\ H_2O(l)$

Total ionic equations:

a. $H^+(aq) + NO_3^-(aq) + K^+(aq) + OH^-(aq) \longrightarrow K^+(aq) + NO_3^-(aq) + H_2O(l)$

b. $H^+(aq) + Cl^-(aq) + NH_4^+(aq) + OH^-(aq) \longrightarrow NH_4^+(aq) + Cl^-(aq) + H_2O(l)$

c. $2\ H^+(aq) + 2\ Br^-(aq) + Ba^{2+}(aq) + 2\ OH^-(aq) \longrightarrow Ba^{2+}(aq) + 2\ Br^-(aq) + 2\ H_2O(l)$

Net ionic equations:

a. $H^+(aq) + \cancel{NO_3^-(aq)} + \cancel{K^+(aq)} + OH^-(aq) \longrightarrow \cancel{K^+(aq)} + \cancel{NO_3^-(aq)} + H_2O(l)$

$H^+(aq) + OH^-(aq) \longrightarrow H_2O(l)$

b. $H^+(aq) + \cancel{Cl^-(aq)} + \cancel{NH_4^+(aq)} + OH^-(aq) \longrightarrow \cancel{NH_4^+(aq)} + Cl^-(aq) + H_2O(l)$

$H^+(aq) + OH^-(aq) \longrightarrow H_2O(l)$

c. $2\ H^+(aq) + \cancel{2Br^-(aq)} + \cancel{Ba^{2+}(aq)} + 2\ OH^-(aq) \longrightarrow \cancel{Ba^{2+}(aq)} + \cancel{2Br^-(aq)} + 2\ H_2O(l)$

$2\ H^+(aq) + 2\ OH^-(aq) \longrightarrow 2\ H_2O$

8.42 **a.** $[H^+] = 1 \times 10^{-4}\ M$

$$[OH^-] = \frac{1 \times 10^{-14}}{1 \times 10^{-4}} = 1 \times 10^{-10}\ M$$

Acidic

b. $[OH^-] = 1 \times 10^{-6}$

$$[H^+] = \frac{1 \times 10^{-14}}{1 \times 10^{-6}} = 1 \times 10^{-8}\ M$$

Basic

c. $[H^+] = 1 \times 10^{-10}\ M$

$$[OH^-] = \frac{1 \times 10^{-14}}{1 \times 10^{-10}} = 1 \times 10^{-4}\ M$$

Basic

8.43 **a.** Basic, $[OH^-] > [K^+] > [H^+]$

b. Acidic, $[H^+] > [NO_3^-] > [OH^-]$

c. Acidic, $[H^+] > [HSO_4^-] > [SO_4^{2-}] > [OH^-]$

d. Basic, $[OH^-] > [Ca^{2+}] > [H^+]$

8.44 Answers will vary, based on the Connect Interactive Activity.

8.45 **a.** Pure water, milk, tomato juice, lemon juice

b. Student answers will vary.

c. Students should answer yes and then describe the pH values. The values will vary depending on their water logs.

8.46 **a.** The lake water with a pH of 4.0 is more acidic. There are 10 times more hydrogen ions in lake water than rainwater.

b. The tap water with a pH of 5.3 is more acidic. There are 1,000 times more hydrogen ions in the tap water than the ocean water.

c. The tomato juice with a pH of 4.5 is more acidic. There are 100 times more hydrogen ions in the tomato juice than the milk.

8.47 Student answers will vary, but should include the fact that a pH of 0 would be extremely acidic and bad for the environment. The senator likely means a pH of 7, to be completely neutral.

8.48 Student answers will vary depending on their water diaries.

8.49 Student answers will vary. For example, in 2012 scientists met at the Third Symposium on the Ocean in a High-CO_2 World.

8.50 **a.** $HCO_3^-(aq) + H^+(aq) \longrightarrow H_2CO_3(aq)$

b. The bicarbonate ion is functioning as a base by accepting H^+.

8.51 **a.** Sulfate: SO_4^{2-}; Hydroxide: OH^-; Calcium: Ca^{2+}; Aluminum: Al^{3+}

b. $CaSO_4$, $Ca(OH)_2$, $Al_2(SO_4)_3$, $Al(OH)_3$, among other compounds that include additional species. Make sure students DO NOT include alcohols in this list.

c. Sodium hypochlorite: NaClO; Calcium hypochlorite: $Ca(ClO)_2$

8.52 **a.** Student answers will vary, but two examples are:

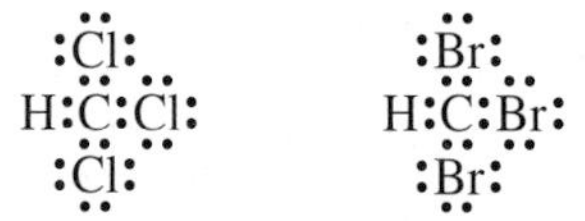

```
  :Cl:          :Br:
H:C:Cl:       H:C:Br:
  :Cl:          :Br:
```

$CHCl_3$	$CHBr_3$
Chloroform	Bromoform

b. THMs have three halogens and one hydrogen. CFCs contain only the chlorine and fluorine halogens and no hydrogens.

c. THMs have higher boiling points than CFCs.

8.53 **a.** Oral health has improved because tooth decay has been prevented. This also saves money for communities no matter their socioeconomic status.

b. People in low-income communities may not have the same access to dental care or expensive fluoridated toothpastes as higher-income communities.

c. Student answers may vary. Answers may include: Could cause other health issues, no advantages, too expensive, etc. There is a discussion about individual rights vs. community benefits that is similar to the vaccination controversy.

8.54 Student answers will vary.

8.55 Student answers will vary depending on their water diary. Conservation efforts can include: not letting the tap run while brushing teeth, collecting rainwater, running water appliances on eco-friendly.

8.56 Student answers will vary. One answer may include: Desalination is an energy-intensive process and would result in the burning of fossil fuels. The burning of fossil fuels would then add to greenhouse gases. This would violate the "Design for Energy Efficiency" and "Less Hazardous Chemical Syntheses" key ideas.

8.57 Fluoride is not a heavy metal.

8.58 **a.** Student answers will vary. One answer may include: Communities can pick up garbage from streams. Farmers can plant plants near river banks to prevent fertilizer runoff.

b. Student answers will vary. One answer may include: Water shortages in third-world countries. It is important because the water is dirty and there isn't an abundance of it. Diseases are spread and much time is spent trying to find clean water. Two ways to address this are to collect rainwater and to dig wells.

8.59 Student answers will vary. A suggestion for a different method of tracking water usage could be to view the meter to see actual usage. Students can use timers to better estimate how long they took a shower, etc.

Chapter 9

REFLECTION

a. Answers will vary. For instance: milk jugs, water bottles, or shampoo bottles.

b. milk jugs: 2 (high-density polyethylene), water bottle: 1 (polyethylene terephthalate), shampoo bottle: 3 (polyvinyl chloride).

9.1 Student answers will vary. Answers may include tennis racquet, tennis court flooring, and clothing. Polymers are flexible, which is good for a tennis racquet frame/strings because they affect how the ball comes off the racquet after being hit. Stretchiness is a good feature for tennis clothing, as the athletes are running and moving around and need flexibility. The tennis court flooring can absorb impact, which is good for athletes' joints!

9.2 Polymerization reactions can be stopped if the monomers are completely used up or through a termination reaction.

9.3 **a.**

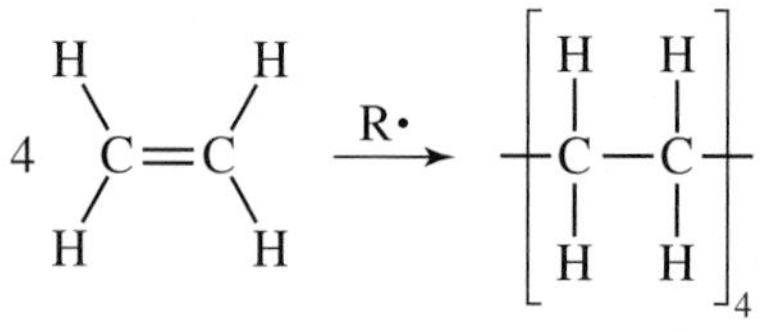

```
  H     H             [ H   H ]
   \   /         R·   [ |   | ]
4   C=C        ----> -[-C---C-]-
   /   \              [ |   | ]
  H     H             [ H   H ]4
```

b.

```
   H   H   H   H   H   H   H   H
   |   |   |   |   |   |   |   |
R--C---C---C---C---C---C---C---C--R
   |   |   |   |   |   |   |   |
   H   H   H   H   H   H   H   H
```

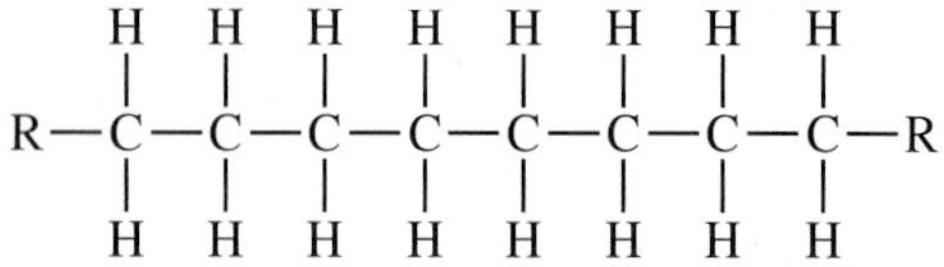

c. Octane is C_8H_{18}. Although the product molecule similarly has eight carbon atoms, it has two fewer hydrogen atoms and two R groups at the ends of the molecule, so its chemical formula is $C_8H_{16}R_2$.

9.4 Student answers will vary. An example of LDPE is plastic wrap. An example of HDPE is plastic bags. LDPE is more flexible than HDPE. LDPE is more translucent than HDPE. HDPE is more often colored with a pigment than LDPE.

9.5 **a.** Necking does not change the number of monomer units.

b. Necking does not affect the bonding between the monomer units with the polymer chain; it's the intermolecular forces that are agitated when necking, not the intramolecular forces (bonds).

9.6 Student answers will vary. Pros may include the energy needed to recycle plastic vs. paper and durability. Cons may include how long plastic takes to biodegrade in the environment, the number of bags used from one shop, the fact that most people won't recycle their single use bags, etc.

9.7 LDPE does not line up in rows because of its branching. HDPE is able to stack rows, adding to its density.

9.8 **a.** Student answers will vary.

b. *Macro* means large. The site is a gallery of large molecules. Similarly, macronutrients are foods that we eat in large quantities.

9.9 Student answers may vary. One answer may include: MDPE is medium-density polyethylene. It is more resistant to stress-cracking than HDPE. LLDPE is linear low-density polyethylene. It is more resistant to punctures than LDPE.

9.10 **a.** Polystyrene, because it is degraded in many organic nonpolar solvents.

b. LDPE if not pigmented, polystyrene if in crystal form, polyethylene terephthalate. Soft drink bottles are made from polyethylene terephthalate.

c. Bottle caps are made from polypropylene. Answers may vary, but may include toughness is also important in luggage material.

d. Polyethylene terephthalate, polyvinyl chloride, HDPE, and LDPE. However, LDPE is typically used for flexible materials. The most likely material would be HDPE.

9.11 Answers will vary, based on the interactive activity on **Connect**.

9.12 **a.** Phenyl group Benzene

$-C_6H_5$ C_6H_6

R—(phenyl ring) R—(phenyl ring) (benzene resonance structures)

b. The two possible resonance structures indicate that the electrons uniformly distribute themselves around the ring. All the C–C bonds are of equal strength and length. The circle inside indicates the uniformity of these six bonds in the ring.

9.13 **a.** Student answers will vary. An example is that polypropylene fibers are used in rope and as an additive to concrete to reduce cracking in case of earthquakes.

b. Polypropylene is resistant to oils.

9.14

H H H H H H
—C—C—C—C—C—C— Head-to-tail
H (phenyl) H (phenyl) H (phenyl)

Head-to-tail is favored because the phenyl groups are large and would have strain if they were immediately next to one another.

9.15 **a.**

$CH_3C(=O)O\text{—}H + H\text{—}OCH_2CH_3$

↓

$H_3C\text{—}C(=O)\text{—}O\text{—}CH_2\text{—}CH_3$

b. No, the reaction ends after the carboxylic acid and alcohol react. In the final product, there are no carboxylic acid or alcohol groups to continue reacting to make another ester group.

9.16

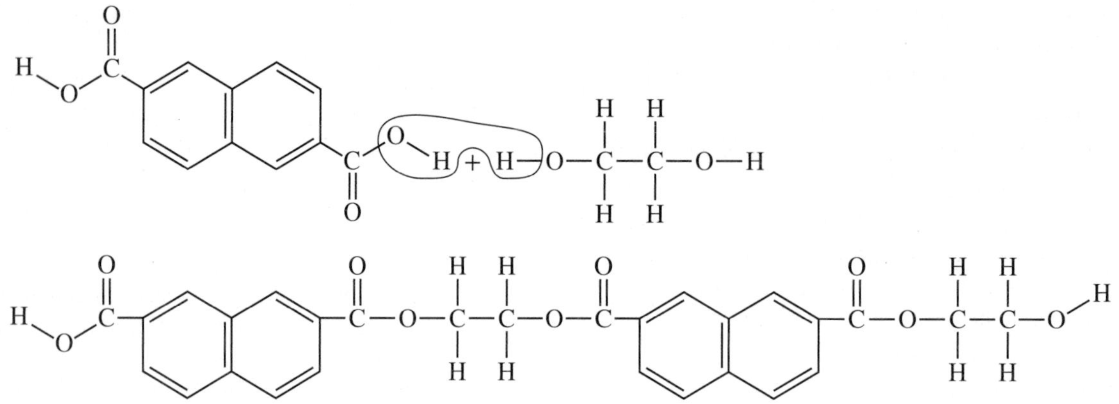

9.17

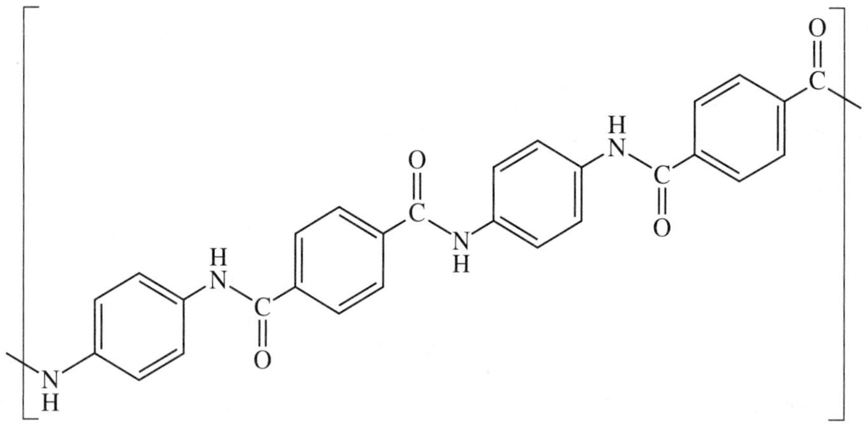

9.18 **a.** Student answers will vary. Some less obvious sources of plastic include Styrofoam or reusable grocery bags composed of polyesters.

b. Student answers will vary (but hopefully students are recycling more than they throw away!).

9.19 Student answers will vary. Many different types of stores offer programs where someone could bring their own reusable bag. Student answers will vary in their willingness to change their "bag habits."

9.20 **a.** Durable goods: 6.9%; Nondurable goods: 2.0%; Containers/Packaging: 15%

b. Student answers will vary. Durable goods may include: computers, TV sets, office chair, carry-on suitcase, lawn furniture. Nondurable goods may include: Disposable silverware, mechanical pencils, Post-it notes. People may not recycle durable items because they are large and difficult to dispose of. People also may not know that they are recyclable. People may not recycle nondurable goods because they seem flimsy or not worth the effort because they are small. Some products have a coating or are painted, which hides the recyclable parts of the product.

9.21 Student answers will vary depending on their plastic-use journals.

9.22 **a.** $2{,}500\ C_2H_4 + 7{,}500\ O_2 \longrightarrow 5{,}000\ CO_2 + 5{,}000\ H_2O$

b. Incomplete combustion produces CO and particulate matter (soot)—both air pollutants.

c. Student answers will vary. Biodegradable plasticizers can be added so that the plastic will biodegrade. Fillers can be added to reduce the total cost of the material. Pigments can be added to polymers to achieve desired colors. Additives could be released as harmful environmental pollutants during combustion.

9.23 **a.** Nonpolar chemicals such as oils can soften HDPE.

b. Examples: cooking oil, shoe polish, alcohols, vinegar, and lighter fluid.

9.24 **a.** 1798 + 1045 + 0.4 + 62 + 0.3 = 2906 million pounds recycled, or 1.45 million tons. This is in the ballpark of the 2.04 million tons quoted by the EPA, especially considering that data for polystyrene are not included in the ACC set, and it does not include plastic containers that are not bottles.

b. Student answers will vary. The metric ton could have been used for international comparisons, as it is based on the metric system. ACC is a U.S.-based company, and the U.S. tends to use pounds rather than metric units such as tons.

9.25 Student answers will vary. Purchase and recycle items might include: soda bottles, shampoo bottles, or milk jugs. Recycled-content products might include: notebooks, egg crates, construction paper. An item could fall into both categories.

9.26 The four plastics could be placed into a saturated solution of $MgCl_2$. PET would sink, but all of the PVC, HDPE and PP would float on the surface.

9.27 **a.** Student answers will vary depending on the five items they found.

b. Student answers will vary depending on the five items they found and their own personal preferences and usage.

9.28 Student answers will vary. The assumption is that the use of plastics will double again in the next 20 years and that recycling rates will remain low. Another assumption is that wildlife populations will remain the same over time and not increase. The steps that can be taken to prevent this scenario are described as (1) create an effective after-use plastics economy, (2) drastically reduce the leakage of plastics into natural systems and other negative externalities, and (3) decouple plastics from fossil feedstocks. One threat to realizing this "new plastics economy" is the fact that current plastics are not "bio-benign."

9.29 **a.**

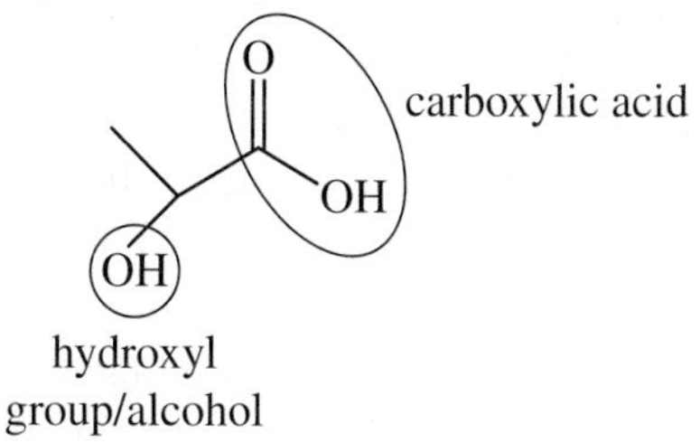

carboxylic acid

O

OH

OH

hydroxyl group/alcohol

b. A condensation reaction because the OH from the carboxylic group and the H from the hydroxyl group form a water molecule, which is a product.

c.

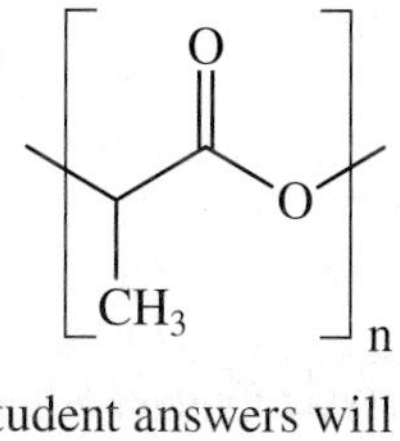

O

O

CH_3

n

9.30 Student answers will vary depending on their campus.

9.31 **a.** Two advantages of plastic jugs: They are lighter and cost less to transport. They are recyclable. Two disadvantages of plastic jugs: Cannot be recycled again into milk jugs because of bacteria from milk. Most milk jugs are not recycled. Two advantages of glass bottles: They can be sterilized to be reused. There is a lot of sand in this world that is available to make more glass. Two disadvantages of glass bottles: It takes a lot of energy to create glass. Glass bottles are heavy and would add to transportation costs.

b. Beer will go flat faster in a plastic bottle than a glass bottle because plastic is more porous than glass is. Chemicals can also leach from the plastic to the beer. Soft drinks are sold in plastic bottles because they are often found in vending machines and they would break if they were glass. The cost of transporting them would also outweigh the money made from selling them.

9.32 **a.** CFCs are linked to the reduction of ozone in the atmosphere; the ozone hole.

b. CFCs break down in the atmosphere to release a chlorine free radical. The free radical acts as a catalyst to break down ozone into oxygen. The reaction propagates and results in another chlorine free radical until a termination step happens. If CFCs stay in the atmosphere for 100 years or more, this process could continue for that long.

c. Until a termination step is executed, a reaction can happen for a long time. In the case of CFCs, this involves breaking down ozone. In the case of polymers, this involves adding monomers together to create long polymer chains.

d. Student answers will vary. One answer could be that plastics are so versatile and are critical for use in existing consumer products. Many industries have evolved and adapted to using plastic. Phasing out plastic would create a huge financial and economic burden to many industries.

e. Student answers will vary. One answer could be that we cannot sustain our current use of plastics because we will run out of space for disposal. Because plastics do not decompose quickly, they will end up stacking up in landfills or floating in the oceans. Fossil fuels are the source of many plastics, and these are not unlimited commodities.

9.33 **a.** An ester is created when an alcohol and a carboxylic acid react. They are characterized by a C–O–C bond and a C=O group on one of the carbons.

b.

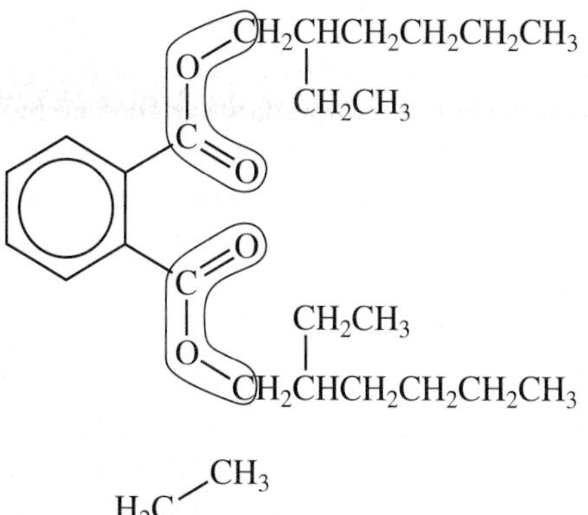

$CH_2CHCH_2CH_2CH_2CH_3$

O

CH_2CH_3

C

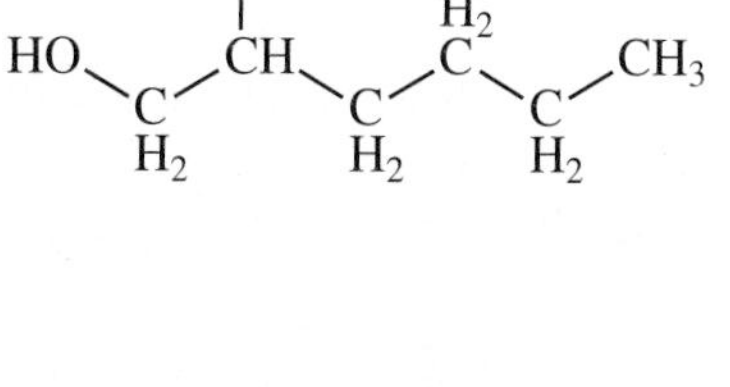

O

O

C

CH_2CH_3

O

$CH_2CHCH_2CH_2CH_2CH_3$

c.

CH_3

H_2C

HO $\quad$ CH $\quad$ H_2 C $\quad$ CH_3

C $\quad$ C $\quad$ C

H_2 $\quad$ H_2 $\quad$ H_2

Chapter 10

REFLECTION

Flavor beads are composed of a gel similar in composition to that of Jell-O®. As the pH of the solution changes, the color of beads changes due to the presence of a pH-sensitive dye.

10.1 Student answers may vary. Some students may respond with the traditional diagram with sour on the sides of the tongue, bitter on the back, and salty/sweet on the front. Some students may have very sensitive palettes and some may have a very dulled sense of taste.

10.2 Recent research published in *Nature* shows that taste buds across the entire tongue can sense all of the tastes and it isn't defined necessarily by regions. There is also an additional taste called umami! Umami is a savory taste. This will serve as a good discussion with "foodie" students about the sensitivity of taste buds and mixing of flavors to achieve delicious results in meals.

10.3 Student answers may vary. Some students may be able to correctly guess the identity of each sample without smelling it, but a majority will be incorrect in their guesses. The students may discuss experience with texture and a difficulty discerning between the tastes they are sensing. For the jelly bean example, the students are able to tell that the samples are overwhelmingly sweet, but without a sense of smell they will report difficulty in discerning flavors.

10.4 White, milk, and dark chocolate each contain different amounts of cocoa beans, thus having different amounts of cocoa butter (fat) and cocoa. This gives them different consistencies/textures and flavors. If the chocolate is allowed to melt in the mouth, the students will notice that each melts differently in the mouth. The higher the percentage of cocoa beans, the lower the amount of sugar and it will taste less sweet.

10.5 Student answers will vary as each person has a different favorite part of a chocolate chip cookie. Crunchy cookies have more flour, sugar and baking soda (and overall a greater mass of ingredients) than chewy cookies. To test the effects of each ingredient or combination of ingredients, a student could make several batches of cookies that vary the individual and combinations of ingredients. The only thing that can be confirmed after the experiment is the effects of each ingredient on the outcome, not which is the best—that is subjective!

10.6 The entire recipe needs to be multiplied by four. This yields 2 cups butter, 4 cups chocolate chips, 2 cups brown sugar, 2 cups white sugar, 4 eggs, 2 teaspoons vanilla, 5 cups flour, 3 teaspoons baking soda, and 1 teaspoon salt.

10.7 If you use all of the cheese and have an unlimited supply of tortilla shells, you could make seven "perfect quesadillas." If you only have eight tortillas, you can make four "perfect quesadillas" and will use 200 g of cheese. The tortillas will be consumed first and there will be 150 g of cheese left over.

10.8 Student answers may vary. The number of grams per cup/teaspoon/tablespoon is dependent on the ingredient, because each ingredient has different densities. 1.5 kilograms of apples is roughly 12.5 cups. 150 grams of sugar is roughly ¾ cup. 25mL of cornstarch is roughly 16 grams and this converts to 2 tablespoons. 4 mL of cinnamon is roughly 1¼ teaspoons. 0.75 mL of salt is between ¼ and 1/8 teaspoons. 0.75 mL of nutmeg is between ¼ and 1/8 teaspoons. 40 grams of butter is roughly 3 tablespoons. This illustrates how using metric units is more accurate and leads to replicable results!

10.9 The average density for flours are as follows: all purpose flour = 0.55 g/mL, cake flour = 0.51 g/mL, whole wheat flour = 0.55 g/mL, potato flour = 0.68 g/mL, corn flour = 0.68 g/mL, and almond flour = 0.38 g/mL. Potato flour and corn flour have the highest densities. (These calculations used 236 mL = 1 cup, and 16 Tbsp = 1 cup.)

10.10 If you do not add salt, it will only change the flavor of the pasta, despite what you may have been told (salt does not make water boil fast, if anything it boils at a higher temperature, but only if you add a significant amount of salt to the water). If you cook the pasta longer than the al dente cook time, it will become mushy. The cooking time will be longer at higher altitudes because the boiling point of water is lower at high altitudes.

10.11 The spaghetti noodles in Denver will take longer to cook than in Los Angeles. This is because the boiling point of water is lower in Denver, thus they need a longer time to cook the noodles.

10.12 *Sous vide* cooking allows for consistent temperature and texture throughout the meat. However, it can take longer to cook this way and would use more electricity. Boiling food is a relatively fast process, but can only be used with foods that aren't messy in water. Boiling points also vary by altitude, so there can be variability between locations. Pressure cookers reduce cooking time and result in less electricity being used. However, the pressure inside a pressure cooker varies with altitude and can result in a variability of outcomes. Cooking with heat/flame allows for the Maillard reaction to occur, which adds to the taste and texture of food. However, some of the by-products of the Maillard reaction or from charred food are carcinogenic. Using coals or flames can also produce by-products that are pollutants. Student answers may vary as to which method is the most environmentally stable depending on the focus of their responses.

10.13 **a.** Modernist Cuisine notes that the vast majority of *sous vide* bags are made with high-density polyethylene (HDPE), low-density polyethylene (LDPE), or polypropylene.

b. According to the Pacific Northwest Pollution Prevention Resource Center, it has not been confirmed that harmful chemicals leech into food from the bags. However, they note that oily and acidic foods could have the potential to increase leeching. The studies they summarize do show the migration of chemicals, but not the exact formulas to find toxicity, nor do they study the *sous vide* specifically. The temperatures of *sous vide* are not high enough to break down the polymer.

c. Inexpensive bags and films in the kitchen can have more harmful leeching than *sous vide* bags because they can contain plasticizers, which aren't healthy to ingest.

10.14 **a.** Microwave regions have the longest wavelength and lowest energy of the three. UV has the shortest wavelength and highest energy of the three. IR falls in between them. Microwaves cause the molecules to rotate. IR waves cause molecules to vibrate and stretch. UV can have enough energy to break bonds.

b. Radio waves have a longer wavelength and even less energy than microwaves. Microwaves already have inconsistent heating because they cannot penetrate food all the way. Radio waves would be able to penetrate even less. This would result in even more inconsistent heating than microwaves. Additionally, it would take longer to cook, and thinner foods would need to be used.

10.15 A microwave oven uses waves to penetrate into the food and rotate the molecules. A conventional oven heats the air around the food and the heated air cooks the food through conduction. Microwaves do no cook food from the inside out, but the molecules at a certain depth inside the food are excited by the waves, which makes it seem like they are cooking from inside out.

10.16 Student answers will vary here. One good resource to drive the conversation is a blog post from *Do the Math* by Tom Murphy: http://physics.ucsd.edu/do-the-math/2012/05/burning-desire-for-efficiency/. Another good resource for this conversation is a post by Pablo Paster on the TreeHugger website: http://www.treehugger.com/clean-technology/ask-pablo-electric-kettle-stove-or-microwave-oven.html. The authors each use different methods for collecting data, but come to the overall similar conclusion that boiling water on a stovetop is a very inefficient way (and thus requires more electricity and produces more associated greenhouse gases) to boil water compared to a microwave. Some students may argue that although boiling water on a stovetop is less efficient, it also acts to heat the kitchen and serves a dual purpose. Others may argue that due to its low efficiency, over the span of a year it would be more sustainable to use a microwave. Students who are especially eco-conscious may point out that the most efficient way to boil water is to actually use a kettle!

10.17 Student answers will vary depending on the foods they ate. They may have had cured meats, smoked fish, pickled gherkins, or dried fruits. This could be a good point of discussion about shelf-life of the different foods and even if different cultures or regions have higher prevalence of different methods.

10.18 **a.** 125 °F; lower than 145 °F on foodsafety.gov

b. 135 °F; lower than 145 °F on foodsafety.gov

c. 145 °F; equal to 145 °F on foodsafety.gov

d. 150 °F; above 145 °F on foodsafety.gov

e. 160 °F; above 145 °F on foodsafety.gov

f. 165 °F; equal to 165 °F on foodsafety.gov

g. 165 °F; equal to 165 °F on foodsafety.gov

h. 140 °F; lower than 145 °F on foodsafety.gov

10.19 17.8 °Bx is equal to 17.8 %(w/w). This converts to 17.8 grams of sucrose per 100 g of solution. 17.8 grams of sucrose is 0.052 moles of sucrose:

$$17.8\ \text{g} \times \frac{1\ \text{mol}}{342.3\ \text{g}};\ 100\ \text{g of solution} \times \frac{1\ \text{mL}}{1.06\ \text{g}} = 94\ \text{mL}$$

of solution. So, 0.052 moles sucrose / 0.094 L of solution = 0.55 M sucrose solution.

10.20 Some of the cans are floating and some are at the bottom. Regular soda contains sugar and diet soda contains sugar substitutes that are much sweeter than sugar. They require less sweetener because of this. Regular soda is therefore more dense than diet soda and will sink in water. The diet soda cans will float in water. Students may suggest that the amount of carbonation or density of other ingredients in the soda could affect the outcome. Or the amount of air between the soda and can could affect whether the can sinks or floats if two cans of regular soda differ.

10.22 The concentration of CO_2 in the bottle before it is opened is 0.039 M (C = 1.25 atm × 0.031 mol/L·atm). When the bottle is opened, the concentration inside the bottle is 0.031M (C = 1.00 atm/ 0.031 mol/L·atm). There are 0.019 moles of carbon dioxide in the solution before it is opened $\frac{0.039\ \text{mol}}{1\ \text{L}} \times 0.500\ \text{L}$ and 0.015 moles after $\frac{0.031\ \text{mol}}{1\ \text{L}} \times 0.50\ \text{L}$. This means that 0.004 moles of CO_2 escaped. This is 0.2 g $0.004\ \text{mol} \times \frac{44\ \text{g}}{1\ \text{mol}}$ of carbon dioxide that escaped the bottle.

Chapter 11

REFLECTION

Answers will vary. Although fruits and vegetables are healthy choices for our diet, these foods may not be grown in an environmentally sustainable manner.

11.1 Student answers will vary in their calculations and assumptions. Assumptions should include average life span and constant body weight over time. The USDA estimates that the average American eats nearly one ton (2,000 pounds) in one year. Students could use this to continue their calculations.

11.2 **a.**

O
OH

b. carboxylic acid group, –COOH

11.3 **a.** Oleic, linoleic, and linolenic acids all have at least one C=C double bond.

b. It is a saturated fatty acid.

O
OH

11.4 Student answers will vary. A sample saturated and unsaturated fatty acid is shown.

Lauric acid, saturated fatty acid

$CH_3CH_2CH_2CH_2CH_2CH_2CH_2CH_2CH_2CH_2CH_2—C(=O)OH$

Oleic acid, monounsaturated fatty acid

$CH_3CH_2CH_2CH_2CH_2CH_2CH_2CH_2CH{=}CHCH_2CH_2CH_2CH_2CH_2CH_2CH_2—C(=O)OH$

Double bonds change the structure so that they are less able to line up and have intermolecular interactions. Because there are fewer intermolecular forces between the molecules, they are more easily separated, and the boiling points are lower.

11.5 **a.** A molecule of octane has fewer carbon atoms than a fat or oil. Octane contains neither a carboxylic acid functional group nor any C=C double bonds. In contrast, fats and oils may contain these groups.

b. A molecule of biodiesel has a similar number of carbon atoms to fats and oils. Biodiesel contains a carboxyl group (C=O) like fats and oils, but no C=C bonds.

11.6 **a.** It is likely soybean oil because it contains mostly polyunsaturated fats and nearly equal amounts of saturated and monounsaturated fats. This matches the composition of soybean oil having large amounts of linoleic acid (a polyunsaturated fat).

b. Student answers will vary depending on their background knowledge, but the correct answer is that vitamin E is part of the oil itself.

11.7 **a.** I Can't Believe It's Not Butter™ has the highest percent of saturated fat (2g/9g=22%). It is still lower than butter (7g/11g=64%).

b. 9% (1g/11g) of the total fat in butter is polyunsaturated.

c. Student answers will vary. Unsaturated fats are healthier than trans fats and saturated fats. Butter contains 36% unsaturated fats, Land O' Lakes™ 55%, I Can't Believe It's Not Butter™ 61%, and Benecol™ 81%.

d. Student answers will vary. One sample answer is the introduction of more sugar into the diet to cover the taste that is missing from removing fats from the diet, which leads to obesity and health issues.

11.8 **a.** The design reduces or eliminates the use or generation of hazardous substances. It uses fewer resources and uses renewable resources.

b. Recall that an enzyme is a catalyst in living organisms. There are catalysts that can help interesterification, but they produce potentially harmful by-products and can be rather complicated. A recent review of the difference between chemical and enzymatic interesterification processes is found in Green Vegetable Oil Processing: AOCS Press: 2014, 205-224 (DOI:10.1016/B978-0-9888565-3-0.50014-5).

11.9 **a.** Student answers will vary. They may contain descriptions of the hydrogen bonding between the sugar and the receptors on the taste buds. Different chemical formulas mean different properties, including degree of sweetness.

b. They are all broken down the same way in the body and thus provide the same amount of energy per gram. Energy per gram is dependent on the chemical bonds being broken and formed, and each of these compounds can be represented by the same chemical formula, so they release the same amount of energy.

11.10 Student answers will vary. One answer could include: Yes, soda has as much if not more added sugar than a candy bar. A can of soda can contain up to 40 grams of sugar. A candy bar such as Snickers has 47 grams of sugar. They are comparable. A student could also compare diet soda with a candy bar and say that it is not a fair characterization because diet soda does not contain sugar but artificial sweeteners.

11.11 a. The summary of data shows that non-Hispanic black men and women consumed more Calories from added sugar in relation to their total Calories than non-Hispanic white or Mexican-American men and women. The number of Calories from added sugars with respect to the total diet declined as age and income increased in adults. For children, the number of Calories from added sugars with respect to the total diet increased with age.

b. Student answers will vary, but in general if 5–15% of the calories should come from added fats and sugars, the values across all groups for added sugars alone were very high.

c. Student answers will vary. In the United Kingdom, it is recommended that no more than 5% of the Calorie intake be from added sugars each day. The World Health Organization (WHO) recommends less than 10% of the Calorie intake be from added sugars each day. Discrepancies may be due to local food ingredient regulations or lifestyles.

11.12 a. Natural compounds are also chemicals.

b. Artificial sweeteners are artificially created or naturally occurring compounds that offer the sweetness of sugar without as many Calories.

11.13 Gly-Gly-Gly, Gly-Gly-Ala, Gly-Ala-Gly, Ala-Gly-Gly, Gly-Ala-Ala, Ala-Gly-Ala, Ala-Ala-Gly, Ala-Ala-Ala

11.14 Folic acid is water soluble because it is a polar molecule (NH and OH sections of the molecule make it polar).

11.15 The observed skin discoloration is due to the fat-soluble component called carotene. It collects in the body rather than being excreted daily.

11.16 a. Amounts far above the average daily recommended amount (usually considered to be more than twice the recommended daily amount). For adult men, the recommended daily dose is 90 mg and for adult women it is 75 mg.

b. Student answers will vary. According to an NIH study (http://www.ncbi.nlm.nih.gov/pubmed/10796569), long-term daily supplementation in large doses do not appear to prevent colds. It could reduce the duration of cold systems. Taking too much vitamin C can decrease the amount absorbed in the body (https://ods.od.nih.gov/factsheets/VitaminC-HealthProfessional/).

c. Student answers will vary.

11.17 a. Student answers will vary. They may include wrinkle creams, body balms, lotions.

b. It is thought that the antioxidant vitamin E protects the skin from free radicals, which damage the skin.

c. Student answers will vary. An example claim could be that taking antioxidants has no real preventative or therapeutic value unless deficiency is your problem. Likewise, because they are free radicals, they can cause harm rather than fix a problem that is present.

11.18 Student answers will vary depending on the items they chose.

11.19 a. I-131 is absorbed in the bloodstream and collects in the thyroid. Once there, it destroys thyroid cells, thus leading to a reduction in the function of the thyroid gland.

b. Student answers will vary. Risks are that the patient has to take medication for the rest of their life because the cells in the thyroid are permanently damaged. Salivary glands may be permanently damaged from the treatment. Benefits are decreased recurrence of hyperthyroidism and decreased mortality due to hyperthyroidism.

c. The half-life of I-131 is 8.0197 days. If it takes 10 half-lives to be "gone" from the body, that would be 80.197 days.

11.20 The FDA states that "low-fat" should have 3 grams total fat or less per serving. For it to be low in saturated fat, it has to be 1 gram or less and 15% or less of Calories from saturated fat. This would meet the guidelines for total fat, but not saturated fat (10 Calories out of 50 Calories is greater than 15%).

11.21 a. No, males require more Calories than females for all ages for the same level of activity. This is due to the increased mass of males compared to females. The more mass you have, the more energy you need.

b. The estimated Calorie requirement decreases with age.

c. Student answers will vary depending on their countries chosen. For the most part, they will be similar but may be in different units; for example, some countries report food-energy content in joules.

11.22 1524 Calories: $\left[\frac{350\text{ Cal}}{1\text{ burger}} \times 2\text{ burgers}\right]$

$$+ \left[\frac{108\text{ Cal}}{1\text{ oz}} \times 3\text{ oz}\right] + \left[\frac{175\text{ Cal}}{4\text{ oz}} \times 8\text{ oz}\right]$$

$$+ \left[\frac{100\text{ Cal}}{8\text{ oz}} \times 12\text{ oz}\right], \text{ or around 3 hours } \frac{1524\text{ Cal}}{490\text{ Cal/hr}}.$$

11.23 a. The Basic Four were vegetables and fruits, milk, meat, and cereals/breads. The Food Pyramid indicated that we should have 6–11 servings of bread/cereal/rice/pasta, 2–4 servings of fruit, 3–5 servings of vegetables, 2–3 servings of milk/yogurt/cheese, 2–3 servings of meat/poultry/fish/dry beans/eggs/nuts, and to use fats/oils/sweets sparingly. MyPyramid doesn't offer serving suggestions, but rather an abstract view of amounts. The most being grains, then equal amounts of vegetables and milk, then fruit is smaller in size, then protein even smaller, and a very small sliver for oils and sweets. It also includes steps to indicate the importance of exercise in addition to watching what you eat. MyPlate divides a plate into largely vegetables and grains, and smaller amounts of fruits and proteins. There is also a glass to indicate dairy.

b. Student answers will vary. Some claim that the MyPlate has flaws because it is missing fats and oils. Others claim that it is unrealistic to actually have a meal that fits those guidelines. Another claim is that it doesn't highlight healthier choices over others (red meat vs. chicken/fish) and it is missing servings per day. A benefit could include that it is easier to interpret with its pie chart analogy.

11.24 a. Student answers will vary depending on the recent event selected.

b. Student answers will vary depending on the recent event selected.

c. Student answers will vary depending on how fearful the student is about food safety. Analytical techniques can be used to identify foodborne pathogens. Regulations can be tightened to decrease instances of food safety issues.

11.25 a. Student answers will vary depending on the chemicals chosen from this report: https://www.nrdc.org/sites/default/files/safety-loophole-for-chemicals-in-food-report.pdf

b. Student answers will vary depending on the chemicals chosen.

11.26 a. One example is the diversion of water from the Aral Sea for crop use (see Figure 8.18).

b. Animas River contamination (Durango, CO—Figure 8.19); Flint, Michigan, water crisis.

11.27 a. If the yield in grain per acre increases, there will be less land necessary to bring one kilogram of beef to the table. If the yield in grain per acre decreases, there will be more land necessary to bring one kilogram of beef to the table.

b. Early in life, beef cattle may graze before heading to the feedlot. Estimates for land use are higher if more of a cow's lifespan is included.
Note: Depending on the land quality and the practices of the farmer or rancher, a cow may require from a few acres to more than 30 acres of grazing land.

c. The breeds of livestock were likely combined to find an average for this data. There will be no measurable effect on the estimate.

11.28 a. Student answers will vary depending on the statement chosen.

b. Student answers will vary.

c. Student answers will vary.

11.29 Pros of eating local foods: Food is picked at its peak freshness and when it is best in season; food does not need to be transported as far (less pollutants, less cost to end user), and it can have more nutrients; local foods support local farmers. Cons of eating local foods: Food can be more expensive because farms are smaller; food can spoil more quickly because it lacks preservatives. Seasonality means that you won't have the same selection year round.

Pros of eating non-local foods: Food is cheaper; lasts longer because it has preservatives; ability to have foods year round. Cons of eating non-local foods: Transportation costs and effects to the environment; can have fewer nutrients because it isn't as fresh; doesn't support local businesses.

Pros of eating organic foods: Food can have more nutrients; no less harmful/toxic pesticides; some feel it tastes better; better for the environment. Cons of eating organic foods: More expensive to purchase; more expensive to farm; strict guidelines to adhere to as a farmer.

Pros of eating non-organic foods: Food is cheaper and more food is produced per acre; doesn't spoil as quickly. Cons of eating non-organic foods: Food might contain less nutrients; risk of toxic pesticides or additives; pesticides can harm the environment.

11.30 a.

$$5 \text{ lbs C} \times \frac{453.591 \text{ g}}{1 \text{ lb}} \times \frac{1 \text{ mol C}}{12.01 \text{ gC}} = 188.84 \text{ moles of C}$$

$$188.84 \text{ moles of } CO_2 \times \frac{44.01 \text{ g } CO_2}{1 \text{ mol } CO_2} \times \frac{1 \text{ lb}}{453.591 \text{ g}}$$

$= 18.32$ lbs CO_2 20 lbs to 1 sig fig

b. $1000 \text{ miles} \times \frac{1 \text{ gallon}}{30 \text{ miles}} = 33.33 \text{ gallons}$

$$\frac{33.33 \text{ gallons}}{x \text{ lbs } CO_2} = \frac{1 \text{ gallon}}{18.32 \text{ lbs } CO_2}$$

$x = 610.67$ lbs CO_2 in one year saved
600 lbs CO_2 to 1 sig fig

Assumptions are how many miles per gallon are used. A sample answer is to use 30 miles per gallon.

11.31 a. Answers will vary but may include: Reduce the amount of transportation necessary, reduce the packaging of materials and make the remaining packing recyclable.

b. Answers will vary but may include: Carbon footprints calculations assume that all resources that are consumed can be tracked/quantified. It assumes that all acres are equivalent. A sample website for this could be: http://rprogress.org/ecological_footprint/footprint_FAQs.htm

11.32 The bond energy for the triple bond in N_2 is very high (946 kJ, nearly double) compared to the O=O bond in O_2 (498 kJ) and the O–H bond in water (467 kJ). O=O and O–H bonds have similar bond energies to one another.

11.33 Student answers will vary depending on the form of nitrogen chosen. A sample answer could be:

$$NH_{3(g)} + HCl_{(aq)} \longrightarrow NH_4Cl_{(aq)}$$

11.34 **a.** Ammonia is soluble in water because it is very polar and can form hydrogen bonds.

b. When ammonia is mixed with water, the ammonium ion (NH_4^+) and a hydroxide ion (OH^-) are formed ($NH_3 + H_2O \longrightarrow NH_4OH$). The NH_4OH is soluble in water, so the $[OH^-]$ increases in the solution, the solution is basic.

c. Nitrates (NO_3^-)

d. Nitrites (NO_2^-)

11.35 Student answers will vary depending on the favorite food chosen.

11.36

```
      Br
      |
  H—C—H
      |
      H
```

When the molecule is in the stratosphere, UV light causes methyl bromide to break apart into $\cdot CH_3$ and a bromine radical. The bromine radical reacts with ozone to produce oxygen similar to how chlorine reacts, and thus it depletes the ozone layer.

11.37 Many possible answers exist, such as acts of terrorism, financial collapse, climate change, rising oil prices, herbicide-resistant weeds, etc.

11.38 **a.** Assuming an annual ethanol production of 13 billion gallons (data from *Your Turn 5.27*), here is the math:

$$13{,}000{,}000{,}000 \text{ gallons ethanol} \times \frac{2000 \text{ lb corn}}{100 \text{ gallons ethanol}} \times \frac{1 \text{ ton corn}}{2000 \text{ lb corn}} = 130 \text{ million tons of corn.}$$

b. Using the same ratio of ethanol to corn calculated in part a, producing 36 billion gallons of ethanol would require 360 million tons of corn, essentially all of the current U.S. harvest of feed corn.

11.39 **a.** Student answers will vary depending on their initial rankings.

b. Student answers will vary depending on the changes they are planning to make, if any.

c. Answers will vary, but according to a 2012 report by the NRDC (https://www.nrdc.org/sites/default/files/wasted-food-IP.pdf) 40% of food in the USA is unused. Some practices to reduce food waste may include: cutting irregularly shaped products into more desirable products (ugly carrots vs. baby carrots), selling at farmers markets, donating food to food banks, re-evaluating sell-by/best-by dates to be more accurate, having a bargain-shelf for nearly expired food, etc.

Chapter 12

REFLECTION

Answers will vary. Some possibilities include: medication (treated a symptom), surgery (cured a problem), pacemaker implant (prevented a future problem).

12.1 A healthy thyroid produces thyroxine. It is converted into triiodothyronine when it gets to the liver and kidneys. It helps the body regulate rate of metabolism and maintenance of bones among other roles.

12.2 **a.** The balanced reaction must follow the language of the sentence closely and show glucose as the reactant and fructose as the product:

glucose $\longrightarrow$ fructose

The two compounds can be written in chemical formulas, but (as this is a structural rearrangement only) the difference between reactant and product is unclear.

$$C_6H_{12}O_6 \longrightarrow C_6H_{12}O_6$$

b. The equilibrium constant must follow the format of concentration of product over concentration of reactant. Again, as the chemical formulas do not show us the important structural differences between glucose and fructose, we should use the chemical names.

$$K_{eq} = \frac{[\text{fructose}]}{[\text{glucose}]}$$

c. Here, we are given the concentrations of our product and reactant already at equilibrium. We input these values into the formula we have generated above to calculate our equilibrium constant.

$$K_{eq} = \frac{[\text{fructose}]}{[\text{glucose}]} = \frac{4.45 \text{ mM}}{6.02 \text{ mM}} = 0.739$$

12.3 In this question, we are discussing the release of epinephrine from the receptor similar to what we have seen in Figure 12.2. The reactant side is the complex with the free receptor and epinephrine as the two products. The K_{eq} constant (or given its special name in biochemistry, the K_d) value is very low (and much lower than 1). Referring back to Figure 12.1, this means that the reactants are heavily favored so the complex must be favored over the free state. This means that the epinephrine prefers to stay bound.

12.4 From the context given, we must decide what is necessary for the relative K_{eq} values for the *release of oxygen* from hemoglobin versus myoglobin. For this transfer to happen, hemoglobin must prefer to release oxygen at the same conditions that myoglobin prefers to stay in complex with oxygen. This means that hemoglobin must have a higher K_{eq} than myoglobin.

12.5 Answers will vary but may include:

Carbonic acid, hydrochloric acid, nitric acid are all acids. We have seen these three in the contexts of ocean acidification due to rising CO_2 levels (see Chapters 4 and 8), tests for the purity of Si for electronics (see Chapter 1), and in the formation of acid rain from byproducts of burning coal (see Chapter 5), respectively.

Ammonia, sodium hydroxide, and calcium hydroxide are all bases. We have seen these three in the contexts of common refrigerant gases before CFCs (see Chapter 3), in the production of biodiesel (see Chapter 5), and as flocculating agents in the water treatment process (see Chapter 8), respectively.

12.6 **a.** $CH_3COOH + H_2O \rightleftharpoons H_3O^+ + \boxed{CH_3COO^-}$

b. $H_2CO_3 + H_2O \rightleftharpoons H_3O^+ + \boxed{HCO_3^-}$

c. $H_3PO_4 + H_2O \rightleftharpoons H_3O^+ + \boxed{H_2PO_4^-}$

12.7 **a.** Phosphoric acid and its conjugate base, dihydrogen phosphate, is an excellent mixture for stabilizing a solution at pH 2.5 because it is within one pH unit of its pKa at 2.1.

b. The dihydrogen phosphate ion, with its conjugate base the hydrogen phosphate ion, is an excellent buffer for stabilizing a solution at pH 7.4 because it is within one pH unit of its pKa at 6.9.

c. Carbonic acid or dihydrogen phosphate ion would be a good buffer for stabilizing a solution at pH 6.8 because they are within one unit of their pKa at 6.3 or 6.9, respectively.

d. Acetic acid or carbonic acid would be a good buffer for stabilizing a solution at pH 5.4 because they are within one unit of their pKa at 4.8 and 6.3, respectively.

12.8 Answers will vary but may include: Octane, pentane, methane, ethene, and ethanoic acid are organic compounds. Sodium chloride, copper(II) nitrate, hydrogen bromide, calcium carbonate, and nitric acid are inorganic compounds. Contexts will vary depending on student experiences.

12.9 **a.** In all of the common arrangements presented, each carbon makes four bonds and contains no lone pairs. In some cases, it has four single bonds (involving eight electrons), in others, it has a double bond and two single bonds (involving eight electrons), and in some cases it has a single bond and a triple bond (involving eight electrons). They all follow the octet rule.

b. Carbon monoxide does not form four bonds (but still follows the octet rule).

12.10 **a.**

```
    H   H   H   H
    |   |   |   |
H — C — C — C — C — H
    |   |   |   |
    H   |   H   H
    H — C — H
        |
        H
```

b.

```
    H   H   H
    |   |   |
H — C — C — C — H
    |   |   |
    H   |   H
    H — C — H
        |
    H — C — H
        |
        H
```

c.

```
            H
            |
        H — C — H
    H   H   |   H
    |   |   |   |
H — C — C — C — C — H
    |   |   |   |
    H   H   |   H
        H — C — H
            |
            H
```

12.11 **a.**

b.

c.

12.12 a. Yes, *n*-butane and iso-butane are structural isomers that have the same chemical formula C_4H_{10}.

b. No, *n*-hexane and cyclohexane are not isomers. They do not have the same chemical formula. The ring structure eliminates two of the hydrogens that hexane has.

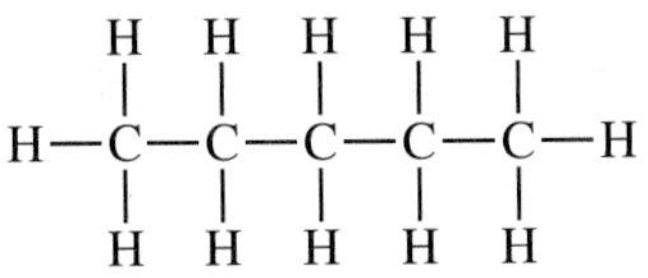

$CH_3CH_2CH_2CH_2CH_3$

c.

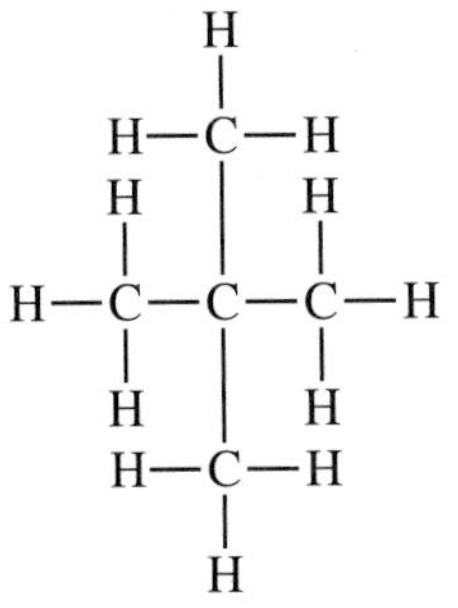

$CH_3C(CH_3)_3$

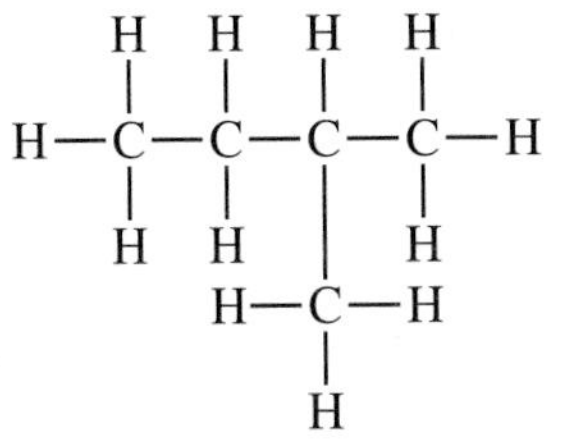

$CH_3CH(CH_3)CH_2CH_3$

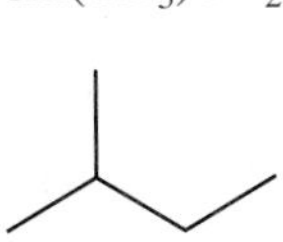

12.13 a. 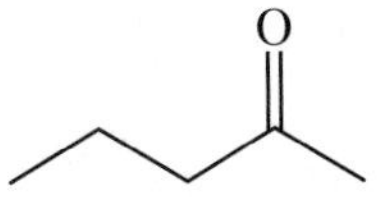 ketone

b. alcohol

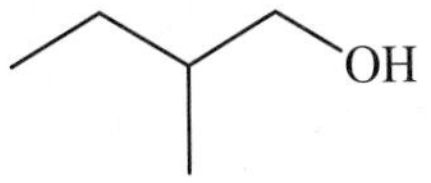

c. amine

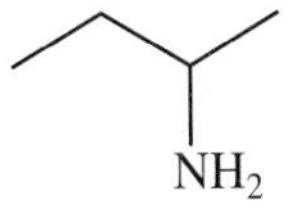

d. ester

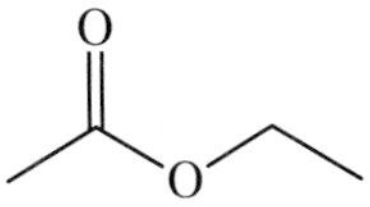

e. aldehyde

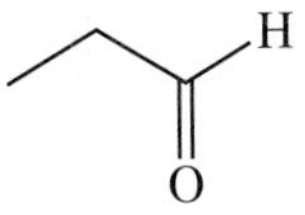

12.14

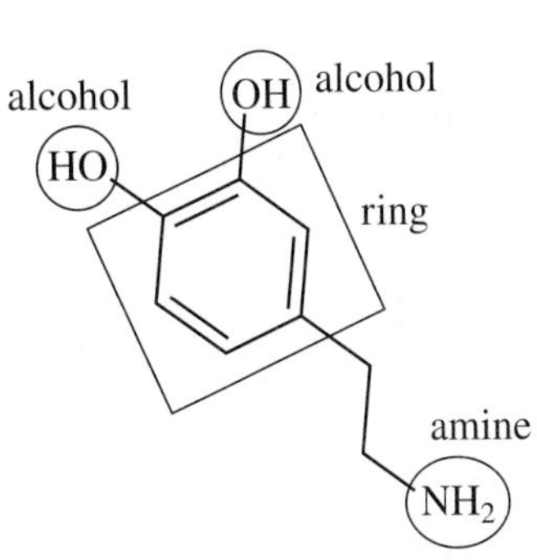

alcohol alcohol ring amine

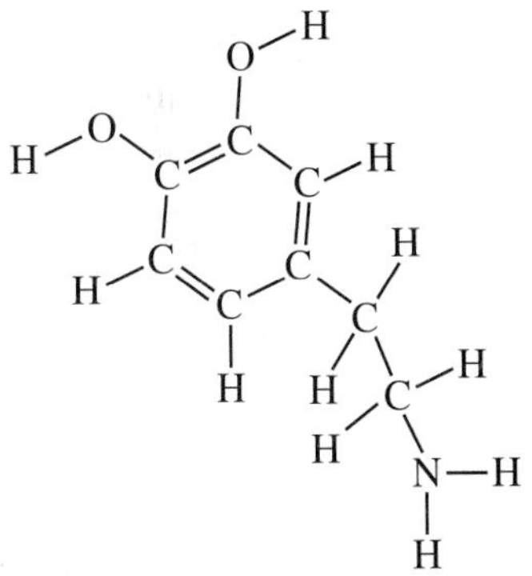

Dopamine differs from epinephrine in that there is an additional -OH on the carbon chain and only an NH (not NH_2) followed by another carbon. Their chemical formulas only differ by one carbon, two hydrogens, and one oxygen.

12.15 Vitamins A, D, and K are all fat soluble. They are nonpolar and can collect in the body. The dosages of these vitamins need to be monitored. Vitamin C is water soluble because it is polar and tends to be flushed out of the body when it is present in excess. The dosages of water-soluble vitamins still need to be monitored if consumed in megadoses.

12.16 a.

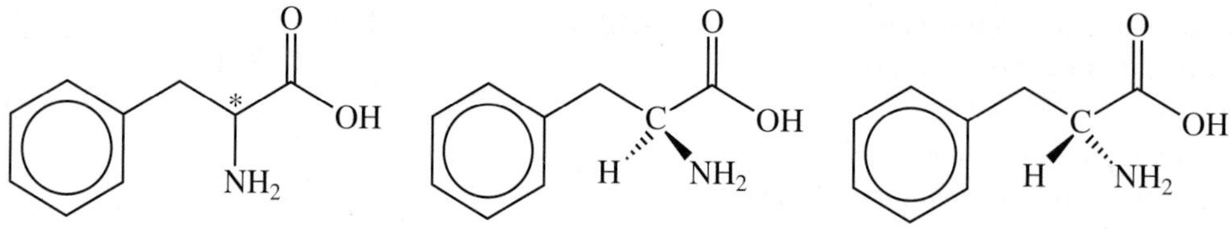

b.

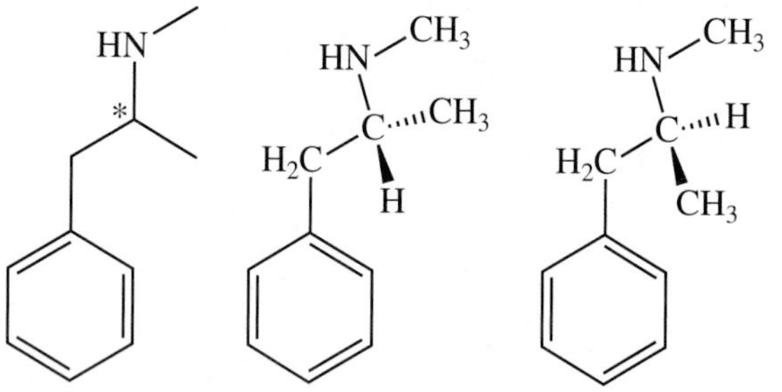

c.

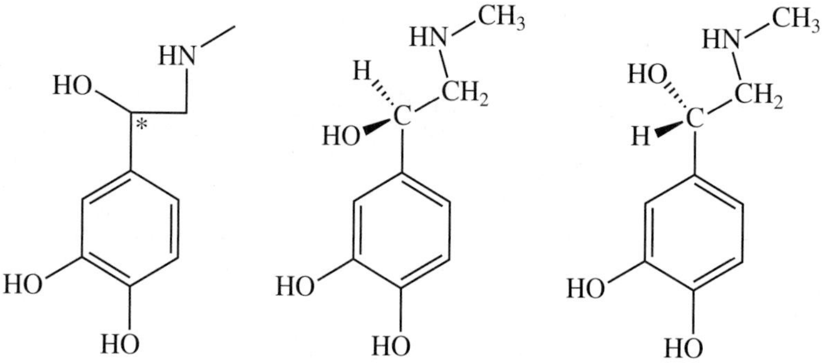

d.

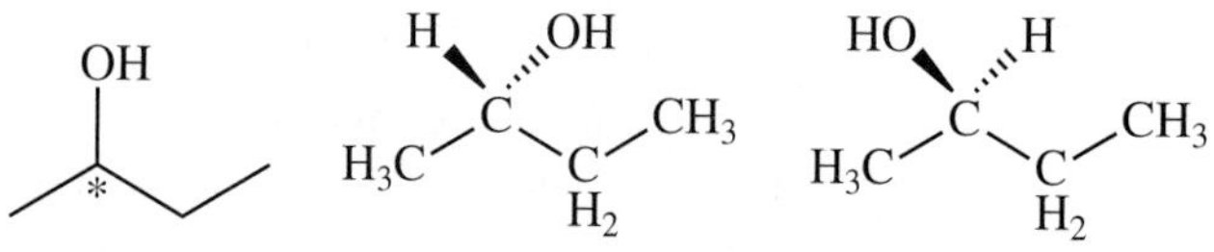

12.17 a.

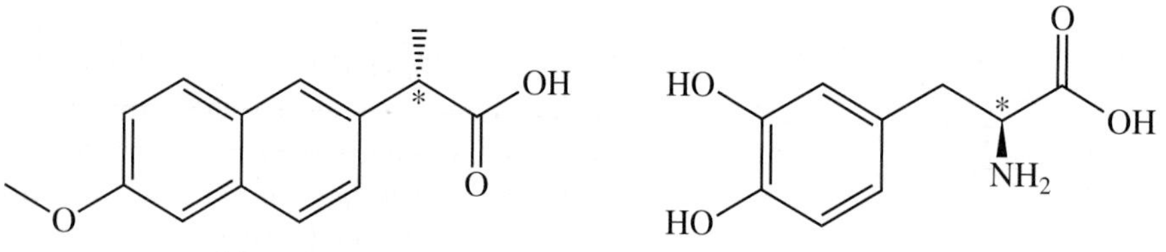

Naproxen

L-DOPA

b. Both drugs contain benzenes and a carboxylic acid functional group. Naproxen also contains an ether functional group. L-DOPA also contains two alcohol (hydroxyl) functional groups and an amine functional group.

c.

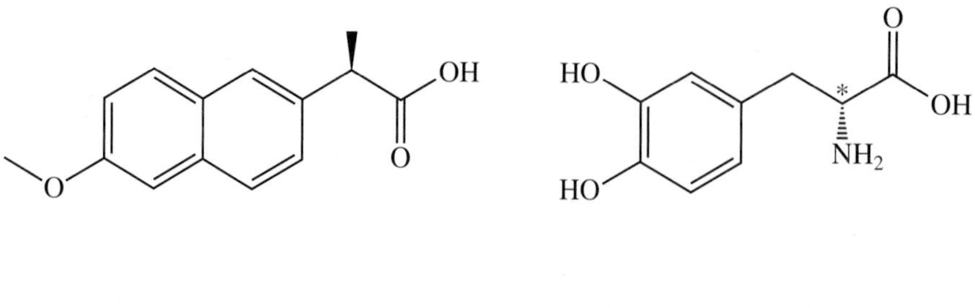

12.18 a. $C_6H_{12}O_6 + 6\ O_2 \longrightarrow 6\ CO_2 + 6\ H_2O$ (already in the answers in book)

b. 10 grams of glucose = 0.055 moles:

$10\ g \times \left(\frac{1\ mol}{180.2\ g}\right)$. To find the number of moles necessary of oxygen:

$$0.055\ \text{mol glucose} \times \frac{6\ \text{mol}\ O_2}{1\ \text{mol}} = 0.33\ \text{mol}\ O_2.$$

$$0.33\ \text{mol of}\ O_2 \times \frac{32\ g\ O_2}{1\ \text{mol}\ O_2} = 11\ \text{grams of}\ O_2.$$

12.19 As seen in Chapter 5, the First Law of Thermodynamics states that energy cannot be created nor destroyed. This statement only considers glucose and ATP. As we convert glucose through the full system, energy can go into producing ATP, as well as other parts of the system. The total amount of energy is conserved through the process.

12.20 Student answers will vary depending on their selection within the Molecule of the Month archives. Options range widely and can be tailored to student interest.

12.21 Previously, we have seen discussions of polyamides as polymers in Chapter 9—as artificial polymers like Kevlar® in bullet-proof vests, and nylon as well as natural polymers like the silk spun by spiders. Finally, we have also seen proteins discussed as a vital macronutrient in Chapter 11, as well as more specific food examples in the white of an egg or the collagen melted during the smoking of meats in Chapter 10.

12.22 Student answers will vary depending on two amino acids chosen. A sample answer is: glycine and alanine. It is nonpolar because glycine and alanine both have nonpolar side chains.

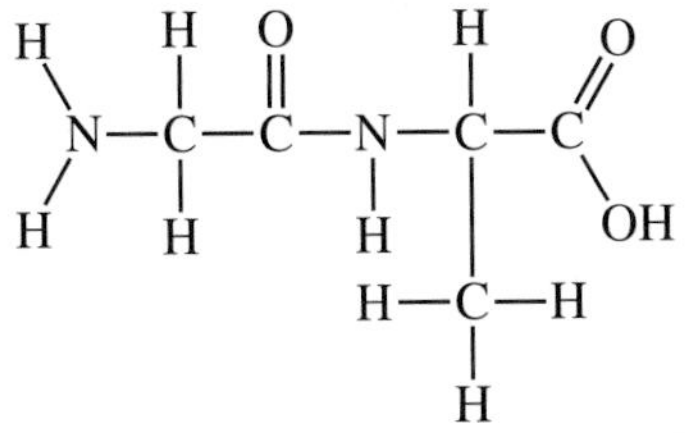

```
H     H  O      H     O
 \    |  ||     |    //
  N—C—C—N—C—C
 /    |      |  |     \
H     H      H  |      OH
             H—C—H
                |
                H
```

12.23 A key concept in this chapter is that molecules interact at multiple points. A strong and selective interaction should take advantage of all of the epinephrine's functional groups: three hydroxyls, an amine, and the planar benzene ring. We could expect that the negatively charged side chains of acidic amino acids would interact strongly with the amine group because it can act as a base and become positively charged. Polar (neutral as well as charged) amino acids should be able to make hydrogen bonds with the two hydroxyl groups. Nonpolar amino acids would interact with the benzene ring.

12.24 Student answers will vary. The structures may differ in only one carbon and four hydrogens, but the structures are very different in terms of functional groups. Estradiol contains a benzene ring with one alcohol group attached, whereas testosterone contains a cyclohexene with a ketone and an additional methyl. Because of these structural differences, they have different properties that change which proteins they interact with specifically. In addition, these are not the only sex hormones that differ between men and women. Men also have another major sex hormone called androsterone. Women also have estrone and progesterone.

12.25 a. Estradiol: $C_{18}H_{24}O_2$; Progesterone: $C_{21}H_{30}O_2$. Both have ring structures: three six-carbon rings and one five-carbon ring.

b. Corticosterone: $C_{21}H_{30}O_4$; Cortisone: $C_{21}H_{28}O_5$. Both have ring structures: three six-carbon rings and one five-carbon ring. They both also contain ketone and alcohol groups.

c. Cholic acid: $C_{24}H_{40}O_5$; Cholesterol: $C_{27}H_{46}O$. Both have ring structures—three six-carbon rings and one five-carbon ring. They both contain alcohol functional groups.

12.26 a.

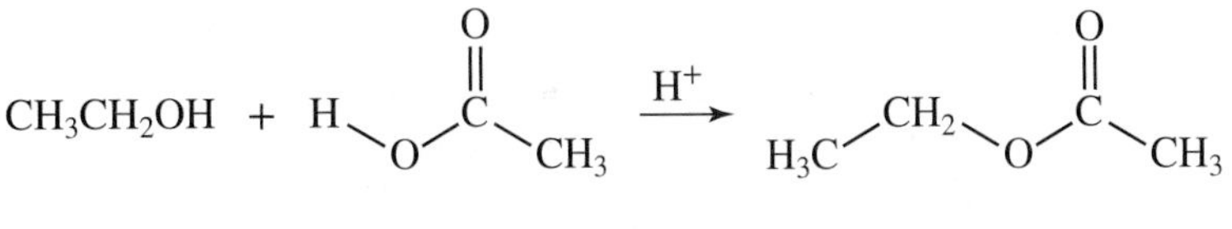

```
                         O                               O
                         ||        H+                    ||
CH3CH2OH  +  H     C      ——→        CH2     C
               \  / \                  /    \   / \
                O    CH3           H3C       O     CH3
```

b.

```
   H        O                                 H       O       H
   |        ||                          H     |       ||      |   H
H—C       C       H  +  CH3OH   H+      \  C       C       C /
  /  \   /  \  /            ——→        /   \   /   \   /  \
 H     C      O                        H      C      O      H
      / \                                    / \
     H   H                                  H   H
```

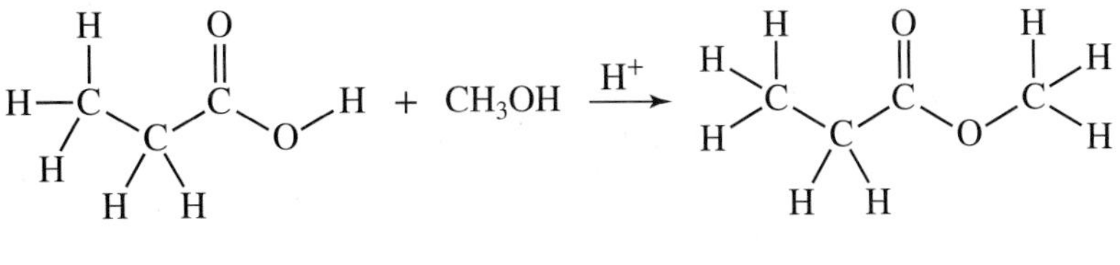

12.27 a. All three molecules have a benzene ring and a C=O or carbonyl. All three also have two (and only two) groups attached to the benzene ring. As differences, only aspirin has an ester and only acetaminophen has an amide. In addition, only acetaminophen has an alcohol group and only ibuprofen has a nonpolar carbon chain.

b. Aspirin: pain, fever, inflammation. Ibuprofen: inflammation and fever. Acetaminophen: pain and fever. They all treat fever.

c. Aspirin: rash, nausea, stomach ache. Ibuprofen: nausea, bloating, head ache, rash, ear ringing. Acetaminophen: nausea, dark urine, jaundice. Nausea is the common side effect.

12.28 a. You will not use the entire bottle before it expires.

b. Aspirin breaks down over time. It undergoes hydrolysis to form salicylic acid and acetic acid, which don't treat the symptoms you take aspirin for. Salicylic acid, remember, led to some concerning side-effects.

12.29 The most popular and common example of a chance drug is Viagra®: It was researched to treat high blood pressure initially. Botox is an injection used for cosmetic purposes, but it also can treat excessive sweating and mirgranes. A reference website that contains many examples is ncbi.nlm.nih.gov/pmc/articles/PMC3181823

12.30 Student answers will vary but may include a variety of antifungals, antivirals, and antimalarial drugs as drugs that kill foreign invaders. Drugs that bring about a desired physiological response could include inhalers, insulin, and cold medications.

12.31 Student answers will vary depending on the properties they think a drug should have. Answers may include targeted treatment, no side effects, no accumulation in the body, no unwanted chemicals when it is metabolized, etc. Their motivations, similarities, and difference on how they might change will vary.

12.32 a. These models allow the viewer to see the orientation of the atoms in three dimensions rather than as dashes and wedges. The elements are also color coded. For some, this may dramatically change how they consider the position of functional groups relative to each other.

b. It helps explain the shape of the molecule and where points of the molecule have space for a reaction to occur and where it is too crowded for a reaction to occur. It also allows for the view to rotate the molecules in different directions to see how it might interact with other molecules. Disadvantages may include that you can't physically hold the molecules and that specific technology is necessary to view it. It may also lead the viewer to believe that a molecule's shape is very static even when it is not. Both 2-D and 3-D models have limitations as compared to a real molecule. They do not have all of the information contained in a molecule, but rather the information that is chosen to be portrayed in the model. The scale may also be a limitation, but that is also a benefit because we cannot see atoms normally.

12.33 Gamma radiation is the best to use for detection through tissues and skin because the rays can pass through the layers of tissue necessary to generate images of deep inside the body. Beta emissions can similarly penetrate the skin, but will not go as deeply into tissues. Positron emissions are similar in size to beta emissions. Beta and positron emitting isotopes are often used due to a concurrent gamma emission. The worst to use for detection through tissues and skin would be alpha radiation because it doesn't have much penetrating power. It can be stopped by a simple sheet of paper.

12.34 a. A salt containing iodate or iodide.

b. Thyroid cancer, lower heart rates, goiters, etc. Students may list any of the symptoms of hypothyroidism because the thyroid would be damaged.

12.35 a. It slows metabolism and induces fatigue. Without enough iodide, the thyroid gland can't produce enough thyroxine.

b. If Megan doesn't keep up with her hormone therapy, she runs the risk of being tired.

c. It's important to make sure the levels of hormone therapy are closely monitored to best mimic the naturally occurring gland.

Chapter 13

REFLECTION

Student answers will vary based on family traits of eye color, hair color, tongue rolling, ear lobe connectivity, or others.

13.1 a.

Thymine

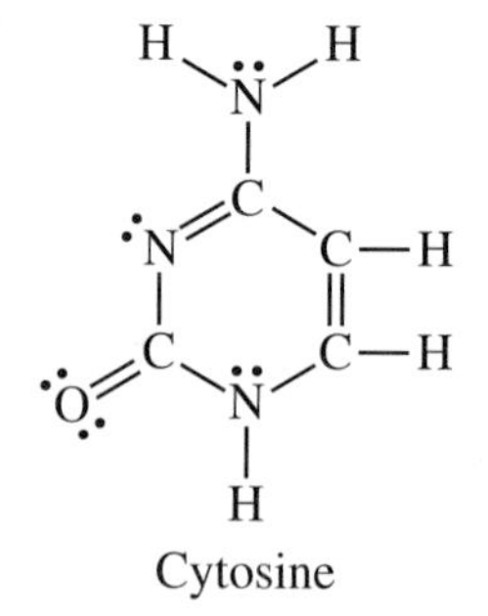

Cytosine

Guanine

Adenine

b. Any H atom attached to an N atom could form a hydrogen bond.

c. Any O and N atom could participate in hydrogen bonding.

13.2 a. Deoxyribose formula: $C_5H_{10}O_4$. Ribose formula: $C_5H_{10}O_5$.

b. Ribose has three carbons in the ring with an alcohol functional group. In deoxyribose, the middle carbon of those three has only an H atom instead.

c. Ribose fits this pattern but deoxyribose does not.

d. The oxygen atoms that are part of the alcohol functional group can participate in hydrogen bonds.

13.3 a. Very close to 0.9 (90%) of H_3PO_4 and 0.1 (10%) of $H_2PO_4^-$.

b. About 0.99 (99%) HPO_4^{2-} and 0.01 (1%) PO_4^{3-}.

c. About 0.99 (99%) $H_2PO_4^-$ and 0.01 (1%) of HPO_4^{2-}.

d. Very close to 0.1 (10%) HPO_4^{2-} and 0.9 (90%) of PO_4^{3-}.

e. Very close to 0.5 (50%) of HPO_4^{2-} and 0.5 (50%) of $H_2PO_4^-$.

13.4

NH_2
HC C N
HC C O
O^-
$O{=}P{-}O{-}CH_2$ O N
O^- C H H C
H H
OH H

13.5 a. TATGGACG

b. CTAGGAT

13.6 a. This representation does not show the phosphate or sugar backbone. It also oversimplifies how the strands are a double helix, making it more circular.

b. This representation highlights the base pairs that consistently complement one another, how they fit together, and how the pairs appear to rotate as the helix turns. The color coding of the base pairs is a beneficial visual aid.

c. Student answers will vary.

13.7 a. The eigth base (G) is where the error occurs; the complement shows a T when it should be a C.

b. The mismatched pair will be less stable than a correct pair because there would not be as much hydrogen bonding. Additionally, the sizes of T and C are different, which would cause a disorganized orientation of base pairs.

13.8 a.
$$\frac{0.34\ \cancel{\text{nm}}}{1\ \text{base}\ \cancel{\text{pair}}} \times \frac{1\ \cancel{\text{m}}}{1 \times 10^9\ \cancel{\text{nm}}} \times \frac{1 \times 10^2\ \text{cm}}{1\ \cancel{\text{m}}}$$
$$\times \frac{1.35 \times 10^8\ \cancel{\text{base pairs}}}{\text{Chromosome 11}} = \frac{4.6\ \text{cm}}{\text{Chromosome 11}}$$

b.
$$4\ \mu\text{m} = 0.0004\ \text{cm}$$
$$4.6\ \text{cm} = 46000\ \mu\text{m}$$
$$\frac{46000\ \mu\text{m}}{4\ \mu\text{m}} = 11{,}500$$

c. Humans have 23 pairs of chromosomes, for a total of 46 chromosomes. The compaction needs to occur, otherwise they wouldn't fit.

13.9 $4 \times 4 \times 4 \times 4 = 4^4 = 256$ different four-base sequences.

13.10 This allows for possible mistakes in the genetic code to still code for the same amino acid.

13.11 a. Both contain C=O and N-H groups.

b. Typically, nylon is synthesized by the reaction of a monomer containing two carboxylic acids, with a monomer containing two amine groups. For example, refer back to Equation 9.7. In contrast, proteins are synthesized from a monomer (an amino acid) that contains one carboxylic acid and one amine group.

c. They are both condensation polymers.

13.12 Student answers will vary. An example is that a student may draw hydrogen bonding between polyamide molecules or Kevlar® to illustrate intermolecular hydrogen bonding (hydrogen bonds that take place between two separate molecules). Students may draw proteins that form an α-helix or β-pleated sheets to illustrate intramolecular hydrogen bonds (hydrogen bonds that take place within a molecule).

13.13 a. Valine and glutamic acid differ in their side chains. Although they both contain three carbons, glutamic acid also has a carboxylic acid group at the end of that carbon chain.

b. Glutamic acid has a polar side chain, and valine has a nonpolar side chain. Glutamic acid would be predicted to have a higher solubility in water.

c. Valine is nonpolar and thus hydrophobic. This change to the surface of the molecule reduces its solubility, and causes the sickle shape to result.

13.14 a. An ion is an atom (or molecule) that has a charge on it. A cation has a positive charge because it has lost electrons and an anion has a negative charge because it has gained electrons.

b. Student answers will vary. Examples include: hydrogen atoms, chlorine atoms, and hydroxyl free radicals.

c. Radiation triggers the dissociation of otherwise stable molecules. In the example of H_2O, a photon dissociates the molecule into a hydrogen free radical and hydroxyl free radical.

d. Free radicals can penetrate the body. Hydroxyl radicals can attack deoxyribose and base-pairs in DNA, which alters the function of these molecules.

13.15 Student answers will vary. As of summer 2016, USDA has until 2018 to write rules on how to communicate whether a food product contains GMOs. Most likely, it will be a symbol or a code readable by cell phones.

Chapter 14

REFLECTION

a. Survey the situation and secure the area; ensure the scene is safe by extinguishing the fire or apprehending the criminal.

b. Carefully document the crime scene and interview witnesses.

c. Carefully collect evidence using appropriate techniques; proper attire must be worn to prevent contamination of the crime scene.

14.1 a. The benefits of "smart" cancer-treatment drugs have fewer side effects because they target the cancer cells without also targeting healthy cells.

b. Phase I studies have healthy volunteers to determine the side effects of the drug. This phase gives information on metabolism and excretion of the drug. Phase II studies are conducted only if Phase I doesn't show high death rates or serious side effects. This phase is focused on determining whether the drug is effective in people who have a certain condition. Phase III studies only begin if sufficient evidence is shown that the drug works in Phase II. This phase is focused on studying effects of varied dosages, being taken with other medications, and in different populations.

14.2 a. The Hazmat crew should use either carbon dioxide or dry-chemical extinguishers (*e.g.*, ammonium phosphate, sodium bicarbonate, potassium chloride), because the solvents are considered to be Class B and flammable. However, the sodium metal is considered to be Class D and also requires a dry-powder extinguisher (graphite or sodium chloride powders pressurized with nitrogen gas). They do not simply use water because it can spread the flammable liquid and fire or produce a violent exothermic reaction (explosion!) when it comes into contact with sodium.

b. Sodium is a strong reducing agent. Each sodium atom donates an electron to a benzophenone molecule, resulting in a sodium ion and a ketyl radical anion. Using sodium metal to purify a chlorinated solvent such as dichloromethane will result in the formation of products that are shock-sensitive explosives!

14.3 A good resource for this question is www.femalifesafety.org. Dry-powder extinguishers are used to either remove the heat from the fire or separate the fuel from the oxygen. Dry-chemical extinguishers stop the chemical reaction and can act as a barrier between the fuel and the oxygen. Wet-chemical extinguishers remove heat from the fire and act as a barrier between the fuel and the oxygen. When barriers are created between the fuel and the oxygen, this can prevent the fire from being reignited.

14.4

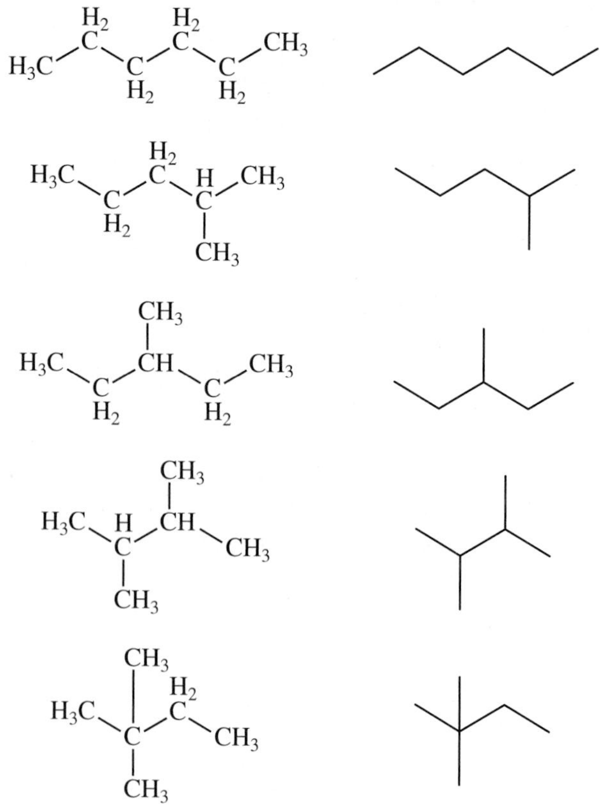

H_3C–CH_2–CH_2–CH_2–CH_2–CH_3 (printed as: H_2 C, H_2 C, CH_3 / H_3C, C H_2, C H_2)

H_3C–CH_2–CH_2–CH(CH_3)–CH_3 (printed as: H_2 C / H_3C, C H_2, H C, CH_3 / CH_3)

H_3C–CH_2–CH(CH_3)–CH_2–CH_3 (printed as: CH_3 / H_3C, C H_2, CH, C H_2, CH_3)

H_3C–CH(CH_3)–CH(CH_3)–CH_3 (printed as: CH_3 / H_3C, H C, CH, CH_3 / CH_3)

H_3C–C(CH_3)$_2$–CH_2–CH_3 (printed as: CH_3, H_2 C / H_3C, C, CH_3 / CH_3)

14.5 **a.** At the flash point, a chemical change occurs. At the boiling point, a physical change takes place.

b. Combustible and flammable liquids both can burn. However, it isn't the liquid itself that is burning but the vapor and air. They differ in the temperature at which they burn. Flammable liquids can ignite at normal temperatures, while combustible liquids ignite at temperatures above normal. The flash point is lower for flammable liquids than it is for combustible liquids.

c. A good reference for the answer to this question is Experimental study of the flash point of flammable liquids under different altitudes in the Tibet plateau. *Fire Materials* **2014**, *38,* 241–246 (DOI: 10.1002/fam.2177). Their study found that flash points decrease with altitude, thus making substances more flammable at normal temperatures. This would mean that flammable liquids would need to be handled with more care and stored at lower temperatures when used at high altitudes.

14.6 Toluene: Has a flash point of 6 °C and boiling point of 111 °C; Class IB

Hexane: Has a flash point of –26 °C and boiling point of 69 °C; Class IB

Tetrahydrofuran: Has a flash point of –14 °C and boiling point of 66 °C; Class IB

Diethyl ether: Has a flash point of –45 °C and boiling point of 35 °C; Class IA

Dichloromethane: Has a flash point of 100 °C and boiling point of 40 °C; Class IIIA

Acetonitrile: Has a flash point of 2 °C and boiling point of 82 °C; Class IB

Ethanol: Has a flash point of 17 °C and boiling point of 78 °C; Class IB

Acetone: Has a flash point of –20 °C and boiling point of 56 °C; Class IB

14.7 **a.** Diethyl ether, acetone, tetrahydrofuran, hexane, ethanol, acetonitrile, and toluene will all catch fire at room temperature with the aid of a match. They have flash points that are less than room temperature.

b. None of the solvents found in Dr. Thompson's laboratory could combust without the presence of an outside source of ignition. The autoignition temperatures are all well above room temperature.

14.8 **a.** Some examples include:

Toluene

$C_7H_8 + 9\ O_2 \longrightarrow 7\ CO_2 + 4\ H_2O$

Hexane

$2\ C_6H_{14} + 19\ O_2 \longrightarrow 12\ CO_2 + 14\ H_2O$

Tetrahydrofuran

$2\ C_4H_8O + 11\ O_2 \longrightarrow 8\ CO_2 + 8\ H_2O$

Diethyl ether

$(C_2H_5)_2O + 4\ O_2 \longrightarrow 4\ CO_2 + 5\ H_2O$

Dichloromethane

$2\ CH_2Cl_2 + 3\ O_2 \longrightarrow 2\ CO_2 + 2\ H_2O + 2\ Cl_2$

Acetonitrite

$4\ C_2H_3N + 15\ O_2 \longrightarrow 8\ CO_2 + 6\ H_2O + 4\ NO_2$

Ethanol

$C_2H_6O + 3\ O_2 \longrightarrow 2\ CO_2 + 3\ H_2O$

Acetone

$C_3H_6O + 4\ O_2 \longrightarrow 3\ CO_2 + 3\ H_2O$

b. Particulate matter such as soot is likely to exceed air quality standards. Nitrogen oxides will also form and likely exceed air quality standards.

c. Compact fluorescent light bulbs (CFLs) contain mercury vapor if cracked during a fire. The mercury contamination lasts many hours after the fire has been extinguished.

14.9 Two good resources for this discussion are www.interfire.org and www.tcforensic.com.au/docs/article2.html. If the materials are easy to cut away, like a piece of wet carpet, it is cut away and stored. If the material is difficult to remove from the scene of the crime, an absorbent can be applied to the surface and the absorbent collected for analysis. Metal paint cans that are unlined are best for storage because they can be sealed. Plastic bags can be punctured and glass jars can be shattered. The samples are stored in cool places because of unknown flash points.

14.10 The highest temperatures of the sensors near the laboratory notebooks were about 190 °C. The autoignition temperature of paper, cotton, and PVC are all higher than this value. It is not likely that these items burned without the use of an accelerant.

14.11

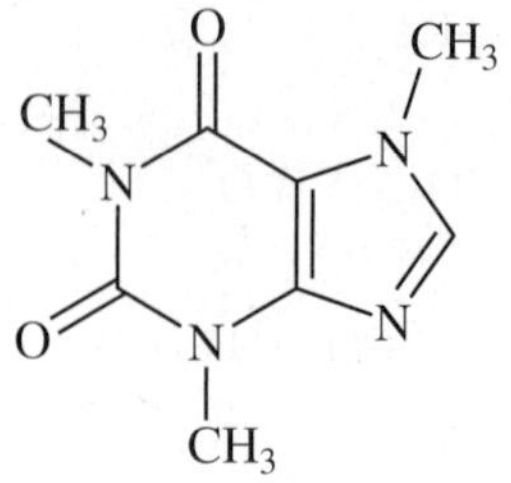

Caffeine

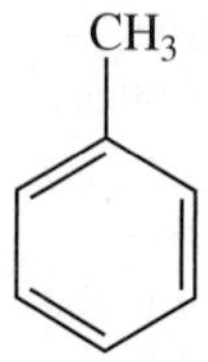

Toluene

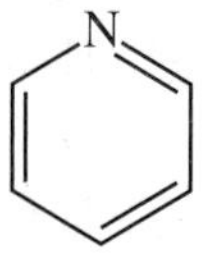

Pyridine

n-heneicosane

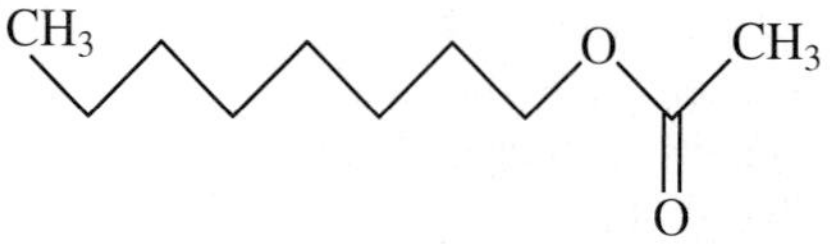

n-octyl acetate

Least polar to most polar: n-heneicosane, toluene, n-octyl acetate, pyridine, caffeine.

The least polar molecules take longer to make it through the column. Thus, the first peak (2.5 mins) is caffeine, the second peak (2.75 mins) is pyridine, the third peak (8.5 mins) is n-octyl acetate, the fourth peak (16.75 mins) is toluene, and the final peak (19 mins) is n-heneicosane.

14.12 The isotopic ratios for wood from matchsticks found in Dr. Thompson's laboratory and those found at the hit-and-run crime scene were similar and within the uncertainty ranges. This does not prove that the same person was involved but rather the same type of matchsticks was used by a person at both scenes. If the same person was at both scenes, it does add evidence that could place that person at the crime scene, but it is not proof alone.

14.13 His stomach was nearly empty indicating a long period since he had eaten his last meal. The contents that were there included starch grains and caffeine. Student answers will vary, but a potential meal could have been a piece of toast and coffee earlier that day. According to www.intox.com (http://www.intox.com/t-AboutAlcohol.aspx), the ethanol would have reached his bloodstream in approximately 30 minutes. The proportions of fat and water in his body, the alcohol concentration in the wine, whether he drank the wine in one gulp or in small sips, and whether he had eaten prior to drinking could have an effect on the metabolism of alcohol in the body.

14.14 Dr. Thompson's total cholesterol was borderline high. His HDL cholesterol was near the risk-factor range and not yet in the desirable range. His LDL cholesterol was high. His triglycerides were borderline high. Overall, he was at risk for heart disease.

14.15 Peak at 3350 is from OH stretching (hydroxyl functional group)

Peaks between 2870–2989 are from C–H bond stretching (alkane functional groups)

Peak at 1730 is a C=O bond (carbonyl functional group)

Peak at 1163 is due to N–H bending (amine functional group)

Broad band at around 1500 is due to aromatic C=C (phenyl functional group)

14.16 **1-** 2972: C–H

2- 1780.58: C=O

3- 1500.96: C=C aromatic

4- 1189.47: C–O

5- 1004.29: C–O

6- 554.28: C–H bending or C=C ring torsion

14.17 The EPA states the range as 265 miles for an 85 kWh battery and the NEDC reports 310 miles. According to Tesla's website, a 90 kWh battery should power a 294-mile trip. However, the speed of the car (faster speed = fewer miles), the outside temperature (higher temperature = more miles), and whether the air conditioning or heat is on (on = fewer miles) affect the mileage. It is unlikely that a driver could maintain exactly the speed and temperature conditions when driving on a highway for 140 miles each way to yield the ideal battery range. Assuming they left with a perfect charge, they would barely make the round trip. The driver would probably have needed to stop to recharge.

14.18 Student answers will vary. One possibility is:

a 60W solar charging kit found on Amazon.com ($195). If the Tesla battery is 90 kWh, it would take 1500 hours (!) to fully charge the battery using this solar charger.

14.19 Density of water = 1 g/cm^3; density of phenol = 1.07 g/cm^3; density of chloroform = 1.49 g/cm^3.

a. When shaken, the water rises to the top and the phenol and chloroform become a mixture together in another layer. Water is polar, phenol and chloroform are nonpolar and mix. The densities of the nonpolar substances are greater than water, therefore that layer sinks.

b. The DNA will be found in the water layer because it is polar.

14.20 Student answers will vary. The evidence suggests that Dr. Thompson was in the lab and handled the acetone containers. There is also evidence that he touched Dr. Thompson's wrists. What we do not know is when his wrists were touched and if it occurred at the crime scene. The DNA on the cigarette butts place Dr. Littleton near the scene of the crime, but do not unequivocally prove that he was the murderer. The chemical evidence does point to him, but alibis and other evidence must be considered before concluding he is guilty.

Appendix 5

Answers to Selected End-of-Chapter Questions Indicated in Blue in the Text

Chapter 1

1. a. Compound (two molecules of one compound made up of two different elements).

b. Mixture (two atoms of one element plus two atoms of another).

c. Mixture (three different substances, two elements and one compound).

d. Element (four atoms of the same element).

3. Exact answer will vary depending on viewing size of text. An approximate measurement for the period could be 0.25 mm. Converting this to nanometers:

$$0.25 \text{ mm} \times \frac{10^{-3} \text{ m}}{1 \text{ mm}} \times \frac{1 \text{ nm}}{10^{-9} \text{ m}}$$

$= 2.5 \times 10^5$ nm or 250,000 nm.

5. 1×10^2 cm, 1×10^6 μm, 1×10^9 nm.

7. a.

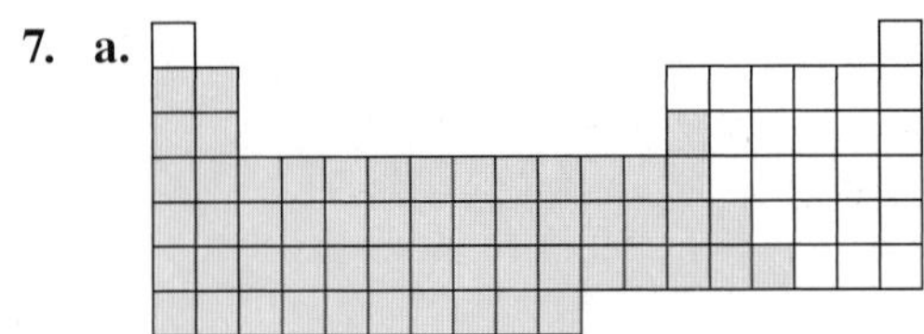

b. iron, Fe; magnesium, Mg; aluminum, Al; sodium, Na; potassium, K; silver, Ag.

c. sulfur, S; oxygen, O, carbon, C, chlorine, Cl, fluorine, F (and others).

9. There are several allotropes of sulfur. The most stable and common allotrope consists of eight atoms in a ring. Other common allotropes include rings of 5, 6, 7, 10, and larger number of atoms. Most allotropes are yellow solids, although they can be found in liquid or gaseous forms at appropriate temperatures. Most of these allotropes are created by heating sulfur of the eight-membered ring form.

11. a. ionic **b.** molecular

c. molecular **d.** ionic

e. molecular

13. a. 1 carbon atom, 2 oxygen atoms

b. 2 hydrogen atoms, 1 sulfur atom

c. 1 nitrogen atom, 2 oxygen atoms

d. 1 silicon atom, 2 oxygen atoms

15. a. copper has a +2 oxidation state;

b. aluminum has a +3 oxidation state;

c. iron has a +3 oxidation state;

d. manganese has a +7 oxidation state

17. As of 2016, the smallest transistors used in electronics devices have features of 14 nm. This is equivalent to

$$14 \text{ nm} \times \frac{10^{-9} \text{ m}}{1 \text{ nm}} \times \frac{1 \text{ km}}{10^3 \text{ m}} = 1.4 \times 10^{-11} \text{ km}.$$

19. There are many steps involved during this process. Many resources are available on the Internet that describe this process (*e.g.*, http://www.australianminesatlas.gov.au/education/fact_sheets/aluminium.html).

21. A thin layer of a transparent conducting material is deposited on the surface of glass. The most common material used for this purpose is indium tin oxide (ITO), which is used in applications ranging from LED displays to solar cells. Do-it-yourself enthusiasts have posted to the Internet a way to coat glass using stannous chloride ($SnCl_2$) to create a layer of conductive tin oxide.

23. Answers will vary depending on device and component choice, but typical components and sizes include length (*e.g.*, 14.5 cm, 145 mm, 145,000 μm, 145,000,000 nm); width (*e.g.*, 7.5 cm, 75 mm, 75,000 μm, 75,000,000 nm); thickness (*e.g.*, 1 cm, 10 mm, 10,000 μm, 10,000,000 nm); camera lens (*e.g.*, 0.3 cm, 3 mm, 3,000 μm, 3,000,000 nm); and speaker hole diameter (*e.g.*, 0.03 cm, 0.3 mm, 300 μm, 300,000 nm).

25. The steps involved to convert silica sand, mostly composed of SiO_2, to high-purity Si are outlined in Section 1.8. In contrast, sea sand contains many impurities such as metals (*e.g.*, Fe, Al, Mg, K, Na, Ca, Zn, Ni, etc.) and nonmetals (*e.g.*, B, P) that would require extensive pre-processing of the sand via chemical reactions, involving acids and high temperatures. Furthermore, the use of sea sand for industrial processes would not be sustainable and would cause a variety of environmental consequences. By disturbing sea sand, an area could be changed physically, biologically, and chemically.

27. While individual cell phones may become smaller and take up less space in, say, a landfill, there may be a much higher impact on the environment, depending on the materials required for making the cell phone and the energy needed for manufacturing the components. To make the devices smaller and less expensive, different materials may be used, which may require more invasive mining or greater waste production in the manufacturing process.

29. The colors of many gemstones comes from impurities in the crystal structure. For example, the purple color of amethyst comes from Fe^{3+} ions in a SiO_2 crystal, and the red color of rubies comes from Cr^{3+} in an Al_2O_3 crystal.

31. A thin layer of material is sandwiched between two pieces of glass. When an electrical current is passed through the glass, the material will line up according to the direction of the current, similar to the liquid crystal display (LCD) of common calculators.

33. Apple has removed mercury and arsenic from screens and lead from solder in their electronics.

35. Old electronic devices could be disassembled and either parts could be reused as is or could be mechanically or chemically separated into raw materials to be used for manufacturing new devices.

37. A good web resource that compares CZ growth *vs.* the float zone technique is http://www.siliconsultant.com/SIcrysgr.htm. The Si ingots are sliced into an appropriate thickness and a variety of processing steps are then used to fabricate a processor chip. Intel provides details regarding these steps (http://download.intel.com/newsroom/kits/chipmaking/pdfs/Sand-to-Silicon_32nm-Version.pdf).

39. Some considerations include toxicity of materials used in the device, availability of raw materials, amount of energy required by the device, and number of uses for the device and disposable components (such as batteries).

41. The increase in demand for rare earth metals amounts to about 40%. Rare earth metals are used in a wide variety of products such as cell phones, computers, rechargeable batteries, wind turbines, speakers, and fluorescent lights. It is doubtful that the U.S. could meet its demand for rare earth elements even with 100% recovery of the metals. The growth of the market for this wide variety of products outpaces the retirement of old products, many of which may not use much, if any, rare earth metals that could be recycled.

Chapter 2

1. **a.**

$$\frac{0.5\text{ L}}{1\text{ breath}} \times \frac{10\text{ breaths}}{1\text{ min}} \times \frac{60\text{ min}}{1\text{ hr}} \times 7.5\text{ hr} = 2250\text{ L}$$

(*i.e.,* 2×10^3 L to one significant figure)

b. Possibilities include burning less (wood, vegetation, cooking fuels, gasoline, incense), using products that pollute less (low-emission paints), and using motor-less appliances and tools (hand lawnmower, egg beater, broom, rake)

2. **a.** $Rn < CO < CO_2 < Ar < O_2 < N_2$

b. CO and CO_2

c. CO. By the time this book is in print, CO_2 may be regulated as well.

d. Rn (radon) and Ar (argon)

4. **a.** 0.934 parts per hundred

$$\times \frac{1{,}000{,}000\text{ parts per million}}{100\text{ parts per hundred}}$$

= 9340 parts per million
(Move the decimal 4 places to the right.)

b. $$2\text{ ppm} \times \frac{100\text{ parts per hundred}}{1{,}000{,}000\text{ ppm}}$$

= 0.0002 parts per hundred or 0.0002%
(Move the decimal 4 places to the left.)
20 ppm is equivalent to 0.0020%. 50 ppm is equivalent to 0.0050%.

c. $$8{,}500\text{ ppm} \times \frac{100\text{ parts per hundred}}{1{,}000{,}000\text{ ppm}}$$

= 0.85 parts per hundred or 0.85%
Be careful not to confuse the absolute humidity calculated in this problem with relative humidity, which is the amount of water vapor in the air compared to the maximum possible amount of water vapor that the air can hold at a particular temperature. For example, in a rainforest, the relative humidity is usually between 75 and 95%.

d. 8 ppm is 0.0008% (move the decimal 4 places to the left).

5. **a.** The chemical formula tells the elements present in a compound and the atomic ratio of the elements.

b. Xe (xenon), N_2O (dinitrogen monoxide or nitrous oxide), CH_4 (methane)

8. Nitrogen is 78.0% of the air, meaning that out of 100 air particles, 78 of them are nitrogen molecules.

$$500\text{ air particles} \times \frac{78\text{ nitrogen molecules}}{100\text{ air particles}} = 390\text{ nitrogen molecules}$$

Oxygen is 21.0% of the air, meaning that out of 100 air particles, 21 of them are oxygen molecules.

$$500\text{ air particles} \times \frac{21\text{ oxygen molecules}}{100\text{ air particles}} = 105\text{ oxygen molecules}$$

Argon is 0.9% of the air, meaning that out of 100 air particles, 0.9 of them are argon atoms.

$$500 \text{ air particles} \times \frac{0.9 \text{ argon atoms}}{100 \text{ air particles}}$$
$= 4.5$ argon atoms (or between 4–5 argon atoms)

10. **a.** Yes, the mass of the reactants equals the mass of the products. The Law of Conservation of Mass applies.

b. No, the numbers of molecules are not the same (four reactant molecules vs. two product molecules).

c. Yes, the numbers of each type of atom present as reactants and products are the same.

12. **a.** $C_3H_8(g) + 5\ O_2(g) \longrightarrow 3\ CO_2(g) + 4\ H_2O(g)$

b. $2\ C_4H_{10}(g) + 13\ O_2(g) \longrightarrow 8\ CO_2(g) + 10\ H_2O(g)$

c. $2\ C_3H_8(g) + 7\ O_2(g) \longrightarrow 6\ CO(g) + 8\ H_2O(g)$
$2\ C_4H_{10}(g) + 9\ O_2(g) \longrightarrow 8\ CO(g) + 10\ H_2O(g)$

13. **a.** $2\ C_2H_6(g) + 3\ O_2(g) \longrightarrow 4\ C(s) + 6\ H_2O(g)$

b. $2\ C_2H_6(g) + 5\ O_2(g) \longrightarrow 4\ CO(g) + 6\ H_2O(g)$

c. $2\ C_2H_6(g) + 7\ O_2(g) \longrightarrow 4\ CO_2(g) + 6\ H_2O(g)$

d. The balanced equations show that complete combustion requires the highest ratio of oxygen to ethane (7:2). If a 5:2 ratio is present, carbon monoxide is formed instead of carbon dioxide. If only a 3:2 ratio is available, then carbon (soot and particulate matter) is formed. *Note:* With less oxygen, the products are likely to be mixed, rather than pure CO or pure soot.

15. In respiration, inhaled oxygen reacts with sugar in your body to produce carbon dioxide and water vapor to produce energy. Therefore, exhaled air has a decreased percentage of oxygen, and increased percentage of carbon dioxide. Oxygen is used to metabolize the food we eat.

16. The troposphere is the layer of air closest to Earth the place where we live. It contains 75% of the air, by mass, and is where air currents and storms occur that mix the air in our atmosphere.

18. NO_2 = nitrogen dioxide
N_2O = dinitrogen monoxide
NO = nitrogen monoxide
NCl = nitrogen trichloride
N_2O_4 = dinitrogen tetroxide

20. **a.** $$400 \text{ parts per million} \times \frac{100 \text{ parts per hundred}}{1{,}000{,}000 \text{ parts per million}}$$
$= 0.04$ parts per hundred $= 4\%$

b. Carbon monoxide is an air pollutant because when breathed into the lungs, CO can be hazardous to human health.

c. Carbon monoxide interferes with the ability of hemoglobin to carry oxygen throughout your body. If you are exposed to CO in high enough concentrations, it can cause a person to die due to lack of oxygen. Shorter-term exposure leads to dizziness or a headache.

22. $6\text{ m} \times 5\text{ m} \times 3\text{ m} = 90\text{ m}^3$

$$3{,}600 \text{ mg acetone} \times \frac{1000\ \mu g}{1 \text{ mg}} \times \frac{1}{90 \text{ m}^3} = 40{,}000 \frac{\mu g}{\text{m}^3}$$

24. Carbon monoxide: Mild CO poisoning makes you feel crummy, causing headache, dizziness, or nausea. You will not be able to exert yourself in your normal manner. More severe poisoning may cause unconsciousness.

Particulate matter: Mild PM poisoning will cause lung and cardiovascular distress. Again, you will not have your normal energy level. More severe poisoning can cause a heart attack.

Ozone: Mild ozone poisoning will cause your eyes and throat to burn. It will aggravate your breathing and asthma.

26. In respiration, inhaled oxygen reacts with substances in your body to produce carbon dioxide and water vapor. Therefore, exhaled air has a decreased percentage of oxygen, and an increased percentage of carbon dioxide.

28. Here are some possibilities:
- Iron and steel would rust more slowly, prolonging the useful life of many objects made from these materials.
- Fires would burn less vigorously and produce more CO and soot. Logs in your fireplace might last longer, putting out heat more slowly.
- Your body can adapt (just as it does at higher elevations) to lower levels of oxygen. But in this case, the level may be too low for metabolic processes involving oxygen to occur at fast enough rates for life as we currently know it.

30. **a.** To convert from percent to ppm, move the decimal point 4 places to the right. Alternatively:

$3\% = 3$ pph

$$3 \text{ pph} \times \frac{1{,}000{,}000 \text{ ppm}}{100 \text{ pph}} = 30{,}000 \text{ ppm}$$

$$3 \text{ pph} \times \frac{1{,}000{,}000{,}000 \text{ ppb}}{100 \text{ pph}} = 30{,}000{,}000 \text{ ppb}$$

b. The NAAQS for CO in an 8-hour period is 9 ppm. The concentration of CO in cigarette smoke is over three thousand times the 8-hour standard. The NAAS for CO in a 1-hour period is 35 ppm. The concentration in cigarette smoke is almost nine hundred times the 1-hour standard.

c. Smokers do not die from CO poisoning because they breathe mainly air, not pure cigarette smoke.

32. Reporting the absolute difference, 0.01 ppm, seems to minimize the amount by which the standard is exceeded, at least in the eyes of the general public. Unless the standard is reported as well, there is no way to compare the magnitude of the difference to the magnitude of the standard. Calculating the percentage by comparing the difference (0.01 ppm) to the standard (0.12 ppm) gives 8%, which may give people a better understanding of the amount by which the standard was exceeded.

34. **a.** The elderly, the young, and people with respiratory diseases such as asthma and emphysema are most affected by PM.

b. December 21–22, December 27, December 31

c. Although PM varies in composition, most of it is less chemically reactive than ozone. It typically is removed from the air by rain or wind.

d. Possibilities include smoke blowing in from a wildfire outside the city, an air inversion, large industrial releases of soot, and a volcanic eruption somewhere in the region that released the ash and soot.

35. **a.** 15 ppm is 0.0015% and 2% is 20,000 ppm. 20,000 ppm is roughly 1300 times larger than 15.

b. $2\ SO_2 + O_2 \longrightarrow 2\ SO_3$

c. $2\ C_{12}H_{26}(l) + 37\ O_2(g) \longrightarrow 24\ CO_2(g) + 26\ H_2O(g)$

d. Ultimately, burning diesel which is derived from the fossil fuel petroleum is not sustainable. In the short term, diesel engines also are old and have high emissions. So these have a high cost in terms of public health. However, the ultra-low sulfur diesel fuel is definitely a step in the right direction.

38. **a.** Reducing the number of cars in use will directly and indirectly reduce the concentrations of NO_x, SO_x, CO, CO_2, and ozone in the air.

b. Geographical features that lead to stagnant air, such as being situated in a valley or surrounded by mountains, may contribute to the higher ozone levels.

43. This phrase refers to what can happen when individuals use a natural resource (*e.g.*, air we breathe, water we drink) that is shared by all for their own interests and then lowers the quality of this resource. This is not in the best interest of a larger group of people. Air pollution is a classic example—people add waste to the air, which in turn affects the health and well-being of others. A case in point would be an industry (group of people) that burns coal to produce electricity. In the process of doing this, oxides of nitrogen and sulfide are released into the air. Other waste products include mercury and the greenhouse gas carbon dioxide. Clearly some people benefit, perhaps even those using the electricity. But all breathe the dirty air. Depending on the concentrations of pollutants, some people may get sick or die.

46. **a.** The rubber may have come from tires abrading as they roll along the highways. Other sources of PM include soot from incomplete combustion and dirt picked up and blown by the wind.

b. Iron, aluminum, and calcium also are commonly present. Other possibilities include sodium, potassium, magnesium, and sulfur.

c. The edges of the particles appear to be irregular and jagged, thus likely to cause inflammation.

48. **a.** This graph clearly indicates that exposure to higher carbon monoxide concentrations over longer time periods becomes increasingly life-threatening.

b. CO poses a serious health threat. This gas is colorless and odorless, making it impossible to detect without a monitor or kit. Furthermore, the initial symptoms of carbon monoxide poisoning are not unique, and those suffering from the associated headaches and nausea could easily presume the symptoms are due to a flu-like illness. Untreated, individuals will ultimately lapse into a coma, after which point they will be unable to call for assistance. For reasons such as these, carbon monoxide detectors are life-saving devices.

50. **a.** The health hazards associated with isocyanate include irritation of the mucous membrane and skin, tightness in the chest, and difficulty breathing. Isocyanate is also a potential carcinogen for humans and is known to cause cancer in animals.

b. Instead of using non-renewable, petroleum-based feedstocks to create adhesives, composites, and foams, Professor Wool's processes use feedstocks from bio-based sources. These renewable sources include flax, chicken feathers, and vegetable oils. In addition to being renewable, their production uses less water and energy and is not as toxic as the petroleum-based counterparts.

Chapter 3

1. The chemical formulas of ozone and oxygen are O_3 and O_2, respectively. Both are gases, but they differ in their properties. Oxygen has no odor; ozone has a very sharp odor. Although both are reactive, ozone is much more highly so. Oxygen is necessary for many forms of life; in contrast, ozone is a harmful air pollutant in the troposphere. However, ozone in the stratosphere helps to protect us from the harmful ultraviolet rays of the sun.

3. **a.** The size of the ozone "hole" varies each year, but has been estimated to be as large as 28 million km^2 in area.

$$10\text{ miles} \times \frac{\text{km}}{0.621\text{ miles}} = 16.1\text{ km}$$

b. Yes, the figure is correct, as the stratosphere extends between 15 and 30 km above Earth's surface.

c. Ozone absorbs UVB and UVC radiation.

6. **a.** The Dobson unit (DU) measures the ozone in a column above a specific location on Earth. If this ozone were compressed at specified conditions of temperature and pressure, it would form a layer. A layer 3-mm thick corresponds to 300 DU. Similarly, a 1-mm layer corresponds to 100 DU.

b. 320 DU > 275 DU. Thus, 320 DU indicates more total ozone overhead.

7. **a.** A neutral atom of oxygen has eight protons and eight electrons.

b. A neutral atom of magnesium has 12 protons and 12 electrons.

c. A neutral atom of nitrogen has seven protons and seven electrons.

d. A neutral atom of sulfur has 16 protons and 16 electrons.

9. **a.** helium, He

b. potassium, K

c. copper, Cu

10. **a.** ·Ca·   **b.** :C̈l·

c. ·N̈·   **d.** He:

12. ·Ö·   Ö=Ö   (ozone structure)   ·Ö—H

The Lewis structures for the oxygen molecule and the ozone molecule both follow the octet rule. In contrast, the oxygen atom has only six outer electrons and does not follow the octet rule. The hydroxyl radical also does not follow the octet rule and has an unpaired electron. Another resonance structure for the ozone molecule may be drawn; the other molecules do not have resonance structures.

14. **a.** This wavelength is in the microwave region of the spectrum.

b. This wavelength is in the infrared region of the spectrum.

c. This wavelength is in the range of violet light in the visible region.

d. This wavelength is in the UHF/microwave region of the spectrum.

16. *Note:* $c = 3.0 \times 10^8$ m/s and $E = h\nu$, in which $h = 6.63 \times 10^{-34}$ J·s.

a. $E = (6.63 \times 10^{-34} \text{ J·s})(1.5 \times 10^{10} \text{ s}^{-1}) = 1.0 \times 10^{-24} \text{ J}$

b. $E = (6.63 \times 10^{-34} \text{ J·s})(8 \times 10^{14} \text{ s}^{-1}) = 5 \times 10^{-19} \text{ J}$

c. $E = (6.63 \times 10^{-34} \text{ J·s})(6 \times 10^{12} \text{ s}^{-1}) = 4 \times 10^{-21} \text{ J}$

d. $E = (6.63 \times 10^{-34} \text{ J·s})(2.0 \times 10^{9} \text{ s}^{-1}) = 1.3 \times 10^{-24} \text{ J}$

The most energetic photon corresponds to the shortest wavelength, 400 nm.

19. $c = \nu\lambda$ and $\lambda = \frac{c}{\nu}$; $c = 3.0 \times 10^8$ m/s

$$\lambda = \frac{3.0 \times 10^8 \text{ m/s}}{2.45 \times 10^9\text{/s}} = 1.2 \times 10^{-1} \text{ m}$$

At 1.2×10^{-1} m, this microwave radiation has a longer wavelength (and lower energy) than X-rays (at $\sim 10^{-10}$ m), but a shorter wavelength (and higher energy) than radio waves (at $\sim 10^3$ m).

23. Answers will vary. To qualify as CFCs, the compounds should contain only carbon, chlorine, and fluorine. Possibilities include:

(Lewis structures of CCl_3F and CCl_2F_2)

CCl_3F trichlorofluoromethane Freon 11

CCl_2F_2 dichlorodifluoromethane Freon 12

25. **a.** No, a CFC molecule can contain only chlorine, fluorine, and carbon atoms.

b. HCFC molecules must contain hydrogen, carbon, fluorine, and chlorine atoms, and no other atoms. In order for a molecule to be classified as an HFC, it must contain hydrogen, fluorine, and carbon (but no other atoms).

27. **a.** Cl· has 7 outer electrons. Its Lewis structure is :C̈l·

$\cdot NO_2$ has 5 + 2(6) = 17 outer electrons.
Its Lewis structure is

Ö::N̈:Ö: or Ö=N̈—Ö:

ClO· has 7 + 6 = 13 outer electrons.
Its Lewis structure is

:C̈l:Ö· or :C̈l—Ö·

·OH has 6 + 1 = 7 outer electrons.
Its Lewis structure is:

·Ö:H or ·Ö—H

b. They all contain an unpaired electron.

29. The first graph is a more realistic representation of the relationship between the percent *reduction* in the concentration of ozone and the percent *increase* in UVB radiation. As the ozone layer is depleted, the concentration of UVB that can penetrate into the atmosphere rises. The second graph shows a type of inverse relationship, which is not substantiated by experimental facts.

30. The message is that ground-level ozone is a harmful air pollutant. Ozone in the stratosphere, on the other hand, is beneficial because it can absorb harmful UV-B before it reaches the surface of Earth.

32. **a.** The most energetic fraction of solar UV radiation is the UVC light.

b. Up in the stratosphere where the air is very thin, UVC splits oxygen molecules, O_2, into two oxygen atoms, O. These in turn react with other oxygen molecules to produce ozone, O_3. See Equation 3.4. Without the UVC light (which does not reach the surface of our planet), the ozone layer would not form.

34. **a.** HFCs are being used to replace HCFCs.

b. HFCs are greenhouse gases!

36. Here are the resonance structures for ozone:

$$:\ddot{O}=\ddot{O}-\ddot{O}: \longleftrightarrow :\ddot{O}-\ddot{O}=\ddot{O}$$

Both contain one double bond (expected length of 121 pm) and one single bond (expected length of 132 pm). But, in reality, the bonds are neither single nor double. Rather, the length of each bond is intermediate between the single and double bond lengths. A reasonable prediction would be 126 or 127 pm for both O-to-O bonds, midway between the two lengths.

39. With respect to valence electron distribution, the Lewis structures of SO_2 and ozone are identical. This should not be surprising, as sulfur and oxygen are in the same group on the periodic table, and thus have the same number of outer electrons. However, the atoms present in the two Lewis structures differ:

$$\ddot{O}::\ddot{O}:\ddot{O}: \longleftrightarrow :\ddot{O}:\ddot{O}::\ddot{O} \quad \text{and} \quad \ddot{O}::\ddot{S}:\ddot{O}: \longleftrightarrow :\ddot{O}:\ddot{S}::\ddot{O}$$

41. The UV Index, typically a number between 1 and 15, helps people to gauge how intense the sunlight is predicted to be on a particular day. A value of 6.5 (color-coded orange) indicates that there is high risk of harm and that you should protect your eyes and skin. A value of 8–10 indicates a very high risk, and above 11 is an extreme risk.

43. Stratospheric ozone is both formed and broken down in a dynamic system. Unless there are disturbances to this system, the system remains in balance and there is no net change in the concentration of stratospheric ozone.

44. These compounds are useful because they are colorless, odorless, tasteless, and generally inert. However, compounds such as these have long atmospheric lifetimes. They persist in the environment and make their way up to the stratosphere where they cause harm to the ozone layer.

46. Cl· acts as a catalyst in the series of reactions in which stratospheric O_3 molecules react to produce O_2 molecules. Because it is not consumed in the reaction, Cl· can continue to catalyze the breakdown of O_3.

50. **a.** These compounds once were manufactured as fire suppressants. They are not water-based, so are excellent for specialized uses such as libraries, aircraft, and electronics. However, their production has been halted because they have high ozone depleting potentials (ODPs).

b. The two halons have different atmospheric lifetimes. According to data from the U.S. EPA, http://www.epa.gov/Ozone/science/ods/classone html (accessed August 2013), the values are 65 years and 16 years, respectively, for Halon-1301 and Halon-1211. A more interesting question is why the different lifetimes, which is beyond the scope of this text.

c. At the time this graph was drawn, it was thought that no substitutes would be found for some uses of methyl bromide. However, it now is looking more likely that substitutes will be found.

51. $5 + 3(6) + 1 = 24$ electrons

$$\left[\begin{array}{c} :\ddot{O}: \\ | \\ \ddot{O}=N-\ddot{O}: \end{array}\right]^- \longleftrightarrow \left[\begin{array}{c} :\ddot{O}: \\ | \\ :\ddot{O}-N=\ddot{O} \end{array}\right]^- \longleftrightarrow \left[\begin{array}{c} :O: \\ \| \\ :\ddot{O}-N-\ddot{O}: \end{array}\right]^-$$

Each atom in all three resonance structures satisfy the octet rule.

54. O_2, O_3, and N_2 all have an even number of valence electrons. In contrast, N_3 would have 15 valence electrons. Molecules with odd numbers of electrons cannot follow the octet rule, making them free radicals and more reactive.

56. Ozonators typically produce ozone either via an electrical discharge or with UV light. The former is similar to the process that produces ozone in a thunderstorm. A bolt of lightning can split O_2 molecules to form O atoms. The latter uses UVC light to split O_2 molecules. In either case, the O atoms then react with another oxygen molecule to produce ozone. The ozone produced works as an effective disinfectant. It can react with many biological molecules, thereby being effective against undesired microbes and viruses. It also can react with many molecules that produce odors.

a. Search the web for examples. Claims include that ozonators can:

- destroy odors from tobacco, smoke, pets, cooking, and chemicals
- kill bacteria and airborne viruses
- remove allergy causing pollen and microbes
- prevent mold and mildew, the leading cause of Legionnaires disease
- eliminate toxic fumes from printing, plating processes, and hair and nail salons
- purify water in holding tanks and emergency storage water tanks
- purify drinking water from well sources or city water supplies
- remove undesirable tastes, odors, and colors

b. Ozone can be a harmful pollutant causing damage to both plants and animals. Any device that creates the gas must carefully contain it.

60. **a.** See Figure 3.22. Most months of the year, it is not cold enough in the Arctic for PSCs to form.

b. $HCl + ClONO_2 \longrightarrow Cl_2 + HNO_3$

The nitric acid remains bound to the ice, but the chlorine gas is released to the atmosphere.

c. In the atmosphere in the presence of sunlight, $Cl_2 \longrightarrow 2\ Cl\cdot$

61. **a.** This is a possible Lewis structure.

$$:\ddot{Cl}-\ddot{O}-\ddot{O}-\ddot{Cl}:$$

b. If Cl_2O_2 is the actual molecule, then it will have to be broken down by UV photons to ClO· free radicals before it can react with oxygen atoms as shown in Equations 3.7 and 3.8. This would add one additional decomposition reaction in the catalytic destruction of ozone.

Chapter 4

2. These two planets are warmer than would be expected because they have atmospheric gases that produce a "greenhouse effect." Sunlight enters the atmosphere of both Earth and Venus, warming the surfaces of the planet. The atmospheric gases are able to trap some of the heat radiated by the planet surfaces. Without these gases, the planets would be the temperatures expected as a result of their distance from the Sun. *Note:* The high concentration of CO_2 in the Venusian atmosphere (98% CO_2) has led to a "runaway greenhouse effect" and resulted in a surface temperature of about 450 °C!

6. a. The rest of the Sun's energy is either absorbed or reflected by the atmosphere. For example, Chapter 3 pointed out that oxygen and stratospheric ozone absorb certain wavelengths of UV light. This chapter points out that clouds may reflect incoming radiation back into space.

b. Under steady-state conditions, 29 MJ/m^2 would leave the atmosphere each day.

7. a. As of 2016, the atmospheric concentration of CO_2 was a little above 400 ppm; however, 20,000 years ago the concentration was only about 190 ppm. Looking back to 120,000 years ago, the concentration was about 270 ppm, still ~40% below current levels.

b. The mean atmospheric temperature at present is somewhat above the 1950–1980 mean atmospheric temperature. 20,000 years ago, the mean atmospheric temperature was lower by about 9 °C. However, 120,000 years ago the mean atmospheric temperature was lower than the present temperature by only about 1 °C.

c. Although there appears to be a *correlation* between mean atmospheric temperature and CO_2 concentration, this graph does not prove *causation* of either by the other.

9. a. Visible light can enter through the glass, but infrared radiation cannot leave through the glass. There also is little exchange of air with the outside, so the heat cannot be dissipated and the temperature inside the car increases.

b. On clear nights, the heat from Earth can radiate through the atmosphere and out into space. When it is cloudy, the water vapor in the clouds absorbs some of the heat, thus retaining it.

c. In the desert, the temperature swings between night and day tend be more pronounced. Clouds and humid air make the temperatures more uniform, because they tend to block or scatter incoming solar radiation and trap outgoing heat. *Note:* If the desert contains large urban areas, the pavement and buildings absorb heat during the day. This heat is released at night and thus can keep the temperature high even when the Sun has set.

d. Dark clothing absorbs much of the light that strikes it; in contrast, lighter clothing reflects most of it. The light energy absorbed converts to heat energy, which can increase the risk of heatstroke.

11. a. H–S̈–H (with two lone pairs on S) bent

b. :C̈l–Ö–C̈l: bent

c. :N̈=N=Ö: or :N≡N–Ö: (with three lone pairs on O) or :N̈–N≡O: (with three lone pairs on terminal N) linear

13. a. 3(1) + 4 + 6 + 1 = 14 outer electrons. This is the Lewis structure.

```
      H
      |    ..
  H — C — O — H
      |    ..
      H
```

b. The geometry around the C atom is tetrahedral, and there are no lone pairs. A H–C–H bond angle of about 109.5° is predicted.

c. There are four pairs of electrons around the O atom, two of which are bonding pairs, while the other two are nonbonded pairs. Repulsion between the two nonbonded electron pairs and their repulsion of the bonding pairs is predicted to cause the H–O–C bond angle to be slightly less than 109.5°, about 104.5°.

15. All can contribute to the greenhouse effect. In each case, the atoms move as the bond stretches or bends, and therefore the charge distribution changes. Unlike the linear CO_2 molecule, the water molecule is bent and so its polarity changes with each of these modes of vibration.

16. a. Use $E = \frac{hc}{\lambda}$ to calculate the energies.

$$E = \frac{(6.63 \times 10^{-34}\ \text{J·s}) \times (3.00 \times 10^8\ \text{m/s})}{4.26\ \mu\text{m} \times \frac{1\ \text{m}}{10^6\ \mu\text{m}}}$$

$$= 4.67 \times 10^{-20}\ \text{J}$$

$$E = \frac{(6.63 \times 10^{-34}\ \text{J·s}) \times (3.00 \times 10^8\ \text{m/s})}{15.00\ \mu\text{m} \times \frac{1\ \text{m}}{10^6\ \mu\text{m}}}$$

$$= 1.33 \times 10^{-20}\ \text{J}$$

b. If the vibrating molecule CO_2 collides with another molecule, such as N_2 or O_2, the energy can be transferred to the second molecule. The energy can also be spontaneously emitted back to the atmosphere or into space.

19. a. $C_6H_{12}O_6 \longrightarrow 3\ CH_4 + 3\ CO_2$

b. In one day:

$$1.0 \text{ mg glucose} \times \frac{1 \text{ g}}{1000 \text{ mg}} \times \frac{1 \text{ mol glucose}}{180 \text{ g glucose}} \times \frac{3 \text{ mol } CO_2}{1 \text{ mol glucose}} \times \frac{44 \text{ g } CO_2}{1 \text{ mol } CO_2} = 7.3 \times 10^{-4} \text{ g}$$

in one year:

$$\frac{7.3 \times 10^{-4} \text{ g } CO_2}{\text{day}} \times \frac{365 \text{ days}}{\text{year}} = 0.27 \text{ g } CO_2/\text{year}$$

21. a. A neutral atom of Ag-107 has 47 protons, 60 neutrons, and 47 electrons.

b. A neutral atom of Ag-109 has 47 protons, 62 neutrons, and 47 electrons. Only the number of neutrons is different.

23. a. $$\frac{107.87 \text{ g}}{1 \text{ mole}} \times \frac{1 \text{ mole}}{6.02 \times 10^{23} \text{ atoms}} = \frac{1.79 \times 10^{-22} \text{ g}}{\text{atom}}$$

b. $$\frac{1.79 \times 10^{-22} \text{ g}}{\text{atom}} \times (10 \times 10^{12} \text{ atoms}) = 1.79 \times 10^{-9} \text{ g}$$

c. $$5.00 \times 10^{45} \text{ atoms} \times \frac{1.79 \times 10^{-22} \text{ g}}{\text{atom}} = 8.95 \times 10^{23} \text{ g}$$

25. a. The mass percent of Cl in CCl_3F (Freon-11) is:

$$\frac{3 \times (35.5 \text{ g/mol})}{12.0 \text{ g/mol} + 3 \times (35.5 \text{ g/mol}) + 19.0 \text{ g/mol}} \times 100 = 77.5\%$$

b. The mass percent of Cl in CCl_2F_2 is 58.7%.

c. Freon-11: 77.5 g; Freon-12: 58.7 g

d. Freon-11: $$77.5 \text{ g Cl} \times \frac{1 \text{ mol Cl}}{35.5 \text{ g Cl}} \times \frac{6.02 \times 10^{23} \text{ atoms Cl}}{1 \text{ mol Cl}} = 1.31 \times 10^{24} \text{ Cl atoms;}$$

Freon-12: 9.95×10^{23} Cl atoms.

26. Concentration of carbon atoms =

$$\frac{7.5 \times 10^{17} \text{ g}}{7.5 \times 10^{22} \text{ g}} \times 100 = 0.001\%$$

$$\frac{0.001 \text{ parts C in living systems}}{100 \text{ parts C on Earth}} = \frac{x \text{ parts C in living systems}}{1{,}000{,}000 \text{ parts C on Earth}}$$

$x = 10$ ppm

29. a. ${}^{19}_{9}F$ **b.** ${}^{56}_{26}Fe$ **c.** ${}^{222}_{86}Rn$

31. a. Before sophisticated analytical instruments were developed, miners would take caged canaries into the mines to warn them if they encountered any toxic gases (quite prevalent in mine shafts). The canaries were more sensitive to gases like CO. If the canary died, the miners knew they had to get out and into better air quickly.

b. The changes that are occurring in the Arctic may be an early warning sign for the rest of the planet in terms of potential consequences of warmer global temperatures.

c. Significant amounts of methane are trapped in the frozen tundra. If the tundra thawed and released this methane into the atmosphere, this would further accelerate global warming elsewhere because methane is a greenhouse gas.

34. One initial reaction would be that the newspaper reporter has confused "the greenhouse effect" with "global warming." The greenhouse effect is necessary for life on Earth to exist; without it, the average temperature would be −15 °C. Global warming, or the "enhanced greenhouse effect," is what is being blamed for the rise in average global temperatures and the consequences for humans that may result.

36. Substances that absorb visible light have observable colors. For example, if the wavelengths associated with red light are absorbed, the object appears green. Because we cannot see any color associated with either carbon dioxide gas or water vapor, we conclude they do not absorb a significant amount of visible light.

37. The energy required would be smaller for each of the IR-absorbing vibrations if single bonds were present. In general, single bonds between atoms are weaker than double bonds, and therefore less energy will be required to cause stretching and bending.

39. a. $C_2H_5OH + 3\ O_2 \longrightarrow 3\ H_2O + 2\ CO_2$

b. Two moles of CO_2 is produced for each mole of ethanol burned.

c. Thirty mol O_2, because for every 1 mole of C_2H_5OH burned, 3 moles of O_2 burn.

41. The main chemical species involved in ozone depletion are O_3 and CFCs, while for climate change, CO_2, CH_4, and N_2O are the main greenhouse gases. Ultraviolet radiation breaks covalent bonds in CFCs, leading to ozone depletion, while infrared radiation is absorbed and trapped by atmospheric gases, causing the greenhouse and enhanced greenhouse effects. Predicted consequences of ozone depletion include increased UV exposure at Earth's surface, increased occurrences of skin cancer in humans, and damage to other biological organisms. Climate change consequences include rising sea level, stresses on freshwater resources, loss of biodiversity, and ocean acidification, among others.

43. $73 \times 10^6 \text{ metric tons } CH_4 \times \frac{12 \text{ metric tons C}}{16 \text{ metric tons } CH_4}$

$= 5.5 \times 10^7$ metric tons C

46. a. On a per capita basis, the United States would rank first. The population of the United States is smaller than that of China.

b. The value for metric tons of CO_2 would be higher than that for tons carbon. The former includes the mass of the oxygen; the latter does not.

48. Arrhenius overestimated the temperature increase caused by a doubling of the atmospheric CO_2 concentration by about a factor of two, compared to the IPCC models. The models project a range of 2–4.5 °C increase for doubling of the CO_2 concentration.

50. a. Burning coal with high sulfur content introduces more SO_x to the atmosphere. This reduces air quality, increases acid precipitation, and speeds the degradation of the environment due to acidification.

b. Sulfur aerosols reflect incoming solar radiation back toward space. They also serve as nuclei around which water vapor can condense to form sunlight-reflecting cloud particles.

c. India

51. a. The area in Mexico previously inhabited is becoming too hot and dry for the butterfly.

b. Land development is the main cause of the parts of California being uninhabitable.

c. Answers can include prohibiting further development in the region, or developing breeding programs to increase the species population.

53. a. The balanced equation for the complete combustion of methane is:

$CH_4 + 2\,O_2 \longrightarrow CO_2 + 2\,H_2O$

Therefore, burning 1196 mol of methane could produce 1196 mol of CO_2.

b. $1196 \text{ mol} \times \frac{44 \text{ g } CO_2}{\text{mol}} = 52{,}624 \text{ g } CO_2$, or 52.62 kg

c. $52.62 \text{ kg} \times \frac{1 \text{ metric ton}}{1000 \text{ kg}}$

$= 0.05262$ metric tons

Chapter 5

2. a. CO_2, carbon dioxide.

b. SO_2 is an air pollutant. Although sulfur is present in low concentrations in coal, large amounts of coal are burned, and collectively large amounts of SO_2 are released.

c. Nitrogen is present in the air (~80% of atmospheric gases). The nitrogen present in air reacts with O_2 (also $N_2 + O_2 \xrightarrow{\text{high temperature}} 2\text{ NO}$ present in air) at high temperatures to form NO:

d. Revisit Chapter 2 for the details. From the EPA website: "Particle exposure [of any size] can lead to a variety of health effects. For example, numerous studies link particle levels to increased hospital admissions and emergency room visits and even to death from heart or lung diseases. Both long- and short-term particle exposures have been linked to health problems. Long-term exposures, such as those experienced by people living for many years in areas with high particle levels, have been associated with problems such as reduced lung function and the development of chronic bronchitis and even premature deaths."

4. a. The fuel in the burner is a source of potential energy. When burned, some of its potential energy is converted to heat through combustion. The heat is converted into kinetic energy of the vaporized water molecules (steam).

b. The kinetic energy of the steam is converted to mechanical energy by spinning a turbine.

c. The mechanical energy generated from the spinning turbine is converted to electrical energy by rotating a wire in a magnetic field.

d. The electrical energy, carried to the city by the power lines, lights bulbs and heats homes.

5. a. $\frac{5.00 \times 10^8 \text{ J}}{\text{s}} \times \frac{3600 \text{ s}}{\text{h}} \times \frac{24 \text{ h}}{\text{day}} \times \frac{365 \text{ days}}{\text{year}}$

$= 1.58 \times 10^{16}$ J of electricity generated per year

$\frac{1.58 \times 10^{16} \text{ J}}{0.375} = 4.2 \times 10^{16}$ J of heat for electricity generated per year

b. $1.58 \times 10^{16} \text{ J produced} \times \frac{1 \text{ kJ}}{1000 \text{ J}} \times \frac{1 \text{ g}}{30 \text{ kJ}}$

$= 5.3 \times 10^{11}$ g per year

$5.3 \times 10^{11} \text{ g} \times \frac{1 \text{ metric ton}}{10^6 \text{ g}} = 5.3 \times 10^5$ metric tons

9. A typical power plant burns 1.5 million tons of coal each year. The first calculation is for coal with 50 ppb mercury; the second is for 200 ppb.

$\frac{x \text{ ton Hg}}{1.5 \times 10^6 \text{ ton coal}} = \frac{50 \text{ ton Hg}}{1 \times 10^9 \text{ ton coal}} \quad x = 0.075 \text{ ton Hg}$

$\frac{x \text{ ton Hg}}{1.5 \times 10^6 \text{ ton coal}} = \frac{200 \text{ ton Hg}}{1 \times 10^9 \text{ ton coal}} \quad x = 0.30 \text{ ton Hg}$

Assuming mercury concentrations in the range of 50–200 ppb, the plant releases between 0.075 and 0.30 ton of Hg per year.

11. a. CH_3CH_3: ethane, $CH_3(CH_2)_2CH_3$: butane

b. C_2H_6 and C_4H_{10}

```
    H  H                 H  H  H  H
    |  |                 |  |  |  |
 H—C—C—H     and     H—C—C—C—C—H
    |  |                 |  |  |  |
    H  H                 H  H  H  H
```

c. Chemical formulas such as C_4H_{10} are compact and easy to write. The same is true for condensed structural formulas, at least in this particular case. Although structural formulas take longer to draw and take up more space, they clearly reveal the arrangement of all the bonds and atoms.

13. a.

```
     H  H  H  H
     |  |  |  |
  H—C—C—C—C—H
     |  |  |  |
     H  H  H  H
```

b. There is only one other isomer:

```
        H
        |
     H—C—H
     H  |  H
     |  |  |
  H—C—C—C—H
     |  |  |
     H  H  H
```

14. Pentane should be a liquid because room temperature (20 °C) is below its boiling point (36 °C) but above its melting point (–130 °C). Triacontane should be solid at room temperature because room temperature is below its melting point (66 °C). Propane should be a gas because room temperature is above its boiling point (–42 °C).

17. a. $C_7H_{16} + 11\ O_2 \longrightarrow 7\ CO_2 + 8\ H_2O$

b. $2.50\ \text{kg} \times \dfrac{10^3\ \text{g}}{\text{kg}} \times \dfrac{1\ \text{mol}\ C_7H_{16}}{100.2\ \text{g}}$

$\times \dfrac{4817\ \text{kJ}}{1\ \text{mol}\ C_7H_{16}} = 1.2 \times 10^7\ \text{kJ}$

19. $92\ \text{kcal} \times \dfrac{4.184\ \text{kJ}}{1\ \text{kcal}} = 380\ \text{kJ}$

20. a. exothermic **b.** endothermic **c.** endothermic

21. a. Bonds broken in the reactants
1 mol N≡N triple bonds = 1(946 kJ) = 946 kJ
3 mol H–H single bonds = 3(436 kJ) = 1308 kJ
Total energy *absorbed* in breaking bonds = 2254 kJ

Bonds formed in the products
6 mol N–H single bonds = 6(391 kJ) = 2346 kJ
Total energy *released* in forming bonds = 2346 kJ
Net energy change is (+2254 kJ)
+ (–2346 kJ) = –92 kJ
The overall energy change is negative, characteristic of an exothermic reaction.

b. Bonds broken in the reactants
1 mol H–H single bonds = 1(436 kJ) = 436 kJ
1 mol Cl–Cl single bonds = 1(242 kJ) = 242 kJ
Total energy *absorbed* in breaking bonds = 678 kJ

Bonds formed in the products
2 mol H–Cl single bonds = 2(431 kJ) = 862 kJ
Total energy *released* in forming bonds = 862 kJ
Net energy change is (+678 kJ)
+ (–862 kJ) = –184 kJ
The overall energy change is negative, characteristic of an exothermic reaction.

24. a. None of these are isomers. All have different chemical formulas.

b. No, no other isomers are possible for ethene.

c. One other isomer is possible, although it contains a very distinct functional group, called an ether. Here is its condensed structural formula: $CH_3–O–CH_3$.

25. a.

```
     H  H  H  H  H  H  H  H
     |  |  |  |  |  |  |  |
  H—C—C—C—C—C—C—C—C—H
     |  |  |  |  |  |  |  |
     H  H  H  H  H  H  H  H

        CH3                  CH3
        |                    |
  CH3—C—CH3                CH—CH3
        |                    |
        CH2                  CH2
        |                    |
        CH—CH3         CH3—C—CH3
        |                    |
        CH3                  CH3
```

b. The second and third are identical.

c. Several other isomers are possible. Here are two.

```
     H  H   H  H  H  H  H
     |  |   |  |  |  |  |
  H—C—C—C—C—C—C—C—H
     |  |   |  |  |  |  |
     H  CH3 H  H  H  H  H

     H  H  H   H  H  H  H
     |  |  |   |  |  |  |
  H—C—C—C—C—C—C—C—H
     |  |  |   |  |  |  |
     H  H  CH3 H  H  H  H
```

28. a. $C_{16}H_{34} \longrightarrow C_5H_{12} + C_{11}H_{22}$

```
     H  [H]  H  H  [H]  H
     |  [|]  |  |  [|]  |
  H—C—[C]—C—C—[C]—C—H  ⟶
     |  [|]  |  |  [|]  |
     H  [H]3 H  H  [H]9 H

     H  [H]  H             H   H  [H]  H
     |  [|]  |              \  |  [|]  |
  H—C—[C]—C—H   +           C=C—[C]—C—H
     |  [|]  |              /      [|]  |
     H  [H]3 H             H       [H]8 H
```

The C–C single bond in the center of the molecule and one of the C–H single bonds must be broken. A second C–C single bond must be broken so that a C=C double bond can be formed in its place. A new C–H single bond must form on the shorter product.

b. Bonds broken in the reactants
2 mol C–C single bonds = 2(356 kJ) = 712 kJ
1 mol C–H single bonds = 1(416 kJ) = 416 kJ
Total energy *absorbed* in breaking bonds = 1128 kJ

Bonds formed in the products
1 mol C–H single bonds = 1(416 kJ) = 416 kJ
1 mol C=C double bonds = 1(598 kJ) = 598 kJ
Total energy *released* in forming bonds = 1014 kJ
Net energy change is (+1128 kJ) + (–1014 kJ) = 114 kJ

The overall energy change has a positive sign, characteristic of an endothermic reaction.

30. a. The hydroxyl functional group, –OH. All of the compounds are alcohols.

b. carbon dioxide, CO_2, and water, H_2O.

c. These three compounds are similar in chemical composition and differ only in number of CH_2 groups. Each have hydrogen bonding as their predominant intermolecular force, but methanol would have the lowest boiling point because it has the lowest molecular mass of the group.

d. *n*-propanol, an alcohol with three carbon atoms, is the most similar to glycerol (which also is an alcohol and has three carbon atoms). However, glycerol contains three –OH groups (one on each carbon) in comparison to the one in *n*-propanol.

32. a. $C_6H_{12}O_6 + 6\ O_2 \longrightarrow 6\ CO_2 + 6\ H_2O + \text{energy}$

b. Cellulose is one of the primary components of wood. Cellulose is a polymer made up of glucose building blocks. As a result, burning cellulose gives products comparable to burning glucose.

35. Figure 5.6 gives the energy released per gram for the combustion of several fuels. Assuming the densities of octane and ethanol are similar (a good assumption), one gallon of gasoline releases more energy than one gallon of ethanol (44.4 kJ/g of gasoline vs. 26.8 kJ/g of ethanol). This makes sense, because ethanol is an oxygenated fuel; it contains oxygen and thus is already "partially burned."

38. A primary component of wood is cellulose, with a chemical formula that can be approximated with that of glucose, $C_6H_{12}O_6$. Given that the ratio of carbon-to-oxygen in glucose is 1:1, the chemical formula for this soft coal most likely would contain much more oxygen than common types of coal. The same is true for hydrogen, because the ratio of carbon-to-hydrogen in glucose is 1:2.

41. Answers may vary. Here is one example: Wouldn't you rather spill a drop of hot coffee on you than the whole cup at the same temperature? Although the drop and the cup full of coffee may initially have the same temperature, you will receive a bigger burn from the bigger volume of coffee because it has the higher heat content. Heat is a form of energy. In contrast, temperature is a measurement that indicates the direction heat will flow. Heat always flows from an object at high temperature to an object at lower temperature. This means that if hot coffee is added to cold coffee, heat will flow from the hot liquid to the cold liquid, and the final temperature of the mixture will be between the original temperatures of the two individual solutions. Heat depends on the temperature and on how much material is present.

44. $H_2CO(g) + O_2(g) \longrightarrow CO_2(g) + H_2O(g)$

Let x represent the C=O bond energy in H_2CO.

Bonds broken in the reactants:
2 mol C–H single bonds = 2(416 kJ) = 832 kJ
1 mol C=O double bonds = 1(x kJ) = x kJ
1 mol O=O double bonds = 1(498 kJ) = 498 kJ
Total energy *absorbed* by breaking bonds = (1330 + x) kJ

Bonds formed in the products
2 mol O–H single bonds = 2(467 kJ) = 934 kJ
2 mol C=O double bonds = 2(803 kJ) = 1606 kJ
Total energy *released* in forming bonds = 2540 kJ
Net energy change:

$$(1330 + x\text{ kJ}) + (-2540\text{ kJ}) = -465\text{ kJ}$$

Rearranging the equation:

$$x\text{ kJ} = -465 + 2540 - 1330\text{ kJ} \quad x = 745\text{ kJ}$$

This value is less than the bond energy for C=O double bonds in carbon dioxide reported in Table 5.1. The C=O double bonds in carbon dioxide are stronger than the C=O double bond in formaldehyde.

45. CFCs are stable because the bond energies for C–Cl and C–F are large compared to other bond energies. It takes less energy to release Cl atoms from CFCs because the C–Cl bond energy (327 kJ/mol) is lower than the C–F bond energy (485 kJ/mol). HFCs, with their C–F bonds and (no C–Cl bonds) release no Cl atoms.

47. From Figure 5.6, fuels containing oxygen have lower energy content per gram than those without oxygen. For example, ethanol and glucose have proportionately more oxygen than the other fuels listed. In essence, they are already "partially burned" (oxidized) and thus their energy content is lower when you look at the values per gram of fuel.

Here are the values in kilojoules per mole.

$$\text{methane } (CH_4)\text{: } \frac{50.1\text{ kJ}}{1\text{ g}} \times \frac{16.05\text{ g}}{1\text{ mol}} = 802\text{ kJ/mol}$$

$$\text{octane } (C_8H_{18})\text{: } \frac{44.4\text{ kJ}}{1\text{ g}} \times \frac{114\text{ g}}{1\text{ mol}} = 5.06 \times 10^3\text{ kJ/mol}$$

coal ($C_{135}H_{96}O_9NS$):

$$\frac{32.8\text{ kJ}}{1\text{ g}} \times \frac{1908\text{ g}}{1\text{ mol}} = 6.26 \times 10^4\text{ kJ/mol}$$

ethanol (C_2H_6O):

$$\frac{26.8\text{ kJ}}{1\text{ g}} \times \frac{46\text{ g}}{1\text{ mol}} = 1.23 \times 10^3\text{ kJ/mol}$$

glucose: ($C_6H_{12}O_6$):

$$\frac{14.1\text{ kJ}}{1\text{ g}} \times \frac{180\text{ g}}{1\text{ mol}} = 2.54 \times 10^3\text{ kJ/mol}$$

With kilojoules per mole, though, the trend observed is different. Here, the fuels with higher numbers of carbon atoms in their chemical formulas (and hence higher molar masses) release more energy when burned. Compare methane and *n*-octane to see the contrast clearly.

49. **a.** When $n = 1$, the balanced equation is

$$CO + 3\ H_2 \longrightarrow CH_4 + H_2O$$

Bonds broken in the reactants
1 mol C≡O triple bonds = 1(1073 kJ) = 1073 kJ
3 mol H–H single bonds = 3(436 kJ) = 1308 kJ
Total energy *absorbed* in breaking bonds = 2381 kJ

Bonds formed in the products
4 mol C–H single bonds = 4(416 kJ) = 1664 kJ
2 mol O–H single bonds = 2(467 kJ) = 934 kJ
Total energy *released* in forming bonds = 2598 kJ

Net energy change is (+2381 kJ) + (–2598 kJ) = –217 kJ

b. Reactions with *n* greater than 1 will release more energy as *n* becomes larger, assuming that we are viewing the energy per mole of the hydrocarbon formed (not per gram). There will always be *n* C≡O triple bonds to break and $(2n + 1)$ H–H single bonds to break. The number of C–H bonds forming will be $(2n + 2)$, the number of O–H bonds forming will be $2n$, and the number of C–C single bonds forming will be $n - 1$. Combining these terms shows that as *n* becomes larger, more and more energy will be released.

50. **a.** The Lewis structure for dimethyl ether is:

```
   H    H
   ..  ..  ..
H:C:O:C:H
   ..  ..  ..
   H    H
```

b. The structural formula for diethyl ether is:

```
    H   H       H   H
    |   |       |   |
H—C—C—O—C—C—H
    |   |       |   |
    H   H       H   H
```

c. The common structural feature is an oxygen atom between two carbon atoms.

51. **a.** The compounds *n*-octane and iso-octane have nearly identical heats of combustion. This makes sense because they have the same number and types of bonds. However, they have very different octane ratings. Therefore, the octane rating is not a measure of the energy content of a substance.

b. Knocking produces an objectionable pinging sound, reduced engine power, overheating, and possible engine damage.

c. The higher octane blends are more expensive to produce because they require more "processing," including energy-intensive cracking reactions that convert lower octane fuels into higher octane ones.

d. The octane ratings tell you nothing about whether or not oxygenates are present. Although oxygenates are one way to improve the octane rating, they are not the only way.

54.

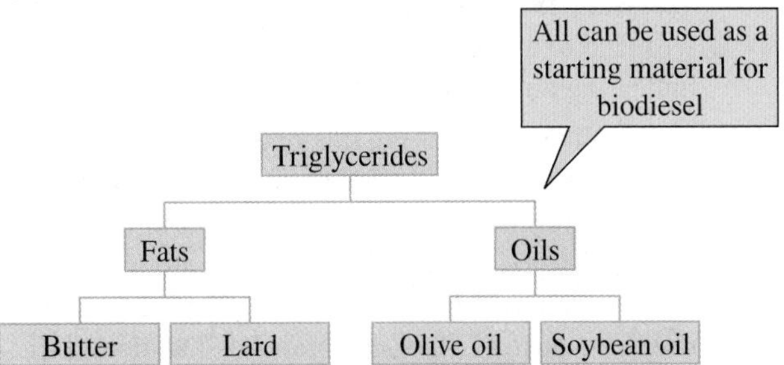

59. The price of electricity varies by locality, but according to www.electricchoice.com, the average cost of electricity in the U.S. in 2015 was 12.5 cents per kWh.

Using a 75 watt incandescent bulb for 10,000 hours would consume 75,000 Wh of electricity. Using an 18 watt compact fluorescent bulb for 10,000 hours would consume 18,000 Wh of electricity.

75 watt incandescent bulb:

$$75{,}000\text{ Wh} \times \frac{1\text{ kWh}}{1000\text{ Wh}} \times \frac{\$0.125}{\text{kWh}} = \$9.37$$

18 watt compact fluorescent bulb:

$$18{,}000\text{ Wh} \times \frac{1\text{ kWh}}{1000\text{ Wh}} \times \frac{\$0.125}{\text{kWh}} = \$2.25$$

These calculations indicate an electricity savings of \$9.37 – \$2.25 = \$7.12 over the life of a compact fluorescent bulb.

To be most accurate, the lifetime of incandescent bulbs should also be taken into account. Incandescent bulbs burn out after ~750 hours, so 10,000 hours of use would require the use of 10,000/750 = 13.33 incandescent bulbs.

At a popular online retailer, a six-pack of 75 W incandescent bulbs costs \$7.97 (or \$1.33 each) and a four-pack of 18 W compact fluorescent bulbs costs \$5.97 (or \$1.49 each). Therefore, the cost of 13.33 incandescent bulbs would be \$17.71. When compared to the cost of one compact fluorescent bulb (\$1.49), it can be seen that an additional savings of \$16.22 can be realized by using the longer lasting compact fluorescent bulbs. *Note:* The price of compact fluorescents has been dropping, so it might make sense to recheck the prices.

62. Consider a natural gas (methane) explosion that releases energy:

$$CH_4 + 2\,O_2 \longrightarrow CO_2 + 2\,H_2O$$

The bond energies involved are: C–H single bonds, 416 kJ/mole; O=O double bonds, 498 kJ/mole; H–O single bonds, 467 kJ/mole; C=O double bonds, 803 kJ/mole. The bond energies of the products are larger than those of the reactants, and thus they release more energy when forming than was necessary to break the bonds of the reactants. This will lead to a large negative net energy change indicating an exothermic reaction.

65. As gasoline additives go, this one ranks high. It is more than double the value of any listed in Table 5.4. The structural formula of TEL shows four ethyl groups ($-C_2H_5$) around a central lead atom. It is highly branched! Just as iso-octane has a high octane rating because of all of its "branches," so does TEL.

66. **a.** The sketch shows that the catalyzed pathway requires less activation energy than the uncatalyzed pathway.

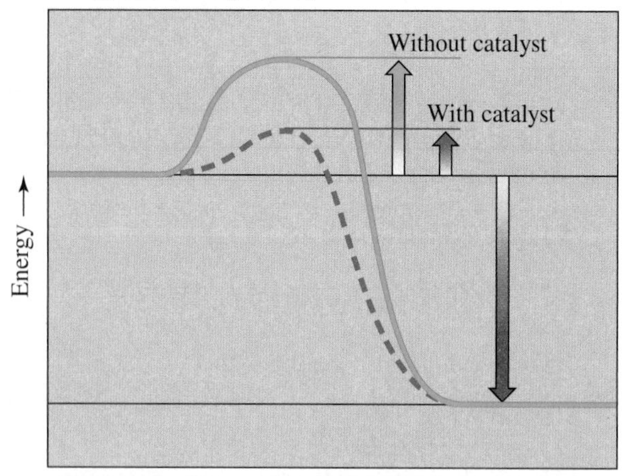

b. In Chapter 2, catalysts were discussed in connection with removing NO from automobile exhaust. Nitrogen oxide can react with oxygen to form NO_2, a criteria pollutant. NO is also involved in forming ozone in the troposphere and contributes to acid rain. To reduce pollution, it is important to reduce NO emissions.

67. Recall that in an endothermic reaction, the potential energy of :N̈=Ö: the products is greater than the potential energy of the reactants. The opposite is true for an exothermic reaction.

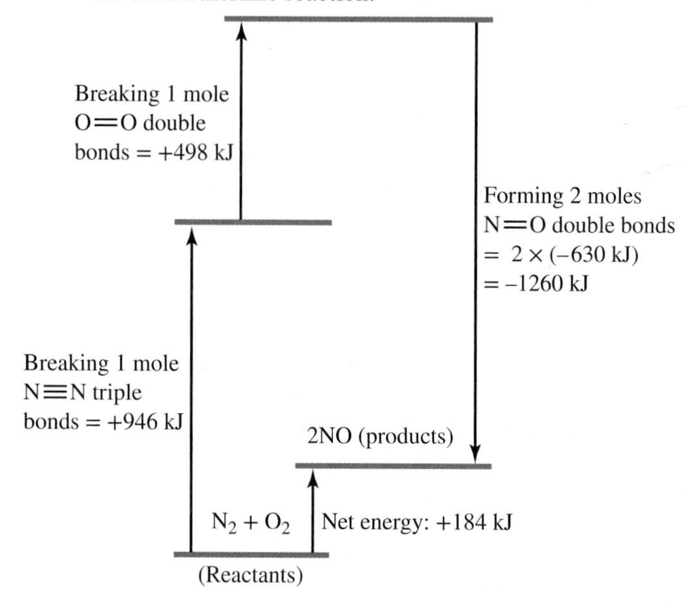

Chapter 6

1. One carbon atom can differ from another in the number of neutrons (such as C-12 and C-13) and in the number of electrons (carbon ions do exist, but we do not discuss them in this text). All carbon atoms differ from all uranium atoms in the number of protons, neutrons, and electrons. Carbon atoms also differ from uranium atoms in their chemical properties.

3. **a.** 94 protons

b. 93 protons = Np (neptunium), 94 protons = Pu (plutonium)

c. 86 protons

5. E represents energy, *m* represents mass lost in a nuclear transformation, and *c* represents the speed of light.

9. **a.** The alpha particle may have come from the radioactive decay of another radioisotope.

b. ${}^{1}_{0}n$ represents a neutron.

c. Curium-243 represents an unstable intermediate in the nuclear reaction. This isotope has an extremely short lifetime, decomposing immediately upon formation into Cm-242 with an accompanying neutron.

12. **a.** ${}^{1}_{0}n + {}^{235}_{92}U \longrightarrow [{}^{236}_{92}U] \longrightarrow {}^{146}_{57}La + {}^{87}_{35}Br + 3\,{}^{1}_{0}n$

b. ${}^{1}_{0}n + {}^{235}_{92}U \longrightarrow [{}^{236}_{92}U] \longrightarrow {}^{140}_{56}Ba + {}^{94}_{36}Kr + 2\,{}^{1}_{0}n$

13. **A** represents the control rod assembly, **B** represents the cooling water out of the core, **C** represents the control rods, **D** represents the cooling water into the core, and **E** represents the fuel rods.

15. The primary coolant is the liquid surrounding the fuel bundles and control rods, a liquid that comes in direct contact with the nuclear reactor to carry away heat. The heat from the primary coolant is transferred to the secondary coolant, water in the steam generators that does not come in contact with the reactor. The steam generators are separated from the nuclear reactor, so the secondary coolant is not housed in the containment dome.

16. a. $^{1}_{0}n + ^{10}_{5}B \longrightarrow [^{11}_{5}B] \longrightarrow ^{4}_{2}He + ^{7}_{3}Li$

b. Boron can be used in control rods because it is a good neutron absorber.

17. a. $^{239}_{94}Pu \longrightarrow ^{235}_{92}U + ^{4}_{2}He$

$^{131}_{53}I \longrightarrow ^{131}_{54}Xe + ^{0}_{-1}e + ^{0}_{0}\gamma$

b. In a particulate form such as a powder or a dust, plutonium can be inhaled. If the plutonium particles become lodged in the lungs, the ionizing radiation they emit (alpha particles) can damage lung cells. The decay products from U-235 also are radioactive and can damage tissue.

c. Iodine accumulates in the thyroid gland.

d. After about 10 half-lives, samples have decayed to very low levels. The half-life of Pu-239 is about 24,000 years, so the timescale for a decrease to background level is on the order of hundreds of thousands of years. The half-life of I-131 is 8.5 days, so 10 half-lives is 85 days or about 3 months. A sample of I-131 will decay to low levels on a timescale of months.

19. $^{235}_{92}U \longrightarrow ^{231}_{90}Th + ^{4}_{2}He$

$^{231}_{90}Th \longrightarrow ^{231}_{91}Pa + ^{0}_{-1}e$

$^{231}_{91}Pa \longrightarrow ^{227}_{89}Ac + ^{4}_{2}He$

$^{227}_{89}Ac \longrightarrow ^{227}_{90}Th + ^{0}_{-1}e$

$^{227}_{90}Th \longrightarrow ^{223}_{88}Ra + ^{4}_{2}He$

$^{223}_{88}Ra \longrightarrow ^{219}_{86}Rn + ^{4}_{2}He$

20. For this type of question, it is helpful to construct a chart.

# of half-lives	% remaining	% decayed
0	100	0
1	50	50
2	25	75
3	12.5	87.5
4	6.25	93.75
5	3.12	96.88
6	1.56	98.44

22. Perhaps someday it can. However, solar energy is diffuse, unequally distributed over Earth's surface, and still presents us with challenges to economically capture and store it.

25. Alchemists were perhaps the first practical chemists, but they did not have the advantage of knowing anything about atomic structure or nuclear reactions. No chemical reaction can produce gold from another element; a nuclear reaction is required. Even if they had envisioned a nuclear reaction that would produce gold from another isotope, they clearly did not have the means to accomplish this. The situation has indeed changed, and modern-day chemists could design experiments to change lead into gold. The question now is why anyone would want to, as the cost would be prohibitive.

27. a. All means of separation depend on the tiny mass difference between U-235 and U-238. For example, it is possible to separate them by converting the uranium sample to gaseous UF_6 and then use gas diffusion. A large high-speed centrifuge also can be used to separate these gas molecules.

b. The uranium must be enriched to provide a critical mass of U-235 to sustain the chain reaction responsible for energy production in the reactor.

c. First, the enrichment procedure is both expensive and energy intensive, so the minimum enrichment level capable of sustaining a chain reaction is preferred. Second, reactors using 80–90% fuels would have safety concerns due to the increased possibility of an uncontrolled chain reaction. Third, such reactors would also have significant security issues. The highly enriched fuel can be used directly in nuclear weapons, making the reactors potential terrorist targets.

d. The difference in the isotopes of uranium is in their nuclear masses. This difference is not enough to significantly affect the chemical reactivity of the two isotopes. For chemical separation, the isotopes of uranium would need to behave differently in a chemical reaction of one sort or another.

29. The Palo Verde power plant produces energy through the process of nuclear fission. Coal and oil burning plants generate energy by burning fossil fuels.

31. a. The subscript for each element is its atomic number or number of protons, which can be found in the periodic table. The subscript for the neutron is zero, which requires knowing or finding the charge of a neutron in a reference table.

b. The superscripts cannot be omitted because nuclear equations must specify a specific isotope and this is something that cannot be determined by looking at the periodic table or another reference.

33. After seven half-lives, 99% of a sample has decayed which is a reasonable approximation of being "gone." However, actually the radioactivity is not gone, as 0.78% of the radioactive sample still remains. Thus, if you start with a large amount of a radioactive substance (for example, 2000 pounds), after seven half-lives you still have about 10 pounds of radioactive substance left!

34. **a.** Bananas are rich in potassium (K). One of the naturally occurring isotopes of potassium (K-40) is radioactive, thereby adding to the radioactivity in bananas.

b. Because the natural abundance of K-40 is only 0.01%, the vice president could have stated that bananas are weakly radioactive and that this radioactivity is natural. Potassium-40 has a long half-life (on the order of billions of years), so it is undergoing radioactive decay very slowly.

c. No. Bananas are a good source of potassium, an essential nutrient. The amount of K-40 in bananas is not significant enough to consider eliminating bananas from your diet on the basis of their radioactivity. Furthermore, any potassium (radioactive or not) that you ingest is not retained. Potassium is lost through sweat and in urine.

37. PV devices have demonstrated their practical utility for satellites, highway signs, security and safety lighting, navigational buoys, and automobile recharging stations.

39. **a.** In the equation $E = mc^2$, the speed of light (c) is 3.00×10^8 meters/second. In order to have joules (J) as the unit of energy (E), the mass (m) must be in kilograms. In addition, use the conversion factor of 1 joule = 1 $kg \cdot m^2/s^2$.

$$50.1 \text{ kJ} \times \frac{10^3 \text{ J}}{1 \text{ kJ}} = \text{m} \times \frac{1 \text{ kg}}{10^3 \text{ g}} \times \left(\frac{3.00 \times 10^8 \text{ m}}{\text{s}}\right)^2 \times \frac{1 \text{ J}}{\text{kg} \cdot \text{m}^2/\text{s}^2}$$

The mass loss is 5.57×10^{-10} g

b. To produce 50.1 kJ of energy, the ratio of masses is 1.00 g of methane burned to 5.57×10^{-10} g methane converted to energy, or about 1.80×10^9 to 1.

c. In a chemical reaction, mass is conserved, so $E = mc^2$ doesn't apply. The energy released in a chemical reaction is a result of potential energy stored in bonds. In a nuclear reaction, a small amount of mass is lost from reactants to products, which is converted to energy.

40. In the equation $E = mc^2$, the speed of light (c) is 3.00×10^8 meters/second. In order to have joules (J) as the unit of energy (E), the mass (m) must be in kilograms. In addition, use the conversion factor of 1 joule = 1 $kg \cdot m^2/s^2$.

$$E = 0.0265 \text{ g} \times \frac{1 \text{ kg}}{10^3 \text{ g}} \times \left(\frac{3.00 \times 10^8 \text{ m}}{\text{s}}\right)^2 \times \frac{1 \text{ J}}{\text{kg} \cdot \text{m}^2/\text{s}^2}$$

$$E = 2.39 \times 10^{12} \text{ J}$$

43. Tritium, H-3, is a radioisotope of hydrogen. Hydrogen is a gas at room temperature, so it is unlikely that the gas itself is contained in the watches. Several descriptions of this watch mention the stainless steel screw-on back, again making it unlikely that tritium gas was present inside the watch. Most likely, the tritium is in a compound in the luminous paint. The paint also contains a phosphor; that is, a compound that glows (is phosphorescent) when hit by ionizing radiation such as the beta particles emitted by tritium. Indeed, these watches can glow brightly as claimed by the advertisement.

45. Similarities between a coal-fueled power plant and nuclear-fueled power plant include:

Both contain a steam generating loop where liquid water is turned to steam. The gaseous water turns a turbine to create electricity, and then is re-condensed to form liquid water again.

The turbine that produces electricity is the same.

Each power plant includes a cooling water loop with an external body of water as the cooling source.

Differences between a coal-fueled and nuclear-fueled power plant include:

The source of the energy to heat the water which turns the turbine is from nuclear fission reactions in the nuclear power plant and from the burning of coal in the coal power plant

The nuclear-fueled power plant has an additional cooling loop (the primary coolant) to cool the reactor core. This coolant is a closed loop so that the secondary coolant is not contaminated with radioactive material.

48. Crystalline silicon is used for the production of photovoltaic cells. Two of the common methods for synthesizing crystalline silicon are Czochralski Crystal Growth and Float Zone Crystal Growth, two different processes that involve manipulating silicon in different physical states. Source: http://www.siliconfareast.com/crystal.htm (accessed November 2016).

To cope with the shortage of silicon, the photovoltaic industry is undergoing a series of changes. Companies are pursuing business models to increase productivity and secure silicon supply. Researchers are developing and testing a number of different synthetic molecules for the next generation of photovoltaic collectors.

50. **a.** As of 2016, the largest PV power plant in the United States is the Solar Star power station near Rosamond, California.

b. As of 2016, Qinghai Province, China, and Kamuthi, India, have two of the world's largest PV power plants.

c. Factors include: (1) having land available for the array, (2) living in a climate with weather favorable to solar collection, (3) having economic conditions that promote an investment with a long-term payback, (4) having the infrastructure to transmit the electricity to population centers.

Chapter 7

2. **a.** Oxidation. Iron loses two electrons to form the Fe^{2+} ion.

b. Reduction. The Ni^{4+} ion gains two electrons to form the Ni^{2+} ion.

c. Oxidation. Each chloride ion loses an electron to form a neutral chlorine atom. These atoms combine to form a chlorine molecule (Cl_2).

3. $Zn(s)$ is oxidized to Zn^{2+} in zinc oxide. $O_2(g)$ is reduced to O^{2-} in zinc oxide.

5. Electric current (an amount of charge per second) is measured in amps (A). In contrast, the electrical potential or the "pressure" behind this current is measured in volts (V).

7. **a.** oxidation half-reaction: $Li(s) \longrightarrow Li^+ + e^-$
reduction half-reaction: $I_2(s) + 2\ e^- \longrightarrow 2\ I^-$

b. overall cell equation: $2\ Li(s) + I_2(s) \longrightarrow 2\ LiI(s)$

c. The oxidation half-reaction $Li(s) \longrightarrow Li^+(aq) + e^-$ occurs at the anode.

The reduction half-reaction $I_2(s) + 2\ e^- \longrightarrow 2\ I^-(aq)$ occurs at the cathode.

8. **a.** The voltage from both kinds of cells is the same (1.5 V) because voltage depends on the chemical reaction that is producing the electrical energy and not on the size of the electrodes.

b. The amount of current (A) produced by a cell depends on the size of the cell. Larger cells contain more materials and can sustain the transfer of electrons over a longer period. For example, the larger D alkaline cell can generate 120 A/hr of current, whereas a tiny AAA alkaline cell generates less current at 12 A/hr.

10. **a.** Oxidation half-reaction: $Zn \longrightarrow Zn^{2+} + 2\ e^-$

b. Reduction half-reaction: $Hg^{2+} + 2\ e^- \longrightarrow Hg$

c. Mercury was once widely used in batteries. By 1990, an awareness of the dangers of mercury in urban trash had grown. Mercury is a toxic metal and (in some forms) can accumulate in the biosphere. Safer batteries and the need to recycle batteries led to the passage of the Mercury-Containing and Rechargeable Battery Management Act (The Battery Act) in 1996.

12. **a.** $Pb(s) + SO_4^{2-}(aq) \longrightarrow PbSO_4(s) + 2\ e^-$
$PbO_2(s) + 4\ H^+(aq) + SO_4^{2-}(aq) + 2\ e^- \longrightarrow PbSO_4(s) + 2\ H_2O(l)$

b. The first half-reaction shows electrons being lost, so it represents oxidation. The second half-reaction shows electrons being gained, so it represents reduction.

c. Lead is oxidized, so lead is the anode. While not as obvious, the lead dioxide or lead(IV) oxide is reduced (note the electrons on the left side of the half-reaction). Thus, lead(IV) oxide is the cathode.

13. It represents the reduction half-reaction. The conversion of O_2 to H_2O requires a supply of electrons.

15. The first half-reaction takes place at the anode (hydrogen is oxidized). The second half-reaction takes place at the cathode (oxygen is reduced).

16. PEM stands for proton exchange (or polymer electrolyte) membrane. As in the fuel cell shown in Question 15, $H_2(g)$ is oxidized to form $H^+(aq)$. In the PEM fuel cell $H^+(aq)$ moves through the membrane to react with $O_2(g)$ (which is reduced) to form water. There is no membrane in the fuel cell in Question 15. In addition, the electrodes differ between the two types of fuel cells as do the electrolytes. The PEM has a solid polymer electrolyte membrane coated with a Pt-based catalyst and the other fuel cell uses a KOH(*aq*) solution as its electrolyte. Finally, the PEM fuel cell operates at room temperature and the one used for the space mission does not.

18. Oxidation half-reaction:

$$CH_4 + 8\ OH^- \longrightarrow CO_2 + 6\ H_2O + 8\ e^-$$

Reduction half-reaction:

$$2\ O_2 + 4\ H_2O + 8\ e^- \longrightarrow 8\ OH^-$$

Overall reaction:

$$CH_4(g) + 2\ O_2(g) \longrightarrow CO_2(g) + 2\ H_2O(l)$$

20. **a.** $K(s) + \frac{1}{2}\ H_2(g) \longrightarrow KH(s)$ or
$2\ K(s) + H_2\ (g) \longrightarrow 2\ KH(s)$

b. $KH(s) + H_2O(l) \longrightarrow H_2(g) + KOH(aq)$

c. Because lithium is less dense than potassium, a given mass will weigh less. This offers an advantage in the handling and transportation of a storage cell. In addition, lithium metal is less reactive than potassium metal, which makes it safer to use in the manufacturing of metal hydride storage systems.

21. Before we can use hydrogen fuel cells more widely, we will need to meet the challenges of safely producing, transporting, and storing large quantities of hydrogen.

24. These batteries derive their voltage from different sets of chemical reactions. A rechargeable battery (such as a Ni–Cd battery) is one in which the oxidation-reduction reaction can be reversed with the input of energy (such as plugging the battery into an electrical outlet.) This recharges the battery. The oxidation-reduction reactions in a non-rechargeable battery, such as an alkaline battery, cannot easily be reversed. Since no simple way exists to recharge alkaline batteries, they are discarded once they stop producing electrical energy.

26. The primary difference is that these produce electricity using different chemical reactions. In addition, a lead–acid storage battery converts chemical energy into electrical energy by means of a reversible reaction. No reactants or products leave the "storage" battery and the reactants can be reformed during the recharging cycle. A fuel cell also converts chemical energy into electrical energy but the reaction is not reversible. A fuel cell continues to operate only if fuel and oxidant are continuously added, which is why it is classified as a "flow" battery.

28. a. "ZPower is taking the leading role in launching the next generation of rechargeable, silver-zinc batteries for microbattery applications. This advanced battery offers superior performance over traditional microbattery technologies—delivering 40% more energy than lithium-ion and 2-3 times the energy of nickel metal-hydride. ZPower batteries also offer a green solution since 95% of the battery elements can be recycled and reused. The mercury-free design is inherently safe due to its water-based electrolyte which is not susceptible to thermal runaway." http://www.zpowerbattery.com/ (Accessed November 2016)

b. oxidation half-reaction (zinc is oxidized):

$$Zn(s) \longrightarrow Zn^{2+}(aq) + 2\,e^-$$

reduction half-reaction (silver is reduced):

$$2\,Ag^+(aq) + 2\,e^- \longrightarrow 2\,Ag(s)$$

32. A Toyota gasoline-battery hybrid car (for example, the Prius) has a gasoline engine sitting side-by-side with a nickel–metal hydride battery and an electric motor. The engine drives the motor to recharge the battery. In addition, regenerative braking recharges the cells in the battery, during which the kinetic energy of the car is stored as electrical energy.

33. All EVs (electric vehicles) share a common technology: the use of electricity to power the vehicle and rechargeable batteries as an energy storage device. If the power goes out, you could drive your vehicle only as far as the energy stored in its batteries would allow. Gasoline-powered vehicles also would be limited. Given that an electrical outage would also render inoperable the pumps at a gasoline filling station, your driving would be limited by the fuel left in your gas tank. Note that in a severe storm, which may topple trees and down power lines, nobody is likely to be driving anywhere!

36. The bond energies listed in Table 5.1 for bonds breaking and forming in gases can be used to calculate the following heats of combustion. These differ somewhat from the values given in the beginning of this chapter, where the product water is given in the *liquid* state.

$H_2(g) + \frac{1}{2}\,O_2(g) \longrightarrow H_2O(g)$
heat of combustion = 249 kJ/mol

$CH_4(g) + 2\,O_2(g) \longrightarrow CO_2(g) + 2\,H_2O(g)$
heat of combustion = 814 kJ/mol

In each case, the units of the calculated heat of combustion can be changed to kJ/gram by dividing by the molar mass of the fuel.

$$\frac{249\text{ kJ}}{\text{mol } H_2} \times \frac{1\text{ mol } H_2}{2.01\text{ g } H_2} = \frac{124\text{ kJ}}{\text{g } H_2}$$

38. a. When water boils, the hydrogen bonds *among* water molecules (intermolecular forces) are disrupted. No bonds are broken within the water molecules.

b. When water is electrolyzed, the covalent bonds *within* water molecules are broken.

41. An energy input of 249 kJ/mol is required in the electrolysis of water. Most of this energy comes from the burning of fossil fuels in conventional power plants. The inherent inefficiency associated with transforming heat into work limits the usefulness of large-scale electrolysis and makes the process energy-intensive. Although not technically feasible yet, using the power of the sun to produce hydrogen from water could be more thermodynamically efficient and would certainly be a far more sustainable method.

43. a. $1.0\text{ mol } H_2 \times \dfrac{2\text{ mol Na}}{1\text{ mol } H_2} \times \dfrac{23.0\text{ g Na}}{1\text{ mol Na}} = 46.0\text{ g Na}$

b. $1.1 \times 10^6\text{ kJ} \times \dfrac{1\text{ mol } H_2}{249\text{ kJ}} \times \dfrac{2\text{ mol Na}}{1\text{ mol } H_2} \times \dfrac{23.0\text{ g Na}}{1\text{ mol Na}}$
$= 2.10 \times 10^5\text{ g Na}$

c. Using the result from part **a**:

$$46.0\text{ g Na} \times \frac{1\text{ kg Na}}{10^3\text{ g Na}} \times \frac{\$165}{1\text{ kg Na}} = \$7.59$$

46. a. Hydrogen is oxidized as it gained an oxygen atom to become water. Oxygen gained hydrogen atoms, so it got reduced.

b. In the first equation, carbon is oxidized as it gained oxygen atoms to become carbon dioxide. In the second more complex equation, oxygen is reduced when it gained hydrogen atoms to become water. The carbon in octane lost hydrogen atoms to become a product, which gained an oxygen. The change to both reactants fits this non-electron definition of oxidation and reduction.

49. Here are ways that some of the principles of green chemistry might apply:

1. *"It is better to prevent waste than to treat or clean up waste after it is formed."* For example, stations are now set up to receive and recycle lead storage batteries. These batteries are no longer sent to landfills or junk yards.

2. *"It is better to minimize the amount of materials used in the production of a product."* For example, find ways to use less silicon in a PV cell, find ways to reduce the packaging for batteries (for example, "button" batteries often come in a plastic package to prevent point of purchase theft). Find another way to prevent theft.
3. *"It is better to use and generate substances that are not toxic."* For example, stop using or minimize the use of batteries that contain toxic metals. This already has been done for mercury batteries. Also develop new batteries that don't contain cadmium, lead, and other toxic materials.

51. Candlelight

Origin	The hot gases that burn and emit light.
Immediate energy source	The hydrocarbon wax, either produced by bees or from petroleum.
Original energy source	Sunlight that drove photosynthesis, which in turn produced the plants from which bees gathered their food (or years ago died and formed fossil fuels).
Products	CO_2, H_2O, and small amounts of soot and CO.
Environmental costs	Primarily the health effects of the particulate matter and soot produced. Also the greenhouse gas produced, CO_2.
Advantages Disadvantages	Convenient, pleasing to view. Produce dirty soot and may cause a fire if unattended.

Light in a battery-powered flashlight

Origin	A wire that glows when it is heated to a high temperature.
Immediate energy source	A chemical reaction in the battery.
Original energy source	Several possibilities, depending on what energy source was used to produce the battery. Could have been fossil fuel consumption (originally solar energy) or nuclear power plant (nuclear fission).
Products	The end products are different chemicals in the battery, while the byproducts are those that are produced during the manufacture of the battery, bulb, and flashlight.
Environmental costs	All those associated with the production and disposal of the battery materials, as well as the side products during the combustion of fossil fuels (or nuclear fission) used in its manufacture.
Advantages Disadvantages	Portable, convenient, clean for the user. Somewhat expensive, becomes waste when energy is spent.

Light from an electric light bulb

Origin	A wire that glows when it is heated to a high temperature.
Immediate energy source	Several possibilities, depending on what energy source was used to produce the electricity (*e.g.*, burning coal or nuclear power plant).
Original energy source	The Sun or the ancient stellar synthesis that produced the uranium and other metals on our planet.
Products	The light bulb is very clean at the site where it is used, but produces pollutants such as NO_x, SO_2, CO_2, and particulate matter at the power plant (if coal, natural gas, or fuel oil combustion) or spent nuclear fuel (if nuclear).
Environmental costs	All those associated with the production and disposal of the light bulb, as well as the side products during the combustion of fossil fuels (or nuclear fission) used to provide the electricity.
Advantages Disadvantages	Convenient, safe, inexpensive. Few to the user, except that the energy costs of incandescent light bulbs are relatively high in comparison to a fluorescent bulb or a light-emitting diode (LED).

Chapter 8

1. **a.** A compound is a pure substance made up of two or more different elements in a fixed, characteristic chemical combination. Water is a compound rather than an element because it contains both the elements H and O in a 2:1 ratio, as evidenced by the chemical formula H_2O.

   b. The Lewis structure for water is:

   H–Ö–H (with two lone pairs on O)

   The molecule is "bent" because the two lone pairs (nonbonding pairs) of electrons on the oxygen atom occupy space (as do the two bonding pairs of electrons on the O atom). The shape of the water molecule maximizes the space between all these electron pairs.

3. **a.** Because water has such a high specific heat, it can moderate climate by capturing and absorbing heat from surrounding land and air.

b. If ice were denser than liquid water, it would sink as it forms. As a result, lakes would freeze from the bottom up, killing forms of life that could not tolerate freezing temperatures.

5. **a.** N and C, 3.0 – 2.5 = 0.5
O and S, 3.5 – 2.5 = 1.0
N and H, 3.0 – 2.1 = 0.9
F and S, 4.0 – 2.5 = 1.5

b. N more strongly attracts electrons than C.
O more strongly attracts electrons than S.
N more strongly attracts electrons than H.
F more strongly attracts electrons than S.

c. N–C < N–H < S–O < S–F

7. **a.** Yes, this is true for hydrocarbons. Distillation towers at a petroleum refinery separate hydrocarbons of different sizes by their boiling point. CH_4 boils at –161.5 °C, while C_4H_{10} boils at –1 °C, and C_8H_{18} boils at 125 °C.

b. Based only on molar mass, you would expect H_2O to have the lowest boiling point because its molar mass is the lowest at 18.0 g/mol.

c. Water is a polar molecule while the rest are nonpolar. Both its geometry and its polar covalent bonds contribute to the formation of strong intermolecular forces. Thus, molar mass is not the only factor that contributes to the boiling point of a substance.

9. The arrow points to a hydrogen bond, an example of an attraction force *between* water molecules and not *within* each water molecule (as is the case for the polar covalent O–H bond).

11. **a.** Here is the Lewis structure $H—\ddot{\underset{\cdot\cdot}{O}}—H$.

b. Lewis structures: $[H]^+$ and $\left[:\ddot{\underset{\cdot\cdot}{O}}—H\right]^-$.

c. $H^+(aq) + OH^-(aq) \longrightarrow H_2O(l)$

13. **a.** Add the liquids in this order: maple syrup, dishwashing detergent and then vegetable oil (most to least dense). Three factors need to be considered: solubility, density, and the care with which each liquid is poured. Maple syrup will probably slowly dissolve in dishwashing liquid; likewise vegetable oil may be slightly soluble in the detergent. But with careful pouring, these three liquids should not easily mix and probably could be added in any order.

b. After vigorous mixing, a cloudy emulsion most likely will form. Over time, it will separate into two layers: one with maple syrup and some of the detergent dissolved in water, one with the rest of the detergent dissolved in oil. You may want to try this experiment and observe the results!

15. **a.** Partially soluble. Orange juice concentrate contains some solids (pulp) that do not dissolve in water. Over time, some of the concentrate may separate from the water. Before drinking, you should give the container a shake or a stir.

b. Very soluble. Note that ammonia is a gas. Ammonia will dissolve in water in any proportion like it does in cleaning products.

c. Not soluble. Sometimes you can see chicken fat floating on top of aqueous chicken soup.

d. Very soluble. When you add laundry detergent to your load of wash, it dissolves in the water.

e. Partially soluble if the chicken broth contains fat or suspended solids, neither of which will dissolve in water. Very soluble if the broth is clear and fat-free.

17. **a.** For Cl and Na, the electronegativity difference is 3.0 – 0.9 = 2.1. For Cl and Si, the electronegativity difference is 3.0 – 1.8 = 1.2.

b. Larger differences in electronegativity are associated with ionic bonds; smaller differences with covalent bonds.

c. When electronegativity differences are relatively large, one or more electrons are transferred, forming ions. When electronegativity differences are smaller, neither atom is able to release its outer electrons to the other, so the outer electrons are shared, resulting in the formation of covalent bonds. In the case of $SiCl_4$, Si and Cl form a polar covalent bond.

19. No, it exceeds the acceptable limit by 35 times. A concentration of 10 ppm is equivalent to 10 mg/L, so 350 mg/L is 350 ppm.

21. **a.** The solution will conduct electricity and the bulb will light. Based on Table 8.6, $CaCl_2$ is a soluble salt and therefore releases ions (Ca^{2+} and Cl^-) when it dissolves. These ions carry the current.

b. The solution will not conduct electricity. Although ethanol (C_2H_5OH) is soluble in water, it is a covalent compound and does not form ions.

c. The solution will conduct electricity and the bulb will light. Sulfuric acid (H_2SO_4) releases ions when it dissolves: H^+ and SO_4^{2-}.

23. All of these compounds are water-soluble.

25. The chemical formula for calcium carbonate is $CaCO_3$. This salt is insoluble in water according to the solubility rules.

27. **a.** HI(*aq*) is acidic. $[H^+] = [I^-] > [OH^-]$

b. NaCl(*aq*) is neutral. $[Na^+] = [Cl^-]$ and $[H^+] = [OH^-]$

c. NH_4OH(*aq*) is basic. $[NH_4^+] = [OH^-] > [H^+]$

d. Basic; $[OH^-] > [H^+]$ (remember, $1 \times 10^{-14} = [H^+] \times [OH^-]$).

e. Basic; $[OH^-] > [H^+]$

f. Acidic; $[H^+] > [OH^-]$

g. Acidic; $[H^+] > [OH^-]$

29. **a.** A solution of pH = 6 has 100 times more $[H^+]$ than a solution of pH = 8.

b. A solution of pH = 5.5 has 10 times more $[H^+]$ than a solution of pH = 6.5.

c. The solution with $[H^+] = 1 \times 10^{-6}$ M has 100 times more $[H^+]$ than a solution with $[H^+] = 1 \times 10^{-8}$ M.

d. Using Equation 8.12, the solution with $[OH^-] = 1 \times 10^{-2}$ M has $[H^+] = 1 \times 10^{-12}$ M. The solution with $[OH^-] = 1 \times 10^{-3}$ M has $[H^+] = 1 \times 10^{-11}$ M. Thus, in the second solution ($[OH^-] = 1 \times 10^{-3}$ M) the $[H^+]$ is higher by a factor of 10.

31. **a.** nitrate = NO_3^-, sulfate = SO_4^{2-}, carbonate = CO_3^{2-}, ammonium = NH_4^+

b. For the nitrate ion, one possibility is nitric acid neutralizing sodium hydroxide.

$H^+(aq) + NO_3^-(aq) + Na^+(aq) + OH^-(aq) \longrightarrow Na^+(aq) + NO_3^-(aq) + H_2O(l)$

For the sulfate ion, one possibility is sulfuric acid neutralizing sodium hydroxide.

$2\ H^+(aq) + SO_4^{2-}(aq) + 2\ Na^+(aq) + 2\ OH^-(aq) \longrightarrow 2\ Na^+(aq) + SO_4^{2-}(aq) + 2\ H_2O(l)$

For the carbonate and ammonium ions, one possibility is carbonic acid neutralizing ammonium hydroxide.

$H_2CO_3(aq) + 2\ NH_4OH(aq) \longrightarrow 2NH_4^+(aq) + CO_3^{2-}(aq) + 2\ H_2O(l)$

Note: Ammonium hydroxide is written in its undissociated form, as explained in the text; see Equation 8.7b. Similarly, carbonic acid is written in its undissociated form.

33. **a.** To prepare 2 liters of 1.50 M KOH, weigh 168 g of KOH:

$$2\ \text{L} \times \frac{1.50\ \text{mol}}{\text{L}} = 3.0\ \text{mol KOH}$$

3.0 mol × 56 g/mol = 168 g.

Place the 168 g KOH into a 2-liter volumetric flask. Add distilled (or deionized) water to fill the flask to the mark. *Note:* If you don't have a 2-L volumetric flask, you will need to repeat the procedure twice with a 1-L flask.

b. To prepare one liter of 0.050 M NaBr, weigh 5.2 g of NaBr and place it into a 1-liter volumetric flask. Add water to fill the flask to the mark.

c. This should be done with a 100-mL volumetric flask. Weigh 7.0 g of $Mg(OH)_2$ and place it into a 100 mL volumetric flask. Add water to the mark.

35. **a.** carbon dioxide, CO_2 ($CO_2 + H_2O \longrightarrow H_2CO_3$)

b. sulfur dioxide, SO_2 ($SO_2 + H_2O \longrightarrow H_2SO_3$)

37. **a.** Possibilities include sodium hydroxide (NaOH), potassium hydroxide (KOH), ammonium hydroxide (NH_4OH), magnesium hydroxide ($Mg(OH)_2$), and calcium hydroxide ($Ca(OH)_2$).

b. In general, bases taste bitter, turn litmus paper blue (and have characteristic color changes with other indicators), have a slippery feel in water, and are caustic to your skin and eyes.

39. Chocolate requires 1,700 liters of water to produce a 100-gram bar. This includes the water necessary to grow and process both the cacao and the sugar used to sweeten it. A pint of beer requires about 140 L of water, mainly for growing and producing the malted barley. Processing the cacao and sugar is highly water-intensive compared to processing barley. These are average global values given on The Water Footprint Network (http://www.waterfootprint.org-accessed November 2016).

41. "Pure" water is usually interpreted as meaning water that has no dissolved impurities, something that is very difficult to achieve. But in reality, water is never pure. For example, if rain has fallen through the atmosphere, it will have picked up carbon dioxide, which explains why all rainwater is slightly acidic. Groundwater can easily pick up water-soluble ions, and even ice may contain suspended particulate matter or gases. Even bottled water usually contains dissolved minerals.

43. A mercury concentration of 1.5 ppb means there are 1.5 parts of mercury for every 10^9 parts fish. The caution sign is necessary because mercury is toxic and capable of causing severe neurological effects in humans. The EPA has set the Maximum Contaminant Level for mercury in drinking water at 2 ppb. This is below that limit, but the caution sign is necessary because mercury is a cumulative poison.

45. The diatomic molecule (XY) with a polar covalent bond *must* be polar because the molecule is linear. An example is HCl. The H–Cl bond is polar, and so is the molecule. If the triatomic molecule contains polar bonds, the geometry of the molecule will determine whether the molecule is polar or nonpolar. For example, although CO_2 has polar C=O double bonds, the molecule is linear and, as a result, nonpolar. The H_2O molecule has polar H–O bonds, but is bent. This geometry causes the molecule to be polar.

47. Like water, NH_3 is a polar molecule. It has polar N–H bonds and a trigonal pyramidal geometry. Therefore, despite its low molar mass, considerable energy must be added to liquid NH_3 to overcome the intermolecular forces (hydrogen bonding) among NH_3 molecules.

49. **a.** The Lewis structure for ethanol is

```
     H   H
     |   |   ..
  H—C—C—O—H
     |   |   ..
     H   H
```

b. The cube sinks because, as is the case for most substances, the density of the solid phase is greater than the density of the liquid phase. Unlike water molecules, the ethanol molecules are closer together in solid ethanol than liquid ethanol. Therefore, solid ethanol has a greater density than liquid ethanol and it sinks.

51. For a given contaminant, the MCLG (a goal) and the MCL (a legal limit) are usually the same. However, the levels may differ when it is not practical or possible to achieve the health goal as set by the MCLG. This sometimes is the case for carcinogens, for which the MCLG is set at zero (under the assumption that any exposure presents a cancer risk).

53. a. Nitrate ion (NO_3^-) and nitrite ion (NO_2^-).

b. In the body, oxygen is needed to metabolize ("burn") glucose to produce energy.

c. The nitrate ion is not volatile. It is a solute that does not evaporate or decompose with heat. Instead, the water in the nitrate-containing solution will evaporate leaving behind NO_3^-.

55. The two most common desalination techniques are distillation and reverse osmosis. Both of these require energy to remove salts from seawater or brackish water, and thus inherently are expensive. If a less expensive option is available, such as hauling fresh water from a distance, then this option is used.

57. Coffee beans that have been soaked in water are placed in a container in which liquid carbon dioxide is injected. The nonpolar solvent, liquid CO_2, attracts caffeine based on the generalization "like dissolves like." This caffeine extract can then be removed from the coffee bean mixture, allowing further processing to the final product.

59. The bond energy of an O–H bond is 467 kJ/mole, or about 10 times greater than the maximum energy of a hydrogen bond. Actually, the hydrogen bonds between water molecules is 20 kJ/mole, this means that the bond energy of an O–H bond is about 20 times greater than the bond energy of a hydrogen bond in water.

60. a. Mining waste, gas and oil operations, cement production.

b. Organic mercury is a carbon compound that contains mercury. These compounds tend to be nonpolar, so they accumulate in fatty tissue that also is comprised of nonpolar molecules (like dissolves like).

61. a. Glycine contains several polar bonds and has several polar areas in its molecule (everything but the $-CH_2$ region).

Bond	Electronegativity Difference
O–H	$3.5 - 2.1 = 1.4$
O–C	$3.5 - 2.5 = 1.0$
N–H	$3.0 - 2.1 = 0.9$
N–C	$3.0 - 2.5 = 0.5$

b. Yes, hydrogen bonding is possible when O–H and N–H bonds are present, both of which are in glycine.

c. Because glycine has polar bonds located in several areas of the molecule and has a relatively small molar mass, glycine should be soluble in water.

63. A complex question! First, you would need to determine the environmental rules and regulations in your region. Most likely these would apply to releases of chemicals into the soil, air, and water. Then, you would need to monitor what is being released by the industry, in what amounts, and with what occurrence. Compliance with environmental controls, economic factors, and community acceptance of the plant all will affect the success of this plant.

65. a. The PUR Purifier of Water is made by Proctor & Gamble and sold as a packet of chemicals that can be added to a sample of nonpotable water. Each packet contains a powdered flocculent, iron(II) sulfate, and a disinfectant, calcium hypochlorite. The contents are added to 10 liters of non-potable water, the water is stirred for 5 minutes, and the solids are allowed to settle. The water is then poured into another container by filtering it through a cotton cloth. After 20 minutes, the disinfectant inactivates any microbes present (including viruses) and the water is ready for consumption.

b. Here are a few comparisons. Both systems offer comparable water disinfection, with one catch. The personal Lifestraw does not protect against viruses, but the PUR system does. Both have different uses: one is portable (Lifestraw) and can be used immediately. The other processes a larger volume of water and filters water by gravity instead of by mouth suction. Thus, a larger quantity of water can be purified in a shorter time. Finally, the water filtered with a personal Lifestraw lacks the chemical aftertaste that occurs following treatment with the PUR system.

67. a. Currently, over 90 substances have health-based standards established under the Safe Drinking Water Act. The EPA uses the Unregulated Contaminant Monitoring program every 5 years to establish a list of contaminants and collect data for those contaminants suspected to be present in drinking water.

b. The EPA uses a CCL to determine what unregulated contaminants could be researched and possibly added as substances to be regulated. This follows the precautionary principle that "stresses the wisdom of acting, even in the absence of full scientific data, before the adverse effects to human health or the environment become significant or irrevocable."

c. The EPA lists "pesticides, disinfection by-products, chemicals used in commerce, waterborne pathogens, pharmaceuticals, and biological toxins" in its CCL. An example of a specific substance from the CCL-3 list is Halon-1011 (bromochloromethane), used as a fire-extinguishing fluid and as a solvent to make pesticides.

69. **a.** As of May 2016, the approximate atmospheric concentration of CO_2 was 404 ppm.

b. The concentration of carbon dioxide in the atmosphere is increasing because humans are burning fossil fuels and cutting down forests which absorb CO_2.

c. Here is the Lewis structure. $\ddot{\underset{\cdot\cdot}{O}}=C=\ddot{\underset{\cdot\cdot}{O}}$.

d. No, you would not. Carbon dioxide is is a non-polar compound, and sea water is a polar solution of water and dissolved ions. "Like dissolves like." Even so, carbon dioxide is slightly soluble in seawater and dissolves to form carbonic acid, H_2CO_3.

Chapter 9

1. Cotton, silk, rubber, wool, and DNA are examples of natural polymers. Synthetic polymers include Kevlar®, polyvinyl chloride (PVC), Dacron™, polyethylene, polypropylene, and polyethylene terephthalate.

5. **a.** At the molecular level, increasing the length of the polymer chain would increase its molar mass and the extent of its interactions with neighboring chains. This would be expected to somewhat increase the polymer's rigidity, strength, and melting point.

b. At the molecular level, aligning polyethylene chains with one another means that the structure is more crystalline and highly ordered. This would be expected to give the polymer slightly more density, more rigidity, and more strength. The melting point would also increase.

c. At the molecular level, this would be the opposite of the previous answer. The structure would be less crystalline, less ordered, and possibly somewhat tangled. This would be expected to make the polymer slightly less dense, less rigid, and not as strong. The melting point would decrease.

7. To serve as monomers, hydrocarbons must have a C=C double bond (and contain no elements other than C and H). Here are two possibilities other than ethylene.

$$H_2C=CH(CH_3) \qquad H_2C=CH(C_6H_5)$$

Propylene Styrene

9. Each ethylene monomer has a molar mass of 28 g/mol. To determine the number of monomers in the polymer, divide 40,000 (the molar mass of the polymer) by 28 (the molar mass of the monomer). The result is 1428 monomers, or 1400 to two significant figures. To determine the number of carbon atoms present in the polymer, note that each monomer contains two carbon atoms ($H_2C=CH_2$). Accordingly, the polymer contains $2 \times 1{,}428$ carbon atoms, or 2856 carbon atoms. In round numbers, there are roughly 3000 carbon atoms.

11. This is the tail-to-tail, head-to-head arrangement of PVC formed from three monomer units.

12. These are different. The top segment represents the "head-to-tail, head-to-tail" arrangement. The Cl atoms in each case are on alternate carbon atoms. It makes no difference if the atom is on the "top" or on the "bottom" of the chain (these positions are equivalent). In contrast, the bottom segment is "head-to-head, tail-to-tail." The Cl atoms are on adjacent carbon atoms.

14. Note that the question asks "most likely." For plastic use, "most likely" changes over time so the answers are a moving target. For example, PVC is being phased out in many uses.

a. PET, polyethylene terephthalate

b. HDPE, high-density polyethylene

c. PVC, polyvinyl chloride ("vinyl")

d. PP, polypropylene

e. LDPE, low-density polyethylene

f. PS, polystyrene

g. HDPE, high-density polyethylene

17. **a.** phenyl group, alkene

b. hydroxyl group (or alcohol)

c. phenyl group, carboxylic acid

d. amine, carboxylic acid

e. amine

f. carboxylic acid

19. Terephthalic acid contains two carboxylic acid groups. Phenylenediamine contains two amine groups. These two monomers react in a condensation polymerization to form amide linkages between the phenyl groups.

22. **a.** A blowing agent is a gas (or a substance capable of producing a gas) used to manufacture a foamed plastic. For example, a blowing agent is used to produce Styrofoam™ from PVC.

b. Carbon dioxide can replace the CFCs or the HCFCs that once were used as blowing agents. Although CO_2 is a greenhouse gas, it still is preferable because CFCs and HCFCs both deplete the ozone layer as well as being potent greenhouse gases.

24. **a.** Post-consumer content includes all types of waste: newspapers, cardboard, foam cups, packing peanuts, 2-liter bottles, and plasticware. Pre-consumer content includes waste created in the manufacturing process, such as scraps of fabric, plastics, paper, wood, and food.

b. Not necessarily. For example, most PET soda bottles are made from petroleum products and do not include recycled PET.

26. Polypropylene (PP) is a tough plastic and bottle caps need to be tough, standing up to repeated use and not losing their shape or their threads. However, PP melts at a higher temperature than PET and has different properties. So somewhere in the recycling process, PP needs to be separated from PET. In essence, in most cases, the caps need to be removed from the bottles. Beverage companies are currently seeking alternatives.

28. Factors other than the chemical composition of the monomer(s) influence the properties of the polymer. These include length of the chain (the number of monomer units), the three-dimensional arrangement of the chains, the degree of branching in the chain, and orientation of monomer units within the chain (such as head-to-tail).

30. For addition polymerization, the monomer must have a C=C double bond. Although some monomers have benzene rings as part of their structures (styrene, for example), the double bond involved in addition polymerization must not be in the ring—the double bonds in the ring are part of the resonance structure and not a true double bond. An example is the formation of PP from propylene.

For condensation polymerization, each monomer must have two functional groups that can react and eliminate a small molecule such as water. For example, an alcohol and a carboxylic acid can react to eliminate water. An example is the formation of PET from ethylene glycol and terephthalic acid.

31. In vinyl chloride, each carbon atom has three bonds (two single bonds and one double bond). These three bonds form an equilateral triangle (trigonal) and the Cl–C–H bond angle is about 120°. In the polymer, each carbon atom is connected to other atoms by four single bonds. The double bond is no longer present, and the four bonds point to the corners of a tetrahedron, with a bond angle of about 109.5°.

34. From the bond energies in Table 5.1, it requires 598 kJ/mol to break C=C double bonds. The formation of C–C single bonds releases 356 kJ/mol. If we consider the reaction of two ethylene monomers, two double bonds are broken and replaced with four single bonds (two bonds between the C atoms of the monomers, one between the first monomer and the second, and a bond extending to what would be the third ethylene monomer). Here is the calculation:

$(2 \times 598 \text{ kJ/mol}) + (4 \times -356 \text{ kJ/mol}) = -228 \text{ kJ/mol}.$

Thus, the reaction is exothermic.

36. The heat of combustion of polyethylene would be most similar to that of octane. Both are hydrocarbons consisting of carbon-carbon single bonds. The other fuels contain different atoms (and thus different bonds would be broken and formed).

38. **a.** PLA stands for polylactic acid, a polymer.

b. The monomer of PLA is lactic acid. In the United States, lactic acid is produced from corn.

c. Reasons include that (1) corn is a renewable resource, (2) PLA is compostable, and (3) PLA is not a petroleum-based polymer.

d. (1) Although corn is a renewable resource, corn is a crop with its share of controversies. These include the degradation of the land on which it is grown and the runoff of fertilizers and pesticides into nearby waterways. (2) Although PLA is compostable, this is true only in industrial composters that most communities do not have. It degrades slowly if at all in a landfill. (3) Although no oil is used in its production, fuels such as petroleum are nonetheless used in the growing of corn and its transportation.

39. **a.** From the bond energies in Table 5.1, it requires 598 kJ/mol to break C=C double bonds. The formation of C–C single bonds releases 356 kJ/mol. If we consider the reaction of two ethylene monomers, two double bonds are broken and replaced with four single bonds (two bonds between the C atoms of the monomers, one between the first monomer and the second, and a bond extending to what would be the third ethylene monomer). Here is the calculation:

$(2 \times 598 \text{ kJ/mol}) + (4 \times -356 \text{ kJ/mol}) = -228 \text{ kJ/mol}.$

With 1000 monomers joining, we multiply the −228 kJ/mol by 500 and the heat released will be 114,000 kJ or 1.14×10^5 kJ.

b. Overall, heat is released, but some energy must be inputted to overcome the activation energy. The reaction is so exothermic that, in the early days of polymer manufacture, polymerization vessels exploded. Manufacturers realized that conditions needed to be carefully controlled to avoid this.

44. The Big Six polymers are generally large (in fact huge) molecules with few polar groups. Furthermore, many are hydrocarbons (HDPE, LDPE, PS, PP) and therefore would not be expected to dissolve in polar solvents such as water. "Like dissolves like," so many polymers, including HDPE and LDPE, soften in hydrocarbons or chlorinated hydrocarbons because these nonpolar solvents interact with the nonpolar polymeric chains.

46. **a.** Starch is a polymer of glucose. Many foods are a source of starch, including corn, potatoes, and rice.

b. Advantages of starch packing peanuts: they are lightweight, compostable, and made from a renewable material. Disadvantages: they will degrade if the package gets wet, they are made from what could be eaten as a food, and they are less "springy" than polystyrene foam peanuts.

c. Composting is a good option. Although some can be washed down the drain, this results in the starch needing to be removed later at a water treatment plant.

48. **a.** A plasticizer is a compound added to a hard or rigid plastic in order to soften it.

b. DEHP was added to PVC in order to make soft vinyl products such as boots, shower curtains, clothing items, and flexible tubing.

c. DEHP has been banned for items that infants repeatedly put in their mouths, such as pacifiers. It has also been banned in some medical devices and children's toys. The use of DEHP remains controversial and different restrictions are in place in different countries.

52. **a.** The two main properties are (1) stable over time of intended use and (2) nontoxic. Other factors to consider are low cost, lack of solubility in body fluids, lack of reactivity in body fluids, and the ease of implantation.

b. Several different types of contact lenses are on the market and each uses a different type of polymer. Polymethyl methacrylate (PMMA), one of the earliest polymers used for rigid gas permeable lenses, is structurally similar to Lucite and plexiglas. Silicone-acrylate materials now are more commonly used under trade names such as Kolfocon. Newer rigid gas permeable (RGP) polymers tend to contain fluorine. Manufacturers' websites are good sources of information.

Desirable properties include being nontoxic, permeable to oxygen, comfortable to wear, and inexpensive. Also desirable is the ability to conform to the shape of the eye and to be easily cleaned (if necessary).

56. **a.** With its high molar mass, this polymer should not dissolve in water. Like many polymers, it would be expected to be an electrical insulator. With its methyl groups, it should be somewhat flammable (but the presence of silicon reduces the flammability and gives it good stability at high temperature).

b. "Silly Putty" bounces, but it breaks when pulled sharply. When it is formed into a shape, it slowly loses this shape over time and flattens out. Ink will stick to Silly Putty and so it can "lift" images from newsprint. These properties are rather unique, and led to the popular acceptance of Silly Putty as a toy long before it was widely used for other products. Answers.com has great information about Silly Putty.

c. Silicone rubber is used for flexible bakeware because of its resistance to high temperatures. It also is used in greases, caulking, and tubing.

59. As you might suspect from the –ol ending, polyols are alcohols. The "poly" means they have multiple hydroxyl (–OH) groups. Many types of polyols exist, some with only two hydroxyl groups and others with many more. For example, in this chapter, you met ethylene glycol, a polyol with two hydroxyl groups (also called a diol). It served as one of the two monomers for PET. In Chapter 5 in the section on biofuels, you met ethylene glycol and propylene glycol, two polyols. In Chapter 11, in regard to fats and triglycerides, you will read about glycerol (glycerin), a polyol that contains three hydroxyl groups (also called a triol).

Polyols can serve as monomers for any condensation polymers such as polyesters made from a "double acid" and a "double alcohol" (diol). An Internet search for soybean plastics or soybean-based polymers should turn up many examples.

Chapter 10

1. For g/ml, $5\,\frac{g}{dm^3} \times \left(\frac{1\ dm^3}{1 \times 10^3\ cm^3}\right) = 5 \times 10^{-3}\ g/cm^3$

For mg/L, $5\,\frac{g}{dm^3} \times \left(\frac{1\ dm^3}{1\ L}\right) \times \left(\frac{1 \times 10^3\ mg}{1\ g}\right) = 5 \times 10^3\ mg/L$

For mg/m^3, $5\,\frac{g}{dm^3} \times \left(\frac{1 \times 10^3\ dm^3}{1\ m^3}\right) \times \left(\frac{1 \times 10^3\ mg}{1\ g}\right) = 5 \times 10^6\ mg/m^3$

3. $1\,\frac{g}{mL} \times \left(\frac{1000\ mg}{1\ g}\right) \times \left(\frac{1000\ mL}{1\ L}\right) = 1 \times 10^6\ mg/L$

In terms of how does a solute affect the density of water, their presence increases the density of the liquid water. Because of the nature of solutions, solid solutes in liquid solvents add mass while not adding appreciable volume, and so, density increases.

5. Recall for gases, $\frac{P_1}{T_1} = \frac{P_2}{T_2}$ or $P_1T_2 = P_2T_1$
Temperature must be expressed in Kelvin, and so, recall K = °C + 273.

(1 atm)(x) = (2 atm)(373 K), x = 746 K, or 473 °C
With a 100 °C increase in temperature, the reaction rate would double 10 times (100/10 = 10). At 2 atm, the rate of reaction would be over 1000 times the original rate at 1 atm ($2^{10} = 1024$). This emphasizes how pressure cookers cook food so much faster than an open pot of boiling water.

7. How strongly molecules stick together via intermolecular forces determine how easily they will transition from the solid or liquid phase to the gas phase. To be considered volatile, these molecules must make this transition fairly easily and not be held too tightly to other molecules. Weaker intermolecular forces among molecules tend to come from nonpolar features in the molecule, as well as the molecule being compact and not made of *many* branched, but single long chains of nonpolar groups. Low molecular masses can also be a factor, but recall nonpolar versus polar molecules tend to be the bigger factor. Numerous molecules that have properties of smells and/or tastes have small rings of nonpolar groups. See examples below:

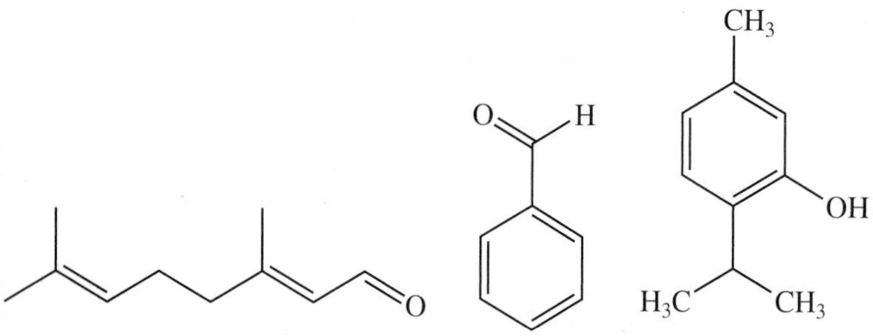

CH_3 O H OH O H_3C CH_3

Lemon Almond Thyme

8. The browning of cut fruit is the result of a reaction of an enzyme (polyphenol oxidase), which is exposed when the fruit is cut, and oxygen from the air. It is an oxidation process much like the formation of rust on iron structures. These products create an unpleasant sour taste and are arguably a defense mechanism so that the fruit is not eaten by animals or even invaded by germs. This is not the same process as carmelization and the Maillard reaction, which occur at high temperature and with no moisture present.

9. Oxygen prefers to react with acidic vitamin C before reacting with the polyphenol oxidase in fruit, and so, the browning product from the oxidase reaction does not form. Thus, the juice of lemons, limes, or even cranberries, which contain high amounts of vitamin C, when squeezed on fruits are effective at preventing the browning process.

12. All green plants (including vegetables) contain chlorophyll, which also includes a magnesium atom attached to it. This gives them their characteristic color. When exposed to heat (or acid) long enough, the magnesium is replaced by hydrogen atoms. This change in composition alters the color property and these vegetables turn from bright green to a more dull, olive green. In terms of texture, long exposure to heat and acid breaks down the fibrous, crunchy structural molecules of the vegetable and they become softer.

14. Heated and usually boiling water is used to make a cup of tea. In order to extract the molecules from the tea leaves, sufficient heating of the water solvent is necessary. Recall that achieving the boiling point of a liquid is actually more related to overcoming the external, atmospheric pressure rather than just merely heating to a particular temperature. Water boils at a lower temperature when the atmospheric pressure is lower, such as at higher altitudes. These lower temperatures may not be sufficient for extracting all the molecules one might enjoy from tea leaves, and so, the brewed tea may not be as flavorful.

16. Baking soda and baking powder are both used in baking recipes. They are different mixtures, but both produce the same result. They are leavening agents. Baking soda is a solid basic salt, $NaHCO_3$. Baking powder is a solid mixture of baking soda and a solid acid of some sort, usually, cream of tartar, an acidic salt, $KHC_4H_4O_6$. Both leavening agents react to produce carbon dioxide, CO_2, gas which is what causes the airy texture in many baked goods. Baking soda is used with the other ingredients, including some significant acid already, like vinegar or buttermilk. The reaction with vinegar is: $HCO_3^-(aq) + HC_2H_3O_2(aq) \longrightarrow C_2H_3O_2^-(aq) + H_2O(l) + CO_2(g)$. In baking powder, once the mixture is in contact with a water solvent, as is the case in many baking recipes, the two acid and base components react: $HCO_3^-(aq) + HC_4H_4O_6^-(aq) \longrightarrow C_4H_4O_6^-(aq) + H_2O(l) + CO_2(g)$. Baking powder is often used in batters that do not have acid in them already.

18. The terms "strong" and "weak" are terms used to describe the chemical properties of an acid (or base). In this case, these describe the degree to which a molecule of a particular acid dissociates in water. Strong acids mean that nearly every molecule of the acid donates its proton when places in water, *e.g.*, hydrochloric acid is a strong acid: $HCl + H_2O \longrightarrow Cl^- + H_3O^+$. For weak acids, only some of their molecules dissociate and donate the protons. For these substances there remains some undissociated molecules and is represented with an equilibrium reaction, *e.g.*, acetic acid is a weak acid: $HC_2H_3O_2 + H_2O \leftrightarrow C_2H_3O_2^- + H_3O^+$.

The terms "concentrated" and "dilute" do not refer to the nature of the acid itself, but to the relative amount of the acid in the water solvent. These terms are independent of the particular acid, so that a solution could be a dilute strong acid solution, *e.g.*, 0.001 M $HCl(aq)$, or a concentrated weak acid solution, 18 M $HC_2H_3O_2(aq)$.

19. Microwave radiation causes changing electric fields. Because water is polar, is responds to the changing negative and positive conditions, alternating between repulsions and attractions that ultimately increase the motion of the water molecules. It is this motion (friction) that heats up the food. Other polar molecules are likely to respond in this manner. Among the three molecules listed, only ammonia (NH_3) is polar and is likely to respond like water to microwaves.

20. Capsaicin is the common name for the chemical responsible for the chemical heat in peppers. It has a chemical formula ($C_{18}H_{27}NO_3$), and its molecules structure is:

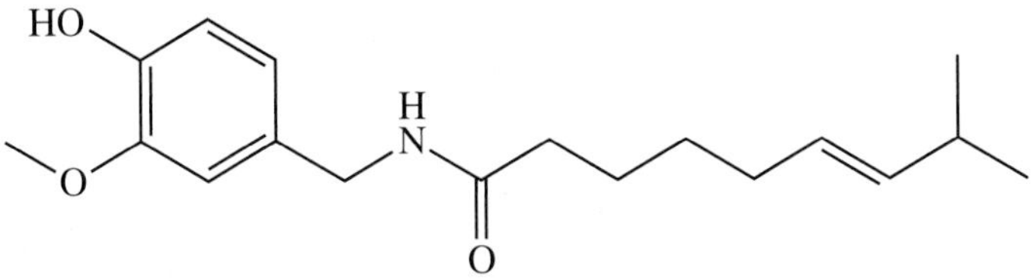

Drinking water is not useful to cool the heat because capsaicin is insoluble in water; it is mostly a nonpolar molecule. The water merely spreads the capsaicin more around your mouth and across your tongue. Drinking milk or beverages with more fat that can dissolve the capsaicin may be more beneficial for cooling down the heat.

23. Flavor tripping is a nickname for the concept in the culinary world that deceives the other senses for what you are about to taste. For example, most people see and touch a lemon wedge and are ready for a very sour taste, but after eating a miracle berry, the raw lemon wedge tastes like the sweetest lemonade. Miracle berries contain a particular molecule called miraculin, which is a glycoprotein. It interacts with particular receptors on your tongue, the ones that detect sour, and block them from detecting the sour part, so the lemon tastes sweet.

This concept of blocking flavors might be interesting because it could make things that are perfectly nutritious but taste dreadful more palatable, which could expand diet options for feeding a growing world population.

25. A food mile is a measure of the energy needed just to transport a particular food item from the point of production to the point of consumption. For example, to someone in the Midwestern United States, obtaining a pineapple will use more food miles than obtaining an ear of corn. Pineapples are grown far from the Midwest, but corn is grown directly there.

We can reduce food miles by eating foods that are produced relatively close to where we consume them—the "eat local" movement. Another option is connected to the flavor tripping notion from Question 23. If we can make local, nutritious foods that don't naturally taste good more palatable, we may have more options for eating locally!

No matter your opinion of the importance of reducing food miles, the concept of energy use and its ongoing availability is a factor in this analysis.

27. The Maillard reaction only takes place when there is no measurable water in the cooking environment. On a hot grill, meat will brown because the grill temperatures cause the water on the surface of the meat to evaporate. The meat can then "brown" on the surface. Done properly, this browning seared the surface of the meat and locks moisture inside the meat so it does not dry out. (With moisture, though, the inside will not brown.) Food items do not brown in boiling water because of the abundance of water in the environment, and food only cooks in the microwave if water is present. Recall earlier questions about how a microwave cooks food.

30. **a.** Teflon™ is actually a long-chain polymer of repeating units of the monomer, C_2F_4. It is mostly non-reactive due to the strong carbon-to-fluorine bonds. Even though Teflon™ contains fluorine, a highly electronegative element, the arrangement of the fluorine atoms in the polymer cancel each other out and the polymer is nonpolar, making it not want to interact with water. Because nearly all the foods we consume contain some degree of water, cooking with Teflon™ still allows the heat to transmit but does not cause the food to stick, because the Teflon™ repels the water in the food.

b. Ultimately, the side of Teflon™ that will "stick" to the pan has to be chemically altered to get it to bind to the other pan materials. This is done by breaking many of the carbon-to fluorine bonds on the side that will stick, making it more chemically conducive to binding to the other pan materials.

31. Gelatin, the primary component in Jell-O®, is a protein that provides structure to the liquid mixture, such that once cooled and/or some water evaporates, the web-like structure of the gelatin molecules allow the material to hold a relatively solid shape. (This is similar to how gluten provides structures for many baked breads and goods.)

Pineapple, kiwi, and papaya contain an enzyme called bromelain, which catalyzes the breakdown of the gelatin protein molecules into its smaller components—its amino acids. These smaller building blocks, when broken apart, cannot provide enough structure to allow the liquid and the Jell-O® mixture does not solidify. (Ironically, extracts of these fruits are used as digestive aids, to help us when our bodies are not able to do an effective job of breaking down the proteins we consume.)

Chapter 11

1. Malnutrition is caused by a diet lacking the proper mix of nutrients, even though the energy content of the food eaten may be adequate. Undernourishment is caused by the insufficient energy content of the food eaten.

3. **a.** The three different types of macronutrients are fats, carbohydrates, and proteins.

   b. Fats are the highest in energy content, almost a factor of two higher than carbohydrates and proteins, which are similar in energy content.

5. The pie chart indicates more carbohydrate is present than would be found in steak, and more protein than would be found in chocolate chip cookies. Of the choices given, the pie chart is likely to be a representation of peanut butter, which has a high oil content. (See Table 11.1 for confirmation.)

7. A steak is 28% protein, 15% fat, and 57% water.

   18 oz × 0.28 = 5.0 oz of protein

   18 oz × 0.15 = 2.7 oz of fat

   18 oz × 0.57 = 10 oz of water

9. **a.** Here is the structural formula for lactic acid:

```
    H   OH   O
    |   |   //
H — C — C — C
    |   |    \
    H   H     OH
```

   b. As a fatty acid, lactic acid would be saturated because the hydrocarbon chain contains only single bonds between the carbon atoms.

   c. No, lactic acid is not a fatty acid. Although it has the carboxylic acid group characteristic of a fatty acid, it lacks the long hydrocarbon chain (12–24 carbon atoms). Lactic acid also has a hydroxyl group (–OH) that is not found in fatty acids.

11. **a.** flaxseed oil

    b. safflower oil

    c. safflower oil

    d. coconut oil

13. **a.** Milk (lactose is "milk sugar")

    b. Many fruits contain fructose ("fruit sugar"). So does honey.

    c. Sucrose ("table sugar") originates from both beets and sugarcane.

    d. Starch is a primary component of potatoes, rice, tapioca, taro root, wheat, and corn.

15. Both starch and cellulose are polymers in which the monomer is glucose. But the glucose units are hooked together in a different manner. Our bodies possess an enzyme that can digest starch. In contrast, we lack the enzyme that would enable us to digest cellulose. In essence, we can derive nutritional value from a potato but not from a piece of paper.

17. See Figure 11.11 for the chemical structures of fructose and glucose. Observe that the chemical structure of fructose is based on a five-membered ring composed of four C atoms and one O atom. In contrast, the chemical structure of glucose is based on a six-member ring composed of five C atoms and one O atom. Glucose has one $-CH_2OH$ side chain, and fructose has two.

19. The "amino" in *amino acid* indicates that there is an amine functional group present. The "acid" indicates there is an acidic functional group; in this case a carboxylic acid.

21. Analogous to Equation 11.4:

```
 H         O               H         O
  \        ||               \        ||
   N      C     H            N      C      H
  / \   /   \  /            / \   /   \   /
 H    C       O      +     H    C       O        →
     / \                       / \    H
    H   H                     H   H  /
                                    C
                                   / \
                               H  /   H
                                C=C
                               /   \
                          H — C     C — H
                              \\   //
                               C — C
                              /     \
                             H       H
```

```
 H        O
  \       ||         O
   N      C          ||
  / \   /   \        C      H
 H    C       N     / \    /
     / \     / \   /   O         + H2O
    H   H   H   C        
               / \    H
              H   H  /
                    C
                   / \
               H  /   H
                C=C
               /   \
          H — C     C — H
              \\   //
               C — C
              /     \
             H       H
```

23. Phenylketonuria (PKU) is a disease in which people lack the enzyme necessary to metabolize phenylalanine, an amino acid. Without the enzyme, phenylalanine accumulates in the body and eventually causes problems in brain development. People who are phenylketonurics must carefully limit their intake of foods rich in proteins. Although aspartame is a sweetener, it is a dipeptide of aspartic acid and phenylalanine, as shown in Figure 11.18. Sucralose, with a chemical structure not related to amino acids, does not pose any risk to people with this disease.

25. The two methods are increasing crop yields (for example, by using fertilizers) and devoting more land to agriculture (for example, deforestation).

27. Several answers are possible. For example, food production may diminish water quality through the amount of: (1) water required for irrigation that may deplete aquifers, causing more salty water to seep in, (2) fertilizers used, which may run off into

waterways providing excess nutrients that promote algal blooms, (3) herbicides and insecticides used that may contaminate local streams and rivers, and (4) draining of wetlands that serve as natural filters for water.

29. **a.** *Reactive nitrogen* refers to the chemical species of nitrogen that cycle relatively quickly through the biosphere and interconvert via several pathways.

b. Atmospheric nitrogen gas (N_2) is an unreactive form of nitrogen.

c. A natural source of reactive nitrogen is nitrogen-fixing bacteria in soils. An unnatural source of reactive nitrogen is synthetic fertilizers from the Haber-Bosch process.

31. To a chemist, all food is composed of chemical compounds ("chemicals"). These include fats, carbohydrates, proteins, minerals, and water. As a result, it is impossible to go on a "chemical-free" diet. Admittedly, though, the term "chemicals" is commonly taken to mean "added chemicals" or perhaps even "bad chemicals." An "all organic" diet is possible, though, and most organic foods strictly limit what chemicals can be used in raising the crops and can be added during food processing.

33. In terms of food chemistry, hydrogenation is the process of "adding hydrogen" to an unsaturated molecule to make it saturated (or more highly saturated). On a molecular level, in hydrogenation a molecule of H_2 is "added" to a C=C double bond to form two new C–H bonds and a C–C bond. Partial hydrogenation is the case in which some of the C=C bonds are hydrogenated, but not all of them. The manufacturers are required to report this, because these are two different chemical substances (*i.e.,* partially hydrogenated soybean oil is different from soybean oil). These substances have different properties and different effects on your health.

35. The process of hydrogenating an oil to "add H atoms" converts some of the C=C double bonds in the oil to C–C single bonds. This is desirable in that it improves the shelf-life of the product (and sometimes the spreadability as well). However, hydrogenation produces *trans* fats as a side-product, a drawback, because *trans* fats both increase the "bad" cholesterol and decrease the "good" cholesterol. Interesterification also reduces the number of C=C double bonds, but accomplishes this in a way that does not produce *trans* fat.

37. Based only on the percent of saturated fat in coconut oil relative to the butterfat in cream, this is not a good plan. Coconut oil is 87% saturated fat, but butterfat is only 63% saturated fat. However, if a person uses a smaller quantity of the nondairy creamer than of cream, his or her total amount of saturated fat consumption may be reduced.

39. According to Table 11.1, peanut butter is a good source of protein but is high in fat/oil, which in this case is peanut oil. On the positive side, unless the peanut butter has been hydrogenated, peanut oil is largely unsaturated, as shown in Figure 11.7.

41. **a.** $1.5 \text{ g fat} \times \frac{9 \text{ Cal}}{\text{g fat}} = 10 \text{ Calories from fat}$

$17 \text{ g carbohydrate} \times \frac{4 \text{ Cal}}{\text{g carbohydrate}}$

$= 70 \text{ Calories from carbohydrate}$

$3 \text{ g protein} \times \frac{4 \text{ Cal}}{\text{g protein}} = 10 \text{ Calories from protein}$

The total number of Calories is 90 Calories per slice of bread.

b. The percent Calories from fat is

$\frac{10 \text{ Cal}}{90 \text{ Cal}} \times 100 = 11\%$

Note: To one significant figure, this is 10%. This is something worth keeping an eye on, but not something we have emphasized in this textbook.

c. One slice of whole wheat bread qualifies as a nutritious food because it provides a serving of whole grains with few additional Calories from sugars and fats. Remember that if one slice of bread counts for one serving, then a sandwich counts as two servings of bread.

43. **a.** Carbohydrates

b. Possibilities include corn, sugarcane, and tapioca.

c. Digesting the starch to make glucose, fermentation of the glucose to produce ethanol, and distillation of the ethanol. See Section 5.15 for more details.

d. One controversy involves the energy costs. The energy gained by burning ethanol in an engine may be less than the energy inputs to produce it. Another controversy involves the environmental costs of growing corn and cutting down forests to produce sugarcane.

45. Answers will vary, but could include eating lower Calorie food, eating a greater variety of food, buying locally grown food, and eating less resource-intensive food such as red meats.

47. Assumptions made in your estimate might include (1) the number of days a year you drink soft drinks (non-diet), (2) the volume you drink or the packets of sugar you add to coffee or tea, and (3) the type of soft drink and grams of sugar it contains.

49. **a.** There are zero Calories in a packet of Splenda.

b. Splenda is made from sucrose by selectively replacing three of the –OH groups with –Cl groups to produce a molecule that is 600 times sweeter than sucrose. It is made from sugar but is a different chemical compound.

51. **a.** As you can see from the pie charts, soybeans are higher in protein (~35%) than wheat (~13%). Soybeans can be used to produce many different food products, including soy milk, textured vegetable protein, and tofu.

b. People in some cultures, particularly in Asian countries, are accustomed to obtaining their protein from soy. The taste of soy-based beverages or other foods is familiar and in widespread use, which would increase its appeal and acceptance.

53. For updates of U.S. ethanol production, the Renewable Fuels Association web page is a useful place to start (http://www.ethanolrfa.org/pages/statistics#A, accessed July 2016). Ethanol is produced primarily from corn in the United States. Keep an eye out for any evidence that cellulosic ethanol (such as switchgrass or another non-food feedstock) has begun commercial production. Look for new estimates and carefully assess the assumptions behind these estimates. Finally, the Triple Bottom Line judges whether a business operation has benefits to the economy, the environment, and to the society. These benefits (or lack thereof) are relevant to U.S. ethanol production.

Chapter 12

2. a. As the K < 1, this means that the reactant (here, glucose) is favored over the product (here, fructose).

b. For a system at equilibrium, the ratio of $K = \frac{[\text{fructose}]}{[\text{glucose}]} = 0.74$. As we know the concentration of glucose, we can solve for the concentration of fructose using the formula:

$$\frac{[\text{fructose}]}{0.22\text{ mM}} = 0.74$$

$$[\text{fructose}] = 0.74 \times 0.22\text{ mM}$$

$$[\text{fructose}] = 0.16\text{ mM}$$

4. In the figure, the AB complex is in the position of the reactants leading to A free and B free as products.

$$K = \frac{[\text{A free}][\text{B free}]}{[\text{AB complex}]}$$

7. There are five different isomers. Here are the structural formulas and line-angle drawings. The hydrogen atoms in the structural formulas have been omitted for clarity.

9. a. ether **b.** carboxylic acid **c.** ketone **d.** amide **e.** ester

11. a. The compound is an alcohol (ethanol). An isomer with a different functional group (an ether) is:

b. Aldehyde. An isomer with a different functional group (a ketone) is:

c. Ester. An isomer with a different functional group (a carboxylic acid) is:

12. **a.** The chemical formula is $C_5H_9N_3$.

b. The amine functional group, $-NH_2$, is present in histamine.

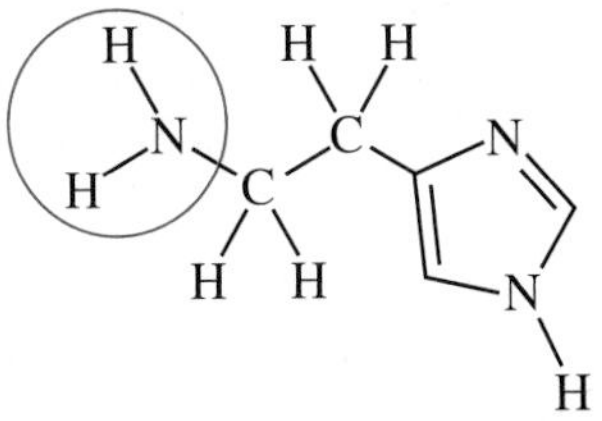

c. The molecule's most polar region will be around the amine group. This can easily take on a proton in water and (particularly when charged) will be the principal part of the molecule that interacts with polar water molecules.

15. **a.** This compound cannot exist in chiral forms. The central carbon atom is bonded to two equivalent $-CH_3$ groups.

b. This compound can exist in chiral forms. The four groups attached to the central carbon atom are all different.

c. chiral

d. not chiral

17. Hormones are chemical signals produced in the body to regulate many physiological events, such as the metabolism of food (insulin), our response to sudden or dangerous events (adrenaline), and many others. Receptors are the proteins that bind to these hormones to transfer the information in that chemical signal. A chemical without a switch to flip will not serve a purpose. Similarly, a protein receptor without a chemical to activate it will be passive.

20.

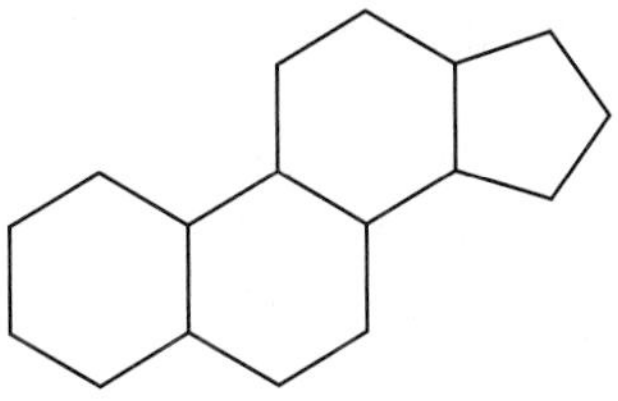

21. Acetic acid is CH_3COOH.

a.

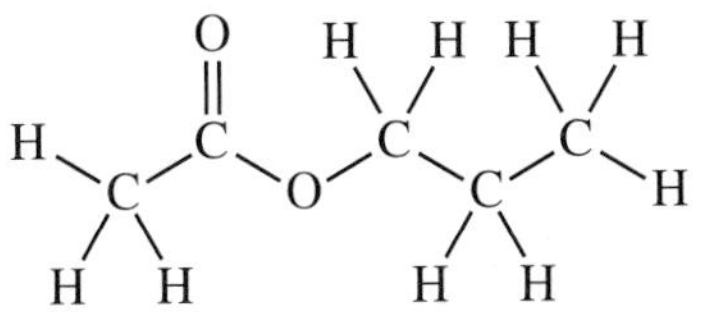

b.

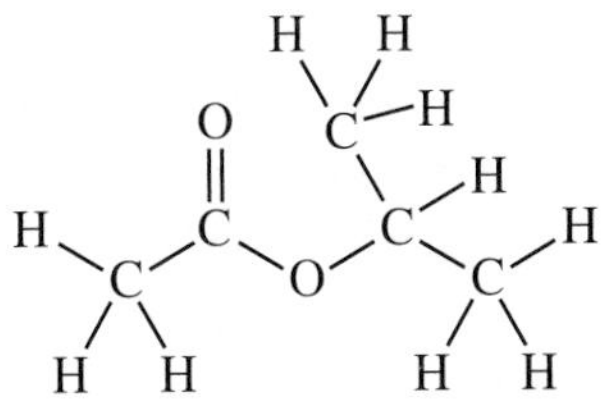

c.

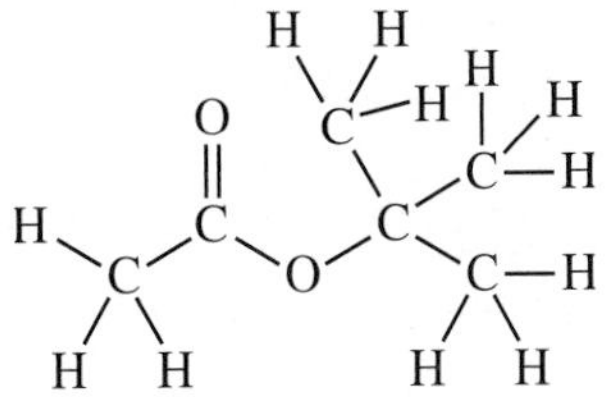

23. A pharmacophore is the three-dimensional arrangement of atoms, or groups of atoms, responsible for the biological activity of a drug molecule.

25. The equilibrium constant shows you whether the reactants or products are at a lower energy level. That constant does not provide any information about the barrier in between the two. To be more technical, the equilibrium constant provides information about the thermodynamics not kinetics.

27. **a.** Four single bonds, one triple bond

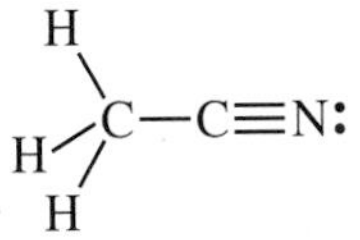

b. Six single bonds, one double bond

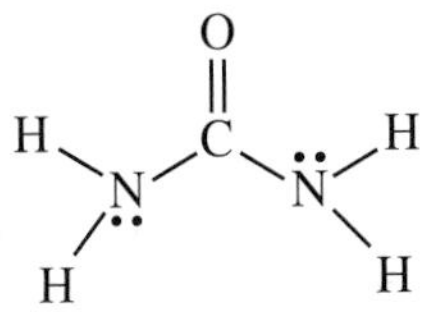

c. 11 single bonds, four double bonds

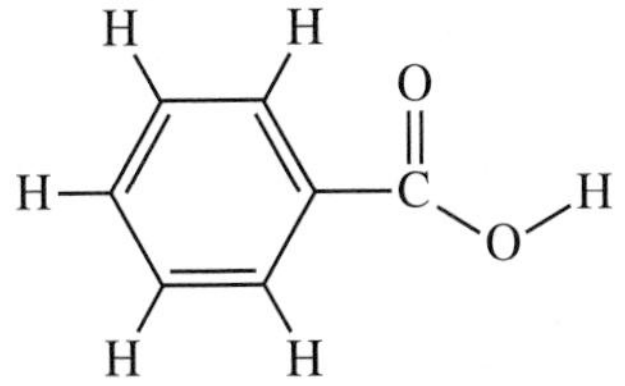

28. Only three isomers are shown here; some structures are duplicates. #1 and #5 are different representations of the *same* isomer. #2, #3, and #4 are all representations of the *same* isomer. #6 is an isomer *different* from #1 and #5, and from numbers #2–4.

30. **a.** The chemical formula is $C_{16}H_{21}N_3$.

b. Both compounds have nitrogen-containing rings, but they are not the same size rings. The major structural similarity is the presence of the $—CH_2\,CH_2N$ group, which is likely the part of each molecule that competes to attach to the receptor site.

33. **a.** The cellular membrane is primarily composed of lipids.

b. Glycogen is a polymer of glucose, specifically a polysaccharide or, more commonly, a carbohydrate.

c. Enzymes are protein-based catalysts.

36. The analogy compares the receptor site on the surface of a cell to a lock that can only be opened by a unique key. Drug molecules can only bind to receptor sites that match the molecule's geometry.

38. **a.** Aspirin produces a physiological response in the body.

b. Antibiotics kill or inhibit the growth of bacteria that cause infections.

c. Morphine induces a physiological response.

d. Estrogen causes a physiological response.

e. Penicillin inhibits the growth of bacterial infections.

40. **a.** The codeine structure has two ether groups, an alcohol group, and an amine group.

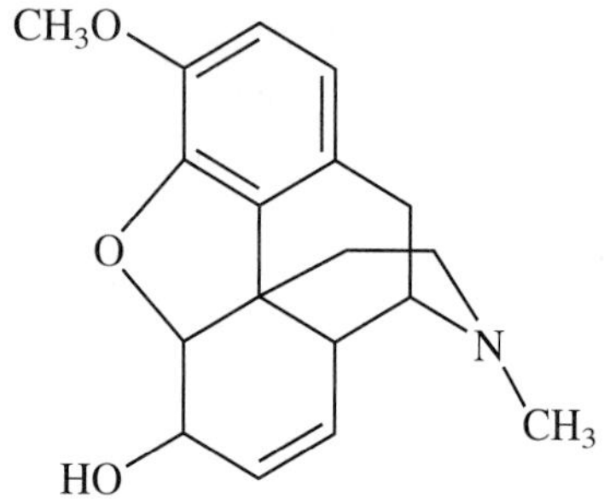

CH_3O O N CH_3 HO

Codeine

b. No, comparison of only two drugs is not enough evidence to draw general conclusions about the role structural changes play in determining drug effectiveness and addictiveness.

43. **a.** There are 6(4) + 6(1) or 30 electrons available. Here is one possible structural formula of a linear isomer for benzene:

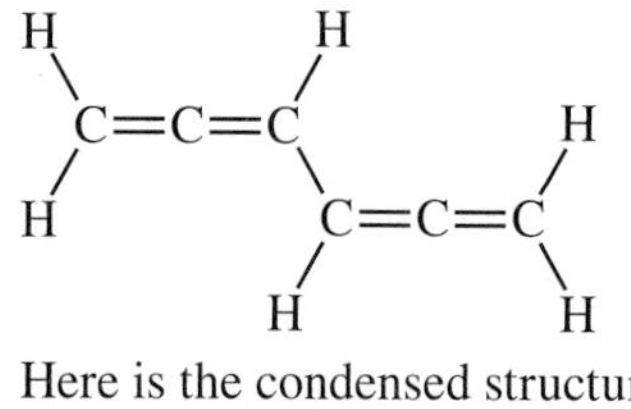

H H H C=C=C H C=C=C H H

b. Here is the condensed structural formula:

$CH_2{=}C{=}CH{-}CH{=}C{=}CH_2$

c. First check to see if all of the structures correctly represent C_6H_6 and that each carbon has four bonds. If these conditions are met, the structures with double bonds should differ only in the placement of the lone C–C single bond. However, structures including carbon-carbon triple bonds can also be drawn; these would be distinctly different.

44. Thalidomide's two optical isomers have very different effects in the body. One isomer treats nausea, while the other produces mutations in babies born to women who take the drug early in pregnancy. Unfortunately, the body can convert each isomer into the other. When the German maker of thalidomide applied to the FDA for approval to market the drug in the U.S. in the 1960s, their application was rejected repeatedly due to a lack of data proving the safety of the drug. In 1998, FDA approved thalidomide for the treatment of skin lesions caused by leprosy, provided that patients are not pregnant and take precautionary measures to avoid becoming pregnant while on the medication.

47. Information on Presidential Green Chemistry Challenge Awards are collected online. The companies collaborated to develop a novel synthesis incorporating an evolved enzyme. The prior synthetic route required high-pressure, expensive metals, and a costly purification step that now can be eliminated. This matches our understanding of the utility of enzymes as biological replacements for organic synthetic steps. If selected correctly, they can do the reaction at lower temperature and with less solvent waste. See EPA award site for more information: https://www.epa.gov/greenchemistry/presidential-green-chemistry-challenge-winners (accessed July 2016).

51. The life of Dorothy Crowfoot Hodgkin (1910-1994) is documented in several biographies. An interesting website is http://www.sdsc.edu/ScienceWomen/hodgkin.html (accessed July 2016). Dorothy was born in Cairo, Egypt, where her father was in the Ministry of Education and administered archaeological sites. Her mother was a self-taught botanist. Dorothy was well-educated at Oxford and together with her mentor, J. D. Bernal, she first applied X-ray diffraction to crystals of biological substances, including pepsin, penicillin, cholesterol, and later insulin. Hodgkin was elected a Fellow of the Royal Society in 1947 after publishing the structure of penicillin and was awarded the Nobel Prize in chemistry in 1964 for solving the structures of important biomolecules, such as vitamin B_{12}. In the words of colleague Max Perutz (Nobelist for his solution of the hemoglobin molecule structure), she was "a great chemist, a saintly, gentle, and tolerant lover of people, and a devoted protagonist of peace."

Chapter 13

2. Farming and medicine are the most obvious industries that have changed, as seen throughout this chapter, but other options are available. Plastics and recycling are also changing due to new methods for making materials with enzymes. Proposed answers can include the fuel industry with increasing biofuels development.

3. A genome is made up of all the genetic information in a cell, while a gene is a small subsection of the genome that codes for a single protein.

5. **a.** A nucleotide links a nitrogen-containing base, a sugar, and a phosphate group.

b. Covalent bonds hold the units together.

8.

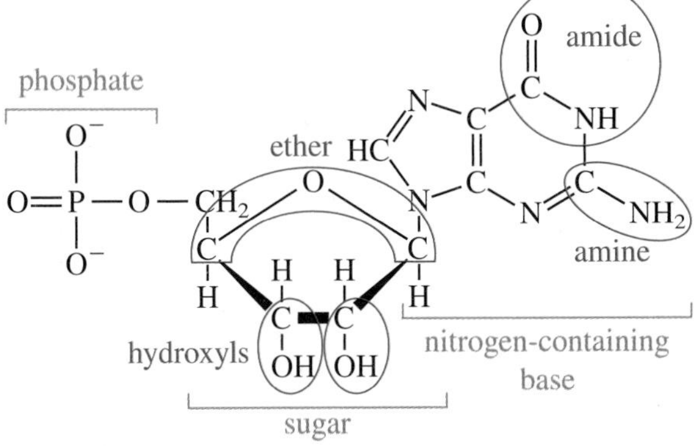

9. a. Nucleotides polymerize to form DNA when the phosphate group from one nucleotide reacts with the hydroxyl group on another nucleotide. The nucleotide shown has two hydroxyl groups. The one closer to the phosphate group is the correct site for polymerization, similar to DNA. The second hydroxyl group results in the significant chemical instability seen in this DNA-related polymer.

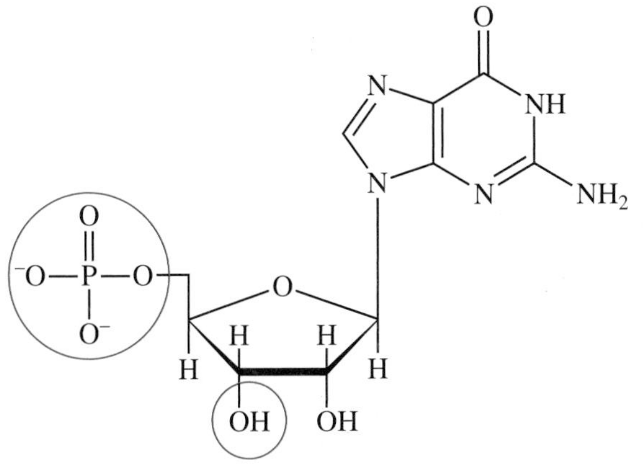

b. In the name DNA, the D stands for the deoxyribose in deoxyribonucleic acid. The name for the polymer built with ribose instead is RNA for ribonucleic acid.

12. Sequences are always read in one direction. On paper, we read each sequence from left to right to determine which amino acid it codes for. Chemically, the directionality comes from the specific order of bonds connecting the alternating sugars and phosphates in the backbone. Although the second sequence in this question is simply the reverse of the first base sequence, it would code for a different amino acid. (ATG codes for methionine; GTA codes for valine.)

13. a. The complementary base sequence is ATAGATC.

b. Your answer should have two lines between each A and T, and three lines between each C and G in the sequence.

TATCTAG
|| || || ||| || || |||
ATAGATC

15. Each codon consists of a three-nucleotide sequence that is specific for an amino acid or the start/stop of protein synthesis. All of the codons together make up the code for translating a sequence of DNA into the amino acid sequence of a protein.

16. a. All amino acids must follow the same general structure (see image to the right). Only the R group varies between acidic, basic, and neutral amino acids. Correct example R groups for each category include:

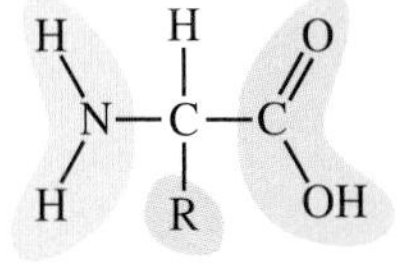

Acidic = glutamic acid ($-CH_2CH_2COOH$)

Basic = lysine ($-CH_2CH_2CH_2CH_2NH_2$)

Polar neutral = serine ($-CH_2OH$)

b. All three categories can make hydrogen bonds, but only acidic and basic amino acids can also make ionic bonds.

c. Acidic amino acids have carboxylic acids within their R groups, while basic amino acids often have amine groups. Neutral amino acids may contain hydroxyl groups or amides in their side chains.

18. Only a minor change in the amino acid composition of human hemoglobin leads to sickle cell anemia. In the hemoglobin S chain, a nonpolar valine replaces a specific charged glutamic acid in the sequence. This seemingly innocuous change in the protein's primary structure dramatically affects the shape of the protein. The tertiary structure must change to accommodate the change in the side chain; the nonpolar valine cannot positively interact with water or other polar groups in the same ways as glutamic acid. The change in protein shape leads to sickled red blood cells (under certain conditions) and a number of health problems.

21. The steps for both actually have a number of parallels. Both require a source that is grown in the lab or in the animal and at least some purification steps before the final medicine. (See diagram below for more information.)

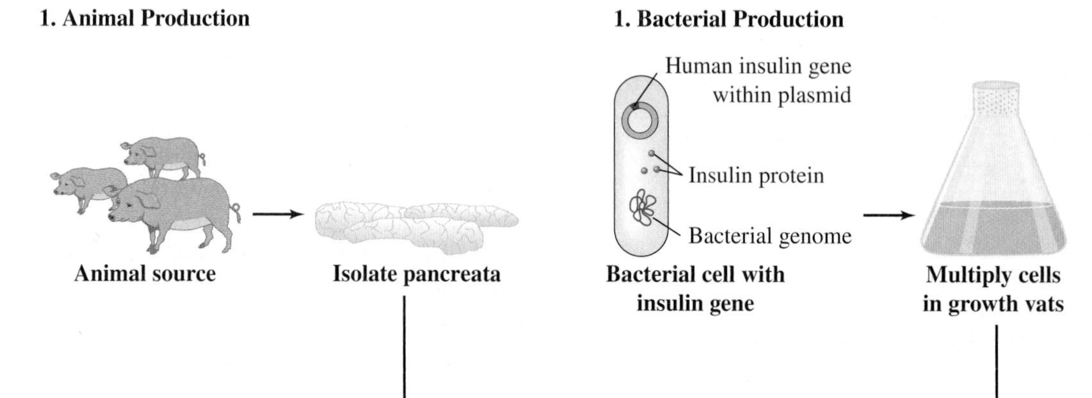

2. Purification stages

The cells or tissue must be broken open. Then, the cell contents, a mixture of blomolecules, are separated by size, charge, or other characteristics. Typlcally, separation occurs by passing through a serles of columns.

Injectible purified insulin

23. Adenine-thymine base pairs are less stable than cytosine-guanine base pairs because they have fewer hydrogen bonds holding them together.

24. **a.** In the electromagnetic spectrum, X-rays have high energies and wavelengths from 0.01 to 10 nm.

b. A beam of X-rays is directed at an unknown substance. The nuclei in the substance scatter the X-rays. A detector measures the intensity and pattern of scattered X-rays. If the atoms in the substance are arranged in a regular pattern, the diffracted X-rays can be used to calculate the distance between atoms.

c. One reason is that salts, such as sodium chloride, easily form crystals. In contrast, nucleic acids and proteins are much larger and do not easily form crystals. Another reason is that when nucleic acids and proteins do crystallize, their X-ray diffraction patterns are far more complex and difficult to interpret.

26. The two DNA strands are complementary, meaning that the sequence on one strand can be reconstructed from the other. If a single strand breaks, the other strand can be used by repair enzymes to correctly replace any lost nucleotides and reassemble the backbone. If both strands break, the information on how to repair the break is lost and the correct sequence of DNA will be permanently altered.

28. Many amino acids are represented by more than one DNA codon. For example, GTT, GTC, GTA, or GTG all translate to valine. If an error in the base sequence of DNA did not change the meaning of the codon, the protein would contain the correct amino acid. This redundancy makes the genome more resilient to change.

30. Reasons for the low public concern include (1) insulin treatment is not new and has been seen to be safe before it was generated by genetically engineered bacteria, and (2) genetically engineered bacteria are kept within laboratories and industrial facilities and are therefore easier to control, while plants spread more easily.

33. Rosalind Franklin was born in 1920, the daughter of a prominent London banking family. All of the children in the family were given encouragement to pursue an education. Both her undergraduate and graduate degrees were from Cambridge University. During World War II, she suspended her graduate research to contribute to the war effort by studying the properties of coal and graphite. After the war, she completed her PhD in physical chemistry and joined a prominent laboratory in Paris where she was introduced to the technique of X-ray crystallography. She soon became an expert in the field, moved back to London in 1951 to work at King's College, and in 1952 produced X-ray photographs of DNA. Before publishing the images, she showed them to Maurice Wilkins, another scientist studying DNA. Without Franklin's knowledge, Wilkins shared the photographs with James Watson, a molecular biologist who was working with Frances Crick to describe the structure of DNA. The X-ray data changed Watson and Crick's hypothesis about the structure, and they published a paper describing DNA as a double helix in 1953. Watson, Crick, and Wilkins shared the Nobel Prize for physiology or medicine in 1962, but Franklin did not live to be considered for the prize. Franklin died of ovarian cancer in 1958. For more information, good references are: http://www.sdsc.edu/ScienceWomen/franklin.html and *Chemical Achievers: The Human Face of the Chemical Science*, a publication of the Chemical Heritage Foundation, Philadelphia, Pennsylvania, 1997.

34. The most popular explanation for the persistence of the sickle cell trait is that carrying the trait protects against death from malaria. Certainly, the regions and climates where the trait is historically common overlap closely with people from those regions affected by malaria. The reason that the trait provides resistance to malaria is not fully understood.

36. Patenting GM plants will give distinct advantages to industrialized nations who have the funds to conduct further research on GM organisms. The patenting process may be advantageous because (1) it provides financial incentive to perform the extremely expensive research required, and (2) it increases the control and decreases the spread of artificial GM crops. The process may be a disadvantage because it (1) prevents the creative and traditional breeding process by farmers and smaller scale plant breeders, and (2) halts further research by academic scientists into a specific gene. Also, more importantly, patenting may make the technology too expensive for the underdeveloped nations that many GM crops are being developed to help.

38. **a.** An excellent source of information specific to the EU is http://www.gmo-compass.org/eng/home/ (Accessed November 2016). Examples include rapeseed, soybean, carnations, and corn. New traits include herbicide tolerance and insect resistance. Some have been approved for import and processing; very few have been accepted for cultivation. Typically, those allowed were cultivated before the onset of regulations.

b. Examples of biotech timeline:

1973: The first genetically engineered bacteria producing a human protein is created.

1997: Two scientists at Harvard University, Walter Gilbert and Allan Maxam, created a method for sequencing DNA using chemicals instead of enzymes. This technique was an important start to identifying DNA sequences efficiently.

1980: The U.S. Supreme Court ruled that genetically engineered organisms can be patented. The ruling allowed Exxon to patent an oil-eating microorganism. More importantly, it opened the door for commercialization of genetically engineered technologies.

1980: Kary Mullis and others at Cetus Corporation in Berkeley, California, created a method for copying target sequences of DNA outside of cells. The technique is now commonly used in labs across the globe to copy, alter, or create DNA sequences.

1982: Genentech, Inc. received FDA approval for the first genetically engineered drug: human insulin protein produced by bacteria.

40. Gene therapy involves introducing normal genes into patients lacking them. It would allow us to treat a specific subset of diseases caused by clear genetic alterations. For even a small subset of medical conditions, high-risk clinical trials have seen both disappointing results as well as some promise. Other diseases might be amenable to treatment with gene therapy, but developing appropriate protocols is costly, both in time and money. The Human Gene Therapy Subcommittee of the Recombinant DNA Advisory Committee of the National Institutes of Health must give approval to all proposed uses of gene therapy.

Example histories: The first person treated was a four-year-old girl suffering from severe combined immunodeficiency disease (SCID) in which a genetic defect prevents the formation of a specific enzyme necessary for the health of white blood cells. Children with SCID have extremely weak immune systems and often die before adulthood. Initial results showed promise, but in January 2003, the FDA temporarily halted all gene therapy trials using retroviral vectors in blood stem cells after two children developed a leukemia-like condition. (The first case of leukemia development in a child participating in a clinical trial occurred in 2002.)

More recently, gene therapy has been used with far more success to treat neural degenerative disorders and blindness. Here, the key to this success is that the genes were more carefully controlled and more specifically placed. At the time of this writing, long-term effects are still unknown in these studies. See *Science* **2009**, *326*, 818–823, and http://sciencenow.sciencemag.org/cgi/content/full/2009/1024/1 for more information.

Chapter 14

1. **a.** Class K (wet chemical, such as aqueous potassium carbonate that includes a detergent)

b. Class C (carbon dioxide or dry chemical such as ammonium phosphate or sodium bicarbonate)

c. Class A (water only)

3. Ester, alkene, and nitrile ("cyano") functional groups.

5. Many possible examples, such as accidentally stepping in a pool of blood in a dimly lit house, inadvertently removing or smearing fingerprints when opening doors to secure the scene, or maybe knocking something over as they move through the scene. They may also be tempted to pick up gauze, latex gloves, etc., left by emergency medical personnel or to "tidy up" before more personnel arrive. They may also not wear protective outer clothing and therefore may leave fibers from their own clothing at the crime scene.

8. Many possible examples, such as no sign of forced entry, or the entry is forced beyond what would be required to gain access; only one specific item was stolen; no search for any valuables in an apparent burglary, or no items have been stolen; excessive ransacking, or too careful of specific items (some items set aside to protect them); victim is posed to suggest or cover up a sexual assault; survivor of an attack has minor wounds only on the side of the body opposite their own handedness (self-inflicted); wounds are consistent with being self-inflicted

10. The order of places where, and the persons with whom, physical evidence was located from the time it was collected to its submission at trial.

12. Any solvents that have a flash point less than 25 °C would ignite in contact with a match. These include diethyl ether, acetone, tetrahydrofuran, hexanes, ethanol, acetonitrile, and toluene.

14. 2,500 nm to 25,000 nm

16. This can be solved by taking the natural log (ln) of both sides of the equation:

$2^n = 1{,}000{,}000$. That is, $n \ln(2) = \ln(1{,}000{,}000)$ or $n = \ln(1{,}000{,}000) / \ln(2)$. This will equal 19.9 or 20 cycles. Refer to Appendix 3 for more information about the math behind logarithms.

18. The spectrum shows the presence of alkyl C–H groups (strong C–H stretching peaks around 3000 cm^{-1} and C–H bending peaks around 1380-1400 cm^{-1}) a C=O carbonyl group (peak around 1700 cm^{-1}), and perhaps an ester C–O peak (1170 cm^{-1}). The absence of a medium-strong, broad peak around 3500 cm^{-1} indicates that an alcohol or carboxylic acid –OH is not present. You are not able to determine the concentration of functional groups based on their peak heights; IR spectroscopy is only used to determine whether certain functional groups are present.

19. No. All of these compounds would show the presence of C–H, C=O, C–O, and O–H groups. In order to distinguish between these compounds, mass spectrometry or nuclear magnetic resonance (NMR) spectroscopy would need to be performed.

22. Yes. Depending on the alloy comprising the gun, various acid mixtures may be applied to the surface, which will etch the metal and reveal the serial number that lies beneath the scratch marks.

23. In addition to fingerprints, earprints (*e.g.,* left behind in a crime scene by someone listening through a door), or recognizing people by their gait—the way they walk.

24. The mass of an electron is 9.11×10^{-31} kg; by plugging these values into the equation, the best resolution would be 1.2×10^{-11} m. This corresponds to 0.012 nm or 0.12 Å.

26. **a.** Not necessarily. Luminol also exhibits bright chemiluminescence upon contact with other substances such as some paints, varnishes, fruit and vegetable juices, and iron-containing compounds.

b. Although luminol may have an effect on the typing of bloodstains using conventional serological testing, it has been found that luminol *does not interfere* with DNA analyses.

29. Visual inspection of the tungsten filament can be used to determine whether headlights were illuminated. The tungsten filament inside sealed headlights gets extremely hot when headlights are turned on. After a crash with an impact speed of >20 km/h, a hot filament will be significantly deformed, whereas a cold filament will resemble a spring with little/no deformation observed. Often, the glass bulb will shatter during impact, which will allow oxygen to react with the hot filament. This will result in a yellowish-white powder that will coat the surface of the filament, which is readily observable by using scanning electron microscopy. Furthermore, small glass particles and grains will implode into the bulb, which will adhere to the filament as it cools down. In contrast, a cold filament will not react with oxygen and will remain clean and shiny. Glass particulates will not adhere to a cold filament.

Glossary

The numbers at the end of each term indicate the section(s) where the term is defined and explained in the text.

A

acid a compound that releases hydrogen ions, H^+, in aqueous solution (8.8)

acid-neutralizing capacity the capacity of a lake or other body of water to resist a decrease in pH (8.12)

activation energy the energy necessary to initiate a chemical reaction (5.14, 12.6, and 14.4)

active site the catalytic region, often a crevice, in an enzyme that binds only specific reactants and accelerates the desired reaction (12.6 and 13.5)

addition polymerization A type of polymerization in which the monomers add to the growing chain in such a way that the polymer contains all the atoms of the monomer. No other products are formed. (9.3)

aerosols liquid or solid particles that remain suspended in the air rather than settling out (4.10)

albedo a measure of the reflectivity of a surface; the ratio of electromagnetic radiation reflected from a surface relative to the amount of radiation incident on it (4.10)

alcohol a hydrocarbon substituted with one or more –OH groups (hydroxyl groups) bonded to its carbon atoms (5.15)

alkane a hydrocarbon composed of only single bonds between neighboring carbon atoms (5.12)

allotrope different structural forms of the same element (1.6 and 2.14)

ambient air the air surrounding us, usually meaning the outside air (2.9)

amino acid Monomer from which our body builds proteins. Each amino acid molecule contains two functional groups: an amine group ($-NH_2$) and a carboxylic acid group (–COOH). (9.7)

amino acid residues amino acids which have been incorporated into the peptide chain (11.6 and 13.4)

amorphous regions a solid in which the constituent atoms, ions, or molecules are arranged in a random, disordered array (1.9 and 9.5)

amphiphilic a molecule that has both nonpolar and polar groups that make it both lipophilic (fat-soluble) and hydrophilic (water-soluble) (10.10)

anode the electrode at which oxidation takes place (7.2)

anthropogenic caused or produced by human activities, such as industry, transportation, mining, and agriculture (4.6)

anthropogenic forcings Man-made factors that influence Earth's energy balance (4.10)

aqueous solution a solution in which water is the solvent (8.6)

aquifer an underground permeable rock formation from which groundwater may be extracted using a well (8.4)

atom the smallest unit of an element that can exit as a stable, independent entity (1.1)

atomic mass the mass (in grams) of the same number of atoms that are found in exactly 12 g of carbon-12 (4.3)

atomic number the number of protons in the nucleus of an atom (1.4)

autoignition temperature the minimum temperature at which the vapor of a substance spontaneously ignites, even in the absence of an ignition source (14.4)

B

basal metabolic rate (BMR) the minimum amount of energy required per day to support basic body functions (11.8)

base a compound that releases hydroxide ions, OH^-, in aqueous solution (8.8)

battery an energy-storage device that converts the energy released from spontaneous chemical reactions into electrical energy (7.1)

biofuel a generic term for a renewable fuel derived from a biological source, such as trees, grasses, animal waste, or agricultural crops (5.15)

biological oxygen demand (BOD) A measure of the amount of dissolved O_2 that microorganisms use up as they decompose the organic wastes found in water. A low BOD is one indicator of good water quality. (8.13)

biomagnification the increase in concentration of certain persistent chemicals in successively higher levels of a food chain (8.7)

biomimetic materials materials that try to replicate specific properties of biological materials for use in human applications (9.7)

blowing agent either a gas or a substance capable of producing a gas used to manufacture a foamed plastic (9.5)

boiling point the temperature at which the vapor pressure of a liquid equals the surrounding atmospheric pressure (5.12)

bond dipole The difference in electronegativity between two atoms in a polar covalent bond, which gives rise to partial positive/negative charges on the atoms. A convention to indicate the bond dipole uses an arrow to point in the direction of the more negatively charged end of the covalent bond (8.2)

bond energy the amount of energy that must be absorbed to break a specific chemical bond (5.5)

Brix scale A unit used to quantitatively express the sugar content of a solution, based on measurements provided by a refractometer. One degree Brix (°Bx) is equal to 1 g of sucrose per 100 g of solution (*i.e.*, 1%(w/w)). (10.9)

buffer a system that responds only gradually or slightly to an external influence (8.11 and 12.2)

C

calorie (cal) the amount of heat necessary to raise the temperature of one gram of water by 1 °C (5.4)

calorimeter a device used to experimentally measure the quantity of heat energy released in a combustion reaction (5.4)

capacitance the ability of a material to store an electrical charge (7.7)

capacity The specific energy of an energy storage device given in ampere-hours (Ah). This represents the discharge current a battery can deliver over time. (7.7)

carbohydrate a compound that contains carbon, hydrogen, and oxygen, with H and O atoms found in the same 2:1 ratio as in H_2O (11.4)

carbon footprint an estimate of the amount of CO_2 and other greenhouse gas emissions in a given time frame, usually a year (4.10)

carbon neutral a situation in which the CO_2 added to the atmosphere is balanced by the CO_2 removed by photosynthesis, sequestration, a carbon offset, or some other process (5.17)

carcinogenic capable of causing cancer (2.15)

catalyst a chemical substance that participates in a chemical reaction and influences its rate, without itself undergoing permanent change. (2.13 and 5.14)

catalytic cracking a process in which catalysts are used to crack larger hydrocarbon molecules into smaller ones at relatively low temperatures (5.13)

catalytic reforming a process in which the atoms within a molecule are rearranged, usually starting with linear molecules and producing ones with more branches (5.13)

cathode The electrode at which reduction takes place. The cathode receives the electrons produced at the anode. (7.2)

cellular membrane the dynamic yet protective outer casing of a cell (12.6)

cellulose A naturally occurring compound composed of C, H, and O that provides structural rigidity in plants, shrubs, and trees. Cellulose is a natural polymer of glucose. (5.15)

cellulosic ethanol ethanol produced from any plant containing cellulose, typically cornstalks, switchgrass, wood chips, and other materials that are nonedible by humans (5.15)

chain reaction a term that generally refers to any reaction in which one of the products becomes a reactant and thus makes it possible for the reaction to become self-sustaining (6.1)

chemical equation a representation of a chemical reaction using chemical formulas (2.11)

chemical formula A symbolic way to represent the elementary composition of a substance. It reveals both the elements present (by chemical symbols) and the atomic ratio of those elements (by the subscripts). (1.3)

chemical reaction a process whereby substances described as reactants are transformed into different substances called products (2.11)

chiral (optical) isomers compounds with the same chemical formula but different three-dimensional molecular structures and different interaction with plane polarized light (12.5)

chlorofluorocarbons (CFCs) compounds composed of the elements chlorine, fluorine, and carbon (but do not contain the element hydrogen) (3.9)

chromosomes rod-shaped, compact coils of DNA and specialized proteins packed in the nucleus of cells (13.3)

climate A term that describes regional temperatures, humidity, winds, rain, and snowfall over decades, not days. Contrast this with *weather.* (4.9)

climate adaptation the ability of a system to adjust to climate change (including climate variability and extremes), to moderate potential damage, to take advantage of opportunities, or to cope with the consequences (4.12)

climate mitigation any action taken to permanently eliminate or reduce the long-term risk and hazards of climate change to human life, property, or the environment (4.12)

codon a sequence of three adjacent nucleotides that either guides the insertion of a specific amino acid or signals the start or end of protein synthesis (13.4)

coenzymes molecules that work in conjunction with enzymes to enhance the enzyme's activity (11.7)

combustion the chemical process of burning; the rapid reaction of fuel with oxygen to release energy in the form of heat and light (2.11)

compostable under the conditions of either a home composter or an industrial composter, the ability for an item to undergo biological decomposition to form a material (compost) that contains no materials toxic to plant growth (9.10)

composition A description of the identity and structure of the subunits that comprise a substance or material. (1.1)

compound a pure substance that is comprised of two or more different types of atoms in a fixed, characteristic chemical combination (1.1)

concentration the ratio of the amount of solute to the amount of solution (8.6)

condensation polymerization a type of polymerization in which a small molecule such as water is split out (eliminated) when the monomers join to form a polymer (9.6)

condensed structural formula a structural formula in which some bonds are not shown; rather, the structural formula is understood to contain an appropriate number of bonds (5.2)

conjugate acid the species formed by adding a proton to a base (8.8)

conjugate base the species formed by removing a proton from an acid (8.8)

copolymer a polymer formed by the combination of two or more different monomers (9.6)

covalent bond a bond formed when electrons are shared between two atoms (3.7)

cradle-to-cradle a term coined in the 1970s that refers to a regenerative approach to the use of things in which the end of the life cycle of one item dovetails with the beginning of the life cycle of another, so that everything is reused rather than disposed of as waste (1.10 and 9.9)

critical mass the amount of fissionable fuel required to sustain a chain reaction (6.1)

crystal a solid-state material that consists of long-range 3-D ordering of its constituent atoms, ions, or molecules (1.9)

crystalline regions in a polymer, a region in which the long polymer molecules are arranged neatly and tightly in a regular pattern (9.5)

crystallization the process of dropping a solid out of a solution in a controllable manner in order to form a solid with an ordered structural array (4.4)

current (electrical) the rate of electron flow through a circuit (7.2)

D

degasification the process of a gas escaping from a liquid or solid (4.2)

denitrification the process of converting nitrates to nitrogen gas (11.12)

density the mass per unit volume (1.9 and 8.3)

deoxyribonucleic acid (DNA) the biological polymer that carries genetic information in all species (13.2)

desalination any process that removes sodium chloride and other minerals from salty water (8.14)

diffusion the net movement of molecules or atoms from an area of higher concentration to a region of lower concentration (5.7)

dipeptide a compound formed from two amino acids (11.6)

disaccharide a "double sugar" formed by joining two monosaccharide units, such as sucrose (table sugar) (11.4)

distillation a separation process in which a liquid solution is heated to its boiling point and the vapors are condensed and collected (5.12, 8.14, and 14.2)

distributed generation generating electricity on-site where it is used (*i.e.,* with a fuel cell), thus avoiding the losses of energy that occur over long electric transmission lines (7.9)

doping the process of intentionally adding small amounts of other elements to pure silicon to modify its semiconductor properties (6.8)

double bond a covalent bond consisting of two pairs of shared electrons (3.7)

double helix a spiral consisting of two strands that coil around a central axis (13.3)

E

electricity the flow of electrons from one region to another that is driven by a difference in potential energy (1.5 and 7.1)

electrochemistry The branch of chemistry that deals with the transformation between chemical and electrical energies. (7.1)

electrodes the electrical conductors (anode and cathode) in an electrochemical cell that serve as sites for chemical reactions (7.2)

electrolysis The process of passing a direct current of electricity of sufficient voltage to cause a chemical reaction to occur. For example, the electrolysis of water decomposes it into H_2 and O_2. (7.10)

electrolyte a solute that conducts electricity in an aqueous solution (7.4 and 8.7)

electrolytic cell a type of electrochemical cell in which electrical energy is converted to chemical energy (7.10)

electromagnetic spectrum continuum of waves that ranges from short, high-energy X-rays and gamma rays to long, low-energy radio waves (3.1)

electronegativity a measure of the attraction of an atom for an electron in a chemical bond (8.2)

element One of the 100 or so pure substances in our world from which compounds are formed. Elements contain only one type of atom. (1.1 and 1.3)

emulsion A mixture of two or more liquids that are normally immiscible. (10.10)

endocrine disrupter a compound that affects the human hormone system, including hormones for reproduction and sexual development (9.11)

endothermic a term applied to any chemical or physical change that absorbs energy (5.4)

energy the ability or capacity of matter to do work, or to produce change (1.10)

energy density the amount of potential energy stored in a given system per unit volume (volumetric energy density) or mass (gravimetric energy density, also referred to as specific energy density) (7.6)

enhanced greenhouse effect the process in which atmospheric gases trap and return *more than* 80% of the heat energy radiated by Earth (4.6)

entropy a measure of how much energy gets dispersed in a given process (5.7)

enzymes proteins that act as biochemical catalysts, influencing the rates of chemical reactions (5.15 and 12.6)

equilibrium the state at which the concentration of products and reactants are equal during a reversible chemical reaction (12.1)

equilibrium constant The concentration of products divided by the concentration of reactants for a reversible reaction that is at equilibrium. (12.1)

equilibrium reaction a reaction that proceeds in both directions in which reactants form products, and products may re-form reactants (8.8)

essential amino acids those amino acids required for protein synthesis that must be obtained from the diet because the body cannot synthesize them (11.6)

exothermic a term to describe any chemical or physical change accompanied by the release of heat (5.4)

exposure the amount of a substance encountered (2.9)

F

fats triglycerides that are solids at room temperature (5.16 and 11.2)

First Law of Thermodynamics Also called the Law of Conservation of Energy, this law states that energy is neither created nor destroyed during any process or transformation. (5.6)

Fischer-Tropsch process A method of producing a variety of liquid hydrocarbons from a series of catalyzed reactions using carbon monoxide and hydrogen gases as reactants. (5.14)

flash point The temperature at which combustion is possible for a solvent. This occurs when the vapor pressure of a solvent is equal to its lower flammability limit (LFL). (14.4)

flashover The dangerous condition during a fire when the majority of exposed surfaces in a room are heated to their autoignition temperatures. This causes the materials to emit flammable gases, which provide additional fuel to the fire. (14.5)

fossil fuel Combustible substances derived from the remnants of prehistoric organisms. The most common examples are coal, petroleum, and natural gas. (5.1)

fracking a controversial method of extracting natural gas from underground rock formations through the injection of high-pressure fluids (5.12)

free radical a highly reactive chemical species with one or more unpaired electrons (3.8)

frequency the number of waves passing a fixed point in 1 second (3.1)

fuel any solid, liquid, or gas that may be burned to provide energy in the form of heat or work (5.1)

fuel cell an electrochemical cell that produces electricity by converting the chemical energy of a fuel directly into electricity without burning the fuel (7.9)

functional group a distinctive arrangement of a group of atoms that imparts characteristic properties to the molecules that contain this group (5.15, 9.6, and 12.4)

G

galvanic cell a type of electrochemical cell that converts the energy released in a spontaneous chemical reaction into electrical energy (7.1)

gas chromatography (GC) a simple and rapid analytical technique commonly used to identify separate components in mixtures of liquids or gases (14.6)

genes short pieces of the genome that code for the production of proteins (13.1)

genetic engineering the direct manipulation of DNA in an organism 13.6

genome the primary route for inheriting biological information required to build and maintain an organism 13.1

global warming a popular term used to describe the increase in average global temperatures that results from an enhanced greenhouse effect 4.6

glycolysis a complex biological process that breaks down glucose in order to provide energy for each living cell (12.6)

green chemistry the design of chemical products and processes that reduce or eliminate the use and generation of hazardous substances (2.16 and 7.11)

greenhouse effect the natural process by which atmospheric gases trap a major portion (about 80%) of the infrared radiation radiated by Earth (4.6)

greenhouse gases Gases capable of absorbing and emitting infrared radiation, thereby warming the atmosphere. Examples include water vapor, carbon dioxide, methane, nitrous oxide, ozone, and chlorofluorocarbons. (4.6)

groundwater fresh water found in underground reservoirs also known as aquifers (8.4)

group A column on the periodic table that organizes elements according to the important properties they have in common. Groups are numbered left to right. (1.1)

H

half-life ($t_{1/2}$) the time required for the level of radioactivity to fall to one half of its initial value (6.4)

half-reaction a type of chemical equation that shows the electrons either lost or gained by the reactants (1.7 and 7.1)

halons Inert, nontoxic compounds that contain chlorine or fluorine (or both, but no hydrogen). In addition, they contain bromine. (3.9)

heat the kinetic energy that flows from a hotter object to a colder one (5.3)

heat of combustion the quantity of heat energy given off when a specified amount of a substance burns in oxygen (5.4)

hemoglobin an iron-containing metalloprotein found within red blood cells used for oxygen transport (12.6)

Henry's Law a formula that describes the concentration of dissolved gas in a solution is proportional to its partial pressure in the gas phase (10.10)

heterogeneous mixture A combination of solids, liquids, or gases that are not uniformly distributed throughout the substance. (1.1 and 1.6)

homogeneous mixture A single-phase combination of solids, liquids, or gases with a uniform distribution of its constituents throughout the substance. (1.1 and 1.6)

hormones chemical messengers produced by the body's endocrine glands (12.6)

hybrid electric vehicle (HEV) a vehicle propelled by a combination of a conventional gasoline engine and an electric motor run by batteries (7.8)

hydrocarbon organic compounds comprised entirely of carbon and hydrogen (2.7, 5.2, and 5.12)

hydrogen bond an electrostatic attraction between a H atom bonded to a highly electronegative atom (O, N, or F) and a neighboring O, N, or F atom, either in another molecule or in a different part of the same molecule (8.3)

hydrogenation a process in which hydrogen gas, in the presence of a metallic catalyst, adds to a C=C double bond and converts it to a single bond (11.3)

hydrometer a device used to measure the density of a liquid (10.9)

I

interesterification any process in which the fatty acids on two or more triglycerides are scrambled to produce a mixture of different triglycerides (11.3)

intermolecular force a force that occurs between molecules (5.12, 8.3, and 9.4)

ionic compound a compound composed of ions that are present in fixed proportions and arranged in a regular, geometric structure (1.3)

isomers molecules with the same chemical formula, but with different structures and properties (5.13 and 12.3)

J

joule (J) a unit of energy equal to 0.239 cal (5.4)

K

kinetic energy the energy of motion (5.3)

kinetics the branch of science that deals with the rates of reactions (7.6)

L

latent heat a measure of the heat absorbed by a substance to induce a phase change (10.5)

latent prints fingerprints left by the natural oils from one's skin, which result in prints left on a surface that are not visible to the naked eye (14.6)

Law of Conservation of Matter and Mass a law stating that in a chemical reaction, matter and mass are conserved (3.6)

Lewis structure a representation of an atom or molecule that shows its outer electrons (3.7)

limiting reagent the reactant that is totally consumed during a chemical reaction, hence limiting the amount of product that may be formed (10.4)

line-angle drawing simplified version of a structural formula that is most useful for representing larger molecules (12.3)

lipids a class of compounds that includes not only all triglycerides, but also related compounds such as cholesterol and other steroids (11.2)

London dispersion forces attractive forces between nonpolar molecules such as hydrocarbons (5.12 and 9.4)

lower flammability limit (LFL) the lowest concentration of solvent vapor that can ignite in air (14.4)

M

macrominerals elements that are necessary for life (Ca, P, Cl, K, S, Na, and Mg) but not nearly as abundant in the body as O, C, H, and N (11.7)

macronutrient the fats, carbohydrates, and proteins that provide essentially all of the energy and most of the raw material for body repair and synthesis (11.1)

Maillard reaction A reaction that occurs at high temperatures involving functional groups present in sugars and proteins within foods. This reaction results in a browned crust that forms on cooked foods such as eggs, meats, breads, etc. (10.5)

malnutrition caused by a diet lacking in proper nutrients, even though the energy content of the food may be adequate (11.1)

mass number the sum of the number of protons and neutrons in the nucleus of an atom (1.4)

mass spectrometry an analytical technique in which a sample vapor is ionized and the resulting ions are separated according to their mass-to-charge ratios (14.6)

matter any solid, liquid, gas, or plasma that occupies space and has a mass (1.1)

medicinal chemistry the branch of chemistry that deals with the discovery or design of new therapeutic chemicals and their development into useful medicines (12.0)

metabolism the complex set of chemical processes that are essential in maintaining life (11.1 and 12.6)

microminerals nutrients that the body requires lesser amounts of, such as Fe, Cu, and Zn (11.7)

micronutrients substances such as vitamins and minerals that are needed only in miniscule amounts, but remain essential for the body to produce enzymes, hormones, and other substances needed for proper growth and development (11.7)

minerals ions or ionic compounds that, like vitamins, have a wide range of physiological functions (1.6 and 11.7)

mixture a physical combination of two or more pure substances present in variable amounts (2.2)

molar mass the mass of Avogadro's number, or one mole, of whatever particles are specified (4.4)

molarity (M) a unit of concentration represented by the number of moles of solute present in one liter of solution (8.6)

mole (mol) an Avogadro's number of objects (1.3 and 4.4)

molecular compound a pure substance that contains two or more atoms from nonmetallic elements. (1.3)

molecule two or more atoms held together by chemical bonds in a certain spatial arrangement (1.3 and 2.6)

monomer a small molecule used to synthesize a larger polymer (from *mono* meaning "one" and *meros* meaning "unit") (9.2)

monosaccharide a single sugar, such as fructose or glucose 11.4

municipal solid waste (MSW) Garbage; that is, everything you discard or throw into your trash, including food scraps, grass clippings, and old appliances. MSW does not include all sources, such as waste from industry, agriculture, mining, or construction sites. (9.9)

N

n-type semiconductor a semiconductor in which there are freely moving negative charges (electrons) (6.8)

nanotechnology the manipulation of matter with at least one dimension sized between 1–100 nanometers (1.2)

neutral solution a solution that is neither acidic nor basic; that is, it has equal concentrations of H^+ and OH^- (8.9)

neutralization reaction a chemical reaction in which the hydrogen ions from an acid combine with the hydroxide ions from a base to form water molecules (8.9)

nitrification the process of converting ammonia in the soil to the nitrate ion (11.12)

nitrogen cycle a set of chemical pathways whereby nitrogen moves through the biosphere (11.12)

nitrogen-fixing bacteria bacteria that remove nitrogen from the air and convert it to ammonia (11.12)

nonpolar covalent bond a covalent bond in which the electrons are shared equally or nearly equally between atoms (8.2)

nuclear fission the splitting of a large nucleus into smaller ones with the release of energy (6.1)

nuclear radiation radiation emitted by a nucleus, such as alpha, beta, or gamma radiation (6.3)

nucleotide covalently bonded combination of a base, a deoxyribose molecule, and a phosphate group (13.2)

Nuffield Council report Nuffield Council on Bioethics, *Biofuels: Ethical Issues* **2011**, *84*. (5.17)

O

ocean acidification the lowering of the ocean pH due to increased atmospheric carbon dioxide (8.12)

octet rule A generalization that electrons are arranged around atoms so that these atoms have a share in eight electrons. Hydrogen is an exception. (3.7 and 5.2)

oils triglycerides that are liquids at room temperature (5.16 and 11.2)

organic chemistry the branch of chemistry devoted to the study of carbon compounds (12.3)

organic compound a compound that always contains carbon, almost always contains hydrogen, and may contain other elements such as oxygen and nitrogen (2.13)

osmosis the passage of water through a semipermeable membrane from a solution that is less concentrated to a solution that is more concentrated (8.14)

oxidation a process in which a chemical species loses electrons (1.7 and 7.1)

oxygenated gasoline a blend of petroleum-derived hydrocarbons with added oxygen-containing compounds such as MTBE, ethanol, or methanol (5.13)

ozone layer a designated region in the stratosphere of maximum ozone concentration (2.8)

P

p-type semiconductor a semiconductor in which there are freely moving positive charges, or holes (6.8)

parts per billion (ppb) one part out of one billion, or 1000 times less concentrated than 1 part per million (8.6)

parts per million (ppm) A concentration of one part out of a million. One ppm is a unit of concentration 10,000 times smaller than 1% (one part per hundred). (8.6)

patent prints fingerprints left by someone who is using a substance such as grease, paint, blood, etc., which results in a visible print on a surface (14.6)

peptide bond the covalent bond that forms when the $–COOH$ group of one amino acid reacts with the $–NH_2$ group of another, thus joining the two amino acids (9.7)

percent (%) Parts per hundred. For example, 15% is 15 parts out of 100. (2.3 and 8.6)

pharmaceuticals therapeutic substances intended to prevent, moderate, or cure illnesses (12.0)

pharmacophore the three-dimensional arrangement of atoms or groups of atoms responsible for the biological activity of a drug molecule (12.10)

photon a way of conceptualizing light as a particle that has energy but no mass (3.2)

photosynthesis the process by which green plants (including algae) and some bacteria capture the energy of sunlight to produce glucose and oxygen from carbon dioxide and water (3.4)

plasmids rings of DNA (13.6)

plasticizer a compound added in small amounts to a polymer to make the polymer softer and more pliable (9.5)

polar covalent bond a covalent bond in which the electrons are not equally shared, but rather are closer to the more electronegative atom (8.2)

polar stratospheric clouds (PSCs) thin clouds composed of tiny ice crystals formed from the small amount of water vapor present in the stratosphere (3.9)

polyamide a condensation polymer that contains the amide functional group (9.7)

polyatomic ion two or more atoms covalently bound together that have an overall positive or negative charge (4.1)

polymer a large molecule built from smaller ones (monomers) that consists of a long chain or chains of atoms covalently bonded together (9.0)

polymerase chain reaction (PCR) a technique that is used to amplify a single copy or a few copies of DNA across several orders of magnitude, generating thousands to millions of copies of a particular DNA sequence (14.14)

polysaccharide A condensation polymer made up of thousands of monosaccharide units. Examples include starch and cellulose. (11.4)

post-consumer content material used by a consumer that would otherwise have been discarded as waste (9.9)

potable water water safe for drinking and cooking (8.4)

potential energy Energy of position, or stored energy. In chemistry, we refer to this energy as that stored in the chemical bonds within a molecule that may be released during a chemical reaction (5.3)

power The rate at which work is performed or energy is converted. For electric power, this corresponds to the rate at which electrical energy is transported through a circuit (voltage × current) (7.6)

power density The amount of power (voltage × current) per unit of volume. This refers to the ability of an energy-storage device to take on or deliver power. That is, a battery with a high power density will charge faster than one with a lower power density. (7.7)

pre-consumer content waste left over from the manufacturing process itself, such as scraps and clippings (9.9)

precautionary principle stresses the wisdom of acting, even in the absence of full scientific data, before the adverse effects on human health or the environment become significant or irrevocable (5.17 and 9.11)

precipitate The solid deposited during a precipitation event, when a solid drops out of a homogeneous solution. This generally refers to an amorphous solid, with no long-range structural order, but it can also be used to describe a crystalline solid deposited slowly from a solution. (4.2)

precipitation The process of a solid dropping out of a homogeneous solution. Usually this refers to rapid deposition of a solid, which forms an amorphous solid with a disordered structural array. (4.2)

primary structure the unique sequence of the amino acids that make up each protein (13.5)

processed foods foods that have been altered from their natural state by techniques such as canning, cooking, freezing, or adding chemicals such as thickeners or preservatives (11.1)

products the substances listed on the right-hand side of a chemical equation, representing the materials that are formed during a standard chemical reaction (1.8)

protein a polyamide or polypeptide; that is, a polymer built from amino acid monomers (11.6)

protein complementarity combining foods that complement essential amino acid content so that the total diet provides a complete supply of amino acids for protein synthesis (11.6)

proton a subatomic positively charged particle with approximately the same mass as a neutron (8.8)

Q

quantized an energy distribution that is not continuous, but rather consists of many individual steps (3.2)

R

racemic mixture mixture consisting of equal amounts of each optical isomer of a compound (12.5)

radiation the emission of energy as electromagnetic waves or as moving subatomic particles (3.1)

radiative forcings factors (both natural and anthropogenic) that influence the balance of Earth's incoming and outgoing radiation (4.10)

radioactive decay series a characteristic pathway of radioactive decay that begins with a radioisotope and progresses through a series of steps to eventually produce a stable isotope (6.3)

radioactivity the spontaneous emission of radiation by certain elements (6.3)

radioisotope an isotope that spontaneously emits nuclear radiation (6.3)

radiopharmaceuticals organic molecules that carry radioactive isotopes to specific regions in the body in order to create contrast between tissue areas (12.10)

reactants the substances listed on the left-hand side of a chemical equation, representing the starting materials for a standard chemical reaction (1.8)

reaction quotient The product concentrations divided by the reactant concentrations during a reversible chemical reaction that may or may not be at equilibrium. (12.1)

reactive nitrogen the compounds of nitrogen that cycle through the biosphere and interconvert with each other relatively quickly (11.12)

receptor a biomolecule that is typically embedded within the cellular membrane that binds with specific molecules, thus producing some effect in the cell (12.6)

recycled-content products products made from material that otherwise would have been in the waste stream (9.9)

reduction a process in which a chemical species gains electrons (1.7 and 7.1)

reflux the repeated process of heating a solvent to boiling and condensing its vapor (14.2)

reformulated gasoline (RFG) an oxygenated gasoline that also contains a lower percentage of certain more volatile hydrocarbons found in nonoxygenated conventional gasoline (5.13)

refractive index a description of how light propagates through a material or liquid, versus through a vacuum. For instance, water has a refractive index of 1.33, which indicates that light propagates through water 1.33 times slower than in a vacuum due to refraction among the constituent water molecules.

refractometry the study of how light propagates through a material or liquid (10.9)

renewable resources those resources that are replenished more quickly over time than they are being consumed (6.0)

replication the process of cell reproduction in which the cell must copy and transmit its genetic information to its progeny (13.3)

residual chlorine The name given to chlorine-containing chemicals that remain in the water after the chlorination step. These include hypochlorous acid (HClO), the hypochlorite ion (ClO^-), and dissolved elemental chlorine (Cl_2). (8.13)

resistor an electrical component that impedes the flow of electrons within an electrical circuit (7.2)

resonance forms Lewis structures that represent hypothetical extremes of electron arrangements in a molecule (3.7 and 12.3)

respiration the process of metabolizing the foods we eat to produce carbon dioxide and water and to release the energy that powers other chemical reactions in our bodies (2.1 and 12.6)

reverse osmosis a process that uses pressure to force the movement of water through a semipermeable membrane from a solution that is more concentrated to a solution that is less concentrated (8.14)

risk assessment the process of evaluating scientific data and making predictions in an organized manner about the probabilities of an outcome (2.9)

rocks heterogeneous solid-state mixtures that contain a variety of ionic compounds (1.6)

S

saturated fatty acid a hydrocarbon containing only single bonds between the carbon atoms (11.2)

scanning electron microscopy (SEM) an analytical technique used to image the surface of a sample by using a high-energy beam of electrons (14.6)

scientific notation a system for writing numbers as the product of a number and 10 raised to the appropriate power (1.8 and 3.1)

second law of thermodynamics a law that can be stated in many ways, including that the entropy of the universe is constantly increasing (5.7)

secondary pollutant a pollutant produced from chemical reactions involving one or more other pollutants (2.14)

secondary structure the folding pattern within a segment of the protein chain (13.5)

self-discharge when an energy storage device loses its charge over time without being connected to an external circuit (7.7)

semiconductor a material that does not normally conduct electricity well, but can do so under certain conditions, such as with exposure to sunlight (6.8)

shifting baseline the idea that what people expect as "normal" on our planet has changed over time, especially with regard to ecosystems (2.16 and 9.11)

single covalent bond a bond formed when two electrons (one pair) are shared between two atoms (3.7)

solute the solid, liquid, or gas that dissolves in a solvent (4.2 and 8.6)

solution a homogeneous (of uniform composition) mixture of a solvent and one or more solutes (1.1 and 8.6)

solvent a substance, often a liquid, capable of dissolving one or more pure substances (8.6)

solvent still a laboratory apparatus used to remove oxygen and moisture from organic solvents (14.2)

specific gravity the ratio of density of a solution to the density of pure solvent (without any dissolved solutes) (10.9)

specific heat the quantity of heat energy that must be absorbed to increase the temperature of one gram of a substance by 1 °C (8.3)

spherification a culinary process of shaping a liquid into spheres (10.10)

stable isotope analysis a technique that compares the concentration ratio of various isotopes for a sample in order to determine its origin (14.6)

standard temperature and pressure (STP) an ambient temperature of 25 °C and pressure of 1 atm (760 Torr) (8.2)

starch A carbohydrate found in many grains, including corn and wheat. Starch is a natural polymer of glucose. (5.15)

steady state a condition in which a dynamic system is in balance so that there is no net change in concentration of the major species involved (3.7)

steroid a class of naturally occurring or synthetic fat-soluble organic compounds that share a common carbon skeleton arranged in four rings (12.8)

strong acid an acid that dissociates completely in water (8.8 and 12.2)

strong base a base that dissociates completely in water (8.8)

structural formula A representation of how the atoms in a molecule are connected. It is a Lewis structure from which the nonbonding electrons have been removed. (3.7)

structure-activity relationship (SAR) study a study in which systematic changes are made to a drug molecule followed by an assessment of the resulting changes in activity (12.10)

sublimation the direct conversion of a solid directly into a gas, without proceeding through a melting transition (1.8)

supercapacitor a device that stores a significant amount of energy by means of a static charge rather than an electrochemical reaction used by batteries (7.7)

surface water fresh water found in lakes, rivers, and streams (8.4)

surfactant a molecule that has both polar and nonpolar regions that allows it to help solubilize different classes of molecules (8.7)

sustainability "meeting the needs of the present without compromising the ability of future generations to meet their own needs" (from *Our Common Future,* a 1987 report by the United Nations) (2.16)

sustainable packaging the design and use of packaging materials to reduce their environmental impact and improve the sustainability of all practices (9.8)

T

temperature a measure of the average kinetic energy of the atoms and/or molecules present in a substance (5.3)

tertiary structure the overall molecular shape of the protein defined by the interactions between amino acids far apart in sequence, but close in space (13.5)

thermal cracking a process that breaks large hydrocarbon molecules into smaller ones by heating them to a high temperature (5.13)

thermodynamics the branch of science that deals with the energies and relative spontaneity associated with chemical reactions or processes (12.1)

thermoplastic polymer a plastic that can be melted and reshaped over and over again (9.5)

three pillars of sustainability Sustainability is meeting the needs of the present without compromising the ability of future generations to meet their own needs. There are three considerations that are equally important for a sustainable practice, which include environmental, social, and economic factors. (1.10)

toxicity the intrinsic health hazard of a substance (2.9)

trace mineral an element present in the body, usually at microgram levels, such as I, F, Se, V, Cr, Mn, Co, Ni, Mo, B, Si, and Sn (11.7)

tragedy of the commons The situation in which a resource is common to all and used by many, but has no one in particular who is responsible for it. As a result, the resource may be destroyed by overuse to the detriment of all that use it. (2.14 and 8.5)

***trans* fat** a triglyceride that is composed of one or more *trans* fatty acids (11.3)

transgenic an organism resulting from the transfer of genes across species (13.6)

triglycerides A class of compounds that includes both fats and oils. Triglycerides contain three ester functional groups and are formed from a chemical reaction with three fatty acids and the alcohol glycerol. (5.16 and 11.2)

trihalomethanes (THMs) compounds such as $CHCl_3$ (chloroform), $CHBr_3$ (bromoform), $CHBrCl_2$ (bromodichloromethane), and $CHBr_2Cl$ (dibromochloromethane) that form from the reaction of chlorine or bromine with organic matter in drinking water (8.13)

triple bond a covalent linkage made up of three pairs of shared electrons (3.7)

triple bottom line a three-way measure of the success of a business based on its benefits to the economy, to society, and to the environment (5.17)

troposphere the lowest region of the atmosphere in which we live that lies directly above the surface of Earth (2.5)

U

undernourishment a condition in which a person's daily caloric intake is insufficient to meet metabolic needs (11.1)

unsaturated fatty acid a fatty acid in which the hydrocarbon chain contains one or more double bonds between carbon atoms (11.2)

V

vapor deposition the conversion of a gas into a solid, commonly used to form thin films from a gas-phase reaction (1.8)

vapor pressure the pressure exerted by gaseous molecules, as a result of vaporization of a liquid or solid (5.12)

vaporization the process of transferring molecules from the liquid to gaseous state (5.12)

vector a modified plasmid used to carry DNA back into the bacterial host (13.6)

vitamin an organic compound, with a wide range of physiological functions, that is essential for good health, proper metabolic functioning, and disease prevention (11.7)

volatile organic compounds (VOCs) carbon-containing compounds that pass easily into the vapor phase (2.13 and 8.7)

volatility the ease at which molecules of a liquid overcome their intermolecular forces to be released into the gaseous phase (5.12)

voltage the difference in electrochemical potential between two electrodes (7.2)

volumetric flask a type of glassware that contains a precise amount of solution when filled to the mark on its neck (8.6)

W

water footprint an estimate of the volume of fresh water used to produce a particular good or to provide a service (8.4)

watt The SI unit of power, equal to 1 J/s (6.2)

wavelength the distance between successive peaks in a sequence of waves (3.1)

weak acid an acid that dissociates only to a small extent in aqueous solution (8.8 and 12.2)

weak base a base that dissociates only to a small extent in aqueous solution (8.8)

weather Conditions that include the daily high and low temperatures, the drizzles and downpours, the blizzards and heat waves, and the fall breezes and hot summer winds, all of which have relatively short durations. Contrast this with *climate*. (4.9)

X

X-ray diffraction an analytical technique in which a crystal is hit by a beam of X-rays to generate a pattern that reveals the positions of the atoms in the crystal (13.3)

Index

Note: Page references followed by *f* or *t* indicate material in figures or tables, respectively. Page numbers preceded by A indicate appendix pages.

A

B

C

G

H

I

N

O

P

T

X

Y

Z